Acknowledgements

Orders: You can order through the website www.tlmaths.com

ISBN: 978-1-3999-9572-6
First Edition: published 2024
This edition is 1.02

Special thanks to Adam Brown in the provision of artwork, and most importantly Laura and Paul for patience and their on-going support of our unnecessarily oversized projects. Thanks also to Mandy Taylor for doing some triple-checks!

Artwork with thanks to Desmos
Published by TLMaths Publications
Typeset: TLMaths Publications, Hampshire
Editors: Jack Brown, Danielle Hosford

Copyright © 2024, TLMaths Publications

All rights reserved. No part of this publication may be reproduced or transmitted in any form or by any means, electronical, mechanical, including photocopy, recording or any information storage and retrieval system, without permission in writing from the publisher or under license from the Copyright Licensing Agency Limited.

Meet the Authors

Jack Brown and Dani Hosford are both 6th form college maths teachers in Hampshire. They have a combined total of over 30 years of experience teaching A Level Maths and A Level Further Maths to year 12 and year 13 students.

Jack Brown first began his TLMaths YouTube channel in 2013. In that time, he has created over 4000 videos for teaching and learning mathematics and has accumulated more than 130 000 subscribers with over 50 million views.

Dani Hosford is a maths textbook author and examiner of A Level Maths and A Level Further Maths. She is more commonly known by students as the teacher who wears hats on Fridays.

Dani first started working with Jack in 2019 and after four years of building and redesigning resources, and un-italicising ds in differentials, they realised that what they really needed to do was write a great textbook. After all, how hard could it be?

 Artwork: Adam Brown

About this Book

Answers and worked solutions

Answers to the exercises are included in this book, however for full worked solutions please visit www.tlmaths.com. Whilst the questions and worked solutions may provide the required support, further guidance from qualified teachers or instructors for clarification on mathematical concepts and techniques should be sought as necessary.

Exam board specifications

We have designed this textbook on England's AS and A-Level Maths specifications, namely Edexcel (8MA0 & 9MA0), AQA (7356 & 7357), OCR A (H230 & H240), and OCR B MEI (H630 & H640). Whilst every effort has been made to ensure this book aligns with all these exam board specifications, the official specifications and guidance materials are the definitive sources of information. For exact guidance, please refer to those official resources.

There are subtle differences between the exam board specifications, such as OCR B MEI being the only specification to mention the midrange as a statistical measure, and rank correlation. If you are unsure of whether a topic is in a part of your specification, please ask your teacher and/or refer to the official specification documents found online.

Answers and fact checks

We have gone through a rigorous editorial process with every effort made to ensure that this publication is free from errors. However, occasionally errors may have occurred. If you spot an error, please let us know by filling in the form at www.tlmaths.com.

The Large Data Set

For access to the large data set for your exam board, please refer to the appropriate website. These websites provide the data sets and guidance, as well as possibly having other additional resources. It is important to ensure that the data set you are using matches the specification of your exam board and in some cases, that you have selected the correct one.

The purpose of the A Level Maths large data set is to provide students with practical experience in handling and analysing real-world data. They are designed for students to develop essential skills in problem solving, and proficiency in the use of appropriate software and data analysis. Students should explore the large data set to understand its structure, perform statistical calculations using it and interpret the results.

However, as part of the process of familiarisation with the large data set, our exercises provide revision of statistical analysis and interpretation framed in the context of the large data set. They may be used as a familiarisation or revision exercise, or otherwise.

Disclaimers

All names used in this book are fictitious. Any resemblance to actual persons, living or dead, is purely coincidental. The characters and events depicted are the product of the author's imagination and are not intended to represent real individuals or events.

Using this Book

Our mission?

To create a textbook that saves teachers time and improves student outcomes.

Traditional textbooks are often weighed down with long-winded introductions, followed by several worked examples, and then a short exercise of questions that aren't challenging enough or escalate in difficulty too quickly. And very few of these questions we would consider to be "exam-style". The exercises often feel like an afterthought, and it is these exercises that will get used the most in class.

Our textbook has stripped away the lengthy introductions and worked examples. This is what the teacher will cover in class, and it is also all contained within Jack's TLMaths videos online. This allowed us to focus on designing exercises that will scaffold students' learning and then push further, helping students master the material rather than skimming through it.

Individual questions can be used as lesson starters, or a set of questions can be used as an exercise in class, or as some homework, or as revision closer to the exam.

We've organised the topics so that AS material comes first (marked by section 1), and within that we have decided on an order that we feel makes sense and is accessible. But we know that Maths departments will likely already have an order of delivery, so we've included **Prerequisite knowledge** markers. If a question requires knowledge that is found elsewhere in that section, this will be identified.

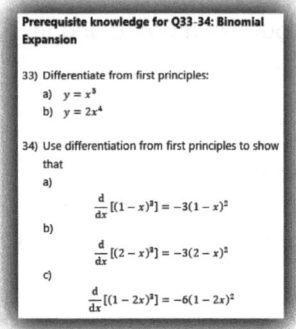

So, if you are a teacher diving into Differentiation from First Principles but haven't covered Binomial Expansion yet, you can tell your students to skip Q33 and Q34 for the time being and return to them later. We haven't included any Prerequisite knowledge markers in section 2 for topics in section 1, otherwise there would be markers all over the place! But hopefully you find these markers useful, and they are searchable in the index too.

As for answers, they're all at the back of the textbook, but they are *just* the answers. This is so students can quickly check their work without having to lug around a textbook that would be three times the size! However, full worked solutions for every question do exist and they are available through www.TLMaths.com. The QR-code below will take you straight there:

Jack & Dani

Contents

	Acknowledgements	i
	Meet the Authors	ii
	About this book	iii
	Using this book	iv
	Contents	v

Pure Mathematics

1.01	Indices	1
1.02	Surds	3
1.03	Coordinate Geometry: Linear Graphs	5
1.04	Graph Sketching: Linear Graphs	7
1.05	Linear Modelling	10
1.06	Quadratics: Introduction	12
1.07	Algebra: Simultaneous Equations	15
1.08	Graph Sketching: Quadratic Graphs	16
1.09	Quadratics: Problem Solving	20
1.10	Quadratics: The Discriminant	21
1.11	Quadratic Modelling	22
1.12	Coordinate Geometry: Circles	24
1.13	Algebra: Algebraic Fractions	28
1.14	Polynomials	29
1.15	Graph Sketching: Cubic Graphs	31
1.16	Graph Sketching: Quartic Graphs	34
1.17	Proportion	37
1.18	Graph Sketching: Rational Functions	38
1.19	Graph Sketching: Radical Functions	40
1.20	Inequalities	41
1.21	Inequalities: Identifying Regions	44
1.22	Exponentials and Logarithms	45
1.23	e and Natural Logarithms	49
1.24	Graph Sketching: Exponentials and Logarithms	52
1.25	Graph Sketching: e and Natural Logarithms	57
1.26	Exponential Growth and Decay	62
1.27	Reduction to Linear Form	66
1.28	Graph Transformations: Single Transformations	71
1.29	Binomial Expansion	77
1.30	Differentiation from First Principles	79
1.31	Differentiation: Introduction	82
1.32	Differentiation: Tangents and Normals	84
1.33	Differentiation: Graphs of Gradient Functions	87
1.34	Differentiation: Stationary Points	90
1.35	Differentiation: Optimisation	91
1.36	Integration: Indefinite Integrals	94
1.37	Integration: Definite Integrals and Areas	97
1.38	Applications of Calculus	104
1.39	Trigonometry: Introduction	105
1.40	Trigonometry: sin(x), cos(x), tan(x) in Degrees	108
1.41	Graph Sketching: sin(x), cos(x), tan(x) in Degrees	112
1.42	Trigonometry: Trigonometric Identities in Degrees	120
1.43	Trigonometry: Modelling in Degrees	125
1.44	Hidden Polynomials	128
1.45	Proof	129
1.46	2D Vectors	130

Statistics

1.47	Sampling Methods	132
1.48	Measures of Central Tendency and Variation	134
1.49	Representing Data	140
1.50	Bivariate Data: PMCC and Linear Regression	144
1.51	Bivariate Data: Association and Spearman's Rank	149
1.52	Probability: Introduction	150
1.53	Probability: Regions of Venn Diagrams	159
1.54	Probability: Independence and Mutually Exclusive	162
1.55	Discrete Random Variables	164
1.56	The Binomial Distribution	168
1.57	Binomial Hypothesis Testing	173
1.58	Large Data Set: Edexcel	176
1.59	Large Data Set: AQA	178
1.60	Large Data Set: OCR MEI (Health)	180
1.61	Large Data Set: OCR MEI (World Bank)	182
1.62	Large Data Set: OCR MEI (London)	184
1.63	Large Data Set: OCR A	186

Mechanics

1.64	Graphs of Motion	188
1.65	Constant Acceleration	193
1.66	Forces: Equilibrium	196
1.67	Forces: Particles in Motion	198
1.68	Forces: Vectors	200
1.69	Forces: Connected Particles	202
1.70	Forces: Pulleys	205
1.71	Variable Acceleration	209

Contents

Pure Mathematics

2.01	Trigonometry: Radians, Arcs and Sectors	213
2.02	Trigonometry: sin(x), cos(x), tan(x) in Radians	217
2.03	Graph Sketching: sin(x), cos(x), tan(x) in Radians	221
2.04	Trigonometry: Trigonometric Identities in Radians	229
2.05	Trigonometry: Modelling in Radians	231
2.06	Sequences and Series: Introduction	234
2.07	Sequences and Series: Arithmetic Sequences	238
2.08	Sequences and Series: Geometric Sequences	241
2.09	Binomial Series	245
2.10	Domain, Range and Composite Functions	247
2.11	Graph Transformations: Combinations	253
2.12	Inverse Functions	256
2.13	Modulus Functions	260
2.14	Trigonometry: Inverse Trigonometric Functions	267
2.15	Trigonometry: Reciprocal Trigonometric Functions	270
2.16	Trigonometry: Compound Angle Formulae	273
2.17	Trigonometry: Double Angle Formulae	276
2.18	Trigonometry: Harmonic Forms	278
2.19	Trigonometry: Small Angle Approximations	280
2.20	Differentiation: Points of Inflection and Applications	282
2.21	Differentiation: Standard Functions	284
2.22	Differentiation: The Chain Rule	287
2.23	Differentiation: Connected Rates of Change	289
2.24	Differentiation: The Product Rule	291
2.25	Differentiation: The Quotient Rule	293
2.26	Differentiation: Implicit Differentiation	297
2.27	Numerical Methods: Change of Sign Method	300
2.28	Numerical Methods: $x = g(x)$ Method	301
2.29	Numerical Methods: Newton-Raphson Method	304
2.30	Integration: The Trapezium Rule	308
2.31	Integration: Areas between Curves	312
2.32	Integration: Standard Functions	314
2.33	Integration: Reversing the Chain Rule Part 1	319
2.34	Integration: Reversing the Chain Rule Part 2	321
2.35	Integration by Substitution	324
2.36	Integration by Parts	328
2.37	Partial Fractions	330
2.38	Integration: Partial Fractions	333
2.39	Integration: Double Angles	334
2.40	Differential Equations	335
2.41	Parametric Equations: Introduction	340
2.42	Parametric Equations: Differentiation	342
2.43	Parametric Equations: Integration	345
2.44	Proof by Contradiction	349
2.45	3D Vectors	350

Statistics

2.46	Bivariate Data: PMCC Hypothesis Testing	353
2.47	Bivariate Data: Spearman's Rank Hypothesis Testing	355
2.48	PMCC and Spearman's Critical Tables	356
2.49	Probability: Conditional Probability	357
2.50	The Normal Distribution	361
2.51	Sample Means Hypothesis Testing	367

Mechanics

2.52	Forces: Equilibrium	371
2.53	Forces: Coefficient of friction	376
2.54	Forces: Connected Particles	381
2.55	Forces: Inclined Planes	383
2.56	Forces: Connected Particles on Inclined Planes	388
2.57	Constant Acceleration: Vectors	392
2.58	Projectiles	395
2.59	Variable Acceleration: Vectors	398
2.60	Moments	401
2.61	Moments: Ladders and Hinges	405

	Answers	408
	Index	559

1.01 Indices (answers on page 408)

1) Without using a calculator, evaluate:
   a) $4^{\frac{1}{2}}$
   b) $9^{-\frac{1}{2}}$
   c) $7^3 \times 7^{-1}$
   d) $2 \times 3^3 \times 3^{-2}$
   e) $3^5 \div 3^{-1}$
   f) $\left(64^{\frac{1}{3}} \times 16^{\frac{1}{4}}\right)^{-1}$
   g) $\left(\frac{1}{16}\right)^{-\frac{1}{2}}$
   h) $\left(\frac{27}{8}\right)^{-\frac{1}{3}}$
   i) $\dfrac{64^{\frac{2}{3}}}{8^{\frac{5}{3}}}$

2) Express the following as a single power:
   a) $x\sqrt{x}$
   b) $x^2\sqrt{x}$
   c) $\dfrac{x}{\sqrt{x}}$
   d) $\dfrac{x^2}{\sqrt{x}}$
   e) $(\sqrt{x})^3$
   f) $(\sqrt[3]{x})^2$
   g) $\dfrac{1}{\sqrt{x}}$
   h) $\dfrac{4}{(\sqrt{x})^3}$
   i) $\dfrac{x\sqrt{x}}{\sqrt[3]{x}}$

3) Express the following in the form $\sqrt[a]{x^b}$, x^b, $\dfrac{1}{\sqrt[a]{x^b}}$ or $\dfrac{1}{x^b}$ as appropriate, where a and b are integers to be found:
   a) $x^{\frac{1}{2}}$
   b) $x^{-\frac{1}{2}}$
   c) x^{-5}
   d) $x^{\frac{2}{3}}$
   e) $x^{-\frac{3}{4}}$
   f) $\dfrac{1}{x^{-\frac{3}{5}}}$

4) Simplify each of the following:
   a) $(4a^{10})^{\frac{3}{2}}$
   b) $(a^3b^2c)^{\frac{1}{3}} \times (4a^4b^2c)^{\frac{1}{2}}$
   c) $(16ab^2)^{\frac{1}{2}} \times (27a^6b^3)^{-\frac{1}{3}}$
   d) $(20a^8bc) \div (5a^2b^2c^{-1})$
   e) $(3a^2b^3c)^2 \div (9a^2b^{-2}c^4)^{\frac{1}{2}}$

5) Simplify each of the following:
   a) $\dfrac{(2a)^3 \times (3a^2)^5}{(6a^5)^2}$
   b) $\dfrac{(3a)^3 \times (4a^2)^2}{(2a^4)^3}$
   c) $\left(\dfrac{a^{\frac{1}{2}} \times a^{\frac{5}{2}}}{(a^3)^4}\right)^{-1}$
   d) $\left(\dfrac{32}{a^8}\right)^{\frac{3}{5}} \times \left(\dfrac{a^7}{4}\right)$

6) Simplify each of the following:
   a) $\sqrt{x^3 \times x^5}$
   b) $\sqrt{x^{\frac{3}{5}} \times x^{\frac{1}{5}}}$
   c) $\sqrt{49x^6y^4} \times (27x^3y^9)^{-\frac{1}{3}}$
   d) $\sqrt{16x^4y^6} \times (9x^2y^3)^{-\frac{1}{2}}$

7) Give each of the following as sums of terms in the form ax^n:
   a) $\sqrt{x}(2x+1)$
   b) $(\sqrt{x}+1)(\sqrt{x}-2)$
   c) $\dfrac{x+1}{\sqrt{x}}$
   d) $\dfrac{x^2-1}{\sqrt{x}}$
   e) $\dfrac{4}{x} + \dfrac{1}{\sqrt{9x}}$
   f) $\dfrac{\sqrt{x}-3}{x\sqrt{x}}$

1.01 Indices (answers on page 408)

8) A right-angled triangle has perpendicular sides $7a^{\frac{1}{2}}$ and $24a^{\frac{1}{2}}$. Find
   a) the length of the hypotenuse side.
   b) the area of the triangle.

9) Solve each of the following:
   a) $x^{\frac{1}{2}} = 2$
   b) $x^{-\frac{1}{2}} = 3$
   c) $x^{\frac{1}{3}} = \frac{1}{2}$
   d) $x^{-\frac{1}{2}} = 11$
   e) $x^{-\frac{3}{2}} = \frac{8}{27}$
   f) $2x^{-\frac{1}{3}} = 3$
   g) $x^{-\frac{2}{3}} = 2\frac{7}{9}$

10) Mo was asked to find the solutions to
$$x^{\frac{7}{4}} - 2187 = 0$$
His solution is shown below.
$$x^{\frac{7}{4}} = 2187$$
$$x^{\frac{1}{4}} = \sqrt[7]{2187} = \pm 3$$
$$x = 3^4 \text{ or } (-3)^4$$
$$x = 81 \text{ or } -81$$
Find his mistake(s) and write a correct solution.

11) Write each of these in the form 8^k:
   a) $\sqrt{8}$
   b) $8^{10} \times 8^{-12}$
   c) $\frac{1}{64}$
   d) $\sqrt[3]{64}$
   e) $\frac{2}{\sqrt{8}}$
   f) $\frac{2^6}{\sqrt{8}}$
   g) $\sqrt[3]{16} \times (8^3)^{\frac{1}{9}}$
   h) $2 \times \sqrt{64} \times \sqrt[3]{16}$

12) Solve each of the following equations:
   a) $2^x \times 2^{\frac{1}{2}} = 2$
   b) $5^x \times \sqrt{5} = \frac{1}{25}$
   c) $3^{x+1} = 9^x$
   d) $4^{2x} = \sqrt{2}$
   e) $2^{3x-1} = 32$
   f) $2^{3x+2} = 4^{x+5}$
   g) $9^{2x-1} = \frac{1}{\sqrt{3}}$
   h) $\frac{9^x}{3^{x+2}} = 27$
   i) $3^{x^2-x} = 729$

13) Simplify each of the following into a single term (selecting the smaller base):
   a) $4^x + 4^x + 4^x + 4^x$
   b) $2^{2x} + 4^x$
   c) $3 \times 2^{3x} + 5 \times 8^x$
   d) $5 \times 16^{2x} - 3 \times 2^{8x}$
   e) $3^{2x+1} + 3^{2x}$
   f) $8 \times 2^x + 3 \times 2^{x+1}$
   g) $2^{2x+4} + 4^x$

14) Simplify each of the following so that there are only positive exponents in the numerator and denominator:
   a) $$\frac{x + x^{-2}}{x^2 y + x^{-1}}$$
   b) $$\frac{x^2 y^{-1} + x^{-1}}{xy^2 + x^{-2}}$$
   c) $$\frac{x^4 - x^{-2} yz^3}{x^2 + z^{-1}\sqrt{y}}$$
   d) $$\frac{x^5 - x^{-1} yz^4}{x^{-3} + z^2 y^{-\frac{1}{2}}}$$

15) Solve each of the following equations:
   a) $4^{2x} \times 3^{2x} = 12$
   b) $5^x \times 2^{2x} = 20$
   c) $3^{2x} \times 9^x = 9$
   d) $\frac{2^x}{2^{\frac{x}{2}} \times 2^4} = 1$

16) Find the solutions to
$$4^{x+2} - 2^{2x+3} - 16 = 0$$

17) Find the value of k such that
$$\frac{2^{501} - 2^{497}}{30} = 64^k$$

18) Solve
$$\frac{2^x}{32 \times 2^{\sqrt{x}}} = 2$$

19) Find one solution to
$$(x^2 - 4x - 1)^{3x+2} = \frac{1}{4}$$

1.02 Surds (answers on page 409)

We would expect completion of these questions without the use of a calculator.

1) Simplify
   a) $\sqrt{32}$
   b) $2\sqrt{12}$
   c) $\sqrt{45}$
   d) $\sqrt{200}$
   e) $-4\sqrt{50}$

2) Simplify
   a) $\sqrt{5} \times \sqrt{2}$
   b) $\sqrt{5}\sqrt{15}$
   c) $\sqrt{8} \times 3\sqrt{2}$
   d) $\sqrt{\frac{1}{9}}$
   e) $\frac{\sqrt{28}}{\sqrt{14}}$
   f) $\sqrt{25x^2}$

3) Simplify
   a) $2\sqrt{5} - \sqrt{5}$
   b) $\sqrt{8} + 2\sqrt{2}$
   c) $\sqrt{12} - \sqrt{3}$
   d) $\sqrt{128} - \sqrt{72}$
   e) $3\sqrt{12} + \sqrt{27}$
   f) $\sqrt{150} - \sqrt{96}$
   g) $\sqrt{3} + \sqrt{27} + \sqrt{12}$
   h) $\sqrt{350} + \sqrt{24}$
   i) $\sqrt{32a} + 2\sqrt{a} - \sqrt{28a}$

4) Simplify
   a) $\sqrt{2} \times \sqrt{7} \times \sqrt{3}$
   b) $\sqrt{18} \times \sqrt{2}$
   c) $2\sqrt{3} \times 2\sqrt{3}$
   d) $5\sqrt{3} \times 2\sqrt{5}$
   e) $-3\sqrt{3} \times 2\sqrt{3}$
   f) $(2\sqrt{6})^2$
   g) $(3\sqrt{8})^2$
   h) $(2\sqrt{5})^3$

5) Express each of the following in the form $\sqrt{a}$ where a is a positive integer:
   a) $2\sqrt{5}$
   b) $4\sqrt{3}$
   c) $3\sqrt{2}$
   d) $2\sqrt{7}$
   e) $2\sqrt{43}$

6) Express the following as a single square root:
   a) $a\sqrt{b}$
   b) $2\sqrt{a}$
   c) $(\sqrt{a})^5$
   d) $(a\sqrt{b})^3$

7) Express $\sqrt{525}$ in the form $p\sqrt{q}\sqrt{r}$ where p, q and r are prime numbers.

8) Write this ratio in its simplest form:
   $\sqrt{300} : \sqrt{48} : \sqrt{27}$

9) Rationalise the following surds:
   a) $\frac{3}{\sqrt{2}}$
   b) $\frac{6}{\sqrt{3}}$
   c) $\frac{6}{\sqrt{10}}$
   d) $\frac{2}{\sqrt{6}}$
   e) $\frac{10}{\sqrt{10}}$
   f) $\frac{4\sqrt{2}}{\sqrt{8}}$

10) Expand and simplify the following:
    a) $\sqrt{5}(1 + \sqrt{5})$
    b) $\sqrt{11}(2 + \sqrt{11})$
    c) $4(\sqrt{3} + 2\sqrt{5})$
    d) $(\sqrt{3} + 1)(\sqrt{3} - 1)$
    e) $(\sqrt{6} + 1)(2 + \sqrt{6})$
    f) $(\sqrt{8} - 1)(2 + \sqrt{8})$
    g) $(\sqrt{24} + 1)(\sqrt{2} - 3)$

11) Rationalise the following surds:
    a) $\frac{1}{\sqrt{2} + 1}$
    b) $\frac{3}{\sqrt{5} - 2}$
    c) $\frac{2}{1 - \sqrt{6}}$
    d) $\frac{1}{a + \sqrt{b}}$
    e) $\frac{1}{a - \sqrt{2b}}$

1.02 Surds (answers on page 409)

12) Simplify the following, giving your answers in the form $p\sqrt{q}$ where p and q are non-zero integers:
   a) $\sqrt{27} + \frac{15}{\sqrt{3}}$
   b) $\frac{\sqrt{54}}{2} - \frac{10}{\sqrt{6}}$
   c) $\frac{\sqrt{150}}{5} + \frac{30}{\sqrt{24}}$

13) Solve
$$x(4 - \sqrt{6}) = 10$$
giving your answer in the form $a + b\sqrt{6}$, where a and b are non-zero integers to be found.

14) Rationalise the denominator and simplify the following:
   a) $\frac{\sqrt{3}}{\sqrt{2} - 2}$
   b) $\frac{2\sqrt{5} - 1}{\sqrt{5} + 4}$
   c) $\frac{5\sqrt{3} - 2}{2 - \sqrt{7}}$
   d) $\frac{2\sqrt{3} - 7}{2\sqrt{3} - 3}$
   e) $\frac{1 - \sqrt{k}}{2 + \sqrt{2k}}$

15) Solve
$$x(1 - \sqrt{2}) + \sqrt{2} = 5$$
giving your answer in the form $a + b\sqrt{2}$, where a and b are non-zero integers to be found.

16) Show that
$$\frac{3\sqrt{10}}{\sqrt{2} + \sqrt{5}}$$
may be written in the form $p\sqrt{q} - q\sqrt{p}$ where p and q are non-zero integers to be found.

17) Show that
$$\frac{\sqrt{3} + \sqrt{7}}{\sqrt{3} - \sqrt{7}}$$
may be written in the form $a + b\sqrt{21}$ where a and b are non-zero constants to be found.

18) The expression
$$\frac{a + \sqrt{6}}{1 - \sqrt{6}}$$
may be written in the form $p + q\sqrt{6}$. Find the values of p and q in terms of a

19) Fully simplify the expression
$$\frac{12}{\frac{1}{\sqrt{3}} + \sqrt{3}}$$

20) A cylinder has a height of $(\sqrt{3} + 1)$cm and a radius of $\left(\frac{1}{\sqrt{3} - 2}\right)$cm.
Show that the volume of the cylinder is exactly $\pi(19 + 11\sqrt{3})$cm³

21) The expression
$$\frac{-k + 2\sqrt{3}}{\sqrt{27} + 1}$$
may be written in the form $a + b\sqrt{3}$. Find the values of a and b in terms of k

22) The expression
$$\frac{2 - \sqrt{k}}{1 + \sqrt{k}}$$
may be written in the form $a + b\sqrt{k}$. Find the values of a and b in terms of k

23)
   a) Simplify
   $$\frac{\sqrt{a+1} - \sqrt{a}}{\sqrt{a+1} + \sqrt{a}}$$
   b) Hence write down the solution to
   $$\frac{\sqrt{7} - \sqrt{6}}{\sqrt{7} + \sqrt{6}}$$

24) Evaluate the sum
$$\frac{1}{\sqrt{1} + \sqrt{2}} + \frac{1}{\sqrt{2} + \sqrt{3}} + \cdots + \frac{1}{\sqrt{99} + \sqrt{100}}$$

25) By using algebraic methods, show that
$$\sqrt{2} + \sqrt{3} > \sqrt{5}$$

26) Show that
$$\frac{2}{\sqrt{5} + \sqrt{2} + 1} = 3 + 2\sqrt{2} - \sqrt{5} - \sqrt{10}$$

1.03 Coordinate Geometry: Linear Graphs (answers on page 410)

1) Find the line passing through each of the following pairs of points giving your answer in the form $ax + by = c$ where a, b and c are integers:
   a) $(2, 5)$ and $(1, 7)$
   b) $(1, 4)$ and $(4, 13)$
   c) $(-1, -3)$ and $(5, 1)$
   d) $(-4, -6)$ and $(-2, -17)$
   e) (p, q) and $(2p, 4q)$
   f) $\left(p, \frac{1}{p}\right)$ and $\left(\frac{1}{p}, p\right)$ where $p \neq -1, 0, 1$

2) State both the gradient of the line and the gradient of a line which is perpendicular to each of the following:
   a) $3x + 6y = 7$
   b) $3y - 2x = 1$
   c) $2x + 8y = 11$
   d) $7y - 5x - 2 = 0$
   e) $20x + 17y = 3$
   f) $ax + by = c$

3) For each of the following pairs of lines, determine if they are perpendicular to each other:
   a) $2x + 3y = 1$ and $y = -\frac{3}{2}x + 5$
   b) $x - 5y + 10 = 0$ and $y = 5x - 4$
   c) $24x + 12y = 16$ and $12y + 24x = 1$
   d) $y = -4x + 8$ and $4y - x = 28$
   e) $ax + by = p$ and $bx - ay = q$

4) For the lines $ax + by = c$ and $px + qy = r$ where a, b, c, p, q and r are integers find a relationship between a, b, p and q such that the two lines are
   a) Parallel.
   b) Perpendicular.

5) Find the midpoint between each of these pairs of points:
   a) $(-4, 3)$ and $(4, 9)$
   b) $(1, 7)$ and $(6, -5)$
   c) $(-8, 10)$ and $(-2, -1)$
   d) $(-7, -2)$ and $\left(\frac{7}{2}, 12\right)$
   e) (p, q) and $(2p, 3q)$
   f) $\left(p, \frac{1}{p}\right)$ and $\left(\frac{1}{p}, p\right)$ where $p \neq -1, 0, 1$

6) For each of the following, find the value of k given the point which lies on the line:
   a) $(k, 4)$ and $2y - 5x = 16$
   b) $(2k, k)$ and $4x - 3y + 20 = 0$

7) Find the perpendicular bisector of the each of the following points giving your answer in the form $ax + by = c$ where a, b and c are integers:
   a) $(-2, 5)$ and $(6, 1)$
   b) $(3, 5)$ and $(6, 8)$
   c) $(1, 2)$ and $(4, 10)$
   d) $(-6, 2)$ and $(-1, -5)$
   e) $(-9, 11)$ and $(-1, -2)$
   f) (p, q) and $(2p, 3q)$

8) Find the equation of the straight line which is parallel to the line $2y - x = 5$ and passes through the point $(-1, 4)$
   Give your answer in the form $ax + by = c$ where a, b and c are integers.

9) Find the equation of the straight line which is perpendicular to the line $6y + 2x - 11 = 0$ and passes through the point $(2, 3)$
   Give your answer in the form $ax + by = c$ where a, b and c are integers.

10) Find the angle (in degrees to 1 decimal place) between each of the following pairs of lines:
    a) $y = 0$ and $y = 2x$
    b) $y = 3x + 3$ and $y = 4x - 1$
    c) $y = 5x + 8$ and $y = -x$
    d) $y = -2x - 4$ and $y = -5x + 1$

11) Find the area of the triangle bounded by the coordinate axes and the line $2y + x - 11 = 0$

12) Find the area in terms of k of the triangle bounded by the coordinate axes and the line $ky + 3x = 2$ where k is a positive constant.

13) The three points $(3, 11)$, $(4, -2)$ and $(-2, 1)$ are all corners of a triangle. By considering the gradients of the lines passing through each of the pairs of points, show that it is a right-angled triangle.

1.03 Coordinate Geometry: Linear Graphs (answers on page 410)

14) The three points $(5, 10)$, $(-1, 8)$ and $(4, -7)$ are all corners of a triangle. By considering the distances between each of the pairs of points, show that it is a right-angled triangle and hence find the area of the triangle.

15) Find the value of k such that $2y - 7x = 0$ and $ky + 6x = 10$ are perpendicular to each other.

16) A line perpendicular to $y + 2x = 9$ passes through $(p, 6)$ and $(1, 7)$. Find the value of p

17) Two points $A(p, -2)$ and $B(5, q)$ have a perpendicular bisector with equation $3y + x = 15$. Find the values of p and q.

18) Show that $(-5, 7)$, $(10, 10)$ and $(-15, 5)$ are collinear.

19) Find the value(s) of k such that
    a) $(-15, 9)$, $(3, 3)$ and $(k, -5)$ are collinear.
    b) $(1, k)$, $(-3k, 3)$ and $\left(7, \frac{3}{2}\right)$ are collinear.

20) Three points A, B and C have coordinates $(-3, 10)$, $(5, 2)$ and $(9, 6)$ respectively.

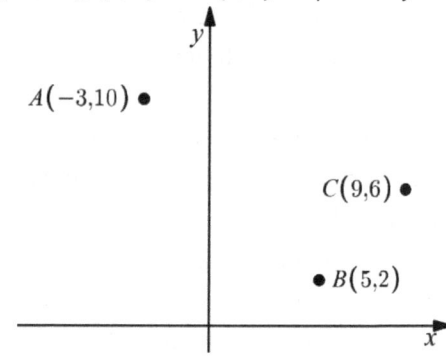

   a) Find the equation of the perpendicular bisector of A and B in the form $y = mx + c$
   b) Find the equation of the perpendicular bisector of B and C in the form $y = mx + c$
   c) Find the point P, where the two perpendicular bisectors intersect.
   d) Find the distance between the point P and each of the three points.

21) By considering the line that is perpendicular to $3y - 4x = 5$ which passes through the point $P(11, 8)$, find the shortest distance between $P(11, 8)$ and the line $3y - 4x = 5$

22) Three points $A(10, 10)$, $B(5, 4)$ and $C(20, 1)$ and $D(15, k)$ form the vertices of a trapezium.

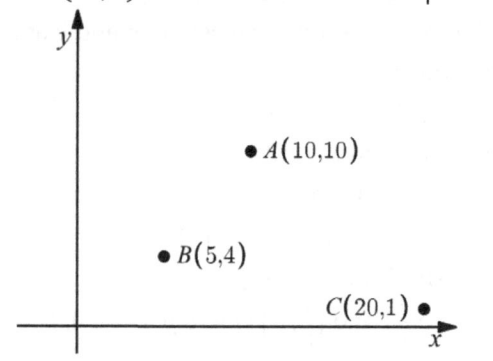

   a) Show that AB is not perpendicular to BC

   Find
   b) the equation of the line passing through B and C in the form $y = mx + c$
   c) the equation of the line passing through A and D in the form $y = mx + c$ given that it is parallel to the line passing through points B and C
   d) the value of k
   e) the distances between A & D, and B & C
   f) the equation of the line which is perpendicular to BC and passes through A
   g) the shortest distance from A to the line BC
   h) the area of the trapezium $ABCD$

23) The points $A(-6, 2)$, $B(1, 3)$, $C(-4, 8)$ and D form a quadrilateral.

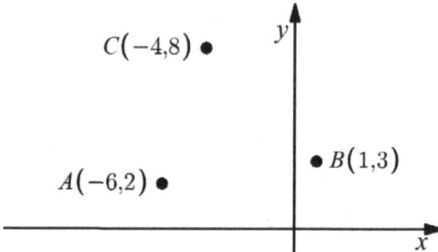

   a) Show that the distances between A and B, and between B and C are the same.

   The line AC is perpendicular to the line BD.
   b) Show that the possible coordinates of the point D satisfy the line $x + 3y = 10$

24) Four points $A(-3, -1)$, $B\left(2, -\frac{3}{5}\right)$, $C(4, 4)$ and $D\left(-1, \frac{18}{5}\right)$ form a quadrilateral. By considering the diagonals only, AC and BD, determine its type.

1.04 Graph Sketching: Linear Graphs (answers on page 411)

1) Find the equation of each of the linear graphs below:

   a)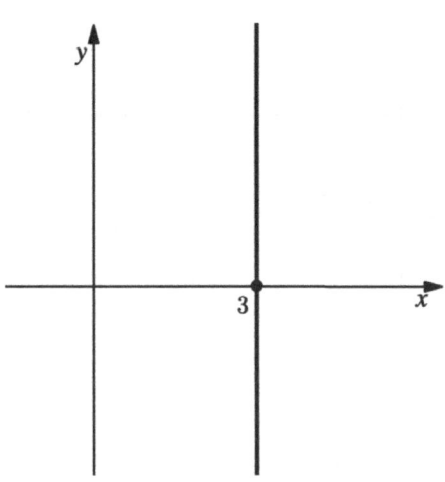

   b)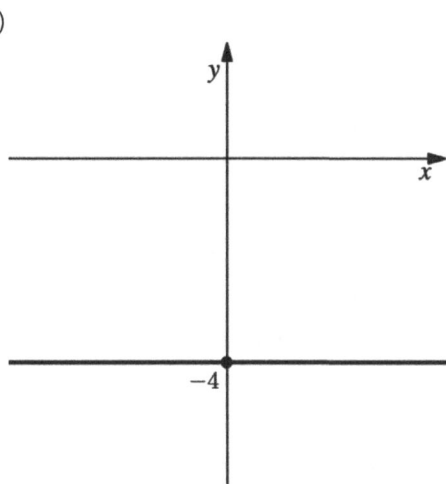

2) Find the equation of each of the following graphs in the form $ax + by = c$, where a, b and c are integers.

   a)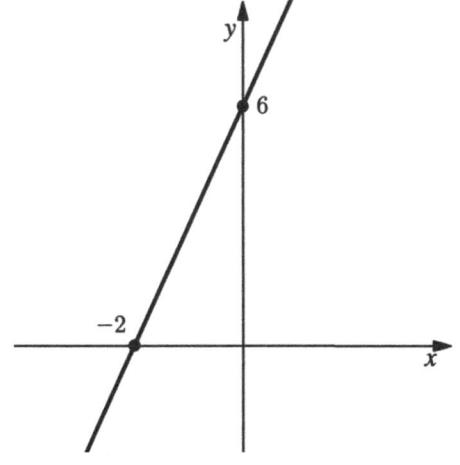

   b)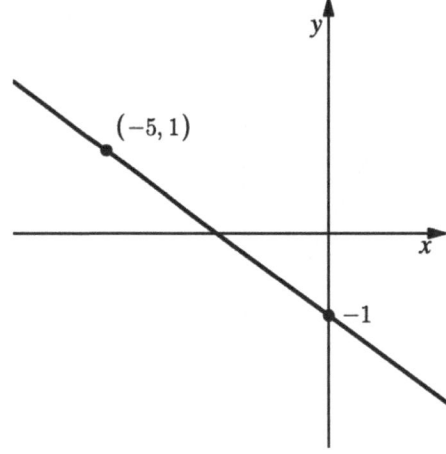

   c)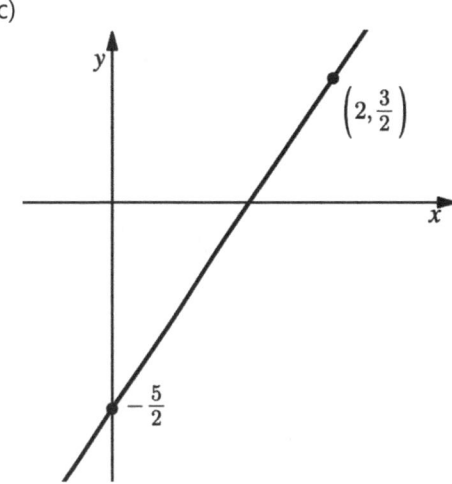

   d)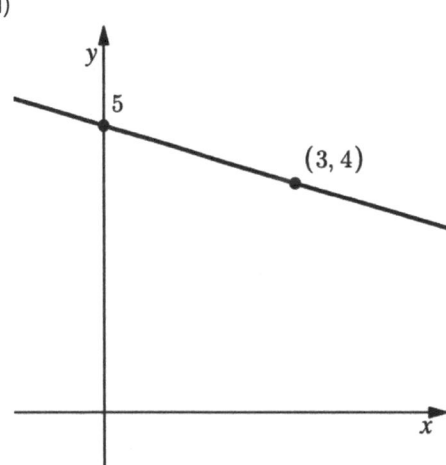

3) Sketch the following linear graphs, identifying where they intersect the coordinate axes:
   a) $y = 2x - 7$
   b) $y = 6 - 10x$
   c) $y = \frac{1}{2}x + 5$
   d) $y = 2 - \frac{1}{3}x$

1.04 Graph Sketching: Linear Graphs (answers on page 411)

4) Find the equation of each of the following graphs in the form $ax + by = c$, where a, b and c are integers.

a)
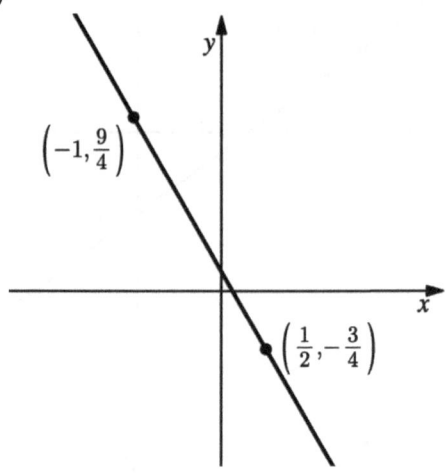

b)
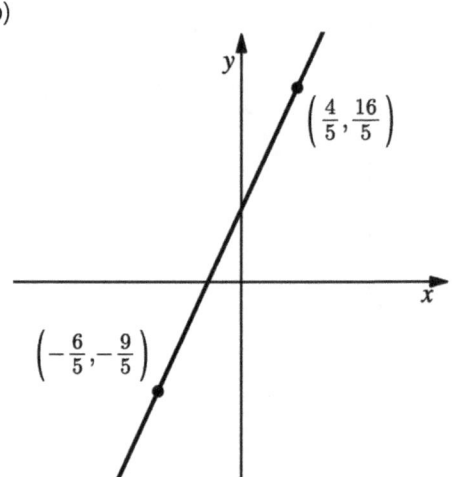

c)
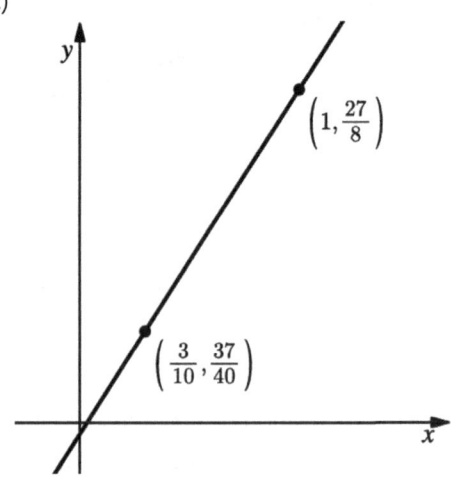

d)
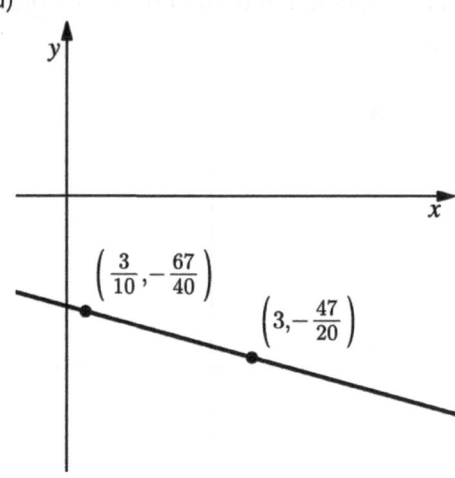

5) Sketch the following linear graphs, identifying where they intersect the coordinate axes:
   a) $2x + 5y = 10$
   b) $4x - 9y = 36$
   c) $-7x + 2y = 28$
   d) $6x + 11y = -2$

6)
   a) Sketch the graphs of $y = 2x + 8$ and $2y = 10 - x$ on the same axes, identifying where the graphs cross the coordinate axes.
   b) Determine where the two lines intersect.
   c) Find the area of the triangle formed by the two lines and the x-axis.
   d) Find the area of the triangle formed by the two lines and the y-axis.

7) Find the equation of each of the following graphs in the form $x + ay = b$, where a and b are integers.

a)
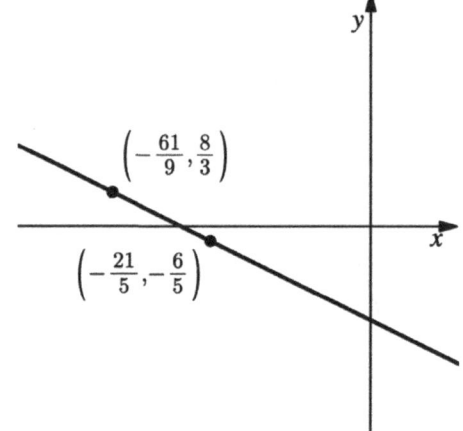

1.04 Graph Sketching: Linear Graphs (answers on page 411)

b)

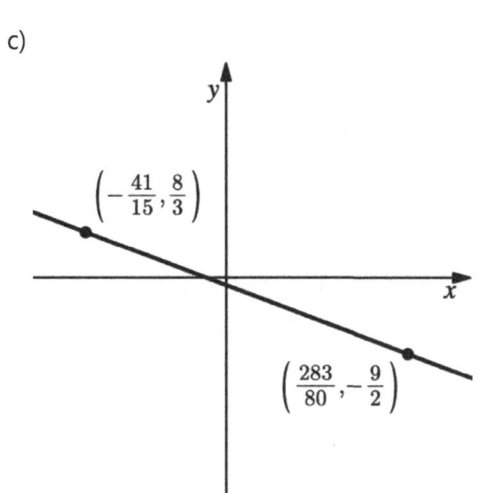

c)

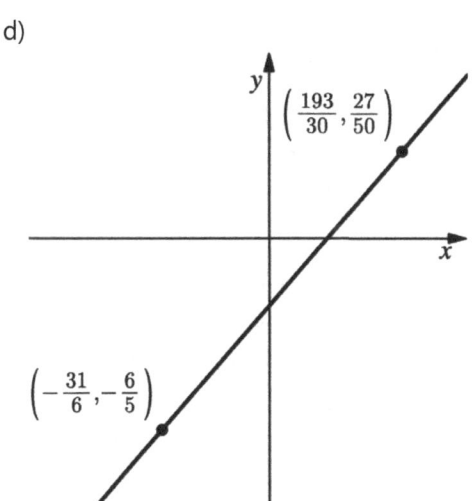

d)

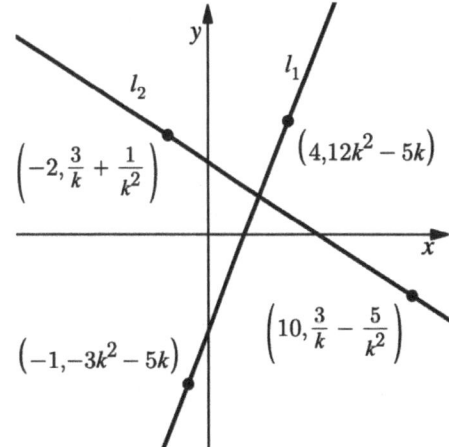

8) Sketch the following linear graphs, identifying where they intersect the coordinate axes:
   a) $2y = 3x - 7$
   b) $5y + 1 = 6x$
   c) $3y - 3 = 8x$
   d) $10y = 4 - 7x$

9) The diagram below shows a linear function. Find the equation of the line in the form $y = ax + b$, where a and b are given in terms of k, where $k > 1$

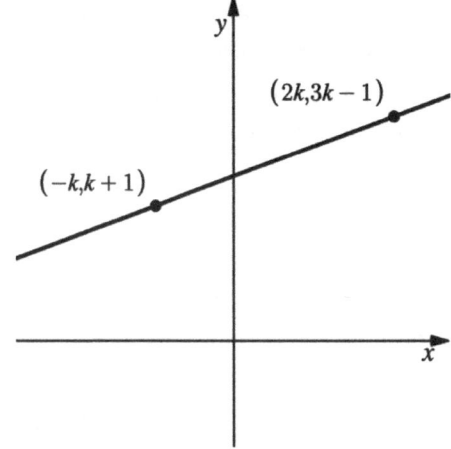

10) Sketch the following linear graphs, identifying where they intersect the coordinate axes. It is given that $k > 0$
    a) $y = kx + 2$
    b) $y = 8 - 2kx$
    c) $ky = 3 - x$
    d) $2x + 7y = k$
    e) $3x + ky = 9$
    f) $6kx - y = -3$
    g) $\frac{x}{k} + \frac{y}{2k} = 4$

11) The diagram below shows two lines, l_1 and l_2.
    a) Find the equation of the line l_1
    b) Find the equation of the line l_2
    c) Determine the coordinates, in terms of k, of where l_1 and l_2 intersect.

1.05 Linear Modelling (answers on page 412)

1) For each of the following situations, write an equation to model the information:
   a) When making pastry, the weight of fats (y) needed is half the weight of the flour (x)
   b) A business adds a tax of 20% to all items it sells (x) before calculating its sales price (y)
   c) For each £1 converted into US dollars (x) you receive \$1.29 (y)

2) Three phone companies charge mobile phone tariffs usually known as 'pay-as-you-go' which charge £C the user per t minutes and have no monthly cost.

Phone Company	Standard Charge (per minute)
H_2O	25p
UU	30p
PiffPaff	24p

   a) On the same axes, sketch a graph representing the cost over time for making calls using each of the companies. Label the point at which £3 has been spent.
   b) In each case write down an equation for C in terms of t

3) The graph below shows a straight-line graph which goes through the point A

   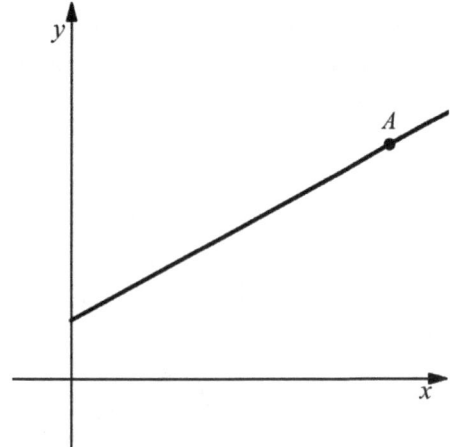

   Find the equation of the line given
   a) Point A is $(1, 9)$ and the y-intercept is 8
   b) Point A is $(5, 4)$ and the y-intercept is 2
   c) Point A is $(3, 7)$ and the y-intercept is 5

4) The graph below shows a straight-line graph going between two points A and B

   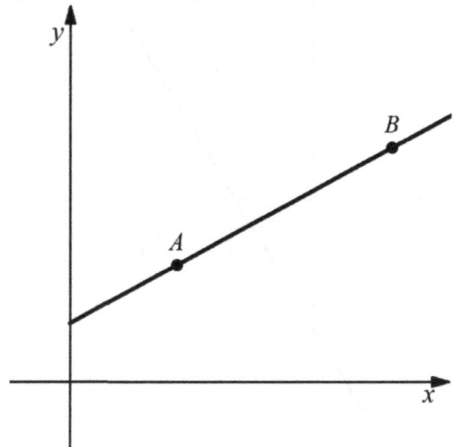

   Find the equation of the line in the form $y = mx + c$ given points A and B are
   a) $(2, 10)$ and $(5, 19)$
   b) $(1, 2.5)$ and $(7, 5.5)$
   c) $(3, 7)$ and $(15, 12)$

5) Data has been collected on the dive of penguins. The depth of their dives d in feet and duration of their dive duration in time t minutes is plotted below.

   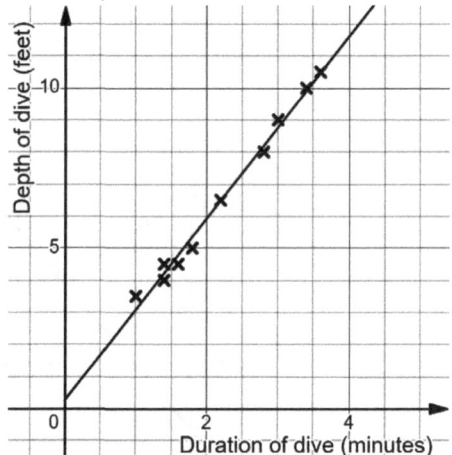

   a) Use the graph to write an equation of d in term of t
   b) Use your equation to predict how deep a penguin will dive in a 2.5 minute dive.
   c) What does the gradient of the graph represent? State the units.
   d) What does the vertical intercept represent in context?
   e) Comment on the validity of the vertical intercept.

1.05 Linear Modelling (answers on page 412)

6) The graph below shows the times of the gold medallists, T seconds, in the men's 100 metre event in the Olympics against the years since 1960, x years. The line is given by the equation
$$T = -0.0082x + 10.2$$

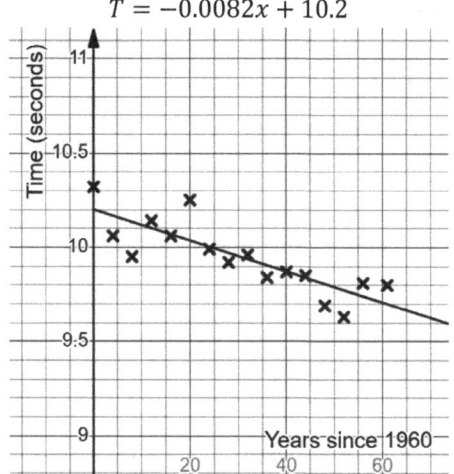

a) According to the model, what time will someone in the 2028 Olympics achieve?
b) Explain what the 10.2 and the -0.0082 represents in context.
c) Explain why a linear model is not valid in the long term.

7) The graph below shows the cost £C of hiring a coach for the number of miles travelled m from two different companies, Stagebus and Zelacoach.

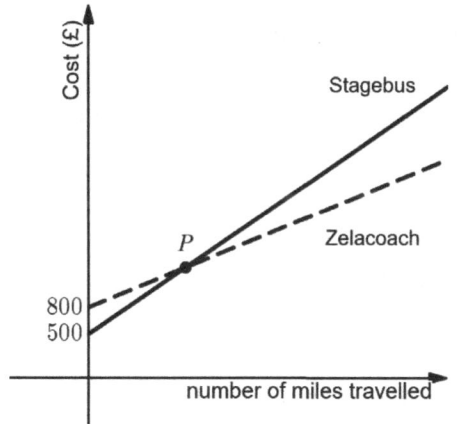

For a distance of 220 miles, Stagebus quotes a cost of £2000 whilst Zelacoach charges £1700
a) Explain what the values on the vertical axis represent in context for each company.
b) Find an equation that models the costs of hiring coaches from the two companies.
c) Identify the coordinates of the point P labelled on the graph and state the relevance of this point in context.

8) The graph below shows the link between the price of an ice cream £P and the number of them sold x in the shop.

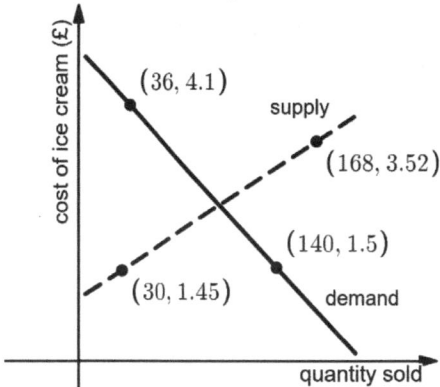

State the point at which the two lines intersect and interpret what that point means in context.

9) Two companies charge different costs for electrical work at their business premises.
Electrosolutions: A call out charge plus £30 an hour after that.
Ohms Electrical: Starting at £90 covering the first 2 hours and then following a linear relationship with an hourly rate.
Rebecca uses both companies for her business and finds that they both charged £170 for 4 hours work.
a) Find the call out charge of Electrosolutions.
b) Find the hourly rate of Ohms Electrical.
c) Ohms Electrical want to change to a call out charge with their current hourly rate. What would be their initial call out charge?

10) A bakery makes muffins. They sell the muffins at £2.00 each.
There is an upfront cost in making the muffins, plus a fixed cost per muffin.
When 500 muffins are made and sold, a profit of £300 is made, but when 100 muffins are made and sold, there is a loss of £50.
a) Show that the cost £C in terms of the number of muffins x, may be modelled as
$$C = 1.125x + 137.5$$
b) Find the minimum number of muffins that need to be made and sold for the bakery to not make a loss.

1.06 Quadratics: Introduction (answers on page 412)

Algebraic Manipulation and Solving Equations

1) Fully factorise the following:
   a) $x^2 + 14x + 45$
   b) $x^2 + 3x - 18$
   c) $x^2 - 7x$
   d) $2x^2 + 11x - 6$
   e) $2x^2 - 17x + 35$
   f) $8x^2 - 14x + 3$
   g) $6x^2 + 17x + 5$
   h) $-2x^2 - x + 3$
   i) $x^2 - 9$
   j) $4x^2 - 5$

2) Fully factorise the following where a and b are constants:
   a) $x^2 - a^2$
   b) $(ax)^2 - b^2$
   c) $x^2 - a$

3) Fully factorise the following:
   a) $8(x + 2) - x(x + 2)$
   b) $-(x - 2) + 2x(x - 2)$
   c) $12(x - 5)^2 + (x - 5)$
   d) $(x - 7)^2 - 4$
   e) $\frac{2}{3}(3x - 10)^2 - \frac{32}{3}$

4) Solve the following:
   a) $x^2 = x$
   b) $x^2 - 4 = 0$
   c) $x^2 - 8 = 0$
   d) $x^2 + 2x = 0$
   e) $(x - 5)^2 = 0$
   f) $x^2 - 12x + 36 = 0$
   g) $\frac{1}{5}x^2 - \frac{1}{5}x = 4$
   h) $(x + 3)^2 = 6$
   i) $(x - 1)(x - 3) = 8 - 2x$

5) Solve the following, giving your answers in exact form:
   a) $x^2 - 8x + 5 = 0$
   b) $2x^2 - 40 = 0$
   c) $2x^2 + 16x + 24 = 0$
   d) $3x^2 - 5x - 1 = 0$
   e) $ax^2 - b = 0, a \neq 0$
   f) $ax^2 - bx = 0, a \neq 0$

6) Two students are trying to find the solutions to
   $(x - 5)^2 = (x - 5)$
   In each case, explain the mistake they made and then show your workings for correctly solving it.

   Student 1:
   When $x = 5$, $(5 - 5)^2 = (5 - 5)$
   so $0 = 0$ which means $x = 5$

   Student 2:
   Divide both sides by $x - 5$: $x - 5 = 1$
   so $x = 6$

7) Solve $x - 3 = 2\sqrt{x}$

8) Solve $2x + 2 = 5\sqrt{x}$

9) By using the substitution $p = x^2$ find the solutions to $2x^4 - 11x^2 + 5 = 0$

10) By using the substitution $p = x^3$ find the solutions to
    $$3x^3 + 19 = \frac{40}{x^3}$$

11) By using the substitution $p = 2x^2 - x$, find the solutions to $12x^4 - 12x^3 + 5x^2 - x - 2 = 0$

Completing the Square

12) Complete the square on each of the following:
    a) $y = x^2 - 8x$
    b) $y = x^2 + 6x$
    c) $y = x^2 - 10x + 40$
    d) $y = x^2 + 8x - 20$
    e) $y = x^2 + 5x + 7$
    f) $y = 2x^2 - 16x - 20$
    g) $y = 4x^2 - 24x + 25$
    h) $y = 3x^2 - 7x + 1$

13) Complete the square on each of the following:
    a) $y = -x(x - 2)$
    b) $y = (x + 3)(x + 2)$
    c) $y = -2x^2 - 4x + 5$
    d) $y = -2x^2 - 7x + 1$
    e) $y = -3x^2 + 5x + 10$
    f) $y = 25 - 4x^2$
    g) $y = x^2 + px + r$

1.06 Quadratics: Introduction (answers on page 412)

14) Explain the graphical relevance of a and b in the curve $y = (x - a)^2 + b$

15) For each of the following curves, find the coordinates of the vertex and state the type of vertex:
    a) $y = (x - 2)^2 - 12$
    b) $y = (x + 2)^2 - 8$
    c) $y = -(x - 1)^2 + 12$
    d) $y = -(x + 2)^2$
    e) $y = 3(x + 4)^2 + 5$
    f) $y = 2\left(x + \frac{5}{2}\right)^2 + \frac{1}{2}$
    g) $y = -10(x - 4)^2 - 100$
    h) $y = x^2 - 6x - 2$
    i) $y = 3x^2 + 6x + 5$
    j) $y = -2x^2 - 4x + 3$
    k) $y = 2x^2 - 5$
    l) $y = -5x^2 + 12$
    m) $y = a(x - b)^2, a < 0$
    n) $y = x^2 + px + r$

16) State the line of symmetry in the following curves:
    a) $y = (x - 4)^2 + 5$
    b) $y = 2(x + 3)^2 + 7$
    c) $y = -3(x - 1)^2 - 6$
    d) $y = (x - 2)(x + 4)$
    e) $y = 2(x - 1)(x - 6)$
    f) $y = 5(x + 2)(x - 8)$
    g) $y = (x - a)(x - b)$
    h) $y = (x - p)^2 + q$

17) In each case, find the equation of the curve in the form $y = (x - a)(x - b)$ and state the coordinates of the vertex:
    a)
    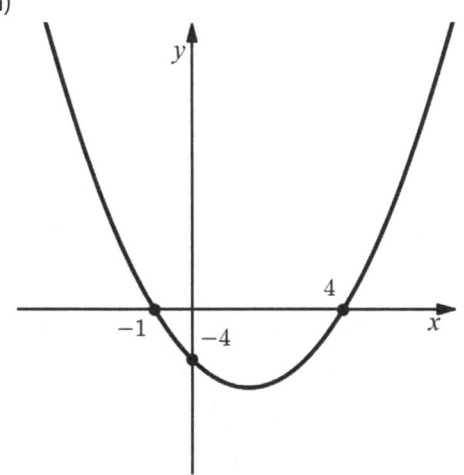

b)
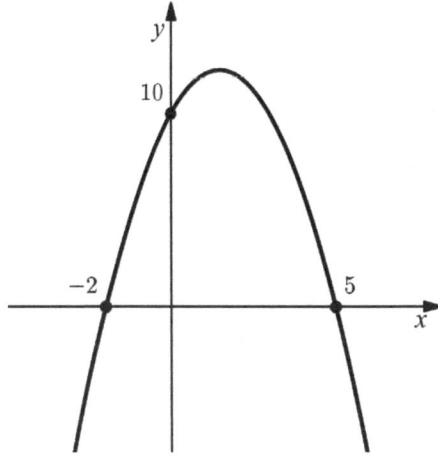

18) For each of the following functions, find the equation of the graph in the form
$$y = a(x - b)^2 + c$$
where a, b and c are values to be found:

a)
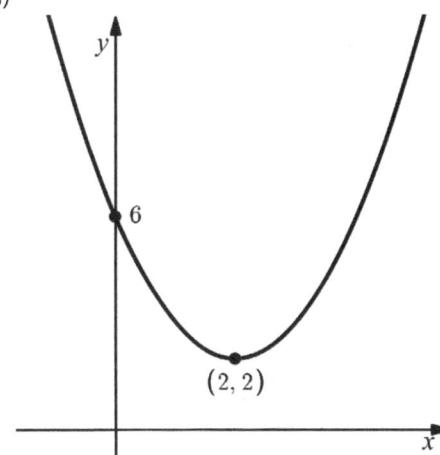

b)
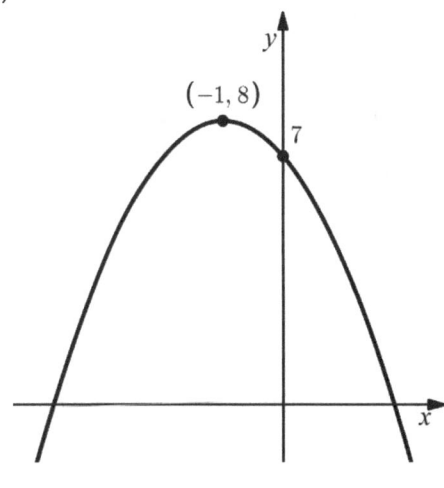

1.06 Quadratics: Introduction (answers on page 412)

c)

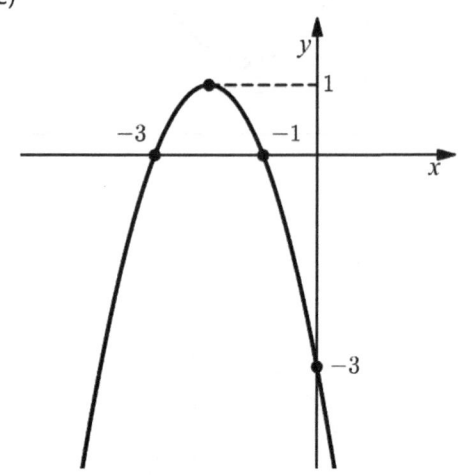

d)

e)

19) Find the equation of the curve in the form
$$y = ax^2 + bx + c$$
for the curves described below:
 a) Minimum point at $(1, 5)$, intersecting the y-axis at $y = 6$
 b) Minimum point at $(3, -1)$, intersecting the y-axis at $y = 17$
 c) Maximum point at $(-2, 4)$ intersecting the y-axis at $y = -76$

Simultaneous Equations

We suggest that Algebra: Simultaneous Equations is helpful with Q20-Q23

20) Find the solutions to following pairs of simultaneous equations:
 a) $y = 2x + 3$, $y = x^2 + x + 1$
 b) $y = -2x + 3$, $y = x^2 - 2x + 2$
 c) $y = 2 - x$, $y = 6 - x^2 - x$
 d) $y = -2x^2 + x + 4$, $y = -x^2 + x + 3$
 e) $y = x^2 - 2x$, $y = 2x^2 + x - 4$

21) Find the point(s) of intersection of the following curves and straight lines:
 a) $y = x^2 + 1$, $y = \frac{1}{2}x + 1$
 b) $y = -x^2 + 5$, $y = -x + 3$
 c) $y = x^2 - 4x + 5$, $y = 2x - 4$
 d) $y = 2x^2 - 4x + 3$, $y = 4x - 5$

22) Find the points of intersection of the quadratic curve and the straight line below:

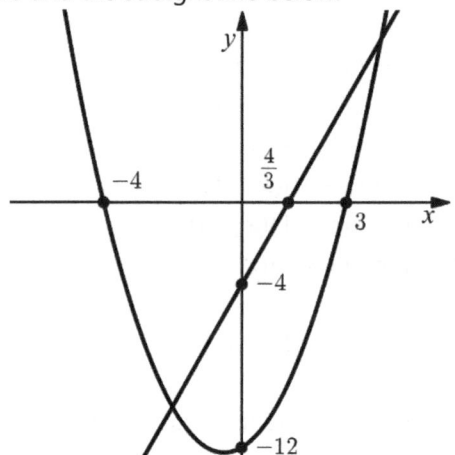

23) Find the points of intersection of the quadratic curves shown in the graph below:

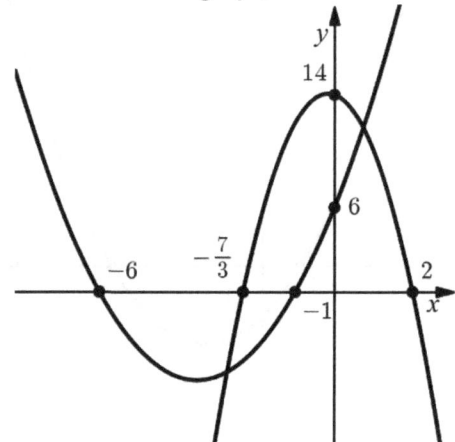

1.07 Algebra: Simultaneous Equations (answers on page 414)

1) Solve the following sets of simultaneous equations:
   a) $2x + 3y = 5$ and $7x - y = 6$
   b) $6x = 5 - 2y$ and $x - 2y = 2$
   c) $\frac{x-9}{3} + y = 0$ and $2(x + y) = 2$
   d) $3x - y = 4$ and $\frac{y}{2} + \frac{x}{4} = 5$

2) It is given that $y - kx = 10$ and $y = 2x - 10$, where k is an unknown constant, intersect when $y = 15$. Find
   a) the x-coordinate of the point of intersection,
   b) the value of k

3) Two lines are given to be
   $$y + ax = 5 \text{ and } \frac{y}{b} + \frac{11x}{80} = 1$$
   where a and b are unknown constants
   The two lines never intersect, and one line passes through the y-axis at $y = 8$
   a) the value of b
   b) the value of a

4) The diagram below shows two lines, which intersect at the point $(12,5)$

   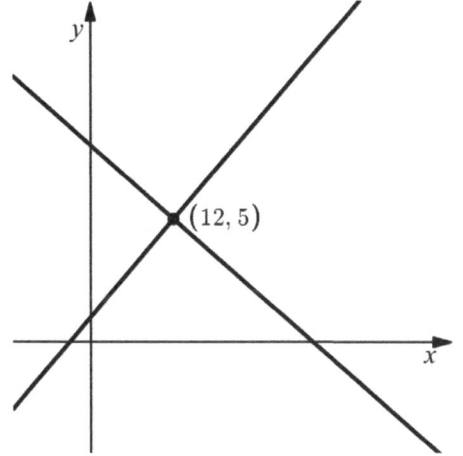

   Given one line has gradient $-\frac{1}{4}$ and the other line intersects the y-axis at $y = 1$, find the equations of both lines in the form $ay + bx = c$ where a, b and c are integers.

5) Solve the following sets of simultaneous equations:
   a) $y = 2x + 4$ and $y = x^2 + 1$
   b) $y = 3x - 2$ and $y = x^2 - 4x - 10$
   c) $y = \frac{1}{2}x$ and $y = x^2 + 2xy$

6) Two equations are given by
   $3y - x + 10 = 0$ and $x^2 + y^2 = 10$
   a) Show that the two equations may be satisfied by
   $$x^2 - 2x + 1 = 0$$
   b) Hence solve the simultaneous equations
   $3y - x + 10 = 0$ and $x^2 + y^2 = 10$

7) Solve the following sets of simultaneous equations:
   a) $y + 3x = 15$ and $y^2 = 9(x - 3)$
   b) $y - x = 4$ and $x^2 + y = 10$
   c) $y - 2x = 12$ and $x^2 + 2xy = 5$

8) Solve the following sets of simultaneous equations:
   a) $y + x = 4$ and $x^2 + y^2 = 16$
   b) $y + 4x = 6$ and $x^2 + 2x + y^2 - 3y = 1$
   c) $2y + x = 8$ and $x^2 + y^2 - xy = 12$
   d) $x = 2y + 1$ and $y^2 + 8 = 3xy$

9) Two equations are given by
   $x^2 + 2x + y^2 = 24$ and $2y = x - 12$
   Show that there are no solutions when solving simultaneously.

10) The curve
    $$y = px^2 - qx + 1$$
    where p and q are positive constants, passes through the points $(5, 36)$ and $(-2, 15)$
    Find the values of p and q

11) The curve
    $$y = px^3 - 5x^2 + x + q$$
    where p and q are positive constants, passes through the points $(1, 6)$ and $\left(-\frac{1}{2}, \frac{21}{4}\right)$
    Find the values of p and q

12) Find the two positive real numbers with a product of 20 and a difference of $\frac{8}{3}$

13) Solve the simultaneous equations
    $2^{x-1} = 4^{2y}$ and $9^{x-2} = 3^{y+1}$

14) Solve the simultaneous equations
    $5^{2x-3} = 25^{y-1}$ and $16^{x-1} = 2^{3y}$

1.08 Graph Sketching: Quadratic Graphs (answers on page 414)

1) Find the equation of each of the following graphs in the form $y = (x + a)(x + b)$

a)
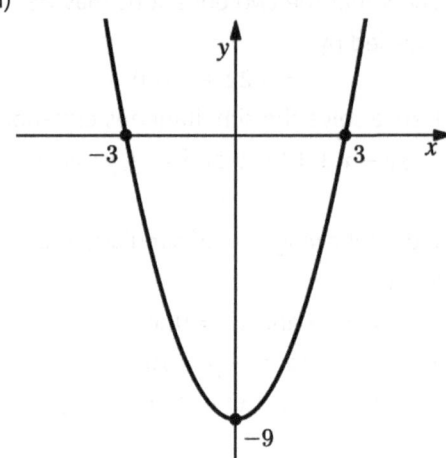

b)
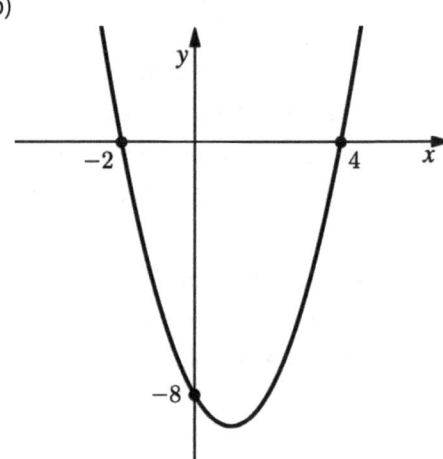

c)
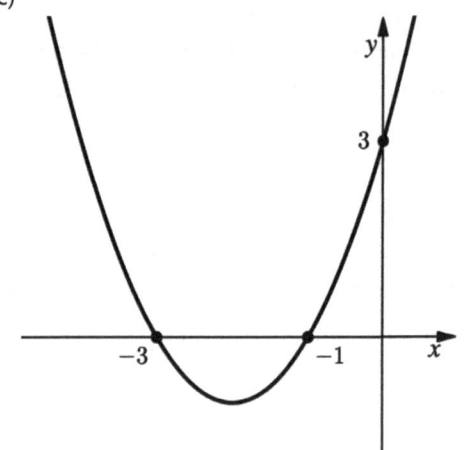

2) Sketch the following graphs, showing where they intersect the coordinate axes:
   a) $y = (x - 5)(x - 7)$
   b) $y = (x + 4)(x - 2)$

3) Find the equation of each of the following graphs in the form $y = -(x + a)(x + b)$

a)
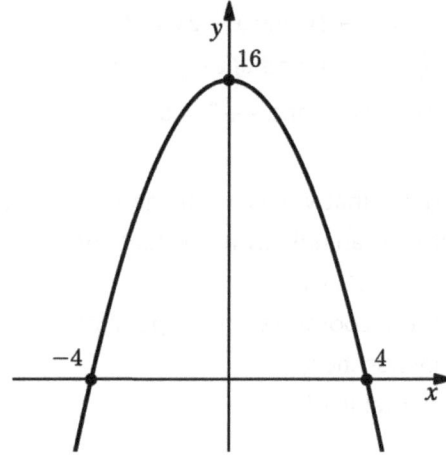

b)
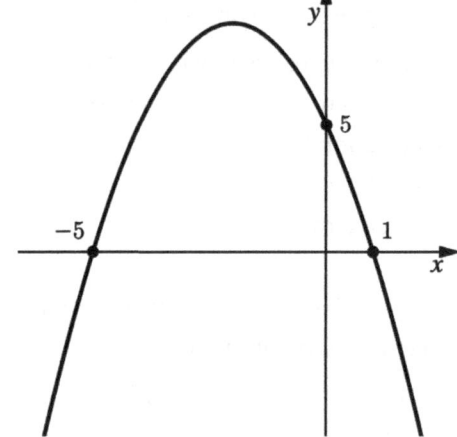

c)
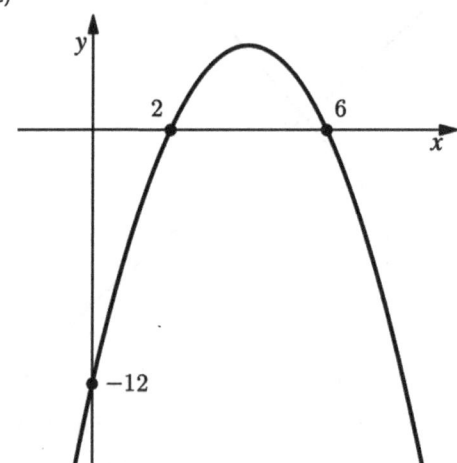

4) Sketch the following graphs, showing where they intersect the coordinate axes:
   a) $y = -(x + 2)(x - 5)$
   b) $y = -(x + 8)(x + 1)$

1.08 Graph Sketching: Quadratic Graphs (answers on page 414)

5) Find the equation of each of the following graphs in the form $y = (x + a)^2 + b$

a)
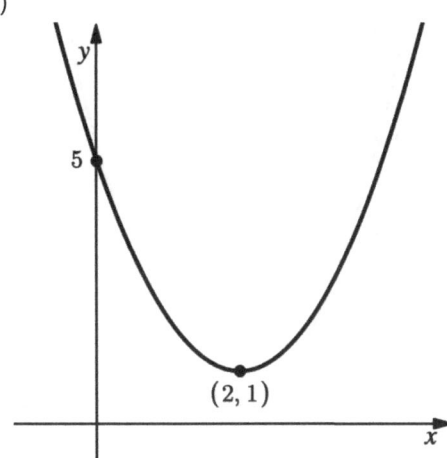

b)
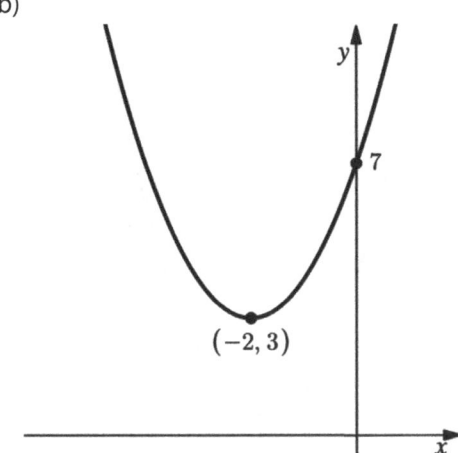

c)
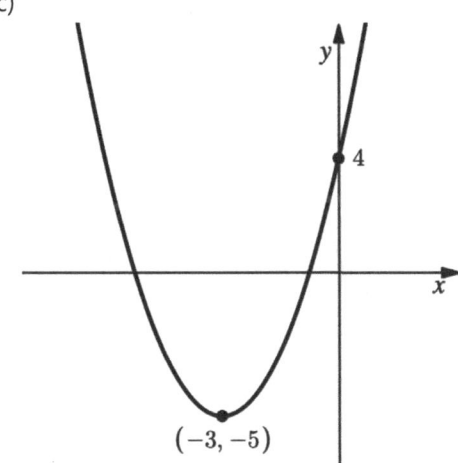

6) Sketch the following graphs, showing the coordinates of the turning point and where they intersect the y-axis:
   a) $y = (x - 4)^2 + 5$
   b) $y = (x - 5)^2 - 6$

7) Find the equation of each of the following graphs in the form $y = -(x + a)^2 + b$

a)
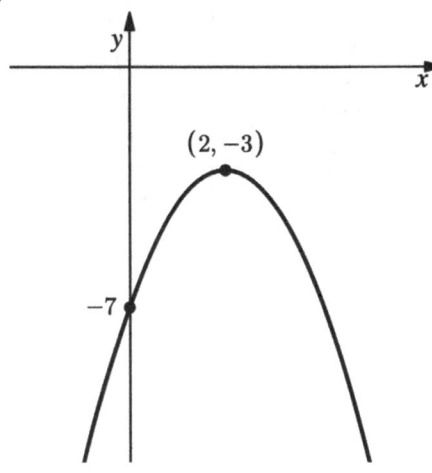

b)
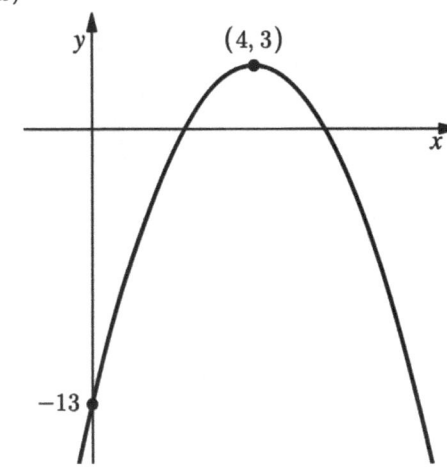

c)
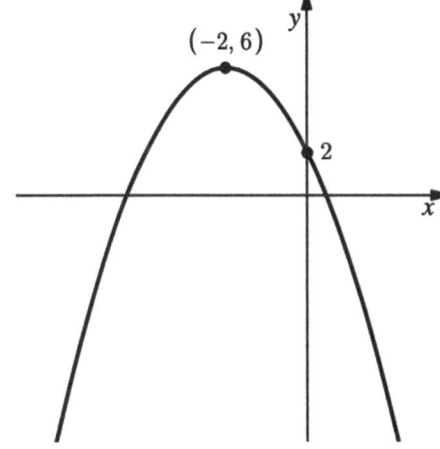

8) Sketch the following graphs, showing the coordinates of the turning point and where they intersect the y-axis:
   a) $y = -(x + 6)^2 + 2$
   b) $y = -(x - 3)^2 - 4$

1.08 Graph Sketching: Quadratic Graphs (answers on page 414)

9) Find the equation of each of the following graphs in the form $y = a(x + b)(x + c)$
   a)
   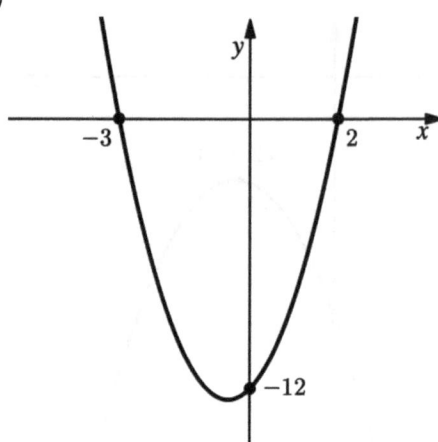

   b)
   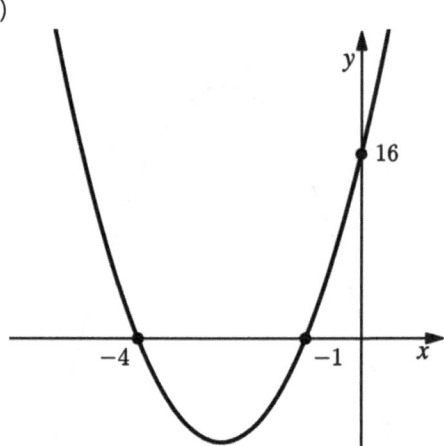

   c)
   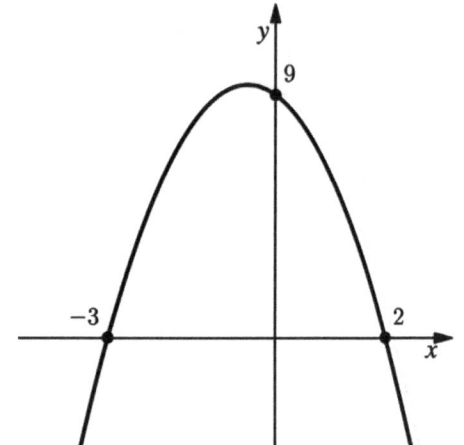

10) Sketch the following graphs, showing where they intersect the coordinate axes:
    a) $y = 3(x - 1)(x + 2)$
    b) $y = -2(x + 4)(x + 3)$

11) Find the equation of each of the following graphs in the form $y = a(x + b)^2 + c$
    a)
    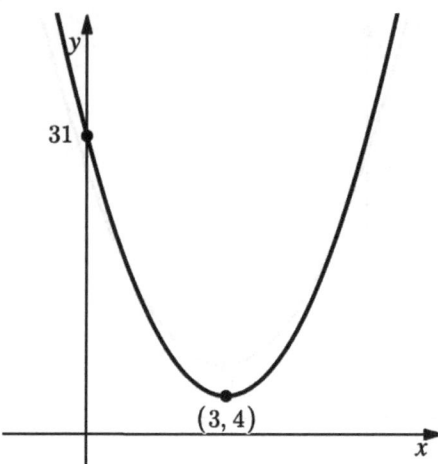

    b)
    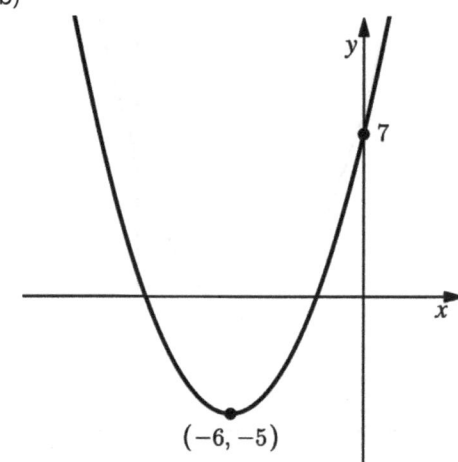

    c)
    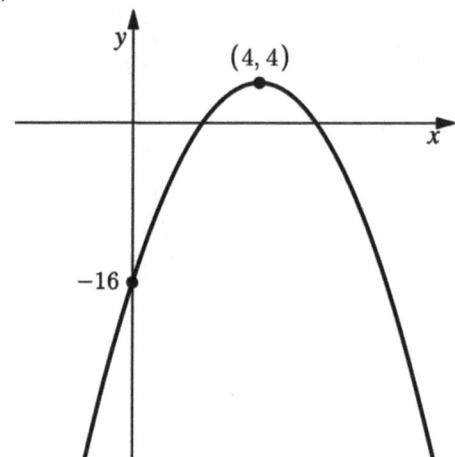

12) Sketch the following graphs, showing the coordinates of the turning point and where they intersect the y-axis:
    a) $y = 5(x + 1)^2 - 2$
    b) $y = -\frac{3}{2}(x - 6)^2 - 12$

1.08 Graph Sketching: Quadratic Graphs (answers on page 414)

13) Find the equation of each of the following graphs in the form $y = ax^2 + bx + c$

  a)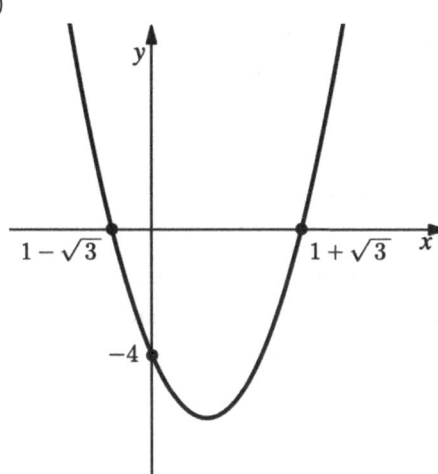

  b)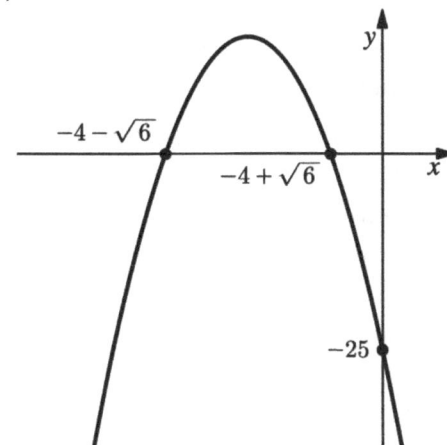

  c)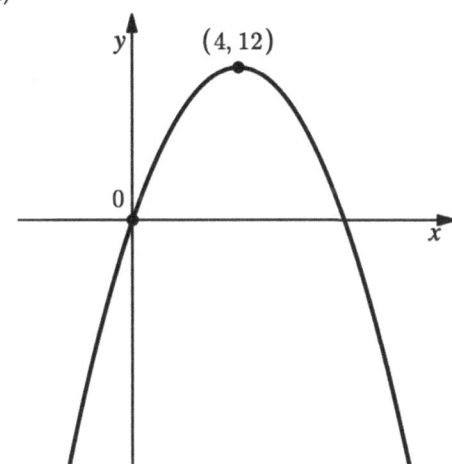

14) Sketch the following graph, showing the coordinates of the turning and where it intersects the y-axis:

$$y = \frac{1}{2}(x+8)^2 - 32$$

15) Sketch the following graphs, showing the coordinates of the turning point and where they intersect the coordinate axes:

  a) $y = (x+2)(x-8)$
  b) $y = (x-6)(x+4)$
  c) $y = 3(x+6)(x+2)$
  d) $y = -4(x-2)(x-4)$

16) Sketch the following graphs, showing the coordinates of the turning point and where they intersect the coordinate axes:

  a) $y = (x-3)(x-7)$
  b) $y = \frac{1}{3}(x+2)(x+6)$
  c) $y = -(x+9)(x+5)$
  d) $y = -\frac{7}{4}(x-8)(x+4)$

17) Sketch the following graphs, showing the coordinates of the turning point and where they intersect the coordinate axes:

  a) $y = (x-6)^2 - 2$
  b) $y = 8 - (x+2)^2$
  c) $y = \frac{4}{3}(x+6)^2 - 9$
  d) $y = -\frac{8}{3}(x-3)^2 + 12$

18) Sketch the following graphs, showing the coordinates of the turning point and where they intersect the coordinate axes:

  a) $y = 3x^2 - 9x + 2$
  b) $y = -2x^2 + 10x + 2$
  c) $y = \frac{1}{2}x^2 - 6x + 9$
  d) $y = -\frac{1}{5}x^2 + 2x + 10$

19) Given that $k > 0$, sketch the following graphs, identifying the turning point and where they cross the coordinate axes in terms of k:

  a) $y = k(2x+k)(x-2k)$
  b) $y = k - \frac{1}{3}(x-3k)^2$
  c) $y = 10 + \frac{2}{k}(kx+1)^2$
  d) $y = (3kx+2)(3k-x)$

20) Given that $k > 0$, sketch the graphs of $y = (k-x)(3x+k)$ and $y = k - x$ on the same axes, showing clearly where the two graphs cross the coordinate axes and where they intersect in terms of k

1.09 Quadratics: Introduction (answers on page 416)

1) A rectangular garden is 30m by 40m. The path surrounding a rectangular pond has a consistent width of x and total area of 864m². Find the width and length of the pond.

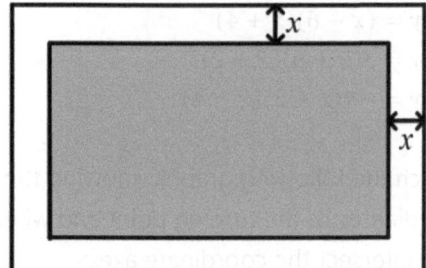

2) A room is comprised of two areas; one square with length x m, and one rectangular with width $(x + 2)$ m and length 10 m. It has a total area of 44 m². Find the value of x

3) For the equations below, without manipulating the equation, state the number of real roots:
   a) $y = (x + 1)^2 + 2$
   b) $y = -(x + 1)^2 + 2$
   c) $y = (x - 4)^2 - 3$
   d) $y = -(x - 4)^2 - 3$
   e) $y = (x + 2)^2$

4)
   a) Sketch the graph of
   $$y = \frac{3}{4}x(x - 1)$$
   for $0 \leq x \leq 4$
   b) Find the minimum of the curve.

5)
   a) Sketch the graph of
   $$y = \frac{22}{25}x(200 - x)$$
   for $0 \leq x \leq 200$
   b) Find the maximum of the curve.

6) Sketch the graph of
$$y = a(x - b)^2 + c$$
for constants a, b and c in each of the cases below. Label where the curve intersects with the axes.
   a) $a > 0, b > 0, c > 0$
   b) $a > 0, b < 0, c > 0$
   c) $a < 0, b > 0, c < 0$
   d) $a < 0, b < 0, c < 0$

7) For the graph of
$$y = a(x - b)^2 + c$$
where a, b and c are constants, state the conditions for the graph to have a minimum in the top left quadrant and the y-intercept to be positive.

8) For the graph of
$$y = a(x - b)^2 + c$$
where a, b and c are constants, state the conditions for the graph to have a maximum in the top right quadrant and the y-intercept to be negative.

9) The graph of
$$y = x^2 + 3x + k$$
where k is a constant has a minimum with coordinates $\left(-\frac{3}{2}, -\frac{1}{2}\right)$
Find the coordinates of where the curve intersects
   a) the y-axis.
   b) the x-axis.

10) The graph of
$$y = -x^2 + 4kx + k$$
where k is a constant has a maximum with coordinates $(2, 5)$
Find the coordinates of the points where the curve intersects
   c) the y-axis.
   d) the x-axis.

11) By completing the square, show that the solutions to $ax^2 + bx + c = 0$ are given by
$$x = \frac{-b \pm \sqrt{b^2 - 4ac}}{2a}$$

12) A curve has equation
$$y = 2x^2 + x + k$$
where $k \neq 0$.
Given that the roots of the equation have a difference of 2.5, find the value of k

13) A curve has equation
$$y = -5x^2 - kx + 3$$
where $k \neq 0$.
Given that the roots of the equation have a difference of 1.6, find the possible values of k

1.10 Quadratics: The Discriminant (answers on page 417)

1) Calculate the discriminant:
   a) $x^2 - 2x - 6$
   b) $x^2 - 9$
   c) $-x^2 + x + 5$
   d) $-0.5x^2 + 12$
   e) $2x^2$
   f) $2x^2 - 3x + 2$
   g) $-3x^2 + 3x - 1$
   h) $x^2 - kx + 2$

2) State the number of roots for each of the following functions:
   a) $f(x) = x^2 - 4x + 5$
   b) $f(x) = x^2 + x - 6$
   c) $f(x) = 2x^2 - 2x - 4$
   d) $f(x) = x^2 - 3x + 8$
   e) $f(x) = x^2 + 8x + 16$
   f) $f(x) = -2x^2 + 5x + 3$
   g) $f(x) = -4x^2 + 12x - 9$

3) Show that $3x = 3x^2 + 1$ has no real solutions.

4) Find the values of k for each of the following equations with equal roots:
   a) $4x^2 + kx + 9 = 0$
   b) $3x^2 + kx + 12 = 0$
   c) $x^2 - 3x + 2 + \frac{1}{k} = 0$
   d) $x^2 + (2k + 10)x + k^2 + 5 = 0$
   e) $16x^2 + 4x + k^2 = 0$

5) The equation
$$x^2 + 3px + p = 0$$
where p is a non-zero, has equal roots.
Find the value of p

6) Show that the discriminant of
$$(x - p)(x - q)$$
is $(p - q)^2$

7) Show that the discriminant of
$$k(x - p)^2 + q$$
is $-4kq$

8) The curve
$$y = -\frac{1}{2}x^2 + kx + k$$
touches the line $2y - 2x = 5$ at one point.
Find the possible value(s) of k

9) The line $5y + x = k$ touches the curve $2x^2 + 2y = 1$ once.
   a) Find the value of k
   b) Hence find the values at which $5y + x = k$ intersects the x-axis and the y-axis.

10) A function
$$f(x) = ax^2 + bx + c$$
has constants a, b and c such that $a < 0$, $b > 0$ and $c < 0$. The discriminant of $f(x)$ is 0
Sketch the graph of $f(x)$. Label the intersections with the axes.

11) Show that
$$y = x^2 + kx - x$$
for $k > 0$, always has two real roots.

12) For the equations
$$2x + y = 1$$
and
$$x^2 - 4ky + 5k = 0$$
where k is non-zero constant,
   a) Show that $x^2 + 8kx + k = 0$
   b) Given that $x^2 + 8kx + k = 0$ has equal roots, find the value of k
   c) Hence for this value of k find the solution of the simultaneous equations.

13) Find the relationship between non-zero constants p and q, such that the equation
$$y = px^2 + (p + q)x + q$$
has equal roots.

14) The curves
$$f(x) = x^2 - kx - 4$$
and
$$g(x) = -x^2 + x + 4k$$
intersect once.
   a) Find the possible value(s) of k
   b) Hence find the coordinates of the point(s) of intersection of the two curves.

15) The equation
$$x^4 - 6x^2 + k = 0$$
has two pairs of equal roots.
Find the value of k

1.11 Quadratics: Modelling (answers on page 417)

1) A ball is thrown vertically upwards with a speed of 19.6ms⁻¹.
   The height of the ball, h metres, may be modelled by the equation
   $$h = 19.6t - 4.9t^2$$
   where t is the time given in seconds.
   a) Find the height of the ball after 2 seconds.
   b) Find the times at which the ball is at a height of 14.7 metres above the ground.
   c) Explain what your two solutions in the previous part represent.

2) A ball is hit into the air, the graph below shows its vertical height above the ground before it hits the ground.

   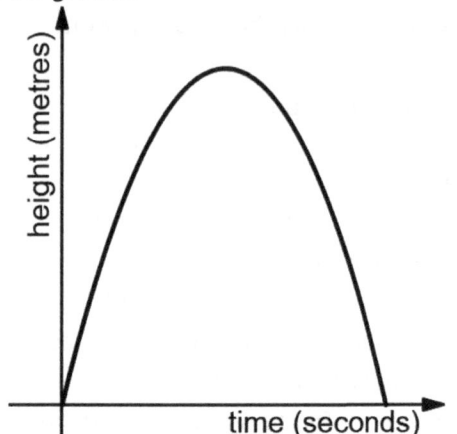

   A model for the height of the ball, h metres at time t seconds is given by
   $$h(t) = 10t - 2t^2$$
   Find
   a) the time when the ball hits the ground.
   b) $h(t)$ in the form
   $$k(t - p)^2 + q$$
   where p, q and k are constants to be found.
   c) the maximum height the ball reaches and the time at which it reaches the maximum height.

   An improved model starts the ball at $h = 1$
   d) State the new model in the form
   $$h(t) = at^2 + bt + c$$
   e) State the coordinate of the maximum height of the curve of this new model.

3) Two identical fence posts are 2.5 metres apart and a rope is hung between them.
   The curve
   $$y = \frac{1}{100}x^2 - \frac{1}{40}x + \frac{1}{2}$$
   may be used to model the rope, where y is the vertical distance in metres above the ground and x is the horizontal distance in metres from the left post.
   The curve is shown in the diagram below:

   a) State the height of the posts.
   b) Rearrange the model into the form
   $$a(x - b)^2 + c$$
   where a, b and c are constants to be found.
   c) State the horizontal and vertical position of the minimum height of the rope.
   d) State why the model is not effective for $x > 2.5$

4) The path of a swimmer may be modelled by
   $$d = 0.1x^2 - x + 0.9$$
   for $x \geq 0$ where d is the depth of the swimmer in relation to the water surface and x metres is the distance from the poolside.
   a) Sketch the graph of the model.

   According to the model:
   b) Find the depth that the swimmer swims below the surface of the water.
   c) Determine the horizontal distance the swimmer swam under the water.
   d) State how far from the pool edge the swimmer is when they first contact the water.
   e) From what value of x does the function become an unrealistic model for the motion of the swimmer?

1.11 Quadratics: Modelling (answers on page 417)

5) The income £I made by a company on various cleaning services they provide may be modelled by the equation
$$I = \frac{5}{2}x(40 - x)$$
where £x is the price charged for each individual service.
   a) Sketch the model for $0 \leq x \leq 40$
   b) Rearrange the model into the form
   $$I = p - \frac{5}{2}(x - q)^2$$
   where p and q are constants to be found.

   According to the model:
   c) Find the maximum income that is made.
   d) Find the price of the service that gives the maximum income.
   e) Find the prices which give an income of £600

6) The profit of a business selling books depends on the price at which each book is sold. The equation
$$y = \frac{15}{2}x - \frac{25}{2} - \frac{5}{8}x^2$$
is used to model the relationship between the profit and the price where £x is the selling price of each book and y pence is the profit achieved at that price.
A sketch of the curve is shown in the diagram below:

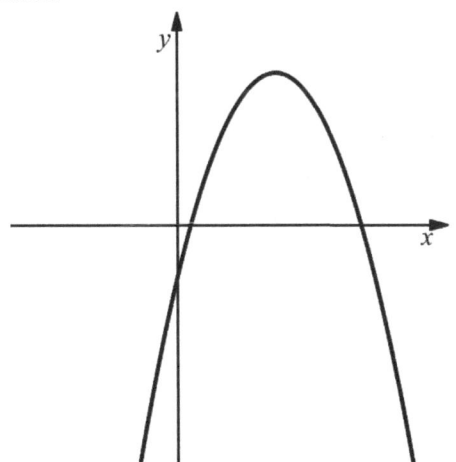

Find
   a) the selling price that generates the maximum profit.
   b) the value of the maximum profit.
   c) the selling price that gives no profit.
   d) the loss if the books are given away.

7) The growth of two flowers may be modelled by quadratic curves as follow:
Flower A: height y cm after x days is given by
$$y = -\frac{2}{45}x^2 + \frac{8}{3}x$$
Flower B: height y cm after x days is given by
$$y = -\frac{1}{10}x^2 + 3x + \frac{15}{2}$$
   a) Calculate the number of days which have passed when they are the same height.
   b) Find the maximum height of the curve representing flower B.
   c) Explain why the growth of the flowers may not best be modelled by these models.

8) A ball is projected from the ground. It reaches a maximum height of 10 metres at 2 seconds since being projected.
Find a model for the trajectory of the ball in the form
$$h = at^2 + bt$$
where h is height (metres), t is time (seconds) and a and b are constants to be found.

9) A stone is projected from the ground. It starts at a horizontal position of a metres from a wall. It is modelled by
$$y = -\frac{1}{8}x^2 + \frac{11}{8}x - \frac{57}{32}$$
where x and y are the horizontal and vertical distance in metres respectively from the bottom of the wall.
   a) Find the value of a
   b) Find the total horizontal distance travelled by the ball before it hits the ground.
   c) Find the maximum height the ball reaches.
   d) A tree of height 1.2 m is 7.5 m from the first wall, does the stone hit the tree?

10) An arch has a span of 350m and a maximum vertical height of 140m above ground level. Let x be the horizontal distance from the left-most point of the arch in metres and y be the vertical distance above the ground.
Find a model in the form
$$y = \frac{p}{q}x(r - x)$$
where p, q and r are integers to be found.

1.12 Coordinate Geometry: Circles (answers on page 418)

1) State the centre and radius of the following circles:
   a) $(x-2)^2 + (y-7)^2 = 3^2$
   b) $x^2 + (y+2)^2 = 3.5^2$
   c) $(x+8)^2 + (y-5)^2 = 16$
   d) $(x+12)^2 + (y+9)^2 = 19$
   e) $(x-3)^2 + (y+16)^2 = 28$
   f) $\left(x-\frac{1}{2}\right)^2 + y^2 = \frac{9}{4}$
   g) $(x+2k)^2 + (y-k+1)^2 = 100k$, $k > 0$

2) Write down the equation of the circle in the form $(x+a)^2 + (y+b)^2 = c$, where $c > 0$, given the following centre and radius:
   a) Centre $(4, -3)$, Radius $= 5$
   b) Centre $(-3, -8)$, Radius $= 9$
   c) Centre $(0, 0)$, Radius $= 100$
   d) Centre $(6, 1)$, Radius $= \sqrt{17}$
   e) Centre $(-11, 6)$, Radius $= 2\sqrt{2}$
   f) Centre $\left(-\frac{2}{3}, -\frac{3}{2}\right)$, Radius $= \frac{4}{9}$
   g) Centre $(k-5, 7-k)$, Radius $= (k+2)$, $k > 0$

3) Write down the exact area of these circles:
   a) $(x+4)^2 + (y-1)^2 = 5^2$
   b) $x^2 + \left(y+\frac{3}{8}\right)^2 = \frac{49}{9}$
   c) $3(x+7)^2 + 3y^2 = 363$

4) Write the equations for each of these circles in the form $x^2 + ax + y^2 + by + c = 0$:
   a) $x^2 + (y+6)^2 = 2$
   b) $(x-10)^2 + y^2 = 130$
   c) $(x+2)^2 + (y-8)^2 = 66$
   d) $(x-11)^2 + (y-15)^2 = 1600$
   e) $(x+4k)^2 + (y-2k)^2 = 20k^2$

5) Write the equations for each of these circles in the form $(x+a)^2 + (y+b)^2 = c$, where $c > 0$:
   a) $x^2 + 2x + y^2 + 6y - 8 = 0$
   b) $x^2 - 8x + y^2 - 20y - 784 = 0$
   c) $x^2 + y^2 + 6x - 14y - 82 = 0$
   d) $x^2 + y^2 - 5x - 9y - \frac{57}{2} = 0$
   e) $x^2 - \frac{2}{3}x + y^2 + \frac{4}{9}y - \frac{230}{81} = 0$
   f) $x^2 + y^2 - \frac{3}{2}x - \frac{5}{4}y - \frac{1091}{64} = 0$
   g) $x^2 + y^2 + \frac{14}{5}x + \frac{5}{2}y + \frac{1009}{400} = 0$
   h) $x^2 + y^2 + kx - 6ky + 3k = 0$, $k > \frac{12}{37}$

6) Determine whether the following equations define a circle:
   a) $x^2 - 6x + y^2 + 16y + 83 = 0$
   b) $x^2 + 10x + y^2 - y + \frac{101}{4} = 0$
   c) $x^2 - 4x - y^2 - 10y - 46 = 0$
   d) $x^2 + 12x + 3y^2 - 6y + 3 = 0$
   e) $x^2 + \frac{2}{3}x + y^2 - \frac{4}{3}y - \frac{1}{3} = 0$

7) Determine the possible values of k such that the equation $x^2 + y^2 + 8x - 12y - k = 0$ represents a circle.

8) Determine the possible values of k such that the equation $x^2 + y^2 - 3x + 9y + \frac{k}{6} = 0$ represents a circle.

9) Determine whether these circles intersect with either the x-axis or the y-axis:
   a) $(x-4)^2 + (y-5)^2 = 9$
   b) $(x+1)^2 + (y-6)^2 = 25$
   c) $(x+3)^2 + (y-4)^2 = 8$
   d) $(x+8)^2 + (y+3)^2 = 10$

10) By considering the centre and radius, determine the values of k, where $k > 0$, such that these circles intersect the x-axis at two distinct points:
    a) $(x-2)^2 + (y-k)^2 = 36$
    b) $(x+5)^2 + (y-k)^2 = 10$
    c) $(x-1)^2 + (y-4)^2 = k^2$
    d) $(x+7)^2 + (y+8)^2 = (k+1)^2$

11) By considering the centre and radius, determine the values of k, where $k > 0$, such that these circles intersect the y-axis at two distinct points:
    a) $(x-k)^2 + y^2 = 49$
    b) $(x+k)^2 + (y-4)^2 = 9$
    c) $(x-6)^2 + (y+3)^2 = k^2$

12) Sketch the following circles, identifying their centre and the coordinates of the point on the circle that is closest to the x-axis:
    a) $(x-3)^2 + (y+5)^2 = 4$
    b) $(x-6)^2 + (y-3)^2 = 4$
    c) $(x+7)^2 + (y+5)^2 = 16$
    d) $(x+6)^2 + (y-6)^2 = 10$

1.12 Coordinate Geometry: Circles (answers on page 418)

13) Sketch the following circles, identifying their centre and where they intersect the coordinate axes:
    a) $x^2 + (y-3)^2 = 9$
    b) $x^2 + (y+2)^2 = 16$
    c) $(x-5)^2 + y^2 = 81$
    d) $(x+5)^2 + y^2 = 25$

14) Sketch the following circles, identifying their centre and where they cross the coordinate axes:
    a) $(x-4)^2 + (y-4)^2 = 16$
    b) $(x+2)^2 + (y+2)^2 = 13$
    c) $(x-1)^2 + (y+3)^2 = 17$
    d) $(x+7)^2 + (y-2)^2 = 20$

15) Sketch the following circles, identifying their centre and the exact points at which they cross the coordinate axes:
    a) $(x+3)^2 + (y-8)^2 = 16$
    b) $(x+7)^2 + (y+5)^2 = 36$
    c) $(x-8)^2 + (y-2)^2 = 81$
    d) $(x-1)^2 + \left(y - \frac{11}{16}\right)^2 = \frac{697}{256}$

16) Determine whether the point $(3,7)$ is inside, on, or outside of these circles:
    a) $(x-1)^2 + (y-5)^2 = 9$
    b) $(x-3)^2 + (y-9)^2 = 4$
    c) $(x-5)^2 + (y-6)^2 = 1$

17) A circle has centre $(2,-2)$ and the point $(14,14)$ lies on the circle. Find the equation of the circle.

18) The points A and B lie on a circle. The line AB passes through the centre of the circle. Find the equation of the circle in each case:
    a) $A = (-5,-7)$ and $B = (9,-5)$
    b) $A = (-8,7)$ and $B = (6,-1)$
    c) $A = (4,7)$ and $B(-4,-9)$
    d) $A = (-2,-1)$ and $B(12,13)$

19) A circle passes through the points $A(6,2)$ and $B(8,4)$, and the centre of the circle lies on the x-axis. Find the equation of the circle.

20) A circle passes through the points $A(-5,1)$ and $B(3,11)$, and the centre of the circle lies on the y-axis. Find the equation of the circle.

21) A circle passes through the points $A(1,-6)$ and $B(17,-10)$, and the centre of the circle lies on the line $y = x$. Find the equation of the circle.

22) Show that the shortest distance from $A(-3,8)$ to the circumference of the circle with equation $(x-8)^2 + (y+3)^2 = 9$ is $a\sqrt{2} + b$, where a and b are integers.

23) The diagram below shows two circles of radius 10 that pass through the points $(1,0)$ and $(9,0)$. Find the equations of the two circles.

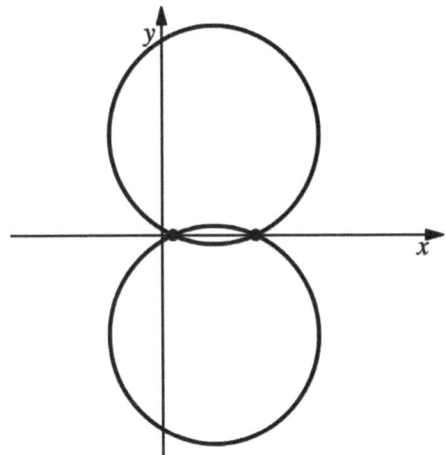

24) A circle passes through the points $A(7,2)$, $B(3,2)$ and $C(3,-6)$. Find the equation of the circle.

25) A circle passes through the points $A(-1,-12)$, $B(0,-13)$ and $C(3,-4)$. Find the equation of the circle.

26) A circle passes through the points $A(-8,1)$, $B(-9,2)$ and $C(0,11)$. Find the equation of the circle.

27) A circle passes through the points $A(1,1)$ and $B(2,6)$. A diameter of the circle lies on the y-axis. Find the equation of the circle.

1.12 Coordinate Geometry: Circles (answers on page 418)

28) The points $A(-6,-1)$, $B(2,7)$ and $C(4,5)$ lie on a circle. It is given that AC is a diameter of the circle. Find the area of the triangle ABC.

29) A circle passes through the points $A(-12,11)$, $B(-8,-1)$ and $C(-2,1)$
    a) Calculate the lengths of the sides of the triangle ABC.
    b) Hence show that AC is a diameter of the circle.
    c) Find the area of triangle ABC.

30) The points $A(2,6)$ and $B(6,k)$ lie on a circle such that AB is a diameter of the circle. The line through A and B also passes through $(10,20)$. Find the equation of the circle.

Lines intersecting Circles

31) Determine whether the following lines intersect the circle with equation
    $$(x+4)^2 + (y-6)^2 = 20$$
    once, twice or not at all.
    a) $y = 5x + 1$
    b) $y = 1 - 4x$
    c) $y = 2x + 24$

32) The line with equation $y = k$ is a tangent to the circle $x^2 + y^2 - 3x + 10y + \frac{427}{16} = 0$
    Find the possible values of k.

33) The vertices of an equilateral triangle lie at $(0,0)$, $(-10, 10\sqrt{3})$ and $(10, 10\sqrt{3})$. A circle is drawn within the triangle so that the three sides are all tangents to the circle. Find the area of the circle.

34) The circle with equation
    $$(x-3)^2 + (y+5)^2 = 34$$
    passes through the origin. Find the gradient of the tangent to the circle at the origin.

35) The point $A(1,2)$ lies on the circle with equation $(x+2)^2 + (y-3)^2 = 10$
    Find the equation of the tangent to the circle at A in the form $y = ax + b$, where a and b are integers.

36) The point A has coordinates $\left(\frac{5}{2}, -\frac{1}{2}\right)$ and lies on the circle with equation
    $$\left(x - \frac{1}{2}\right)^2 + \left(y + \frac{3}{2}\right)^2 = 5$$
    Find the equation of the tangent to the circle at A in the form $y = ax + b$

37) The point A has coordinates $(1,1)$ and lies outside of the circle $(x+3)^2 + (y-9)^2 = 10$. The point B lies on the circle such that AB is a tangent to the circle. Find the length of AB.

38) Find the two equations of the tangents to the circle $(x-3)^2 + (y-4)^2 = 100$ when $x = -5$ in the form $y = ax + b$, where a and b are rational numbers.

39) Find the two equations of the tangents to the circle $(x+5)^2 + (y-9)^2 = 10$ when $y = 10$ in the form $y = ax + b$, where a and b are integers.

40) The line $x + 7y = 82$ is a tangent to a circle with centre $\left(-\frac{5}{2}, \frac{17}{2}\right)$. Find the equation of the circle.

41) A circle has equation
    $$(x-11)^2 + (y+4)^2 = 16$$
    The straight line with equation $y = kx$, where k is a constant, meets the circle.
    a) Show that the coordinates of any points of intersection between the line and the circle satisfies the equation
    $$(1+k^2)x^2 + 2(4k-11)x + 121 = 0$$
    b) Hence, find the two values of k for which the line $y = kx$ is a tangent to the circle.

42) A circle has equation
    $$(x+3)^2 + (y-3)^2 = 80$$
    The straight line with equation $y = 2x + k$, where k is a constant, meets the circle.
    a) Show that the coordinates of any points of intersection between the line and the circle satisfies the equation
    $$5x^2 + 2(2k-3)x + k^2 - 6k - 62 = 0$$
    b) Hence, find the two values of k for which the line $y = 2x + k$ is a tangent to the circle.

1.12 Coordinate Geometry: Circles (answers on page 418)

43) The diagram below shows the circle with equation $(x-10)^2 + y^2 = 17$. The line passing through A and B is a tangent to the circle at the point where $x = 9$. Find the area of the shaded region.

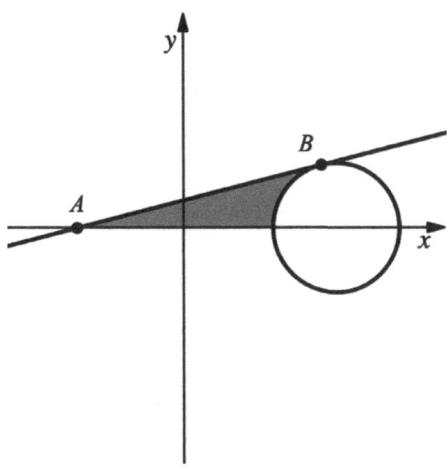

44) The diagram below shows the circle with equation $(x-25)^2 + (y-25)^2 = 1250$. The two lines are both tangents to the circle. The tangent at A has equation $y = -\frac{1}{7}x + a$ and the tangent at B has equation $y = -7x + b$. Show that the two lines intersect at $\left(\frac{225}{4}, \frac{225}{4}\right)$

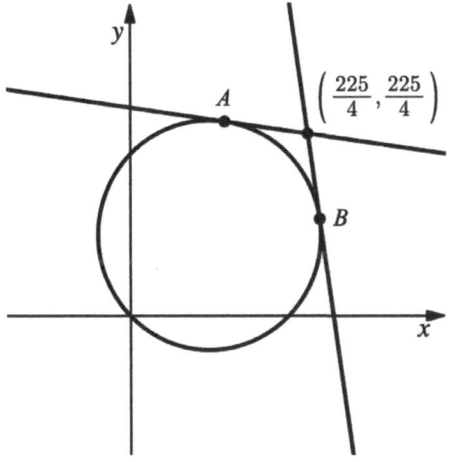

45) Find the exact shortest distance from the line $y = 2x + 1$ to the circle $x^2 + (y-7)^2 = 3$

46) A circle has equation
$$x^2 - 14x + y^2 + 18y - 40 = 0$$
The points A and B lie on the circle such that $(2, 1)$ is the midpoint of A and B. Find the shortest distance around the circumference of the circle from A to B.

47) A circle, P, has equation $x^2 + y^2 + 6y = 7$. A second circle, Q, has its centre on the x-axis and has the same radius as P. Q touches P at one point. Find the possible equations of Q.

48) A circle centred at $(8, k)$ has $3x - 2y = 12$ and $3x - 2y = 24$ as tangents. Find the equation of the circle.

49) A circle with radius k has both the x-axis and the y-axis as tangent lines. $2x + y = 50$ is also a tangent line to the circle. Find the two possible exact values of k and sketch both situations on the same graph.

50) A circle of area 25π square units is drawn that has both the x-axis and $y = \frac{1}{2}x$ as tangents, as shown in the diagram below. Find the equation of the circle.

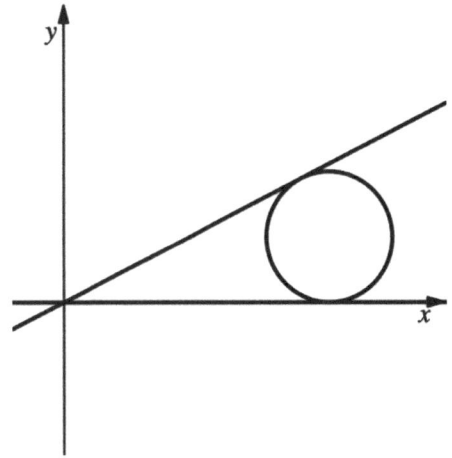

51) For the circle shown in the diagram below, determine the percentage of the circle's area that is below the x-axis.

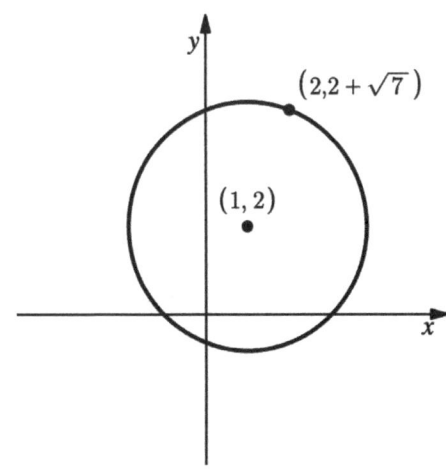

1.13 Algebraic fractions (answers on page 420)

1) Simplify the following as far as possible:

   a) $$\frac{(x+4)(x+3)}{(x+3)}$$

   b) $$\frac{x^2+11x+30}{x+5}$$

   c) $$\frac{x^2+5x}{x^2+7x+10}$$

   d) $$\frac{x^2+2x-35}{x^2+4x-21}$$

   e) $$\frac{2x^2+x-1}{2x^2+15x-8}$$

   f) $$\frac{3x^2+7x+4}{9x^2-16}$$

   g) $$\frac{2x^2+9x+4}{4x^3-8x^2-5x}$$

   h) $$\frac{2x^2+17x+30}{2x^3+11x^2+15x}$$

2) Write as a single fully simplified fraction:

   a) $$\frac{1}{x}+\frac{3}{2x}$$

   b) $$\frac{1}{x^2}+\frac{1}{x}+\frac{3}{5}$$

   c) $$\frac{1}{x}+\frac{1}{x+1}$$

   d) $$\frac{1}{x+2}+\frac{2}{x+4}$$

   e) $$\frac{3}{x+1}-\frac{1}{x-3}$$

   f) $$\frac{5}{x+2}+\frac{7}{2x-1}$$

   g) $$\frac{3}{2(x+5)}+\frac{1}{4(x-3)}$$

   h) $$\frac{x}{x^2-4}+\frac{5}{x+2}$$

3) Simplify and fully factorise the following:

   a) $$\frac{x^2-5x+4}{x^2-x}\times\frac{2x+3}{5x-20}$$

   b) $$\frac{x-3}{3x^2-6x}\times\frac{3x^2-12}{x^2-9}$$

   c) $$\frac{16x^2-4}{3x-1}\times\frac{3x^2+14x-5}{4x+2}$$

   d) $$\frac{x+2}{x+3}\div\frac{x^2+3x+2}{2x^2-2x-12}$$

4) Fully simplify the following:

   a) $$\frac{\frac{1}{x}}{\frac{1}{x+1}}$$

   b) $$\frac{\frac{1}{x-2}}{\frac{1}{2x}}$$

   c) $$\frac{\frac{1}{x+3}+\frac{1}{x-5}}{\frac{x-5}{x+3}}$$

   d) $$\frac{\frac{1}{x+4}+\frac{1}{x-1}}{\frac{x+2}{x-1}}$$

   e) $$\frac{(3x+1)(5x-1)+4(5x-1)}{\frac{5x-1}{3x+1}}$$

5) Prove each of the following:

   a) $$\frac{3}{x+2}+\frac{x-2}{(x+2)(x+3)}\equiv\frac{4x+7}{(x+2)(x+3)}$$

   b) $$\frac{2x^2-x-1}{x^2-1}-\frac{1}{x^2+x}\equiv\frac{2x-1}{x}$$

   c) $$1-\frac{1}{x-3}+\frac{2}{x^2-9}\equiv\frac{x^2-x-10}{(x-3)(x+3)}$$

   d) $$\frac{9-4x}{2(x-1)}+\frac{1}{x+3}+2\equiv\frac{7x+13}{2(x-1)(x+3)}$$

1.14 Polynomials (answers on page 420)

1) Expand each of the following:
   a) $(x-1)(2x+1)(7x+3)$
   b) $(2x-1)(3x-4)(5x-2)$
   c) $(4x+3)(2x+1)(3x-2)$

2) Fully factorise each of the following:
   a) $8x^3 - 12x^2 - 28x$
   b) $10x^4 - 20x^3 - 30x^2$
   c) $6x^4 - 45x^3 - 24x^2$

3) Fully factorise each of the following:
   a) $2x^2y^3z - 6xy^4z^2$
   b) $4x^2y^3z^4 + 14x^3y^2z^5$
   c) $3(x-1)^3y^2z - 9(x-1)^2yz^3$
   d) $9x^2(2y-1)^3z - 15x^3(2y-1)z^2$
   e) $8x^2(3y-2)^3(z+1) - 4x^3(3y-2)(z+1)^2$

4) State the degree of the following polynomials:
   a) $6x^4 - 9x^2 + 2$
   b) $4x^2 - 3x^5 + 1$
   c) $8 + 4x^3 - x^6$

5) Expand each of the following:
   a) $(x+2)(x^2 - 3x + 1)$
   b) $(2x-1)(x^3 + 2x^2 - x + 3)$
   c) $(x^2 + 2x + 3)(3x^2 - x + 2)$
   d) $(2x^2 - 4x + 1)(x^2 + 3x - 2)$

6) Simplify these fractions:
   a) $$\frac{3x^3 + 5x^2}{x}$$
   b) $$\frac{7x^3 + 9x^{10}}{3x}$$
   c) $$\frac{(x+1)(x-5)}{x+1}$$
   d) $$\frac{(x-2)^3}{x-2}$$
   e) $$\frac{x+4}{(3x-1)(x+4)}$$
   f) $$\frac{x-5}{2x^2 - 9x - 5}$$
   g) $$\frac{2x^2 - 15x + 7}{2x^2 + 5x - 3}$$

7) Divide
   a) $4x^3 - 22x^2 + 34x - 12$ by $2x - 4$
   b) $x^4 + 6x^3 + 7x^2 + x + 6$ by $x + 2$
   c) $6x^5 - 15x^4 + 9x^3 - 36x^2 + 72x - 24$ by $3x - 6$
   d) $32x^4 - 12x^3 + x^2 + 20x - 5$ by $4x - 1$
   e) $2x^6 + x^5 - 3x^4 + 4x + 6$ by $2x + 3$
   f) $x^5 + 3x^4 + 4x^3 - 2x^2 - 4x - 5$ by $x^2 + x + 1$
   g) $x^7 - 5x^6 + x^5 - 4x^4 - 8x^3 + 13x^2 - 8x + 12$ by $x^2 + 1$

8) Divide
   a) $x^3 + 9x^2 + 19x + 6$ by $x + 1$
   b) $4x^3 - 10x^2 - 12$ by $2x - 1$
   c) $x^4 + 2x^3 - x^2 + x + 16$ by $x + 2$
   d) $x^5 - 4x^3 + x^2 + 10$ by $x^2 - 1$

9) Given $(x-5)$ is a factor of $f(x)$, where
   $$f(x) = x^3 - 3x^2 - 8x - 10$$
   a) Write $f(x)$ as a product of a quadratic and a linear expression.
   b) Show that $f(x) = 0$ only has one real solution.

10) Given $(x+2)$ is a factor of $f(x)$, where
    $$f(x) = 2x^3 - 7x^2 - 10x + 24$$
    a) Write $f(x)$ as a product of a quadratic and a linear expression.
    b) Fully factorise $f(x)$
    c) Find the solutions to $f(x) = 0$
    d) Write down the roots of $f(x-2) = 0$

11) Given $(2x+1)$ is a factor of $f(x)$, where
    $$f(x) = 2x^3 - x^2 + 35x + 18$$
    a) Write $f(x)$ as a product of a quadratic and a linear root.
    b) Show that $f(x) = 0$ only has one real solution.
    c) State the one real solution to $f(x) = 0$
    d) Write down the real root of $f(x-4) = 0$

12) Given $f(x)$ and $g(x)$ both have a common factor of $(x-1)$ where
    $f(x) = x^3 + 4x^2 - 5x$ and $g(x) = x^3 - 3x + 2$
    a) Fully factorise $f(x)$ and $g(x)$
    b) Show that $(x-1)$ is also a factor of $f(x) - g(x)$

1.14 Polynomials (answers on page 420)

The Factor Theorem

13) Use the factor theorem to prove that
    a) $(x-3)$ is a factor of $x^3 - 10x^2 + 13x + 24$
    b) $(x+5)$ is a factor of $x^3 - 5x^2 - 10x + 200$
    c) $(4x-1)$ is a factor of
       $4x^3 - 21x^2 + 29x - 6$

14) In each case, use the factor theorem to prove the given factor is a factor and hence fully factorise the expression:
    a) $(x-6)$ is a factor of $2x^3 - 11x^2 - 9x + 18$
    b) $(2x+5)$ is a factor of $6x^3 + 19x^2 + 8x - 5$

15) It is given that $(x+1)$ is a factor of $(x+3)(x^2 - kx + 4) - 20$ where k is a positive constant. Find the value of k

16) It is given that $(x+3)$ is a factor of
    $$f(x) = 2x^3 + 7x^2 - px - 18$$
    where p is a positive constant.
    a) Find the value of p
    b) Fully factorise $f(x)$

17) It is given that $(8x + 15)$ is a factor of
    $$f(x) = 32x^3 + px^2 + 219x + 45$$
    where p is a positive constant.
    a) Find the value of p
    b) Fully factorise $f(x)$

18) It is given that $(x+1)$ is a factor of
    $$f(x) = 2x^3 + 11x^2 - px - 2p$$
    where p is a positive constant.
    a) Find the value of p
    b) Fully factorise $f(x)$

19) Given that $(x+2)$ and $(x-5)$ are factors of
    $$f(x) = px^3 - 5x^2 - 23x - q$$
    where p and q are positive constants.
    a) Find the values of p and q
    b) Fully factorise $f(x)$

20) Given that $(x+3)$, $(x-1)$ and $(2x-1)$ are factors of
    $$f(x) = px^4 + qx^3 + rx^2 - 37x + 15$$
    where p, q and r are constants. Find the values of p, q and r

21) It is known that $(x+9)$ is a factor of
    $$f(x) = 3x^4 + 59x^3 + px^2 + 369x - 162$$
    a) Find the value of p
    b) Fully factorise $f(x)$

22) It is given that $x^2 + bx + c$ has a factor of $(x-1)$ and $x^2 + px + q$ has a factor of $(x+2)$. Show that $5 + b + c - 2p + q = 0$

23) Show that $135k^3 + 270k^2 + 165k + 30$ is always a multiple of 30 when k is a positive, whole number.

Prerequisite knowledge for Q24-Q28: Graph Sketching

24)
    $$f(x) = x^3 + 2x^2 + 2x - 20$$
    a) Use the factor theorem to show that $(x-2)$ is a factor of $f(x)$
    b) Prove that the graph of $f(x)$ intersects the x-axis only once and state its coordinates.
    c) Given that the point of inflection is the left of the root of the curve, sketch $y = f(x)$

25) It is given that $(x+3)$ is a factor of
    $$f(x) = x^3 - 7x^2 - px + 75$$
    where p is a positive constant.
    a) Find the value of p
    b) Find the solutions to $f(x) = 0$
    c) Sketch the graph of $y = f(x)$

26) Given that $(x-3)$ and $(x+1)$ are factors of
    $$f(x) = x^3 - (2+k)x^2 + (2k-3)x + 3k$$
    where $0 < k < 3$. Sketch the graph of $f(x)$

27) It is given that
    $$f(x) = x^3 - (2k+4)x^2 + (k^2 + 8k)x - 4k^2$$
    where k is a positive constant.
    a) Prove that $(x-4)$ is a factor of $f(x)$
    b) Given $0 < k < 4$, sketch the graph of $f(x)$

28)
    a) Divide $x^3 - x^2 + 5$ by $x - 1$
    b) Hence, sketch the graph of
    $$y = \frac{x^3 - x^2 + 5}{x - 1}$$

1.15 Graph Sketching: Cubic Graphs (answers on page 421)

1) The following graphs are all functions that can be written in the form $y = (x - p)^3 + q$. Find the equation of the curve in the form
$$y = x^3 + bx^2 + cx + d$$

a) [Graph showing point of inflection at $(0, -4)$]

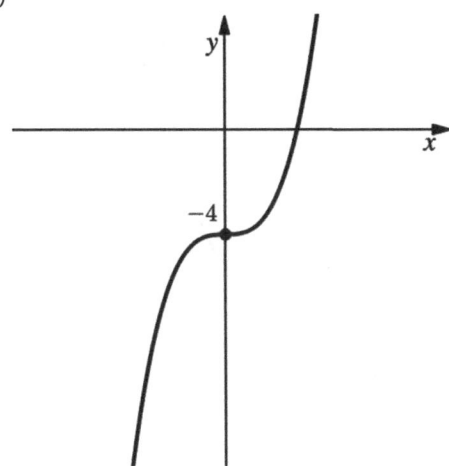

b) [Graph showing point of inflection at $(2, 1)$ and y-intercept -7]

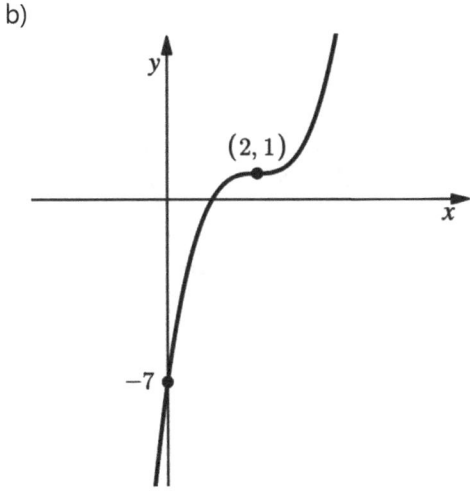

c) [Graph showing point of inflection at $(-3, -20)$ and y-intercept 7]

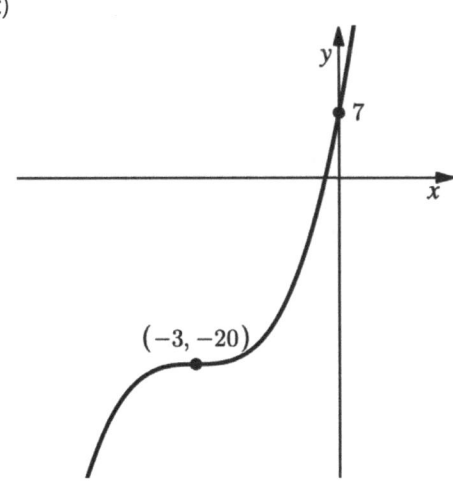

2) Sketch the following graphs:
   a) $y = (x + 5)(x + 3)(x + 1)$
   b) $y = (x - 5)(x - 3)(x - 1)$

3) The following graphs are all functions that can be written in the form $y = q - (x - p)^3$. Find the equation of the curve in the form
$$y = -x^3 + bx^2 + cx + d$$

a) [Graph showing point of inflection at $(0, -3)$]

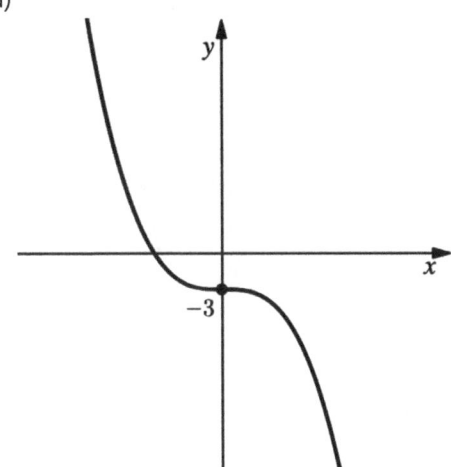

b) [Graph showing point of inflection at $(-4, 0)$ and y-intercept -64]

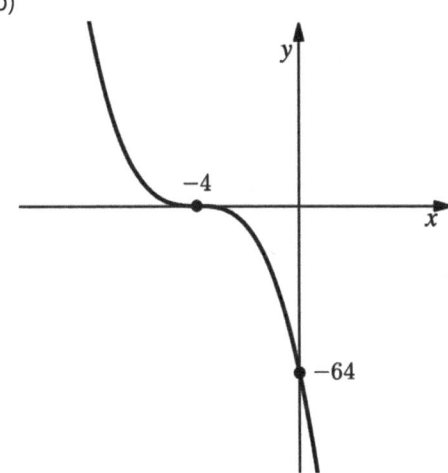

c) [Graph showing point of inflection at $(2, 20)$ and y-intercept 28]

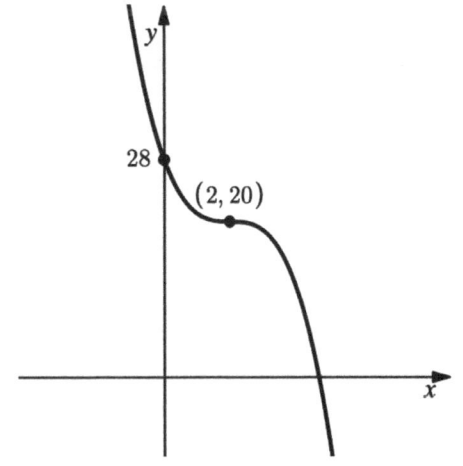

4) Sketch the following graphs:
   a) $y = -(x + 5)(x + 3)(x + 1)$
   b) $y = -(x - 5)(x - 3)(x - 1)$

1.15 Graph Sketching: Cubic Graphs (answers on page 421)

5) Find the equation of each of the following graphs in the form
$$y = (x-a)(x-b)(x-c)$$

a)

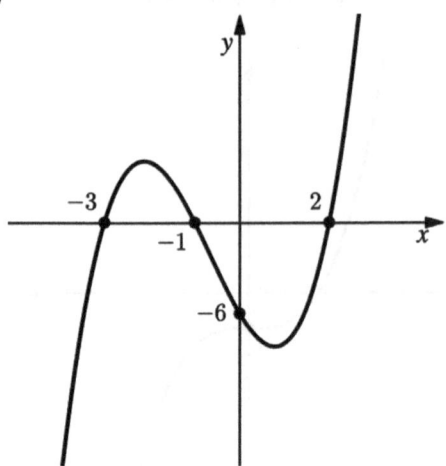

b)

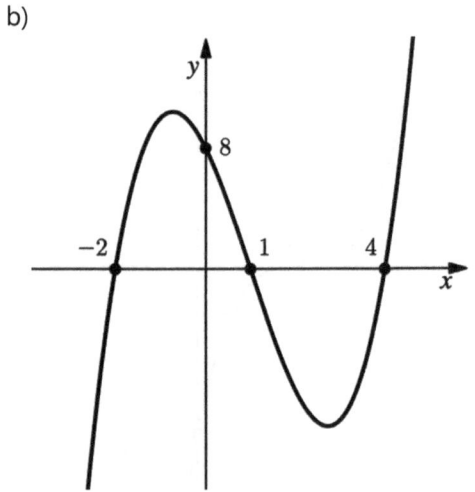

c)

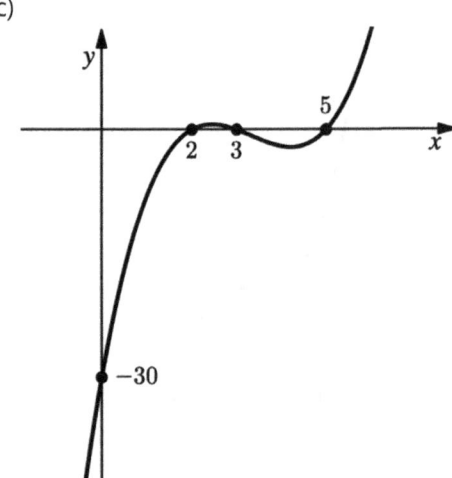

6) Sketch the following graphs:
   a) $y = (x+2)(x-2)^2$
   b) $y = (x-2)(x+2)^2$

7) Find the equation of each of the following graphs in the form
$$y = (x-a)(x-b)^2$$

a)

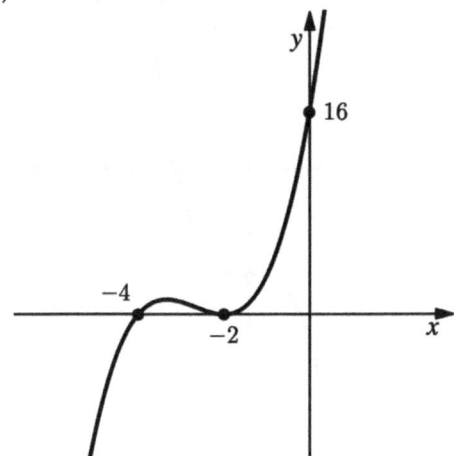

b)

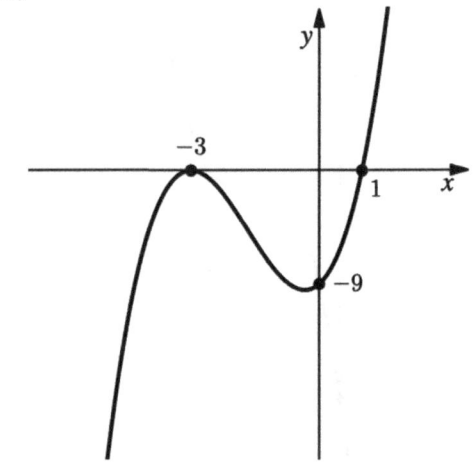

c)

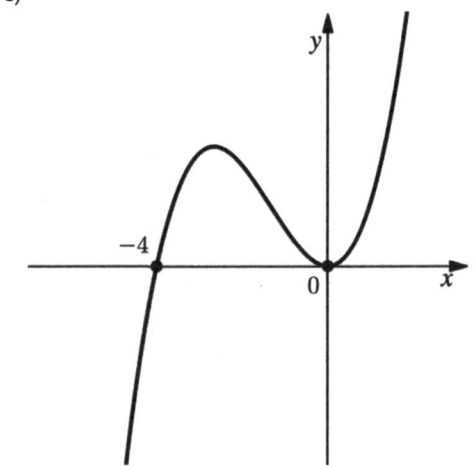

8) Sketch the following graphs:
   a) $y = -(x+3)(x-3)^2$
   b) $y = (3-x)(x+3)^2$

1.15 Graph Sketching: Cubic Graphs (answers on page 421)

9) Find the equation of each of the following graphs in the form
$$y = k(x-a)(x-b)(x-c)$$
   a) [graph with x-intercepts at -3, 2, 4 and y-intercept at -48]

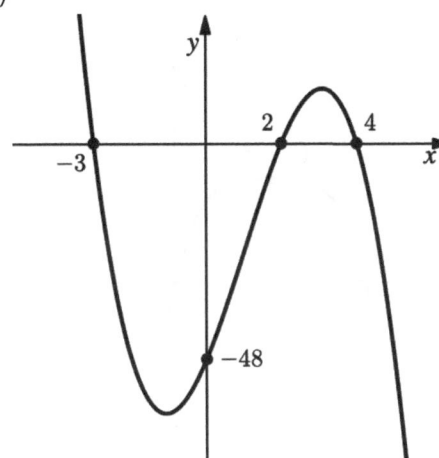

   b) [graph with x-intercepts at -6, -3, 2 and y-intercept at 24]

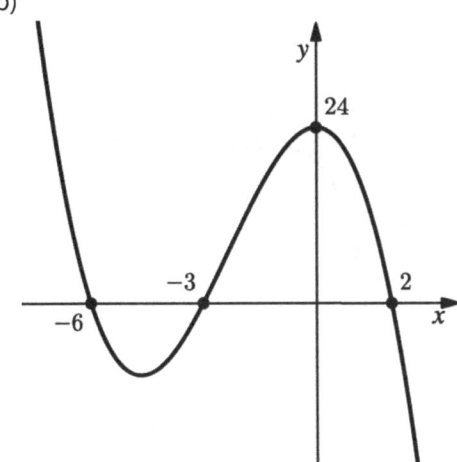

   c) [graph with x-intercepts at -2, 0, 4 and minimum at $(2, -8)$]

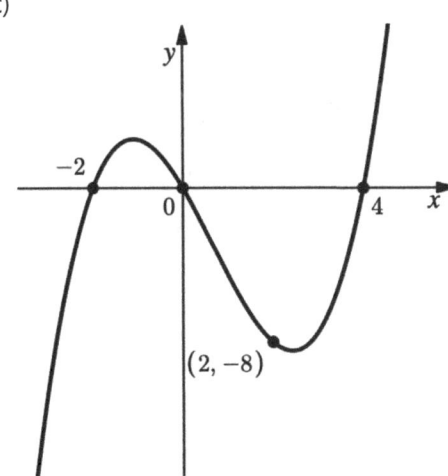

10) Sketch the following graphs:
    a) $y = \frac{1}{3}(x-6)(x-3)(x+3)$
    b) $y = -\frac{3}{4}(x+8)(x+4)(x-4)$

11) Sketch the following graphs, showing where they intersect the coordinate axes:
    a) $y = (2x+1)(x-2)(2x-7)$
    b) $y = (3x-1)(2x-3)^2$
    c) $y = x(4x-1)^2$
    d) $y = x^2(4x-1)$

12) Sketch the following graphs, showing where they intersect the coordinate axes:
    a) $y = -(x+3)(2x-1)(2x-7)$
    b) $y = (4-x)(3x+2)^2$
    c) $y = \frac{1}{5}(x+5)^2(5x-4)$
    d) $y = -\frac{5}{3}(x+9)(x-3)(x-6)$

13) Given that $k > 0$, sketch the following graphs:
    a) $y = (x-k)(x-2k)(x-3k)$
    b) $y = (x+k)(x+2k)(x-3k)$
    c) $y = (k-x)(x-2k)(x+3k)$
    d) $y = (k-x)(2k-x)(x+3k)$

14) Given that k is a positive constant, find the equation of the following graph in the form
$$y = a(bx+c)(dx+e)(fx+g)$$
[graph with x-intercepts at $-\frac{5k}{2}$, $\frac{6}{k}$, $6k$ and y-intercept at $60k$]

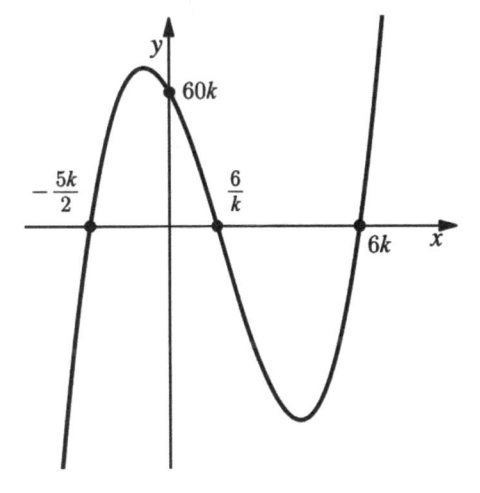

15) Given that $k > 0$, sketch the following graph:
$$y = k(2k-x)(x-k)^2$$

16) Given that $k > 0$, sketch the graphs of
$$y = (x-2k)(3kx+20)$$
and
$$y = (x-2k)(3kx+20)x$$
on the same axes, showing clearly where the two graphs cross the coordinate axes and each other.

1.16 Graph Sketching: Quartic Graphs (answers on page 423)

1) The following graphs are all functions that can be written in the form $y = (x-p)^4 + q$. Find the equation of the curve in the form
$$y = x^4 + bx^3 + cx^2 + dx + e$$

a) [graph showing curve with y-intercept at 1 and minimum at (1, 0)]

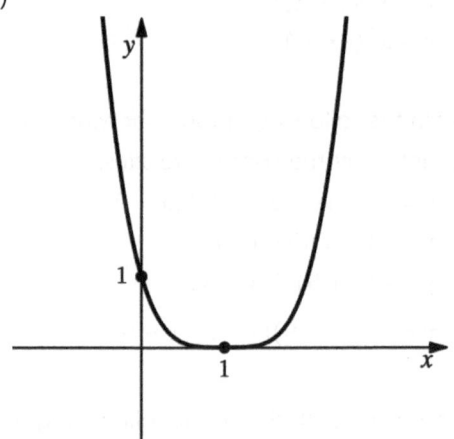

b) [graph showing curve with y-intercept at 18 and minimum at (−2, 2)]

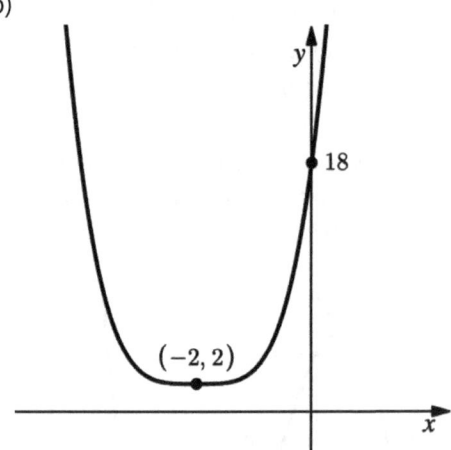

c) [graph showing curve passing through (0, 0) and (6, 0) with minimum at (3, −81)]

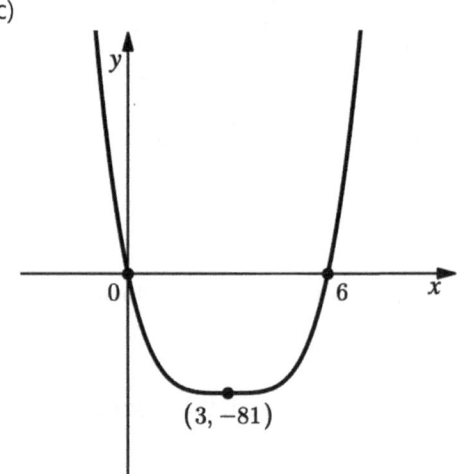

2) Sketch the following graphs:
   a) $y = (x-2)(x-1)(x+1)(x+2)$
   b) $y = (x-3)(x+1)(x-1)(x+4)$

3) The following graphs are all functions that can be written in the form $y = q - (x-p)^4$. Find the equation of the curve in the form
$$y = -x^4 + bx^3 + cx^2 + dx + e$$

a) [graph with maximum at (−1, 1), passing through −2 and 0]

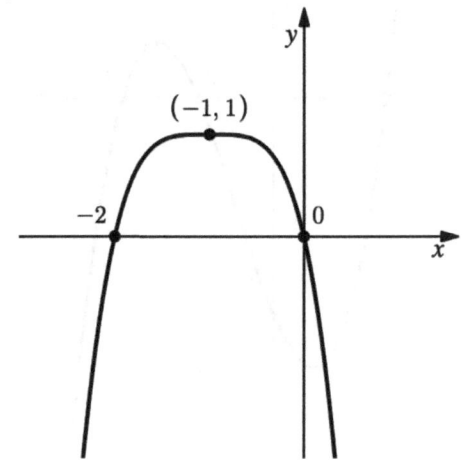

b) [graph with maximum at (1, 2), y-intercept at 1]

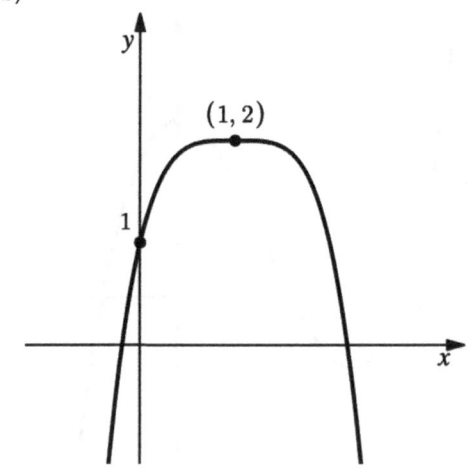

c) [graph with maximum at (−3, −19), y-intercept at −100]

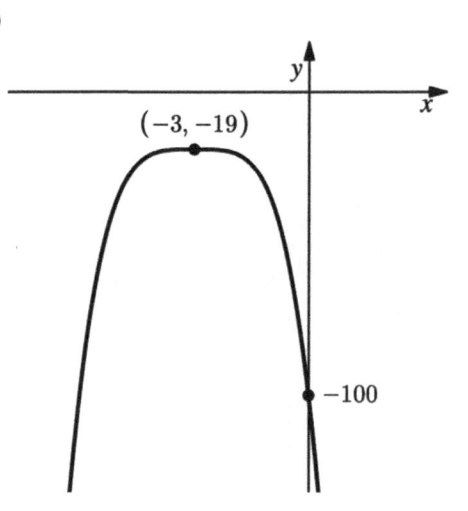

4) Sketch the following graphs:
   a) $y = -(x-3)(x+1)(x+2)(x+3)$
   b) $y = -(x+4)(x+3)(x+2)(x+1)$

1.16 Graph Sketching: Quartic Graphs (answers on page 423)

5) Find the equation of each of the following graphs in the form
$$y = (x-a)(x-b)(x-c)(x-d)$$
   a) 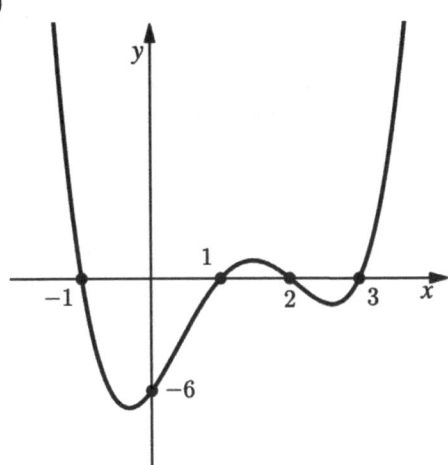
   b) 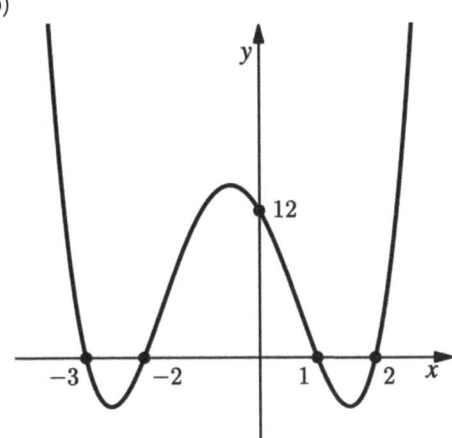
   c)

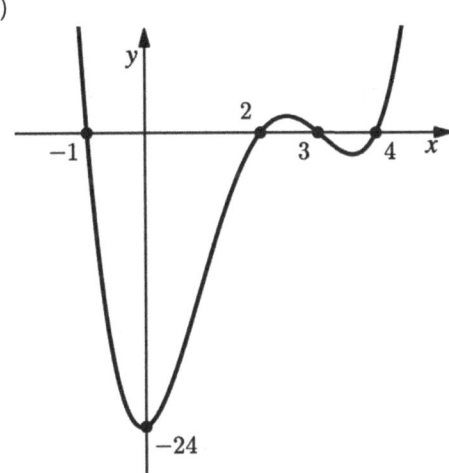

6) Sketch the following graphs:
   a) $y = (x-1)(x+3)(x-2)^2$
   b) $y = (x+1)^2(x+2)(x-3)$
   c) $y = (x+1)^2(x-2)^2$

7) Find the equation of each of the following graphs in the form
$$y = (x-a)(x-b)(x-c)^2$$
   a) 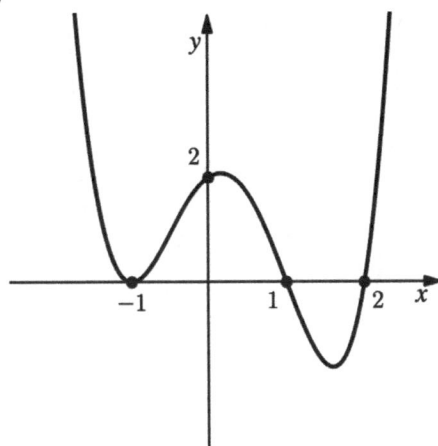
   b) 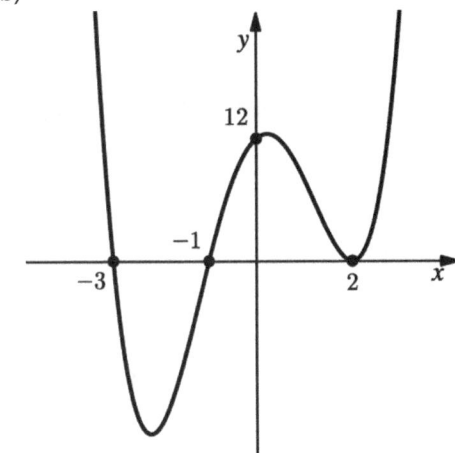
   c)

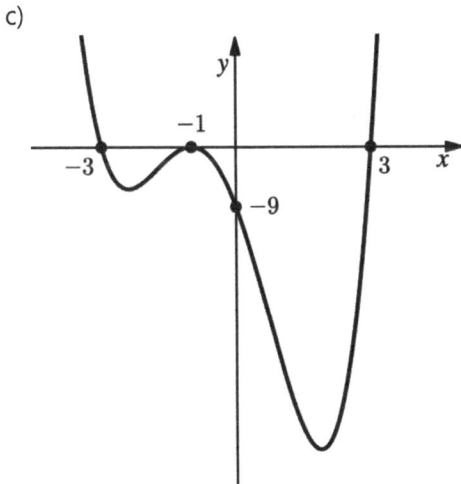

8) Sketch the following graphs:
   a) $y = -(x-4)(x+1)(x+2)^2$
   b) $y = (x-1)(3-x)(x+1)^2$
   c) $y = -x^2(x-3)^2$

1.16 Graph Sketching: Quartic Graphs (answers on page 423)

9) Find the equation of each of the following graphs in the form
$$y = k(x-a)(x-b)(x-c)(x-d)$$
   a) [graph with y-intercept 24, x-intercepts at −2, −1, 2, 3]

   b) [graph with y-intercept −72, x-intercepts at 1, 2, 3, 4]

   c) [graph with y-intercept −15, x-intercepts at −4, −1, 3, 5]

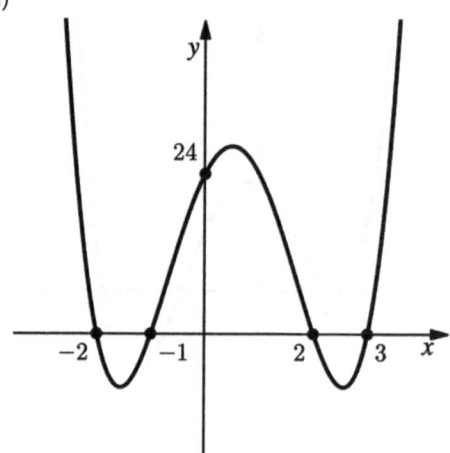

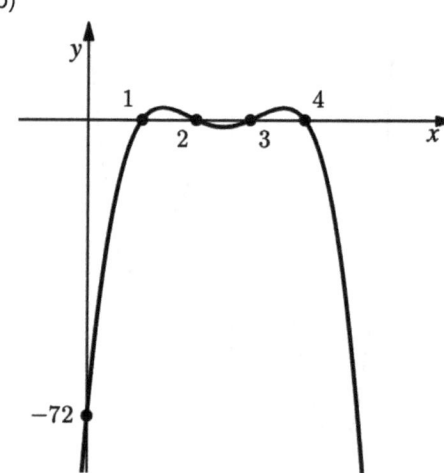

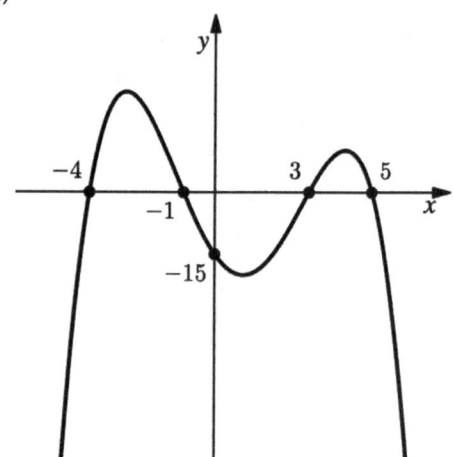

10) Sketch the following graphs:
    a) $y = \frac{1}{2}(x-4)(x+2)(x-2)(x-6)$
    b) $y = -\frac{2}{3}(x+3)(x-3)(x-6)^2$

11) Sketch the following graphs, showing where they intersect the coordinate axes:
    a) $y = (2x-1)(x+1)(2x+1)(3x-1)$
    b) $y = (x+2)(2x-1)(3x-5)^2$
    c) $y = x(3x+4)(5x+2)^2$
    d) $y = -x^2(4x-3)(3x-4)$
    e) $y = (2x-1)^2(2x+3)^2$
    f) $y = -\frac{1}{6}(6x-1)^2(12x+5)^2$

12) Sketch the following graphs, showing where they intersect the coordinate axes:
    a) $y = (x-2)(x+3)^3$
    b) $y = (2x+3)(4x-3)^3$
    c) $y = (3-2x)x^3$
    d) $y = -\frac{1}{4}(3x+4)(3x-1)^3$

13) Given that $k > 0$, sketch:
    a) $y = x(x-k)(x-2k)(x-3k)$
    b) $y = (x+k)(x+3k)(x-2k)(x+2k)$
    c) $y = (k-x)(2k+x)(3k-x)^2$
    d) $y = (4k-x)(2k+x)^3$

14) Given that k is a positive constant, find the equation of the following graph in the form
$$y = a(bx+c)(dx+e)(fx+g)(hx+i)$$
[graph with x-intercepts at $-\frac{k}{2}$, $\frac{k}{3}$, $\frac{2}{k}$, $4k$, y-intercept $-8k$]

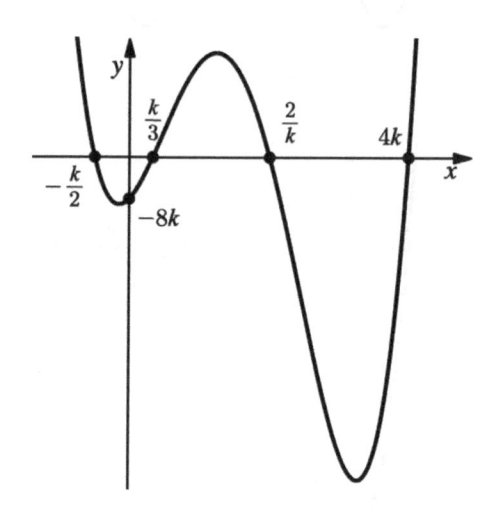

15) Given that $k > 0$, sketch the graphs of
$$y = (x-3k)(x+3k)(x-k)^2$$
and
$$y = (3k-x)(3k+x)(k-x)$$
on the same axes, showing clearly where the two graphs cross the coordinate axes and where they intersect.

1.17 Proportion (answers on page 424)

1) It is given that y is inversely proportional to x^3. State which of the following diagrams could represent the graph of y against x.

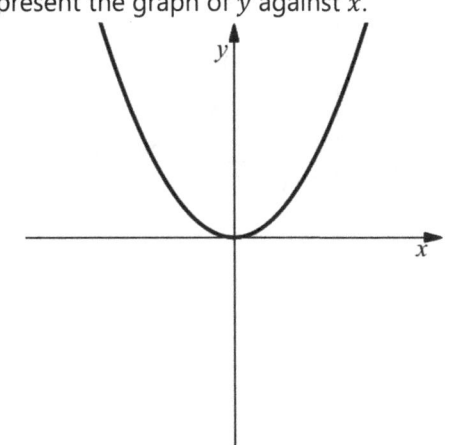

A

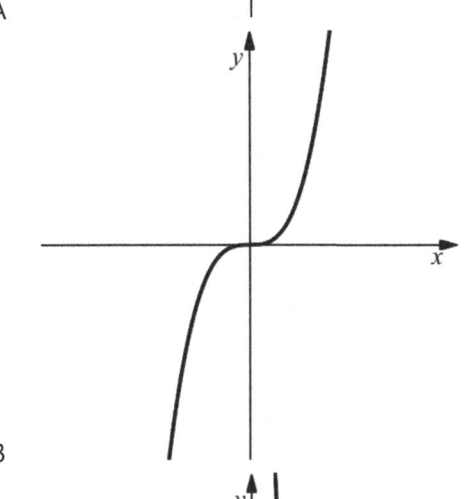

B

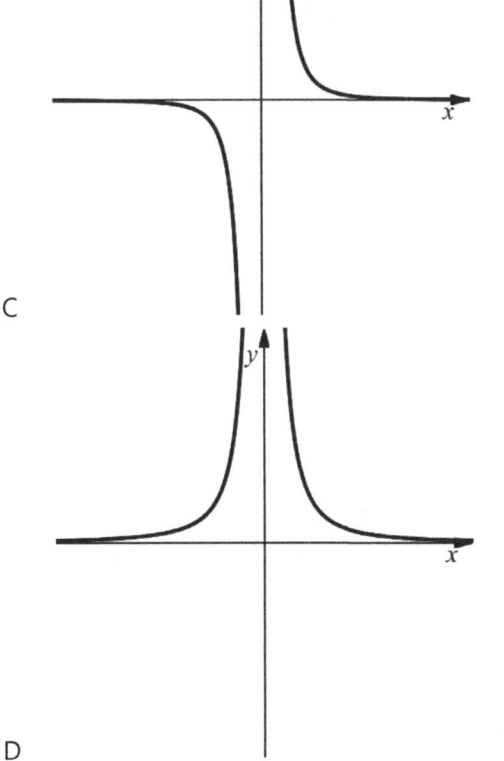

C

D

2) In each of these cases, write down an equation for y in terms of x and k, where k is a constant of proportionality:
   a) y is directly proportional to the square of x
   b) y is inversely proportional to x
   c) y is directly proportional to the cube root of x
   d) y is inversely proportional to the cube of $x+1$

3) Given that y is directly proportional to x^2, and when $x = 5$, $y = 100$, find y when $x = 10$.

4) Given that y is inversely proportional to $\sqrt[4]{x}$, and when $x = 81$, $y = 18$, find x when $y = 180$.

5) The volume of gas, V litres, that a gas canister can hold is directly proportional to the temperature, T kelvin. It is given that when $T = 300$, $V = 24$. Determine the volume of gas when the temperature is 400 Kelvin.

6) The time taken, t hours, to complete a job is inversely proportional to the number of workers, n, employed. It is given that when $n = 5$, $t = 8$. Determine the number of workers required to complete the job in 4 hours.

7) The rate of reaction, R mol/s, of a certain chemical reaction is directly proportional to the concentration C mol/L of a reactant. It is given that when $C = 0.5$, $R = 2$. Determine the concentration that will give a reaction rate of 6 mol/s.

8) The intensity of light, I lumens, from a light source is inversely proportional to the square of the distance d metres from the source. It is given that when $d = 2$, $I = 400$. Determine the distance that will give a light intensity of 100 lumens.

9) It is given that y is directly proportional to $\sqrt{x}$, and that z is inversely proportional to y^{10}. Given that $z = 2$ when $x = 4$, find x when $z = \frac{1}{16}$.

1.18 Graph Sketching: Rational Functions (answers on page 425)

1) Find the equation of the following graph in the form
$$y = \frac{k}{x}$$

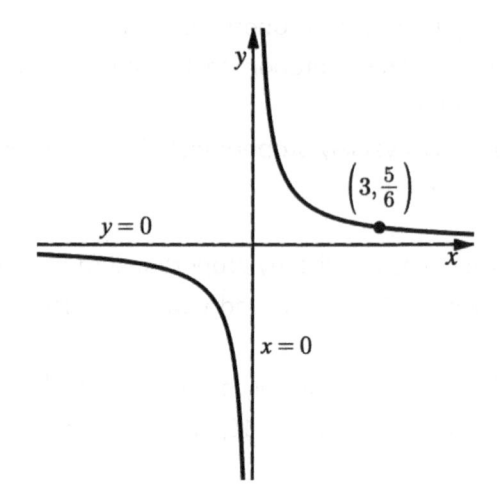

2) Find the equation of each of the following graphs in the form
$$y = \frac{1}{x-a} + b$$

a)

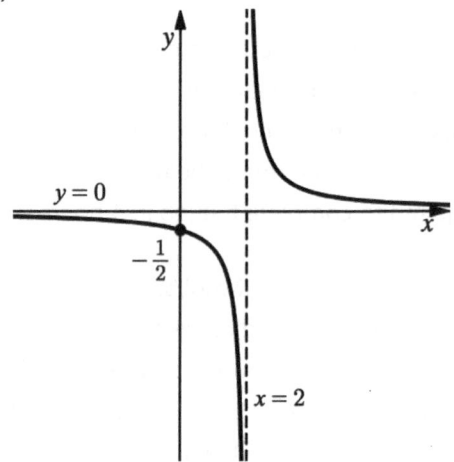

b)

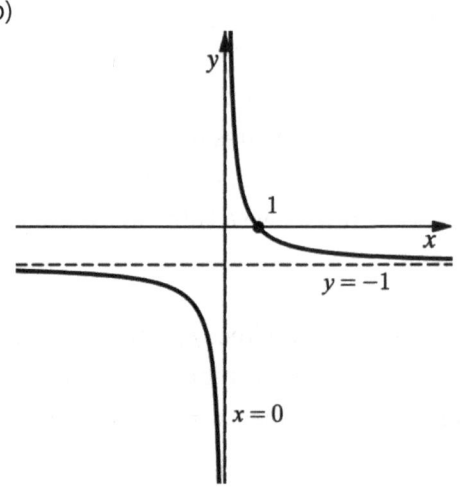

c)

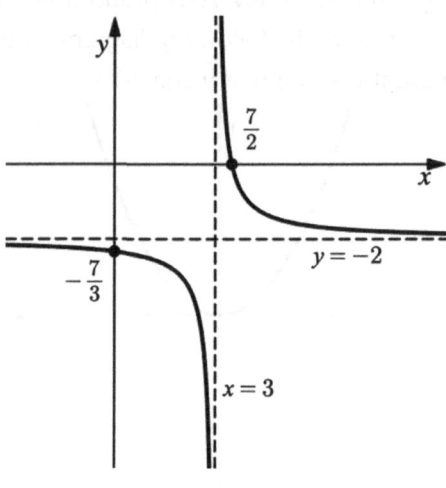

d)

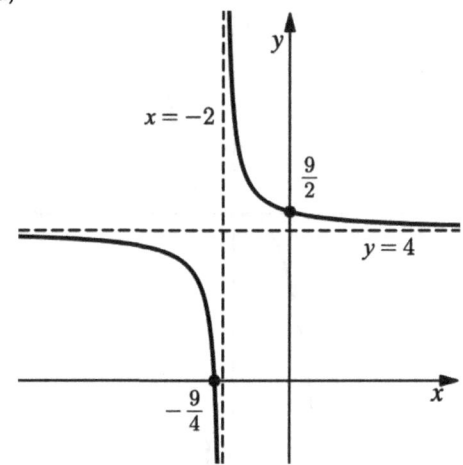

3) State the equations of the asymptotes of the curve $y = \frac{2}{x-1} + 3$

4) Sketch the following graphs, identifying where they intersect the coordinate axes and the equations of any asymptotes:
   a) $y = \frac{1}{x} - 5$
   b) $y = \frac{1}{x+3} - 1$
   c) $y = \frac{4}{x-4} + 2$
   d) $y = 3 - \frac{1}{x}$
   e) $y = 5 - \frac{1}{x+3}$

5) It is given that $k > 0$. Sketch the following graphs, identifying where they intersect the coordinate axes and the equations of any asymptotes:
   a) $y = \frac{k}{x} - 2k$
   b) $y = k - \frac{1}{x-2k}$

1.18 Graph Sketching: Rational Functions (answers on page 425)

6) Find the equation of the following graph in the form
$$y = \frac{k}{x^2}$$

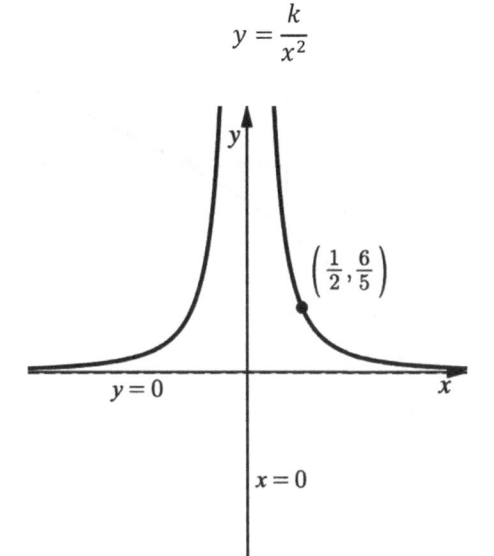

7) Find the equation of each of the following graphs in the form
$$y = \frac{1}{(x-a)^2} + b$$

a)

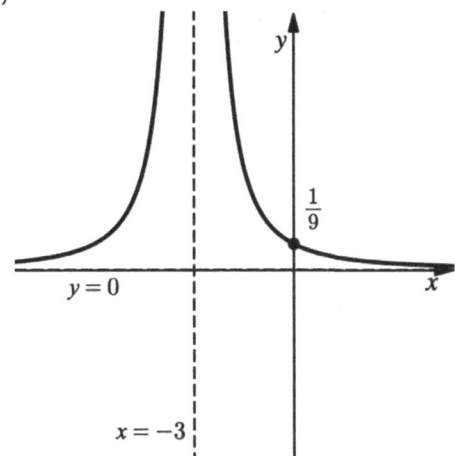

b)

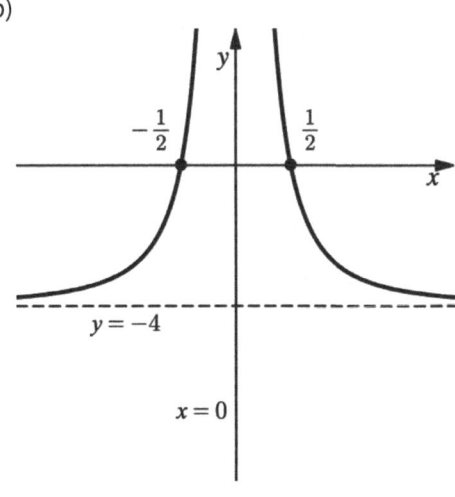

c)

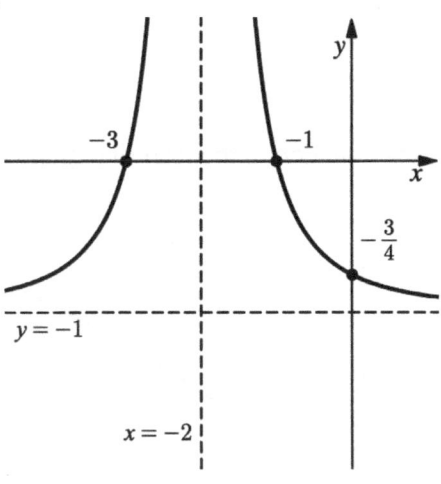

d)

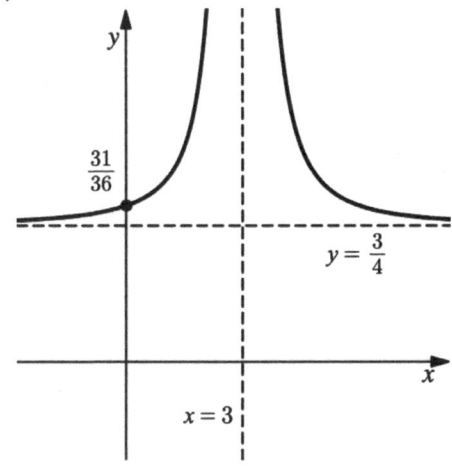

8) State the equations of the asymptotes of the curve $y = \frac{1}{(x+5)^2} - 2$

9) Sketch the following graphs, identifying where they intersect the coordinate axes and the equations of any asymptotes:
   a) $y = \frac{1}{x^2} + 2$
   b) $y = \frac{1}{(x-1)^2} - 4$
   c) $y = \frac{4}{(x-4)^2} + 9$
   d) $y = 4 - \frac{1}{x^2}$
   e) $y = 9 - \frac{1}{(x+2)^2}$

10) It is given that $k > 0$. Sketch the following graphs, identifying where they intersect the coordinate axes and the equations of any asymptotes:
   a) $y = \frac{1}{(x-k)^2} + 3k$
   b) $y = 4k^2 - \frac{9}{(x+k)^2}$

1.19 Graph Sketching - Roots (answers on page 426)

1) Sketch the following graphs:
   a) $y = \sqrt{x}$
   b) $y = -\sqrt{x}$

2) Find the equation of the following graph in the form
$$y = k\sqrt{x}$$

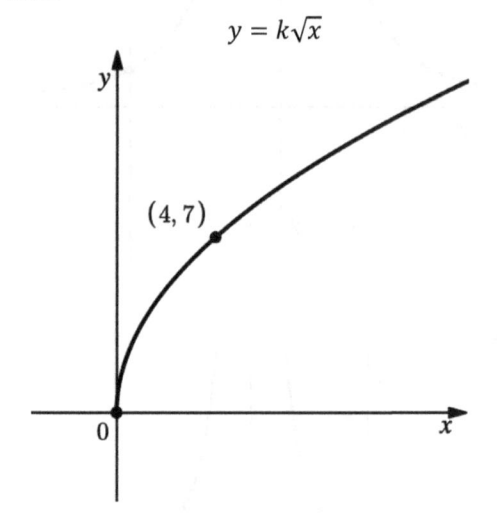

   (graph passes through $(4, 7)$)

3) Find the equation of each of the following graphs in the form
$$y = \sqrt{x - a}$$
   a) (graph starts at $x = 2$)
   b) (graph starts at $x = -3$, passes through $\sqrt{3}$ on y-axis)

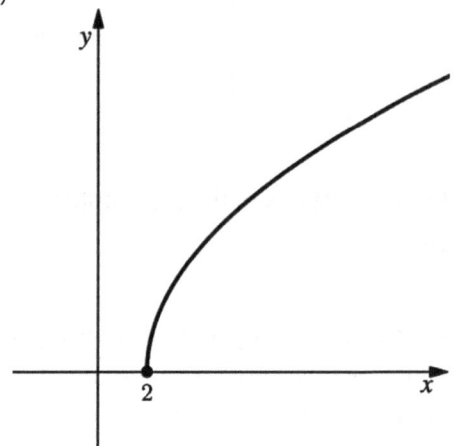

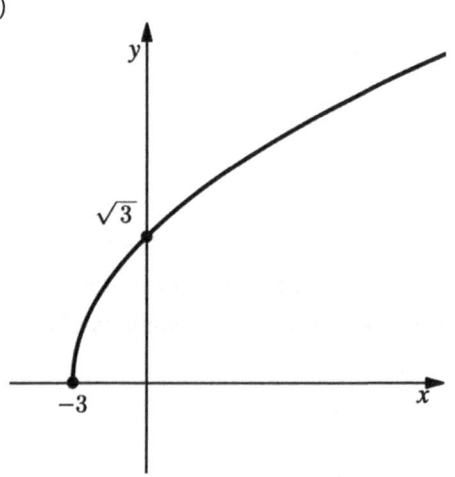

4) Find the equation of each of the following graphs in the form
$$y = \sqrt{x - a} + b$$
   a) (graph passes through $(\sqrt{5} - 3, 0)$ and 4 on x-axis, starts at $(-5, -3)$)
   b) (graph passes through origin, starts at $(-2, b)$)

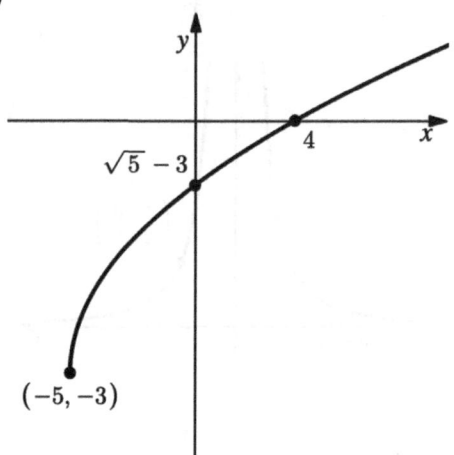

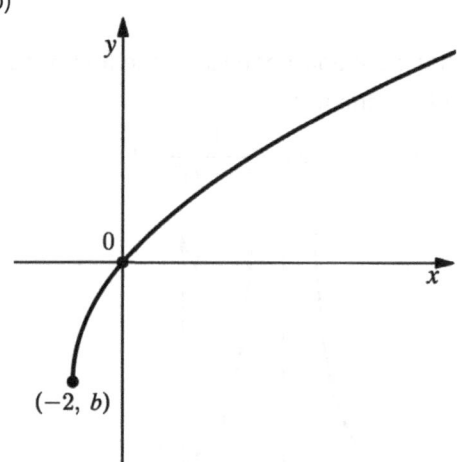

5) Sketch the following graphs:
   a) $y = \sqrt{x - 3} - 2$
   b) $y = 4 - \sqrt{x + 2}$

6) The diagram below shows the graphs $y = 4 - \sqrt{x}$ and $y = \sqrt{x + 4}$. Find the exact coordinates of the points of intersection.

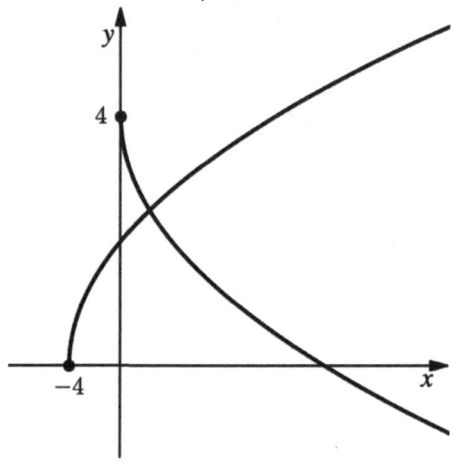

1.20 Inequalities (answers on page 426)

1) For each of the following find the values which have been identified, using set notation:
   a) [number line: -4 to 4, filled from 2 to 3, closed at 2, open at 3]
   b) [number line: -4 to 4, open at -2, closed at 1]
   c) [number line: -4 to 4, closed at -1, closed at 1]
   d) [number line: -4 to 4, closed at 2, open at 3]

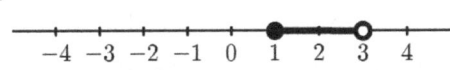

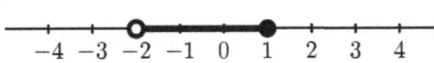

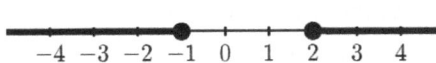

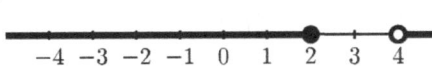

2) For each of the following find the values which have been identified, using interval notation:
   a) [number line: open at -3, closed at 1]
   b) [number line: closed at -1, extending right]
   c) [number line: open at -3, open at 2]
   d) [number line: open at 0, closed at 3]

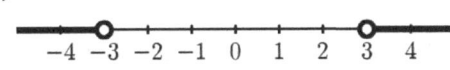

3) Draw a number line to shows the values of real number x which belong in the following sets:
   a) $\{x: x \leq -3\} \cup \{x: x > 2\}$
   b) $\{x: x < 1\} \cup \{x: x \geq 2\}$
   c) $\{x: x \geq -2\} \cap \{x: x < 3\}$
   d) $\{x: x > -1\} \cap \{x: x \leq 2\}$
   e) $\{x: x > 1\} \cap \{x: x < -3\}$

4) Solve the following inequalities giving your answer in set notation:
   a) $3x - 8 < 4$
   b) $6 - 2x \leq 15$
   c) $6 + 5x > 2 - 3x$
   d) $1 + 9x \leq 6 - x$
   e) $-2 \leq 3x + 4 \leq 10$
   f) $-5 < 5x - 8 < 22$
   g) $-1 < 2x + 1 < 1$
   h) $-1 < ax + b < 1$ where $a > 0$
   i) $-1 < ax + b < 1$ where $a < 0$

5) Solve the following inequalities, writing your answer in interval notation:
   a) $\frac{x+14}{3} < 2(x-1)$
   b) $\frac{4(x+1)}{3} \geq \frac{x-6}{2}$
   c) $\frac{4x+1}{3} - \frac{2x-1}{2} \leq 1$

6) Solve the following inequalities giving your answer in set notation:
   a) $(x+2)(x-4) < 0$
   b) $(x+1)(x+7) \geq 0$
   c) $x^2 + 3x - 10 \leq 0$
   d) $x^2 + 5x + 6 > 0$
   e) $-(x+3)(x-8) < 0$
   f) $(2-x)(4+x) \geq 0$
   g) $x^2 + 3x - 1 > 0$
   h) $x^2 - 2x + 1 \geq 0$
   i) $-x^2 + x - 3 > 0$

7) Solve the following inequalities giving your answer in interval notation:
   a) $x^2 - x > 20$
   b) $2x^2 < 10 - 8x$
   c) $4x^2 + 4x \geq 35$
   d) $6x^2 \leq 7x + 5$

8) Freddie tries to solve the inequality
   $$2x - x^2 + 8 \leq 0$$
   His workings and answer are as follows:
   $$x^2 - 2x - 8 \leq 0$$
   $$(x+4)(x-2) \leq 0$$
   $$x = -2, x = 4$$
   $$\{x: x < -2\} \cup \{x: x > 4\}$$
   a) Identify all his mistakes.
   b) Fully correct his workings.

9) Find the set of values of x for which the line
   $$y = 2x + 3$$
   lies below the curve
   $$y = x^2$$
   giving your answer in set notation.

10) Find the set of values of x for which the line
    $$y = 19 - 10x$$
    lies above the curve
    $$y = (2x+1)^2$$
    giving your answer in interval notation.

1.20 Inequalities (answers on page 426)

11) Solve
$$\frac{1}{x^2} < 529$$
giving your answer in interval notation.

12) Solve
$$\frac{1}{x^2} \geq a^2$$
where $a \neq 0$. Give your answer in terms of a and in set notation.

13) Re-write the following phrases into mathematical inequalities:
   a) x is more than 12
   b) x is less than 100
   c) x is at most 5
   d) x is at least 10
   e) x is at least 50% more than y
   f) x is at most 20% more than y

14) A picture frame of height x cm is 6 cm wider than it is tall. The width needs to be at least 75% more than its height.

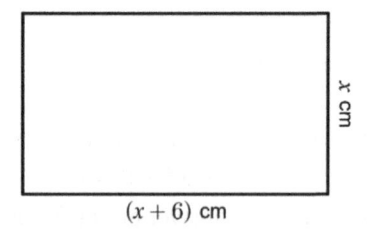

   a) Show that $x \leq 8$

   The back of the picture frame has an area of at most 160 cm².
   b) Find the possible values of x giving your answer in set notation.
   c) Hence state its maximum area.

15) Joe wants to install a rectangular swimming pool in his garden.
   He would like it to be 2 metres longer than it is wide but at least 8 metres long. He wants a pool that is less than 80 m² in area.
   Given the width is x metres, find the possible values of x giving your answer in set notation.

16) Prove that $x^2 + 3x > -4$ for all real values of x

17) The equation $x^2 + kx + 9 = 0$ where k is a constant has real roots.
   a) Show that $k^2 - 36 \geq 0$
   b) Hence find the set of possible values of k giving your answer in set notation.

18) Find the values of k for which each of the following equations have no real roots, giving your answer in set notation:
   a) $x^2 + 3kx + 1 = 0$
   b) $x^2 - kx + 5 - 2k = 0$
   c) $(2k-1)x^2 + (2k+3)x + k = 0$, $k \neq \frac{1}{2}$

19) Find the set of values of k such the curve
$$y = x^2 - 3kx + 8k$$
lies entirely above the x-axis.
Write your answer in set notation.

20) Find the set of values of k such that the curve
$$y = (k+2)x^2 + kx - 2k$$
lies entirely above the x-axis.
Write your answer in interval notation.

21) Find the set of values of k such that the curve
$$y = (k+1)x^2 - kx + k$$
lies entirely below the x-axis.
Write your answer in set notation.

Prerequisite knowledge for Q22: Circles

22) Find the values of k such that the line
$$y = x + k$$
intersects the circle with equation
$$x^2 + y^2 = 16$$
Write your answer in set notation.

Prerequisite knowledge for Q23: Cubic Curves

23) By first sketching the curve, solve the following inequalities, writing your answer in set notation:
   a) $(x-3)(x+2)(x+1) > 0$
   b) $(1-x)(x-5)(1+x) > 0$
   c) $(x-6)(x-1)(2-x) \leq 0$
   d) $x(5-x)(3+x) \geq 0$

1.20 Inequalities (answers on page 426)

24)
    a) Sketch the curve
$$f(x) = x^2(x-2)$$
labelling the points where the curve and the axes meet.
    b) Hence find the set of values of x such that $f(x) \leq 0$ giving your answer in set notation.

25)
    a) Sketch the curve
$$f(x) = (x+1)(x-3)^2$$
labelling the points where the curve and the axes meet.
    b) Hence find the set of values of x such that $f(x) > 0$ giving your answer in set notation.

26)
    a) Sketch the curve
$$f(x) = x^2(x-a)$$
where $a > 0$, labelling the points where the curve and the axes meet.
    b) Hence find the set of values of x such that $f(x) > 0$ giving your answer in set notation.

27)
    a) Sketch the curve
$$f(x) = (x+2)(x-a)^2$$
where $a > 0$, labelling the points where the curve and the axes meet.
    b) Hence find the set of values of x such that $f(x) \leq 0$ giving your answer in set notation.

28)
    a) Sketching the graph of
$$y = (a-x)(x-b)^2$$
where $a > 0$ and $0 < b < a$, labelling the points where the curve and the axes meet.
    b) Find the values of x such that
$$(a-x)(x-b)^2 \geq 0$$
giving your answer in set notation.
    c) Find the values of x such that
$$(a-x)(x-b)^2 \leq 0$$
giving your answer in interval notation.

Prerequisite knowledge for Q29: Factor Theorem

29) Given
$$f(x) = 4x^3 - 24x^2 + 21x - 5$$
    a) Use the factor theorem to show that $(x-5)$ is a factor of $f(x)$
    b) Hence fully factorise $f(x)$
    c) Sketch $y = f(x)$
    d) Find the values for which $f(x) \leq 0$ giving your answer in interval notation.

Prerequisite knowledge for Q30-Q31: Graph Transformations

30) Given
$$f(x) = x^3 + 4x^2 - 3x - 18$$
    a) Show that $(x-2)$ is a factor of $f(x)$
    b) Hence fully factorise $f(x)$
    c) Sketch $y = f(x)$
    d) Solve $f(2x) = 0$
    e) Hence, find the values for which $f(2x) > 0$ giving your answer in set notation.

31) A curve is given by
$$f(x) = -2x^3 + 18x^2 - 28x - 48$$
    a) Fully factorise $f(x)$
    b) Hence, find the values for which $f\left(\frac{1}{2}x\right) > 0$ giving your answer in set notation.

Prerequisite knowledge for Q32-Q34: Exponentials and Logarithms

32) Solve the following inequalities, writing your answer in set notation:
    a) $e^x - 10 \leq 0$
    b) $4^x - 5 \geq 0$
    c) $\left(\frac{1}{3}\right)^x > 9$
    d) $\left(\frac{1}{5}\right)^x < 1$
    e) $0.4^x > 6.25$

33) Giving your answer in set notation, find the set of values of x for which $2 \times 3^{2x} - 3^x - 28 < 0$

34) Giving your answer in set notation, find the set of values of x for which
$$e^x - \frac{4}{e^x} > -3$$

1.21 Inequalities (answers on page 427)

1) Sketch and shade the region satisfied by
   a) $y > 3x + 1$
   b) $y \leq 2x - 5$

2) Sketch and shade the region satisfied by both
   $$y \leq 2x - 1 \text{ and } x + 2y > 8$$

3) Sketch and shade the region satisfied by both
   $$y > 3x + 8 \text{ and } x + 3y < 4$$

4) In each case, sketch a diagram and shade the region which satisfies the inequality:
   a) $y \leq (x - 5)(x + 3)$
   b) $y \geq (2x + 1)(x - 4)$
   c) $y < x^2 + x - 2$
   d) $y > 2x^2 - 7x - 15$

5) In each case, sketch a diagram and shade the region which satisfies the inequalities:
   a) $y + x < -5$ and $y > x^2 + x - 20$
   b) $y - 2x < 6$ and $y \leq -x^2 - x + 6$

6) Sketch and shade the region satisfied by
   $$3y \geq x, y + x < 4 \text{ and } y \geq (x - 2)^2$$

7) Sketch and shade the region satisfied by
   $$y \geq x^2, x + y \leq 6 \text{ and } x \leq 0$$

8) The diagram shows the line $y = x + 1$ and the curve $y = 6 + 5x - x^2$

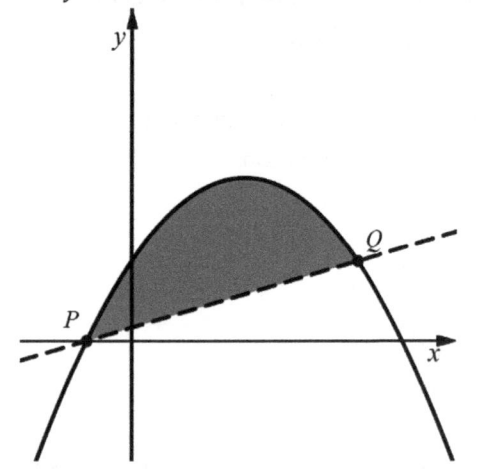

   a) Write down the inequalities that define the shaded region.
   b) Find the coordinates of the points P and Q
   c) Solve the inequality
   $$6 + 5x - x^2 > x + 1$$
   giving your answer in set notation.

9) The diagram shows two curves of the form
   $$y = ax^2 + bx + c$$
   where a, b, c are non-zero constants.

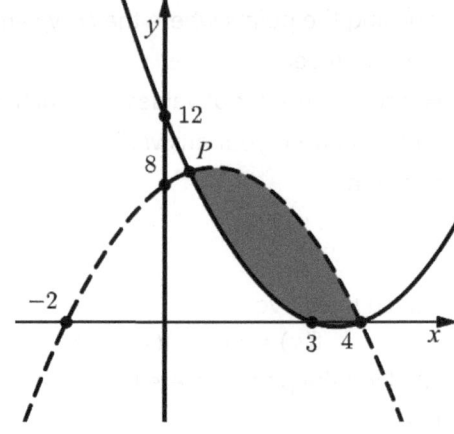

   a) Write down the inequalities that define the shaded region.
   b) Find the coordinates of the points P

10)
   a) Sketch and shade the region which satisfies both
   $$y + 3 \geq x^2 \text{ and } y + 2x^2 < x + 1$$
   b) Hence, solve the inequality
   $$3x^2 - x - 4 < 0$$
   giving your answer in set notation.

Prerequisite knowledge for Q11-Q12: Cubic Curves

11) Sketch and shade the region which satisfies
    $$y < x(x - 2)(x + 4)$$

12) Sketch and shade the region which satisfies
    $$y \geq x^3 - 6x^2 + 11x - 6 \text{ and } y < 2x - 2$$

Prerequisite knowledge for Q13-Q14: Circles

13)
   a) Show that $x^2 + 2x + y^2 = 3$ is an equation of a circle, stating the centre and the radius.
   b) Hence sketch a diagram and shade the region which satisfies
   $$x^2 + 2x + y^2 < 3$$

14) Sketch a diagram and shade the region which satisfies both
    $$(x - 3)^2 + (y - 1)^2 \leq 25 \text{ and } y > 3 - x$$

1.22 Exponentials and Logarithms (answers on page 428)

1) Solve the following equations, writing your answer in logarithmic form:
   a) $4^x = 3$
   b) $3^x = 4$
   c) $2^x = \frac{1}{3}$
   d) $9^x = \frac{1}{100}$

2) Solve the following equations, writing your answer in logarithmic form:
   a) $2^{x+1} = 3$
   b) $3^{x-5} = \frac{1}{5}$
   c) $5^{2x-1} = 7^{500}$
   d) $6^{3x+4} = 9^{250}$

3) Solve the equation
   $$6^{x^2-5} = \frac{1}{216}$$
   giving your answers in exact form.

4) Solve the equation
   $$8^{10-x^2} = 32768$$
   giving your answers in exact form.

5) Solve the following equations, writing your answer in the form $x = a^b$:
   a) $\log_3 x = 8$
   b) $\log_8 x = 3$
   c) $\log_5 x = 7$
   d) $\log_7 x = 5$

6) Write each of these as a single logarithm:
   a) $\log_3 4 + \log_3 5$
   b) $\log_5 3 + \log_5 6$
   c) $\log_6 10 - \log_6 2$
   d) $\log_{10} 4 - \log_{10} 8$

7) Write each of these as a single logarithm:
   a) $\log_{10} 4 + \log_{10} 3 - \log_{10} 2$
   b) $\log_2 20 - \log_2 6 + \log_2 3$
   c) $\log_5 9 - \log_5 10 - \log_5 3$
   d) $\log_9 8 - \log_9 10 + \log_9 4$

8) Solve the equation
   $$\log_3 x - \log_3 5 = \log_3 8$$

9) Solve the equation
   $$\log_7 9 - \log_7 x = \log_7 18$$

10) Evaluate each of these:
    a) $\log_3 3$
    b) $\log_8 8$
    c) $2\log_5 5$
    d) $-4\log_{10} 10$

11) Write each of these in the form $\log_a b$:
    a) $2\log_3 5$
    b) $3\log_4 3$
    c) $4\log_9 2$
    d) $5\log_7 10$

12) Write each of these in the form $\log_a b$, where b is a rational number:
    a) $-\log_2 3$
    b) $-\log_3 2$
    c) $-2\log_5 2$
    d) $-3\log_8 3$

13) Write each of these as a single logarithm:
    a) $\log_b 2 + 2\log_b 5$
    b) $\log_b 5 + 3\log_b 3$
    c) $\log_b 10 - 4\log_b 2$
    d) $\log_b 4 - 2\log_b 10$

14)
    a) Express $3\log_2 x + \log_2 k$ as a single logarithm.
    b) Given that $3\log_2 x + \log_2 k = 3$, express x in terms of k

15)
    a) Express $5\log_3 x - \log_3 m$ as a single logarithm.
    b) Given that $5\log_3 x - \log_3 m = 2$, express x in terms of m

16) Write each of these as a single logarithm:
    a) $\log_5 30 - 2\log_5 2$
    b) $\log_2 9 - 3\log_2 3$
    c) $\log_9 100 + 2\log_9 5$
    d) $\log_3 20 + 3\log_3 10$

17) Given that $\log_5 y^5 - \log_5 x^3 = 10$, show that $y = mx^n$, where m is an integer and n is a rational number.

1.22 Exponentials and Logarithms (answers on page 428)

18) Write each of these as a single logarithm:
    a) $2\log_3 4 + 4\log_3 2$
    b) $3\log_5 4 - 2\log_5 2$
    c) $3\log_8 5 + 2\log_8 10$
    d) $4\log_4 3 - 3\log_4 9$

19) Write each of these as a single logarithm:
    a) $\frac{1}{2}\log_7 4 + \frac{1}{3}\log_7 8$
    b) $\frac{1}{2}\log_{10} 81 - \frac{1}{3}\log_{10} 125$
    c) $\frac{2}{3}\log_2 27 + \frac{3}{2}\log_2 9$
    d) $\frac{4}{5}\log_9 32 - \frac{5}{4}\log_9 16$

20) Write each of these as a single logarithm:
    a) $4\log_3 2 + 1$
    b) $3\log_2 5 - 1$
    c) $2\log_5 6 + 2$
    d) $3 - 2\log_4 10$

21) Write each of these as a single logarithm:
    a) $2\log_b 3 + 2$
    b) $4\log_b 2 - 1$
    c) $3\log_b 4 + 1$
    d) $1 - 2\log_b 20$

22) It is given that
    $$\log_b p - 2\log_b q = \log_b (p - 2q)$$
    Show that
    $$p = \frac{2q^3}{q^2 - 1}$$
    where $p > 2q > 0$

23) It is given that
    $$3\log_b q - \frac{1}{2}\log_b p = \log_b (3q - p)$$
    Show that
    $$p = \frac{q^6}{9q^2 - 6pq + p^2}$$
    where $3q > p > 0$

24) Find the value of:
    a) $\log_b b^2 + \log_b b^5$
    b) $\log_b b^4 - \log_b \frac{1}{b}$
    c) $\log_b \frac{1}{b^2} + \log_b b^3$
    d) $\log_b \frac{1}{b} - \log_b \frac{1}{b^3}$

25) Show that, for $x > 0$,
    $$\log_3 \frac{x^3}{9} + \log_3 27x - \log_3 x^5 \equiv \log_3 \frac{3}{x}$$

26) Given that $m = \log_4 x$, where x is positive, write each of these in their simplest form in terms of m:
    a) $\log_4 \sqrt{x}$
    b) $\log_4 \frac{x}{4}$
    c) $\log_4 \frac{16}{x^2}$

27) It is given that
    $$\log_b y = 3\log_b 2 + \log_b 5 - \frac{3}{2}$$
    Write y in terms of b without any logarithms.

28) It is given that
    $$\log_k y = \log_k 8 - 2\log_k \frac{1}{2} + \frac{3}{4}$$
    Write y in terms of k without any logarithms.

29) Write each of these as a single logarithm:
    a) $2\log_b p + 3\log_b q + 2\log_b r$
    b) $3\log_b p + 5\log_b q - 3\log_b r$
    c) $5\log_b p - 2\log_b q + 4\log_b r$
    d) $8\log_b p - 6\log_b q - 9\log_b r$

30) Write each of these as a single logarithm:
    a) $\log_7 30 - (\log_7 2 + \log_7 3)$
    b) $\log_8 15 - 2(\log_8 6 - \log_8 2)$
    c) $\log_6 60 - 3(\log_6 5 + \log_6 2)$
    d) $\log_{10} 50 - 2(\log_{10} 80 - \log_{10} 16)$

31) Write each of these as a single logarithm:
    a) $\log_b p - (\log_b q + \log_b r)$
    b) $\log_b p - 2(\log_b q - \log_b r)$
    c) $\log_b p - 3(\log_b q + \log_b r)$
    d) $2\log_b p - 2(\log_b q - 2\log_b r)$

32) It is given that $x = \log_b 3$, $y = \log_b 5$ and $z = \log_b 7$. Write each of these in terms of x, y and z:
    a) $\log_b 15$
    b) $\log_b 9$
    c) $\log_b 21$
    d) $\log_b \frac{49}{25}$

1.22 Exponentials and Logarithms (answers on page 428)

33) By writing $\log_3 4 \times \log_3 64$ in terms of $\log_3 4$, show that
$$\log_3 4 \times \log_3 64 > 3$$

34) Showing your working, write each of these in a form not involving logarithms:
   a) $\log_{100} 10 + \log_{10} 100$
   b) $\log_2 8 - \log_8 2$
   c) $\log_9 81 + \log_3 \frac{1}{3}$
   d) $\log_{25} \frac{1}{625} + \log_5 625$

35) A right-angled triangle ABC has hypotenuse AC. The length of AB is $(\log_5 16)$ cm and the length of BC is $(\log_5 64)$ cm. Without the use of a calculator, show that the area of the triangle is larger than 0.75 cm².

36) Using $y = 2^x$ as a substitution, show that
$$4^x - 7 \times 2^{x+1} + 45 = 0$$
can be written as
$$y^2 - 14y + 45 = 0$$
Hence show that $4^x - 7 \times 2^{x+1} + 45 = 0$ has two solutions of the form $\log_2 k$.

37) By writing $\log_4 3 \times \log_4 9 \times \log_4 81$ in terms of $\log_4 3$, show that
$$\log_4 3 \times \log_4 9 \times \log_4 81 < 8$$

38) Using $y = 2^{2x}$ as a substitution, show that
$$16^x + 2^{2x+1} - 99 = 0$$
can be written as
$$y^2 + 2y - 99 = 0$$
Hence show that $16^x + 2^{2x+1} - 99 = 0$ has only one solution of the form $\log_2 k$.

39) Solve the following equations, writing your answers in logarithmic form:
   a) $2^{2x} - 8 \times 2^x + 15 = 0$
   b) $3^{2x} - 9 \times 3^x + 20 = 0$
   c) $4^{2x} - 6 \times 4^x - 7 = 0$
   d) $3^{2x} - 6 \times 3^x - 40 = 0$

40) Solve the following equations, writing your answers in logarithmic form:
   a) $2^{2x} - 5 \times 2^{x+1} + 21 = 0$
   b) $5^{2x} - 3 \times 5^{x+1} + 50 = 0$
   c) $4^{2x} + 2^{2x+1} - 15 = 0$
   d) $4^{2x} - 5 \times 2^{2x+1} + 2^{2x+3} - 80 = 0$

41) Solve the equation
$$8^x - 2^{2x+2} + 3 \times 2^x = 0$$

42) Solve the following equations, writing your answers in logarithmic form:
   a) $3^{2x+1} - 13 \times 3^x - 10 = 0$
   b) $16^{x+1} - 139 \times 4^x - 45 = 0$
   c) $4^{2x+3} - 9 \times 2^{2x+5} = 0$
   d) $2^{4x+7} + 2^{2x+3} - 3 \times 4^{x+2} - 3 = 0$

43) Show that the solution to the equation
$$5^{x-3} = 2^{x+4}$$
can be written in the form
$$\frac{4\log_5 2 + 3}{1 - \log_5 2}$$

44) Show that the solution to the equation
$$8^{x+2} = 3^{2-x}$$
can be written in the form
$$\frac{2 - 2\log_3 8}{\log_3 8 + 1}$$

45) Show that the solution to the equation
$$3^{x+1} = 8^{x-1}$$
can be written in the form
$$\frac{\log_{10} 8 + \log_{10} 3}{\log_{10} 8 - \log_{10} 3}$$

46) Solve each of these equations by taking logarithms of base 3 of both sides:
   a) $3^x = 4^{x-1}$
   b) $3^{x+5} = 5^{2-x}$
   c) $8^{x-3} = 3^{x+8}$
   d) $10^{5-x} = 3^{x-6}$

47) Solve each of these equations by taking logarithms of base 2 of both sides:
   a) $5^x = 9^{x+2}$
   b) $3^{x-2} = 7^{3-x}$
   c) $9^{x+4} = 10^{4-x}$
   d) $5^{8-x} = 6^{4-x}$

48)
   a) Given that
   $$2\log_{10}(2-x) = \log_{10}(28-9x)$$
   show that $x^2 + 5x - 24 = 0$
   b) Hence show that
   $$2\log_{10}(2-x) = \log_{10}(28-9x)$$
   has only one solution.

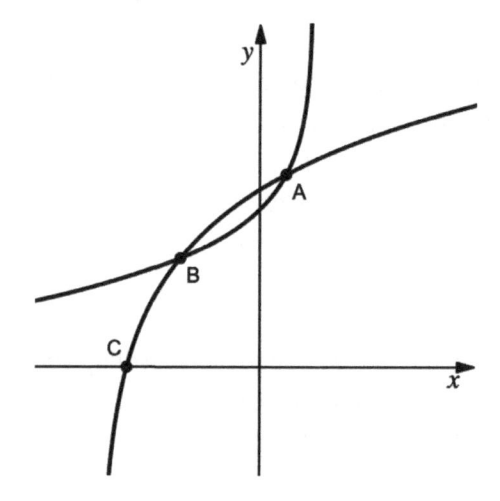

1.22 Exponentials and Logarithms (answers on page 428)

49) Solve the equation
$$\log_5(8x - 1) - \log_5(2 - x) = 2$$

50) Julia is trying to solve the equation
$$3(\log_8 x)^2 + 11 \log_8 x - 20 = 0$$
Julia's attempt is shown below:

$3(\log_8 x)^2 + 11 \log_8 x - 20 = 0$ Line 1
$6 \log_8 x + 11 \log_8 x - 20 = 0$ Line 2
$17 \log_8 x - 20 = 0$ Line 3
$\log_8 x = \frac{20}{17}$ Line 4
$x = \left(\frac{20}{17}\right)^8$ Line 5

a) Identify the two errors made by Julia.
b) Solve the equation
$$3(\log_8 x)^2 + 11 \log_8 x - 20 = 0$$
giving your answers in exact form.

51) George is trying to solve the equation
$$\log_3 \frac{1}{x} + \frac{1}{4 \log_3 x} = 0$$
George's attempt is shown below:

$\log_3 \frac{1}{x} + \frac{1}{4 \log_3 x} = 0$ Line 1
$\log_3 \frac{1}{x} + \frac{1}{4} \log_3 \frac{1}{x} = 0$ Line 2
$\frac{5}{4} \log_3 \frac{1}{x} = 0$ Line 3
$-\frac{5}{4} \log_3 x = 0$ Line 4
$x = 3$ Line 5

a) Identify the two errors made by George.
b) Solve the equation
$$\log_3 \frac{1}{x} + \frac{1}{4 \log_3 x} = 0$$
giving your answers in exact form.

52)
a) Show that the equation
$$2 \log_3 x = \log_3(mx - 4) + 4$$
where m is a positive constant, can be expressed in the form
$$x^2 - 81mx + 324 = 0$$
b) Given that $x^2 - 81mx + 324 = 0$ has only one root, find the value of m

53) Solve $2^{2^x} = 5$, writing your answer in an exact form.

54) The graph below shows two curves that intersect at the point C. The curve that passes through A has equation
$$y = \log_5(6x - 15) + 1$$
and the curve that passes through B has equation
$$y = 2 \log_5(x - 3)$$
Determine the exact coordinates of the points A, B and C.

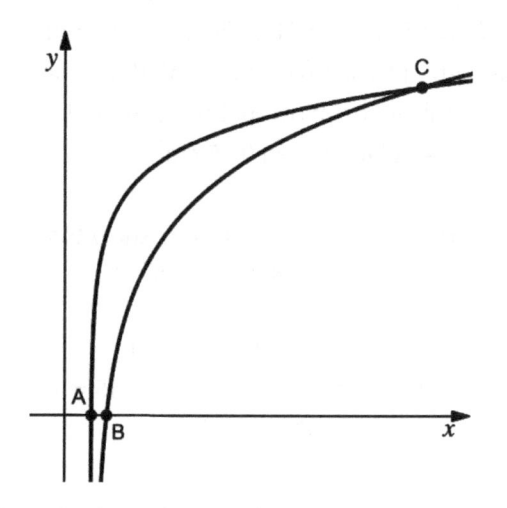

55) The graph below shows two curves that intersect at the points A and B. The curve that passes through C has equation
$$y = 2 \log_4(x + 6)$$
and the other curve has equation
$$y = \log_4(x + 48) - \log_4(2 - x)$$
Determine the exact coordinates of the points A, B and C.

1.23 e and Natural Logarithms (answers on page 430)

1) Solve the following equations, writing your answer in logarithmic form:
   a) $e^x = 8$
   b) $e^{2x} = 4$
   c) $e^{3x} = 8$
   d) $e^{5x} = \frac{1}{100}$

2) Solve each of these equations:
   a) $e^{4x+3} = 9$
   b) $e^{5x-2} = 10$
   c) $e^{9-2x} = 4$
   d) $e^{1-5x} = \frac{1}{5}$

3) Solve the following equations, writing your answer in the form $x = e^a$.
   a) $\ln x = 7$
   b) $\ln x = -5$
   c) $\ln x = \frac{1}{3}$
   d) $\ln x = -\frac{2}{5}$

4) Write each of these as a single logarithm:
   a) $\ln 4 + \ln 5 + \ln 6$
   b) $\ln 8 + \ln 3 - \ln 4$
   c) $\ln 12 - \ln 4 - \ln 6$
   d) $\ln 40 - \ln 8 + \ln 5$

5) Solve the following equations:
   a) $\ln x + \ln 2 = \ln 8$
   b) $\ln x - \ln 5 = \ln 3$
   c) $\ln 8 + \ln x = \ln \frac{1}{3}$
   d) $\ln 50 - \ln x = \ln \frac{3}{2}$

6) Evaluate each of these:
   a) $\ln e$
   b) $4 \ln \frac{1}{e}$
   c) $\ln \frac{1}{e^5}$
   d) $\ln \sqrt[3]{e}$

7) Write each of these as a single logarithm:
   a) $2 \ln 2 + 2 \ln 4$
   b) $3 \ln 2 - 2 \ln 3$
   c) $2 \ln 9 + 3 \ln 3 - 2 \ln 6$
   d) $2 \ln 10 - 4 \ln 2 - 2 \ln 5$

8) Write $3 \ln x - 4 \ln y$ as a single logarithm.

9) Write each of these as a single logarithm:
   a) $4 \ln p + 2 \ln q + 3 \ln r$
   b) $3 \ln p + 5 \ln q - 4 \ln r$
   c) $6 \ln p - 2 \ln q - 8 \ln r$
   d) $9 \ln p - 5 \ln q + 7 \ln r$

10) Write each of these as a single logarithm:
    a) $2 \ln p - (\ln q + 3 \ln r)$
    b) $\ln p - (2 \ln q - \ln r)$
    c) $3 \ln p - 2(\ln q + 2 \ln r)$
    d) $4 \ln p - 3(2 \ln q - 3 \ln r)$

11) Write each of these equations in a form not involving logarithms:
    a) $\ln x + \ln y = 5$
    b) $\ln y - \ln x = 12$
    c) $2 \ln x - \ln y = 4$
    d) $4 \ln x + 3 \ln y = 8$

12) The diagram below shows the straight-line graph $7y = x + 10$ and the curve with equation $\ln x = 2 \ln y$. Determine the exact coordinates of where the two graphs intersect.

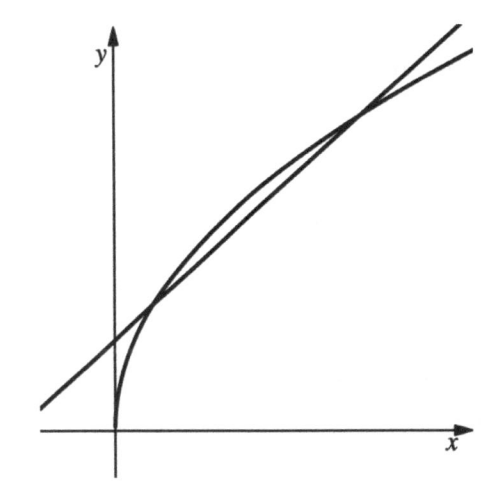

13) Solve each of these equations:
    a) $e^{2x} - 16 e^x = 0$
    b) $e^{3x} - 16 e^x = 0$
    c) $e^{3x} - 16 e^{2x} = 0$
    d) $e^{4x} - 16 e^{2x} = 0$

14) Solve each of these equations:
    a) $e^x = e^{2x+1}$
    b) $e^{2x} = e^{3x-5}$
    c) $e^x = 2 e^{4x}$
    d) $e^{2x} = 3 e^{x-1}$

1.23 e and Natural Logarithms (answers on page 430)

15)
   a) Using $y = e^x$ as a substitution, show that
   $$4e^x + 27e^{-x} - 39 = 0$$
   can be written as
   $$4y^2 - 39y + 27 = 0$$
   b) Hence solve the equation
   $$4e^x + 27e^{-x} - 39 = 0$$
   giving your answers as exact values.

16) Solve the following equations, writing your answers in logarithmic form:
   a) $e^{2x} - 6e^x + 8 = 0$
   b) $e^{2x} - 13e^x + 40 = 0$
   c) $e^x + 4 - 45e^{-x} = 0$
   d) $e^x + 5 - 150e^{-x} = 0$

17) Solve the following equations, writing your answers in logarithmic form:
   a) $2e^{2x} - 7e^x + 6 = 0$
   b) $4e^{2x} - 28e^x + 45 = 0$
   c) $6e^x - 23 - 18e^{-x} = 0$
   d) $8e^x + 14 - 15e^{-x} = 0$

18) Show that the solution to the equation
   $$4^{x-3} = 5^{x+6}$$
   can be written in the form
   $$\frac{6\ln 5 + 3\ln 4}{\ln 4 - \ln 5}$$

19) By writing $\ln 5 \times \ln 125 \times \ln 625$ in terms of $\ln 5$, show that
   $$\ln 5 \times \ln 125 \times \ln 625 > 12$$

20) Solve each of these equations by taking natural logarithms of both sides:
   a) $2^x = 5^{x+4}$
   b) $4^{x-1} = 6^{x+2}$
   c) $5^{3-x} = 7^{x+1}$
   d) $8^{9-x} = 6^{8-x}$

21) Solve each of these equations, writing your answers in exact form:
   a) $2\ln(3x - 5) - 4 = 0$
   b) $2\ln(3 - 2x) - 9 = 0$
   c) $2\ln(x + 4) - \ln 5 = 0$
   d) $5\ln 2x - 3\ln 3x = 1$

22) Solve the equation $2\ln x - \ln(x + 12) = 0$

23) Solve each of these equations, writing your answers in exact form:
   a) $\ln x = 2 + \ln(x + 3)$
   b) $\ln(x - 1) = 3 + \ln(x + 1)$
   c) $\ln 2x = 4 - \ln x$
   d) $\ln 4x = 5 - \ln(x + 1)$

24) The diagram below shows the curves $y = \ln(x^2 + 2)$ and $y = 4\ln x$. Find the exact coordinates of where the two curves intersect.

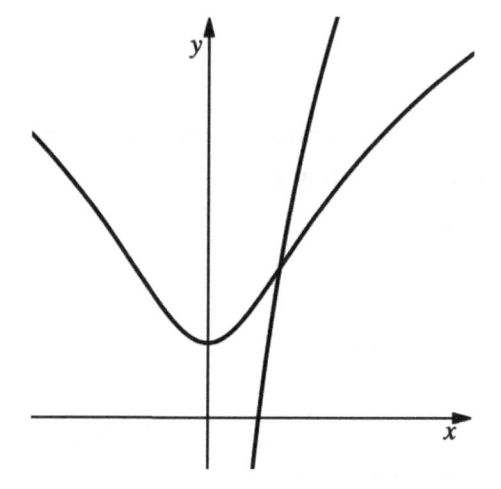

25) Solve each of these equations, writing your answers in exact form:
   a) $2\ln x = 2 + \ln 2x$
   b) $2\ln x = 3 - \ln 8x$
   c) $2\ln x = \ln 4x - 1$
   d) $2\ln x = \ln 10x - 4$

26) Solve the equation
   $$3\ln x = 3 - \ln 16x$$
   writing your answer in exact form.

27) Solve each of these equations, writing your answers in exact form:
   a) $\ln x + \frac{32}{\ln x} = 12$
   b) $2\ln x + \frac{6}{\ln x} = 7$
   c) $6\ln x + 11 = \frac{10}{\ln x}$
   d) $10\ln x - 33 = \frac{7}{\ln x}$

28) Solve the equation
   $$2\ln(x + 10) = \ln x + \ln(2x - 1)$$

29) Solve the equation
   $$\ln(5 - 3x) + \ln(8 + 4x) = 2\ln(x + 2)$$

1.23 e and Natural Logarithms (answers on page 430)

30) Solve the equation
$$2\ln(5-2x) = \ln 2 + \ln\left(\frac{233}{8} - 2\left(x - \frac{5}{4}\right)^2\right)$$

31)
   a) Using $y = e^x$ as a substitution, show that
   $$1 - 28e^{-3x} - 4e^{-2x} + 7e^{-x} = 0$$
   can be written as
   $$y^3 + 7y^2 - 4y - 28 = 0$$
   b) Hence solve the equation
   $$1 - 28e^{-3x} - 4e^{-2x} + 7e^{-x} = 0$$
   giving your answers as exact values.

32) Given that $y = \ln x$ and $y^2 - 3y + 2 = 0$, find the possible values of x

33) Solve the equation
$$6(\ln x)^2 - 19\ln x + 15 = 0$$

34) Solve the equation
$$\ln 2x + \ln 2x^2 + \ln 2x^3 = 6$$

35) Solve the equation
$$\frac{e^x}{3^x} = e$$
writing your answer in the form
$$\frac{p}{q + \ln r}$$
where p, q and r are integers.

36) Solve the equation
$$3^x e^{2x+3} = 2$$
writing your answer in the form
$$\frac{p + \ln q}{r + \ln s}$$
where p, q, r and s are integers.

37) Solve the equation
$$5^x e^{8x} = e^{10}$$
writing your answer in the form
$$\frac{p}{q + \ln r}$$
where p, q and r are integers.

38) Fully simplify
$$\frac{3x^2 - 19x + 6}{x^2 - x - 30}$$
Hence, given that
$$\ln(3x^2 - 19x + 6) - 1 = \ln(x^2 - x - 30)$$
where $x \neq 6$, find x in terms of e.

39) Determine the exact coordinates of where the curves $y = 2e^x - 11$ and $y = 40e^{-x}$ intersect.

40) A curve has equation $y = ae^x + b$ and passes through the points $(\ln 2, 12)$ and $\left(-3\ln 2, \frac{87}{16}\right)$. Find the values of a and b

41) A curve has equation $y = p\ln x + q$ and passes through the points $\left(e^{12}, -\frac{55}{2}\right)$ and $\left(e, \frac{11}{6}\right)$. Find the values of p and q

42) Solve the simultaneous equations
$$e^x + y^2 = 4$$
$$2\ln y = x$$

43) Determine the exact coordinates of where the curves $x + e^y = 4$ and $\ln((x-4)^2) - y = 2$ intersect.

44) The water level, h m, in a large tank over time, t hours, is modelled by the equation
$$h = 20 - 5\ln(t + 1)$$
   a) What is the initial height of the water in the tank?
   b) How high is the water in the tank after 1 hour?
   c) After how many hours will the water level be 10 metres?

45) A population of hawks, P, can be modelled by the equation
$$P = 100\ln\left(\frac{t+3}{3}\right) + 500$$
where t is the number of years since a conservation project started.
   a) What was the initial population of hawks when the conservation project started?
   b) What does the model suggest the size of the population will be in 5 years after the conservation project started?
   c) After how many years does the model suggest the population will reach 700 hawks?

46) The temperature, T°C, of the water in a lake can be modelled by the equation
$$T = 2\ln(t+3) - \ln(t+2) + \ln 100$$
where t is the number of days after the 1st March.
   a) What was the temperature of the lake on the 1st March?
   b) How many days after the 1st March will the temperature of the lake be $(\ln 3000)$°C ?

1.24 Graph Sketching: Exponentials and Logarithms (answers on page 432)

1) Find the equation of each of the following graphs in the form
$$y = 2^x + b$$
a)

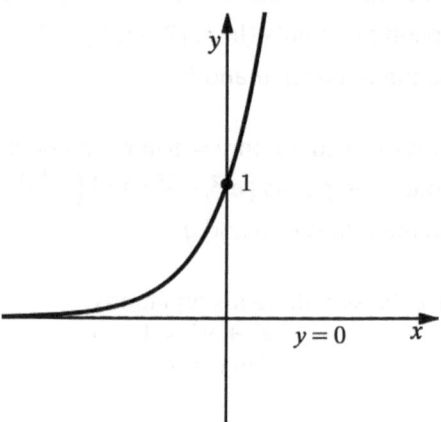

b)

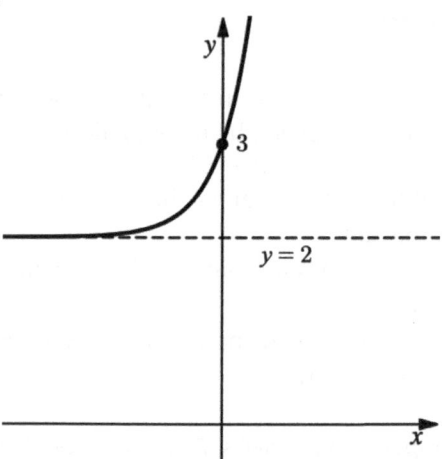

c)

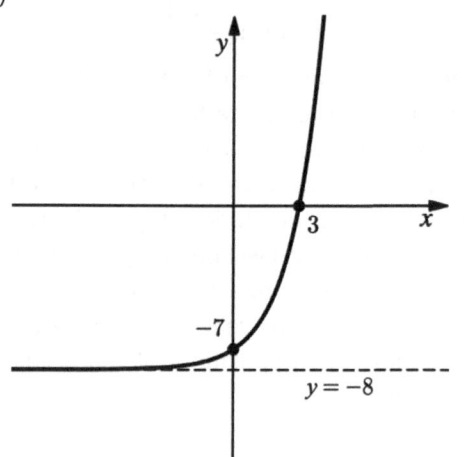

2) Sketch the following graphs, identifying where they intersect the coordinate axes and the equations of any asymptotes:
a) $y = 3^x - 9$
b) $y = 4^x + 6$

3) Find the equation of each of the following graphs in the form
$$y = 2^{x-a} + b$$
a)

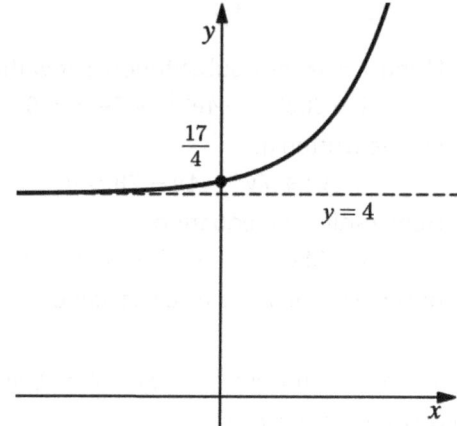

b)

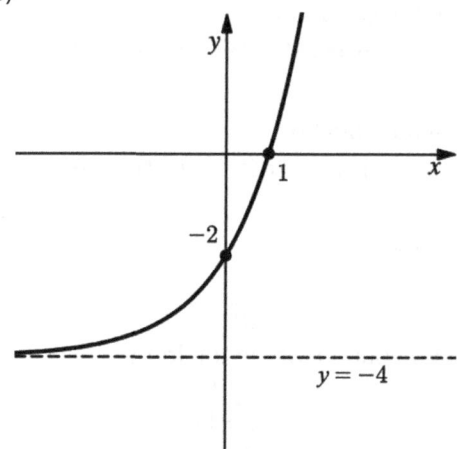

c)

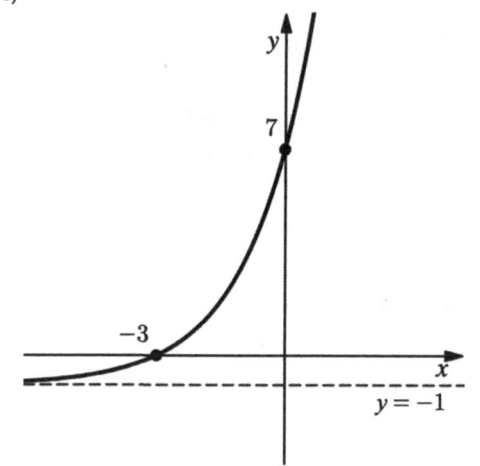

4) Sketch the following graphs, identifying where they intersect the coordinate axes and the equations of any asymptotes:
a) $y = 2^{x-4} - 4$
b) $y = 3^{x+2} + 2$

1.24 Graph Sketching: Exponentials and Logarithms (answers on page 432)

5) Find the equation of each of the following graphs in the form
$$y = 2^{a-x} + b$$
a)

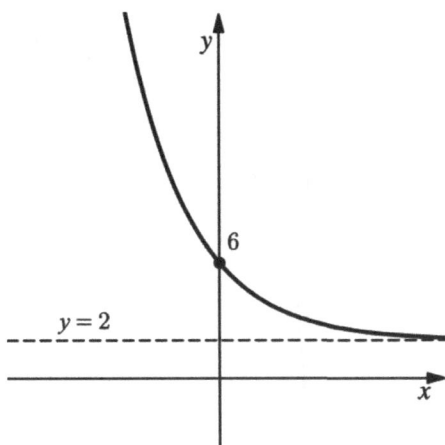

b)

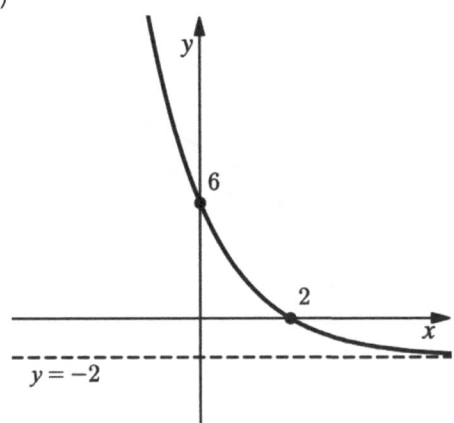

c)

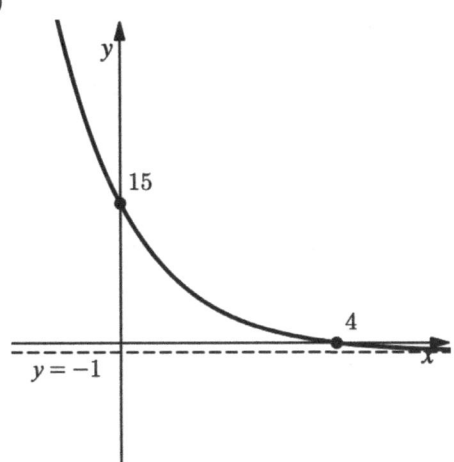

6) Sketch the following graphs, identifying where they intersect the coordinate axes and the equations of any asymptotes:
a) $y = 3^{2-x} + 4$
b) $y = 5^{-1-x} - 1$

7) Find the equation of each of the following graphs in the form
$$y = b - 2^{x-a}$$
a)

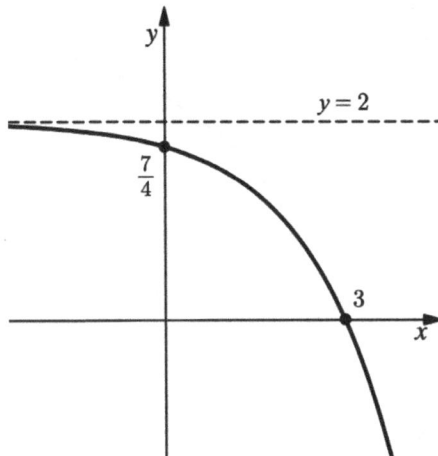

b)

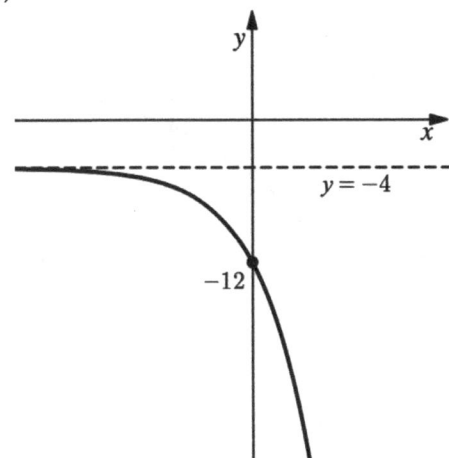

c)

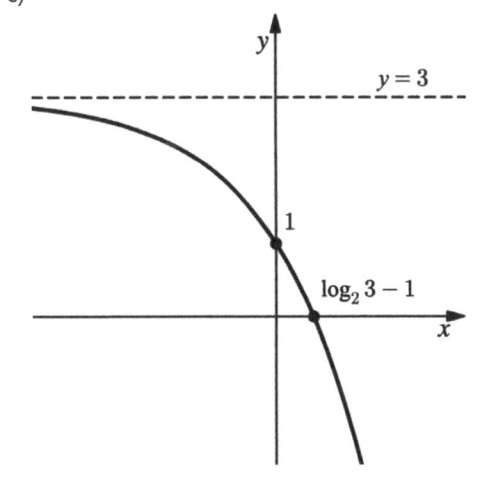

8) Sketch the following graphs, identifying where they intersect the coordinate axes and the equations of any asymptotes:
a) $y = 2 - 3^{x-1}$
b) $y = 3 - 2^{1-x}$

1.24 Graph Sketching: Exponentials and Logarithms (answers on page 432)

9) Find the equation of each of the following graphs in the form
$$y = \log_2(x - a)$$

a) [graph with asymptote $x = 2$, passing through $(3, 0)$]

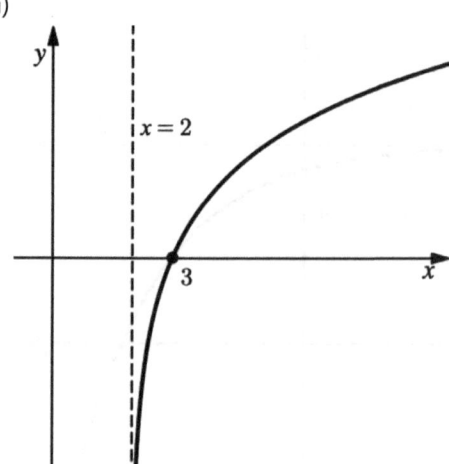

b) [graph with asymptote $x = -4$, passing through $(-3, 0)$ and $(0, 2)$]

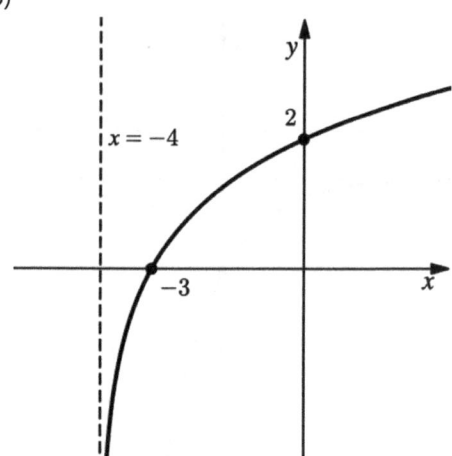

c) [graph with asymptote $x = 10$, passing through $(11, 0)$]

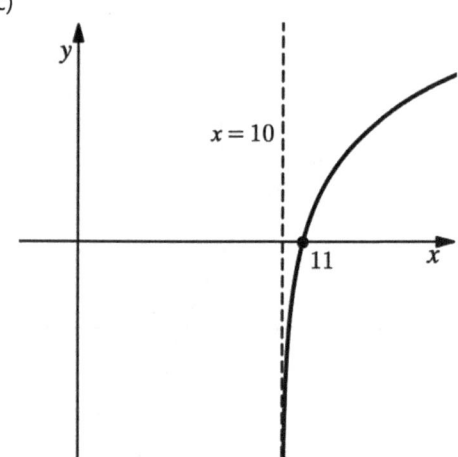

10) Sketch the following graphs, identifying where they intersect the coordinate axes and the equations of any asymptotes:
a) $y = \log_3(x - 2)$
b) $y = \log_5(x + 4)$

11) Find the equation of each of the following graphs in the form
$$y = \log_2 x + b$$

a) [graph with asymptote $x = 0$, passing through $(\tfrac{1}{2}, 0)$]

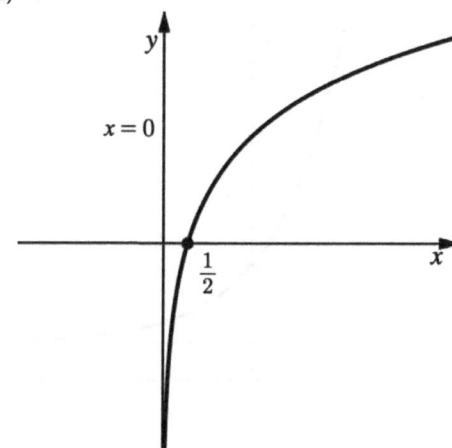

b) [graph with asymptote $x = 0$, passing through $(4, 0)$]

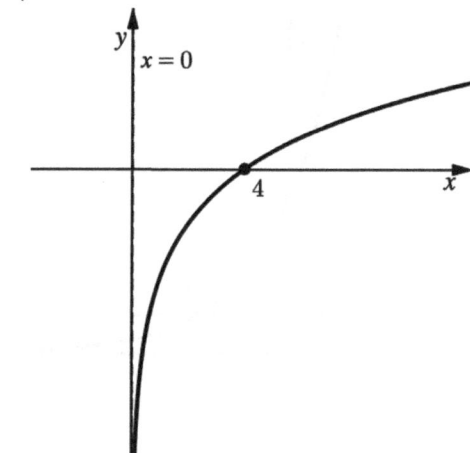

c) [graph with asymptote $x = 0$, passing through $(\tfrac{\sqrt{2}}{2}, 0)$]

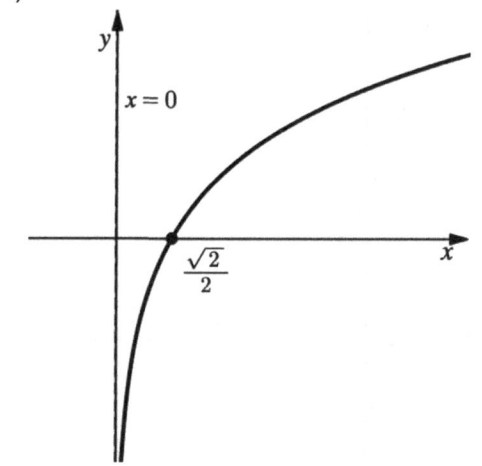

12) Sketch the following graphs, identifying where they intersect the coordinate axes and the equations of any asymptotes:
a) $y = \log_5 x - 5$
b) $y = \log_3 x + 4$

1.24 Graph Sketching: Exponentials and Logarithms (answers on page 432)

13) Find the equation of each of the following graphs in the form
$$y = \log_2(x - a) + b$$

a)

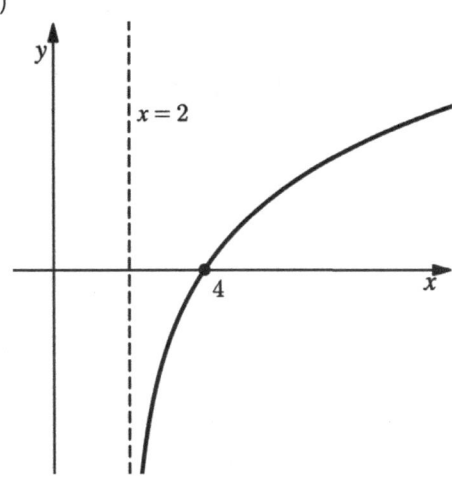

b)

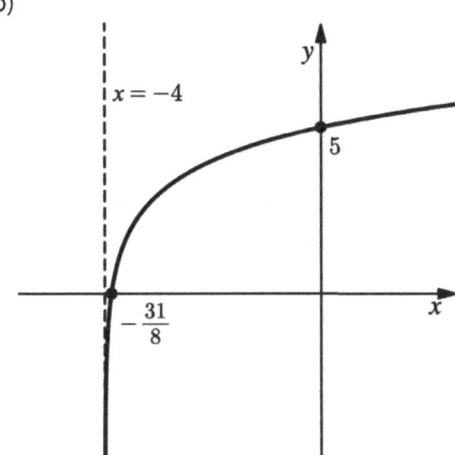

c)

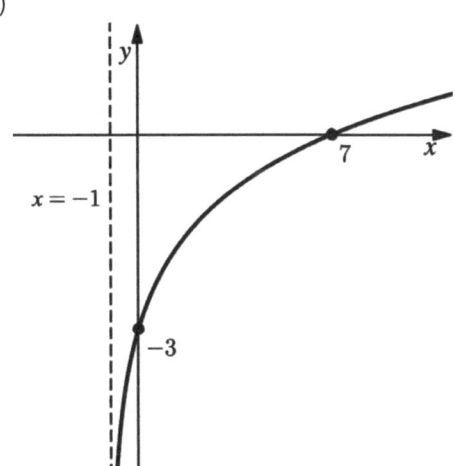

14) Sketch the following graphs, identifying where they intersect the coordinate axes and the equations of any asymptotes:
   a) $y = \log_4(x - 2) + 1$
   b) $y = \log_3(x + 3) - 3$

15) Find the equation of the following graph in the form
$$y = \log_2(a - x) + b$$

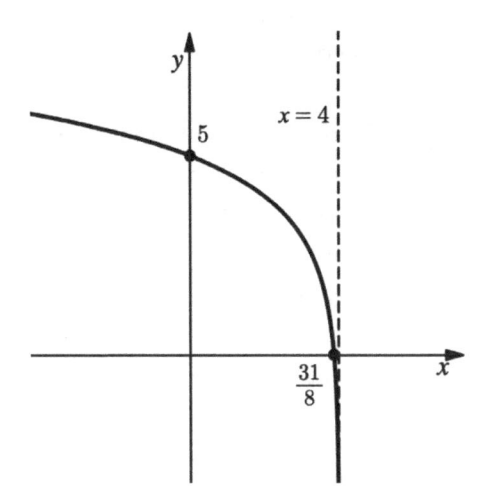

16) Sketch the following graphs, identifying where they intersect the coordinate axes and the equations of any asymptotes:
   a) $y = \log_4(5 - x) + 3$
   b) $y = \log_5(10 - x) - 2$
   c) $y = \log_3(-3 - x) + 1$
   d) $y = \log_4(-8 - x) - 2$

17) Find the equation of the following graph in the form
$$y = b - \log_2(x - a)$$

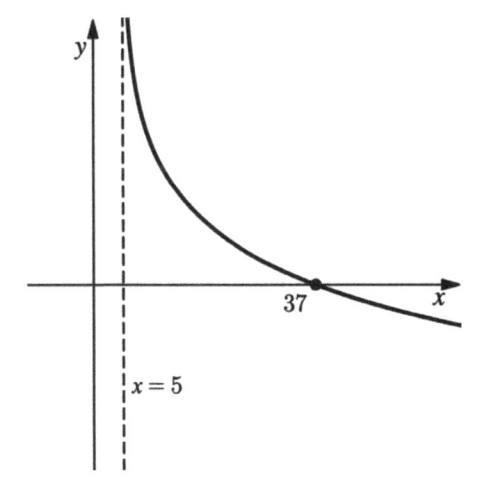

18) Sketch the following graphs, identifying where they intersect the coordinate axes and the equations of any asymptotes:
   a) $y = 3 - \log_3(x - 3)$
   b) $y = 2 - \log_4(x + 2)$
   c) $y = 1 - \log_5(3 - x)$
   d) $y = 2 - \log_3(9 - x)$

1.24 Graph Sketching: Exponentials and Logarithms (answers on page 432)

19) Sketch the following pairs of graphs on the same axes, clearly identifying which is which:
   a) $y = 2^x$ and $y = 4^x$
   b) $y = 3^x$ and $y = 4^{-x}$
   c) $y = 2^{-x}$ and $y = 3^{-x}$

20) Sketch the following pairs of graphs on the same axes, clearly identifying which is which:
   a) $y = -5^x$ and $y = -3^x$
   b) $y = -2^x$ and $y = 3^x$
   c) $y = -4^{-x}$ and $y = -2^{-x}$

21) Sketch the following pairs of graphs on the same axes, clearly identifying which is which:
   a) $y = \log_2 x$ and $y = \log_4 x$
   b) $y = \log_5 x$ and $y = \log_3(-x)$
   c) $y = \log_6(-x)$ and $y = \log_2(-x)$

22) Sketch the following pairs of graphs on the same axes, clearly identifying which is which:
   a) $y = -\log_4 x$ and $y = -\log_3 x$
   b) $y = -\log_8 x$ and $y = \log_4 x$
   c) $y = -\log_2(-x)$ and $y = -\log_4(-x)$

23) It is given that $k > 1$. Sketch the following graphs, identifying where they intersect the coordinate axes and the equations of any asymptotes:
   a) $y = k^x$
   b) $y = k^x - k$
   c) $y = k^{x-k} + k$

24) It is given that $k > 1$. Write down a suitable equation for each of the following curves.
   a)

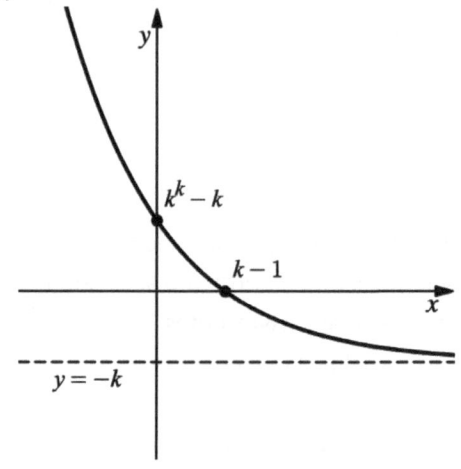

   b)

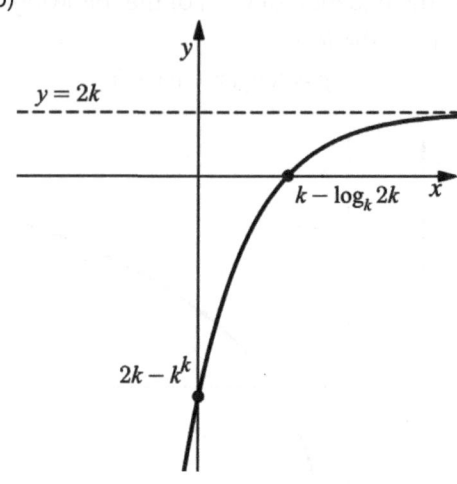

25) It is given that $k > 1$. Sketch the following graphs, identifying where they intersect the coordinate axes and the equations of any asymptotes:
   a) $y = \log_k x + k$
   b) $y = \log_k(x + k) - k$

26) It is given that $k > 1$. Write down a suitable equation for each of the following curves.
   a)

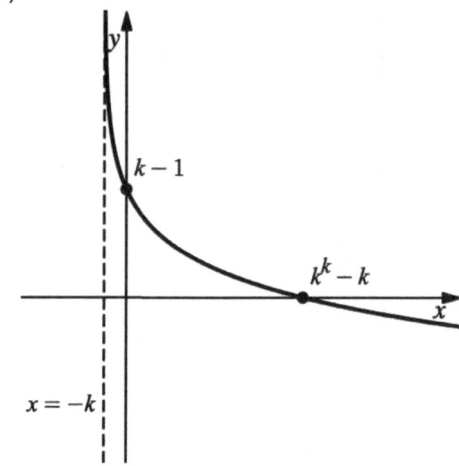

   b)

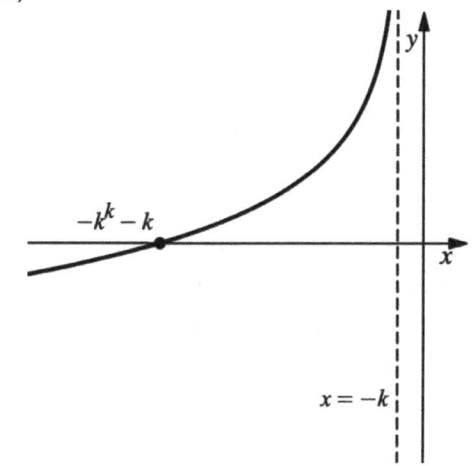

1.25 Graph Sketching: e and Natural Logarithms (answers on page 434)

1) Find the equation of each of the following graphs in the form
$$y = e^x + b$$

a)

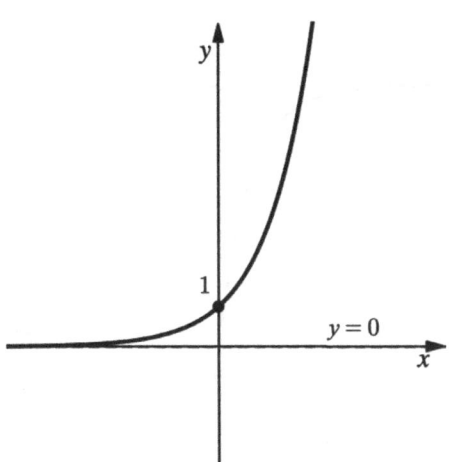

b)

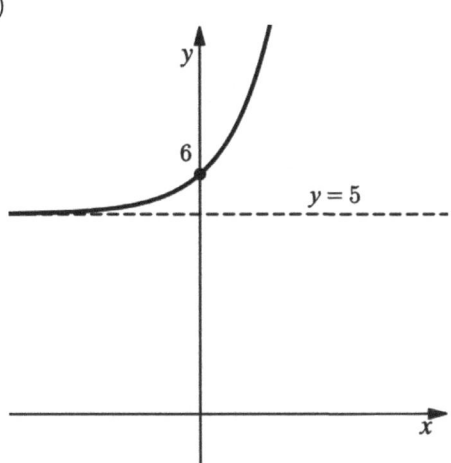

c)

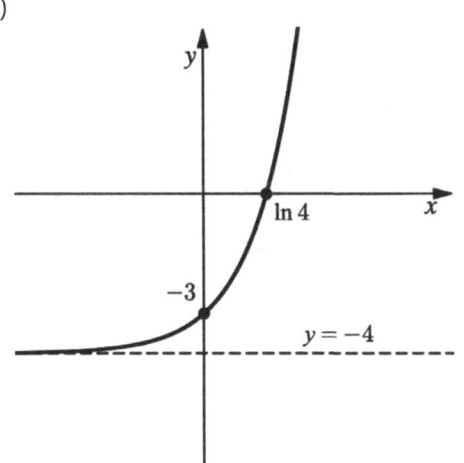

2) Sketch the following graphs, identifying where they intersect the coordinate axes and the equations of any asymptotes:
   a) $y = e^x - 6$
   b) $y = e^x + e$

3) Find the equation of each of the following graphs in the form
$$y = e^{x-a} + b$$

a)

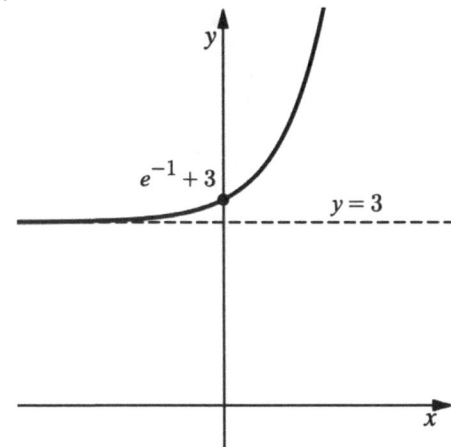

b)

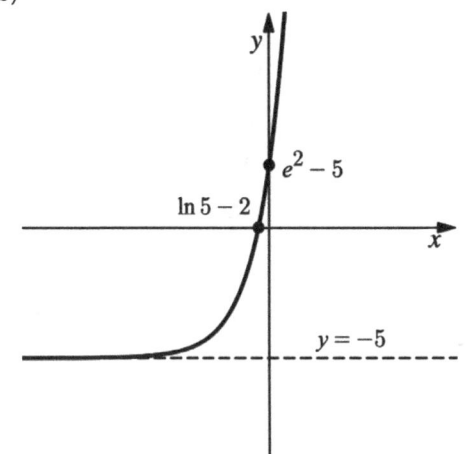

c)

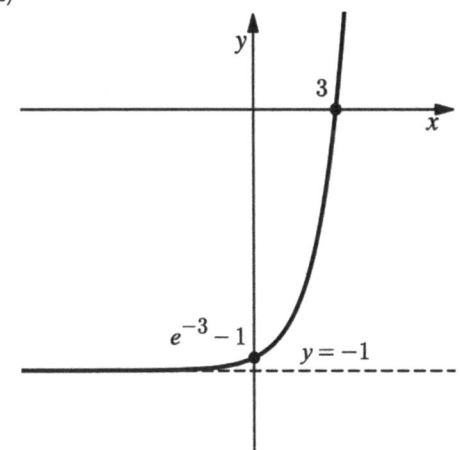

4) Sketch the following graphs, identifying where they intersect the coordinate axes and the equations of any asymptotes:
   a) $y = e^{x+1} - 10$
   b) $y = e^{x-e} - e$

1.25 Graph Sketching: e and Natural Logarithms (answers on page 434)

5) Find the equation of each of the following graphs in the form
$$y = e^{a-x} + b$$

a) [graph: horizontal asymptote $y=2$, y-intercept $e+2$]

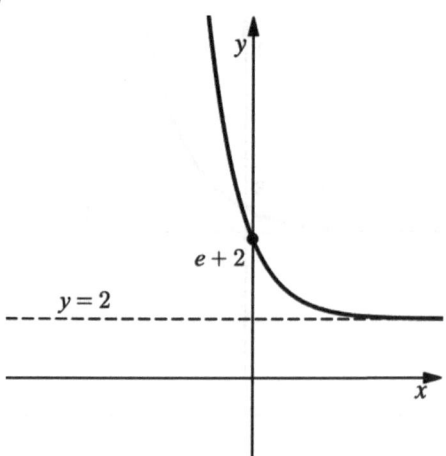

b) [graph: horizontal asymptote $y=-3$, y-intercept $e^2 - 3$, x-intercept $2 - \ln 3$]

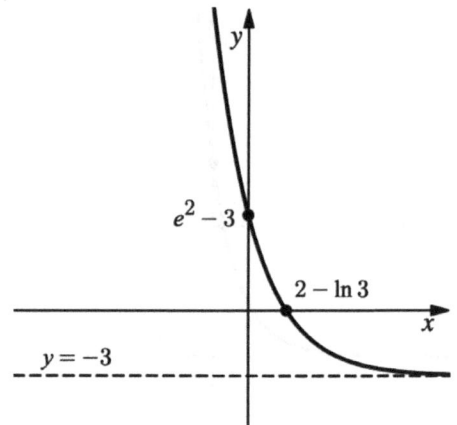

c) [graph: horizontal asymptote $y=-9$, x-intercept $e^{-3} - 9$, y-intercept $-3 - \ln 9$]

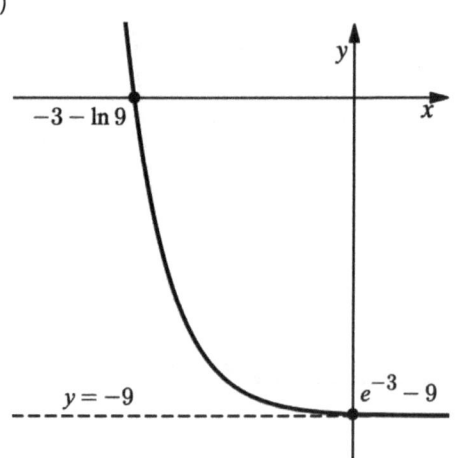

6) Sketch the following graphs, identifying where they intersect the coordinate axes and the equations of any asymptotes:
   a) $y = e^{5-x} - 20$
   b) $y = e^{-1-x} + 20$

7) Find the equation of each of the following graphs in the form
$$y = b - e^{x-a}$$

a) [graph: horizontal asymptote $y=1$, passing through 0]

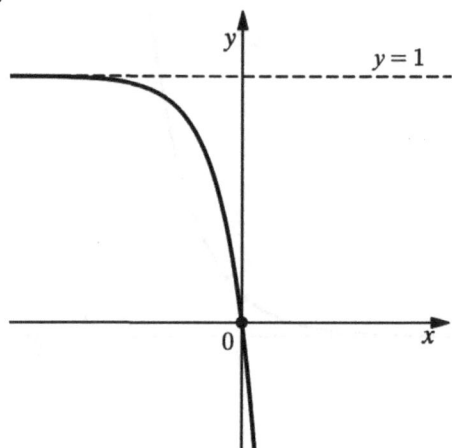

b) [graph: horizontal asymptote $y=-2$, y-intercept $-2 - e^{-1}$]

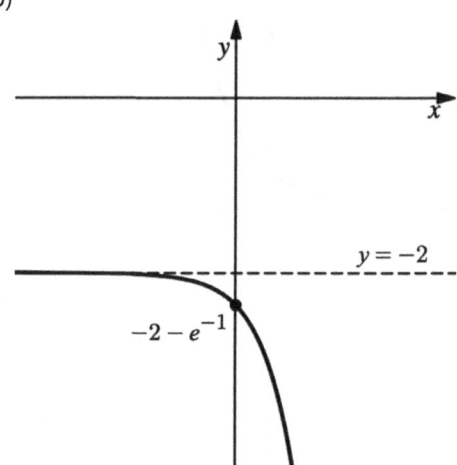

c) [graph: horizontal asymptote $y=4$, x-intercept $\ln 4 - 2$, y-intercept $4 - e^2$]

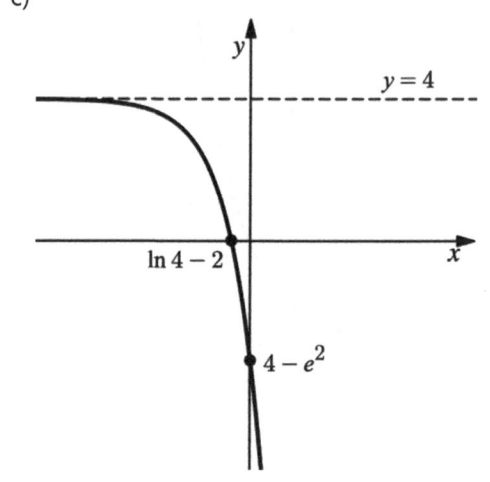

8) Sketch the following graphs, identifying where they intersect the coordinate axes and the equations of any asymptotes:
   a) $y = 10 - e^{x+4}$
   b) $y = 5 - e^{4-x}$

1.25 Graph Sketching: e and Natural Logarithms (answers on page 434)

9) Find the equation of each of the following graphs in the form
$$y = \ln(x - a)$$

a) [Graph showing asymptote $x = 3$, passing through $(4, 0)$]

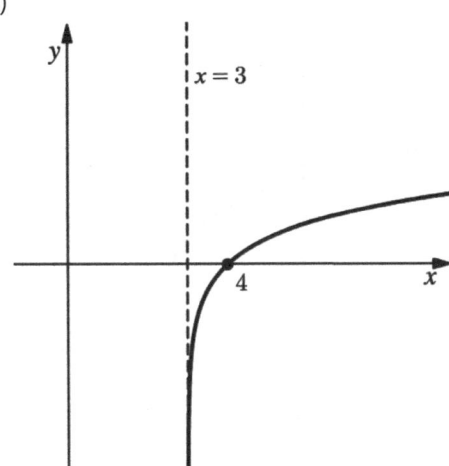

b) [Graph showing asymptote $x = -5$, passing through $(-4, 0)$ and $(0, \ln 5)$]

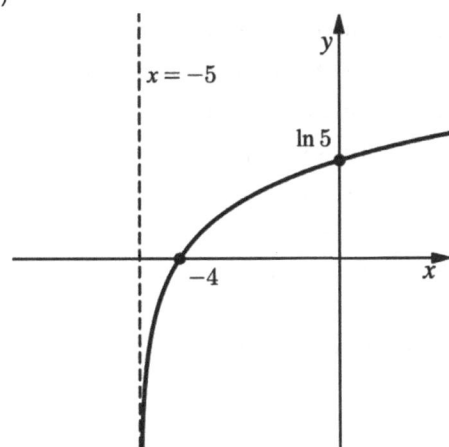

c) [Graph showing asymptote $x = -1$, passing through $(0, 0)$]

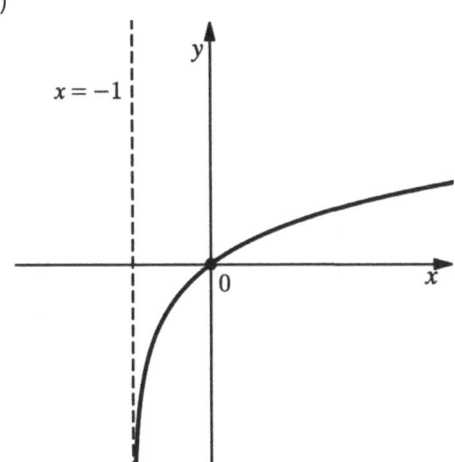

10) Sketch the following graphs, identifying where they intersect the coordinate axes and the equations of any asymptotes:
   a) $y = \ln(x - 2)$
   b) $y = \ln(x + 2)$

11) Find the equation of each of the following graphs in the form
$$y = \ln x + b$$

a) [Graph showing asymptote $x = 0$, passing through $(e^{-1}, 0)$]

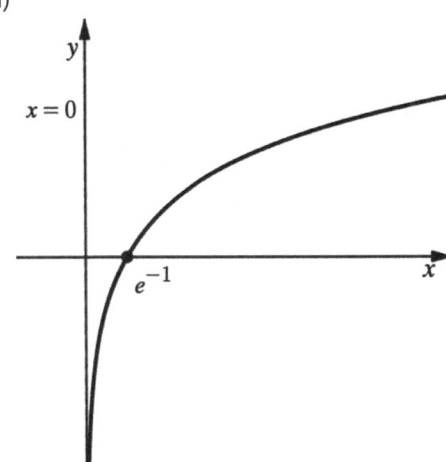

b) [Graph showing asymptote $x = 0$, passing through $(e^2, 0)$]

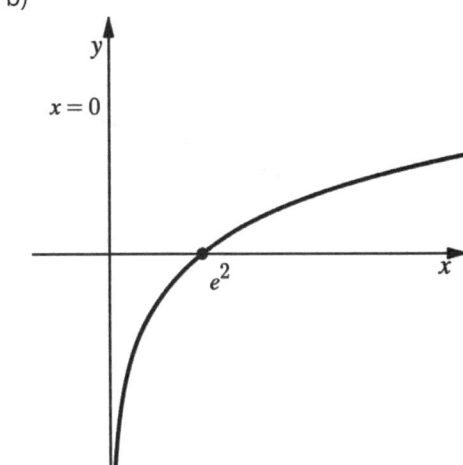

c) [Graph showing asymptote $x = 0$, passing through $(e^{-\frac{1}{2}}, 0)$]

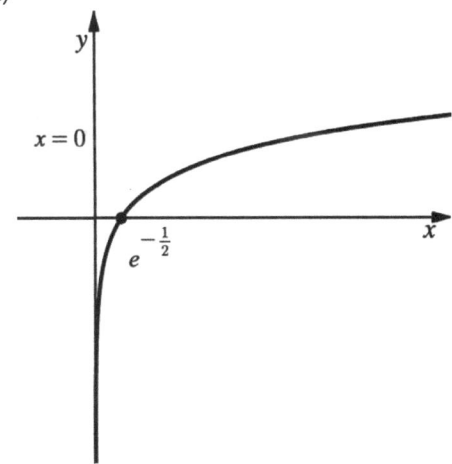

12) Sketch the following graphs, identifying where they intersect the coordinate axes and the equations of any asymptotes:
   a) $y = \ln x + 3$
   b) $y = \ln x - e$

1.25 Graph Sketching: e and Natural Logarithms (answers on page 434)

13) Find the equation of each of the following graphs in the form
$$y = \ln(x - a) + b$$

a) [Graph showing vertical asymptote $x = 1$ and x-intercept at $e^{-1} + 1$]

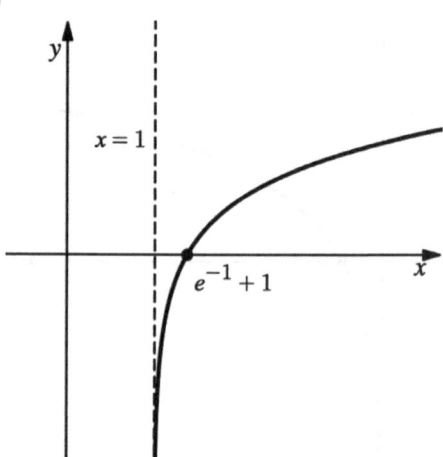

b) [Graph showing vertical asymptote $x = -2$, y-intercept at $\ln 2 + 3$, and x-intercept at $e^{-3} - 2$]

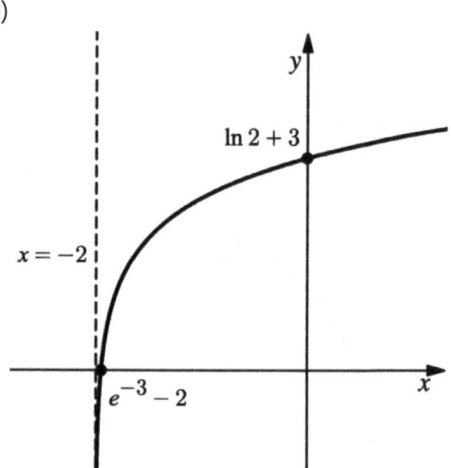

c) [Graph showing vertical asymptote $x = -1$, x-intercept at $e^2 - 1$, and y-intercept at -2]

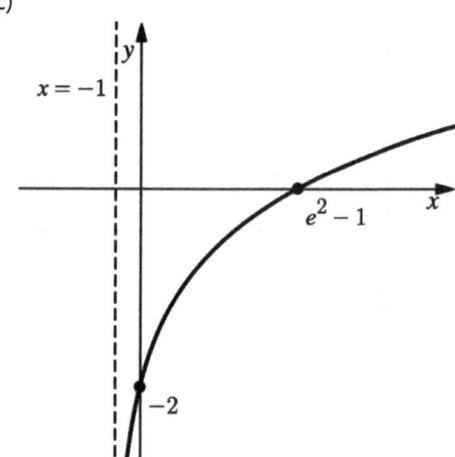

14) Sketch the following graphs, identifying where they intersect the coordinate axes and the equations of any asymptotes:
   a) $y = \ln(x - 3) - 3$
   b) $y = \ln(x - e) + e$

15) Find the equation of the following graph in the form
$$y = \ln(a - x) + b$$

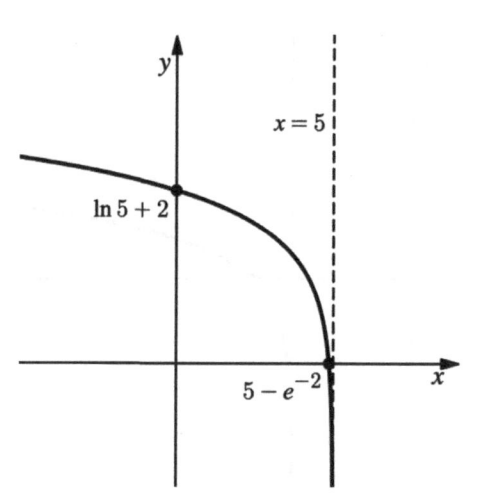

[Graph showing vertical asymptote $x = 5$, y-intercept at $\ln 5 + 2$, and x-intercept at $5 - e^{-2}$]

16) Sketch the following graphs, identifying where they intersect the coordinate axes and the equations of any asymptotes:
   a) $y = \ln(8 - x) - 1$
   b) $y = \ln(1 - x) + 3$
   c) $y = \ln(e - x) - e$
   d) $y = \ln(-e - x) + 2e$

17) Find the equation of the following graph in the form
$$y = b - \ln(x - a)$$

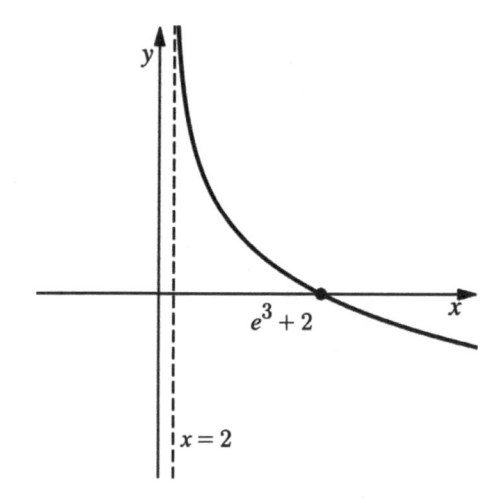

[Graph showing vertical asymptote $x = 2$ and x-intercept at $e^3 + 2$]

18) Sketch the following graphs, identifying where they intersect the coordinate axes and the equations of any asymptotes:
   a) $y = 1 - \ln(x - 1)$
   b) $y = e - \ln(x + e)$
   c) $y = 1 - \ln(2 - x)$
   d) $y = \frac{e}{2} - \ln(3e - x)$

1.25 Graph Sketching: e and Natural Logarithms (answers on page 434)

Prerequisite knowledge for Q19-Q22:
Exponentials and Logarithms

19) Sketch the following pairs of graphs on the same axes, clearly identifying which is which:
   a) $y = 2^x$ and $y = e^x$
   b) $y = 3^x$ and $y = e^x$
   c) $y = e^x$ and $y = e^{-x}$

20) Sketch the following pairs of graphs on the same axes, clearly identifying which is which:
   a) $y = -2^x$ and $y = -e^x$
   b) $y = 3^x$ and $y = -e^x$
   c) $y = -e^{-x}$ and $y = -e^x$

21) Sketch the following pairs of graphs on the same axes, clearly identifying which is which:
   a) $y = \log_2 x$ and $y = \ln x$
   b) $y = \log_3 x$ and $y = \ln x$
   c) $y = \log_2(-x)$ and $y = \ln(-x)$

22) Sketch the following pairs of graphs on the same axes, clearly identifying which is which:
   a) $y = -\log_3 x$ and $y = -\ln x$
   b) $y = -\log_2 x$ and $y = \ln x$
   c) $y = -\log_3(-x)$ and $y = -\ln(-x)$

23) Given that $k > 1$, sketch the following graphs, identifying where they intersect the coordinate axes and the equations of any asymptotes:
   a) $y = e^{kx} - 2k$
   b) $y = k - e^{-kx}$

24) It is given that $k > 1$. Write down a suitable equation for each of the following curves.
   a)

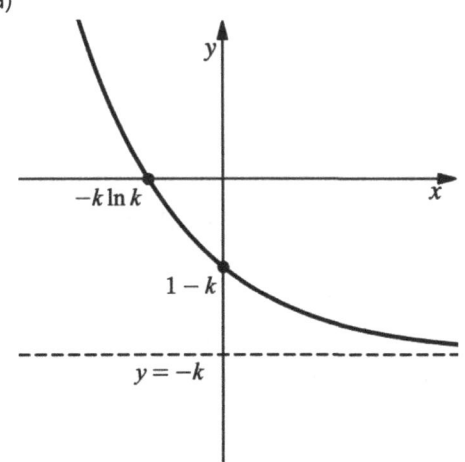

   b)

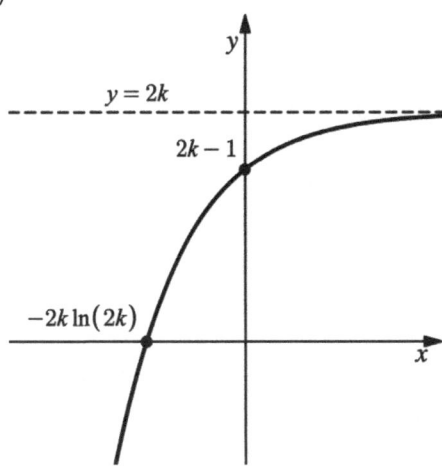

25) Given that $k > 1$, sketch the following graphs, identifying where they intersect the coordinate axes and the equations of any asymptotes:
   a) $y = k - \ln x$
   b) $y = k \ln(x + k) - k$

26) Given that $k > 1$, write down a suitable equation for each of the following curves.
   a)

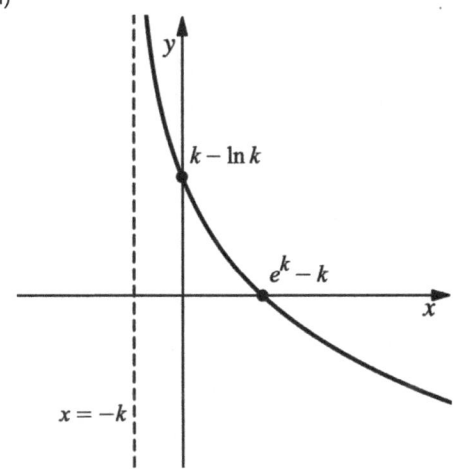

   b)

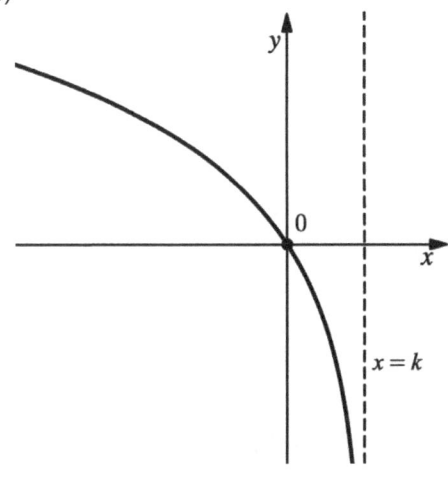

1.26 Exponential Growth and Decay (answers on page 436)

1) For each of the equations below find the value of y when $x = 0$:
   a) $y = 1 + 2^{-x}$
   b) $y = 20 - 2^{0.1x}$
   c) $y = 10 + e^{-0.1x}$
   d) $y = 5 - e^{0.001x}$
   e) $y = 21 + 5e^{-0.5x}$
   f) $y = 3 - 5 \times \left(\frac{1}{2}\right)^{-\frac{1}{4}x}$

2) Find the value of P when $t = 0$ for the following:
   a)
   $$P = \frac{100}{1 + e^{-\frac{1}{2}t}}$$
   b)
   $$P = \frac{18}{1 + e^{-0.1t}}$$

3) A population of bacteria is modelled with the equation
   $$P = 250e^{0.15t}$$
   where P is the population size and t is the time in hours.
   a) State the size of the initial population.
   b) Find the size of the population after 20 hours.
   c) When does the population reach 1000 bacteria?

4) For each of the models below, find the time t (hours) it takes for an amount of a substance y (grams) to double from its initial amount, giving your answers in exact form.
   a) $y = 3 \times 2^t$
   b) $y = 4e^{1.5t}$
   c) $y = 150 \times 1.6^t$
   d) $y = 20 + 4e^{0.1t}$

5) The quantity of radioactive isotope Uranium-235 y grams over time may be modelled by the equation
   $$y = 1000 \times 2^{-0.00142t}$$
   where t is millions of years.
   a) State the initial mass of the Uranium-235
   b) Show that the half-life of Uranium-235 is about 704 million years.

6) For each of the models below, find the initial value of y and state what is happening to the value of x as $x \to \infty$
   Hence sketch each of the models.
   a) $y = 10 + 2e^{-0.3x}$
   b) $y = -5 + 7e^{-2x}$
   c) $y = 1.2 - 0.6e^{-7.6x}$
   d) $y = 0.2 + 1.8e^{2.3x}$

7) The graph below shows the temperature of a cup of coffee over time.
   The room temperature is 22°C.
   The temperature, T°C, after t minutes may be modelled by the equation
   $$T = A + Be^{-\frac{1}{5}t}$$

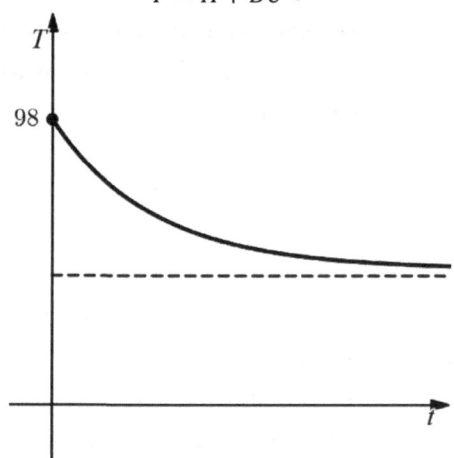

   a) State the initial temperature of the coffee.
   b) Write down the equation of the horizontal asymptote.
   c) Hence find the values of A and B
   d) Suggest how the model may be refined for a cup of coffee that is initially 100°C.
   e) Suggest values of A and B for a cup of coffee cooling at the same rate but from initial temperature 97°C in a room at 18°C.

8) The height of a plant is given by the equation
   $$h = \frac{50}{1 + 4e^{-\frac{1}{5}t}}$$
   where h is the height in cm and t is the number of weeks that have passed.
   a) According to this model state the initial height of the plant.
   b) What height does this model predict in the long term for this plant?

1.26 Exponential Growth and Decay (answers on page 436)

9) The height of a child was recorded for the first seven years of their life. A model suggested for the child's height, h cm, at x years old is given by
$$h = 130 - 80 \times 1.45^{-x}$$
where $0 \leq x \leq 7$. Using the model:
   a) State the height of a newborn baby.
   b) Find the height of a child after 18 months, giving your answer to the nearest millimetre.
   c) Estimate the age of the child in months (to the nearest month) when their height is 1 m.
   d) Explain why this model might be unsuitable for predicting the child's height as a teenager.

10) An investment of £1200 is made at 5% compound interest per annum, with interest added to the balance only at the end of each complete year.
    a) Create an equation to model the value of the investment over time.
    b) Sketch the curve for your model, including labelling any points of intersections with axes.
    c) Find the value of the investment after 4 years giving your answer to the nearest penny.
    d) Find when the investment reaches more than £2000.

11) A fine for speeding is initially £A. Each week it is not paid, the fees increase exponentially. After 1 week the fine becomes £150 and after 2 weeks it becomes £225.
    a) Calculate the initial fine, £A
    b) Find a model which for this data, giving your answer in the form
    $$C = A \times r^t$$
    where C is the cost, t is the time (in weeks) and A and r are constant values to be found.
    c) Use your model to find the cost of the fine in week 4.

12) The number of people using a social media platform is declining exponentially since its peak of popularity (at $t = 0$).
    At 3 years and 6 years it had 180 and 144 million active users respectively.
    a) Find the number of active users at its peak.
    b) Hence find a model which fits the data, giving your answer in the form
    $$N = A \times r^{\frac{1}{3}t}$$
    where N is the number of active users, in millions, t is the time, in years, and A and r are constant values to be found.
    c) Use your model to predict the number of users after 10 years has passed.

13) A population of insects is known to be growing exponentially. The population 2 and 4 months into the data collection is 1440 and 3240 insects respectively.
    A model for the data may be given by
    $$P = A \times r^t$$
    where P is population, t is time (in months) and A and r are constant values to be found.
    a) Find the values of A and r in the model.
    b) Interpret the value of the constants A and r in context.

14) The amount of stock in a warehouse of a particular product P, is known to be 12 000 units. After 5 months this is expected to have reduced by 9500.
    a) Find a model of the form
    $$P = ae^{kt}$$
    where t is the number of months passed, where a and k are constant values to be found.
    b) How much stock is available after 2 months?
    c) The company orders a delivery from the manufacturers when the stock drops to 500 units. After how many months should they make this order?
    d) What does this model predict in the long term for the stock of this product? Comment on its reliability.

1.26 Exponential Growth and Decay (answers on page 436)

15) The amount, C milligrams, of a particular drug in a person's body after t hours may be given by a function of the form
$$C = Ae^{-kt}$$
   a) Given that the initial dose taken is 200mg and in 4 hours the amount of drug has halved once, find the values of A and k giving your answer in exact form.
   b) Hence, find the number of hours it takes for the drug to reduce to 50 mg in a person's body.

16) The number P of restaurants in a particular fast food chain in their nth year may be modelled by the function
$$P = Ae^{kn}$$
In the first year ($n = 1$), there were 1000 restaurants and in the second year there were 1400.
   a) Show that $e^k = 1.4$
   b) Find the exact value of k and A.
   c) How many restaurants does the model predict in the 4th year?
   d) How many whole years is it before there are 100 000 restaurants?
   e) Explain why the model is invalid in the long-term.

17) A substance p is known to decay exponentially according to the model
$$p = p_0 e^{-kt}$$
where t is time in years, p_0 is the initial quantity and k is an unknown constant. Find the amount of time needed for the quantity to have reduced by 99%, giving your answer in the form $\frac{1}{k}\ln A$ where A is a positive integer.

18) The amount of a drug y mg in the blood stream is modelled by the equation
$$y = Ae^{-kt}$$
where t is the time in hours since a tablet was taken. After 2 hours there was 500mg in the blood stream, and after 7 hours there was 88mg in the blood stream.
   a) Find the exact value of k
   b) Find the value of A to 1 decimal place.

19) The populations of two types of fish in a lake may be modelled with the following equations:
   Fish A:
$$P_A = ae^{0.08t}$$
   Fish B:
$$P_B = be^{0.12t}$$
where P_A and P_B are the respective populations of fish A and fish B and t is the number of months passed since the start of the year. The total number of fish in the lake at the start of the year is 85, then after 6 months passes the total population is 153.
   a) Find the values of a and b giving your answers to 1 decimal place.
   b) Interpret what your values for a and b mean in context.
   c) Find the number of months passed when the population of fish B first exceeds the population of fish A.
   d) Explain why the models are invalid in the long-term.

20) A beaker of liquid is cooling. Its initial temperature is 65°C.
After 3 minutes its temperature is 35°C. The temperature is modelled using the equation
$$T = 17 + ae^{-kt}$$
where T is the temperature in °C, t is the time in minutes and a and k are constants.
   a) Find the exact values of a and k.
   b) What is the temperature of the liquid after 5 minutes?
   c) What is the temperature of the liquid in the long term?
   d) State how the model could be modified if the room temperature is reduced.

The temperature of a second liquid, y°C, cooling from 80°C at the same point in time in the same room is modelled by the equation
$$y = A + be^{-1.5kt}$$
where k has the same value from before.
   e) Find the time at which both liquids have the same temperature.

1.26 Exponential Growth and Decay (answers on page 436)

**Prerequisite knowledge for Q21-Q26:
Differentiation of the form ae^{kx}**

21) For each of the equations calculate the rate of change in y when $x = 2$:
    a) $y = 20 + 2e^{-0.3x}$
    b) $y = 4(5 + e^{0.2x})$
    c) $y = 70 - 60e^{-0.8x}$

22) The population of red squirrels in the UK can be modelled by the equation
$$P = 3500e^{-\frac{1}{225}\ln\left(\frac{700}{57}\right)t}$$
where P is their population (in thousands) and t is the number of years since 1800.
Show that in 2000, according to the model, the amount of squirrels is decreasing at a rate of approximately 4.2 thousand per year.

23) A volume V (ml) of liquid being poured from a bottle may be modelled by the equation
$$V = 330(1 - e^{-0.1t})$$
after t seconds.
   a) State the theoretical maximum amount of liquid the bottle is holding, according to the model.
   b) Find the rate at which the liquid is being poured when $t = 5$ seconds.

24) The amount of stock in a shop of two products P may be modelled by the following equations where t is the number of months since the start of the year.
   Product A: $P = 100(1 + e^{-kt})$
   Product B: $P = 400e^{-kt}$
   where $t \geq 0$ and k is a positive constant.
   a) Sketch the two models on the same axes, include labelling points where the curves meet the axes and any asymptotes.
   b) State the amount of each product in the long term according to the model.
   c) Explain the limitations of the model.
   d) Show that the rate of decrease of product B is always 4 times that of product A.
   e) Find the exact value of k given there is the same amount of each product after two months have passed.

25) The height of a tree is given by the equation
$$h = \frac{18e^{\frac{1}{2}t}}{e^{\frac{1}{2}t} + 3}$$
where h is the height in metres and t is the number of years that have passed since it is planted. The graph of height against time is shown below.

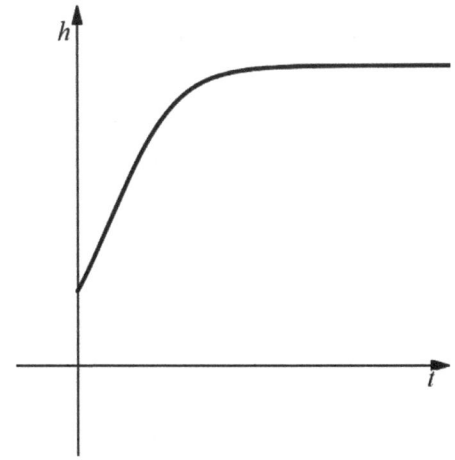

   a) According to this model state the initial height of the tree.
   b) Show that
$$h = \frac{18}{1 + 3e^{-\frac{1}{2}t}}$$
   c) What height does this model predict in the long term for this tree?
   A second tree was planted at the same time, and it may be modelled with equation
$$h = a - 14e^{-\frac{1}{2}t}$$
   where a is an unknown constant.
   d) Find the value of a given this tree has a height of 2 m when planted.
   e) Find the rate of growth of the second tree at 5 years.
   f) State which of the two trees is expected to be taller.

26) Two models of mass m kg over an exponentially decaying substance over time t days are given by
$$m = 100e^{-0.05t}$$
and
$$m = 10 + 90e^{-kt}$$
where k is a positive constant.
Find the ranges of values of k for which the second model is initially decaying at a faster rate than the first.

1.27 Reduction to Linear Form (answers on page 438)

Answer sheets for questions marked with a (*) can be found at TLMaths.com

1) Show that $y = ax^b$ may be written in the form
$$\log_{10} y = b \log_{10} x + \log_{10} a$$

2) Show that $y = a \times b^x$ may be written in the form
$$\log_{10} y = x \log_{10} b + \log_{10} a$$

3) Show that $y = ae^{bx}$ may be written in the form
$$\ln y = bx + \ln a$$

4) A model for some data is given as
$$y = 2 \times 1.05^x$$
Show that the graph of $\log_{10} y$ against x will be a straight line.

5) A model for some data is given as
$$y = 0.5x^{1.2}$$
Show that the graph of $\ln y$ against $\ln x$ will be a straight line.

6) A model for some data is given as
$$y = 0.4 \times 0.88^{-x}$$
Show that the graph of $\log_{10} y$ against x will be a straight line.

7) For each of the following, rearrange to find an equation of y in terms of x, giving your answer in the form $y = ax^b$ where a and b are values given to 2 significant figures:
   a) $\log_{10} y = 0.478 + 5 \log_{10} x$
   b) $\ln y = 3 \ln x + 0.815$
   c) $\log_{10} y = -1.347 - 0.89 \log_{10} x$
   d) $\ln y = 0.51 \ln x - 1.15$

8) For each of the following, rearrange to find an equation of y in terms of x, giving your answer in the form $y = a \times b^x$ where a and b are values given to 2 significant figures:
   a) $\log_{10} y = 0.214x - 1.54$
   b) $\ln y = 0.555x + 0.444$
   c) $\ln y = -0.235x + 0.159$
   d) $\log_2 y = 0.895x + 2.03$

9) Convert each of the following into the form $\ln y = a + bx$ where a and b are constants to be found:
   a) $y = 40e^{-0.2x}$
   b) $y = 0.5 \times 1.04^{0.5x}$
   c) $y = 1.62 \times 2.13^{-0.2x}$

10) A data set (x, y) has been plotted on a graph of $\log_{10} y$ against x

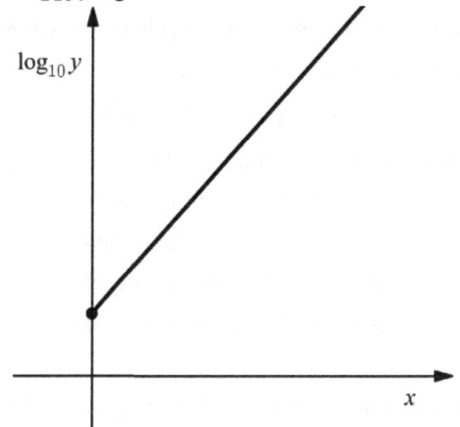

A linear equation is found to best fit the data. In each of the following cases, find a model in the form
$$y = a \times b^x$$
where a and b are values to be found:
   a) gradient of 3 and an intercept of 0.5
   b) gradient of 0.25 and an intercept of 0.0714

11) A data set (x, y) has been plotted on a graph of $\ln y$ against x

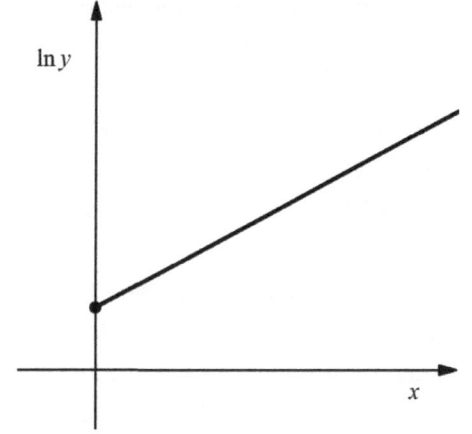

The linear equation found to best fit the data. In each of the following cases, find a model in the form
$$y = ae^{bx}$$
where a and b are values to be found:
   a) gradient of 4 and an intercept of 2
   b) gradient of 0.75 and an intercept of 0.1

1.27 Reduction to Linear Form (answers on page 438)

12) A data set (x, y) has been plotted on a graph of $\log_{10} y$ against $\log_{10} x$

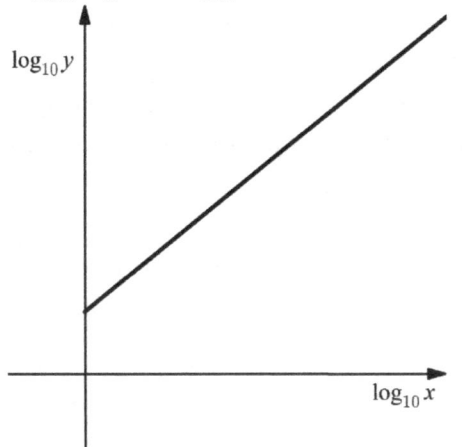

The linear equation found to best fit the data. In each of the following cases, find a model in the form
$$y = ax^n$$
where a and n are values to be found:
a) gradient of 4 and an intercept of 0.1
b) gradient of 0.5 and an intercept of 4

13) A data set (x, y) has been plotted on a graph of $\log_{10} y$ against $\log_{10} x$ The linear equation found to best fit the data.

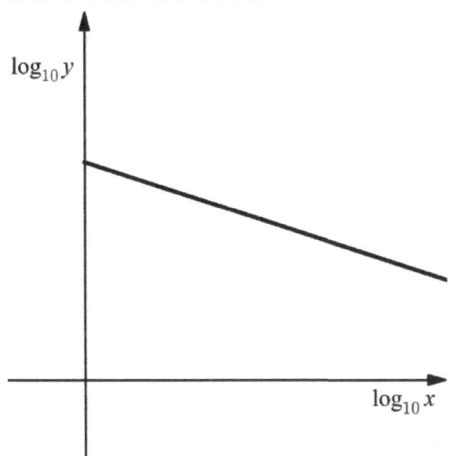

In each of the following cases, find a model in the form
$$y = ax^n$$
where a and n are values to be found:
a) gradient of -5 and an intercept of 0.12
b) gradient of -0.4 and an intercept of 1.17

14) A straight-line graph of $\ln y$ against $\ln x$ passes through the y-axis at 2 and the point with coordinates $(1.8, 2.9)$. Find the equation which represents the equation of the line in terms of y and x

15) A straight-line graph of $\log_{10} y$ against $\log_{10} x$ passes through the y-axis at 3 and the point with coordinates $(1.6, 3.704)$. Find the equation which represents the equation of the line in terms of y and x

16) A straight-line graph of $\log_{10} y$ against x passes through the y-axis at 0.5 and the point with coordinates $(1.01, 0.803)$. Find the equation which represents the equation of the line in terms of y and x

17) A straight-line graph of $\ln y$ against x passes through the points $(1.1, 3.06)$ and $(2.4, 4.12)$. Find the equation which represents the equation of the line in terms of y and x

18) (*) The number of visits to a website y since its launch x days before is recorded in the following table:

Days since launch, x	Number of visits, y
2	5
30	15
60	24
100	30
150	40

It is thought that the data may be modelled in the form
$$y = ax^b$$
a) Complete a table of values of $\log_{10} x$ and $\log_{10} y$, giving your answers to 3 decimal places.
b) Plot a graph of $\log_{10} y$ against $\log_{10} x$ and draw a line of best fit on your plot.
c) Find the y-intercept and the gradient of your line of best fit.
d) Show that $y = ax^b$ may be written in the form
$$\log_{10} y = \log_{10} a + b \log_{10} x$$
e) Use your graph to estimate the values of a and b to 1 decimal place.

1.27 Reduction to Linear Form (answers on page 438)

19) (*) The number of app downloads, N millions at time t months after launch may be modelled with the equation $N = at^b$. The data collected within the first 24 months is shown in the table below:

Time, t months	Downloads, N
3	12.1
5	24.7
8	50.3
10	65.9
18	170.2
24	249.1

a) Complete a table of values of $\log_{10} x$ and $\log_{10} y$, giving your answers to 3 decimal places.
b) Plot a graph of $\log_{10} y$ against $\log_{10} x$ and draw a line of best fit.
c) Use your graph to estimate the values of a and b to 1 decimal places.

Use your model to find
d) The amount of app downloads at 12 months.
e) The time it took to reach 200 million downloads.

20) (*) The mass of radioactive iodine, m grams after t days can be modelled with the exponential equation $m = ae^{bt}$ where a and b are constants. Data is collected every 10 days and is recorded in the table below:

Time, t days	Mass, m grams
0	50
10	20.9
20	8.9
30	3.6
40	1.7

a) Complete a table of values of $\ln m$ and t, giving your answers to 3 decimal places.
b) Plot $\ln m$ against t and draw a line of best fit on your graph.
c) Find the values of a and b
d) Use your model to predict the mass of iodine after 50 hours since the start.

21) (*) Morgana has recorded how many minutes T she has spent exercising per week over the first 10 weeks of the year. She thinks the data may be modelled in the form $T = a \times t^b$ where t is the number of weeks since the beginning of the year. The data she collected is given in the table below:

Time, t weeks	Time spent exercising, T minutes
1	90
2	82
4	77
6	76
8	73
10	71

a) Plot a graph of $\log_{10} T$ against $\log_{10} t$ and draw a line of best fit on your graph.
b) Find the values of a and b and hence use your values of a and b to create a model in the form
$$T = a \times t^b$$
c) Interpret the meaning of the value a in the context of the model.
d) Use your model to find the amount of exercise she did 3 weeks after starting.
e) Comment on the validity of the model when $t = 0$
f) Comment on the long-term reliability of the model.

22) Matt is trying to find a model to represent the price of an item every month. He thinks it will be modelled in the form $y = ax^b$ where y is the price and x is the time passed in months that has passed.
He plotted $\log_{10} y$ against x, and found the y-intercept to be 2.4 and calculated the gradient as 0.6
He is unsure what to do next and suspects he has gone wrong.
a) State the mistake in his method.
b) Working with the graph he plotted, write an equation of y in terms of x with the information you have been given.

1.27 Reduction to Linear Form (answers on page 438)

23) Dev has found a model for data that she has collected, but when she uses her model to make predictions the results are not what she expects.

x	y	$\ln x$
2	2300	7.74
4	2125	7.66
6	1975	7.59
8	1825	7.51
10	1675	7.24
12	1550	7.34

Gradient is -0.0395, intercept is 7.82
$$\ln y = -0.0395x + 7.82$$
$$y = 10^{-0.0395x + 7.82}$$

a) Find the mistake in her working.
b) Find the equation of y in terms of x in the form
$$y = e^{px+q}$$
where p and q are constants to be found.

24) The number of cinema admissions N millions in the UK, may be modelled by the equation
$$N = a \times t^b$$
where t is the number of years since 1954. The plot below shows the graph of $\ln y$ against $\ln t$, with two of points on the line labelled.

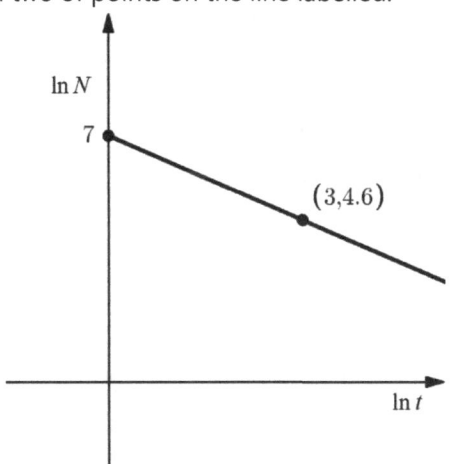

a) Find an equation for the line of best fit in terms of $\ln N$ and $\ln t$
b) Express N in terms of t
c) Use your model to predict the cinema admissions in 2020
d) Comment on the suitability of the model for predicting the number of cinema admissions in 2020

25) The semi-conductor industry uses Moore's Law to predict the number of transistors y in a microprocessor, modelled by
$$y = a \times 2^{bx}$$
where x is the number of years since 1975. The plot below shows the graph of $\log_2 y$ against x, with two points on the line labelled.

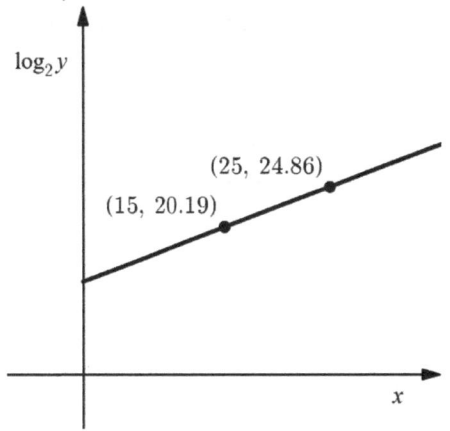

Points: $(15, 20.19)$ and $(25, 24.86)$

a) Find an equation for the line of best fit in terms of $\log_2 y$ and x
b) Express y in terms of x
c) With reference to the model, explain the meaning of the value of a in context
d) Use your model to predict the number of transistors in 2030

26) A content creator is monitoring their number of subscribers N to their channel, which may be modelled by
$$\log_{10} N = 0.238t + 1.975$$
where t is the number of subscribers t years after when the channel went live.
a) Show that $N = ab^t$ where a and b are constants to be found.
b) Interpret the meanings of the values a and b in the context of the model.
c) Use the model to calculate the total number of subscribers 10 years after the channel went live.

27) Zain leaves a cup of hot water to cool, he collects his data, and establishes a model for the temperature $T°C$ over time t minutes
$$\ln T = -0.0701t + 4.585$$
a) Determine the initial temperature of the hot water
b) State an equation for T in the form $T = ae^{bt}$ where a and b are constants.

1.27 Reduction to Linear Form (answers on page 438)

28) It is found that the time it takes T days for a planet to orbit the sun may be given by
$$\ln T = 1.48 \ln d - 1.6$$
where d million kilometres is average distance of a planet from the Sun.
   a) Find an equation for T in terms of d

   Mars has an average distance of 228 million kilometres from the sun.
   b) Using the model, find the number of days it will take Mars to orbit the sun, giving your answer to the nearest day.

   It actually takes 687 days for Mars to orbit the sun.
   c) Find the percentage error in the prediction made by the model.

29) The planet Saturn has many moons. The radius of each of the moons is given by r million km. The number of days it takes a moon to complete one orbit of Saturn is given by T days. They may be modelled by
$$r = a \times T^b$$

	Radius, r million km	Period, T days
Titan	1.22	15.9
Rhea	0.53	4.5

   a) Show that the equation may be written in the form $\ln r = a + b \ln T$
   b) Given the data in the table, find the values of a and b
   c) Sketch the graph of r against T, labelling points which represent Titan and Rhea.

   Another moon of Saturn, Dione has a period of 2.7 days and a radius of 3.8×10^5 kilometres.
   d) Use your model to predict the radius of the moon giving your answer to 3 significant figures.
   e) Find the % error of the prediction by the model.

30) The life expectancy of a person aged y years, may be defined to be a function of the number of years they have left to live, x years. A model for this may be given by
$$\ln y = \ln b - ax$$
It is known that when $x = 5$, $y = 71$ and when $x = 55$, $y = 24$
   a) Find the value of a
   b) Find the value of b
   c) Use the model to find the life expectancy of someone aged 40
   d) Find the age corresponding to a life expectancy of 25

31) The diagram below shows a straight-line graph of y against x^2 passing through $P(0.3, 5.41)$ and $Q(4.7, 8.49)$

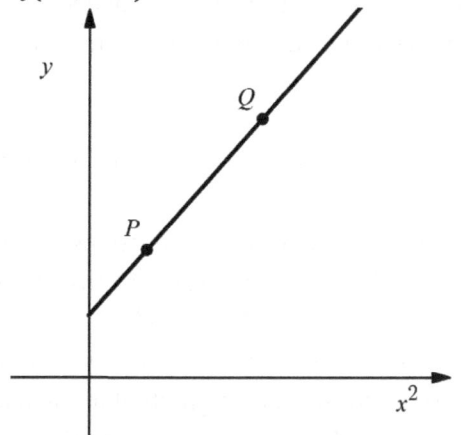

   Find an expression for y in the form $ax^b + k$ where a, b and k are constants to be found.

32) The diagram below shows a straight-line graph of y against $\sin x$ passing through $P\left(\frac{\pi}{2}, 5.3\right)$ and $Q\left(\frac{3\pi}{5}, 7.8\right)$

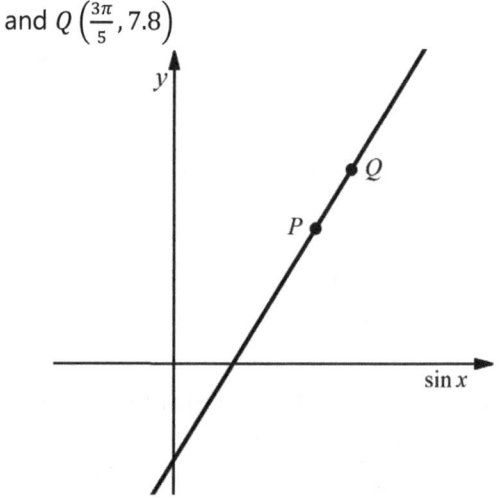

   Write y in the form $a \sin(x) + b$ where a and b are constants to be found.

1.28 Graph Transformations: Single Transformations (answers on page 439)

Translations

1) Describe the transformation that maps $y = f(x)$ onto each of these:
   a) $y = f(x - 6)$
   b) $y = f(x) + 8$
   c) $y = f(x + 4)$
   d) $y = f(x) - 11$

2) Describe the transformation that maps $y = x^2$ onto each of these:
   a) $y = (x - 9)^2$
   b) $y = (x + 5)^2$
   c) $y = x^2 + 2$
   d) $y = x^2 - 10$
   e) $y + 4 = x^2$
   f) $y - 8 = x^2$
   g) $y = (x + 7)^2 - 3$
   h) $y = (x - 1)^2 + 7$

3) Describe the transformation that maps $y = x^3$ onto each of these:
   a) $y = (x + 4)^3$
   b) $y = (x - 2)^3$
   c) $y = x^3 - 7$
   d) $y = x^3 + 15$
   e) $y + 9 = x^3$
   f) $y = (x - 3)^3 + 8$

4) Write down the image of $y = 9x + 2$ after a translation by the vector $\begin{pmatrix} -2 \\ 5 \end{pmatrix}$

5) The following functions are to be translated by the vector $\begin{pmatrix} 4 \\ -7 \end{pmatrix}$. Write down the image in each case.
   a) $y = 4 - 6x$
   b) $y = x^2$
   c) $y = x^2 + 3$
   d) $y = (x + 2)(x - 8)$
   e) $y = (x - 9)(x + 1)$
   f) $y = 3^x$
   g) $y = e^x - 5$
   h) $y = \frac{1}{x}$
   i) $y = \frac{1}{x^2}$
   j) $y = \sqrt{x}$
   k) $y = \cos x$

6) The point P(7, −5) lies on the graph with equation $y = f(x)$. State the coordinates of the image of P after $y = f(x)$ is mapped to each of these:
   a) $y = f(x - 8)$
   b) $y = f(x) + 3$
   c) $y = f(x + 12)$
   d) $y = f(x + 1) - 5$

7) The graph of $y = f(x)$ is shown in the diagram below.

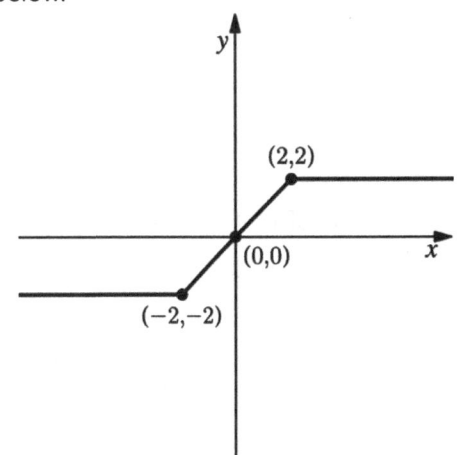

   Write down the equation of the graph shown in the diagram below.

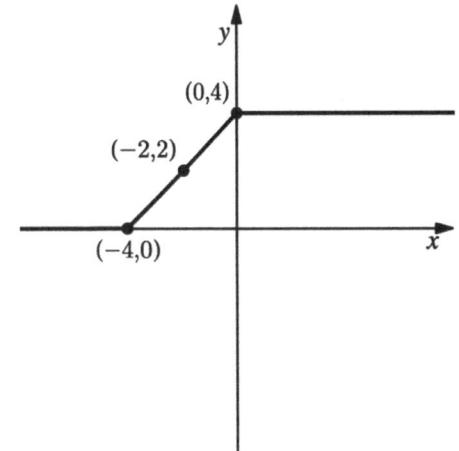

8) Describe the transformation that maps $y = 3x^4$ onto the image $y = 3(x + 4)^3 - 10$

9) Describe the transformation that maps $y = x^8 + 7$ onto the image $y = (x - 8)^8$

10) Describe the transformation that maps $y = \sqrt{x} + 2$ onto the image $y = \sqrt{x + 2}$

1.28 Graph Transformations: Single Transformations (answers on page 439)

Prerequisite knowledge for Q11-Q13:
Logarithms

11) Find the translation vector that maps $y = \log_5 x$ onto the image $y = \log_5\left(\frac{x-4}{25}\right)$

12) Find the translation vector that maps $y = \log_3 x$ onto the image $y = \log_3(9(x-2))$

13) Fully describe the transformation that maps $y = \ln x$ onto the image $y = \ln\left(\frac{x+2e}{e^5}\right)$

14) Find the image of $x^2 + y^2 = 9$ after a translation by the vector $\begin{pmatrix} -7 \\ 5 \end{pmatrix}$

15) Find the image of $x^2 + y^2 = 3$ after a translation by the vector $\begin{pmatrix} 2 \\ -8 \end{pmatrix}$

16) Find the image of $(x-2)^2 + y^2 = 10$ after a translation by the vector $\begin{pmatrix} -9 \\ -3 \end{pmatrix}$

17) Find the image of $(x+10)^2 + (y-3)^2 = 16$ after a translation by the vector $\begin{pmatrix} 2 \\ -1 \end{pmatrix}$

18) Find the image of $xy + x^2 - y = 10$ after a translation by the vector $\begin{pmatrix} -3 \\ 4 \end{pmatrix}$

19) Find the image of $y = x^3 - 2x^2 + 8x + 2$ after a translation by the vector $\begin{pmatrix} -2 \\ 3 \end{pmatrix}$

20) Find the image of $y = 8x^4 - 4x^8$ after a translation by the vector $\begin{pmatrix} 4 \\ -8 \end{pmatrix}$

21) Find the image of $y = \frac{3}{x^2}$ after a translation by the vector $\begin{pmatrix} -2 \\ -6 \end{pmatrix}$

22) Find the image of $y = -\frac{4}{x-7}$ after a translation by the vector $\begin{pmatrix} 7 \\ 0 \end{pmatrix}$

23) Fully describe the transformation that maps $y = 4 \times 4^x$ onto the image $y = 4^{x-5}$

24) The diagram below shows the graph of $y = f(x)$. It is given that $f(x)$ is symmetric about $x = -1$.

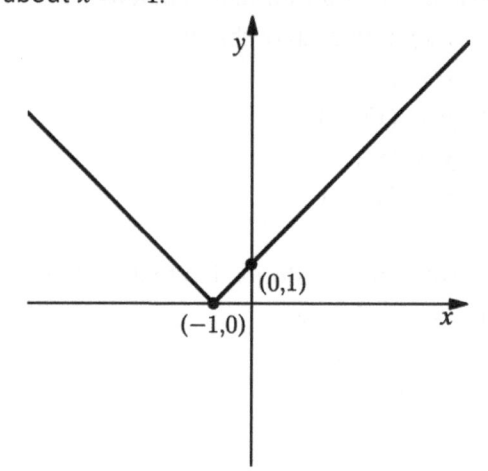

On separate diagrams, draw a sketch of each of these:
a) $y = f(x + 1)$
b) $y = f(x) - 2$
c) $y = f(x - 4) + 1$
d) $y - 2 = f(x + 2)$

Stretches

25) Describe the transformation that maps $y = f(x)$ onto each of these:
a) $y = f(5x)$
b) $y = \frac{1}{6}f(x)$
c) $y = f\left(\frac{2}{3}x\right)$

26) Describe the transformation that maps $y = x^2 - 3$ onto each of these:
a) $y = 3x^2 - 9$
b) $y = 9x^2 - 3$

27) Describe the transformation that maps $y = x^3 + 2$ onto each of these:
a) $y = 8x^3 + 2$
b) $y = 8x^3 + 16$
c) $y = \frac{1}{8}x^3 + \frac{1}{4}$
d) $y = \frac{1}{8}x^3 + 2$

28) Write down the image of $y = 5x + 3$ after:
a) a stretch, parallel to the x-axis, scale factor $\frac{1}{4}$
b) a stretch, parallel to the y-axis, scale factor $\frac{1}{5}$

1.28 Graph Transformations: Single Transformations (answers on page 439)

29) The following functions are to be stretched, parallel to the x-axis, scale factor 3. Write down the image in each case.
    a) $y = 4 - 6x$
    b) $y = x^2$
    c) $y = x^2 + 3$
    d) $y = (x+2)(x-8)$
    e) $y = (x-9)(x+1)$
    f) $y = 3^x$
    g) $y = e^x - 5$
    h) $y = \frac{1}{x}$
    i) $y = \frac{1}{x^2}$
    j) $y = \sqrt{x}$
    k) $y = \cos x$

30) The following functions are to be stretched, parallel to the y-axis, scale factor 3. Write down the image in each case.
    a) $y = 4 - 6x$
    b) $y = x^2$
    c) $y = x^2 + 3$
    d) $y = (x+2)(x-8)$
    e) $y = (x-9)(x+1)$
    f) $y = 3^x$
    g) $y = e^x - 5$
    h) $y = \frac{1}{x}$
    i) $y = \frac{1}{x^2}$
    j) $y = \sqrt{x}$
    k) $y = \cos x$

31) The point P$(-2, 10)$ lies on the graph with equation $y = f(x)$. State the coordinates of the image of P after $y = f(x)$ is mapped to each of these:
    a) $y = 6f(x)$
    b) $y = f\left(\frac{1}{2}x\right)$
    c) $y = \frac{2}{5}f(x)$
    d) $y = f(6x)$

32) Describe the transformation that maps $y = \sin x$ onto the image $y = \sin 5x$

33) Describe the transformation that maps $y = \cos 8x$ onto the image $y = 4\cos 8x$

34) The graph of $y = f(x)$ is shown in the diagram below.

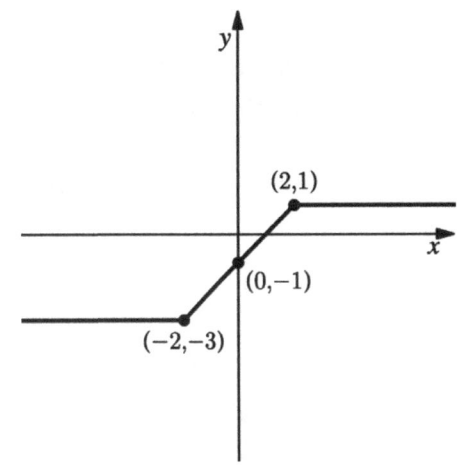

Points: $(-2, -3)$, $(0, -1)$, $(2, 1)$

Write down the equation of the graph shown in the diagram below.

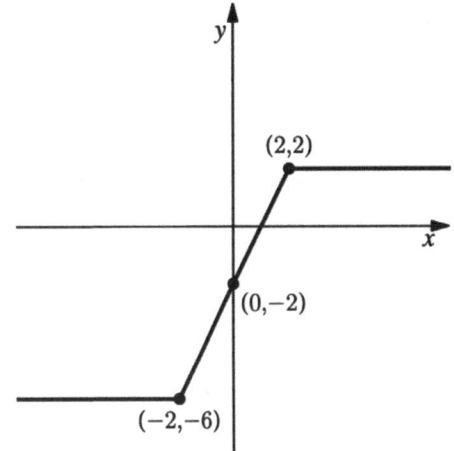

Points: $(-2, -6)$, $(0, -2)$, $(2, 2)$

35) Describe the transformation that maps $y = (x+3)^3$ onto the image $y = \left(\frac{1}{4}x + 3\right)^3$

36) Describe the transformation that maps $y = x^3$ onto the image $y = 27x^3$ as:
    a) a stretch parallel to the y-axis,
    b) a stretch parallel to the x-axis.

37) Describe the transformation that maps $y = \frac{1}{x}$ onto the image $y = \frac{8}{x}$ as:
    a) a stretch parallel to the y-axis,
    b) a stretch parallel to the x-axis.

38) Describe the transformation that maps $y = \frac{1}{x^2}$ onto the image $y = \frac{100}{x^2}$ as:
    a) a stretch parallel to the y-axis,
    b) a stretch parallel to the x-axis.

1.28 Graph Transformations: Single Transformations (answers on page 439)

39) Describe the transformation that maps $x^2 + y^2 = 16$ onto the image $x^2 + 9y^2 = 16$

40) Describe the transformation that maps $12x^2 + y^2 = 9$ onto the image $x^2 + y^2 = 9$

41) The graph of $y = f(x)$ is shown in the diagram below.

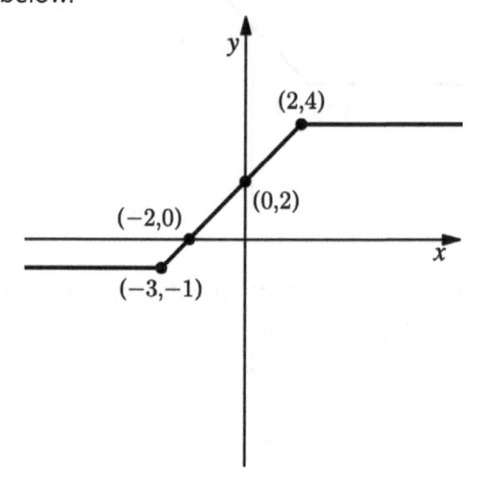

Points shown: $(-3,-1)$, $(-2,0)$, $(0,2)$, $(2,4)$

Write down the equation of the graph shown in the diagram below.

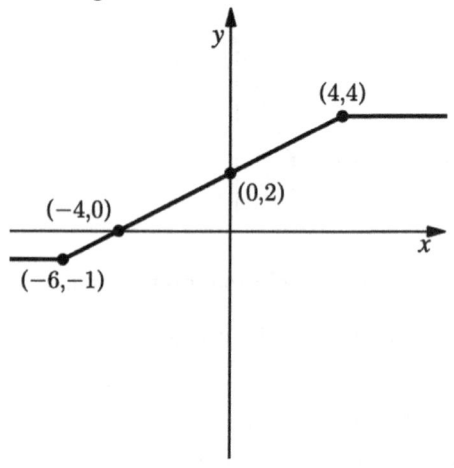

Points shown: $(-6,-1)$, $(-4,0)$, $(0,2)$, $(4,4)$

42) Describe the single transformation that maps $y = 3(x - 2)^2 + 5$ onto the image $y = 12(x - 1)^2 + 5$

43) Describe the single transformation that maps $y = 4(x + 8)^2 - 2$ onto the image $y = 64(x + 2)^2 - 2$

44) Describe the single transformation that maps $y = 5\left(x - \frac{1}{4}\right)^2 + 1$ onto the image $y = \frac{5}{64}(x - 2)^2 + 1$

45) The diagram below shows the graph of $y = f(x)$. It is given that $f(x)$ is symmetric about $x = -1$

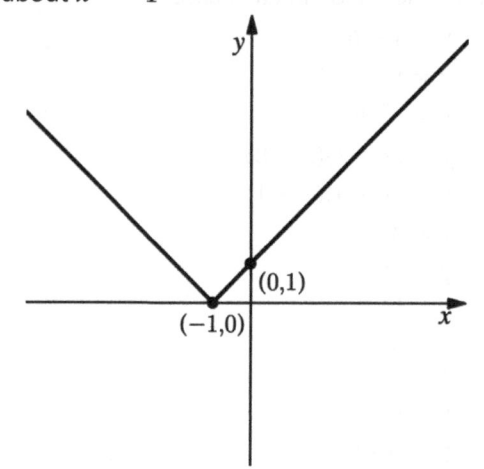

Points shown: $(-1,0)$, $(0,1)$

On separate diagrams, draw a sketch of each of these:
a) $y = f(2x)$
b) $y = 4f(x)$
c) $3y = f(x)$
d) $y = f\left(\frac{1}{3}x\right)$

Reflections

46) Describe the transformation that maps $y = f(x)$ onto each of these:
a) $y = f(-x)$
b) $y = -f(x)$

47) Describe the transformation that maps $y = (x + 3)^2 - 4$ onto each of these:
a) $y = 4 - (x + 3)^2$
b) $y = (3 - x)^2 - 4$

Prerequisite knowledge for Q48: e^x & $\ln x$

48) Describe the transformation that maps $y = e^x - 2$ onto each of these:
a) $y = e^{-x} - 2$
b) $y = 2 - e^x$
c) $y = \frac{1 - 2e^x}{e^x}$

49) Write down the image of $y = 8x - 9$ after:
a) a reflection in the x-axis,
b) a reflection in the y-axis.

50) Write down the image of $y = 3 - 6x$ after:
a) a reflection in the x-axis,
b) a reflection in the y-axis.

1.28 Graph Transformations: Single Transformations (answers on page 439)

51) The following functions are to be reflected in the y-axis. Identify which (if any) of these functions will remain unchanged:
    a) $y = 5 - x$
    b) $y = x^2 + 2$
    c) $y = x^3$
    d) $y = 4^x$
    e) $y = \frac{1}{x}$
    f) $y = \frac{1}{x^2}$
    g) $y = \sin x$
    h) $y = \cos x$
    i) $y = \frac{e^x + e^{-x}}{2}$

52) The following functions are to be reflected in the x-axis. Identify which (if any) of these functions will remain unchanged:
    a) $y = 5 - x$
    b) $y = x^2 + 2$
    c) $y = x^3$
    d) $y = 4^x$
    e) $y = \frac{1}{x}$
    f) $y = \frac{1}{x^2}$
    g) $y = \sin x$
    h) $y = \cos x$
    i) $y = \frac{e^x + e^{-x}}{2}$

53) The diagram below shows the graph of $y = f(x)$. It is given that $f(x)$ is symmetric about $x = -1$.

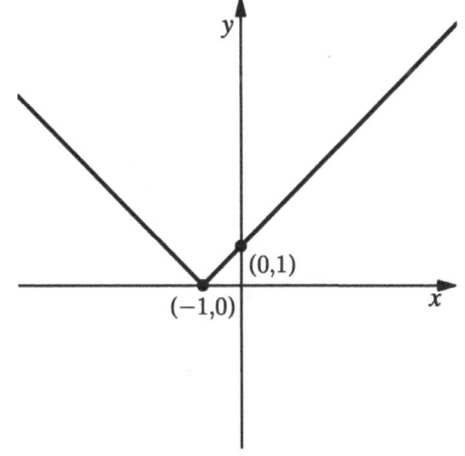

    On separate diagrams, draw a sketch of each of these:
    a) $y = f(-x)$
    b) $y = -f(x)$

54) The point $P(4, -1)$ lies on the graph with equation $y = f(x)$. State the coordinates of the image of P after $y = f(x)$ is mapped to each of these:
    a) $y = -f(x)$
    b) $y = f(-x)$

Mix

55) Describe the transformation that would map $y = \sqrt{x}$ onto each of these:
    a) $y = \sqrt{x} - 3$
    b) $y = \sqrt{2x}$
    c) $y = \sqrt{x + 9}$
    d) $y = -\sqrt{x}$
    e) $y = 8\sqrt{x}$

56) It is given that $f(x) = 2x^2$ and $g(x) = x - 1$. Show that $y = f(x) + 1$ and $y = g(3x)$ do not intersect.

57) It is given that $f(x) = (x - 1)^2 - 1$ and $g(x) = 3x$. Show that $y = 2f(x)$ and $y = g(x - 2)$ intersect at two distinct points.

58) Describe a single transformation that will map $y = \frac{16}{x+1}$ onto the image $y = \frac{9}{x+1}$

59) It is given that $f(x) = \frac{1}{x^2}$. State the asymptotes of the following graphs:
    a) $y = f(x + 2)$
    b) $y = -f(x)$
    c) $y = f(x - 5) + 4$
    d) $y = f\left(\frac{1}{3}x\right)$

60) It is given that $y = f(x)$ is a polynomial of degree 3. It is also given that $f(-2) = 0$, $f(5) = 0$ and $f(7) = 0$. $y = f(x)$ is mapped to the image $y = g(x)$. It is now given that $g(a) = 0$, $g(b) = 0$ and $g(c) = 0$. State the values of a, b and c when:
    a) $g(x) = f(x - 2)$
    b) $g(x) = f(x + 10)$
    c) $g(x) = 4f(x)$
    d) $g(x) = f(2x)$
    e) $g(x) = f\left(\frac{1}{3}x\right)$
    f) $g(x) = f(-x)$

1.28 Graph Transformations: Single Transformations (answers on page 439)

61) It is given that $f(x) = (x-2)(x+4)(x+5)$
   a) Solve $f(x) = 0$
   b) Solve $f(x-2) = 0$
   c) Solve $f(2x) = 0$
   d) Solve $f\left(\frac{1}{3}x\right) = 0$
   e) Solve $f(-x) = 0$
   f) Solve $f(x+15) = 0$

62) It is given that $f(x) = \frac{8}{x-6}$. State the equation of the image of $f(x)$ after each of these transformations:
   a) A translation by the vector $\begin{pmatrix} -3 \\ 0 \end{pmatrix}$
   b) A stretch, parallel to the y-axis, scale factor $\frac{1}{2}$
   c) A reflection in the y-axis
   d) A stretch, parallel to the x-axis, scale factor $\frac{1}{4}$

63) It is given that $y = f(x)$ is a polynomial of degree 3. It is also given that $f(-3) = 0$, $f(6) = 0$ and $f(8) = 0$. $y = f(x)$ is mapped to the image $y = g(x)$, where $g(x) = f(x-k)$ and k is a constant. It is also given that $y = g(x)$ passes through $(1,0)$. State the possible values of k.

64) It is given that $y = f(x)$ is a polynomial of degree 3. It is also given that $f(-8) = 0$, $f(-2) = 0$ and $f(3) = 0$. $y = f(x)$ is mapped to the image $y = g(x)$, where $g(x) = f(kx)$, where k is a constant. It is also given that $y = g(x)$ passes through $(-1,0)$. State the possible values of k.

65) It is given that $f(x) = 4^x$. The graph of $y = f(x)$ is mapped onto the image $y = g(x)$, where $g(x) = 4^{x-3}$.
   a) Fully describe the transformation using a translation.
   b) Fully describe the transformation using a stretch.

66) Fully describe the single transformation that maps $y = x^2$ onto the curve $y = x^2 + 8x - 3$

67) Fully describe the single transformation that maps $y = 2x^2$ onto the curve $y = 2x^2 - 8x - 5$

68) It is given that $f(x) = 3x^2 - 2x + 1$. Find the new equation of the curve in the form $y = ax^2 + bx + c$ after each of these transformations:
   a) A reflection in the y-axis
   b) A translation by the vector $\begin{pmatrix} 3 \\ -6 \end{pmatrix}$
   c) A stretch, parallel to the x-axis, scale factor $\frac{1}{3}$

69) It is given that $f(x) = 2x^4 - 7x^3 - 4x^2$.
   a) Solve $f(x) = 0$

   It is given that
   $g(x) = 2(x+3)^4 - 7(x+3)^2 - 4(x+3)^2$
   b) Use your answer to a) to solve $g(x) = 0$

   It is given that
   $h(x) = 162x^4 - 189x^3 - 36x^2$
   c) Use your answer to a) to solve $h(x) = 0$

   It is given that
   $m(x) = 2x^4 + 7x^3 - 4x^2$
   d) Use your answer to a) to solve $m(x) = 0$

70) The diagram below shows the graph of $y = f(x)$. It is given that $f(x)$ is symmetric about $x = -2$

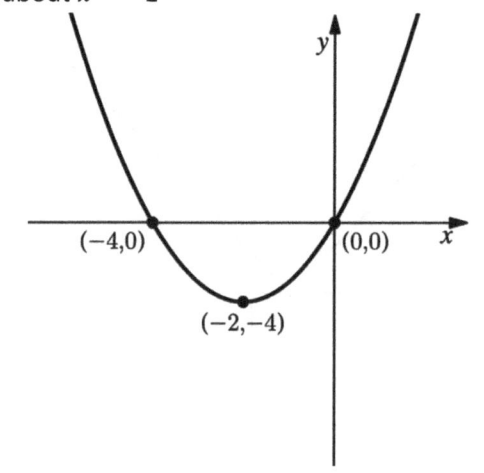

   On separate diagrams, draw a sketch of each of these:
   a) $y = -f(x)$
   b) $y = f(x-3)$
   c) $y = \frac{5}{2}f(x)$

1.29 Binomial Expansion (answers on page 442)

1) Without using your calculator, find the value of
   a) $\frac{8!}{6!}$
   b) $\frac{50!}{47!3!}$
   c) $\frac{50!}{47!4!}$
   d) $\binom{7}{6}$
   e) $\binom{100}{98}$
   f) $\binom{100}{99}$
   g) $^{50}C_{49}$
   h) $^{102}C_{102}$

2) Simplify
   a) $\frac{n!}{(n-1)!}$
   b) $\frac{n!}{2!(n-2)!}$
   c) $\binom{n}{1}$
   d) $\binom{n+1}{n}$
   e) ^nC_n
   f) $\binom{2n-1}{2}$
   g) $\binom{2n-4}{2n-6}$

3) Show that $10\binom{n}{5} = 21\binom{n}{3}$ simplifies to $n^2 - 7n - 30$ and hence find the value of n

4) Solve the following equations:
   a) $3 \times {}^nC_6 = 11 \times {}^nC_4$
   b) $28 \times {}^nC_4 = 15 \times {}^nC_6$
   c) $33 \times {}^nC_2 = 2 \times {}^nC_5$

5) Fully expand in ascending powers of x:
   a) $(1+x)^6$
   b) $(2+x)^5$
   c) $(x+3)^6$
   d) $(x-1)^4$

6) Fully expand in descending powers of x:
   a) $(x-3)^5$
   b) $(5-x)^6$
   c) $(2x+1)^4$
   d) $(3x-3)^5$
   e) $\left(1+\frac{1}{2}x\right)^4$
   f) $\left(2-\frac{x}{4}\right)^5$
   g) $\left(2x-\frac{1}{x}\right)^4$

7) In ascending order, fully expand and simplify, $-5(3-2x)^4$

8) In ascending order, fully expand and simplify, $(1+2x)(2-x)^4$

9) In ascending order, fully expand and simplify, leaving your answers in exact form:
   a) $(1+\sqrt{3})^4$
   b) $(\sqrt{2}+\sqrt{5})^4$
   c) $(\sqrt{7}-\sqrt{2})^5$

10) Find the x^2 coefficient in the expansion of:
    a) $(x+5)^4$
    b) $(2x-1)^5$
    c) $(8+x)^5$
    d) $(1-3x)^6$
    e) $(1+2x)^{78}$

11) Find the x^2 coefficient in the expansion of:
    a) $\left(x-\frac{1}{x}\right)^4$
    b) $\left(3x-\frac{2}{x}\right)^6$
    c) $\left(x-\frac{1}{x}\right)^{18}$

12) By using the definition of $\binom{n}{r}$, show that
    $$\binom{n}{5} = \binom{n-1}{4} + \binom{n-1}{5}$$

13) By using the definition of $\binom{n}{r}$, show that
    $$\binom{n}{r} = \binom{n-1}{r-1} + \binom{n-1}{r}$$

14) Explain why in the expansion of $(1-3x)^9$, the terms with odd powers of x have negative coefficients.

15) Find the value of n in the expansion of
    $$(1+x)^n$$
    where $n \geq 5$, given that the coefficient of x^2 and the coefficient of x^5 is the same.

16) Find the value of n in the expansion of
    $$(1+x)^n$$
    where $n \geq 4$, given that the coefficient of x^4 is 13 times the coefficient of x^2

1.29 Binomial Expansion (answers on page 442)

17) Find the value of n in the expansion of
$$(1 + 2x)^n$$
where $n \geq 2$, given that the coefficient of x^2 is 10 times the coefficient of x

18) Find the value of n in the expansion of
$$\left(1 + \frac{x}{2}\right)^n$$
where $n \geq 5$, given that the coefficient of x^4 is 2 times the coefficient of x^5

19) Find the x^2 coefficient in the expansion of:
   a) $x^2 \left(x + \frac{1}{x}\right)^{12}$
   b) $(1 - x)(1 + 4x)^{10}$
   c) $(1 - 2x)(3 - x)^9$

20)
   a) Fully expand $(3 + x)^5$
   b) Hence simplify $\left(3 - \sqrt{2}\right)^5$ giving your answer in the form $a + b\sqrt{2}$ where a and b are constants.

21)
   a) Fully expand $(2 + 5x)^4$
   b) State the full expansion of $(2 - 5x)^4$
   c) Hence fully simplify
   $$\left(2 + \sqrt{75}\right)^4 + \left(2 - \sqrt{75}\right)^4$$

22)
   a) Find the first 4 terms in ascending order of $(1 + ax)^8$ where a is a positive constant.
   b) Given that the coefficient is x^3 is 3 times the coefficient of x^2, find the value of a

23) Find the possible values of a in the expansion of $(a + 2x)^6$ given the x^2 term has coefficient 4860

24) Find the values of a and n such that
$$(1 + ax)^n = 1 + 40x + 700x^2 + \cdots$$
where $a, n \in \mathbb{Z}$.

25) Find the value of a given in the expansion of
$$\left(2x - \frac{1}{ax}\right)^6$$
the coefficient of x^4 is -12

26)
   a) In ascending order, find the expansion of
   $$(2 - 7x)^5$$
   b) Given that in the expansion of
   $$(1 + ax)(2 - 7x)^5$$
   the x term has coefficient of -464, find the value of a

27)
   a) Find the first 3 terms in ascending order of
   $$(1 - 2x)^9$$
   b) Hence find the value of 0.98^9 giving your answer to 3 decimal places.

28) Explain how you can use the expansion of
$$\left(3 + \frac{2x}{5}\right)^5$$
to find the value of 2.96^5

29)
   a) Find the first 3 terms in ascending order of the expansions of
   $$\left(1 - \frac{x}{2}\right)^5 \text{ and } \left(2 - \frac{3x}{4}\right)^6$$
   b) Hence, estimate $0.995^5 + 1.9925^6$ giving your answer to 3 decimal places.

30) Without fully expanding $\left(x^2 + \frac{1}{x^4}\right)^{12}$ explain why all the indices of the powers of x are multiples of 6

31)
   a) Fully expand $(1 + 3x)^4$
   b) Hence state the full expansion of $(1 - 3x)^4$
   c) State the minimum value of
   $$f(x) = (1 + 3x)^4 + (1 - 3x)^4$$

32)
   a) Find the first 3 terms in ascending order in the expansion of
   $$(5 - 2x)^6$$
   b) Hence determine the coefficient of the quadratic term in
   $$(5 - 2x - 6x^2)^6$$
   c) Use your answer to part a) to find the coefficient of the x^2 term of
   $$(5 + 3x - 2x^2)^6$$

1.30 Differentiation from First Principles (answers on page 443)

1) For $f(x) = 2x^2 + 1$, find
    a) $f(1)$
    b) $f(-1)$
    c) $f(3)$
    d) $f(-4)$
    e) $f\left(\frac{1}{2}\right)$

2) For each of the following curves, the points A and B lie on the curve when $x = 0$ and $x = 1$ respectively. Find the gradient between these two points on:
    a) $y = 3x^2 + x + 1$
    b) $y = 2^x + 1$
    c) $y = \frac{1}{x-2}$
    d) $y = \frac{x}{1+x}$

3) The table below shows some calculations made with the curve
$$f(x) = 7x^2$$

x	$f(x)$	h	$x + h$	$f(x+h)$
2	28	1	3	63
2	28	0.1	2.1	30.87
2	28	0.01	2.01	28.2807
2	28	0.001	2.001	
2	28	0.0001		

Complete the table and hence state the value of
$$\lim_{h \to 0} \frac{f(x+h) - f(x)}{h}$$

4) The table below shows some calculations made with the curve $y = f(x)$

x	$f(x)$	h	$x + h$	$f(x+h)$
1	4	1	2	6
1	4	0.1	1.1	4.143546925
1	4	0.01	1.01	4.0139111
1	4	0.001	1.001	4.001386775
1	4	0.0001	1.0001	4.000138634

a) Calculate
$$\frac{f(x+h) - f(x)}{h}$$
for each line of the table, giving your answers to 5 decimal places.
b) State the value of
$$\lim_{h \to 0} \frac{f(x+h) - f(x)}{h}$$
to 2 decimal places.

5) The diagram below shows
$$y = -5 \times (0.4)^x + 10$$

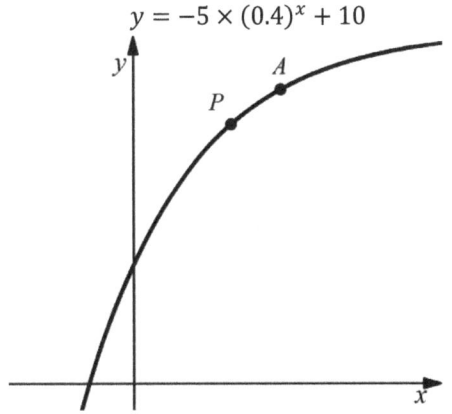

The point P lies on the curve when $x = 2$ and the point A lies on the curve when $x = 3$
a) Find the gradient of the chord connecting the two points A and P.

A third point B lies on the curve when $x = 2.1$
b) Find the gradient of the chord connecting the two points B and P giving your answer to 3 significant figures.
c) Explain how you could improve on the approximation to the gradient of the tangent at the point P.

6) A curve has equation
$$y = \frac{1}{x - 3}$$
The point $P(2, -1)$ lies on the curve. Find the gradient of the chord connecting P with each of the following values of x and where necessary explain why this does not approximate the gradient at $x = 2$
a) $x = 2.5$
b) $x = 3$
c) $x = 3.5$

1.30 Differentiation from First Principles (answers on page 443)

7) Expand the following:
   a) $(1+h)^2$
   b) $5(1+h)^2$
   c) $(1+h)^3$
   d) $-3(1+h)^3$

8) Expand the following:
   a) $(2+h)^2$
   b) $7(3+h)^2$
   c) $(-2+h)^3$
   d) $-3(10+h)^3$

9) Expand and simplify the following:
   a) $(3+h)^2 - 9$
   b) $6(1+h)^2 - 6$
   c) $(-1+h)^3 + 1$
   d) $-2(4+h)^3 + 128$

10) Find f(2) and f(2 + h) for each of the following:
    a) $f(x) = x^2$
    b) $f(x) = x^2 - x$
    c) $f(x) = x^2 - 2x + 5$
    d) $f(x) = x - 3x^2$
    e) $f(x) = 10 - x - 2x^2$

11) Find the value of each of the following:
    a) $\lim_{h \to 0}(2+h)$
    b) $\lim_{h \to 0}(h-5)$
    c) $\lim_{h \to 0}(2h+3)$
    d) $\lim_{h \to 0}\left(\frac{5h^2+h}{h}\right)$
    e) $\lim_{h \to 0}\left(\frac{6h^2-4h}{h}\right)$

12) Explain why
$$\lim_{h \to 0}\left(\frac{h^2-2}{h}\right)$$
cannot be found.

13)
   a) Expand
   $$(1+h)^3$$
   b) Simplify
   $$\frac{8(1+h)^3 - 8}{h}$$
   c) Hence find the gradient of the curve $y = 8x^3$ when $x = 1$

14) Using differentiation from first principles, find the gradient of the curve of f(x) at the point given.
    a) $f(x) = x^2$ at $x = 2$
    b) $f(x) = 3x^2$ at $x = 1$
    c) $f(x) = x^2 - 5x$ at $x = -1$

15) A curve is given as $f(x) = x^2 + 5$
    a) Calculate the gradient of the chord joining the two points on the curve when $x = 4$ and $x = 4.1$
    b) Find, in its simplest form,
    $$\frac{f(4+h) - f(4)}{h}$$
    c) Hence, state the gradient of f(x) when $x = 4$

16) The point (3, 45) lies on the curve $y = 5x^2$ and at this point has the gradient m.
    a) Show that
    $$m = \lim_{h \to 0}(30 + 5h)$$
    b) Deduce the value of m

17) Using differentiation from first principles, show that the derivative of $5x^2$ is $10x$

18) Differentiate from first principles:
    a) $y = 2x^2 + 1$
    b) $y = 5x^2 + x$
    c) $y = x - 4x^2$
    d) $y = x^2 - 2x + 1$
    e) $y = x^3$

19) Use differentiation from first principles to find the exact value of $f'(2)$ for each of the following:
    a) $f(x) = 3x^2$
    b) $f(x) = 1 - 2x^2$

20) Show from first principles that the gradient function of kx^3 is $3kx^2$ where k is a constant.

21) A curve has equation $y = 12x - x^3$. It passes through the point $P(2,16)$
    a) Find the gradient between point P and point Q which has x-ordinate $2 + h$
    b) Using your answer to part a), find the gradient at the point P

1.30 Differentiation from First Principles (answers on page 443)

22) James is attempting to use differentiation from first principles to prove that the gradient of $f(x) = 2x^3 - 9$ at $x = 3$ is 54
    Find the two mistakes in his workings below and correct his proof:
    $$f(3) = 2 \times 3^3 - 9 = 45$$
    $$f(3 + h) = 2(3 + h)^3 - 9$$
    $$= 2h^3 + 9h^2 + 54h + 45$$

    $$\frac{dy}{dx} = \lim_{h \to 0} \frac{2h^3 + 9h^2 + 54h + 45 - 45}{h}$$

    $$= 2h^2 + 9h + 54 = 54$$

23) Using differentiation from first principles find the derivative of $y = ax + b$, where a and b are constants.

24) Using differentiation from first principles find the derivative of $y = ax^2 + bx + c$, where a, b and c are constants.

25) When $x = a$, where $a > 3$, the curve $y = f(x)$ is 8 units higher than when $x = 3$
    Given that $f(x) = x^2 + 2x - 7$, find the exact value of a

26) It is known that two functions
    $$f(x) = px^3 \text{ and } g(x) = qx^2 - 5px + 1$$
    are related by
    $$f'(2) = g'(2) = 24$$
    By using differentiation from first principles, find the values of p and q

27)
    a) Using differentiation from first principles, find the slope of the tangent to the curve $f(x) = 4x^2 - 7$ at the point $P(2, 9)$
    b) Calculate the positive angle which the tangent makes with the positive x-axis at P giving your answer in degrees to 1 decimal place.

28) Using differentiation from first principles, for the curve $y = 4x^2 - x + 1$ find:
    a) The gradient of the curve when $x = 1$
    b) The point on the curve where the gradient is 15

29) Use differentiation from first principles to find $\frac{dy}{dx}$ for each of the following:
    a) $y = \frac{1}{x}$
    b) $y = \frac{1}{1-x}$
    c) $y = \sqrt{x+1}$

30) By differentiating
    $$f(x) = \frac{1}{7x + 2}, \quad x \neq -\frac{2}{7}$$
    from first principles show that
    $$f'(x) = -\frac{7}{(7x + 2)^2}$$

31) By differentiating
    $$f(x) = \frac{2}{3x - 1}, \quad x \neq \frac{1}{3}$$
    from first principles show that
    $$f'(x) = -\frac{6}{(3x - 1)^2}$$

Prerequisite knowledge for Q32: Stationary points

32) Two points A and B lie curve with equation
    $$y = x^3 - 12x$$
    Point A is when $x = 2$ and point B is when $x = 2 + h$
    a) Show that the gradient of the chord AB is $h^2 + 6h$
    b) How does your answer to part a) demonstrate point A is a stationary point?

Prerequisite knowledge for Q33-Q34: Binomial Expansion

33) Differentiate from first principles:
    a) $y = x^5$
    b) $y = 2x^4$

34) Use differentiation from first principles to show that
    a) $$\frac{d}{dx}[(1 - x)^3] = -3(1 - x)^2$$
    b) $$\frac{d}{dx}[(2 - x)^3] = -3(2 - x)^2$$
    c) $$\frac{d}{dx}[(1 - 2x)^3] = -6(1 - 2x)^2$$

1.31 Differentiation: Introduction (answers on page 445)

1) Find $\frac{dy}{dx}$ for each of these:
   a) $y = 4x^3 + 5x^2$
   b) $y = 2x - 9x^2$
   c) $y = x^6 - 18x^2 + 3$
   d) $y = x^{-1} + x^4$
   e) $y = x^3 - x^{-2}$
   f) $y = x^2 + x^{-2} + x + 1$
   g) $y = 3x^5 - 2x^{-2} - x + 6$
   h) $y = 2x^a - x^b$

2) Find $f'(t)$ for each of these:
   a) $f(t) = \frac{1}{2}t^2 + \frac{3}{5}t^{-5}$
   b) $f(t) = \frac{1}{4}t - \frac{3}{4}t^4$
   c) $f(t) = \frac{1}{12}t^6 + 10t^2 + 7$
   d) $f(t) = \frac{1}{3}t^{-1} + \frac{2}{3}t^6$
   e) $f(t) = \frac{7}{8}t^2 - \frac{1}{4}t^{-3} - 4$
   f) $f(t) = \frac{2}{7}t^8 + \frac{1}{7}t^{-3} + \frac{1}{5}t + \frac{4}{7}$
   g) $f(t) = -\frac{1}{4}t^6 - \frac{1}{2}t^{-2} + 2t + \frac{5}{9}$
   h) $f(t) = -\frac{1}{2}t^{\frac{a}{b}}, b \neq 0$

3) Find $\frac{dy}{dt}$ for each of these:
   a) $y = 8t^{\frac{1}{2}} + 6t^{\frac{1}{3}}$
   b) $y = 5t - 12t^{\frac{1}{4}}$
   c) $y = t^{\frac{2}{3}} + 7t^{\frac{3}{5}}$
   d) $y = t^{-\frac{1}{2}} + t^{\frac{1}{4}} + \frac{1}{4}$
   e) $y = -t^{\frac{5}{8}} - t^{-\frac{3}{8}} - \frac{1}{8}t + \frac{1}{8}$
   f) $y = t^{\frac{1}{3}} + t^{-\frac{1}{3}} + 3t + 3$
   g) $y = -t^{\frac{3}{4}} + t^{-\frac{8}{9}} - \frac{1}{12}t + 1$

4) Find $f'(x)$ for each of these:
   a) $f(x) = x(x + 3)$
   b) $f(x) = (x + 1)(x + 5)$
   c) $f(x) = (x - 2)(2x + 5)$
   d) $f(x) = (x + 4)^2$
   e) $f(x) = x^2(x - 1)$
   f) $f(x) = (x + 2)^2(5x - 1)$
   g) $f(x) = x(7x - x^3)$
   h) $f(x) = x^2(9x - x^{-1})$
   i) $f(x) = x(7x^{-1} - x^{-2} + 8)$
   j) $f(x) = \left(x + \frac{1}{x}\right)^3$
   k) $f(x) = (x + p)(x + q)$
   l) $f(x) = (x - a)\left(x + \frac{1}{a}\right), a \neq 0$

5) Write each of the following in the form x^k:
   a) $\sqrt{x}$
   b) $\sqrt{x^5}$
   c) $\frac{1}{x^4}$
   d) $(x\sqrt{x})^3$
   e) $\sqrt[3]{x}$
   f) $\frac{1}{\sqrt{x}}$
   g) $\frac{x}{\sqrt{x}}$
   h) $\frac{1}{x\sqrt{x}}$
   i) $\frac{1}{x \times \sqrt[a]{x}}$ where $a > 0$

6) Write each of the following in the form ax^b:
   a) $5\sqrt[3]{x}$
   b) $\sqrt{4x^3}$
   c) $\frac{3}{x^2}$
   d) $(2x\sqrt{x})^5$
   e) $\frac{\sqrt[8]{x}}{3}$
   f) $\frac{4}{\sqrt{x}}$
   g) $\frac{3x}{\sqrt{9x^4}}$
   h) $\frac{5}{x\sqrt{16x}}$
   i) $\frac{2}{\sqrt{25ax}}$ where $a > 0$

7) Find $\frac{dy}{dx}$ for each of these:
   a) $y = \sqrt[4]{x}$
   b) $y = \sqrt{9x^5}$
   c) $y = \frac{5}{x^3}$
   d) $y = (x\sqrt{x})^3$
   e) $y = \frac{\sqrt[3]{x}}{5}$
   f) $y = \frac{3}{\sqrt{x}}$
   g) $y = \frac{x}{\sqrt{x^4}}$
   h) $y = \frac{2x}{\sqrt{16x^4}}$

8) Find $f'(x)$ for each of these:
   a) $f(x) = \sqrt{x}(x + 2)$
   b) $f(x) = \sqrt{x}(x^2 - 1)$
   c) $f(x) = \frac{x^3 + 4x}{x}$
   d) $f(x) = \frac{2x^3 + 4x^{-2}}{x}$
   e) $f(x) = \frac{9x - 1}{x^2}$
   f) $f(x) = \frac{x^2 + 2x - 1}{\sqrt{x}}$
   g) $f(x) = \frac{x^2 - 5x^{-\frac{3}{2}} + 1}{\sqrt{x}}$

1.31 Differentiation: Introduction (answers on page 445)

9) For each of the following, find $\frac{dy}{dx}$ and $\frac{d^2y}{dx^2}$:
   a) $y = x^3 + x^{-3} + 2x^2 - 5x + 8$
   b) $y = \frac{1}{12}x^4 + \frac{1}{8}x^2 + 7x$
   c) $y = -x^3 - 2x^{-\frac{1}{2}} + 15x$
   d) $y = -x^{\frac{1}{7}} - x^{-\frac{1}{7}} - \frac{1}{9}x + \frac{1}{9}$
   e) $y = x^3(2x + x^{-2})$
   f) $y = \sqrt{16x^3}$
   g) $y = \frac{2}{\sqrt{x}}$
   h) $y = \frac{x^5 + 4x^{-2}}{x}$

10) Find the value of the gradient of the curve at the given points:
    a) $y = x^2 + 5x - 1$ at $x = 1$
    b) $y = x^3 - 2x + 4$ at $x = -2$
    c) $y = \frac{1}{\sqrt{x}} - 2x^2$ at $x = 4$
    d) $y = \sqrt{x} - x$ at $(4, -2)$
    e) $y = \frac{x^4 - 3x^3}{x}$ at $(2, -4)$

11) Find $f''(1)$ for each of these:
    a) $f(x) = x^3 - 5x^2 + 2$
    b) $f(x) = \frac{1}{x^2} - x^2$
    c) $f(x) = \sqrt{x} + 2x^2$
    d) $f(x) = x\sqrt[3]{x} - 4x$
    e) $f(x) = p\sqrt{x} - \frac{4}{qx}$ where $q \neq 0$

Prerequisite knowledge for Q12-Q17: Differentiation of the form ae^{kx}

12) Find $f'(x)$ for each of these:
    a) $f(x) = e^{2x}$
    b) $f(x) = e^{13x}$
    c) $f(x) = e^{-8x}$
    d) $f(x) = e^{\frac{1}{5}x}$
    e) $f(x) = e^{-\frac{4}{7}x}$
    f) $f(x) = 2e^{3x}$
    g) $f(x) = 5e^{-x}$
    h) $f(x) = \frac{1}{4}e^{-2x}$
    i) $f(x) = \frac{1}{e^{5x}}$
    j) $f(x) = \sqrt{e^{3x}}$
    k) $f(x) = \frac{e^x}{\sqrt{e^x}}$
    l) $f(x) = \frac{e^x}{\sqrt{e^{6x}}}$
    m) $f(x) = \sqrt{e^{ax}}$
    n) $f(x) = \frac{e^x}{\sqrt{e^{ax}}}$

13) Find $\frac{dy}{dx}$ for each of these:
    a) $y = 2e^{2x} - 5e^{3x}$
    b) $y = 2x - e^{3x}$
    c) $y = e^{-x} + x^5$
    d) $y = x^5 - \frac{1}{4}x^4 + 3e^{-\frac{1}{5}x}$
    e) $y = \sqrt{x} + \frac{3}{e^{2x}}$
    f) $y = \sqrt{ax} + \frac{5}{e^{bx}}$

14) Find $f'(x)$ and $f''(x)$ for each of these:
    a) $f(x) = e^{3x} + 8e^{-x}$
    b) $f(x) = 5x^2 - e^{5x}$
    c) $f(x) = 2e^{-x} + x^2$
    d) $f(x) = x^4 - 8e^{-\frac{1}{2}x} - \frac{1}{2}x^{-3}$

15) Find the value of the gradient of the curve at the given points, giving your answers in exact form:
    a) $y = e^x$ at $x = 2$
    b) $y = 5e^x$ at $x = -1$
    c) $y = x^2 - e^x$ at $(0, -1)$
    d) $y = x - e^x$ at $(1, 1 - e)$
    e) $y = 2(x + e^{2x})$ at $\left(\frac{1}{2}, 1 + 2e\right)$
    f) $y = -3(x + e^{ax})$ at $(1, -3 - 3e^a)$

16) The graph below shows the curve
    $$y = 3(x - e^x) + 5$$
    Find the value of the gradient at each of the points where the curve meets the axes giving your answers to 2 decimal places.

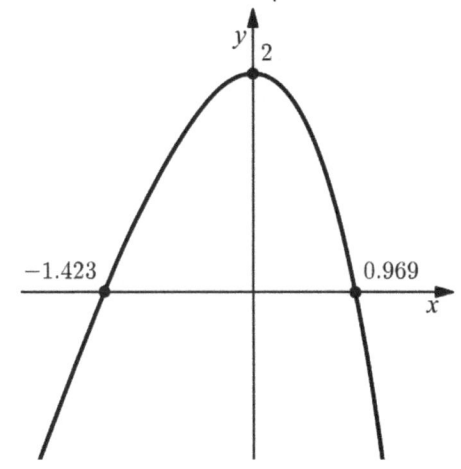

17) Find the values of k and p such that the second derivative of
    $$y = \frac{k}{e^{px}}$$
    is always positive.

1.32 Differentiation: Tangents and Normals (answers on page 446)

1) Evaluate $f'(2)$ for each of the following, leaving your answers in exact form:
   a) $f(x) = x^3 - 2x^{-1}$
   b) $f(x) = \sqrt{x} + 2x^4$
   c) $f(x) = \frac{x^5 - 1}{x^2}$
   d) $f(x) = \sqrt{x}(1 - x)$

2) For each of the following curves, find the equation of the tangent line which passes through the given point.
   Give your answers in the form $ax + by + c = 0$ where a, b, c are integers and $a > 0$:
   a) $y = x^2$ at $(2, 4)$
   b) $y = x^3 - x + 5$ at $(1, 5)$
   c) $y = \frac{1}{x}$ at $\left(-2, -\frac{1}{2}\right)$
   d) $y = 2\sqrt{x}$ at $x = 1$

3)
   a) Show that the tangent to the curve $y = 3 - x^2$ when $x = 0$ may be written in the form $y - k = 0$ where k is a value to be found.
   b) Explain the significance of the gradient of this tangent.

4) For each of the following curves, find the equation of the normal line which passes through the given point.
   Give your answers in the form $ax + by + c = 0$ where a, b, c are integers and $a > 0$:
   a) $y = 3x^2$ at $(1, 3)$
   b) $y = 5x^4 - x^2$ at $(-1, 4)$
   c) $y = \frac{1}{x^2}$ at $\left(2, \frac{1}{4}\right)$
   d) $y = \sqrt{x}$ at $(1, 1)$
   e) $y = x - x^4$ when $x = -1$

5) For the equation $y = 5x^2 + 1$, find the value of x such that $\frac{dy}{dx} = 30$

6) For the equation $y = x^3 - 2x$, find the values of x such that $\frac{dy}{dx} = 10$

7) For the equation $y = x^{\frac{1}{2}}$, find the value of x such that $\frac{dy}{dx} = \frac{1}{4\sqrt{2}}$

8) The graph of $y = \frac{1}{x} + 2$ is shown below.

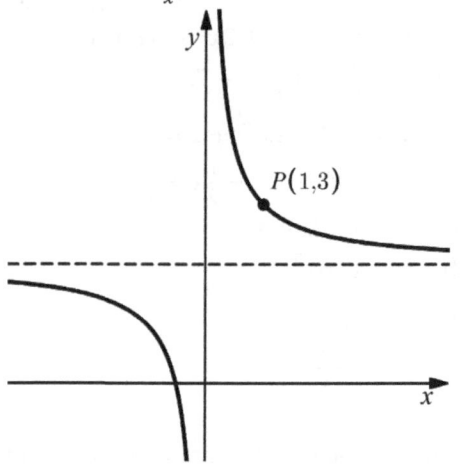

   The point P has coordinates $(1, 3)$
   a) Find the equation of the tangent line at P in the form $y = mx + c$ where m and c are integers to be found.
   b) Find the equation of the normal line at P in the form $y = mx + c$ where m and c are integers to be found.
   c) Find the coordinates where the tangent and normal line at P meets the coordinate axes.

9) The graph of $y = x^3 - 3x^2 + 2$ is shown below.

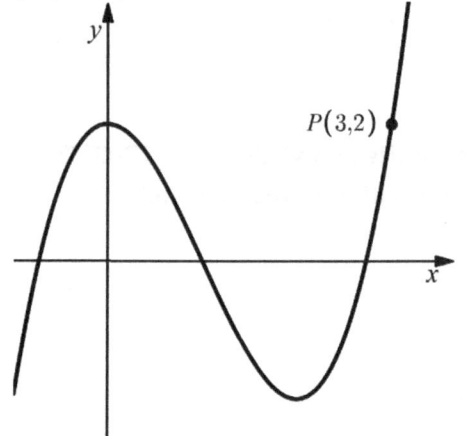

   The point P has coordinates $(3, 2)$
   a) Find the equation of the normal line at P in the form $ax + by = c$ where a, b and c are integers to be found and $a > 0$
   b) Find the coordinates where the normal line crosses the x-axis and y-axis.
   c) Hence calculate the area of the triangle bounded by the normal line at P and the coordinate axes.

1.32 Differentiation: Tangents and Normals (answers on page 446)

10) The graph of $y = \frac{1}{x} + x^3 + 1$ is shown below.

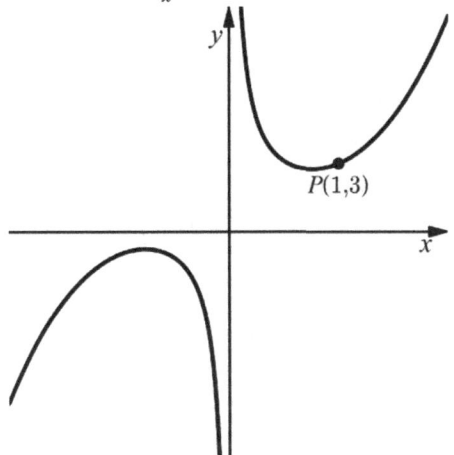

The point P has coordinates $(1, 3)$
   a) Find the equation of the tangent line at P in the form $y = mx + c$ where m and c are integers to be found.
   b) Find the coordinates where the tangent line crosses the x-axis and y-axis.
   c) Find the coordinates where the tangent meets the curve again.

11) The point A lies on the curve
$$y = 5x^2 - \frac{1}{x}$$
The tangent line to the curve at point A which lies on the curve is parallel to the line $y + 9x = 5$
   a) Find the coordinates of the point A.
   b) Find the equation of the tangent line to the curve at the point A in the form $y = mx + c$ where m and c are integers to be found.
   c) Hence find the other point where the tangent intersects the curve.

12) A curve has equation
$$y = \frac{k}{x}$$
where k is a positive constant.
Find $\frac{dy}{dx}$ at $\left(\frac{1}{2}, 2k\right)$ giving your answer in terms of k

13) A curve has equation
$$y = x^3 + px^2 + qx + 5$$
It passes through the point $P(2, 1)$. The gradient of the curve at P is 10
Find the value of p and q

14) The graph below shows the curve
$$f(x) = -x^2 + 2x + 8$$
which meets the x-axis at points A and B
The tangent lines to the curve at A and B meet at point C. Find the area of the triangle ABC.

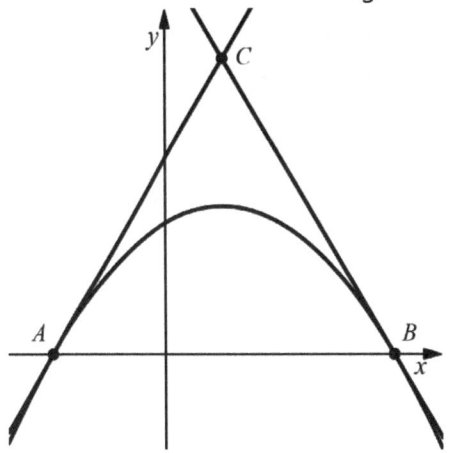

15) A curve has equation
$$y = x^3 + (1 + p)x^2 + qx + 5$$
It passes through the point $P(-1, 6)$. The gradient of the normal to the curve at P is $\frac{1}{2}$
Find the value of p and q

16) A curve has equation
$$y = 3x^3 - 5x^5 + 1$$
At $x = -1$ the tangent line has the same gradient as another tangent line to the curve. Find the equation of this other tangent line.

17) Show that the equation of the tangent to the curve $y = \frac{1}{x^2}$ at the point $x = a$ is given by $a^3y + 2x = 3a$

18)
   a) Show that the equation of the tangent to the curve $y = x(x^2 + 2)$ at the point $x = a$ is given by $y = (3a^2 + 2)x - 2a^3$
   b) Hence explain why the gradient of the curve $y = x(x^2 + 2)$ is always positive.

Prerequisite knowledge for Q19: Second derivatives

19) A curve has equation
$$y = x^2(1 - kx)$$
where k is a positive constant.
Find $\frac{d^2y}{dx^2}$ when $x = 2$ giving your answer in terms of k

1.32 Differentiation: Tangents and Normals (answers on page 446)

Prerequisite knowledge for Q20-Q27:
Differentiation of the form ae^{kx}

20) Evaluate $f'(1)$ for each of the following, leaving your answers in exact form:
    a) $f(x) = e^{6x}$
    b) $f(x) = \sqrt{x} - e^{2x}$
    c) $f(x) = x^{\frac{4}{3}} + e^{2x}$

21) Find the equation of the tangent line to the curve
$$y = e^x$$
at $(0, 1)$ in the form $y = mx + c$ where m and c are integers to be found.

22) Find the equation of the normal line to the curve
$$y = x^2 - e^{2x}$$
at $(0, -1)$ in the form $y = mx + c$ where m and c are integers to be found.

23) The curve of
$$y = e^x - 2x^2$$
is shown on the diagram below. The point P has coordinates $(1, e - 2)$

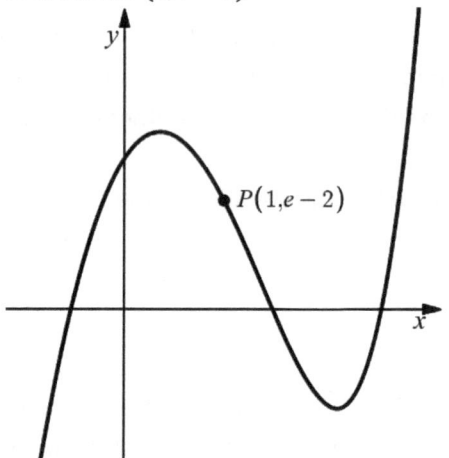

a) Find the equation of the tangent line at P, giving your answer in the form $y = mx + c$ where m and c are constants to be found.
b) Find the coordinates where the tangent line crosses the x-axis and y-axis.
c) Hence calculate the area of the triangle bounded by the tangent line at P and the coordinate axes giving your answer in the form
$$\frac{a}{b - e}$$
where a and b are integers to find.

24) A curve has equation
$$y = 4e^{-kx}$$
where k is a positive constant.
a) Find $\frac{dy}{dx}$ when $x = 2$ giving your answer in terms of k
b) Show that
$$\frac{d^2y}{dx^2} = k^2 y$$

25) A curve has equation
$$y = e^{5kx}(1 - k)$$
where k is a positive constant. Given that at $x = \frac{1}{k'}$
$$\frac{d^2y}{dx^2} = -100e^5$$
find the value of k

26) The curve of
$$y = 5x - e^x$$
is shown on the diagram below.
The point P has coordinates $(2, 10 - e^2)$

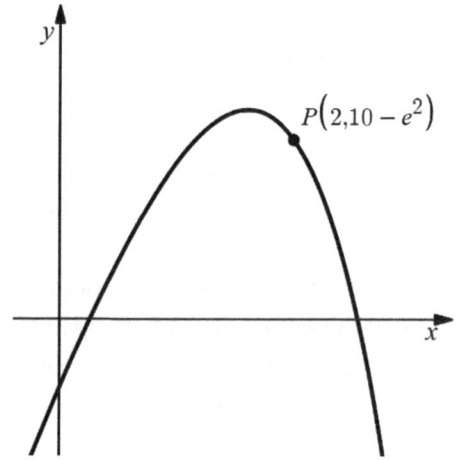

a) Find the equation of the normal line at P, giving your answer in the form $y = mx + c$ where m and c are constants to be found.
b) Find the coordinates where the normal line meets the x-axis, giving your answer in the form $(e^4 + ae^2 + b, 0)$ where a and b are integers to be found.

27) The normal to the curve
$$f(x) = e^x - 2x^2 - x + e^{-x}$$
at $x = 0$ is the same line as the tangent to the curve
$$g(x) = \frac{1}{2}x^2 + (5 + k)x + p$$
at the point $(7, 9)$
Find the values of p and k

1.33 Differentiation: Graphs of Gradient Functions (answers on page 447)

1) Draw the gradient function on copies of each of the following graphs:

a)

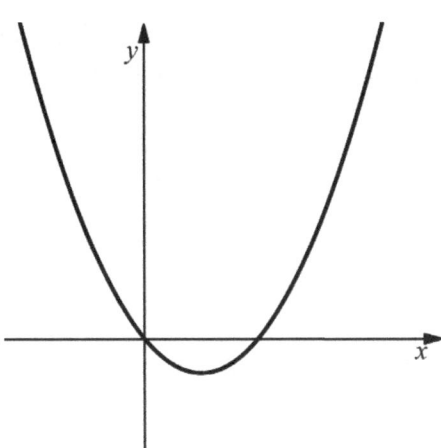

b)

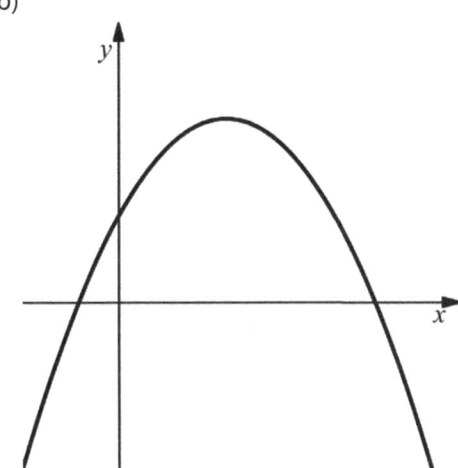

c)

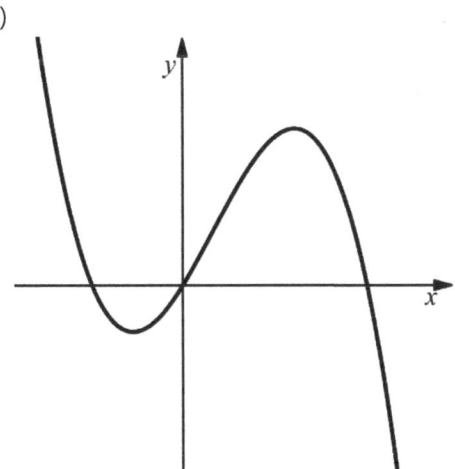

d)

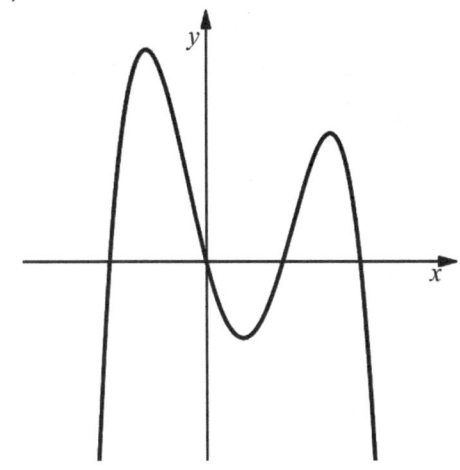

e)

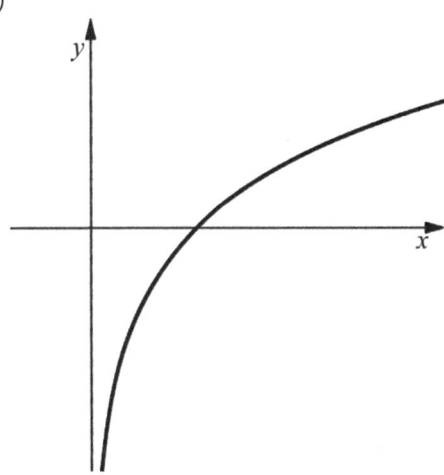

f)

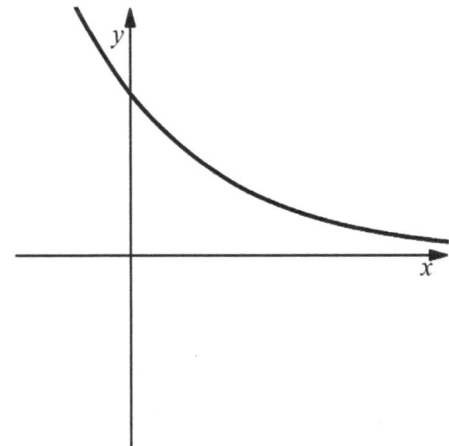

1.33 Differentiation: Graphs of Gradient Functions (answers on page 447)

2) Given that in each of the following functions, their maximum and minimum gradient is the same magnitude as the maximum and minimum values respectively, draw the gradient function on copies of each of the following graphs:

a)

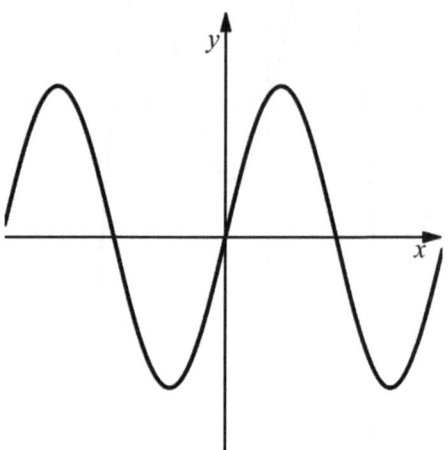

b)

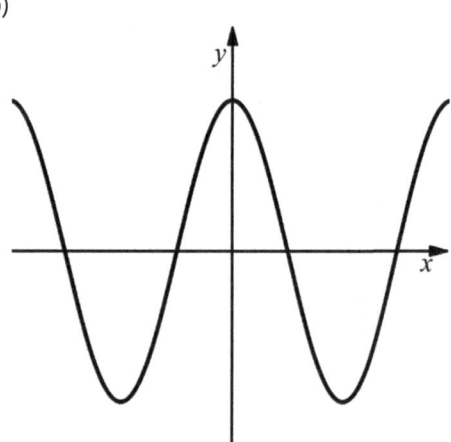

c)

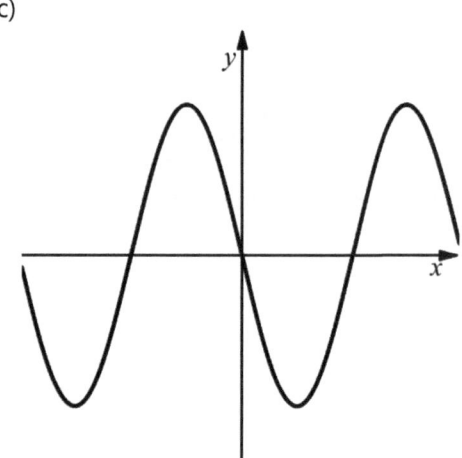

3) The gradient function of the curve
$$f(x) = ax^3 + bx^2 + cx$$
is sketched.
In each case state whether each of a, b, and c are greater than, equal to or less than 0

a)

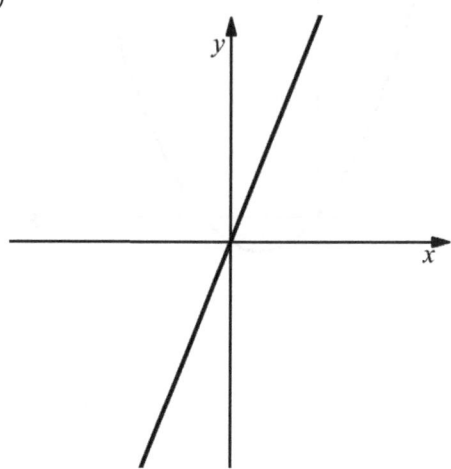

b)

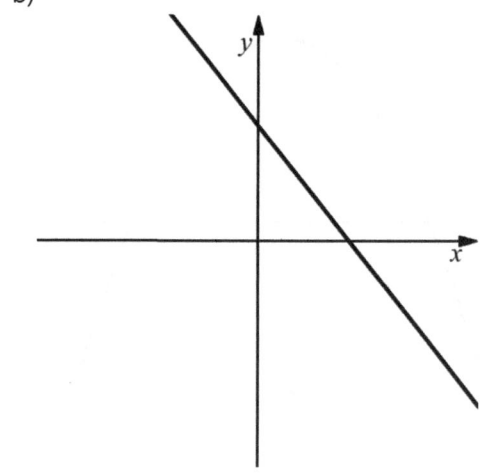

c)

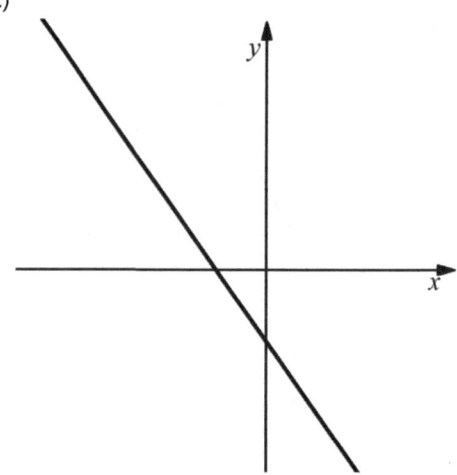

1.33 Differentiation: Graphs of Gradient Functions (answers on page 447)

d)

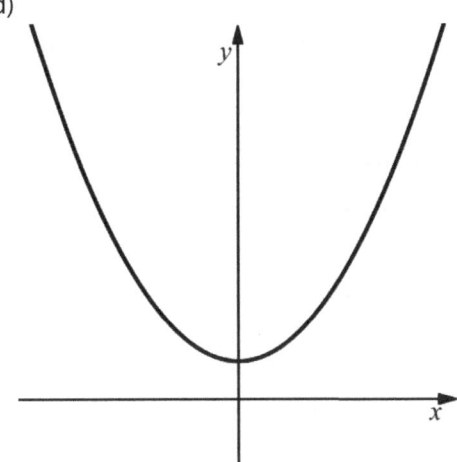

e)

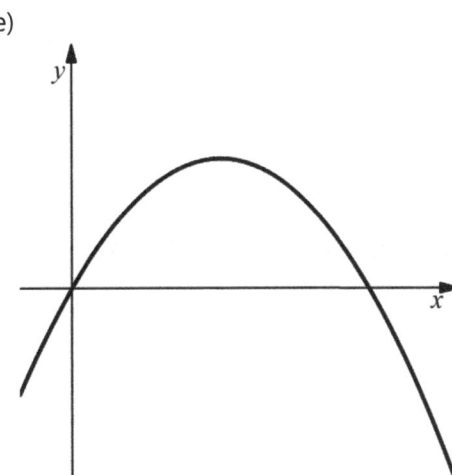

f)

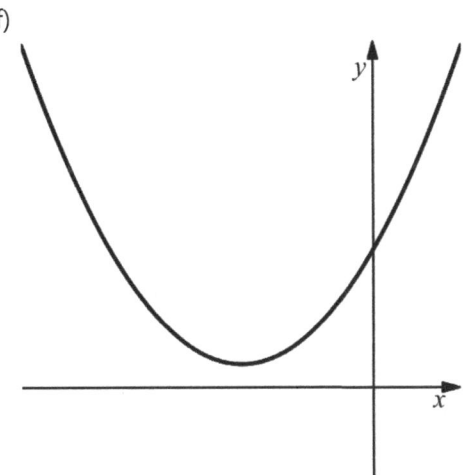

g)

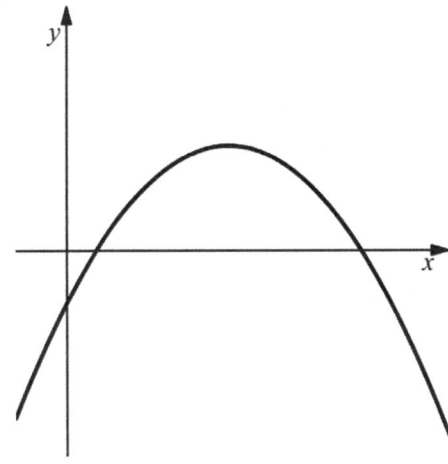

h)

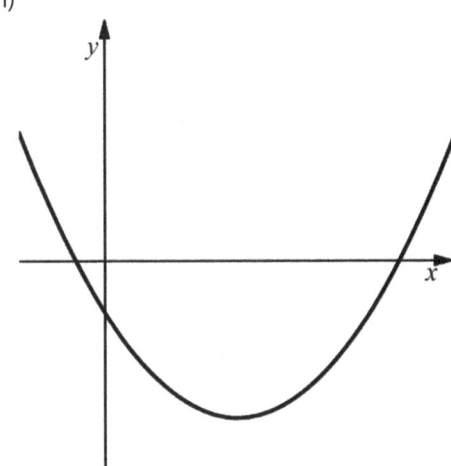

4) For the curve shown on the graph below, the two points of inflection are labelled A and B. Sketch the gradient function on a copy of the graph.

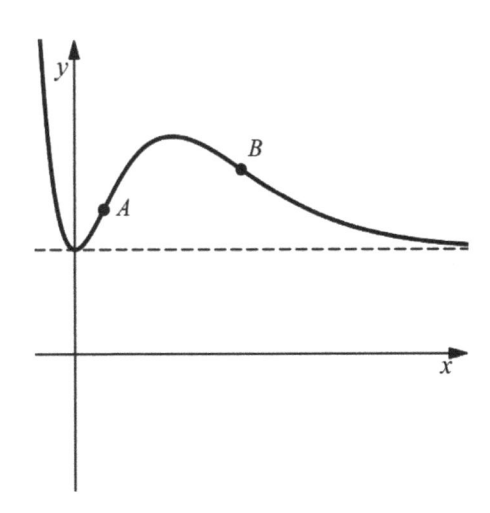

1.34 Differentiation: Stationary Points (answers on page 448)

1) For each of following, find the value(s) of x for which $\frac{dy}{dx} = 0$
   a) $y = x^2 - 2x$
   b) $y = -2x^2 + 18x - 37$
   c) $y = x^3 - 3x^2 + 3x - 4$
   d) $y = x - \frac{4}{x^2}$
   e) $y = x + \frac{4}{x}$
   f) $y = \frac{x^3+2}{x}$
   g) $y = \frac{x^4+3}{3x^3}$

2) Find the interval on which the function each of the following functions are decreasing, giving your answer in set notation:
   a) $f(x) = x^2 + 4x$
   b) $f(x) = 6x^2 - 10x + 5$
   c) $f(x) = x^3 - 3x$
   d) $f(x) = 2x^3 + 3x^2 - 36x - 10$
   e) $f(x) = -2x^3 - 18x^2 - 48x - 1$

3) Find the interval on which the function
   $$f(x) = -x^3 - 6x^2 + 15x + 1$$
   is increasing, giving your answer in interval notation.

4) Show that the function
   $$f(x) = x^3 - 3x^2 + 4x - 8$$
   is always increasing.

5) Show that the function
   $$f(x) = -3x^3 + 3x^2 - 8x + 2$$
   is always decreasing.

6) Find the coordinates of the stationary points of the following curves and determine their types:
   a) $y = -x^2 + 5x - 6$
   b) $y = 2x^3 - 15x^2 + 24x$
   c) $y = 2x^3 - 3x^2 - 12x + 7$
   d) $y = -x^3 + 3x^2 + 72x$
   e) $y = 2x^4 - 4x^2$
   f) $y = 4x + \frac{1}{x}$
   g) $y = \frac{1}{x^2} + 2x$
   h) $y = 3x^6 - 6x^3$

7) Find the coordinates of the maximum of the gradient function of
   $$y = -3x^3 - 2x^2 + 1$$

8) Find the coordinates of the stationary points, determine their types and hence sketch each of the following curves:
   a) $y = x^3 - 3x^2 - 45x$
   b) $y = 16x^4 - 32x^2 + 7$
   c) $y = 5x - 2x^{\frac{1}{2}}$ for $x \geq 0$

9) Find the values of a, b and c such that
   $$y = -x^3 + ax^2 + bx + c$$
   has stationary points when $x = 1$ and $x = 3$, and passes through the point $(2, 4)$

10) Find the values of a, b and c such that
    $$y = ax^4 + bx^2 + c$$
    passes through the points $(1, 0)$ and $(0, -3)$, and has a stationary point at $(\sqrt{2}, 1)$. Find all the stationary points and determine their types.

Prerequisite knowledge for Q11-Q13: Differentiation of the form ae^{kx}

11) Giving your answer in set notation, find the interval on which $f(x) = e^{2x} - x$ is increasing.

12) The diagram shows the curve
    $$y = 5e^x - e^{2x}$$

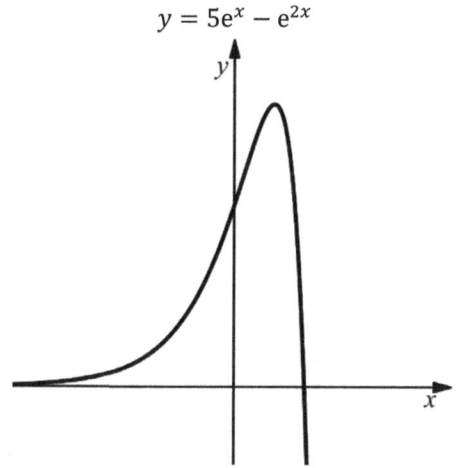

    a) Find the coordinates at which the curve crosses the x-axis and the y-axis.
    b) Find the coordinates of the stationary point, giving your answer in exact form.
    c) Show that the stationary point is a maximum.

13) Find the coordinates of the stationary point of
    $$y = e^{3x} - e^{2x} - e^x$$
    and determine its type.

1.35 Differentiation: Optimisation (answers on page 448)

1) For each of the following shapes, write an algebraic expression for the area, A and the perimeter, P:
   a) Square with sides of length x
   b) Rectangle with sides of length x and $2x$
   c) Circle with radius x
   d) Isosceles triangle with base of x and perpendicular height of $2x$

2) For each of the following 3D shapes, write an algebraic expression for the volume, V and the total surface area, A:
   a) Cube with sides of length x
   b) Cuboid with sides of length x, x and $3x$
   c) Cylinder with radius x and a length h
   d) Isosceles triangular based prism of length h, with the triangle surface having base x and a perpendicular height of x

3) For each of the following, find:
   a) $\frac{dV}{dr}$ for $V = 20r - \frac{1}{4}r^2$
   b) $\frac{dP}{dt}$ for $P = \frac{1}{t} + \frac{t}{2}$
   c) $\frac{dA}{dr}$ for $A = 2r - \frac{10}{r^2}$
   d) $\frac{dC}{dx}$ for $C = \frac{10}{x} + \frac{x^2}{4}$

4) Solve each of the following equations, leaving your answer in exact form:
   a) $25\pi - 4\pi r^2 = 0$, $r > 0$
   b) $-\frac{144}{x^2} + \frac{1}{4} = 0$, $x > 0$
   c) $-\frac{10}{x^2} + \frac{1}{2}x = 0$, $x > 0$
   d) $\frac{1}{4}\pi r - \frac{27\pi^2}{r^2} = 0$, $r > 0$
   e) $\frac{1}{5}\pi r - \frac{3}{r^2} = 0$, $r > 0$

5) A rectangular enclosure is created within a field, with a wall forming one side with a width of x metres. A rope of length 30 metres is used to form the remaining three sides.

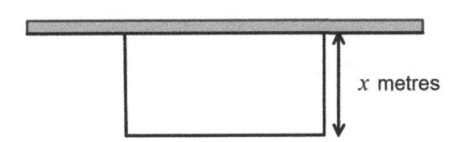

   a) Show that the area A may be written as
   $$A = 30x - 2x^2$$
   b) Hence, find the maximum area of the enclosure.

6) The cost of a journey £C when a van when driving at a steady speed in v kilometres per hour may be modelled by
$$C = \frac{1156}{v} + \frac{v}{9} + 50$$
where $v > 0$
   a) Find the speed at which the minimum cost occurs.
   b) Find the minimum cost.

7) The diagram shows a solid cuboid with lengths x cm, $4x$ cm and y cm.

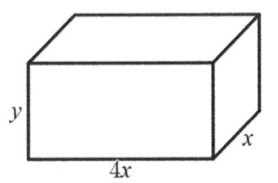

   The total surface area of the cuboid is 100 cm².
   a) Show that the volume may be written as
   $$V = 40x - \frac{16}{5}x^3$$
   b) Hence, find the maximum volume of the cuboid giving your answer to 1 decimal place.
   c) Find the exact dimensions of the box for this maximum volume to occur.

8) The diagram shows an open-topped cuboid tank.
   Its base has width x metres and length $2x$ metres and has a height of y metres.

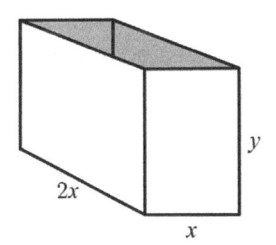

   It is made from 48 m² of sheet metal.
   a) Show that the volume may be written as
   $$V = 16x - \frac{2}{3}x^3$$
   b) Hence find the maximum volume giving your answer to 1 decimal place.
   c) Find the exact dimensions of the box for this maximum volume to occur.

1.35 Differentiation: Optimisation (answers on page 448)

9) The diagram shows a plant pot which is a cylindrical container of radius r cm and height h cm.
   The plant pot has an open top.
   The total surface area is 48π cm².

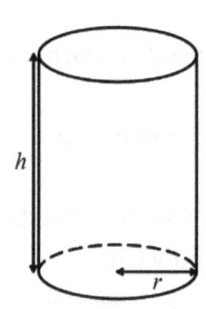

   a) Show that the volume may be written as
   $$V = 24\pi r - \frac{\pi}{2}r^3$$
   b) Hence find the maximum volume giving your answer in exact form.

10) The diagram shows a piece of square cardboard with width 40 cm. Four squares are cut from the corners each of width x cm. The cardboard is then folded into an open tray.

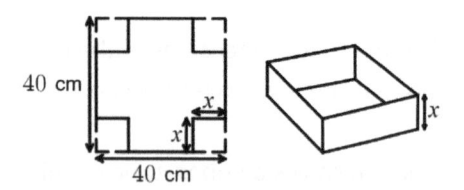

    Find the maximum capacity of the tray. Give your answer to 1 decimal place.

11) A container is created as a closed cylinder with capacity 432π cm³.

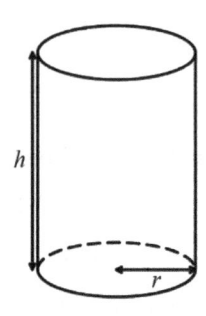

    a) Find the dimensions of the container when its surface area is minimised.
    b) For sealing the side of the container, a line of fluid is dropped along the length. Find the cost given it costs £0.04 per cm.

12) The curve
    $$f(x) = 10 - x^2$$
    and rectangle ABCD is shown in the diagram below.

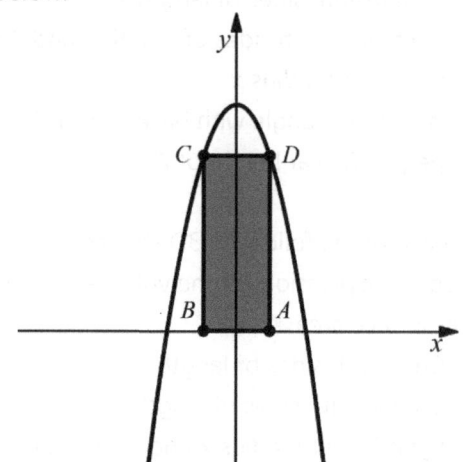

    Given that the rectangle is bounded by the curve and the x-axis, find the maximum area that the rectangle can have.

13) A container for ice cream desserts is being designed to use as little plastic as possible. It needs to have a volume of 144π cm³. The vendors of the ice cream would like it to be comprised of an open top semi-sphere with a cylinder with the same radius attached, and that it should have no base or top.
    The radius of the cylinder is r cm and the length of the cylinder is h cm.

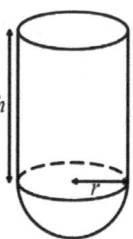

    A sphere of radius r has volume $\frac{4}{3}\pi r^3$ and surface area $4\pi r^2$

    a) Show that the area of plastic used is given by
    $$A = \frac{2}{3}\pi r^2 + \frac{288\pi}{r}$$
    b) Hence find the radius which minimises the area.
    c) By calculating the value of h for this to be minimised, state an issue with the vendor's requirement.

1.35 Differentiation: Optimisation (answers on page 448)

14) A cylindrical metal container which is closed at either end, has a capacity of 500cm³.
The metal used for the curved side costs 0.03 pence/cm³. The cost for the circular bottom and top costs 0.07 pence/cm³.
   a) Show that the cost of the metals for making the container may be given by
   $$C = \frac{7}{50}\pi r^2 + \frac{30}{r}$$
   where r is the radius of the container and C is the cost in pence.
   b) Find the radius for the minimum cost and height for that cost.
   c) Hence calculate the minimum cost of making one of these containers to the nearest pence.

15) A solid triangular based prism is made from wood with a volume of $\left(500(\sqrt{2} - 1)\right)$ cm³
The two triangular ends are isosceles triangles with base $2x$ cm and perpendicular height x cm. The length of the prism is y cm.
The rectangular sides are painted with paint that costs £0.02 per cm² and the triangular ends are painted with paint that costs £0.04 per cm².
   a) Show that the slanted sides (not the base) have a total area of $2\sqrt{2}xy$
   b) Show that the cost of painting the prism is given by
   $$C = \frac{20}{x} + \frac{2}{25}x^2$$
   c) Hence calculate the minimum cost of painting one of these prisms.

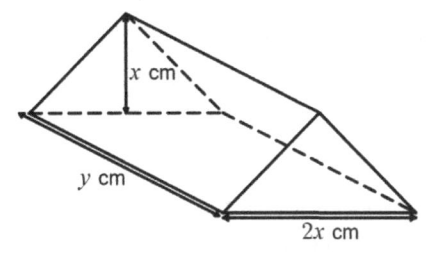

16) A part of a garden is going to be filled with two different types of decorative stones.
The design is a quarter circle with radius x.
Each rectangle has width y metres and length x metres and is laid out as shown in the diagram.

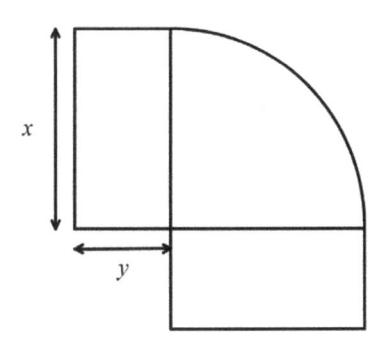

For the quarter circle, the stones will be 80% of the cost of the stones for the rectangular areas. The total perimeter is 16 metres.
   a) Show that the cost may be given by
   $$C = 8px - \left(1 + \frac{\pi}{20}\right)px^2$$
   for some cost per £p per metre².
   b) Show that the length of x for the maximum cost is occurs when
   $$x = \frac{80}{20 + \pi}$$
   c) Calculate the area of the quarter circle and the rectangles separately for this value of x, giving your answer to 3 significant figures.

Prerequisite knowledge for Q17: Arc Length and Sector Area in Radians

17) A toy slice of cake is being manufactured and can be modelled by a sector of a circle with radius r
The volume is to be 350cm³.
The angle of the sector is 0.7 radians.

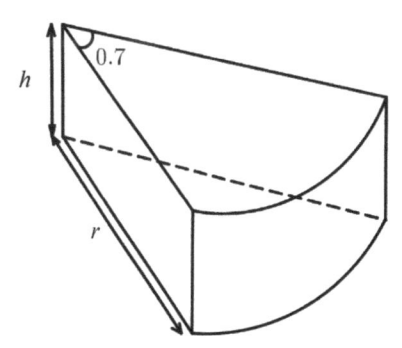

   a) Show that the surface area may be written in the form
   $$A = \frac{2700}{r} + 0.7r^2$$
   b) Hence find the minimum surface area and the values of r and h which obtain this, giving your answer to 3 significant figures.

1.36 Integration: Indefinite Integrals (answers on page 449)

1) Find
   a) $\int (8x + 1) \, dx$
   b) $\int (6x^2 - 3) \, dx$
   c) $\int (10x^3 + 9) \, dx$
   d) $\int (3x + 5)(4x^2 - 1) \, dx$

2) Find
   a) $\int \left(\frac{2}{3}x + \frac{5}{3}x^4\right) dx$
   b) $\int \left(\frac{3}{4}x^2 - \frac{14}{3}x^6\right) dx$
   c) $\int \left(\frac{2}{5}x^7 - \frac{3}{8}x^5\right) dx$
   d) $\int \left(\frac{1}{3}x^2 + 4\right)\left(\frac{2}{3}x^2 + 6\right) dx$

3) The curve y has gradient function
$$\frac{dy}{dx} = 2 - 2x$$
Sketch the family of solutions that solve this differential equation.

4) The curve y has gradient function
$$\frac{dy}{dx} = \left(\frac{x}{2} - 1\right)\left(\frac{x}{3} + 1\right)$$
and passes through the point (6,1).
Show that y can be written in the form $y = ax^3 + bx^2 + cx + d$, where a, b, c and d are rational numbers.

5) Find the equation of the curve y given the gradient function is:
   a) $\frac{dy}{dx} = x^{\frac{2}{3}}$
   b) $\frac{dy}{dx} = 6x^{\frac{1}{5}}$
   c) $\frac{dy}{dx} = 21x^{\frac{3}{4}}$
   d) $\frac{dy}{dx} = \left(2x^{\frac{1}{6}} + 3\right)^2$

6) Find
   a) $\int \left(x^3 + x^{\frac{3}{2}}\right) dx$
   b) $\int \left(5x^{-2} - 2x^{\frac{2}{3}}\right) dx$
   c) $\int \left(8x^{-3} - 4x^{\frac{6}{5}}\right) dx$
   d) $\int \left(2x^{-4} - 3x^{\frac{4}{3}}\right)^2 dx$

7) Find
$$\int (1 - x^{-2})(1 - x^{-4}) \, dx$$

8) The curve y has gradient function
$$\frac{dy}{dx} = 40x^{\frac{1}{3}} - 4x$$
and passes through the point (1, −1).
Show that y can be written in the form $y = ax^{\frac{4}{3}} + bx^2 + c$, where a, b and c are integers.

9) Find the equation of the curve y given the gradient function is:
   a) $\frac{dy}{dx} = 5 - 2x^{-4}$
   b) $\frac{dy}{dx} = 6x^{-3}(x - 2)$
   c) $\frac{dy}{dx} = x^{\frac{1}{2}}(8x - 3)$
   d) $\frac{dy}{dx} = \left(2x^{\frac{3}{2}} - 3\right)\left(2x^{-\frac{7}{2}} + 7\right)$

10) Find
    a) $\int 4\sqrt{x}(x - 1) \, dx$
    b) $\int \sqrt[4]{x} \, dx$
    c) $\int \sqrt[5]{32x} \, dx$
    d) $\int \sqrt[4]{81x^{\frac{1}{4}}} \, dx$

11) The curve y has gradient function
$$\frac{dy}{dx} = x^2(7\sqrt{x} - 3)$$
and passes through the point (1, −4).
Show that y can be written in the form $y = ax^{\frac{7}{2}} - bx^3 + c$, where a, b and c are integers.

12) Find the equation of the curve y given the gradient function is:
    a) $\frac{dy}{dx} = \frac{4}{x^2}$
    b) $\frac{dy}{dx} = \frac{6}{x^3}$
    c) $\frac{dy}{dx} = \frac{9}{5x^4}$
    d) $\frac{dy}{dx} = \frac{100}{9x^{10}}$

13) Find
    a) $\int \frac{4}{3\sqrt{x}} \, dx$
    b) $\int \frac{6}{\sqrt[3]{x}} \, dx$
    c) $\int \left(\frac{8}{\sqrt[5]{x}} - \frac{100}{\sqrt[3]{8x}}\right) dx$
    d) $\int \left(\frac{1}{\sqrt[3]{x}} - \frac{3}{x\sqrt{x}}\right)^2 dx$

1.36 Integration: Indefinite Integrals (answers on page 449)

14) The curve y has gradient function
$$\frac{dy}{dx} = \frac{8}{\sqrt{x}}$$
and passes through the point $(16, -8)$.
Show that y can be written in the form
$y = a(2\sqrt{x} + b)$, where a and b are integers.

15) The curve y has gradient function
$$\frac{dy}{dx} = \frac{750}{x^4}$$
and passes through the point $(5, 248)$.
Show that y can be written in the form
$y = a\left(1 - \frac{1}{x^3}\right)$, where a is an integer.

16) The curve y has gradient function
$$\frac{dy}{dx} = \frac{6}{x\sqrt{x}}$$
and passes through the point $(8, 0)$.
Show that y can be written in the form
$y = a\left(\sqrt{b} + \frac{c}{\sqrt{x}}\right)$, where a, b and c are integers.

17) The curve y has gradient function
$$\frac{dy}{dx} = \frac{12}{\sqrt[3]{x}}$$
and passes through the point $(27, 0)$.
Show that y can be written in the form
$y = a\left(x^{\frac{2}{3}} + b\right)$, where a and b are integers.

18) The curve y has gradient function
$$\frac{dy}{dx} = \frac{1}{x^3\sqrt{x}}$$
and passes through the point $\left(1, \frac{3}{5}\right)$.
Show that y can be written in the form
$y = a - \frac{b}{cx^2\sqrt{x}}$, where a, b and c are integers.

19) Find
$$\int \left(\frac{4}{x^2\sqrt{x}} - \frac{\sqrt{x}}{4x^2}\right) dx$$

20) Find
$$\int \left(\frac{6}{5x^3\sqrt{x}} - \frac{\sqrt[3]{x}}{6x^3}\right) dx$$

21) Find
$$\int \left(\frac{3\sqrt{x}}{4x^4} - \frac{x\sqrt{x}}{\sqrt[3]{x}}\right) dx$$

22) It is given that
$$f'(x) = \left(3x - \frac{4}{x}\right)^2$$
and $f(2) = -2$. Show that
$$f(x) = ax^3 + bx + \frac{c}{x} + d$$
where a, b, c and d are integers.

23) It is given that $k > 0$. Find the equation of the curve y given the gradient function is:
 a) $\frac{dy}{dx} = kx^4$
 b) $\frac{dy}{dx} = (kx)^4$
 c) $\frac{dy}{dx} = \frac{1}{kx^4}$
 d) $\frac{dy}{dx} = \frac{18}{(2kx)^5}$

24) Find
 a) $\int \frac{5+x^3}{x^2} dx$
 b) $\int \frac{3x-2x^3}{x^5} dx$
 c) $\int \frac{8x^2-9}{x^4} dx$
 d) $\int \frac{30x^3-7}{14x^5} dx$

25) Find the equation of the curve y given the gradient function is:
 a) $\frac{dy}{dx} = \frac{5x^4-7}{5x^3}$
 b) $\frac{dy}{dx} = \frac{3-8x^3}{4x^5}$
 c) $\frac{dy}{dx} = \frac{6x^7-8}{12x^4}$
 d) $\frac{dy}{dx} = \frac{8x-3x^2-6}{24x^5}$

26) It is given that
$$f'(x) = \frac{x-8}{x^3}$$
and $f(1) = 0$. Show that
$$f(x) = \frac{ax^2 + bx + c}{x^2}$$
where a, b and c are integers.

27) It is given that
$$f'(x) = \left(\frac{14 - 7x}{\sqrt[3]{x}}\right)^2$$
and $f(1) = 462$. Show that
$$f(x) = 21x^{\frac{1}{3}}(ax^2 + bx + c)$$
where a, b and c are integers.

1.36 Integration: Indefinite Integrals (answers on page 449)

Prerequisite knowledge for Q28: Stationary Points

28) A curve has gradient function
$$\frac{dy}{dx} = 3x^2 + kx - 21$$
The curve has two turning points. One of the turning points is at $(1, -6)$. Find the coordinates of the other turning point.

Prerequisite knowledge for Q29-Q30: The Factor Theorem

29) A curve has gradient function
$$f'(x) = kx^2 + 74x + 11$$
where $f(-1) = -100$. It is also given that $(x + 5)$ is a factor of $f(x)$. Find $f(x)$ in its simplest form.

30) A curve has gradient function
$$f'(x) = 576x^3 + ax^2 - 208x + 40$$
It is given that $(6x - 5)$ and $(2x + 1)$ are factors of $f(x)$. Find $f(x)$ in its simplest form.

Prerequisite knowledge for Q31: Binomial Expansion

31) Find
   a) $\int [(5+x)^4 - (5-x)^4]\, dx$
   b) $\int [(3+2x)^5 - (3-2x)^5]\, dx$
   c) $\int [(1-3x)^6 - (1+3x)^6]\, dx$
   d) $\int [(1+2x)^7 - (1-2x)^7]\, dx$

32) The curve y has gradient function
$$\frac{dy}{dx} = (4+2x)^3 - (4-2x)^3$$
and passes through the point $(-2, 1024)$. Show that y can be written in the form $y = 4(x^2 + a)^2$, where a is an integer.

Prerequisite knowledge for Q33: Second Derivatives

33) Find the equation of the curve y given:
   a) $\frac{d^2y}{dx^2} = 3$
   b) $\frac{d^2y}{dx^2} = 4x - 5$
   c) $\frac{d^2y}{dx^2} = 6x^2 - 8x + 9$
   d) $\frac{d^2y}{dx^2} = 8x^3 - 7x^5$

34) Find the equation of the curve y given:
   a) $\frac{d^2y}{dx^2} = \sqrt{x}$
   b) $\frac{d^2y}{dx^2} = \frac{5}{\sqrt{x}}$
   c) $\frac{d^2y}{dx^2} = \frac{3}{x^3} - \frac{x^3}{3}$
   d) $\frac{d^2y}{dx^2} = \frac{8}{x^4\sqrt{x}}$

35) Find the equation of the curve y given
$$\frac{d^3y}{dx^3} = \frac{10}{x\sqrt{x}}$$

36) It is given that the curve $y = f(x)$, where $f(-3) = 2$ and $f'(3) = 2$. It is also given that
$$f''(x) = 14 - 6x$$
Find the equation of the curve $y = f(x)$ in its simplest form.

37) It is given that the curve $y = f(x)$, where $f(1) = 3$ and $f'(1) = 5$. It is also given that
$$f''(x) = 36x^2 + 24x + 12$$
Find the equation of the curve $y = f(x)$ in its simplest form.

38) It is given that the curve $y = f(x)$, where $f(2) = 5$ and $f'(1) = -2$. It is also given that
$$f''(x) = \frac{18}{x^3}$$
Find the equation of the curve $y = f(x)$ in its simplest form.

39) It is given that the curve $y = f(x)$, where $f(1) = 11$ and $f'(0) = 0$.
It is also given that
$$f''(x) = 210(x - \sqrt{x} + x\sqrt{x})$$
Find the equation of the curve $y = f(x)$ in its simplest form.

Prerequisite knowledge for Q40: Stationary Points

40) It is given that the curve $y = f(x)$, where $f(2) = -37$ and $f'(1) = -15$.
It is also given that
$$f''(x) = 12(kx - 1)$$
and the curve has a stationary point at $x = \frac{1}{2}$.
Find the equation of the curve $y = f(x)$ in its simplest form.

1.37 Integration: Definite Integrals and Areas (answers on page 450)

1) Evaluate the following definite integrals:
   a) $$\int_2^6 (8x+1)\,dx$$
   b) $$\int_3^4 (6x^2-1)\,dx$$
   c) $$\int_4^8 (500+4x-x^3)\,dx$$
   d) $$\int_{-1}^1 (14x-9x^5)\,dx$$

2) The diagram below shows the curve with equation $y = 10x - x^4$. Find the exact value of the shaded area.

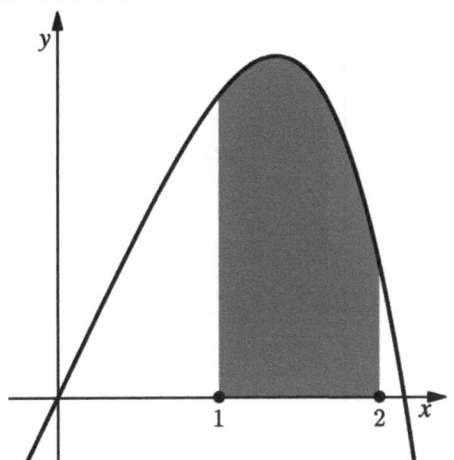

3) The diagram below shows the curve with equation $y = -2 - x^2 + x^3 - x^4$. Find the exact value of the shaded area.

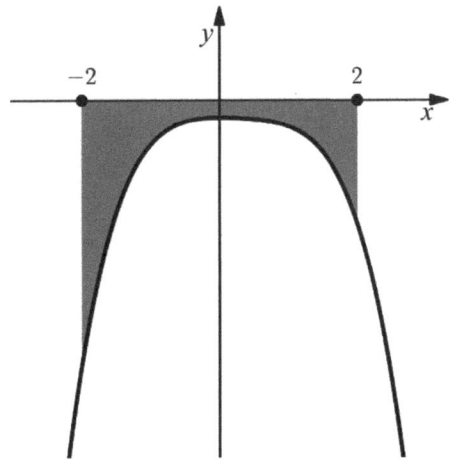

4) It is given that $\int_1^3 (f(x)+2)\,dx = 50$. Find the value of $\int_1^3 f(x)\,dx$

5) The diagram below shows the curve with equation $y = 10 - x^5 + 4x^3$. Find the exact value of the shaded area.

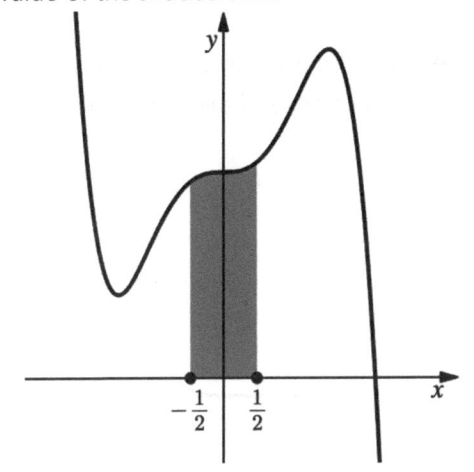

6)
   a) Sketch the graph of $y = (x-1)(4-x)$
   b) Explain why
      $$\int_1^5 (x-1)(4-x)\,dx$$
      does not give the area between the curve, the x-axis and the lines $x = 1$ and $x = 5$

7) The diagram below shows the curve with equation $y = 4x - 2x^3$

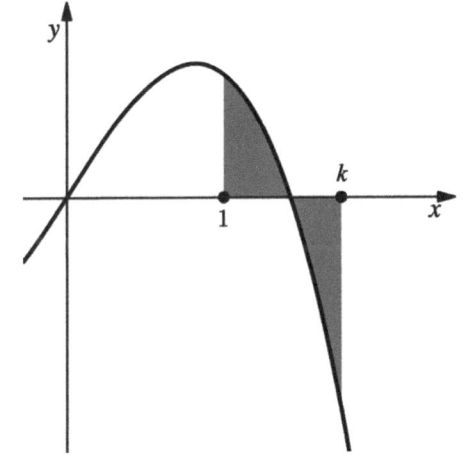

   It is given that
   $$\int_1^k (4x - 2x^3)\,dx = 0$$
   where k is a positive constant and $k > 1$
   Find the value of k

8) A quadratic curve has the equation $y = ax^2 + bx + c$. It is given that the curve passes through $(-2, 0)$, $(6, 0)$ and $(0, 240)$. Find the exact finite area between the quadratic curve and the x-axis.

1.37 Integration: Definite Integrals and Areas (answers on page 450)

9) Find the value of the constant k given that
$$\int_{-3}^{3}(kx^2 - 5x^4)\,dx = 18$$

10) The diagram below shows the curve $y = x^5$. Find the exact value of the shaded area.

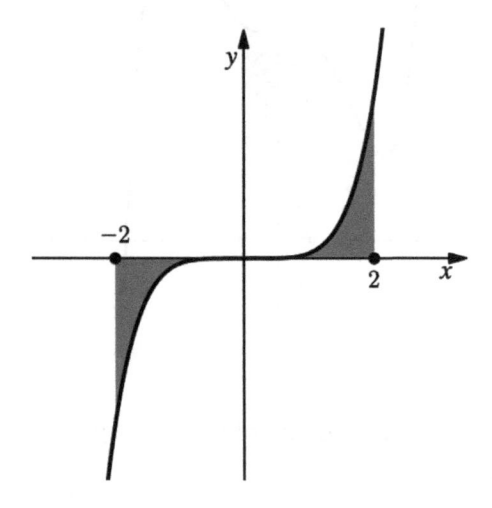

11) The diagram below shows the curve $y = 4 - x^2$. Find the exact value of the shaded area.

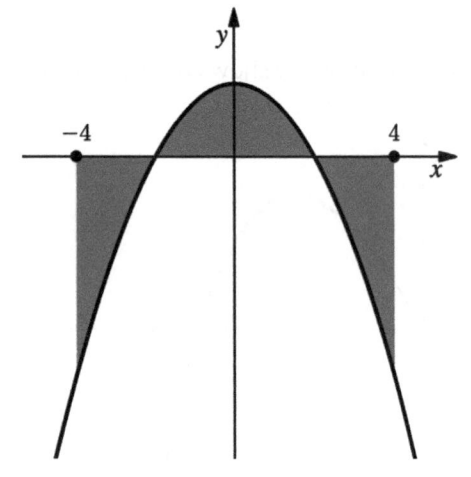

12) Show that
$$\int_{\sqrt{2}}^{\sqrt{3}}(3x^2 + 1)\,dx = 4\sqrt{3} - 3\sqrt{2}$$

13) Show that
$$\int_{1}^{1+\sqrt{2}}(50 - 6x^2)\,dx = 40\sqrt{2} - 12$$

14) Show that
$$\int_{1+\sqrt{3}}^{2+\sqrt{3}} 3x(x+2)\,dx = 25 + 15\sqrt{3}$$

15) Find the exact value of the following definite integrals:

a)
$$\int_{4}^{16} \sqrt{x}\,dx$$

b)
$$\int_{-2}^{-1} \frac{6}{x^2}\,dx$$

c)
$$\int_{1}^{9} \frac{4}{\sqrt{x}}\,dx$$

d)
$$\int_{1}^{64}\left(\frac{1}{\sqrt[3]{x}} - \frac{1}{\sqrt[4]{x}}\right)dx$$

16) The diagram below shows the curve with equation $y = \sqrt[4]{x}$. Find the exact value of the shaded area.

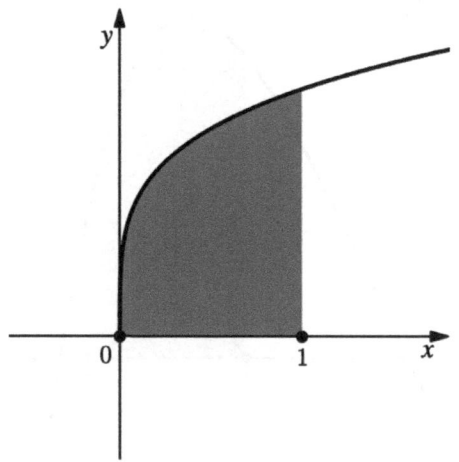

17) The diagram below shows the curve with equation $y = \frac{12}{\sqrt[3]{x}}$. Find the exact value of the shaded area.

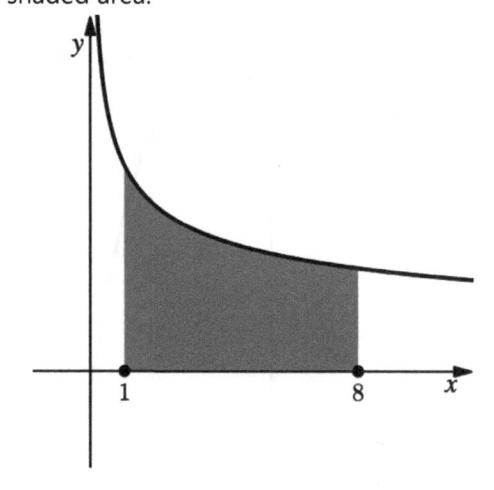

1.37 Integration: Definite Integrals and Areas (answers on page 450)

18) The diagram below shows the curve with equation $y = \frac{9}{x^2}$. Find the exact value of the shaded area.

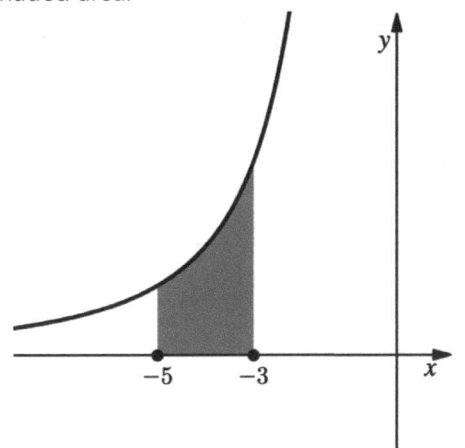

19) The diagram below shows the curve $y = x^3 + 4x^2 + x - 6$. Find the exact value of the shaded area.

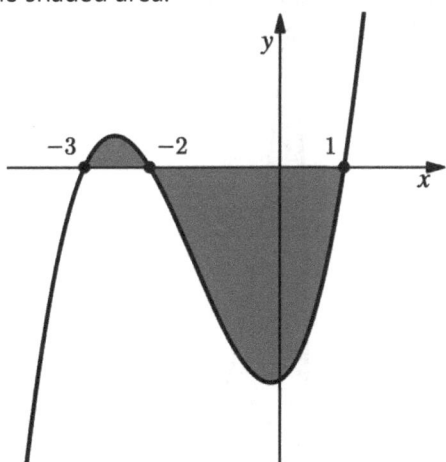

20) The diagram below shows the curve $y = \frac{1}{2}x^2 + 3$. Find the area bounded by the curve, the x-axis, and the lines $x = k$ and $x = 2k$, giving your answer in terms of k, where k is a non-zero constant.

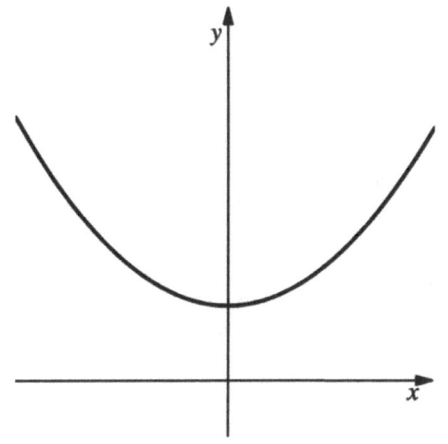

21) The diagram below shows the curve with equation $y = x(5 - x)$ and the line with equation $y = x$. Find the exact value of the shaded region.

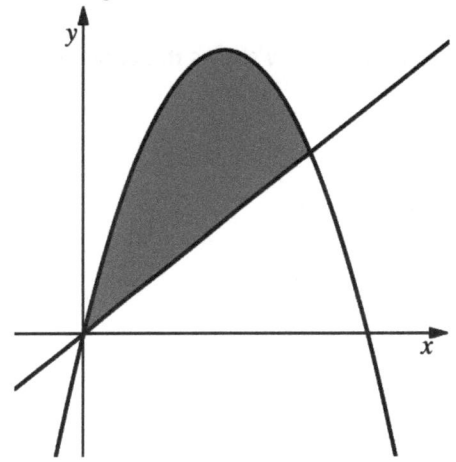

22) The diagram below shows the curve with equation $y = 20 - x^2(x - 3)$ and the line with equation $y = 20 - 4x$. Find the exact value of the shaded region.

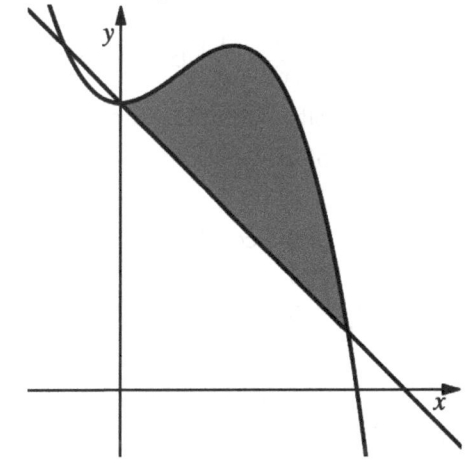

23) Find the value of the constant k given that
$$\int_1^2 \left(\frac{k}{x^3} + 2\right) dx = 5$$

24) Find the value of the constant k given that
$$\int_1^9 \left(2kx - 3\sqrt{x}\right) dx = 108$$

25) Find the possible values of the constant k given that
$$\int_1^{16} \left(k\sqrt{x} + 2x + 2\right) dx = 3k^2$$

26) Find the exact finite area enclosed between the curve $y = 6x^2 - 5x + 1$ and the x-axis.

1.37 Integration: Definite Integrals and Areas (answers on page 450)

27) Find the maximum value of
$$\int_2^4 (3p - 2p^2 x)\, dx$$
where p is a positive constant.

28) The diagram below shows the curve with equation
$$y = \frac{5}{x^3} - \frac{1}{x^2}$$

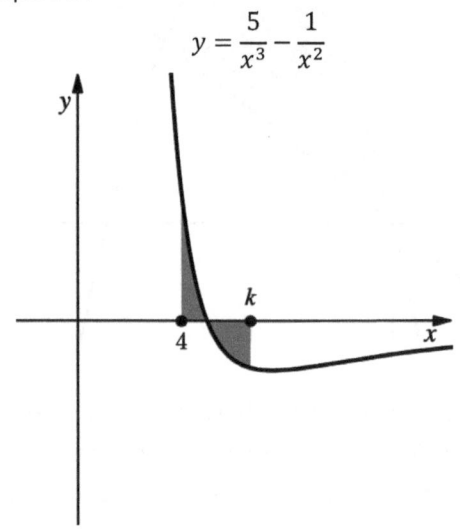

It is given that
$$\int_4^k \left(\frac{5}{x^3} - \frac{1}{x^2}\right) dx = 0$$
where k is a positive constant with $k > 4$
Find the value of k

29) Find the value of the constant k given that
$$\int_1^k \left(\frac{30}{\sqrt{x}} + 10\right) dx = 6090$$

30) The diagram below shows the curve with equation
$$y = x\sqrt{x} - 8$$
Find the exact value of the shaded area.

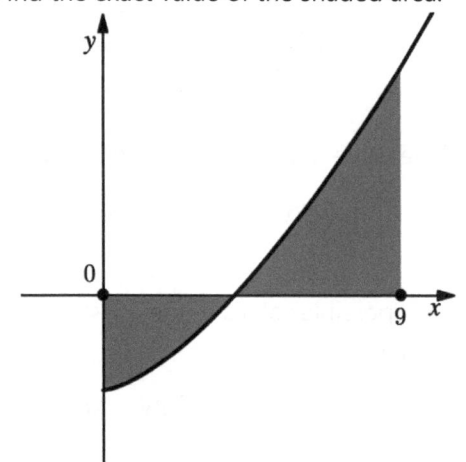

31) The diagram below shows the curve with equation $y = x + \frac{1}{x^2}$. Find the exact value of the shaded area.

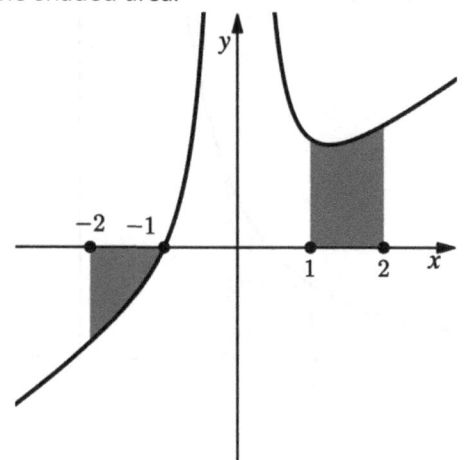

32) The diagram below shows the curve with equation $y = \frac{1}{x^2} - \frac{6}{x^3} + \frac{11}{x^4} - \frac{6}{x^5}$. Find the exact value of the shaded area.

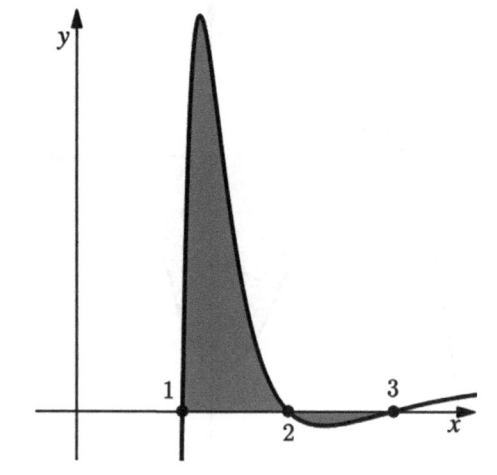

33) The diagram below shows the curve with equation $y = 35(x-3)(x-5)\sqrt{x}$ and the line $y = k$. The shaded area is $300\sqrt{3}$ units². Find the value of k

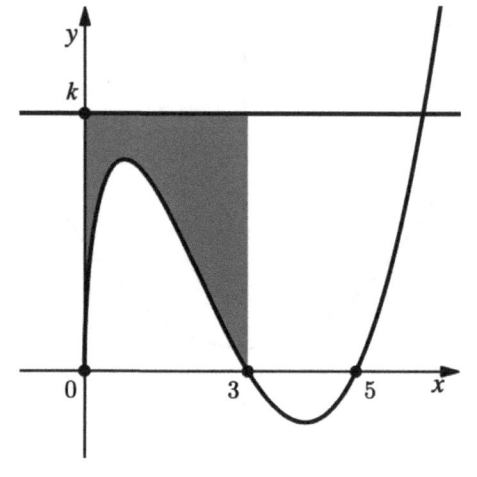

1.37 Integration: Definite Integrals and Areas (answers on page 450)

34) The diagram below shows the curve with equation $y = \frac{2}{\sqrt{x}} + x$. Find the area bounded by the curve, the x-axis, and the lines $x = k$ and $x = 9k$, giving your answer in terms of k, where k is a positive constant.

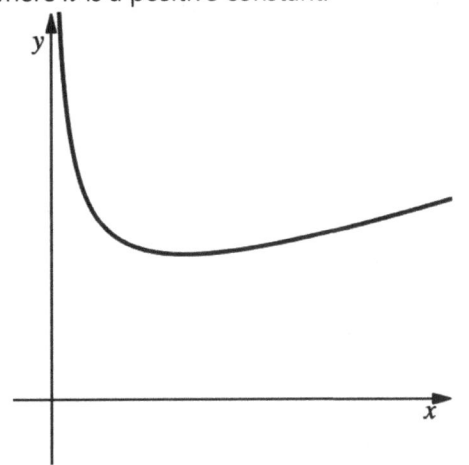

35) The diagram below shows the curve with equation $y = \frac{2}{x^2}$ and the line with equation $y = 3 - x$
Find the exact value of the shaded region.

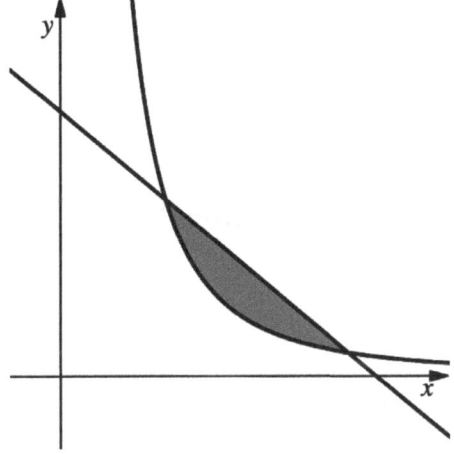

36) Find the exact value of the following definite integrals:

a) $$\int_1^3 \frac{x+2}{x^3} dx$$

b) $$\int_2^4 \frac{3x+8}{x^4} dx$$

c) $$\int_{-3}^{-1} \frac{4-5x}{2x^3} dx$$

d) $$\int_{-4}^{-2} \frac{3x^2-2}{6x^4} dx$$

37) The diagram below shows the curve with equation $y = \frac{x^2+10}{x^2}$
Find the exact value of the shaded area.

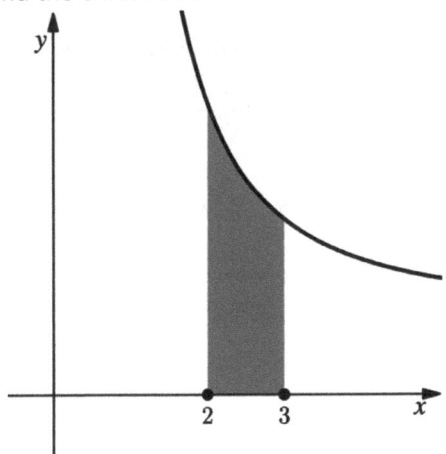

38) The diagram below shows the curve with equation $y = \frac{\sqrt{x}-2}{\sqrt{x}}$
Find the exact value of the shaded area.

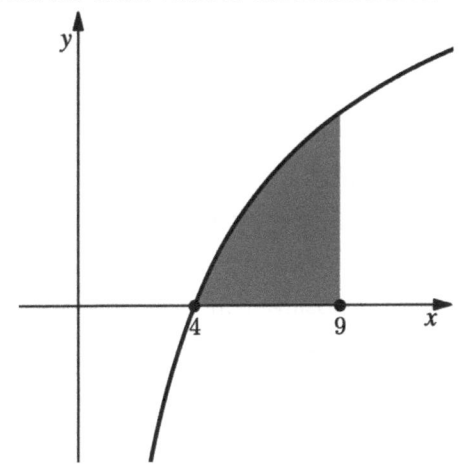

39) The diagram below shows the curve with equation $y = \frac{x+3}{x^3}$
Find the exact value of the shaded area.

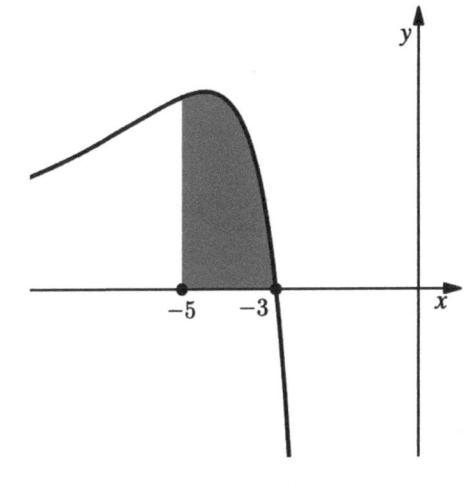

1.37 Integration: Definite Integrals and Areas (answers on page 450)

40) Find the value of the constant k given that
$$\int_1^2 \frac{kx+1}{x^4} dx = \frac{11}{12}$$

41) Find the value of the constant k given that
$$\int_4^9 \frac{4-k\sqrt{x}}{6\sqrt{x}} dx = 1$$

42) The diagram below shows the curve with equation
$$y = \frac{x^2-1}{x^2}$$

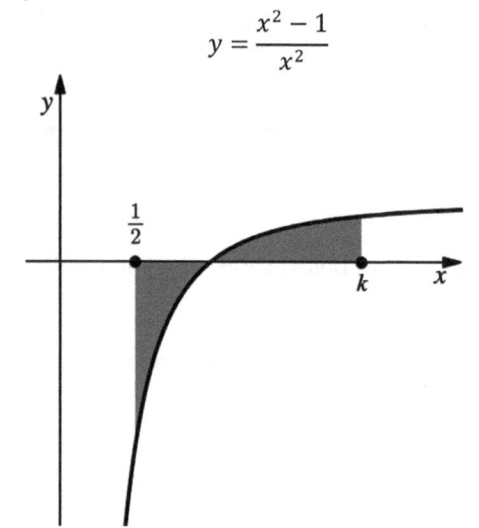

It is given that
$$\int_{\frac{1}{2}}^k \frac{x^2-1}{x^2} dx = 0$$
where k is a positive constant and $k > \frac{1}{2}$
Find the value of k

43) The diagram below shows the curve with equation $y = \frac{(3-x)(5-x)}{x^4}$
Find the exact value of the shaded area.

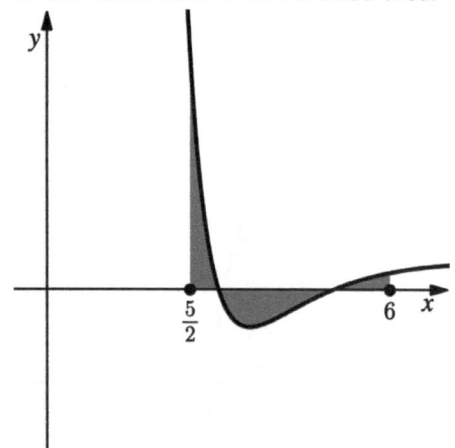

44) Find the exact area enclosed between the curve $y = 84x^{\frac{5}{2}}$, the x-axis, and the line $x = 2$

45) The diagram below shows the curve with equation $y = \frac{3x-1}{x\sqrt{x}}$
Show that the area bounded by the curve, the x-axis, and the lines $x = k$ and $x = 4k$, where k is a positive constant, is $\frac{6k-1}{\sqrt{k}}$

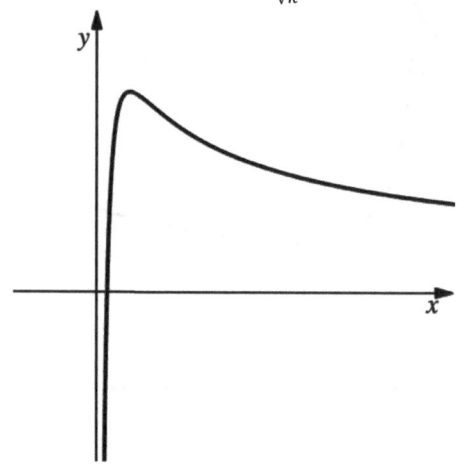

46) The diagram below shows the curve with equation $y = \frac{x+1}{\sqrt{x}}$ and the line with equation $3x + 14y = 47$
Find the exact value of the shaded region.

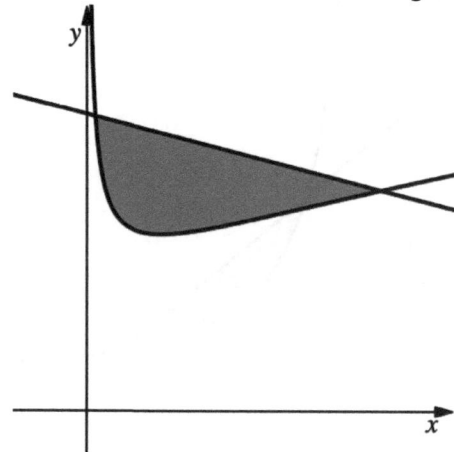

Prerequisite knowledge for Q47: Graph Sketching: Radical Functions

47) Find the exact finite area enclosed between the curve $y = \sqrt{x} - 16$, the x-axis and the y-axis.

Prerequisite knowledge for Q48: Graph Sketching: Cubic Graphs & Quartic Graphs

48) Find the exact finite area enclosed between the following curves and the x-axis:
   a) $y = (x-1)(x-2)(x-3)$
   b) $y = (x-1)^2(x-2)$
   c) $y = (x-1)(x-2)^2$
   d) $y = (x-1)^2(x-2)^2$

1.37 Integration: Definite Integrals and Areas (answers on page 450)

49) A puzzle piece is designed using a 6 cm by 6 cm square with the origin as its centre, as shown in the diagram below. Two identical curved shapes are cut out of the square to allow other pieces to be joined on the sides. The designer uses the equation $y = x^4 + 2$ to create one of the curved sections. Find the exact area of the puzzle piece.

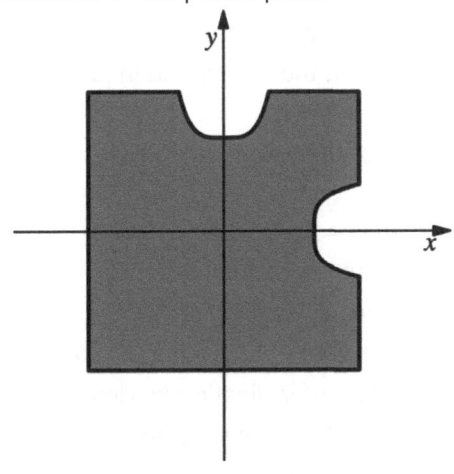

50) The diagram below shows the circle with equation $x^2 + y^2 = 2$ and the curve with equation $y = x^2$. Find the exact value of the shaded area.

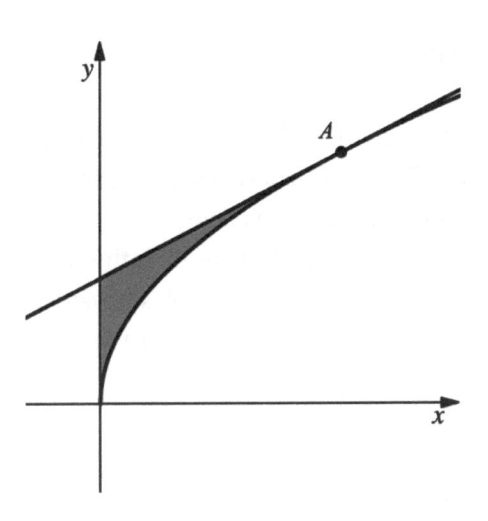

52) The diagram below shows the curve with equation $y = (3 - x)(x + 2)$. The points A and B lie on the curve and have x-coordinates 0 and 2 respectively. The shaded region is formed by the curve, the line $y = 7$, the normal to the curve at A and the normal to the curve at B. Find the exact shaded area.

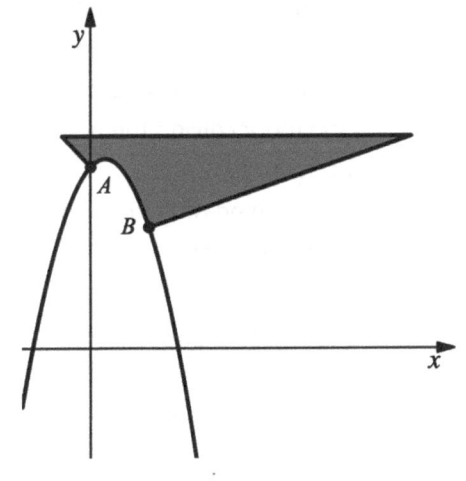

53) The diagram below shows the curve with equation $y = 2x^3 + 15x^2 - 144x + 500$. The points A and B lie on the turning points of the curve. Find the exact value of the shaded area.

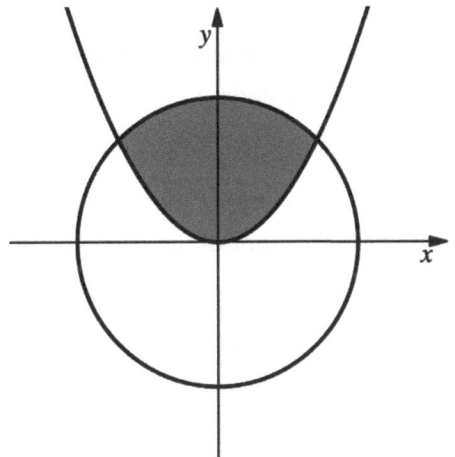

Prerequisite knowledge for Q51-Q53: Applications of Differentiation

51) The diagram below shows the curve with equation $y = 2\sqrt{x}$. The tangent to the curve at $x = 4$ is also shown in the diagram. The point A is where the tangent meets the curve. The shaded area is the region bounded by the curve, the tangent and the y-axis. Find the exact value of the shaded area.

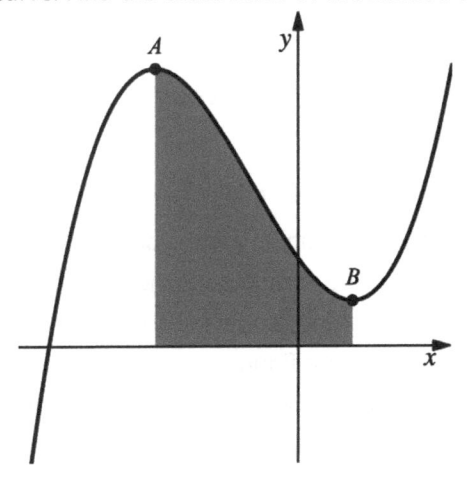

1.38 Applications of Calculus (answers on page 451)

Applications of Differentiation

1) For each of the following situations, find the first differential, explain what it means in context and specify its units:
   a) $C = 5.8x + 14.2$ where £C is the cost of the transport and x is the miles travelled.
   b) $h = -t^2 + 6t + 135$ at $t = 10$ where h is the height of a ball (in metres) and t is the number of seconds passed.
   c) $T = 0.01t^3 - 0.2t^2 + 20$ at $t = 5$ where T is the temperature (°C) of a substance and t is the number hours that have passed.

2) The spread of a virus over a period of 22 days may be modelled by the equation
$$N = -0.49t^2 + 10t + 3.54$$
where N is the number of people with the virus and t is the number of days passed. Find
   a) the rate of spread of the virus on day 5
   b) the rate of spread of the virus on day 15
   c) the coordinate of the maximum of the curve of the model giving the points to 3 significant figures.
   d) explain the meaning of the maximum point you found in part c)

3) The height h of a seedling in mm is given by
$$h = -0.028n^2 + 1.95n + 6.2$$
where n is the number of days passed.
   a) Find $\frac{dh}{dn}$ and state its physical relevance.
   b) Use this model to predict the day at which the seedling reaches maximum height and state its height.
   c) How effective is the model in the long term?

4) The velocity of an object over t seconds may be given by
$$v = \frac{7}{3}t^3 - 8t^2 + 4t + 10, \quad 0 \le t \le 3$$
where v is velocity in metres per second.
   a) Find the coordinates of the stationary points of the model giving your answer to 3 significant figures.
   b) Explain the meaning of stationary points in the context of the object.

5) The depth of water y cm below a model boat at time t seconds is modelled by
$$y = t^3 - 8t^2 + 15t + 48, \quad 0 \le t \le 5$$
   a) Find the greatest depth of water below the model boat.
   b) Find the minimum value of $\frac{dy}{dt}$, the time at which it occurs and the depth of the water at this time.
   c) State the vertical direction that the boat is moving at the time found in part b)

6) A model for the height h of a roller coaster in feet is
$$h(x) = 2x^3 - 30x^2 + 94x + 150$$
where x is the horizontal distance in hundreds of feet. Find
   a) the gradient of h(x) when $x = 0$
   b) the angle of inclination (in degrees) to the ground of the rollercoaster initially.
   c) the highest point of the roller coaster and where they occur, giving your answer to the nearest feet.

Applications of Integration

7) The height h of a pile of sand above ground level can be modelled by
$$h(x) = -10x^3 + 14x^2 + \frac{3}{2}x$$
where x is the number of metres from one side. The mean height of the pile of sand may be given by
$$A = \frac{1}{2}\int_0^2 h(x)\ dx$$
Find the value of A

8) The value of some shares £V during the first 12 months of the year may be modelled by
$$V(t) = 0.5t^3 + 3.5t^2 - 12t + 50$$
where t is the number of months passed.
   a) Use this model to predict the value of the shares when 12 months have passed.

The mean share of the shares value may be given by
$$A = \frac{1}{12}\int_0^{12} V(t)\ dt$$
   b) Find the value of A

1.39 Trigonometry: Introduction (answers on page 452)

1) Find all the missing angles and sides in the following right-angled triangles, giving your answers to three significant figures.

   a) [Triangle ABC with right angle at B, AB = 19 cm, angle at C = 50°]

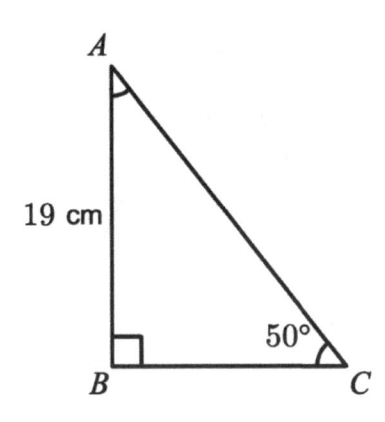

   b) [Triangle with right angle at C, AB = 26 cm, BC = 30 cm]

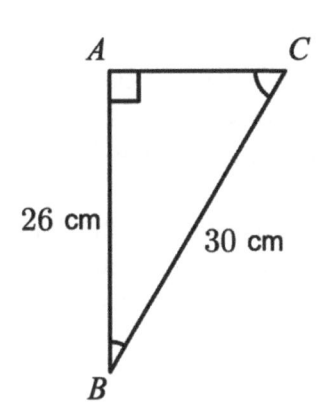

2) The diagram below shows the kite $ABCD$, where angle $DAC = 60°$ and angle $ACD = 30°$. It is also given that the length of AD is 1 m. Find the exact area of the kite.

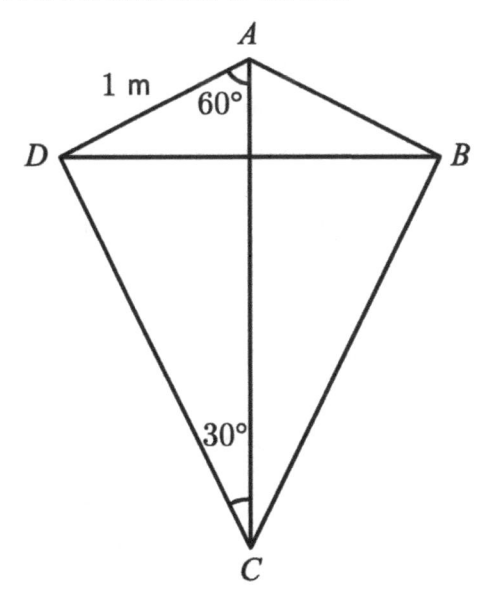

   [Kite ABCD with AD = 1 m, angle DAC = 60°, angle ACD = 30°]

3) In the diagram below, the angles ABC and CDE are both 90°, and the length DE is 1 m. It is given that BCD is a straight line. The area of triangle ABC is one third of the area of triangle CDE. Find the exact length of AB.

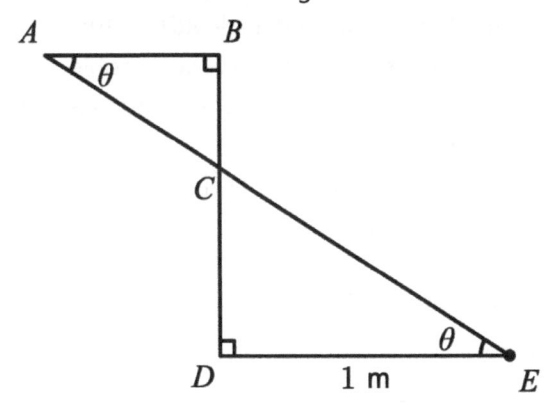

   [Diagram with triangles ABC and CDE, angle θ marked at A and E, DE = 1 m]

4) In the diagram below, the angles OAB, OBC and OCD are all 90°, and the lengths AB, BC and CD are all 1 unit.

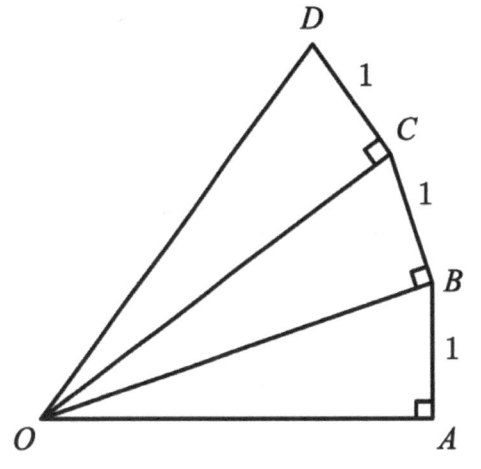

   [Diagram showing point O with successive right triangles, each with outer side length 1]

   Find the length OA given that
   $$\frac{OD}{OA} = \sqrt{2}$$

5) Find the area of the triangle shown in the diagram below, giving your answer to three significant figures.

   [Triangle ABC with right angle at B, angle A = 75°, AB = 23 m, AC = 44 m]

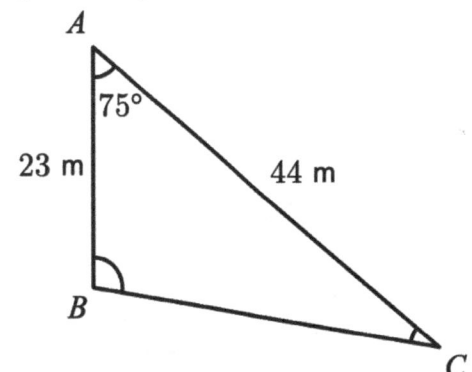

1.39 Trigonometry: Introduction (answers on page 452)

6) It is given that ABC is a triangle where angle ABC is 45°, side $AB = 20$ cm, and side $BC = 15$ cm. Find the length of side AC.

7) It is given that ABC is a triangle where AB is 9 cm, BC is 12 cm, and angle $ACB = 40°$.
   a) Use the sine rule to find two possible values for angle BAC, giving your answers to one decimal place.
   b) Hence draw and label two possible triangles ABC.

8) Find all the missing angles and sides in the following triangles, giving your answers to three significant figures.
   a) 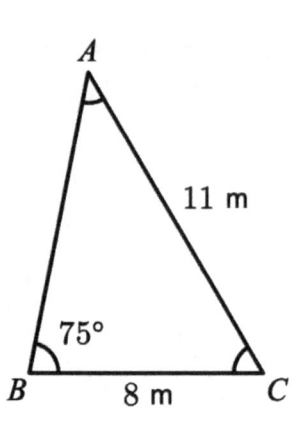
   b) 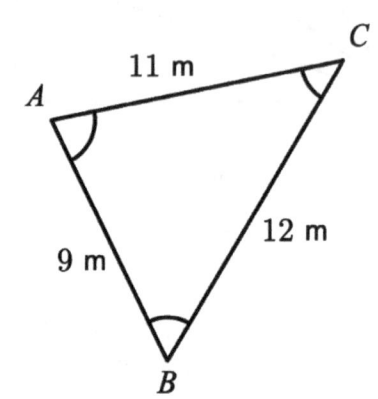
   c)

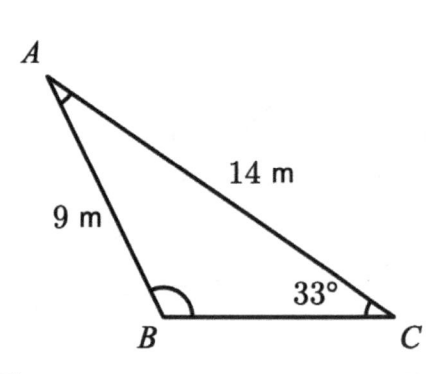

9) Find the area of the triangle shown in the diagram below, giving your answer to three significant figures.

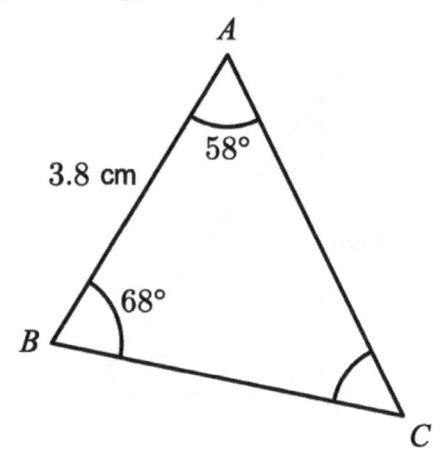

10) It is given that ABC is a triangle where AB is 16 cm, AC is 9 cm and angle BAC is $x°$. The area of the triangle is 25 cm². Find the two possible values of x to one decimal place.

11) The diagram below shows the triangle ABC. It is given that angle ACB is obtuse. Find the length of XC given that BX bisects the triangle.

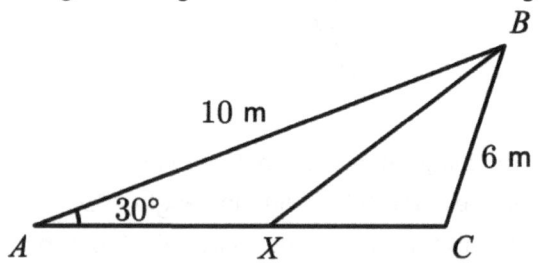

12) It is given that ABC is a triangle where AB is $5m$ cm, AC is $2m^2$ cm and angle BAC is 45°. The area of the triangle is 640 cm². Show that $m = k\sqrt{2}$ where k is an integer to be found.

13) The diagram below shows the triangle ABC.
   a) Show that $11x - 4x^2 = 0$
   b) Find the value of x
   c) Find the largest angle in the triangle.

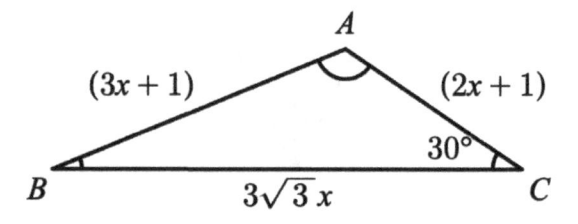

1.39 Trigonometry: Introduction (answers on page 452)

14) The diagram below shows a parallelogram with sides of length 86 cm and 140 cm. It is given that angle $ABC = 120°$. Show that the exact area of the parallelogram is $k\sqrt{3}$ m² where k is a rational number.

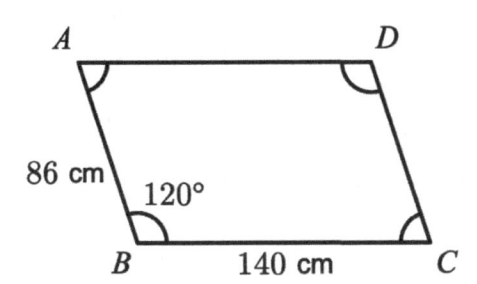

15) It is given that a triangle has sides of length b, $b + 1$ and $b + 2$, where $b > 0$. The largest angle in the triangle is labelled x. Use the cosine rule to show that $\cos x = \dfrac{b-3}{2b}$

16) Ben measures the angle of elevation from where he is standing to the top of a tall building to be $30.5°$. He then walks straight towards the building for 20 metres, stops, and measures the angle of elevation again. This time it is $42.7°$. Determine the height of the building.

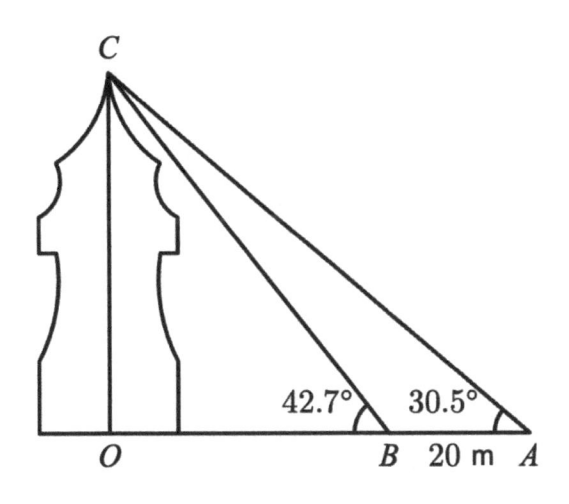

17) Three archaeological digs, A, B and C, are drawn on a map. B is 180 metres due South of A. C is 320 metres from A on a bearing of $075°$. Find
   a) the bearing of B from A
   b) the bearing of A from C
   c) the bearing of B from C

18) The diagram below shows a triangle, ABC, with sides of length p, $p + 5$ and 10. $\angle ABC = 60°$. Find the exact value of p

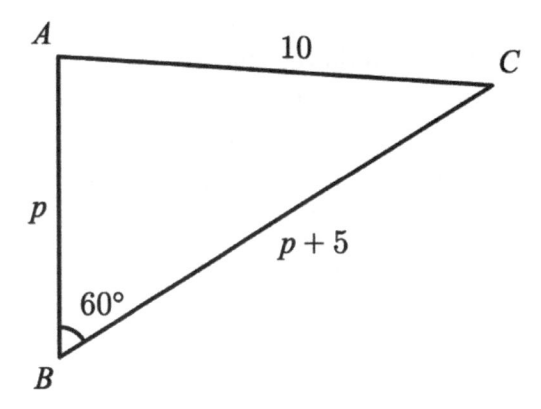

19) The area of a regular pentagon is 60 cm². Calculate the length of one of its sides.

20) The area of a regular octagon is 450 cm². Calculate the length of one of its sides.

21) The diagram below shows a square, $ABCD$, with area 100 units² drawn on the coordinate axes. The square has been divided up into four triangles, XBC, XAB, XCD, and XAD, with areas 10 units², 20 units², 30 units², and 40 units² respectively.

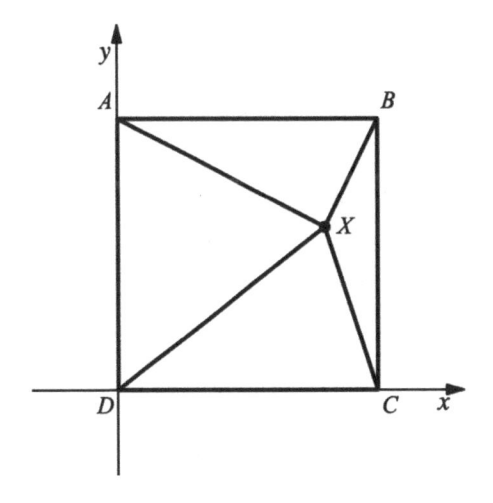

   a) X has exact coordinates (p, q). Find the value of p and the value of q
   b) Y has exact coordinates (r, q), where $0 < r < p$. The triangle XYD has an area that is at least half of the area of triangle XAD. Find the range of possible values of r

1.40 Trigonometry: sin x, cos x, tan x in Degrees (answers on page 452)

1) Sketch the following graphs for $0° \leq x \leq 360°$:
   a) $y = \sin x$
   b) $y = \cos x$
   c) $y = \tan x$

2) Solve the following equations:
   a) $\sin \theta = 0$, $0° \leq \theta < 360°$
   b) $\sin \theta = 1$, $0° \leq \theta < 360°$
   c) $\sin \theta = -1$, $0° \leq \theta < 360°$

3) Solve the following equations:
   a) $\cos \theta = 0$, $0° \leq \theta < 360°$
   b) $\cos \theta = 1$, $0° \leq \theta < 360°$
   c) $\cos \theta = -1$, $0° \leq \theta < 360°$

4) Solve the following equations:
   a) $\tan \theta = 0$, $0° \leq \theta < 360°$
   b) $\tan \theta = 1$, $0° \leq \theta < 360°$
   c) $\tan \theta = -1$, $0° \leq \theta < 360°$

5) Find the values of x for which $y = \sin x$ is an increasing function, for $0° \leq x \leq 360°$

6) Find the values of x for which $y = \cos x$ is a decreasing function, for $0° \leq x \leq 360°$

7) Solve the following equations:
   a) $\sin \theta = \frac{1}{3}$, $0° \leq \theta < 360°$
   b) $\cos \theta = \frac{2}{3}$, $0° \leq \theta < 360°$
   c) $\tan \theta = \frac{4}{3}$, $0° \leq \theta < 360°$

8) Solve the following equations:
   a) $\sin \theta = 0.4$, $-360° \leq \theta < 0°$
   b) $\cos \theta = \frac{3}{7}$, $-360° \leq \theta < 0°$
   c) $\tan \theta = \frac{4}{9}$, $-360° \leq \theta < 0°$

9) Solve the following equations:
   a) $\sin \theta = -0.2$, $0° \leq \theta < 360°$
   b) $\cos \theta = -\frac{4}{7}$, $0° \leq \theta < 360°$
   c) $\tan \theta = -4$, $0° \leq \theta < 360°$

10) Solve the following equations:
    a) $\sin \theta = -0.9$, $-360° \leq \theta < 0°$
    b) $\cos \theta = -\frac{3}{5}$, $-360° \leq \theta < 0°$
    c) $\tan \theta = -\frac{9}{2}$, $-360° \leq \theta < 0°$

11) Solve the following equations:
    a) $\sin \theta = \frac{3}{8}$, $-360° \leq \theta < 360°$
    b) $\cos \theta = 0.7$, $-360° \leq \theta < 360°$
    c) $\tan \theta = 6$, $-360° \leq \theta < 360°$
    d) $\sin \theta = -\frac{6}{7}$, $-360° \leq \theta < 360°$
    e) $\cos \theta = -\frac{2}{9}$, $-360° \leq \theta < 360°$
    f) $\tan \theta = -1.3$, $-360° \leq \theta < 360°$

12) Find the values of x for which $y = \sin x$ is a decreasing function, for $-720° \leq x \leq -360°$

13) Find the values of x for which $y = \cos x$ is an increasing function, for $180° \leq x \leq 540°$

14) Solve the following equations:
    a) $\sin \theta = \frac{4}{7}$, $0° \leq \theta < 720°$
    b) $\cos \theta = -\frac{8}{9}$, $720° \leq \theta < 1080°$
    c) $\tan \theta = -9$, $-720° \leq \theta < -540°$

15) Solve the following equations:
    a) $\frac{1}{\sin \theta} = 18$ for $0° \leq \theta < 180°$
    b) $\frac{5}{\tan \theta} = 13$ for $180° \leq \theta < 540°$
    c) $\frac{0.1}{\cos \theta} = 2$ for $-180° \leq \theta < 180°$

16) A student is solving a problem with a triangle. They correctly use the sine rule to find $\frac{\sin A}{18} = \frac{\sin 20°}{15}$. Find the two possible values of A

17) The graph below shows the curve $y = \sin x$ for $-360° \leq x \leq 360°$, and the line $y = k$ where $0 < k < 1$. The point A has x-coordinate p, where $p > 0$. Write down the x-coordinates of B, C and D in terms of p

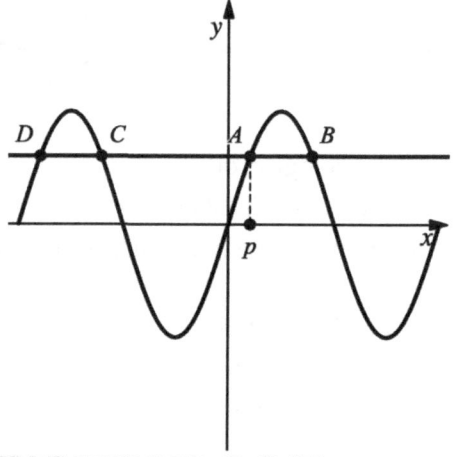

1.40 Trigonometry: $\sin x$, $\cos x$, $\tan x$ in Degrees (answers on page 452)

18) It is given that $\sin 60° = \frac{\sqrt{3}}{2}$

    Explain why $\sin 480° = \frac{\sqrt{3}}{2}$

19) It is given that $\cos 45° = \frac{\sqrt{2}}{2}$

    Explain why $\cos 495° = -\frac{\sqrt{2}}{2}$

20) It is given that $\tan(-210°) = -\frac{\sqrt{3}}{3}$

    Explain why $\tan 390° = \frac{\sqrt{3}}{3}$

21) The graph below shows the curve $y = \sin x$ for $-360° \leq x \leq 360°$, and the line $y = k$ where $0 < k < 1$. The point A has x-coordinate p, where $p > 0$. Write down the x-coordinates of B, C and D in terms of p

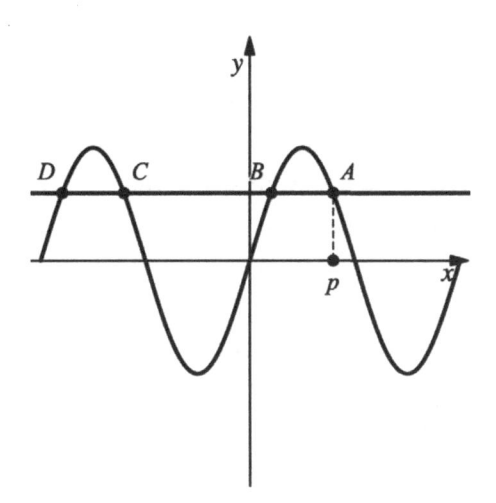

22) The graph below shows the curve $y = \sin x$ for $-360° \leq x \leq 360°$, and the line $y = k$ where $-1 < k < 0$. The point A has x-coordinate $-p$, where $p > 0$. Write down the x-coordinates of B, C and D in terms of p

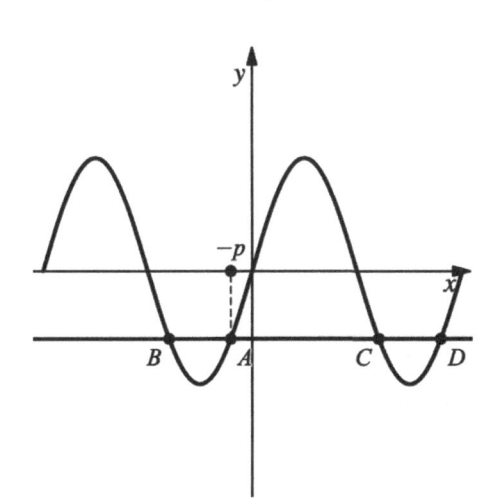

23) Solve the following equations:
    a) $\cos^2 \theta = \frac{1}{16}$ for $0° \leq \theta < 360°$
    b) $3\tan^2 \theta = 0.8$ for $-180° \leq \theta < 180°$
    c) $10\sin^2 \theta = 7$ for $0° \leq \theta < 540°$

24) Solve the following equations:
    a) $\cos^2 \theta = 2\cos \theta$ for $0° \leq \theta < 360°$
    b) $7\tan^2 \theta + 4\tan \theta = 0$ for $0° \leq \theta < 360°$
    c) $2\sin \theta = 5\sin^2 \theta$ for $0° \leq \theta < 360°$

25) The graph below shows the curve $y = \cos x$ for $-360° \leq x \leq 360°$, and the line $y = k$ where $0 < k < 1$. The point A has x-coordinate p, where $p > 0$. Write down the x-coordinates of B, C and D in terms of p

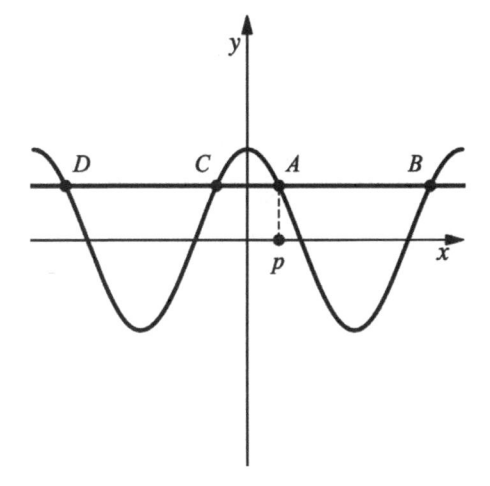

26) The graph below shows the curve $y = \cos x$ for $-360° \leq x \leq 360°$, and the line $y = k$ where $-1 < k < 0$. The point A has x-coordinate p, where $p > 0$. Write down the x-coordinates of B, C and D in terms of p

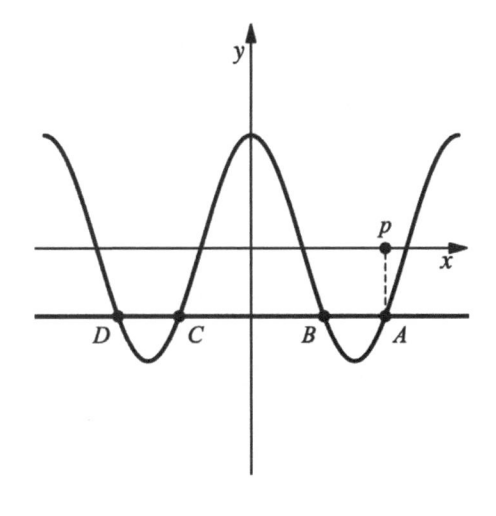

1.40 Trigonometry: sin x, cos x, tan x in Degrees (answers on page 452)

27) The graph below shows the curve $y = \tan x$ for $-360° \leq x \leq 360°$, and the line $y = k$ where $0 < k < 1$. The point A has x-coordinate p, where $p > 0$. Write down the x-coordinates of B, C and D in terms of p

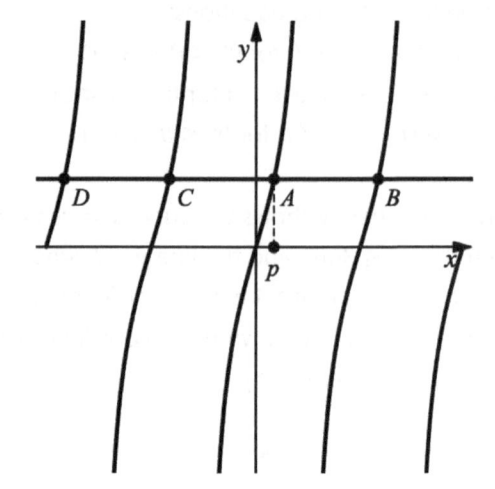

28) Solve the following equations:
   a) $3 \sin \theta \cos \theta = \cos \theta$ for $0° \leq \theta < 360°$
   b) $8 \sin \theta \tan \theta - 9 \sin \theta = 0$ for $0° \leq \theta < 360°$
   c) $6 \cos \theta - 5 \sin \theta \cos \theta = 0$ for $0° \leq \theta < 360°$

29) Solve the following equations:
   a) $2 \sin^2 \theta + \sin \theta - 1 = 0$ for $0° \leq \theta < 360°$
   b) $3 \cos^2 \theta - \cos \theta - 2 = 0$ for $0° \leq \theta < 360°$
   c) $4 \tan^2 \theta + 5 \tan \theta - 6 = 0$ for $0° \leq \theta < 360°$

30) Solve the equations for $0° \leq \theta < 360°$:
   a) $6 \sin \theta = 1 + \frac{1}{\sin \theta}$
   b) $9 \cos \theta = \frac{16}{\cos \theta} - 32$
   c) $30 \tan \theta = 23 + \frac{40}{\tan \theta}$

Prerequisite knowledge for Q31: Quadratics

31) It is given that $f(x) = 3x^2 - 12x + 17$
   a) Write $f(x)$ in the form $a(x + b)^2 + c$, where a, b and c are integers.
   b) Write down the minimum point of the curve with equation $y = f(x)$

   It is given that $g(\theta) = 3 \sin^2 \theta - 12 \sin \theta + 17$
   c) Using your answers to a) and b), find the minimum value of $g(\theta)$ and the smallest positive value of θ, in degrees, for which this minimum occurs.

32) Write down the period of the following functions in degrees:
   a) $y = \sin x$
   b) $y = \cos x$
   c) $y = \tan x$
   d) $y = \sin 2x$
   e) $y = \cos 3x$
   f) $y = \tan 4x$
   g) $y = \sin \frac{1}{3}x$
   h) $y = \cos \frac{2}{3}x$
   i) $y = \tan \frac{4}{5}x$

33) State the number of solutions to the following equations. You are not required to solve them.
   a) $\sin 2x = 0.3$ for $0° \leq x < 360°$
   b) $\sin 2x = 0.3$ for $0° \leq x < 180°$
   c) $\sin 2x = 0.3$ for $0° \leq x < 720°$

34) State the number of solutions to the following equations. You are not required to solve them.
   a) $\cos 2x = 0.8$ for $0° \leq x < 180°$
   b) $\cos 3x = 0.8$ for $0° \leq x < 360°$
   c) $\cos 4x = 0.8$ for $0° \leq x < 720°$

35) State the number of solutions to the following equations. You are not required to solve them.
   a) $\tan 2x = 0.8$ for $0° \leq x < 1080°$
   b) $\tan 3x = 0.8$ for $0° \leq x < 720°$
   c) $\tan 4x = 0.8$ for $-270° \leq x < 360°$

36) Write down the period of the following functions in degrees:
   a) $y = \sin(x - 15°)$
   b) $y = \cos(x + 80°)$
   c) $y = \tan(x - 300°)$
   d) $y = \sin(60° - x)$
   e) $y = \cos(45° - x)$
   f) $y = \tan(5° - x)$
   g) $y = \sin(2x + 50°)$
   h) $y = \cos(4x - 90°)$
   i) $y = \tan(9x + 35°)$
   j) $y = \sin\left(100° - \frac{1}{5}x\right)$
   k) $y = \cos\left(\frac{315° - x}{8}\right)$
   l) $y = \tan\left(\frac{28° - 5x}{9}\right)$

1.40 Trigonometry: $\sin x$, $\cos x$, $\tan x$ in Degrees (answers on page 452)

37) Solve the following equations:
    a) $\sin(\theta + 10°) = 0.3$ for $0° \leq \theta < 360°$
    b) $\sin(\theta - 50°) = \frac{2}{11}$ for $0° \leq \theta < 360°$
    c) $\sin(\theta + 205°) = -0.4$ for $0° \leq \theta < 360°$

38) Solve the following equations:
    a) $\cos(\theta - 60°) = \frac{4}{11}$ for $0° \leq \theta < 360°$
    b) $\cos(\theta + 25°) = 0.85$ for $0° \leq \theta < 360°$
    c) $\cos(\theta - 195°) = -\frac{7}{12}$ for $0° \leq \theta < 360°$

39) Solve the following equations:
    a) $\tan(\theta - 45°) = 2.5$ for $0° \leq \theta < 360°$
    b) $\tan(\theta + 80°) = \frac{5}{6}$ for $0° \leq \theta < 360°$
    c) $\tan(\theta - 215°) = -4.2$ for $0° \leq \theta < 360°$

40) Find the values of x for which $y = \sin(x - 30°)$ is an increasing function, for $0° \leq x \leq 360°$

41) Solve the following equations:
    a) $\sin 2\theta = \frac{\sqrt{2}}{2}$ for $0° \leq \theta < 360°$
    b) $\cos 2\theta = \frac{\sqrt{2}}{4}$ for $0° \leq \theta < 360°$
    c) $\tan 2\theta = \frac{\sqrt{3}}{3}$ for $0° \leq \theta < 360°$

42) Solve the following equations:
    a) $\sin 3\theta = 0.8$ for $0° \leq \theta < 360°$
    b) $\cos 3\theta = -0.3$ for $0° \leq \theta < 360°$
    c) $\tan 3\theta = -3$ for $0° \leq \theta < 360°$

43) Solve the following equations:
    a) $\sin \frac{1}{2}\theta = 0.2$ for $0° \leq \theta < 720°$
    b) $\cos \frac{1}{4}\theta = -\frac{4}{5}$ for $0° \leq \theta < 3600°$
    c) $\tan \frac{2}{3}\theta = -5$ for $0° \leq \theta < 270°$

44) Solve the following equations:
    a) $\sin(2\theta - 25°) = \frac{1}{2}$ for $0° \leq \theta < 360°$
    b) $\cos(3\theta + 50°) = \frac{1}{2}$ for $0° \leq \theta < 360°$
    c) $\tan(2\theta + 65°) = 1$ for $0° \leq \theta < 360°$

45) Solve the following equations:
    a) $\sin(4\theta + 161°) = \frac{5}{6}$ for $0° \leq \theta < 180°$
    b) $\cos(6\theta - 94°) = 0.03$ for $-90° \leq \theta < 90°$
    c) $\tan(5\theta - 213°) = 8$ for $100° \leq \theta < 200°$

46) Find the values of x for which $y = \cos 2x$ is a decreasing function, for $0° \leq x \leq 180°$

47) Find the values of x for which $y = \sin 3x$ is an increasing function, for $0° \leq x \leq 360°$

48) Solve the following equations:
    a) $\sin(84° - 3\theta) = \frac{1}{2}$ for $0° \leq \theta < 360°$
    b) $\cos(115° - 2\theta) = \frac{1}{2}$ for $0° \leq \theta < 360°$
    c) $\tan(5° - 4\theta) = 1$ for $0° \leq \theta < 360°$

49) Solve the following equations:
    a) $\sin(28° - 5\theta) = 0.1$ for $0° \leq \theta < 144°$
    b) $\cos(77° - 3\theta) = 0.99$ for $0° \leq \theta < 300°$
    c) $\tan(174° - 8\theta) = -0.56$ for $0° \leq \theta < 55°$

50) Solve the following equations:
    a) $\sin^2 2\theta - \frac{1}{4} = 0$ for $0° \leq \theta < 360°$
    b) $\cos^2 3\theta - \frac{1}{16} = 0$ for $0° \leq \theta < 180°$
    c) $\tan^2 5\theta - 121 = 0$ for $-40° \leq \theta < 40°$

51) The graph below shows the curve $y = \sin 2x$ for $-360° \leq x \leq 360°$, and the line $y = k$ where $0 < k < 1$. The point A has x-coordinate p. Write down the x-coordinates of B, C and D in terms of p

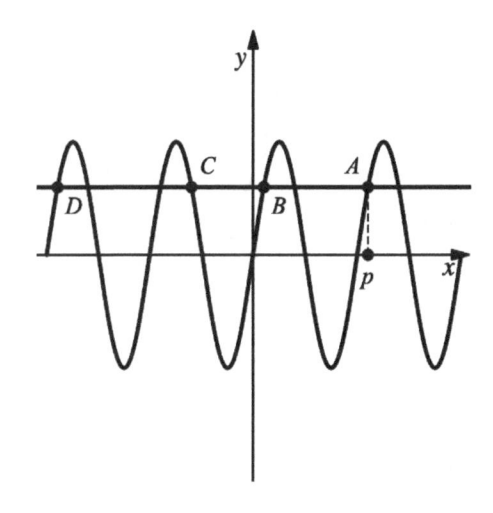

52) Solve the following equations:
    a) $\sin(\theta - 40°) = \sin 80°$ for $0° \leq \theta < 360°$
    b) $\sin(\theta + 35°) = \sin 280°$ for $0° \leq \theta < 360°$
    c) $\cos(\theta + 15°) = \cos 50°$ for $0° \leq \theta < 360°$
    d) $\tan(\theta - 70°) = \tan 20°$ for $0° \leq \theta < 360°$

53) Solve the following equations:
    a) $\sin(35° - \theta) = \sin 75°$ for $0° \leq \theta < 360°$
    b) $\cos(2\theta + 80°) = \cos 55°$ for $0° \leq \theta < 360°$
    c) $\tan(20° - 3\theta) = \tan 140°$ for $0° \leq \theta < 360°$

Prerequisite knowledge for Q54: Polynomials

54) It is given that $f(x) = 45x^3 - 63x^2 - 2x + 8$
    a) Show that $(5x - 2)$ is a factor of $f(x)$
    b) Hence solve the equation
       $45\cos^3 \theta - 63\cos^2 \theta - 2\cos \theta + 8 = 0$
       for $0° \leq \theta \leq 360°$

1.41 Graph Sketching: $\sin x$, $\cos x$, $\tan x$ in Degrees (answers on page 454)

1) Find the equation of each of the following graphs in the form
$$y = a \sin x$$
   a) 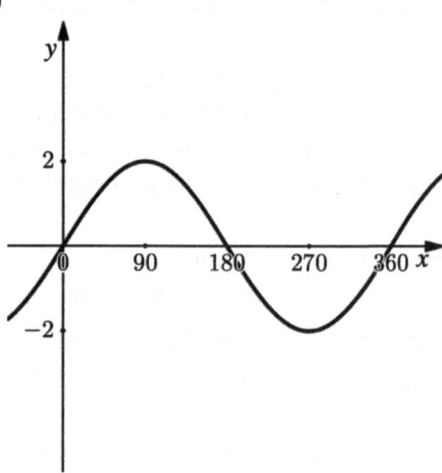

   b)

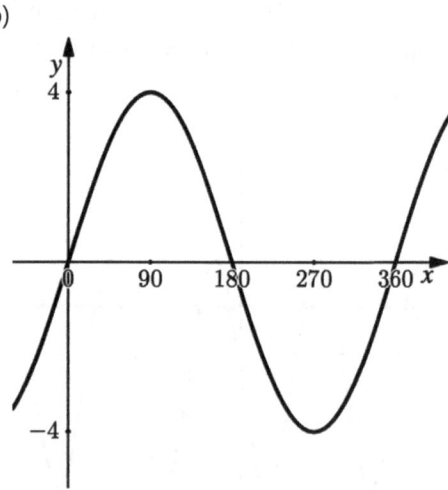

2) Find the equation of each of the following graphs in the form
$$y = a \cos x$$
   a) 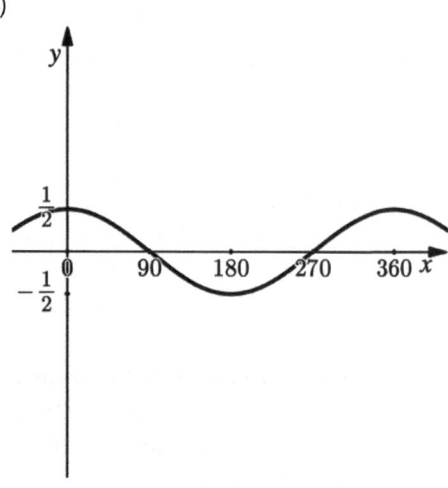

   b)

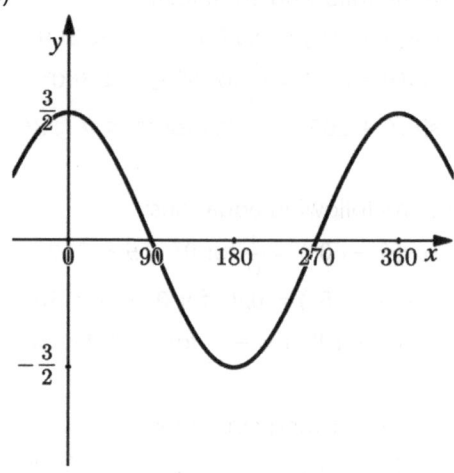

3) Sketch the following graphs for $0° \leq x \leq 360°$:
   a) $y = \frac{6}{5} \sin x$
   b) $y = \frac{2}{3} \cos x$

4) Find the equation of each of the following graphs in the form
$$y = \sin ax$$
   a) 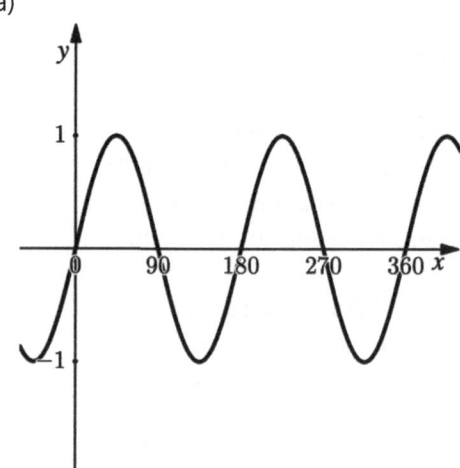

   b)

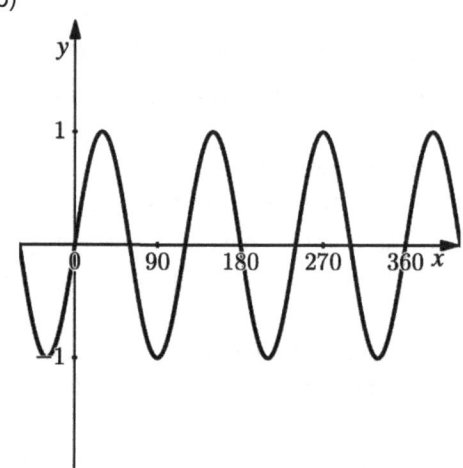

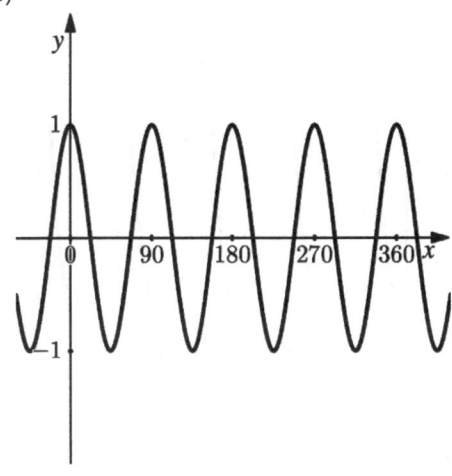

1.41 Graph Sketching: $\sin x$, $\cos x$, $\tan x$ in Degrees (answers on page 454)

c)

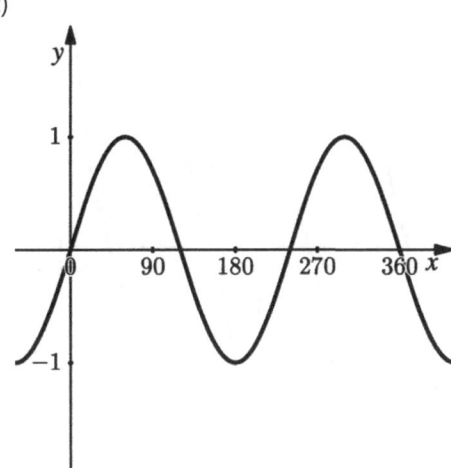

b)

d)

c)

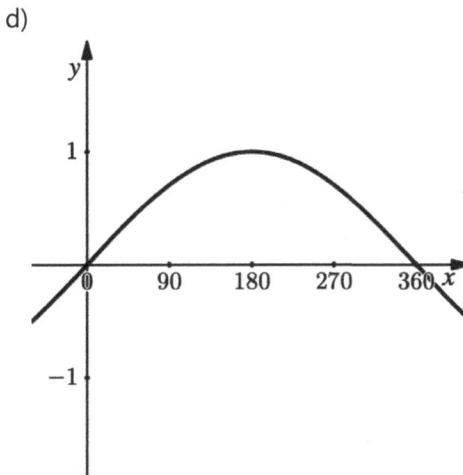

5) Sketch the graph of $y = -2\sin x$ for $0° \leq x \leq 360°$

6) Find the equation of each of the following graphs in the form
$$y = \cos ax$$

a)

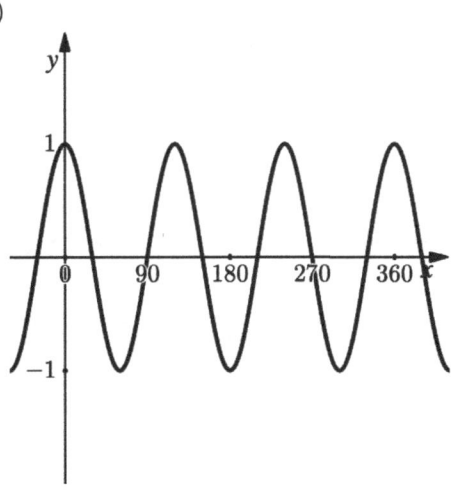

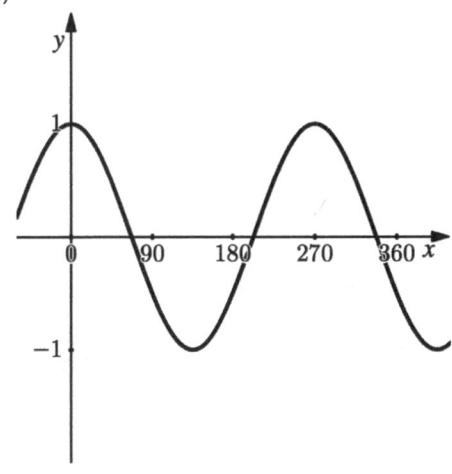

d)

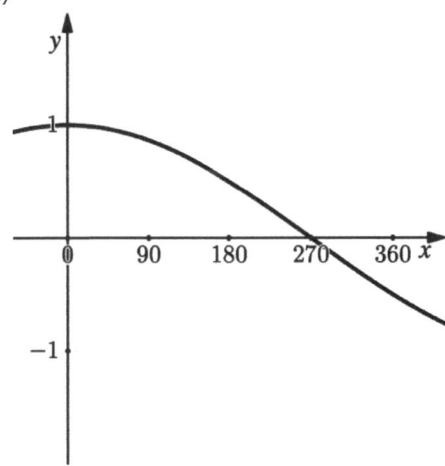

7) Sketch the graph of $y = -3\cos x$ for $0° \leq x \leq 360°$

8) Sketch the graph of $y = \sin(-2x)$ for $0° \leq x \leq 360°$

1.41 Graph Sketching: $\sin x$, $\cos x$, $\tan x$ in Degrees (answers on page 454)

9) Find the equation of each of the following graphs in the form
$$y = \tan ax$$

a)

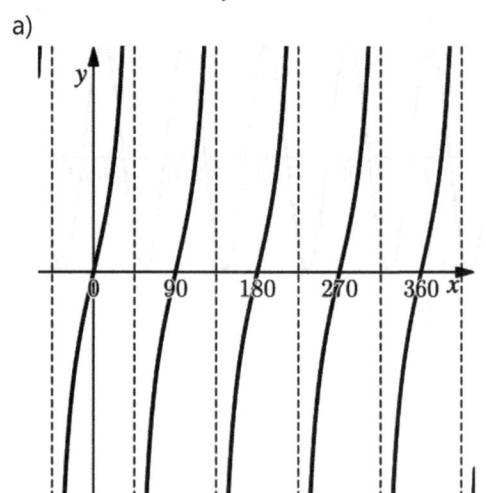

b)

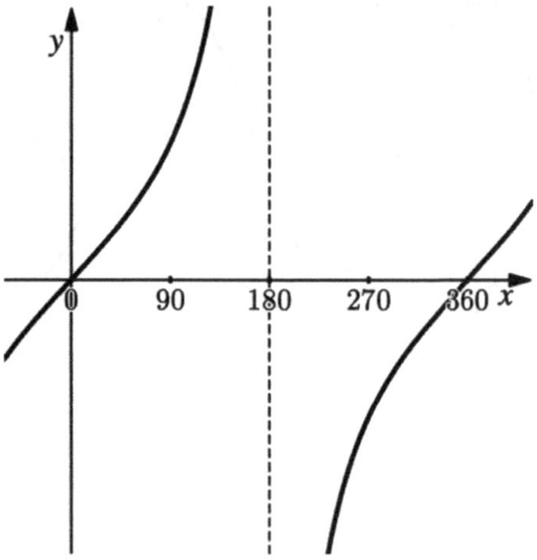

c)

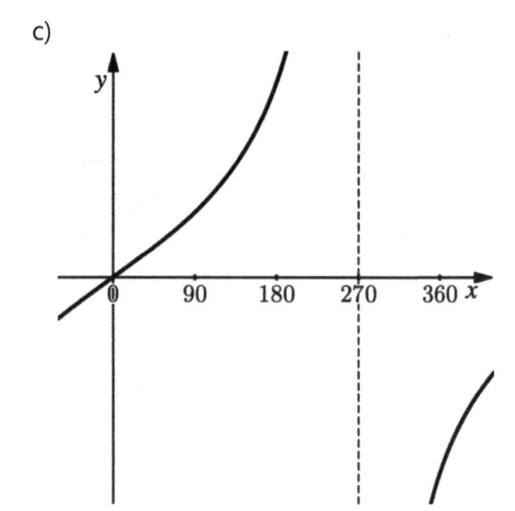

d)

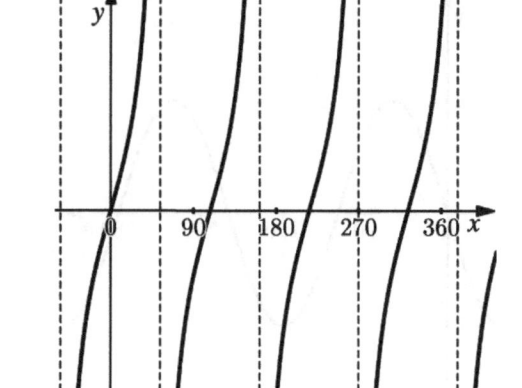

10) Sketch the following graphs for $0° \leq x \leq 360°$:
   a) $y = \tan \frac{3}{2} x$
   b) $y = \tan \frac{2}{3} x$

11) Find the equation of each of the following graphs in either the form $y = a + \sin x$ or the form $y = a + \cos x$

a)

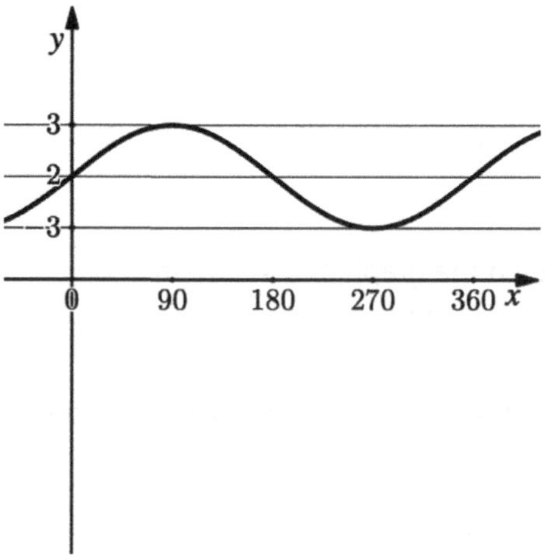

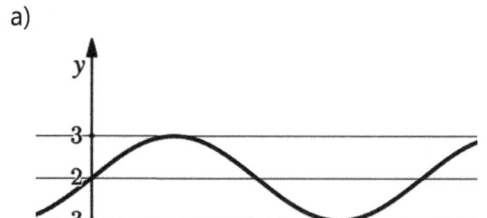

b)

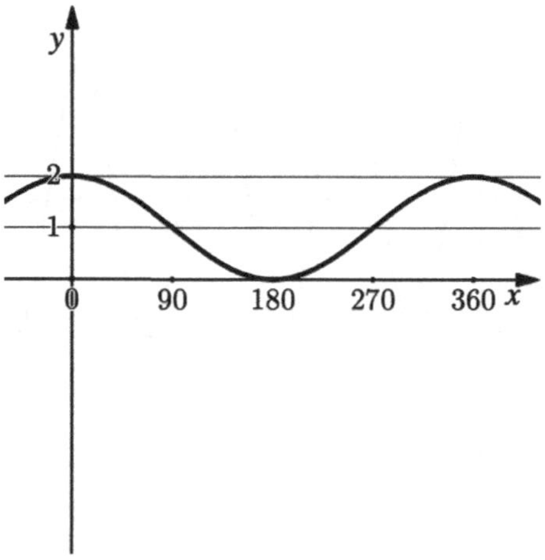

1.41 Graph Sketching: $\sin x$, $\cos x$, $\tan x$ in Degrees (answers on page 454)

c)

b)

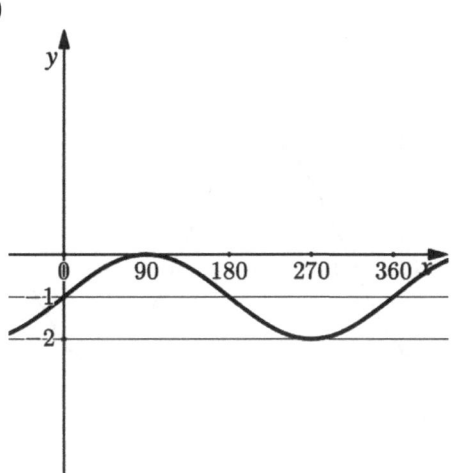

d)

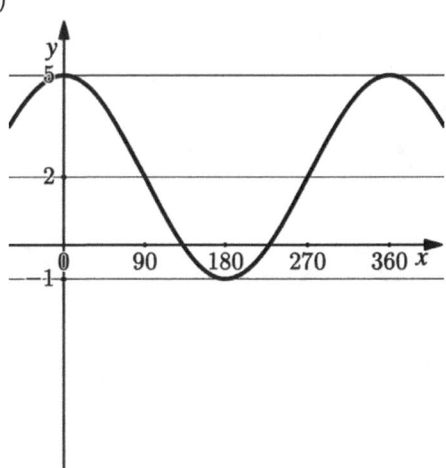

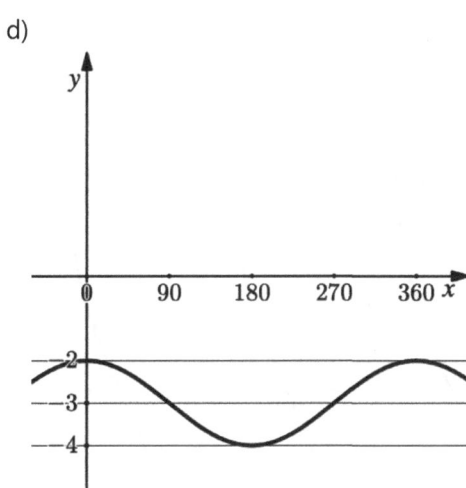

c)

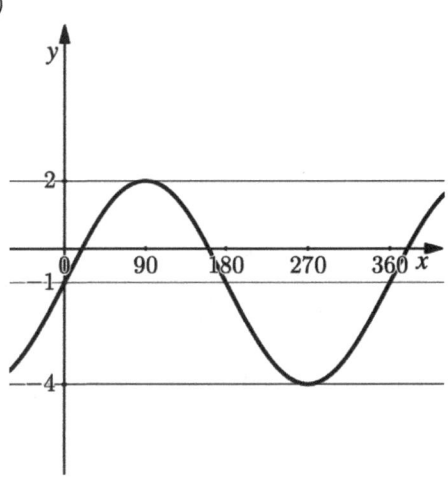

Prerequisite knowledge for Q12-Q13:
Combinations of Graph Transformations

12) Find the equation of each of the following graphs in the form $y = a + b \sin x$ or the form $y = a + b \cos x$

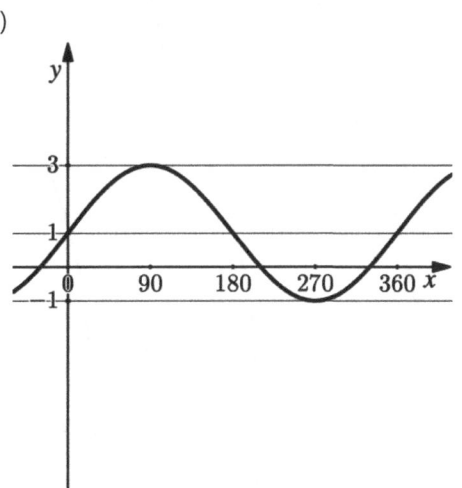

a)

d)

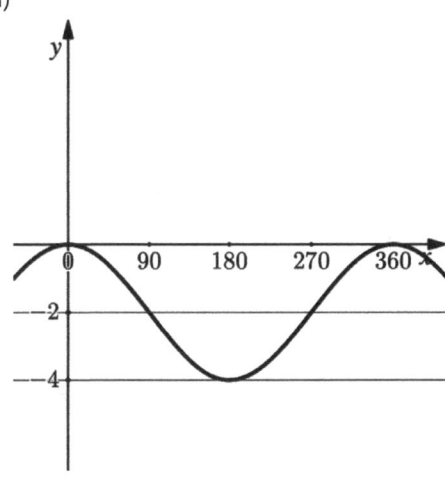

13) Sketch the following graphs for $0° \leq x \leq 360°$:
   a) $y = 1 + 3\sin x$
   b) $y = -1 + 4\cos x$
   c) $y = 2 - 2\sin x$
   d) $y = -3 - 2\cos x$

1.41 Graph Sketching: $\sin x$, $\cos x$, $\tan x$ in Degrees (answers on page 454)

14) Find the equation of each of the following graphs in the form
$$y = \sin(x + a)$$

a)

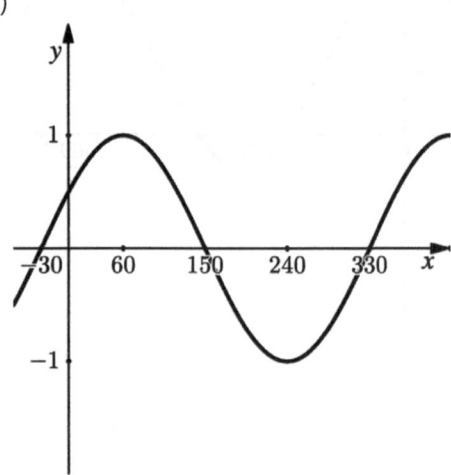

b)

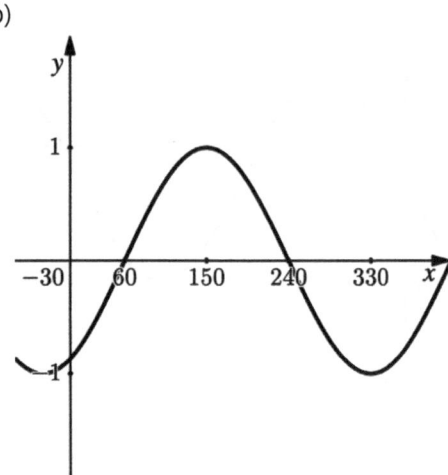

15) Find the equation of each of the following graphs in the form
$$y = \cos(x + a)$$

a)

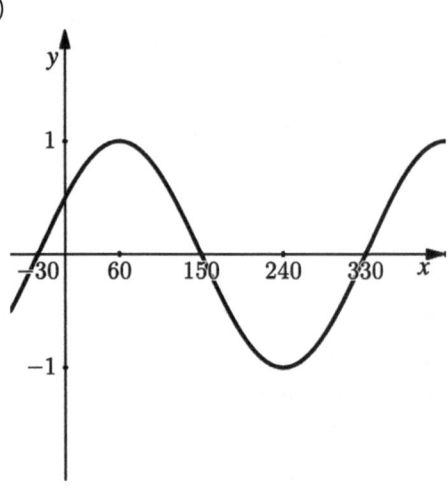

b)

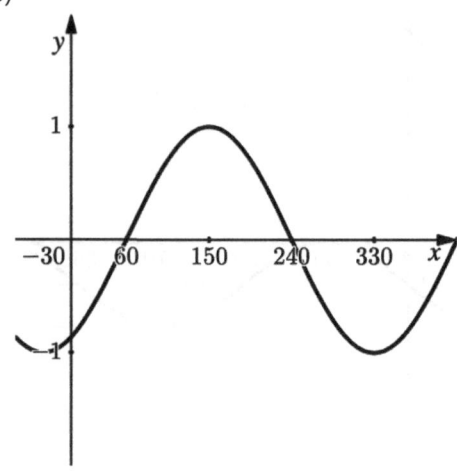

16) Find the equation of each of the following graphs in the form
$$y = \tan(x + a)$$

a)

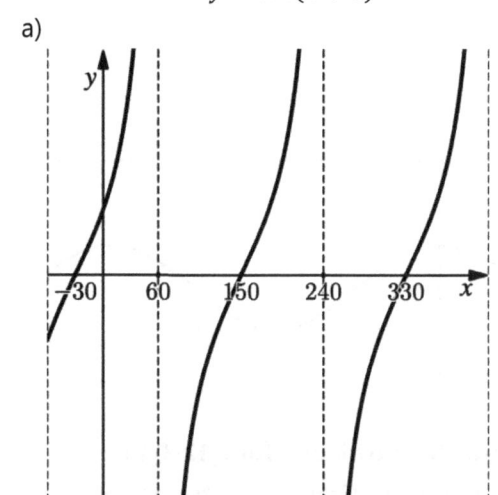

b)

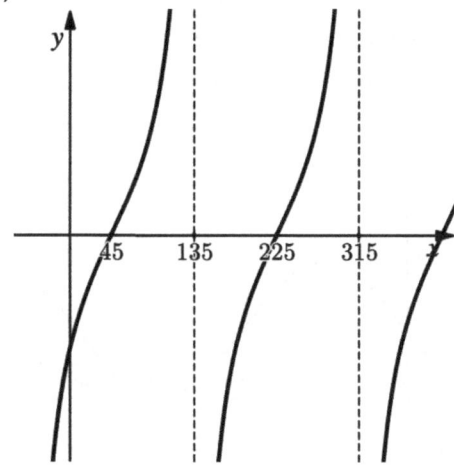

17) It is given that k is a constant where $k > 1$. Sketch the following graphs for $0° \leq x \leq 360°$:

a) $y = k + \sin x$
b) $y = -k + \cos x$

1.41 Graph Sketching: $\sin x$, $\cos x$, $\tan x$ in Degrees (answers on page 454)

18) Sketch the following graphs for $0° \leq x \leq 360°$:
    a) $y = \sin(x - 90°)$
    b) $y = \cos(x + 60°)$
    c) $y = \tan(x + 90°)$
    d) $y = \sin(x + 120°)$
    e) $y = \cos(x - 40°)$
    f) $y = \tan(x - 50°)$

19) It is given that k is a constant where $0° < k < 90°$. Sketch the following graphs for $0° \leq x \leq 360°$:
    a) $y = \sin(x + k)$
    b) $y = \cos(x - k)$

Prerequisite knowledge for Q20-Q25: Combinations of Graph Transformations

20) Find the equation of each of the following graphs in the form
$$y = \sin(ax + b)$$
    a) 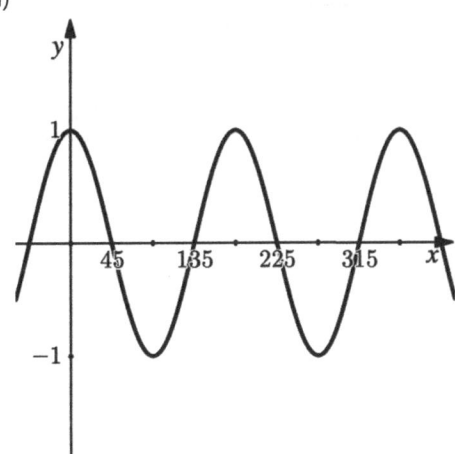
    b)

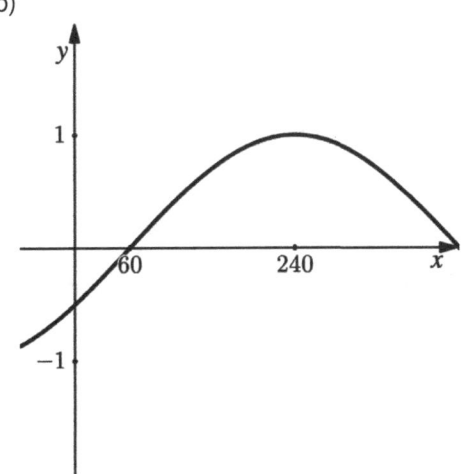

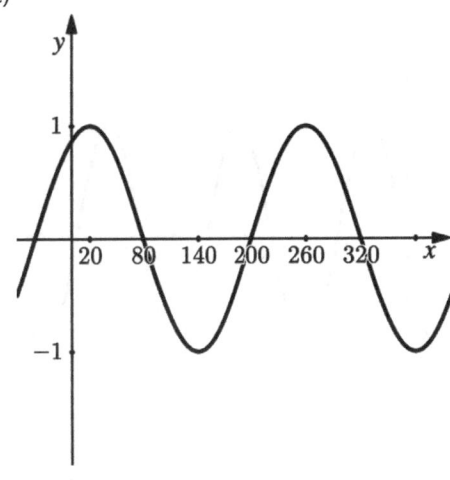

21) Find the equation of each of the following graphs in the form
$$y = \cos(ax + b)$$
    a) 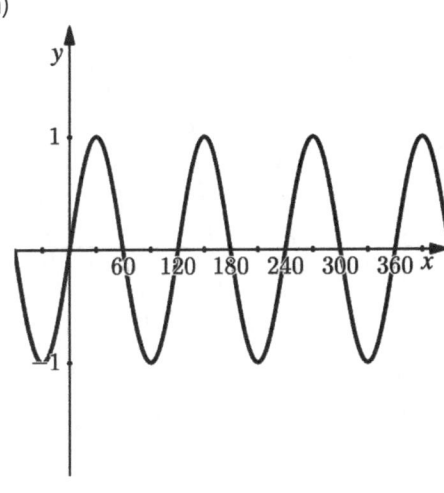
    b)
    c)

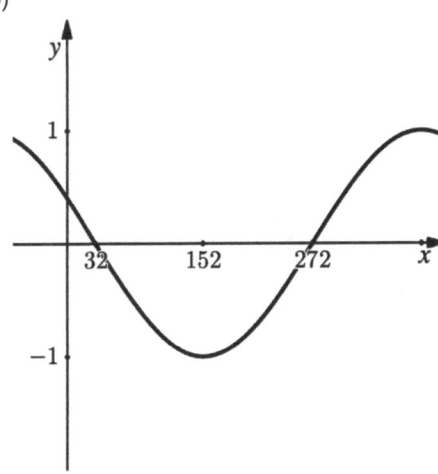

1.41 Graph Sketching: $\sin x$, $\cos x$, $\tan x$ in Degrees (answers on page 454)

c)

c)

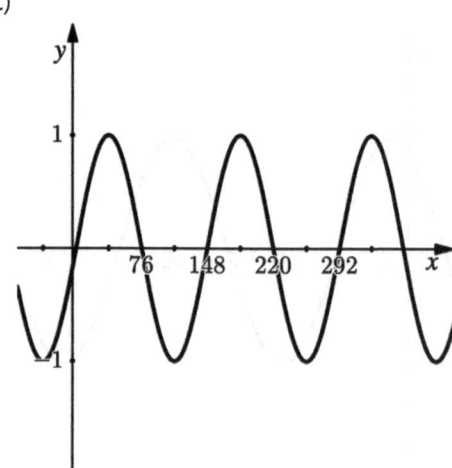

22) Find the equation of each of the following graphs in the form
$$y = \tan(ax + b)$$

a)

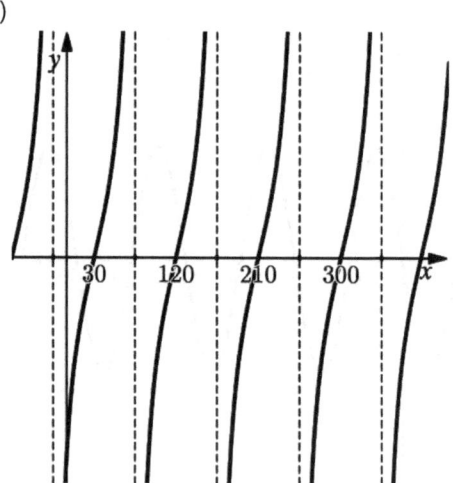

b)

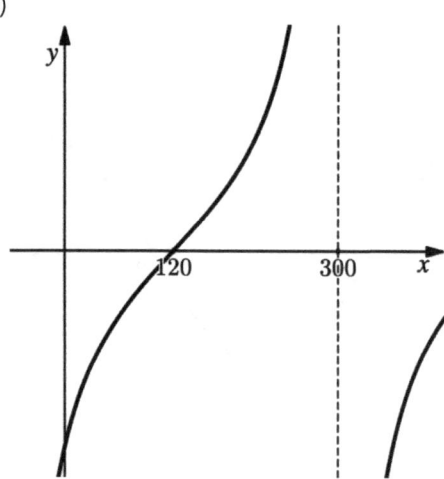

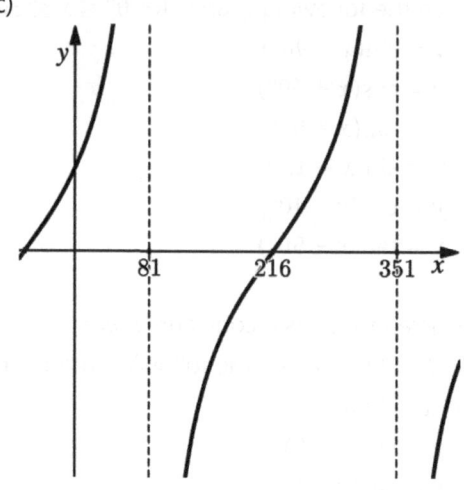

23) Find the equation of each of the following graphs in the form
$$y = a + b\sin(cx + d)$$
where $-180° < d < 180°$

a)

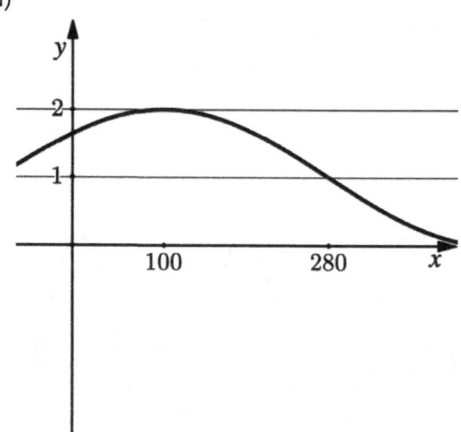

b)

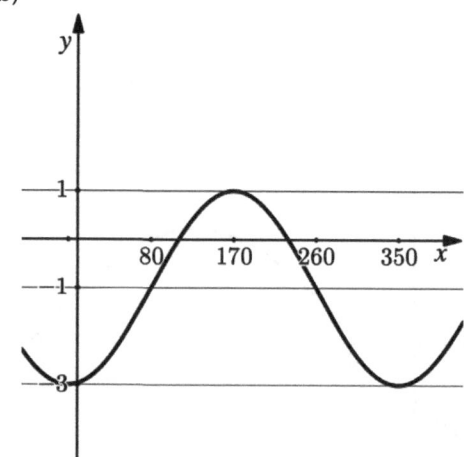

1.41 Graph Sketching: $\sin x$, $\cos x$, $\tan x$ in Degrees (answers on page 454)

c)

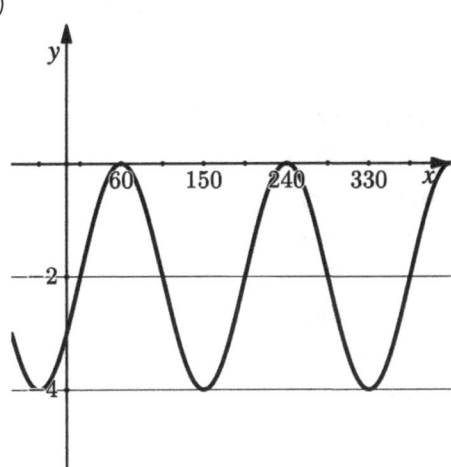

b)

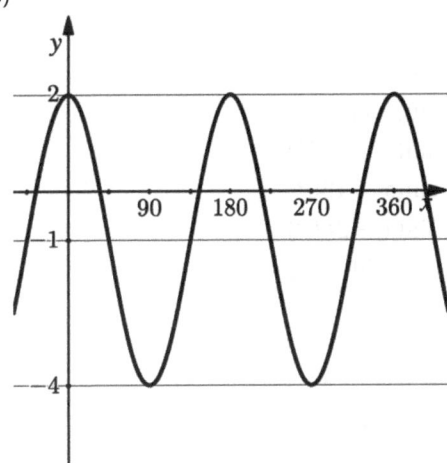

d)

c)

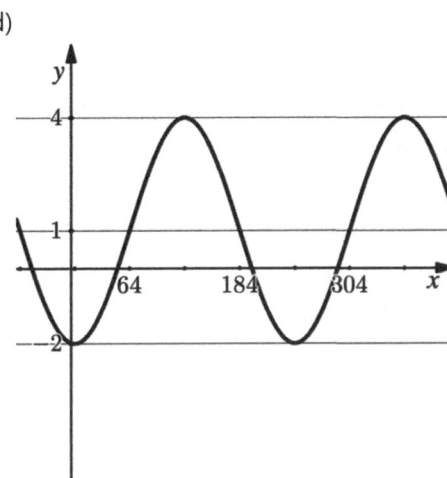

24) Find the equation of each of the following graphs in the form
$$y = a + b\cos(cx + d)$$
where $-180° < d < 180°$

a)

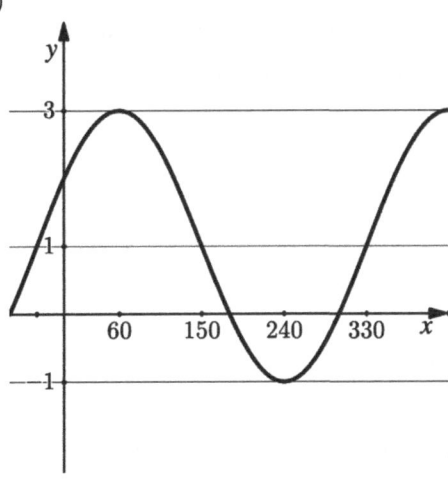

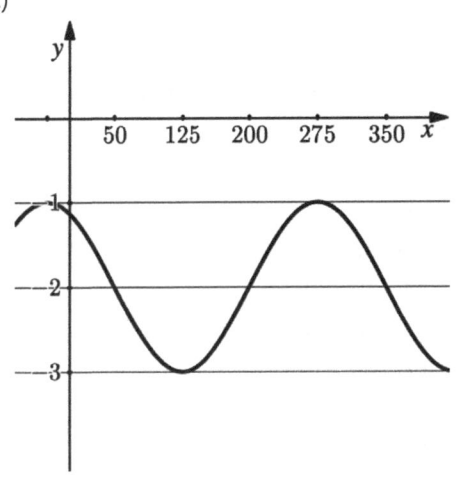

d)

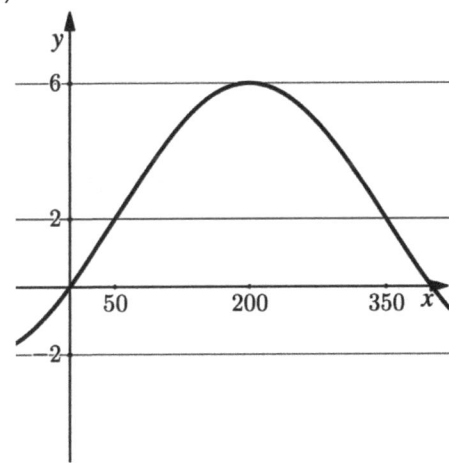

25) It is given that a and b are positive constants. Sketch the following graphs for $0° \leq x < 360°$:
a) $y = a + b\sin x$ where $a > b$
b) $y = a + b\sin x$ where $a = b$
c) $y = a + b\sin x$ where $a < b$

1.42 Trigonometry: Trigonometric Identities in Degrees (answers on page 456)

1) Write each of the following expressions in terms of $\sin x$:
   a) $\sin(x + 180°)$
   b) $\sin(-x)$
   c) $\sin(x + 360°)$
   d) $\sin(x - 180°)$
   e) $\sin(x - 360°)$
   f) $\sin(180° - x)$
   g) $\sin(360° - x)$
   h) $\sin(540° - x)$
   i) $\sin(x - 540°)$
   j) $\sin(x + 540°)$
   k) $\sin(x + 720°)$
   l) $\sin(900° - x)$
   m) $\sin(3600° - x)$

2) Write each of the following expressions in terms of $\cos x$:
   a) $\cos(x - 180°)$
   b) $\cos(-x)$
   c) $\cos(180° - x)$
   d) $\cos(x + 180°)$
   e) $\cos(x + 360°)$
   f) $\cos(360° - x)$
   g) $\cos(x - 360°)$
   h) $\cos(540° - x)$
   i) $\cos(x + 540°)$
   j) $\cos(x - 540°)$
   k) $\cos(x - 720°)$
   l) $\cos(900° - x)$
   m) $\cos(3600° - x)$

3) Write each of the following expressions in terms of $\tan x$:
   a) $\tan(x + 180°)$
   b) $\tan(-x)$
   c) $\tan(x + 360°)$
   d) $\tan(x - 180°)$
   e) $\tan(360° - x)$
   f) $\tan(180° - x)$
   g) $\tan(x - 360°)$
   h) $\tan(540° - x)$
   i) $\tan(x - 540°)$
   j) $\tan(x + 540°)$
   k) $\tan(x + 720°)$
   l) $\tan(900° - x)$
   m) $\tan(3600° - x)$

4) Write each of the following expressions in terms of $\cos x$
   a) $\sin(x + 90°)$
   b) $\sin(x - 90°)$
   c) $\sin(x + 270°)$
   d) $\sin(90° - x)$
   e) $\sin(270° - x)$
   f) $\sin(x + 450°)$
   g) $\sin(x - 450°)$
   h) $\sin(450° - x)$
   i) $\sin(x - 630°)$
   j) $\sin(630° - x)$
   k) $\sin(x + 810°)$
   l) $\sin(990° - x)$
   m) $\sin(9090° - x)$

5) Write each of the following expressions in terms of $\sin x$
   a) $\cos(x + 90°)$
   b) $\cos(x - 90°)$
   c) $\cos(x + 270°)$
   d) $\cos(90° - x)$
   e) $\cos(270° - x)$
   f) $\cos(x + 450°)$
   g) $\cos(x - 450°)$
   h) $\cos(450° - x)$
   i) $\cos(x - 630°)$
   j) $\cos(630° - x)$
   k) $\cos(x + 810°)$
   l) $\cos(990° - x)$
   m) $\cos(9450° - x)$

6) It is given that $\sin x = \frac{1}{3}$ and x is acute.
   Find the exact values of:
   a) $\cos x$
   b) $\tan x$

7) It is given that $\cos x = \frac{2}{7}$ and x is acute.
   Find the exact values of:
   a) $\sin x$
   b) $\tan x$

8) It is given that $\tan x = \frac{3}{10}$ and x is acute.
   Find the exact values of:
   a) $\sin x$
   b) $\cos x$

1.42 Trigonometry: Trigonometric Identities in Degrees (answers on page 456)

9) It is given that $\sin x = \frac{2}{9}$ and x is obtuse.
   Find the exact values of:
   a) $\cos x$
   b) $\tan x$

10) It is given that $\cos x = -\frac{3}{7}$ and x is obtuse.
    Find the exact values of:
    a) $\sin x$
    b) $\tan x$

11) It is given that $\tan x = -\frac{11}{20}$ and x is obtuse.
    Find the exact values of:
    a) $\sin x$
    b) $\cos x$

12) It is given that $\sin x = -\frac{5}{6}$ and $180° < x < 270°$.
    Find the exact values of:
    a) $\cos x$
    b) $\tan x$

13) It is given that $\cos x = -\frac{1}{3}$ and $180° < x < 270°$. Find the exact values of:
    a) $\sin x$
    b) $\tan x$

14) It is given that $\tan x = \frac{7}{2}$ and $180° < x < 270°$.
    Find the exact values of:
    a) $\sin x$
    b) $\cos x$

15) It is given that $\sin x = -\frac{1}{10}$ and $270° < x < 360°$. Find the exact values of:
    a) $\cos x$
    b) $\tan x$

16) It is given that $\cos x = \frac{14}{15}$ and $270° < x < 360°$.
    Find the exact values of:
    a) $\sin x$
    b) $\tan x$

17) It is given that $\tan x = -\frac{20}{3}$ and $270° < x < 360°$. Find the exact values of:
    a) $\sin x$
    b) $\cos x$

18) It is given that
    $$\cos 75° = \frac{\sqrt{2-\sqrt{3}}}{2}$$
    a) Find the value of $\sin 75°$
    b) Using your answer to a), show that $\tan 75° = \sqrt{7 + 4\sqrt{3}}$

19) Simplify the following trigonometric expressions as far as possible, writing your answers in terms of $\sin x$ and $\cos x$ only:
    a) $\cos x \tan x$
    b) $\sin x \tan x$
    c) $\frac{\tan x}{\sin x}$
    d) $\frac{\tan x}{\cos x}$

20) Simplify the following trigonometric expressions as far as possible, writing your answers in terms of $\sin x$ and $\cos x$ only:
    a) $\sin x \cos x \tan x$
    b) $\sin x \cos^2 x \tan^2 x$
    c) $\frac{\sin x}{\cos x \tan^2 x}$
    d) $\frac{\tan^2 x}{\sin x \cos x}$

21) Simplify the following trigonometric expressions:
    a) $1 - \cos^2 x$
    b) $\sqrt{1 - \sin^2 x}$
    c) $9 \cos^2 x - 9$
    d) $8 \cos^2 x + 8 \sin^2 x$

22) Simplify the following trigonometric expressions:
    a)
    $$\frac{1 - \sin^2 x}{\cos^2 x}$$
    b)
    $$\frac{1 - \cos^2 x}{\sin^2 x}$$
    c)
    $$\frac{4 - 4\cos^2 x}{5 - 5\sin^2 x}$$
    d)
    $$\frac{\tan x}{6 - \frac{6}{\cos^2 x}}$$

1.42 Trigonometry: Trigonometric Identities in Degrees (answers on page 456)

23) Solve the following equations:
   a) $\sin\theta = \cos\theta$, $0° \leq \theta < 360°$
   b) $2\sin\theta = 3\cos\theta$, $0° \leq \theta < 360°$
   c) $8\sin\theta + 5\cos\theta = 0$, $0° \leq \theta < 360°$
   d) $7\cos\theta - 2\sin\theta = 0$, $0° \leq \theta < 360°$

24)
   a) Sketch the graphs of $y = \cos x$ and $y = 2\sin x$ on the same axes between $0°$ and $360°$.
   b) Determine the coordinates of where the two graphs intersect.

25) Solve the following equations:
   a) $6\sin 3\theta = 5\cos 3\theta$, $0° \leq \theta < 240°$
   b) $7\sin\frac{1}{2}\theta + 9\cos\frac{1}{2}\theta = 0$, $0° \leq \theta < 720°$
   c) $11\cos\frac{1}{3}\theta - 25\sin\frac{1}{3}\theta = 0$, $0° \leq \theta < 2160°$

26) Solve the following equations:
   a) $4\tan\theta + \cos\theta = 0$, $0° \leq \theta < 360°$
   b) $\sin\theta\tan\theta + 6 = 0$, $0° \leq \theta < 360°$
   c) $8\tan 2\theta + \cos 2\theta = 0$, $0° \leq \theta < 360°$

27) Solve the following equations:
   a) $2\cos^2\theta = 3\sin\theta$, $0° \leq \theta < 180°$
   b) $4\sin^2\theta = 15\cos\theta$, $0° \leq \theta < 540°$
   c) $7\cos^2\theta = 48\sin\theta$, $0° \leq \theta < 360°$
   d) $10\sin^2\theta = 99\cos\theta$, $-180° \leq \theta < 360°$

28) Write the expression $2\cos^4 x + \cos^2 x$ in terms of $\sin x$ only.

29) Solve the following equations:
   a) $2\cos^2\theta + 3\sin\theta - 3 = 0$, $0° \leq \theta < 360°$
   b) $3\sin^2\theta - 5\cos\theta - 5 = 0$, $0° \leq \theta < 360°$
   c) $6\cos^2\theta + \sin\theta - 4 = 0$, $0° \leq \theta < 360°$
   d) $20\sin^2\theta - 7\cos\theta - 14 = 0$, $0° \leq \theta < 360°$

30) Solve the following equations:
   a) $3\sin^2\theta + \sin\theta - 3\cos^2\theta + 2 = 0$, $0° \leq \theta < 360°$
   b) $4\cos^2\theta - 16\cos\theta + 3\sin^2\theta - 19 = 0$, $0° \leq \theta < 360°$
   c) $10\sin^2\theta - 11\cos^2\theta + 46\sin\theta + 35 = 0$, $0° \leq \theta < 360°$

31)
   a) Solve the equation
   $5\cos^2 x + 3\cos x + 2 = 4\sin^2 x$
   for $-360° \leq x < 360°$
   b) Hence find the smallest positive solution to the equation
   $5\cos^2(2\theta + 50°) + 3\cos(2\theta + 50°) + 2$
   $= 4\sin^2(2\theta + 50°)$

32) Solve $2\sin^2 x + 3 = \tan 5x - 2\cos^2 x$ for $0° \leq x < 72°$

33) Solve the following equations:
   a) $10\sin\theta\tan\theta + 21 = 0$, $0° \leq \theta < 360°$
   b) $2\sin\theta\tan\theta + 20\cos\theta + 15 = 0$, $0° \leq \theta < 360°$
   c) $\frac{2\cos\theta}{\tan\theta} = 18\sin\theta - 3$, $0° \leq \theta < 360°$

34) The graph below shows the curves with equation $y = 12\sin^2 x - 28\sin x\cos x$ and $y = 5\cos^2 x$ from $0°$ to $360°$. Determine the coordinates of where the two graphs intersect.

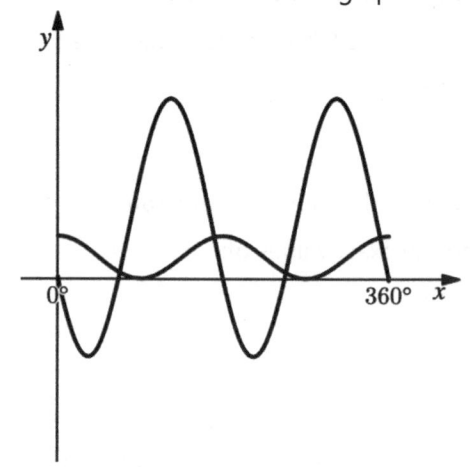

35) Solve the following equations:
   a) $\sin\theta\tan\theta + 4\cos\theta = 5\sin\theta$, $0° \leq \theta < 360°$
   b) $5\sin\theta\tan\theta = 3\cos\theta + 14\sin\theta$, $0° \leq \theta < 360°$
   c) $10\sin\theta\tan\theta + 3\sin\theta = 18\cos\theta$, $0° \leq \theta < 360°$

36) Solve the following equations:
   a) $2\cos\theta + 13\cos\theta\tan\theta + 6\sin\theta\tan\theta = 0$, $0° \leq \theta < 360°$
   b) $15\cos\theta + 6\sin\theta\tan\theta = 19\cos\theta\tan\theta$, $0° \leq \theta < 360°$

1.42 Trigonometry: Trigonometric Identities in Degrees (answers on page 456)

37) Prove the following trigonometric identities:

   a) $(\sin x + \cos x)^2 \equiv 1 + 2\sin x \cos x$

   b) $\sin^3 x + \cos^3 x \equiv (\sin x + \cos x)(1 - \sin x \cos x)$

   c) $1 + \tan^2 x \equiv \dfrac{1}{\cos^2 x}$

   d) $\dfrac{1}{\tan^2 x} + 1 \equiv \dfrac{1}{\sin^2 x}$

38) Prove the following trigonometric identities:

   a) $\dfrac{\cos x}{\sin x} + \dfrac{\sin x}{\cos x} \equiv \dfrac{1}{\sin x \cos x}$

   b) $\dfrac{1}{\cos x + 1} - \dfrac{1}{\cos x - 1} \equiv \dfrac{2}{\sin^2 x}$

   c) $\dfrac{\cos x}{1 - \sin x} + \dfrac{\cos x}{1 + \sin x} \equiv \dfrac{2}{\cos x}$

   d) $\dfrac{\cos x}{\sin x - 1} + \dfrac{\cos x}{\sin x + 1} \equiv -2\tan x$

39) Prove the following trigonometric identities:

   a) $\dfrac{1}{1 + \tan^2 x} \equiv \cos^2 x$

   b) $\dfrac{\tan x}{1 + \tan^2 x} \equiv \sin x \cos x$

   c) $\dfrac{1}{1 - \tan^2 x} \equiv \dfrac{\cos^2 x}{1 - 2\sin^2 x}$

   d) $\dfrac{\cos^2 x}{1 - \tan^2 x} \equiv \dfrac{1}{1 - \tan^4 x}$

40) Prove the following trigonometric identities:

   a) $\dfrac{1}{\frac{1}{\sin^2 x} - 1} \equiv \tan^2 x$

   b) $\dfrac{1}{\frac{1}{\cos^2 x} - 1} \equiv \dfrac{1}{\sin^2 x} - 1$

   c) $\dfrac{1}{\frac{1}{\tan^2 x} + 1} \equiv \sin^2 x$

   d) $\dfrac{1}{\frac{1}{\sin^2 x} + \frac{1}{\cos^2 x}} \equiv (\sin x \cos x)^2$

41) Prove the following trigonometric identities:

   a) $\sin^4 x - \cos^4 x \equiv \sin^2 x - \cos^2 x$

   b) $\sin^2 x + \dfrac{\sin^4 x}{\cos^2 x} \equiv \tan^2 x$

   c) $\cos^4 x + \sin^2 x + \sin^2 x \cos^2 x + \sin^2 x \tan^2 x \equiv \cos^2 x + \tan^2 x$

   d) $\sin^2 x \cos^2 x - 1 + \sin^4 x - \tan^2 x \equiv \sin^2 x - \dfrac{1}{\cos^2 x}$

42) Prove the following trigonometric identities:

   a) $\dfrac{\sin x}{1 - \tan x} + \dfrac{\cos x}{1 - \frac{1}{\tan x}} \equiv 0$

   b) $\dfrac{\cos x}{1 - \tan x} + \dfrac{\sin x}{1 - \frac{1}{\tan x}} \equiv \sin x + \cos x$

   c) $\dfrac{\sin x}{1 - \tan x} - \dfrac{\cos x}{1 - \frac{1}{\tan x}} \equiv \dfrac{2\sin x - 2\cos x}{2 - \frac{1}{\sin x \cos x}}$

   d) $\dfrac{\cos x}{1 - \tan x} - \dfrac{\sin x}{1 - \frac{1}{\tan x}} \equiv \dfrac{1}{\cos x - \sin x}$

43) Prove the following trigonometric identities:

   a) $\dfrac{\sin x}{1 - \cos x} \equiv \dfrac{1 + \cos x}{\sin x}$

   b) $\dfrac{\cos x}{1 - \sin x} \equiv \dfrac{1 + \sin x}{\cos x}$

   c) $\dfrac{\sin x}{1 - \tan x} \equiv \dfrac{\sin x \cos x}{\cos x - \sin x}$

   d) $\dfrac{1 - \sin x}{1 - \cos x} \equiv \dfrac{1 + \cos x}{\sin^2 x} - \dfrac{1 + \cos x}{\sin x}$

1.42 Trigonometry: Trigonometric Identities in Degrees (answers on page 456)

44) Prove the following trigonometric identities:

   a) $$\frac{\tan x + \frac{1}{\tan x}}{\frac{1}{\cos x} + \frac{1}{\sin x}} \equiv \frac{1}{\sin x + \cos x}$$

   b) $$\frac{\cos x - \frac{1}{\cos x}}{\frac{1}{\sin x} - \frac{1}{\cos x}} \equiv \frac{\sin^2 x}{1 - \frac{1}{\tan x}}$$

   c) $$\frac{\sin x - \frac{1}{\sin x}}{\frac{1}{\sin x} - \frac{1}{\cos x}} \equiv \frac{\cos^3 x}{\sin x - \cos x}$$

45) Prove the following trigonometric identities:

   a) $$\frac{\sin x \cos x}{1 + \sin x + \cos x - \cos^2 x} \equiv \frac{1}{\frac{1}{\sin x} + \frac{1}{\cos x} + \tan x}$$

   b) $$\frac{\sin x \cos^2 x}{\cos^3 x - 2\cos^2 x - \cos x + 1} \equiv \frac{1}{\frac{\tan x}{\cos x} - \frac{1}{\sin x} - \tan x}$$

46) Prove the following trigonometric identities:

   a) $\left(\frac{1}{\cos x} - \sin x\right)\left(\frac{1}{\sin x} - \cos x\right) \equiv \frac{(\sin x \cos x - 1)^2}{\sin x \cos x}$

   b) $\left(\frac{1}{\sin x} - \sin x\right)\left(\frac{1}{\cos x} - \cos x\right) \equiv \frac{1}{\tan x + \frac{1}{\tan x}}$

47) Prove the following trigonometric identity:
$$\frac{2\cos^2 x - 1}{\cos x - \sin x} \equiv \frac{1}{\sin x - \cos x + \frac{2\cos^2 x}{\sin x + \cos x}}$$

48) Prove the following trigonometric identity:
$\sin^2 A + \sin^2 B - 1 \equiv \sin^2 A \sin^2 B - \cos^2 A \cos^2 B$

49) Isla is attempting to solve the equation
$\sqrt{2}\cos^2 x + \sin x - \sqrt{2} = 0$ for $180° < x < 360°$.
Her attempt is shown below:
$\sqrt{2}\cos^2 x + \sin x + \sqrt{2} = 0$
$\Rightarrow \sqrt{2}(\sin^2 x - 1) + \sin x + \sqrt{2} = 0$
$\Rightarrow \sqrt{2}\sin^2 x - \sqrt{2} + \sin x + \sqrt{2} = 0$
$\Rightarrow \sqrt{2}\sin^2 x + \sin x = 0$
$\Rightarrow \sqrt{2}\sin x + 1 = 0$
$\Rightarrow \sin x = -\frac{1}{\sqrt{2}}$
$\Rightarrow x = -45°$
Explain the two errors that Isla has made.

50)
   a) Show that the equation
   $20\sin x \tan x + 31\sin x = 7\cos x$
   can be written in the form
   $(4\tan x + 7)(5\tan x - 1) = 0$

   b) Hence solve the equation
   $20\sin 4\theta \tan 4\theta + 31\sin 4\theta = 7\cos 4\theta$
   for $0° \le \theta < 45°$

51)
   a) Show that
   $$\frac{\tan x}{\frac{1}{\cos x} - \cos x} \equiv \frac{1}{\sin x}$$

   b) Using your answer to a), solve the equation
   $$\frac{\tan x}{\frac{1}{\cos x} - \cos x} = 6$$
   for $0° \le x < 360°$

   c) Hence solve the equation
   $$\frac{\tan(2\theta - 55°)}{\frac{1}{\cos(2\theta - 55°)} - \cos(2\theta - 55°)} = 6$$
   for $0° \le \theta < 180°$

52)
   a) Show that
   $$\frac{\cos^3 x}{\sin x} + \sin x \cos x \equiv \frac{1}{\tan x}$$

   b) Using your answer to a), solve the equation
   $$\frac{\cos^3 x}{\sin x} + \sin x \cos x = \frac{1}{20}$$
   for $0° \le x < 360°$

   c) Hence solve the equation
   $$\frac{\cos^3(2\theta + 9°)}{\sin(2\theta + 9°)} + \sin(2\theta + 9°)\cos(2\theta + 9°) = \frac{1}{20}$$
   for $0° \le \theta < 180°$

53) Solve the equation
$$\sin^4 \theta - \cos^4 \theta = \frac{1}{2}$$
for $0° \le \theta \le 360°$

54) Solve the equation
$2\cos^4 \theta + \cos^2 \theta \sin \theta - 2\cos^2 \theta = 0$
for $0° \le \theta \le 360°$

1.43 Trigonometry: Modelling in Degrees (answers on page 459)

1) State the period of each of the following functions:
   a) $f(x) = \sin 2x$
   b) $f(x) = \cos 6x$
   c) $f(x) = \tan x$
   d) $f(x) = \sin\left(\frac{1}{2}x\right)$
   e) $f(x) = \cos\left(\frac{1}{4}x\right)$
   f) $f(x) = \tan(2x - 45)$
   g) $f(x) = \sin\left(\frac{360}{7}x\right)$
   h) $f(x) = \cos\left(\frac{90}{13}x + \frac{360}{13}\right)$
   i) $f(x) = \sin\left(\frac{72}{73}(x - 30)\right)$
   j) $f(x) = \cos(15(x + 7))$

2) In each of the following cases, the curve
$$y = \sin x$$
has been stretched parallel to the x-axis. State the equation of each curve.

   a)

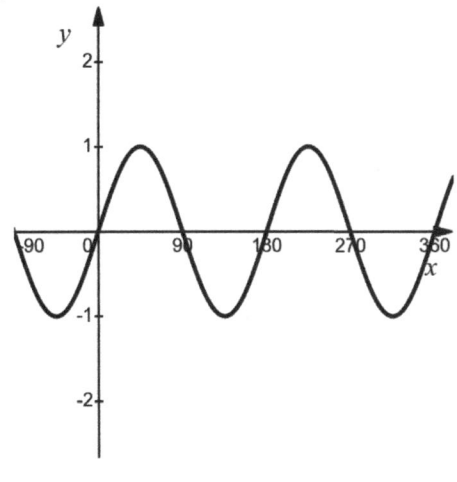

   b)

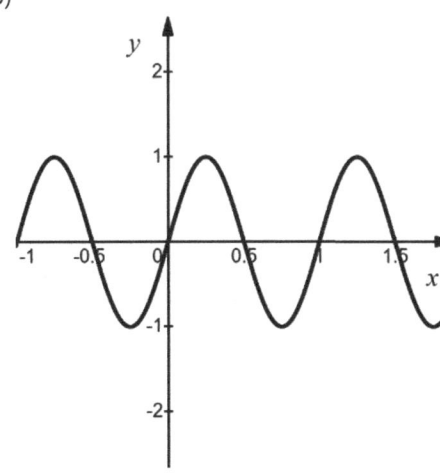

   c)

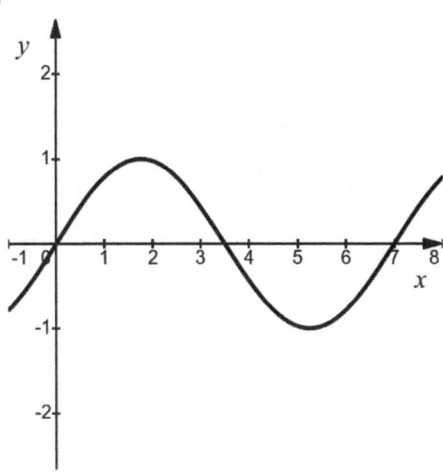

   d)

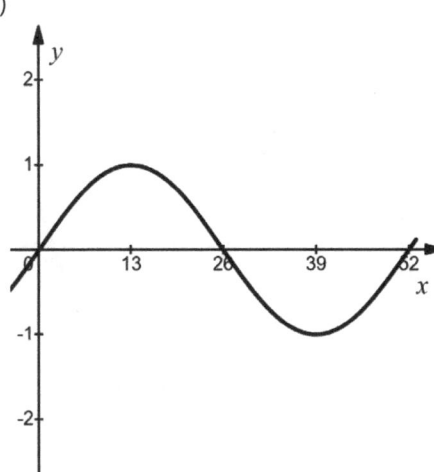

3) All the following graphs are transformations of
$$y = \sin x$$
State the equation of each:

   a)

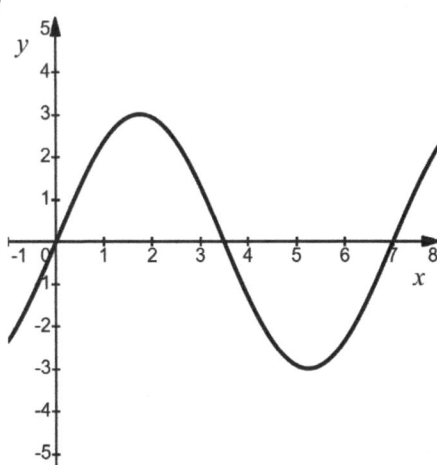

1.43 Trigonometry: Modelling in Degrees (answers on page 459)

b)

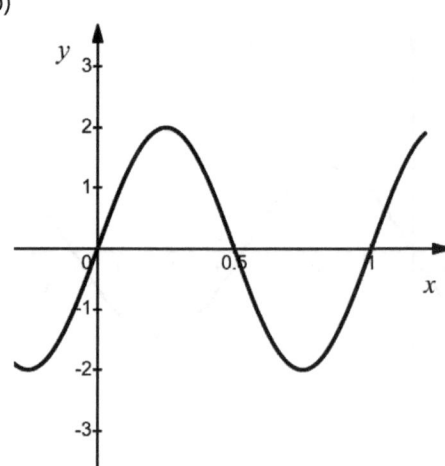

c)

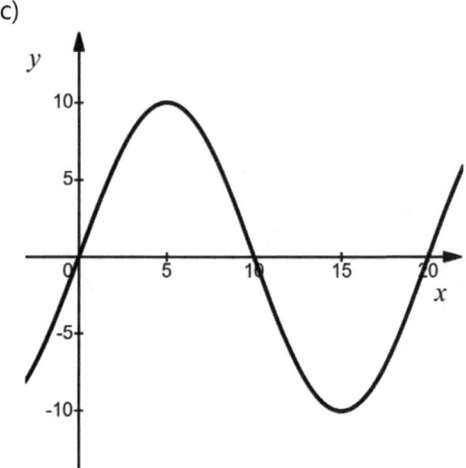

d)

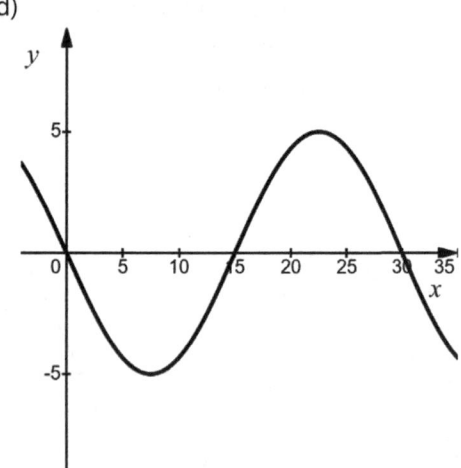

4) State the amplitude of each of the following periodic functions:
   a) $y = \sin 2x$
   b) $y = 2 + \cos(15x + 3)$
   c) $y = 40 + 18\sin(12x - 4)$
   d) $y = 3 - 5\sin(0.5x + 1.1)$

5) For each of the following functions state the minimum and maximum of the function:
   a) $y = \sin 3x$
   b) $y = 4 + \cos(12x° - 7)$
   c) $y = 25 + 15\sin(30x + 2)$
   d) $y = 3 - 8\sin(0.5x + 0.2)$

6) All the following graphs are of the form
   $$y = a + b\sin(cx)$$
   In each case, find the equation of the graph:
   a)

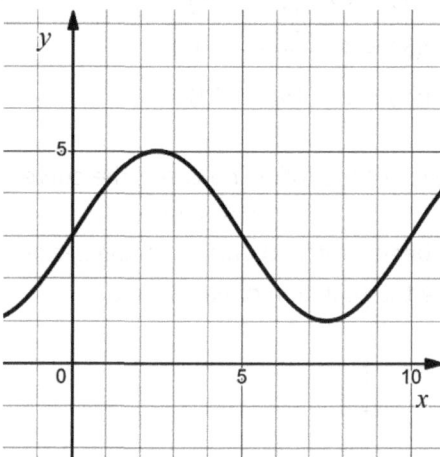

   b)

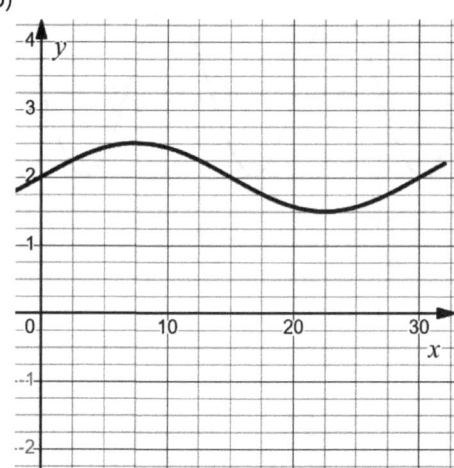

   c)

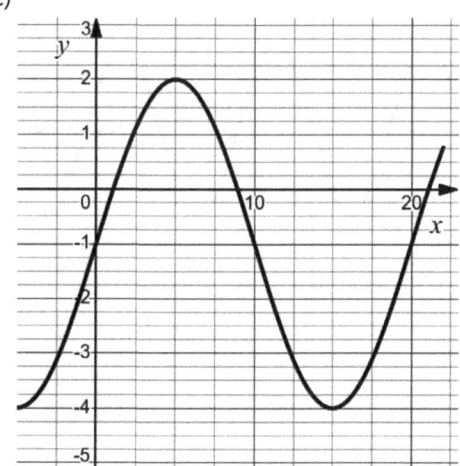

1.43 Trigonometry: Modelling in Degrees (answers on page 459)

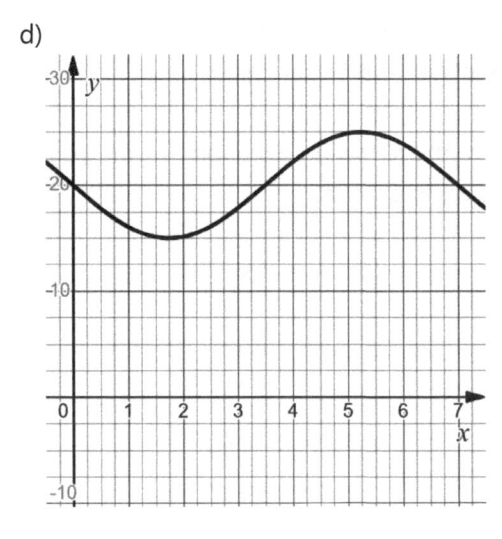

d) [Graph showing a sinusoidal curve]

7) The time of sunrise, y hours since midnight for each day x of a year may be modelled by the equation
$$y = a + b \cos x°$$
It is known that the latest sunrise is at 8:00 am on January 1st and the earliest is at 4:30 am, six months later.
   a) Find the values of a and b
   b) Suggest how the model may be improved.

8) A person's blood pressure, y, oscillates between 120 and 80 with their heartbeat occurring every 1 second.
   a) Find a model in the form
$$y = a + b \sin kt°$$
   where t is the time that has passed in seconds and a, b, k are constants.
   b) Explain why this model will not always be accurate for predicting blood pressure

9) The size of a population, P, of wild mice is recorded over time, and may be modelled by a formula of the form
$$P = 20 + 6\sin(3t - 10)°, \quad 0 \leq t \leq 121$$
where t is the number of days since the start of the year. According to the model, find
   a) the number of mice in the population after 5 days have passed.
   b) the maximum and minimum number of mice in the population.
   c) the days on which the maximum and the minimum number of mice in the population occurs.

10) The profits on the sales of ice cream for a shop, £P, may be given by the equation
$$P = 55 - 50\cos\left(\frac{72}{73}(x - 45)\right)°$$
where x is the number of days since the 1st January. According to the model,
   a) write down the difference between the largest and smallest values of profits.
   During the 136th day of the year, the rate of change of the profits is at its maximum.
   b) Find the next day when the model predicts the rate of change will be the same.

11) Laiyla is monitoring the height of the tide, h metres, in her local harbour. At various points of time, she records the height along with the time t hours after midnight.

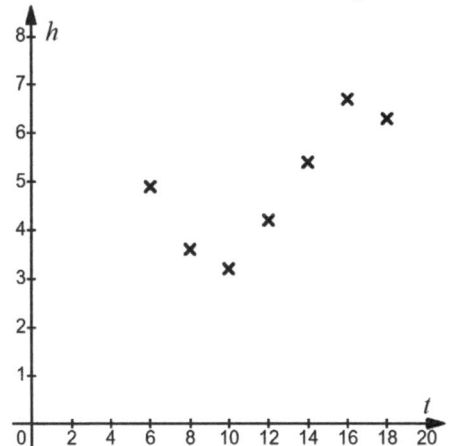

[Scatter plot of h vs t]

Laiyla models the data with the equation
$$h = 5 + 1.7\sin(25t - 330)°$$
where t is the number of hours since midnight.
   a) Find the minimum height of the tide predicted by Laiyla's model.
   b) Find the maximum number of consecutive hours that Laiyla's model predicts that the height of the tide will exceed 6 metres.
   c) Find the second time after midnight at which the tide has a height of 6 metres.

Further along the coast, a second location's tide has the same times for its high and low tides, varying in depth by 3.3 metres. At high tide, it is 2.55 metres less deep than the high tide found by Laiyla's model.
   d) Make appropriate refinements to Laiyla's equation to create a model for the height of the tide.

1.44 Hidden Polynomials (answers on page 459)

1) Solve the following equations, giving your answer in exact form:
   a) $$5x^4 = 3 - 2x^2$$
   b) $$3x - 16\sqrt{x} = 35$$
   c) $$2x^{\frac{2}{3}} + 7x^{\frac{1}{3}} = 15$$
   d) $$\frac{6}{x^2} + \frac{7}{x} = 5$$
   e) $$7^x + 7^{2x} = 2$$
   f) $$4^x - 3 \times 2^{x-1} + \frac{1}{2} = 0$$
   g) $$3^{3x+1} + 20 \times 3^{2x} - 7 \times 3^x = 0$$
   h) $$2^{x+1} + 11 = \frac{6}{2^x}$$

2) By first factorising, express $5^8 - 3^8$ as a product of its prime factors.

3) Find the solutions to
$$\frac{2x^4 + 2}{5} = x^2$$

4) Given
$$f(x) = 8x^3 + \frac{1}{x^3}$$
solve $f(x) = -9$

5) Given
$$f(x) = 2x^6 + \frac{1}{x^3} \text{ and } g(x) = x^3 + 2$$
solve $f(x) = g(x)$

6) Find the solutions to
$$\left(x - \frac{4}{x}\right)^2 + 8 = 6\left(x - \frac{4}{x}\right)$$

7) Find the solutions to
$$\frac{5x^2}{\sqrt{x}} + 22 + \frac{24}{\sqrt{x}} = 12x + \frac{39x}{\sqrt{x}}$$

8) Find the solutions to
$$3^{2x+1} + 20 \times 3^x + \frac{75}{3^{2x}} = 22 + \frac{220}{3^x}$$

Prerequisite knowledge for Q9-Q12: Exponentials and Logarithms

9) Solve the following equations, giving your answer in exact form:
   a) $$3 \times 2^{2x} + 7 \times 2^x = 6$$
   b) $$2e^{3x} + 3e^{2x} - 17e^x + 12 = 0$$
   c) $$4 \times 3^x + 13 = \frac{2}{3^x} + \frac{15}{3^{2x}}$$

10)
   a) Solve
   $$\frac{4}{(x-2)^2} - \frac{7}{x-2} = 2$$
   b) Hence, solve
   $$\frac{4}{(5^x - 2)^2} - \frac{7}{5^x - 2} = 2$$
   giving your answer in exact form.

11) Solve the following equations, giving your answer in exact form:
   a) $$\log_2(x+3) = 2\log_2 x - \log_2 4$$
   b) $$\log_4(5x+2) = \log_4(13x-6) - 2\log_4 x$$
   c) $$-1 + \log_5(x^4 + 4) = \log_5 x + \log_5(3x + 14 - 5x^2) - \log_5 6$$

12) Solve for the following equation giving your answer in exact form
$$\ln x + \frac{4}{\ln x} = \frac{\ln(x^4)}{\ln x}$$

Prerequisite knowledge for Q13-Q14: Solving Trigonometric Equations

13) Solve
$$\sin^2 x = \frac{7 - 13\sin x}{2}$$
for $0 \leq x \leq 360°$

14) Solve
$$\frac{\cos^2 x (3\cos x + 11)}{20} = \cos x$$
for $0 \leq x \leq 360°$

1.45 Proof (answers on page 460)

1) In each case, choose one of the statements
$$A \Rightarrow B \quad A \Leftarrow B \quad A \Leftrightarrow B$$
to describe the relationship between A and B
   a) A: The shape has 4 equals sides.
      B: The shape is a square.
   b) A: A quadrilateral has perpendicular diagonals.
      B: The quadrilateral is a kite.
   c) A: A quadrilateral has one pair of parallel sides.
      B: The quadrilateral is a trapezium.
   d) A: $PQRS$ is a rhombus.
      B: $PQRS$ is a parallelogram.

2) In each case, choose one of the statements
$$A \Rightarrow B \quad A \Leftarrow B \quad A \Leftrightarrow B$$
to describe the relationship between A and B
   a) A: $x = 3$ B: $(x-3)(x+2) = 0$
   b) A: $x > 6$ B: $x = 10$
   c) A: $x = 4$ B: $x^2 = 16$
   d) A: $x^3 = -1$ B: $x = -1$

3) For each of the following determine which of these statements are true, where it is false explain your answer:
   a) If n is an even integer, n^2 is even
   b) The product of 5 consecutive integers is a multiple of 5
   c) $x^2 > 0$ for all $x \in \mathbb{R}$
   d) The square root of a positive number is always positive
   e) The product of any two odd integers is always even
   f) If $\cos x° = 0.5$, then $x = 60$

4) Find the mistake in the proof below:
$$2 = \frac{20}{10} = \frac{10+10}{10} = \frac{(10-10)(10+10)}{10(10-10)}$$
$$= \frac{10^2 - 10^2}{100 - 100} = \frac{100 - 100}{100 - 100} = 1$$

5) For each statement, find an example that disproves it:
   a) All prime numbers are odd.
   b) All squares are even numbers.
   c) All multiples of 3 are also multiples of 6
   d) All real numbers have a square root.
   e) All perfect squares end in an odd number.

6) For each statement, find an example that disproves it:
   a) If $x^4 > x$, then $x > 1$
   b) If n is a prime number, then $n^2 - n + 13$ is prime.
   c) The value of $n^2 + n + 17$ is a prime number for all positive integer values of n

7) Use proof by exhaustion to prove that
   a) the sum of the digits of any 2-digit number is less than or equal to 18
   b) if a 2-digit number is divisible by 3, then the sum of its digits is divisible by 3
   c) 89 is a prime number.

8) Prove that no square number ends in a 3

9) Show that
   a) the sum of 2 consecutive numbers is always odd.
   b) the sum of any two consecutive odd numbers is a multiple of 4
   c) the square of an odd number is always odd.
   d) the sum of the squares of 2 consecutive odd numbers is always even.

10) Prove
    a) the product of any three consecutive positive integers is a multiple of 6
    b) the product of any four consecutive positive integers is a multiple of 24

11) Prove that for any even number, n, the expression $6n^2 + 10n$ has a factor of 4

12) Prove that $n^4 - n^2$ is a multiple of 6 for all positive integer values of n

13) Prove that $n^3 - n$ is a multiple of 3 for all positive integer values of n, using
    a) Proof by deduction
    b) Proof by cases, by considering $n = 3k$, $n = 3k + 1$ and $n = 3k + 2$

14) For a positive integer number n, prove that $n^4 - 4$ is not divisible by 8

1.46 2D Vectors (answers on page 460)

1) Given $\mathbf{a} = 2\mathbf{i} + \mathbf{j}, \mathbf{b} = 3\mathbf{j}$ and $\mathbf{c} = -\mathbf{i} - 4\mathbf{j}$ then find the following, giving your answer both in component and column vector form:
   a) $3\mathbf{a}$
   b) $2\mathbf{b} + \mathbf{c}$
   c) $\mathbf{a} - 2\mathbf{c}$

2) For the quadrilateral shown, $\overrightarrow{OA} = \mathbf{a}, \overrightarrow{OB} = \mathbf{b}$ and $\overrightarrow{OC} = \mathbf{c}$

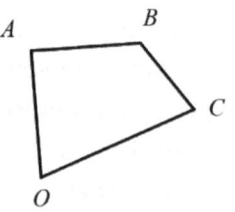

   find the following in terms of **a**, **b** and **c**:
   a) $\overrightarrow{AB}$
   b) $\overrightarrow{CB}$
   c) $\overrightarrow{AC}$

3) For the parallelogram shown, $\overrightarrow{OA} = \mathbf{a}$ and $\overrightarrow{OB} = \mathbf{b}$

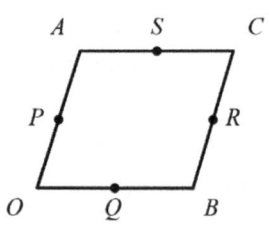

   The midpoints of OA, OB, BC and AC are P, Q, R and S respectively. Find in terms of **a** and **b** each of the following:
   a) $\overrightarrow{OP}$
   b) $\overrightarrow{OQ}$
   c) $\overrightarrow{OR}$
   d) $\overrightarrow{OS}$
   e) $\overrightarrow{BQ}$
   f) $\overrightarrow{SR}$
   g) $\overrightarrow{RP}$
   h) $\overrightarrow{AQ}$

4) Given two non-parallel vectors **a** and **b**, state whether each of the following pairs of vectors are parallel:
   a) $4\mathbf{a}$ and $-2\mathbf{a}$
   b) $2\mathbf{a} + \mathbf{b}$ and $\mathbf{a} + \frac{1}{2}\mathbf{b}$
   c) $3\mathbf{a} - \mathbf{b}$ and $-3\mathbf{a} - \mathbf{b}$
   d) $-2\mathbf{a} - 4\mathbf{b}$ and $8\mathbf{a} + 24\mathbf{b}$

5) Find the magnitude of the following vectors:
   a) $\mathbf{a} = 4\mathbf{i} + 5\mathbf{j}$
   b) $\mathbf{b} = \begin{pmatrix} 7 \\ 1 \end{pmatrix}$
   c) $\mathbf{c} = -3\mathbf{i} + \mathbf{j}$
   d) $\mathbf{d} = \begin{pmatrix} p \\ q \end{pmatrix}$

6) Find the magnitude and the bearing of each of the following vectors, where vectors **i** and **j** are unit vectors due east and north respectively:
   a) $\mathbf{a} = 2\mathbf{i} + 12\mathbf{j}$
   b) $\mathbf{b} = -5\mathbf{i} + 3\mathbf{j}$
   c) $\mathbf{c} = -10\mathbf{i} - 5\mathbf{j}$
   d) $\mathbf{d} = 8\mathbf{i} - 15\mathbf{j}$

7) Write the following in the form $p\mathbf{i} + q\mathbf{j}$, giving p and q to 3 significant figures:
   a)
   b)
   c)
   d)

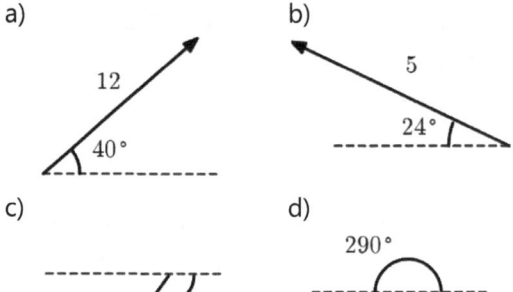

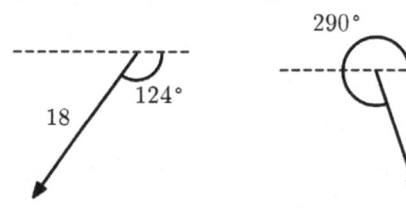

8) Find the unit vector in the direction of the following vectors:
   a) $\mathbf{a} = 2\mathbf{i} + 3\mathbf{j}$
   b) $\mathbf{b} = \begin{pmatrix} -1 \\ 4 \end{pmatrix}$
   c) $\mathbf{c} = 5\mathbf{i} - 8\mathbf{j}$

9) The vector **a** is parallel to the given vector **p**. In each case, given the magnitude of **a**, find the vector **a**:
   a) $\mathbf{p} = -2\mathbf{i} + 6\mathbf{j}$ and $|\mathbf{a}| = 3\sqrt{10}$
   b) $\mathbf{p} = 2\mathbf{i} + 7\mathbf{j}$ and $|\mathbf{a}| = \frac{1}{2}\sqrt{53}$
   c) $\mathbf{p} = \mathbf{i} - 8\mathbf{j}$ and $|\mathbf{a}| = 3\sqrt{65}$

10) State the position vector of each of the following coordinates:
    a) $(1, 2)$
    b) $(-1, 6)$
    c) (p, q)

1.46 2D Vectors (answers on page 460)

11) It is given that A, B and C have the coordinates $(1,2)$, $(-2,6)$ and $(3,-10)$ respectively. Find
    a) $\overrightarrow{AB}$
    b) $\overrightarrow{BC}$
    c) $\overrightarrow{AC}$
    d) $\overrightarrow{BA}$

12) Given each of the following position vectors, find the magnitude of the vector $\overrightarrow{AB}$:
    a) $\overrightarrow{OA} = \begin{pmatrix} 3 \\ -1 \end{pmatrix}$ and $\overrightarrow{OB} = \begin{pmatrix} -5 \\ 2 \end{pmatrix}$
    b) $\overrightarrow{OA} = 2\mathbf{i} - 7\mathbf{j}$ and $\overrightarrow{OB} = 9\mathbf{i} + 4\mathbf{j}$

13) Given each of the following position vectors, find the unit vector in the direction of the vector $\overrightarrow{AB}$:
    a) $\overrightarrow{OA} = \begin{pmatrix} 1 \\ -5 \end{pmatrix}$ and $\overrightarrow{OB} = \begin{pmatrix} 3 \\ 6 \end{pmatrix}$
    b) $\overrightarrow{OA} = 5\mathbf{i} + 2\mathbf{j}$ and $\overrightarrow{OB} = -3\mathbf{i} + 5\mathbf{j}$

14) Find the angle between each of the following pair of vectors, giving your answer to 1 decimal place:
    a) $\begin{pmatrix} 0 \\ 5 \end{pmatrix}$ and $\begin{pmatrix} 5 \\ 0 \end{pmatrix}$
    b) $2\mathbf{i} + 7\mathbf{j}$ and $3\mathbf{i} + 3\mathbf{j}$
    c) $2\mathbf{i} - 10\mathbf{j}$ and $8\mathbf{i} - 3\mathbf{j}$
    d) $\begin{pmatrix} 0 \\ -4 \end{pmatrix}$ and $\begin{pmatrix} 3 \\ 5 \end{pmatrix}$
    e) $\mathbf{i} - 2\mathbf{j}$ and $-\mathbf{i} + 7\mathbf{j}$

15) It is given that
    $$\overrightarrow{AB} = \begin{pmatrix} 1 \\ -2 \end{pmatrix} \text{ and } \overrightarrow{BC} = \begin{pmatrix} 3 \\ -1 \end{pmatrix}$$
    and the angle between $\overrightarrow{AB}$ and $\overrightarrow{BC}$ is θ. Show that $\cos\theta = -\frac{\sqrt{2}}{2}$.

16) It is given that
    $$\overrightarrow{OA} = \begin{pmatrix} 3 \\ 2 \end{pmatrix}, \overrightarrow{OB} = \begin{pmatrix} 4 \\ 8 \end{pmatrix} \text{ and } \overrightarrow{OC} = \begin{pmatrix} -5 \\ 1 \end{pmatrix}$$
    It is known that $\overrightarrow{BC} = \overrightarrow{AD}$, find the position vector of the point D.

17) In each case, determine whether each set of points A, B and C with the given coordinates are respectively collinear:
    a) $(-8,2)$, $(0,7)$ and $(8,12)$
    b) $(-5,4)$, $(4,-5)$ and $(7,-8)$
    c) $(-10,17)$, $(-3,13)$ and $(5,8)$

18) The midpoint of A and B is M. Point A has position vector $-5\mathbf{i} + 13\mathbf{j}$ and the position vector of M is $-\mathbf{i} + 4\mathbf{j}$.
    Find the position vector of the point B.

19) The position vectors of A, B and C respectively are $\mathbf{a} = \begin{pmatrix} -3 \\ 7 \end{pmatrix}, \mathbf{b} = \begin{pmatrix} 4 \\ 3 \end{pmatrix}$ and $\mathbf{c} = \begin{pmatrix} 1 \\ k \end{pmatrix}$. There is a further point $D(p,8)$ such that $\overrightarrow{AB} = \overrightarrow{CD}$
    a) Find the value of p and k
    b) Determine the shape of $ABDC$.

20) The points A and B have position vectors $\mathbf{a}$ and $\mathbf{b}$ respectively. A third point C forms the parallelogram $OABC$.
    a) In terms of $\mathbf{a}$ and $\mathbf{b}$,
        i) find $\overrightarrow{OC}$
        ii) find $\overrightarrow{OP}$, where P is the midpoint of AB
        iii) find $\overrightarrow{OQ}$, where Q is the midpoint of OC
    b) Show that the midpoint of $\overrightarrow{PQ}$ is at the centre of the parallelogram.

21) A quadrilateral is given by $ABCD$. It is known that $\overrightarrow{AB} = 4\mathbf{i} + 5\mathbf{j}$ and $\overrightarrow{DC}$ is the same length and direction as $\overrightarrow{AB}$. Two points C and D have position vectors $\mathbf{c} = 5\mathbf{i} + \mathbf{j}$ and $\mathbf{d} = k\mathbf{i} - 4\mathbf{j}$ respectively. Find
    a) the value of k
    b) the coordinates of A and B given $ABCD$ is a square.

22) The position vectors of A, B and C respectively are $\mathbf{a} = \begin{pmatrix} 5 \\ 2 \end{pmatrix}, \mathbf{b} = \begin{pmatrix} 1 \\ 6 \end{pmatrix}$ and $\mathbf{c} = \begin{pmatrix} -5 \\ 4 \end{pmatrix}$. The midpoint of A and B is P and the midpoint of B and C is Q
    a) Find the coordinates of the points P and Q
    b) Show that $\overrightarrow{PQ} = \frac{1}{2}\overrightarrow{AC}$
    A further point D exists such that $|\overrightarrow{DB}| = \sqrt{74}$ and $\overrightarrow{BC}$ is parallel to $\overrightarrow{AD}$.
    c) find the possible coordinates of D

23) The points A and B have position vectors $\mathbf{a}$ and $\mathbf{b}$ relative to the origin O where $\mathbf{a} = \begin{pmatrix} 1 \\ 2 \end{pmatrix}$ and $\mathbf{b} = \begin{pmatrix} 2 \\ 10 \end{pmatrix}$. Find the possible coordinates of a third point C such that $\overrightarrow{AB}$ is the hypotenuse of a right-angled isosceles triangle.

1.47 Sampling Methods (answers on page 461)

1) A by-election is due to take place in Wellingborough. Amelia is a politics student who is interested in the election and decides to survey local residents. She stands in the main high street and surveys the first 500 people she meets. Identify the name of this type of sampling.

   simple random   opportunity

   cluster   quota

2) In a small village of 156 households, Jason wants to investigate how many use the local post office. He decides to number the households from 001 to 156, and then use a random number generator to obtain 30 different three-digit numbers between 001 and 156. He will then select the corresponding households and include them in the sample. Identify the name of this type of sampling.

   convenience   self-selection

   simple random   systematic

3) The table below shows the share price of HSBC on the 1st January, 1st April, 1st July and 1st October across a sample of five years.

	01/01	01/04	01/07	01/10
1993	156.57	173.13	189.10	221.34
1998	464.77	548.13	442.40	406.67
2003	549.88	597.37	671.01	771.22
2008	654.88	766.43	731.13	664.03
2013	716.70	703.50	747.80	682.10

   Identify the sampling procedure that most likely chose the years selected in the sample.

   quota   systematic

   simple random   stratified

4) In a secondary school, there are 12 classes in year 9 and the school wants to collect some feedback on a recent futures fair that visited the school. Sandra, the head of year 9, decides to select three classes at random and then get feedback from every student in those classes. Identify the name of this type of sampling.

   stratified   quota

   opportunity   cluster

5) Ibrahim works for a local charity and is asked to interview members of the public in the high street on a Saturday afternoon. He is asked to interview 30 people under the age of 40, and 30 people who are aged 40 or over. Identify the name of this type of sampling.

   simple random   stratified

   systematic   quota

6) A large retailer of flat-pack furniture employs 7200 members of staff across 24 stores in the UK. The retailer wants to investigate their staff's job satisfaction through a survey. The HR team suggest four possible sampling methods:

   Method A   Randomly select 120 people from the whole list.
   Method B   Randomly select five stores and then randomly select 24 people from each of these stores.
   Method C   Randomly select 120 people from the London store.

   a) Identify the population in the retailer's investigation.
   b) Give one advantage of using Method B over Method C.
   c) Give one advantage of using Method B over Method A.

7) The Office for National Statistics conducts a census for England and Wales every 10 years. Describe one disadvantage with a census.

1.47 Sampling Methods (answers on page 461)

8) In 2019, there were estimated to be approximately 20,850 breeding pairs of African penguins. The health of the population is to be investigated by sampling 100 breeding pairs. Explain why a simple random sample cannot be used.

9) A national union of care workers is holding its annual conference. Chloe wants to sample the views of care workers on conditions in their place of work. She decides to set up a stall near the entrance to the conference and ask attendees who visit her stall to fill in a questionnaire. State two disadvantages with this sampling method.

10) Clive runs a takeaway business. He wants to revise the menu down to reduce costs. He has recorded the choices made by each customer since the business opened and now has 680 customers and their choices in a spreadsheet. Describe how Clive could obtain a simple random sample of size 40 from his list of customers.

11) Lundi supermarkets split their workforce into those who work on the shop floor and those who work in the warehouse. Across all of their stores, 3265 work in the shops and 2435 work in the warehouses. The company wants to conduct a survey of 150 members of staff, and they decide to use a stratified sample. Calculate how many from the shop floor and how many from the warehouses should be included in the sample.

12) Harry works in a large office block. Near his office, there is a kitchen where employees store their mug in a cupboard. The mugs are all different, but his is often used by somebody else accidentally. Harry wants to label all the mugs with their owner's name. Suggest why an opportunity sample conducted near the kitchen would be more appropriate to gather opinions than a simple random sample to the whole workforce.

13) Susan is the Subject Leader of Maths at a large sixth form college. She wants to investigate the number of hours it takes for the students taking A-Level Maths to complete their homework each week. There are 1200 students in 50 classes. Describe how Susan could obtain a cluster sample of three classes of students for her survey.

14) A company wishes to survey its employees. It decides to send out an online survey to all of their employees' work emails and then analyse the results they get within a 24-hour window. Suggest two disadvantages with using this method.

15) A work-place choir is made up of four companies: Aurora, BrightByte, Cosmic and Dreamscape. The number of people from each company who take part in the choir are 25, 30, 50 and 45 respectively. A stratified sample is taken of size k to gain feedback on the choir, and the samples taken from Aurora, BrightByte and Cosmic are 8, 10 and 17 respectively. In the calculations to determine how many people must be sampled from each company, the number was always rounded to the nearest person.
   a) Determine the value of k
   b) Determine how many people who work at Dreamscape should be sampled.

16) A sixth form college employs 120 members of staff. A survey is to be sent out to 20 members of staff using the following method:
   ➢ List all of the members of staff in alphabetical order by surname and assign each of them a number from 1 to 120.
   ➢ Choose a random number from 1 to 6 and choose this person.
   ➢ Select every 6th person on the list from there until 20 people are selected.
   a) Identify this sampling method.
   b) Give a reason why this method does not provide a random sample.

1.48 Measures of Central Tendency and Variation (answers on page 462)

1) For the two data sets below, find the mean, median, mode, range and interquartile range, and hence compare the two sets of data.

| Set A | 3 | 3 | 6 | 7 | 8 | 12 | 15 | 16 | 20 | 25 | 28 |
| Set B | 3 | 3 | 6 | 9 | 10 | 12 | 13 | 16 | 20 | 23 | 28 |

2) Give one reason why:
   a) the mean would be preferable to use as an average instead of the median,
   b) the median would be preferable to use as an average instead of the mean.

3) The following data was recorded:

| 1 | 4 | 8 | 8 | 9 | 10 |
| 11 | 14 | 15 | 15 | 25 | |

   a) Find the mean and standard deviation.
   b) Values that are less than two standard deviations below or above the mean are considered outliers. Determine whether there are any outliers in the data.
   c) Find the median and interquartile range.
   d) Values that are less than $LQ - 1.5 \times IQR$ or more than $UQ + 1.5 \times IQR$ are considered outliers. Determine whether there are any outliers in the data.

4) The following data was recorded:

| 3 | 8 | 9 | 10 | 15 | 17 |
| 21 | 25 | 26 | 28 | 29 | 55 |

   a) Find the mean and standard deviation.
   b) Values that are less than two standard deviations below or above the mean are considered outliers. Determine whether there are any outliers in the data.
   c) Find the median and interquartile range.
   d) Values that are less than $LQ - 1.5 \times IQR$ or more than $UQ + 1.5 \times IQR$ are considered outliers. Determine whether there are any outliers in the data.

5) The following data was recorded:

| 35 | 36 | 45 | 48 | 52 | 59 |

   a) Find the mean.
   b) Without performing any calculations, explain what would happen to the mean if 59 was removed.

6) The following data was recorded:

| 21 | 25 | 40 | 61 | 90 | 102 |

   a) Find the mean.
   b) Without performing any calculations, explain what would happen to the mean if 40 was changed to 56.5

7) Find the mean, $\bar{x}$,
   a) given that $\sum x = 860$, $n = 5$
   b) given that $\sum x = 86753$, $n = 35$

8) A data set, x, contains n values. Given that $\sum x = 51324$ and $\bar{x} = 52$, determine the value of n

9) Given that $\sum x = 613$, $\sum x^2 = 56905$, $n = 10$, find the standard deviation, σ, to 3 significant figures.

10) Given that $\sum x = 412$, $\sum x^2 = 7322$, $n = 25$, find the standard deviation, σ, to 3 significant figures.

11) Given that $\sum x = 3305$, $\sum x^2 = 183075$, $n = 60$, find the variance, σ^2, to 3 significant figures.

12) Given that $\sum x = 1354$, $\sum x^2 = 212182$, $n = 9$, find the variance, σ^2, to 3sf.

13) The following data was recorded:

| 109 | 112 | 113 | 117 | 118 | 125 |

   a) Find the standard deviation.
   b) Without performing any calculations, explain what would happen to the standard deviation if 125 was removed.

1.48 Measures of Central Tendency and Variation (answers on page 462)

14) The following data was recorded:

| 37.4 | 39.2 | 40.1 | 40.3 | 42.1 | 43.6 |

    a) Find the standard deviation.
    b) Without performing any calculations, explain what would happen to the standard deviation if 43.6 was changed to 48.6

15) The following data was recorded:

| 8 | 11 | 13 | 16 | 18 | 21 |

    a) Find the mean.
    b) Find the standard deviation.
    c) Two more data points are added, x and y, such that $x < y$. It is given that the mean remains the same but the standard deviation is reduced. Determine possible values for x and y

16) The following data is recorded:

| 21 | 25 | 32 | 34 | 38 | 42 |

    a) Find the mean.
    b) Find the median.
    c) Two more data points are added, x and y, such that $x < y$. It is given that the mean remains the same but the median is reduced. Determine possible values for x and y

17) The following data is recorded:

| 8 | 12 | 15 | x | 22 | 28 |

    a) Given that the mode is 22, find the value of x
    b) Given instead that the mean is 17.05, find the value of x
    c) Given instead that the median is 16.3 and $15 < x < 22$, find the value of x

18) Chrissy grows a certain variety of sunflowers. One day, she measures the heights, in cm, of a random sample of 12 of the sunflowers in her garden and her collected data is shown below.

| 85 | 102 | 121 | 127 | 132 | 133 |
| 140 | 145 | 149 | 160 | 162 | 198 |

    a) Find the mean height of the sunflowers to 3 significant figures.
    b) Find the sample standard deviation of the heights of the sunflowers to 3 significant figures.
    c) Any sunflower with a height of less or more than 2 standard deviations from the mean may be regarded as an outlier. Using your answers to a) and b), determine if there are any outliers.

19) Collin sampled 80 students at his school about the number of pets their household had. This data is shown in the following table:

Number of pets	Frequency
0	7
1	31
2	21
3	8
4	6
5	5
6	2

Calculate the mean number of pets per household, and the standard deviation.

20) Alexis surveyed 95 colleagues at her company about the number of projects they were working on. This data is shown in the following table:

Number of projects	Frequency
0	3
1	34
2	48
3	10

Calculate the mean number of projects per employee, and the standard deviation.

1.48 Measures of Central Tendency and Variation (answers on page 462)

21) Sofia sampled 60 properties listed on an estate agent's website and recorded the number of bedrooms the properties had. This data is shown in the following table:

Number of bedrooms	Frequency
1	12
2	18
3	21
4	6
5 to 7	2
8 to 10	1

a) Calculate an estimate for the mean number of bedrooms.
b) Find the median number of bedrooms.

22) The following data was recorded:

x	Frequency
$14 \leq x < 18$	100
$18 \leq x < 22$	200
$22 \leq x < 30$	250
$30 \leq x \leq 32$	50

a) Calculate an estimate for the mean of x
b) Calculate an estimate for the standard deviation of x
c) State the assumption that explains why the midpoints of the class intervals are used to estimate the mean and standard deviation.

23) The weights, in kg, of 30 swans kept on a nature reserve was recorded:

Weight	Frequency
5 to 8	5
9 to 11	15
12 to 13	8
14 to 15	2

a) State the upper class boundary of the "9 to 11" class.
b) Calculate an estimate for the mean weight of the swans.
c) Calculate an estimate for the standard deviation of the weights of the swans.

24) The ages, in years, of 90 teachers working at a sixth form college was recorded:

Age	Frequency
22 to 26	12
27 to 34	30
35 to 45	22
46 to 54	19
55 and over	7

a) State the upper class boundary of the "22 to 26" class.
b) Explain why changing the "55 and over" class to "55 to 66" would be reasonable.
c) Using the class interval suggested in b), calculate estimates for the mean and standard deviation of the ages of the teachers.

25) A publishing company based in London employs 17 members of staff. Their salaries, in thousands of pounds, are recorded in the table below.

Salary	Frequency
35 –	3
40 –	7
45 –	4
50 –	1
55 – 75	2

a) Estimate the mean of the company's salaries.
b) Estimate the standard deviation of the company's salaries.
c) Explain why your answers to a) and b) are only estimates.
d) Explain why your answer to a) should not be publicised as the average wage of people who work in publishing across the UK.

26) Given that $\bar{x} = 9$, $\sigma^2 = 2$, $n = 33$, find $\sum x$ and $\sum x^2$

27) Given that $\bar{x} = 90.2$, $\sigma^2 = 10.96$, $n = 100$, find $\sum x$ and $\sum x^2$

1.48 Measures of Central Tendency and Variation (answers on page 462)

28) Ebba takes a contract with a company that makes A-Level Maths resources. Her contract says that the average number of hours working per week should be approximately 4, although they will pay her for each hour worked up to a maximum of 8 hours a week. Some projects take Ebba longer than this. Ebba records the number of hours, t, she works each week in the table below.

Hours Worked (t)	Frequency
$0 \leq t < 2$	12
$2 \leq t < 4$	17
$4 \leq t < 6$	21
$6 \leq t < 8$	7
$8 \leq t < 12$	2

Ebba is paid £15.60 an hour for her work. Estimate the mean gross pay she earns from the company each week.

29) Given that $\sum x = 40000$, $\sum x^2 = 80000$, $\sigma = 1$, find n

30) The following data has a mean of 30 and a sample standard deviation of 8.70 to 3 significant figures.

| 19 | 21 | 26 | 27 | 29 | 35 | 39 | 44 |

One piece of data is removed and replaced with a new value. The mean increases to 32 but the sample standard deviation decreases. Identify which of the pieces of data could have been removed and find the values they would need to be replaced with.

31) It is given that $\bar{x} = k$, $\sigma = 5$ and $n = 10$. Two more pieces of data are added, $x = k - 1$ and $k + 1$. Determine the standard deviation of the 12 pieces of data.

32) It is given that $\bar{x} = 200$, $\sigma = k$ and $n = 10$. Two more pieces of data are added, $x = 2 + k$ and $2 - k$ so that σ is now 99. Determine the value of k

33) On a rainy day, 23 students sit a test. Their teacher marks their papers and finds there to be a mean score of 30 and a standard deviation of 5. Due to the rain, Tom was late to school and missed the test, and so had to complete it later. Tom scored 35. Determine the mean and standard deviation of the scores of all 24 students.

34) On a sunny day, 25 students sit a test. Their teacher marks their papers and finds there to be a mean score of 40 and a standard deviation of 8. The teacher finds that one of his students has cheated and copied another's work. The student's score was 52 and the teacher wants to remove this score from the analysis. Determine the mean and standard deviation of the scores of the remaining 24 students.

Linear Coding

35) The data set, x, shown below is coded using
$$y = \frac{x + 5}{4}$$
Write down the data set y

| x | 15 | 85 | 155 | 295 | 455 |

36) The values x_i have a mean of $\bar{x} = 85$ and standard deviation $\sigma_x = 12$. The data is coded to the values y_i such that $y = 10x$
Find $\bar{y}$ and σ_y, the mean and standard deviation respectively of the values y_i

37) The values x_i have a mean of $\bar{x} = 120$ and standard deviation $\sigma_x = 5$. The data is coded to the values y_i such that $y = 3x + 50$
Find $\bar{y}$ and σ_y, the mean and standard deviation respectively of the values y_i

38) The values x_i have a mean of $\bar{x}$ and standard deviation σ_x. The data is coded to the values y_i such that $y = 8x - 10$. It is given that $\bar{y} = 500$ and $\sigma_y = 120$
Find $\bar{x}$ and standard deviation σ_x

1.48 Measures of Central Tendency and Variation (answers on page 462)

39) The values x_i have a mean of $\bar{x}$ and standard deviation σ_x. The data is coded to the values y_i such that $y = \frac{20-3x}{9}$. It is given that $\bar{y} = 250$ and $\sigma_y = 45$. Find $\bar{x}$ and standard deviation σ_x

40) The temperatures (°C) at midday for a week in May in Valletta, Malta are recorded in the table below.

| 24 | 23 | 25 | 27 | 29 | 26 | 24 |

a) Find the mean and standard deviation of the temperatures in °C
b) The formula to convert from °C to °F is given by $F = \frac{9C}{5} + 32$
Use this formula to determine the mean and standard deviation of the temperatures in °F

41) The temperatures (°F) at midday for a week in May in Southampton, UK are recorded in the table below.

| 61 | 63 | 59 | 59 | 64 | 63 | 61 |

a) Find the mean and standard deviation of the temperatures in °F
b) The formula to convert from °F to °C is given by $C = \frac{5(F-32)}{9}$
Use this formula to determine the mean and standard deviation of the temperatures in °C

The Mid-Range

42) For the following data, find the mid-range.

| 28 | 35 | 42 | 59 | 60 | 75 |

43) Colm teaches Biology to four classes. He set some homework for his students on an online platform and his dashboard recorded how often his students logged in during the week to work on it. This data is shown in the following table:

Log-ins	Frequency
0	12
1	42
2	27
3	10
4	8
5	3
6	1

a) State the mode of the data.
b) State the mid-range of the data.
c) State the median of the data.

Linear Interpolation

44) The following data was recorded:

x	Frequency
$0 \leq x < 5$	35
$5 \leq x < 10$	60
$10 \leq x < 15$	15
$15 \leq x \leq 20$	10

Use linear interpolation to calculate an estimate for the median of x

45) The following data was recorded:

x	Frequency
$0 \leq x < 8$	82
$8 \leq x < 12$	74
$12 \leq x < 20$	62
$20 \leq x \leq 24$	44

Use linear interpolation to calculate an estimate for the median of x

46) The following data was recorded:

x	Frequency
$10 \leq x < 30$	55
$30 \leq x < 70$	95
$70 \leq x < 80$	150
$80 \leq x < 110$	75
$110 \leq x \leq 140$	25

Use linear interpolation to calculate an estimate for the median of x

1.48 Measures of Central Tendency and Variation (answers on page 462)

47) The following data was recorded:

x	Frequency
$50 \leq x < 60$	12
$60 \leq x < 70$	16
$70 \leq x < 80$	28
$80 \leq x < 100$	32
$100 \leq x \leq 120$	14

Use linear interpolation to calculate an estimate for the interquartile range of x

48) The following data was recorded:

x	Frequency
$0 \leq x < 15$	48
$15 \leq x < 40$	93
$40 \leq x < 55$	117
$55 \leq x < 70$	152
$70 \leq x \leq 95$	146
$95 \leq x < 110$	121
$110 \leq x < 135$	94
$135 \leq x < 150$	29

a) Use linear interpolation to calculate an estimate for 20th to the 80th interpercentile range of x
b) Estimate the number of data values that fall within the 20th to the 80th interpercentile range.

49) Andrew measures the acidity of the soil in several randomly selected locations at a nursery. The following data was recorded:

x (pH)	Frequency
$6 \leq x < 6.2$	12
$6.2 \leq x < 6.4$	27
$6.4 \leq x < 6.6$	38
$6.6 \leq x < 6.8$	22
$6.8 \leq x < 7.0$	9

Use linear interpolation to calculate an estimate for the upper quartile of the pH levels recorded.

50) Freya measures the heights, in metres, of a random sample of oak trees in a large forest. The following data was recorded:

Height (m)	Frequency
13 to 15	7
16 to 18	15
19 to 21	23
22 to 24	9
25 to 26	2

Use linear interpolation to calculate an estimate for the lower quartile of the heights of the oak trees recorded.

51) Lisa runs a carp farm that has over 50 outdoor ponds dedicated to stocking fisheries in the UK. In one pond, there are a large number of young carp. Lisa collects a sample and measure their lengths in centimetres. The following data was recorded:

Length (cm)	Frequency
6 to 7	56
8 to 9	84
10 to 11	129
12 to 13	76
14 to 15	43

Use linear interpolation to calculate an estimate for the length 70% of the fish are longer than.

52) Gerald records the ages, in years, of the members of a local choir. The following data was recorded:

Age (years)	Frequency
25 to 30	9
31 to 40	32
41 to 50	19
51 to 65	26
66 to 80	12

Use linear interpolation to calculate an estimate for the median age of the choir.

1.49 Representing Data (answers on page 463)

1) Find the mean from the dotplot:

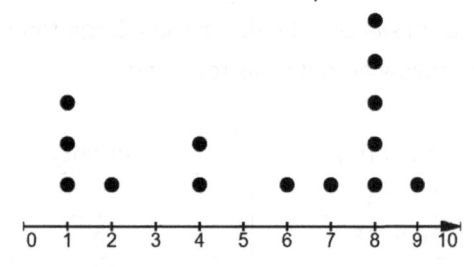

2) Use the terms: uniform, unimodal, bimodal, continuous, discrete, symmetric, positively skewed, and/or negatively skewed to describe:

a)

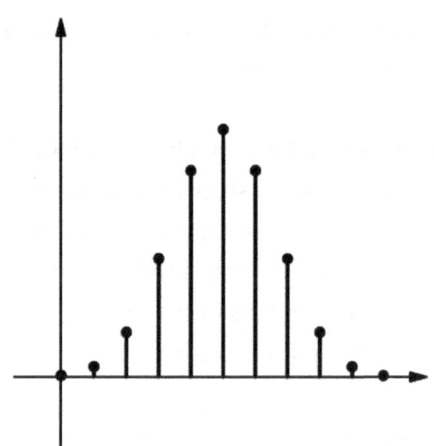

b)

c)

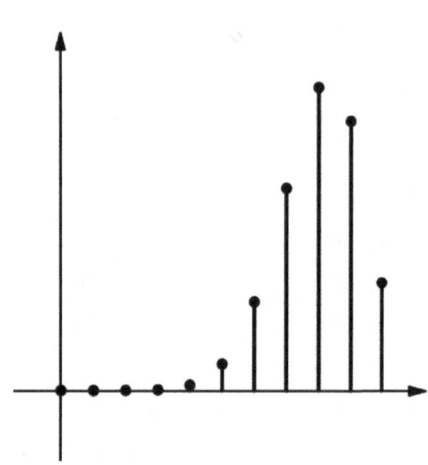

d)

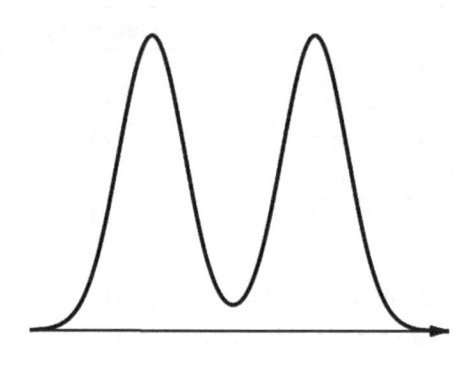

e)

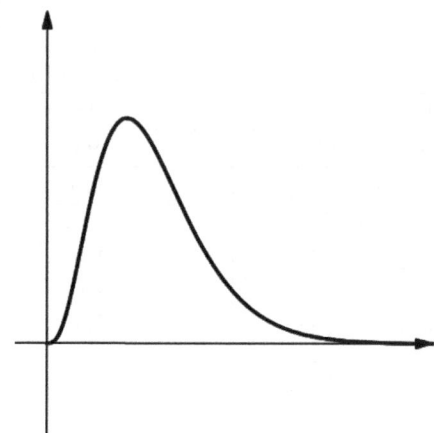

f)

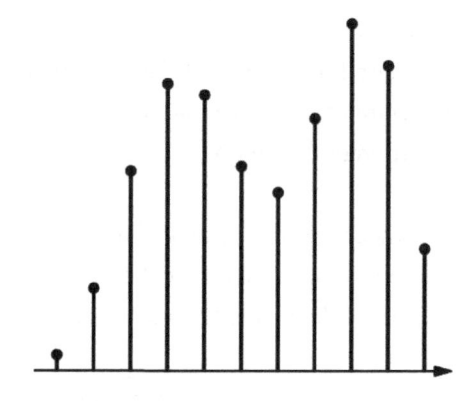

g)

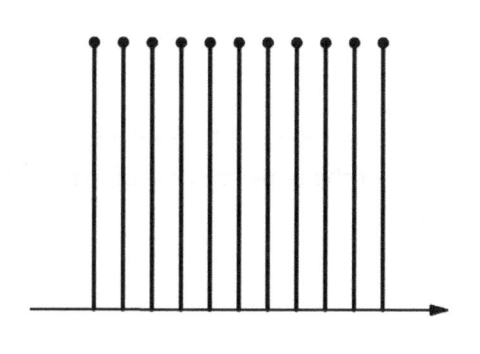

1.49 Representing Data (answers on page 463)

3) The stem-and-leaf diagram shows the weights of a sample of 29 male moose measured to the nearest kg.

38	1	2	5	6	6		
39	3	3	4	7	8	9	9	9
40	0	0	2	2	3	8	
41	0	1	2	3	9		
42	2	2	2				
43	0	1					

Key: 38|1 represents 381 kg

a) Describe the shape of the distribution.
b) Find the median.
c) Find the interquartile range.

4) The stem-and-leaf diagram shows the heights of a sample of 13 red kangaroos measured to the nearest cm.

7	2	3	3			
8	1	2	3	6	7	8
9	5	5				
10	0					
11						
12	9					

Key: 7|4 represents 74 cm

a) Find the median and interquartile range of the red kangaroo heights.
b) Calculate the mean and standard deviation of the red kangaroo heights.
c) Explain why the median would be a better measure of central tendency than the mean in this case.

5) The stem-and-leaf diagram shows the wingspans of a sample of 23 Arctic terns measured to the nearest cm.

31	0					
32	8	9				
33	0	1	2	4		
34	0	0	4	4	8	
35	3	3	7	8	8	9
36	0	0	0	1	1	

Key: 31|0 represents 31.0 cm

a) Describe the shape of the distribution.
b) Alan looks at the data in the stem-and-leaf diagram and determines that 36.0 cm is the average wingspan of an Arctic tern. Explain how Alan can justify this result.
c) Brian looks at the data in the stem-and-leaf diagram and determines that 34.8 cm is the average wingspan of an Arctic tern. Explain how Brian can justify this result.
d) Draw a boxplot to represent the data.

6) Draw a boxplot with negative skew.

7) For the following boxplots, identify whether they are roughly symmetric, positively skewed or negatively skewed:

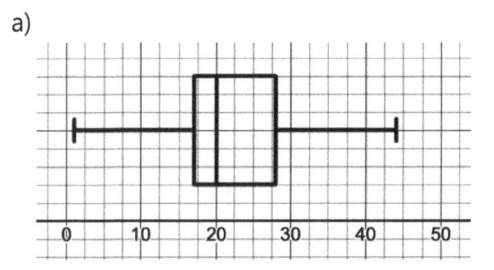

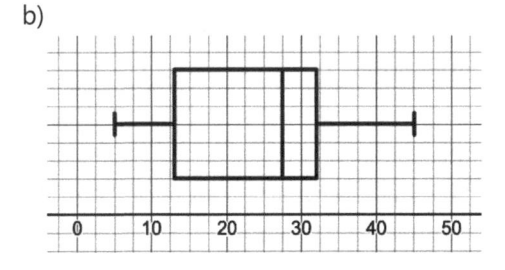

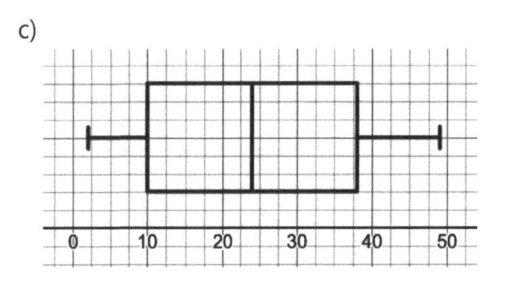

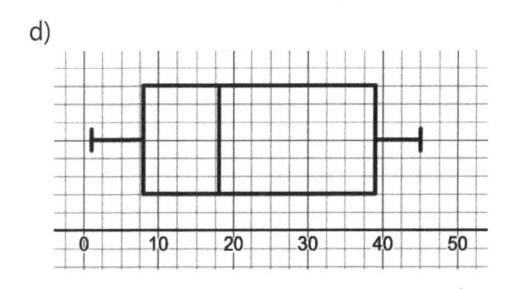

a)

b)

c)

d)

1.49 Representing Data (answers on page 463)

8) The boxplot below represents the commute times, in minutes, of employees at a large company.

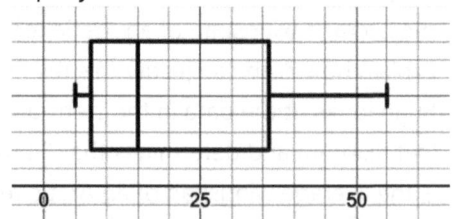

   a) State the name of the measure of central tendency identified by the box plot, and its value.
   b) Determine whether the box plot is symmetric, negatively skewed or positively skewed.

9) The boxplot below represents the number of dogs that attended puppy training each Thursday evening for 20 weeks.

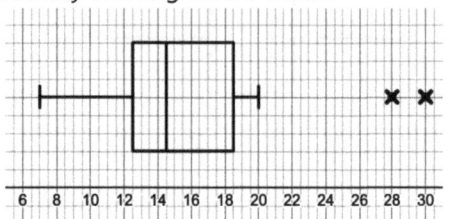

   a) Find the median number of dogs that attended.
   b) Find the range of the number of dogs that attended.
   c) Find the interquartile range of the number of dogs that attended.

10) Two classes sat a test where the maximum score was 100. The results are displayed in the two boxplots below.

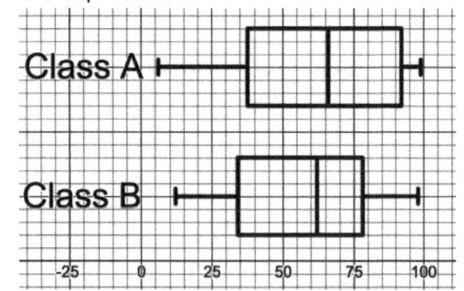

    a) Estimate the median of Class A and the median of Class B.
    b) Estimate the interquartile range of Class A and the interquartile range of Class B.
    c) Use your answers to compare the results of the two classes.

11) The histogram below shows the times, x minutes, that employees at a company take to walk to work. Estimate the proportion of employees that take between 9 minutes and 25 minutes to walk to work.

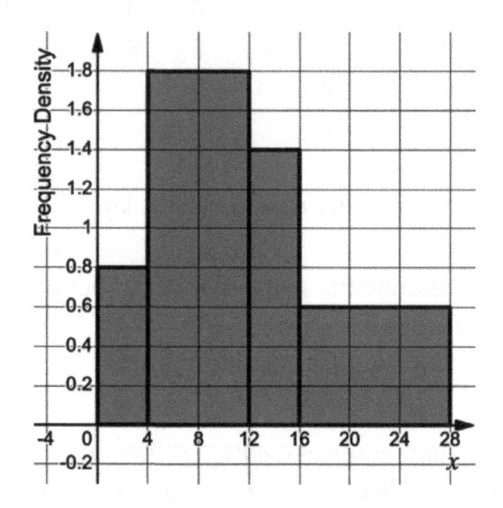

12) Aminah is investigating the length of time, t minutes, meetings take at her place of work. She is partway through drawing a histogram from the following table of data:

Meeting Length (t)	Frequency
$0 \leq t < 30$	30
$30 \leq t < 60$	45
$60 \leq t < 120$	20
$120 \leq t < 150$	15
$150 \leq t < 180$	

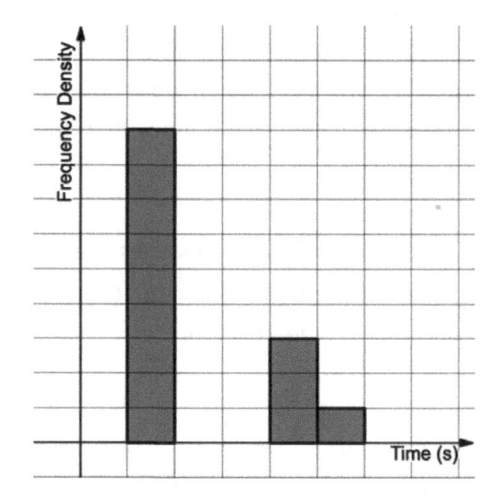

   a) Copy and complete the table and histogram.
   b) Estimate the number of meetings that last between 40 and 99 minutes.

1.49 Representing Data (answers on page 463)

13) The histogram below shows the times, in seconds, for a sample of adults to complete a puzzle.

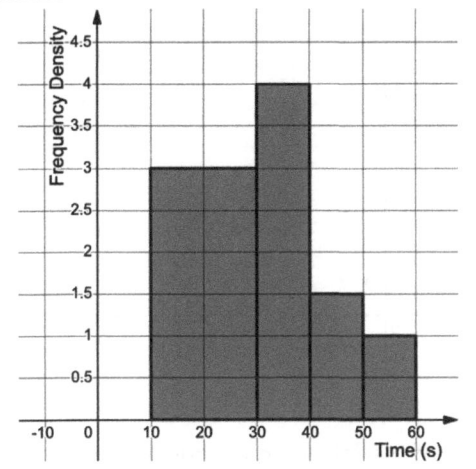

   a) Which class interval contains the median?
   b) Estimate the mean.
   c) Estimate the standard deviation to 3sf.
   d) An adult is selected at random. Find the probability that they took between 28 and 32 seconds to complete the puzzle.

14) Umar took a random sample of 24 apples and they weighed between 160 and 180 grams. He drew a histogram to represent the data and the bar representing 160 to 180 grams has a width of 2 cm and a height of 6 cm. There are 16 apples that weigh between 180 and 210 grams. Determine the width and height of the bar on the histogram representing 180 to 210 grams.

15) Use the following cumulative frequency graph to estimate the median and interquartile range.

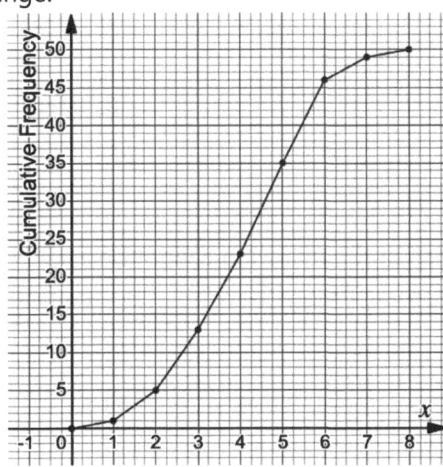

16) Amber sells ice-creams. One year, she counts the number of customers she sells ice creams to over ten weeks of Sundays in the summer. Her sales are shown in the table below:

Week	0 −	1 −	2 −	3 −	4 −
Sales	40	56	82	105	129
Week	5 −	6 −	7 −	8 −	9 −
Sales	131	95	87	51	29

   a) Amber draws two different cumulative frequency graphs. These are shown below. Which of the two should she use to represent her data? Explain your reasoning.

   Graph A:

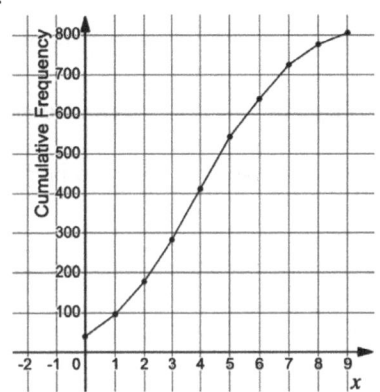

   Graph B:

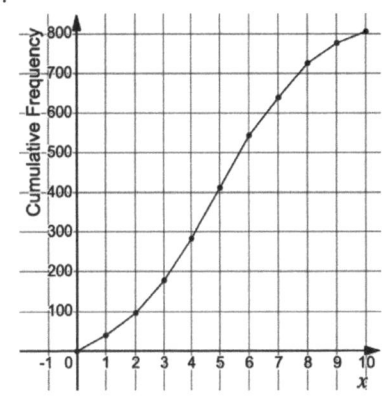

   b) Amber shows the correct graph to Terry who says that it shows that week on week her sales are increasing. With reference to the graph, explain whether Terry is correct.
   c) Terry goes on to say that Amber can use the correct graph to estimate the exact time during a week when Amber had sold half of the ice creams. Explain why this would be inappropriate.

1.50 Bivariate Data: PMCC & Linear Regression (answers on page 464)

1) The diagram below shows bivariate data.

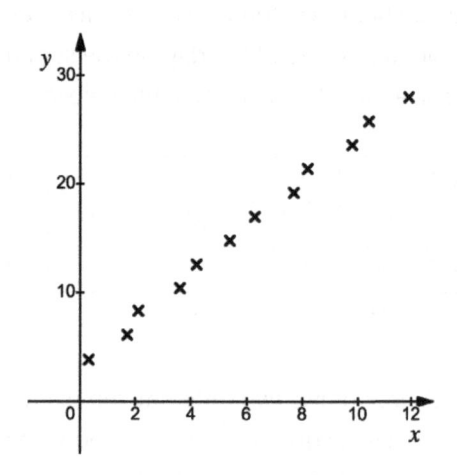

Which of the following best describes the correlation shown in the diagram?

Strong Positive   Moderate Positive

Weak Positive   Weak Negative

Moderate Negative   Strong Negative

2) The diagram below shows bivariate data.

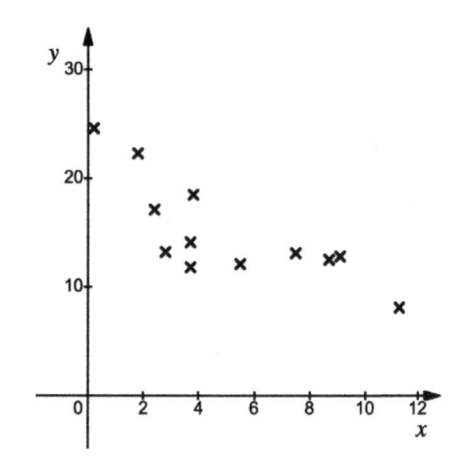

Which of the following best describes the correlation shown in the diagram?

Strong Positive   Moderate Positive

Weak Positive   Weak Negative

Moderate Negative   Strong Negative

3) The diagram below shows bivariate data.

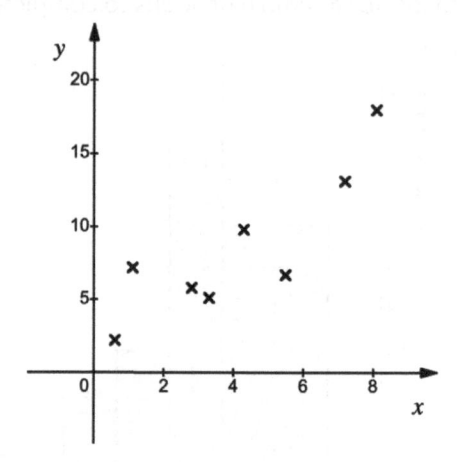

One of the following product moment correlation coefficient is correct for the above data. Which is it?

0.864   0.164

-0.164 -0.864

4) The diagram below shows bivariate data.

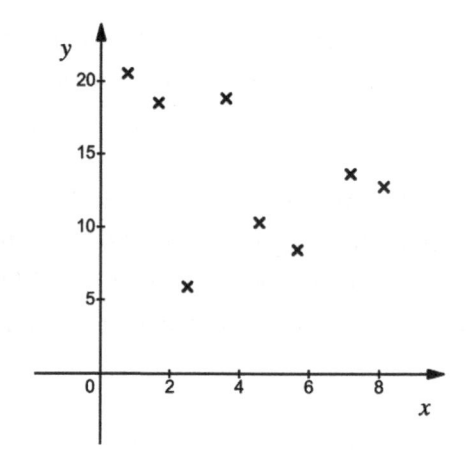

One of the following product moment correlation coefficient is correct for the above data. Which is it?

0.891   0.391

-0.391 -0.891

5) Which of the following cannot be a value for the product moment correlation coefficient?

0 $\dfrac{3}{4}$ $-\dfrac{7}{3}$ -1 $\dfrac{8}{7}$

1.50 Bivariate Data: PMCC & Linear Regression (answers on page 464)

6) A researcher conducts a study to understand the impact of physical exercise on mental health among college students. The researcher collects data on the number of hours students exercise per week and their levels of stress, measured on a standardised scale. Identify which of these is the explanatory variable and which is the response variable.

   Age of students   Hours of exercise per week

   Diet quality   Levels of stress

7) Asher collects data on a person's age and their reaction time as measured in a standardised test. He calculates the product moment correlation coefficient to be 0.79. Which of the following best describes the pmcc found?

   Definitely Correct   Probably Correct

   Probably Incorrect   Definitely Incorrect

8) Jane wants to determine if there is any correlation between a person's salary and how much they spend at a large supermarket. Jane collects a random sample of adults and calculates the product moment correlation coefficient to be 1.13. Which of the following best describes the pmcc found?

   Definitely Correct   Probably Correct

   Probably Incorrect   Definitely Incorrect

9) Franklin has two data points: $(13, 7)$ and $(20, 11)$. He calculates the product moment correlation coefficient between x and y to be 1. Which of the following best describes the pmcc found?

   Definitely Correct   Probably Correct

   Probably Incorrect   Definitely Incorrect

10) Ayaan works in an office in a busy city centre. He collects a random sample of his colleagues, and his results are shown in the table below.

Distance from workplace (miles)	Commute Time (minutes)
1.2	14
2.5	21
4.1	24
5.8	31
7.3	36
9.0	39
10.7	44
12.4	51

    Use your calculator to calculate the product moment correlation coefficient between the distances from the workplace and the commute times.

11) Grace asks a random sample of her the students in her year group how many hours they studied in a week (to the nearest hour) in preparation for a test, and their test score out of 100. The results are shown in the table below.

Hours studied in a week	Test Score (/ 100)
3	65
6	75
9	77
4	62
8	80
10	85
3	57
7	78

    a) Use your calculator to calculate the product moment correlation coefficient between the hours studied in a week by the students and their test score.

    b) Describe what the PMCC shows in this case, in context.

1.50 Bivariate Data: PMCC & Linear Regression (answers on page 464)

12) A scientist is investigating the effects of smoking on the respiratory system in youths. A random sample of 15-18 year olds found a PMCC of −0.76 between the frequency of smoking and their respiratory function. Describe what this shows in context.

13) Angela is investigating the average price of a house compared with the distance the house is away from London. She decides to consider areas in a South-Westerly direction from the centre of London. Her results are shown in the table below.

Location	Dist. from London (miles)	Av. House Price (£ thousands)
Richmond	8.1	1100
Twickenham	10.5	753
Hampton	12.1	616
Chertsey	18.3	502
Bagshot	26.2	448
Farnborough	30.7	372
Winchester	59.8	582
Tidworth	68.7	269

Angela plotted the data in a scatter graph, as shown below.

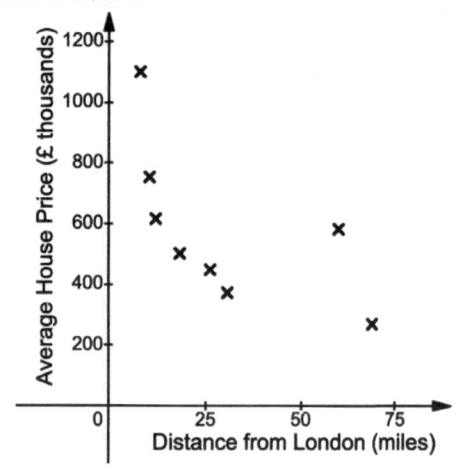

a) Identify a possible outlier and give a possible reason for it being an outlier.
b) If the outlier is discarded, determine if the scatter diagram shows any correlation for the remaining points.
c) Interpret the scatter diagram in context.

14) Charlene organises a running club. In week 10, she plots the data of her running club members with the amount of exercise they do the previous week in hours along the horizontal axis, and their endurance, measured by how far they can run on the day they meet in kilometres along the vertical axis, as shown in the graph below.

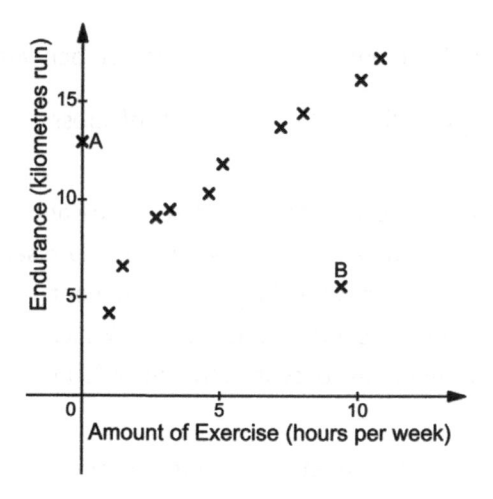

a) Explain why data point A may **not** have been inputted correctly.
b) Explain why data point B may have been inputted correctly.

15) A teacher plots the scatter diagram below showing the arithmetic and spelling scores from a recent test sat by their students.

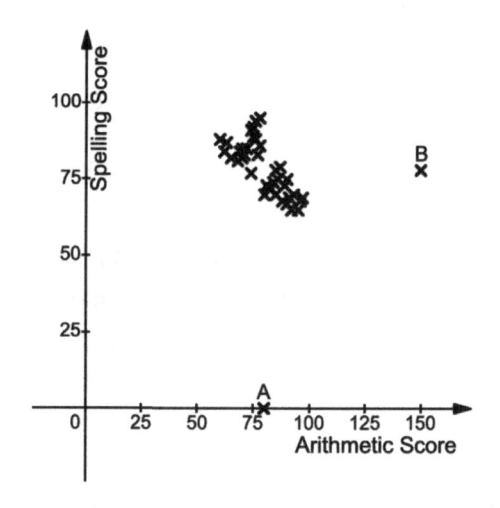

a) Explain why data point A may have been inputted correctly.
b) Explain why data point B may **not** have been inputted correctly.

1.50 Bivariate Data: PMCC & Linear Regression (answers on page 464)

16) Tristan is doing some research into how sleep cycles affect productivity.
In his trial, he decides to randomly sample twelve adults and when they wake up they must complete a task.
The results of his experiment are shown in the diagram below.

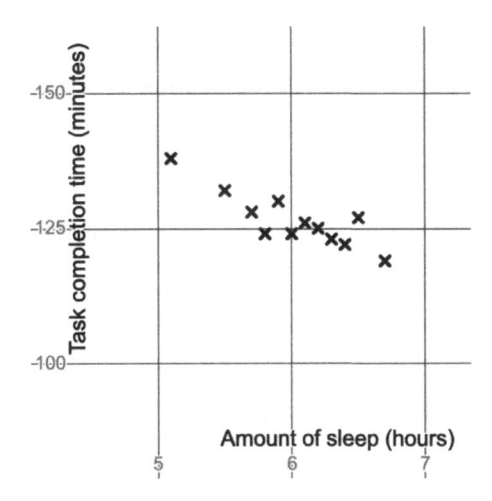

Tristan claims that the scatter diagram proves that increasing the number of hours of sleep reduces the amount of time it takes to complete the task.
Comment on the validity of this claim.

17) The diagram below shows bivariate data.

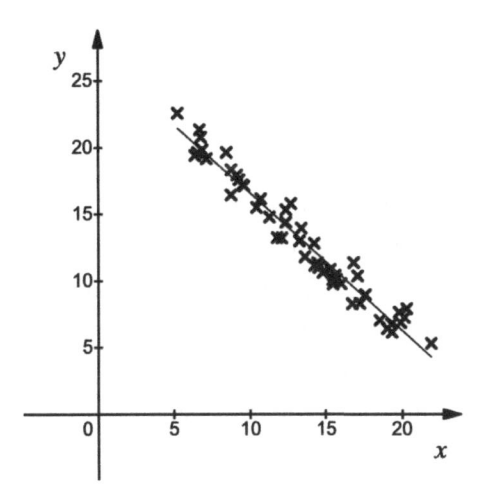

The product moment correlation coefficient for the data is -0.979. Explain why it would be reasonable to model the relationship between x and y as a linear graph.

18) Maria wants to compare students' scores from a practice paper, where the maximum score was 100, against their total final exam score, where the maximum score was 300. She takes a random sample of students and the data is shown in the table below.

Score from Practice Paper (/ 100)	Total Final Exam Score (/ 300)
72	217
55	168
83	267
46	121
79	202
63	199
88	243
91	265

Use your calculator to calculate the equation of the regression line, where x is the score on the practice paper and y is the total final exam score.

19) Chris is investigating how the amount of debt an adult has affects their credit score. He takes a random sample of adults with debts and collects data on the amount of debt they have, in thousands of pounds, and their credit scores. The data is shown in the table below.

Amount of Debt (in thousands of dollars)	Credit Score
12	723
18	716
21	663
27	668
37	672
38	650
41	565
49	575

Use your calculator to calculate the equation of the regression line, where x is the amount of debt and y is the credit score.

1.50 Bivariate Data: PMCC & Linear Regression (answers on page 464)

20) The regression line between a company's monthly advertising expenditure (£ a) and its monthly sales revenue (£ s) is given by $s = 1.96a + 2920$. Describe in context what the 1.96 represents.

21) A company makes cans of soup that are sold at different price points across the country. They want to investigate the relationship between how much the soup costs and how much they sell. The regression line between the price of a can of soup (p pence) and the number of cans sold in a week (N) is given by $N = 502 - 2.22p$. Describe in context what the -2.22 represents.

22) Osian is investigating how sleep affects sprinters' performance. He takes a random sample of runners and asks them how much sleep they had the night before, in hours, and then records their times running 100 metres. The regression line between the amount of sleep (s hours) and their running times (t seconds) is given by $t = 12.3 - 0.15s$. Describe a limitation with this model.

23) Throughout the year, Peter runs a drop-in lunchtime workshop for his students that works on exam questions and exam technique. He maintains a register throughout the year and records the total number of hours each student attended, and their scores in the final exam (which is out of a maximum score of 300). The regression line between the number of hours they attended the workshop (t) and their final exam score (q) is given by $q = 98 + 1.12t$.
   a) State the units of the value 1.12 in the regression line equation.
   b) With reference to the regression line, describe what effect, on average, attending 2.5 hours of the lunchtime workshop throughout the year would have on a student's final exam score.

   The one-hour lunchtime workshop runs Monday to Friday every week throughout the academic year for 38 weeks.
   c) Peter wants to use the model to estimate the value of q when $t = 200$. Explain why this would not be appropriate.

24) Amelia is investigating the growth of a fungus. She measures the width of the fungus (P mm) every twelve hours (t). She models the growth of the fungus using the regression line given by the equation $P = 3.1 + 0.67t$.
   a) With reference to the regression line, describe what effect, on average, three days will have on the width of the fungus.
   b) Assuming that the model continues to be accurate, use the regression line to determine an estimate for how many days it would take the fungus to reach a width of 5 cm.

25) A random sample of farms report the amount of Nitrogen Fertiliser (kilogram per hectare) they used and their Crop Yield (tonnes per hectare). The data is shown in the diagram below.

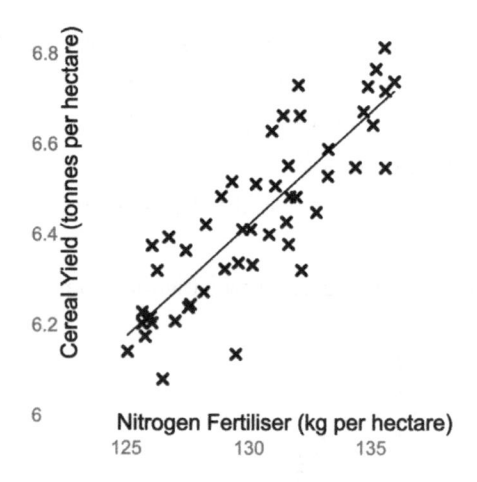

   The equation of the regression line is $y = 0.0497x - 0.0277$, where y is the cereal yield and x is the amount of nitrogen fertiliser.
   a) Farm A reports that it uses 132 kg per hectare of nitrogen fertiliser. Use the equation of the regression line to estimate Farm A's cereal yield.
   b) Farm B reports that it uses 121 kg per hectare of nitrogen fertiliser. Use the equation of the regression line to estimate Farm B's cereal yield.
   c) Explain why the estimate for Farm A's cereal yield is more reliable than the estimate for Farm B's cereal yield.

1.51 Bivariate Data: Association & Spearman's Rank (answers on page 465)

1) The diagram below shows bivariate data.

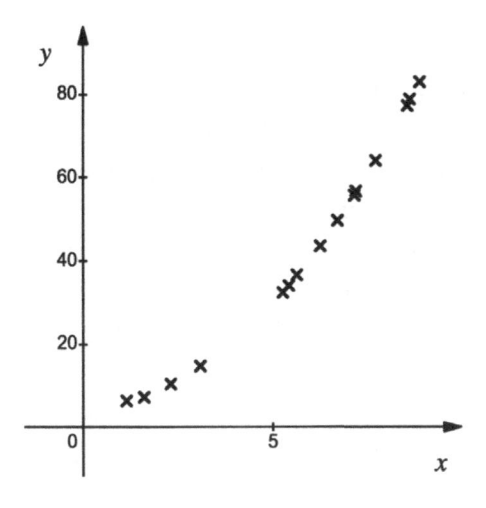

   Which of the following best describes the data shown in the diagram?

   Positive correlation   Negative correlation

   Positive nonlinear association   Negative nonlinear association

2) The diagram below shows bivariate data.

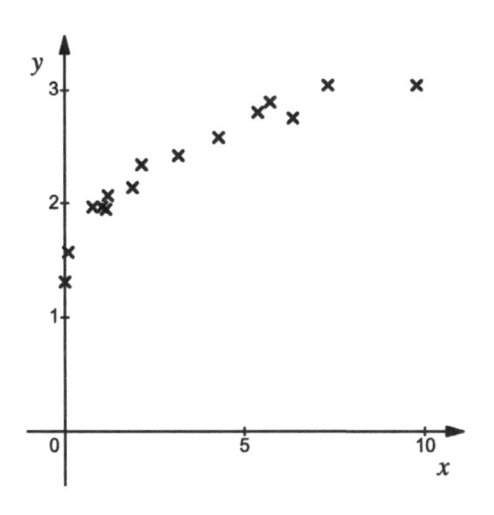

   Which of the following best describes the data shown in the diagram?

   Positive correlation   Negative correlation

   Positive nonlinear association   Negative nonlinear association

3) Barnaby collects the bivariate data shown in the table below.

x	0.8	1.2	2.6	2.9	1.9	1.6	2.5	2.1
y	2.5	1.9	5.8	8.4	4.3	0.9	5.9	4.0

   He ranks the data from smallest to largest and correctly finds the differences, d, and the squares of the differences, d^2, as shown in the table below.

Rank x	1	2	7	8	4	3	6	5
Rank y	3	2	6	8	5	1	7	4
d	-2	0	1	0	-1	2	-1	1
d^2	4	0	1	0	1	4	1	1

   a) Given that $\sum d_i^2$ is the sum of the squares of the differences, and n is the number of pairs of bivariate data in the original table, calculate Spearman's Rank correlation coefficient using the formula
   $$r_s = 1 - \frac{6 \sum d_i^2}{n(n^2 - 1)}$$
   b) Which of the following best describes the association found in a)?

   Strong Positive   Moderate Positive

   Weak Positive   Weak Negative

   Moderate Negative   Strong Negative

4) Calculate Spearman's Rank correlation coefficient for the following bivariate data:

x	4.3	3.8	8.6	8.0	1.1	5.5	4.8	6.5
y	7.9	8.1	7.3	7.2	9.3	7.7	7.8	7.6

5) Uzair wants to investigate whether there is an association between the number of hours students study, x, and their scores on a maths test, y. The data collected from a random sample of 8 students is shown in the table below.

Student	A	B	C	D	E	F	G	H
x	2	3	4	5	1	6	3	2
y	49	58	62	60	59	71	46	66

   Calculate Spearman's Rank correlation coefficient for the data.

1.52 Probability: Introduction (answers on page 465)

1) Yusuf flips two fair coins simultaneously.
   a) Draw a sample space diagram to show all of the possible outcomes.
   b) Find the probability that there is at least one tails.

2) Sylvia rolls two fair dice, both of which have four faces numbered 1 to 4. The product of the two scores is to be investigated.
   a) Draw a sample space diagram to show all of the possible outcomes.
   b) Find the probability that Sylvia scores a product greater than 10

3) Joanne has two spinners. The first spinner has four equal sections numbered with the first four prime numbers, and the second spinner has three equal sections numbered with the first three square numbers. Joanne spins both spinners simultaneously and adds the scores together.
   a) Draw a sample space diagram to show all of the possible outcomes.
   b) Find the probability that Joan's score is an even number.

4) Matt has two packs of cards. Each pack contains three cards, one labelled A, one labelled B and one labelled C. Matt randomly selects a card from the first pack, and then randomly selects a card from the second pack.
   a) Draw a sample space diagram to show all of the possible outcomes.
   b) Find the probability that Matt selects two cards that are both consonants.

5) Kenneth has a covered box that contains 3 red balls, 2 green balls and 1 blue ball. He draws two balls from the box, one after the other, without replacement.
   a) Write down a list of all the possible outcomes.
   b) Find the probability that Kenneth does not draw the blue ball.

6) Halima has two bags of marbles. The first bag has 3 red marbles and 3 blue marbles. The second bag has 4 red marbles and 4 blue marbles. She draws a marble from each of the bags without replacement.
   a) Write down a list all the possible outcomes.
   b) Find the probability that Janie draws two red marbles.
   c) Find the probability that the two marbles are different colours.
   d) Find the probability that one bag now has more red marbles than blue marbles, and the other bag now has more blue marbles than red marbles.

7) A lottery is played in a town hall. There are two buckets, both of which contain balls numbered 0, 1, 2, 3, 4 and 5. A ball is randomly selected from the first bucket and this becomes the first digit of the winning lottery number. A ball is then randomly selected from the second bucket and this becomes the second digit of the winning lottery number.
   a) Draw a sample space diagram to show all of the possible outcomes.
   b) Find the probability that the winning lottery number is a nonzero multiple of 11
   c) Find the probability that the winning lottery number is a nonzero multiple of 4

8) Victor has a bag of marbles. The bag contains 2 red marbles, 2 blue marbles, and 1 green marble. Victor draws three marbles from the bag without replacement.
   a) Write down a list of all the possible outcomes.
   b) Find the probability that Victor draws two red marbles.
   c) Find the probability that Victor draws marbles of all three colours.
   d) Find the probability that the bag now only contains marbles of one colour.

1.52 Probability: Introduction (answers on page 465)

9) Events A and B are shown in the Venn diagram below.

   [Venn diagram: A only = 0.25, $A \cap B$ = 0.3, B only = 0.35, outside = 0.1]

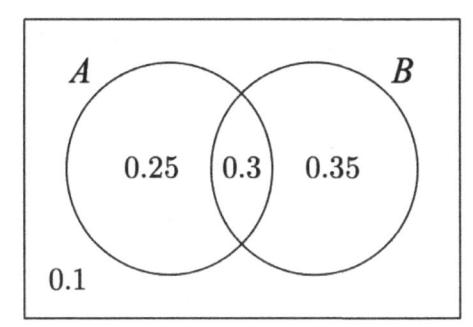

   Find
   a) $P(A)$
   b) $P(A \cap B)$
   c) $P(B')$
   d) $P(A \cup B)$
   e) $P(A \cap B')$
   f) $P(A \cup B')$

10) Events A and B are shown in the Venn diagram below.

    [Venn diagram: A only = 0.03, $A \cap B$ = 0.6, B only = 0.23, outside = 0.14]

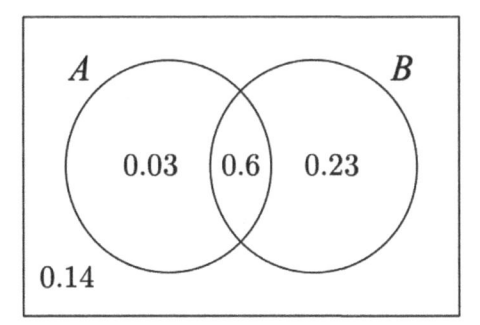

    Find
    a) $P(B)$
    b) $P(A' \cap B)$
    c) $P(A' \cup B)$
    d) $P((A \cup B)')$
    e) $P((A \cap B)')$
    f) $P(A' \cup B')$

11) Draw a Venn diagram for events A and B, given that $P(A \cap B) = 0.2$, $P(A) = 0.3$ and $P(B) = 0.4$

12) Draw a Venn diagram for events C and D, given that $P(C \cap D) = 0.7$, $P(C) = 0.75$ and $P(D) = 0.85$

13) Events A and B are shown in the Venn diagram below.
    a) Find the value of k
    b) Find $P(A)$

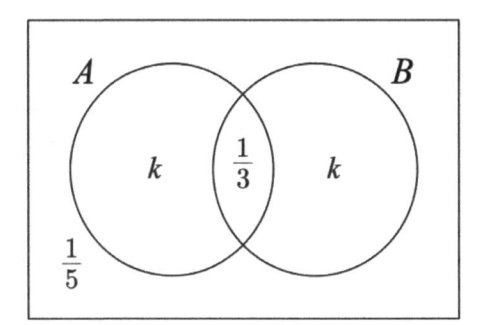

    [Venn diagram: A only = k, $A \cap B$ = $\frac{1}{3}$, B only = k, outside = $\frac{1}{5}$]

14) Events A and B are shown in the Venn diagram below. It is given that $P(A) = 2P(B)$. Find the value of p and the value of q

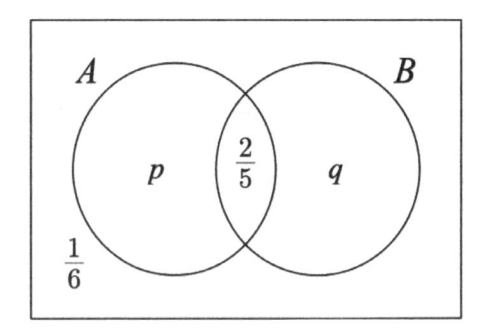

    [Venn diagram: A only = p, $A \cap B$ = $\frac{2}{5}$, B only = q, outside = $\frac{1}{6}$]

15) Events A and B are shown in the Venn diagram below. It is given that $2P(A) = P(B)$. Find the value of r and the value of s

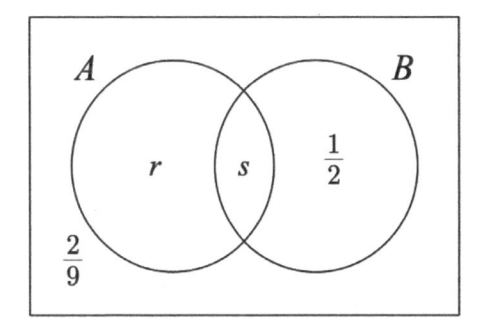

    [Venn diagram: A only = r, $A \cap B$ = s, B only = $\frac{1}{2}$, outside = $\frac{2}{9}$]

16) Events A and B are such that $P(A \cap B) = \frac{1}{10}$, $P(A) = v$ and $P(B) = w$. It is also given that $P(A \cup B) = 4P(A)$ and that $P((A \cup B)') = \frac{1}{5}$.
    a) Find the value of v and the value of w.
    b) Draw a fully labelled Venn diagram showing the events A and B.

1.52 Probability: Introduction (answers on page 465)

17) 150 adults were surveyed and they were asked which, if either, of the following takeaways they have eaten in the last month: Fish and Chips (F), and Pizza (P). The results are recorded in the table below:

F	P	F ∩ P
67	108	52

Copy and complete the following Venn diagram to display this information.

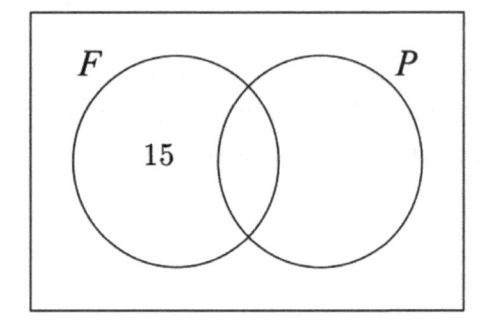

18) Events A, B and C are shown in the Venn diagram below. Find the value of k

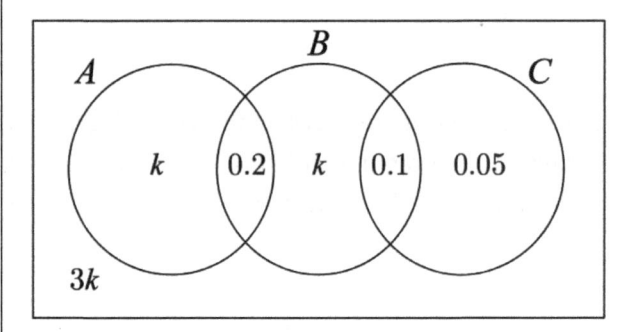

19) Events A, B and C are shown in the Venn diagram below. Find the value of q

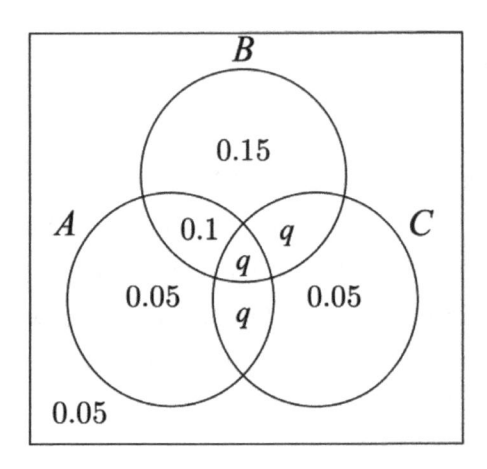

20) 580 adults were surveyed and they were asked which, if any, of the following non-religious holidays they celebrated: April Fools' Day (A), Valentine's Day (V), and Earth Day (E). The results are recorded in the table below:

A	V	E	A ∩ V	A ∩ E	V ∩ E	A ∩ V ∩ E
80	410	30	63	5	23	4

Copy and complete the following Venn diagram to display this information.

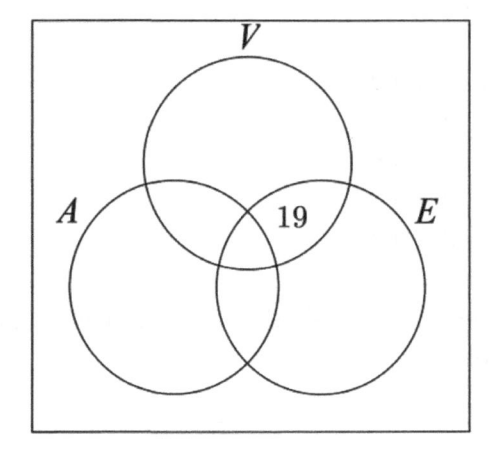

21) Events A, B and C are shown in the Venn diagram below.

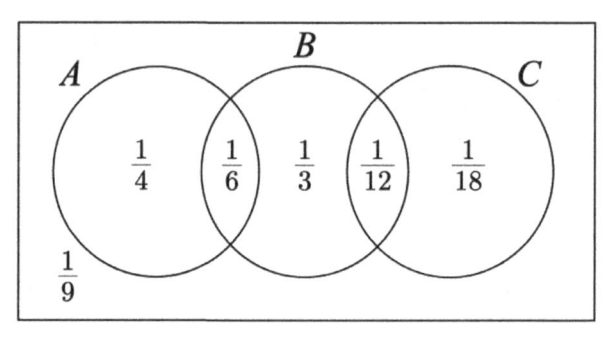

Find
a) $P(B)$
b) $P(A \cup B)$
c) $P(B \cup C)$
d) $P(A \cap B)$
e) $P(A \cap C)$
f) $P(A \cup B \cup C)$
g) $P(C')$
h) $P(A \cap C')$
i) $P(B \cap C')$
j) $P((A \cup B) \cap C)$

1.52 Probability: Introduction (answers on page 465)

22) Events A, B and C are shown in the Venn diagram below.

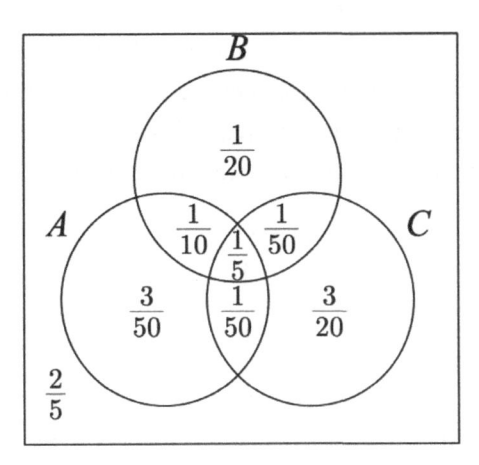

Venn diagram values: B only: $\frac{1}{20}$; $A \cap B$ only: $\frac{1}{10}$; $B \cap C$ only: $\frac{1}{50}$; $A \cap B \cap C$: $\frac{1}{5}$; A only: $\frac{3}{50}$; $A \cap C$ only: $\frac{1}{50}$; C only: $\frac{3}{20}$; outside: $\frac{2}{5}$.

Find
a) $P(A)$
b) $P(A \cap B)$
c) $P(B \cup C)$
d) $P(A' \cap C)$
e) $P(A \cap B \cap C)$
f) $P(A \cup B \cup C)$
g) $P(A' \cup B')$
h) $P((A \cup B) \cap C)$
i) $P((A \cap B) \cup C)$
j) $P(A' \cup B' \cup C')$

23) In a sixth form college, there are 252 students in Year 12. 85 study both Physics and Chemistry, while 35 do not study either Physics or Chemistry. There are twice as many students who study Physics but not Chemistry, as there are who study Chemistry and not Physics. A student is randomly selected from Year 12. Find the probability that they study Physics.

24) A company employs 360 people. There are certain areas of the company that require a Level 1 or Level 2 passkey. Some employees have both passkeys, some have just one, and others have neither. 105 employees just have a Level 1 passkey and 145 just have a Level 2 passkey. Four times as many employees do not have either passkey as have both. Two members of staff are picked at random. Find the probability that they both have a Level 2 passkey.

25) A company has 150 employees. One day, they have a 'Health and Wellbeing Day' where each employee much choose at least one activity to take part in. There are three activities to choose from: Yoga, Pottery, and a Hike.
 77 choose Pottery,
 59 choose Yoga,
 68 choose a Hike,
 22 choose Pottery and Yoga,
 19 choose Yoga and a Hike,
 21 choose Pottery and a Hike.
An employee is chosen at random. Find the probability that the employee chose all three activities.

26) At a Primary school, there are three after school clubs that students can optionally attend: Art Club (A) on a Monday, Book Club (B) on a Wednesday, and Creative Writing (C) on a Thursday. There are 120 students at the school, and the Venn diagram below shows the number of students who regularly attend each after school club.

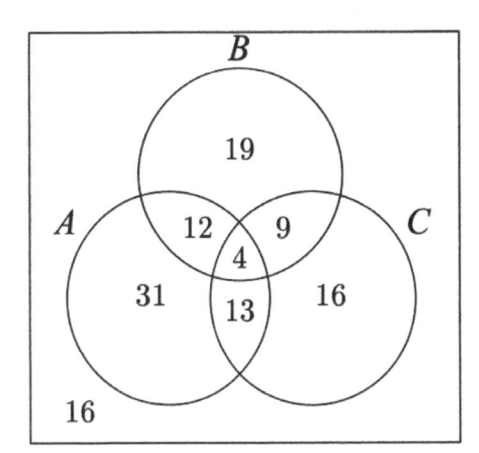

Venn diagram values: B only: 19; $A \cap B$ only: 12; $B \cap C$ only: 9; $A \cap B \cap C$: 4; A only: 31; $A \cap C$ only: 13; C only: 16; outside: 16.

Two students are randomly selected.
a) Find the probability that they both attend Art Club and Book Club but not Creative Writing.
b) Find the probability that the first attends Creative Writing and the second attends all three.
c) Find the probability that the two students both attend the same two clubs.

1.52 Probability: Introduction (answers on page 465)

27) In a survey that can be assumed to be a random sample of people living in a large town, 150 adults were asked whether they owned a dog and whether they were under 40 years old. Let A be the event that they owned a dog, and let B be the event that they were under 40 years old. The results are shown in the two-way table below.

	A	A'	Total
B	20	35	55
B'	40	55	95
Total	60	90	150

Find:
a) $P(A)$
b) $P(A \cap B)$
c) $P(B')$
d) $P(A \cup B)$
e) $P(A \cap B')$
f) $P(A \cup B')$

28) In a survey that can be assumed to be a random sample of people living in a large town, 300 adults were asked whether they were vegetarian and whether they were under 40 years old. Let A be the event that they were vegetarian and let B be the event that they were under 40 years old. The results are shown in the two-way table below.

	A	A'	Total
B	20	155	175
B'	15	110	125
Total	35	265	300

Find:
a) $P(B)$
b) $P(A' \cap B)$
c) $P(A' \cup B)$
d) $P((A \cup B)')$
e) $P((A \cap B)')$
f) $P(A' \cup B')$

29) The employees of a large company were surveyed about how they get to and from work. Let A be the event that they travel to and from work by bicycle and let B be the event that they had worked at the company for more than 5 years. The probabilities of these events are shown in the two-way table below.

	A	A'
B	$\frac{2}{19}$	$\frac{7}{19}$
B'	$\frac{1}{38}$	$\frac{1}{2}$

a) Write down fully what is meant by the following events:
   i) $A \cap B'$
   ii) $A' \cap B$
   iii) $A' \cup B'$
b) Find:
   i) $P(A)$
   ii) $P(B)$
   iii) $P(A \cap B)$
   iv) $P(A \cup B)$
   v) $P(A' \cap B')$
   vi) $P(A' \cup B')$

30) A doughnut stall offers doughnuts that are dipped in white chocolate, in milk chocolate, both or neither. Let A be the event that the customer orders white chocolate and let B be the event that the customer orders milk chocolate. The percentage of customers ordering each of these are shown in the two-way table below.

	A	A'
B	23%	36%
B'	14%	27%

Find
a) $P((A' \cup B)')$
b) $P((A \cap B')')$

1.52 Probability: Introduction (answers on page 465)

31) The two-way table below shows the events A and B. Copy and complete the table, given that $P(A) = 3P(B)$.

	A	A'
B	0.19	
B'	0.47	

32) Students at a university must choose one of three science courses: Biology, Chemistry or Physics, and they can opt to complete an additional Maths course if they choose. Let A be the event they choose Biology, let B be the event they choose Chemistry, let C be the event they choose Physics, and let D be the event they choose to study the additional Maths course. The two-way table below shows the percentages that choose each option.

	A	B	C
D	0.1	0.25	0.3
D'	0.2	0.1	0.05

Find
a) $P(C \cap D')$
b) $P(B \cup D)$
c) $P(A \cup B)$
d) $P((A \cup C) \cap D')$
e) $P((A \cap D) \cup B)$

33) A supermarket buys onions from two farmers to sell on their shelves. 40% of the onions come from farm A and the rest come from farm B. A certain percentage from each of the farms is spoilt in the transportation process. Overall, 2.1% of the onions are spoilt. The table below collects this information together.

	A	B
supplied	40%	
spoilt	1.5%	

Determine the percentage of onions bought from farm B that is spoilt.

34) A school trip is offered to three classes of students. 25% of the students on the trip come from class A and 45% come from class B. Of the students from class A who are on the trip, 85% are right-handed, and of the students from class B who are on the trip, 70% are right-handed. It is given that 81.25% of the students overall on the trip are right-handed. The table below collects this information together.

	Class A	Class B	Class C
Students on the trip	25%	45%	
Right-handed	85%	70%	

Determine the percentage of students from class C who are on the trip who are right-handed.

35) The two-way table below shows the events A and B and their respective probabilities. It is given that $3P(A) = 5P(B)$. Find the value of p and the value of q

	A	A'
B	p	$3p$
B'	q	$5q$

36) The two-way table below shows the events A, B, C and D and their respective probabilities. It is given that the probability of event B occurring is 26%, and the probability of event A **not** occurring is 44%. Find the value of r, the value of s and the value of t

	A	A'
B	$2r$	t
C	$3s$	$3r$
D	$2t$	s

1.52 Probability: Introduction (answers on page 465)

37) The probability tree below shows the events A and B and their respective probabilities.

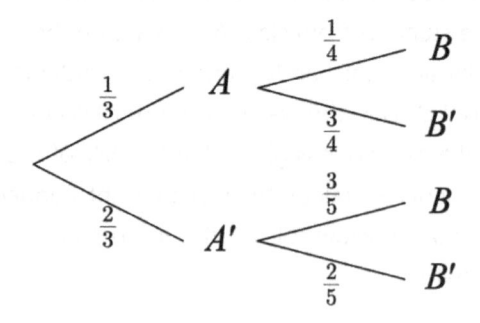

Find
a) $P(A)$
b) $P(A')$
c) $P(B)$
d) $P(B')$

38) The probability tree below shows the events A, B and C and their respective probabilities.

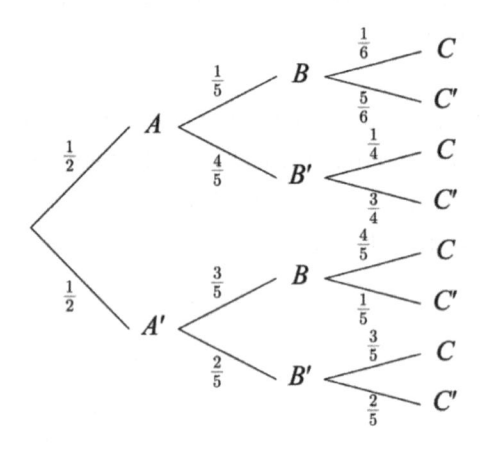

Find
a) $P(A$ then B then $C)$
b) $P(A'$ then B then $C')$
c) $P(C)$

39) In a pack of ten playing cards, 8 are blue and 2 are red. One card is randomly chosen from the pack and then replaced. Another card is then randomly chosen from the pack and then replaced.
a) Draw a probability tree to display this information.
b) Find the probability that the two cards that are chosen are the same colour.

40) In a pack of twelve playing cards, 8 are blue and 2 are red. One card is randomly chosen from the pack and is not replaced. Another card is then randomly chosen from the pack and not replaced.
a) Draw a probability tree to display this information.
b) Find the probability that the two cards that are chosen are the same colour.

41) The probability tree below shows the events A, B, C, D and E and their respective probabilities.

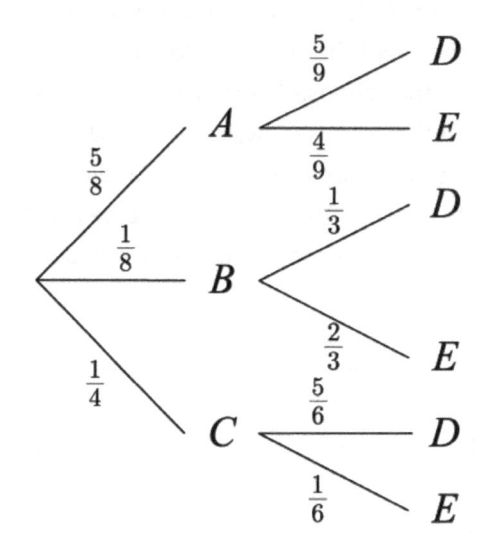

Find
a) $P(B$ then $D)$
b) $P(E)$
c) $P(A$ or E or both)

42) Daniel usually walks to work and the probability of him arriving before 8:30am is 0.95. If it rains, Daniel catches the bus and then the probability of him arriving before 8:30am is 0.75. The probability on any given day of it raining is 0.2
a) Draw a probability tree to display this information.
b) Find the probability that, on any randomly chosen day, Daniel arrives at work after 8:30am.
c) Five days throughout the year are randomly selected. Find the probability that Daniel arrives at work before 8:30am on all five of these days.

1.52 Probability: Introduction (answers on page 465)

43) In the UK in 2022, 17.47% of the population was aged 14 or under, 63.36% was between the ages of 15 to 64 inclusive, and 19.17% were aged 65 or over. Two people are picked at random from the population.
    a) Find the probability that both of them are aged 65 or over.
    b) Find the probability that one is aged 65 or over, and the other is aged 14 or under.

44) In a village in the UK in 2022, 46 of the population was aged 14 or under, 194 was between the ages of 15 to 64 inclusive, and 140 were aged 65 or over. Two people are picked at random from the population.
    a) Find the probability that both of them are aged 65 or over.
    b) Find the probability that one is aged 65 or over, and the other is aged 14 or under.

45) Aisha has six cards numbered 1 to 6 lying face-down on a table. She randomly selects each card and turns it over.
    a) Find the probability that she picks the cards in perfect ascending order.
    b) Find the probability that the last two cards to be turned over are 5 and then 6
    c) Find the probability that the first card to be turned over is 1 and the last is 6
    d) Find the probability that the second card to be turned over is 2, the fourth card to be turned over is 4 and the sixth card to be turned over is 6

46) Haroon has eight cards. Five of them are blue and three of them are red, and they are lying face-down on a table. He randomly selects each card and turns it over.
    a) Find the probability that he picks all the red cards first, followed by all the blue cards.
    b) Find the probability that he picks all the blue cards first, followed by all the red cards.
    c) Find the probability that he picks a blue card first and then the colours alternate until there are no more red cards.

47) Pete and Harvey have ten cards numbered 1 to 10 lying face-down on a table. They decide to take it in turns to randomly select cards and turn them over. If someone selects a multiple of 4, then that person automatically wins. If someone selects a prime number, then that person automatically loses. Pete picks first.
    a) Find the probability that Pete loses on his first card.
    b) Find the probability that Harvey wins on his first card.
    c) Find the probability that the maximum number of cards are drawn before someone wins or loses.
    d) Find the probability that Pete wins after either of the first two cards being drawn.
    e) Find the probability that Pete wins.
    f) Find the probability that Harvey wins.

48) Bag A contains 3 red counters and 1 blue counter. Bag B is initially empty. One by one, the counters in Bag A are randomly chosen and placed into Bag B. Find the probability that the first and last counters placed into Bag B are both red.

49) Bag A contains 2 red counters and 1 blue counter. Bag B contains 1 red counter and 3 blue counters. One by one, the counters in Bag A are randomly chosen and placed into Bag B. Find the probability that there are always fewer red counters than blue counters in Bag B.

50) Bag A contains 3 red counters, 10 blue counters and 5 yellow counters. Bag B contains 4 red counters and 5 blue counters. One counter is drawn from Bag A and one counter is drawn from Bag B.
    a) Find the probability that both counters are the same colour.

    The counters are placed back in their original bags. A counter is now drawn at random from Bag B and placed into Bag A. One counter is then drawn from Bag A and one counter is drawn from Bag B.
    b) Find the probability that both counters are the same colour.

1.52 Probability: Introduction (answers on page 465)

51) Bag A contains 8 red counters, 7 yellow counters and 3 green counters. Bag B contains 3 red counters, 2 yellow counters and k green counters. One counter is drawn from Bag A and one counter is drawn from Bag B. The probability of the two colours being different is $\frac{7}{9}$. Find the value of k

52) Bag A contains 4 red counters, 5 yellow counters and 3 green counters. Bag B contains 3 red counters, 2 yellow counters and 1 green counter. One counter is drawn at random from Bag B and placed into Bag A. Then a counter is drawn at random from Bag A and placed into Bag B. Find the probability that Bag B still contains 3 red counters, 2 yellow counters and 1 green counter.

53) Bag A contains 6 red counters, 3 yellow counters and 2 green counters. Bag B contains 5 red counters, 4 yellow counters and 2 green counters. One counter is drawn at random from Bag B and placed into Bag A. Then a counter is drawn at random from Bag A and placed into Bag B. Find the probability that Bag B now has an equal number of red and yellow counters.

54) Janice and Eric are playing with marbles. There are 10 blue marbles and 10 red marbles in a bag. They take it in turns to pick a marble from the bag. If Janice picks a blue marble, or if Eric picks a red marble, the game continues. If Janice picks a red marble, or if Eric picks a blue marble, the game is over. Janice plays first.
   a) Draw a probability tree to show the possible results from the first three picks.
   b) Find the probability that six marbles are picked from the bag and the game is not over.

55) A sixth form college has 48 students who study A-Level Further Maths across 3 classes. Each student is given the option of teaching a short revision lesson to their peers on one of three topics. The table below shows the number of students from each class who chose each topic.

	Class A	Class B	Class C	Total
Complex Loci	5	6	9	20
Matrices	5	7	6	18
Series	5	3	2	10
Total	15	16	17	48

Three students were randomly selected from the 48 students.
Find the probability that:
   a) one taught a lesson on 'Complex Loci', one taught a lesson on 'Matrices', and one taught a lesson on 'Series',
   b) all three taught a lesson on 'Complex Loci',
   c) one student from each class is picked,
   d) two taught a lesson on 'Matrices' and one taught a lesson on 'Series'.

56) Erin and Molly are playing a game where in each round they must complete a puzzle as quickly as possible. They play four rounds, and they decide that someone can only win the game if someone wins two rounds in a row. The probability of Erin winning a round is 0.75 and the probability of Molly winning a round is 0.25
   a) Find the probability that Erin wins the game.
   b) Find the probability that neither of them wins the game.

57) Cecil and Hilda are to play a game of netball and to practice they make up a short game. They will both get three chances of trying to get the ball in the basket. They must get at least two balls into the basket to win, and a draw is considered a loss for both players. The probability of Cecil scoring is 0.35 and the probability of Hilda scoring is 0.55
   a) Find the probability that Cecil wins the game.
   b) Find the probability that neither wins the game.

1.53 Probability: Regions of Venn Diagrams (answers on page 468)

1) State, in terms of A and B, what each of these regions represent:
   a) 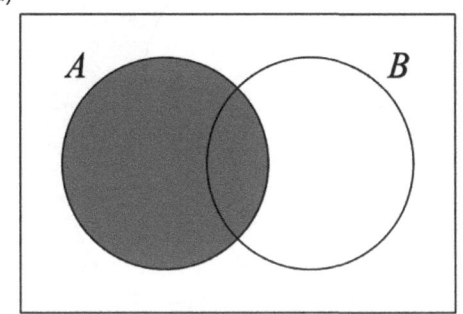
   b) 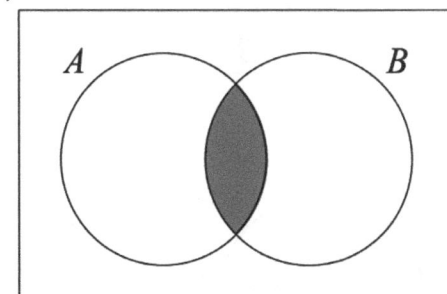
   c) 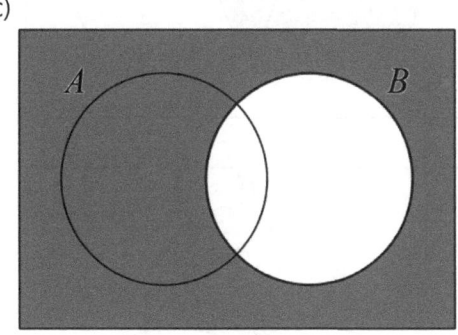
   d) 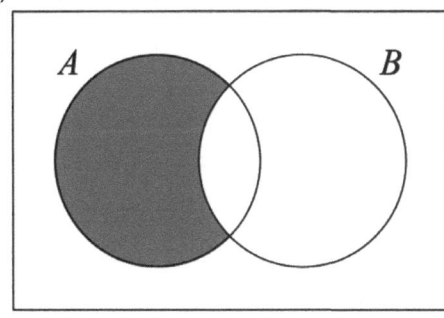
   e)

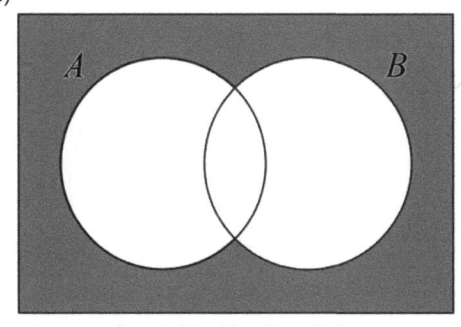

2) State, in terms of A, B and C, what each of these regions represent:
   a) 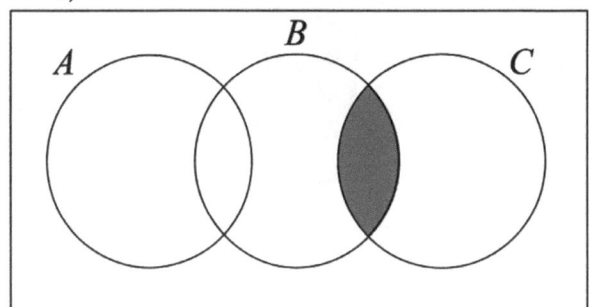
   b) 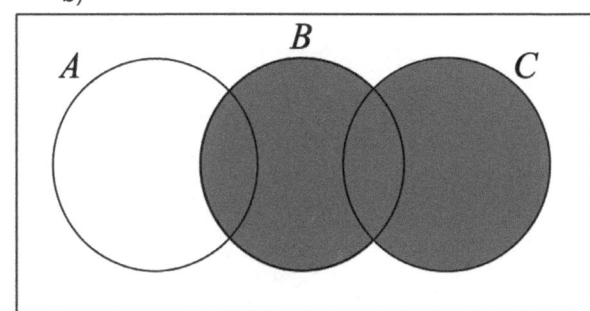
   c) 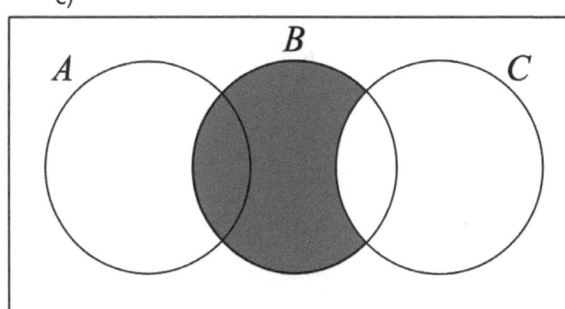
   d) 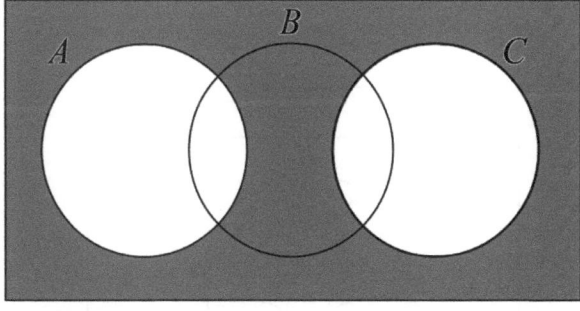
   e)

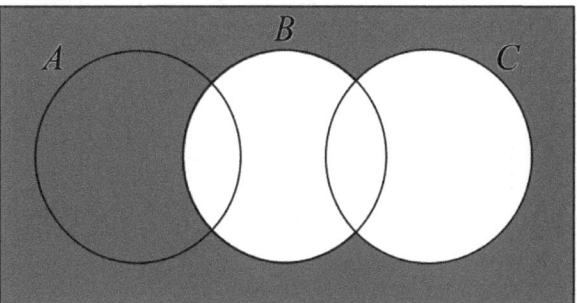

1.53 Probability: Regions of Venn Diagrams (answers on page 468)

f)

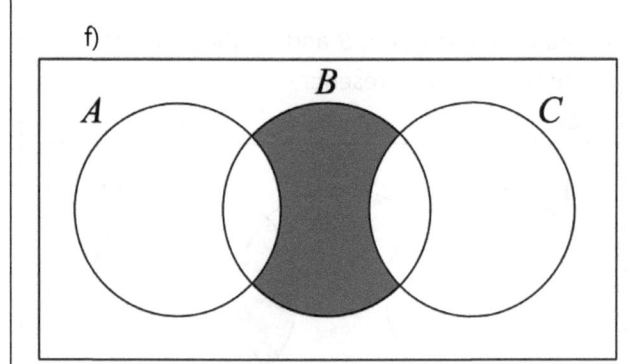

g)

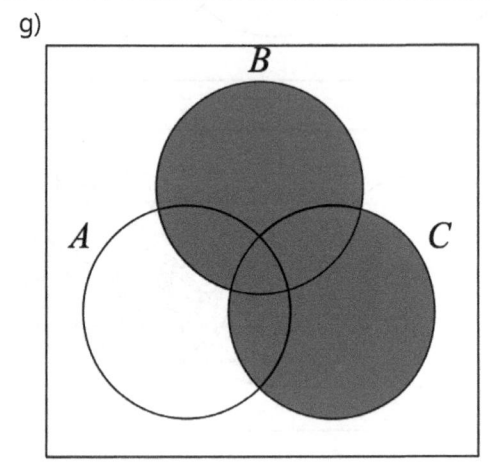

h)

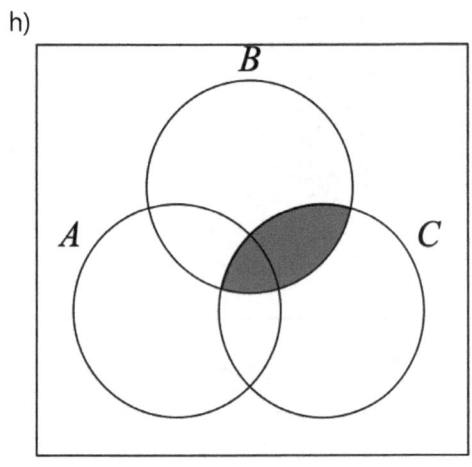

i)

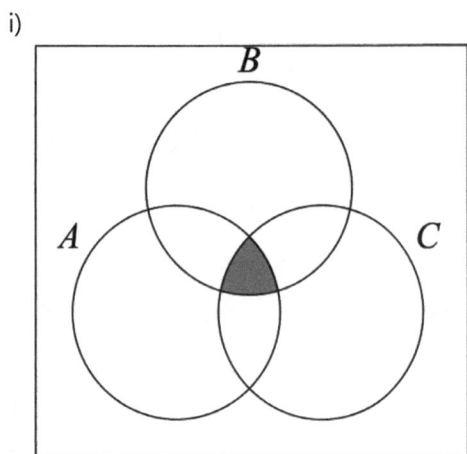

j)

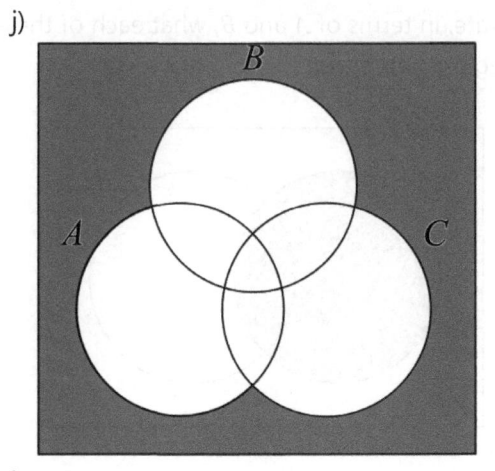

k)

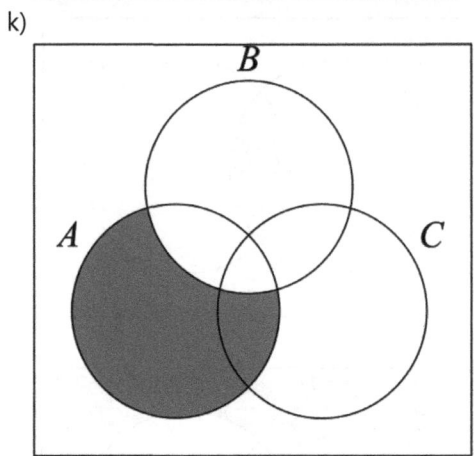

l)

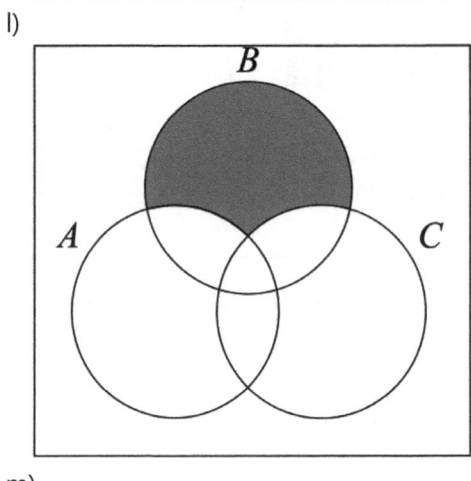

m)

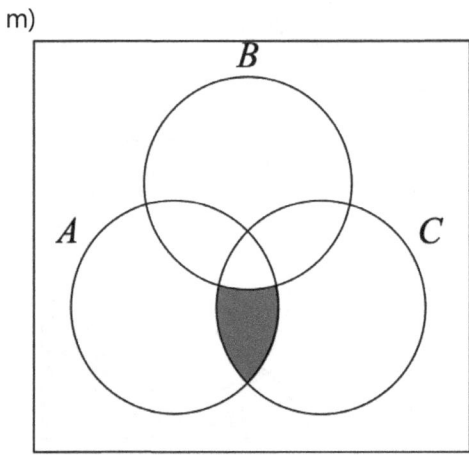

1.53 Probability: Regions of Venn Diagrams (answers on page 468)

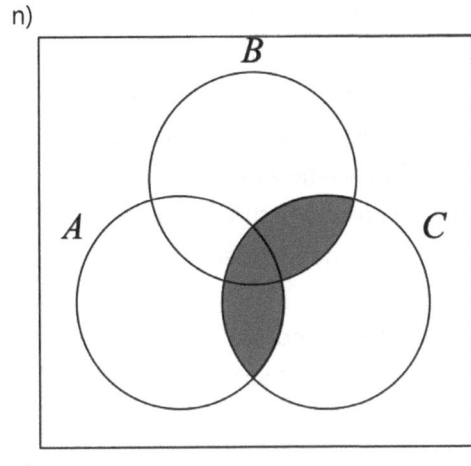

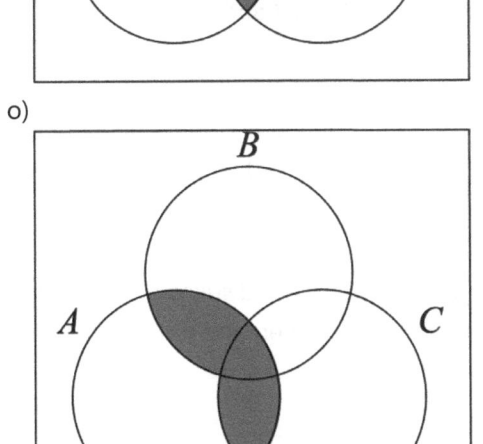

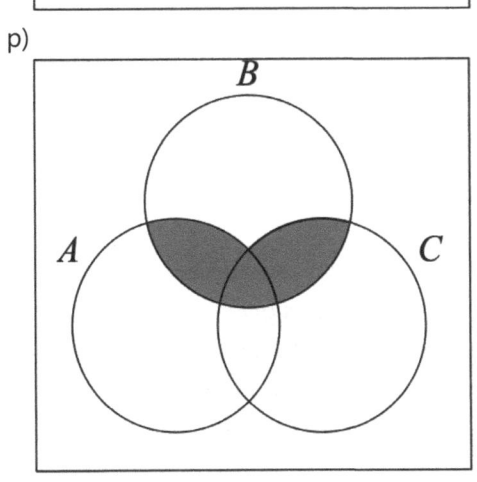

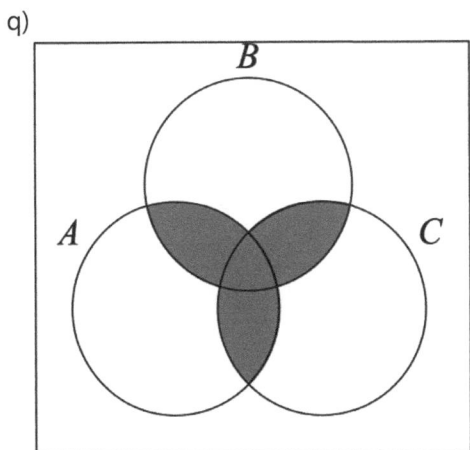

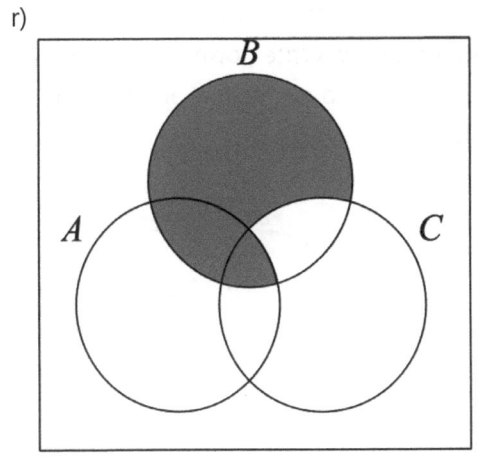

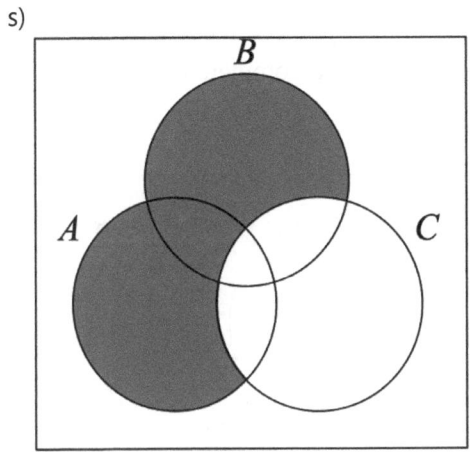

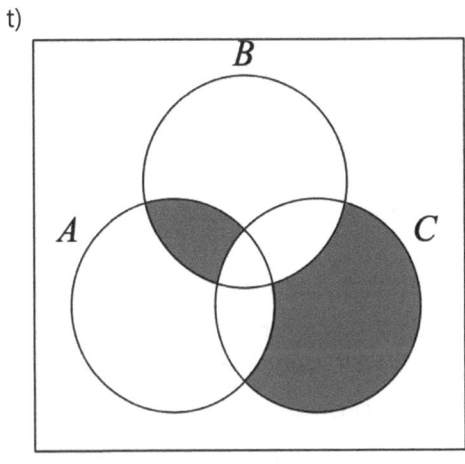

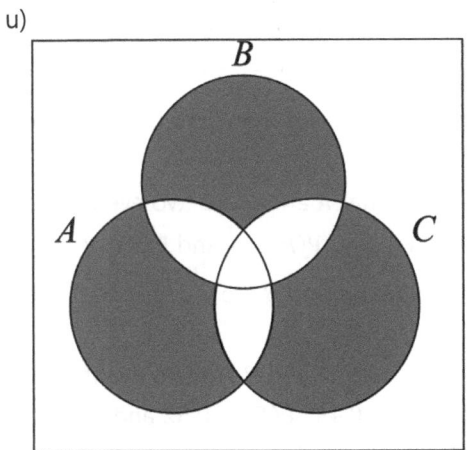

n)

o)

p)

q)

r)

s)

t)

u)

1.54 Probability: Independence and Mutually Exclusive (answers on page 468)

1) Events A, B, C and D are displayed in the Venn diagram below. Write down three different examples of two events that are mutually exclusive.

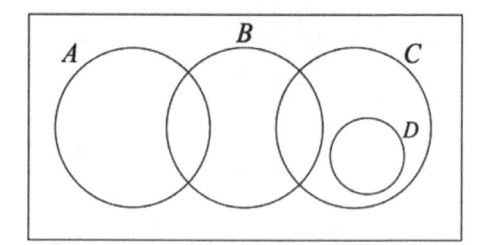

2) It is given that A and B are independent events. Copy and complete the following Venn diagram.

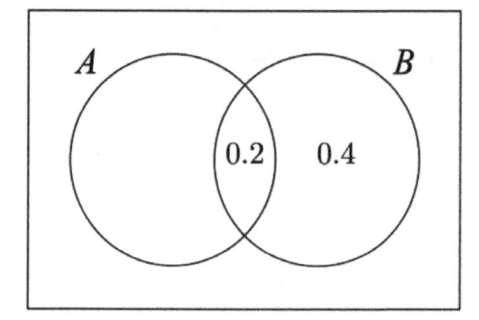

3) It is given that A and B are two events such that $P(A) = 0.4$, $P(B) = 0.55$ and $P(A \cup B) = 0.95$. Determine if A and B are mutually exclusive.

4) It is given that A and B are two events such that $P(A) = 0.32$, $P(B) = 0.46$ and $P(A \cup B) = 0.7$. Determine if A and B are mutually exclusive.

5) It is given that A and B are two events such that $P(A) = \frac{3}{7}$, $P(B) = \frac{2}{3}$ and $P(A \cap B) = \frac{2}{7}$. Determine if A and B are independent events.

6) It is given that A and B are two events such that $P(A) = \frac{3}{10}$, $P(B) = \frac{5}{6}$ and $P(A \cap B) = \frac{1}{2}$. Determine if A and B are independent events.

7) It is given that A and B are two events such that $P(A) = 0.45$, $P(B) = 0.52$ and $P(A \cap B) = 0.23$. Determine if A and B are independent events.

8) An estate agent has 30 flats advertised on their website. 16 have an ensuite bathroom, 8 have a balcony, and 6 have both an ensuite bathroom and a balcony. A flat is chosen at random. Determine whether the events 'the flat has an ensuite bathroom' and 'the flat has a balcony' are independent.

9) It is given that A and B are independent events such that $P(A) = 0.3$ and $P(A \cup B) = 0.75$. Find $P(B)$

10) It is given that A and B are independent events such that $P(A \cap B) = 0.85$ and $P(A) = 0.95$. Find $P(A \cup B)$

11) Events A, B and C are displayed in the Venn diagram below. Determine if A and B are independent events.

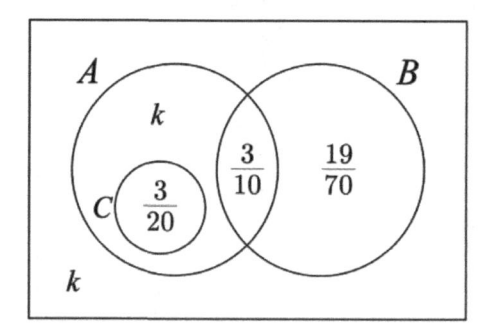

12) Events A, B and C are displayed in the Venn diagram below. It is given that A and C are independent events, and that A and B are mutually exclusive and B and C are also mutually exclusive. Find the value of k and the value of p

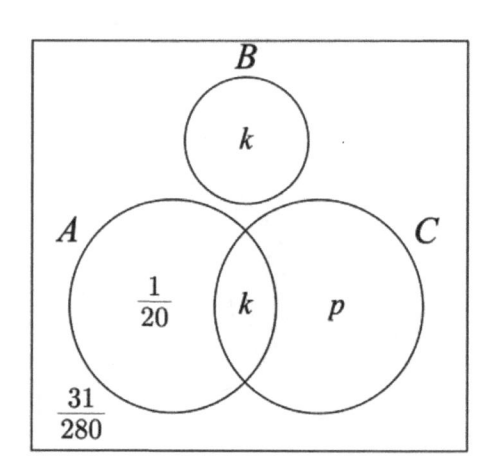

1.54 Probability: Independence and Mutually Exclusive (answers on page 468)

13) Events A, B and C are displayed in the Venn diagram below. It is given that B and C are mutually exclusive, and that A and B are independent. Find the possible values of x, y and z

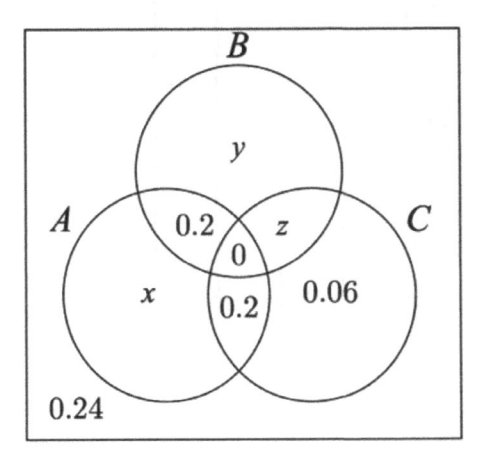

(Venn diagram: Three circles labelled A, B, C. Values shown: y in B only, 0.2 in $A \cap B$, z in $B \cap C$, 0 in centre, x in A only, 0.2 in $A \cap C$, 0.06 in C only, 0.24 outside.)

14) Events A and B are independent. It is given that $P(A \cap B) = \frac{1}{6}$ and $P(A \cup B) = \frac{3}{4}$. Find the possible values of $P(A)$ and $P(B)$

15) Events A, B and C have probabilities 0.05, 0.4 and 0.1 respectively.
    A and C are independent events.
    A and B are mutually exclusive.
    B and C are also mutually exclusive. Draw a fully labelled Venn diagram showing the events A, B and C.

16) Events A, B and C have probabilities 0.05, 0.1 and 0.4 respectively.
    A and B are independent events.
    B and C are also independent events.
    A and C are mutually exclusive. Draw a fully labelled Venn diagram showing the events A, B and C.

17) The two-way table below shows the events A and B. Determine whether events A and B are independent.

	A	A'
B	0.0522	0.0928
B'	0.3078	0.5472

18) Three classes, A, B and C in a sixth form college have 75 students, who either study three A-Levels, three A-Levels and one AS-Level, or four A-Levels. The table below shows the number and type of courses the students in each class are studying.

	Class A	Class B	Class C	Total
Three A-Levels	20	14	18	52
Three A-Levels + One AS	4	7	4	15
Four A-Levels	1	3	4	8
Total	25	24	26	75

A is the event 'the student is studying three A-Levels'.
B is the event 'the student is from Class B'.
A student is selected at random.
a) Find $P(B)$
b) Find $P(A \cup B)$
c) Find $P(A \cap B')$
d) Determine if A and B are independent events.

19) Events A, B and C are displayed in the Venn diagram below. Determine whether $A' \cap B$ and C are independent.

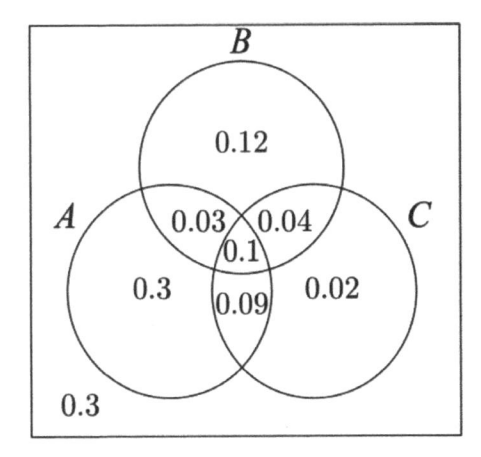

(Venn diagram: Three circles labelled A, B, C. Values: 0.12 in B only, 0.03 in $A \cap B$, 0.04 in $B \cap C$, 0.1 in centre, 0.3 in A only, 0.09 in $A \cap C$, 0.02 in C only, 0.3 outside.)

20) Events A, B and C are such that $P(A) = 0.2$, $P(B) = 0.3$ and $P(C) = k$. It is given that A and B are independent, and A and C are mutually exclusive. It is also given that B and C are independent and $P((A \cup B \cup C)') = 0$. Find the value of k

1.55 Discrete Random Variables (answers on page 469)

1) The table below shows the probability distribution for a discrete random variable X.

x	1	2	3	4	5
$P(X = x)$	0.11	0.13	0.29	0.30	0.17

Find $P(2 \leq X < 5)$

2) A probability distribution is given by
$$P(X = x) = \frac{x}{100} + \frac{1}{5}$$
for $x = 2, 4, 6, 8$. Draw a table showing the probability distribution of X.

3) A probability distribution is given by
$$P(X = x) = \frac{x + 6}{40}$$
for $x = 1, 3, 5, 7$. Find $P(X > 4)$

4) The table below shows the probability distribution for a discrete random variable X.

x	0	1	2	3	4 or more
$P(X = x)$	0.05	0.1	k	0.35	0.32

Find the value of k

5) The table below shows the probability distribution for a discrete random variable X.

x	1	2	3	4	5
$P(X = x)$	k	$3k$	$5k$	k	k

Find the value of k

6) A probability distribution is given by
$$P(X = x) = k(10 - x)$$
for $x = 2, 4, 6, 8$ where k is a constant.
a) Show that $k = \frac{1}{20}$
b) Calculate $P(X \geq 6)$

7) A probability distribution is given by
$$P(X = x) = \frac{x + k}{10x + 10}$$
for $x = 1, 2, 3$. Find the value of k

8) The stick graph below shows the probability distribution for the discrete random variable X.

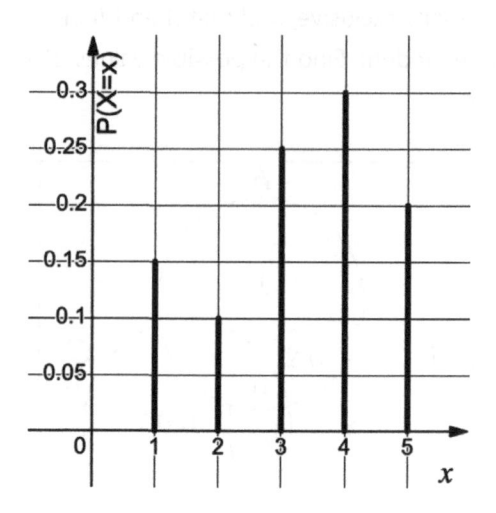

Find $P(1 < X \leq 4)$

9) The stick graph below shows the probability distribution for the discrete random variable X.

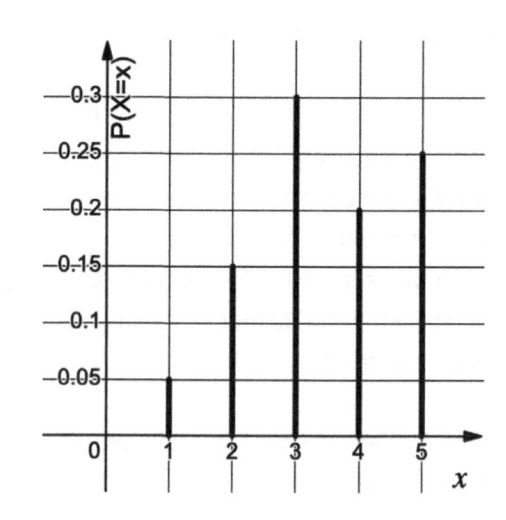

Two values of X are selected at random. Assume these events are independent. Find the probability that both values are greater than or equal to 4

10) The table below shows the probability distribution for a discrete random variable X.

x	1	2	3
$P(X = x)$	0.71	0.19	0.1

Two values of X are chosen at random. Assume these events are independent. Find the probability that both values are the same.

1.55 Discrete Random Variables (answers on page 469)

11) The table below shows the probability distribution for a discrete random variable X.

x	1	2	3
P($X = x$)	0.9	0.09	0.01

Two values of X are chosen at random. Assume these events are independent. Find the probability that the second value is less than the first.

12) The table below shows the probability distribution for a discrete random variable X.

x	1	2	3
P($X = x$)	0.24	0.36	0.4

Two values of X are chosen at random. Assume these events are independent. Find the probability that the first value is odd and the second value is prime.

13) The table below shows the probability distribution for a discrete random variable X.

x	5	8	11
P($X = x$)	0.15	0.25	0.6

Two values of X are chosen at random. Assume these events are independent. Find the probability that the sum of the two scores is less than 17

14) In a game, one spin of a biased spinner can score 1, 2, 3 or 4 points. The random variable X represents a player's score, and this has the following probability distribution:

x	1	2	3	4
P($X = x$)	0.02	0.14	0.26	0.58

The spinner is spun twice and the scores added. It is given that each game is independent. Find the probability of getting a total of 4 points.

15) The table below shows the probability distribution for a discrete random variable X.

x	0	1	2
P($X = x$)	0.33	0.22	0.45

Two values of X are chosen at random. Find the probability that the product of these two values is zero.

16) The discrete random variable X is modelled by the probability distribution:
$$P(X = x) = \frac{x}{20}$$
for $x = 3, 4, 6$ and 7. The independent random variables X_1 and X_2 have the same distribution as X. Find $P(X_1 = X_2)$

17) A spinner can be modelled by X with the probability distribution:
$$P(X = x) = \frac{x + 1}{14}$$
for $x = 1, 2, 3$ and 4. The spinner is spun three times. Find the probability that the first value is greater than the sum of the next two values.

18) It is given that a discrete random variable X can take on the values 1, 2, 4 and 5, and that $P(X = 1) = 0.2$. When $r = 1$ and $r = 2$, it is also given that $P(X = r) = P(X = r + 3)$ Draw a table to show the probability distribution of X.

19) It is given that a discrete random variable X can take on the values 4, 5, 9 and 10, and that $P(X = 5) = 0.18$. When $r = 4$ and $r = 5$, it is also given that $P(X = r) = P(X = r + 5)$ Draw a table to show the probability distribution of X.

20) It is given that a discrete random variable X can take on the values 2, 3, 4, 5 and 6, and that $P(X = 2) = 0.11$. When $r = 2$ and $r = 3$, it is also given that $P(X = r) = P(X = r + 3)$ It is also given that $P(X = 4) = P(X = 3) + P(X = 5)$. Draw a table to show the probability distribution of X.

1.55 Discrete Random Variables (answers on page 469)

21) The discrete random variable X is modelled by the following probability distribution:

$$P(X = x) = \begin{cases} \frac{1}{5} \times \frac{2^{x-1}}{3^{x-1}} & x = 1,2 \\ k & x = 3,4 \\ 0 & \text{otherwise} \end{cases}$$

where k is a constant.
a) Find $P(X = 2)$.
b) Find the value of k
c) Find $P(2 \leq X \leq 3)$

22) The discrete random variable X is modelled by the following probability distribution:

$$P(X = x) = \begin{cases} ax^2 & x = 1,2 \\ bx + 0.05 & x = 3,4 \\ 0 & \text{otherwise} \end{cases}$$

where a and b are constants.
a) Find $P(X = 2)$ in terms of a
b) Given that $P(X = 2) = P(X \geq 3)$, find the exact values of a and b

23) The discrete random variable X is modelled by the following probability distribution:

$$P(X = x) = \begin{cases} a(15 - 3x) & x = 1,2,3,4 \\ bx & x = 5,6 \\ 0 & \text{otherwise} \end{cases}$$

where a and b are constants.
a) Show that $30a + 11b = 1$
b) Given that $P(X \geq 4) = \frac{1}{4}$ find the exact values of a and b

24) The table below shows the probability distribution for a discrete random variable X.

x	0	1	2	3
$P(X = x)$	$3k^2$	$3k$	$\frac{9}{2}k^2$	$\frac{1}{2}k$

Find the value of k and hence find $P(X > 1)$

25) The discrete random variable X is modelled by the following probability distribution:

$$P(X = x) = \begin{cases} k & x = 1,2,3,4 \\ \frac{1}{3}P(X = x - 1) & x = 5,6 \\ 0 & \text{otherwise} \end{cases}$$

where k is a constant.
Show that $k = \frac{9}{40}$

26) The discrete random variable X is modelled by the following probability distribution:

$$P(X = x) = \begin{cases} \frac{1}{5}P(X = x + 1) & x = 1,2,3 \\ k & x = 4,5 \\ 0 & \text{otherwise} \end{cases}$$

where k is a constant. Find the value of k and hence find $P(X \leq 3)$

27) The table below shows the probability distribution for a discrete random variable X.

x	0	1	2	3
$P(X = x)$	$2k$	0.42	0.37	k

Two values of X are chosen at random. Find the probability that the product of these two values is greater than their sum.

28) The table below shows the probability distribution for a discrete random variable X.

x	0	1	2	3	4
$P(X = x)$	$\frac{1}{9}$	$\frac{1}{6}$	$\frac{1}{3}$	$\frac{1}{9}$	$\frac{5}{18}$

Two independent values, X_1 and X_2, of X are chosen. Find $P(X_1 + X_2 = 5)$

29) The table below shows the probability distribution for a discrete random variable X.

x	2	3	4	5
$P(X = x)$	0.8	0.09	0.06	0.05

It is given that $Y = X^2$. Find $P(Y + 2X \leq 15)$

30) The table below shows the probability distribution for a discrete random variable X.

x	4	5	9	13
$P(X = x)$	0.7	0.2	0.05	0.05

It is given that $Y = \frac{15}{X}$. Find $P(X + Y \geq 8)$

1.55 Discrete Random Variables (answers on page 469)

31) Tom models the number of games of pool he will win in a match by the random variable X with the following probability distribution:

x	0	1	2	3	More than 3
$P(X = x)$	$\frac{1}{15}$	$\frac{1}{10}$	$\frac{1}{3}$	$\frac{1}{2}$	0

The number of games of pool won in a match is independent of any other match. Tom chooses three pool matches at random. Use the model to find the probability of:
 a) Tom winning the same number of games of pool in all three matches;
 b) Tom winning exactly two games of pool in total in the three matches;
 c) Tom winning fewer games of pool than the preceding match across the three matches.

32) The table below shows the probability distribution for a discrete random variable X.

x	1	2	3	4
$P(X = x)$	a	b	c	0.45

It is given that $P(X \leq 2) = \frac{1}{3} P(X \geq 3)$.
Find $P(X \geq 3)$

33) A probability distribution is given by
$$P(X = x) = \begin{cases} \frac{k}{x+1} & x = 1,3,7 \\ 0 & \text{otherwise} \end{cases}$$
where k is a constant. Three independent values, X_1, X_2 and X_3, of X are chosen.
Find $P(X_1 + X_2 > X_3)$

34) The table below shows the probability distribution for a discrete random variable S.

s	6	12	18	24
$P(S = s)$	$3k$	$4k$	$5k$	$6k$

S_1 and S_2 have the same distribution as S and are independent. Find $P(S_1 + S_2 = 36)$

35) Two bias spinners, X and Y, can be represented by the probability distributions given by $P(X = x) = \frac{x}{25}$ for $x = 12,13$, and $P(Y = y) = \frac{y}{27}$ for $x = 13,14$ respectively. Spinner X is spun first and then spinner Y is spun. Find the probability that the sum of the scores is 26

36) The random variable X is a discrete uniform distribution where every outcome is equally likely. X can take on the values 1, 2, 3, 4 and 5. Two values, X_1 and X_2, of X are chosen at random. Find $P(X_1 = X_2)$

37) The random variable X is a discrete uniform distribution where every outcome is equally likely. X can take on the values 1, 2, 3, ..., n. Two values, X_1 and X_2, of X are chosen at random. Find $P(X_1 = X_2)$

38) The random variable X is a discrete uniform distribution where every outcome is equally likely. X can take on the values 1, 2, 3, ..., n. Two values, X_1 and X_2, of X are chosen at random. It is given that $P(X_1 = X_2) = 0.1$. Find $P(X_1 < X_2)$

Prerequisite knowledge for Q39-Q40: Conditional Probability

39) A probability distribution is given by
$$P(X = x) = \begin{cases} 2kx & x = 3,4,5,6 \\ 0 & \text{otherwise} \end{cases}$$
where k is a constant. Two independent values, X_1 and X_2, of X are chosen.
 a) Find $P(X_1 + X_2 = 9)$
 b) It is given that $X_1 + X_2 = 9$. Find the probability that one of X_1 or X_2 is 4

40) A probability distribution is given by
$$P(X = x) = \frac{2^x}{k}$$
for $x = 2, 3, 4, 5$ where k is a constant. Three independent values, X_1, X_2 and X_3, of X are chosen.
 a) Find $P(X_1 + X_2 + X_3 = 9)$
 b) It is given that $X_1 + X_2 + X_3 = 9$. Find the probability that at least one of X_1, X_2 or X_3 is 4

1.56 The Binomial Distribution (answers on page 470)

1) The four diagrams below show $X \sim B(6,0.2)$, $X \sim B(6,0.5)$, $X \sim B(6,0.7)$ and $X \sim B(6,0.89)$. Which is which?

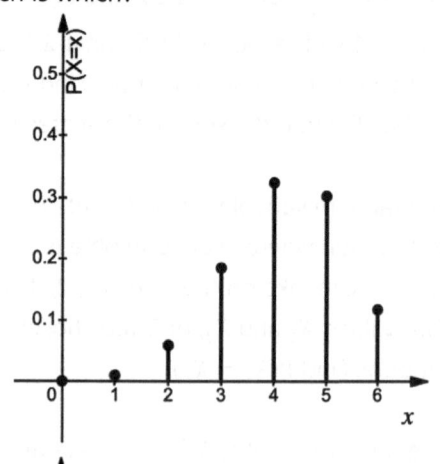

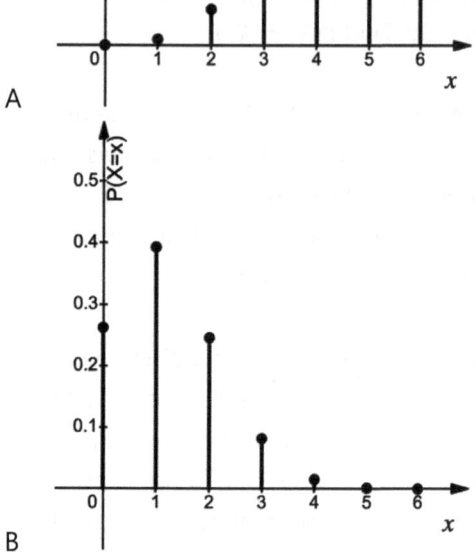

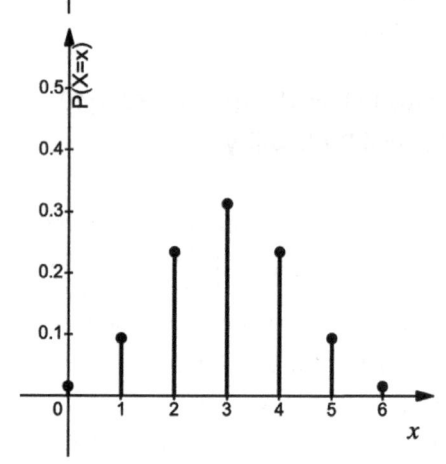

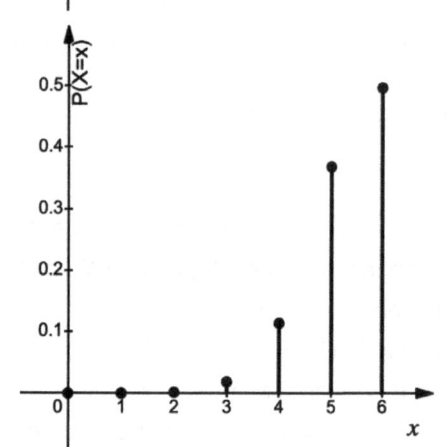

A

B

C

D

2) Consider the stick graph of the following distributions. Do they show a positive skew, a negative skew, or are they symmetric?
   a) $X \sim B(45,0.3)$
   b) $X \sim B(32,0.5)$
   c) $X \sim B(24,0.9)$
   d) $X \sim B(60,0.51)$

3) Given $X \sim B(3,0.2)$, copy and complete the table shown below and draw a stick graph to represent the distribution.

x	0	1	2	3
$P(X = x)$				

4) Given $X \sim B(3,0.6)$, copy and complete the table shown below and draw a stick graph to represent the distribution.

x	0	1	2	3
$P(X = x)$				

5) Given $X \sim B(4,0.5)$, copy and complete the table shown below and draw a stick graph to represent the distribution.

x	0	1	2	3	4
$P(X = x)$					

6) The diagram below shows $X \sim B(n, p)$. Write down n and suggest a value for p

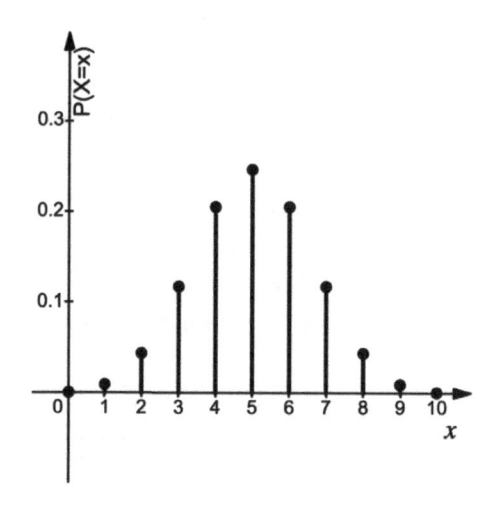

1.56 The Binomial Distribution (answers on page 470)

7) Given $X \sim B(15, 0.3)$, **use the formula** to find:
   a) $P(X = 3)$
   b) $P(X = 4)$
   c) $P(X = 5)$
   d) $P(X = 9)$

8) **Use the formula** in each case.
   a) Given $X \sim B(10, 0.6)$, find $P(X = 6)$
   b) Given $X \sim B(25, 0.2)$, find $P(X = 4)$
   c) Given $X \sim B(60, 0.9)$, find $P(X = 57)$
   d) Given $X \sim B(105, 0.56)$, find $P(X = 52)$

9) Given $X \sim B(10, 0.7)$, use your calculator's binomial probability function to find:
   a) $P(X = 5)$
   b) $P(X = 7)$
   c) $P(X = 10)$

10) Given $X \sim B(20, 0.2)$, use your calculator's binomial probability function to find:
    a) $P(X = 1)$
    b) $P(X = 3)$
    c) $P(X = 5)$

11) Given $X \sim B(17, 0.16)$, **use the formula** to find:
    a) $P(X \leq 1)$
    b) $P(X \leq 2)$
    c) $P(X \leq 3)$

12) Given $X \sim B(22, 0.35)$, use your calculator's binomial probability function to find:
    a) $P(X \leq 2)$
    b) $P(X \leq 5)$
    c) $P(X \leq 8)$

13) Use your calculator's binomial probability function in each case.
    a) Given $X \sim B(18, 0.24)$, find $P(X \leq 5)$
    b) Given $X \sim B(27, 0.36)$, find $P(X \leq 15)$
    c) Given $X \sim B(39, 0.49)$, find $P(X \leq 20)$
    d) Given $X \sim B(47, 0.61)$, find $P(X \leq 25)$

14) Given $X \sim B(18, 0.68)$, use your calculator's binomial probability function to find:
    a) $P(X < 9)$
    b) $P(X < 11)$
    c) $P(X < 14)$

15) Given $X \sim B(17, 0.49)$, use your calculator's binomial probability function to find:
    a) $P(X \geq 8)$
    b) $P(X \geq 11)$
    c) $P(X \geq 12)$

16) Use your calculator's binomial probability function in each case.
    a) Given $X \sim B(16, 0.37)$, find $P(X \geq 4)$
    b) Given $X \sim B(38, 0.02)$, find $P(X \geq 2)$
    c) Given $X \sim B(71, 0.77)$, find $P(X \geq 52)$

17) Given $X \sim B(23, 0.66)$, use your calculator's binomial probability function to find:
    a) $P(9 \leq X \leq 12)$
    b) $P(10 \leq X \leq 14)$
    c) $P(12 \leq X \leq 14)$

18) Given $X \sim B(18, 0.24)$, use your calculator's binomial probability function to find:
    a) $P(2 < X \leq 3)$
    b) $P(3 < X \leq 7)$
    c) $P(4 < X \leq 10)$

19) Given $X \sim B(29, 0.49)$, use your calculator's binomial probability function to find:
    a) $P(14 \leq X < 17)$
    b) $P(13 \leq X < 18)$
    c) $P(10 \leq X < 13)$

20) Given $X \sim B(13, 0.45)$, use your calculator's binomial probability function to find:
    a) $P(4 < X < 6)$
    b) $P(5 < X < 10)$
    c) $P(6 < X < 9)$

21) In a large population of a certain type of tropical bird, 20% are known to have yellow feathers. A random sample of 30 of these birds are selected. Calculate the probability that the sample contains exactly 5 with yellow feathers.

22) A factory produces blades for pencil sharpeners. The probability of one of these blades being misshapen is 0.008. In a random sample of 200 blades, find the probability of less than 3 being misshapen.

1.56 The Binomial Distribution (answers on page 470)

23) Write down the mean, E(X), and variance, Var(X) of the following distributions.
   a) $X \sim B(8, 0.5)$
   b) $X \sim B(20, 0.3)$
   c) $X \sim B(16, 0.9)$
   d) $X \sim B(25, 0.2)$
   e) $X \sim B(40, 0.6)$

24) Given $X \sim B(22, 0.6)$ and the mean, $E(X) = \mu$,
   a) find $P(X \leq \mu)$,
   b) find $P(X > \mu)$.

25) Given $X \sim B(n, p)$ and
$$P(X = 10) = \binom{45}{10} \times 0.25^{10} \times 0.75^{35}$$
write down the mean, E(X), and variance, Var(X).

26) Given $X \sim B(450, p)$ and the mean, E(X), is 100, find the value of p

27) Given $X \sim B(1560, p)$ and the mean, E(X), is 120, find the value of p

28) Given $X \sim B(n, 0.3)$ and the variance, Var(X), is 42, find the value of n

29) Given $X \sim B\left(n, \frac{1}{6}\right)$ and the variance, Var(X), is 375, find the value of n

30) Given $X \sim B\left(n, \frac{1}{8}\right)$ and the standard deviation is 35, find the value of n

31) Given $X \sim B\left(n, \frac{5}{8}\right)$ and the standard deviation is 105, find the value of n

32) Given $X \sim B(n, p)$, with $E(X) = 25.2$ and Var(X) = 2.52, find the values of n and p

33) Given $X \sim B(n, p)$, with $E(X) = 14.4$ and Var(X) = 8.64, find the values of n and p

34) Given $X \sim B(n, p)$, with $E(X) = 5$ and $Var(X) = \frac{60}{13}$, find the values of n and p

35) Given $X \sim B(100, 0.2)$ and $Y \sim B(20, k)$, and that Var(X) = E(Y), find the value of k

36) It is given that $X \sim B(n, p)$ and $Y \sim B(k, p)$, where $0 < p < 1$. It is also given that $E(X) = 3E(Y)$ and $Var(X) = 2E(Y)$. Find the value of p

37) It is given that $X \sim B(n, p)$ and $Y \sim B(k, p)$, where $0 < p < 1$. It is also given that $E(X) = 5Var(Y)$ and $E(Y) = 3E(X)$. Find the value of p

38) State the conditions under which a situation can be modelled by a binomial distribution.

39) A fair six-sided die is rolled until a 6 appears. Explain why we cannot use a binomial distribution to model this.

40) In a small village, the probability of it raining on any one day of the year is known to be 0.22. Nico has sixteen days between breaking up and going back to school for his Easter holidays and wants to determine the probability that for at most 3 days it will be raining. Explain why it would be inappropriate to use a binomial distribution in this situation.

41) Georgia attempts a crossword puzzle on the puzzle page of a newspaper every day. Over several months, Georgia has found that she is able to successfully complete the crossword puzzle 82% of the times.
   a) Using the binomial distribution, find the probability that, out of a random sample of 21 days, Georgia successfully completes at least 18 crossword puzzles.
   b) Using the same binomial distribution in part a), find the probability that Georgia successfully completes less than 16 crossword puzzles.

42) A fair 8-sided spinner has sides numbered 1, 2, 3, 4, 5, 6, 7 and 8. The spinner is spun 35 times. The random variable X represents the number of times the spinner lands on 8.
   a) Find the variance, Var(X), of the number of times the spinner lands on 8
   b) Find the probability that the spinner lands on an 8 at most 6 times.
   c) Find $P(4 < X < 8)$.

1.56 The Binomial Distribution (answers on page 470)

43) Emma plays snooker and is practicing a trick-shot. Each evening, she attempts the trick-shot 50 times. Every time she attempts the trick-shot, there is a probability of 0.1 that she miss-cues. The situation is modelled as a binomial distribution.
    a) Find the mean number of times that Emma miss-cues in one evening.
    b) Find the variance of the number of times that Emma miss-cues in one evening.
    c) Find the probability that, on a particular evening, she miss-cues exactly 5 times.
    d) Find the probability that, on a particular evening, she miss-cues at most 6 times.
    e) Emma practices the trick-shot 50 times each evening for one week. Find the probability that she miss-cues at most 6 times on each of the 7 days.

44) Curtis plays darts. He has determined the probability of him hitting the bullseye with one dart is 0.15. Curtis throws twelve darts. The situation is modelled as a binomial distribution.
    a) Find the probability that Curtis will hit the bullseye more than twice.
    b) Curtis throws twelve darts at three different randomly selected times. Find the probability that Curtis will hit the bullseye more than twice on **one** of these three times.
    c) State two assumptions, in context, that would be necessary for the binomial distribution to be valid in this case.

45) In a survey of 3500 A-Level Maths students across several sixth form colleges, 2240 use graphical calculators.
    a) Using the binomial distribution, find the probability that, out of a random sample of 24 students, exactly 15 students have a graphical calculator.
    b) Explain why the answer to part a) may not be valid for 24 students from a single class.

46) A shop sells interlocking floor tiles. Tara visits the shop to purchase thirty of these tiles. It is given that 2.5% of all interlocking floor tiles of these style are faulty. By modelling the situation as a binomial distribution,
    a) Find the probability that none of the tiles Tara buys are faulty.
    b) Find the probability that at least three of the tiles Tara buys are faulty.
    c) State two assumptions, in context, that would be necessary for the binomial distribution to be valid in this case.

47) At a small sixth form college, 12% of the students belong to the cinema club. A random sample of 40 students are selected from the college. By modelling the situation as a binomial distribution,
    a) Find the probability that exactly 4 students belong to the cinema club.
    b) Find the probability that at least 3 but at most 8 belong to the cinema club.
    c) Of the students who belong to the cinema club, 30% also belong to the book club. Assuming that these events are independent, find the probability that a randomly chosen student belongs to both the cinema club and the book club.
    d) In the random sample of 40 students, find the probability that exactly one student belongs to both the cinema club and the book club.

48) A type of lightbulb has a $\frac{3}{16}$ chance of being faulty. In a random sample of 15 lightbulbs, show that the probability of 2 lightbulbs being faulty is the same as the probability of 3 lightbulbs being faulty.

49) Every day, Kai takes a walk in a park near his house. The probability that he spots a squirrel in the park on any particular day is 0.13
    a) Find the probability that in a week Kai spots a squirrel on exactly 3 days.
    b) Find the probability that Kai spots a squirrel on exactly 3 days in a week, during at most 1 of 7 randomly chosen weeks.

1.56 The Binomial Distribution (answers on page 470)

Prerequisite knowledge for Q50: Conditional Probability

50) A fair die has eight faces numbered 1 to 8. The die is rolled 40 times and the number of times an 8 is rolled is recorded. Find the probability that at least eight 8s are rolled,
    a) given that fewer than eleven 8s are rolled.
    b) given that more than five 8s are rolled.
    c) given that more than five but fewer than eleven 8s are rolled.

51) On each of the twelve floors of a large office building, there is a printer. Four of these are designated to only print colour, and the other eight only print black and white. The colour printers are in use 15% of the time and the black and white printers are in use 35% of the time. Find the probability that at any one time,
    a) one of the colour printers is in use;
    b) at most two black and white printers are in use;
    c) none of the printers are in use;
    d) exactly two of the printers are in use.

52) Given $X \sim B(150, 0.25)$, find the largest possible value of k such that $P(X \leq k) < 0.1$

53) Given $X \sim B(500, 0.15)$, find the largest possible value of k such that $P(X < k) < 0.1$

54) In a town, it is known that 10% of local households use online food shopping. A random sample of 300 households are selected. The random variable X is defined as the number of households in the sample who use online food shopping. Calculate the largest possible value of k such that $P(X < k) < 0.01$

55) In a constituency, it is known that 35% of constituents traditionally vote for the independent candidate. A random sample of 400 adults in the constituency are selected. The random variable X is defined as the number of adults in the sample who voted for the independent candidate. Calculate the least possible value of k such that $P(X > k) < 0.05$

Prerequisite knowledge for Q56-Q59: Logarithms

56) It is given that $X \sim B(n, 0.07)$. Estimate the value of n given that $P(X = 0) = 0.00465$

57) It is given that $X \sim B(n, 0.03)$. Estimate the value of n given that $P(X = 0) = 0.02921$

58) It is given that $X \sim B(n, 0.995)$. Estimate the value of n given that $P(X = n) = 0.03983$

59) It is given that $X \sim B(n, 0.974)$. Estimate the value of n given that $P(X = n) = 0.01557$

Use your calculator's SOLVE function for Q60 and Q61

60) It is given that $X \sim B(n, 0.023)$. Estimate the value of n given that $P(X \leq 1) = 0.39021$

61) It is given that $X \sim B(n, 0.998)$. Estimate the value of n given that $P(X \geq n - 1) = 0.72913$

Prerequisite knowledge for Q62-Q63: Logarithms

62) In a town, it is known that one person in every 15 people, on average, has the blood type B positive. A local hospital is looking for potential donors of this blood type. Determine the least number of blood donors that must be selected, at random, in order for the probability of including at least one donor that is type B positive is 0.95 or more.

63) A factory has been made aware that there is a fault that affects one in every 70 components that it manufactures. To investigate, an employee is tasked with taking a random sample of components from the production line and finding a faulty component. Determine the least number of components that must be randomly sampled in order for the probability of including at least one faulty component is 0.99 or more.

1.57 Binomial Hypothesis Testing (answers on page 472)

1) Describe what is meant by the significance level.

2) In each case, find the p-value and determine whether you reject H_0 or fail to reject H_0
   a) $H_0: p = 0.7$, $H_1: p < 0.7$
      $X \sim B(20, 0.7)$
      Observed 10, 5% significance level
   b) $H_0: p = 0.3$, $H_1: p < 0.3$
      $X \sim B(30, 0.3)$
      Observed 5, 5% significance level
   c) $H_0: p = 0.6$, $H_1: p < 0.6$
      $X \sim B(27, 0.6)$
      Observed 12, 10% significance level
   d) $H_0: p = 0.9$, $H_1: p < 0.9$
      $X \sim B(52, 0.9)$
      Observed 40, 1% significance level

3) In each case, find the p-value and determine whether you reject H_0 or fail to reject H_0
   a) $H_0: p = 0.4$, $H_1: p > 0.4$
      $X \sim B(15, 0.4)$
      Observed 9, 5% significance level
   b) $H_0: p = 0.85$, $H_1: p > 0.85$
      $X \sim B(45, 0.85)$
      Observed 41, 5% significance level
   c) $H_0: p = 0.55$, $H_1: p > 0.55$
      $X \sim B(36, 0.55)$
      Observed 26, 10% significance level
   d) $H_0: p = 0.12$, $H_1: p > 0.12$
      $X \sim B(90, 0.12)$
      Observed 18, 1% significance level

4) In each case, find the p-value and determine whether you reject H_0 or fail to reject H_0
   a) $H_0: p = 0.8$, $H_1: p \neq 0.8$
      $X \sim B(25, 0.8)$
      Observed 15, 5% significance level
   b) $H_0: p = 0.72$, $H_1: p \neq 0.72$
      $X \sim B(33, 0.72)$
      Observed 29, 5% significance level
   c) $H_0: p = 0.38$, $H_1: p \neq 0.38$
      $X \sim B(67, 0.38)$
      Observed 33, 10% significance level
   d) $H_0: p = 0.05$, $H_1: p \neq 0.05$
      $X \sim B(180, 0.05)$
      Observed 3, 1% significance level

5) Describe what is meant by:
   a) the critical region,
   b) the acceptance region.

6) In each case, find the critical region and write down the actual significance level.
   a) $H_0: p = 0.5$, $H_1: p < 0.5$
      $X \sim B(18, 0.5)$
      5% significance level
   b) $H_0: p = 0.55$, $H_1: p < 0.55$
      $X \sim B(38, 0.55)$
      5% significance level
   c) $H_0: p = 0.35$, $H_1: p < 0.35$
      $X \sim B(29, 0.35)$
      10% significance level
   d) $H_0: p = 0.86$, $H_1: p < 0.86$
      $X \sim B(65, 0.86)$
      1% significance level

7) In each case, find the critical region and write down the actual significance level.
   a) $H_0: p = 0.2$, $H_1: p > 0.2$
      $X \sim B(40, 0.2)$
      5% significance level
   b) $H_0: p = 0.33$, $H_1: p > 0.33$
      $X \sim B(49, 0.33)$
      5% significance level
   c) $H_0: p = 0.11$, $H_1: p > 0.11$
      $X \sim B(65, 0.11)$
      10% significance level
   d) $H_0: p = 0.74$, $H_1: p > 0.74$
      $X \sim B(125, 0.74)$
      1% significance level

8) In each case, find the critical region and write down the actual significance level.
   a) $H_0: p = 0.4$, $H_1: p \neq 0.4$
      $X \sim B(60, 0.4)$
      5% significance level
   b) $H_0: p = 0.9$, $H_1: p \neq 0.9$
      $X \sim B(44, 0.9)$
      5% significance level
   c) $H_0: p = 0.28$, $H_1: p \neq 0.28$
      $X \sim B(55, 0.28)$
      10% significance level
   d) $H_0: p = 0.09$, $H_1: p \neq 0.09$
      $X \sim B(135, 0.09)$
      1% significance level

1.57 Binomial Hypothesis Testing (answers on page 472)

9) Fran is a florist. She regularly makes the most sales the day before Mother's Day each year, and over several years 25% of customers are found to purchase a chrysanthemum bouquet on that day. Fran decides to make the chrysanthemum bouquets larger for the following year and consequently they become more expensive. She believes this will result in fewer bouquets of this variety being sold. The following year, in a random sample of 34 customers, 5 customers bought a chrysanthemum bouquet. Fran carries out a hypothesis test, at the 10% significance level, to test her belief. Fran's hypothesis test is shown below.

Let X be the number of customers purchasing a chrysanthemum bouquet.
$H_0: p = 0.25$
$H_1: p \leq 0.25$
$X \sim B(34, 0.25)$
$P(X = 5) = 0.0647$
$0.0647 > 0.1$ so we fail to reject H_0
There is insufficient evidence to suggest that the proportion of customers purchasing a chrysanthemum bouquet has decreased.

  a) Identify the first two errors made by Fran in her hypothesis test.
  b) Correct her test.

10) Chaisai is playing a game of chance that is designed to have a probability of winning of $\frac{1}{5}$. Chaisai believes he has found a way to improve his chances and wishes to conduct a hypothesis test to test this belief. He plays the game 45 more times and wins 15 times. Carry out a hypothesis test at the 5% significance level to test Chaisai's claim.

11) The National Health and Medical Research Council (NHMRC) 2013 Australian Dietary Guidelines set out the recommended servings of fruit and vegetables to be eaten per day. In 2022, it was reported that 44.1% of adults met the fruit recommendation of 2 servings. Amy believes that this proportion is too high. She takes a random sample of 60 adults and 20 of them reported they met the fruit recommendation of 2 servings a day. Carry out a hypothesis test at the 5% significance level to investigate whether Amy's belief is supported.

12) The UK employment rate for Sep-Nov 2023 was 75.8%. Daniel lives in Brighton and believes the employment rate there is significantly lower than the UK average. He takes a random sample of 110 adults between the ages of 16 to 64 and finds that 70 of them are employed. Carry out a hypothesis test at the 1% significance level to investigate whether Daniel's belief is supported.

13) Energee lithium-ion batteries have a 77% chance of lasting longer than 20 hours on a single charge. Energee releases a new version of the battery which they hope will on average last longer. A random sample of 150 of the new batteries is taken and 126 last longer than 20 hours. Carry out a hypothesis test at the 5% level of significance to investigate Energee's claim about the new batteries.

14) Thomas works for a company that designs and produces television screens. Thomas leads the team that develops the remote control, and it has been found that 4% of the remote controls produced have had a fault. A change is made to the manufacturing process and Thomas wants to determine if this has had any effect on changing the percentage that have a fault. A random sample of 60 brand new remote controls are checked and six are found to have the fault. Carry out a hypothesis test at the 5% significance level to investigate whether the new manufacturing process has changed the percentage that have a fault.

15) It is known that 35% of roses grown by a horticulturist suffer from powdery mildew. The horticulturist separates the roses from their other plants and grows them in a large new greenhouse instead. They want to conduct a hypothesis test to determine whether this has

1.57 Binomial Hypothesis Testing (answers on page 472)

had any change on the proportion that suffer from powdery mildew. A random sample of 300 roses from the new greenhouse are collected and 81 are found to suffer from powdery mildew. Test, at the 3% significance level, whether there is any evidence to suggest the proportion has changed.

16) Chloe and Ajay are playing a board game with a single six-sided die. After several turns, Chloe believes the die is bias in favour of scoring a 1. They decide to pause the board game and roll the die 35 times to test this theory.
   a) Write down suitable hypotheses to test Chloe's belief.
   b) Using a 10% significance level, find the critical region for the test.
   c) Of the 35 rolls, 1 appears seven times. Complete the test.

17) The probability of having to wait longer than three hours at a hospital emergency unit is known to be 76%. Another doctor is employed in the hope that waiting times will be reduced. A hypothesis test is to be conducted to test this belief.
   a) Write down the hypotheses that should be used to test whether there has been a reduction in the number of patients waiting longer than three hours.
   b) A random sample of 70 patients is taken. Using a 1% level of significance, find the critical region for the test.
   Of the 70 patients in the sample, 43 waited longer than three hours to be seen.
   c) Complete the test.

18) A large supermarket sells soup by the can or in packs of six. Of those customers who bought soup in 2022, it was recorded that 32% bought packs of six. During the following year, Becky, the manager of the supermarket, believes that the percentage has changed. Using the 2023 records, she collects a random sample of 80 customers who bought cans of soup.
   a) Write down the hypotheses that should be used to test Becky's belief.
   b) Using a 5% significance level, find the critical region for a two-tailed test.
   c) Find the actual significance level of a test based on your critical region.
   Of the 80 customers who were selected in the sample, 45 bought cans of soup in packs of six.
   d) Complete the test.

19) Aaron Farms grows grain for stockfeed. Richard, a farmer, receives bags of grain from Aaron Farms and knows from experience that 5% will be spoilt. Richard decides to test whether another supplier, Baker Farms, will deliver fewer spoilt bags of grain. Richard makes several orders with Baker Farms and takes a random sample of 550 bags of grain.
   a) Write down suitable hypotheses to test whether the number of bags of spoilt grain has been reduced.
   b) Using a 10% significance level, find the critical region such a test.
   c) Of the 550 bags of grain, 13 are found to be spoilt. Complete the test.

20) John collects dinosaur models that are sold in sealed egg containers that the collector cannot see through. On average, 8% of the egg containers contain a limited edition gold dinosaur. John purchases 25 of the eggs. This may be considered to be a random sample.
   a) Find the probability that
      i) exactly one of the eggs contains a gold dinosaur,
      ii) at most two eggs contain gold dinosaurs.
   b) John buys 30 more eggs from a website. This may also be considered to be a random sample. John believes there is a greater chance of obtaining gold dinosaurs online. John decides to conduct a hypothesis test at the 5% level.
      i) Write down suitable hypotheses for the test.
      ii) Find the critical region for the test.
      iii) John opens the eggs to find there are four golden dinosaurs. Complete the test.

1.58 Large Data Set: Edexcel (answers on page 473)

1) State the following facts about the Large Data Set:
   a) The 5 locations in the UK.
   b) The 3 locations not in the UK.
   c) The years in which the data was collated.
   d) The months which are included and the maximum total number of days in those months.

2) Which of the locations
   a) are costal?
   b) are south of the equator?

3) Why does Perth have a lower temperature than Jacksonville for the data in the Large Data Set?

4) Which of the UK locations has the highest windspeed?

5) State the units of the following:
   a) Daily mean temperature.
   b) Daily total rainfall.
   c) Daily total sunshine.
   d) Daily maximum relative humidity.
   e) Daily mean windspeed.
   f) Cloud cover.
   g) Visibility.
   h) Pressure.

6) Convert the following:
   a) 3 knots into mph.
   b) 5.75 mph into knots.
   c) 30 mph into knots.

7) Explain what it means for cloud coverage to be
   a) 1 oktas
   b) 4 oktas
   c) 8 oktas

8) According to the Large Data Set, on 29th May 1987, the wind in Heathrow was blowing with bearing 290°. In which direction was the wind blowing as a compass direction?

9) In the Large Data Set, is a location more likely to be high cloud coverage or low cloud coverage?

10) Zara has calculated the mean and standard deviation of a sample of rainfall in Leeming in September 2015.
    She used only the data which has a value, obtaining $\bar{x} = 1.6$ and $\sigma = 4.2$ where $n = 24$
    a) Comment on the validity of her results.
    b) Comment on how the mean and standard deviation will change after a correction.

11) Two months' worth of data for hours of daily sunshine, x has been summarised for Hurn for July & August 2015 below:
    $\Sigma x = 321.2, \Sigma x^2 = 2692.66, n = 62$
    a) Calculate the mean and the standard deviation, where $s_x = \sqrt{\dfrac{\Sigma x^2 - \frac{(\Sigma x)^2}{n}}{n-1}}$, giving your answer to 2 decimal places.

    In Hurn for the same months in 1987 gives
    $\bar{x} = 6.42$ and $s_x = 4.62$
    b) Compare these to the statistics calculated for 2015 above.

12) The daily temperature for the same two consecutive months in both 1987 and 2015 in Camborne are shown in the boxplots below:

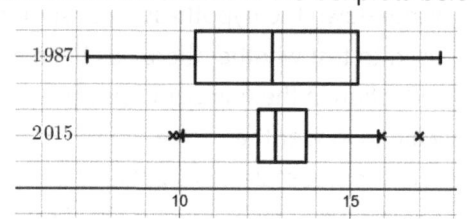

    a) Which are these months likely to be?

    It is claimed based on these data that the mean temperature is becoming more consistent across the UK.
    b) Explain the validity of this claim.

13) Jeremy is investigating the daily total rainfall in Jacksonville in October 2015. He selects data by selecting the first data point, then picks every 6th one after that until he reaches the end of the data set. State
    a) State the type of sampling Jeremy used.
    b) State a mistake with his application of this sampling method.

1.58 Large Data Set: Edexcel (answers on page 473)

14) The table below shows the probability distribution for c, the cloud coverage in Leeming in 2015.

c	1	2	3	4
$P(X=c)$		0.04	0.09	0.14
c	5	6	7	8
$P(X=c)$	0.17	0.22	0.23	

a) Find the missing probabilities given it is known that
$$10P(c=1) = P(c=8)$$
b) Find the expected number of days that the cloud coverage is 3 oktas or less.

It is claimed that this data could be modelled by $X \sim B\left(54, \frac{1}{9}\right)$

c) Using this model, find the number of days that cloud coverage is 3 oktas or less.
d) Comment on the validity of this claim.

15) A sample of 100 data points has been taken from Beijing in 1987. The windspeed w (knots) has been plotted against pressure p (hPa) and is shown in the diagram below. A regression line has been found, given by
$$w = 0.151p - 148$$

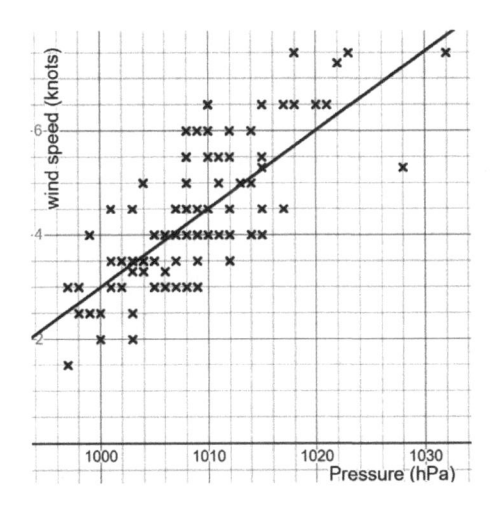

a) Use the model to predict the windspeed when there is a pressure of 1026 hPa.

The PMCC is found to be $r = 0.734$

b) Comment on why there is reason to believe that that data may not have been sampled randomly.

Prerequisite knowledge for Q16: Normal Distribution

16) The daily windspeed in Perth in 2015 is shown on the histogram below:

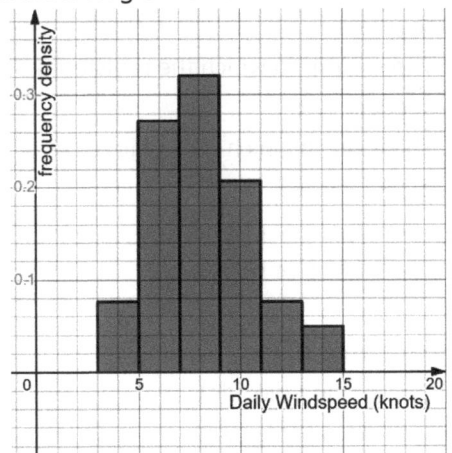

The heights of the bars are
$\{0.076, 0.272, 0.321, 0.206, 0.076, 0.049\}$

a) Estimate the mean and the standard deviation of the data.
b) Estimate the number of days the windspeed is above 12 knots.
c) By modelling the data using a normal distribution with your mean and standard deviation, find the probability of the windspeed being over 12 knots.
d) Is the normal distribution a suitable model for these data?

Prerequisite knowledge for Q17: PMCC Hypothesis Testing

17) Adrian believes that there is a negative correlation between the windspeed and relative humidity in Leeming in 2015.
He collects a random sample of 20 dates and calculates the sample's correlation coefficient to be -0.3613
The critical value for $n = 20$ for a one-tailed test at a 5% significance level is 0.3783.

a) Carry out a hypothesis test at the 5% significance level to investigate whether Adrian's belief is supported.
b) A second sample of 20 dates is taken, and the correlation coefficient is -0.6234. Write the conclusion of the hypothesis test for this sample.

1.59 Large Data Set: AQA (answers on page 473)

1) State the following facts about the Large Data Set:
   a) The five makes of vehicle.
   b) The years in which the vehicles were registered.
   c) The regions in which the registered keepers live.
   d) The years in which the vehicles were manufactured.
   e) The 5 types of propulsion.
   f) The 9 types of body type.

2) State the units of the following:
   a) CO_2 (carbon dioxide).
   b) CO (carbon monoxide).
   c) NOX (oxides of nitrogen).
   d) part (particulate emissions).
   e) hc (hydrocarbon emissions).
   f) mass.
   g) engine size.

3) Explain why, when sampling from the Large Data Set, that blanks may be found in some the data.

4) Kerry samples a car from the Large Data Set and notes down that its mass, particulate emissions, and engine size are all 0.
   In each case, state whether Kerry may have made a mistake or has correctly noted the details of the car.

5) For each of the following, given a sample is being taken from the Large Data Set, write down the name of the sampling method used.
   a) The order of the cars is randomised, then every 100th is noted.
   b) The order of the cars is randomised, then the first 10 of each type of make is noted to obtain 50 cars.
   c) The data set is sorted by makes and the proportions of each type calculated. Using a random number generator and the reference number, each make is selected to the same proportion as the full data set.

6) Martha uses the complete Large Data Set and finds that the mean mass of a car in 2002 is 1278.7 kg and in 2016 is 1416.6 kg. She finds out that the average mass of a human in 2002 and 2016 was 75.9 kg and 78.0 kg respectively.
   a) She says the actual mass of the cars is 1202.8 kg and 1338.6 kg. Explain why Martha is incorrect.
   b) She claims that all vehicles in the UK are heavier in 2016 than in 2002. Comment on the validity of Martha's claim.

7) The table below shows a sample of 10 vehicles from the Large Data Set.

Make	Region	Engine Size
Vauxhall	London	1398
Volkswagen	London	1198
Ford	North West	999
Volkswagen	London	1598
Vauxhall	South West	1398
Ford	London	998
Toyota	London	1398
Vauxhall	South West	1398
BMW	London	2171
Ford	North West	1498

   a) Find the mean and the standard deviation of the engine size of this sample.
   b) Give a reason why the sample used above may have biased data.

Any value more than 2 standard deviations from the mean can be identified as an outlier.
   c) Find any outliers in this sample of engine size.

The mean engine size for the whole of the Large Data Set is 1646.2 cm^3 and the standard deviation is 456.0 cm^3
   d) Give one comparison of central tendency for this sample vs the full data set.
   e) Give one comparison of spread of this sample vs the full data set.

1.59 Large Data Set: AQA (answers on page 473)

8) The table below shows the CO_2 emissions (g/km) in 2002 and 2016, for all the diesel cars in the Large Data Set.

	2002	2016
$\sum x$	46 335	127 096
Number of cars	302	1 097

   a) Find the mean of the CO_2 emissions for 2002 and for 2016

   The mean CO_2 emissions for all the cars registered in 2016 in the Large Data Set is 120.37. There are 2529 total cars in the Large Data Set in the year 2016.

   b) Find the mean of all the CO_2 emissions for all the vehicles in 2016 which are not diesel.

   c) It is claimed that the increase in CO_2 emissions is caused by the electric cars which feature solely in the 2016 data. Explain, with reference to the Large Data Set, why this cannot be the case.

9) A set of cars were randomly sampled from the Large Data Set and the masses have been displayed in the histogram below:

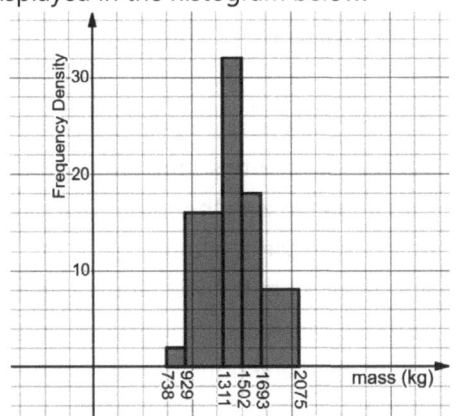

   It is known that the bar between 1693 kg and 2075 kg represents 8 cars.

   a) Calculate how many cars have been sampled altogether.
   b) Estimate the mean mass of a car in this sample.
   c) Find the probability of picking a car of mass between 1120 kg and 1693 kg.

10) The Keeper Title of a sample of 250 cars are taken from the Large Data Set and a probability distribution is given:

Keeper Title	Probability
Male	$\frac{2}{5}$
Female	k
Unknown	$\frac{4}{125}$
Company	$\frac{9}{25}$

   a) Find the value of k
   b) State the number of each of the Keeper Title sampled.
   c) Give one reason why the sample is representative of the Large Data Set.

11) A set of 100 cars were sampled from the Large Data Set and the engine sizes of these cars are displayed in the boxplot below:

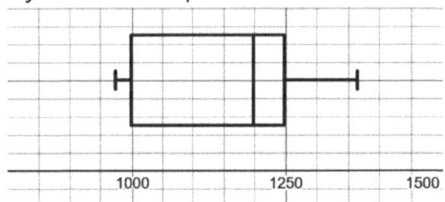

   Using your knowledge of the Large Data Set, explain why it is possible that it was not sampled randomly.

Prerequisite knowledge for Q12: Normal Distribution

12) The CO_2 emissions, x g/km, from a sample of 40 cars are taken from the Large Data Set. The summary statistics are as follows:
$$\sum x = 5592, \quad \sum x^2 = 802\,846$$

   a) Calculate the mean and the standard deviation, where $s_x = \sqrt{\frac{\sum x^2 - \frac{(\sum x)^2}{n}}{n-1}}$, giving your answer to 4 significant figures.

   This data is believed to be normally distributed. Find the probability that the CO_2 emissions are
   b) greater than 150 g/km.
   c) exactly 114 g/km.

1.60 Large Data Set: OCR MEI (Health) (answers on page 474)

These exercises were written using LDS6 but should continue to be of use for future Health Large Data Set samples.

1) State the following facts about the Large Data Set:
   a) The age of the youngest person.
   b) The age of the oldest person.
   c) The six categories of marital status.
   d) What the category "Food30" is recording.
   e) What the category "Arm" is recording.
   f) The country in which the original survey is conducted.

2) State the units of the following:
   a) Weight.
   b) Height.
   c) Body mass index.
   d) Lengths of body parts.
   e) Pulse rate.
   f) Blood pressure.

3) What is the data set taken from?

4) Explain the difference between the systolic blood pressure and diastolic blood pressure.

5) Explain why there are columns in the Large Data Set labelled with a 1, 2, 3, 4, in the context of what was happening on the day that the data was collected.

6) What data do the columns labelled Systolic and Diastolic represent in the Large Data Set?

7) Mustafa wants a sample of 20 people from the Large Data Set.
   He claims that all he needs to do is pick 20 random numbers from between 1 and 200 and use those rows from the spreadsheet.
   a) State what type of sampling he has chosen.
   b) Explain why this might not end up with 20 useful sets of data.

8) Kristina would like to clean the data set before she samples from it. She decides to remove any rows which contain a #N/A except the 'Systolic4' and the 'Distolic4' columns. Explain the effect of this on the data set.

9) Katie would like to take a sample to do a comparison of body mass index for different categories of marital status. She would like a sample of 50 with each marital status proportionally represented.
   a) State the type of sampling she should use.
   b) Explain how she would do this.

10) The table below shows a sample of 10 people from the Large Data Set.

Sex	Height
Female	173.4
Female	171.4
Male	183.5
Female	174.7
Male	180.1
Male	184.9
Male	186.2
Male	173.3
Male	159.4
Male	188.4

   a) Find the mean and the standard deviation of the height of this sample.
   b) Give a reason why the sample used above may have biased data.
   c) Any value more than 2 standard deviations from the mean can be identified as an outlier. Find any outliers in this sample of heights.

   The mean height for the whole of the Large Data Set is 168.0 cm^3 and the standard deviation is 9.9 cm^3.
   d) Give one comparison of central tendency for this sample vs the full data set.
   e) Give one comparison of spread of this sample vs the full data set.

1.60 Large Data Set: OCR MEI (Health) (answers on page 474)

11) The scatter graph below shows the height against thigh length of the 179 data points in the Large Data Set. A regression line has been found, given by
$$h = 1.95t + 88.70$$
where h is the height of a person in cm and t is the thigh length of a person in cm.

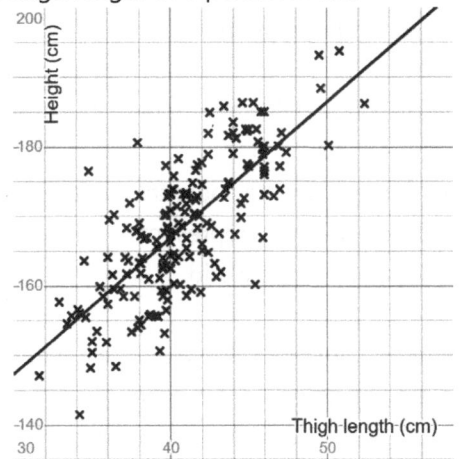

One of the people not included in the data set is 116.3 cm tall.
   a) Assuming you can make a prediction by using the regression line, calculate the thigh length of this person, giving your answer to 1 decimal place.

The product moment correlation coefficient for the data is 0.7518
   b) Explain why it is reasonable to model the relationship between t and h as a linear graph.

12) The average diastolic blood pressure for males and females is shown in the boxplots below.

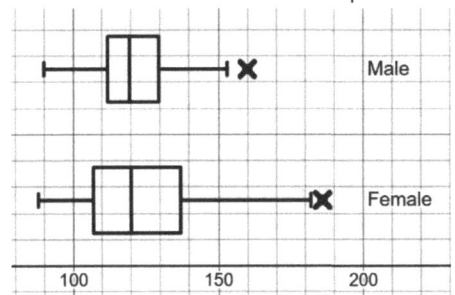

   a) Estimate the median of males and the median of females.
   b) Estimate the interquartile range of males and the interquartile range of females.
   c) Use your answers to compare the average diastolic blood pressure for males and females in the Large Data Set.

13) The histogram below shows the ages, in years, of the people in the Large Data Set.

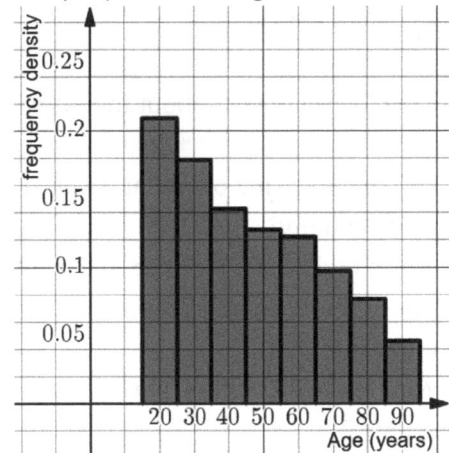

The heights of the bars are
{0.209, 0.179, 0.143, 0.128, 0.122, 0.097, 0.076, 0.046}
   a) Which class interval contains the median?
   b) Estimate the mean.

An adult is selected at random.
   c) Find the probability that they are younger than 45 years old.
   d) Explain why the final bar may not be a suitable width to represent the data.

Prerequisite knowledge for Q14: PMCC Hypothesis Testing

14) Emily believes that there is a positive correlation between BMI and average systolic blood pressure. She collects a random sample of 20 people from the Large Data Set and calculates the sample's correlation coefficient to be 0.8159. The critical value for $n = 20$ for a one-tailed test at a 5% significant level is 0.3783
   a) Carry out a hypothesis test at the 5% significance level to investigate whether Emily's belief is supported.

Emily's teacher is suspicious of these results, and they create a scatter graph for the whole data set, drawing a vertical and horizontal line going through the mean point. They state that "there are approximately the same number of data points in each quarter of the graph."
   b) Explain what the teacher's graph shows and why they believe Emily selected the data with bias.

1.61 Large Data Set: OCR MEI (World Bank) (answers on page 475)

These exercises were written using LDS7 but should continue to be of use for future World Bank Large Data Set samples.

1) State the following facts about the Large Data Set:
   a) The country is the official name used by which country?
   b) Which regions are included in the data?
   c) How is the population of each country found?
   d) What is the unit of the total area of a country?

2) Explain what a birth or death rate per 1000 is and what point of the year it is based on.

3)
   a) Explain what GDP is.
   b) What does 'real GDP per capita' mean?

4) Explain what is meant by 'physician density' and what definition a physician has for this data set.

5)
   a) The World Health Organisation categorises healthcare workers as what?
   b) According to the World Health Organisation, fewer than how many healthcare workers per 1000 would be insufficient to achieve coverage of primary healthcare needs?

   Using your answer to part b)
   c) Identify one of the 3 countries in Europe that do not meet the World Health Organisation value for coverage of primary healthcare needs.
   d) Identify the region of the world in which no countries meet the World Health Organisation value for coverage of primary healthcare needs.

6) Explain what is meant by 'current health expenditure' and why it is a useful piece of data.

7) Caspian claims that the world has more mobile phones than people.
   a) Explain how the Large Data Set indicate that this may be the case in some countries.
   b) In which region are there the fewest mobile phone subscribers per 100 people in the population?

   Caspian would like a sample of 46 countries to test his theory.
   c) Explain how he could take a sample which proportionally represents each of the regions in the world.

8) Jamie is comparing the life expectancy in 2020 to the life expectancy in other years. Are there any countries in which life expectancy went down in the time frame from
   a) 1960 to 2020?
   b) 1990 to 2020?

9) The median age of countries in Europe and Asia are shown in the boxplots below:

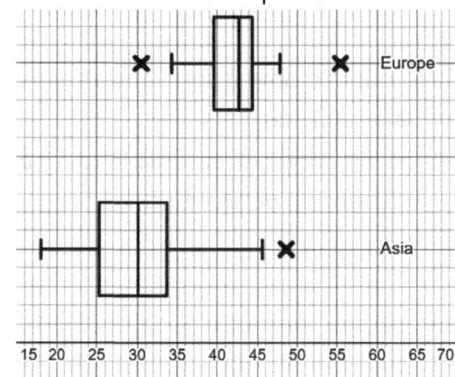

   a) Identify the outliers in each of the boxplots.
   b) The Europe data had to be cleaned before it could be used, explain why this is the case and how it could be identified in the Large Data Set.
   c) Make a comparison of the two data sets.

   Oscar states that there is a country missing from the boxplots.
   d) Which country was not included in either boxplot, but lies in the correct part of the world?

1.61 Large Data Set: OCR MEI (World Bank) (answers on page 475)

10) The table below shows the current health expenditure for the Caribbean after it has been cleaned:

Country	CHE (%)
Antigua and Barbuda	4.4
The Bahamas	5.8
Barbados	6.3
Cuba	11.3
Dominica	5.5
Dominican Republic	5.9
Grenada	5
Haiti	4.7
Jamaica	6.1
Saint Kitts and Nevis	5.4
Saint Lucia	4.3
Saint Vincent and the Grenadines	4.8
Trinidad and Tobago	7

a) Find the mean and the standard deviation of the current health expenditure for the Caribbean.

Any value more than 2 standard deviations from the mean can be identified as an outlier.

b) Identify any countries which are outliers.

The current health expenditure in Oceania is $\bar{x} = 8.79\%$ and $s_x = 5.87\%$

c) Compare these to the statistics calculated for the Caribbean above.
d) Give two reasons why it may be fair to compare the Caribbean with Oceania.
e) Give a reason why it may not be fair to compare the Caribbean with Oceania.

11) The life expectancy at birth, L, between 1960 and 2020 can be modelled using a straight line where t is the number of decades since 1960.

Botswana: $L = 2.01t + 50.1$
Chile: $L = 3.90t + 55.5$

a) Predict the life expectancy for 1995 for both countries.
b) Explain what each of the values in the models mean for each country.
c) Compare the change in life expectancy of the two countries.

12) The median age of a country is plotted against the physician density (physicians/1000 population) of a country, creating the graph below. The product moment correlation coefficient is $r = 0.7949$ and the least squares regression line is given by
$$y = 4.37x + 22.6$$

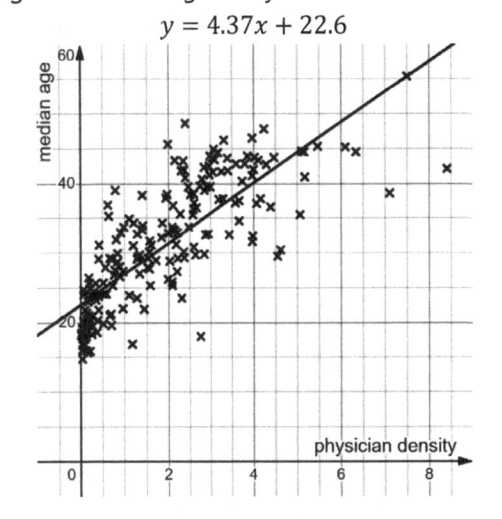

a) Comment on the value of r and what it means for the association between median age and physician density.

It is known that Cuba has a very high physician density but not a high median age.

b) Identify which is likely to be the point which represents Cuba.

A country which was not included in the graph decides to manage their physician density, so it is 7 per 1000 people with the aim to increase the median age of people in their country.

c) Use the regression line to determine the median age for this country.
d) Comment on your result in part c) with reference to the aim of the country.

Prerequisite knowledge for Q13: Normal Distribution

13) The death rate per 1000 people across the world may be considered normally distributed, where $X \sim N(7.5, 6.92)$. Find
a) the probability that the death rate per 1000 people of a country is higher than 6
b) $P(\mu - 2\sigma < X < \mu + 2\sigma)$
c) the death rate per 1000 for the 10% of the countries with the highest death rate.

1.62 Large Data Set: OCR MEI (London) (answers on page 476)

These exercises were written using LDS8 but should continue to be of use for future London Large Data Set samples.

1) State the following facts about the Large Data Set:
   a) How many boroughs of London are listed.
   b) How many areas of England are recorded (ignoring categories, Wales, England, England and Wales, and UK).
   c) Who defines which boroughs are Inner or Outer London for this data set?

2) Explain what each of the following categories in the Large Data Set are:
   a) Employment Rate (age 16 to 64).
   b) Mean and median Income of Taxpayers (£).
   c) All pupils at end of KS4 achieving 5+ A*-C including English and Maths.
   d) Household Waste Recycling Rate.

3) Which properties are excluded from the median house price category?

4) Why should the data related to the mean and median income of taxpayers be treated with caution?

5) A sample of 5 London boroughs are being sampled from the Large Data Set. For each of the following, write down the name of the sampling method used.
   a) Each borough is given a number between 1 and 33 then a random number generator is used to select 5, until 5 unique boroughs have been identified.
   b) The order of the boroughs is randomised, then every 6th is noted.
   c) The order of the boroughs is randomised, then the first 5 are sampled.
   d) The data is sorted into inner and outer London and then 1 inner London and 4 outer London boroughs are randomly selected using a similar approach to in part a).

6) The table below a sample of 10 boroughs in London and their median house price in 2021:

Area	Inner or Outer London	Median House Price (£)
Barnet	Outer	575 500
Brent	Outer	525 000
Camden	Inner	765 000
Ealing	Outer	522 000
Hammersmith and Fulham	Inner	780 000
Harrow	Outer	535 000
Kensington and Chelsea	Inner	1 250 000
Merton	Outer	545 000
Richmond upon Thames	Outer	715 000
Wandsworth	Inner	656 178

   a) Find the mean and the standard deviation of the median house price.
   b) Give a reason why the sample used above may have biased data.

   Any value more than 2 standard deviations from the mean can be identified as an outlier.
   c) Identify any boroughs which are outliers.

   The same boroughs in 2007 for median house price £x gives
   $\bar{x} = £566\,727.50$ and $s_x = £176\,435.80$
   d) Compare these to the statistics calculated above.

7) The mean income of taxpayers and median income of taxpayers in 2018/19 for Lewisham and for Kensington and Chelsea are shown in the table below:

	Mean (£)	Median (£)
Kensington and Chelsea	173 000	39 300
Lewisham	38 000	27 900

Compare the distributions of the income of taxpayers for both areas.

1.62 Large Data Set: OCR MEI (London) (answers on page 476)

8) The mean income I in £1000s has been recorded against time t for 12 consecutive years for both Hackney and the North West of England.

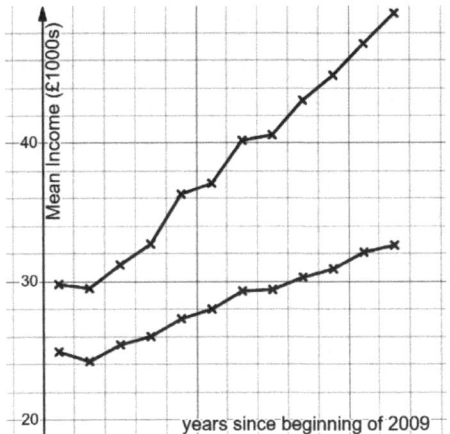

   a) Identify which of the graphs is Hackney.

   The two lines may be modelled by:
   $I = 1.88t + 26.3$ and $I = 0.781t + 23.3$

   b) Predict the mean income for 2021/2022 according to these models.
   c) Explain what the values 1.88 and 26.3, and 0.781 and 23.3, within the models above mean in context.
   d) Compare the change in mean income of the two areas.

9) The percentage of all pupils at the end of KS4 achieving 5 or more A*- C grades including English and Maths in the academic year 2007/8 for Inner London, Outer London and the rest of England are shown in the boxplots below.

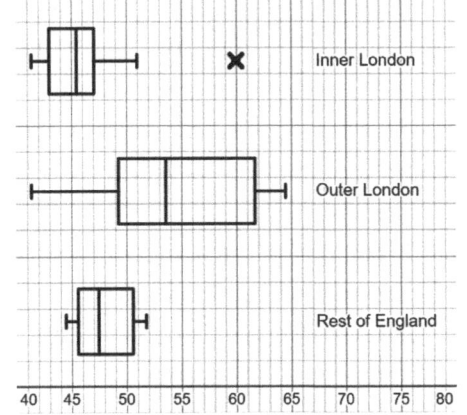

   a) Using your knowledge of the Large Data Set, identify the area of the outlier in Inner London.
   b) Compare the 3 data sets.

10) The scatter graph below shows the household recycling rate (%) against the employment rate for 16-64 year olds for 20 of the London boroughs in 2011. The product moment correlation coefficient is $r = 0.7227$ and the least squares regression line is given by
    $$y = 0.979x - 33.0$$

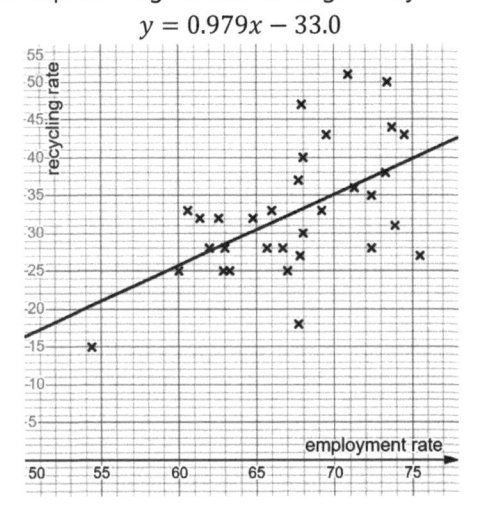

    a) What does the value of r mean for the association between the household recycling rate and the employment rate?

    One of the boroughs not included in the graph has an employment rate of 73.4%.
    b) Find the recycling rate of this borough according to the regression line.

    Lewisham was not included in the sample. It had an employment rate of 67.7% and a recycling rate of 18% in 2011.
    c) If this data point were to be included, explain how the regression line and the value of r would change.

Prerequisite knowledge for Q11: Normal Distribution

11) The percentage of all pupils achieving 5 or more A*- C grades across the whole of England in 2011/2 may be considered normally distributed, where $X \sim N(59.4, 4)$. Find
    a) the probability that the percentage is higher than 58%.
    b) the probability that the percentage is less than 55%.
    c) $P(\mu - 2\sigma < X < \mu + 2\sigma)$

1.63 Large Data Set: OCR A (answers on page 477)

1) State the following facts about the Large Data Set:
   a) the 10 different categories of method of travel to work (excluding 'other').
   b) the extra category to ensure that all people are included in the travel data collection.
   c) the 2 different years from which the data was collected.

2) What happened to the authorities between 2001 and 2011 which has an impact when comparing the two years with this data?

3) Kyra states that if she wants to find all the people who come to work by park and ride the only category she needs to consider is "bus, minibus or coach". Explain why Kyra is wrong.

4) In 2011, Abdullah was unemployed the week before the census, but started a new job close to his house the day he filled it in. Which category should he have identified as?

5) Toby is considering the 2011 age data by authority. He notices that Exeter has 14155 people aged 20-24 whilst Ashford has 6245 aged 20-24 despite a similar total number of residents. Explain a reason why this might be the case.

6) Jakub states that in 2001, the number of people using the Underground, metro, light rail and tram in the South East and the South West can be summarised using the following statistics:
   $\Sigma x = 1915$, $\Sigma x^2 = 196\,261$, and $n = 36$
   a) Calculate the mean and the standard deviation, where $s_x = \sqrt{\frac{\Sigma x^2 - \frac{(\Sigma x)^2}{n}}{n-1}}$, giving your answer to 1 decimal place.

   Jakub's teacher explains he is wrong as he has missed a data point.
   b) Calculate the new mean and standard deviation which includes this data point.

7) State how a baby which is less than 1 year is categorised in the Large Data Set.

8) For each of the following, given a sample of number of cyclists per authority is being taken from the Large Data Set, write down the name of the sampling method used.
   a) The order of the authorities is randomised, then every 100th is noted.
   b) The order of the authorities is randomised, then the first 5 from each region is noted to obtain 50 authorities.
   c) The data set is sorted by authorities and the proportion of each authority is calculated. Using a random number generator and the position in the spreadsheet, each authority is selected to the same proportion as the full data set.
   d) The order of the authorities is randomised, then the first 20 authorities are chosen.

9) The table below summarises some age data about London.

	2001	2011
Age 0 to 4	478187	591495
Age 5 to 14	887190	939674
Age 15 to 24	947810	1101631
Age 25 to 44	2533089	2903920
Age 45 to 64	1434225	1732472
Age 65 to 74	468067	473058
Age 75 to 84	310553	308661
Age 85 to 94	112970	123030
	7172091	8173941

   a) Is there a significant increase in the proportion of 25–44-year-olds living in London?
   b) By comparing the number of 15-24-year-olds living in London in 2011, to the 5–14-year-olds living in London in 2001, comment on the change in this demographic.
   c) Explain why the final category includes data outside of the category range.

1.63 Large Data Set: OCR A (answers on page 477)

10) The scatter graph below shows the proportion of working people "traveling to work by any means which is not driving" against people "driving a car" in the North West in 2001. The product moment correlation coefficient, r, is -0.899 to 3 significant figures.

   a) One of the points is identified with a circle rather than a cross.
   By considering this point, explain why all the points do not lie exactly in a line with $r = 1$.

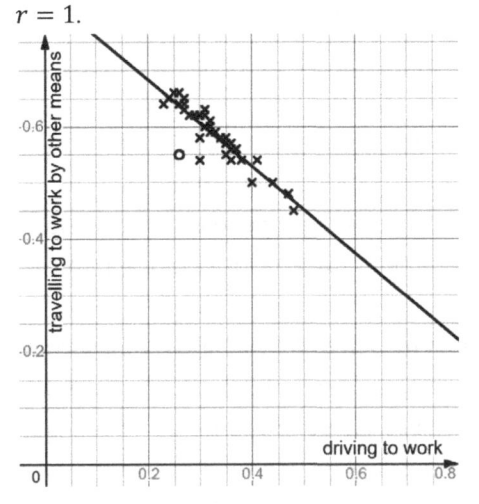

   The least squares regression line is given by
   $$y = -0.7718x + 0.8374$$
   where y is the proportion of people not travelling to work by driving and x is the proportion driving to work.

   b) By using the regression line, calculate how many of the people in the North West would typically drive to work if they were in a local authority where 50% of people drive to work.

11) The proportion of people travelling by bus, minibus or coach in 2001 and 2011 is shown in the boxplots below

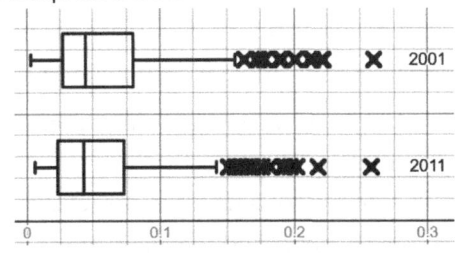

    Compare the two data sets.

12) Consider the two pie charts below showing data from 2001. One is in London and the other is in a rural location.

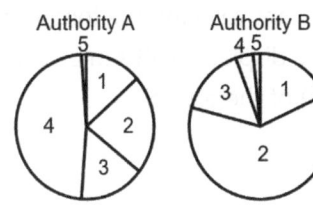

    The categories are 'any form of public transport', 'working from home', by 'bicycle or on foot', 'car as a passenger or a driver' and 'other' but not in that order.
    Identify which is which and make suggestions on which categories 1-5 will be, given they represent the same type of data in both pie charts.

Prerequisite knowledge for Q13: Normal Distribution

13) Alex creates a histogram to show the distribution of the age of people in the South East in 2011.
    There are 8 634 750 people living in this region in 2011.

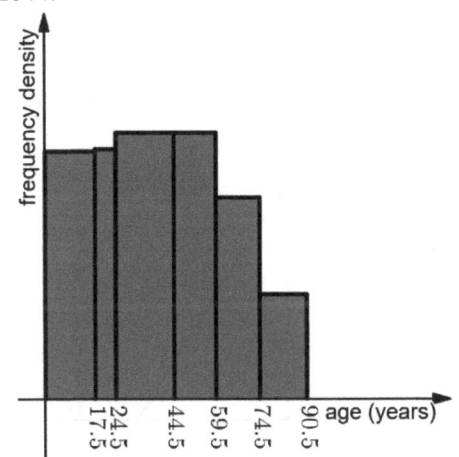

    The heights of the bars are
    $\{213, 215, 229, 229, 173, 90\}$

    a) Estimate the mean and the standard deviation of the data.
    b) Estimate the number of people aged 68 or above. Explain why this is an estimation.
    c) Would it be reasonable to assume that this data is normally distributed? Fully justify your answer.
    d) Explain how the histogram created by Alex could be improved to better represent the data.

1.64 Graphs of Motion (answers on page 478)

1) A battery-powered toy car is placed on a straight track. The car is switched on, so that when it contacts the track, it instantly moves with constant velocity. The following graph shows the displacement s (metres), over time t (seconds) of the toy car:

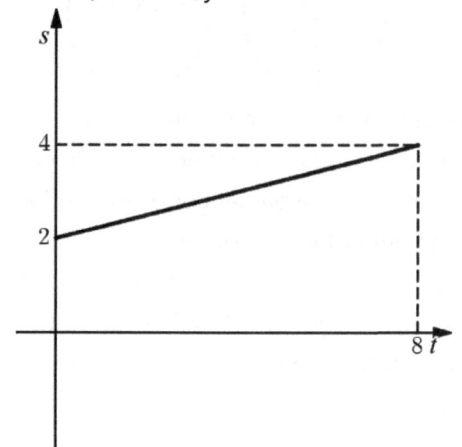

   a) How far along the track is the car initially placed?
   b) How far along the track is the car when 8 seconds have passed?
   c) Find the velocity of the car.
   d) How far along the track is the car after 5 seconds?

2) Displacement s (metres) over time t (seconds) for a particle is given in the graph below:

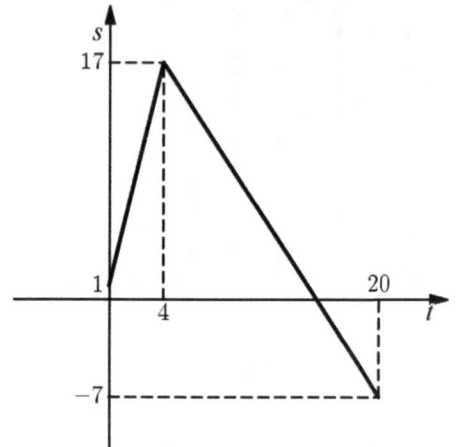

   Find
   a) the distance travelled over the first 20 seconds.
   b) the displacement over the first 20 seconds.
   c) the average speed of the particle.
   d) the velocity for the first part of the motion.
   e) the velocity for the second part of the motion.
   f) the average velocity of the particle.

3) Displacement s (metres) over time t (seconds) for a particle is given in the graph below:

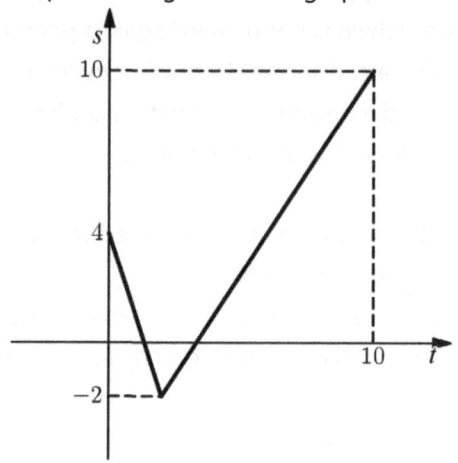

   Find, after the first 10 seconds,
   a) the distance travelled by the particle.
   b) the displacement of the particle.
   c) the average speed of the particle.
   d) the average velocity of the particle.

4) Lisa runs in a straight line for 60 seconds at a constant speed of 3 ms^{-1}. She then pauses for 20 seconds before turning and running in the opposite direction for 45 seconds at 4 ms^{-1}. Draw a displacement-time graph to show the athlete's motion.

5) A child is running along a straight, flat path. He runs for 6 seconds at 2 ms^{-1}. He then stops and waits for 5 seconds, before running back along the path for 9 seconds, returning to his original starting point.
   a) Sketch the displacement-time graph for the motion of the child.
   b) Find the exact velocity for the final stage of the motion.

6) An object travels 20 metres at constant velocity v ms^{-1} taking t seconds. It then pauses and remains stationary for 5 seconds before returning to its original position in $\frac{1}{2}t$ seconds.
   a) Sketch a displacement-time graph for the motion of the object.
   b) Given the total time the object is in motion is 6 seconds, find the value of t

188

1.64 Graphs of Motion (answers on page 478)

7) Describe what is happening to the objects in each of the following velocity-time graphs for velocity v, in metres per second and time t, in seconds. For each stage of motion comment on the speed and where possible its position:

   a)

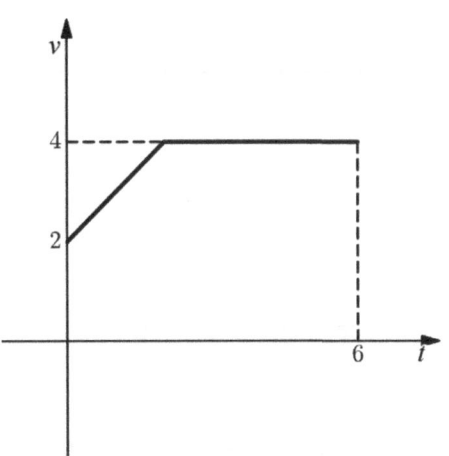

   b)

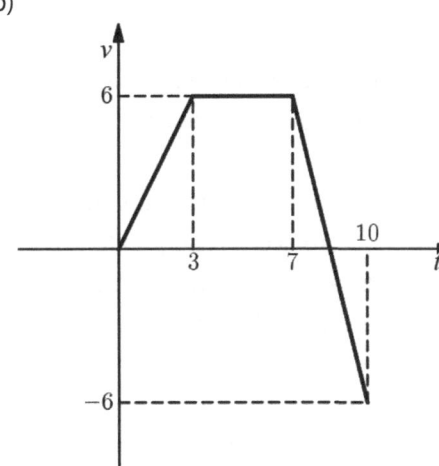

8) For each of the following velocity-time graphs below, sketch the corresponding acceleration-time graph:

   a)

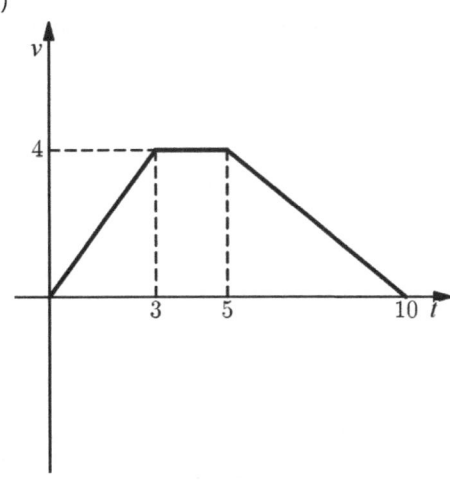

   b)

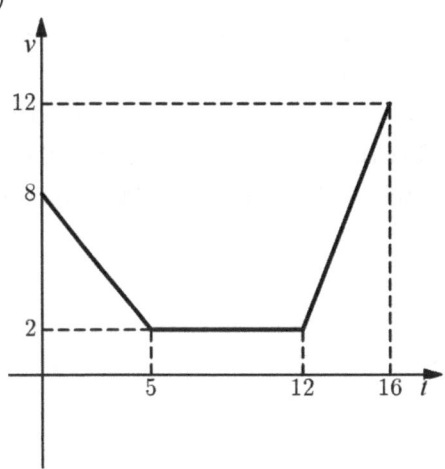

9) For each of the following acceleration-time graphs below, sketch the corresponding velocity-time graph given that the object is initially at rest:

   a)

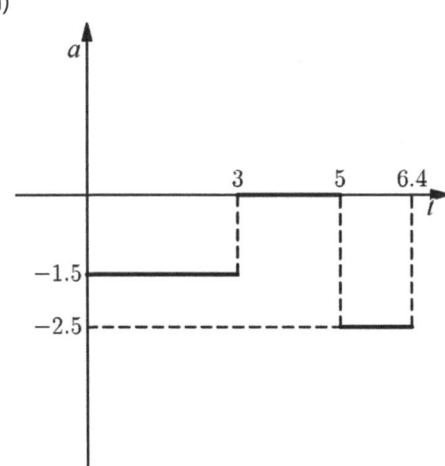

   b)

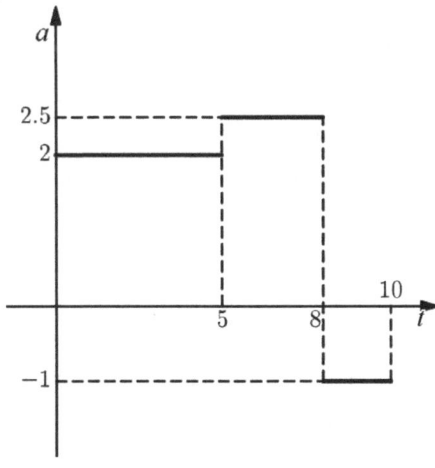

1.64 Graphs of Motion (answers on page 478)

10) A car which is initially at rest, accelerates at a constant rate of 4.5 ms^{-2} for 3 seconds and then remains at this constant speed.
    a) Sketch a velocity-time graph to show the first 10 seconds of the motion of the car.
    b) Find the distance travelled by the car in the first 10 seconds of motion.

11) A cyclist is timed along a section of track, with velocity v recorded in metres per second at $t = 10$, $t = 14$ and $t = 20$ seconds. The velocity-time graph for the cyclist is shown below.

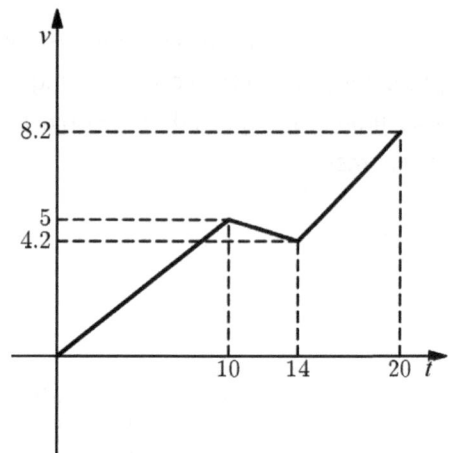

    Find the
    a) acceleration of the cyclist at each stage.
    b) total distance travelled by the cyclist.
    c) time when the cyclist is travelling at exactly 4 ms^{-1}.
    d) speed the cyclist is travelling at 17 seconds.

12) Sketch a velocity-time graph which represents a car initially at rest, with:
    a) constant acceleration for the first 6 seconds before reaching a fixed velocity of 10 ms^{-1} which it remains at for a further T seconds.
    b) increasing acceleration for the first 2 seconds, constant acceleration for the next 3 seconds, then decreasing acceleration for the next 1 second, before reaching a fixed velocity of 10 ms^{-1} which it remains at for a further T seconds.

13) The velocity v ms^{-1} at a given time t seconds of a van in motion on a straight flat road is modelled using the velocity-time graph below:

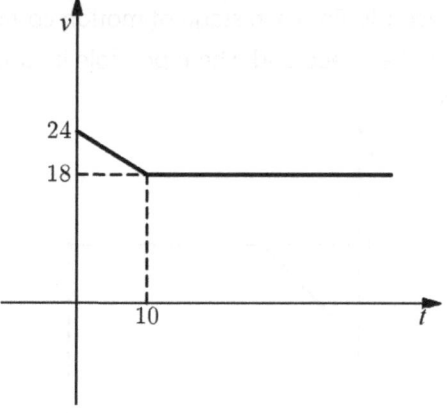

    a) Find the deceleration of the van in the first 10 seconds.
    b) At 10 seconds, the van continues to travel at a constant speed of 18 ms^{-1} for a distance of 1026 m. Find the time it spends at this speed.
    c) Find the total distance travelled by the van at the time given in your part b).

14) An object is moving in a straight line. It starts at 4 ms^{-1}, before accelerating uniformly to 10 ms^{-1} in 8 seconds, moving at 10 ms^{-1} for 5 seconds before finally decelerating to rest in 7 seconds.
    a) Sketch a velocity-time graph which represents the motion of the object.
    b) Find the distance travelled by the object.
    c) Find the average velocity of the object.

15) A van is driving along a straight stretch of motorway. It starts at 33 ms^{-1}, before decelerating to 22 ms^{-1} uniformly over 1.1 km.
    a) Sketch a velocity-time graph which represents the motion of the van.
    b) Find the time the van spends decelerating according to this model.

    The van maintains this speed for a further 400 metres.
    c) Find the average speed of the van.
    d) Two speed cameras, one at the start and one at the end of this 1.5 km stretch of road, measure the speed of the van and finds its average. Find this speed.

1.64 Graphs of Motion (answers on page 478)

16) A vehicle is being driven along a flat, straight road. Its motion is modelled using the velocity-time graph below where velocity v is given in metres per second and time t is given in seconds.

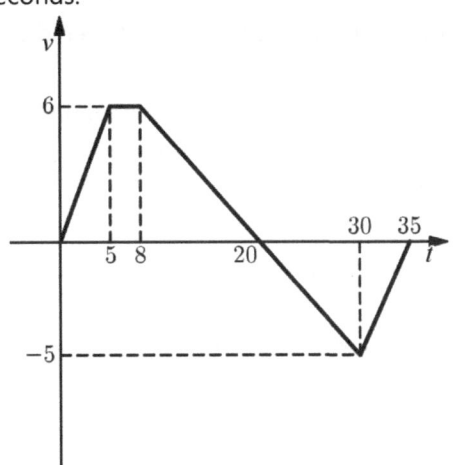

   a) Find the acceleration at each stage of motion.
   b) Find the total distance driven by the vehicle in the first 20 seconds.
   c) Find the total distance driven by the vehicle in the final 15 seconds.
   d) State the displacement of the vehicle at $t = 20$ seconds.
   e) State the displacement of the vehicle at $t = 35$ seconds.
   f) State the total distance driven by the vehicle.
   g) Find the average speed of the vehicle over the 35 seconds.

17) A cyclist starts from rest and is cycling along a flat straight road. He accelerates at a constant acceleration of 0.5 ms^{-2} for the first 5 seconds, before accelerating at a constant acceleration of 0.3 ms^{-2} for the next 5 seconds. He then remains at a constant speed for a further 20 m.
   a) Sketch a velocity-time graph to model the motion of the bicycle.
   b) State the velocity of the cyclist in the third stage of the motion.
   c) Find how far the cyclist has cycled since the initial start point.

18) The velocity v ms^{-1} at a given time t seconds of a car in motion is modelled using the velocity-time graph below.

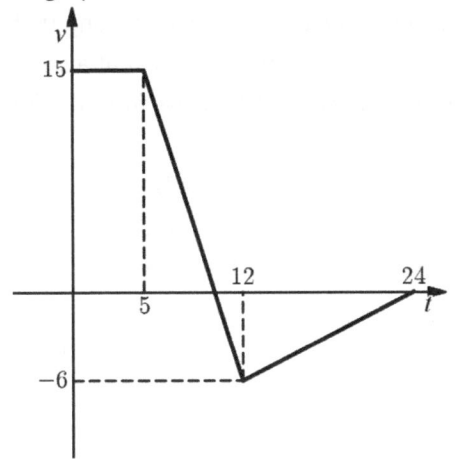

   Find
   a) the acceleration for all three stages of the motion.
   b) the exact times at which the car, according to the graph, has instantaneously stopped.
   c) the total distance travelled by the car.
   d) the displacement of the car from its initial position.

19) A ride at a theme park with a 35 second duration has a carriage that travels along a flat, straight track.
   The velocity-time graph below shows the motion of the carriage in time t in seconds.

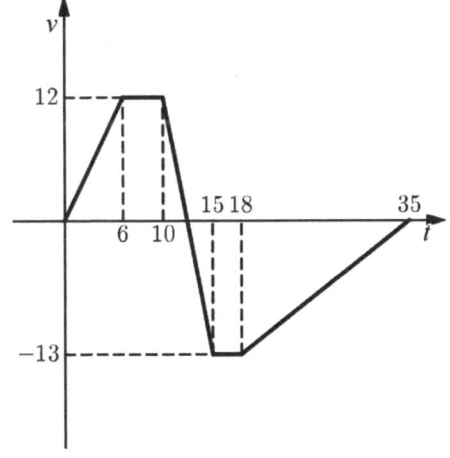

   a) Find the maximum magnitude of acceleration over the 35 second ride.
   b) Find the times at which the carriage is instantaneously stationary.
   c) If the passengers disembark at 35 seconds, where are they in relation to the start point?

1.64 Graphs of Motion (answers on page 478)

20) A bus travels from rest from a bus stop, on a straight, flat road. It drives with acceleration 0.9 ms^{-2} for 15 seconds. It is then driven at constant velocity before decelerating uniformly at 1 ms^{-2} where it comes to rest at a second bus stop which is 918 metres from the first bus stop.
    a) Find the velocity of the bus when it is moving with constant velocity.
    b) Sketch a velocity-time graph to represent the motion of the bus.
    c) Find the time taken to get between the bus stops.

21) The velocity-time graph shows the motion of a runner running along a straight flat track of length 200 metres before coming to a stop.

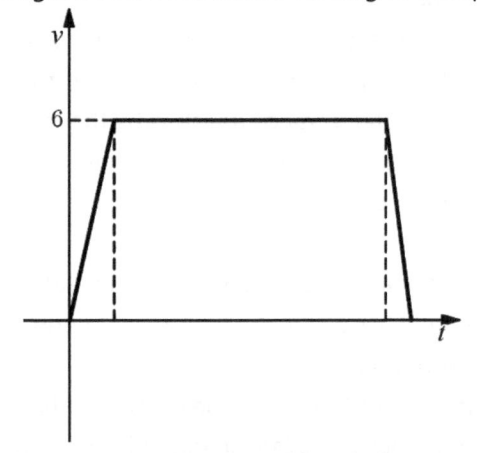

    The runner accelerates uniformly from rest over the first 15 metres until reaching a speed of 6 ms^{-1} before continuing to run at 6 ms^{-1} until they complete the 200m. A further 3 seconds later they come to rest. Find
    a) the exact time the runner completes the 200 metres.
    b) the total distance they run.

22) An object starts from rest and is travelling in a straight line with uniform acceleration for T seconds, travelling 24 metres. It then moves with constant velocity for 9 seconds before coming to rest uniformly in a further $\frac{3}{5}T$ seconds.
    a) Sketch a velocity-time graph for the motion of the object.
    b) Given the displacement of the object is 146.4 metres, find the value of T

23) Two cars, A and B are driving along a straight, flat road.
    Car A is driving with constant velocity 15 ms^{-1}.
    Car B, which is initially driving at 12 ms^{-1}, is driving with constant acceleration until 32 seconds have passed at which it then drives at 20 ms^{-1}.
    The velocity-time graph demonstrating the motion of the two cars is shown below:

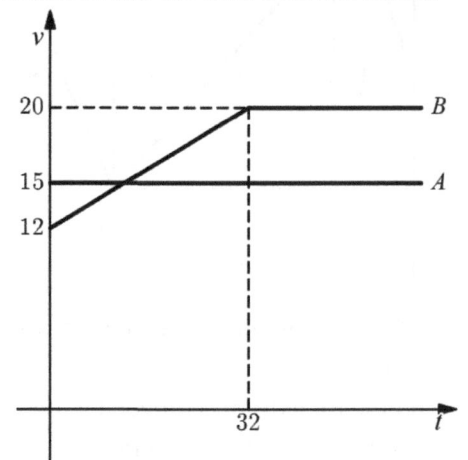

    a) Find the acceleration of car B for the first 32 seconds of the motion.
    b) Find the equation that gives the velocity of car B for the first 32 seconds of motion.
    c) Find the exact time at which car B overtakes car A according to the model.
    d) Find the distance the cars have travelled when car B overtakes car A.
    e) State what the point of intersection of the two lines represents.

24) Two cars, A and B are driving along a straight, flat road. Car A is driving with constant velocity u ms^{-1}. Car B, which is initially driving at $(u-5)$ ms^{-1}, is driving with constant acceleration until T seconds have passed at which it then drives at a constant velocity. Car B overtakes car A at exactly T seconds.
    a) Sketch the velocity-time graph for the two cars.
    b) State the velocity of car B when it overtakes car A.
    c) Given that the cars are going the same speed exactly when $t = 20$ and car B overtakes car A after driving 1000 m, find the values of T and u

1.65 Constant Acceleration (answers on page 479)

1) Using the following velocity-time graph

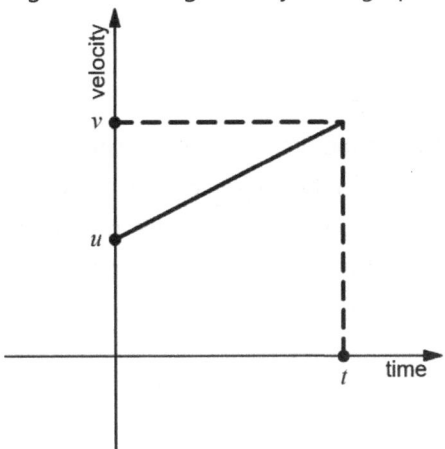

   show that
   a) $v = u + at$
   b) $s = \frac{1}{2}(u + v)t$
   c) $s = ut + \frac{1}{2}at^2$
   d) $s = vt - \frac{1}{2}at^2$

2) By substituting $v = u + at$ into $s = vt - \frac{1}{2}at^2$ show how $v^2 = u^2 + 2as$ may be derived.

3) In each case, use the constant acceleration formulae with the given values:
   a) $a = 2, u = 4, v = 6$, to find t
   b) $a = -2, t = 3, s = 5$, to find u
   c) $s = 12, t = 4, v = 5$, to find a
   d) $u = 2, a = 1.5, v = 3$, to find s
   e) $s = 20, t = 8, u = 5$, to find v

4) If $s = 20, t = 6, v = 2.5$, using the constant acceleration formulae, find u and a

5) If $s = 15, v = 4.5, u = 1.5$, using the constant acceleration formulae, find a and t

6) A particle is moving in a straight line from A to B with constant acceleration 2 ms^{-2}. Its speed at A is 6 ms^{-1} and it takes 7 seconds to travel between A and B. Find
   a) the speed of the particle at B
   b) the distance from A to B

7) Katie is walking along a straight path of length 100 m initially at 1.1 ms^{-1} and has constant acceleration 0.05 ms^{-2}. Find the time it takes for Katie to travel the path.

8) Fasil is running along the road and for this part he can be modelled as running with constant acceleration of 0.1 ms^{-2}. He is running at 5 ms^{-1} when he passes a lamp post. A further 10 seconds later he passes a litter bin. Find
   a) the distance between the lamp post and the litter bin.
   b) the speed at which Fasil passes the litter bin.

9) Find the acceleration and the displacement given each of the following scenarios, where the object starts from rest:
   a) $v = 4$ ms^{-1}, $t = 3$ s
   b) $v = 1.4$ cms^{-1}, $t = 2$ s
   c) $v = 10$ ms^{-1} giving your answer in terms of t

10) A cyclist is cycling on a straight flat road. Her initial velocity is u ms^{-1}, then she accelerates at a constant rate of a ms^{-2} for 12 seconds before reaching a speed of $3u$ ms^{-1}. Express a in terms of u

11) In perfect conditions, it takes 38 metres to stop a car driving at a speed of 80 kmh^{-1}.
    a) Find the time it takes to stop, assuming the deceleration is constant.
    b) Comment on why this may not be an accurate answer for modelling the stopping of a car in a real-world scenario.

12) Ceri cycles 2 km straight horizontal section of a race. She is cycling 28 kmh^{-1} initially and finishes the section at 32 kmh^{-1}. Find her acceleration for this section, assuming she accelerates at a constant rate.

13) A car travels along a straight horizontal road with constant acceleration, passing three points A, B and C. It takes 15 s to travel from B to C. It is moving at 12ms^{-1} and 16ms^{-1} at B and C respectively.
    a) Find the acceleration of the car
    b) Find the speed of the car at A given the distance between A and B is 150 metres.

1.65 Constant Acceleration (answers on page 479)

14) A car is travelling along a straight horizontal road with a constant acceleration, passing three points A, B and C.
It takes 6 seconds to travel from A to B and a further 4 seconds to travel 10.4 metres from B to C.
It is travelling at 5 ms^{-1} as it passes point B.
Find
   a) the acceleration of the car.
   b) the speed of the car at A.
   c) how far the car travels beyond point C before it comes to rest.

15) A vehicle travels with constant acceleration along a straight flat road.
It passes three checkpoints A, B and C and in each case is travelling at 15 ms^{-1}, 30 ms^{-1} and 50 ms^{-1}.
Find
   a) the ratio of the times taken to drive the distances AB and BC.
   b) the distance between AB given the distance AC is 150 metres.

16) A truck is travelling along a straight horizontal road with constant acceleration.
A set of traffic lights is 25 metres away from the truck. A junction is a further 110 metres on from the traffic lights.
The truck takes 2 seconds to travel to the traffic lights and 5 seconds to travel between the traffic lights the junction.
Find
   a) the acceleration of the truck.
   b) the speed of the truck in its initial position.

17) Two cars, A and B, are travelling along a flat, straight road.
Car A is driving at u ms^{-1} throughout its motion.
Initially, car B is 15 metres behind car A and is driving 0.2 ms^{-1} slower than Car A but is accelerating at 0.8 ms^{-2}.
Find the time it takes for Car B to catch up Car A.

18) A car is travelling on a straight flat road.
Initially it has velocity of 6 ms^{-1} and drives with constant acceleration, a ms^{-2}, past two checkpoints.
It reaches the first checkpoint which is at 48 metres from the initial point, in k seconds.
   a) Find a giving your answer in terms of k
   b) Find the initial speed for the second phase of the motion in terms of k

The next checkpoint is a further 204 metres away and it reaches it in 10 seconds.
   c) Find the value of k
   d) Find the value of a

19) A stone is dropped on a planet from a height of 3 metres. It takes 2.2 seconds to hit the surface. Find the acceleration due to gravity on this planet.

The following questions use $g = 9.81$ ms^{-2}

20) A ball is dropped, and it takes 0.8 seconds to reach the ground. Find
   a) the height above the ground from which the ball is dropped.
   b) the speed in terms of t
   c) the speed when the ball hits the ground.

The following questions use $g = 9.8$ ms^{-2}

21) An object, which is initially at rest is dropped from a height of 8m above the ground.
   a) Calculate the speed of the object when it hits the ground.
   b) Calculate the time it takes to hit the ground.
   c) State a refinement to the model, apart from air resistance, which would make the model more realistic.

22) A ball is projected upwards from the ground with speed 14.7 ms^{-1}. Find
   a) the time it takes to reach ground again.
   b) the total distance travelled in the first 2 seconds of motion.

1.65 Constant Acceleration (answers on page 479)

23) A pebble is thrown upwards in the air from the edge of a bridge with a velocity of 24.5 ms^{-1}.
    a) Find the time it takes for the ball to reach its maximum height.

    After 6.5 seconds from being thrown, the pebble lands in the river below. Find
    b) the height of the bridge at the point where the pebble was thrown.
    c) the total distance travelled by the pebble.

24) A stone is thrown vertically upwards in the air starting 1.65 m above the ground being projected with initial speed 6.25 ms^{-1}. The stone moves freely under gravity.
    a) Find the time at which the stone hits the ground.
    b) Find the times at which the stone is precisely 2 m above the ground.
    c) Hence find the total time the stone spends more than 2 metres above the ground.

The following questions use g = 10 ms^{-2}

25) A ball is thrown vertically upwards with initial speed u ms^{-1} from a height of 15m above the ground. It hits the ground with speed 22ms^{-1}
    a) Find the initial speed.
    b) What is the greatest height it reaches?
    c) Find the time it takes to hit the ground.
    d) Given the ball should be subject to air resistance, explain the how this would affect your answers to the previous parts of the question.

26) A parachutist steps off a plane at a height of 2800 m. They free-fall for 4 seconds before opening their parachute.
    a) Find the speed of the parachutist the moment the parachute opens.

    It is given that from the point the parachute opens, they decelerate at 12 ms^{-2} for 3 seconds before reaching a constant speed.
    b) Find the total time it takes to reach the ground from initially stepping off the plane.

27) A ball is thrown vertically upwards from the ground at 20 ms^{-1} whilst simultaneously a second identical ball is thrown vertically downwards at 5 ms^{-1} from a height of 10 metres above the ground. Find
    a) the time at which both balls are at the same height.
    b) the distance from the ground at which both balls are at the same height.

28) A ball A is thrown vertically upwards from the ground at 20 ms^{-1} whilst simultaneously a second identical ball B is thrown vertically upwards at 12 ms^{-1} from a height of 5 metres above the ground. Find
    a) the amount of time that both balls are above the height of the initial start point of ball B.
    b) the time at which both balls are at the same height.
    c) the distance from the ground at which both balls are at the same height.

The following questions use g in exact form

29) An object is dropped from a height of h metres. Show that $v = \sqrt{2gh}$

30) A stone is dropped from a height of h metres above horizontal ground. It falls freely downwards. Its speed when reaching the ground is 8 ms^{-1}. Show that $h = \frac{32}{g}$

31) A ball is projected directly upwards at u ms^{-1}. Find an expression for
    a) the maximum height in terms of u
    b) the time it takes to return to its start point in terms of u

32) Two particles A and B are released from rest. A is released from k metres above a horizontal surface. B is released 0.5 seconds later from $\frac{10}{k}$ metres above the horizontal surface. They land simultaneously on the surface at t seconds after the release of A.
    Find the value of t and show that $k \approx 0.267g$

1.66 Forces: Equilibrium (answers on page 480)

1) Find the resultant force in each of the following cases, stating both the value and direction:

a) Up: 25N, Down: 10N, Left: 8N, Right: 8N

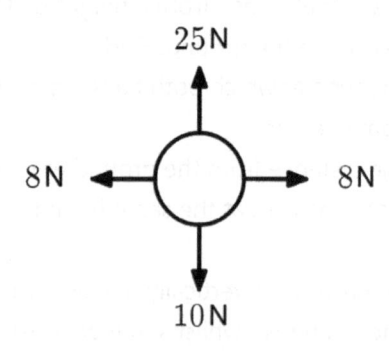

b) Up: 11N, Down: 11N, Left: 11N, Right: 4N

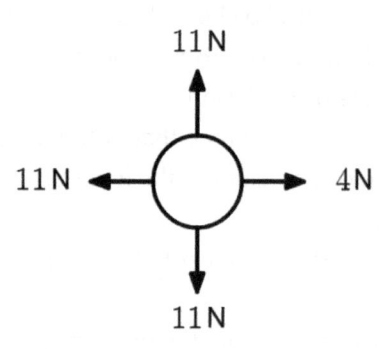

c) Up: 22N and (P+4)N, Down: 60N, Left: 5N, Right: 5N

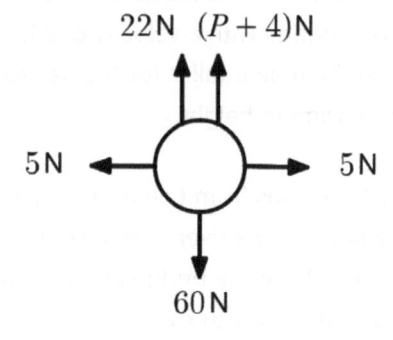

d) Up: 18N, Down: 18N, Left: 40N and PN, Right: 3PN

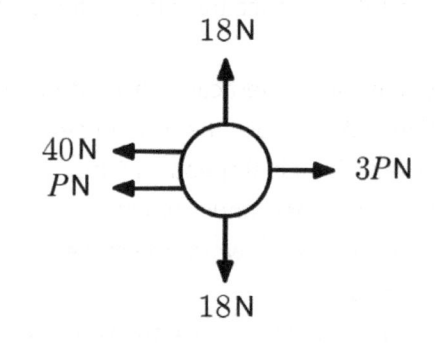

2) Given that each of these particles is stationary, find the value of P:

a) Up: 6N, Down: 6N, Left: 2N, Right: PN

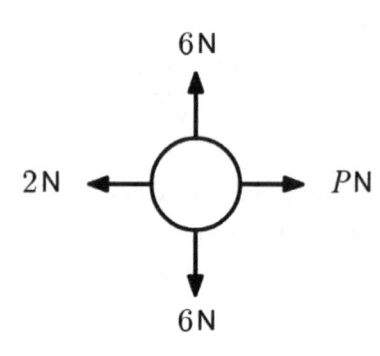

b) Up: 7N, Down: 7N, Left: 8N, Right: PN and 2N

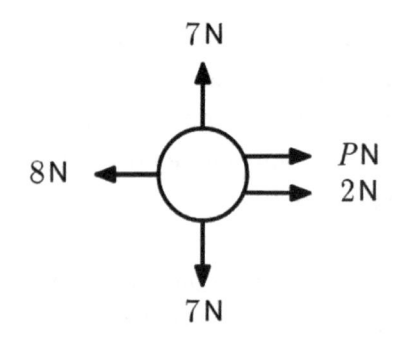

3) Find the value of P and Q such that the object is at rest.

a) Up: 62N, Down: PN, Left: 0.5PN, Right: QN

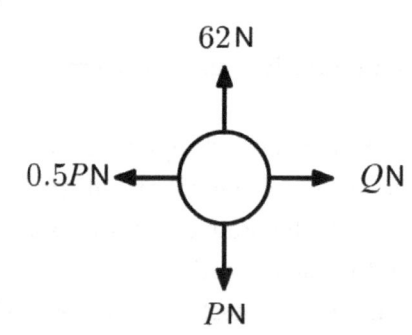

b) Up: 16N and (P−3)N, Down: 40N, Left: 2QN, Right: 4N and 12N

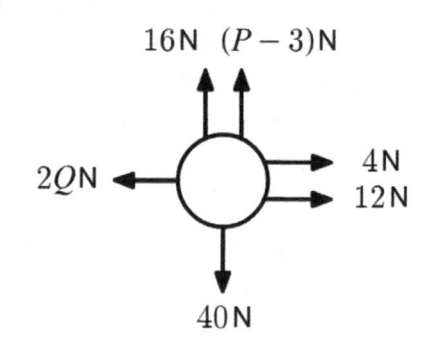

c) Up: 12N and PN, Down: 24N, Left: 3QN, Right: PN

196 THE A-LEVEL MATHS TEXTBOOK

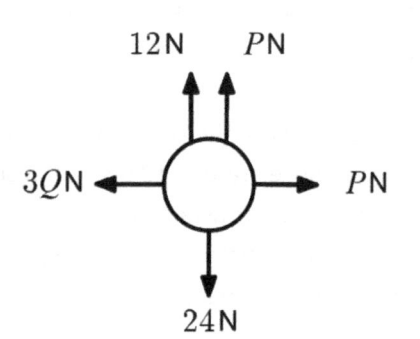

1.66 Forces: Equilibrium (answers on page 480)

4) Each of the following diagrams shows an object acted on by a set of forces. Find the value of P and Q such that the object is moving with constant velocity.

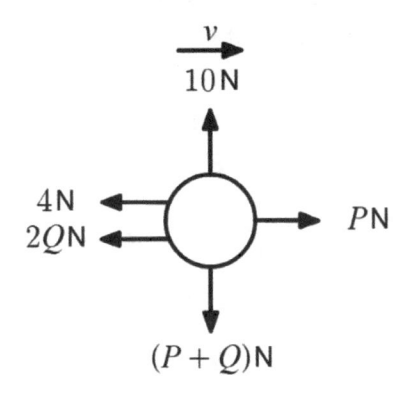

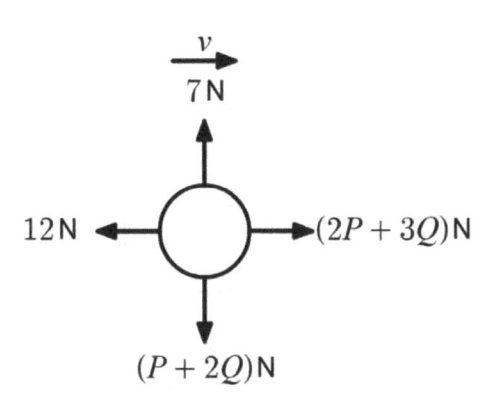

   a) Forces on object: 10 N upwards (velocity v to the right), 4 N left, 2Q N left, P N right, $(P+Q)$ N downwards.

   b) Forces on object: 7 N upwards (velocity v to the right), 12 N left, $(2P+3Q)$ N right, $(P+2Q)$ N downwards.

5) An object of weight 50 N is hanging by a string in equilibrium as shown in the diagram.

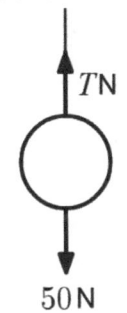

   Forces: T N upwards, 50 N downwards.

   a) State the value of T.
   b) By using $g = 10$ ms^{-2}, find the mass of the object.

   The string snaps and the object drops.
   c) State the size and direction of the resultant force acting on the object.

6) State the weight of an object of mass 50 kg when
   a) on earth given $g = 9.81$ ms^{-2}.
   b) on the moon given $g = 1.63$ ms^{-2}.

7) An object has weight 4000 N on Earth where $g = 9.8$ ms^{-2}. Find the weight of the same object on the moon where the acceleration due to gravity is 1.6 ms^{-2}.

8) An object of mass 0.4 kg is being held in equilibrium by an upwards force of P N. Using $g = 9.8$ ms^{-2}
   a) Find the weight of the object.
   b) Hence, state the value of P

   The upwards force is removed.
   c) State the size and direction of the resultant force acting on the object.

9) A small body is hanging in equilibrium at the end of a vertical string.
   a) Draw a force diagram to represent the forces acting on the body.

   Find, using $g = 9.8$ ms^{-2}, the
   b) weight of the body if the mass is 2 kg.
   c) mass of the body if its weight is 34.3 N.
   d) tension in the string if the mass is 0.5 kg.
   e) tension in the string if the weight of the body is 14.7 N.

10) A box is at rest on a horizontal table.
    a) Draw a force diagram to represent the forces acting on the box.

    Find, using $g = 10$ ms^{-2}, the
    b) weight of the box if the box is 1.5 kg.
    c) mass of the box if its weight is 60 N.
    d) mass of the box if the normal reaction of the table on the box is 5 N.
    e) normal reaction of the table on the box if the mass of the box is 1.2 kg.
    f) normal reaction of the table on the box if the weight of the body is 25 N.

11) A book of mass 400 g is being pushed along with a force of 2 N on a smooth horizontal table.
    Draw a force diagram to represent the forces acting on the book.

1.67 Forces: Particles in Motion (answers on page 481)

1) A particle mass m kg is acted upon by a resultant force F N has acceleration a ms^{-2}. In each case, find the unknown variable:
   a) $m = 0.8$ kg, $a = 1.6$
   b) $a = 0.8$, $F = 20$
   c) $m = 5$ kg, $F = 15$
   d) $m = 200$ g, $a = 0.8$

The following questions use g = 9.8 ms^{-2}

2) A particle of mass m kg with forces acting on it as shown in the diagram below, is accelerating at a ms^{-2}. In each case, state the value of m and find the values of P and Q
   a)

   Forces: $a = 2$ (right), Q N (up), 10 N (left), P N (right), $5g$ N (down)

   b)

   Forces: $a = 1.5$ (right), Q N (up), 8 N (left), P N (right), $3g$ N (down)

   c)

   Forces: $a = 3$ (left), Q N (up), 40 N (left), P N (right), $6.5g$ N (down)

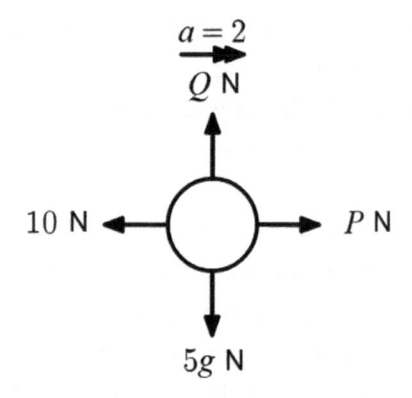

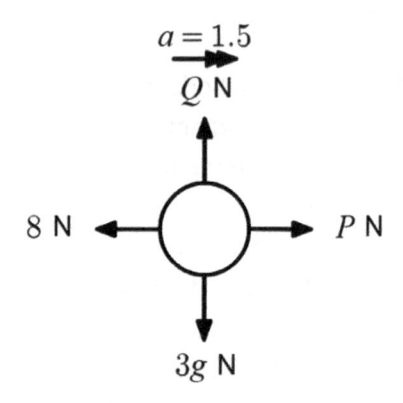

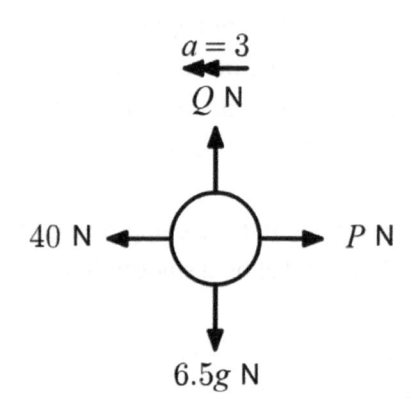

The following questions use g = 10 ms^{-2}

3) A particle of mass m kg with forces acting on it as shown in the diagram below, is accelerating at a ms^{-2}. In each case, state the value of m and find the value of P
   a)

   Forces: P N (up), $a = 2$ (up), $3g$ N (down)

   b)

   Forces: P N (up), $a = 1.5$ (up), $7.5g$ N (down)

   c)

   Forces: P N (up), $a = 0.8$ (down), $1.2g$ N (down)

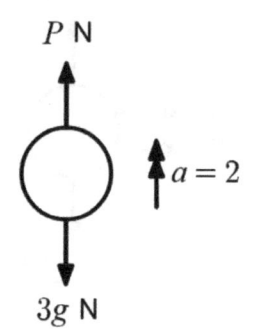

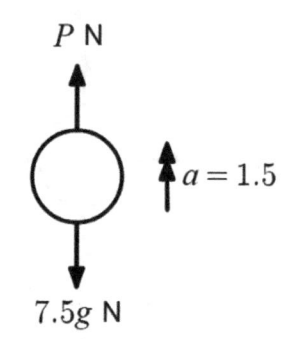

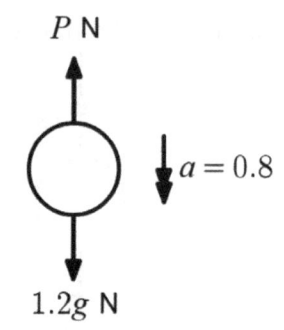

4) A particle of mass 2 kg is attached to the end of a light, vertical string. The particle is accelerating vertically at a ms^{-2}. In each case find the tension in the string when:
   a) $a = 3$ ms^{-2} upwards.
   b) $a = 1.5$ ms^{-2} downwards.
   c) $a = -2.5$ ms^{-2} upwards.
   d) moving with constant velocity.

5) A particle of mass 0.5 kg is attached to the end of a light, vertical string. The particle is decelerating upwards at 0.2 ms^{-2}. Find the tension in the string.

1.67 Forces: Particles in Motion (answers on page 481)

The following questions use $g = 9.8$ ms^{-2}

6) A block of mass 8 kg is moving in a straight line with constant acceleration 0.1 ms^{-1}. The force driving it forward is 12 N. Find the total resistance acting against the motion of the block.

7) A box is being pushed on a horizontal surface such that it is moving with constant velocity. The pushing force is 100 N. Find the total resistive force.

8) A box is being pushed on a horizontal smooth surface by a 24 N force so that it is accelerating at 1.5 ms^{-1}.
   a) Find the weight of the box.
   b) State a modelling assumption used in your calculation.

9) A lorry of mass 3 tonnes is driving along a straight road. The lorry is experiencing resistance forces of 900 N. The engine is driving forward with a force of 2100 N. Find the acceleration of the lorry.

10) A pebble of mass 50 g is falling to solid ground through a still body of water. The pebble is experiencing a downwards acceleration of 2.5 ms^{-2}. Find the total resistive forces acting on the pebble.

11) A car of mass 2000 kg reduces its speed from 18 ms^{-1} to rest in 5 seconds, along a straight flat road.
    a) Find the breaking force required to achieve this.
    b) State the modelling assumption you have made in calculating the breaking force.

12) A vehicle of mass 900 kg reduces its speed from 20 ms^{-1} to rest in 6 seconds, along a straight flat road. It experiences a resistive force of 450 N during this time. Find the breaking force of the vehicle.

Prerequisite knowledge for Q13-Q18: Constant Acceleration formulae

13) A stone of mass 400 g is dropped from a bridge and falls vertically. It hits the ground 1.5 seconds after being released. A constant resistance to motion of the stone is given to be 2 N. Find the height from which the stone is dropped.

14) A body of weight 20 N is accelerating along rough horizontal ground, being pushed by a force of 7 N acting parallel to the ground. A constant resistance force of 5.5 N acts against the particle.
    a) Find the mass of the particle.
    b) Given the particle starts at rest, find its speed after 10 seconds.

15) A block of mass 3 kg starts at rest and is accelerating across a smooth horizontal ground by a force of 1.5 N acting parallel to the ground. Find how far it has travelled in the first 4 seconds of motion.

16) A car of mass 1650 kg is driving with constant speed 10 ms^{-1} when it begins accelerating. After 30 m it reaches a speed of 18 ms^{-1}. Find the resultant force acting on the car during the period of acceleration.

17) A car of mass 2000 kg is driving at 47 km h^{-1} when the driver notices an accident ahead. When the car is 20 metres away from the accident, the driver applies the brakes creating a total resistance force of 10 000 N. Determine whether the car stops before the accident.

18) A van of mass 2250 kg drives a distance of 40 metres along a straight road, accelerating uniformly from rest to 25 ms^{-1}. Find
    a) the acceleration of the van.
    b) the magnitude of the force exerted by the engine, stating any modelling assumptions you have made in this calculation.
    c) the magnitude of the normal reaction between the van and the road.

1.68 Forces: Vectors (answers on page 481)

1) Two forces **P** and **Q** act on a particle. The vectors **i** and **j** are unit vectors due east and north respectively. Find the magnitude and direction (anticlockwise from the positive x-axis) of the resultant force in each case, giving your answers to 3 significant figures:
    a) $\mathbf{P} = (2\mathbf{i} - \mathbf{j})$ N and $\mathbf{Q} = (-\mathbf{i} + 4\mathbf{j})$ N
    b) $\mathbf{P} = \binom{3}{2}$ N and $\mathbf{Q} = \binom{5}{-4}$ N
    c) $\mathbf{P} = (3\mathbf{i} + \mathbf{j})$ N and $\mathbf{Q} = (-5\mathbf{i} + 8\mathbf{j})$ N
    d) $\mathbf{P} = \binom{-5}{-9}$ N and $\mathbf{Q} = 4\binom{1}{2}$ N
    e) $\mathbf{P} = (-0.5\mathbf{i} + 2\mathbf{j})$ N and $\mathbf{Q} = 3(\mathbf{i} - 6\mathbf{j})$ N

2) Two forces **P** and **Q** act on a particle. The vectors **i** and **j** are unit vectors due east and north respectively. Find the magnitude and the bearing of the resultant force in each case:
    a) $\mathbf{P} = (5\mathbf{i} + 3\mathbf{j})$ N and $\mathbf{Q} = (-\mathbf{j})$ N
    b) $\mathbf{P} = \binom{-5}{2}$ N and $\mathbf{Q} = \binom{0}{1}$ N
    c) $\mathbf{P} = (2\mathbf{i} - \mathbf{j})$ N and $\mathbf{Q} = (3\mathbf{i} - 11\mathbf{j})$ N
    d) $\mathbf{P} = \binom{5}{-1.2}$ N and $\mathbf{Q} = \binom{-6}{1}$ N
    e) $\mathbf{P} = (1.25\mathbf{i} - 0.75\mathbf{j})$ N and $\mathbf{Q} = (-2.25\mathbf{i} - 1.75\mathbf{j})$ N

3) The force $\binom{k}{5}$ N where k is a positive constant, has a magnitude of $\sqrt{29}$ N. Find the value of k

4) The force $(2k\mathbf{i} - 2k\mathbf{j})$ N where k is a constant, has a magnitude of 12 N and a bearing of 315°. Find the value of k

5) Two forces **P** and **Q** act on a particle. The vectors **i** and **j** are unit vectors due east and north respectively. Find a third force **R** such that the three forces are in equilibrium:
    a) $\mathbf{P} = (-2\mathbf{i} + \mathbf{j})$ N and $\mathbf{Q} = (-3\mathbf{i} + 11\mathbf{j})$ N
    b) $\mathbf{P} = \binom{-1}{0}$ N and $\mathbf{Q} = \binom{-4}{\sqrt{3}}$ N
    c) $\mathbf{P} = (\sqrt{2}\mathbf{i} - \mathbf{j})$ N and $\mathbf{Q} = (-3\mathbf{j})$ N
    d) $\mathbf{P} = \binom{k}{1}$ N and $\mathbf{Q} = \binom{-k}{1}$ N
    where k is a constant.
    e) $\mathbf{P} = (a\mathbf{i} + b\mathbf{j})$ N and $\mathbf{Q} = (-a\mathbf{i} + 2b\mathbf{j})$ N where a and b are constants.
    f) $\mathbf{P} = \binom{2a}{6a}$ N and $\mathbf{Q} = \binom{b}{6b}$ N where a and b are constants.

6) The forces $\binom{a}{3}$ N, $\binom{2b}{-a}$ N and $\binom{-6b}{b}$ N act on an object which is in equilibrium. Find the values of a and b

7) Three forces, $(6\mathbf{i} + 10\mathbf{j})$ N, $(-3.5\mathbf{i} - 12\mathbf{j})$ N and $(-2.5\mathbf{i})$ N act on a particle. It begins at rest at the origin. The vectors **i** and **j** are unit vectors due east and north respectively.
    a) Find the resultant force in the form $(a\mathbf{i} + b\mathbf{j})$ N where a and b are constants to be found.
    b) Find the magnitude and the bearing of the resultant force.
    c) Describe the motion of the particle.
    d) State the additional force that would be needed for the particle to instead be at rest.

8) Three forces, $(12\mathbf{i} + 11\mathbf{j})$ N, $(3\mathbf{i} - 15\mathbf{j})$ N and $(a\mathbf{i} + b\mathbf{j})$ N act on a particle. The vectors **i** and **j** are unit vectors due east and north respectively. State the values of a and b such that the particle is moving due East.

9) Two forces **P** and **Q** act on a particle. Find the angle between the two forces:
    a) $\mathbf{P} = (3\mathbf{i} + 4\mathbf{j})$ N and $\mathbf{Q} = (5\mathbf{i} + 12\mathbf{j})$ N
    b) $\mathbf{P} = \binom{2}{3}$ N and $\mathbf{Q} = \binom{5}{1}$ N

10) Two forces are given by $\binom{1}{-7}$ N and $\binom{k}{3k}$ N, where k is a positive constant. Given that the resultant of these two forces is parallel to $\binom{-4}{7}$, find the value of k

11) Three forces, $\binom{1}{2}$ N, $\binom{-3}{8}$ N and $\binom{a}{b}$ N act on a particle. The vectors **i** and **j** are unit vectors due east and north respectively.
    a) State the values of a and b such that the particle is moving due North.

    It is given that $a = 8$
    b) Find the value of b such that the particle is moving due North-East.

1.68 Forces: Vectors (answers on page 481)

12) Find the unit vector of each of the following vectors:
    a) $\begin{pmatrix} 3 \\ 4 \end{pmatrix}$
    b) $3\mathbf{i} - \mathbf{j}$
    c) $\begin{pmatrix} -1 \\ 4 \end{pmatrix}$
    d) $-8\mathbf{i} - 6\mathbf{j}$

13) The force **F** is parallel to $\overrightarrow{AB}$. Given the coordinates of A and B and the magnitude of **F**, find **F** in each case:
    a) $A(0,0), B(-3,-3), |\mathbf{F}| = 5$ N
    b) $A(0,-2), B(3,0), |\mathbf{F}| = 4$ N
    c) $A(1,2), B(5,6), |\mathbf{F}| = 8$ N

Prerequisite knowledge for Q14-Q23: Constant Acceleration

14) A resultant force of $(3\mathbf{i} + 4\mathbf{j})$ N is pushing against a particle of mass 0.5 kg. The particle is initially at rest.
    a) Find the magnitude of the acceleration of the mass
    b) Hence find the time it takes for the particle to move 80 metres.

15) A resultant force of $\begin{pmatrix} 1 \\ 2 \end{pmatrix}$ N is pushing against a particle of mass 2 kg. The particle is initially at rest.
    a) Find the magnitude of the acceleration of the mass.
    b) Hence find its speed of motion after 4 seconds.

16) Two forces given by $(5\mathbf{i} + \mathbf{j})$ N and $(-2\mathbf{i} + 3\mathbf{j})$ N are pushing against a particle of mass 4 kg. The particle is initially at rest.
    a) Find the magnitude of the acceleration of the mass.
    b) Hence find the time it takes for the particle to move 10 metres.

17) A force of $\begin{pmatrix} 5 \\ 11.6 \end{pmatrix}$ N is pulling on an object of mass 2kg which is falling under the influence of gravity, where $g = 9.8$ ms^{-2}. Find the magnitude of the acceleration of the object.

18) An object of mass 2 kg is dropped. A force of $(4.5\mathbf{i} + 13.6\mathbf{j})$ N is pulling on the object as it falls under the influence of gravity, where $g = 9.8$ ms^{-2}.
    a) Find the magnitude of the object's acceleration.
    b) Another force of $(-4.5\mathbf{i} + 5.5\mathbf{j})$ N is applied to the object. Find the magnitude of the object's acceleration now.

19) A resultant force **F** N parallel to $4\mathbf{i} - 3\mathbf{j}$ is pulling a box of mass 4 kg. The particle is initially at rest. It takes 2 seconds to travel 5 metres. Find the vector form of **F**

20) A resultant force **F** N parallel to $8\mathbf{i} + 4\mathbf{j}$ is pulling a box of mass 3 kg. Initially it is moving at 1 ms^{-1} but reaches 4 ms^{-1} in 5 seconds. Find the vector form of **F**

21) Two vectors $\begin{pmatrix} 3 \\ -2 \end{pmatrix}$ N and $\begin{pmatrix} p \\ q \end{pmatrix}$ N are acting on a box of mass 5 kg which is initially at rest. The box moves in the direction $\begin{pmatrix} 2 \\ 1 \end{pmatrix}$
    a) Show that $p - 2q + 7 = 0$

    Given that $p = 5$
    b) Find the distance the box has travelled in 5 seconds.

22) Two vectors $3\mathbf{i} - \mathbf{j}$ N and $p\mathbf{i} + q\mathbf{j}$ N are acting on a box of mass 3 kg which is initially at rest. The box moves in the direction $\mathbf{i} + 5\mathbf{j}$
    a) Show that $5p - q + 16 = 0$

    Given that $q = 6$
    b) Find the speed of the box after 3 seconds.

23) Two vectors $\begin{pmatrix} -1 \\ 4 \end{pmatrix}$ N and $\begin{pmatrix} p \\ q \end{pmatrix}$ N are acting on a box of mass 2 kg which is initially at rest. The box moves in the direction $\begin{pmatrix} 3 \\ 4 \end{pmatrix}$
    a) Show that $4p - 3q - 16 = 0$

    It begins at position A with a speed of 2 ms^{-1}. Given that $p = -2$
    b) Find the speed of the box after 3 seconds.

1.69 Forces: Connected Particles (answers on page 482)

1) By modelling each object as a particle and ignoring air resistance, draw a diagram showing all of the forces acting on:
   a) A car of mass M kg towing a caravan of mass m kg on a rough surface.
   b) A train engine of mass M kg pulling two trucks of mass m kg on a rough surface.
   c) A helicopter of mass M kg with a person of mass m kg hanging directly below.

2) A car of mass 2000 kg is towing a caravan of mass 500 kg along a straight flat road. The driving force produced by the engine is 600 N. By modelling both the car and caravan as having no resistance to motion, find
   a) the acceleration of the car.
   b) the tension in the tow bar.

3) A car of mass 2200 kg is towing a trailer of mass 400 kg along a straight flat road. The engine of the car is creating a driving force of 1100 N. The resistance force acting on the car is 72 N and the resistance force acting on the trailer is 16 N. Find
   a) the acceleration of the car and trailer.
   b) the tension in the tow bar.

4) Two particles, P and Q which are connected by a light, inextensible string have mass 6 kg and 1.5 kg respectively. Particle P is pulled along by a horizontal force of 60 N. Particles P and Q have resistances of 12 N and 3 N acting against them respectively.
   a) Find the acceleration of the particles.
   b) Find the tension in the string.
   c) Explain how the modelling assumptions that the string is light and inextensible have been used.

5) Two particles, P and Q which are connected by a light, inextensible rod have mass 10 kg and 2 kg respectively. Particle P is pushed along a rough horizontal surface by a horizontal force of 30 N. Particles P and Q both experience resistances which are proportional to their masses. Given that they are accelerating at 0.5 ms^{-2}, find the thrust in the rod.

6) Two particles, P and Q which are connected by a light, inextensible string have mass 10 kg and m kg respectively. Particle P is pulled along a smooth horizontal surface by a horizontal force of 120 N which creates an acceleration of 3 ms^{-2}. Find
   a) the value of m
   b) the tension in the string.

7) Two particles, P and Q which are connected by a light, inextensible string have mass $5m$ kg and $2m$ kg respectively. Particle P is pulled along by a horizontal force of 37.8 N. Particles P and Q have resistances of 6 N and 2.4 N acting against them respectively. Find
   a) the acceleration in terms of m
   b) the size of the tension in the string.
   c) Explain how the modelling assumptions that the string is light and inextensible have been used.

8) A train engine of mass 10000 kg is towing 2 identical trucks of mass 3500 kg along a flat straight track which are experiencing the same resistive forces. The driving force produced by the engine is 25 000 N. They are moving with constant acceleration 1.2 ms^{-2}. The engine has a resistance of 2400 N acting against it. Find
   a) the total resistance acting against each of the trucks.
   b) the tension in each of the couplings.

9) A car is towing a caravan. The driving force created by the car is 300 N. A constant resistance force of 180 N acts against the caravan. They move at constant velocity. Find the total resistance acting on the car.

10) A car of mass 1800 kg is towing a caravan of mass 800 kg along a straight flat road. They start at rest and then reach 5 ms^{-1} after having travelled 20 metres. The resistance force acting on the car is 90 N and the resistance force acting on the caravan is 40 N. Find
    a) the acceleration of the car and caravan.
    b) the driving force by the engine of the car.
    c) the tension in the tow bar.

1.69 Forces: Connected Particles (answers on page 482)

11) Two particles, P and Q which are connected by a light, inextensible string have mass 20 kg and 6 kg respectively. Particle P is pulled by a force of 145.6 N. Particles P and Q have resistances of 60 N and 18 N acting against them respectively. The system starts from rest. Find
   a) the acceleration of P.
   b) the distance P moves over the first 5 s.
   c) the tension in the string between P and Q.

12) A train engine of mass 7000 kg is towing a carriage of mass 2000 kg along a straight flat track. The driving force produced by the engine is 12 600 N. The resistance forces on the train engine and carriage are $700k$ N and $200k$ N respectively.
   a) Show that $a = \frac{7}{5} - \frac{k}{10}$
   b) Find the tension in the coupling between the train engine and the carriage.

   Given they start at rest and reach 3 ms^{-1} in 5 s,
   c) find the resistance on the carriage.
   d) find the resistance on the train engine.

13) A van of mass 2500 kg is towing a trailer of mass 1000 kg along a straight flat road, with a tow bar. The driver reduces the speed of the van from 81 km h^{-1} to 63 km h^{-1} by using the van's brakes, creating a total force of 2590 N against the motion. The resistance force acting on the van is 150 N and the resistance force acting on the trailer is 60 N throughout the motion.
   a) find the deceleration of the van.
   b) find the time taken to slow to 63 km h^{-1}.
   c) find the thrust in the tow bar.
   d) State a modelling assumption you have made.

   Whilst travelling at 63 km h^{-1} the tow bar breaks. The trailer slows to a rest without a collision. The total resistance to motion to the trailer is now 400 N.
   e) find how far the trailer travels before coming to a stop.
   f) Explain what the driver of the van would have noticed with the motion of the van.

The following questions use g = 9.8 ms^{-2}

14) An object P of mass 1.2 kg is suspended by a light inextensible string. A second object Q of mass 0.8 kg is suspended by a second light inextensible string onto object P, as shown in the diagram. The first string has a force of 15 N applied, and the system moves upwards.

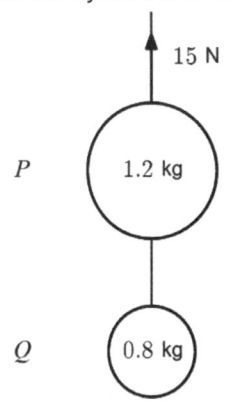

   Find:
   a) the acceleration of the system.
   b) the tension in the lower string.
   c) the value the 15 N needs to be increased by to achieve constant velocity.

For each of the following questions containing a crate in a lift, the following diagram may be used

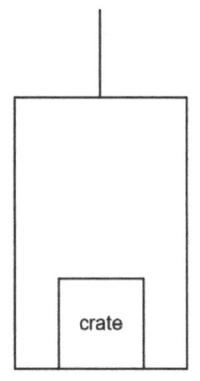

15) A crate of mass m is in a lift which is travelling upwards. Find the magnitude of the force exerted on the crate by the lift when:
   a) $m = 10$ kg and an acceleration 0.5 ms^{-2}
   b) $m = 100$ g and an acceleration 0.7 ms^{-2}
   c) $m = 2$ kg and a deceleration of 1.2 ms^{-2}
   d) $m = 450$ g and a deceleration of 0.8 ms^{-2}

1.69 Forces: Connected Particles (answers on page 482)

16) A lift of mass 800 kg is attached to a light, inextensible cable. It contains a crate of mass 100 kg.
Find the tension in the cable when the lift is:
a) Accelerating downwards at 0.5 ms^{-2}
b) Accelerating upwards at 0.5 ms^{-2}
c) Decelerating downwards at 0.4 ms^{-2}
d) Decelerating upwards at 0.4 ms^{-2}

Find the acceleration upwards of the lift when the tension in the cable is:
e) 2700 N
f) 8100 N
g) 9450 N

The following questions use g = 10 ms^{-2}

17) A lift of mass 600 kg is attached to a light, inextensible cable. It contains a crate of mass 50 kg. The lift starts from rest and moves upwards, reaching the required floor which is 10 metres above in 15 seconds. Find
a) the acceleration of the lift.
b) the tension in the cable.
c) the magnitude of the force exerted on the crate by the floor of the lift.

18) A lift of mass 100 kg accelerates upwards at 0.4 ms^{-2}, being pulled by a light inextensible vertical cable. The lift contains two boxes, P and Q of masses 5 kg and 10 kg respectively. Box P sits directly on top of box Q as shown in the diagram below:

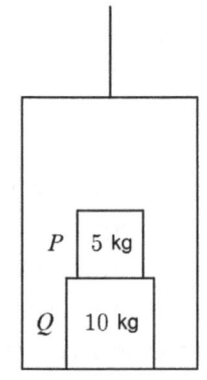

Find
a) the tension in the cable
b) the force exerted on Q by P
c) the force exerted by Q on the lift.

19) A lift of mass 120 kg decelerates upwards at 0.25 ms^{-2}, being pulled by a light inextensible vertical cable. The lift contains two crates, P and Q of masses 3 kg and 15 kg respectively. Crate P sits directly on top of crate Q

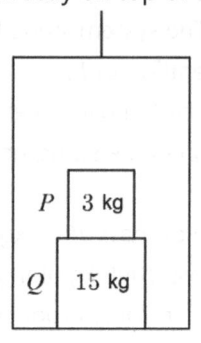

Find
a) the tension in the cable.
b) the force exerted on Q by P.
c) the force exerted by Q on the lift.

20) A lift of mass 120 kg accelerates upwards at 0.1 ms^{-2} pulled by a light inextensible vertical cable, carrying three identical crates, A, B and C, each with a mass of 2 kg.

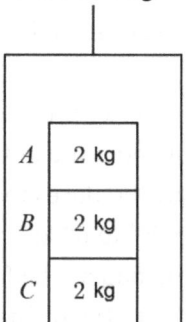

Find
a) the tension in the cable.
b) the force exerted on the lift by C.

Prerequisite knowledge for Q21: Variable Acceleration

21) Two particles P and Q which are connected by a light, inextensible string have mass 8 kg and 5 kg respectively. P is being pulled along a smooth horizontal surface with velocity v ms^{-1} at time t seconds given by
$$v = -\frac{1}{5}t^2 + 6t, 0 \leq t \leq 30$$
Find, at 10 seconds into the motion:
a) the acceleration of the particles.
b) the tension in the string.
c) the force pulling P.

1.70 Forces: Pulleys (answers on page 483)

The following questions use $g = 9.8$ ms^{-2}

1) Two particles A and B of mass m_A and m_B are attached to the end of a light inextensible string which passes over a small smooth fixed pulley. The masses hang with the string taut, and the system is released from rest.

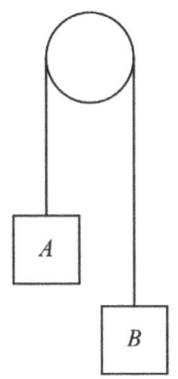

   The system is released from rest. In each case find the tension in the string and the acceleration of the system:
   a) $m_A = 9$ kg, $m_B = 3$ kg
   b) $m_A = 2$ kg, $m_B = 8$ kg
   c) $m_A = 200$ g, $m_B = 500$ g

2) Two particles A and B of masses $5m$ kg and $3m$ kg respectively are attached to the end of a light inextensible string which passes over a small smooth fixed pulley. The masses hang with the string taut, and the system is released from rest. Find
   a) the acceleration of the system.
   b) the tension in the string in terms of m
   c) the magnitude of the force exerted on the pulley while A and B are in motion.

The following questions use exact values of g

3) Two particles A and B of masses m kg and M kg respectively, where $M > m$, are attached to the end of a light inextensible string which passes over a small smooth fixed pulley. The masses hang with the string taut, and the system is released from rest. The tension in the string is T N and the acceleration is a ms^{-2}. Show that
$$a = \left(\frac{M - m}{M + m}\right)g$$

4) Two particles A and B of masses km kg and m kg respectively where $k > 1$. They are attached to the end of a light inextensible string which passes over a small smooth fixed pulley. The masses hang with the string taut, and the system is released from rest. The acceleration of the system is a ms^{-2}. Show that
$$a = \left(\frac{k - 1}{k + 1}\right)g$$

The following questions use $g = 9.81$ ms^{-2}

5) Two particles A and B are attached to the end of a light inextensible string which passes over a small smooth fixed pulley. The masses hang with the string taut, and the system is released from rest. A has mass 2 kg and descends 0.8 metres in the first 4 seconds after the system is released. Find
   a) the acceleration of the system.
   b) the tension in the string.
   c) the mass of B.

6) Two particles A and B of masses 4 kg and 3 kg respectively are attached to the end of a light inextensible string which passes over a small smooth fixed pulley.

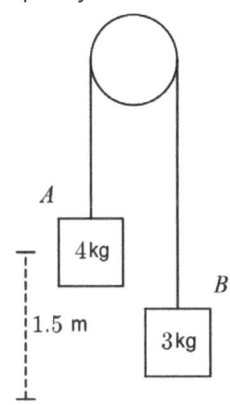

   The masses hang with the string taut, and the system is released from rest. Initially, particle A begins 1.5 metres above the ground.
   Find
   a) the acceleration of the system.
   b) the tension in the string.
   c) the time it takes for A to hit the ground.
   d) the greatest height of B above the floor, stating any modelling assumptions you have made.

1.70 Forces: Pulleys (answers on page 483)

The following questions use g = 9.8 ms⁻²

7) Two boxes A and B of masses 5 kg and m kg respectively are attached to the end of a light inextensible string which passes over a smooth small fixed pulley. The masses hang with the string taut, and the system is released from rest.

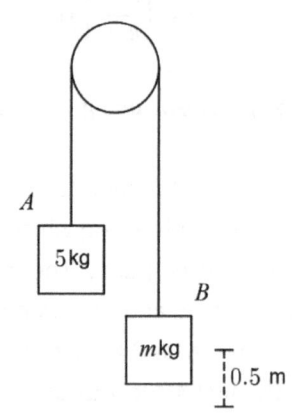

Box B starts 0.5 metres above the ground. Box A ascends 0.2 metres in the first 0.5 seconds after the system is released.
a) Find the acceleration of the system.
b) Find the tension in the string.
c) Show that the mass of B is approximately 7 kg.
d) Find the velocity at which B hits the ground.

Once Box B hits the ground, the string holding up box A becomes slack.
e) Given box A starts 1 metre above the ground, find the greatest height that box A reaches above the ground.
f) Find the time box A spends with the string not taut.

A different box (box C) is attached to the string, in replacement of box B. The system is re-set so that Box C starts 0.5 m above the ground.
Explain what box A will experience with its acceleration (including the direction) and the tension in the string in each of the following scenarios:
g) Box C has a larger mass than box B.
h) Box C has the same mass as box A.
i) Box C has a lower mass than box A.

8) Two boxes A and B of masses 3 kg and 0.5 kg respectively are attached to the end of a light inextensible string which passes over a small smooth fixed pulley. The system is released from rest. Box A begins $5h$ metres from the ground and box B begins h metres from the ground.

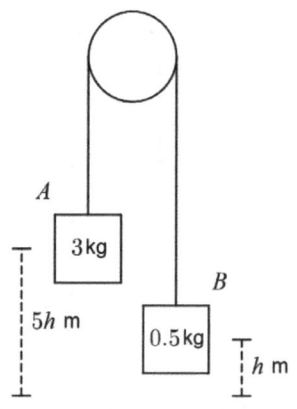

Find
a) the acceleration of the system.
b) the tension in the string.
c) Show that the time after release that both boxes are at the same height is given by
$$t = \sqrt{\frac{4h}{7}}$$

9) A scale pan A of mass m kg hangs freely and is attached by a light inextensible string to a mass B which is 8 kg. The string passes over a small smooth fixed pulley. The masses hang with the string taut.

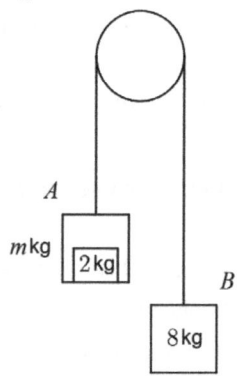

An object of mass 2 kg is placed into the scale pan and the system is released from rest. The scale pan A descends but doesn't exceed 1.96 ms⁻². Find
a) the range of values of T where T N is the tension in the string.
b) the range of values of m

1.70 Forces: Pulleys (answers on page 483)

10) Two particles, A and B have masses 1.2 kg and 0.5 kg respectively. They are connected by a light, inextensible string. Particle A lies on a smooth horizontal table and the string passes over a small round smooth pulley which is fixed at the edge of a table. Particle B hangs freely.

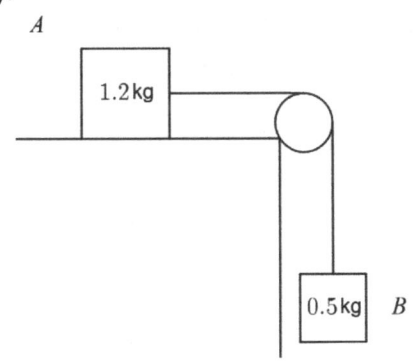

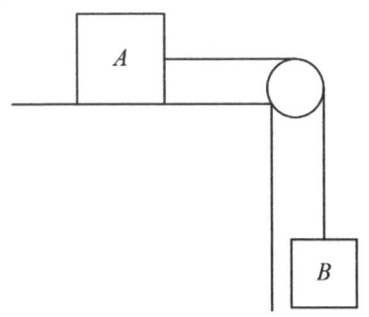

Find
   a) the acceleration of the system.
   b) the tension in the string.

11) Two particles, A and B have masses m_A and m_B respectively. They are connected by a light, inextensible string. Particle A lies on a rough horizontal table and the string passes over a small round pulley which is fixed at the edge of a table. Particle B hangs freely. The resistance force acting on particle A is F N.

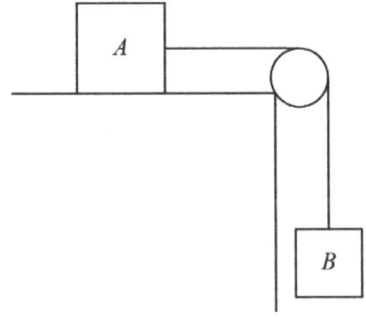

The system is released from rest.
Find, in each case, the acceleration of the system and the tension in the string.
   a) $m_A = 2$ kg, $m_B = 3$ kg, $F = 20$
   b) $m_A = 5$ kg, $m_B = 7$ kg, $F = 12$
   c) $m_A = 200$ g, $m_B = 500$ g, $F = 2$
   d) $m_A = 8$ kg, $m_B = 6$ kg, $F = 58.8$

12) For each of the scenarios in question 11), find the time it takes for particle B to travel 1 m, assuming particle B begins over 1 metre from the floor.

The following questions use $g = 10$ ms^{-2}

13) A van A of mass 2000 kg is attached with a cable to a pallet of building materials B which has mass 100 kg. The cable passes over a smooth fixed pulley, and B hangs freely.

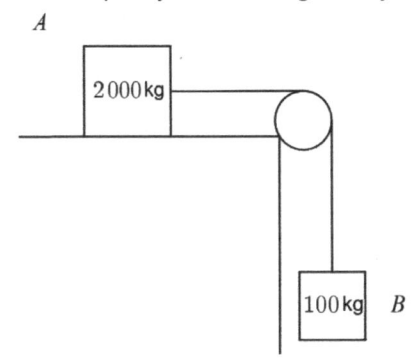

Initially the system is at rest and the cable is taut. The van begins to drive away from the pulley with driving force 2500 N. The resistance force acting on the van is 450 N.
   a) Find the acceleration of the system.
   b) Find the tension in the cable.
   c) State two modelling assumptions that you have made.

The van drives forward for 10 seconds before the pallet is getting close to reaching the pulley.
   d) Find the height the pallet has been lifted.
   e) State any assumptions in you have used.

14) Two particles, A and B have masses M kg and m kg respectively. They are connected by a light, inextensible string. Particle A lies on a smooth horizontal table and the string passes over a small round smooth pulley which is fixed at the edge of a table. Particle B hangs freely.

Show that
$$a = \frac{10m}{m + M}$$

1.70 Forces: Pulleys (answers on page 483)

The following questions use g = 9.8 ms⁻²

15) A wooden block A with mass 300 g sits on a rough table, attached to a string which passes over a small, smooth pulley which is fixed to the edge of the table. A plastic block B of mass m kg hangs freely at the end of the string.

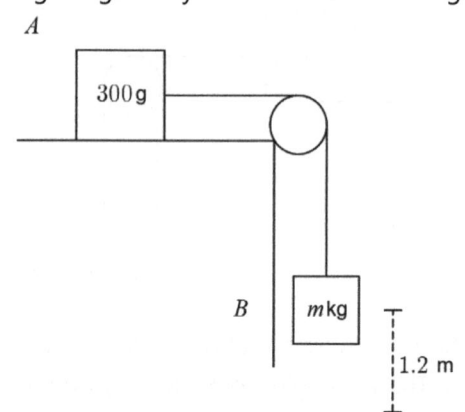

    The system is released from rest and block B descends. The resistance to motion on block A is 5 N.
    a) Find the acceleration of the system in terms of m

    The block B initially starts 1.2 metres above the ground. It takes 2 seconds for B to reach the ground.
    b) Find the value of m
    c) State any modelling assumptions you have made.

16) Particles A, B and C, have masses 2 kg, 8 kg and 3 kg respectively. A and B are attached to the ends of a light inextensible string. B and C are attached to the ends of another light inextensible string. B rests on a smooth table. A and C hang freely. Both strings pass over small smooth fixed pulleys at the edge of the table with the string taut.

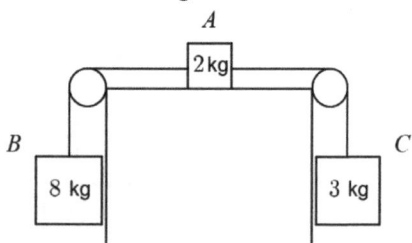

    The system is released from rest. Find
    a) the acceleration of the system.
    b) the tension in the two strings.

The following questions use exact values of g

17) Two particles, A and B have masses m kg and km kg respectively where $k > 1$. They are connected by a light, inextensible string. Particle A lies on a rough horizontal table and the string passes over a small round smooth pulley which is fixed at the edge of a table. Particle B hangs freely.

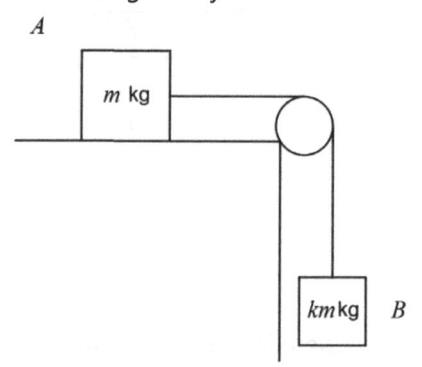

    The system is released from rest. The resistance to motion on particle A is given as $\frac{1}{2}kmg$ N. Show that
    $$a = \frac{kg}{2(1+k)}$$

18) Two particles, A and B have masses $3m$ kg and km kg respectively where $k > 0$. They are connected by a light, inextensible string. Particle A lies on a rough horizontal table and the string passes over a small round smooth pulley which is fixed at the edge of a table. Particle B hangs freely.

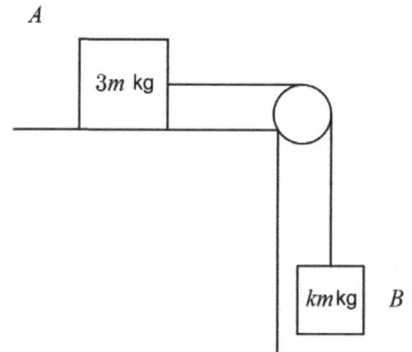

    A horizontal force of $20mg$ N is applied to A away from the pulley. The resistance to motion on particle A is given as $10mg$ N. Initially at rest, particle A travels 0.5 m in 10 s.
    a) Find the acceleration of the system
    b) Show that
    $$k = \frac{1000g - 3}{100g + 1}$$

1.71 Variable Acceleration (answers on page 484)

1) A particle moves along a straight line so that its displacement s metres at time t seconds, where $t \geq 0$, is given by each of the following functions. Find expressions for the velocity and the acceleration of the object at time t for:
   a) $s = 4t^3 - 5t$
   b) $s = t^5 + 2t^2$
   c) $s = t(t - 1)$
   d) $s = 2t - \frac{1}{t}$

2) A particle starts at rest before moving along a straight line. Its velocity v ms^{-1} at time t seconds is given by each of the following functions, where $t \geq 0$. In each case find the displacement in terms of t:
   a) $v = 8t^3 - 6t$
   b) $v = 6t^2 + 2t^3$
   c) $v = \frac{1}{10}t^2(t - 5)$
   d) $v = \frac{1}{2}t(t^2 + 1)$

3) A particle is moving along a straight line. Its acceleration a ms^{-2} at time t seconds is given by each of the following functions, where $t \geq 0$.
   In each case find the velocity at in terms of t, given that is known that the particle is moving with velocity v_1 at time t_1
   a) $a = 15t^4 - 4t, t_1 = 1, v_1 = 2$
   b) $a = 9t^2 + 5t^4, t_1 = 2, v_1 = 40$
   c) $a = -0.09t(t - 30), t_1 = 1, v_1 = 3$
   d) $a = \sqrt{t} - \frac{1}{t^3}, t_1 = 1, v_1 = 2$

4) A car drives between two sets of traffic lights, stopping at both. Its speed v ms^{-1} at time t s is modelled by
$$v = -\frac{1}{50}t^2 + \frac{6}{5}t, \quad 0 \leq t \leq 60$$
   a) Find the times at which the car is stationary.
   b) Sketch the velocity-time graph for the motion of the car.
   c) Find the distance between the two sets of traffic lights.
   d) Find an expression for acceleration a ms^{-2} in terms of t
   e) Explain why the model is no longer appropriate after 60 seconds.

5) A particle's acceleration a ms^{-2} with time t seconds is given by
$$a = 15\sqrt{t} + 2, \quad t \geq 0$$
The initial velocity is 1 ms^{-1} and when $t = 1$ the displacement is of the particle is 6 m.
   a) Find an expression for the velocity v with time t
   b) Find an expression for the displacement s with time t

6) A particle's acceleration a ms^{-1} at any given time t seconds is given by
$$a = -30t^2 + 25t^3, \quad t \geq 0$$
The particle initially starts at position O, with a velocity of 5 ms^{-1}.
   a) Find an expression for the velocity given the particle initially starts at position O.
   b) Find the displacement of the particle when 2 seconds have passed.

7) The displacement s (metres) over time t (seconds) for a particle is given by
$$s = -\frac{3}{8}t^2 + kt + 2, \quad t \geq 0$$
where k is a positive constant and is shown in the diagram below.

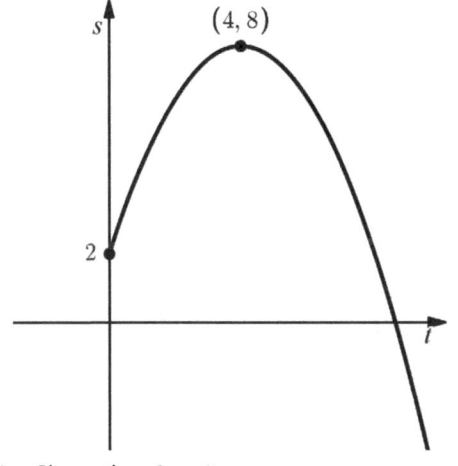

   a) Show that $k = 3$
   b) State the distance travelled by the particle when $t = 4$
   c) Find the time at which the particle returns to its initial position.
   d) Find an expression for the velocity v in terms of t
   e) Find the speed of the particle when $t = 10$ seconds.

1.71 Variable Acceleration (answers on page 484)

Prerequisite knowledge for Q8: $F_{net} = ma$

8) An object of mass 0.4 kg is moving in a straight line under the action of a resultant force F N. Its velocity v ms^{-1} at any given time t seconds may be given by
$$v = 8t - 3t^2, \quad t \geq 0$$
   a) Find an expression for the object's displacement, s metres, in terms of t
   b) Find an expression for the object's acceleration, a ms^{-2}, in terms of t
   c) Find an expression for F in terms of t

9) The velocity, v ms^{-1} at any given time t seconds of a car may be given by
$$v = kt - 3 + \frac{1}{t^2}, \quad \frac{1}{2} \leq t \leq 10$$
   where k is an unknown positive constant. It is known that the acceleration when $t = 2$ is 1.75 ms^{-2}.
   a) Find the value of k
   b) Find the total distance travelled by the car for $\frac{1}{2} \leq t \leq 10$

10) Ellis is going to the bus stop. His displacement, s metres at any given time t seconds is
$$s = -\frac{1}{k}t(t - 240), \quad 0 < t \leq 120$$
    where k is an unknown positive constant. It is known that the velocity when $t = 20$ is 4 ms^{-1}.
    a) Show that $k = 50$
    b) Find the speed at which Ellis is travelling when $t = 60$ seconds
    c) Find the speed at which Ellis is travelling when $t = 120$ seconds.
    d) Explain in context what happens to Ellis' velocity and displacement if this same model is used for $t \geq 120$

11) A particle's velocity, v ms^{-1} with time t seconds is given by
$$v = -t + 10, \quad t \geq 0$$
    Find when the particle will return to its original position.

12) An object moves along a straight line so that its acceleration a ms^{-2} at time t seconds is given by
$$a = 4 + 3t - 6t^2, \quad t \geq 0$$
    a) Given that the object is initially stationary, find the velocity, v, in terms of t
    b) Given that when 1 second has passed, the object has moved in a positive direction of 6 metres from its initial position, find the displacement, s, in terms of t
    c) Hence, find the time at which the object returns to its original position, giving your answer to 3 significant figures.

13) A particle's displacement, s metres with time t seconds is given by
$$s = t^3 - 6t^2 + 12t, \quad 0 \leq t \leq 10$$
    a) Find the distance travelled by the particle over the 10 seconds.
    b) Find the time when the particle is at rest.
    c) Show that, according to this model, the particle will never reverse direction.
    d) Show that it does not have a constant acceleration.

14) A particle's velocity, a ms^{-2} with time t seconds is given by
$$a = -1.8t + 18$$
    where $t \geq 0$
    Find the time it takes for the particle to travel 600 metres given it starts from rest.

15) A car is being driven on a straight flat road, with velocity v ms^{-1} at time t seconds is given by
$$v = 2t - 0.001t^3, \quad 0 \leq t \leq 25$$
    a) Find the distance travelled by the car in the first 20 seconds.
    b) Show that for $0 \leq t \leq 25$, the car is always accelerating (and not decelerating).

16) The displacement, s metres at any given time t seconds of a car may be given by
$$s = \frac{4}{15}t^2 - \frac{1}{675}t^3, \quad 0 \leq t \leq 120$$
    a) Find the velocity of the car v in terms of t
    b) Find the maximum speed for $0 \leq t \leq 120$

1.71 Variable Acceleration (answers on page 484)

Prerequisite knowledge for Q17-Q18: Constant Acceleration

17) A car is being tested on a straight, flat road. Its velocity is found to be 12.5 ms^{-1} at 5 seconds and 25 ms^{-1} at 10 seconds. A model for its velocity v ms^{-1} against time t seconds is
$$v = at^2 + bt^3, \quad t \geq 0$$
where a and b are unknown constants.
   a) Find the values of a and b

   A second model suggests that the cars motion may be modelled with constant acceleration.
   b) Show that the distance travelled for both these two models in the first 10 seconds is the same.

18) A runner having started from rest reaches a velocity of 3.5 ms^{-1} in 7 seconds.
   a) Assuming constant acceleration, find a model for the velocity v ms^{-1} at any given time t
   b) Find the distance travelled by the runner in the 7 second duration using this model.

   An alternative model is given by
   $$v = -\frac{1}{k}(t-7)^2 + 3.5, \quad t \geq 0$$
   where k is a positive constant.
   c) Find the value of k
   d) Find the distance travelled by the runner in the 7 second duration according to this second model.
   e) By first sketching both models on a graph, explain why the second model gives a larger distance than the first.

19) The velocity, v ms^{-1} at any given time t seconds of a cyclist for a 6 second section of their journey may be given by
$$v = \frac{2}{5}t^3 - 4t^2 + 10t, \quad 0 \leq t \leq 6$$
   a) Find the distance the cyclist has travelled in the first 5 seconds.
   b) Find the maximum speed of the cyclist and the time at which it occurs.
   c) Find the minimum acceleration of the cyclist and state what its value represents in context of the cyclist.
   d) Explain why this model would not be valid in the long term.

20) The velocity, v ms^{-1} at any given time t seconds of a person walking along a flat, straight path may be given by
$$v = \frac{k}{\sqrt{t}} + \frac{t}{80}, \quad 0 < t \leq 120$$
where k is an unknown positive constant. It is known that the acceleration when $t = 25$ is 0.0065 ms^{-2}.
   a) Show that $k = \frac{3}{2}$
   b) Find the distance travelled by the person for $60 < t \leq 120$
   c) Comment on the validity of the model for small values of t

21) The velocity, v ms^{-1} at any given time t seconds of a particle may be given by
$$v = pt^2 + qt + r, \quad t \geq 0$$
where p, q and r are unknown constants. Initially, it has velocity $v = 7.8$ ms^{-1}. It reaches its minimum velocity of 3 ms^{-1} when $t = 4$
Find the values of p, q and r

22) The acceleration, a ms^{-2} at any given time t seconds of a particle may be given by
$$a = kt - 4, \quad t \geq 0$$
where k is an unknown constant. The initial velocity is twice the magnitude of the minimum velocity which occurs when $t = 2$
Find the value of k and hence an expression for the velocity, v in terms of t only.

23) The velocity, v ms^{-1} at any given time, t seconds, of a particle may be given by
$$v = \frac{3}{2}t^2 - \frac{9}{2}t + 3, \quad t \geq 0$$
The particle is initially at the origin. Determine the displacement of the particle from the origin when it has covered a total distance of 8.5 metres.

1.71 Variable Acceleration (answers on page 484)

24) The velocity v ms^{-1} at any given time t seconds of a particle starting at initial position O may be given by
$$v = -t^2 + kt, \qquad 0 < t \leq 9$$
where k is an unknown positive constant. Initially, the acceleration is 6 ms^{-2}.
   a) Show that $k = 6$
   b) Show that the maximum velocity occurs when $t = 3$
   c) Find the distance it has travelled when it has reached its maximum velocity.
   d) Find the displacement of the particle over the 9 second duration.
   e) Find the time at which the particle is not moving.
   f) Find the distance travelled over the 9 second duration.

25) The velocity v ms^{-1} at any given time t seconds of a cyclist may be given by
$$v = \begin{cases} 0.81t^2 - 0.09t^3 & 0 \leq t < 6 \\ 9.72 & 6 \leq t \leq 20 \end{cases}$$
   a) State the initial velocity.
   b) Describe the motion of the cyclist.
   c) Find the maximum acceleration of the cyclist.

26) Filip is walking to a shop which is 800 m from his house, but on arriving he realises he left his phone at home and returns to retrieve it. His displacement from his house s m at any given time t seconds may be given by
$$s = \begin{cases} \dfrac{3}{200}t^2 - \dfrac{1}{40000}t^3 & 0 \leq t \leq t_1 \\ -3.2t + 2080 & t_1 \leq t \leq t_2 \end{cases}$$
where t_1 is the time at which Filip arrives at the shop, and t_2 is the time at which Filip arrives home. According to the model,
   a) find the time t_1 it takes to reach the shop.
   b) find the time t_2 taken for the whole journey, giving your answer to the nearest second.
   c) find the maximum velocity during his journey to the shop.
   d) find the maximum speed for his full journey.
   e) Comment on the validity of the model for the return journey.

Prerequisite knowledge in Q27-Q28: Differentiation of the form ae^{kx}

27) A school is holding a 20 m race. The displacement s m at any given time t seconds of the fastest child running the race is given by
$$s = 6t - 15 + 15e^{-0.4t}, \qquad 0 \leq t \leq k$$
   a) Show that the time k seconds according to the model, at which the fastest child finishes the 20 m is approximately 5.56 s.
   b) Find an expression for the velocity v ms^{-1} in terms of t
   c) Find the initial velocity of the child.
   d) Find the maximum acceleration a ms^{-2} of the child during the race.

28) The vertical displacement s metres above ground level of a carriage in "The Drop", a ride at an amusement park, at any given time t in the first 40 seconds may be given by
$$s = 0.8e^{0.1t} + 8e^{-0.2t}, \qquad 0 \leq t \leq 40$$
The ride stops at 40 seconds before dropping.
   a) State the height above the ground at which the ride begins.
   b) Find the time at which the ride changes direction, giving your answer to the nearest second.
   c) Find the distance the ride drops before travelling upwards again.
   d) Find an expression for the vertical acceleration a ms^{-2} in terms of t
   e) Find the maximum vertical speed of the ride according to the model.
   In the second phase of the ride at $t = 40$, the ride drops in free-fall for 2 seconds before applying a brake.
   f) Using g = 9.8 ms^{-2} determine the distance the carriage drops before the brake is applied.

Prerequisite knowledge for Q29: $F_{\text{net}} = ma$

29) An object of mass 0.2 kg is moving in a straight line under the action of a resultant force F N. Its displacement s m at any given time t seconds may be given by
$$s = -\dfrac{1}{10}t^2(t - 4), \qquad t \geq 0$$
Find an expression for F in terms of t

2.01 Trigonometry: Radians, Arcs and Sectors (answers on page 485)

1) Copy and complete the following table:

Degrees	Radians
180°	
90°	
	$\frac{\pi}{3}$
	$\frac{\pi}{4}$
	$\frac{\pi}{5}$
30°	
20°	
	$\frac{8\pi}{9}$
1.62°	
	1.62

2) Find the length of each of these sector's arc with angle θ radians and radius r:
   a) $\theta = 0.5$, $r = 10$ cm
   b) $\theta = \frac{\pi}{4}$, $r = 20$ m
   c) $\theta = 0.75$, $r = 15$ cm
   d) $\theta = \frac{3\pi}{8}$, $r = 120$ m

3) Find the area of each of these sectors with angle θ radians and radius r:
   a) $\theta = 0.4$, $r = 20$ cm
   b) $\theta = \frac{\pi}{3}$, $r = 9$ m
   c) $\theta = 1.15$, $r = 64$ cm
   d) $\theta = \frac{\pi}{20}$, $r = 18.5$ m

4) Find the area and perimeter of the following sectors, given that the angle shown is measured in radians:

   a) [sector with angle 0.8 and side 8 cm]

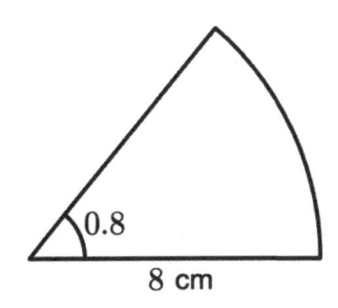

   b) [sector with 14 m and angle $\frac{\pi}{5}$]

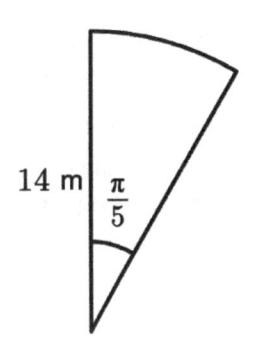

   c) [sector with 6.5 cm and angle 2.95]

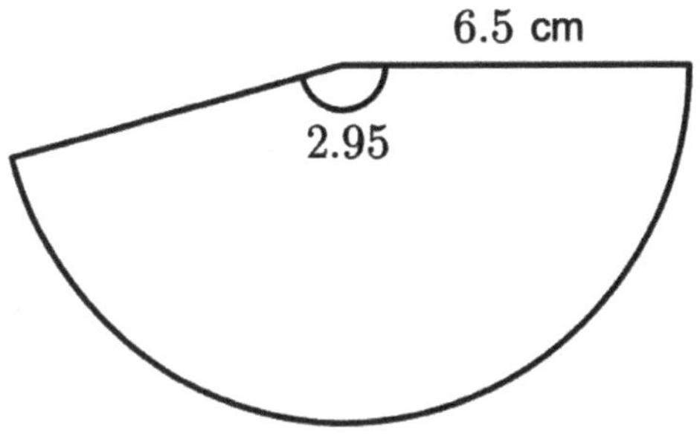

   d) [sector with angle $\frac{11\pi}{18}$ and 360 m]

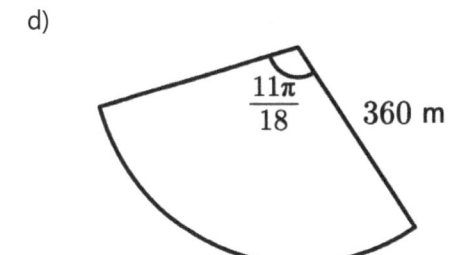

5) A sector of a circle, OAB, with radius 8 cm has perimeter 25 cm. The angle AOB is θ radians. Find the value of θ

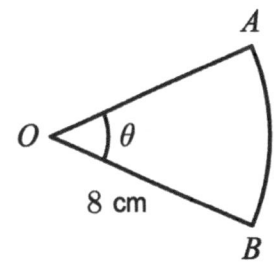

6) A sector of a circle with radius r cm and angle θ radians has an area of 50 cm². Write down an equation for r in terms of θ

7) A sector of a circle with radius r cm and angle θ radians has a perimeter of 84 m. Write down an equation for θ in terms of r

2.01 Trigonometry: Radians, Arcs and Sectors (answers on page 485)

8) The diagram below shows two sectors, OAB and OCD. OAC and OBD are straight lines. The angle AOB is $\frac{\pi}{6}$ radians. The length of OC is 9 cm. It is given that the area of sector OCD is three times larger than the area of OAB. Find the exact length of OA.

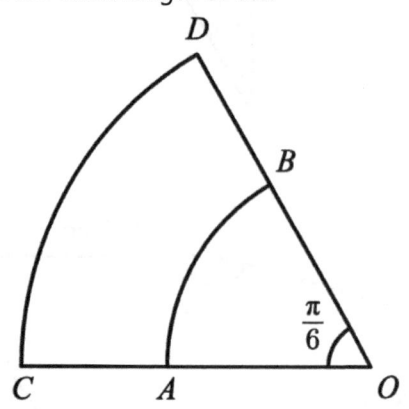

9) The diagram below shows a sector of a circle with radius 6 cm and angle 0.9 radians.

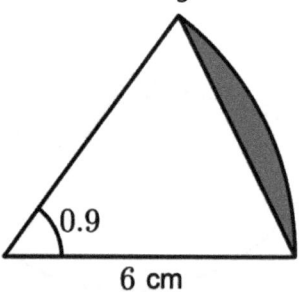

   a) Find the area of the shaded segment.
   b) Find the perimeter of the shaded segment.

10) The diagram below shows the circle with equation $x^2 + y^2 - 4x + 14y + 19 = 0$. It is given that OAB is a sector with angle 2.25 radians.

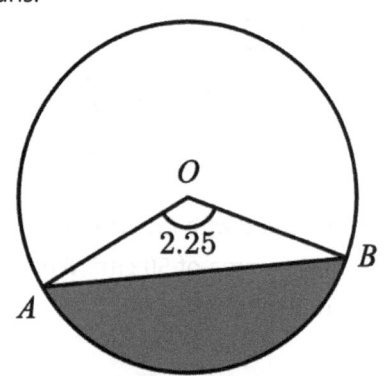

    a) Find the area of the shaded segment.
    b) What percentage of the circle is shaded in the diagram?

11) The diagram below shows a sector, OAB, of a circle with radius 10 m and angle $\frac{2\pi}{7}$ radians. It is given that C is the midpoint of O and B. Find the shaded area.

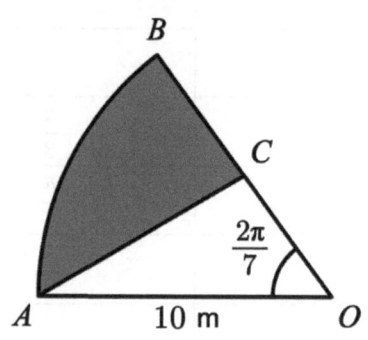

12) The diagram below shows a sector, OAB, of a circle with radius 9 cm and angle θ radians. It is given that AC has length 8 cm, and that OC has length 7 cm.

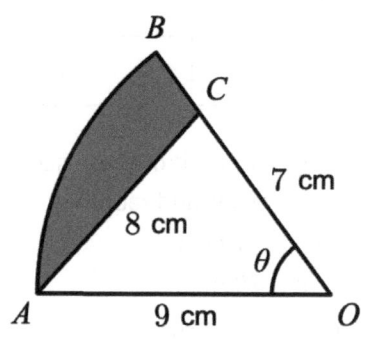

    a) Show that $\cos\theta = \frac{11}{21}$
    b) Hence show that $\sin\theta = \frac{8\sqrt{5}}{21}$
    c) Find the area of the shaded region.

13) The diagram below shows the triangle OAB. Within the triangle is a sector of a circle, where angle AOB is 1.1 radians. It is given that $OA = 15$ cm and $OB = 5\sqrt{5}$ cm. The area of the shaded region is 20 cm². Find the perimeter of the shaded region to 3 significant figures.

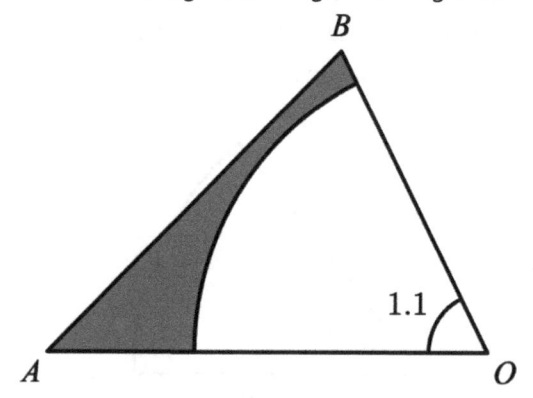

2.01 Trigonometry: Radians, Arcs and Sectors (answers on page 485)

14) The diagram below shows a sector of a circle with radius r m and angle θ radians. It is given that the area of the sector is 2 m² and the perimeter is 6 m.

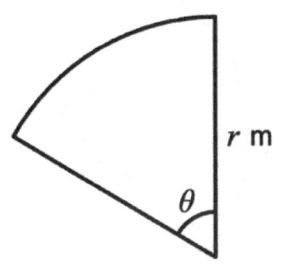

   a) Show that $r^2 - 3r + 2 = 0$
   b) Hence find the two possible values of θ

15) The diagram below shows a sector of a circle with radius r m and angle θ radians. It is given that the area of the sector is 45 m² and the perimeter is six times the length of the arc AB. Find the values of r and θ

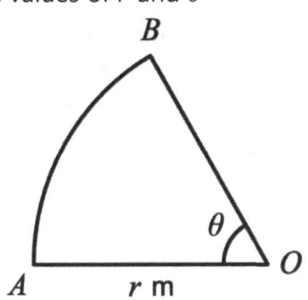

16) The diagram below shows the cross-section of a design for a new spinning toy.

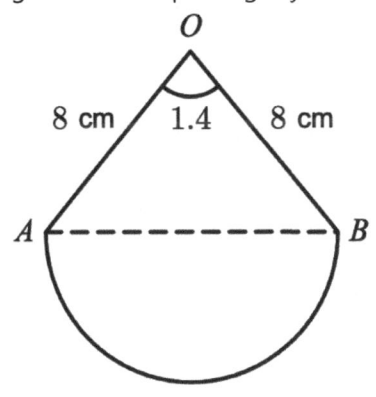

   The shape is constructed from an isosceles triangle OAB and a semi-circle with diameter AB. The toy is redesigned so that the cross-section is instead a sector of a circle with the same area as the original design. The sector has an angle of 1.4 radians and radius r cm. Determine the value of r correct to 3 significant figures.

17) A company logo is made from two similar sectors, as shown in the diagram below, where OA is twice the length of OD. The length of OA is r and the angle AOB is θ radians. It is given that the perimeter and area of the whole logo is 9.2 and 2 respectively. Show that there is only one valid pair of values for r and θ

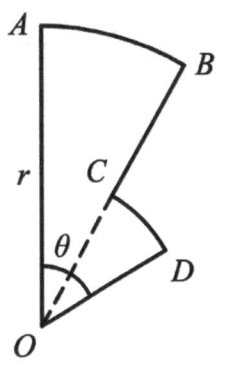

18) It is given that OAC is an equilateral triangle with side length r, and the angle BOC is θ radians. It is also given that OBF and ODE are sectors. B is the midpoint of AC.

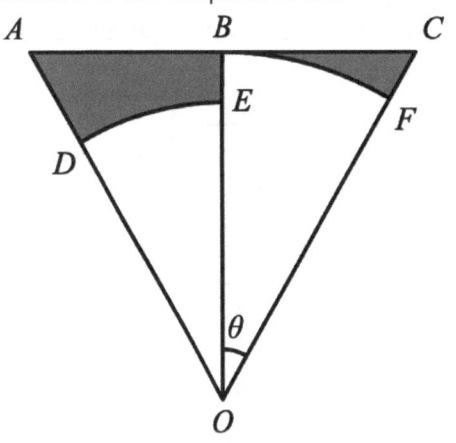

   Given that $OB = 5BE$, show that the shaded area can be written in the form $(a\sqrt{3} + b\pi)r^2$, where a and b are rational numbers.

19) The diagram below shows an equilateral triangle, ABC, with side length 5 cm. Glued to the edge of each side of the triangle is the segment of a circle centred at the opposite corner. Find the total exact area of the badge.

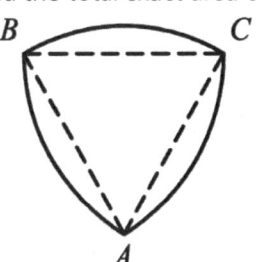

2.01 Trigonometry: Radians, Arcs and Sectors (answers on page 485)

20) The diagram below shows part of a design for a playground. It is made from three sectors. The length of the arc PQ is 12π m. The two sectors OAC and OBD are precisely the same size, and their combined area is equal to the area of the sector OPQ. It is given that AOB is a straight line. Find the length of AB.

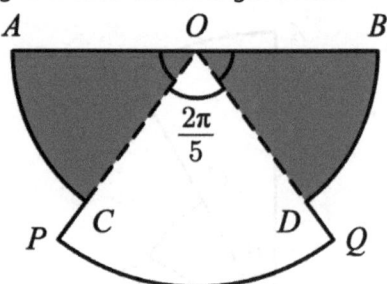

21) The diagram below shows the circle with equation $x^2 + y^2 - 6x + 12y - 4 = 0$. It is given that AB is a diameter of the circle, and both C and D lie on the circle. The angle CAB and the angle DAE are both equal to 0.6 radians. ADE is a sector of a circle centred at A. Find the area of the sector ADE to 3 significant figures.

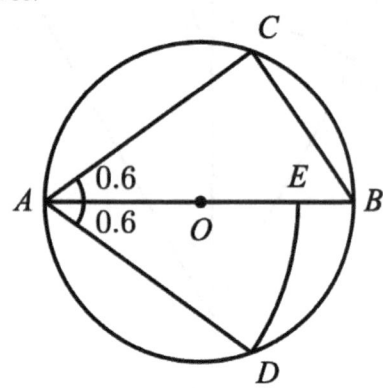

22) In the diagram below, OAB is a right-angled triangle and OBC is a sector of a circle centred at O. Triangle OAB and sector OBC have the same area. The angle BOC is θ radians. Find the shortest distance from C to the line AO

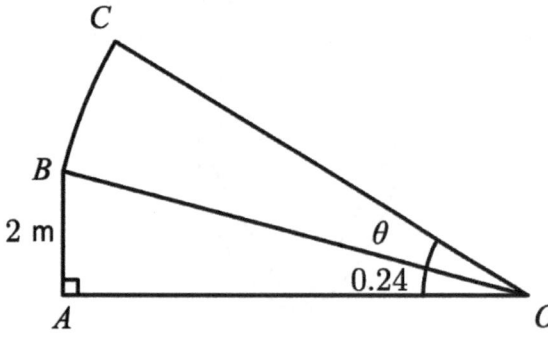

23) The diagram below shows the two circles with equations $(x - 5)^2 + (y - 5)^2 = 25$ and $(x - 17)^2 + (y - 14)^2 = 100$

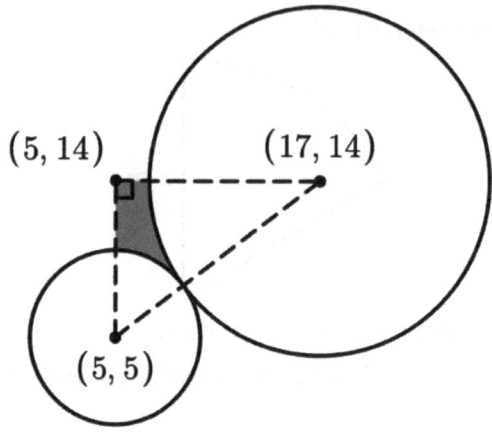

a) Show that the circles intersect at a single point.
b) Find the shaded area.

24) A square is placed over the top of a circle, as shown in the diagram below, to create a new shape. The circle, centred at O, has equation $x^2 + y^2 = 36$. The point O is the midpoint of one of the sides of the square. The square has an area of 64 units². Find the perimeter of the new shape, ignoring any dashed lines.

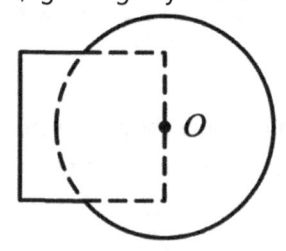

25) The diagram below shows the two circles $x^2 + y^2 = 16$ and $(x - k)^2 + y^2 = 16$. It is given that the perimeter of the shape formed by the overlapping circles is 10π. Find the exact value of k, given that $k > 0$

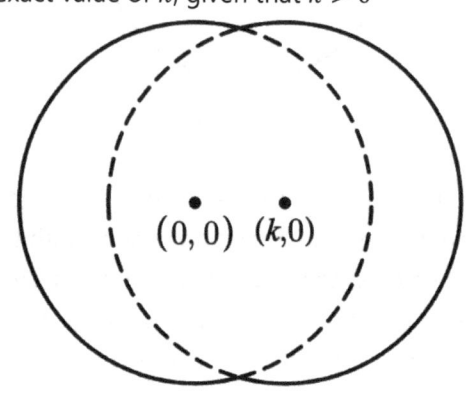

2.02 Trigonometry: $\sin x$, $\cos x$, $\tan x$ in Radians (answers on page 486)

1) Sketch the following graphs for $0 \leq x \leq 2\pi$:
   a) $y = \sin x$
   b) $y = \cos x$
   c) $y = \tan x$

2) Solve the following equations:
   a) $\sin \theta = 0$, $0 \leq \theta < 2\pi$
   b) $\sin \theta = 1$, $0 \leq \theta < 2\pi$
   c) $\sin \theta = -1$, $0 \leq \theta < 2\pi$

3) Solve the following equations:
   a) $\cos \theta = 0$, $0 \leq \theta < 2\pi$
   b) $\cos \theta = 1$, $0 \leq \theta < 2\pi$
   c) $\cos \theta = -1$, $0 \leq \theta < 2\pi$

4) Solve the following equations:
   a) $\tan \theta = 0$, $0 \leq \theta < 2\pi$
   b) $\tan \theta = 1$, $0 \leq \theta < 2\pi$
   c) $\tan \theta = -1$, $0 \leq \theta < 2\pi$

5) Find the values of x for which $y = \sin x$ is a decreasing function, for $0 \leq x < 2\pi$

6) Find the values of x for which $y = \cos x$ is an increasing function, for $0 \leq x < 2\pi$

7) Solve the following equations:
   a) $\sin \theta = \frac{1}{4}$, $0 \leq \theta < 2\pi$
   b) $\cos \theta = \frac{3}{4}$, $0 \leq \theta < 2\pi$
   c) $\tan \theta = \frac{5}{4}$, $0 \leq \theta < 2\pi$

8) Solve the following equations:
   a) $\sin \theta = 0.9$, $-2\pi \leq \theta < 0$
   b) $\cos \theta = \frac{1}{9}$, $-2\pi \leq \theta < 0$
   c) $\tan \theta = \frac{6}{5}$, $-2\pi \leq \theta < 0$

9) Solve the following equations:
   a) $\sin \theta = -\frac{1}{6}$, $0 \leq \theta < 2\pi$
   b) $\cos \theta = -\frac{8}{9}$, $0 \leq \theta < 2\pi$
   c) $\tan \theta = -\frac{1}{3}$, $0 \leq \theta < 2\pi$

10) Solve the following equations:
    a) $\sin \theta = -0.66$, $-2\pi \leq \theta < 0$
    b) $\cos \theta = -0.55$, $-2\pi \leq \theta < 0°$
    c) $\tan \theta = -8.6$, $-2\pi \leq \theta < 0$

11) Solve the following equations:
    a) $\sin \theta = \frac{4}{9}$, $-2\pi \leq \theta < 2\pi$
    b) $\cos \theta = 0.11$, $-2\pi \leq \theta < 2\pi$
    c) $\tan \theta = 3.2$, $-2\pi \leq \theta < 2\pi$
    d) $\sin \theta = -0.83$, $-2\pi \leq \theta < 2\pi$
    e) $\cos \theta = -\frac{10}{99}$, $-2\pi \leq \theta < 2\pi$
    f) $\tan \theta = -6.6$, $-2\pi \leq \theta < 2\pi$

12) Find the values of x for which $y = \sin x$ is an increasing function, for $-5\pi \leq x \leq 0$

13) Find the values of x for which $y = \cos x$ is a decreasing function, for $7\pi \leq x \leq 10\pi$

14) Solve the following equations:
    a) $\sin \theta = \frac{3}{16}$, $0 \leq \theta < 4\pi$
    b) $\cos \theta = -\frac{5}{11}$, $3\pi \leq \theta < 6\pi$
    c) $\tan \theta = -0.94$, $-7\pi \leq \theta < -5\pi$

15) Solve the following equations:
    a) $\frac{1}{\sin \theta} = 16.5$ for $0 \leq \theta < \pi$
    b) $\frac{14}{\tan \theta} = 19$ for $\pi \leq \theta < 4\pi$
    c) $\frac{0.02}{\cos \theta} = 0.3$ for $-\frac{5\pi}{2} \leq \theta < -\frac{3\pi}{2}$

16) A student is solving a problem with a triangle. They correctly use the sine rule to find $\frac{\sin A}{150} = \frac{\sin 1^c}{135}$. Find the two possible values of A

17) The graph below shows the curve $y = \sin x$ for $-2\pi \leq x \leq 2\pi$, and the line $y = k$ where $0 < k < 1$. The point A has x-coordinate $-p$, where $p > 0$. Write down the x-coordinates of B, C and D in terms of p

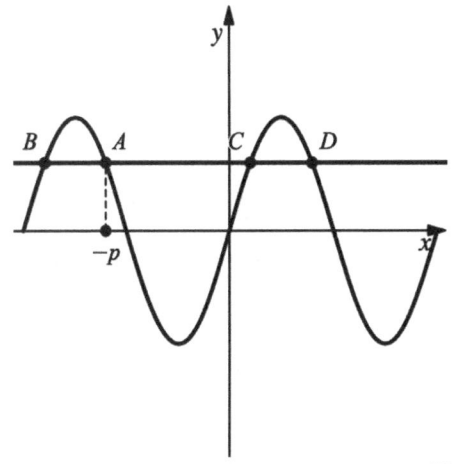

2.02 Trigonometry: $\sin x$, $\cos x$, $\tan x$ in Radians (answers on page 486)

18) It is given that $\sin\frac{\pi}{6} = \frac{1}{2}$

    Explain why $\sin\frac{17\pi}{6} = \frac{1}{2}$

19) It is given that $\cos\frac{\pi}{4} = \frac{\sqrt{2}}{2}$

    Explain why $\cos\left(-\frac{29\pi}{4}\right) = -\frac{\sqrt{2}}{2}$

20) It is given that $\tan\left(-\frac{39\pi}{4}\right) = 1$

    Explain why $\tan\frac{19\pi}{4} = -1$

21) The graph below shows the curve $y = \sin x$ for $-2\pi \leq x \leq 2\pi$, and the lines $y = k$ and $y = -k$ where $0 < k < 1$. The point A has x-coordinate $-p$, where $p > 0$. Write down the x-coordinates of B, C and D in terms of p

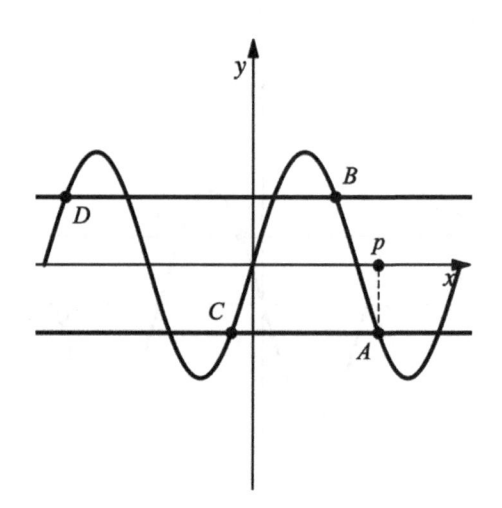

23) Solve the following equations:
    a) $\sin^2\theta = \frac{1}{25}$ for $0 \leq \theta < 3\pi$
    b) $16\tan^2\theta = 19$ for $-3\pi \leq \theta < -\pi$
    c) $25\cos^2\theta = \frac{1}{4}$ for $\pi \leq \theta < 4\pi$

24) Solve the following equations:
    a) $\sin^2\theta = 8\sin\theta$ for $0 \leq \theta < 2\pi$
    b) $9\tan^2\theta - 5\tan\theta = 0$ for $0 \leq \theta < 2\pi$
    c) $-6\cos\theta = 11\cos^2\theta$ for $0 \leq \theta < 2\pi$

25) The graph below shows the curve $y = \cos x$ for $-2\pi \leq x \leq 2\pi$, and the lines $y = k$ and $y = -k$ where $0 < k < 1$. The point A has x-coordinate p, where $p > 0$. Write down the x-coordinates of B, C and D in terms of p

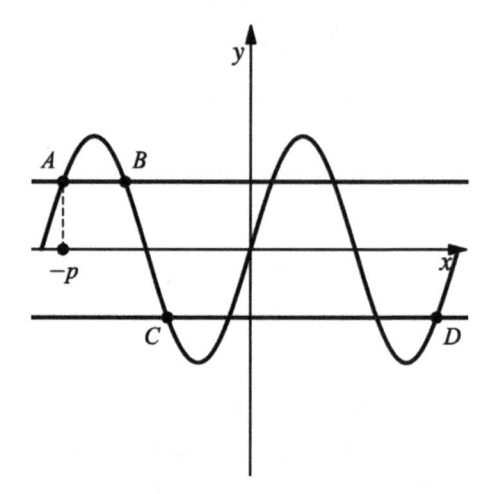

22) The graph below shows the curve $y = \sin x$ for $-2\pi \leq x \leq 2\pi$, and the lines $y = k$ and $y = -k$ where $0 < k < 1$. The point A has x-coordinate p, where $p > 0$. Write down the x-coordinates of B, C and D in terms of p

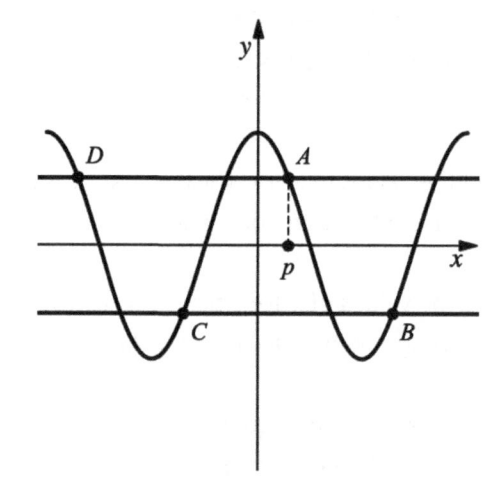

26) The graph below shows the curve $y = \cos x$ for $-2\pi \leq x \leq 2\pi$, and the lines $y = k$ and $y = -k$ where $0 < k < 1$. The point A has x-coordinate $-p$, where $p > 0$. Write down the x-coordinates of B, C and D in terms of p

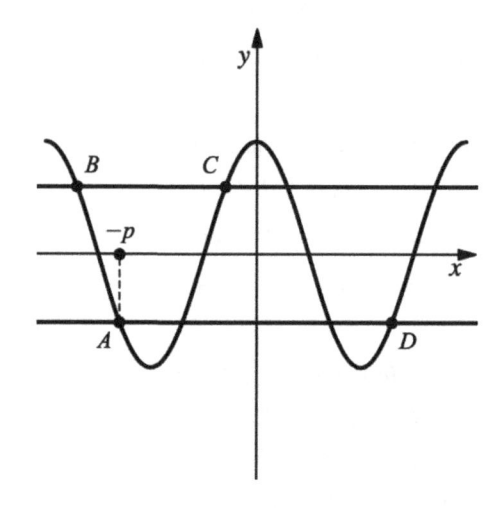

2.02 Trigonometry: $\sin x$, $\cos x$, $\tan x$ in Radians (answers on page 486)

27) The graph below shows the curve $y = \tan x$ for $-2\pi \leq x \leq 2\pi$, and the lines $y = k$ and $y = -k$ where $0 < k < 1$. The point A has x-coordinate $-p$, where $p > 0$. Write down the x-coordinates of B, C and D in terms of p

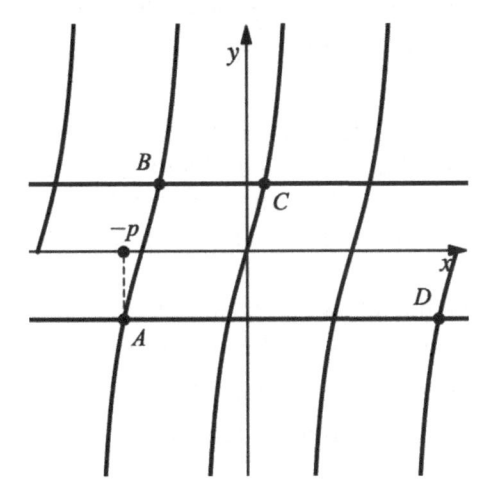

28) Solve the following equations:
    a) $5 \sin \theta \cos \theta = 3 \cos \theta$ for $0 \leq \theta < 2\pi$
    b) $\frac{2}{5} \sin \theta \tan \theta + 3 \sin \theta = 0$ for $0 \leq \theta < 2\pi$
    c) $29 \sin \theta - 31 \sin \theta \cos \theta = 0$ for $0 \leq \theta < 2\pi$

29) Solve the following equations:
    a) $6 \sin^2 \theta - \sin \theta - 1 = 0$ for $0 \leq \theta < 2\pi$
    b) $15 \cos^2 \theta - 2 \cos \theta - 1 = 0$ for $0 \leq \theta < 2\pi$
    c) $18 \tan^2 \theta + 15 \tan \theta + 2 = 0$ for $0 \leq \theta < 2\pi$

30) Solve the equations for $0 \leq \theta < 2\pi$:
    a) $32 \sin \theta = \frac{9}{\sin \theta} - 12$
    b) $54 \cos \theta = 51 + \frac{14}{\cos \theta}$
    c) $20 \tan \theta = 48 + \frac{77}{\tan \theta}$

31) It is given that $f(x) = 5x^2 + 70x + 219$
    a) Write $f(x)$ in the form $a(x + b)^2 + c$, where a, b and c are integers.
    b) Write down the minimum point of the curve with equation $y = f(x)$

    It is given that $g(\theta) = 5 \cos^2 \theta - 70 \cos \theta + 219$
    c) Using your answers to a) and b), find the minimum value of $g(\theta)$ and the smallest positive value of θ, in radians, for which this minimum occurs.

32) Write down the period of the following functions in radians:
    a) $y = \sin x$
    b) $y = \cos x$
    c) $y = \tan x$
    d) $y = \sin 3x$
    e) $y = \cos 2x$
    f) $y = \tan 6x$
    g) $y = \sin \frac{1}{5} x$
    h) $y = \cos \frac{8}{3} x$
    i) $y = \tan \frac{9}{16} x$

33) State the number of solutions to the following equations. You are not required to solve them.
    a) $\sin 2x = 0.7$ for $0 \leq x < 2\pi$
    b) $\sin 2x = 0.7$ for $0 \leq x < 3\pi$
    c) $\sin 2x = 0.7$ for $-4\pi \leq x < 8\pi$

34) State the number of solutions to the following equations. You are not required to solve them.
    a) $\cos 2x = 0.2$ for $-\pi \leq x < \pi$
    b) $\cos 3x = 0.2$ for $0 \leq x < 5\pi$
    c) $\cos 4x = 0.2$ for $-6\pi \leq x < \frac{3\pi}{2}$

35) State the number of solutions to the following equations. You are not required to solve them.
    a) $\tan 2x = 2.1$ for $-2\pi \leq x < 3\pi$
    b) $\tan 3x = 2.1$ for $-5\pi \leq x < 10\pi$
    c) $\tan 4x = 2.1$ for $-25\pi \leq x < 40\pi$

36) Write down the period of the following functions in radians:
    a) $y = \sin(x + \pi)$
    b) $y = \cos\left(x - \frac{2\pi}{3}\right)$
    c) $y = \tan(x - 0.1)$
    d) $y = \sin(4\pi - x)$
    e) $y = \cos(0.9 - x)$
    f) $y = \tan(1.3 - x)$
    g) $y = \sin\left(3x + \frac{\pi}{9}\right)$
    h) $y = \cos(6x - 0.34)$
    i) $y = \tan\left(12x + \frac{2\pi}{5}\right)$
    j) $y = \sin\left(0.4 - \frac{1}{6}x\right)$
    k) $y = \cos\left(\frac{0.56 - x}{0.16}\right)$
    l) $y = \tan\left(\frac{5\pi - 15x}{12}\right)$

2.02 Trigonometry: sin x, cos x, tan x in Radians (answers on page 486)

37) Solve the following equations:
   a) $\sin(\theta + 0.32) = 0.45$ for $0 \leq \theta < 2\pi$
   b) $\sin\left(\theta - \frac{8\pi}{9}\right) = \frac{6}{7}$ for $0 \leq \theta < 2\pi$
   c) $\sin(\theta - 1.2) = -0.33$ for $0 \leq \theta < 2\pi$

38) Solve the following equations:
   a) $\cos(\theta - 0.97) = 0.2$ for $0 \leq \theta < 2\pi$
   b) $\cos(\theta + 2.3) = 0.49$ for $0 \leq \theta < 2\pi$
   c) $\cos\left(\theta - \frac{5\pi}{3}\right) = -0.89$ for $0 \leq \theta < 2\pi$

39) Solve the following equations:
   a) $\tan(\theta + 2.01) = 6$ for $0 \leq \theta < 2\pi$
   b) $\tan\left(\theta - \frac{\pi}{5}\right) = \frac{3}{8}$ for $0 \leq \theta < 2\pi$
   c) $\tan(\theta + 0.99) = -8.2$ for $0 \leq \theta < 2\pi$

40) Find the values of x for which $y = \sin\left(x - \frac{\pi}{4}\right)$ is a decreasing function, for $0 \leq x \leq 2\pi$

41) Solve the following equations:
   a) $\sin 2\theta = \frac{\sqrt{2}}{2}$ for $0 \leq \theta < 2\pi$
   b) $\cos 2\theta = \frac{\sqrt{3}}{2}$ for $0 \leq \theta < 2\pi$
   c) $\tan 2\theta = \sqrt{3}$ for $0 \leq \theta < 2\pi$

42) Solve the following equations:
   a) $\sin 3\theta = 0.76$ for $0 \leq \theta < 2\pi$
   b) $\cos 3\theta = -0.95$ for $0 \leq \theta < 2\pi$
   c) $\tan 3\theta = -0.32$ for $0 \leq \theta < 2\pi$

43) Solve the following equations:
   a) $\sin\frac{1}{2}\theta = -0.28$ for $0 \leq \theta < 4\pi$
   b) $\cos\frac{1}{5}\theta = \frac{2}{9}$ for $0 \leq \theta < 10\pi$
   c) $\tan\frac{7}{5}\theta = -2.4$ for $0 \leq \theta < \pi$

44) Solve the following equations:
   a) $\sin(3\theta - 0.31) = \frac{2}{7}$ for $0 \leq \theta < \pi$
   b) $\cos(4\theta + 0.97) = \frac{6}{11}$ for $0 \leq \theta < \pi$
   c) $\tan(5\theta + 1.45) = \frac{7}{8}$ for $0 \leq \theta < \frac{\pi}{2}$

45) Solve the following equations:
   a) $\sin(5\theta + 3.2) = -0.16$ for $-\pi \leq \theta < 0$
   b) $\cos(7\theta - 4.2) = -0.12$ for $\pi \leq \theta < \frac{3\pi}{2}$
   c) $\tan(3\theta - 6.4) = -12$ for $2\pi \leq \theta < 4\pi$

46) Find the values of x for which $y = \cos 3x$ is a decreasing function, for $0 \leq x \leq \pi$

47) Find the values of x for which $y = \sin\frac{1}{4}x$ is an increasing function, for $0 \leq x \leq 10\pi$

48) Solve the following equations:
   a) $\sin\left(\frac{\pi}{8} - 2\theta\right) = \frac{\sqrt{2}}{2}$ for $0 \leq \theta < 2\pi$
   b) $\cos\left(\frac{3\pi}{5} - 3\theta\right) = \frac{\sqrt{3}}{2}$ for $0 \leq \theta < 2\pi$
   c) $\tan\left(\frac{3\pi}{10} - 4\theta\right) = 1$ for $0 \leq \theta < 2\pi$

49) Solve the following equations:
   a) $\sin(1.2 - 4\theta) = -0.07$ for $0 \leq \theta < 2$
   b) $\cos(2.1 - 5\theta) = 0.93$ for $0 \leq \theta < 1.5$
   c) $\tan(4.3 - 7\theta) = -0.16$ for $0 \leq \theta < 1$

50) Solve the following equations:
   a) $\sin^2 4\theta - \frac{1}{25} = 0$ for $-1.4 \leq \theta < 1.4$
   b) $\cos^2 3\theta - \frac{1}{81} = 0$ for $-2 \leq \theta < 2$
   c) $\tan^2 9\theta - 169 = 0$ for $-0.5 \leq \theta < 0.5$

51) The graph below shows the curve $y = \cos 3x$ for $-2\pi \leq x \leq 2\pi$, and the lines $y = k$ and $y = -k$ where $0 < k < 1$. The point A has x-coordinate $-p$, where $p > 0$. Write down the x-coordinates of B, C and D in terms of p

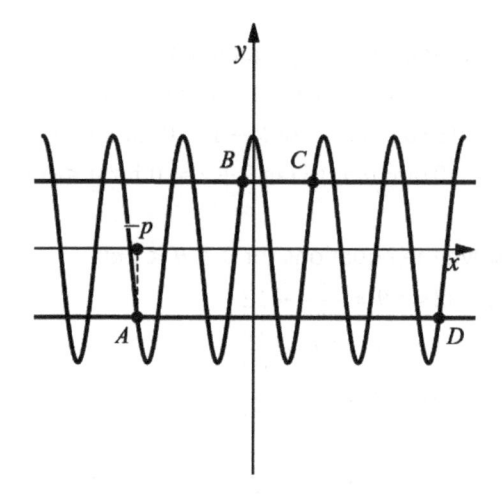

52) Solve the following equations:
   a) $\sin\left(\theta - \frac{\pi}{3}\right) = \sin\frac{\pi}{6}$ for $0 \leq \theta < 2\pi$
   b) $\sin\left(\theta + \frac{2\pi}{5}\right) = \sin\frac{11\pi}{10}$ for $0 \leq \theta < 2\pi$
   c) $\cos\left(\theta + \frac{3\pi}{7}\right) = \cos\frac{4\pi}{3}$ for $0 \leq \theta < 2\pi$
   d) $\tan\left(\theta - \frac{5\pi}{8}\right) = \tan\frac{\pi}{16}$ for $0 \leq \theta < 2\pi$

53) Solve the following equations:
   a) $\sin\left(\frac{\pi}{3} - \theta\right) = \sin\frac{46\pi}{9}$ for $0 \leq \theta < 2\pi$
   b) $\cos\left(2\theta - \frac{5\pi}{6}\right) = \cos\frac{11\pi}{12}$ for $0 \leq \theta < 2\pi$
   c) $\tan\left(\frac{7\pi}{6} - 4\theta\right) = \tan\frac{8\pi}{3}$ for $0 \leq \theta < 2\pi$

2.03 Graph Sketching: $\sin x$, $\cos x$, $\tan x$ in Radians (answers on page 488)

1) Find the equation of each of the following graphs in the form
$$y = a \sin x$$

a)

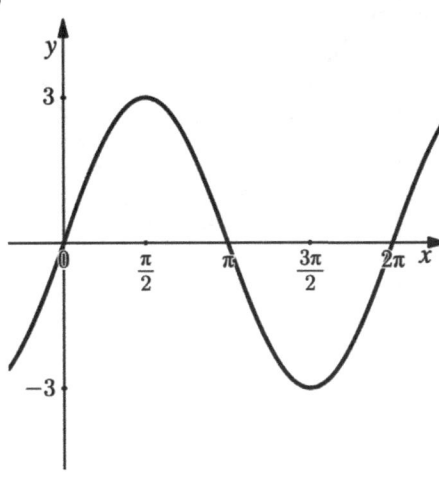

b)

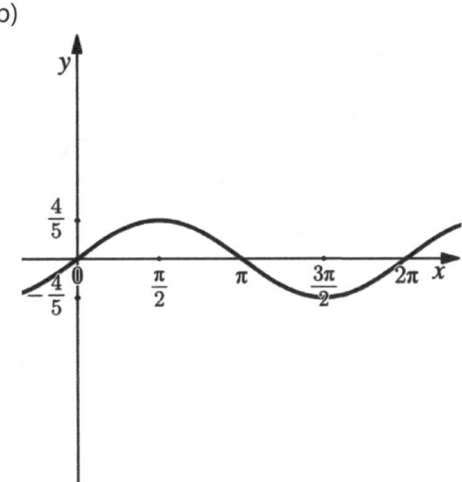

2) Find the equation of each of the following graphs in the form
$$y = a \cos x$$

a)

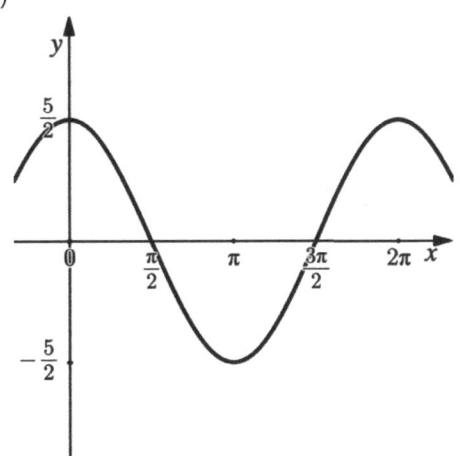

b)

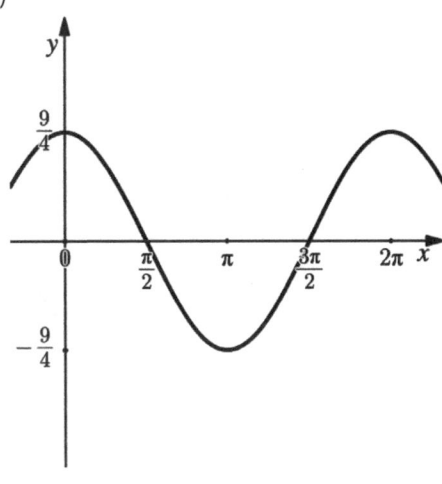

3) Sketch the following graphs for $0 \leq x \leq 2\pi$:
   a) $y = \frac{1}{3} \sin x$
   b) $y = \frac{8}{7} \cos x$

4) Find the equation of each of the following graphs in the form
$$y = \sin ax$$

a)

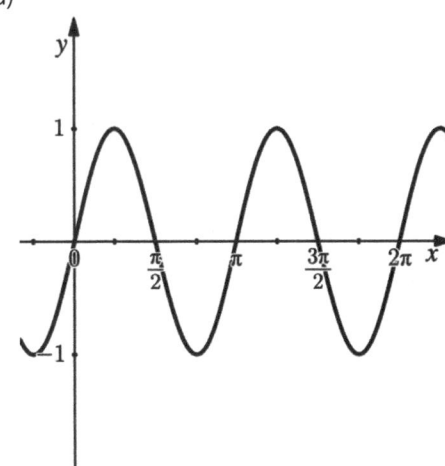

b)

221

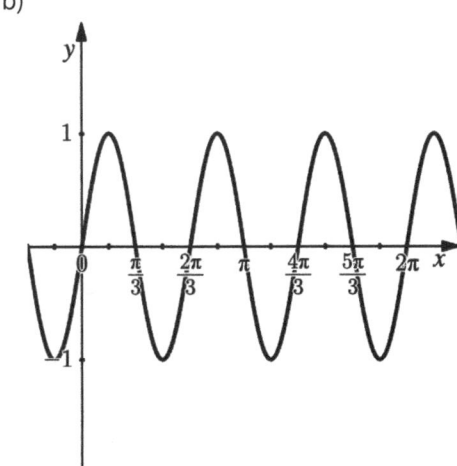

2.03 Graph Sketching: $\sin x$, $\cos x$, $\tan x$ in Radians (answers on page 488)

c)

b)

d)

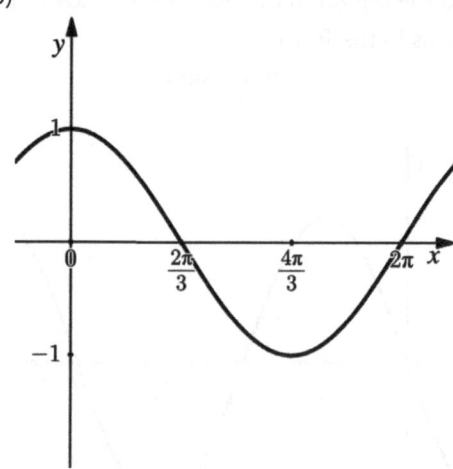

c)

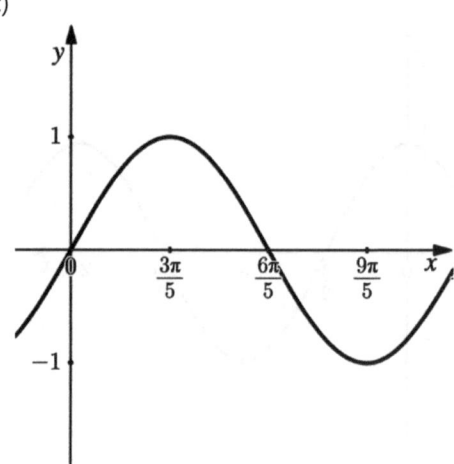

d)

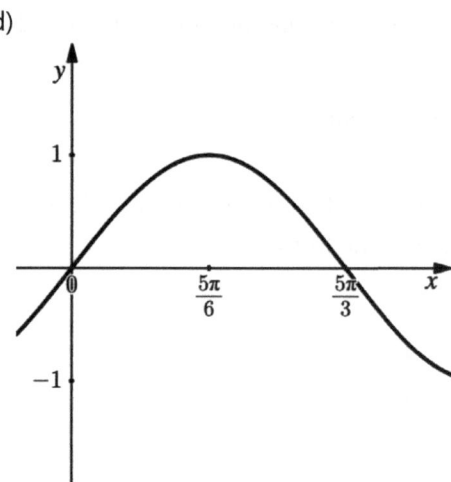

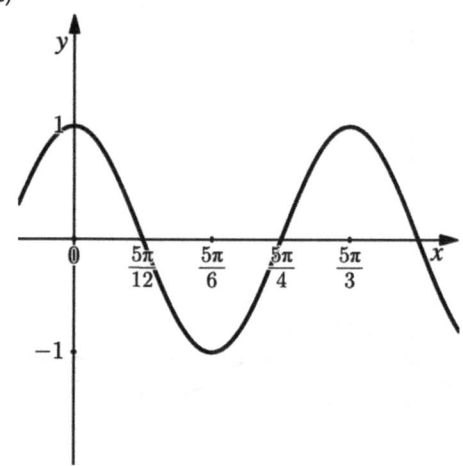

5) Sketch the graph of $y = -4 \sin x$ for $0 \leq x \leq 2\pi$

6) Find the equation of each of the following graphs in the form
$$y = \cos ax$$

a)

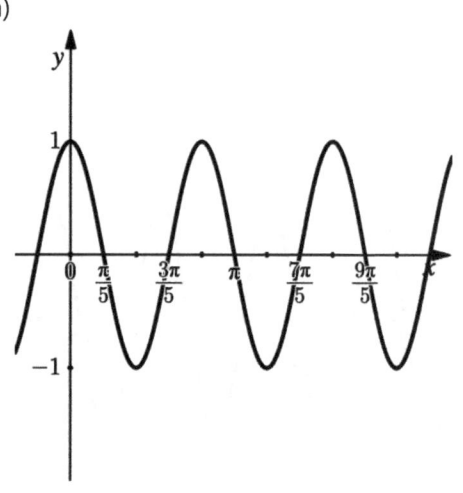

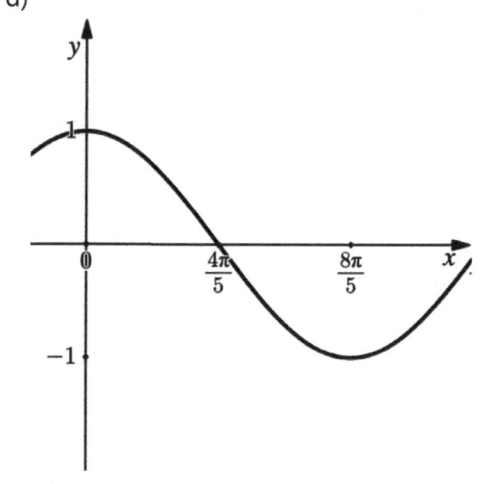

7) Sketch the graph of $y = -\frac{3}{2} \cos x$ for $0 \leq x \leq 2\pi$

8) Sketch the graph of $y = \sin\left(-\frac{2}{3}x\right)$ for $0 \leq x \leq 2\pi$

2.03 Graph Sketching: $\sin x$, $\cos x$, $\tan x$ in Radians (answers on page 488)

9) Find the equation of each of the following graphs in the form
$$y = \tan ax$$

a)

b)

c)

d)

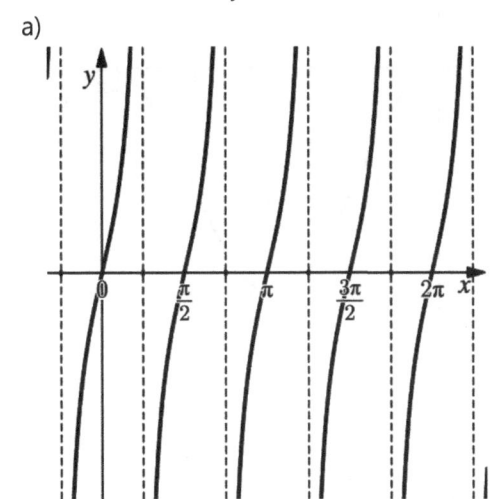

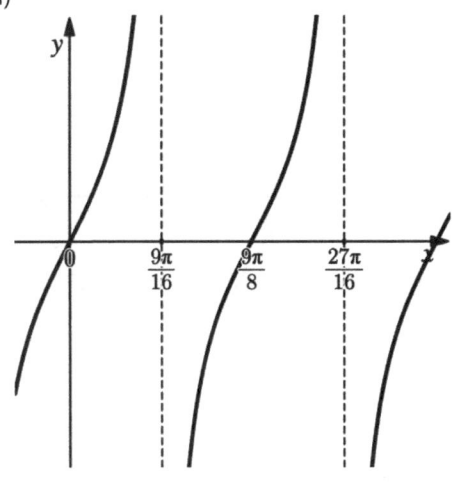

10) Sketch the following graphs for $0° \leq x \leq 2\pi$:
   a) $y = \tan \frac{2}{5}x$
   b) $y = \tan \frac{10}{9}x$

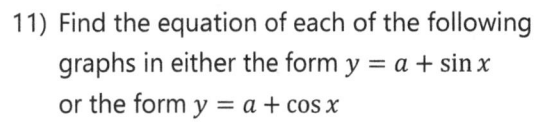

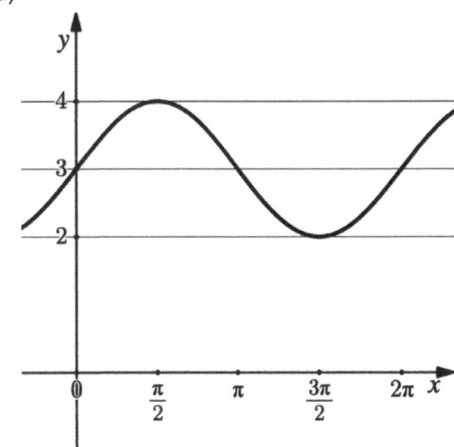

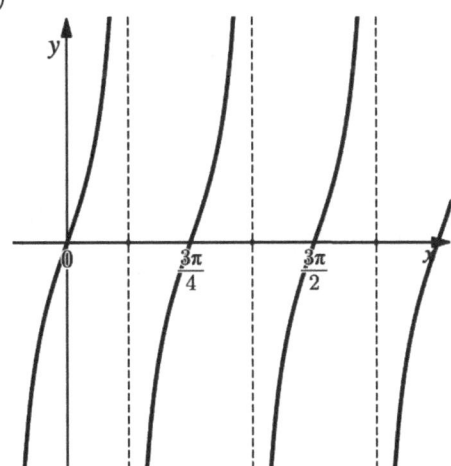

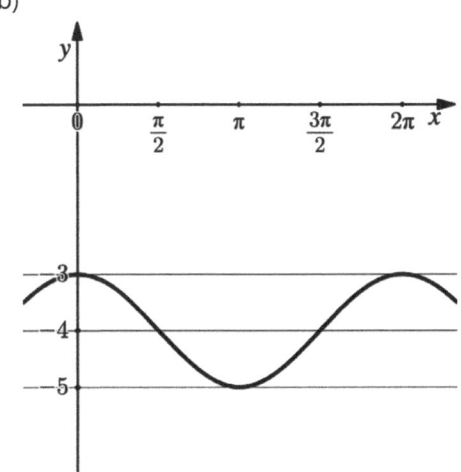

11) Find the equation of each of the following graphs in either the form $y = a + \sin x$ or the form $y = a + \cos x$

a)

b)

2.03 Graph Sketching: $\sin x$, $\cos x$, $\tan x$ in Radians (answers on page 488)

c)

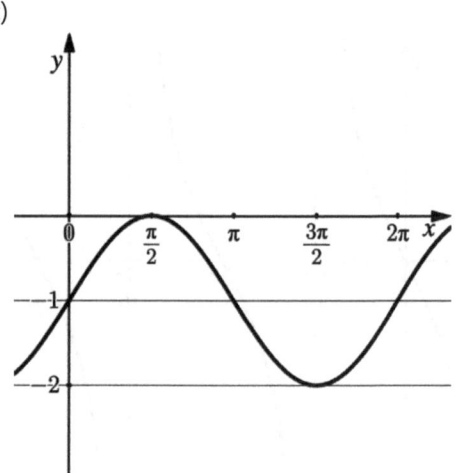

b)

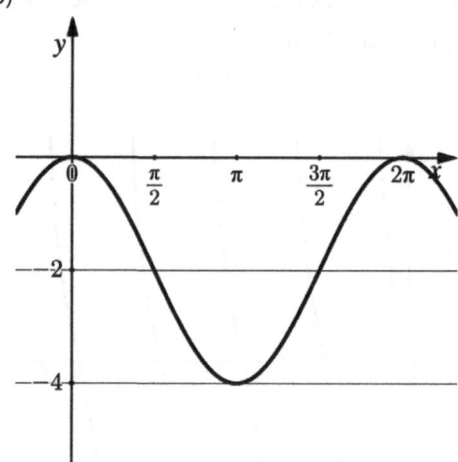

d)

c)

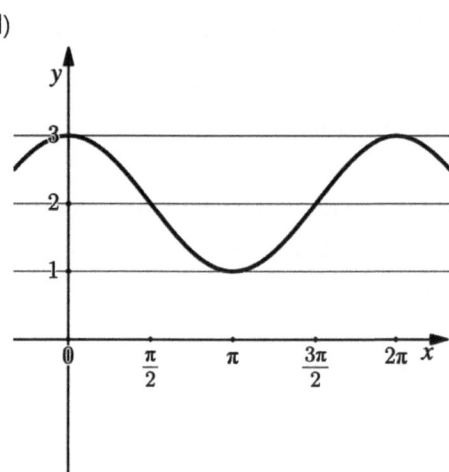

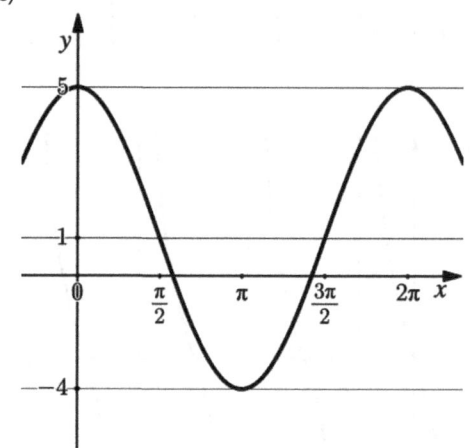

Prerequisite knowledge for Q12-Q13:
Combinations of Graph Transformations

d)

12) Find the equation of each of the following graphs in the form $y = a + b \sin x$ or the form $y = a + b \cos x$
   a)

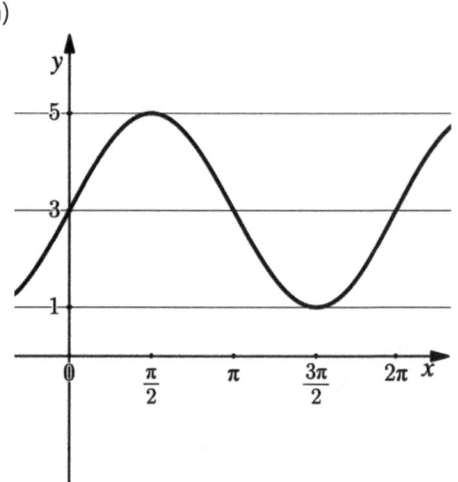

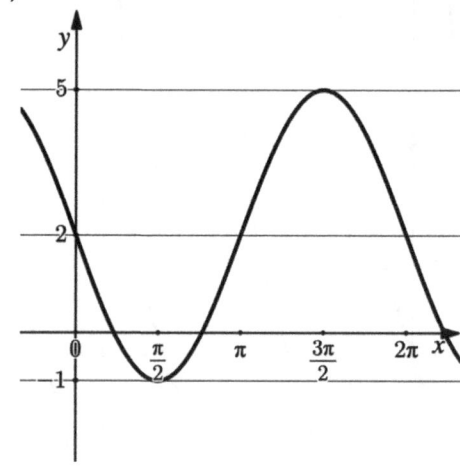

13) Sketch the following graphs for $0 \leq x \leq 2\pi$:
   a) $y = 1 + 3 \sin x$
   b) $y = -2 + 3 \cos x$
   c) $y = \frac{3}{2} - 4 \sin x$
   d) $y = -\frac{2}{3} - 2 \cos x$

2.03 Graph Sketching: $\sin x$, $\cos x$, $\tan x$ in Radians (answers on page 488)

14) Find the equation of each of the following graphs in the form $y = \sin(x + a)$ where $-\pi < a < \pi$

   a)

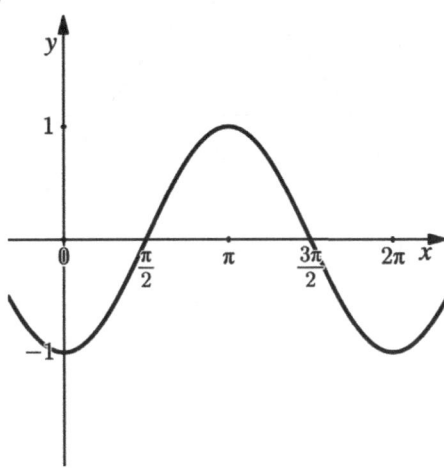

   b)

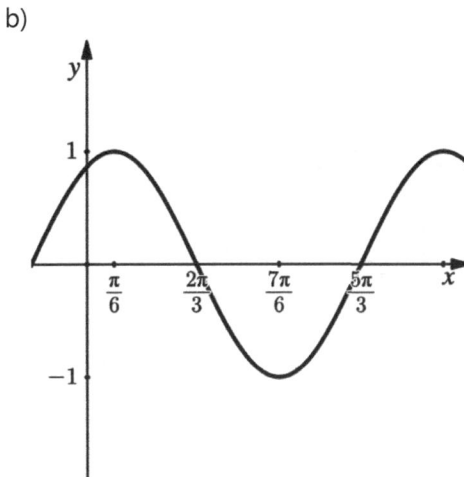

15) Find the equation of each of the following graphs in the form $y = \cos(x + a)$ where $-\pi < a < \pi$

   a)

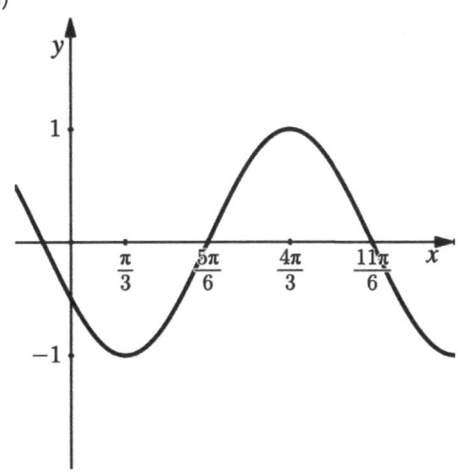

   b)

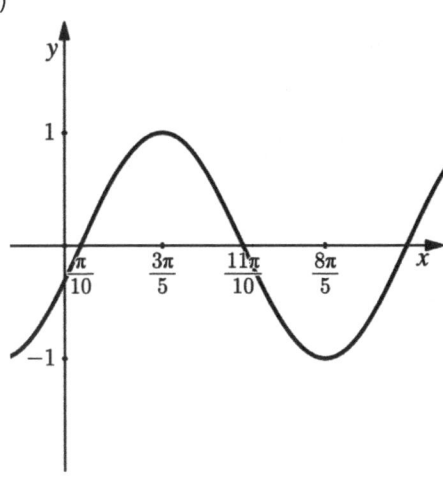

16) Find the equation of each of the following graphs in the form $y = \tan(x + a)$ where $-\frac{\pi}{2} < a < \frac{\pi}{2}$

   a)

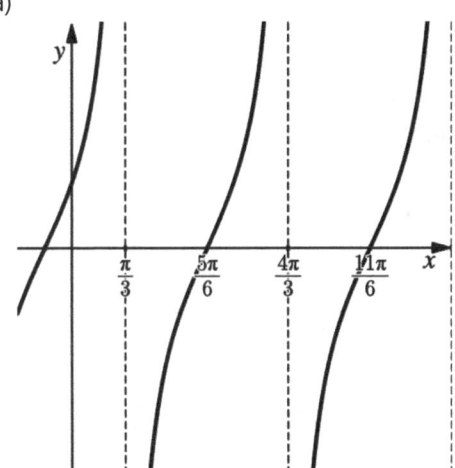

   b)

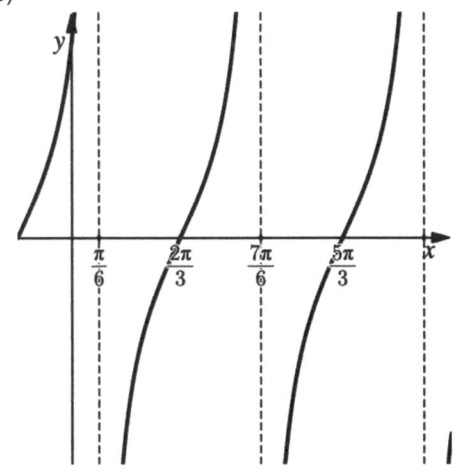

17) It is given that k is a constant where $k > 1$. Sketch the following graphs for $0 \leq x \leq 2\pi$:

   a) $y = -k + \sin x$
   b) $y = k + \cos x$

2.03 Graph Sketching: sin x, cos x, tan x in Radians (answers on page 488)

18) Sketch the following graphs for $0 \leq x \leq 2\pi$:
    a) $y = \sin\left(x - \frac{\pi}{3}\right)$
    b) $y = \cos\left(x + \frac{\pi}{3}\right)$
    c) $y = \tan\left(x - \frac{\pi}{5}\right)$
    d) $y = \sin\left(x + \frac{4\pi}{5}\right)$
    e) $y = \cos\left(x - \frac{\pi}{6}\right)$
    f) $y = \tan\left(x + \frac{3\pi}{8}\right)$

19) It is given that k is a constant where $0 < k < \frac{\pi}{2}$.
    Sketch the following graphs for $0 \leq x \leq 2\pi$:
    a) $y = \sin(x - k)$
    b) $y = \tan(x + k)$

Prerequisite knowledge for Q20-Q25:
Combinations of Graph Transformations

20) Find the equation of each of the following graphs in the form $y = \sin(ax + b)$ where $-\pi < b < \pi$
    a) 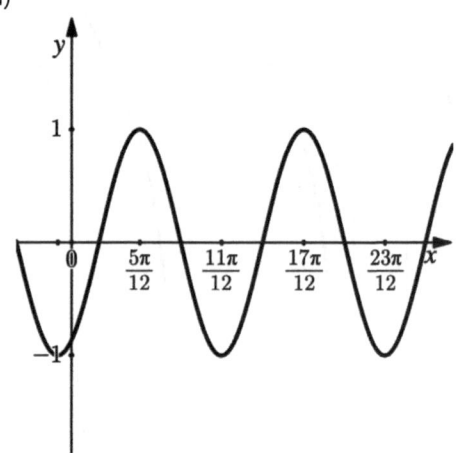
    b) 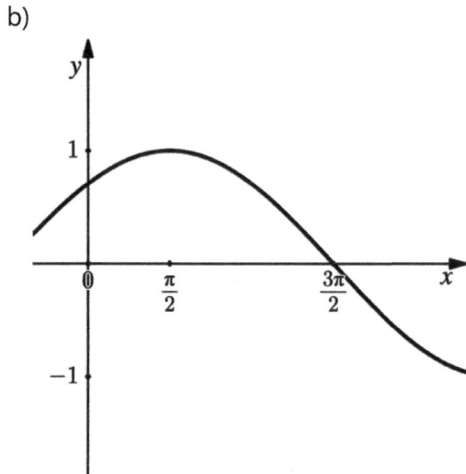
    c)

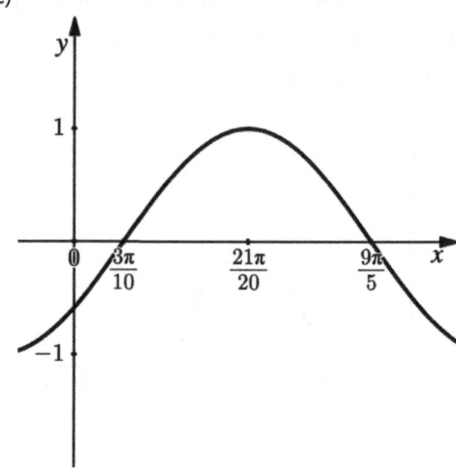

21) Find the equation of each of the following graphs in the form $y = \cos(ax + b)$ where $-\pi < b < \pi$
    a) 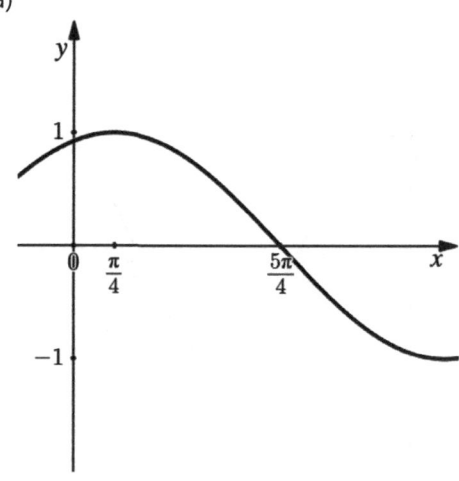
    b)

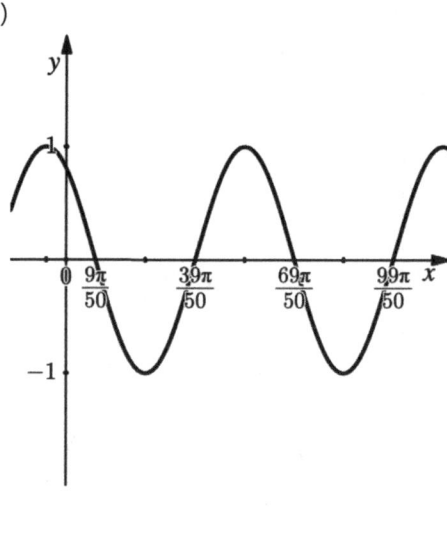

2.03 Graph Sketching: $\sin x$, $\cos x$, $\tan x$ in Radians (answers on page 488)

c) [Graph showing a sine-like curve passing through origin, with x-axis markings at $\frac{7\pi}{9}$ and $\frac{133\pi}{90}$, y-range from -1 to 1]

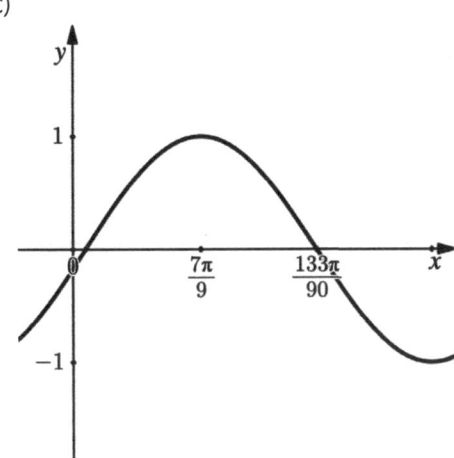

c) [Graph showing a tangent-like curve with vertical asymptote at $\frac{3\pi}{4}$ and x-axis marking at $\frac{15\pi}{8}$]

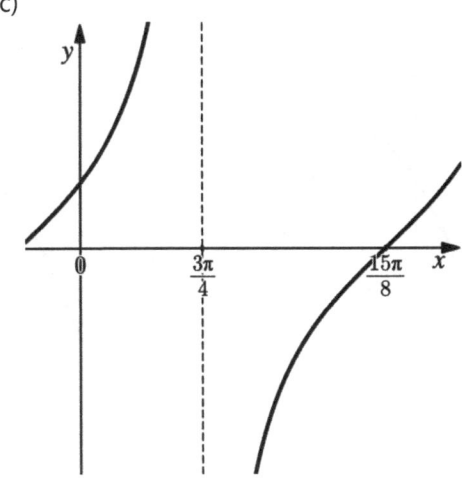

22) Find the equation of each of the following graphs in the form $y = \tan(ax + b)$ where $-\frac{\pi}{2} < b < \frac{\pi}{2}$

a) [Graph with vertical asymptotes and x-axis markings at $\frac{2\pi}{5}$, π, $\frac{8\pi}{5}$]

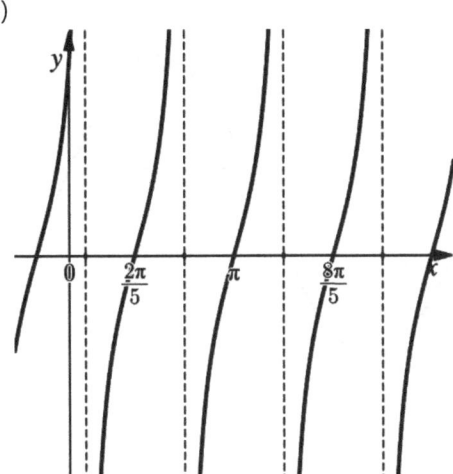

b) [Graph with vertical asymptotes and x-axis markings at $\frac{2\pi}{7}$, $\frac{6\pi}{7}$, $\frac{10\pi}{7}$, 2π]

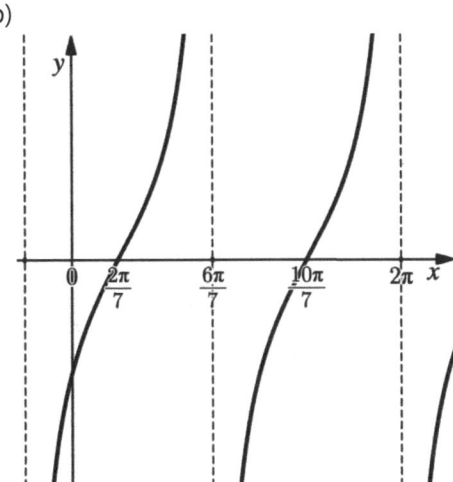

23) Find the equation of each of the following graphs in the form
$$y = a + b\sin(cx + d)$$
where $-\pi < d < \pi$

a) [Graph of sinusoidal curve with x-axis markings at $\frac{3\pi}{8}$, $\frac{7\pi}{8}$, $\frac{11\pi}{8}$, $\frac{15\pi}{8}$, with maximum 5 and minimum -1]

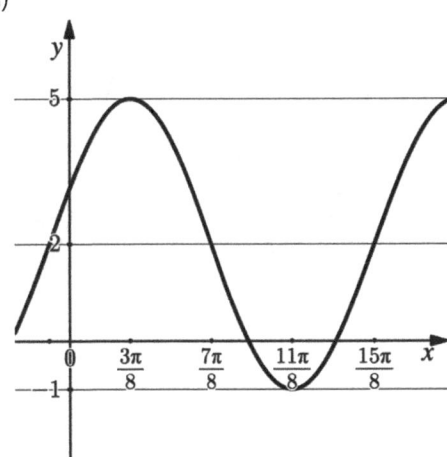

b) [Graph of sinusoidal curve with x-axis markings at $\frac{2\pi}{5}$, $\frac{11\pi}{15}$, $\frac{16\pi}{15}$, $\frac{7\pi}{5}$, $\frac{26\pi}{15}$, with maximum -2 and minimum -4]

227

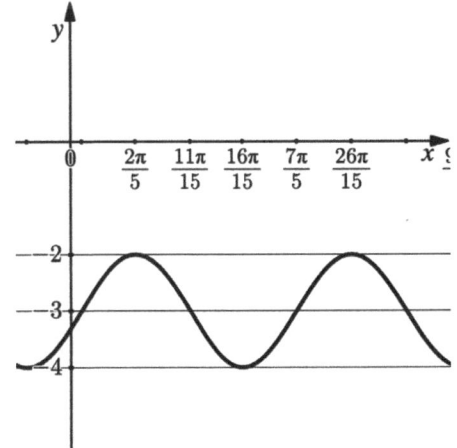

2.03 Graph Sketching: $\sin x$, $\cos x$, $\tan x$ in Radians (answers on page 488)

c)

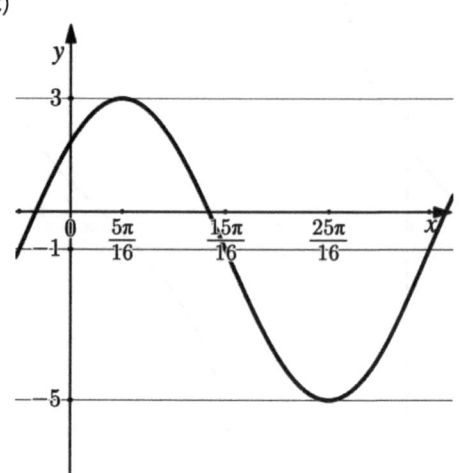

b)

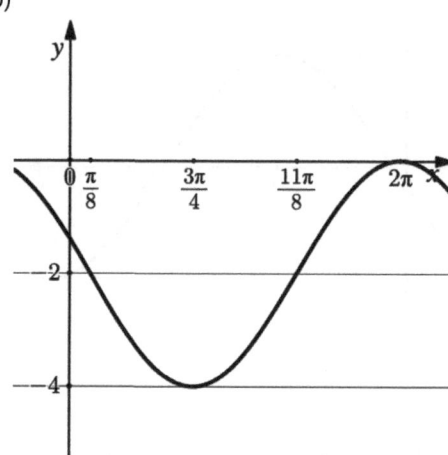

d)

c)

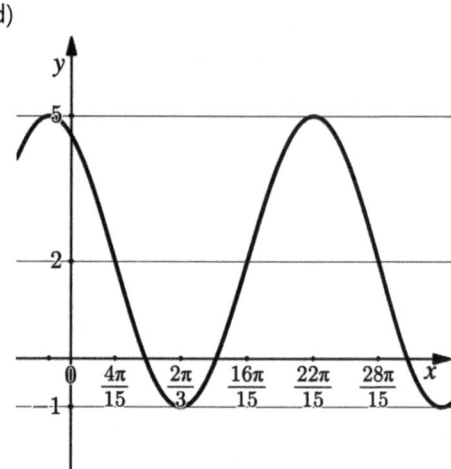

24) Find the equation of each of the following graphs in the form
$$y = a + b\cos(cx + d)$$
where $-\pi < d < \pi$

a)

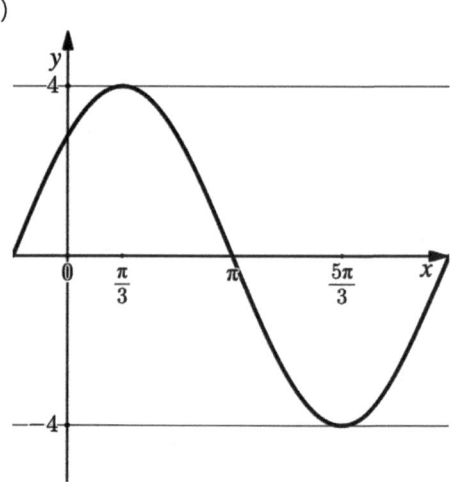

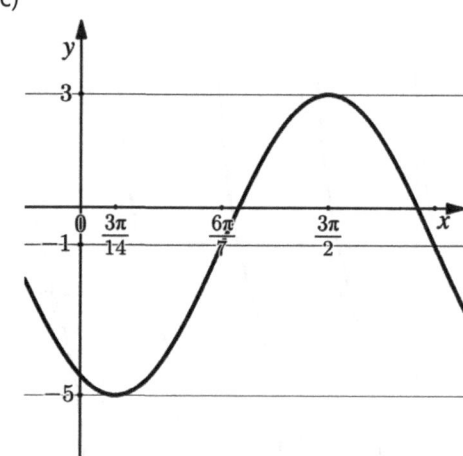

d)

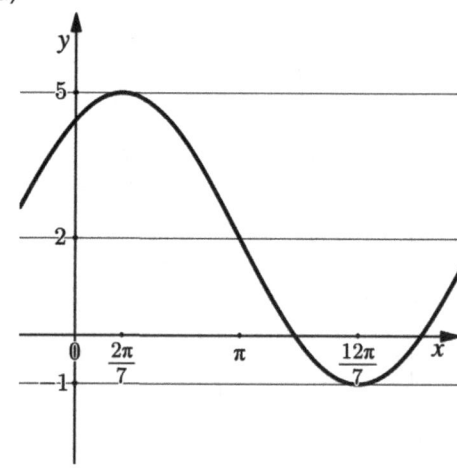

25) It is given that a and b are positive constants. Sketch the following graphs for $0 \leq x < 2\pi$:
   a) $y = a + b\cos x$ where $a > b$
   b) $y = a + b\cos x$ where $a = b$
   c) $y = a + b\cos x$ where $a < b$

2.04 Trigonometry: Trigonometric Identities in Radians (answers on page 490)

1) It is given that $\sin^2 x = \frac{4}{5}$ and x is acute. Find the exact values of:
   a) $\cos x$
   b) $\tan x$

2) It is given that $\cos^2 x = \frac{8}{9}$ and x is acute. Find the exact values of:
   a) $\sin x$
   b) $\tan x$

3) It is given that $\tan^2 x = \frac{25}{6}$ and x is acute. Find the exact values of:
   a) $\sin x$
   b) $\cos x$

4) It is given that $\sin^2 x = \frac{5}{12}$ and x is obtuse. Find the exact values of:
   a) $\cos x$
   b) $\tan x$

5) It is given that $\cos^2 x = \frac{12}{13}$ and x is obtuse. Find the exact values of:
   a) $\sin x$
   b) $\tan x$

6) It is given that $\tan^2 x = \frac{30}{7}$ and x is obtuse. Find the exact values of:
   a) $\sin x$
   b) $\cos x$

7) It is given that $\sin^2 x = \frac{3}{16}$ and $\pi < x < \frac{3\pi}{2}$. Find the exact values of:
   a) $\cos x$
   b) $\tan x$

8) It is given that $\cos^2 x = \frac{11}{12}$ and $\pi < x < \frac{3\pi}{2}$. Find the exact values of:
   a) $\sin x$
   b) $\tan x$

9) It is given that $\tan^2 x = \frac{18}{11}$ and $\pi < x < \frac{3\pi}{2}$. Find the exact values of:
   a) $\sin x$
   b) $\cos x$

10) It is given that $\sin^2 x = \frac{24}{25}$ and $\frac{3\pi}{2} < x < 2\pi$. Find the exact values of:
    a) $\cos x$
    b) $\tan x$

11) It is given that $\cos^2 x = \frac{24}{49}$ and $\frac{3\pi}{2} < x < 2\pi$. Find the exact values of:
    a) $\sin x$
    b) $\tan x$

12) It is given that $\tan^2 x = \frac{60}{17}$ and $\frac{3\pi}{2} < x < 2\pi$. Find the exact values of:
    a) $\sin x$
    b) $\cos x$

13) It is given that
$$\cos\frac{\pi}{10} = \sqrt{\frac{5}{8} + \frac{\sqrt{5}}{8}}$$
   a) Find the exact value of $\sin\frac{\pi}{10}$
   b) Using your answer to a), show that
$$\tan\frac{\pi}{10} = \sqrt{1 - \frac{2}{\sqrt{5}}}$$

14) Serkan is attempting to solve the equation $3\sin x = \tan x$ for $0° \leq x < 360°$. His attempt is shown below:
$$3\sin x = \tan x$$
$$\Rightarrow 3\sin x = \frac{\sin x}{\cos x}$$
$$\Rightarrow 3\sin x \cos x = \sin x$$
$$\Rightarrow 3\cos x = 1$$
$$\Rightarrow \cos x = \frac{1}{3}$$
$$\Rightarrow x = 1.23 \text{ or } 5.05 \text{ to 3sf}$$
Explain the two errors that Serkan has made.

15) Solve the following equations:
    a) $\sin\theta = \cos\theta$, $0 \leq \theta < 4\pi$
    b) $20\sin\theta + 3\cos\theta = 0$, $0 \leq \theta < 2\pi$
    c) $11\cos\theta - 4\sin\theta = 0$, $0 \leq \theta < 2\pi$

16) Solve the following equations:
    a) $\sin 2\theta = \cos 2\theta$, $0 \leq \theta < 3\pi$
    b) $7\sin 4\theta = 2\cos 4\theta$, $0 \leq \theta < \frac{\pi}{2}$
    c) $10\sin\frac{1}{2}\theta + 3\cos\frac{1}{2}\theta = 0$, $0 \leq \theta < 5\pi$
    d) $19\cos\frac{1}{4}\theta - 16\sin\frac{1}{4}\theta = 0$, $0 \leq \theta < 20\pi$

2.04 Trigonometry: Trigonometric Identities in Radians (answers on page 490)

17) Solve the following equations:
   a) $6 \tan \theta + \cos \theta = 0$, $0 \leq \theta < 2\pi$
   b) $\sin \theta \tan \theta + 16 = 0$, $0 \leq \theta < 2\pi$
   c) $12 \tan \frac{1}{2}\theta + \cos \frac{1}{2}\theta = 0$, $0 \leq \theta < 8\pi$

18) Solve the following equations:
   a) $3 \cos^2 \theta = 8 \sin \theta$, $0 \leq \theta < \pi$
   b) $9 \sin^2 \theta = 80 \cos \theta$, $0 \leq \theta < 3\pi$
   c) $12 \cos^2 \theta = 143 \sin \theta$, $0 \leq \theta < 2\pi$
   d) $50 \sin^2 \theta = 2499 \cos \theta$, $-\pi \leq \theta < \pi$

19) Solve the following equations:
   a) $4 \cos^2 \theta + 9 \sin \theta - 6 = 0$, $0 \leq \theta < 2\pi$
   b) $9 \sin^2 \theta + 3 \cos \theta - 7 = 0$, $0 \leq \theta < 2\pi$
   c) $15 \cos^2 \theta - 31 \sin \theta - 25 = 0$, $0 \leq \theta < 2\pi$
   d) $27 \sin^2 \theta + 12 \cos \theta - 7 = 0$, $0 \leq \theta < 2\pi$

20) Solve the following equations:
   a) $8 \sin^2 \theta - 26 \sin \theta - 16 \cos^2 \theta + 21 = 0$, $0 \leq \theta \leq 2\pi$
   b) $13 \cos^2 \theta + 13 \cos \theta - 22 \sin^2 \theta + 10 = 0$, $0 \leq \theta < 2\pi$
   c) $20 \sin^2 \theta - 4 \cos^2 \theta + 38 \sin \theta - 51 = 0$, $0 \leq \theta < 2\pi$
   d) $50 \sin^2 \theta + 28 = 30 \cos^2 \theta + 31 \cos \theta$, $0 \leq \theta < 2\pi$

21)
   a) Solve the equation
   $20 \sin^2 x - 14 \sin x - 1 = 4 \cos^2 x$
   for $-2\pi \leq x < 2\pi$
   b) Hence find the smallest positive solution to the equation
   $20 \sin^2 \left(3\theta + \frac{\pi}{3}\right) - 14 \sin \left(3\theta + \frac{\pi}{3}\right) - 1$
   $= 4 \cos^2 \left(3\theta + \frac{\pi}{3}\right)$

22) Solve the following equations:
   a) $36 \sin \theta + \frac{10}{\cos \theta} = 24 \tan \theta + 15$, $0 \leq \theta < 2\pi$
   b) $3 \sin \theta \tan \theta = 21 \cos \theta + 14$, $0 \leq \theta < 2\pi$
   c) $\frac{6 \cos \theta}{\tan \theta} = 29 \sin \theta - 1$, $0 \leq \theta < 2\pi$
   d) $10 \sin \theta \cos \theta - 2 \cos^2 \theta + 5 \cos \theta$
   $= 2 \sin \theta (2 + \sin \theta)$, $0 \leq \theta < 2\pi$

23) Solve the following equations:
   a) $3 \sin \theta \tan \theta = 2 \sin \theta + \cos \theta$, $0 \leq \theta < 2\pi$
   b) $8 \sin \theta \tan \theta + 39 \sin \theta = 5 \cos \theta$, $0 \leq \theta < 2\pi$
   c) $74 \cos \theta \tan \theta = 45 \sin \theta \tan \theta + 24 \cos \theta$, $0 \leq \theta < 2\pi$
   d) $40 \sin \theta \tan \theta = 15 \cos \theta + 38 \cos \theta \tan \theta$, $0 \leq \theta < 2\pi$

24)
   a) Solve the equation $18 \sin^2 x + 3 \cos x = 8$, $0 \leq x < 4\pi$
   b) Hence solve the equation
   $18 \sin^2 \frac{1}{3}\theta + 3 \cos \frac{1}{3}\theta = 8$, $0 \leq \theta < 12\pi$

25)
   a) Show that
   $$\frac{2 \sin x - 5 \cos^2 x - 2}{5 \sin x + 7} \equiv \sin x - 1$$
   b) Hence solve the equation
   $$\frac{2 \sin \left(\theta + \frac{\pi}{3}\right) - 5 \cos^2 \left(\theta + \frac{\pi}{3}\right) - 2}{5 \sin \left(\theta + \frac{\pi}{3}\right) + 7}$$
   $$= 10 \cos \left(\theta + \frac{\pi}{3}\right) - 1$$
   for $0 \leq \theta < 2\pi$

26)
   a) Show that
   $$\frac{25 - 40 \cos^2 x - 38 \sin x}{4 \sin x - 5} \equiv 3 + 10 \sin x$$
   b) Hence solve the equation
   $$\frac{25 - 40 \cos^2 2\theta - 38 \sin 2\theta}{4 \sin 2\theta - 5} = 3 - 8 \cos 2\theta$$
   for $0 \leq \theta < \pi$

27) Solve the equation
   $\sqrt{18 \sin^2 \theta - 21 \sin \theta \cos \theta} = 2 \cos \theta$
   for $0 \leq \theta < 2\pi$

28) Prove the following trigonometric identities:
   a)
   $$\frac{\cos^2 \theta - \cos^4 \theta}{2 \cos^2 \theta - 1} \equiv \frac{\sin \theta \cos \theta}{\frac{1}{\tan \theta} - \tan \theta}$$
   b)
   $$\frac{\sin^4 \theta - \sin^2 \theta}{\sin^3 \theta + 2 \sin^2 \theta - \sin \theta - 1} \equiv \frac{\sin \theta \cos \theta}{\frac{1}{\tan \theta} - \tan \theta + \cos \theta}$$

2.05 Trigonometry: Modelling in Radians (answers on page 491)

1) State the period of each of the following functions:
   a) $f(x) = \sin(x)$
   b) $f(x) = \cos(4\pi x)$
   c) $f(x) = \tan(\pi x)$
   d) $f(x) = \sin\left(\frac{\pi}{2}x\right)$
   e) $f(x) = \cos\left(\frac{\pi}{4}x\right)$
   f) $f(x) = \tan\left(\frac{\pi}{2}x - \frac{\pi}{4}\right)$
   g) $f(x) = \sin\left(\frac{2\pi}{52}x\right)$
   h) $f(x) = \cos\left(\frac{2\pi}{7}x + \frac{\pi}{18}\right)$
   i) $f(x) = \sin\left(\frac{2\pi}{365}(x - 30)\right)$
   j) $f(x) = \cos\left(\frac{\pi}{26}(x + 7)\right)$

2) In each of the following cases, the curve
$$y = \sin x$$
has been stretched parallel to the x-axis. State the equation of each curve.
   a) 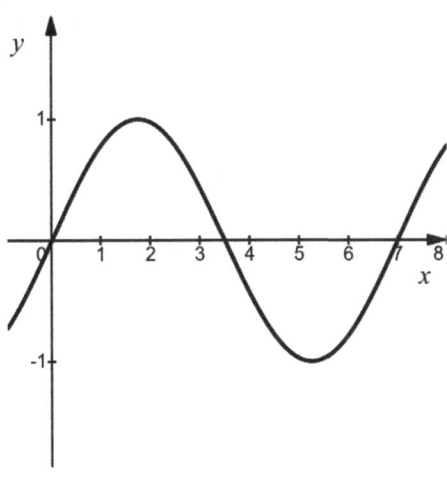
   b) 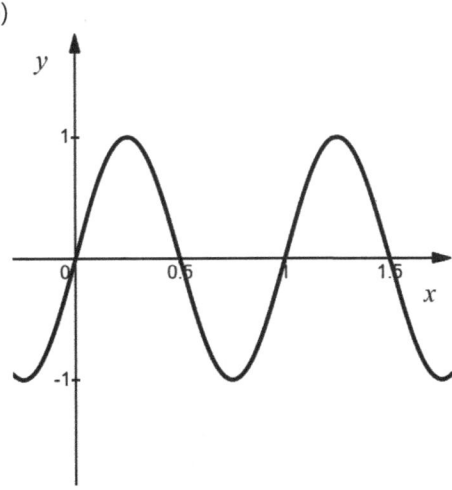
   c) 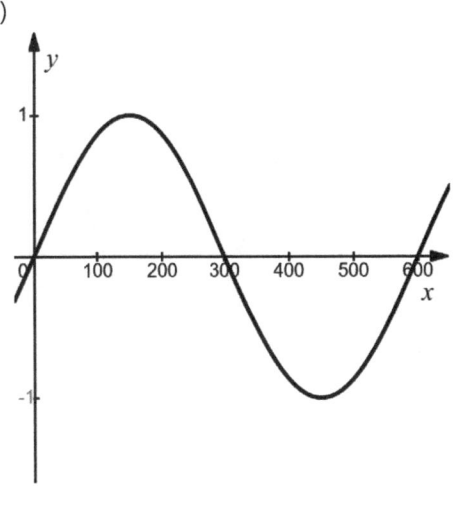
   d)

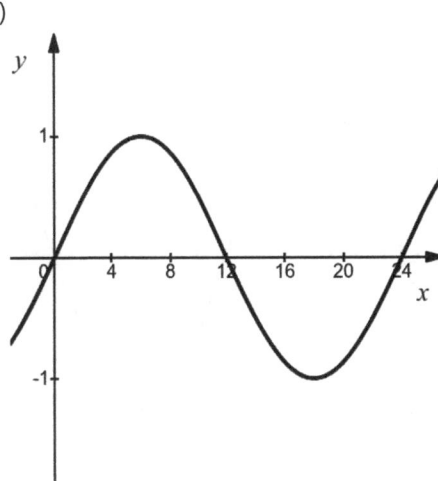

3) All the following graphs are transformations of
$$y = \sin x$$
State the equation of each:
   a)

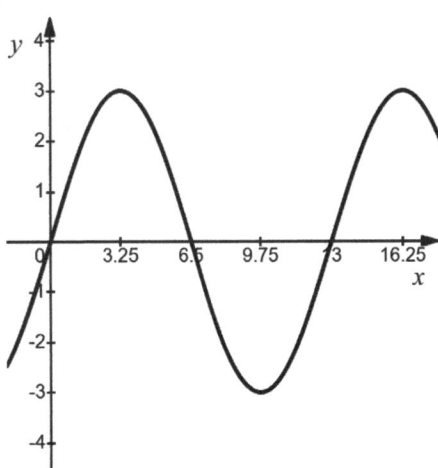

2.05 Trigonometry: Modelling in Radians (answers on page 491)

b) *[graph]*

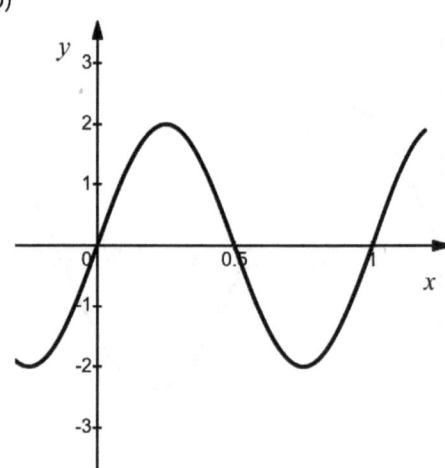

c) *[graph]*

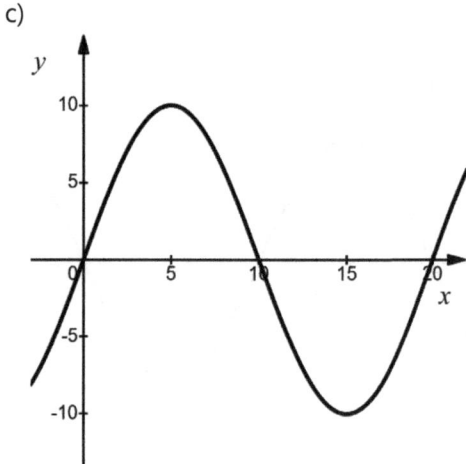

d) *[graph]*

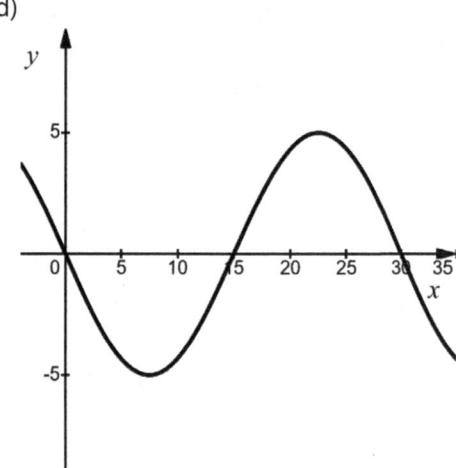

4) State the amplitude of each of the following periodic functions:
   a) $y = \sin(3\pi x)$
   b) $y = 4 + \cos(\pi x + 0.1)$
   c) $y = 55 + 12\sin(\pi x)$
   d) $y = 1 - 2\sin\left(x + \frac{\pi}{2}\right)$

5) For each of the following functions state the minimum and maximum of the function:
   a) $y = \sin(2\pi x)$
   b) $y = 3 + \cos\left(\pi x - \frac{\pi}{2}\right)$
   c) $y = 14 + 6\sin(30\pi x)$
   d) $y = 2 - 7\cos\left(\frac{\pi}{2}x + 0.2\right)$

6) All the following graphs are all the form
$$y = a + b\sin(cx)$$
   In each case, find the equation of the graph:

   a) *[graph]*

   b) *[graph]*

   c) *[graph]*

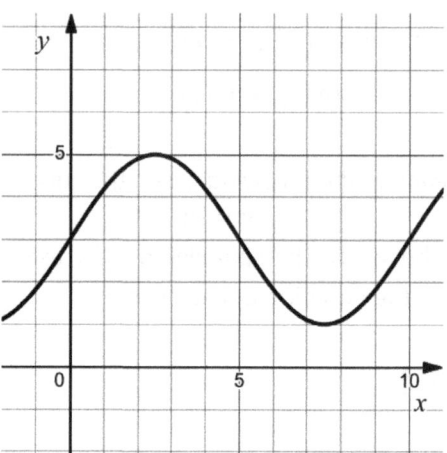

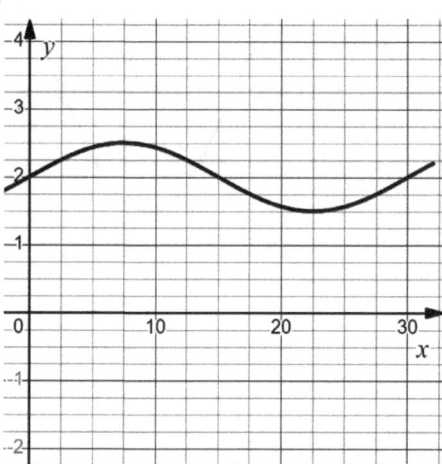

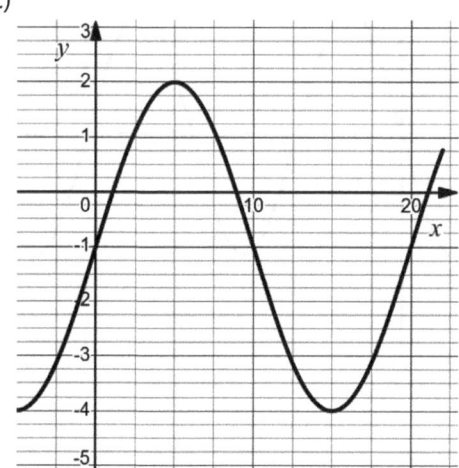

2.05 Trigonometry: Modelling in Radians (answers on page 491)

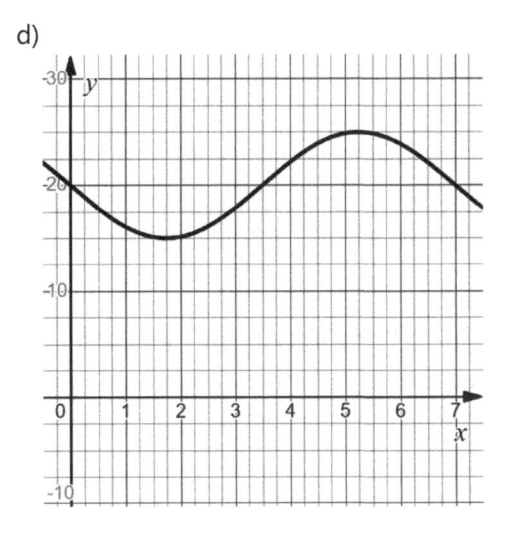

d)

7) The river Hamble is tidal. On one day the depth of the Hamble oscillates between a maximum of 3.6 metres at 8:00 am and a minimum of 2.0 metres at 12:30 pm. Kaelen decides to model the depth of the river, d metres by the equation
$$d = a + b\cos\left(\frac{2}{5}\pi t\right)$$
where t is the number of hours since 8:00 am.
   a) State the values of a and b that Kaelen should use.
   b) State the period of this equation.
   c) Calculate the total time, according to this model, that the Hamble is less than 2.5 metres in depth within the first 7 hours.

8) The temperature, $T°C$ of an average human can be modelled by the equation
$$T = 36.8 - 0.64\sin\left(\frac{\pi}{12}(h+2)\right)$$
where h is the number of hours since midnight.
   a) Find the maximum temperature of an average human.
   b) Find the first time at which this maximum occurs after midnight.

The temperature, $T°C$ of an average dog varies between 38.3°C and 39.2°C
   c) Find an equation to model a dog's temperature in the form
$$T = a - b\sin\left(\frac{\pi}{12}(h+2)\right)$$

Prerequisite Knowledge for Q9: Graph Transformations (combinations)

9) The time of sunset, t hours for the first day of the week x for a year may be modelled by the equation
$$t = 18.5 + 2.4\sin\left(\frac{2\pi}{52}(x+26)\right)$$
for $0 \le x \le 52$
   a) Describe the transformations which map
$$t = \sin\left(\frac{2\pi}{52}(x+26)\right)$$
onto
$$t = 18.5 + 2.4\sin\left(\frac{2\pi}{52}(x+26)\right)$$
   b) Find the time of the sunset in the 14th week of the year.

10) The number of passengers waiting on a platform for a train is recorded over the course of a day, and they are plotted on a graph. Passengers are only counted once as they embark on a train before the next count of passengers takes place.

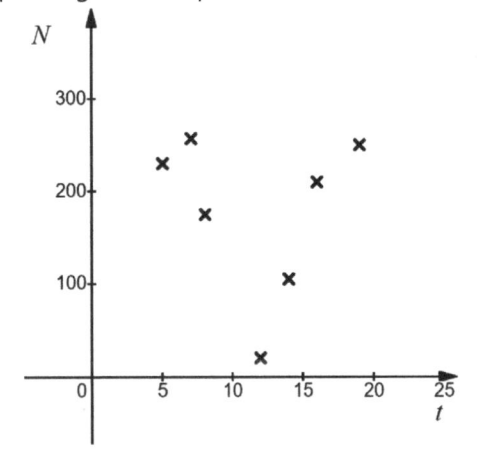

It is believed that the data can be modelled with the equation
$$N = 140 + 120\sin\left(\frac{\pi}{6}(t-3)\right)$$
where N is the number of passengers on the platform and t is the number of hours after midnight. According to the model, find
   a) the difference in the number of passengers between 5 am and 9 am
   b) the times at which the total number of passengers on the platform is over 200 over the 24-hour period.
   c) Explain why it may not be suitable to use this function to model the data.

2.06 Sequences and Series: Introduction (answers on page 491)

1) For each of the following sequences, explain whether they can be described as convergent, divergent, periodic, or none of these.
   a) $1, 2, 4, 8, 16, 32, \ldots$
   b) $1, 5, 1, 5, 1, 5, \ldots$
   c) $1, \frac{1}{2}, \frac{1}{4}, \frac{1}{8}, \frac{1}{16}, \frac{1}{32}, \ldots$
   d) $50, 60, 65, 67.5, 68.75, 69.375, \ldots$

Sigma Notation

2) Evaluate each of these:
   a) $\sum_{n=1}^{4}(5n-1)$
   b) $\sum_{n=1}^{3}(3^n + n^2)$
   c) $\sum_{n=1}^{5}\frac{2}{n}$
   d) $\sum_{n=1}^{4} 4 \times \left(\frac{1}{2}\right)^{n-1}$

3) Evaluate each of these:
   a) $\sum_{n=3}^{6}(n^2 - 10)$
   b) $\sum_{n=99}^{100}\frac{3}{n}$
   c) $\sum_{n=6}^{8} 50 \times 5^{n-5}$

4) Simplify each of these, where k is a constant:
   a) $\sum_{n=1}^{5}(n+k)$
   b) $\sum_{n=1}^{4}(1 - kn^3)$
   c) $\sum_{n=2}^{4}\frac{k}{n+1}$
   d) $\sum_{n=3}^{6}\frac{kn}{k+1}$

5) It is given that $f(n)$ is the nth term of a sequence. Some further information about $f(n)$ is given in the table below.

k	19	20	21	22	23	24	25
$\sum_{n=1}^{k} f(n)$	$\frac{19}{10}$	$\frac{21}{10}$	$\frac{58}{25}$	$\frac{127}{50}$	$\frac{69}{25}$	3	$\frac{163}{50}$

   Find the value of
   $$\sum_{n=20}^{24} f(n)$$

6) It is given that
   $$g(k) = \sum_{n=1}^{k}(n+2)^2$$
   where k is an integer such that $k \geq 1$. Find:
   a) $g(2)$
   b) $g(5)$
   c) $g(7) - g(6)$
   d) $g(k+1) - g(k)$

7) Find
   $$\sum_{i=1}^{100} 1 + \sum_{i=20}^{100} 1 + \sum_{i=50}^{80} 1$$

8) Show that
   $$\sum_{n=1}^{4} \frac{6(n+1)}{\sqrt{n}} = 27 + 9\sqrt{2} + 8\sqrt{3}$$

9) Show that
   $$\sum_{n=1}^{3} \frac{n}{\sqrt{n} + \sqrt{n+1}} = 5 - \sqrt{2} - \sqrt{3}$$

10) Show that
    $$\sum_{n=1}^{3} \ln(n+1) = \ln 24$$

2.06 Sequences and Series: Introduction (answers on page 491)

11) Show that
$$\sum_{n=1}^{4} \ln\left(\frac{1}{n}\right) = -\ln 24$$

12) Show that
$$\sum_{n=1}^{5} \ln\left(\frac{n+1}{n}\right) = \ln 6$$

13) It is given that
$$h(k) = \sum_{n=1}^{k} \frac{n}{n+1}$$
Find the value of k such that
$$h(k+1) - h(k) = \frac{99}{100}$$

Recurrence Relations and Limits

14) It is given that
$$u_{n+1} = 3u_n + 4$$
with $u_1 = 50$.
Find the values of u_2, u_3 and u_4

15) It is given that
$$u_{n+1} = \frac{u_n + 2}{10}$$
with $u_1 = 1000$.
Find the values of u_2, u_3 and u_4

16) It is given that
$$u_{n+1} = 5u_n + k$$
where k is a constant, with $u_1 = 2$ and $u_3 = 4$.
Find the value of k

17) It is given that
$$u_{n+1} = \frac{1}{u_n + 2}$$
with $u_1 = 3$. Calculate
$$\sum_{n=1}^{3} u_n$$

18) It is given that
$$u_{n+1} = \frac{10}{u_n} - k$$
where k is a constant, with $u_1 = 10$.
It is also given that
$$\sum_{n=1}^{3} u_n = 0$$
Find the two possible values of k

19) In each of the following cases, use your calculator to determine if the sequence converges or diverges.
   a) $u_{n+1} = 0.3u_n - 8$, $u_1 = 1$
   b) $u_{n+1} = 1.1u_n - 3$, $u_1 = 2$
   c) $u_{n+1} = 2.5u_n + 2$, $u_1 = -3$
   d) $u_{n+1} = 3 - 1.8u_n$, $u_1 = 4$
   e) $u_{n+1} = 100 - 0.9u_n$, $u_1 = 10$

20) In each of the following cases, determine if the sequence converges or diverges.
   a) $$u_n = 3 + 0.6n$$
   b) $$u_n = 2^n$$
   c) $$u_n = \left(\frac{1}{3}\right)^n$$
   d) $$u_n = \frac{10}{n+1}$$

21) In each of the following cases, it is given that the limit of u_n as n tends to infinity is L
In each case, write down an equation for L and find its value.
   a) $$u_{n+1} = \frac{u_n}{3} + 50$$
   b) $$u_{n+1} = 0.85u_n - 6$$
   c) $$u_{n+1} = \frac{u_n + 15}{30}$$
   d) $$u_{n+1} = \frac{0.2u_n - 100}{25}$$

22) Arnold takes an initial dose of 400 mg of a certain type of painkiller, and every four hours must take another dose. The amount of painkiller in Arnold's system at the end of a four-hour period, just after he has taken the next dose, can be modelled by the recurrence relation $P_{n+1} = 0.25P_n + 400$.
   a) Explain the significance of 0.25 in the recurrence relation.
   b) Show that the amount of painkiller in Arnold's system 12 hours after he took the first dose, is 531 mg to the nearest mg.
   c) Use algebra to determine the long-term amount of painkiller in Arnold's system.

2.06 Sequences and Series: Introduction (answers on page 491)

23) In each of the following cases, determine the possible value(s) of u_1 that will generate a constant sequence.
   a) $$u_{n+1} = \frac{9u_n}{8} + 5$$
   b) $$u_{n+1} = \frac{10 - u_n}{15}$$
   c) $$u_{n+1} = \frac{5}{2} + \frac{21}{u_n}$$
   d) $$u_{n+1} = \frac{55 - 3u_n - 8u_n^2}{10u_n}$$

24) Write down the limit of the following sequences:
   a) $$u_n = \frac{1}{n + 10}$$
   b) $$u_n = \frac{n}{n + 10}$$
   c) $$u_n = \frac{n + 11}{n + 10}$$
   d) $$u_n = \frac{2n + 11}{n + 10}$$
   e) $$u_n = \frac{2n + 11}{n^2 + 10}$$

25) It is given that
$$u_{n+1} = 1 - u_n$$
with $u_1 = 1$
Find the value of u_{155}

26) It is given that
$$u_{n+1} = 8 - u_n$$
with $u_1 = 1$
Find the value of u_{88}

27) It is given that
$$u_{n+1} = -u_n$$
with $u_1 = 1$
   a) Find the value of u_{100}
   b) Find the value of
$$\sum_{n=1}^{100} u_n$$

28) It is given that
$$u_{n+1} = 13 - u_n^2$$
with $u_1 = 4$
   a) Find the value of u_{500}
   b) Find the value of
$$\sum_{n=1}^{500} u_n$$

29) It is given that
$$u_{n+1} = -\frac{1}{u_n}$$
with $u_1 = \frac{5}{3}$
   a) Find the value of u_{99}
   b) Find the value of
$$\sum_{n=1}^{99} u_n$$

30) It is given that
$$u_{n+1} = \frac{u_n - 1}{ku_n}$$
with $u_1 = 3$, where k is a constant, generates a sequence with period 3
   a) Find the value of k
   b) Hence find the value of
$$\sum_{n=1}^{100} u_n$$

31) It is given that
$$u_{n+3} = u_n$$
with $u_1 = 4$, $u_2 = -6$ and $u_3 = 8$
Find the value of
$$\sum_{n=1}^{500} u_n$$

32) It is given that
$$u_{n+4} = u_n$$
with $u_1 = 1$, $u_2 = 4$, $u_3 = -7$ and $u_4 = -2$
Find the value of
$$\sum_{n=1}^{499} u_n$$

33) It is given that
$$u_{n+3} = u_n + 2$$
with $u_1 = 0$, $u_2 = 1$ and $u_3 = 5$
Find k such that $u_k = 200$

2.06 Sequences and Series: Introduction (answers on page 491)

Periodic, Increasing, Decreasing

34) It is given that
$$u_n = 3 + (-1)^n$$
is a periodic sequence.
State the period of the sequence.

35) It is given that
$$u_n = \sin\left(\frac{3\pi n}{2}\right)$$
is a periodic sequence.
State the period of the sequence.

36) It is given that
$$u_n = \cos\left(\frac{2\pi n}{3}\right)$$
is a periodic sequence.
State the period of the sequence.

37) It is given that
$$u_n = \sin\left(\frac{2\pi n}{3}\right) - \cos\left(\frac{5\pi n}{3}\right)$$
is a periodic sequence.
State the period of the sequence.

38) It is given that
$$u_n = \cos\left(\frac{\pi n}{4}\right) - \sin\left(\frac{5\pi n}{2}\right)$$
is a periodic sequence.
State the period of the sequence.

39) It is given that
$$u_n = (-1)^n + \sin\left(\frac{\pi n}{2}\right)$$
is a periodic sequence.
State the period of the sequence.

40) It is given that
$$u_n = 3(-1)^n + \cos\left(\frac{\pi n}{3}\right)$$
is a periodic sequence.
State the period of the sequence.

41) It is given that
$$u_n = 2(-1)^n - \frac{1}{\cos\left(\frac{\pi n}{3}\right)}$$
is a periodic sequence.
State the period of the sequence.

42) Show that
$$u_n = \frac{3n}{2 + n^2}$$
is not an increasing sequence.

43) Show that
$$u_n = \frac{n+1}{n^3 + n - 1}$$
is not an increasing sequence.

44) Show that
$$u_n = n - \frac{1}{n+1}$$
is not a decreasing sequence.

45) Show that
$$u_n = n^2 - 100n$$
is not a decreasing sequence.

46) Prove that
$$u_n = 3^{-n}$$
is a decreasing sequence.

47) Prove that
$$u_n = \frac{1}{n^2}$$
is a decreasing sequence.

48) Prove that
$$u_n = \frac{n+1}{n+2}$$
is an increasing sequence.

49) Prove that
$$u_n = \frac{n+3}{n+2}$$
is a decreasing sequence.

50) Prove that
$$u_n = \frac{5}{n+5}$$
is a decreasing sequence.

51) Prove that
$$u_n = \frac{2-n}{n}$$
is a decreasing sequence.

52) Prove that
$$u_n = (2n+1)^2 - (n-1)^2$$
is an increasing sequence.

53) Prove that
$$u_n = \sqrt{n+1} + \sqrt{n}$$
is an increasing sequence.

2.07 Sequences and Series: Arithmetic Sequences (answers on page 493)

1) Write the nth term, u_n, for each of these arithmetic sequences in the form
$$u_n = a + (n-1)d$$
where a is the first term and d is the common difference.
   a) 3, 8, 13, 18, 23, ...
   b) 25, 35, 45, 55, 65, ...
   c) 12, 6, 0, −6, −12, ...
   d) 150, 135, 120, 105, 90, ...
   e) −44, −41, −38, −35, −32, ...
   f) −100, −121, −142, −163, −184, ...
   g) $\frac{5}{6}, \frac{31}{30}, \frac{37}{30}, \frac{43}{30}, \frac{49}{30}, \ldots$
   h) $\frac{11}{4}, -\frac{7}{4}, -\frac{25}{4}, -\frac{43}{4}, -\frac{61}{4}, \ldots$

2) It is given that 9, 12, 15, 18, 21, ... is an arithmetic sequence. Find the 500th term in the sequence.

3) It is given that $\frac{3}{2}, \frac{11}{6}, \frac{13}{6}, \frac{5}{2}, \frac{17}{6}, \ldots$ is an arithmetic sequence. Find the 240th term in the sequence.

4) It is given that −50, −45, −40, −35, ... , 765 is a finite arithmetic sequence. Determine how many terms there are in the sequence.

5) It is given that 31.3, 31.6, 31.9, 32.2, ... , 117.7 is a finite arithmetic sequence. Determine how many terms there are in the sequence.

6) It is given that 58.2, 57.5, 56.8, 56.1, ... , −25.1 is a finite arithmetic sequence. Determine how many terms there are in the sequence.

7) An arithmetic progression has 3rd term 22 and 7th term 54. Find the first term and the common difference.

8) An arithmetic progression has 5th term 38 and 10th term 50.5. Find the first term and the common difference.

9) An arithmetic progression has 2nd term 71 and 9th term 631. Find the first term and the common difference.

10) An arithmetic progression has 6th term 20 and 12th term 38. Find the first term that exceeds 100

11) An arithmetic sequence has 3rd term $2 + \sqrt{3}$ and 9th term $3 + \sqrt{5}$. Find the common difference in the form $a + b\sqrt{3} + c\sqrt{5}$, where a, b and c are rational numbers.

12) A botanist is studying the growth of a particular species of plant. The height of the plant after 2 weeks is 15 cm, and after 7 weeks it is 55 cm. Given that the height of the plant over the weeks follows an arithmetic sequence, find the height of the plant after 3 weeks.

13) An arithmetic sequence has the three consecutive terms $\ln \frac{1}{4}k$, $\ln \frac{1}{3}k$, $\ln p$, where k, p are positive constants. Find p in terms of k

14) An arithmetic sequence has the three consecutive terms $6k^2$, $5k$, $(21k − 35)$, where k is a constant. Find the possible values of k

15) An arithmetic sequence has the three consecutive terms $\frac{2}{k}$, $20k$, $−(5k + 13)$, where k is a constant. Find the possible values of k

16) An arithmetic sequence has the three consecutive terms $16e^k$, 20, $16e^{-k}$, where k is a constant. Find the possible values of k

17) An arithmetic sequence has the three consecutive terms 28, $\frac{200}{2^k}$, $\left(\frac{240}{2^k} - 2^k\right)$, where k is a constant. Find the possible values of k

18) The first three terms of an arithmetic sequence are $32 + 3k$, $13 − 4k$, $78 + 10k$, where k is a constant.
   a) Find the value of k.
   b) Find the 100th term in the sequence.

19) Find the sum of the arithmetic series
$$3 + 11 + 19 + 27 + \cdots + 155$$

20) Find the sum of the arithmetic series
$$14 + 20 + 26 + 32 + \cdots + 1238$$

2.07 Sequences and Series: Arithmetic Sequences (answers on page 493)

21) Find the sum of the arithmetic series
$$-11 - 20 - 29 - 38 - \cdots - 542$$

22) Find the sum of the arithmetic series
$$\frac{20}{3} + \frac{35}{6} + 5 + \frac{25}{6} + \cdots - 60$$

23) An arithmetic series is given by
$$\sum_{i=7}^{40}\left(6i - \frac{1}{2}\right)$$
   a) Write down the first term and common difference of the series.
   b) Write down how many terms there are in the series.

24) An arithmetic sequence has rth term given by $u_r = 7 - 10r$. Show that
$$\sum_{r=1}^{40} u_r = -7920$$

25) An arithmetic sequence has kth term given by $u_k = -3 + \frac{k}{3}$. Show that
$$\sum_{k=10}^{30} u_k = 77$$

26) Find the sum of the first 150 terms of the arithmetic sequence
$$25 + 26.5 + 28 + 29.5 + \cdots$$

27) Find the sum of the first 80 terms of the arithmetic sequence
$$\frac{1}{5} + \frac{7}{5} + \frac{13}{5} + \frac{19}{5} + \cdots$$

28) Find the sum of the first 185 terms of the arithmetic sequence
$$100 + 99.2 + 98.4 + 97.6 + \cdots$$

29) Find the sum of the first 390 terms of the arithmetic sequence
$$10.3 + 10 + 9.7 + 9.4 + \cdots$$

30) On his 1st birthday, Jaime's parents put £35 into a bank account. On his 2nd birthday, Jaime's parents put £65 into the bank account. The amounts deposited into the bank account form an arithmetic sequence. Determine the total amount of money in the bank account the day after Jaime's 18th birthday.

31) It is given that S_n is the sum of the first n terms of the arithmetic sequence
$$104 + 101 + 98 + 95 + \cdots$$
Find the maximum value of S_n

32) An arithmetic sequence has nth term given by $u_n = \frac{6n}{5} + \frac{2}{3}$. Find the value of k such that
$$\sum_{n=1}^{k} u_n < 200 \text{ and } \sum_{n=1}^{k+1} u_n > 200$$

33) An arithmetic sequence has nth term given by $u_n = 4.1n - 3$. Find the value of k such that
$$\sum_{n=1}^{k} u_n < 3000 \text{ and } \sum_{n=1}^{k+1} u_n > 3000$$

34) Aiden buys a new phone on a contract that is worth £900 at 0% interest. The contract provider requires Aiden to pay back £10.50 at the end of week 1, £10.60 at the end of week 2, £10.70 at the end of week 3, etc, until the £900 is fully paid off, forming an arithmetic sequence. At the end of the final week, Aiden pays off the remaining balance if it is less than the amount he paid the previous week.
   a) Determine how many weeks it will take to pay off the contract.
   b) Determine how much Aiden pays off in the final week.

35) An arithmetic series has 3rd term 46 and 15th term 222. Find the sum of the first 30 terms.

36) An arithmetic series has 1st term 3. The sum of the first 100 terms is 12. Find the common difference.

37) The first three terms of an arithmetic progression are $8k$, $k^2(k-4)$, 10, where k is a real constant. It is given that the sum of the first n terms is -53900. Find the value of n

38) An arithmetic sequence has first term a and common difference d. The sum of the first 15 terms is -465.
   a) Show that $a + 7d = -31$

   It is given that the 50th term is -283
   b) Find the value of a and the value of d

2.07 Sequences and Series: Arithmetic Sequences (answers on page 493)

39) An arithmetic sequence has first term a and common difference d. The sum of the first 10 terms is 410.
   a) Show that $2a + 9d = 82$

   It is given that the sum of the first 20 terms is 1620
   b) Find the value of a and the value of d
   c) Find the sum of the first 30 terms.

40) An arithmetic sequence has first term a and common difference d. The sum of the first 490 terms is 1489 times the sum of the first 10 terms.
   a) Show that $3a = 11d$
   b) Given that the 10th term is 38, find the value of a and the value of d

41) An arithmetic progression has first term a and common difference d. The sum of the first 220 terms is 211 times the sum of the first 20 terms.
   a) Show that $a + 4d = 0$
   b) Given that the 105th term is -200, find the value of a and the value of d

42) An arithmetic sequence has first term a and common difference d. The sum of the first 425 terms is 221 times the square of the sum of the first 5 terms.
   a) Show that
   $221a^2 - 17a + 884d^2 - 3604d + 884ad = 0$
   b) Given that the 2nd term is 4 and that both a and d are integers, find the value of a and the value of d

43) The first three terms of an arithmetic sequence are given by $\frac{1}{k-4}$, $-\frac{1}{k}$ and $\frac{1}{k+4}$, where k is a positive constant. Show that the sum of the first 52 terms is $a + b\sqrt{2}$, where a and b are integers.

44) The first three terms of an arithmetic sequence are 8, 11, and 14. The sum of the first n terms is greater than 1000 but less than 2000. Find the possible values of n

45) An arithmetic series with sum 100 is given by
$$\sum_{n=1}^{100}(pn + q)$$
where p and q are constants. It is also given that $u_{51} + u_{53} + u_{55} = u_{59}$. Find the value of p and the value of q

46) An arithmetic sequence has first term $(3k + 2)$ and last term $(9k - 1)$, where k is a positive constant. The common difference of the sequence is 6. Find the number of terms in the sequence in terms of k

47) It is given that k is a positive integer. A_k is the sum of the first k terms of an arithmetic series with first term 3 and common difference 8. B_k is the sum of the first k terms of an arithmetic series with first term -52 and common difference 18. Find the value of k such that $A_k = B_k$

48) It is given that k is a positive integer. C_k is the sum of the first k terms of an arithmetic series with first term 1500 and common difference -9. D_k is the sum of the first k terms of an arithmetic series with first term 6 and common difference 3. Find the value of k such that $C_k = D_k$

49) An arithmetic sequence has first term a and common difference d. Show that S_n, the sum of the first n terms, is $\frac{1}{2}n(2a + (n-1)d)$

50) Find the sum of all the even numbers from 100 to 500.

51) Find the sum of all the odd numbers from 5 to 555.

52) It is given that $u_1 = p$, $u_2 = (2p - 1)$, $u_3 = (3p + 2)$, $u_4 = \left(15p + \frac{2}{3}\right)$, ... is **not** an arithmetic sequence, where p is a constant such that $p \geq 1$. However, the sequence formed by the ratios $\frac{u_{n+1}}{u_n}$ **is** an arithmetic sequence. Find the value of p and hence find the value of u_5

2.08 Sequences and Series: Geometric Sequences (answers on page 494)

1) Write the nth term, u_n, for each of these geometric sequences in the form
$$u_n = ar^{n-1}$$
where a is the first term and r is the common ratio.
   a) 3, 12, 48, 192, 768, ...
   b) 80, 120, 180, 270, 405, ...
   c) 240, 120, 60, 30, 15, ...
   d) 10, 9, 8.1, 7.29, 6.561, ...
   e) 22, −44, 88, −176, 352, ...
   f) $\frac{1}{3}, -\frac{4}{3}, \frac{16}{3}, -\frac{64}{3}, \frac{256}{3}, \ldots$
   g) $\frac{9}{10}, -1, \frac{10}{9}, -\frac{100}{81}, \frac{1000}{729}, \ldots$
   h) 88, −66, 49.5, −37.125, 27.84375, ...

2) It is given that 1.5, 3, 6, 12, ... is a geometric sequence. Find the 15th term in the sequence.

3) It is given that $6, 8, \frac{32}{3}, \frac{128}{9}, \ldots$ is a geometric sequence. Find the 6th term in the sequence.

4) It is given that 0.004, 0.012, 0.036, 0.108, ..., 57395.628 is a finite geometric sequence. Determine how many terms there are in the sequence.

5) It is given that 1.02, 4.08, 16.32, 65.28, ..., 1069547.52 is a finite geometric sequence. Determine how many terms there are in the sequence.

6) It is given that 10000, 11000, 12100, 13310, ..., 21435.8881 is a finite geometric sequence. Determine how many terms there are in the sequence.

7) A geometric progression has first term 0.2 and common ratio 2. Find the first term in the sequence that exceeds 1000

8) A geometric progression has 4th term 28 and 15th term 57344. Find the first term and the common ratio.

9) A geometric progression has 3rd term 2 and 7th term 13122. Find the possible first terms and their corresponding common ratios.

10) A geometric progression has 2nd term 15 and 5th term −234375. Find the first term and the common ratio.

11) A geometric progression has 6th term 62.208 and 7th term 373.248. Find the first term and the common ratio.

12) A geometric progression has 3rd term 19.8 and 6th term 534.6. Find the first term in the sequence that exceeds ten million.

13) A geometric sequence has 2nd term $1 + 2\sqrt{5}$ and 3rd term $38 + 19\sqrt{5}$
   a) Show that the common ratio is $8 + 3\sqrt{5}$
   b) Find the 1st term in the sequence, writing it in the form $p + q\sqrt{5}$, where p and q are rational numbers.

14) A geometric sequence has the three consecutive terms $(k+1), 5k, \frac{50k}{7}$, where k is a constant. Find the value of k

15) A geometric sequence has the three consecutive terms $\frac{49k}{36}, (k+2), (3k-2)$, where k is a constant. Find the possible values of k

16) A geometric sequence has the three consecutive terms $2^{3k-7}, 2^{2k+1}, 2^{4k-5}$, where k is a constant. Find the value of k

17) A geometric sequence has the three consecutive terms $3e^{\frac{7}{4}+2k}, \frac{15}{2}e^{k+\frac{67}{24}}, \frac{75}{4}e^{\frac{101}{24}-k}$, where k is a constant. Find the value of k

18) Find the sum of the geometric series
$$5 + 10 + 20 + \cdots + 2560$$

19) Find the sum of the geometric series
$$0.13 + 0.52 + 2.08 + \cdots + 2129.92$$

20) Find the sum of the geometric series
$$\frac{1}{5000} + \frac{1}{1250} + \frac{2}{625} + \cdots + \frac{8192}{625}$$

21) Find the sum of the geometric series
$$9000 - 9900 + 10890 - \cdots - 17538.4539$$

22) Find the sum of the geometric series
$$\frac{2}{3} - 1 + \frac{3}{2} - \cdots + \frac{19683}{512}$$

2.08 Sequences and Series: Geometric Sequences (answers on page 494)

23) Aaliyah decides to go on a seven-day walking holiday along a Highland trail, camping at the end of each day. On the first day, she walks 15 miles. On each subsequent day, she walks 8% less than the previous day.
   a) Determine how far she walks on the last day, to two decimal places.
   b) Determine the total distance Aaliyah walks on her holiday, to two decimal places.

24) A geometric series is given by
$$\sum_{i=6}^{25} \frac{4}{3} \times \left(\frac{2}{5}\right)^{i-1}$$
   a) Write down the first term and common ratio of the series.
   b) Write down how many terms there are in the series.

25) Find the sum of the first 8 terms of the following geometric series to 2 decimal places:
$$66 + 88 + \frac{352}{3} + \cdots$$

26) Find the sum of the first 9 terms of the following geometric series to 2 decimal places:
$$5200 + 4160 + 3328 + \cdots$$

27) A geometric sequence has kth term given by $u_k = 128 \times \left(\frac{7}{2}\right)^{k-1}$. Show that
$$\sum_{k=1}^{5} u_k = 26840$$

28) A geometric sequence has kth term given by $u_r = 19683 \times \left(\frac{1}{3}\right)^{r-1}$. Show that
$$\sum_{r=4}^{10} u_r = 1093$$

29) A geometric sequence, u_n, has a common ratio $r > 1$. Find the value of r such that
$$\sum_{n=1}^{4} u_n = 8 \sum_{n=1}^{2} u_n$$

30) A geometric series has first term 100 and common ratio $\frac{9}{10}$. It is given that the sum of the first k terms is greater than 870.
   a) Show that
$$k > \frac{\ln 0.13}{\ln 0.9}$$
   b) Hence find the smallest possible value of k

31) A geometric series has first term 2 and common ratio $\frac{26}{25}$. It is given that the sum of the first k terms is greater than 1000.
   a) Show that
$$k > \frac{\ln 21}{\ln 1.04}$$
   b) Hence find the smallest possible value of k

32) A geometric series has first term 20 and common ratio $\frac{1}{8}$. It is given that the sum of the first k terms is less than 22.85
   a) Show that
$$k > \frac{\ln 3200}{\ln 8}$$
   b) Hence find the smallest possible value of k

33) Find the smallest possible value of k such that the sum to k terms of the series
$$5 + 10 + 20 + 40 + \cdots$$
exceeds 1500

34) Find the smallest possible value of k such that the sum to k terms of the series
$$1.1 + 3.3 + 9.9 + 29.7 + \cdots$$
exceeds 30000

35) A small clothes shop made a net profit of £250 in their first week of trading. The shop's owner predicts that the net profit each week will form a geometric sequence, increasing by 3% each week.
   a) Find the net profit predicted by the owner for the end of the fifth week.
   b) How many weeks will need to have passed since opening for the weekly net profit to be over £1000?
   c) Determine the total amount of net profit in one year (52 weeks) of trading?

36) Explain why the geometric series
$$7 + 14 + 28 + 56 + \cdots$$
is not convergent and therefore has no finite sum to infinity.

37) Determine which of the following geometric series are convergent:
   a) $0.4 + 1.2 + 3.6 + 10.8 + \cdots$
   b) $10 + 5 + 2.5 + 1.25 + \cdots$
   c) $20 + 30 + 45 + 67.5 + \cdots$
   d) $100 + 95 + 90.25 + 85.7375 + \cdots$
   e) $100 - 95 + 90.25 - 85.7375 + \cdots$
   f) $1.6 - 1.44 + 1.296 - 1.1664 + \cdots$

2.08 Sequences and Series: Geometric Sequences (answers on page 494)

38) Find the sum to infinity of the geometric series with the following properties:
    a) $a = 30, r = 0.1$
    b) $a = 800, r = \frac{2}{5}$
    c) $a = 99, r = 0.99$
    d) $a = -8000, r = \frac{3}{8}$
    e) $a = 410, r = -\frac{5}{6}$
    f) $a = -9, r = -\frac{11}{12}$

39) A geometric series has first term 80 and sum to infinity 90. Find the common ratio r

40) Find the sum to infinity of the geometric series
$$1 + \frac{1}{3} + \frac{1}{9} + \frac{1}{27} + \cdots$$

41) Find the sum to infinity of the geometric series
$$100 + 80 + 64 + 51.2 + \cdots$$

42) Find the sum to infinity of the geometric series
$$\frac{1}{\sqrt{2}} + \frac{1}{\sqrt{6}} + \frac{1}{3\sqrt{2}} + \frac{1}{3\sqrt{6}} + \cdots$$

43) A geometric series has first term 50 and the sum to infinity is less than 100. Find the range of possible values of r

44) A geometric series has first term 9 and the sum to infinity is less than 40. Find the range of possible values of r

45) A geometric series has first term 100 and the sum to infinity is greater than 500. Find the range of possible values of r

46) A geometric series has first term 80 and the sum to infinity is greater than 100 but less than 200. Find the range of possible values of r

47) Find the value of
$$\sum_{n=1}^{\infty} 3000 \times \left(\frac{2}{3}\right)^{n-1}$$

48) Find the value of
$$\sum_{n=3}^{\infty} 25350 \times \left(\frac{11}{13}\right)^{n}$$

49) Find the value of
$$\sum_{n=10}^{\infty} 1\,000\,000 \times \left(\frac{1}{5}\right)^{n-1}$$

50) The geometric sequence u_n has terms $18, \frac{27}{2}, \frac{81}{8}, \frac{243}{32}, \ldots$
    a) Find the sum to infinity of the series generated by the terms of u_n
    b) The sequence v_n is found by multiplying each term in the sequence u_n by 32. Write down the first four terms of the sequence v_n
    c) Using just your answer to a), write down the sum to infinity of the series generated by the terms of v_n

51) The 2nd term of a geometric series is 10. The sum to infinity is 40. Find the first term and common ratio.

52) The 2nd term of a geometric series is 10800 and the sum to infinity is 45000. Find the two possible first terms of this series.

53) The 2nd term of a geometric series is 95 and the sum to infinity is $\frac{9500}{9}$. It is given that $r > \frac{1}{2}$. Find the sum of the first 6 terms in the series to two decimal places.

54) A geometric sequence, u_n, has first term a and common ratio r. It is given that
$$\sum_{n=1}^{\infty} u_n = 10 \sum_{n=1}^{2} u_n$$
Find the two possible values of r

55) The first three terms of a geometric series are given by $\frac{5}{k}, \frac{1}{k^4}$ and $\frac{k}{80}$, where k is a positive constant. Show that the sum to infinity can be written in the form $ak + b$, where a and b are rational numbers and k is a value to be found.

56) The first four terms of a geometric series are:
$$10 + \frac{10}{2k} + \frac{10}{(2k)^2} + \frac{10}{(2k)^3} + \cdots$$
where k is a positive constant. It is given that the series has a sum to infinity. State the possible values of k

57) Find the value of
$$\sum_{n=1}^{\infty} \left(\frac{1}{2}\ln e^2 - 2\ln e^{\frac{1}{4}}\right)^n$$

58) Find the exact value of
$$\sum_{n=1}^{\infty} (\sin 45° \cos 30°)^n$$

2.08 Sequences and Series: Geometric Sequences (answers on page 494)

59) Find the exact value of
$$\sum_{n=1}^{\infty} \left(\frac{\sin(90+180n)°}{5}\right)^n$$

60) Find the exact value of
$$\sum_{n=1}^{\infty} \left(\frac{\sqrt{5}}{2}-\frac{\sqrt{3}}{3}\right)^n$$

61) A geometric sequence has nth term $u_n = \frac{100}{9} \times \left(\frac{3}{5}\right)^{n-1}$. Solve the inequality
$$\sum_{n=1}^{\infty} u_n - \sum_{n=1}^{k} u_n > 0.09$$
where k is a constant.

62) A geometric sequence has first term a and common ratio r. Show that S_n, the sum of the first n terms, is $\frac{a(r^n-1)}{r-1}$

63) A geometric sequence has nth term $u_n = 8000 \times \left(\frac{1}{2}\right)^{n-1}$. Solve the inequality
$$\sum_{n=1}^{\infty} u_n - \sum_{n=1}^{k} u_n < 5000$$
where k is a constant.

64) A geometric sequence has nth term $u_n = k \times 0.5^{n-1}$ where k is a positive constant.
   a) Find the value of the 5th term in terms of k
   b) Show that the sum to infinity is $2k$
   c) Explain how the sum to infinity would be affected if the nth term of the sequence was $v_n = 0.5 + k \times 0.5^{n-1}$

65) Jenny is an art student. She decides to paint a series of thin concentric circles on a canvas, with the first being the largest, having a radius of 40 cm, and the next one always has a radius that is 98% the length of the previous one. The total length of all the circumferences of the circles on the canvas will not exceed k cm. Find the value of k

66) A geometric sequence has first term 108 and common ratio $\frac{3}{8}$. Show that the nth term for the sequence can be written as
$$\frac{3^{n+2}}{2^{3n-5}}$$

67) A geometric sequence has first term 400 and common ratio $\frac{5}{4}$. Show that the nth term for the sequence can be written as
$$\frac{5^{n+1}}{2^{2n-6}}$$

68) A geometric sequence has first term $(2k^2 + 5k - 6)$ and second term $\left(2k + 5 - \frac{6}{k}\right)$, where k is a rational constant. It is given that the sum to infinity is 3. Find the value of the k

69) It is given that
$$\sum_{n=1}^{\infty} \left(\frac{k+1}{2k+3}\right)^n = 2$$
Find the value of
$$\sum_{n=1}^{6} \left(\frac{k+1}{2k+3}\right)^n$$

Prerequisite knowledge for Q70-Q72:
Recurrence Relations & Arithmetic Sequences

70) Determine if any of these sequences are arithmetic, geometric, or neither:
   a) $u_{n+1} = u_n + 3, u_1 = 5$
   b) $u_{n+1} = 2u_n, u_1 = 4$
   c) $u_{n+1} = 2u_n + 3, u_1 = 6$

71) It is given that u_n is an arithmetic sequence with first term 3 and common difference 4. It is also given that v_n is a geometric sequence with first term 2 and common ratio 3
   a) Determine whether the following sequence is arithmetic, geometric or neither:
   $(u_1 + v_1), (u_2 + v_2), (u_3 + v_3), ..., (u_n + v_n)$
   b) Find the exact value of
   $$\sum_{n=1}^{8} (u_n + v_n)$$

72) An arithmetic progression has first term 10 and common difference 700, and a geometric progression has first term 5 and common ratio 3. The kth term for both sequences is the first time when the geometric term is greater than the arithmetic term.
   a) Show that $3^k + 414 > 420k$
   b) Determine the value of k

2.09 Binomial Series (answers on page 495)

1) For each of the following, expand up to and including the x^3 term, stating in each case the values of x for which the expansions are valid:
   a) $(1+x)^{\frac{1}{4}}$
   b) $(1+2x)^{-1}$
   c) $\sqrt{1-3x}$
   d) $(2+x)^{-3}$
   e) $(9+3x)^{\frac{1}{2}}$
   f) $\sqrt{16-36x}$
   g) $\sqrt[3]{8-2x}$
   h) $\frac{1}{\sqrt{3+2x}}$

2) Write down the coefficient of the x^4 term in each of the following expansions:
   a) $(1+3x)^{\frac{1}{4}}$
   b) $(1-2x)^{\frac{2}{3}}$
   c) $\left(1+\frac{x}{2}\right)^{-\frac{3}{4}}$

3)
   a) Find the first 3 terms in the expansion of
   $$\frac{1}{\sqrt{9+x}}$$
   b) Hence find the first 3 terms in the expansion of
   $$\frac{1}{\sqrt{9-x^2}}$$

4)
   a) Find the expansion of $(1+x)^{-\frac{1}{2}}$ up to and including the x^2 term.
   b) Hence find the first 3 terms in the expansion of
   $$\left(1+\frac{2}{5}x\right)^{-\frac{1}{2}} \text{ and } \sqrt{\frac{10}{5+2x}}$$

5)
   a) Find the first 3 terms in the expansion of $\sqrt{9-x}$
   b) Hence find an approximation to $\sqrt{8}$

6)
   a) Expand $(1+5x)^{\frac{1}{3}}$ up to and including the term in x^3.
   b) Hence find an approximation to the value of $\sqrt[3]{1.1125}$ to 4 decimal places.

7) By finding the first 3 terms in the expansion of
$$(2-3x)^{\frac{1}{2}}$$
use $x = \frac{1}{10}$ to find an approximation for $\sqrt{170}$ giving your answer in the form $p\sqrt{2}$ where p is a value to be found.

8)
   a) Find the first 3 terms in the expansions of $(1+3x)^{\frac{1}{2}}$ and $(1+x)^{-\frac{1}{2}}$
   b) Hence, show that
   $$\sqrt{\frac{1+3x}{1+x}} \approx 1 + x - \frac{3}{2}x^2$$
   c) State the values of x for which the expansion is valid.

9)
   a) Find the first 3 terms in the expansion of $(1-2x)^{\frac{1}{2}}$
   b) Hence, show that
   $$\frac{\sqrt{1-2x}}{1+x} \approx 1 - 2x + \frac{3}{2}x^2$$
   c) State the values of x for which the expansion is valid.

10) Given the binomial expansion
$$(1+qx)^p = 1 - 2x + 6x^2 + \cdots$$
where p and q are constants.
   a) Find the values of p and q
   b) Hence state the values of x for which the expansion is valid.

11) Given that
$$\sqrt[3]{8-x} \approx 2 - \frac{1}{12}x + kx^2 + \cdots$$
where k is a constant.
   a) Find the value of k.
   b) By using $x = 1$ in the expansion, find an approximation to the value of $\sqrt[3]{7}$.

12)
   a) By using $x = \frac{1}{8}$ in the first three terms in the expansion of $\sqrt{3-8x}$, find an approximation for the value of $\sqrt{2}$
   b) Explain whether your approximation is an under- or over-estimate.

2.09 Binomial Series (answers on page 495)

13)
  a) Find the terms up to and including the x^2 term in the expansion of $\frac{1}{\sqrt{2-x}}$

  Jenna considers using the values $x = -3$, $x = -\frac{6}{5}$ and $x = \frac{1}{5}$.

  b) Explain how she used the values to find $\sqrt{5}$
  c) Which of the three values should not be used in the expansion?

14) By considering the first three terms in the expansion of $\frac{1}{\sqrt{7-x}}$ find an approximation for the value of $\sqrt{3}$

15) The graph below shows $y = \sqrt{1-x^2}$

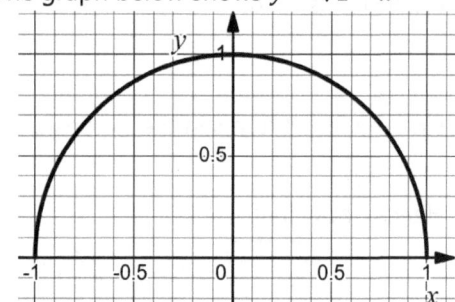

By considering the graph of
$$1 + \frac{1}{2}x^2 + \cdots$$
explain why this is not a correct expansion of $\sqrt{1-x^2}$

16) Find the values of a and b given
$$\frac{a+bx}{\sqrt{9-x}} = 1 - \frac{11}{18}x - \frac{7}{216}x^2$$
where a and b are constants.

17)
  a) Find the first 4 terms in the expansion of $(1+px)^{\frac{1}{2}}$ where p is a constant

  Given for the constant term q
  $$(1+qx)(1+px)^{\frac{1}{2}} = 1 - 4x - 14x^2 + 16x^3 - \cdots$$
  b) find the values of p and q

18)
  a) Find the first 4 terms in the expansion of $(1-x)^{-2}$
  b) Hence find $2 + 3x + 4x^2 + 5x^3 + \cdots$ in the form $\frac{a-x}{(b-x)^2}$ where a and b are integers to be found.

Prerequisite knowledge for Q19-Q21: Partial Fractions

19) It is known that
$$\frac{2-5x}{(x-1)(2x-1)}, \quad x \neq 1, x \neq \frac{1}{2}$$
can be expressed in the form $\frac{A}{x-1} + \frac{B}{2x-1}$
  a) Find the values of A and B
  b) Hence find the first three terms in the expansion of $\frac{2-5x}{(x-1)(2x-1)}$

20) By expressing
$$\frac{11x+5}{(5x+1)(x+3)}, \quad x \neq -\frac{1}{5}, x \neq -\frac{1}{3}$$
in the form $\frac{A}{5x+1} + \frac{B}{x+3}$
  a) Find the first 3 terms in the binomial expansion of $\frac{11x+5}{(5x+1)(x+3)}$
  b) State the values for which the expansion is valid.

21) By expressing
$$\frac{9x^2 - 5x + 2}{(x+1)(3x+1)^2}, \quad x \neq -1, x \neq -\frac{1}{3}$$
as partial fractions,
  a) Find the first 3 terms in the binomial expansion of $\frac{9x^2-5x+2}{(x+1)(3x+1)^2}$
  b) State the values for which the expansion is valid.

Prerequisite knowledge for Q22: Integration of Polynomials

22)
  a) Find the first three terms in the expansion of
  $$\frac{1}{\sqrt{9-x}}$$
  b) Hence find an approximation for
  $$\int_0^1 \frac{1}{\sqrt{9-x^2}} dx$$

Prerequisite knowledge for Q23: Radians

23) A 1 radian sector of a circle has area $(x+2)$ units². Show that for small values of x, the radius is given by
$$r \approx 2 + px - qx^2$$
where p and q are constants to be found.

2.10 Domain, Range and Composite Functions (answers on page 496)

1) Identify which of the following mapping diagrams depict a function:
   a) 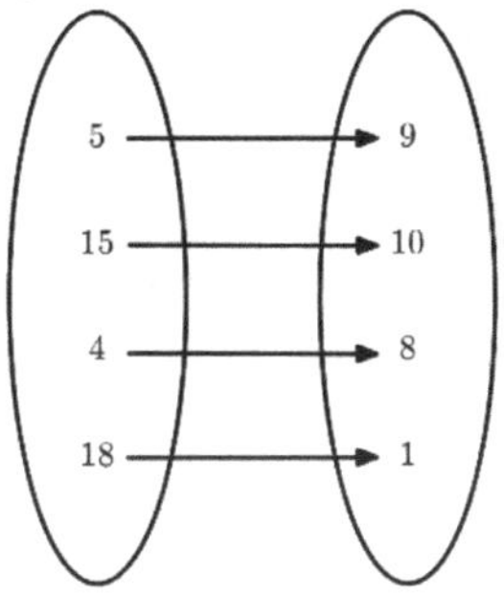
   b) 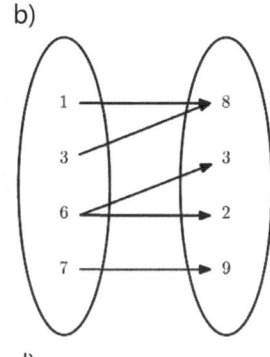
   c) 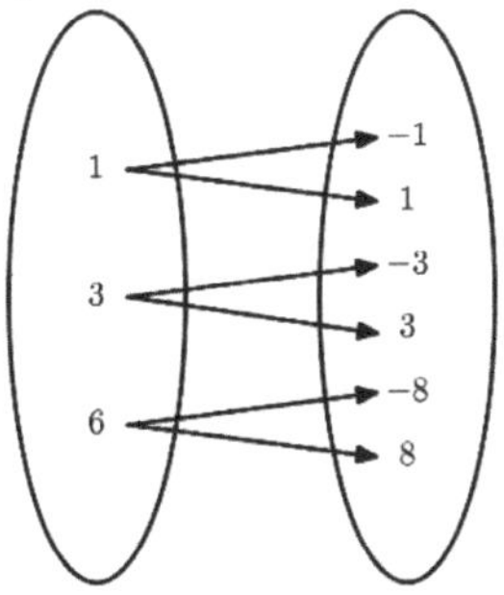
   d)

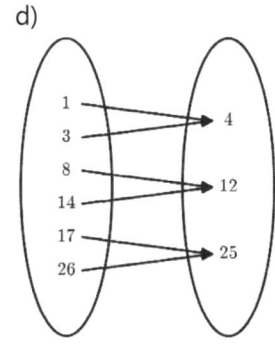

2) For each of the following, state whether it is one-to-one, many-to-one, one-to-many or many-to-many and if it is function or not:
   a) 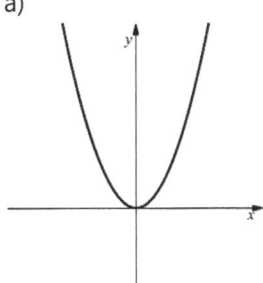
   b) 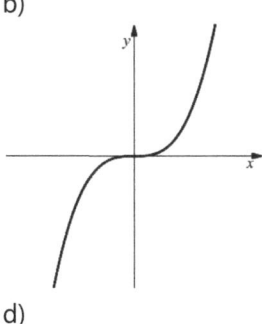
   c) 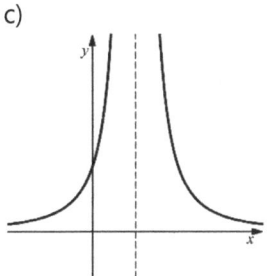
   d) 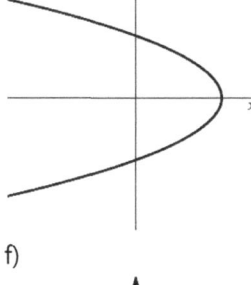
   e) 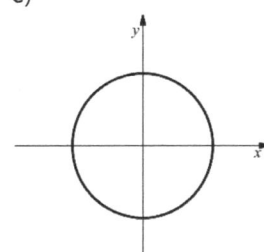
   f)

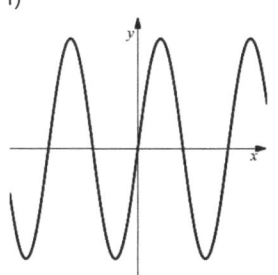

3) It is given that
$$f: x \to 5x - 3$$
where $x \in \{1, 2, 3, 4\}$
   a) Represent f on a mapping diagram.
   b) State whether $f(x)$ is a function and if so, what type of function it is.

4) It is given that
$$f(x) = 3x^2 - 1$$
where $x \in \{-2, -1, 0, 1\}$
   a) Represent $f(x)$ on a mapping diagram.
   b) State whether $f(x)$ is a function and if so, what type of function it is.

5) It is given that
$$f: x \to \sqrt{x - 1}$$
where $x \in \{3, 5, 8, 10\}$
   a) Represent f on a mapping diagram.
   b) State whether f is a function and if so, what type of function it is.

6) Determine which of these functions are one-to-one and which are many to one:
   a) $f(x) = 4x - 3$, $\{x: x \in \mathbb{R}\}$
   b) $f: x \to x^2 + 5$, $\{x: x \in \mathbb{R}\}$
   c) $f(x) = \frac{1}{x}$, $\{x: x \in \mathbb{R}. x \neq 0\}$
   d) $f: x \to 2x^3$, $\{x: x \in \mathbb{R}\}$
   e) $f(x) = x(x + 4)$, $\{x: x \in \mathbb{R}\}$
   f) $f: x \to 3 \cos x$, $\{x: x \in \mathbb{R}, 0° \leq x \leq 180°\}$
   g) $f(x) = (x - 7)^2 + 3$, $\{x: x \in \mathbb{R}, x > 7\}$
   h) $f: x \to x^2(x - 4)$, $\{x: x \in \mathbb{R}, x > 0\}$
   i) $f(x) = 3 \ln x$, $\{x: x \in \mathbb{R}, x > 0\}$
   j) $f: x \to \sqrt{x}$, $\{x: x \in \mathbb{R}, x \geq 4\}$

The answers for the following questions which require a range of values should be given in set notation

7) State the range of the following, given that in each case $x \in \mathbb{R}$:
   a) $f(x) = 1 + 2^{-x}$
   b) $y = (x + 3)^2 - 5$
   c) $f(x) = x(x - 2)^2$
   d) $y = -2e^{-4x}$
   e) $f(x) = 5 - 3^{-x}$

2.10 Domain, Range and Composite Functions (answers on page 496)

8) For each of the following functions $y = f(x)$ below, state the largest possible domain and the corresponding range of f:
   a) $y = \cos x$
   b) $y = \sqrt{x}$
   c) $y = -(x+4)^2 - 9$
   d) $y = 2x^2 - 6x + 9$
   e) $y = \frac{1}{x} + 4$
   f) $y = \frac{1}{x-3}$
   g) $y = 2 + e^x$
   h) $y = 4 \ln x$
   i) $y = 3 \ln(x-2)$

9) For each of the graphs below, state the domain and the range:

   a) endpoints $(-5, 6)$ and $(3, -2)$

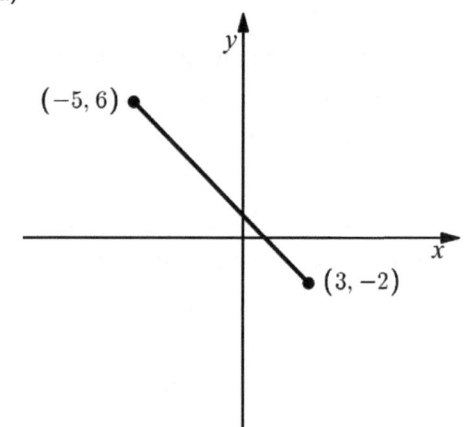

   b) endpoints $(-3, -5)$ and $(4, 9)$

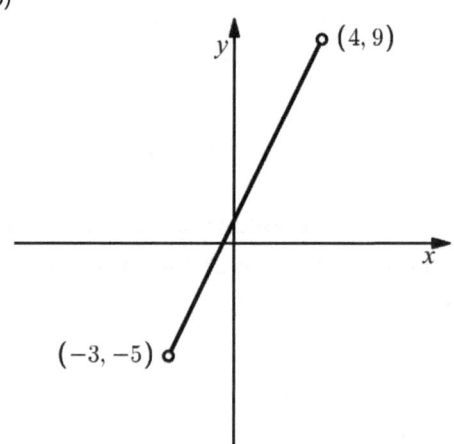

   c) endpoints $(-2, -8)$ and $(2, 8)$

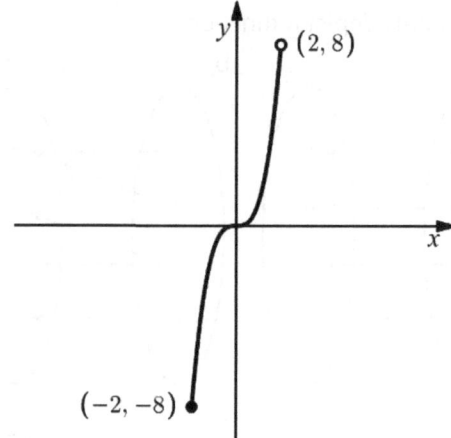

   d) endpoints $(-3, 9)$ and $(2, 4)$

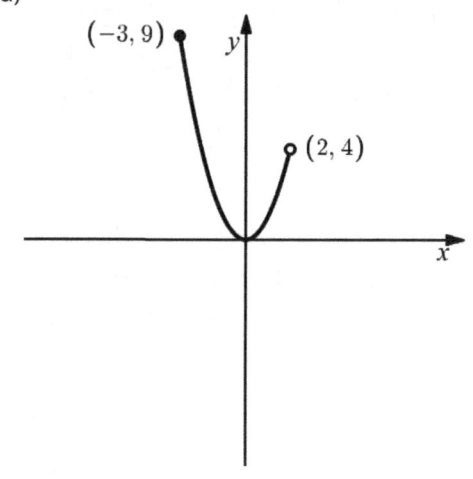

   e) endpoints $(-3, \sqrt{7})$ and $(-3, -\sqrt{7})$, passing through $(4, 0)$

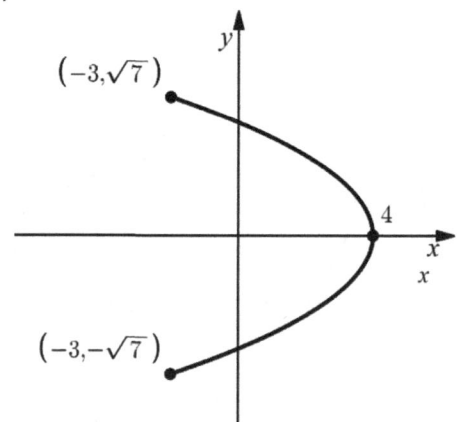

2.10 Domain, Range and Composite Functions (answers on page 496)

f) [Graph showing curve with point (−2,1) and open circle at (3,−6.5), y-intercept at 7]

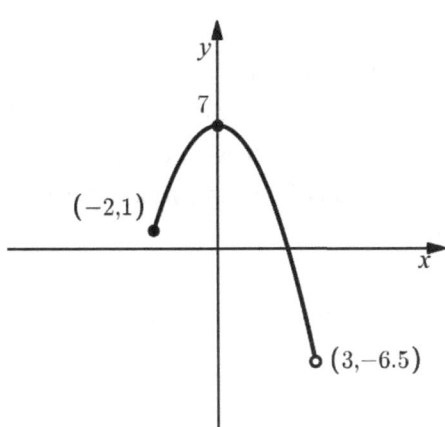

g) [Graph showing curve with horizontal asymptotes at y = 2π and y = −2π]

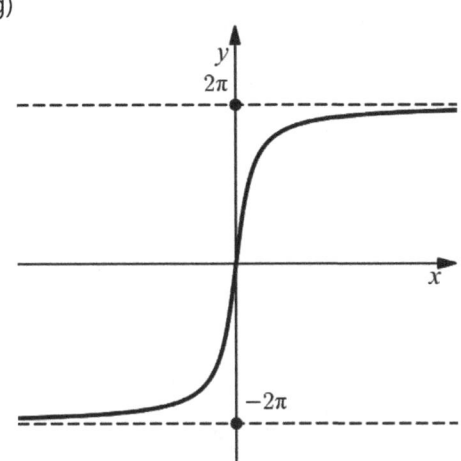

h) [Graph showing curve with vertical asymptotes at x = −π/2 and x = π/2, minimum at y = 4, and branches near x = −π and x = π]

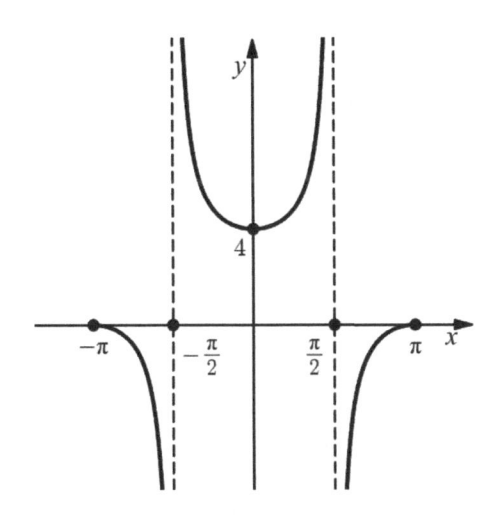

10) Sketch a graph which is many-to-one, has domain $\{x: x \in \mathbb{R}\}$, range $\{y: y \in \mathbb{R}, y \leq 3\}$, the y-intercept is negative, and its only maximum occurs at $x = p$ where $p > 0$

11) Sketch a graph which is one-to-one, has domain $\{x: x \in \mathbb{R}\}$, range $\{y: y \in \mathbb{R}\}$, the y-intercept is positive and it has a single stationary point which occurs at $x = p$ where $p > 0$

12) The graph below shows
$$f(x) = \frac{1}{2}(x-3)^2 - 2$$

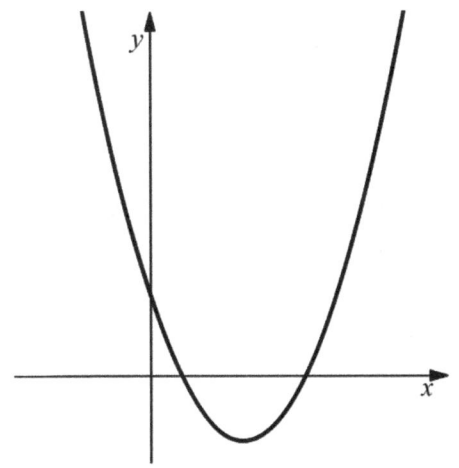

[Graph of upward parabola]

Find the range of $f(x)$ when the domain is restricted as follows:
a) $x \geq 5$
b) $x \leq 2$
c) $0 \leq x \leq 4$

13) The graph below shows
$$f(x) = -0.5x^2 + 5x + 5.5$$

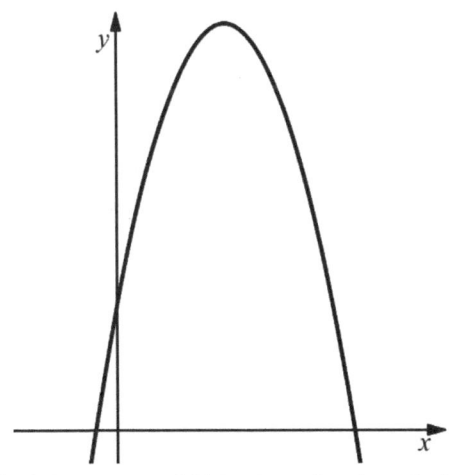

[Graph of downward parabola]

Find the range of $f(x)$ when the domain is restricted as follows:
a) $x \geq 5$
b) $x \leq 2$
c) $0 < x < 4$
d) $3 \leq x \leq 7$

2.10 Domain, Range and Composite Functions (answers on page 496)

14) The graph below shows
$$f(x) = \frac{1}{12}(x-5)^2 e^x$$

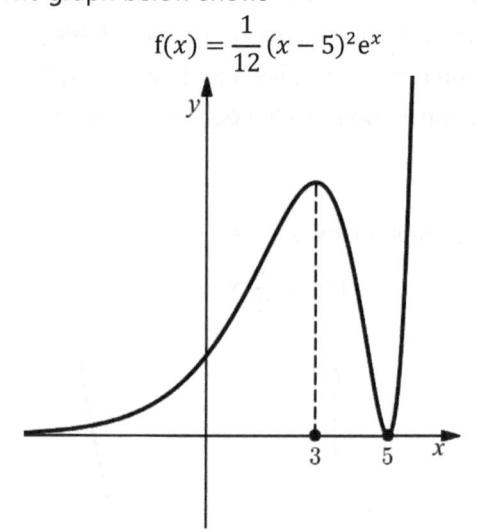

Find the range of $f(x)$ when the domain is restricted as follows:
a) $x < 5$
b) $3 \leq x \leq 5$
c) $x \geq 5$
d) $0 \leq x < 3$

15) Given
$$f: x \to x^2 + 8x + 4$$
a) Sketch the graph of $y = f(x)$, labelling any intersections with the axes and any stationary points.
b) State its range for $x \in \mathbb{R}$

16) Find the range of each of the following functions:
a) $f(x) = 4x + 1, \{x: x \in \mathbb{R}, x \geq 1\}$
b) $f: x \to \frac{1}{3}x - 2, \{x: x \in \mathbb{R}, x < -2\}$
c) $f(x) = 5 - 8x, \{x: x \in \mathbb{R}, -2 \leq x < 4\}$
d) $f(x) = 2\sqrt{x}, \{x: x \in \mathbb{R}, 4 < x \leq 16\}$
e) $f: x \to 3x - x^2, \{x: x \in \mathbb{R}, 0 < x < 3\}$
f) $f(x) = \frac{1}{1+x^2}, \{x: x \in \mathbb{R}\}$

17) Given that for $x \in \mathbb{R}$
$$f(x) = \frac{1}{2x^2 - 4x + 9}$$
By first completing the square on $2x^2 - 4x + 9$ state the range of f

18) The function $f(x) = 2x^2 + kx + 13$, where k is a constant, has domain $x \in \mathbb{R}$. It is known that the range of f is $f(x) \geq -5$
Find the possible value(s) of k

19) The function
$$f(x) = k(x-a)(x-b)$$
where a, b, k are all positive, is restricted in its domain such that it is one-to-one and $x \geq p$ where p gives the largest possible range.
Find the value of p giving your answer in terms of a and b

20) Find the largest possible domain and range of
a) $f: x \to -x^4 + 4x$
b) $f: x \to 4x - e^{2x}$

Prerequisite knowledge for Q21: Differentiation of standard functions and product rule

21) The graph below shows
$$f(x) = x \ln x$$

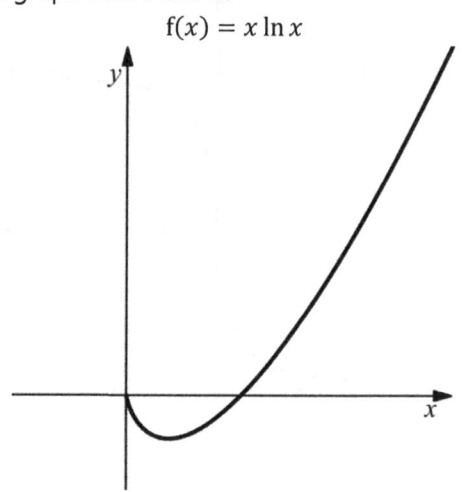

Find the largest possible domain and range of
$$f: x \to x \ln x$$

22) The graph below shows
$$f(x) = x^3 - 3x^2 - 24x$$
for $-2 < x < 5$

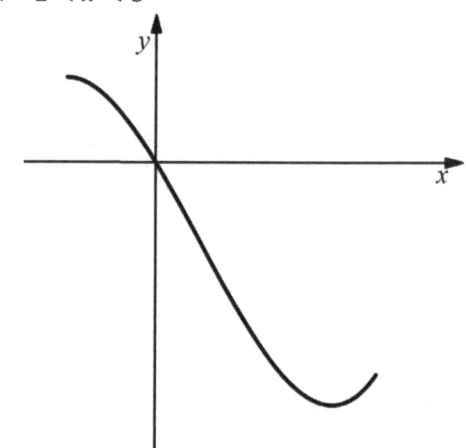

Find the range of
a) f
b) g given that $g(x) = 2f(x)$
c) h given that $h(x) = 3f(x) - 4$

250

2.10 Domain, Range and Composite Functions (answers on page 496)

23) Given f: $x \to 5x - 1$ and g: $x \to x^2 + 2$, find
    a) fg(−1)
    b) gf(2)
    c) ff(4)

24) Given $f(x) = x^2$ and $g(x) = \frac{1}{x-2}$, find fg(4)

25) Given $f(x) = 2x + 1$ and $g(x) = x^2$, find
    a) fg(x)
    b) gf(x)
    c) ff(x)
    d) the value of a such that fg(a) = 33 and a is positive
    e) the values of x such that fg(x) = gf(x)

26) Given $f(x) = \frac{1}{x}$, find
    a) $f^{89}(x)$
    b) $f^{100}(x)$

27) Given $f(x) = 1 + \sqrt{x}$ and $g(x) = 3 - 2^{-x}$
    Solve gf(x) = $\frac{23}{8}$

28) Given f: $x \to 5\log_{10} x$ and g: $x \to 10^x$
    a) find fg(x)
    b) show that fgf(x) = $25 \log_{10} x$

29) A function is given by
    $$f(x) = -kx^2 + 12x - 11$$
    where k is a positive constant.
    It is known that ff(2) = −2
    Find the possible values of k

30) The graph below shows
    $$f(x) = \begin{cases} x + 4, & x \leq 1 \\ 7 - 2x, & x > 1 \end{cases}$$

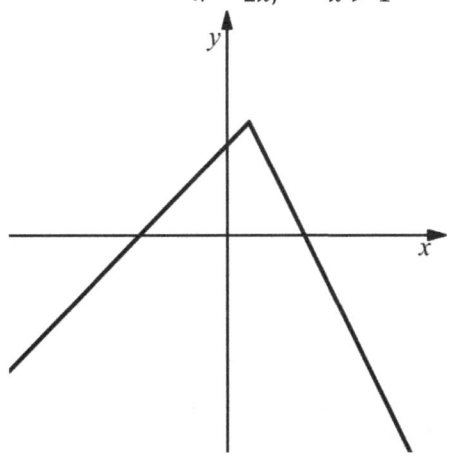

    a) State the value of ff(1)
    b) Find the value of ff$\left(\frac{1}{2}\right)$
    c) Find the range of values of x for which $f(x) < -2$

31) The graph below shows
    $$f(x) = \begin{cases} -x - 3, & x < 1 \\ x^2 - 2x - 3, & x \geq 1 \end{cases}$$

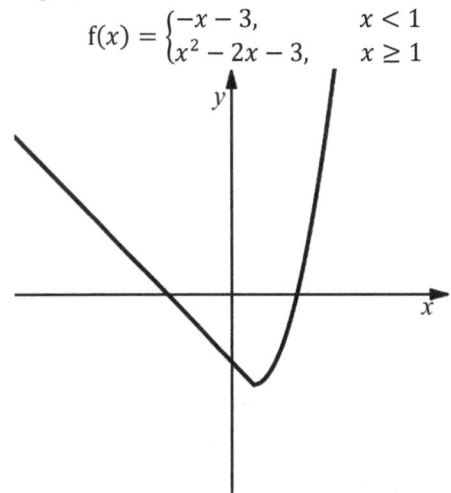

    a) State the value of ff(1)
    b) State the range of f
    c) Find the value of ff(2)
    d) Find the range of values of x for which $f(x) < 0$

32) The graph below shows
    $$f(x) = \begin{cases} x + 1, & x < a \\ 6 - e^{x-4}, & x \geq a \end{cases}$$

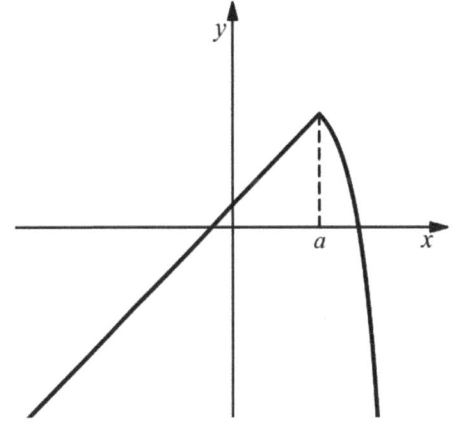

    The range of f is $f(x) \leq 5$. Find
    a) the value of a
    b) the exact value of ff$\left(\frac{7}{2}\right)$
    c) the range of values of x for which $f(x) \geq 0$

33) Given
    $$f: x \to \frac{2x + 1}{x - 5}, \quad x \neq 5$$
    a) Find the range of f
    b) Find ff(x)

2.10 Domain, Range and Composite Functions (answers on page 496)

34) Given
$$f(x) = \frac{ax+1}{x-4}, \quad x \neq 4$$
where a is a constant and $a \neq -\frac{1}{4}$
   a) Find the range of f
   b) Find ff(x)

35) Given
$$f(x) = x^2 - ax$$
and
$$g(x) = a^2 - x$$
where a is a positive constant and $fg(2) = 0$
   a) find $fg(x)$
   b) find the possible values of a giving a in exact form.

The answers for the following questions which require a range of values should be given in interval notation

36) The graph below shows
$$f(x) = 2(x+3)^2 e^{-x}$$

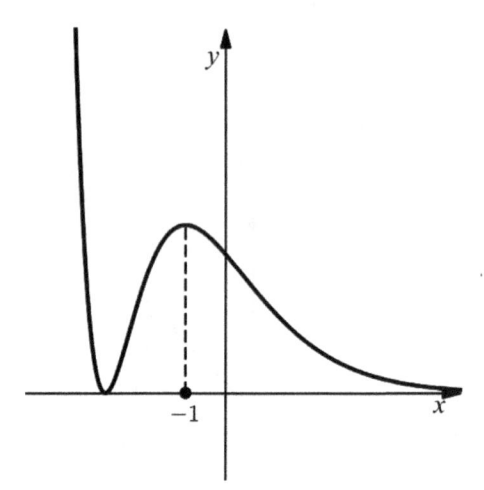

Find the range of $f(x)$ when the domain is restricted as follows:
   a) $x < -3$
   b) $-3 \leq x \leq -1$
   c) $x \geq 0$
   d) $-3 \leq x < 0$

37) Given
$$f(x) = 2x^2 - 12x + 13$$
   a) Sketch the graph of $y = f(x)$, labelling any intersections with the axes and any stationary points.
   b) State its range for $x \in \mathbb{R}$

38) The graph below shows
$$f(x) = 16x^5 - 15x^3 + 1$$
for $-1 < x < 1$

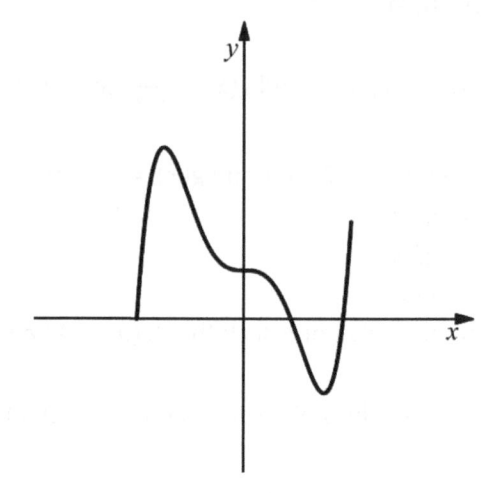

Find the range of
   a) f
   b) g given that $g(x) = 64f(x)$
   c) h given that $h(x) = 8f(x) - 10$

Find the domain of
   d) k given $k(x) = f(2x)$

39) The graph below shows
$$f(x) = \begin{cases} 3 - kx, & x < a \\ e^{x-2} - 2, & x \geq a \end{cases}$$
where k is a positive constant.

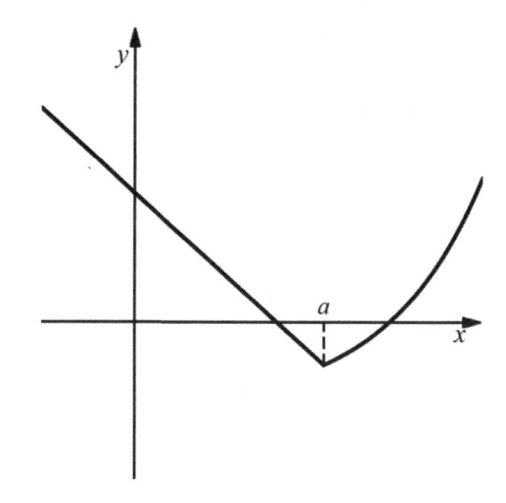

The range of f is $f(x) \geq -1$
   a) Find the value of a
   b) Find the value of k
   c) Find the exact value of $ff(3)$
   d) Find the range of values of x for which $f(x) \leq 0$

2.11 Graph Transformations: Combinations (answers on page 498)

1) Describe a sequence of two transformations that will map:
   a) $y = f(x)$ onto $y = 3f(4x)$
   b) $y = f(x)$ onto $y = \frac{1}{2}f\left(\frac{1}{3}x\right)$

2) Describe a sequence of two transformations that will map:
   a) $y = f(x)$ onto $y = -f(x+6)$
   b) $y = f(x)$ onto $y = f(-x) + 6$

3) Describe a sequence of two transformations that will map:
   a) $y = f(x)$ onto $y = \frac{1}{6}f(x-4)$
   b) $y = f(x)$ onto $y = f(8x) - 5$

Prerequisite knowledge for Q4: Radians

4) Describe a sequence of two transformations that will map $y = \sin x$ onto the image:
   a) $y = 2\sin\left(x - \frac{\pi}{3}\right)$
   b) $y = \sin\left(\frac{1}{5}x\right) + 2$

5) Describe a sequence of two transformations that will map $y = e^x$ onto the image:
   a) $y = 5e^{\frac{x}{3}}$
   b) $y = e^{-x} - 9$

6) Describe a sequence of two transformations that will map $y = \frac{1}{x}$ onto the image:
   a) $y = \frac{2}{x-4}$
   b) $y = 5 - \frac{1}{x}$

7)
   a) Describe a sequence of two transformations that will map $y = \ln x$ onto the image $y = \ln(2x) - 3$
   b) Describe a **single** transformation that will map $y = \ln x$ onto the image $y = \ln(2x) - 3$

8) The point P(−6, 12) lies on the graph with equation $y = f(x)$. State the coordinates of the image of P after $y = f(x)$ is mapped to each of these:
   a) $y = 2f(3x)$
   b) $y = -f(x-5)$

9) The point Q(4, −3) lies on the graph with equation $y = f(x)$. State the coordinates of the image of Q after $y = f(x)$ is mapped to each of these:
   a) $y = \frac{1}{3}f\left(\frac{1}{6}x\right)$
   b) $y = f(-x) + 9$

10) The graph of $y = f(x)$ is shown in the diagram below.

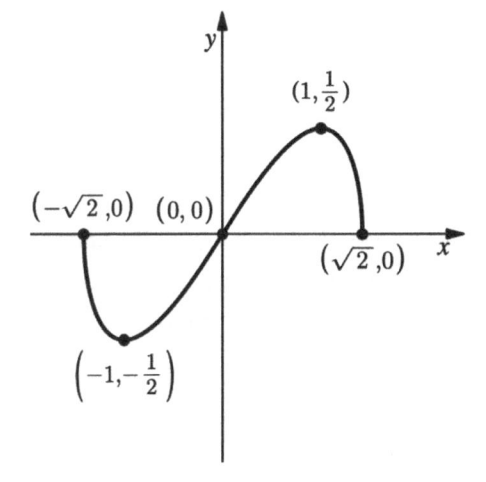

    Points shown: $(1, \frac{1}{2})$, $(-\sqrt{2}, 0)$, $(0, 0)$, $(\sqrt{2}, 0)$, $(-1, -\frac{1}{2})$

    Sketch the graphs of
    a) $y = 2f\left(\frac{1}{2}x\right)$
    b) $y = -f(x + \sqrt{2})$
    c) $y = f(2x) + \frac{1}{2}$
    d) $y = f(-x) - 1$

11) Describe a sequence of two transformations that will map:
    a) $y = f(x)$ onto $y = 2f(x) + 3$
    b) $y = f(x)$ onto $y = 9f(x) - 8$
    c) $y = f(x)$ onto $y = \frac{1}{5}f(x) - 10$
    d) $y = f(x)$ onto $y = \frac{3}{2}f(x) + 20$

12) Describe a sequence of two transformations that will map:
    a) $y = 4f(x) - 1$ onto $y = 5f(x) + 2$
    b) $y = -f(x) - 5$ onto $y = -\frac{1}{2}f(x) + 7$

13) Describe a sequence of two transformations that will map:
    a) $y = f(x)$ onto $y = f(3x + 1)$
    b) $y = f(x)$ onto $y = f(2x - 9)$
    c) $y = f(x)$ onto $y = f(2 - x)$
    d) $y = f(x)$ onto $y = f(-x - 3)$

2.11 Graph Transformations - Combinations (answers on page 498)

14) Describe a sequence of two transformations that will map:
    a) $y = f(3x + 8)$ onto $y = f(6x - 4)$
    b) $y = f(2 - x)$ onto $y = f\left(3 - \frac{1}{5}x\right)$

15) The point $P(10, -3)$ lies on the graph with equation $y = f(x)$. State the coordinates of the image of P after $y = f(x)$ is mapped to each of these:
    a) $y = 2f(x - 5) + 3$
    b) $y = -f(x) - 9$
    c) $y = f(3x + 5)$
    d) $y = f\left(\frac{1}{4}x - 2\right)$

Prerequisite knowledge for Q16: Radians

16) Describe a sequence of two transformations that will map $y = \sin x$ onto the image:
    a) $y = 4\sin(x) - 6$
    b) $y = \sin\left(\frac{x}{6} - \frac{\pi}{9}\right)$
    c) $y = \cos 2x$
    d) $y = 3\cos x$

17) Describe a sequence of two transformations that will map $y = e^x$ onto the image:
    a) $y = e^{6x-10}$
    b) $y = e^{9-x}$
    c) $y = 10e^x + 20$
    d) $y = 18 - e^x$

18) Describe a sequence of two transformations that will map $y = \frac{1}{x}$ onto the image:
    a) $y = \frac{1}{5x-2}$
    b) $y = \frac{8}{x} - 9$
    c) $y = \frac{6-5x}{x}$
    d) $y = \frac{2x+7}{20x}$

19) Describe a sequence of two transformations that will map $y = \frac{1}{x^2}$ onto the image:
    a) $y = -\frac{1}{5x^2}$
    b) $y = 1 + \frac{2}{x^2}$
    c) $y = \frac{2+3x^2}{6x^2}$
    d) $y = \frac{1-6x^2}{8x^2}$

20)
    a) Describe a sequence of two transformations that will map $y = \ln(x + 1)$ onto the image $y = \ln(3x + 3)$
    b) Describe a **single** transformation that will map $y = \ln(x + 1)$ onto the image $y = \ln(3x + 3)$

21) Describe a sequence of two transformations that will map $y = 4x + 3$ onto the image:
    a) $y = 8x + 7$
    b) $y = 2x + 6$

22) Describe a sequence of two transformations that will map $y = 4 - 5x$ onto the image:
    a) $y = -10x - 8$
    b) $y = 50x - 40$

23) Write down the image of $y = 5x + 2$ after a translation by the vector $\begin{pmatrix} 2 \\ -3 \end{pmatrix}$, followed by a stretch, parallel to the x-axis, scale factor 2

24) Write down the image of $y = 8 - 6x$ after a stretch, parallel to the y-axis, scale factor $\frac{1}{3}$, followed by a translation by the vector $\begin{pmatrix} -\frac{2}{3} \\ \frac{4}{3} \end{pmatrix}$

25) Describe a sequence of two transformations that will map $y = x^2$ onto the image:
    a) $y = 4x^2 + 9$
    b) $y = 5 - x^2$
    c) $y = \frac{1}{4}(x - 6)^2 + 3$
    d) $y = 8(x + 1)^2 - 2$
    e) $y = (4x + 1)^2$
    f) $y = \left(\frac{1}{3}x - 9\right)^2$
    g) $y = 2x^2 - 4x - 7$
    h) $y = 3x^2 + 12x + 1$
    i) $y = -(x + 1)(x - 9)$
    j) $y = (6x - 4)(3x + 10)$

26) Describe a sequence of two transformations that will map $y = x^2 + 2$ onto the image:
    a) $y = 9x^2 + 10$
    b) $y = 8x^2$
    c) $y = 7 - x^2$

2.11 Graph Transformations: Combinations (answers on page 498)

27) Write down the image of $y = x^3$ after a stretch, parallel to the x-axis, scale factor $\frac{1}{2}$, followed by a translation by the vector $\begin{pmatrix} -1 \\ 2 \end{pmatrix}$

28) Write down the image of $y = \ln x$ after a stretch, parallel to the x-axis, scale factor 9, followed by a translation by the vector $\begin{pmatrix} 6 \\ -\ln 3 \end{pmatrix}$, showing that it can be written in the form $y = \ln(ax + b)$, where a and b are rational numbers.

29) Find the equation of the curve $y = x^2$ after a reflection in the line:
   a) $y = 2$
   b) $x = 3$

Prerequisite knowledge for Q30: Radians

30) Find the equation of the curve $y = \cos x$ after a reflection in the line:
   a) $y = -3$
   b) $x = -\frac{\pi}{3}$

31) The graph of $y = f(x)$ is shown in the diagram below.

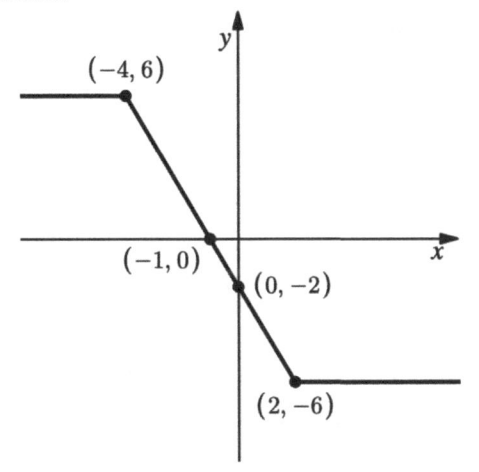

(−4, 6)
(−1, 0)
(0, −2)
(2, −6)

Sketch the graphs of
   a) $y = f\left(\frac{1}{2}x - 2\right)$
   b) $y = f(2x + 5)$
   c) $y = 3f(x) + 1$
   d) $y = \frac{1}{2}f(x) - 4$

32) Describe the transformation that maps $y = f(x)$ onto the image $x = f(y)$

33) Describe the transformation that maps $y = 8 - (x - 3)^2$ onto the image $x = 8 - (y - 3)^2$

34) Sketch the image of $(x - 1)^2 + y^2 = 9$ once it has been reflected in the line $y = x$

35) Describe a sequence of two transformations that will map $y = \ln x$ onto the image $y = e^{3x}$

36) Describe a sequence of two transformations that will map $y = \ln(x - 5)$ onto the image $y = e^x$

37) Describe a sequence of **three** transformations that will map $y = \ln x$ onto the image $y = 2e^x - 7$

38) The graph of $y = f(x)$ is shown in the diagram below.

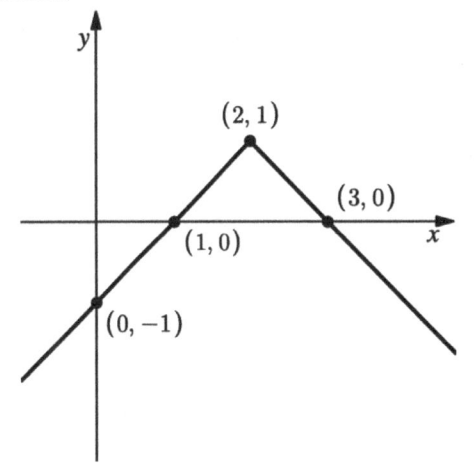

(2, 1)
(1, 0)
(3, 0)
(0, −1)

Sketch the graph of $x = f(y)$

39) Describe a sequence of **four** transformations that will map $y = f(x)$ onto the image $y = 2f\left(\frac{1}{3}x - 9\right) - 3$

40) Describe a sequence of **four** transformations that will map $y = f(x)$ onto the image $y = \frac{3}{5}f\left(\frac{2}{5}x + 4\right) + 2$

41) Find the equation of the curve $y = x^2$ after a reflection in the line $y = -x$

2.12 Inverse Functions (answers on page 501)

1) For each of the following, state whether or not f^{-1} exists:
   a) $f(x) = x^3 + 5$ for $x \in \mathbb{R}$
   b) $f: x \to \cos x$ for $x \in \mathbb{R}$
   c) $f(x) = \tan x$ for $0 \leq x < \pi, x \neq \frac{\pi}{2}$
   d) $f(x) = x^2 + 5$ for $x \in \mathbb{R}$
   e) $f: x \to x^2 + 4x$ for $x \geq -2$

2) Find the inverse of the following functions:
   a) $$f(x) = 1 - 2x, \quad x \in \mathbb{R}$$
   b) $$f(x) = \frac{x+3}{4}, \quad x \in \mathbb{R}$$
   c) $$f(x) = \frac{7}{2x}, \quad x \in \mathbb{R}, x \neq 0$$
   d) $$f(x) = \frac{2x+1}{x-3}, \quad x \in \mathbb{R}, x \neq 3$$
   e) $$f(x) = \frac{4x-9}{2x+1}, \quad x \in \mathbb{R}, x \neq -\frac{1}{2}$$

3) Given
$$f: x \to \frac{1}{2}x - 5$$
for $x \in \mathbb{R}$.
   a) Find $y = f^{-1}(x)$
   b) sketch $y = f(x)$ and $y = f^{-1}(x)$ on the same axes
   c) Find $f(12)$ and $f^{-1}(1)$, comment on your result.

4) For any given one-to-one function, f, state the transformation which maps the graph of $y = f(x)$ onto the graph of its inverse function $y = f^{-1}(x)$

5) It is given that the graph of the one-to-one function $y = f(x)$ passes through the given point. State the coordinates of the point corresponding to the same point on the curve $y = f^{-1}(x)$
   a) $(1, 3)$
   b) $(-5, 4)$
   c) $(3, -3e^2)$
   d) (a, b)

6) For each of the following functions where $x \geq 0$, solve the equation $f(x) = f^{-1}(x)$
   a) $f(x) = 5x - 2$
   b) $f(x) = -x + 4$
   c) $f(x) = x^2 - 6$
   d) $f(x) = -2x^2 - x + 4$
   e) $f(x) = x^3$
   f) $f(x) = \frac{2}{x+1}$

7) Given $f(x) = x^2 + 2x + 5$, explain why f has no inverse.

8) Given
$$f: x \to -x + 12, \quad x \neq 0$$
   a) show that $ff(x) = x$
   b) Hence write down $f^{-1}(x)$

9) Given
$$f(x) = \frac{1}{x}, \quad x \neq 0$$
   a) show that $ff(x) = x$
   b) Hence write down $f^{-1}(x)$

10) Given
$$f(x) = \frac{7x - 48}{x - 7}, \quad x \neq 7$$
   a) show that $ff(x) = x$
   b) Hence write down $f^{-1}(x)$

11) Given
$$f(x) = \frac{ax - 1}{x - a}, \quad a > 1$$
   a) show that $ff(x) = x$
   b) Hence write down $f^{-1}(x)$

The answers for the following questions which require a range of values should be given in set notation

12) Given
$$f: x \to 2x^2 + 1$$
for $x \geq 0$
   a) find $y = f^{-1}(x)$
   b) show that there are no solutions to $f(x) = f^{-1}(x)$
   c) sketch $y = f(x)$ and $y = f^{-1}(x)$ on the same axes
   d) state the domain and range of f^{-1}

2.12 Inverse Functions (answers on page 501)

13) Given
$$f(x) = 2e^{3x} - 1$$
for $x \in \mathbb{R}$
a) find $y = f^{-1}(x)$
b) given $f(x) = f^{-1}(x)$ has two solutions sketch $y = f(x)$ and $y = f^{-1}(x)$ on the same axes
c) state the domain and range of f^{-1}

14) Given
$$f: x \to 3 + 5^{-x}$$
for $x \in \mathbb{R}$
a) find $y = f^{-1}(x)$
b) sketch $y = f(x)$ and $y = f^{-1}(x)$ on the same axes
c) state the domain and range of f^{-1}

15) Given
$$f(x) = \frac{2x}{x+5}$$
for $-4 \leq x \leq 1$
a) find $y = f^{-1}(x)$
b) solve $f(x) = f^{-1}(x)$
c) sketch $y = f(x)$ and $y = f^{-1}(x)$ on the same axes
d) state the domain and range of f^{-1}

16) For each of the following functions, find the domain and range of f^{-1}:
a) $f(x) = 5x - 2$ for $-1 \leq x \leq 5$
b) $f(x) = \ln x$ for $x > 0$
c) $f(x) = e^{x-2}$ for $x \in \mathbb{R}$
d) $f(x) = \sin x$ for $-\frac{\pi}{2} \leq x \leq \frac{\pi}{2}$
e) $f(x) = \cos x$ for $0 \leq x \leq \pi$
f) $f(x) = \tan x$ for $-\frac{\pi}{2} < x < \frac{\pi}{2}$

17) A function is given by
$$f(x) = x^4 - x^3 + 2x$$
where $x \in \mathbb{R}$
a) By using the factor theorem, show that $(x + 1)$ is a factor of $f(x)$
b) Show that $f(x) = 0$ only has two solutions.
c) Explain why an inverse function does not exist.

The domain of f is restricted such that
$$0 \leq x \leq 2$$
d) Find the domain and range of f^{-1}

18) Given
$$f(x) = x^2 - 6x + 5$$
for $x \geq p$ where p is a constant
a) sketch $y = f(x)$ for $x \in \mathbb{R}$
b) state the minimum value that p can be for f to be one-to-one.

For your value of p
c) find the range of f
d) state the domain and range of f^{-1}
e) show that $f(x) = f^{-1}(x)$ only has one solution.
f) sketch $y = f(x)$ and $y = f^{-1}(x)$ on the same axes.

19) Given
$$f(x) = \ln(x - 3)$$
for $x > p$ where p is a constant,
a) state the value of p such that f has the largest possible domain.

For your value of p
b) find the range of f
c) find $y = f^{-1}(x)$
d) state the domain and range of f^{-1}
e) sketch $y = f(x)$ and $y = f^{-1}(x)$ on the same axes.

20) Given
$$f(x) = \sqrt{x - 5}$$
for $x \geq p$ where p is a constant.
a) State the value of p such that f has the largest possible domain.

For your value of p
b) find the range of f
c) find $y = f^{-1}(x)$
d) state the domain and range of f^{-1}

It is known that $x^4 + 10x^2 - x + 30 = 0$ has no solutions.
e) Hence, by considering $f(x) = f^{-1}(x)$, show that there is no intersection between the graphs of $y = f(x)$ and $y = f^{-1}(x)$
f) Sketch $y = f(x)$ and $y = f^{-1}(x)$ on the same axes.

2.12 Inverse Functions (answers on page 501)

21) Given
$$f(x) = 3 - \frac{1}{x+2}$$
for $x \neq p$ where p is a constant,
a) state the value of p

For $x > p$
b) find the range of f
c) find $y = f^{-1}(x)$
d) state the domain and range of f^{-1}
e) find the coordinates of the point of intersection of the graphs of $y = f(x)$ and $y = f^{-1}(x)$
f) sketch $y = f(x)$ and $y = f^{-1}(x)$ on the same axes.

22) Given
$$f(x) = \frac{4x-1}{x-5}$$
for $x \neq p$ where p is a constant,
a) state the value of p
b) find the range of f
c) state the domain and range of f^{-1}

23) For each of the following functions find $f^{-1}(1)$
a)
$$f: x \rightarrow 5x + 3$$
b)
$$f: x \rightarrow x^2 - 8, \quad x \geq 0$$
c)
$$f: x \rightarrow \frac{4x+9}{2x+1}, \quad x \neq -\frac{1}{2}$$

24) For
$$f(x) = ax + b$$
where a and b are positive constants, find
a) the gradient function of f
b) the gradient function of f^{-1}
c) the values of a such that the gradient of $f(x)$ is equal to the gradient of $f^{-1}(x)$

25) For
$$f(x) = x^2, \quad x \geq 0$$
a) find the gradient of $f(x)$ at $x = 2$
b) find the gradient of $f^{-1}(x)$ at $x = 2$
c) find the value of x such that the gradient of $f(x)$ is equal to the gradient of $f^{-1}(x)$

26) Given
$$f(x) = -x + 5$$
and
$$g(x) = 2x - 1$$
for $x \in \mathbb{R}$
a) find $fg(x)$
b) find the value of constant k such that $fg(5) = g^{-1}(k)$

27) Given
$$f(x) = 2x^2 - 8x$$
and
$$g(x) = 5x + 1$$
for $x \in \mathbb{R}$
a) find $fg(x)$
b) find the value of constant k such that $fg(2) = g^{-1}(k)$

28) Given
$$f(x) = -2x + 1, \quad x < 0$$
and
$$g(x) = 3x + 5, \quad x > 1$$
a) find $gf(x)$
b) find the domain of gf
c) state the range of gf
d) find $(gf)^{-1}(x)$
e) state the domain and range of $(gf)^{-1}(x)$

29) Given
$$f: x \rightarrow x + 4, \quad x < 0$$
and
$$g: x \rightarrow x^2 - x, \quad x > 1$$
a) find $gf(x)$
b) find the domain of gf
c) state the range of gf
d) find $(gf)^{-1}(x)$
e) state the domain and range of $(gf)^{-1}(x)$

30) Given
$$f(x) = 3x + 1$$
and
$$g(x) = \frac{2x+2}{x+2}, \quad x \neq -2$$
a) find $fg(x)$
b) find the domain of fg
c) state the range of fg
d) find $(fg)^{-1}(x)$
e) state the domain and range of $(fg)^{-1}(x)$

2.12 Inverse Functions (answers on page 501)

The answers for the following questions which require a range of values should be given in interval notation

31) Given
$$f(x) = x^2 - 6x + 1$$
for $x \leq 1$
a) state the range of f
b) find $y = f^{-1}(x)$
c) find the solution to $f(x) = f^{-1}(x)$
d) sketch $y = f(x)$ and $y = f^{-1}(x)$ on the same axes
e) state the domain and range of f^{-1}

32) Given
$$f: x \rightarrow e^{2x} + 2$$
for $x \in \mathbb{R}$
a) find $y = f^{-1}(x)$
b) sketch $y = f(x)$ and $y = f^{-1}(x)$ on the same axes
c) state the domain and range of f^{-1}

33) Given
$$f(x) = \frac{3x}{x-4}$$
for $-5 \leq x \leq 2$
a) find $y = f^{-1}(x)$
b) solve $f(x) = f^{-1}(x)$
c) sketch $y = f(x)$ and $y = f^{-1}(x)$ on the same axes
d) state the domain and range of f^{-1}

34) For each of the following functions, find the domain and range of f^{-1}:
a) $f(x) = -2x + 3$ for $-2 \leq x \leq 7$
b) $f(x) = 2\ln x$ for $x > 0$
c) $f(x) = 2 + e^{x+5}$ for $x \in \mathbb{R}$

35) Given
$$f(x) = \ln(x+2)$$
for $x > p$ where p is a constant,
a) state the value of p

For your value of p
b) find the range of f
c) find $y = f^{-1}(x)$
d) state the domain and range of f^{-1}

e) Given $f(x) = f^{-1}(x)$ has two solutions, sketch $y = f(x)$ and $y = f^{-1}(x)$ on the same axes.

36) Given
$$f(x) = \frac{5x+2}{x-9}$$
for $x \neq p$ where p is a constant,
a) state the value of p
b) find the range of f
c) state the domain and range of f^{-1}

37) Given
$$f: x \rightarrow 2x^2 + 4x - 5$$
for $x \geq p$ where p is a constant,
a) state the minimum value that p can be for f to be one-to-one.

For your value of p
b) find the domain and range of f^{-1}
c) sketch $y = f(x)$ and $y = f^{-1}(x)$ on the same axes.

38) Given
$$f(x) = 1 + 2x, \quad x < 1$$
and
$$g(x) = x^2, \quad x > 2$$
a) find $gf(x)$
b) find the domain of gf
c) state the range of gf
d) find $(gf)^{-1}(x)$
e) state the domain and range of $(gf)^{-1}(x)$

39) Given
$$f(x) = -2x + 3$$
and
$$g(x) = \frac{8x+21}{x+7}$$
a) find $fg(x)$
b) find the domain of fg
c) state the range of fg

Given fg is now restricted such that $x > -7$
d) find $(fg)^{-1}(x)$
e) state the domain and range of $(fg)^{-1}(x)$

2.13 Modulus Functions (answers on page 504)

1) For the functions $f(x) = |x - 2|$ and $g(x) = |x| - 2$, find
   a) f(1)
   b) f(−3)
   c) f(0)
   d) g(1)
   e) g(0)
   f) g(−4)

2) If $|x| = 4$ find the possible values of $|2x - 5|$

3) If $|x| = 2$ find the possible values of $|4x + 1|$

4) Write down the y-intercept of each of the graphs of the following functions:
   a) $y = |x + 2|$
   b) $y = |x - 1|$
   c) $y = |x - 3| + 5$
   d) $y = 2|x + 4| + 8$
   e) $y = -5|x - 2| + 1$
   f) $y = 6 - 4\left|x + \frac{1}{2}\right|$

5) Write down the vertex of the graph of each of the following functions:
   a) $y = |x + 3|$
   b) $y = |x - 2|$
   c) $y = |x - 1| + 5$
   d) $y = |x + 3| - 8$
   e) $y = 2|x + 4| + 13$
   f) $y = -8|x - 1| + 9$
   g) $y = 2 - 7\left|x + \frac{1}{2}\right|$

6) Sketch each of the following, labelling only the vertex and the y-intercept:
   a) $y = |x + 3|$
   b) $y = |x - 2| + 4$
   c) $y = |x + 1| - 1$
   d) $y = |x - 7| - 7$
   e) $y = 2|x + 1| + 3$
   f) $y = 3|x - 5| + 1$
   g) $y = 2|x + 9|$
   h) $y = -4|x - 1|$
   i) $y = 7 - |x - 6|$
   j) $y = 9 - 2|x + 4|$

7) For each of the following graphs, state the equations:

a)

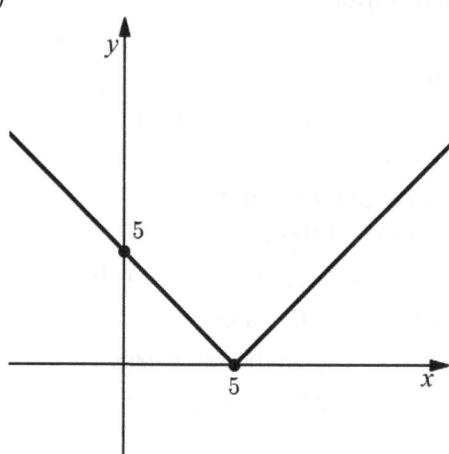

b)

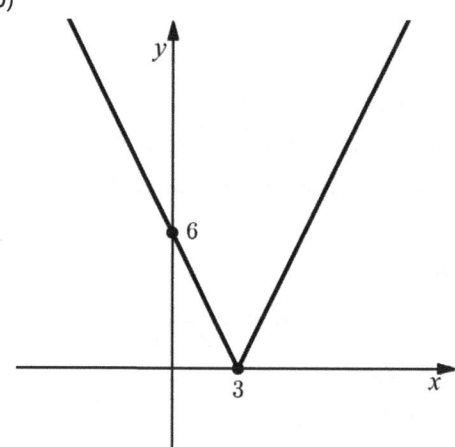

c)

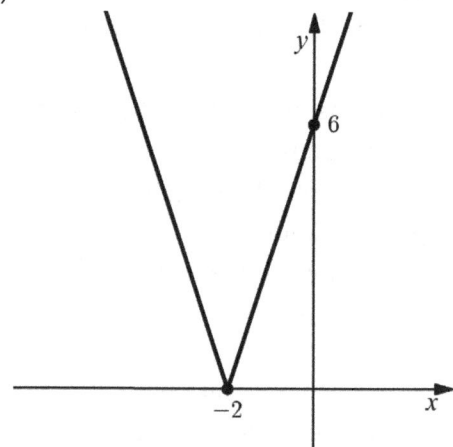

2.13 Modulus Functions (answers on page 504)

d) Graph with vertex at $(-3, 1)$ and y-intercept at 4.

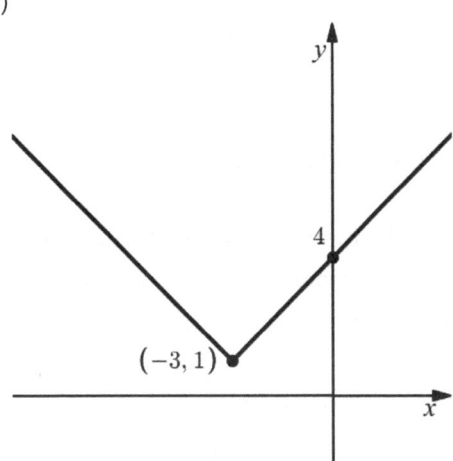

e) Graph with vertex at $(2, -2)$ and x-intercept at 4.

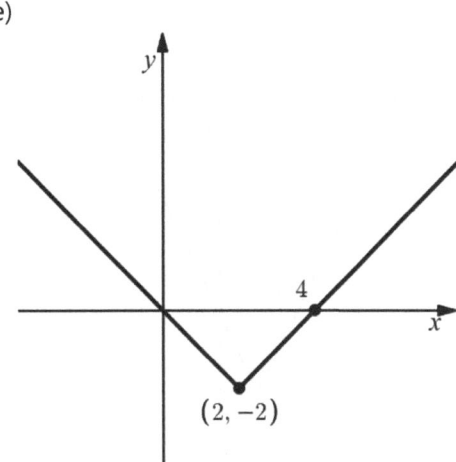

f) Graph with vertex at $(2, -6)$, x-intercepts at -1 and 5, and y-intercept at -2.

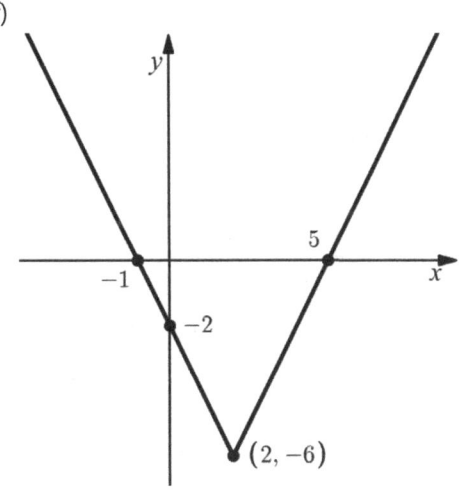

g) Graph with vertex at $(2, 4)$, x-intercepts at -2 and 6, and y-intercept at 2.

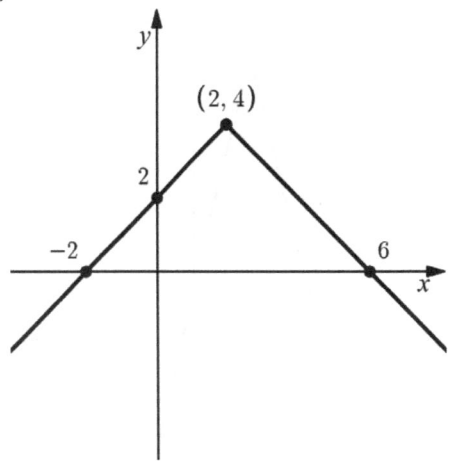

h) Graph with vertex at $(5, -1)$ and y-intercept at -6.

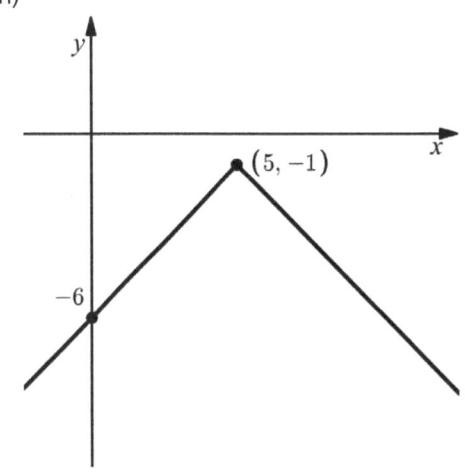

i) Graph with vertex at $(-7, 8)$, x-intercepts at -9 and -5, and y-intercept at -20.

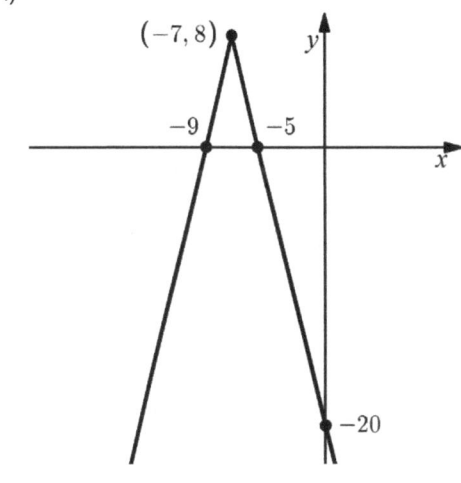

8) Solve each of the following equations:
   a) $|x + 3| = 5$
   b) $|x - 2| = 7$
   c) $|2x + 1| = 4$
   d) $-|x + 5| = -3$
   e) $|x - 5| = -3$

2.13 Modulus Functions (answers on page 504)

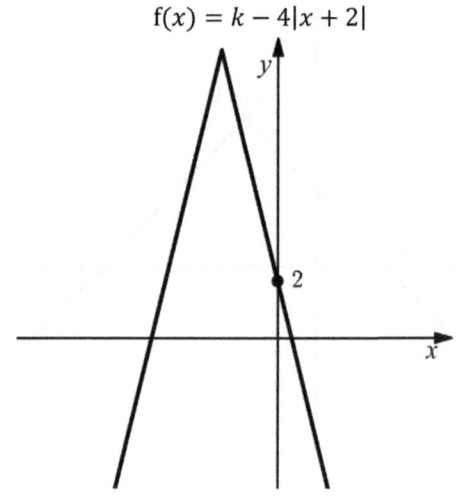

9) The diagram shows a sketch of
$$f(x) = k - 4|x + 2|$$

Find
a) the value of k
b) the values where $f(x)$ intersects with the x-axis.

10) Solve each of the following equations:
a) $|x - 1| = |x + 8|$
b) $|x + 3| = |x - 4|$
c) $|x - 10| = |x - 1|$

11) Solve each of the following equations:
a) $|x + 5| = 2|x + 3|$
b) $|x + 6| = 3|x - 3|$
c) $|x + 2| = 5|x + 8|$
d) $|x + 2| = 10 - |x - 5|$
e) $|x - 4| - 8 = -|x - 3|$
f) $|x + 4| = 4 - |x - 3|$

12) Write the following as a modulus inequality:
a) $-6 < x < 6$
b) $-4 < x < 6$
c) $-3 < x < 25$
d) $x < 1, x > 5$

13) Write $6 \leq x \leq 14$ in the form $|2x - a| \leq b$ where a and b are integers to be found.

14) Sketch the graph of $y = 6 - |2x - 7|$ and $y = 3$ on the same axes, and hence solve the inequality $6 - |2x - 7| > 3$

15) Sketch $y = |2x|$ and $y = |3x - 9|$ on the same axes and hence solve $|2x| \geq |3x - 9|$

16) Find the sets of values of x for which, giving your answer in interval notation:
a) $|x + 3| > 10$
b) $4 - |x| \leq |x - 2|$
c) $|4x| - 2 \geq |7 - 2x|$

17) Sketch the graph of $y = |3x + k|$ where k is a positive constant. Include all points of intersection with the coordinate axes.

18) A function is given to be
$$f(x) = \frac{7}{8}|x + 1| + 4$$
a) Sketch $f(x)$
b) Solve $f(x) = 2x + 3$

19) The diagram shows a sketch of
$$y = k + 2|x + 1|$$

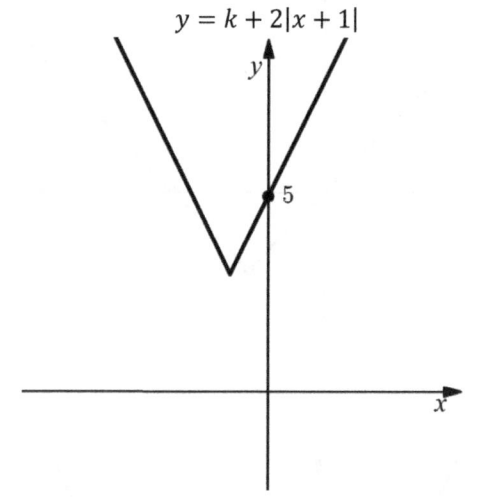

a) Find the value of k
b) For value of k solve
$$k + 2|x + 1| = x + 8$$

20) Find the values of x such that
a) $|x| + 4 \geq |x + 4|$
b) $-2|x| + 8 > -2|x + 4|$

21) For the following, find the possible values of k for which there are two distinct solutions:
a) $2|x - 1| + 6 = k$
b) $3|4 - x| - 4 = k$
c) $8 - 2|x + 1| = k$

22) It is given that $f(x) = |x + 4|$
a) Sketch $f(x)$
b) Find the values of k, such that $f(x) = kx$ has 2 distinct solutions.

2.13 Modulus Functions (answers on page 504)

23) It is given that $f(x) = 4 + |x - 2|$
   a) Sketch $f(x)$
   b) Find the values of k such that
   $$f(x) = 4 + kx$$
   has 2 distinct solutions.

24) The diagram shows a sketch of
   $$y = |2x - 5|$$

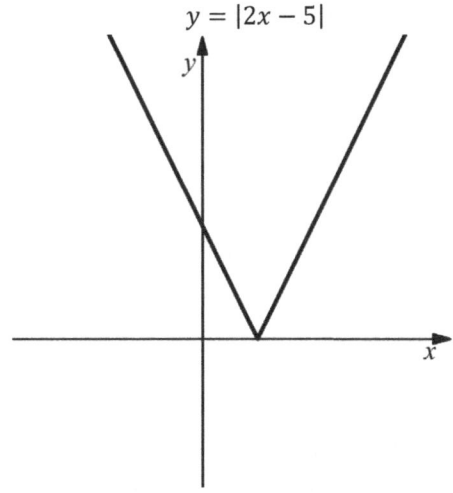

   a) State the coordinates of the points of intersection with the axes.
   b) The graph of
   $$y = kx + 2$$
   has two distinct points of intersection, find the possible values of k
   c) State the solutions of
   $$|2x - 5| = kx + 2$$
   giving your answer in terms of k

25) Describe the transformations that are required to transform the graph of
   $$y = |x + 3|$$
   onto the graph of
   a) $y = 2|x + 3|$
   b) $y = |x - 2|$
   c) $y = |x + 5|$
   d) $y = |2x + 6|$
   e) $y = |2x + 3|$

26) Describe the transformations that are required to transform the graph of $y = |2x - 1|$ onto the graph of
   a) $y = 3|2x - 1|$
   b) $y = |6x - 3|$
   c) $y = |2x - 3|$
   d) $y = |2x + 5| + 4$
   e) $y = |2x - 4| - 1$

27) Describe the transformations that are required to transform the graph of $y = |ax + b|$ onto the graph of
   a) $y = \frac{1}{2}|ax + b|$
   b) $y = \left|\frac{1}{3}ax + b\right|$
   c) $y = |ax + b + 4a| - 5$

28) The diagram shows a sketch of
   $$f(x) = 3|x + 2| - 2$$

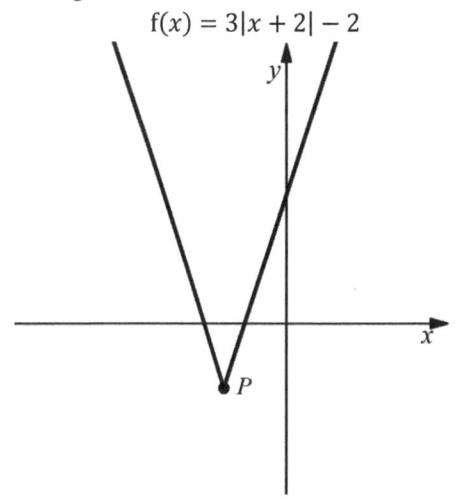

   The point P is at the vertex of $f(x)$
   a) State the coordinates of P
   b) Solve the equation
   $$3|x + 2| - 2 = -3x + 18$$

   A line $y = kx$ where k is a constant.
   c) Find the possible values of k such that it intersects $y = 3|x + 2| - 2$ at least once.

29) A function is given to be
   $$f(x) = k - 3|x + 2|$$
   where $k \geq 6$
   a) Sketch $f(x)$ including labels for all the points of intersection
   b) Find the solutions to
   $$k - 3|x + 2| = x + 4$$
   giving your answer in terms of k

30)
   a) Sketch on a single diagram
   $$y = |x + 2| \text{ and } y = \frac{k}{x}$$
   where k is a positive constant.
   b) Explain why $|x + 2| = \frac{k}{x}$ has exactly one solution for any positive value of k
   c) Find the solution to $x|x + 2| = k$ where $k > 0$, giving your answer in terms of k

263

2.13 Modulus Functions (answers on page 504)

31) The diagram shows a sketch of
$$f(x) = 3|x + 2k|$$
where $k \geq 0$

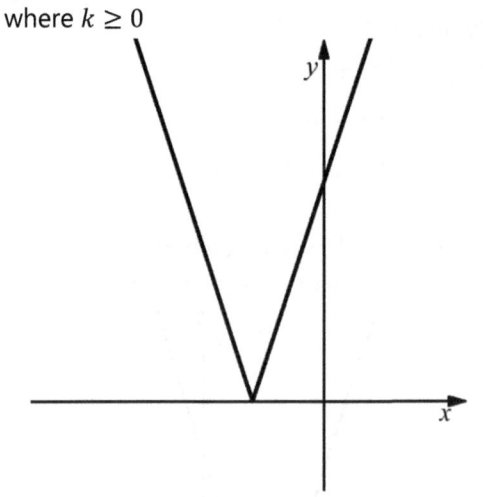

Sketch the graph of each of the following, including the coordinate of the vertex and the y-intercept in each case
a) $y = -3|x + 2k|$
b) $y = 3|x + 2k| + 4$

32) Given that $f(x) = k + |x - 2|$ where $k \geq 0$, sketch each of the following, labelling any intersections with the axes:
a) $y = f(x)$
b) $y = -f(x)$
c) $y = 2f(x)$

33) The diagram shows a sketch of
$$f(x) = |px + q|$$

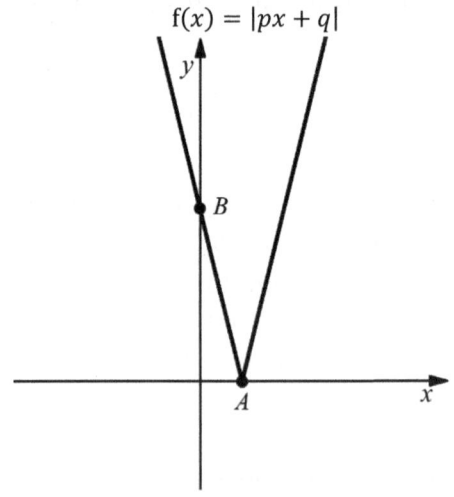

The point A is the minimum of the graph and is on the x-axis. The point B is where the graph intersects with the y-axis.
The line $y = 4x + k$ is parallel to part of the graph and $|px + q| = 4x + k$ has no solutions for $k < -9$
Find the values of p and q

34) State the number of solutions to the equation
$$|x + 2| + |x| = 1$$

Prerequisite knowledge for Q35-Q37: Domain and Range of Functions

35) State the range each of the following, giving your answer in set notation:
a) $f(x) = |x + 1|$
b) $f(x) = -|x + 3|$
c) $f(x) = |x - 1| + 3$
d) $f(x) = 2|x + 4| - 1$
e) $f(x) = 5 - |x - 3|$
f) $f(x) = 1 - 2|x + 7|$

36) It is given that
$$f(x) = 10|x - 2| - 12$$
where $x \geq 0$
a) Sketch $f(x)$
b) State the range of $f(x)$ and domain of $f(x)$
c) Find the values of x such that
$$f(x) \leq -\frac{1}{2}x + \frac{13}{4}$$

37) It is given that
$$f(x) = 3|5 - x| + 5, \qquad x \geq 0$$
a) Sketch $f(x)$
b) State the range of $f(x)$
c) Solve $f(x) = \frac{1}{3}x + 15$

Prerequisite knowledge for Q38-Q39: Composite Functions

38) For $f(x) = 2x + 1$, $g(x) = |x|$ and $h(x) = -|x|$, find:
a) $fg(x)$
b) $gf(x)$
c) $hf(x)$
d) $hfg(x)$

39) For $f(x) = x^2 + 1$, $g(x) = |x|$ and $h(x) = -|x|$, find:
a) $fg(x)$
b) $gf(x)$
c) $hf(x)$
d) $fh(x)$
e) $hg(x)$

2.13 Modulus Functions (answers on page 504)

Prerequisite knowledge for Q40-Q42: The Newton-Raphson Method

40)
  a) Sketch on a single diagram
  $$y = |x+1|$$
  and
  $$y = k \ln x$$
  where k is a negative constant.
  b) Explain why $|x+1| = k \ln x$ has exactly one solution for any negative value of k
  c) Starting with $x_0 = 1$, use the Newton-Raphson method to find the solution to
  $$|x+1| = -5 \ln x$$
  giving your answer to 3 significant figures.

41)
  a) Sketch on a single diagram
  $$y = 2|x+1|$$
  and
  $$y = ke^x$$
  where k is a positive constant and $k > 2$
  b) Explain why $2|x+1| = ke^x$ has exactly one solution.
  c) Starting with $x_0 = -2$, use the Newton-Raphson method to find the solution to
  $$|x+1| = 2e^x$$
  giving your answer to 3 significant figures.

The following questions involve f(|x|)

42) Sketch each of the following:
  a) $y = 2|x| - 5$
  b) $y = 4|x|^2 - 5$
  c) $y = 3|x|^2 - 4|x|$
  d) $y = 2|x|^2 - 3|x| - 5$
  e) $y = -|x|^2 + 4|x| + 5$
  f) $y = \ln|x|$
  g) $y = \frac{1}{|x|^2}$

43) Sketch each of the following:
  a) $y = |(x-2)(x+5)|$
  b) $y = |\ln x|$
  c) $y = |\sin x|$ for $-\pi \leq x \leq 2\pi$
  d) $y = |\cos x|$ for $-\pi \leq x \leq 2\pi$
  e) $y = |x-3| + |x+4|$
  f) $y = |x-2| + |2x+1|$
  g) $y = |x+5| - |2x+7|$

44) Sketch the following for $-360° \leq x \leq 360°$:
  a) $y = \sin|x|$
  b) $y = \tan|x|$

45) It is given that
$$f(x) = x^3 + 6x^2 + 3x - 10$$
Sketch on separate axes:
  a) $y = f(x)$
  b) $y = |f(x)|$
  c) $y = f(|x|)$

46) It is given that
$$f(x) = \cos 2x$$
for $-\pi \leq x \leq \pi$
Sketch on separate axes:
  a) $y = f(x)$
  b) $y = |f(x)|$
  c) $y = f(|x|)$

47) It is given that
$$f(x) = \tan 3x$$
for $-\frac{\pi}{3} \leq x \leq \frac{\pi}{3}$
Sketch on separate axes:
  a) $y = f(x)$
  b) $y = |f(x)|$
  c) $y = f(|x|)$

48) The diagram below shows part of the curve with equation $y = f(x)$

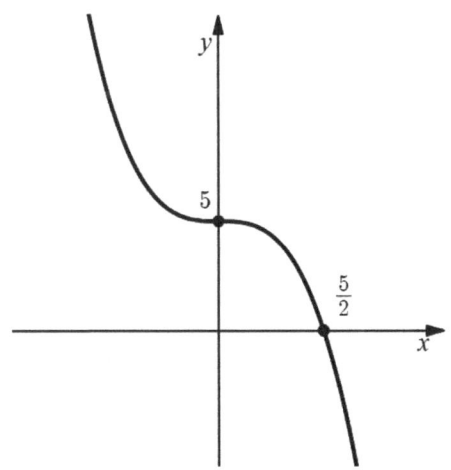

The curve passes through the points $(0, 5)$ and $\left(\frac{5}{2}, 0\right)$ Sketch on separate axes:
  a) $y = |f(x)|$
  b) $y = f(|x|)$
  c) $y = 2f(x)$

2.13 Modulus Functions (answers on page 504)

49) The diagram below shows part of the curve with equation $y = f(x)$

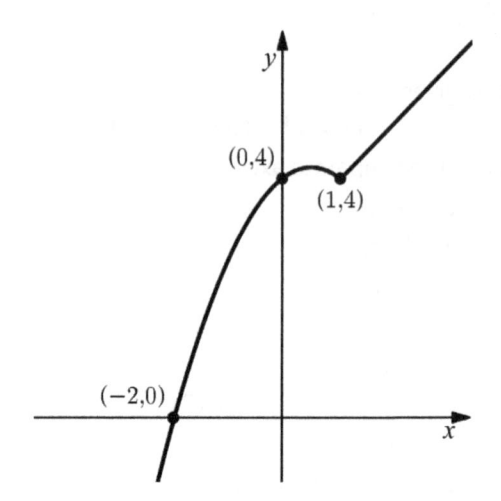

The curve passes through the points $(-2, 0)$, $(0, 4)$ and $(1, 4)$

Sketch on separate axes:
a) $y = |f(x)|$
b) $y = f(|x|)$
c) $y = f(2x)$

50) The diagram below shows part of the curve with equation $y = f(x)$

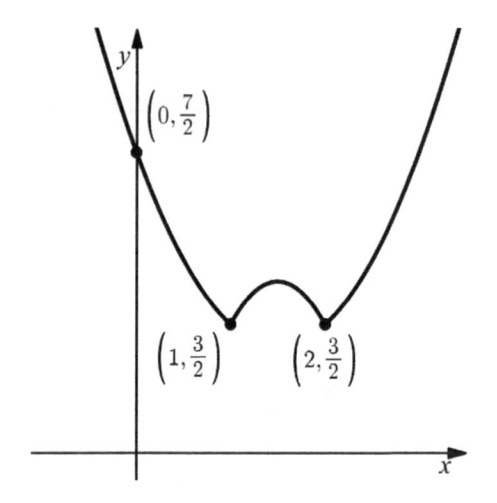

The curve passes through the points $\left(0, \frac{7}{2}\right)$, $\left(1, \frac{3}{2}\right)$ and $\left(2, \frac{3}{2}\right)$

Sketch on separate axes:
a) $y = f(|x|)$
b) $y = f\left(\frac{1}{3}x\right)$
c) $y = 2f(2x)$

51) It is given that
$$f(x) = x^2 + x - 6$$
a) Sketch $y = |f(x)|$
b) Sketch $y = f(|x|)$
c) Find the set of values of x for which $f(|x|) = |f(x)|$

52) It is given that
$$f(x) = \frac{a}{x}$$
for $a > 0$
a) Sketch $y = |f(x)|$
b) Sketch $y = -f(|x|)$
c) Find the set of values of x for which $f(|x|) = |f(x)|$

53) It is given that
$$f(x) = e^x - a$$
for $a > 0$
a) Sketch $y = f(x)$
b) Sketch $y = |f(x)|$
c) Sketch $y = f(|x|)$
d) Find the set of values of x for which $f(|x|) = |f(x)|$

54) Find the set of values for which
a) $x^3 = |x|^3$
b) $x^3 - 4x = |x^3 - 4x|$
c) $\ln x = |\ln x|$
d) $e^{x-1} - 4 = |e^{x-1} - 4|$

55)
a) Write
$$-(x+2)^2 - 1$$
in the form $ax^2 + bx + c$ where a, b and c are integers to be found.
b) Sketch
$$y = -(x+2)^2 - 1$$
and
$$y = -|x|^2 - 4|x| - 5$$
on the same axes
c) Hence find the set of value(s) of x such that
$$-(x+2)^2 - 1 = -|x|^2 - 4|x| - 5$$

56) Solve
$$x^2 - 4x + 5 = -|x|^2 - 4|x| + 6$$

2.14 Trigonometry: Inverse Trigonometric Functions (answers on page 509)

1) Sketch the following graphs, in radians, identifying the coordinates of any endpoints and the equations of any asymptotes.
   a) $y = \arcsin x$
   b) $y = \arccos x$
   c) $y = \arctan x$

2) Sketch the following graphs, in radians, identifying the coordinates of any endpoints.
   a) $y = 3 \arcsin x$
   b) $y = \frac{1}{4}\arcsin x$

3) Sketch the following graphs, in radians, identifying the coordinates of any endpoints.
   a) $y = \arcsin 3x$
   b) $y = \arcsin \frac{2}{5}x$

4) The following are all functions that are single transformations of $y = \arcsin x$. Write down the equation in each case.
   a) [graph with endpoints $(-1, 0)$ and $(1, \pi)$, passing through $(0, \frac{\pi}{2})$]
   b) [graph with endpoints $(-1, \frac{\pi}{2})$ and $(1, -\frac{\pi}{2})$, passing through $(0, 0)$]
   c) [graph with endpoints $(0, -\frac{\pi}{2})$ and $(2, \frac{\pi}{2})$, passing through $(1, 0)$]
   d) [graph with endpoints $(-2, -\pi)$ and $(0, 0)$, passing through $(-1, -\frac{\pi}{2})$]

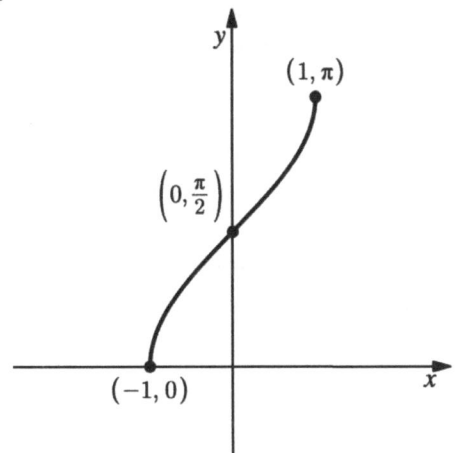

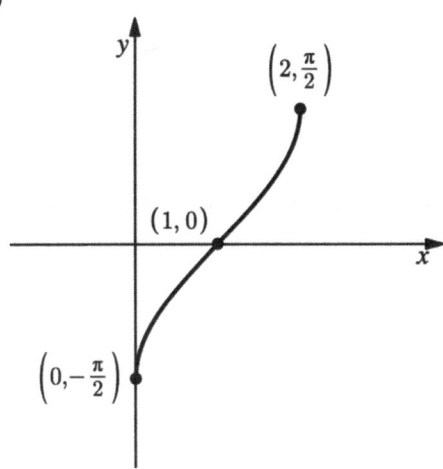

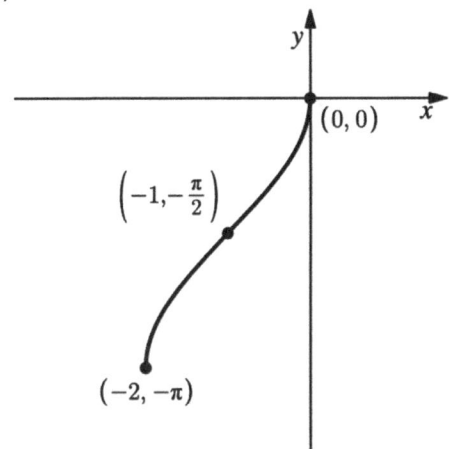

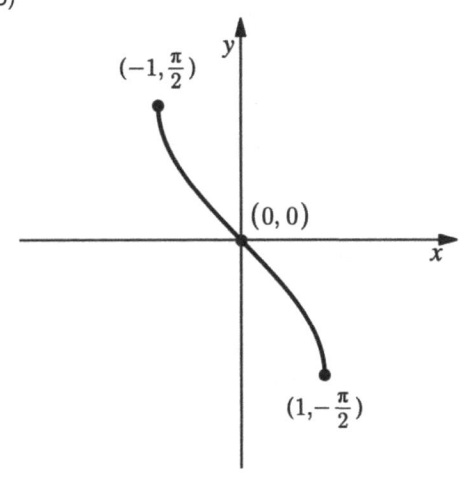

5) Sketch the graph of $y = -\arcsin(x + 1)$ in radians, identifying the coordinates of any endpoints.

6) Sketch the graph of $y = \frac{\pi}{2} - \arcsin x$ in radians, identifying the coordinates of any endpoints.

7) It is given that $f(x) = \arcsin x$, where $-1 \leq x \leq 1$. Find the exact value for which
$$\pi - 3f(2x - 3) = 0$$

Prerequisite knowledge for Q8: Domain and Range and Inverse Functions

8) It is given that $f(x) = 2 - 3\sin x$ for $-\frac{\pi}{2} \leq x \leq \frac{\pi}{2}$
   a) Show that $f^{-1}(x) = \arcsin\left(\frac{2-x}{3}\right)$
   b) State the domain and range of $y = f^{-1}(x)$

9) Sketch the following graphs, in radians, identifying the coordinates of any endpoints.
   a) $y = \frac{5}{2}\arccos x$
   b) $y = \frac{1}{5}\arccos x$

2.14 Trigonometry: Inverse Trigonometric Functions (answers on page 509)

10) Sketch the following graphs, in radians, identifying the coordinates of any endpoints.
    a) $y = \arccos 4x$
    b) $y = \arccos \frac{4}{5}x$

11) The following are all functions that are single transformations of $y = \arccos x$. Write down the equation in each case.
    a) [graph with points $(1, 0)$, $(0, -\frac{\pi}{2})$, $(-1, -\pi)$]
    b) [graph with points $(1, \pi)$, $(0, \frac{\pi}{2})$, $(-1, 0)$]
    c) [graph with points $(-1, \frac{\pi}{2})$, $(0, 0)$, $(1, -\frac{\pi}{2})$]
    d) [graph with points $(-2, 0)$, $(-1, -\frac{\pi}{2})$, $(0, -\pi)$]

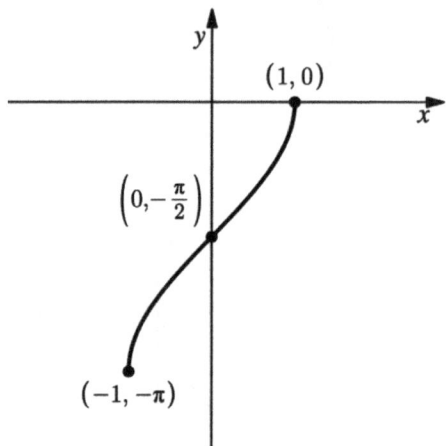

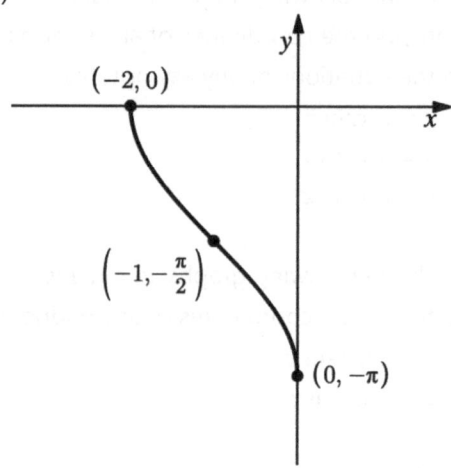

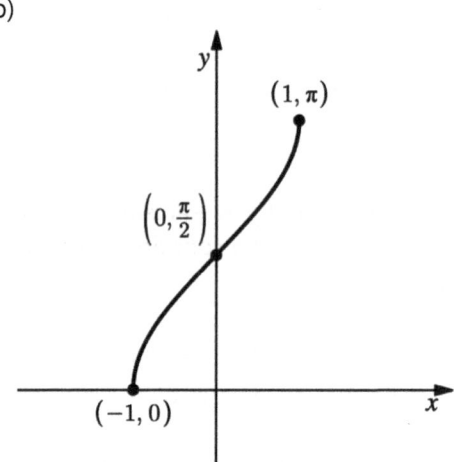

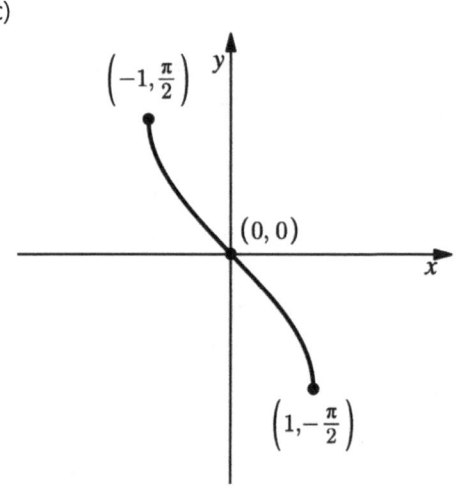

12) Sketch the graph of $y = -\arccos(x - 1)$ in radians, identifying the coordinates of any endpoints.

13) Sketch the graph of $y = \arccos 2x - \pi$ in radians, identifying the coordinates of any endpoints.

14) It is given that $g(x) = \arccos x$, where $-1 \leq x \leq 1$. Find the exact value for which
$$6\pi - 9f(0.2x + 3) = 0$$

15)
    a) Solve the equation $\arccos x = \frac{\pi}{3}$
    b) Hence find the exact value of $\arcsin x$

16) Sketch the following graphs, in radians, identifying the equations of any asymptotes:
    a) $y = 8 \arctan x$
    b) $y = \frac{1}{6} \arctan x$

17) Sketch the following graphs, in radians, identifying the equations of any asymptotes:
    a) $y = 2 + 3 \arctan x$
    b) $y = 5 - 2 \arctan x$

18) Sketch the following graphs, in radians, identifying any endpoints and the equations of any asymptotes:
    a) $y = 3 \arcsin x + \pi$
    b) $y = \arccos\left(\frac{1}{2}x - 2\right)$
    c) $y = 2 \arctan x - \frac{\pi}{3}$
    d) $y = \arcsin(4x + 5)$
    e) $y = \frac{2\pi}{3} - \arccos x$
    f) $y = 3\pi - \arctan x$

2.14 Trigonometry: Inverse Trigonometric Functions (answers on page 509)

19) Write down **two** suitable equations for the following graphs:

a) Graph passing through $\left(1, \frac{5\pi}{4}\right)$, $\left(0, \frac{\pi}{4}\right)$, $\left(-1, -\frac{3\pi}{4}\right)$

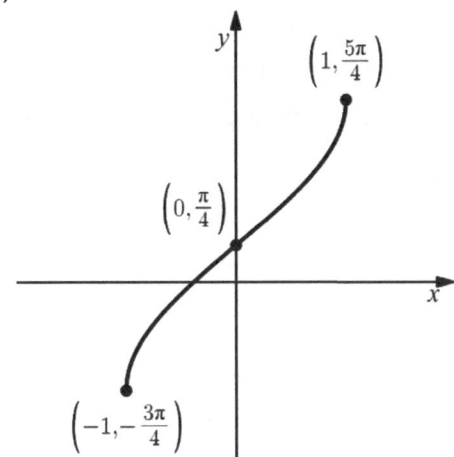

b) Graph passing through $\left(-\frac{3}{8}, \pi\right)$, $\left(\frac{1}{8}, \frac{\pi}{2}\right)$, $\left(\frac{5}{8}, 0\right)$

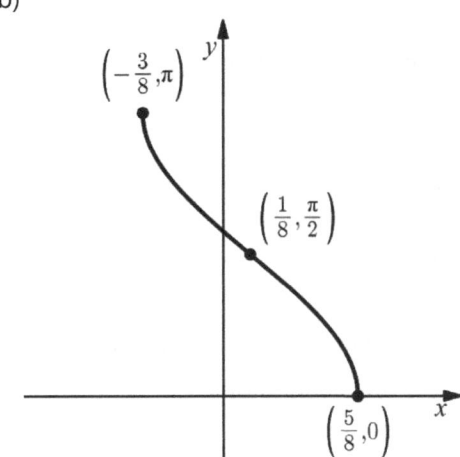

c) Graph passing through $\left(\frac{1}{18}, \frac{\pi}{2}\right)$, $\left(-\frac{5}{18}, 0\right)$, $\left(-\frac{11}{18}, -\frac{\pi}{2}\right)$

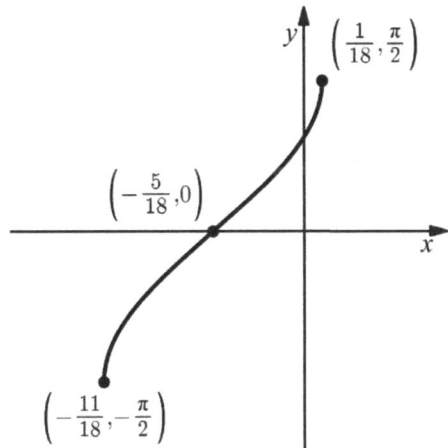

20) Solve the equation $\arcsin x = \arccos x$

Prerequisite knowledge for Q21: Double Angle Formulae

21)
   a) Solve the equation $\arcsin x = 2 \arccos x$
   b) Solve the equation $2 \arcsin x = \arccos x$

22) Given $\arcsin x = \frac{\pi}{5}$, show that $x = \sqrt{\frac{5-\sqrt{5}}{8}}$

The remaining questions in this exercise are all in radians and are to be completed without a calculator.

23) Find
   a) $\arcsin 1$
   b) $\arccos(-1)$
   c) $\arcsin 0$
   d) $\arccos 1$

24) Find
   a) $\arcsin \frac{\sqrt{2}}{2}$
   b) $\arccos \frac{\sqrt{2}}{2}$
   c) $\arcsin\left(-\frac{\sqrt{2}}{2}\right)$
   d) $\arccos\left(-\frac{\sqrt{2}}{2}\right)$

25) Find
   a) $\arctan 1$
   b) $\arctan(-1)$
   c) $\arctan \sqrt{3}$
   d) $\arctan\left(-\frac{\sqrt{3}}{3}\right)$

26) Find
   a) $\arcsin\left(\sin \frac{\pi}{2}\right)$
   b) $\arccos\left(\cos\left(-\frac{\pi}{2}\right)\right)$
   c) $\arcsin\left(\sin \frac{3\pi}{2}\right)$
   d) $\arccos\left(\cos\left(-\frac{3\pi}{2}\right)\right)$
   e) $\arcsin\left(\sin \frac{\pi}{3}\right)$
   f) $\arccos\left(\cos \frac{2\pi}{3}\right)$
   g) $\arcsin\left(\sin \frac{5\pi}{3}\right)$
   h) $\arccos\left(\cos\left(-\frac{\pi}{3}\right)\right)$
   i) $\arctan\left(\tan \frac{\pi}{4}\right)$
   j) $\arctan\left(\tan \frac{2\pi}{3}\right)$

27) Find
   a) $\sin\left(\arcsin \frac{\sqrt{2}}{2}\right)$
   b) $\cos\left(\arccos\left(-\frac{\sqrt{2}}{2}\right)\right)$
   c) $\sin\left(\arccos \frac{1}{2}\right)$
   d) $\cos\left(\arcsin \frac{1}{2}\right)$
   e) $\sin\left(\arccos \frac{\sqrt{3}}{2}\right)$
   f) $\cos\left(\arcsin\left(-\frac{\sqrt{3}}{2}\right)\right)$
   g) $\tan\left(\arcsin\left(-\frac{1}{2}\right)\right)$

2.15 Trigonometry: Reciprocal Trigonometric Functions (answers on page 511)

1) Sketch the following graphs for $0° < x < 360°$:
   a) $y = \csc x$
   b) $y = \sec x$
   c) $y = \cot x$

2) Sketch the following graphs for $0 < x < 2\pi$:
   a) $y = \csc x$
   b) $y = \sec x$
   c) $y = \cot x$

3) Sketch the following graphs for $0° < x < 360°$:
   a) $y = 2\csc x$
   b) $y = \frac{1}{3}\sec x$
   c) $y = -\cot x$
   d) $y = \csc(x - 90°)$
   e) $y = \sec(x + 60°)$
   f) $y = \cot(x - 270°)$

4) Sketch the following graphs for $0 < x < 2\pi$:
   a) $y = \frac{1}{4}\csc x$
   b) $y = \frac{4}{3}\sec x$
   c) $y = \cot(-x)$
   d) $y = \csc\left(x + \frac{\pi}{2}\right)$
   e) $y = \sec\left(x - \frac{\pi}{3}\right)$
   f) $y = \cot\left(x + \frac{2\pi}{3}\right)$

Prerequisite knowledge for Q5-Q6:
Combinations of Graph Transformations

5) Sketch the following graphs for $0° < x < 360°$:
   a) $y = 1 + 3\csc x$
   b) $y = 2 - \sec x$
   c) $y = -\cot(x - 30°)$
   d) $y = \csc(2x - 90°)$
   e) $y = \sec(2x + 60°)$
   f) $y = \cot\left(\frac{1}{2}x - 30°\right)$

6) Sketch the following graphs for $0 < x < 2\pi$:
   a) $y = \frac{1}{2}\csc x - 1$
   b) $y = \frac{1}{3}\sec x - 2$
   c) $y = \cot(-2x)$
   d) $y = \csc\left(\frac{1}{2}x + \frac{\pi}{4}\right)$
   e) $y = \sec\left(\frac{1}{2}x - \frac{\pi}{3}\right)$
   f) $y = \cot\left(2x + \frac{\pi}{3}\right)$

7) Solve the following equations:
   a) $\csc\theta = 2$, $0° < \theta < 360°$
   b) $\sec\theta = 3$, $0 < \theta < 2\pi$
   c) $\cot\theta = 4$, $0° < \theta < 360°$
   d) $\csc\theta = -\frac{5}{4}$, $0 < \theta < 2\pi$
   e) $\sec\theta = -\frac{9}{2}$, $0° < \theta < 360°$
   f) $\cot\theta = -\frac{18}{5}$, $0 < \theta < 2\pi$

8) State the values of x for which $y = \csc x$ is an increasing function, for $0 < x < 2\pi$

9) State the values of x for which $y = \sec x$ is a decreasing function, for $0° < x < 360°$

10) It is given that $\csc x = \frac{7}{2}$ and x is acute. Find the exact values of:
    a) $\sin x$
    b) $\cos x$
    c) $\tan x$
    d) $\sec x$
    e) $\cot x$

11) It is given that $\sec x = \frac{9}{4}$ and x is acute. Find the exact values of:
    a) $\cos x$
    b) $\sin x$
    c) $\tan x$
    d) $\csc x$
    e) $\cot x$

12) It is given that $\cot x = \frac{12}{11}$ and x is acute. Find the exact values of:
    a) $\tan x$
    b) $\sin x$
    c) $\cos x$
    d) $\csc x$
    e) $\sec x$

13) It is given that $\csc x = \frac{10}{9}$ and x is obtuse. Find the exact values of:
    a) $\sin x$
    b) $\cos x$
    c) $\tan x$
    d) $\sec x$
    e) $\cot x$

2.15 Trigonometry: Reciprocal Trigonometric Functions (answers on page 511)

14) It is given that $\sec x = -\frac{12}{5}$ and $180° < x < 270°$. Find the exact values of:
   a) $\cos x$
   b) $\sin x$
   c) $\tan x$
   d) $\cosec x$
   e) $\cot x$

15) It is given that $\cot x = -\frac{2}{7}$ and $270° < x < 360°$. Find the exact values of:
   a) $\tan x$
   b) $\sin x$
   c) $\cos x$
   d) $\cosec x$
   e) $\sec x$

16) It is given that $8\sec^2 x - 14\sec x + 3 = 0$ and x is acute. Find the exact values of:
   a) $\sec x$
   b) $\cos x$
   c) $\sin x$
   d) $\tan x$
   e) $\cosec x$
   f) $\cot x$

17) Without using any calculus, state the maximum value of the curve with equation $y = \frac{2}{\cosec^2 x + 2}$

18) Write the following trigonometric expressions in terms of $\tan x$, $\cosec x$, $\sec x$ and $\cot x$ only:
   a) $\frac{\sin x}{\cos^2 x}$
   b) $\frac{\cos x}{\sin^2 x}$

19) Simplify the following trigonometric expressions as far as possible:
   a) $\cosec x \sin x$
   b) $\sec x \cos x$
   c) $\cot x \tan x$

20) Simplify the following trigonometric expressions as far as possible, writing your answers in terms of $\sin x$ and $\cos x$ only:
   a) $\cosec x \sec x$
   b) $\cosec x \cot x$
   c) $\sec x \cot x$
   d) $\cosec x \sec x \cot x$

21) Simplify the following trigonometric expressions as far as possible:
   a) $\frac{\cosec x}{\sec x}$
   b) $\frac{\cosec x}{\cot x}$
   c) $\frac{\sec x}{\cosec x}$
   d) $\frac{\sec x}{\cot x}$
   e) $\frac{\cot^2 x}{\sec x}$
   f) $\frac{\cot^2 x \sin^3 x}{\cosec x \sec x}$

22) Solve the following equations:
   a) $\cosec \theta = \sec \theta$, $0° < \theta < 360°$
   b) $\cot \theta = -\sec \theta$, $0 < \theta < 2\pi$
   c) $\sec \theta = \cot \theta$, $0° < \theta < 360°$
   d) $\cot \theta = \tan \theta$, $0 < \theta < 2\pi$

23)
   a) Sketch the graphs of $y = \cosec x$ and $y = 2\sec x$ on the same axes in radians between 0 and 2π
   b) Determine the coordinates of where the two graphs intersect.

24) State the values of k for which $\sec x - 3 + k = 0$ has real solutions.

25) Starting from $\sin^2 x + \cos^2 x \equiv 1$
   a) derive an identity for $\cot x$ and $\cosec x$,
   b) derive an identity for $\tan x$ and $\sec x$

26) Simplify the following trigonometric expressions:
   a) $\frac{1+\cot^2 x}{\cosec^2 x}$
   b) $\frac{1+\tan^2 x}{\sec^2 x}$
   c) $\frac{1+\tan^2 x}{1+\cot^2 x}$
   d) $\frac{2+2\cot^2 x}{5+5\tan^2 x}$

27) Solve the following equations:
   a) $\cot^2 \theta + 3\cosec \theta - 3 = 0$, $0° < \theta < 360°$
   b) $\tan^2 \theta - \sec \theta - 11 = 0$, $0 < \theta < 2\pi$
   c) $\cosec^2 \theta + 5\cot \theta - 15 = 0$, $0° < \theta < 360°$
   d) $\sec^2 \theta + 6\tan \theta - 28 = 0$, $0 < \theta < 2\pi$

2.15 Trigonometry: Reciprocal Trigonometric Functions (answers on page 511)

28) Solve the following equations:
   a) $3\csc^2\theta + 2\cot\theta - 4 = 0$,
      $0° < \theta < 360°$
   b) $4\sec^2\theta - 4\tan\theta - 19 = 0$,
      $0 < \theta < 2\pi$
   c) $6\cot^2\theta + 23\csc\theta + 13 = 0$,
      $0° < \theta < 360°$
   d) $18\tan^2\theta + 9\sec\theta - 2 = 0$,
      $0 < \theta < 2\pi$

29) Show that $3\tan^2 x + 4k\sec x + \frac{4k^2}{3} + 3$ can be written in the form $3\left(\sec x + \frac{2k}{3}\right)^2$

30) It is given that $10\tan^2 x + 17\sec x + 16 = 0$ and x is obtuse. Find the exact values of:
   a) $\sec x$
   b) $\cos x$
   c) $\sin x$
   d) $\tan x$
   e) $\csc x$
   f) $\cot x$

31) It is given that $16\cot^2 x + 32\csc x + 31 = 0$ and $\frac{3\pi}{2} < x < 2\pi$. Find the exact values of:
   a) $\csc x$
   b) $\sin x$
   c) $\cos x$
   d) $\tan x$
   e) $\sec x$
   f) $\cot x$

32) Prove there are no real solutions to the equation $12\cot^2\theta - 5\csc\theta + 10 = 0$

33) It is given that $p(x) = 12x^3 + 8x^2 - 3x - 2$
   a) Show that $(3x + 2)$ is a factor of $p(x)$
   b) Write $p(x)$ as a product of three linear factors.
   c) Hence prove that there are no real solutions to the equation
      $12\csc^3\theta + 8\cot^2\theta - 3\csc\theta + 6 = 0$

34) Prove the following trigonometric identities:
   a) $\cot x + \cos x \equiv \cot x (1 + \sin x)$
   b) $\cot x + \sec x \equiv \csc x \sec x (\sin x + \cos^2 x)$
   c) $\cot x + \sin x \equiv \csc x (\sin^2 x + \cos x)$
   d) $\cot x + \tan x \equiv \csc x \sec x$

35) Prove the following trigonometric identities:
   a) $\sec x - \cos x \equiv \sin x \tan x$
   b) $\csc x - \sin x \equiv \cos x \cot x$
   c) $\tan x - \cot x \equiv \csc x \sec x (2\sin^2 x - 1)$
   d) $\tan x - \csc x \equiv \csc x (\sin x \tan x - 1)$

36) Prove the following trigonometric identities:
   a) $\sin^2 x - \cot^2 x \equiv \csc^2 x (\sin^4 x - \cos^2 x)$
   b) $\cos^2 x - \cot^2 x \equiv -\cos^2 x \cot^2 x$
   c) $\sec^2 x - \csc^2 x \equiv \csc^2 x \sec^2 x (2\sin^2 x - 1)$
   d) $\tan^2 x - \cot^2 x \equiv \csc^2 x \sec^2 x (2\sin^2 x - 1)$

37) Prove the following trigonometric identities:
   a) $(\sec x + \cot x)^2 \equiv \tan^2 x + 2\csc x + \csc^2 x$
   b) $(\sec x + \csc x)^2 \equiv \tan^2 x + \cot^2 x + 2(\tan x + \cot x + 1)$
   c) $(\csc x + \cot x)^2 \equiv 2\cot^2 x + 1 + 2\cos x (\cot^2 x + 1)$
   d) $(\tan x + \sec x)^2 \equiv 2\sec^2 x (\sin x + 1) - 1$
   e) $(\tan x + \sec x)^2 \equiv \frac{1 + \sin x}{1 - \sin x}$

38)
   a) Prove the following trigonometric identity:
      $\frac{\sin\theta}{1 - \cos\theta} - \csc\theta \equiv \cot\theta$
   b) Hence solve the equation
      $\frac{\sin\theta}{1 - \cos\theta} = 10 + \cot\theta$
      for $0 < \theta < 2\pi$

Prerequisite knowledge for Q39: Double Angle Formulae

39) Solve the equation $\csc\theta = 10\cos\theta$ for $0 < \theta < 2\pi$

2.16 Trigonometry: Compound Angle Formulae (answers on page 514)

1) The diagram below shows the triangle ABC. It is given that angle ACM is x degrees and angle BCM is y degrees. The length of BC is a and the length of AC is b

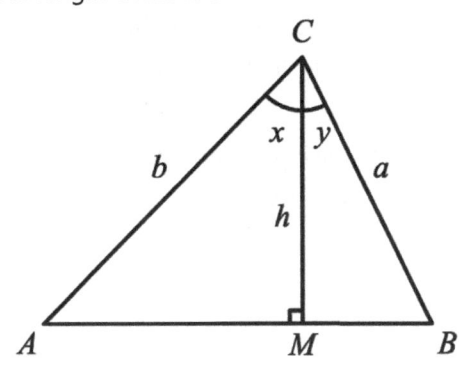

   a) Write h in terms of b and x
   b) Write h in terms of a and y
   c) Find the area of triangles ACM and BCM.
   d) Find the area of triangle ABC.
   e) Hence show that
   $$\sin(x+y) = \sin x \cos y + \cos x \sin y$$

2) The diagram below shows the rectangle $ABDF$. It is given that angle CAE is x degrees and angle EAF is y degrees. It is also given that the length AC is 1 unit.

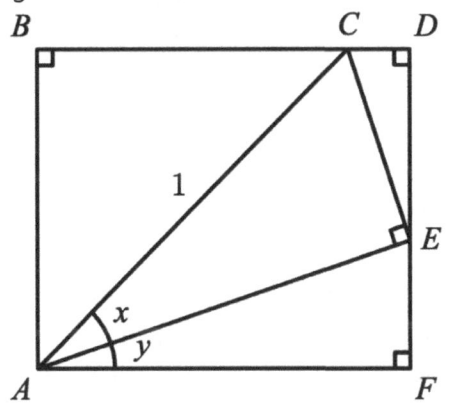

   a) Find the lengths of AE & CE in terms of x
   b) Find the angles ACE, AEF, CED, ECD, BCA and BAC in terms of x and y
   c) Find the lengths of BC and AB in terms of x and y
   d) Find the lengths of AF and EF in terms of x and y
   e) Find the lengths of DE and CD in terms of x and y
   f) Use the fact that $AB = DE + EF$ and $BC = AF - CD$ to write down two compound angle formulae.

3) Given $\sin(x+y) = \sin x \cos y + \cos x \sin y$, show that $\sin(x-y) = \sin x \cos y - \cos x \sin y$

4) Given $\cos(x+y) = \cos x \cos y - \sin x \sin y$, show that $\cos(x-y) = \cos x \cos y + \sin x \sin y$

5) By writing $\tan(x+y)$ as $\frac{\sin(x+y)}{\cos(x+y)}$, show that $\tan(x+y) = \frac{\tan x + \tan y}{1 - \tan x \tan y}$

6) Given $\tan(x+y) = \frac{\tan x + \tan y}{1 - \tan x \tan y}$, show that $\tan(x-y) = \frac{\tan x - \tan y}{1 + \tan x \tan y}$

7) Write the following in the form $\sin k$, where k is an acute angle:
   a) $\sin 30° \cos 45° + \cos 30° \sin 45°$
   b) $\sin 35° \cos 20° - \cos 35° \sin 20°$
   c) $\sin 70° \cos 50° - \cos 70° \sin 50°$
   d) $\sin 70° \cos 50° + \cos 70° \sin 50°$

8) Write the following in the form $\cos k$, where k is an acute angle:
   a) $\cos 50° \cos 30° + \sin 50° \sin 30°$
   b) $\cos 68° \cos 13° + \sin 68° \sin 13°$
   c) $\cos 10° \cos 54° - \sin 10° \sin 54°$
   d) $\cos 300° \cos 50° - \sin 300° \sin 50°$

9)
   a)
      i) Write down the exact values of $\sin 30°$, $\sin 45°$ and $\sin 60°$
      ii) Write down the exact values of $\cos 30°$, $\cos 45°$ and $\cos 60°$
   b) Hence use a compound angle formula to find the exact value of:
      i) $\sin 75°$
      ii) $\sin 15°$
      iii) $\cos 105°$
      iv) $\cos 15°$

10)
   a) Write down the exact values of $\tan 30°$, $\tan 45°$ and $\tan 60°$
   b) Hence use a compound angle formula to find the exact value of:
      i) $\tan 75°$
      ii) $\tan 15°$

2.16 Trigonometry: Compound Angle Formulae (answers on page 514)

11) Use the compound angle formulae to simplify each of these:
    a) $\sin(x + 90°)$
    b) $\cos(x - 90°)$
    c) $\cos(x - 180°)$
    d) $\sin(x - 270°)$

12) Write the following in the form $\sin k$, where k is an acute angle:
    a) $\cos 40° \cos 32° - \sin 40° \sin 32°$
    b) $\cos 128° \cos 125° + \sin 128° \sin 125°$

13) Write the following in the form $\cos k$, where k is an acute angle:
    a) $\sin 96° \cos 5° - \cos 96° \sin 5°$
    b) $\sin 25° \cos 42° + \cos 25° \sin 42°$

14) It is given that $\cos x = \frac{2}{9}$ and $\cos y = \frac{6}{7}$ and that x and y are acute.
    Find the exact value of:
    a) $\cos(x - y)$
    b) $\sin(x - y)$

15) It is given that $\tan 85° = b$. Show that
$$\tan 130° = \frac{1+b}{1-b}$$

16) It is given that $\cos 55° = m$. Show that
$$\cos 5° = \frac{m + \sqrt{3 - 3m^2}}{2}$$

17) Find the minimum value of the following expressions and the positive value of θ, in degrees, for which the minimum first occurs:
    a) $\sin \theta \cos 60° + \cos \theta \sin 60°$
    b) $2 \cos \theta \cos 35° + 2 \sin \theta \sin 35°$
    c) $4 \sin \theta \cos 42° - 4 \cos \theta \sin 42°$

Prerequisite knowledge for Q18-Q19: Radians

18) Find the minimum value of the following expressions and the positive value of θ, in radians, for which the minimum first occurs:
    a) $\sin \theta \cos \frac{\pi}{4} + \cos \theta \sin \frac{\pi}{4}$
    b) $3 \cos \theta \cos \frac{5\pi}{4} + 3 \sin \theta \sin \frac{5\pi}{4}$
    c) $-2 \cos \theta \sin 1.6 + 2 \sin \theta \cos 1.6$

19) Find the maximum value of the following expressions and the positive value of θ, in radians, for which the maximum first occurs:
    a) $\sin 3\theta \cos 2\theta + \cos 3\theta \sin 2\theta$
    b) $9 \cos 5\theta \cos 2\theta + 9 \sin 5\theta \sin 2\theta$
    c) $-\frac{2}{3} \cos 8\theta \sin 2\theta + \frac{2}{3} \sin 8\theta \cos 2\theta$

20) Solve the following equations:
    a) $\sin \theta \cos 20° + \cos \theta \sin 20° = 0.5$ for $0° \leq \theta < 360°$
    b) $\cos \theta \cos 40° + \sin \theta \sin 40° = 0.34$ for $0° \leq \theta < 360°$

21) Write the following in the form $\tan k$, where k is an acute angle:
    a) $$\frac{\tan 200° + \tan 34°}{1 - \tan 200° \tan 34°}$$
    b) $$\frac{\tan 18° - \tan 130°}{1 + \tan 18° \tan 130°}$$

22) Solve the following equations:
    a) $$\frac{\tan \theta + \tan 20°}{1 - \tan \theta \tan 20°} = 4$$ for $0° \leq \theta < 360°$
    b) $$\frac{\tan 3\theta - \tan 33°}{1 + \tan 3\theta \tan 33°} = -6$$ for $0° \leq \theta < 120°$

23) Simplify the following expressions:
    a) $\sin(x + 45°) + \cos(x + 45°)$
    b) $\sin(x + 135°) - \cos(x + 135°)$
    c) $\sin(x - 75°) - \cos(x - 75°)$

Prerequisite knowledge for Q24: Radians

24) Simplify the following expressions:
    a) $\sin\left(x - \frac{\pi}{3}\right) + \cos\left(\frac{\pi}{3} - x\right)$
    b) $\sin\left(\frac{\pi}{3} - x\right) - \cos\left(\frac{4\pi}{3} - x\right)$
    c) $-\sin\left(\frac{\pi}{4} - x\right) + \cos\left(\frac{2\pi}{3} - x\right)$

25) Solve the following equations:
    a) $\sin(\theta + 30°) = \cos \theta$ for $0° < \theta < 360°$
    b) $\sin(\theta + 30°) = \sin \theta$ for $0° < \theta < 360°$
    c) $\cos(\theta + 30°) = \sin \theta$ for $0° < \theta < 360°$

2.16 Trigonometry: Compound Angle Formulae (answers on page 514)

26) Solve the following equations:
    a) $\sin(\theta - 30°) = \cos(\theta + 45°)$
       for $0° < \theta < 360°$
    b) $\cos(\theta - 60°) = \cos(\theta - 45°)$
       for $0° < \theta < 360°$
    c) $\cos(\theta - 120°) = \sin(\theta + 60°)$
       for $0° < \theta < 360°$

27) In the diagram below, angle ABC is x degrees and angle CBD is y degrees. The lengths of AC and CD are $\sqrt{2}$ and $\sqrt{5}$ respectively.
    Show that $DE = \sqrt{2} + \sqrt{5}\cos x$

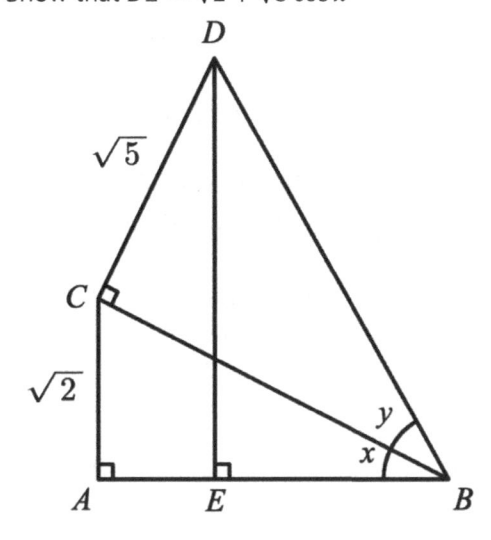

28) Prove the following trigonometric identity
$$\frac{\cos(x+y)}{\cos x \cos y} \equiv 1 - \tan x \tan y$$

29) Solve the equation
$$\frac{\tan\theta - \tan 30°}{1 + \tan\theta \tan 30°} = 2$$
for $0° < \theta < 360°$

30)
   a) Solve the equation
      $3\cos(x - 45°) = 2\sin(x + 60°)$
      for $0° < x < 360°$
   b) Hence solve the equation
      $3\cos(3\theta - 45°) = 2\sin(3\theta + 60°)$
      for $0° < \theta < 120°$

31) It is given that $\sin x = \frac{2}{\sqrt{14}}$ and $\cos y = \frac{3}{7}$ and that x is obtuse and y is acute. Show that the exact value of $\tan(x - y)$ is $\frac{13\sqrt{10}}{5}$

32) It is given that $\tan x = \frac{1}{2}$ and that x is an acute angle. Given that $\tan(x + y) = 13$, find the exact value of $\tan y$

33) Prove the following trigonometric identity
$2\sin(x + 45°)\sin(x - 45°) \equiv \sin^2 x - \cos^2 x$

34) The diagram below shows the rod OAB which is pivoted at O. The length of OA is 10 units and the length of AB is 2 units. Taking O as the point with coordinates $(0,0)$, show that the coordinates of B are
$(9\cos\theta + \sqrt{3}\sin\theta, 9\sin\theta - \sqrt{3}\cos\theta)$

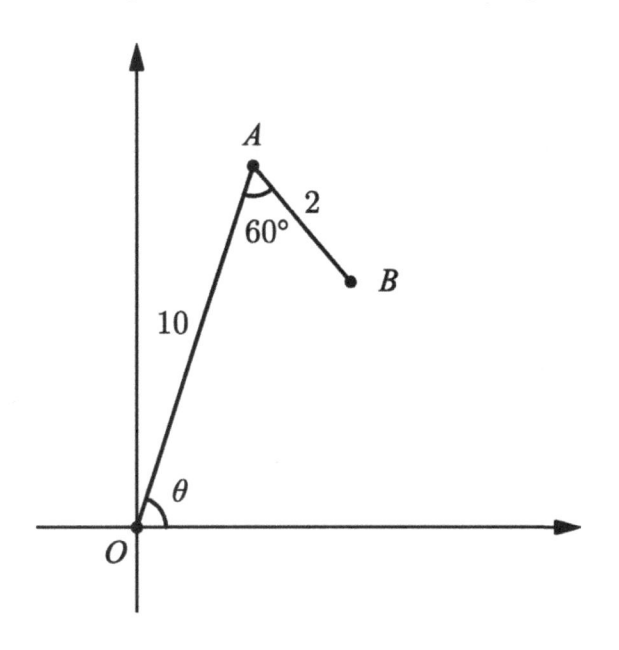

Prerequisite knowledge: Small Angle Approximation

35) By considering compound angle formulae, show that
$$\cos A - \cos B = -2\sin\left(\frac{A+B}{2}\right)\sin\left(\frac{A-B}{2}\right)$$
Hence, prove by first principles that
$$\frac{d}{dx}(\cos x) = -\sin x$$

36) By considering compound angle formulae, show that
$$\sin A - \sin B = 2\cos\left(\frac{A+B}{2}\right)\sin\left(\frac{A-B}{2}\right)$$
Hence, prove by first principles that
$$\frac{d}{dx}(\sin x) = \cos x$$

2.17 Trigonometry: Double Angle Formulae (answers on page 515)

1) Use the Compound Angle Formulae
   $\sin(A \pm B) = \sin A \cos B \pm \cos A \sin B$ to find the Double Angle Formula for $\sin 2x$

2) Use the Compound Angle Formulae
   $\cos(A \pm B) = \cos A \cos B \mp \sin A \sin B$ to find a simplified form for $\cos 2x$

3) Write down $\cos 2x$ just in terms of $\cos x$

4) Write down $\cos 2x$ just in terms of $\sin x$

5) Use the Compound Angle Formulae
   $$\tan(A \pm B) = \frac{\tan A \pm \tan B}{1 \mp \tan A \tan B}$$
   to find a simplified form for $\tan 2x$

6) Use $\tan 2x \equiv \frac{\sin 2x}{\cos 2x}$ to find the same result found in question 5)

7) Simplify the following expressions:
   a) $2 \sin 2x \cos 2x$
   b) $2 \cos^2 3x - 1$
   c) $1 - 2 \sin^2 4x$
   d) $4 \sin \frac{1}{2}x \cos \frac{1}{2}x$
   e) $6 \cos^2 \frac{1}{3}x - 6 \sin^2 \frac{1}{3}x$
   f) $6 \cos^2 \frac{3}{2}x - 3$

8) Solve the following equations:
   a) $\sin 2\theta - 2 \sin \theta = 0$ for $0° \leq \theta < 360°$
   b) $\sin 2\theta + 5 \sin \theta = 0$ for $0 \leq \theta < 2\pi$
   c) $3 \sin 2\theta = 2 \cos \theta$ for $0° \leq \theta < 360°$
   d) $8 \sin 2\theta + 5 \cos \theta = 0$ for $0 \leq \theta < 2\pi$

9) Solve the following equations:
   a) $\cos 2\theta + \cos \theta = 0$ for $0° \leq \theta < 360°$
   b) $3 \cos 2\theta = 7 \cos \theta$ for $0 \leq \theta < 2\pi$
   c) $2 \cos 2\theta + 2 \sin \theta = 0$ for $0° \leq \theta < 360°$
   d) $2 \cos 2\theta = 9 \sin \theta$ for $0 \leq \theta < 2\pi$

Prerequisite knowledge for Q10: Reciprocal Trigonometric Functions

10) Solve the following equations:
    a) $2 \sin 2\theta \csc \theta = 1$ for $0° < \theta < 360°$
    b) $5 \sin 2\theta \sec \theta = 2$ for $0 < \theta < 2\pi$
    c) $\cos 2\theta \csc \theta = 3$ for $0° < \theta < 360°$
    d) $9 \cos 2\theta \sec \theta = 4$ for $0 < \theta < 2\pi$

11) It is given that $\sin x = \frac{2}{5}$ and that x is acute. Find the exact value of $\sin 2x$

12) It is given that $\sin x = \frac{3}{8}$ and that x is acute. Find the exact value of $\cos 2x$

13) It is given that $\cos x = -\frac{2}{9}$ and that x is obtuse. Find the exact value of $\tan 2x$

14) Given that $10 \sin^2 x + 5 \sin 2x = 4$, show that $3 \tan^2 x + 5 \tan x - 2 = 0$

15) Solve the following equations:
    a) $34 \sin^2 \theta - \sin 2\theta = 30$ for $0° \leq \theta < 360°$
    b) $22 \cos^2 \theta - 13 \sin 2\theta = 10$ for $0 < \theta < 2\pi$

Prerequisite knowledge for Q16-Q21: Reciprocal Trigonometric Functions

16) Prove the following trigonometric identities:
    a) $\tan 2x + \cot x \equiv \cot x \sec 2x$
    b) $\tan 2x + \sin 2x \equiv \tan 2x (\cos 2x + 1)$
    c) $\tan 2x + \cot 2x \equiv 2 \csc 4x$

17) Prove the following trigonometric identities:
    a) $\cot 2x - \cos 2x \equiv \cot 2x (1 - \sin 2x)$
    b) $\cot 2x + \csc 2x \equiv \csc 2x (1 + \cos 2x)$
    c) $\cot 2x + \csc 2x \equiv \cot x$

18) Prove the following trigonometric identity:
    $\tan 2x + \csc 2x$
    $\equiv 2 \csc 4x (\sin^2 2x + \cos 2x)$

19) Prove the following trigonometric identities:
    a) $$\frac{2 \tan 2x}{1 + \tan^2 2x} \equiv \sin 4x$$
    b) $(\sin 2x + \cos 2x)^2 \equiv 1 + \sin 4x$
    c) $$\frac{2 \sin x}{\cos 2x + 1} - \frac{2 \sin x}{\cos 2x - 1} \equiv \csc x \sec^2 x$$

20) Prove the following trigonometric identities:
    a) $$\frac{1 - \sin 2x}{1 + \sin 2x} \equiv \frac{2}{(\sin x + \cos x)^2} - 1$$
    b) $$\frac{1 - \cos 2x}{1 + \sin 2x} \equiv \frac{2 \sin^2 x}{(\sin x + \cos x)^2}$$
    c) $$\frac{1 + \cos 2x - \sin 2x}{1 + \sin 2x - \cos 2x} \equiv \frac{\cos x (\cot x - 1)}{\sin x + \cos x}$$

2.17 Trigonometry: Double Angle Formulae (answers on page 515)

21)
a) Prove the following trigonometric identity:
$$\operatorname{cosec}^2 x \equiv \frac{\cos 2x + 3\sin^2 x + \cos^2 x}{2\sin^2 x}$$
b) Hence solve
$$(\cot^2 \theta - 1)\operatorname{cosec}^2 \theta = \frac{\cos 2\theta + 3\sin^2 \theta + \cos^2 \theta}{6\sin^2 \theta}$$
for $0 < \theta < 2\pi$

22)
a) Solve
$$11 \sin 2x = 2\cos x + 6\cos^3 x$$
for $0 < x < 2\pi$
b) Solve
$$11 \sin\left(6\theta - \frac{\pi}{2}\right) = 2\cos\left(3\theta - \frac{\pi}{4}\right) + 6\cos^3\left(3\theta - \frac{\pi}{4}\right)$$
for $0 < \theta < \frac{2\pi}{3}$

23) Prove the following trigonometric identities:
a)
$$\sqrt{2} \sin\left(2x + \frac{\pi}{4}\right) \equiv \cos^2 x + 2\sin x \cos x - \sin^2 x$$
b)
$$2\sqrt{3} \sin\left(\frac{\pi}{3} - 2x\right) \equiv 3\cos^2 x - 2\sqrt{3}\sin x \cos x - 3\sin^2 x$$
c)
$$\sin\left(\frac{\pi}{4} - 2x\right) + \cos\left(2x + \frac{\pi}{4}\right) + 2\sqrt{2}\sin^2 x \equiv \sqrt{2}(\cos x - \sin x)^2$$
d)
$$\frac{1}{2}\cos\left(\frac{\pi}{6} - 2x\right) - \frac{1}{2}\sin\left(\frac{\pi}{3} - 2x\right) \equiv \sin x \cos x$$

24) Prove the following trigonometric identity:
$$\left(\sin\left(x + \frac{\pi}{4}\right) + \cos\left(x + \frac{\pi}{4}\right)\right)^2 \equiv \cos 2x + 1$$

25) Prove the following trigonometric identity:
$$\left(\sin\left(2x + \frac{\pi}{4}\right) - \cos\left(2x + \frac{\pi}{4}\right)\right)^2 \equiv 1 - \cos 4x$$

26) Use the Compound Angle Formulae
$\sin(A \pm B) = \sin A \cos B \pm \cos A \sin B$ to show that $\sin 3x \equiv 3\sin x - 4\sin^3 x$

27) Use the Compound Angle Formulae
$\cos(A \pm B) = \cos A \cos B \mp \sin A \sin B$ to show that $\cos 3x \equiv 4\cos^3 x - 3\cos x$

28) Prove the following trigonometric identity:
$$\sqrt{2} \sin\left(\frac{\pi}{4} - x\right)(2\sin 2x + 1) \equiv \sin 3x + \cos 3x$$

29) Find the period of $f(x) = 5 - \cos^2 x$, $x \in \mathbb{R}$

30) Prove the following trigonometric identity:
$$\frac{\sin 3x}{\sin x} + \frac{\cos 3x}{\cos x} = 4\cos 2x$$

31) Use the Compound Angle Formulae
$$\tan(A \pm B) = \frac{\tan A \pm \tan B}{1 \mp \tan A \tan B}$$
to find $\tan 3x$ in terms of $\tan x$ only

32) Solve the equation $\tan 3\theta + \tan \theta = 0$ for $0° < \theta < 360°$

33) Solve $\sin \frac{2\theta}{3} \cos \frac{2\theta}{3} = \frac{1}{2}$ for $0 < \theta < 2\pi$

34) Solve $\cos^2 \frac{3\theta}{2} - \sin^2 \frac{3\theta}{2} = \frac{1}{3}$ for $0° < \theta < 360°$

35) The diagram below shows the rod OAB which is pivoted at O. The lengths of OA and AB are both k units. Taking O as the point with coordinates $(0,0)$,
a) show that the coordinates of B are $(4k\sin^2\theta\cos\theta, 2k\sin\theta - 4k\sin\theta\cos^2\theta)$
b) Find the distance of B from O when $\theta = 60°$ in terms of k

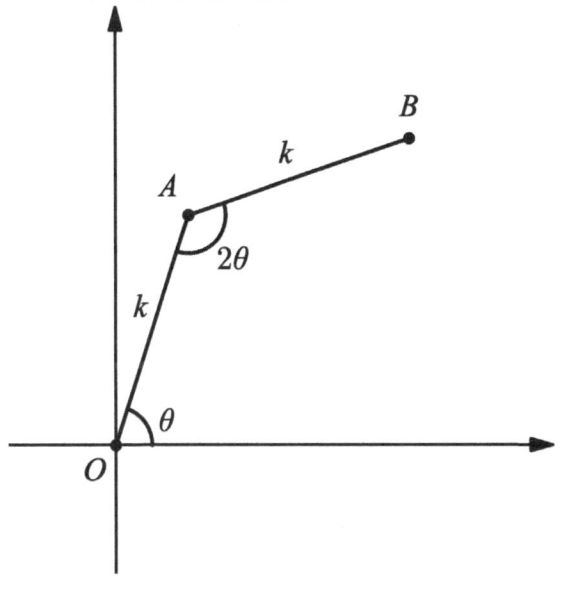

36) Prove the following trigonometric identity:
$$\sin 4x \equiv 4\sin x \cos^3 x - 4\sin^3 x \cos x$$

37) Prove the following trigonometric identity:
$$\cos 4x \equiv 8\cos^4 x - 8\cos^2 x + 1$$

Prerequisite knowledge for Q38: Reciprocal Trigonometric Functions

38) Prove the following trigonometric identity:
$$\frac{\sin 4x}{\cos x} + \frac{\cos 4x}{\sin x} \equiv (2\cos 2x - 1)\operatorname{cosec} x$$

2.18 Trigonometry: Harmonic Forms (answers on page 516)

1) Each of the following can be written in at least one of the forms $R\sin(\theta \pm \alpha)$ or $R\cos(\theta \pm \alpha)$, where $R > 0$ and $0° < \alpha < 90°$. Write down the exact value of R in each case:
   a) $8\sin\theta + 15\cos\theta$
   b) $12\sin\theta - 35\cos\theta$
   c) $14\cos\theta - 20\sin\theta$
   d) $2\sqrt{2}\cos\theta + 5\sqrt{2}\sin\theta$

2) State the maximum and minimum values of each of the following expressions:
   a) $9\sin\theta - 40\cos\theta$
   b) $10\sin\theta + 24\cos\theta$
   c) $6\sqrt{3}\cos\theta + 4\sqrt{3}\sin\theta$
   d) $3\sqrt{5}\cos\theta - 5\sqrt{3}\sin\theta$

3) State the maximum and minimum values of each of the following expressions:
   a) $12 + 220\cos\theta - 21\sin\theta$
   b) $50 - \cos\theta + 18\sin\theta$
   c) $(16\cos\theta + 63\sin\theta)^2$
   d) $(5 + 4\sin\theta - 9\cos\theta)^2$

4) State the maximum and minimum values of each of the following expressions:
   a) $$\frac{1}{14 + 5\sin\theta + 12\cos\theta}$$
   b) $$\frac{15}{90 - 84\sin\theta - 13\cos\theta}$$
   c) $$\frac{100}{(6 + \sin\theta + \cos\theta)^2}$$
   d) $$\frac{625}{90 + (-33\cos\theta + 56\sin\theta)^2}$$

5) The maximum value of the expression $\sqrt{10 + 3\sqrt{3}\cos\theta + k\sin\theta}$ is 4. Find the possible values of k

6) Write the following in the form $R\sin(x + \alpha)$ where $R > 0$ and $0° < \alpha < 90°$:
   a) $3\sin x + 3\cos x$
   b) $5\sin x + 5\sqrt{3}\cos x$
   c) $9\sqrt{3}\sin x + 9\cos x$
   d) $4\sqrt{2}\sin x + 2\sqrt{2}\cos x$

Prerequisite knowledge for Q7: Radians

7) Write the following in the form $R\sin(x - \alpha)$ where $R > 0$ and $0 < \alpha < \frac{\pi}{2}$:
   a) $2\sqrt{2}\sin x - 2\sqrt{2}\cos x$
   b) $3\sin x - 4\cos x$
   c) $12\sin x - 5\cos x$
   d) $11\sin x - \cos x$

8) Write the following in the form $R\cos(x + \alpha)$ where $R > 0$ and $0° < \alpha < 90°$:
   a) $4\cos x - 4\sqrt{3}\sin x$
   b) $5\cos x - \sin x$
   c) $20\cos x - 30\sin x$
   d) $13\cos x - 6\sin x$

Prerequisite knowledge for Q9: Radians

9) Write the following in the form $R\cos(x - \alpha)$ where $R > 0$ and $0 < \alpha < \frac{\pi}{2}$:
   a) $20\cos x + 99\sin x$
   b) $56\cos x + 33\sin x$
   c) $25\sqrt{3}\cos x + 75\sin x$
   d) $2\sqrt{5}\cos x + 5\sqrt{5}\sin x$

10) Write each of the following in an appropriate harmonic form: $R\sin(\theta \pm \alpha)$ or $R\cos(\theta \pm \alpha)$, where $R > 0$ and $0° < \alpha < 90°$
    a) $6\cos\theta - 8\sin\theta$
    b) $7\cos\theta + 24\sin\theta$
    c) $5\sin\theta - 5\cos\theta$
    d) $8\sqrt{2}\sin\theta + 10\sqrt{2}\cos\theta$

11) Find the maximum value of the following expressions, and the smallest positive value of θ, in degrees, for which this maximum occurs:
    a) $\sin\theta + 5\cos\theta$
    b) $9\cos\theta - \sqrt{2}\sin\theta$
    c) $6\sin\theta - 4\cos\theta$
    d) $10\cos\theta + 20\sin\theta$

12) Solve the following trigonometric equations:
    a) $4\sqrt{2}\sin\theta - 4\sqrt{2}\cos\theta = 7$, $0° < \theta < 360°$
    b) $4\sin\theta + 2\cos\theta = 3$, $0 < \theta < 2\pi$
    c) $6\sqrt{3}\cos 3\theta + 5\sin 3\theta = 1$, $0° < \theta < 240°$
    d) $10\sqrt{5}\cos 2\theta - 12\sqrt{5}\sin 2\theta = 5$, $0 < \theta < 2\pi$

2.18 Trigonometry: Harmonic Forms (answers on page 516)

Prerequisite knowledge for Q13: Combinations of Graph Transformations

13) Describe a sequence of two graph transformations, in radians, that would map $y = \sin x$ onto the image
$$y = 4 + 10 \sin x - \sqrt{2} \cos x$$

14) Write $40 \cos 5x - 42 \sin 5x$ in the form $R \cos(5x + \alpha)$ where $R > 0$ and $0 < \alpha < \frac{\pi}{2}$

15) Write $8 \cos \frac{1}{2}x + 3 \sin \frac{1}{2}x$ in the form $R \cos\left(\frac{1}{2}x - \alpha\right)$ where $R > 0$ and $0 < \alpha < \frac{\pi}{2}$

16) The diagram below shows a curve with equation $y = p \sin x + q \cos x$, where p and q are positive constants. The curve passes through the point $\left(-\frac{\pi}{4}, \sqrt{2} - \sqrt{6}\right)$. It is given that the minimum value of the curve is -4. Find the exact values of p and q

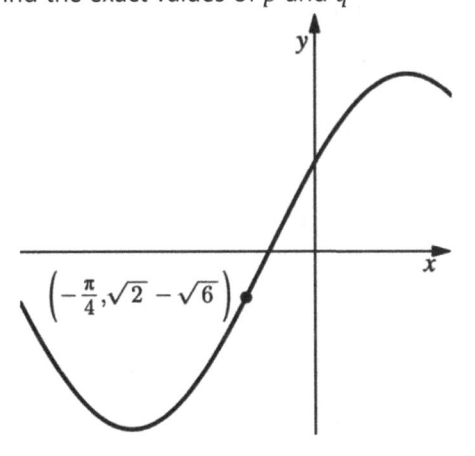

Use Radians for Q17-Q19

17) For a land-based wind turbine, the height above the ground, h metres, of the tip of one of the wind turbine blades after t minutes is modelled by the equation
$$h = k + 28 \cos 4t + 54 \sin 4t$$
a) Before the wind turbine is switched on and starts moving, the blade tip in question is locked into a position 117 metres above the ground. Find the value of k
b) The wind turbine is switched on. Find the maximum height the blade tip reaches above the ground as it rotates, and the time it takes the blade tip to first reach this position.

18) The depth of water, D meters, at a point in a tidal river is modelled by the equation
$$D = 4 + \sin\left(\frac{\pi}{6}t - 2\right) + 1.5 \cos\left(\frac{\pi}{6}t - 2\right)$$
where t is the number of hours after midnight.
a) Find the minimum depth of the river at the point being measured.
b) A ferry can only cross the river when the depth of the water is at least 5 metres. Derek arrives at the ferry at 1pm. How long must he wait before the ferry starts operating again?

19) An engine piston moves up and down in a vertical cylinder. The point A is at the top of the piston. The vertical displacement, h mm, of A from its lowest position is modelled by the equation
$$h = 50 + 40 \cos 20t - 30 \sin 20t$$
where t is the time in seconds after the piston starts moving.
a) Find the maximum displacement of A
b) In a single cycle of the engine, A is more than 80 millimetres above its lowest position for T seconds. Find the value of T
c) The engine's speed is increased so that the piston now completes 15 cycles per second. Explain how the equation of the model should be modified to reflect this change in speed.

20)
a) Express $\sqrt{129} \cos \theta - 20 \sin \theta$ in the form $a \cos(\theta + b)$, where $a > 0$ and $0 < b < 90°$

A meteorologist studies the temperature and humidity in Saskatoon over one year. The temperature, $T°C$, is modelled by the equation
$$T = \sqrt{129} \cos(30t)° - 20 \sin(30t)° + C$$
where t is the time in months after the start of the year.
b) Given that during the year, the maximum temperature in Saskatoon is 25°C, find the value of C
c) The humidity, $H\%$, during the same year is modelled by $H = 10 \cos(30t + 45°) + 72$. Find the temperature in Saskatoon when the humidity is at its maximum.

2.19 Trigonometry: Small Angle Approximations (answers on page 517)

1) The diagram below shows a sector of a circle, ABC. By considering trigonometric ratios, show that as $\theta \to 0$ that $\sin\theta \approx \tan\theta \approx \theta$

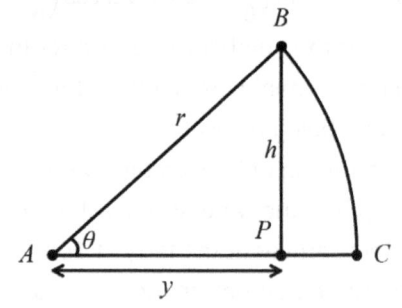

2) Show that if $\sin\theta \approx \theta$ then as $\theta \to 0$
$$\cos\theta \approx 1 - \frac{\theta^2}{2}$$

3) When θ is small, find an approximation for the following:
   a) $\theta \sin\theta$
   b) $\cos\theta - 1$
   c) $\cos 6\theta$
   d) $\tan\theta - 2\sin\theta$
   e) $\theta \tan 3\theta + \cos 3\theta$

4) When θ is small enough for θ^3 to be ignored, find an approximation for the following:
   a) $$\frac{\theta \tan\theta}{1 - \cos\theta}$$
   b) $$\cos\theta \cos 3\theta$$
   c) $$\cos^2\theta$$
   d) $$\sin^2\theta + \cos^2\theta$$
   e) $$\frac{\theta^2 \sin\theta}{1 - \cos 2\theta}$$
   f) $$\frac{\cos 8\theta - \cos 4\theta}{\sin 8\theta - \sin 4\theta}$$
   g) $$\frac{1 - \cos 2\theta}{6 \tan\theta}$$

5)
   a) Find an approximation for $\cos 0.4$ by using $\theta = 0.1$ and small angle approximations.
   b) Find the percentage error of your approximation.

6) When θ is small, use the standard small angle approximations for $\sin\theta$ and $\tan\theta$ find:
   a) $\theta \csc\theta$
   b) $\cot 2\theta$
   c) $\cot 4\theta - \csc 2\theta$

7) $$f(x) = \sec x\,(\sin x + \cos x)$$
   a) By using the small angle approximations for $\tan x$ show that $f(x) \approx 1 + x$
   b) By using the small angle approximation for $\sin x$ and alternative approximation, $\cos x \approx 1$, show that $f(x) \approx 1 + x$
   c) Find the percentage error when approximating $f(0.1)$

8)
   a) When θ is small, show that
   $$-2 + \cos\theta + \sin^2\theta \approx -1 + \frac{1}{2}\theta^2$$
   b) Hence find an approximate solution to
   $$-2 + \cos\theta + \sin^2\theta = 4\tan\theta$$

9) Given x is small, find the approximate solution to
$$\frac{1 - \cos\frac{1}{10}x}{\sin x} = \frac{1}{5} - \tan x$$

10)
   a) Find, for small values of x, an approximation for
   $$f(x) = 2 + 4\tan\frac{x}{2} - 8\sin\frac{x}{3}$$
   b) Use your result in part a) to sketch an approximation for the graph of $f(x)$ for $-0.2 \leq x \leq 0.2$ Include the start and end coordinates in your sketch.

11) Given that the root of the equation is small, find the single solution to
$$2x - \frac{1}{4} + \cos x = 0$$

12) Given x is small enough for x^3 to be ignored, by approximating
$$\tan x + \tan x \cos 4x$$
show that
$$\int_0^{0.3} \sqrt[3]{\tan x\,(1 + \cos 4x)}\,dx \approx 0.190$$

2.19 Trigonometry: Small Angle Approximations (answers on page 517)

13)
   a) When θ is small, find an approximation to
   $$\cos\theta - \cos 2\theta$$
   b) Hence find
   $$\lim_{\theta \to 0} \frac{\cos\theta - \cos 2\theta}{\theta^2}$$

14) Find these limits:
   a)
   $$\lim_{h \to 0} \frac{\cos h - 1}{h}$$
   b)
   $$\lim_{h \to 0} \frac{\sin h}{h}$$

Prerequisite knowledge for Q15-Q18: Differentiation from First Principles and Compound Angle Formulae

15) Differentiate from first principles the following functions by using knowledge about small angle approximations:
   a) $f(\theta) = \sin\theta$
   b) $f(\theta) = \cos\theta$

16) Using differentiation from first principles, for $f(x) = \cos(x)$, find the exact value of $f'\left(\frac{\pi}{3}\right)$

17) Prove, using differentiation from first principles, that
$$\frac{d}{dx}(\tan\theta) = \sec^2\theta$$

18) Prove, using differentiation from first principles, that
$$\frac{d}{dx}(\sec\theta) = \sec\theta\tan\theta$$

Prerequisite knowledge for Q19-Q20: Binomial Series

19)
   a) Find the first three terms, in ascending powers of x of the binomial expansion of
   $$(1-8x)^{\frac{1}{2}}$$
   b) Hence, for small values of x, show that
   $$\sin 2x + \sqrt{\cos 4x} \approx a + bx + cx^2$$
   where a, b and c are constants to be found and x is small enough for x^4 to be ignored.

20)
   a) Find the first two terms, in ascending powers of x of the binomial expansion of
   $$\left(1 - \frac{1}{2}x^2\right)^{\frac{1}{3}}$$
   b) Hence find an approximation for
   $$\int_0^{0.2} \sqrt[3]{\cos x}\ dx$$
   giving your answer to 4 decimal places.

Prerequisite knowledge for Q21-Q22: Differentiation of Trigonometric Functions

21) Find the value of $\frac{dy}{dx}$ at the origin of
$$x = \frac{1}{2}\tan 4y$$
by:
   a) using Calculus
   b) using a small angle approximation and Calculus

22) Find the value of $\frac{dy}{dx}$ when $x = 0$ for
$$x = 3\cos 16y$$
where $0 < y < \frac{\pi}{2}$ by:
   a) using Calculus
   b) using a small angle approximation and Calculus

Prerequisite knowledge for Q23-Q24: Newton-Raphson Method

23) Given x is small, find an approximate solution giving your answers to 3 significant figures for
$$\frac{1}{2} - x - \sin 2x = 0$$
   a) by using small angle approximation
   b) by using a Newton-Raphson approximation starting at $x_0 = 0$

24) Given x is small, find an approximate solution giving your answers to 3 significant figures for
$$2 + 5x + \sin x = \cos x$$
   a) by using small angle approximation
   b) by using a Newton-Raphson approximation starting at $x_0 = 0$

2.20 Differentiation: Points of Inflection and Applications (answers on page 517)

1) For each of the following graphs, identify the points of inflection on a copy of the graph and describe the change in concavity at each point of inflection:
   a) 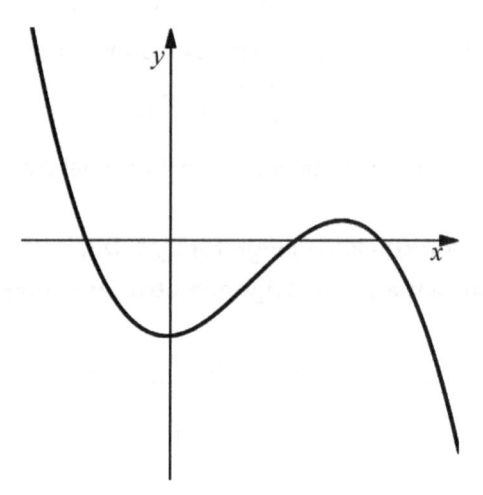

   b) 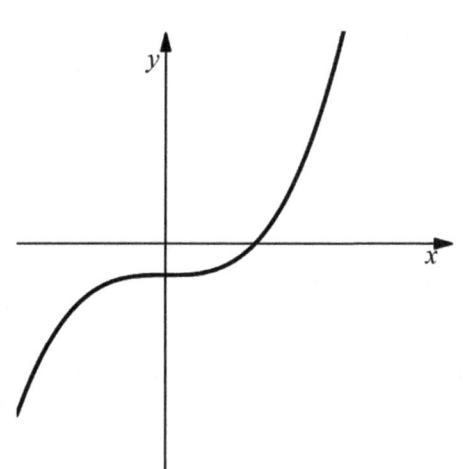

   c)

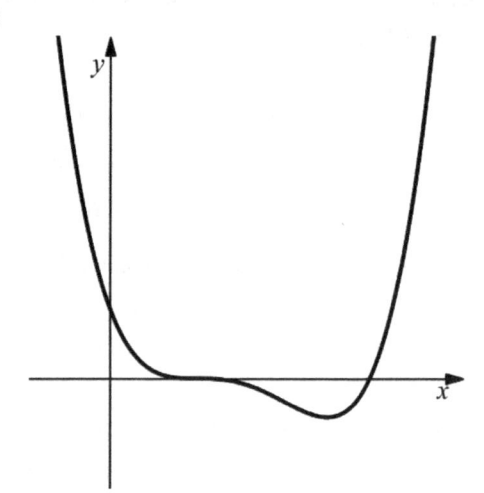

2) Sketch a graph with a single point of inflection at $(0,0)$ which is non-stationary, concave (concave downwards) before $(0,0)$ and convex (concave upwards) after $(0,0)$

3) Sketch a graph which has a stationary point which is a minimum at $x = -2$, a non-stationary point of inflection at $x = 1$ and a stationary point which is maximum at $x = 4$ Label each of the points clearly.

4) Determine whether each of the curves are convex (concave upwards), concave (concave downwards) or neither at the given point:
   a) $y = 3x^3 - x^2 - x - 1$ at $x = 0$
   b) $y = x^4 + 3x^2 + x - 6$ at $x = 1$
   c) $y = 5x^3 - x^2 + x$ at $x = -1$
   d) $y = x^3 - 12x^2 + 48x - 2$ at $x = 4$

5) The curve
   $$y = x^6 - 3x^5 - 10x^4 + 10x^3 + 45x^2$$
   is shown in the graph below. 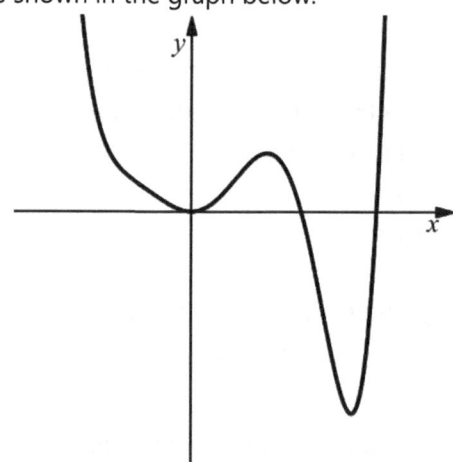

   Find all the points of inflection of the curve.

6) Determine the values of x for which
   $$y = 12x^5 - 20x^4 - 50x^3 + 90x^2$$
   is convex (concave upwards). Give your answer in set notation.

7) Find the coordinates of the points of inflection for each of the following curves. Determine in each case whether they are stationary or not.
   a) $f(x) = x^4 + 9x^3 - 15x^2 + 4$
   b) $f(x) = x^4 + 2x^3 - 2x - 1$

8) Given that the function
   $$h = 2t^3 - 9t^2 + 12t - 3$$
   is used to model the height, h metres, of a passenger on a ride over time t, find the value of t when $\frac{d^2h}{dt^2} = 0$ and state the physical relevance of this point.

2.20 Differentiation: Points of Inflection and Applications (answers on page 517)

9) The value of some shares £V during the first 12 months of the year may be modelled by
$$V = 0.25t^3 - 3.25t^2 + 10t + 20$$
where t is the number of months since the start of the year.
   a) Use this model to predict the value of the shares when 12 months have passed.
   b) Find when $\frac{d^2V}{dt^2} = 0$ and state its physical relevance.

10) Show that
$$y = 2x^3 + 16x^2 + 43x + 38$$
has a single non-stationary point of inflection.

11)
$$f(x) = x^3 + 3x^2 + 3x + 2$$
   a) Find the coordinates of the single stationary point of $y = f(x)$
   b) Show that the stationary point is a point of inflection of $y = f(x)$
   c) Hence, sketch the curve $y = f(x)$

12)
$$f(x) = x^2 - 3x - \frac{1}{x}, \quad x \neq 0$$
   a) Find the stationary points of $y = f(x)$ and determine their nature.
   b) Show that there are no non-stationary points of inflection on the curve $y = f(x)$

13)
$$f(x) = 3x^4 - 4x^3 + 2$$
   a) Show that the graph of $y = f(x)$ only has two stationary points.
   b) Show that the two stationary points of $y = f(x)$ are different types of points.
   c) Hence, sketch the curve $y = f(x)$

14) For the curve $y = 2x^4 + 4x^3$
   a) Find the stationary points on the curve and determine their type.
   b) Find the non-stationary point of inflection.
   c) Hence, sketch the curve.

15) Find the values of a, b and c such that
$$y = x^3 + ax^2 + bx + c$$
has a single stationary point of inflection at the point $(5, 2)$

16) Find the values of a, b and c such that
$$y = -x^3 + ax^2 + bx + c$$
has a single stationary point of inflection at the point $(1, 6)$

Prerequisite knowledge for Q17-Q18: Differentiation of the form ae^{kx} and solving equations involving e^x

17) Determine whether each of the curves are convex (concave upwards) or concave (concave downwards) at the given point.
   a) $y = e^x - x$ at $x = 2$
   b) $y = 2e^x - e^{2x}$ at $x = 0$
   c) $y = e^{2x} + 4$ at $x = -1$
   d) $y = e^{-x} + x^2$ at $x = -1$
   e) $y = e^{-x} - x^2$ at $x = 0$

18)
   a) Show that
$$y = e^x - x^2$$
has a single non-stationary point of inflection.
   b) Find the coordinates of this point.

Prerequisite knowledge for Q19: Differentiation of Standard Functions

19) Show that there is a single non-stationary point of inflection on the curve
   a)
$$y = x + \sin\left(\frac{1}{2}x\right), \quad 0 < x \leq 2\pi$$
   b)
$$y = x^2 + \ln x, \quad x > 0$$

Prerequisite knowledge for Q20: Standard Functions and Product Rule

20) The price per litre (in pence), P, from January to June in 2024 is modelled by function
$$P = 144 + (t - 3)e^{0.5t}$$
where t is the number of months after 1 January 2024.
   a) According to the model, find the minimum price of the fuel in this time frame and the month in which it occurs.
   b) Show that the curve has no points of inflection and explain what this means in context.

2.21 Differentiation: Standard Functions (answers on page 518)

Differentiation of $(ax+b)^n$

1) Differentiate each of the following:
   a) $y = (x+3)^5$
   b) $y = (4x+1)^6$
   c) $y = (6-5x)^4$
   d) $y = (2x-3)^7$
   e) $y = (7-4x)^{\frac{1}{4}}$
   f) $y = \frac{1}{x+2}$
   g) $y = \frac{3}{(x-5)^2}$
   h) $y = \frac{1}{(8-3x)^4}$
   i) $y = \sqrt{2+x}$
   j) $y = \frac{1}{\sqrt{3x-1}}$

2) Find $f''(x)$ for each of the following:
   a) $f(x) = (7x-1)^5$
   b) $f(x) = \sqrt{3-x}$
   c) $f(x) = \frac{1}{\sqrt[3]{2x+9}}$

3) Find coordinates of any stationary points on the following curves:
   a) $y = (2x-7)^5$
   b) $y = 4x + \frac{1}{x-1}$

4) It is given that $f(x) = (3x+1)^4$
   a) Find $f'(x)$ and expand the result fully.
   b) Expand $f(x)$ using binomial expansion, and hence find $f'(x)$ to confirm your answer to part a).

5) Find the tangent line to the curve $y = \sqrt{x+7}$ when $x = 2$ giving your answer in the form $ax + by = c$ where a, b and c are integers to be found.

6) The height of water, h cm in a harbour for a 12-hour period can be modelled by the equation
   $$h = 16(t-6)^2 + 200, \quad 0 \leq t \leq 12$$
   and t is the time in hours after 6 am.
   a) By using this model and calculus, predict the minimum height of the water which occurred that day.
   b) By considering $\frac{d^2h}{dt^2}$, confirm that your answer to a) is a local minimum.

Differentiation of $\sin kx$ and $\cos kx$

7) Differentiate each of the following:
   a) $y = \sin 3x$
   b) $y = \cos 5x$
   c) $y = -7 \sin 2x$
   d) $y = \frac{1}{2} \cos 6x$
   e) $y = -\cos \frac{1}{4} x$
   f) $y = -\frac{1}{5} \sin 10x$
   g) $y = -\cos \pi x$
   h) $y = \cos \frac{x}{\pi}$

8) Find the equation of the tangent to the curve $y = 2 \sin 6x$ when $x = \pi$ giving your answer in the form $y = mx + c$

9) Find the equation for the tangent to the curve $y = 1 + \cos 3x$ when $x = \frac{\pi}{2}$ giving your answer in the form $y = mx + c$

10) Find the equation for the normal to the curve $y = 3 - \sin(\pi x)$ when $x = 2$ giving your answer in the form $y = mx + c$

11) Given that $y = \cos 3x$, show that
    $$\frac{d^2y}{dx^2} = -9y$$

12) Show that the curve $y = \cos x - x$ has a single stationary point in the range $0 < x < 2\pi$. Find the exact coordinates of the stationary point.

13) The number of visitors at a tourist attraction, p, may be modelled by the function
    $$p = 300 - 40 \cos \frac{\pi}{6} t$$
    where t is the number of months since 1 January 2024.
    a) Find the number of visitors predicted on 1 January.
    b) Find an expression for $\frac{dp}{dt}$
    c) Find the value of $\frac{dp}{dt}$ when $t = 6$. What can you deduce in context from its value?
    d) Find an expression for $\frac{d^2p}{dt^2}$. What can you deduce from its value when $t = 6$?

2.21 Differentiation: Standard Functions (answers on page 518)

14) The distance, s cm, of a weight dropped from an equilibrium position may be modelled by the equation
$$s = 3\cos\frac{\pi}{3}t$$
where t is the number of seconds after the weight is released.
   a) Find an expression for $\frac{ds}{dt}$
   b) Find the maximum value of $\frac{ds}{dt}$ giving your answer in exact form and find the first time when this occurs. Explain the relevance of this point in context.

15) Given the gradient of the normal to the curve
$$y = 6\cos 2\pi x, \; 0 < x < \frac{\pi}{3}$$
is $\frac{1}{6\pi}$, find the possible values of x for when this occurs giving your answer in exact form.

16) Differentiate $y = \sin x°$, giving your answer in degrees.

Differentiation of e^{kx}, $\ln kx$, a^x

17) Differentiate each of the following:
   a) $y = \ln 5x$
   b) $y = 3^x$
   c) $y = e^{5x}$
   d) $y = 2\ln\frac{1}{4}x$
   e) $y = -6^x$
   f) $y = (e^x - 1)^2$

18) For each curve, find the coordinates of the stationary points and identify their type:
   a) $y = 2e^x - 4x$
   b) $y = \ln x - 8x$
   c) $y = 2x^2 + 19x - 5\ln x$

19) Find the x-value for which $f'(x) = 1$ giving your answers in exact form:
   a) $f(x) = \ln 2x$
   b) $f(x) = e^{2x} + 4$
   c) $f(x) = 3^x - 5$

Prerequisite knowledge for Q20: Points of Inflection

20) Show that the curve $y = 8\ln x - 5x^2$ is concave (concave downwards) for all values of $x > 0$

21) Find the equation of the tangent to the curve
$$y = \ln 5x$$
when $x = \frac{e}{5}$ giving your answer in the form $y = kx$ where k is a constant.

22) Find the equation of the normal to the curve
$$y = 5^x$$
when $x = 2$ in the form $ay + bx = c$ where a, b and c are constants given in exact form.

23) A population P of bugs has been found and a treatment applied so their population is reduced. The number of bugs after t days can be modelled by
$$P = 500 \times 20^{-\frac{1}{2}t}$$
   a) Determine how many whole hours need to have elapsed for the number of bugs to drop to 50.
   b) Determine the rate of change of bugs after 7 days.

24) Find the coordinates of the single stationary point of the curve
$$y = 9^x - 3^x$$
and determine its type.

25) The graph below shows the curve with equation
$$y = 2 \times 2^x$$

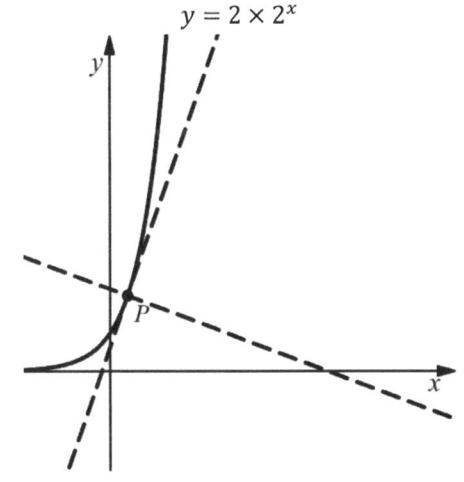

The point P lies on the curve at $x = 1$
By finding the equation of the tangent and normal line at the point P, find the area of the triangle bounded by those two lines and the x-axis. Give your answer in the form
$$p \times \frac{1}{\ln 2} + q\ln 2$$
where p and q are constants to be found.

285

2.21 Differentiation: Standard Functions (answers on page 518)

26) Show that the equation of the tangent to the curve
$$y = \ln x + a$$
when $x = e^b$ is
$$e^b y - x = (a + b - 1)e^b$$
for constants a and b

Differentiation of tan kx

27) Differentiate each of the following:
   a) $y = \tan 3x$
   b) $y = 2 \tan 4x$
   c) $y = -\frac{1}{3} \tan 9x$
   d) $y = -\tan \pi x$
   e) $y = -\frac{1}{12} \tan \frac{6}{\pi} x$

28) Find the equation for the tangent to the curve
$$y = 4 + 2 \tan \frac{\pi}{9} x$$
when $x = \frac{9}{4}$ in the form $y = mx + c$

29) Show that the equation for the normal to the curve
$$y = x - 2 \tan \frac{\pi}{4} x$$
when $x = 4$ crosses the x-axis at $x = 8 - 2\pi$

Mix of Standard Functions

30) Find the gradient function of each of the following:
   a) $y = 3 \sin 7x + 2 \tan 2x$
   b) $y = 15^x - 3 \cos \frac{5}{2} x$
   c) $y = \ln 180x + \tan \sqrt{2} x$
   d) $y = \sqrt{x} - 3 \sin 2x$
   e) $y = e^{3x} + \sin 8x$
   f) $y = \cos \frac{1}{4} x - \cos 2\pi x$
   g) $y = 4^x + \tan \pi x$

31) Find $\frac{d^2 y}{dx^2}$ for each of the following:
   a) $y = 3x^5 - e^{5x}$
   b) $y = x^2 - \sin 2x$
   c) $y = \sqrt{x} + \ln 2x$
   d) $y = \frac{1}{x} - \cos \frac{1}{2} x$
   e) $y = e^{2x} - \sin \pi x$
   f) $y = e^{-\frac{1}{2} x} + \sin 3x - \cos 3x$
   g) $y = \frac{1}{\sqrt{x}} - \ln \frac{1}{2} x$

32) Find the exact value of $f'(x)$ at the given value of x, giving your answer as an exact value:
   a) $f(x) = e^{2x} + x^{\frac{1}{4}}$ when $x = 1$
   b) $f(x) = 5 \ln x + \frac{1}{x^2}$ when $x = 2$
   c) $f(x) = 2^x + \sin 2\pi x$ when $x = \frac{1}{2}$
   d) $f(x) = \cos\left(\frac{\pi}{2} x\right) + \tan\left(-\frac{\pi}{2} x\right)$ when $x = 0$

33) Determine whether the curve
$$y = e^{-2x} - 3 \sin 5x$$
is increasing or decreasing when $x = \pi$

34) Find the equation for the tangent to the curve
$$y = \ln 2x + \sin \pi x, \; x > 0$$
in the form $y = mx + c$ when $x = 3$

35) The curve shown below is
$$y = e^{-2x} + 3 \cos \pi x$$

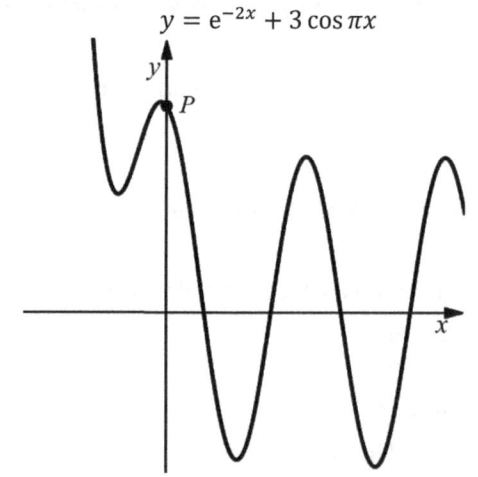

Point P is where the curve meets the y-axis.
   a) Find the tangent line to the curve at P.
   b) Find the normal line to the curve at P.
   c) Find the area bounded by the normal and tangent lines at point P and the x-axis.

Prerequisite knowledge for Q36: Points of Inflection

36) Show that the curve
$$y = 2^x + \cos 3x$$
   a) is convex (concave upwards) at $x = \pi$
   b) is concave (concave downwards) at $x = \frac{2\pi}{3}$

37) Show clearly that the curve
$$y = \sin 3x + \tan 3x$$
has a single stationary point for $0 < x < \pi$ and state the coordinate of this stationary point.

2.22 Differentiation: The Chain Rule (answers on page 520)

Differentiation of $(ax^k + \cdots)^n$

1) Find the gradient function of:
   a) $y = (2x^2 + 1)^3$
   b) $y = (2x^3 - 5x)^4$
   c) $y = (3x^4 - 1)^{\frac{1}{2}}$
   d) $y = (1 - x^2)^{-\frac{3}{2}}$
   e) $y = \frac{1}{5 + \sqrt{x}}$
   f) $y = \left(\frac{1}{x} - x\right)^7$

2) Find the gradient of the curve
$$y = \sqrt{5x^2 - 2x + 1}$$
at $(2, \sqrt{17})$

3) Find the equation of the tangent to the curve
$$y = \sqrt{1 + 7x^2}$$
when $x = 3$ giving your answer in the form $ay + bx = c$ where a, b and c are integers to be found.

4) Find the equation of the normal to the curve
$$y = \frac{1}{\sqrt{4 - 3x^2}}$$
when $x = -1$ giving your answer in the form $ay + bx = c$ where a, b and c are integers to be found.

5) Find the coordinates of all the stationary points of each of the following curves:
   a) $y = (3x^2 - 1)^4$
   b) $y = \frac{1}{\sqrt{1 + x^2}}$

6) For each of the following, find $\frac{dy}{dx}$ giving your answer in terms of y:
   a) $x = 3y^2 - y$
   b) $x = 2y^4 - 1$
   c) $x = (3y^2 - 1)^{\frac{1}{4}}$

7) Find $\frac{dy}{dx}$ of each of the following at the given point:
   a) $x = (3 - y)^5$ when $y = 2$
   b) $x = (y^2 - 2y + 1)^{\frac{1}{2}}$ when $y = 4$
   c) $x = \frac{1}{\sqrt{y^4 + 3}}$ for $y > 0$ when $x = \frac{1}{2}$

8) Find the tangent of the curve
$$y = \frac{1}{3}(x^2 + 5)^{\frac{3}{2}}$$
at $x = 2$ giving your answers in the form $y = mx + c$

9) A curve has equation
$$x = (y - 2)^6$$
   a) Find $\frac{dx}{dy}$ in terms of x
   b) Hence find $\frac{dy}{dx}$

Prerequisite knowledge for Q10: Points of Inflection

10) Find the x-value of the point of inflection on the curve f(x) given that
$$f'(x) = \frac{1}{(1 + 3x - 2x^2)^3}$$

Differentiation of $\sin(f(x))$ **&** $\cos(f(x))$

11) Differentiate each of the following:
   a) $y = \sin^2 x$
   b) $y = \cos^2 x$
   c) $y = 2\sin(3x - 7)$
   d) $y = -\frac{1}{5}\cos\left(\pi x - \frac{\pi}{2}\right)$
   e) $y = \cos\sqrt{x}$
   f) $y = -2\sin(x^3)$
   g) $y = \cos(\sin x)$
   h) $y = \sin(\sin x)$
   i) $y = \sqrt{\sin x}$
   j) $y = \sqrt[3]{\cos x}$
   k) $y = 1 - \cos^3(5x)$

12) Show that
$$\frac{d}{dx}\left[\frac{1}{\sin x}\right] = -\csc x \cot x$$

13) A model for the tide height over a 24-hour period in Southampton is
$$x = 2.85 + 0.75 \sin\left(\frac{17\pi}{120}t + \frac{3\pi}{2}\right)$$
where the depth of the water is x metres and t is the number of hours passed since midnight.
   a) State the maximum and minimum height the water reaches.
   b) By using Calculus, find the two times at which it reaches its maximum depth in the first 24-hour period.

2.22 Differentiation: The Chain Rule (answers on page 520)

14) A curve is given by
$$y = a\sin^2 x + b\cos^2 x$$
where a and b are positive constants and are such that $a \neq b$. Show that the gradient function of the curve at $x = \pi$ is never zero.

15) A model for the number of daylight hours, t (in hours), on the xth day of the year may be given by
$$t = 10 + 2.7\sin\left(\frac{2\pi}{365}(x - 78)\right)$$
Use this model to compare how the number of hours of daylight is changing on day 17 compared to on day 85.

16) Show that the stationary points of
$$f(x) = \sin\left(\frac{\pi}{2}e^x\right)$$
occur when $x = \ln k$ where k is a positive odd integer.

17) Show that
$$f(x) = \frac{1}{2}x + \cos\left(\frac{2}{x}\right)$$
is an increasing function for all $x > 2$.

Differentiation of $e^{f(x)}$, $\ln(f(x))$, $a^{f(x)}$

18) Differentiate each of the following:
   a) $y = (e^{2x} - 1)^5$
   b) $y = (5^x - 1)^4$
   c) $y = (\ln x)^4$
   d) $y = e^{3x^2 - 1}$
   e) $y = 4^{x^2 + 5x + 1}$
   f) $y = -5^{2x^2 + 1}$

19) Find coordinates of the stationary points on the following curves:
   a) $y = 6x - e^{2x}$
   b) $y = 2\ln(x - x^2)$

20) Differentiate each of these:
   a) $y = \ln(2x^3 + x^2 - 1)$
   b) $y = \ln(\sqrt{x})$
   c) $y = \ln\left(\frac{1}{\sqrt{x}}\right)$
   d) $y = \ln(\sin x)$
   e) $y = \ln(e^x + 1)$
   f) $y = \ln(\ln x)$

Differentiation of $\tan(f(x))$

21) Differentiate each of the following:
   a) $y = \tan\sqrt{x}$
   b) $y = \tan(\sin x - x^2)$
   c) $y = e^{\tan x}$
   d) $y = \tan(\ln x + e^{2x})$
   e) $y = \tan(\tan x)$

22) Show that
$$\frac{d}{dx}[\ln(\tan x)] = \operatorname{cosec} x \sec x$$

23) A curve is given by
$$y = \tan(x^2)$$
where $0 < x < \sqrt{\frac{\pi}{2}}$
Show that the curve is always increasing.

24) A curve is given by
$$y = \tan^2(ax)$$
where $0 < a \leq 2\pi$
Given that when $x = 1.5$, $\frac{dy}{dx} = 0$ find the possible values of a

25) An ice cream cone is to be made from 30π cm² of wafer.
The ice cream cone may be modelled with radius r cm, height y cm and slant height h cm and an assumption of negligible thickness.

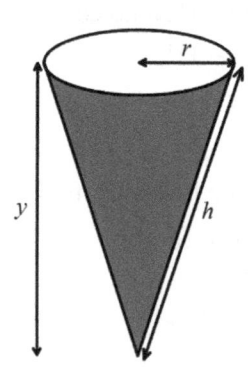

Curved surface area: $A = \pi r h$
Volume: $V = \frac{1}{3}\pi r^2 y$

a) Show that the volume is given by
$$V = \frac{1}{3}\pi\sqrt{900r^2 - r^6}$$

b) Hence, find the maximum volume of the cone.

2.23 Differentiation: Connected Rates of Change (answers on page 521)

1) State each of the following:
   a) The area of a circle in terms of its radius r
   b) The area of an equilateral triangle in terms of its side length x
   c) The volume of a cylinder V in terms of its radius r and its height h
   d) The volume of a sphere in terms of its radius r
   e) The surface area of a sphere in terms of its radius r
   f) The volume of a cone with base of radius r and height h in terms of r and h

2) Given
   a) $A = \frac{1}{2}\pi r^2$ and $\frac{dr}{dt} = 2$, find $\frac{dA}{dt}$ in terms of r
   b) $y = 10x^2$ and $\frac{dx}{dt} = -\frac{1}{2}$, find $\frac{dy}{dt}$ in terms of x
   c) $P = 2x^3 - x$ and $\frac{dx}{dt} = 3$, find $\frac{dP}{dt}$ in terms of x
   d) $M = -2x^5 + x^2$ and $\frac{dx}{dt} = \frac{1}{2}$, find $\frac{dM}{dt}$ in terms of x

3) The side of a square x is increasing at a rate of 10 cm s^{-1}.

   Find the rate at which the area A of the square is increasing when the side of the square is exactly 4 cm.

4) The radius of a circle r is increasing at a rate of 3 cm s^{-1}.

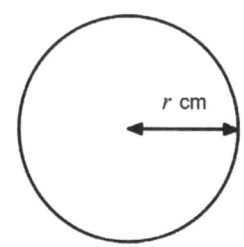

   Find the rate at which the area A of the circle is increasing when the radius of the circle is 5 cm giving your answer in terms of π

5) The radius of a circle r is decreasing at a constant rate of 0.1 cm s^{-1}.

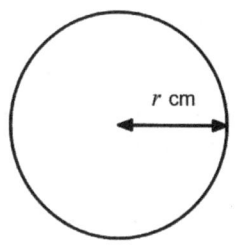

   Find the rate at which the perimeter of the circle is decreasing, giving your answer in terms of π

6) The side of an equilateral triangle x is increasing at a rate of 12 cm s^{-1}.

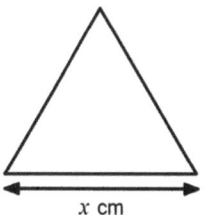

   Find the rate at which the area of the triangle is increasing when the side of the triangle is $2\sqrt{3}$ cm.

7) The surface area of a sphere is increasing at a rate of 12π cm^2 s^{-1}. Find the rate of change of the radius of the sphere exactly when its diameter is 6 cm.

8) A cube is growing so that its volume increases at a rate of 5 cm^3 s^{-1}.

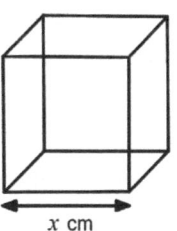

   The length of each of its sides is x cm.
   a) State the volume of the cube V and the surface area of the cube A in terms of x
   b) Show that the rate of increase of the surface area of the cube may be given by
   $$\frac{dA}{dt} = \frac{20}{x}$$
   c) Hence, find the rate of increase of the cube's surface area when it has a length of 4 cm.

2.23 Differentiation: Connected Rates of (answers on page 521)

9) A piston can slide inside a combustion cylinder which is closed at one end. The cylinder is filled with gas whose pressure p in atmospheric units is inversely proportional to the distance of the piston from the closed end, x cm.

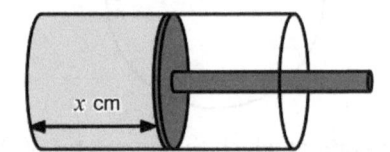

   It is known that the atmospheric pressure is 1.2 when x is 50 cm. At a different point in time, the distance from the closed end is 25 cm and the piston is moving away from the closed end at 5 cm per minute.
   Find the rate at which the pressure is changing at that point in time.

10) A sphere is growing so that its volume increases at a rate of 4 cm³ s⁻¹. Find the rate of increase of the cube's surface area when its diameter is 50 cm.

11) A cylindrical tank with radius r cm has water being poured into it at a rate of 10 cm³ s⁻¹.

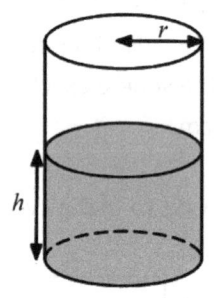

    The height of water after t seconds is given by h cm.
    a) Find an expression for the rate of increase of the water level in the tank in terms of r
    b) Given the tank has diameter 90 cm, find the rate of change of the water level giving your answer to 3 significant figures.

12) A cylindrical tank with radius 10 cm which initially starts full, has water dripping out of a hole in the bottom at a rate of 5 cm³ per minute. The height of water after t seconds is given by h cm.
    Find the rate of change of the water level giving your answer to 3 significant figures.

13) Water is being poured into an empty cylindrical cone of vertical height 10 cm at a constant rate of 6 cm³ s⁻¹. The cylindrical cone has an angle of 30° to the vertical.

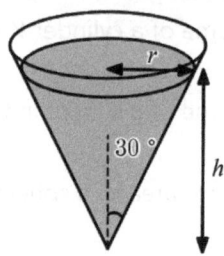

    a) Show that the volume of the water V when it is at height h cm is given by
    $$V = \frac{1}{9}\pi h^3$$
    b) Find $\frac{dV}{dh}$
    c) Find $\frac{dh}{dt}$ when the cone is full.

14) Sand is being poured onto a table and forms a pile in the shape of a cone. The height of the cone is equal to the diameter of the base.

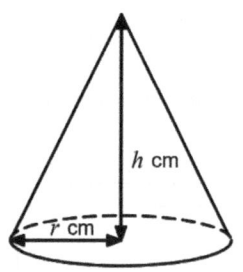

    The sand is poured at 3 cm³ per second. Find the rate at which the height of the pile is increasing when the base has a diameter of 4 cm.

15) The volume of a cube with sides of length x metres is increasing at a constant rate. Show that the rate of increase of the sides, is inversely proportional to the area of one side of the cube.

16) The radius of a circle r changes at a rate which is inversely proportional to the square of r
    a) Given that the rate is 0.5 cm s⁻¹ when the radius is 2 cm, find $\frac{dr}{dt}$
    b) Show that the rate at which the area of the circle, A, changes satisfies
    $$\frac{dA}{dt} = \sqrt{\frac{16\pi^3}{A}}$$

2.24 Differentiation: The Product Rule (answers on page 521)

1) Find $\frac{dy}{dx}$ for each of the following:
   a) $y = x(x-1)^2$
   b) $y = x^2(2x+7)^2$
   c) $y = 4x(7-x)^4$
   d) $y = (x+2)^2(3x+1)^3$
   e) $y = (3-x)^2(2x+9)^3$
   f) $y = \sqrt{2x+5}\,(4x+1)^2$
   g) $y = \left(\frac{1}{x} - 2x\right)\left(\frac{1}{x^3} + \frac{2}{x^5}\right)$

2) Three students attempt to find
$$\frac{d}{dx}\left((x^2-1)(x^3+5)\right)$$
Identify the correct version and spot the error(s) in the other two:

Student A:
$(x^2-1)(x^3+5) = x^5 - x^3 + 5x^2 + 5$
$\frac{dy}{dx} = 5x^6 - 3x^2 + 10x$

Student B:
$\frac{dy}{dx} = (2x)(x^3+5) + (x^2-1)(3x^2)$
$= 5x^4 - 3x^2 + 10x$

Student C:
$\frac{dy}{dx} = 2x \times 3x^2 = 6x^3$

3) Differentiate the following:
   a) $y = x \ln x$
   b) $y = \sqrt{x} \tan x$
   c) $y = \sin x \cos x$
   d) $y = \cos x \ln x$
   e) $y = 2^x \sin x$
   f) $y = (x^2+1) \ln x$

4) Differentiate and fully factorise the following:
   a) $y = xe^x$
   b) $y = x^2 4^x$
   c) $y = e^x \tan x$
   d) $y = 2^x \sin x$
   e) $y = (x^2 - 12)e^{-2x}$
   f) $y = e^{4x} \sin 2x$

5) Find the equation of the tangent line to the curve at the given point, giving your answers in the form $y = mx + c$:
   a) $y = x^3(x+2)^2$ at $x = 1$
   b) $y = x^5 \sqrt{3x+10}$ at $x = 2$
   c) $y = (x^2 - x + 4)e^x$ at $x = 0$

6) Fully factorise each of the following:
   a) $4x^2 + 2x^2(x+1)$
   b) $3(2x-1)^5 - (x+1)(2x-1)^4$
   c) $4(3x+1)^3 + (x+2)(3x+1)^2$
   d) $\frac{1}{3}(6x+5)^3 + (x-1)(6x+5)^2$
   e) $\frac{1}{2}(1-x)^{-5} + 4(1-x)^{-6}$
   f) $4(4x+1)^3(3x-1) + 9(3x-1)^2(4x+1)^2$

7) Solve each of the following:
   a) $2(x+1)^3 + (x-1)(x+1)^2 = 0$
   b) $\frac{1}{3}(2x+1)^{-2} + (2x+1)^{-3} = 0$
   c) $5(7x-1)^3(x-1) - 3(x-1)^2(7x-1)^2 = 0$

8) The curve shown below is
$$y = x^3(3x-5)^2$$

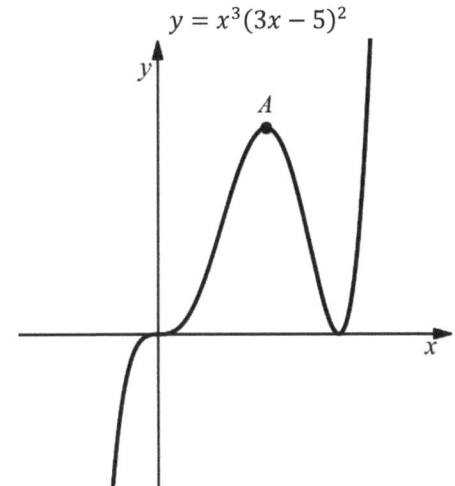

The minimum value is at the point A.
   a) Find the coordinates of A.
   b) State the coordinates of the other stationary points.
   c) State the values of y for which the $y = x^3(3x-5)^2$ has three unique solutions.

9) For each of the following, differentiate and fully factorise:
   a) $y = x(x+3)^7$
   b) $y = (x-3)(2x+1)^4$
   c) $y = 5x(2x+1)^4$
   d) $y = (1+x^2)e^{2x}$

10) A curve is given by
$$y = 4000xe^{-2x}$$
   a) Find the stationary point of the curve.
   b) Determine its type.

2.24 Differentiation: The Product Rule (answers on page 521)

11) The number of employees y, in a company in the first 12 months may be modelled by
$$y = 150 - 3t \sin\left(\frac{\pi}{6}t\right)$$
where t is the number of months after 1st January.
   a) Find the number of employees predicted by the model on 1st January.
   b) Find an expression for $\frac{dy}{dt}$
   c) Find the rate of change in the number of employees predicted by the model on 1st April.
   d) Interpret your result in part c) giving your answer in context.

12) Find the equation of the tangent line to the curve
$$y = x^4 e^{2x}$$
when $x = 1$. Give your answer in the form $y + pe^2 x + qe^2 = 0$.

13) A curve is given by
$$y = x \ln x, \quad x > 0$$
Find the stationary point and determine its type.

14) A curve is given by
$$y = e^{-x} \sin x, \quad x \geq 0$$
   a) Show that the curve has a stationary point at $\left(\frac{\pi}{4}, \frac{\sqrt{2}}{2} e^{-\frac{\pi}{4}}\right)$
   b) Determine its type.

Prerequisite knowledge for Q15: Geometric Sequences

15) A curve is given by
$$y = 10e^x \cos x, \quad x \geq 0$$
   a) Show that the x coordinates of the stationary points of the curve satisfy the equation $\tan x = 1$
   The y-values of the stationary points satisfy a geometric sequence,
   b) find the value of the common ratio r.

16) Find $f'(x)$ for each of the following:
   a) $f(x) = e^{10x^2} \sin(2x^3)$
   b) $f(x) = 5^{x^2} \cos(1 - x^2)$
   c) $f(x) = (x^2 + 1)^3 \ln(x^2)$

Prerequisite knowledge for Q17: Points of Inflection

17) Show that
$$y = 2^{3x^2}$$
is convex (concave upwards) for all values of x

18) Show that that the curve
$$y = (5 - 2x) \ln(x + 1)$$
has a stationary point at $x \approx 1.04$

19) For each of the following curves find $\frac{dy}{dx}$, find the coordinates of any stationary points.
   a) $y = x^2 e^x$
   b) $y = (3x - 1)^4 (3x^2 + 1)^5$
   c) $y = \sqrt{x}(x - 1)^2$, $x > 0$

20) For the curve given by
$$x = (y + 2) \ln(3y - 5), \quad y > \frac{5}{3}$$
   a) Find $\frac{dx}{dy}$ in terms of y
   b) Hence find the value of $\frac{dy}{dx}$ when $x = 0$

21) A curve has the equation
$$y = x^2 \tan^3 x, \quad -\frac{\pi}{2} < x < \frac{\pi}{2}$$
Show that the gradient of the curve at $x = \frac{\pi}{4}$ is $\frac{\pi}{8}(3\pi + 4)$

22) The height, h metres, from the ground of a bungee jumper is modelled by the equation
$$h = 10 + 25e^{-0.05t}\left(1 + \cos\left(\frac{\pi}{3}t\right)\right)$$
where t is the time in seconds after jumping.
   a) State the height above the ground of the start position.
   b) Find $\frac{dh}{dt}$ and the value of $\frac{dh}{dt}$ when $t = 3$
   c) Explain what is happening to the bungee jumper at precisely 3 seconds.

23) Given that
$$f(x) = x\sqrt{4x + 3}, \quad x \geq -\frac{3}{4}$$
   a) Find $f'(x)$
   b) Show that
$$f''(x) = \frac{12(x + 1)}{(3 + 4x)^{\frac{3}{2}}}$$
   c) Find the x-value of the turning point and determine its nature.

2.25 Differentiation: The Quotient Rule (answers on page 522)

Standard Functions

1) Find f'(x) and simplify:
   a) $$f(x) = \frac{3x+1}{2x-1}$$
   b) $$f(x) = \frac{x}{1-4x}$$
   c) $$f(x) = \frac{3x-3}{3x-1}$$
   d) $$f(x) = \frac{e^x}{x-3}$$
   e) $$f(x) = \frac{\ln x}{x^3}$$
   f) $$f(x) = \frac{\sin(10x)}{2x^3}$$
   g) $$f(x) = \frac{5^x}{x^2}$$
   h) $$f(x) = \frac{e^{2x}}{\cos x}$$
   i) $$f(x) = \frac{\ln 5x}{2x+1}$$
   j) $$f(x) = \frac{x^2-1}{\sin x}$$

2) Given that
   $$y = \frac{x-5}{\sqrt{x}}$$
   where $x \neq 0$
   Show that
   $$\frac{dy}{dx} = \frac{x+p}{qx\sqrt{x}}$$
   where p and q are integers to be found.

3) Given that
   $$y = \frac{3^x}{x^2 \ln 3}$$
   where $x \neq 0$
   Show that
   $$\frac{dy}{dx} = \frac{3^x(x \ln 3 - k)}{x^3 \ln 3}$$
   where k is an integer to be found.

4) Simplify each of the following:
   a) $$\frac{(x+1)^2 - 3x(x+1)}{(x+1)^4}$$
   b) $$\frac{(x-4)^2 - 2(x+1)(x-4)}{(x-4)^4}$$
   c) $$\frac{2(x+2)^3 - 4(7x+1)(x+2)^2}{(x+2)^6}$$
   d) $$\frac{4(x+3)^3(5x-1)^3 - 3(x+3)^4(5x-1)^2}{(5x-1)^6}$$
   e) $$\frac{-6(1-2x)^3(2x+5)^3 + 6(1-2x)^4(2x+5)^2}{(1-2x)^8}$$

5) Show that
   $$\frac{(x+1)^{\frac{1}{2}}e^x - \frac{1}{2}(x+1)^{-\frac{1}{2}}e^x}{x+1}$$
   may be rearranged to
   $$\frac{e^x(1+2x)}{2(x+1)^{\frac{3}{2}}}$$

6) Find f'(x) and simplify:
   a) $$f(x) = \frac{4x}{(x+1)^2}$$
   b) $$f(x) = \frac{9x-2}{(1+2x)^3}$$
   c) $$f(x) = \frac{e^{4x}}{(x+1)^2}$$

7) Find the x-value of the stationary points of the following:
   a) $$y = \frac{4x-35}{(2x+5)^4}$$
   b) $$y = \frac{(3-2x)^2}{(1+x)^3}$$
   c) $$y = \frac{e^{\frac{1}{2}x}}{2(x-2)^2}$$
   d) $$y = \frac{e^{ax}}{(x+a)^2}, \quad a \neq 0$$

2.25 Differentiation: The Quotient Rule (answers on page 522)

Reciprocal Trigonometric Functions

8) By writing each first as a quotient, show that:
   a) $$\frac{d}{dx}(\tan x) = \sec^2 x$$
   b) $$\frac{d}{dx}(\operatorname{cosec} x) = -\operatorname{cosec} x \cot x$$
   c) $$\frac{d}{dx}(\sec x) = \sec x \tan x$$

9) Find $\frac{dy}{dx}$ for each of the following:
   a) $y = \tan 3x$
   b) $y = \operatorname{cosec} 2x$
   c) $y = \sec \frac{1}{3}x$
   d) $y = -2 \tan \pi x$
   e) $y = \tan\left(-\frac{1}{5}x\right)$
   f) $y = -\sec 9x$
   g) $y = \operatorname{cosec}\left(-\frac{1}{2}x\right)$

Mix of Functions

10) Find when the curve
$$y = \frac{2x-1}{4x^2+3}$$
is decreasing, giving your answer in interval notation.

11) Find any stationary points of the curve
$$y = \frac{2x-4}{(4x-7)^2}$$
and determine their nature.

12)
$$f(x) = -\frac{1}{2}\sec \pi x, \quad 0 < x < 2$$
   a) Find the stationary point on the curve $y = f(x)$
   b) Determine its type.

Prerequisite knowledge for Q13: Points of Inflection

13) Show that the curve
$$y = \frac{\ln x}{x^2}, \quad x > 0$$
has exactly one single non-stationary point of inflection.

14) The curve
$$y = \frac{2\sqrt{x} + 3x - 9}{x-2}, \quad x > 0$$
is shown on the diagram below.

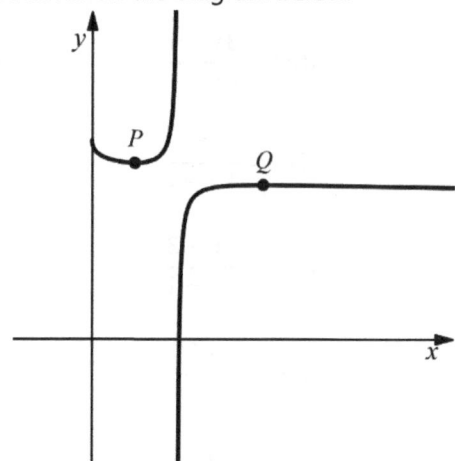

The stationary points are labelled P and Q.
   a) Find the coordinates of P and Q.
   b) Find the equation of the straight line passing through P and Q.

A triangle is bounded by the line passing through the points P and Q, the x-axis and the y-axis.
   c) Find the area of the triangle.

15) Given
$$f(x) = \frac{ax-1}{x^2-a}$$
show that the gradient at $x = 0$ is always -1.

16) A function is defined by
$$f(x) = \frac{3}{k+x^2}, \quad x > 0$$
and k is a positive constant.
   a) Show that f is a decreasing function for all values of k

The point of inflection on the curve is at $x = \frac{2}{\sqrt{3}}$
   b) Find the value of k

17) Given that
$$f(x) = \frac{e^{-x}}{k + 10x^2}$$
where k is a positive constant,
   a) find $f'(x)$
   b) Hence find the range of values of k such that $y = f(x)$ has at least one stationary point.

2.25 Differentiation: The Quotient Rule (answers on page 522)

The Chain Rule

18) Find $\frac{dy}{dx}$ for each of the following:
    a) $y = \tan(x^2 - 1)$
    b) $y = \csc(e^{2x})$
    c) $y = \sec(\sin x)$

19) Show that the curve
$$y = \frac{e^{3x-7}}{(2x-3)^2}$$
has exactly one stationary point.

20)
$$f(x) = \frac{(x^2+1)^2}{10(1-2x)}, \quad x \neq k$$
   a) State the value of k
   b) Show that the x-values of the maximum and minimum point on the curve $y = f(x)$ satisfy the equation $3x^2 - 2x - 1 = 0$
   c) Hence find the exact values of the maximum and minimum points.
   d) Sketch $f(x)$

21) Find $\frac{dy}{dx}$ for each of the following:
    a) $y = \frac{e^{x^2}}{x^2 - 1}$
    b) $y = \frac{\sin(x^3)}{e^{2x+1}}$
    c) $y = \frac{3^{\cos x}}{x^2}$

22)
   a) Find
   $$\frac{d}{dx}\left(\frac{x^2-1}{x-3}\right)^4$$
   b) Hence show that the tangent line the point $x = 2$ is
   $$y - 756x + 1431 = 0$$

23)
   a) Find
   $$\frac{d}{dx}\left(\frac{\ln x}{x^2}\right)^2$$
   b) Hence show that at $x = 1$ there is a stationary point.

24) The sketch below shows the curve
$$y = \frac{5 \sin 3x}{e^{\sqrt{3}x+2}}$$
for $0 \leq x \leq \frac{2\pi}{3}$

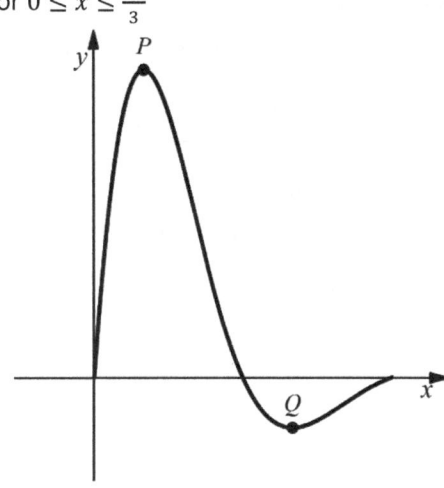

The stationary points are labelled P and Q.
   a) Show that the x coordinates of the stationary points are solutions to the equation
   $$\tan 3x = \sqrt{3}$$
   b) Find the x values of the stationary points of the curve.

25)
$$f(x) = \frac{\ln(x+1)}{(x+1)^2}, \quad x \geq 0$$
   a) Show that
   $$f'(x) = \frac{1 - 2\ln(x+1)}{(x+1)^3}$$
   The diagram below shows the area enclosed by the curve
   $$y = \frac{1 - 2\ln(x+1)}{(x+1)^3}$$
   and the x-axis and y-axis.

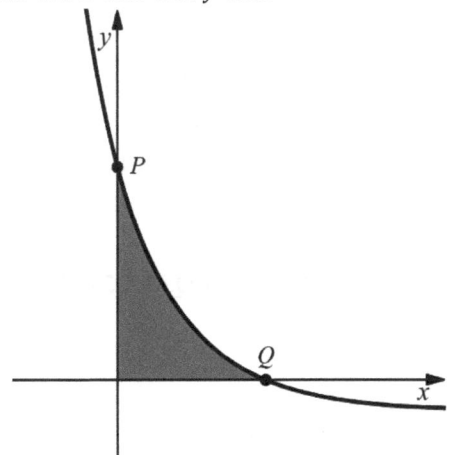

   b) Show that the shaded region is
   $$\frac{1}{2e}$$

295

2.25 Differentiation: The Quotient Rule (answers on page 522)

Prerequisite knowledge for Q26-Q29: The Product Rule

26) Two functions
$$f(x) = (1 - 2x)(2x^2 + ax + b)$$
and
$$g(x) = \frac{9(ax + b)}{x^2 + 2}$$
where a and b are integers.
Given that $f'(1) = g'(1) = -8$ find the values of a and b

27) Find $f'(x)$ for each of the following:
   a) $$f(x) = \frac{x^2 e^x}{2x - 1}$$
   b) $$f(x) = \frac{2^x \sin x}{\sqrt{x}}$$
   c) $$f(x) = \frac{x \ln x}{(x - 1)^2}$$

28) A curve has equation
$$y = \frac{(x^2 - 2x - 3)e^x}{\sqrt{2x + 1}}$$
Show that that equation of the tangent to the curve at the point $x = 0$ is $y = -2x - 3$

29) A curve is given by
$$y = \frac{e^{\frac{1}{2}x}}{x^2 + 2x + 1}$$
for $x \geq 0$
   a) Find the x-value of the single stationary point on the curve.
   b) Show that
   $$\frac{d^2y}{dx^2} = \frac{e^{\frac{1}{2}x}(x^2 - 6x + 17)}{4(x + 1)^4}$$
   c) Hence show why the curve has no points of inflection.

Prerequisite knowledge for Q30-Q31: Double Angle Formulae

30) Show that
$$\frac{d}{dx}\left(\frac{\sin 2x}{\cos x}\right) = 2 \cos x$$

31) Show that
$$\frac{d}{dx}\left(\frac{2 \sin x}{5 \sin x + 5 \cos x}\right) = \frac{p}{q(1 + \sin 2x)}$$
where p and q are integers and $\frac{p}{q}$ is in its most simplified form.

Prerequisite knowledge for Q32-Q34: Domain and Range of Functions

32) The sketch below shows the curve $y = f(x)$ where
$$f(x) = \frac{x - 1}{(x - 3)^2}, \quad x \neq k$$

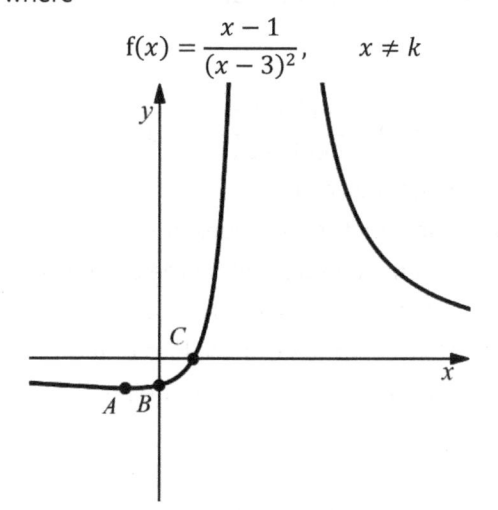

The stationary point is labelled A. The intercepts with the axes are labelled P and Q.
   a) State the value of k
   b) Find the coordinates of the points B and C.
   c) Find the coordinates of A.
   d) Sketch on separate axes, labelling the new values of A, B and C:
      i) $f(x) + 1$
      ii) $f(2x)$

33) Given that
$$f(x) = \frac{2 \ln x - 5}{\ln x - 1}$$
   a) State the maximum domain of f
   b) Show that $f(x)$ is an increasing function.
   c) Find the set of values for which $f(x) < 0$

34) Given that
$$f(x) = \frac{2^x}{x - 1}$$
   a) State the maximum possible domain of f
   b) Find $f'(x)$ and hence find the coordinate of the single stationary point of f giving your answer to 3 significant figures.
   c) Hence find the range of f

2.26 Differentiation: Implicit Differentiation (answers on page 524)

1) Find
   a) $\frac{d}{dx}(x+y^2)$
   b) $\frac{d}{dx}(xy)$
   c) $\frac{d}{dx}(x^2 y)$
   d) $\frac{d}{dx}(e^x \sin y)$
   e) $\frac{d}{dx}(e^y \cos x)$
   f) $\frac{d}{dx}(x \ln y - 5)$
   g) $\frac{d}{dx}(2^x y + 3)$

2) For each of the following, find $\frac{dy}{dx}$:
   a) $y^2 = 8x$
   b) $xy = 12$
   c) $x^3 + xy^2 - y^2 = 0$
   d) $\ln x + y^2 = 2x$
   e) $5x^2 - 2xy + 5y^2 = 25$
   f) $\sin x + xe^y = 1$
   g) $x \cos y - 5 = x$
   h) $x \ln y + y \ln x = 10$

3) By using implicit differentiation, find the tangent line to the circle
   $$x^2 + y^2 = 25$$
   at the point $(4, 3)$

4) Use implicit differentiation on $e^y = x$ to show that
   $$\frac{d}{dx}(\ln x) = \frac{1}{x}$$

5) Use implicit differentiation on $\ln y = \ln a^x$ to show that
   $$\frac{d}{dx}(a^x) = a^x \ln a$$

Differentiation of Inverse Trigonometric Functions

6) Use implicit differentiation on $\sin y = x$ to show that
   $$\frac{d}{dx}(\arcsin x) = \frac{1}{\sqrt{1-x^2}}$$

7) Use implicit differentiation on $\cos y = x$ to show that
   $$\frac{d}{dx}(\arccos x) = -\frac{1}{\sqrt{1-x^2}}$$

8) Use implicit differentiation on $\tan y = x$ to show that
   $$\frac{d}{dx}(\arctan x) = \frac{1}{1+x^2}$$

Mix

9) Find the value(s) of $\frac{dy}{dx}$ at the given point:
   a) $\sin 2y - x = 2$ at $(-2, \pi)$
   b) $x^2 y = x$ when $x = 2$
   c) $xe^y - x^2 + 4 = 0$ at $y = \ln 3$

10) Show that if $x^2 \sin y = 4$ then
    $$\frac{dy}{dx} = -\frac{8}{x\sqrt{16-x^4}}$$

11) Show that if $x^2 \tan y = 1$ then
    $$\frac{dy}{dx} = -\frac{2x}{x^4+1}$$

12) The equation of a curve is given by
    $$x^2 + 2y^3 = 2$$
    a) Find the coordinates of the stationary points on the curve.
    b) Find the coordinates of the points where the curve has an infinite gradient.

Prerequisite knowledge for Q13: Inverse Functions

13)
   a) Sketch the graphs of
      $$y = f(x) = x^2 + 4$$
      for $x \geq 0$ and its inverse $y = g(x)$ on the same axes, labelling any intersections with the axes.
   b) Show that
      $$f'(3) = \frac{1}{g'(13)}$$

14) Given that
    $$f(x) = x(x-3), \quad x \geq 3$$
    calculate the gradient of the graph $y = f^{-1}(x)$ at the point $(10, 5)$

15) The equation of a curve is given by
    $$(x-2)^2 + 2 = e^y$$
    Find the coordinates of the point where the tangent line is parallel to the x-axis.

2.26 Differentiation: Implicit Differentiation (answers on page 524)

16) Show that the curve
$$xy^2 + \sqrt{y} = 5, \quad y > 0$$
has no stationary points.

17) A curve has equation
$$(2x + y)^2 = x - y$$
   a) Find the tangent lines to the curve at $x = \frac{1}{4}$.
   b) Hence, show that the two tangent lines meet at the point $\left(-\frac{5}{12}, \frac{1}{3}\right)$.
   c) Find the exact area of the triangle bound by the two tangent lines and the x-axis.

18) Show that for the curve
$$5x^2 + 12xy - 4y^2 = 10$$
that there are no tangent lines parallel to the line $y = \frac{3}{2}x$.

19) A curve has equation
$$(x + y)^3 = 10x^2 - 2y$$
   a) Show that the point $(1, 1)$ lies on the curve.
   b) Show that the tangent to the curve at the point $(1, 1)$ is given by
   $$7y - 4x = 3$$

20) The curve
$$x^3 + y^2 = 5$$
is shown in the diagram below.

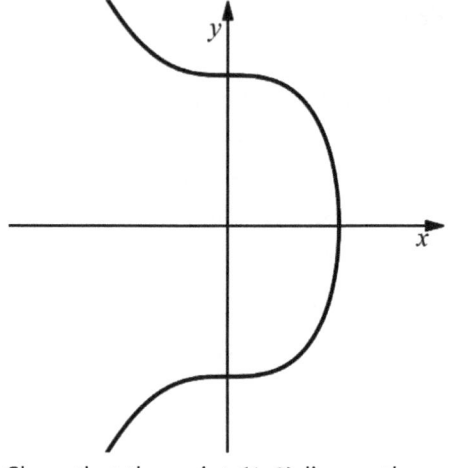

   a) Show that the point $(1, 2)$ lies on the curve.
   b) Find the normal line to the curve at the point $(1, 2)$ in the form $ax + by = c$ where a, b and c are integers to be found.
   c) Show that the normal line found in part b) does not intersect the equation again.

21) The equation of a curve is given by
$$x^2 \sin y + \cos y = \frac{1}{2}x$$
   a) Show that $\left(0, \frac{3\pi}{2}\right)$ lies on the curve.
   b) Show that
   $$\frac{dy}{dx} = \frac{4x \sin y - 1}{2 \sin y - 2x^2 \cos y}$$
   c) Find the equation of the tangent to the curve at the point $\left(0, \frac{3\pi}{2}\right)$ giving your answer in the form $y = mx + c$

22) The equation of a curve is given by
$$x^2 - 2x \ln y = 1$$
   a) Show that $(1, 1)$ lies on the curve.
   b) Show that
   $$\frac{dy}{dx} = \frac{y(x - \ln y)}{x}$$
   c) Find the equation of the normal to the curve at the point $(1, 1)$ giving your answer in the form $y = mx + c$

23) The curve
$$2(x - 5)^2 - 2xy + (y + 1)^2 = 1$$
is shown in the diagram below.

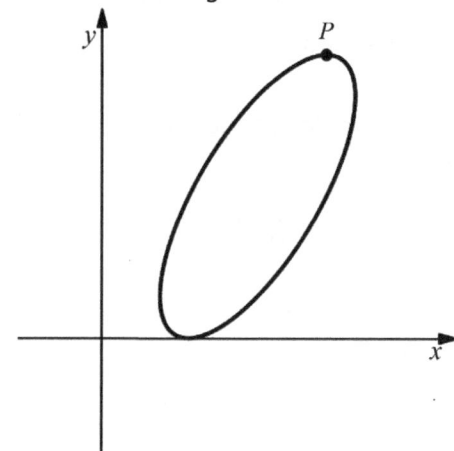

   a) Find the coordinates of the maximum point P of the ellipse.
   b) Show that the distance between where the ellipse meets the x-axis and the point P is $k\sqrt{5}$ where k is an integer to be found.

24) The curve
$$y^2 + \ln x = k, \quad x > 0$$
where $k > 0$ passes through the point $(1, -\sqrt{2})$
   a) Find the value of k.
   b) Find the exact coordinates at which the tangent to the curve is parallel to the y-axis.

2.26 Differentiation: Implicit Differentiation (answers on page 524)

25) The equation of a curve is given by
$$xe^y = 32$$
   a) Show that the point $(2, 4\ln 2)$ lies on the curve.
   b) Find the equation of the tangent line and the normal line at the point $(2, 4\ln 2)$
   c) Find the values where the equation of the tangent line and the normal line meet the x-axis.
   d) Hence show that the triangle bounded by the tangent and normal line at $(2, 4\ln 2)$ and the x-axis has an area of $20(\ln 2)^2$

Prerequisite knowledge for Q26: Domain and Range of Functions

26) A curve is given by
$$x^2 - 2xy + 5y^2 = 20$$
and is shown on the diagram below.

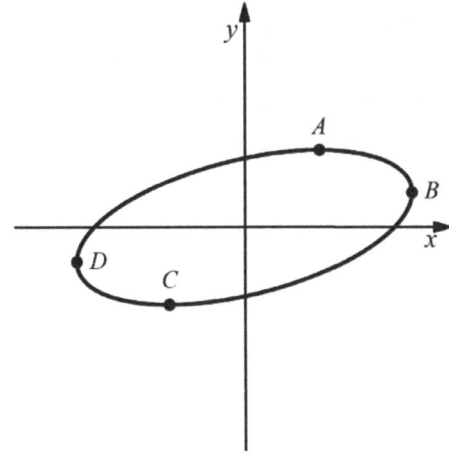

The points A, B, C and D lie on the curve.
The domain is such that the curve lies between the x-values of D and B.
The range is such that the curve lies between the y-values of C and A.
Find the coordinates of A, B, C and D.

27) A curve is given by
$$xy^2 + x^3y = k$$
where $k \neq 0$
   a) Show that the curve never crosses the x-axis.
   b) Show that
$$\frac{dy}{dx} = \frac{-y^2 - 3x^2y}{2xy + x^3}$$
   c) Given that a stationary point occurs when $x = -1$, find the value of k

28) A curve has equation
$$3x + xy + y^2 = k$$
where k is a positive integer such that $k < 9$

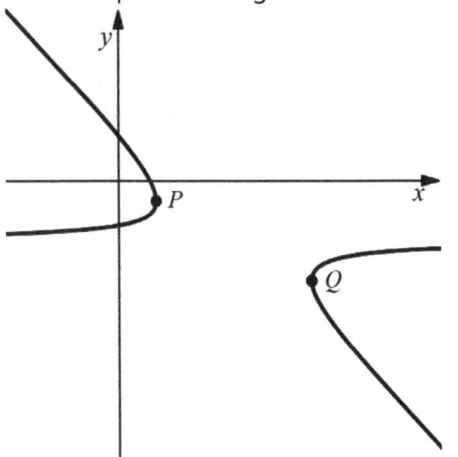

There are two points $P(a, b)$ and $Q(5a, 5b)$ on the curve such that the tangents to the curve are parallel to the y-axis.
   a) Find the values of a, b and k
   b) State the coordinates of the points P and Q

29) A curve has equation
$$(x + y)^4 = 2x - y + 8$$
   a) Show that the point $(-2, 3)$ lies on the curve.
   b) Find the tangent line to the curve at the point $(-2, 3)$ in the form $ax + by = c$ where a, b and c are integers to be found.
   c) Prove that the tangent line found in part b) does not intersect the equation again.

Prerequisite knowledge for Q30: Newton-Raphson Method

30) The percentage of woodland cover in the UK, P, since 1965 may be modelled by the equation
$$P = 8.5 + \arctan(t - 31)$$
where t is the number of years since 1965.
   a) Show that
$$\frac{d}{dx}(\arctan x) = \frac{1}{x^2 + 1}$$
   b) Hence, find
$$\frac{dP}{dt}$$
   c) Use Newton-Raphson's method starting at $t_1 = 30$ to find the year in which, according to the model, the percentage of woodland coverage is 9.95%.

Numerical Methods – Location of roots (answers on page 525)

1) Show that the following equations have a solution in the given interval:
   a) $x^3 - 4x - 1 = 0$ for $x \in [2, 3]$
   b) $2 + x - e^x = 0$ for $x \in [1, 2]$
   c) $x - 4 - 2\ln x = 0$ for $x \in [8, 9]$
   d) $1 - x \times 5^x = 0$ for $x \in [0, 1]$

2)
$$f(x) = \frac{1}{x-3}$$
   a) Find f(2) and f(4)
   b) Sketch f(x)
   c) Explain why a change in sign of f(x) between $x = 2$ and $x = 4$ is not sufficent to show that there is a root.

3) Draw a sketch for each of these situations for a continuous function:
   a) no change of sign, no real roots.
   b) no change of sign, one real root.
   c) no change of sign, two real roots.

4) Draw a sketch for each of these situations for a discontinuous function:
   a) change of sign, no real roots
   b) change of sign, one real root
   c) change of sign, two real roots
   d) no change of sign, no real roots
   e) no change of sign, one real root
   f) no change of sign, two real roots

Prerequisite knowledge for Q5: Radians

5)
$$f(x) = \frac{1}{2x} - \tan x$$
   a) Find f(−1), f(1) and f(2)
   b) Explain if there are any roots between $x = -1$ and $x = 1$, and between $x = 1$ and $x = 2$

6)
   a) Show that the equation $e^x - \frac{1}{x} = 0$ has a root between $x = 0.5$ and $x = 1$
   b) Explain whether or not there is a root between $x = -1$ and $x = 1$

7) Sketch the graphs of $y = e^x$ and $y = 4 - x^2$ showing that $e^x = 4 - x^2$ has two solutions.

8) Sketch the graphs of $y = \sqrt{x-1}$ and $y = \frac{1}{2}x$ showing that $\sqrt{x-1} = \frac{1}{2}x$ has one solution.

9) The diagram shows the graphs of
$$y = \frac{1}{x} \text{ and } y = x^3 + 3$$

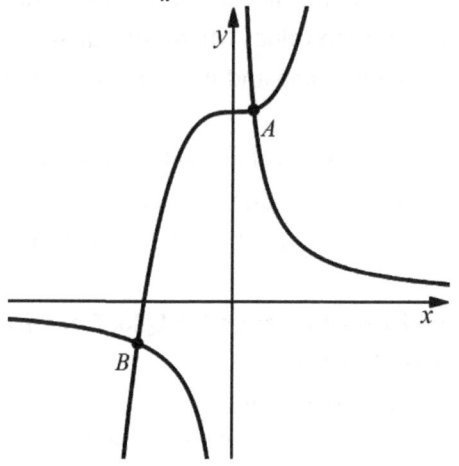

   a) Show that the x-coordinates of point A and point B satisfy
   $$x^4 + 3x - 1 = 0$$
   b) Show that the x-coordinate of point A lies between $x = 0$ and $x = 1$
   c) Show that the x-coordinate of point B lies between $x = -2$ and $x = -1$

10) The diagram shows the graphs of
$$y = x^2 \text{ and } y = -\frac{1}{4}x^3 + 4$$
with point of intersection P.

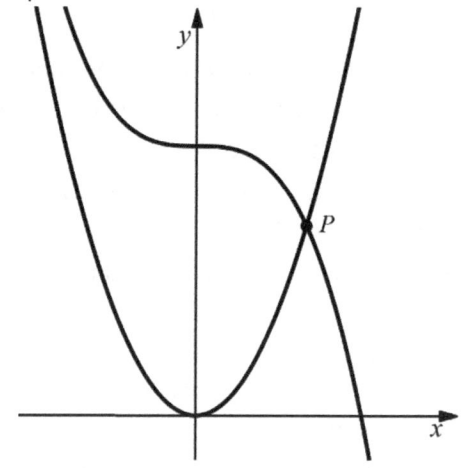

   a) Show that the x-coordinate of P is the solution of
   $$x^3 + 4x^2 - 16 = 0$$
   b) Show that the x-coordinate of point P lies between 1.67855 and 1.67865
   c) Write down the value of α to 4 decimal places.

2.28 Numerical Methods: $x = g(x)$ Method (answers on page 525)

1) Show that the equation
$$x^2 + x - 4 = 0$$
may be arranged in to each of the following:
   a) $x = \sqrt{4 - x}$
   b) $x = \frac{4}{x+1}$

2) Show that the equation
$$x^3 + 2x - 5 = 0$$
may be arranged in to each of the following:
   a) $x = \sqrt[3]{5 - 2x}$
   b) $x = \frac{5}{x^2+2}$

3) Given that
$$f(x) = x^3 - x^2 - 2x - 3$$
   a) Show that the equation $f(x) = 0$ may be written in the form
$$x = \sqrt{\frac{3 + 2x}{x - 1}}$$
   b) Use your answer to part b) to write a recurrence relation for finding a solution to $f(x) = 0$ then starting with $x_1 = 2$, find a solution to 4 decimal places.

4) Given that
$$f(x) = x^2 \ln x - 5x$$
   a) Show that $f(x) = 0$ has a solution between $x = 3$ and $x = 4$
   b) Show that the equation $f(x) = 0$ may be written in the form
$$x = \sqrt{\frac{5x}{\ln x}}$$
   c) Use your answer to part b) to write a recurrence relation for finding a solution to $f(x) = 0$ then starting with $x_1 = 3$, find a solution to 4 decimal places.

Prerequisite knowledge for Q5 b): Radians

5) For each of the following recurrence relations, find x_2, x_3 and x_4 giving your answer to 4 decimal places:
   a) $x_{n+1} = \sqrt{x_n + 3}$ where $x_1 = 5$
   b) $x_{n+1} = 2\sin(x_n)$ where $x_1 = 1$
      (x_n is measured in radians)
   c) $x_{n+1} = 8 - \frac{1}{10} e^{0.4x}$ where $x_1 = 0$
   d) $x_{n+1} = 0.1x_n^3 - 0.5x_n + 2$ where $x_1 = 2$

6) For each graph below, draw a diagram to show the convergence or divergence cobweb or staircase starting at the indicated point, labelling each of x_2, x_3 and x_4.
Identify whether it is a convergent or a divergent cobweb or staircase diagram.

a)

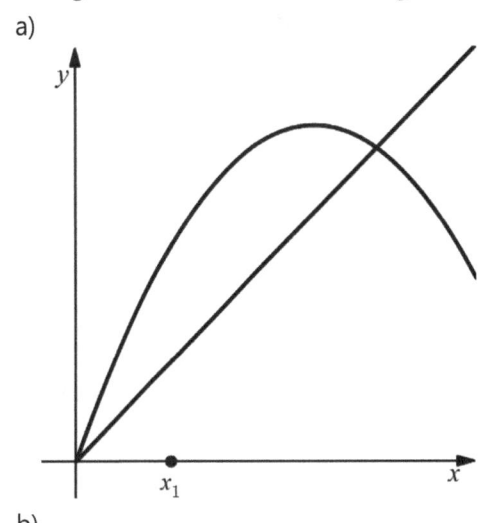

b)

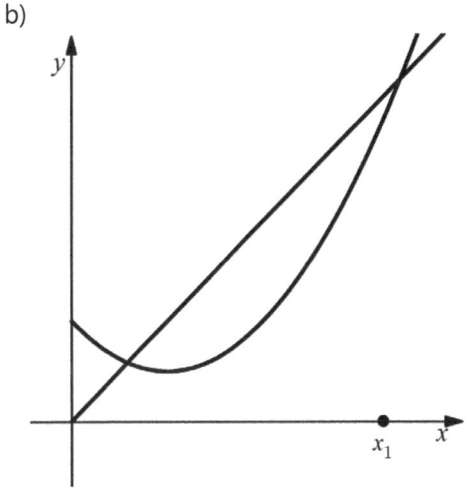

c)

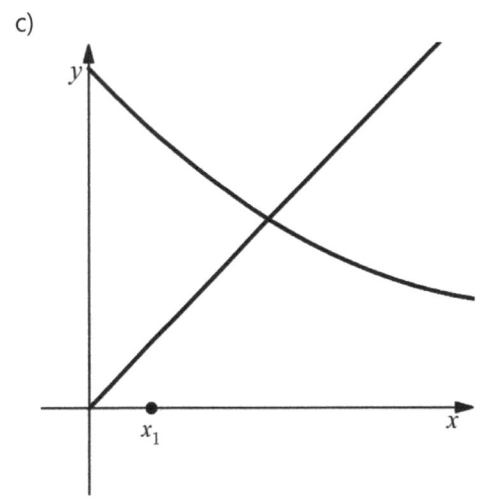

2.28 Numerical methods: $x = g(x)$ Method (answers on page 525)

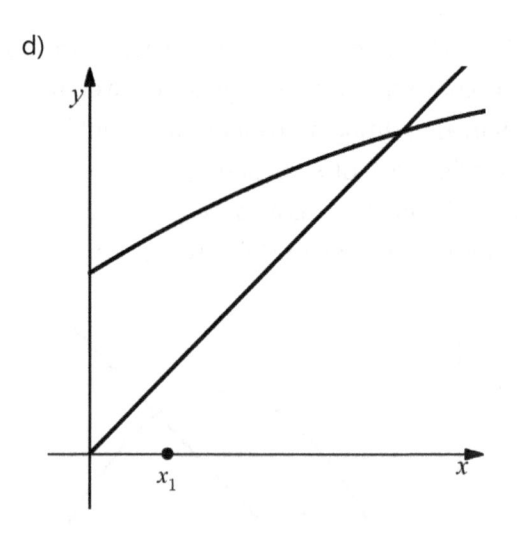

d)

7) The diagram shows
$$y = 2x^2 - 1.5x + 0.5 \text{ and } y = x$$
on the same graph. They intersect when $x = \frac{1}{4}$ and $x = 1$

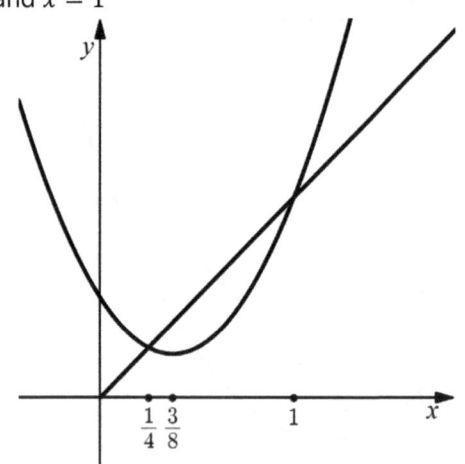

Describe what happens in terms of a cobweb or staircase diagram for each of the following starting points:
a) $-\frac{1}{4} < x_1 < \frac{1}{4}$
b) $x_1 = \frac{1}{4}$
c) $\frac{1}{4} < x_1 < 1$
d) $x_1 = 1$
e) $x_1 > 1$

Prerequisite knowledge for Q8 b): Radians

8) Each of the following iterative formulae has a sequence which converges to a limit $x = \alpha$
In each case, deduce an interval of width 0.001 in which α lies:
a) $x_{n+1} = 2 + \ln x_n$ where $x_1 = 3$
b) $x_{n+1} = \cos^2(x_n) + 1$ where $x_1 = 1.15$
(x_n is measured in radians)
c) $x_{n+1} = \frac{3}{2} x_n - \frac{1}{10} x_n^4$ where $x_1 = 1$

9) An iterative formula for finding root α is given by $x_{n+1} = g(x_n)$. By considering $g'(\alpha)$ in each of the following cases, explain whether the iterative formula will converge or diverge to the given root (given to 3 decimal places) from a point considered close to the root.
a) $x_{n+1} = x_n^5 - 2x_n^4 - 5$ where $\alpha = 2.273$
b) $x_{n+1} = e^{x_n} - 2$ where $\alpha = 1.146$
c) $x_{n+1} = e^{x_n} - 2$ where $\alpha = -1.841$

10) Harry is trying to find the values of the three roots of the equation
$$y = x^3 - 3x^2 - x + 4$$
He rearranges the equation into four of the possible iterative formula $x_{n+1} = g(x_n)$ to find a solution α, including finding an initial point:

A: $x_{n+1} = x_n^3 - 3x_n^2 + 4$ $\quad x_1 = 2$
B: $x_{n+1} = \sqrt[3]{3x_n^2 + x_n - 4}$ $\quad x_1 = 2.8$
C: $x_{n+1} = \dfrac{x_n - 4}{x_n^2 - 3x_n}$ $\quad x_1 = 1.2$
D: $x_{n+1} = -\sqrt{\dfrac{x_n^3 - x_n + 4}{3}}$ $\quad x_1 = -1$

a) Find the three roots using the iterative formulae B, C and D at the given initial points above, giving your answers to 2 decimal places.
b) By considering $g'(\alpha)$ on iterative formula A, explain why this iterative process does not converge to a root for each of the roots that you found in part a).

11)
a) Sketch the graphs $y = \frac{1}{x}$ and $y = \sin x$ to show that the equation
$$\frac{1}{x} = \sin x$$
has exactly one solution for $0 < x < \frac{\pi}{2}$
b) Show that the solution to the equation lies between $x = 1$ and $x = 1.2$
c) By using a suitable iterative formula, obtain an approximation solution to
$$\frac{1}{x} = \sin x$$
to 4 decimal places.

2.28 Numerical Methods: $x = g(x)$ Method (answers on page 525)

12)
  a) Sketch the graphs
  $y = x$, $y = e^{x-2}$ and $y = 2 + \ln x$
  on the same axes, for $0 < x < 4$
  b) Explain why the solutions to the equation
  $$x = e^{x-2}$$
  are also solutions to the equation
  $$e^{x-2} = 2 + \ln x$$

13) The number of tickets sold N for an event in the first 48 hours of sales is modelled by
  $$N = 3t - e^{0.1t}$$
  where t is the time in hours passed. When the number of tickets sold drops to zero and $t = T$
  a) Show that T satisfies $T = 10\ln(3T)$
  b) Write a recurrence relation to find the solutions to $T = 10\ln(3T)$ and starting with $T_1 = 48$, find a solution to the equation to 3 decimal places.
  c) Hence state the number of hours that have passed, to the nearest hour, when the number of tickets being sold is zero.

Prerequisite knowledge for Q14: Radians

14) A circle with centre O has a chord such that the angle $AOB = \theta$ radians where $0 < \theta < \pi$

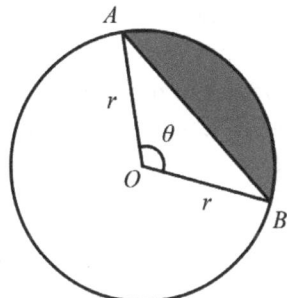

The area of the segment cut off by AB is $\frac{1}{6}$ of the area of the circle.
  a) Show that
  $$\theta - \sin\theta = \frac{\pi}{3}$$
  The solution lies between $\theta = 1$ and $\theta = 3$
  b) Using an iterative method, find the solution correct to 2 decimal places.

15) Giving your answers to 3 decimal places, use an iterative method to find the x-coordinates of the points of intersection of:
  a) $y = x$ and $y = \cos x$
  b) $y = x^2$ and $y = x\log_3(0.5x + 1)$

Prerequisite knowledge for Q16: Inverse Trigonometric Functions

16)
  a) Sketch the graphs of $y = 3\arctan(2x)$ and $y = x^2$ on the same axes for $-4 \leq x \leq 4$
  b) Show that there is a solution to
  $$x^2 - 3\arctan(2x) = 0$$
  between $x = 1.8$ and $x = 2.2$
  c) By using a suitable iterative formula, obtain an approximation solution to $x^2 - 3\arctan(2x) = 0$ to 3 decimal places.

Prerequisite knowledge for Q17: Differentiation of $\ln x$

17) The diagram shows the graph of
$$y = -\frac{2x^2 + 24}{3\sqrt{x}} - 4\ln x$$
The stationary point of the curve is labelled P.

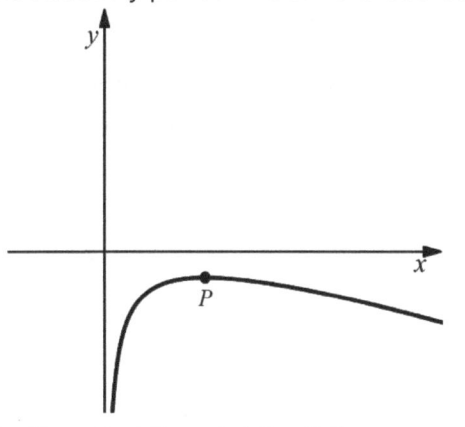

  a) Show that the point P satisfies
  $$0 = \sqrt{x} - \frac{1}{4}x^2\sqrt{x} - x$$
  b) Use the iteration formula
  $$x_{n+1} = \sqrt{x_n} - \frac{1}{4}x_n^2\sqrt{x_n}$$
  starting at $x_1 = 1$ to find the x-coordinate of point P to 2 decimal places.

Prerequisite knowledge for Q18: Double Angle Formulae

18) An approximate solution to an equation can be found using the iterative formula
$$x_{n+1} = \frac{1}{4}\arccos x_n$$
  a) By using $x_1 = 0.3$, find x_2, x_3 and x_4 giving your answer to 4 decimal places.
  b) Show that the iterative formula may be used to find a solution to
  $$x - 1 + 8\sin^2 x \cos^2 x = 0$$

2.29 Numerical Methods: Newton-Raphson Method (answers on page 526)

1) Use the Newton Raphson Method with the value of x_1 given to find the next approximation, x_2, to find a root of each of the following equations, giving your answers to 2 decimal places:
   a) $x^4 + x^2 - 80 = 0$ using $x_1 = 3$
   b) $x^3 + 2x^2 - 5x - 4 = 0$ using $x_1 = 2$
   c) $e^x + x - 4 = 0$ using $x_1 = 1.5$

2) Use the Newton Raphson Method with the value of x_1 given to find the next approximation, x_2, to find a solution of each of the following equations, giving your answers to 2 decimal places:
   a) $2x^4 + x^3 = 5$ using $x_1 = 1.1$
   b) $\frac{1}{6}x^6 - 8x = 4$ using $x_1 = 2$
   c) $e^{2x} - x = 3$ using $x_1 = 0.65$

3) For each graph below, draw a diagram to show the Newton-Raphson starting at the indicated point, labelling each of x_2, x_3 and x_4.
   a) 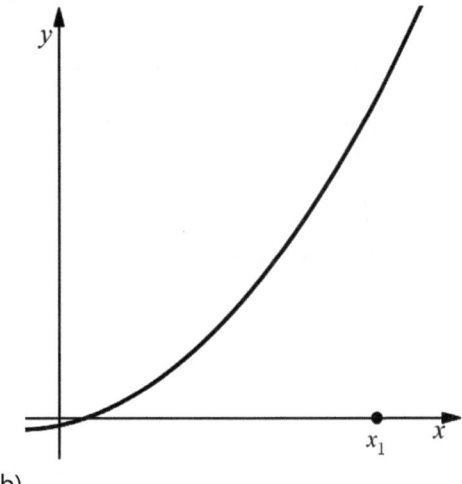
   b) 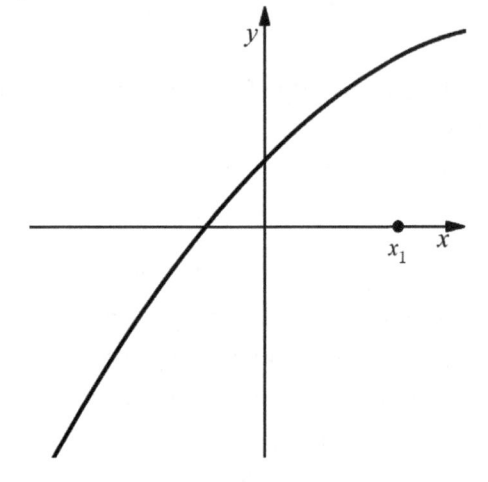
   c) 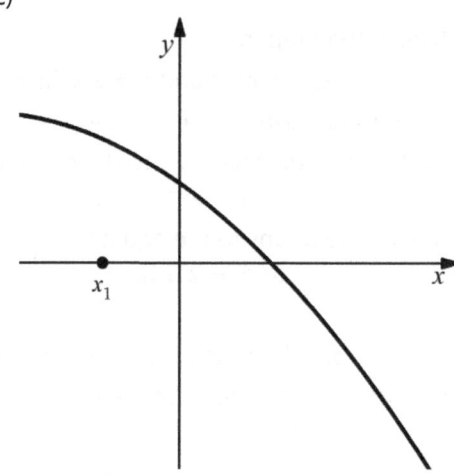
   d)

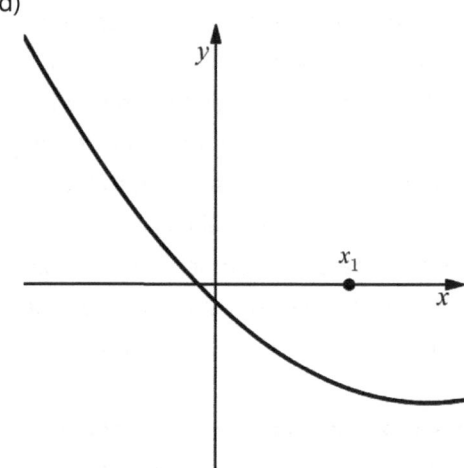

4) The diagram shows
$$y = 2x^2 - 4x - 1$$
with the x-values of the roots and the minimum point labelled on the x-axis.

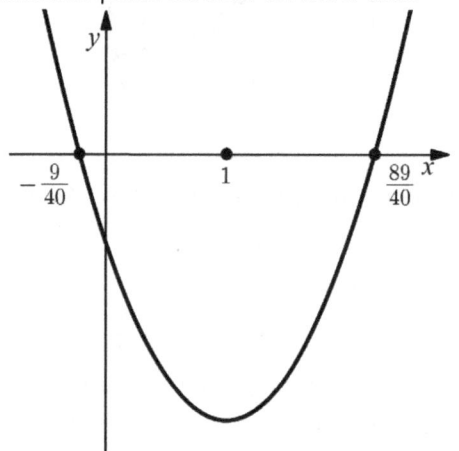

Describe what happens when the Newton-Raphson iterative method is used with each of the following starting points:
   a) $x_1 < 1$
   b) $x_1 = 1$
   c) $x_1 > 1$

2.29 Numerical Methods: Newton-Raphson Method (answers on page 526)

5) The diagram shows
$$y = x^3 + x^2 - 3x + 1$$
with the x-values of the roots and the turning points labelled on the x-axis.

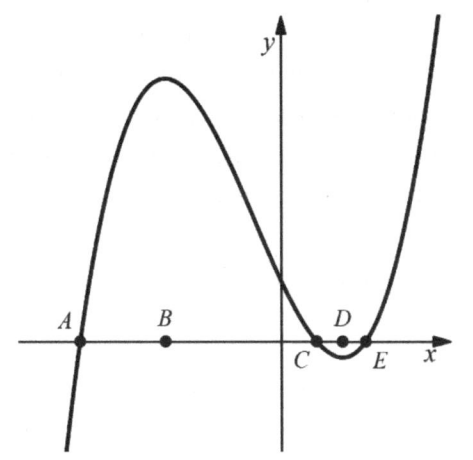

Describe what root is reached when using the Newton-Raphson method for each of the following starting points:
a) $x_1 = A$
b) $x_1 < B$
c) $x_1 = B$
d) $B < x_1 < D$
e) $x_1 = C$
f) $x_1 = D$
g) $x_1 > D$
h) $x_1 = E$

6) By considering the root of the equation
$$x^3 - 20 = 0$$
near $x = 3$, use Newton-Raphson method to find $\sqrt[3]{20}$ correct to 3 decimal places.

7) The equation
$$3x^3 + 4x^2 - 3 = 0$$
has one real root.
   a) By considering a change in sign, show that there is a root between $x = 0$ and $x = 1$
   b) Show that, the Newton-Raphson formula for finding the root can be written in the form:
   $$x_{n+1} = \frac{6x_n^3 + 4x_n^2 + 3}{9x_n^2 + 8x_n}$$
   c) Use the Newton-Raphson method with $x_1 = 1$ to find x_2 and x_3 to 4 decimal places.
   d) Explain why the method fails when $x_1 = 0$ or $x_1 = -\frac{8}{9}$

Prerequisite knowledge for Q8: Arithmetic Series

8) An arithmetic series has first term 2 and common difference 3. The sum of the first n terms is $\frac{12}{n}$
   a) Show that n satisfies the equation
   $$3n^3 + n^2 - 24 = 0$$
   b) Apply the Newton-Raphson method to the equation above using $n_1 = 2$ to obtain an answer correct to 4 significant figures.

Prerequisite knowledge for Q9: Geometric Series

9) A geometric series has first term 2 and common ratio x. The sum of the first 8 terms is 100.
   a) Show that x satisfies the equation
   $$x^8 - 50x + 49 = 0$$
   b) Apply the Newton-Raphson method to the equation above using $x_1 = 1.5$ to obtain an answer correct to 4 significant figures.
   c) Hence determine the sum to infinity of the geometric series giving your answer to 3 significant figures.

Prerequisite knowledge for Q10: Modulus Functions

10)
   a) Sketch the graph of
   $$y = 5 - \frac{1}{x^2} \text{ and } y = 4 - |2x|$$
   on the same axes showing clearly the points of intersection
   b) Show that α, the negative root of
   $$5 - \frac{1}{x^2} = 4 - |2x|$$
   satisfies the equation
   $$2x^3 - x^2 + 1 = 0$$
   c) Hence show that α may be found using a Newton-Raphson iterative formula which may be written in the form
   $$x_{n+1} = \frac{-4x_n^3 + x_n^2 + 1}{2x_n - 6x_n^2}$$
   d) Use the formula starting with $x_1 = -1$ to find an approximation for α to 3 decimal places.
   e) State the starting values at which the Newton-Raphson iterative formula will fail.

2.29 Numerical Methods: Newton-Raphson Method (answers on page 526)

Prerequisite knowledge for Q11-Q14: Differentiation of Standard Functions

Prerequisite knowledge for Q15-Q20: Product Rule

11)
  a) Show that
  $$\sin x = 2 - x$$
  has a solution between $x = 1$ and $x = 2$
  b) Using the Newton-Raphson method and $x_1 = 0.4$ find the solution to $\sin x = 2 - x$ to 3 significant figures.

12)
  a) Draw a sketch of
  $$y = \ln x \text{ and } y = 3 - x$$
  on the same axes for $0 \le x \le 3$
  b) By considering a suitable change in sign, show that the intersection of the two graphs lies between $x = 1$ and $x = 3$
  c) Use the Newton-Raphson method with $x_1 = 2$ to find the solution to 3 significant figures.

13) The equation
$$x^2 + \ln(5x) = 0$$
has one real root.
  a) Show that there is a root between $x = 0.1$ and $x = 0.2$
  b) Show that the Newton-Raphson formula can be written in the form:
  $$x_{n+1} = \frac{x_n^3 + x_n - x_n \ln(5x_n)}{2x_n^2 + 1}$$
  c) Use the Newton-Raphson method starting with $x_1 = 0.1$ to find x_2, x_3 and x_4 to 4 decimal places.
  d) Explain why the Newton-Raphson method can never fail to find the root.

14) A curve has equation
$$y = 3x - x^2 + 10\ln(kx + 1), \qquad x > 0$$
where k is a positive constant. It has a local maximum when $x = 3$.
  a) Show that $k = 3$
  b) Show that the curve intersects the axes between $x = 7$ and $x = 8$
  c) Use the Newton-Raphson method with initial value $x_1 = 7$ to find the x-coordinate of this point of intersection to 3 decimal places.

15)
  a) Show that
  $$x \sin x = 1$$
  has a solution between $x = 1$ and $x = 2$
  b) Using the Newton-Raphson method and $x_1 = 1$ find a solution to $x \sin x = 1$ giving your answer correct to 3 significant figures.

16) Amy is trying to solve
$$2^x x^2 = 3$$
She sets
$$f(x) = 2^x x^2 - 3$$
and decides to use the Newton-Raphson method, starting at $x_1 = 0$
  a) Explain why she will not succeed in finding the solution.
  b) Use the Newton-Raphson method at $x_1 = 1$ to find x_3, an approximation of the root giving your answer to 3 decimal places.

17) The diagram below shows
$$y = x^3 \ln x, \qquad x > 0$$
with the local minimum and the root labelled as A and B respectively.

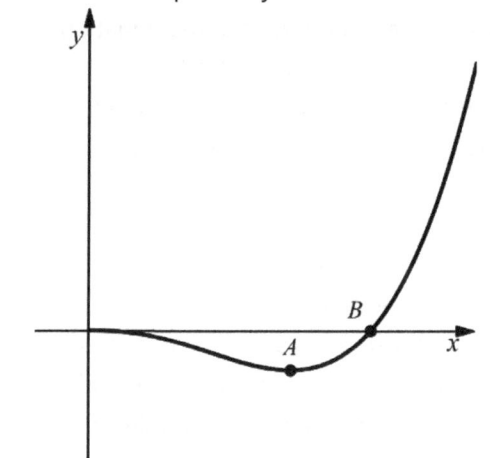

  a) State the coordinates of point B
  b) Show that
  $$\frac{d^2y}{dx^2} = 5x + 6x \ln x$$
  c) By using Newton-Raphson's method and starting at $x_1 = 1$, show that the x-coordinate of the minimum point A is 0.717 to 3 significant figures.

2.29 Numerical Methods: Newton-Raphson Method (answers on page 526)

18) The equation
$$x^2 e^x - 3x + 1 = 0$$
has two roots.
   a) Show that there is a root between $x = 0.4$ and $x = 0.5$
   b) Show that the Newton-Raphson formula to find the roots of $x^2 e^x - 3x + 1 = 0$ can be written in the form:
$$x_{n+1} = \frac{x_n^3 e^{x_n} + x_n^2 e^{x_n} - 1}{x_n^2 e^{x_n} + 2x_n e^{x_n} - 3}$$
   c) Use the above iterative formula with starting point $x_1 = 0.4$ to find x_2 and x_3 giving your answers to 3 significant figures.
   d) Use the Newton-Raphson method starting at $x_1 = 0.6$ to find the root of
$$x^2 e^x + 2x e^x - 3 = 0$$
   giving your answer to 3 significant figures.
   e) State the relevance of the result you found in part d) to with regards the curve
$$y = x^2 e^x - 3x + 1$$

19) The curve
$$y = x \times 3^x - 8x^2 - 2x, \quad x > 0$$
has a single stationary point which is close to $x = 2.2$

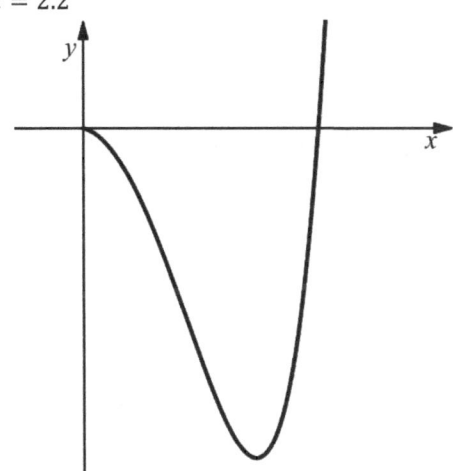

   a) Show that the x-coordinate of this stationary point satisfies the equation
$$3^x + (3^x \ln 3 - 16)x - 2 = 0$$
   b) Show that the Newton-Raphson iterative formula for finding the x-value of the stationary point is
$$x_{n+1} = x_n - \frac{3^{x_n} + (3^{x_n} \ln 3 - 16)x_n - 2}{(2\ln 3 + (\ln 3)^2 x_n)3^{x_n} - 16}$$
   c) Find the x-coordinate of the stationary point giving it correct to 3 significant figures.

20) The number of views per day of a video N on a social media platform may be modelled by
$$N = 8e^{10-0.25t} \sin(3.15 - 0.19t)$$
where t is the number of days passed since the video was shared.
The maximum of the model is found to occur when $t \approx 3.4646$
   a) Show that that the number of days passed when it reaches the maximum number of views per day is approximately 3.4646
   b) Use Newton-Raphson's method to find the number of days passed to the nearest hour when the views per day is zero.

Comment on the validity of the model
   c) when 0 days have passed
   d) after the number of days calculated in part b) has passed.

Prerequisite knowledge for Q21: Trapezium Rule

21) The diagram below shows
$$y = 0.1 + 0.1x^2 \sin(x - 0.5), \quad x > 0$$
with the root labelled as A.

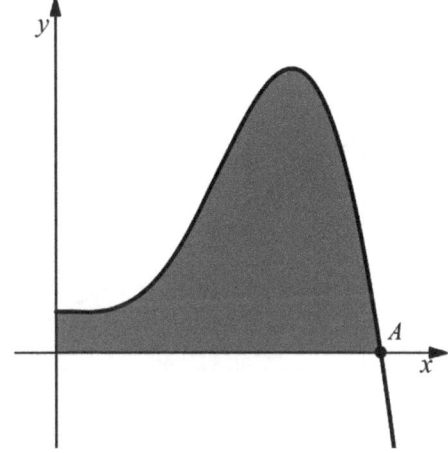

The shaded region is bounded by $x = 0$, $y = 0$ and where the graph meets the x-axis.
   a) Show that the x-coordinate of the point A lies between $x = 3$ and $x = 4$
   b) Use the Newton-Raphson method, to find an approximation for the x-coordinate of A giving your answer to 3 decimal places.
   c) Using your x-coordinate for A, use the trapezium rule with 4 strips to find an approximate value for the shaded area. Give your answer to 1 decimal place.

2.30 Integration: The Trapezium Rule (answers on page 528)

1) The four shaded areas below are to be estimated using the trapezium rule. Which would produce an underestimate?

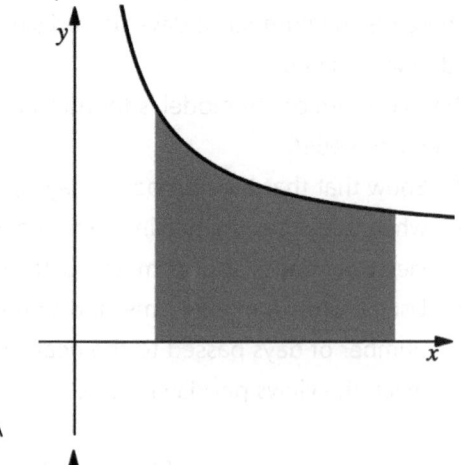

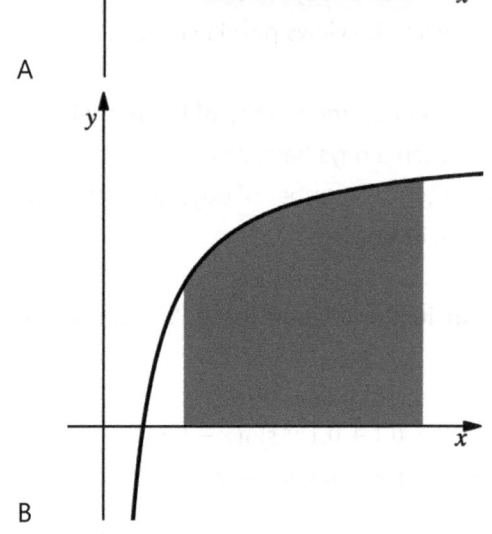

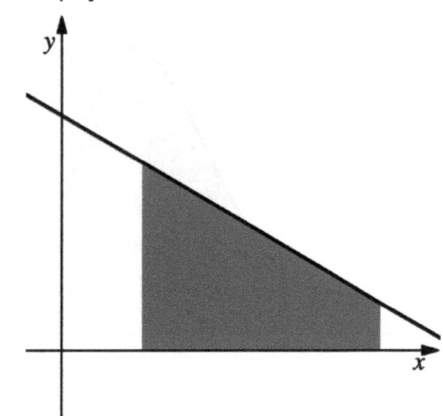

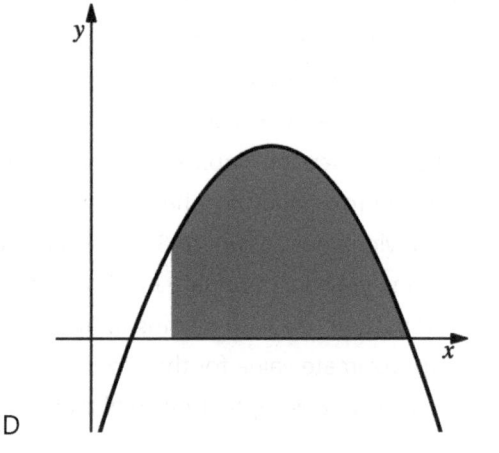

A

B

C

D

2) Amelia uses the Trapezium Rule to estimate an area from $x = 3$ to $x = 8$ using 9 ordinates. What is the horizontal width of each trapezium?

3) Use the Trapezium Rule, with the stated number of strips, to estimate the following integrals. Give your answers in exact form.
   a) $\int_0^3 x^2 \, dx$, 3 strips
   b) $\int_0^4 (x^2 + 1) \, dx$, 4 strips
   c) $\int_0^2 x^3 \, dx$, 4 strips
   d) $\int_0^3 (x^3 + 2) \, dx$, 6 strips

4) Use the Trapezium Rule, with the stated number of ordinates, to estimate the following integrals. Give your answers in exact form.
   a) $\int_2^4 (2x^2 + 3x) \, dx$, 5 ordinates
   b) $\int_1^2 (4 - (x - 1)^2) \, dx$, 5 ordinates
   c) $\int_1^4 (20 - x^2) \, dx$, 7 ordinates
   d) $\int_{-3}^2 ((x + 2)^2 + 3) \, dx$, 6 ordinates

5) Use the Trapezium Rule, with the stated number of strips, to estimate the following integrals. Give your answers in exact form.
   a) $\int_{-3}^1 (x^3 + 2x^2 + 10) \, dx$, 4 strips
   b) $\int_{-2}^2 (5 - (x - 1)(x + 2)) \, dx$, 8 strips
   c) $\int_0^2 \left(3 - \frac{1}{2}x^2\right) dx$, 5 strips
   d) $\int_{-3}^1 \left(2 - \frac{1}{3}x^3\right) dx$, 5 strips

6) Use the Trapezium Rule, with the stated number of ordinates, to estimate the following integrals. Give your answers in exact form.
   a) $\int_2^8 (8 - x)(x - 1)(x + 2) \, dx$, 5 ordinates
   b) $\int_1^9 (3^x + 2) \, dx$, 5 ordinates
   c) $\int_{-7}^{-4} \left(\frac{2}{x+2} + 2\right) dx$, 7 ordinates
   d) $\int_5^7 \frac{x+2}{x-4} \, dx$, 4 ordinates

7) Use the trapezium rule, with four strips, to find an estimate for
$$\int_1^2 \frac{\ln x}{\sqrt{e^x + 1}} \, dx$$

2.30 Integration: The Trapezium Rule (answers on page 528)

8) Given that $k > 0$, use the trapezium rule, with four strips, to show that:
   a) $\int_0^8 \frac{k+x^2}{2} dx = 4k + 88$
   b) $\int_1^5 \frac{k+x}{x} dx = \frac{101k+240}{60}$
   c) $\int_1^5 \ln(kx) dx = \ln(24\sqrt{5}k^4)$
   d) $\int_1^5 \ln\left(\frac{x}{k}\right) dx = \ln 24 + \frac{1}{2}\ln 5 - 4\ln k$

9) The curve shown below has the equation $y = \ln(9-x)$.

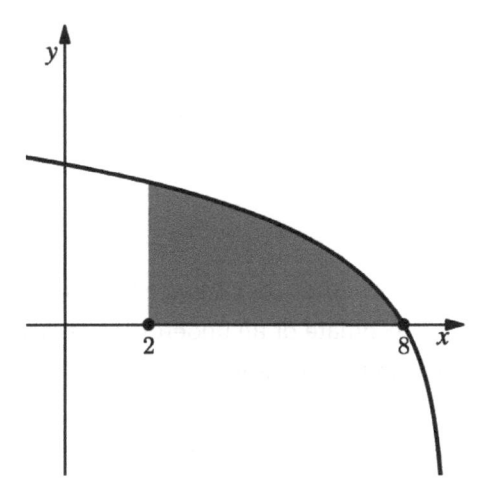

   Use three trapeziums to estimate the shaded region, giving your answer to three decimal places. Explain how you know this will give an underestimate.

10) A particle is moving in a straight line with velocity v ms^{-1} at time t seconds as shown in the graph below. Use the trapezium rule with four strips to estimate the distance travelled by the particle in the first 20 seconds.

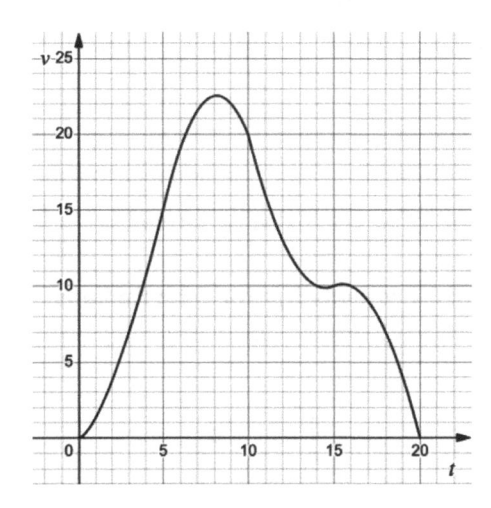

11)
    a) Use the trapezium rule, with four strips, to find an estimate for
    $$\int_0^1 \frac{1}{\sqrt{4+x^2}} dx$$
    b) Explain how the trapezium rule might be used to give a better estimate to the integral given in a).

12) Use the trapezium rule, with six ordinates, to find an estimate for
$$\int_0^1 \frac{x}{\sqrt{x^3+1}} dx$$

13) A particle is moving in a straight line and its speed is measured every 3 seconds. The results are given in the table below with the time in seconds and the speed in ms^{-1}. Use this information with the trapezium rule to estimate the distance travelled by the particle.

Time (s)	0	3	6	9	12	15	18
Speed (ms^{-1})	1	2.8	4.4	6.2	7.7	8.9	10.1

14)
    a) The table below shows the values of x and y for
    $$y = \sqrt{\frac{2+x}{x}}$$

x	2.5	3	3.5	4	4.5
y	1.342	1.291	1.254	1.225	1.202

The values of y are given to 4 significant figures. Use the trapezium rule with all the values of y in the table to estimate
$$\int_{2.5}^{4.5} \sqrt{\frac{2+x}{x}} dx$$

b) Use your answer to deduce an estimate to
$$\int_{2.5}^{4.5} \sqrt{\frac{2+x}{2x}} dx$$
giving your answer to 3 significant figures.

2.30 Integration: The Trapezium Rule (answers on page 528)

15)
  a) Use the trapezium rule, with four strips, to find an estimate for
  $$\int_1^2 e^{3x^2}\,dx$$
  b) Use your answer to deduce an estimate to
  $$\int_1^2 e^{3x^2+1}\,dx$$

16) The screenshot below shows part of a calculation to use the trapezium with five strips to estimate
$$\int_0^2 e^{8-x^3}\,dx$$

	A	B	C	D
1	x	8-x^3	y	
2	0.0000	8.0000	2980.9580	1490.4790
3	0.4000	7.9360	2796.1535	2796.1535
4	0.8000	7.4880	1786.4756	1786.4756
5	1.2000	6.2720	529.5354	529.5354
6	1.6000	3.9040	49.6005	49.6005
7	2.0000	0.0000	1.0000	0.5000
8				6652.7439

Use the values in the spreadsheet to estimate the value of the integral to 3 significant figures.

17)
  a) The table below shows the values of x and y for
  $$y = \log_5 \frac{x}{6}$$

x	8	12	16	20	24
y	0.1787	0.4307	0.6094	0.7481	0.8614

  The values of y are given to 4 decimal places. Use the trapezium rule with all the values of y in the table to estimate
  $$\int_8^{24} \log_5 \frac{x}{6}\,dx$$
  b) Use your answer to deduce an estimate to
  $$\int_8^{24} \log_5 \frac{x^2}{36}\,dx$$

18)
  a) The diagram below shows the curve with equation $y = \sqrt[3]{x^3 + 25}$. Use the trapezium rule with five ordinates to estimate
  $$\int_{-2.5}^{-0.5} \sqrt[3]{x^3 + 25}\,dx$$

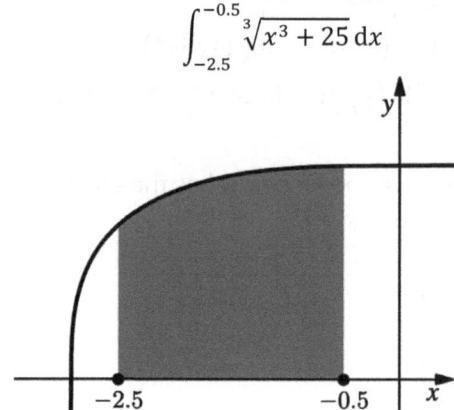

  b) State whether your answer to a) is an overestimate or an underestimate, giving a reason for your answer.

19) The screenshot below shows part of a calculation to use the trapezium with five strips to estimate
$$\int_0^6 \sqrt{e^{-x} - e^{-2x}}\,dx$$

	A	B	C	D	E
1	x	exp(-x)	exp(-2x)	y	
2	0.0000	1.0000	1.0000	0.0000	0.0000
3	1.2000	0.3012	0.0907	0.4588	0.4588
4	2.4000	0.0907	0.0082	0.2872	0.2872
5	3.6000	0.0273	0.0007	0.1630	0.1630
6	4.8000	0.0082	0.0001	0.0903	0.0903
7	6.0000	0.0025	0.0000	0.0497	0.0249
8					1.0242

Use the values in the spreadsheet to estimate the value of the integral to 3 significant figures.

Prerequisite knowledge for Q20: Radians

20) Use the trapezium rule, with six ordinates, to find an estimate for
$$\int_0^{\pi} \sin^2 x\,dx$$

2.30 Integration: The Trapezium Rule (answers on page 528)

21) The table below shows the values of x and y for $y = f(x)$

x	5	5.4	5.8	6.2	6.6
y	p	15	22	p	18

 a) Use the trapezium rule with all the values of y in the table to estimate $\int_5^{6.6} f(x)\,dx$ in terms of p
 b) It is given that the estimate in part a) is 25. Find the value of p

22) The diagram below shows the curve with equation $y = \ln(x^2)$
 a) The area bounded by the curve, the x-axis, and the lines $x = -20$ and $x = -4$, is approximated by a single trapezium. Show that the area of the trapezium is
 $$64 \ln 2 + 16 \ln 5$$

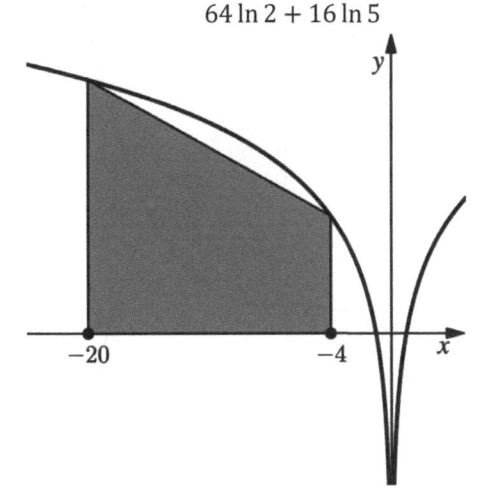

 b) The area is now approximated using eight rectangles of equal width. Show that the total area of the rectangles is
 $$60 \ln 2 + 16 \ln 3 + 4 \ln 5 + 4 \ln 7$$

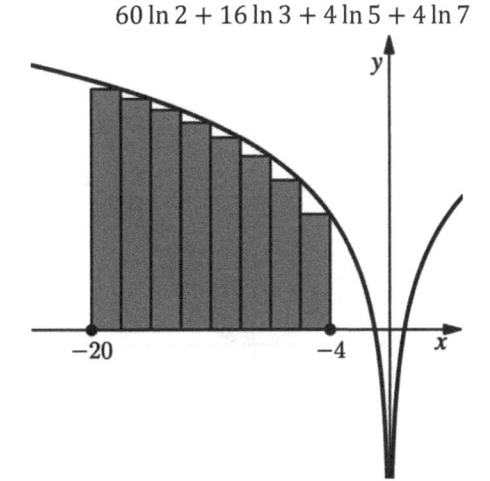

Prerequisite knowledge: Integration by Substitution (for part d)

23) Laura is trying to find an estimate for
$$\int_{-5}^{5} \frac{25}{5 + x^2}\,dx$$

 a) She first uses five rectangles of equal width as shown in the diagram below. Show that this lower bound for the area is given by 18.81 to 4 significant figures.

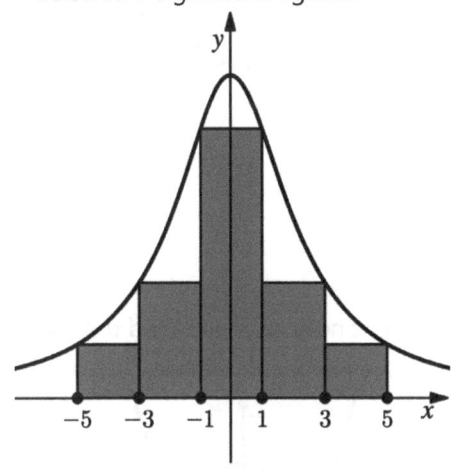

 b) Laura then finds an upper bound for the value of the integral using five rectangles of equal width as shown in the diagram below. Determine the value of this upper bound to 4 significant figures.

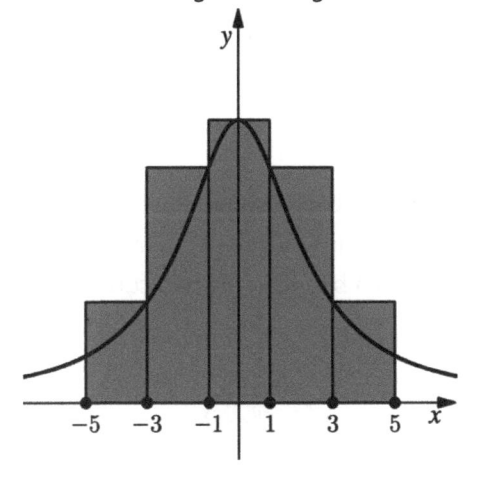

 c) Laura then uses the trapezium rule with five strips to estimate the value of
 $$\int_{-5}^{5} \frac{25}{5 + x^2}\,dx$$
 Find her result to 4 significant figures.

 d) Finally, Laura uses the substitution $x = \sqrt{5} \tan u$ to integrate and find an accurate value for the area. Find her result to 4 significant figures.

2.31 Integration: Area Between Curves (answers on page 528)

1) The diagram below shows the curves $y = x^2$ and $y = x^3$. Find the exact shaded area between the two curves.

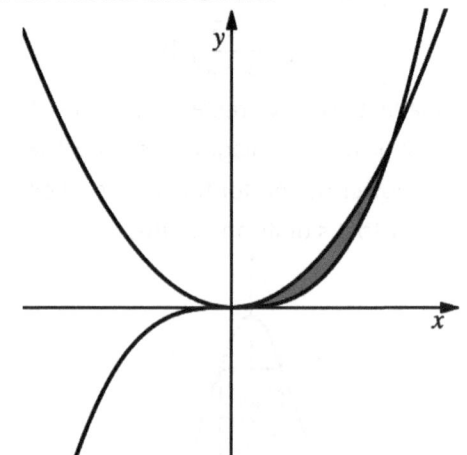

2) The diagram below shows the curves $y = 4 - x^2$ and $y = x^2 + 2$. Find the exact shaded area between the two curves.

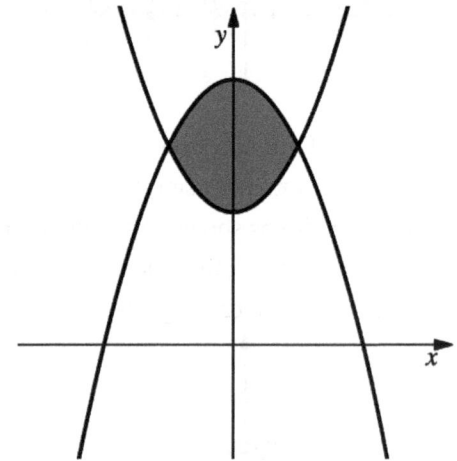

3) The diagram below shows the curves $y = 5 - x^2$ and $y = (x - 1)^2$. Find the exact shaded area between the two curves.

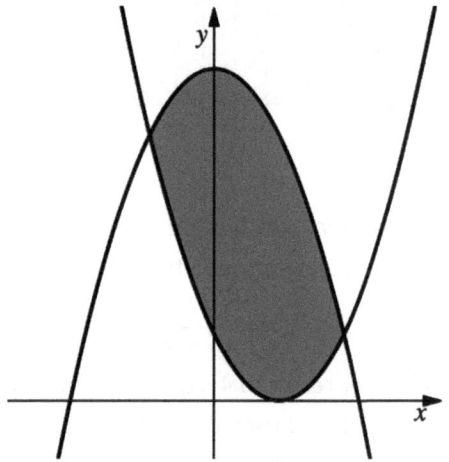

4) The diagram below shows the curves $y = (x - 1)^2 + 2$ and $y = 6 - (x + 1)^2$. Find the exact shaded area between the two curves.

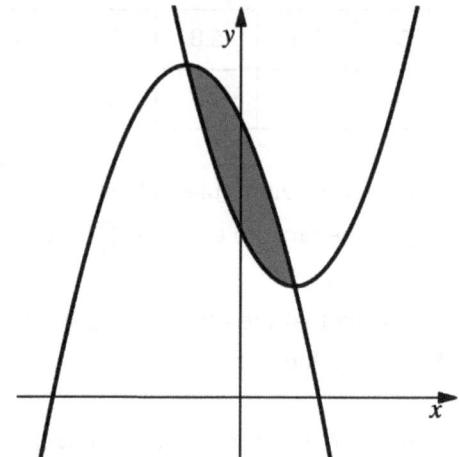

5) The diagram below shows the curves $y = x^3 + 1$ and $y = 3 - x^3$. Find the exact shaded area between the two curves and the y-axis.

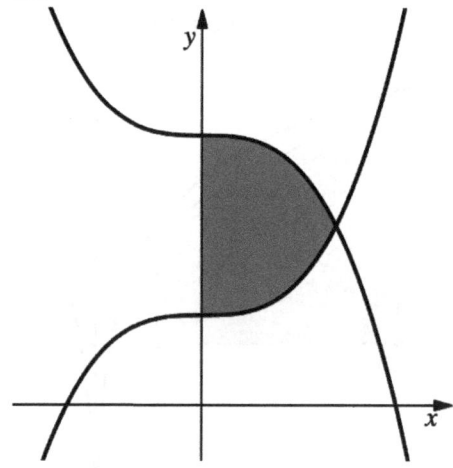

6) The diagram below shows the curves $y = x^2$ and $y = \sqrt{x}$. Find the exact shaded area between the two curves.

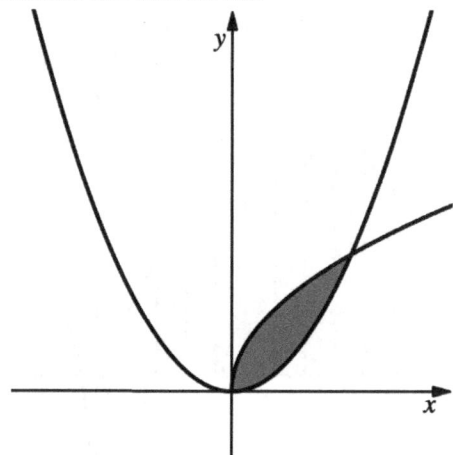

2.31 Integration: Area Between Curves (answers on page 528)

7) The diagram below shows the curves $y = 1 - x^2$ and $y = \frac{4}{3}(1 - x^2)$. Find the exact shaded area between the two curves.

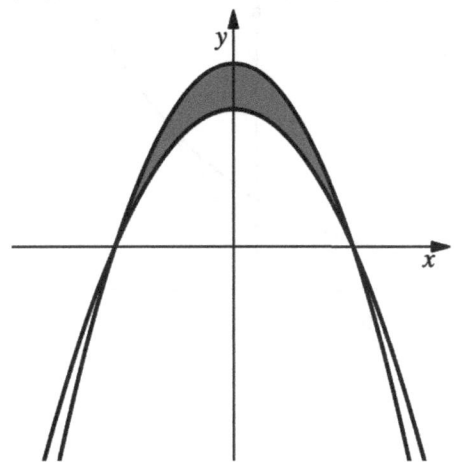

8) The diagram below shows the curves $y = \frac{1}{x^2}$ and $y = \frac{1}{x^3}$, and the line $x = 2$. Find the exact shaded area.

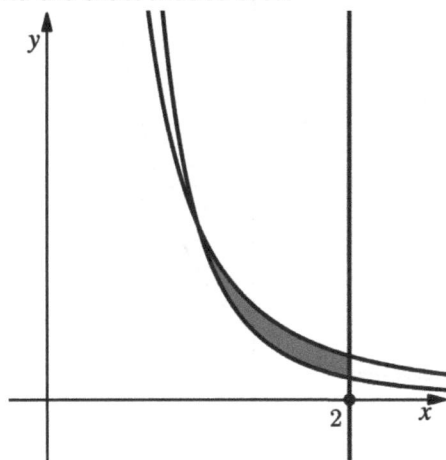

9) The diagram below shows the curves $y = x(x-1)(x+1)$ and $y = 2x(x-1)(x+1)$. Find the exact shaded area between the two curves.

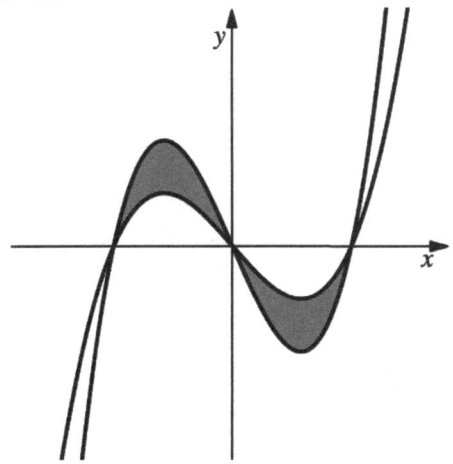

10) The diagram below shows the curves $y = (x+1)(x-2)(x-3)$ and $y = 3(x+1)(x-2)(x-4)$. Find the exact shaded area.

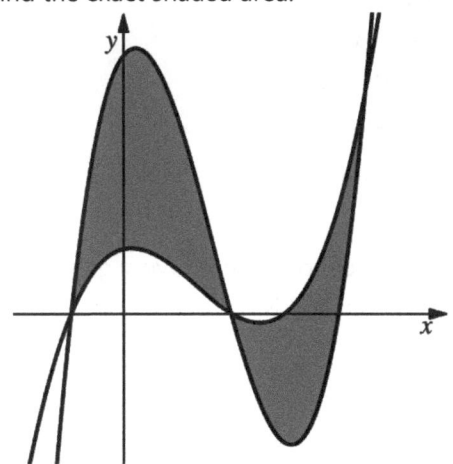

11) It is given that $f(x) = 8x^3 + 6x^2 - 36x$. The diagram below shows the curves $y = f(x)$ and $y = 81 - f(x)$. The curves intersect at precisely two points. Find the exact shaded area.

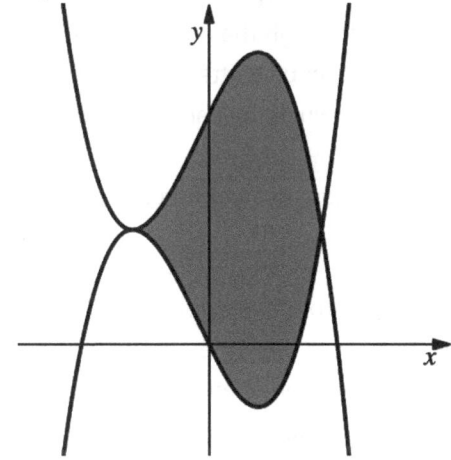

12) The diagram below shows the curves $y = \frac{1}{x^2}$ and $y = \frac{2}{x^2}$, and the lines $y = 1$ and $y = 2$. Find the exact shaded area.

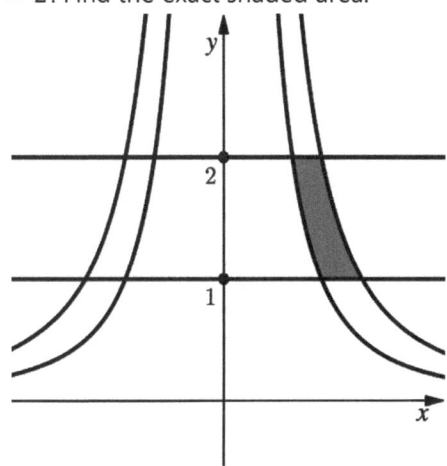

2.32 Integration: Standard Functions (answers on page 529)

Integration of e^{kx}

1) Find
   a) $\int (e^x + x)\, dx$
   b) $\int e^{2x}\, dx$
   c) $\int (2e^{3x} - x^2)\, dx$
   d) $\int \left(\sqrt{e^x} - \dfrac{9}{x^2}\right) dx$
   e) $\int \dfrac{2+e^x}{e^{8x}}\, dx$

2) Find the equation of the curve y given the gradient function is:
   a) $\dfrac{dy}{dx} = e^{-3x}$
   b) $\dfrac{dy}{dx} = 4 - 2e^{-2x}$
   c) $\dfrac{dy}{dx} = \sqrt{x} - \dfrac{1}{3}e^{\frac{1}{6}x}$
   d) $\dfrac{dy}{dx} = x - \dfrac{6}{e^{4x}}$

3) The curve y has gradient function
   $$\dfrac{dy}{dx} = 12e^{6x}$$
   and passes through the point $(\ln 2, 128)$. Show that y can be written in the form $y = ae^{bx} + c$, where a, b and c are integers.

4) The curve y has gradient function
   $$\dfrac{dy}{dx} = \dfrac{24}{\sqrt{e^{3x}}}$$
   and passes through the point $(\ln 4, 1)$. Show that y can be written in the form $y = a + be^{cx}$, where a and b are integers and c is a rational number.

5) Evaluate the following definite integrals:
   a) $\int_0^3 (e^x + 2)\, dx$
   b) $\int_{\ln 2}^{\ln 3} (100 - e^x)\, dx$
   c) $\int_{-1}^{1} \left(24e^{\frac{2}{3}x} - 2x\right) dx$
   d) $\int_0^{\ln 2} \left(\dfrac{5}{e^x} - \dfrac{2}{e^{2x}}\right) dx$

6) Find the exact value of:
   a)
   $$\lim_{\delta x \to 0} \sum_{x=0}^{1} 3e^{18x}\, \delta x$$
   b)
   $$\lim_{\delta x \to 0} \sum_{x=2}^{\ln 8} \dfrac{1}{\sqrt[3]{e^x}}\, \delta x$$

7) The diagram below shows the curve with equation $y = \sqrt{e^x}$. Find the exact value of the shaded area.

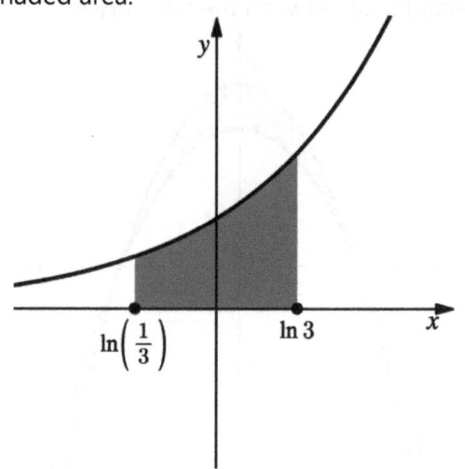

8) The diagram below shows the curves $y = 2 - e^x$ and $y = e^{2x} - 4$. Find the exact shaded area between the two curves.

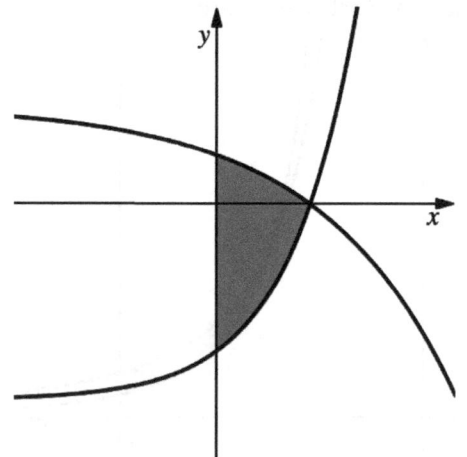

9) The diagram below shows the curve with equation $y = e^2 x - e^x$. The curve has a turning point at A, and AC bisects BD. It is given that $BD = 2$ units. Find the exact shaded area.

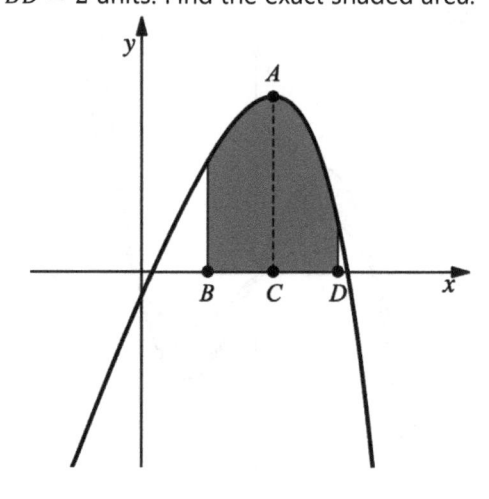

2.32 Integration: Standard Functions (answers on page 529)

Integration of $\sin kx$ and $\cos kx$

10) Find
    a) $\int (\sin x + \cos x)\, dx$
    b) $\int (2\sin x - \cos x)\, dx$
    c) $\int (2\sin 2x + \cos 4x)\, dx$
    d) $\int \left(3\sin\frac{1}{2}x - 6\cos\frac{2}{3}x\right) dx$

11) Find the equation of the curve y given the gradient function is:
    a) $\frac{dy}{dx} = 3\cos 6x$
    b) $\frac{dy}{dx} = \frac{3}{x^2} - 4\sin\frac{1}{4}x$
    c) $\frac{dy}{dx} = \sqrt[3]{x} - \frac{2}{3}\cos\frac{5}{6}x$

12) The curve y has gradient function
    $$\frac{dy}{dx} = -4\sin 3x$$
    and passes through the point $\left(\frac{\pi}{6}, 3\right)$
    Show that y can be written in the form $y = a\cos bx + c$, where b and c are integers, and a is a rational number.

13) The curve y has gradient function
    $$\frac{dy}{dx} = 2\sin\frac{1}{3}x - 3\cos\frac{1}{3}x$$
    and passes through the point $\left(\pi, -\frac{9\sqrt{3}}{2} - 3\right)$
    Show that y can be written in the form $y = a\cos\frac{x}{b} + c\sin\frac{x}{b}$, where a, b and c are integers.

14) Evaluate the following definite integrals:
    a) $\int_0^{\frac{\pi}{2}} \sin x\, dx$
    b) $\int_0^{\frac{\pi}{4}} \cos 2x\, dx$
    c) $\int_{-\frac{5\pi}{6}}^{\frac{5\pi}{3}} 3\sin\frac{1}{5}x\, dx$
    d) $\int_{\frac{\pi}{8}}^{\frac{\pi}{4}} (5\sin 4x - 6\cos 4x)\, dx$

15) Find the exact value of:
    a) $$\lim_{\delta x \to 0} \sum_{x=0}^{\frac{\pi}{12}} (8\sin 4x - 16\cos 4x)\, \delta x$$
    b) $$\lim_{\delta x \to 0} \sum_{x=9\pi}^{10\pi} \left(\frac{1}{8}\sin\frac{1}{4}x - \frac{1}{16}\cos\frac{1}{4}x\right) \delta x$$

16) The diagram below shows the curve with equation $y = 3 + 2\sin\frac{1}{2}x$. Find the exact value of the shaded area.

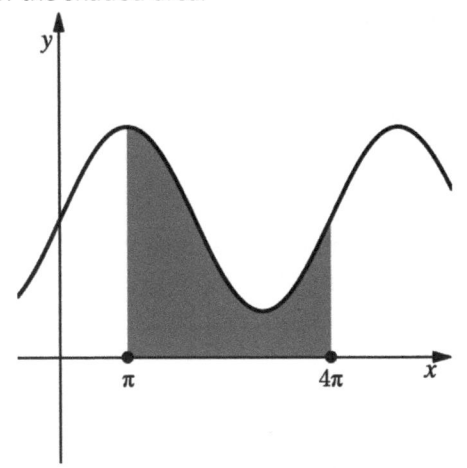

17) The diagram below shows the curves $y = \sin\frac{1}{3}x - 2$ and $y = \cos\frac{1}{3}x + 2$. Find the exact shaded area between the two curves.

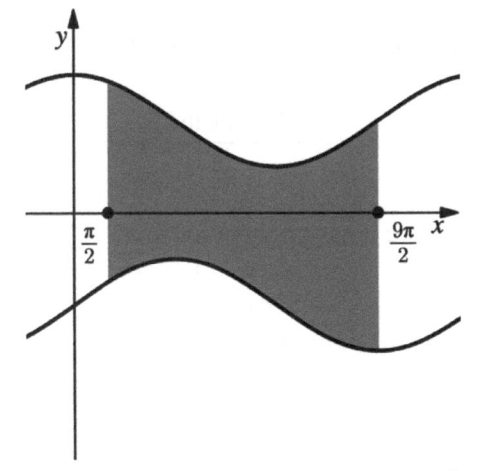

18) The diagram below shows the curve with equation $y = \sin x$ and the curve with equation $y = a + \sin bx$, where x is in radians. Find the value of a and the value of b, and hence find the exact shaded area.

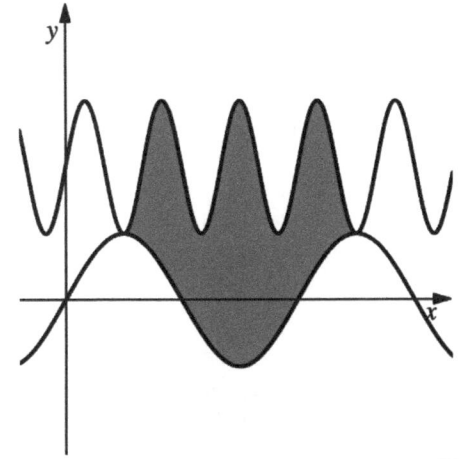

2.32 Integration: Standard Functions (answers on page 529)

Integration of a^x

19) Find
   a) $\int 3^x \, dx$
   b) $\int (3^{2x} + 4^{3x}) \, dx$
   c) $\int \left(\frac{1}{5^x} - \frac{2}{2^x}\right) dx$
   d) $\int \frac{3^x - 4}{9^x} \, dx$

20) Find the equation of the curve y given the gradient function is:
   a) $\frac{dy}{dx} = 8^x$
   b) $\frac{dy}{dx} = x - 3^x \ln 3$
   c) $\frac{dy}{dx} = 6^{3x} - \frac{1}{6^{3x}}$
   d) $\frac{dy}{dx} = \frac{16^x - 4}{2^x}$

21) The curve y has gradient function
$$\frac{dy}{dx} = 25^x$$
and passes through the point $(0, 20)$.
Show that
$$y = \frac{25^x + 20 \ln 25 - 1}{\ln 25}$$

22) The curve y has gradient function
$$\frac{dy}{dx} = \frac{1}{4^x}$$
and passes through the point $\left(0, \frac{1}{\ln 4}\right)$.
Show that
$$y = \frac{1 - 2^{-2x-1}}{\ln 2}$$

23) Evaluate the following definite integrals:
   a) $\int_{-1}^{1} 3^x \, dx$
   b) $\int_{0}^{2} (2^x - 4^x) \, dx$
   c) $\int_{-\frac{1}{3}}^{\frac{1}{3}} (5^{3x} - x) \, dx$
   d) $\int_{1}^{2} \frac{2 - 4^x}{4^x} \, dx$

24) Find the exact value of:
   a)
   $$\lim_{\delta x \to 0} \sum_{x=5}^{10} 5^{0.2x} \, \delta x$$
   b)
   $$\lim_{\delta x \to 0} \sum_{x=\frac{1}{10}}^{\frac{1}{5}} \frac{10}{2^{5x}} \, \delta x$$

25) The diagram below shows the curve with equation $y = 4x - 4^x$. Find the exact value of the shaded area.

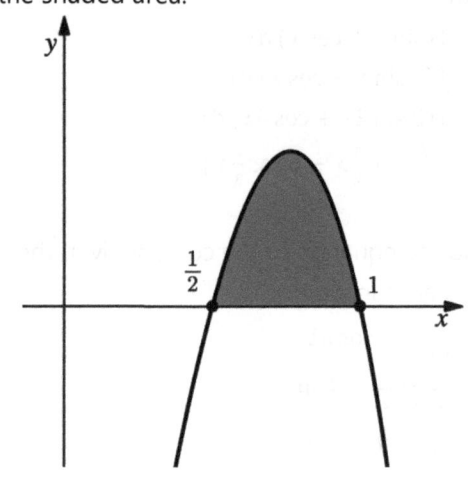

26) The diagram below shows the curves $y = 2^x + 5$ and $y = 4^x - 1$. Find the exact shaded area between the two curves.

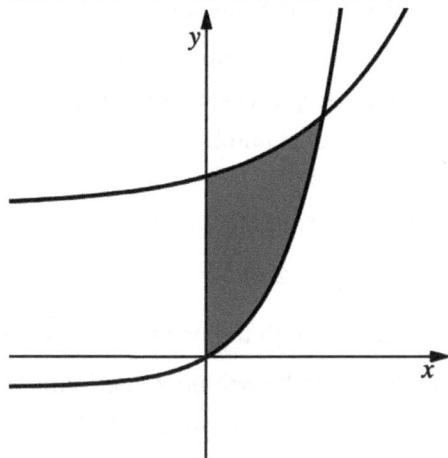

27) The diagram below shows the curve with equation $y = 5^x$. The tangent to the curve at $(1, 5)$ is also shown. Find the exact shaded area.

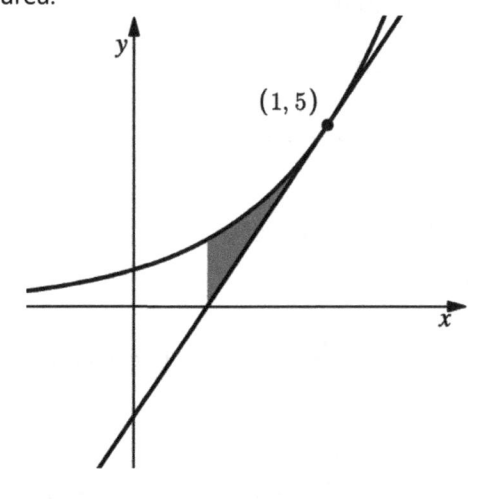

2.32 Integration: Standard Functions (answers on page 529)

Integration of $\frac{1}{x}$

28) Find
   a) $\int \frac{1}{x} dx$
   b) $\int \left(\frac{3}{x} - \frac{x}{3}\right) dx$
   c) $\int \left(\frac{5}{3x} + \frac{2}{5x^2}\right) dx$
   d) $\int \frac{3+8x}{4x} dx$

29) Find the equation of the curve y given the gradient function is:
   a) $\frac{dy}{dx} = \frac{4}{x} - \frac{4}{x^2}$
   b) $\frac{dy}{dx} = 2 - \frac{7}{4x}$
   c) $\frac{dy}{dx} = \frac{3\sqrt{x}+1}{x^{\frac{3}{2}}}$
   d) $\frac{dy}{dx} = \frac{8x-3}{5x^2}$

30) The curve y has gradient function
$$\frac{dy}{dx} = \frac{5}{x}$$
and passes through the point $(e^5, 1)$
Show that y can be written in the form $y = a + b \ln|x|$, where a and b are integers.

31) The curve y has gradient function
$$\frac{dy}{dx} = \frac{3 - 10x}{x}$$
and passes through the point $(1, 1)$
Show that y can be written in the form $y = ax + b + c \ln|x|$, where a, b and c are integers.

32) Evaluate the following definite integrals:
   a) $\int_1^2 \frac{9}{x} dx$
   b) $\int_{-3}^{-1} \frac{8}{x} dx$
   c) $\int_1^4 \frac{2+x}{10x} dx$
   d) $\int_{-3}^{-2} \frac{3-x}{5x} dx$

33) Find the exact value of:
   a) $$\lim_{\delta x \to 0} \sum_{x=-8}^{-4} \left(\frac{4}{x} - 1\right) \delta x$$
   b) $$\lim_{\delta x \to 0} \sum_{x=e}^{e^2} \left(\frac{3}{2x} + 12x\right) \delta x$$

34) The diagram below shows the curve with equation $y = \frac{10}{x}$. Find the exact value of the shaded area.

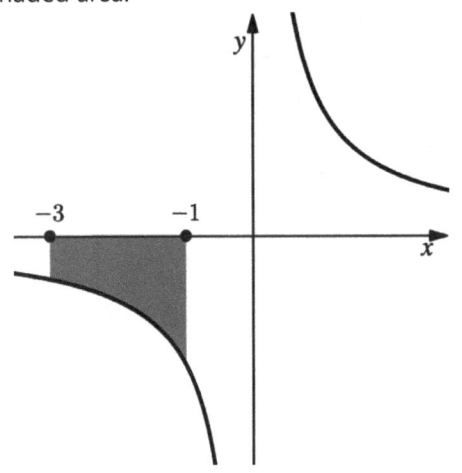

35) The diagram below shows the curves $y = \frac{9-3x}{x}$ and $y = \frac{3-3x}{x}$. Find the exact shaded area between the two curves.

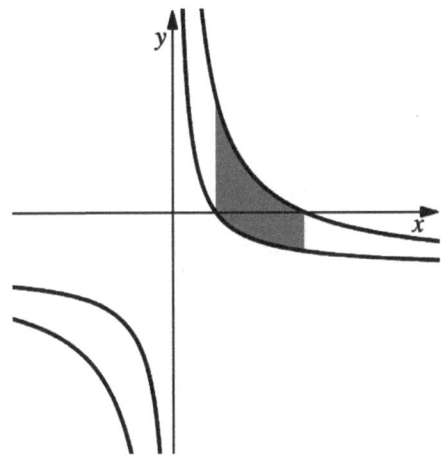

36) The diagram below shows the curve with equation $y = 3 + \frac{2}{x}$. The normal to the curve at $(-1, 1)$ is also shown. Find the exact shaded area, bounded by the curve, the normal and the line $x = 1$

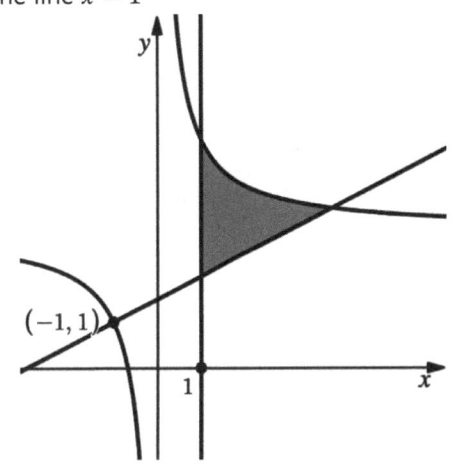

2.32 Integration: Standard Functions (answers on page 529)

Integration of Trigonometric Functions

37) Find
   a) $\int \sec^2 x \, dx$
   b) $\int \csc x \cot x \, dx$
   c) $\int \sec x \tan x \, dx$
   d) $\int \csc^2 x \, dx$
   e) $\int \sec^2 4x \, dx$
   f) $\int \csc^2 \frac{1}{3} x \, dx$

38) Find the equation of the curve y given the gradient function is:
   a) $\frac{dy}{dx} = \csc 2x \cot 2x$
   b) $\frac{dy}{dx} = 8 \csc \frac{1}{4} x \cot \frac{1}{4} x$
   c) $\frac{dy}{dx} = \frac{2}{3} \sec \frac{10}{3} x \tan \frac{10}{3} x$

39) The curve y has gradient function
$$\frac{dy}{dx} = 5 \sec^2 3x$$
and passes through the point $\left(\frac{\pi}{12}, 5\right)$.
Write y in terms of x

40) The curve y has gradient function
$$\frac{dy}{dx} = \sec \frac{1}{2} x \tan \frac{1}{2} x$$
and passes through the point $\left(\frac{\pi}{3}, 0\right)$.
Write y in terms of x

41) Evaluate the following definite integrals:
   a) $\int_0^{\frac{\pi}{6}} 4 \sec^2 2x \, dx$
   b) $\int_{\frac{\pi}{2}}^{\frac{3\pi}{2}} 9 \csc \frac{1}{3} x \cot \frac{1}{3} x \, dx$
   c) $\int_{\frac{\pi}{6}}^{\frac{\pi}{4}} \frac{1}{3} \csc^2 3x \, dx$
   d) $\int_{-\frac{\pi}{4}}^{0} -\sec \frac{2}{3} x \tan \frac{2}{3} x \, dx$

42) Find the exact value of:
   a)
   $$\lim_{\delta x \to 0} \sum_{x=-\frac{\pi}{2}}^{\frac{\pi}{2}} \frac{5}{4} \sec^2 \frac{5}{6} x \, \delta x$$
   b)
   $$\lim_{\delta x \to 0} \sum_{x=\frac{\pi}{9}}^{\frac{2\pi}{9}} \frac{3}{8} \sec \frac{3}{2} x \tan \frac{3}{2} x \, \delta x$$

43) The diagram below shows the curve with equation $y = \sec^2 4x - 2$. Find the exact value of the shaded area.

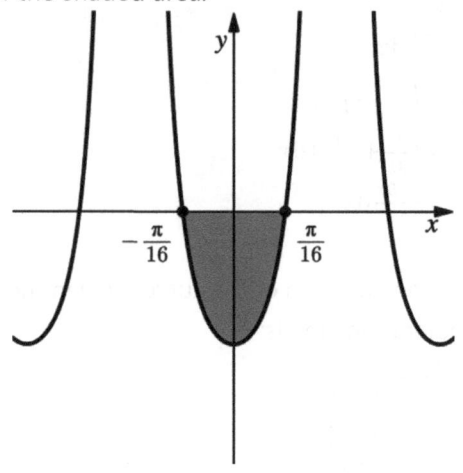

44) The diagram below shows the curves $y = \csc x \cot x$ and $y = \sec x \tan x$. Find the exact shaded area between the two curves.

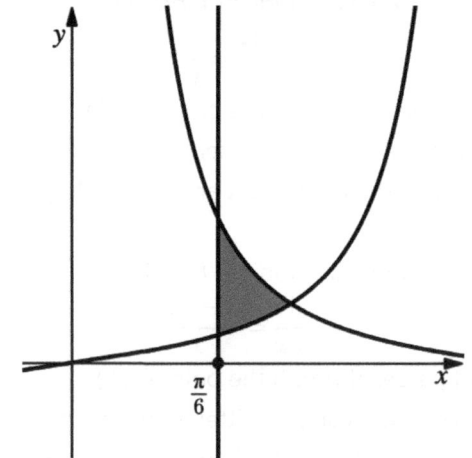

45) The diagram below shows the curve with equation $y = \csc^2 x$. AB is the tangent to the curve at $x = \frac{\pi}{4}$. Find the exact shaded area, bounded by the curve, the tangent, the coordinate axes, and the line $x = \frac{\pi}{2}$

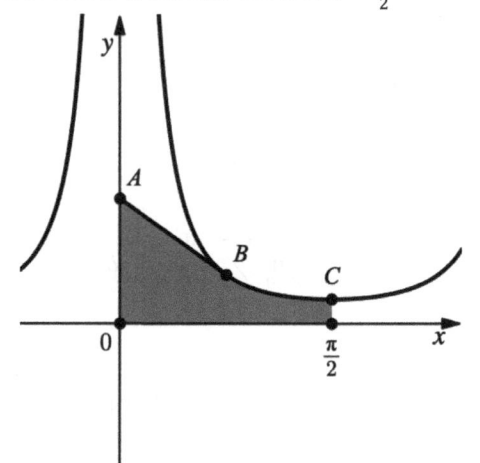

2.33 Integration: Reversing the Chain Rule Part 1 (answers on page 530)

1) Find
   a) $\int (x+1)^3 \, dx$
   b) $\int (x-4)^6 \, dx$
   c) $\int 6(2+x)^7 \, dx$
   d) $\int (10(x-1)^4 - 4(x-3)^5) \, dx$

2) Find the equation of the curve y given the gradient function is:
   a) $\frac{dy}{dx} = (x-3)^{\frac{3}{4}}$
   b) $\frac{dy}{dx} = (9+x)^{-3}$
   c) $\frac{dy}{dx} = \sqrt{x+7}$
   d) $\frac{dy}{dx} = \frac{4}{\sqrt{5+x}}$

3) The curve y has gradient function
   $$\frac{dy}{dx} = \frac{6}{(x-2)^5}$$
   and passes through the point $\left(3, \frac{5}{2}\right)$
   Show that y can be written in the form
   $y = a + \frac{b}{c(x-2)^4}$ where a, b and c are integers.

4) The curve y has gradient function
   $$\frac{dy}{dx} = \sqrt[3]{4+x}$$
   and passes through the point $(4, 3)$
   Show that y can be written in the form
   $y = a(4+x)^b + c$, where a and b are rational numbers, and c is an integer.

5) Evaluate the following definite integrals:
   a) $\int_2^3 (x-2)^8 \, dx$
   b) $\int_{-6}^{-3} 24(6+x)^5 \, dx$
   c) $\int_{-1}^1 \sqrt{x+3} \, dx$
   d) $\int_{-3}^4 \frac{44}{\sqrt[3]{4+x}} \, dx$

6) Find the exact value of:
   a)
   $$\lim_{\delta x \to 0} \sum_{x=-5}^{26} (6+x)^{\frac{3}{5}} \delta x$$
   b)
   $$\lim_{\delta x \to 0} \sum_{x=19}^{84} \frac{4}{\sqrt[4]{x-3}} \delta x$$

7) Find
   a) $\int (2x+1)^4 \, dx$
   b) $\int (2-3x)^5 \, dx$
   c) $\int 4(5+6x)^9 \, dx$
   d) $\int (3(2x-3)^3 - 4(4-5x)^4) \, dx$

8) Find the equation of the curve y given the gradient function is:
   a) $\frac{dy}{dx} = (5x+9)^{\frac{7}{3}}$
   b) $\frac{dy}{dx} = (6-x)^{-5}$
   c) $\frac{dy}{dx} = \sqrt{10+3x}$
   d) $\frac{dy}{dx} = \frac{20}{\sqrt[3]{1-x}}$

9) The curve y has gradient function
   $$\frac{dy}{dx} = \frac{6}{(3x-2)^2}$$
   and passes through the point $(0, 2)$
   Show that y can be written in the form
   $y = \frac{ax+b}{cx+d}$, where a, b, c and d are integers.

10) The curve y has gradient function
    $$\frac{dy}{dx} = \frac{14}{(4x+1)^{\frac{9}{2}}}$$
    and passes through the point $(0, 2)$
    Show that y can be written in the form
    $y = a + \frac{b}{(4x+1)^c}$, where a and b are integers and c is a rational number.

11) Evaluate the following definite integrals:
    a) $\int_2^3 (2x-5)^5 \, dx$
    b) $\int_{-\frac{1}{4}}^{\frac{1}{8}} \sqrt{3+8x} \, dx$
    c) $\int_1^2 \frac{15}{(5-6x)^2} \, dx$
    d) $\int_{\frac{5}{16}}^{\frac{3}{2}} \frac{8}{\sqrt[3]{3+16x}} \, dx$

12) Find the exact value of:
    a)
    $$\lim_{\delta x \to 0} \sum_{x=-\frac{1}{3}}^{10} \frac{2}{5}(2+3x)^{\frac{3}{5}} \delta x$$
    b)
    $$\lim_{\delta x \to 0} \sum_{x=\frac{1}{9}}^{\frac{1}{3}} \frac{243}{(4-9x)^4} \delta x$$

2.33 Integration: Reversing the Chain Rule Part 1 (answers on page 530)

13) Find
   a) $\int 2x(x^2+3)^5\,dx$
   b) $\int 12x^2(4x^3-5)^3\,dx$
   c) $\int x(2x^2-5)^4\,dx$
   d) $\int 2x^3\sqrt{x^4-3}\,dx$

14) Find the equation of the curve y given the gradient function is:
   a) $\frac{dy}{dx} = 9x^2(3x^3+2)^{-2}$
   b) $\frac{dy}{dx} = 10x^3(2-5x^4)^{\frac{5}{2}}$
   c) $\frac{dy}{dx} = x^3\sqrt{3+4x^2}$
   d) $\frac{dy}{dx} = \frac{100x^8}{(2x^9+3)^6}$

15) The curve y has gradient function
$$\frac{dy}{dx} = 5x(5x^2-1)^7$$
and passes through the point $(0, 1)$.
Show that y can be written in the form
$y = a(5x^2-1)^b + c$, where a and c are rational numbers, and b is an integer.

16) The curve y has gradient function
$$\frac{dy}{dx} = \frac{9x^2}{\sqrt{27x^3-1}}$$
and passes through the point $\left(\frac{1}{3}, 0\right)$.
Show that y can be written in the form
$y = a(27x^3-1)^b$, where a and b are rational numbers.

17) Evaluate the following definite integrals:
   a) $\int_0^2 x(10x^2+1)^4\,dx$
   b) $\int_{\frac{\sqrt{2}}{2}}^{\frac{3}{2}} x(4x^2-1)^{\frac{2}{3}}\,dx$
   c) $\int_2^3 \frac{6x^2}{(x^3+2)^2}\,dx$
   d) $\int_0^{\frac{1}{2}} \frac{800x^3}{(1-x^4)^3}\,dx$

18) Find the exact value of:
   a) $\lim_{\delta x \to 0} \sum_{x=0}^{1} 98x(1+14x^2)^{\frac{5}{2}}\,\delta x$
   b) $\lim_{\delta x \to 0} \sum_{x=0}^{9} \frac{\sqrt{\sqrt{x}+1}}{\sqrt{x}}\,\delta x$

19) The diagram below shows the curve with equation $y = \frac{1}{2}x(x^2-1)^5$. Find the exact value of the shaded area.

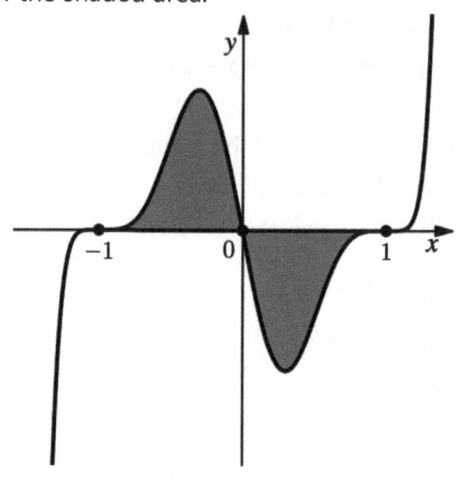

20) The diagram below shows the curves $y = \frac{2}{(x+1)^2}$ and $y = \frac{2}{(2x+1)^2}$. Find the exact shaded area between the two curves.

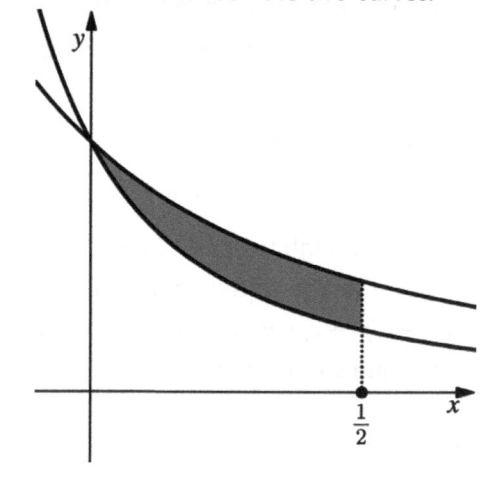

21) The diagram below shows the curve with equation $y = \frac{12x-9x^2}{(2x^2-x^3+1)^3}$. Find the exact value of the shaded area.

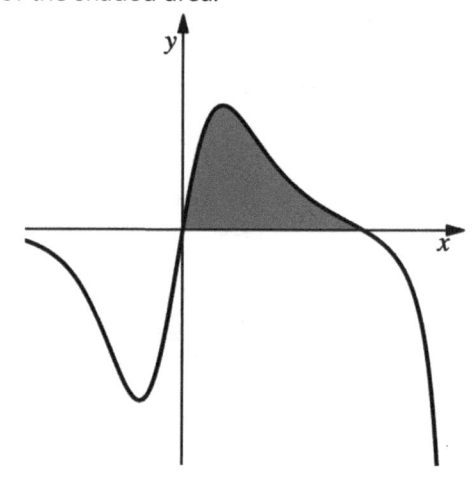

2.34 Integration: Reversing the Chain Rule Part 2 (answers on page 531)

$f'(x)e^{f(x)}$

1) Find
   a) $\int e^{2x+1} \, dx$
   b) $\int e^{5-3x} \, dx$
   c) $\int xe^{x^2} \, dx$
   d) $\int 4x^3 e^{6x^4} \, dx$

2) Find
   a) $\int \cos x \, e^{\sin x} \, dx$
   b) $\int \sin 3x \, e^{6\cos 3x} \, dx$
   c) $\int 4\cos\tfrac{1}{2}x \, e^{\sin\tfrac{1}{2}x} \, dx$
   d) $\int 9\sin\tfrac{1}{3}x \, e^{3\cos\tfrac{1}{3}x} \, dx$

3) Find
   a) $\int (\ln 2) 2^x e^{2^x} \, dx$
   b) $\int 3^x e^{3^x} \, dx$
   c) $\int \frac{1}{\ln 5} 5^x e^{5^x+1} \, dx$
   d) $\int 4^x e^{4^x+1} \, dx$

4) Find the equation of the curve y given the gradient function is:
   a) $\frac{dy}{dx} = \sec^2 x \, e^{\tan x}$
   b) $\frac{dy}{dx} = 2\cosec x \cot x \, e^{4\cosec x}$
   c) $\frac{dy}{dx} = 3\sec 6x \tan 6x \, e^{\sec 6x}$
   d) $\frac{dy}{dx} = -8\cosec^2 \tfrac{1}{2}x \, e^{16\cot\tfrac{1}{2}x}$

5) The curve y has gradient function
$$\frac{dy}{dx} = -\frac{e^{\tfrac{1}{x}}}{x^2}$$
and passes through the point $(1, -e)$.
Write y in terms of x

6) Evaluate the following definite integrals:
   a) $\int_0^1 x^2 e^{x^3} \, dx$
   b) $\int_0^{\tfrac{\pi}{4}} \cos 2x \, e^{\sin 2x} \, dx$
   c) $\int_1^2 (\ln 8) 2^x e^{2^x} \, dx$
   d) $\int_0^{\tfrac{\pi}{16}} 20 \sec^2 4x \, e^{\tan 4x} \, dx$

7) Find the exact value of
$$\lim_{\delta x \to 0} \sum_{x=1}^{4} \frac{e^{2\sqrt{x}}}{\sqrt{x}} \delta x$$

$f'(x)\sin(f(x))$ or $f'(x)\cos(f(x))$

8) Find
   a) $\int \sin\left(5x + \tfrac{\pi}{7}\right) dx$
   b) $\int 8x \cos 2x^2 \, dx$
   c) $\int (2x+1)\sin(x^2+x-1) \, dx$
   d) $\int (x^2-3)\cos\left(6x - \tfrac{2}{3}x^3\right) dx$

9) Find
   a) $\int \cos x \sin(\sin x) \, dx$
   b) $\int \sin x \sin(\cos x) \, dx$
   c) $\int \cos x \cos(\sin x) \, dx$
   d) $\int \sin x \cos(\cos x) \, dx$

10) Find
    a) $\int (\ln 5) 5^x \sin(5^x) \, dx$
    b) $\int 9^x \cos(9^x) \, dx$
    c) $\int \frac{\sin(\ln x)}{x} \, dx$
    d) $\int \frac{\cos(2\ln x)}{x} \, dx$

11) Find the equation of the curve y given the gradient function is:
    a) $\frac{dy}{dx} = \cosec^2 2x \sin(\cot 2x)$
    b) $\frac{dy}{dx} = 4\cosec 3x \cot 3x \cos(8\cosec 3x)$
    c) $\frac{dy}{dx} = 9\sec^2 \tfrac{1}{3}x \sin\left(90\tan\tfrac{1}{3}x\right)$
    d) $\frac{dy}{dx} = \sec\tfrac{2}{3}x \tan\tfrac{2}{3}x \cos\left(18\sec\tfrac{2}{3}x\right)$

12) The curve y has gradient function
$$\frac{dy}{dx} = 12x \sin x^2$$
and passes through the point $\left(\tfrac{\sqrt{\pi}}{2}, 24\right)$.
Write y in terms of x

13) Evaluate the following definite integrals:
    a) $\int_{\ln \tfrac{\pi}{2}}^{\ln \pi} e^x \sin(e^x) \, dx$
    b) $\int_0^{\sqrt{\pi}/4} 40x \cos 8x^2 \, dx$
    c) $\int_{\sqrt{2\pi}}^{\sqrt{3\pi}} 18x \sin x^2 \, dx$
    d) $\int_{\tfrac{\pi}{4}}^{\tfrac{\pi}{3}} (24x - 7\pi)\cos(12x^2 - 7\pi x + \pi^2) \, dx$

14) Find the exact value of
$$\lim_{\delta x \to 0} \sum_{x=\tfrac{16}{\pi}}^{\tfrac{8}{\pi}} \frac{\sin\left(\tfrac{4}{x}\right)}{8x^2} \delta x$$

2.34 Integration – Reversing the Chain Rule Part 2 (answers on page 531)

$\dfrac{f'(x)}{f(x)}$

15) Find
   a) $\int \dfrac{1}{x+1}\,dx$
   b) $\int \dfrac{2}{2x-5}\,dx$
   c) $\int \dfrac{8}{4x+3}\,dx$
   d) $\int \dfrac{9}{5x-7}\,dx$
   e) $\int \dfrac{3}{1-4x}\,dx$
   f) $\int \dfrac{15}{5-3x}\,dx$

16) Find
   a) $\int \dfrac{2x}{x^2+3}\,dx$
   b) $\int \dfrac{3x^2}{5+x^3}\,dx$
   c) $\int \dfrac{15x}{5x^2-4}\,dx$
   d) $\int \dfrac{10x^3}{6-x^4}\,dx$

17) Find
   a) $\int \dfrac{2x+4}{x^2+4x-3}\,dx$
   b) $\int \dfrac{3x^2-3}{x^3-3x+9}\,dx$
   c) $\int \dfrac{x+1}{4x^2+8x+5}\,dx$
   d) $\int \dfrac{6x^2-24x}{x^3-6x^2-1}\,dx$

18) Find
   a) $\int \dfrac{e^x}{e^x+1}\,dx$
   b) $\int \dfrac{(\ln 2)2^x-1}{2^x-x}\,dx$
   c) $\int \tan x\,dx$
   d) $\int \cot x\,dx$

19) Find
   a) $\int \dfrac{\cos 2x}{\sin 2x+1}\,dx$
   b) $\int \dfrac{6\sin 3x}{4-\cos 3x}\,dx$
   c) $\int \dfrac{24\cos 6x}{3\sin 6x-5}\,dx$
   d) $\int \dfrac{18\sin 9x}{15\cos 9x+4}\,dx$

20) Find the equation of the curve y given the gradient function is:
   a) $\dfrac{dy}{dx} = \dfrac{(\ln 3)3^x + e^x}{3^x + e^x}$
   b) $\dfrac{dy}{dx} = \dfrac{8+4\sec^2 2x}{4x+\tan 2x}$
   c) $\dfrac{dy}{dx} = \dfrac{\cos x + \frac{1}{2\sqrt{x}}}{\sin x + \sqrt{x}}$
   d) $\dfrac{dy}{dx} = \dfrac{30e^{3x}-20e^{2x}}{e^{2x}-e^{3x}}$

21) The curve y has gradient function
$$\dfrac{dy}{dx} = \dfrac{2x+1}{(x-1)(x+2)}$$
and passes through the point $(2, 2\ln 2)$
Write y in terms of x

22) Evaluate the following definite integrals:
   a) $\int_0^1 \dfrac{12}{3x+1}\,dx$
   b) $\int_{\sqrt{2}}^{\sqrt{3}} \dfrac{3x}{x^2-1}\,dx$
   c) $\int_\pi^{2\pi} \dfrac{\sin x}{\cos x + 2}\,dx$
   d) $\int_{\ln 2}^{\ln 4} \dfrac{4e^x}{4+e^x}\,dx$

23) Find the exact value of
$$\lim_{\delta x \to 0} \sum_{x=1}^{3} \dfrac{6x+1}{6x^2+2x+1}\,\delta x$$

$f'(x)(f(x))^n$

24) Find
   a) $\int e^x(e^x+5)^3\,dx$
   b) $\int e^{3x}(e^{3x}-3)^4\,dx$
   c) $\int 2e^{4x}(4e^{4x}-1)^5\,dx$
   d) $\int 10e^{\frac{1}{2}x}\sqrt{e^{\frac{1}{2}x}+2}\,dx$

25) Find
   a) $\int \cos x \sin^2 x\,dx$
   b) $\int \cos x \sin^3 x\,dx$
   c) $\int \sin x \cos^4 x\,dx$
   d) $\int \sin 2x \cos^5 2x\,dx$

26) Find
   a) $\int \sec^2 x\,(\tan x - 4)^5\,dx$
   b) $\int \sec^2 3x \tan^3 3x\,dx$
   c) $\int 2^x(2^x+4)^4\,dx$
   d) $\int 3^{x+1}(3^x-2)^5\,dx$

27) Find the equation of the curve y given the gradient function is:
   a) $\dfrac{dy}{dx} = \dfrac{(\ln x + 3)^3}{x}$
   b) $\dfrac{dy}{dx} = \dfrac{(2\ln x + 5)^5}{12x}$
   c) $\dfrac{dy}{dx} = \sec x \tan x\,(\sec x + 2)^3$
   d) $\dfrac{dy}{dx} = \operatorname{cosec}^8 x \cot x$

2.34 Integration: Reversing the Chain Rule Part 2 (answers on page 531)

28) Evaluate the following definite integrals:
   a) $\int_{\ln 2}^{\ln 4} e^x (4e^x - 5)^3 \, dx$
   b) $\int_0^{\frac{\pi}{2}} \cos x \sin^4 x \, dx$
   c) $\int_{2\pi}^{3\pi} \sin \frac{1}{2}x \cos^5 \frac{1}{2}x \, dx$
   d) $\int_e^{e^2} \frac{(\ln x + 3)^5}{2x} \, dx$

Mix

29) Given that k is a positive constant, find
$$\int_k^{2k} \frac{8}{4x + k} \, dx$$

30) Given that k is a positive constant, find
$$\int_{\frac{1}{k}}^{\frac{2}{k}} \frac{k}{kx + 5} \, dx$$

31) Given that k is a positive constant, find
$$\int_{\frac{2}{k}}^{\frac{3}{k}} \frac{k^2 x}{k^2 x^2 - 2} \, dx$$

32) Given that k is a positive constant, show that
$$\int_{\sqrt{k}}^{\sqrt{4k}} 8x e^{x^2} \, dx = 4e^k(e^{3k} - 1)$$

33) Find the exact value of
$$\int_{\ln\left(\frac{\sqrt{\pi}}{4}\right)}^{\ln\left(\frac{\sqrt{\pi}}{2}\right)} 8e^{2x} \sin(4e^{2x}) \, dx$$

34) The diagram below shows the curve with equation $y = \frac{2x^2}{4x^3 + 1}$. Find the exact value of the shaded area.

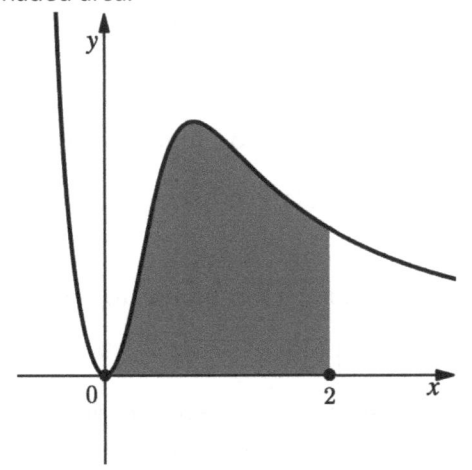

35) The diagram below shows the curves $y = \frac{3}{3x+1}$ and $y = \frac{3}{(3x+1)^2}$. Find the exact shaded area between the two curves.

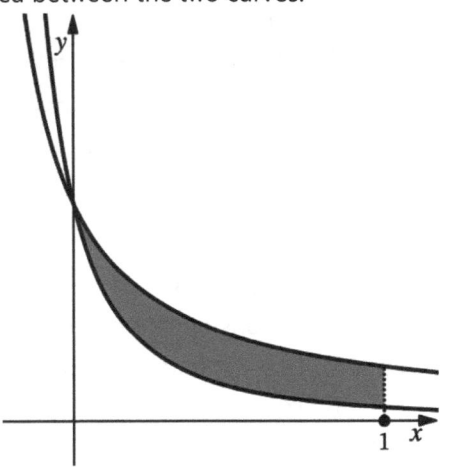

36) The diagram below shows the curve with equation $y = \sin 2x \cos^5 2x$ where C has x-coordinate $\frac{\pi}{12}$. Find the exact shaded area.

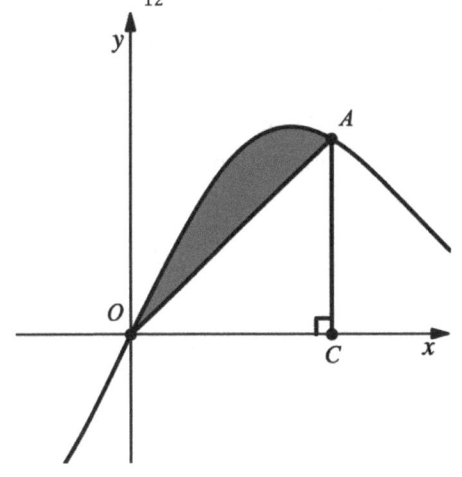

Prerequisite knowledge for Q37: Quotient Rule

37) The diagram below shows $y = \frac{\sin 3x}{\cos 3x + 2}$. A and C are turning points on the curve, and the angle ABC is 90°. Find the exact shaded area.

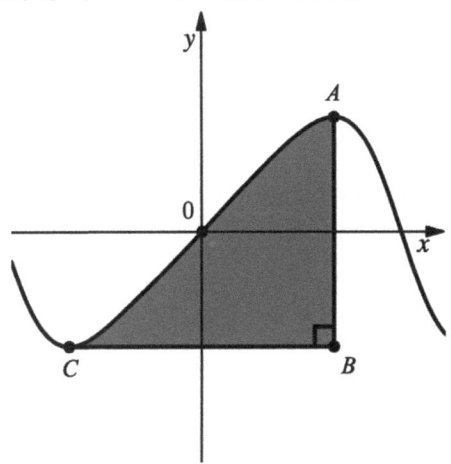

2.35 Integration by Substitution (answers on page 532)

1) Each of the following integrals can be found using reversing the chain rule. However, in each case, evaluate the integrals using the given substitution:
   a) $\int (3x+1)^4 \, dx$ $u = 3x + 1$
   b) $\int \frac{4}{(2x-3)^3} \, dx$ $u = 2x - 3$
   c) $\int \sqrt{8x+5} \, dx$ $u = 8x + 5$
   d) $\int \frac{6}{\sqrt[3]{5-2x}} \, dx$ $u = 5 - 2x$

2) It is given that $y = x(x-3)^5$
   a) Describe the transformation that would map $y = x(x-3)^5$ onto the image $y = (x+3)x^5$
   b) Use the substitution $u = x - 3$ to show that
   $$\int x(x-3)^5 \, dx = \int (u+3)u^5 \, du$$

3) It is given that $y = 2x\sqrt{x+4}$
   a) Describe the transformation that would map $y = 2x\sqrt{x+4}$ onto the image $y = 2(x-4)\sqrt{x}$
   b) Use the substitution $u = x + 4$ to show that
   $$\int 2x\sqrt{x+4} \, dx = \int 2(u-4)\sqrt{u} \, du$$

4) Evaluate the following integrals using the given substitution:
   a) $\int x\sqrt{x+1} \, dx$ $u = x + 1$
   b) $\int 2x\sqrt{2x+3} \, dx$ $u = 2x + 3$
   c) $\int x^3\sqrt{x^2+1} \, dx$ $u = x^2 + 1$

5) Evaluate the following integrals using the given substitution:
   a) $\int \frac{x}{x+1} \, dx$ $u = x + 1$
   b) $\int \frac{2x}{2x+3} \, dx$ $u = 2x + 3$
   c) $\int \frac{x^3}{x^2+1} \, dx$ $u = x^2 + 1$

6) Evaluate the following integrals using the given substitution:
   a) $\int \frac{x^2}{x+1} \, dx$ $u = x + 1$
   b) $\int \frac{4x^2}{2x+3} \, dx$ $u = 2x + 3$
   c) $\int \frac{x^3}{x^2+1} \, dx$ $u = x^2 + 1$

7) Evaluate the following integrals using the given substitution:
   a) $\int x^2\sqrt{x+1} \, dx$ $u = x + 1$
   b) $\int 4x^2\sqrt{2x+3} \, dx$ $u = 2x + 3$
   c) $\int x^3\sqrt{x^2+1} \, dx$ $u = x^2 + 1$

8) Use the substitution $u = 2 - 3x$ to find
   $$\int 189x(2-3x)^5 \, dx$$

9) Use the substitution $u = x + 1$ to find the exact value of
   $$\int_0^1 3x(x+1)^{-\frac{1}{2}} \, dx$$

10) Use the substitution $u = 3x + 4$ to show that
    $$\int_{-2}^{-\frac{4}{3}} (6x+1)(3x+4)^7 \, dx = \frac{3040}{27}$$

11) Use integration by substitution to show that
    $$\int_1^2 \frac{x}{x+2} \, dx = a + \ln\frac{b}{c}$$
    where a, b and c are positive integers.

12) Use integration by substitution to show that
    $$\int_1^3 \frac{9x}{3x+5} \, dx = 6 + 5\ln\frac{4}{7}$$

13) Use the substitution $u = 2x + 3$ to find the exact value of
    $$\int_1^2 \frac{8x-1}{(2x+3)^3} \, dx$$

14) Use integration by substitution to show that
    $$\int_{\frac{3}{2}}^{\frac{19}{2}} 5x\sqrt{2x-3} \, dx = 672$$

15) Use integration by substitution to show that
    $$\int_0^1 140x(4x+1)^{\frac{3}{2}} \, dx = 1 + 225\sqrt{5}$$

16) Use integration by substitution to find
    $$\int (30x - 45)\sqrt{x+2} \, dx$$
    in the form $(x+2)^{\frac{3}{2}}(ax + b) + c$, where a and b are integers, and c is a constant.

2.35 Integration by Substitution (answers on page 532)

17) Use integration by substitution to find the exact value of
$$\int_{\frac{1}{12}}^{\frac{3}{4}} (2-6x)(12x-1)^{\frac{1}{3}} \, dx$$

18) Use the substitution $u = x^2 + 5$ to find
$$\int \frac{2x^3 - x}{\sqrt{x^2 + 5}} \, dx$$

19) Use the substitution $u = 4 - x^4$ to find the exact value of
$$\int_0^1 \frac{x^7}{4 - x^4} \, dx$$

20) Use the substitution $u = x^5 + 3$ to show that
$$\int_0^1 \frac{20x^9}{(x^5 + 3)^2} \, dx = \ln \frac{a}{b} - 1$$
where a and b are positive integers.

21) Use the substitution $u = 2\sqrt{x} - 1$ to find
$$\int \frac{2}{2\sqrt{x} - 1} \, dx$$

22) Evaluate the following definite integral
$$\int_1^4 \frac{2}{\sqrt{x} + 1} \, dx$$
using the substitution $u = \sqrt{x} + 1$

23) Evaluate the following definite integral
$$\int_2^5 \frac{2x^2}{\sqrt{x - 1}} \, dx$$
using the substitution $u = \sqrt{x - 1}$

24) Use integration by substitution to find
$$\int \frac{3x}{1 + \sqrt{x}} \, dx$$

25) Evaluate the following definite integral
$$\int_1^4 \frac{1}{\sqrt{x}(2 + \sqrt{x})} \, dx$$
using the substitution $u = \sqrt{x}$

26) Use the substitution $u = 1 + 2\sqrt{x}$ to show that
$$\int_1^9 \frac{1 + \sqrt{x}}{1 + 2\sqrt{x}} \, dx$$
can be written in the form $a + b \ln c$, where a is an integer, and b and c are rational numbers.

27) Evaluate the following definite integral
$$\int_0^3 \frac{8 - x}{3 + \sqrt{x + 1}} \, dx$$
using the substitution $u = \sqrt{x + 1}$

28) Evaluate the following definite integrals using the given substitutions:
   a) $\int_0^{16} \frac{4}{1 + \sqrt{x}} \, dx$ $\quad u^2 = x$
   b) $\int_1^3 \frac{8x}{\sqrt{4x - 3}} \, dx$ $\quad u^2 = 4x - 3$

29) Evaluate the following definite integral:
$$\int_1^4 \frac{12}{4\sqrt{x} + 3x} \, dx$$
   a) Using the substitution $u = \sqrt{x}$
   b) Using the substitution $u = 3\sqrt{x} + 4$

30) Use the substitution $u^2 = 2x^2 + 1$ to evaluate
$$\lim_{\delta x \to 0} \sum_{x=0}^{2} \frac{x^3}{\sqrt{2x^2 + 1}} \delta x$$

Standard Functions

31) Each of the following integrals can be found using reversing the chain rule. However, in each case, evaluate the integrals using the given substitution:
   a) $\int \cos x \sin^4 x \, dx$ $\quad u = \sin x$
   b) $\int 8x e^{x^2 + 1} \, dx$ $\quad u = x^2 + 1$
   c) $\int \sin x \, e^{\cos x} \, dx$ $\quad u = \cos x$
   d) $\int \frac{\sec^2 x}{\tan x} \, dx$ $\quad u = \tan x$

32) Evaluate the following definite integral
$$\int_0^{\frac{\pi}{2}} \frac{6 \cos x}{9 + \sin x} \, dx$$
using the substitution $u = 9 + \sin x$

33) Use integration by substitution to find
$$\int \frac{5e^x}{5 + e^x} \, dx$$

34) Evaluate the following definite integrals using the given substitutions:
   a) $\int_0^{\ln 2} \frac{3e^{2x}}{3 + e^x} \, dx$ $\quad u = 3 + e^x$
   b) $\int_{\ln 2}^{\ln 3} \frac{8e^x}{(2 + e^x)^2} \, dx$ $\quad u = 2 + e^x$

2.35 Integration by Substitution (answers on page 532)

35) Evaluate the following definite integrals using the given substitutions:
   a) $\int_0^{\frac{\pi}{4}} 8 \cos 2x \sin^5 2x \, dx$ $u = \sin 2x$
   b) $\int_0^{\frac{\pi}{8}} \sin^2 2x \cos^3 2x \, dx$ $u = \sin 2x$

36) Evaluate the following definite integrals using the given substitutions:
   a) $\int_0^{\frac{\pi}{4}} \sec^2 x \, e^{\tan x + 1} \, dx$ $u = \tan x + 1$
   b) $\int_0^{\frac{\pi}{4}} \sec^4 x \sqrt[3]{\tan x} \, dx$ $u = \tan x$

37) Use the substitution $u = 1 + \tan x$ to find
$$\int \frac{1}{\cos^2 x \, (1 + \tan x)^2} \, dx$$

38) Evaluate the following definite integral
$$\int_1^2 \frac{5^x}{(5^x - 1)^3} \, dx$$
   a) Using the substitution $u = 5^x - 1$
   b) Using the substitution $u = 5^x$

39) Evaluate the following definite integral
$$\int_0^1 8^x (8^x + 1)^{\frac{3}{2}} \, dx$$
   a) Using the substitution $u = 8^x + 1$
   b) Using the substitution $u = 8^x$

40) Use the substitution $u = 3 - \ln x$ to find
$$\int \frac{6}{x(3 - \ln x)} \, dx$$

41) Evaluate the following definite integrals using the given substitutions:
   a) $\int_1^{2e} \frac{4}{x(5 + \ln x)^3} \, dx$ $u = 5 + \ln x$
   b) $\int_1^e \frac{2 \ln x}{x(4 - \ln x)^2} \, dx$ $u = 4 - \ln x$

Trigonometric and other substitutions

42) Use the substitution $x = \sin u$ to find:
   a) $\int \frac{1}{\sqrt{1 - x^2}} \, dx$
   b) $\int \frac{1}{(1 - x^2)^{\frac{3}{2}}} \, dx$

43) Evaluate the following definite integral
$$\int_0^{\frac{1}{4}} \frac{1}{\sqrt{1 - 4x^2}} \, dx$$

44) Use the substitution $x = \frac{3}{2} \sin u$ to find
$$\int \frac{1}{\sqrt{9 - 4x^2}} \, dx$$

45) Use the given substitutions to find:
   a) $\int \frac{1}{1 + x^2} \, dx$ $x = \tan u$
   b) $\int \frac{5}{25 + x^2} \, dx$ $x = 5 \tan u$

46) Evaluate the following definite integral
$$\int_0^1 \frac{1}{(1 + x^2)^2} \, dx$$
using the substitution $x = \tan u$

47) Evaluate the following definite integral
$$\int_0^1 \frac{1 - x^2}{1 + x^2} \, dx$$
using the substitution $x = \tan u$

48) Evaluate the following definite integrals using the given substitutions:
   a) $\int_2^3 \frac{1}{(x-2)^2 + 1} \, dx$ $x = 2 + \tan u$
   b) $\int_{-3}^{-1} \frac{1}{(x+3)^2 + 4} \, dx$ $x = 2 \tan u - 3$

49) Use integration by substitution to find
$$\int_5^8 \frac{1}{x^2 - 10x + 34} \, dx$$

50) Use integration by substitution to find
$$\int_5^8 \frac{12x}{x^2 - 10x + 34} \, dx$$

51) Evaluate the following definite integral
$$\int_{\frac{2\sqrt{3}}{3}}^2 \frac{6}{x^2 \sqrt{x^2 - 1}} \, dx$$
using the substitution $x = \text{cosec} \, u$

52) Given that $x > 3$, use the substitution $x = 3 \, \text{cosec} \, u$ to show that
$$\int \frac{1}{x^2 \sqrt{x^2 - 9}} \, dx = \frac{\sqrt{x^2 - 9}}{9x} + c$$

Prerequisite knowledge for Q53: Integration by Parts

53) Use the substitution $2x = e^u$ to find
$$\int (\ln 2x)^2 \, dx$$

2.35 Integration by Substitution (answers on page 532)

54) Use the substitution $x = \sin u$ to find
$$\int \frac{3}{(1-x^2)^{\frac{5}{2}}} dx$$

Prerequisite knowledge for Q55-Q58: Double Angle Formulae

55) Use the substitution $u = \cos x - 2$ to find
$$\int \frac{4 \sin 2x}{\cos x - 2} dx$$

56) Use integration by substitution to find
$$\int \frac{x^2}{\sqrt{1-x^2}} dx$$

57) Evaluate the following definite integral using the given substitution:
   a) $\int_0^{\frac{3}{2}} \sqrt{9-x^2}\, dx \qquad x = 3 \sin u$
   b) $\int_0^{\frac{1}{3}} \sqrt{1-9x^2}\, dx \qquad x = \frac{1}{3}\sin u$

58) Evaluate the following definite integrals using the given substitutions:
   a) $\int_0^1 \frac{1-x^2}{(1+x^2)^2} dx \qquad x = \tan u$
   b) $\int_0^{\frac{5}{2}} \sqrt{\frac{x}{5-x}}\, dx \qquad x = 5\sin^2 u$

Prerequisite knowledge for Q59-Q62: Integration by Parts

59) Use integration by substitution to find:
   a) $\int 2x^3 \ln(x^2 + 4)\, dx$
   b) $\int 3^x \ln(3^x + 1)\, dx$

60) Use the substitution $u = \sqrt{x+2}$ to find
$$\int e^{4\sqrt{x+2}}\, dx$$

61) Evaluate the following definite integral
$$\int_0^4 e^{\sqrt{2x+1}}\, dx$$
using the substitution $u^2 = 2x + 1$

62) Evaluate the following definite integrals using the given substitutions:
   a) $\int_0^1 5x^5 \ln(10x^3 + 1)\, dx \quad u = 10x^3 + 1$
   b) $\int_{-1}^{\frac{\pi^2}{4}-1} \cos(\sqrt{x+1})\, dx \quad u = \sqrt{x+1}$

Prerequisite knowledge for Q63-Q65: Partial Fractions

63) Evaluate the following definite integrals
$$\int_0^{\ln 5} \frac{10}{5 + e^x}\, dx$$
using the substitution $u = e^x$

64) Evaluate the following definite integrals
$$\int_1^4 \frac{2}{x(1 + 3\sqrt{x})}\, dx$$
using the substitution $x = u^2$

65) The diagram below shows the curve with equation
$$y = \frac{1}{x\sqrt{x+1}}$$

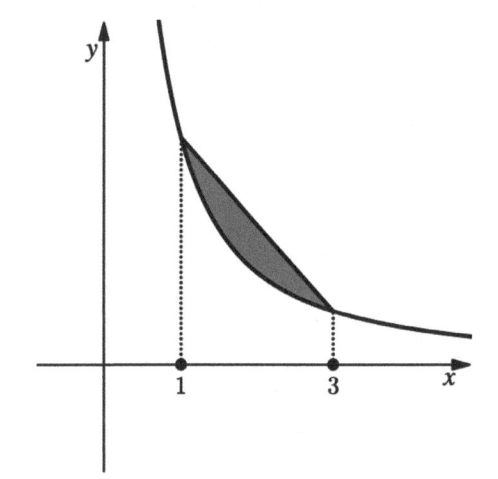

   a) Using the substitution $u = \sqrt{x+1}$, find the exact value of the definite integral
$$\int_1^3 \frac{1}{x\sqrt{x+1}}\, dx$$
   b) Find the exact value of the shaded area.

Prerequisite knowledge for Q66: Implicit Differentiation

66)
   a) Use the substitution $e^x = \sec^2 u$ to show that:
$$\int \sqrt{e^x - 1}\, dx = \int 2 \tan^2 u\, du$$
   b) Hence find the exact value of:
$$\int_0^{\ln \frac{4}{3}} \sqrt{e^x - 1}\, dx$$

2.36 Integration by Parts (answers on page 534)

1) Find
   a) $\int xe^x \, dx$
   b) $\int (x-1)e^x \, dx$
   c) $\int (2x+5)e^x \, dx$

2) Evaluate the following definite integral:
   $$\int_{-\frac{1}{3}}^{\frac{1}{3}} xe^{3x} \, dx$$

3) Evaluate the following definite integral:
   $$\int_{-1}^{0} \frac{x+1}{3e^x} \, dx$$

4) The curve y has gradient function
   $$\frac{dy}{dx} = 2x\sqrt{e^x}$$
   and passes through the point $(\ln 4, 2)$
   Show that y can be written in the form
   $y = a(x-2)\sqrt{e^x} + b \ln 4 + c$, where a, b and c are integers.

5) Find
   a) $\int x \sin x \, dx$
   b) $\int \left(x + \frac{\pi}{3}\right) \sin x \, dx$

6) Evaluate the following definite integral:
   $$\int_{-\frac{\pi}{2}}^{\frac{\pi}{2}} x \sin 2x \, dx$$

7) Find
   a) $\int x \cos x \, dx$
   b) $\int \left(x - \frac{\pi}{6}\right) \cos x \, dx$

8) Evaluate the following definite integral:
   $$\int_{\frac{3\pi}{2}}^{\frac{5\pi}{2}} \left(x - \frac{2\pi}{3}\right) \cos x \, dx$$

9) Evaluate the following definite integral:
   $$\int_{\frac{\pi}{4}}^{\frac{\pi}{2}} \left(2x - \frac{\pi}{4}\right) \sin 2x \, dx$$

10) Evaluate the following definite integral:
    $$\int_{-\frac{\pi}{8}}^{\frac{\pi}{8}} \left(x + \frac{\pi}{8}\right) \cos 4x \, dx$$

11) The curve y has gradient function
    $$\frac{dy}{dx} = 64x \sin 8x$$
    and passes through the point $\left(\frac{\pi}{16}, 1\right)$
    Show that y can be written in the form
    $y = a \sin 8x + bx \cos 8x$, where a and b are integers.

12) The curve y has gradient function
    $$\frac{dy}{dx} = 6x \cos \frac{1}{3}x$$
    and passes through the point $(\pi, -1)$
    Show that y can be written in the form
    $y = ax \sin\frac{1}{3}x + b \cos\frac{1}{3}x + p\sqrt{3}\pi + q$, where a, b, p and q are integers.

13) Find
    a) $\int x \ln x \, dx$
    b) $\int x^2 \ln x \, dx$
    c) $\int (x+4) \ln x \, dx$
    d) $\int (x^3 + 3x) \ln x \, dx$

14) Evaluate the following definite integral:
    $$\int_{1}^{2} \frac{\ln 8x}{x^4} \, dx$$

15) The curve y has gradient function
    $$\frac{dy}{dx} = 4x \ln \frac{1}{2}x$$
    and passes through the point $(2e, 1)$
    Show that y can be written in the form
    $y = ax^2 + bx^2 \ln\frac{1}{2}x + pe^2 + q$, where a, b, p and q are integers.

16) Find
    a) $\int x 2^x \, dx$
    b) $\int (3x+9) 3^x \, dx$

17) Evaluate the following definite integral:
    $$\int_{0}^{1} (x+2) 2^x \, dx$$

18) Evaluate the following definite integral:
    $$\int_{1}^{2} 10x 5^{x-1} \, dx$$

2.36 Integration by Parts (answers on page 534)

19) The curve y has gradient function
$$\frac{dy}{dx} = \frac{x}{4^x}$$
and passes through the point $(\ln 4, 1)$
Write y in terms of x

20) The diagram below shows the curve $y = x\sec^2 x$. Show that the exact shaded area is $\frac{\pi}{\sqrt{3}} - \ln 2$

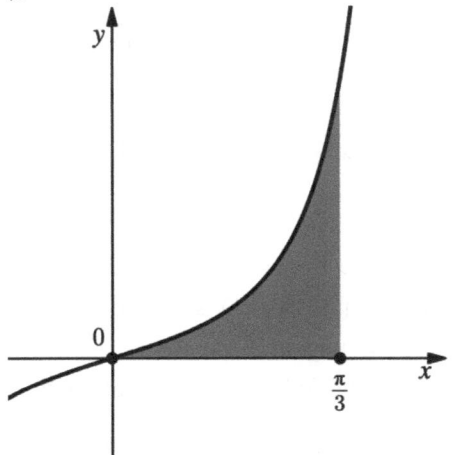

21) Evaluate the following definite integral:
$$\int_{\frac{\pi}{6}}^{\frac{\pi}{3}} 18x \cosec^2 x \, dx$$

22) Find
   a) $\int \ln x \, dx$
   b) $\int (\ln x)^2 \, dx$

Prerequisite knowledge for Q23: Integration by Substitution

23) Evaluate the following definite integral:
$$\int_3^{e^2+2} \ln(x-2) \, dx$$

Integration by Parts Twice

24) Find
   a) $\int x^2 e^x \, dx$
   b) $\int \frac{x^2}{e^x} \, dx$
   c) $\int x^2 \sin x \, dx$
   d) $\int x^2 2^x \, dx$

25) Evaluate the following definite integral:
$$\int_0^{\frac{1}{4}} 20x^2 e^{-4x} \, dx$$

26) The diagram below shows the curves $y = xe^x$ and $y = 2e - x^2 e^x$. Find the exact shaded area between the two curves.

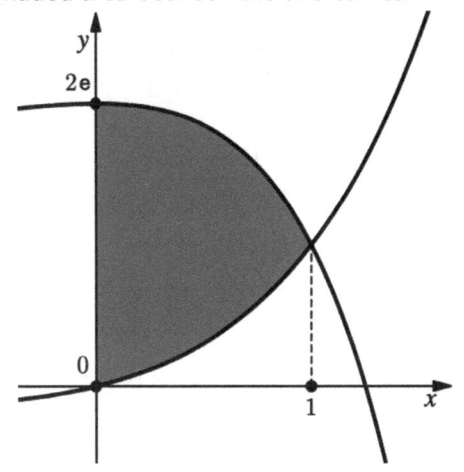

27) Evaluate the following definite integral:
$$\int_1^e (\ln x)^3 \, dx$$

Cyclic Integration

28) Find
   a) $\int e^x \sin x \, dx$
   b) $\int e^x \cos x \, dx$

29) Evaluate the following definite integral:
$$\int_0^{\frac{\pi}{2}} e^{5x} \sin 2x \, dx$$

30) The diagram below shows the curve $y = \ln(2x - k)$. It is given that the curve passes through the origin, O, and the point A with x-coordinate 1. The normal to the curve at A intersects the x-axis at B. Find the exact shaded area of the shape OAB.

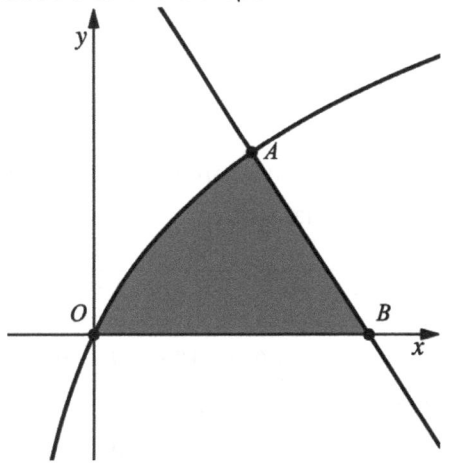

2.37 Partial Fractions (answers on page 535)

1) Express
$$\frac{7x + 17}{(x + 1)(x + 3)}$$
in the form
$$\frac{A}{x + 1} + \frac{B}{x + 3}$$

2) Express
$$\frac{12x - 3}{(x - 2)(x + 1)}$$
in the form
$$\frac{A}{x - 2} + \frac{B}{x + 1}$$

3) Write each of the following as partial fractions:
   a)
   $$\frac{2x + 3}{(x - 2)(x + 5)}$$
   b)
   $$\frac{16x + 1}{(x + 4)(2x + 1)}$$
   c)
   $$\frac{4}{(x + 2)(x + 7)}$$
   d)
   $$\frac{14}{(x + 2)(x + 6)}$$
   e)
   $$\frac{4}{x(x - 2)}$$
   f)
   $$\frac{x + 9}{x^2 + 8x + 15}$$
   g)
   $$\frac{49 - x}{2x^2 + 13x - 24}$$
   h)
   $$\frac{2(11x + 4)}{8x^2 + 6x + 1}$$
   i)
   $$\frac{k}{x^2 + x - 20}$$
   where k is a positive constant.

4) Tom is working on a partial fraction question and so far he has written:
$$\frac{1}{25 - 4x^2} \equiv \frac{A}{5 - 2x} + \frac{B}{5 + 2x}$$
$$25 - 4x^2 \equiv A(5 + 2x) + B(5 - 2x)$$
When $x = -\frac{5}{2}$, $0 = -10B \Rightarrow B = 0$
   a) State what his mistake is.
   b) Find the correct partial fraction.

5) Express
$$\frac{3x^2 - 14x + 31}{(x - 5)(x - 3)(x + 1)}$$
in the form
$$\frac{A}{x - 5} + \frac{B}{x + 1} + \frac{C}{x - 3}$$

6) Write each of the following as partial fractions:
   a)
   $$\frac{4x^2 - 25x + 61}{(x - 5)(x - 2)(x + 1)}$$
   b)
   $$\frac{2}{(x - 4)(x + 3)(2x - 1)}$$

7) Given
$$f(x) = 4x^3 - 8x^2 - 15x + 9$$
   a) show that $(2x - 1)$ is a factor of $f(x)$
   b) Hence fully factorise, $f(x)$
   c) Write the following as partial fractions
   $$\frac{5}{4x^3 - 8x^2 - 15x + 9}$$

8) Lisa is working on a partial fraction question and so far she has written:
$$\frac{1}{(x + 2)(x^2 - 1)} \equiv \frac{A}{x + 2} + \frac{B}{x - 1} + \frac{C}{x + 1}$$
$$1 \equiv A(x + 2) + B(x - 1) + C(x + 1)$$
When $x = -2$, $1 = -3B - C$
When $x = 1$, $1 = 3A + 2C$
When $x = -1$, $1 = A - 2B$
   a) State what her mistake is.
   b) Find the correct partial fractions.

9) Express in the form
$$\frac{7x^2 + 21x + 12}{(x + 1)^2(x + 3)}$$
in the form
$$\frac{A}{x + 1} + \frac{B}{(x + 1)^2} + \frac{C}{x + 3}$$

10) Express in the form
$$\frac{32 - 5x - x^2}{(x - 2)^2(x + 4)}$$
in the form
$$\frac{A}{x - 2} + \frac{B}{(x - 2)^2} + \frac{C}{x + 4}$$

2.37 Partial Fractions (answers on page 535)

11) Write each of the following as partial fractions:
    a) $$\frac{3}{(x-1)(x+2)^2}$$
    b) $$\frac{2x+1}{(2x-1)(x+1)^2}$$
    c) $$\frac{2}{x^2(x-1)}$$
    d) $$\frac{x+3}{(x-1)^2}$$

12) For each of the following select the format you should use to begin the method of finding the partial fraction:

A:	$\frac{A}{x+2} + \frac{B}{x-3}$
B:	$\frac{A}{x+2} + \frac{B}{(x+2)^2} + \frac{C}{x-3}$
C:	$\frac{A}{x+2} + \frac{B}{x-3} + \frac{C}{(x-3)^2}$

    a) $$\frac{5}{(x+2)(x-3)}$$
    b) $$\frac{4x+1}{(x+2)^2(x-3)}$$
    c) $$\frac{3x-2}{x^2-x-6}$$
    d) $$\frac{2x-1}{x^3+x^2-8x-12}$$
    e) $$\frac{5x^2+1}{x^3-4x^2-3x+18}$$

13) Write each of the following as partial fractions:
    a) $$\frac{2x+3}{x^2+3x-10}$$
    b) $$\frac{4}{x^3-4x^2+4x}$$
    c) $$\frac{5x+1}{4x^3+8x^2-x-2}$$
    d) $$\frac{2}{2x^3+x^2-4x-3}$$

14) Given
    $$f(x) = \frac{5x-1}{(x-2)(x+7)}, \quad x \neq 2, x \neq -7$$
    a) Write f(x) as partial fractions.

    Given
    $$g(x) = \frac{3x^2-19x+44}{(x+7)(x-2)^2}, \quad x \neq 2, x \neq -7$$
    b) Write g(x) as partial fractions.
    c) Hence, find the exact solution to the equation f(x) = g(x)

15) It is known that
    $$\frac{5x+19}{(x+3)(x+p)^2} \equiv \frac{1}{x+3} - \frac{1}{x+p} + \frac{3}{(x+p)^2}$$
    for some value of p
    Find the value of p

Prerequisite knowledge for Q16-Q19: Differentiation of $(ax+k)^n$

16) Given
    $$f(x) = \frac{x+2}{(x-4)(2x+1)}, \quad x \neq 4, x \neq -\frac{1}{2}$$
    a) Write f(x) as partial fractions.
    b) Find the exact coordinates of the stationary points of the curve $y = f(x)$

17) Given
    $$f(x) = \frac{3x}{(x-2)^2(5-x)}, \quad x \neq 2, x \neq 5$$
    a) Write f(x) as partial fractions.
    b) Show that the tangent to the curve $y = f(x)$ at the point $x = 3$ has the equation $4y + 21x = 81$

18) Given
    $$f(x) = \frac{2x+1}{(x+3)(2-x)}, \quad x > -3$$
    a) Write f(x) as partial fractions.
    b) Show that f(x) is always increasing.

19) Given
    $$f(x) = \frac{5}{(x+1)(x+2)}, \quad x \neq -1, -2$$
    a) Write f(x) as partial fractions.
    b) Show that f(x) is always decreasing.

2.37 Partial Fractions (answers on page 535)

Prerequisite knowledge for Q20-Q22: Binomial Series

20) Given
$$f(x) = \frac{1}{(1-2x)(1+x)}, \quad x \neq \frac{1}{2}, x \neq -1$$
   a) Write $f(x)$ as partial fractions.
   b) Hence expand $f(x)$ up the term in x^3
   c) State the values for which the expansion is valid.

21) Given
$$f(x) = \frac{1}{(x+1)(x-2)}, \quad x \neq -1, x \neq 2$$
   a) Write $f(x)$ as partial fractions.
   b) Hence expand $f(x)$ up the term in x^3
   c) State the values for which the expansion is valid.

22) Given
$$f(x) = \frac{3x^2 - 5x + 6}{(1-x)^2(3+x)}, \quad x \neq 1, x \neq -3$$
   a) Write $f(x)$ as partial fractions.
   b) Hence expand $f(x)$ up the term in x^3
   c) State the values for which the expansion is valid.

23) Given
$$\frac{8x^2 + 4x + 1}{(2x-1)(x^2+1)} \equiv \frac{A}{2x-1} + \frac{Bx+C}{x^2+1}$$
find the values of A, B and C

The following questions consider improper fractions

24) Given
$$\frac{3x+2}{x-1} \equiv A + \frac{B}{x-1}$$
   a) find the values of A and B
   b) Hence, sketch
$$y = \frac{3x+2}{x-1}$$

25) Given
$$\frac{12 - 5x}{2-x} \equiv A + \frac{B}{x-2}$$
   a) find the values of A and B
   b) Hence, sketch
$$y = \frac{12-5x}{2-x}$$

26) Find the constants in the following identities:
   a) $10 - 4x^2 - 7x \equiv (Ax + B)(1 - 4x) + C$
   b) $2x^2 + 15 \equiv (Ax + B)(3 - x) + C$
   c) $3x^3 + 2x^2 \equiv (Ax + B)(x^2 + 1) + Cx + D$
   d) $x^3 + 2 \equiv (Ax^2 + Bx + C)(x + 1) + Dx$

27) Given
$$\frac{x^2 - 3x - 13}{x - 5} \equiv A + Bx + \frac{C}{x-5}$$
find the values of A, B and C

28) Given
$$\frac{2 + 5x - 5x^2}{1-x} \equiv Ax + \frac{B}{1-x}$$
find the values of A and B

29) Given
$$\frac{x^3}{x^2 + 3x - 4} \equiv A + Bx + \frac{C}{x-1} + \frac{D}{x+4}$$
find the values of A, B and C

30) Given
$$\frac{x^4 - 1}{(x-2)(x-1)(x+4)} \equiv Ax + B + \frac{C}{x-2} + \frac{D}{x+4}$$
find the values of A, B, C and D

31) Given
$$\frac{2x^4 - x^3 - x + 4}{x^2 + 1} \equiv Ax^2 + Bx + C + \frac{D}{x^2+1}$$
find the values of A, B, C and D

Prerequisite knowledge for Q32: Arithmetic Series

32)
   a) Write
$$\frac{4p+5}{(p+1)(p+2)}, \quad p \neq -1, p \neq -2$$
   in partial fractions.

   It is known that $\frac{A}{p+1}$ and $\frac{B}{p+2}$ are the first two terms in an arithmetic sequence
   b) Using your values for A and B find the third term in the arithmetic sequence in terms of p

   The sum of the first 3 terms is 1
   c) Find the value of p and state the first 3 terms of the sequence.

2.38 Integration: Partial Fractions (answers on page 536)

1) It is given that $f(x) = \frac{8x+23}{x+2}$
   a) Write $f(x)$ in the form $A + \frac{B}{x+2}$
   b) Hence find $\int f(x)\,dx$

2) It is given that $f(x) = \frac{2x^2-2x-1}{x+1}$
   a) Write $f(x)$ in the form $Ax + B + \frac{C}{x+1}$
   b) Hence find $\int f(x)\,dx$

3)
   a) Express $\frac{4}{(x-3)(x+1)}$ as partial fractions.
   b) Show that
   $$\int \frac{4}{(x-3)(x+1)}\,dx = \ln\left|\frac{x-3}{x+1}\right| + c$$

4)
   a) Find $\int \frac{10}{25-x^2}\,dx$
   b) Hence show that
   $$\int_{-1}^{1} \frac{10}{25-x^2}\,dx = \ln\frac{a}{b}$$
   where a and b are integers to be found.

5)
   a) Find $\int \frac{2x-1}{(x+5)(x-6)}\,dx$
   b) Hence find the value of k such that
   $$\int_{7}^{k} \frac{2x-1}{(x+5)(x-6)}\,dx = \ln 5$$

6)
   a) Express $\frac{k-4x}{x(2-x)}$ as partial fractions, where k is a positive constant.
   b) Find the value of k such that
   $$\int_{3}^{4} \frac{k-4x}{x(2-x)}\,dx = \ln\frac{128}{27}$$

7)
   a) Express $\frac{23x+11}{(3-x)(x+2)(3x-1)}$ as partial fractions.
   b) Write
   $$\int_{1}^{2} \frac{23x+11}{(3-x)(x+2)(3x-1)}\,dx$$
   in the form $\ln k$, where k is a positive constant.

8)
   a) Express $\frac{4x^2+1}{x^2(2x+1)}$ as partial fractions.
   b) Show that
   $$\int \frac{4x^2+1}{x^2(2x+1)}\,dx = \frac{a}{x} + \ln f(x) + c$$
   where a is an integer to be found.

9)
   a) Write $\frac{4x-2}{(x-2)^2}$ in the form $\frac{A}{x-2} + \frac{B}{(x-2)^2}$ where A and B are constants.
   b) Show that
   $$\int_{3}^{4} \frac{4x-2}{(x-2)^2}\,dx = a + \ln b$$
   where a and b are integers to be found.

10)
    a) Find $\frac{10x^2+6}{(x+1)^2(3x-1)}$ as partial fractions.
    b) Hence show that
    $$\int_{1}^{\frac{5}{3}} \frac{10x^2+6}{(x+1)^2(3x-1)}\,dx = a + b\ln 2 + c\ln 9$$
    where a, b and c are constants to be found.

11) The diagram below shows the curve with equation
    $$y = \frac{x^2+5x+6}{x^2(2-x)}, \quad x \neq 0, x \neq 2$$
    Find the exact value of the shaded area.

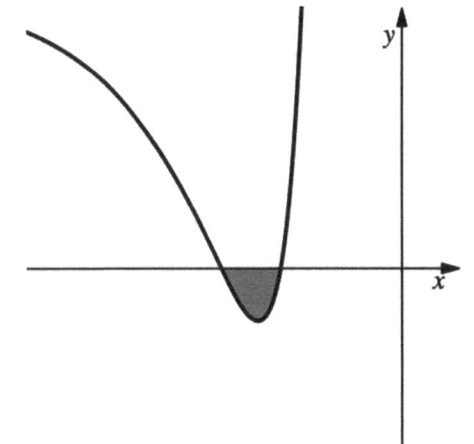

12) Find the exact value of
    $$\lim_{\delta x \to 0} \sum_{x=0}^{\frac{1}{8}} \frac{360x^2 + 157x + 16}{(3x+1)(2-5x)(8x+3)} \delta x$$

Prerequisite knowledge for Q13-Q14: Integration by Substitution

13) Use the substitution $u = \sqrt{x} - 2$ to find
    $$\int \frac{9}{2(\sqrt{x}-2)(\sqrt{x}+1)(\sqrt{x}+4)}\,dx$$

14) Use the substitution $u = \sqrt{x-2}$ to show that
    $$\int_{3}^{6} \frac{6}{(3+2\sqrt{x-2})(x-2)}\,dx = 4\ln\frac{10}{7}$$

Integration – Double Angles (answers on page 536)

1)
   a) Write $\cos 2x$ in terms of $\cos x$
   b) Hence show that $\int_0^{\frac{\pi}{4}} \cos^2 x \, dx = \frac{2+\pi}{8}$

2)
   a) Write $\cos 2x$ in terms of $\sin x$
   b) Hence show that $\int_0^{\frac{\pi}{2}} \sin^2 x \, dx = \frac{\pi}{4}$

3) Find
   a) $\int (1 + \sin x)^2 \, dx$
   b) $\int (1 - \cos x)^2 \, dx$
   c) $\int (\sin x - \cos x)^2 \, dx$
   d) $\int (\sin x - \cos x)(\sin x + \cos x) \, dx$

4) Find
   a) $\int \sin x \cos x \, dx$
   b) $\int 8 \sin 2x \cos 2x \, dx$
   c) $\int 12 \sin 5x \cos 5x \, dx$
   d) $\int \cos^2 6x \tan 6x \, dx$

5) Evaluate the following definite integrals:
   a) $\int_0^{\frac{\pi}{2}} (4 - 3 \sin x)^2 \, dx$
   b) $\int_0^{\frac{\pi}{2}} (1 - 6 \cos x)^2 \, dx$

6) Evaluate the following definite integrals:
   a) $\int_0^{\frac{\pi}{4}} (1 - 8 \sin 2x)^2 \, dx$
   b) $\int_{-\pi}^{\pi} (7 + 9 \cos 3x)^2 \, dx$

7) Find
   a) $\int (2 \cos^2 3x - 1) \, dx$
   b) $\int (1 - 2 \sin^2 5x) \, dx$
   c) $\int (\sin 4x \cot 2x - 1) \, dx$
   d) $\int (1 - \sin 20x \tan 10x) \, dx$

8) Find
$$\int \frac{10 \sin x \cos x}{2 \cos^2 x + 3} \, dx$$

9) The curve y has gradient function
$$\frac{dy}{dx} = \frac{4 \tan x}{\sqrt{2 \cos 2x + 2}}$$
and passes through the point $(\pi, 6)$
Show that y can be written in the form $y = a \sec x + b$, where a and b are integers to be found.

10) Evaluate the following definite integral
$$\int_0^{\frac{\pi}{4}} \frac{\sqrt{1 - \cos 2x}}{\cos 2x + 1} \, dx$$

11) Find the exact value of
$$\lim_{\delta x \to 0} \sum_{x = \frac{\pi}{4}}^{\frac{\pi}{2}} \frac{(\cos 2x + 1) \operatorname{cosec} x}{\sin 2x} \delta x$$

12) The diagram below shows the curve with equation
$$y = \frac{2 \cos x}{\sin 2x \, (\operatorname{cosec} x - \sin x)}$$
Find the exact value of the shaded area.

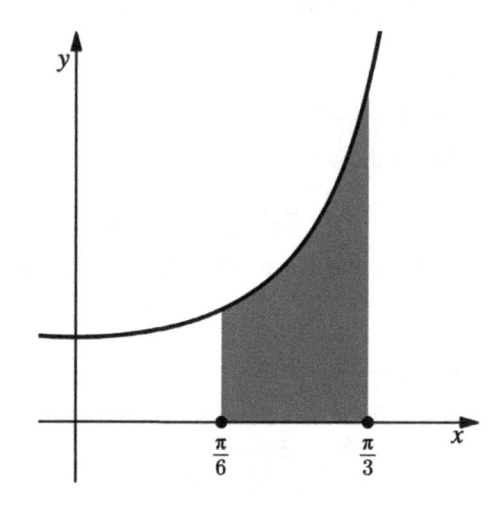

13)
   a) Use the Compound Angle Formulae to show that $\sin 3x \equiv 3 \sin x - 4 \sin^3 x$
   b) Hence find $\int \sin^3 x \, dx$

14)
   a) Use the Compound Angle Formulae to show that $\cos 3x \equiv 4 \cos^3 x - 3 \cos x$
   b) Hence find $\int \cos^3 x \, dx$

Prerequisite knowledge for Q15-Q16: Integration by Substitution

15) Evaluate the following definite integral
$$\int_0^{\frac{1}{2}} \sqrt{\frac{16x}{1 - x}} \, dx$$
using the substitution $x = \cos^2 u$

16) Evaluate the following definite integral
$$\int_{-1}^{5} \sqrt{(x+1)(5-x)} \, dx$$
using the substitution $x = 2 - 3 \cos u$

2.40 Differential Equations (answers on page 537)

1) Solve the following differential equations:
   a) $$\frac{dy}{dx} = y$$
   b) $$\frac{dy}{dx} = \frac{y}{x}$$
   c) $$\frac{dy}{dx} = \frac{y}{x^2}$$
   d) $$\frac{dy}{dx} = \frac{y}{6x^4}$$

2) Solve the following differential equations:
   a) $$\frac{dy}{dx} = \frac{x}{y}$$
   b) $$\frac{dy}{dx} = \frac{x^2}{y^2}$$
   c) $$\frac{dy}{dx} = \frac{x^3}{2y}$$
   d) $$\frac{dy}{dx} = \frac{x^5}{3y^3}$$

3) Given that $x = 1$ when $t = 1$, solve the differential equation
   $$\frac{dx}{dt} = \frac{x^4}{10t}$$

4) Given that $x = 4$ when $t = 0$, solve the differential equation
   $$\frac{dx}{dt} = \frac{t}{\sqrt{x}}$$
   Hence find the value of x when $t = 2\sqrt{39}$

5) Solve the following differential equations:
   a) $$\frac{dy}{dx} = \frac{y^2}{2x}$$
   b) $$\frac{dy}{dx} = \frac{y^2}{x^2}$$
   c) $$\frac{dy}{dx} = \frac{y^3}{3x^3}$$

6) Solve the following differential equations:
   a) $$\frac{dy}{dx} = xy$$
   b) $$\frac{dy}{dx} = x^2 y$$
   c) $$\frac{dy}{dx} = 4x^3 y$$
   d) $$\frac{dy}{dx} = y\sqrt{x}$$

7) Given that $P = 500$ when $t = 0$, solve the differential equation
   $$\frac{dP}{dt} = -100Pt^3$$
   Hence find the value of P when $t = 0.7$, giving your answer to three significance figures.

8) Solve the following differential equations:
   a) $$\frac{dy}{dx} = xy^2$$
   b) $$\frac{dy}{dx} = x^2 y^2$$
   c) $$\frac{dy}{dx} = \frac{2}{3}(xy)^3$$

9) Given that $y(1) = 6$, solve the differential equation
   $$\frac{dy}{dx} = \frac{1}{3}(xy)^3$$
   Hence find the exact possible values of $y(0)$

10) Solve the following differential equations:
    a) $$\frac{dy}{dx} = \frac{\sin x}{y}$$
    b) $$\frac{dy}{dx} = \frac{\cos x}{4y}$$
    c) $$\frac{dy}{dx} = \frac{\sin 2x}{3\sqrt{y}}$$
    d) $$\frac{dy}{dx} = \frac{5\cos 4x}{\sqrt{16y}}$$

2.40 Differential Equations (answers on page 537)

11)
   a) Given that $y(0) = 0$, solve the differential equation:
   $$\frac{dy}{dx} = \frac{32\cos 3x}{\sqrt{8y}}$$
   b) State the exact maximum value of y

12) Solve the following differential equations:
   a)
   $$\frac{dy}{dx} = \frac{\cos 2x}{e^y}$$
   b)
   $$\frac{dy}{dx} = \frac{y+1}{x+1}$$
   c)
   $$\frac{dy}{dx} = \frac{12x^2}{e^{0.2y}}$$
   d)
   $$\frac{dy}{dx} = \frac{4y-3}{e^{0.2x}}$$

13) Solve the following differential equations:
   a)
   $$\frac{dy}{dx} = e^{x-y}$$
   b)
   $$\frac{dy}{dx} = e^{y-x}$$
   c)
   $$\frac{dy}{dx} = 2e^{4x-3y}$$
   d)
   $$\frac{dy}{dx} = 8e^{4y-\frac{1}{2}x}$$

14) Given that $y(0) = 0$, solve the differential equation
   $$\frac{dy}{dx} = e^{y-6x}$$
   Hence find the exact value of $y(0.5)$.

15) Solve the differential equation
   $$\frac{24}{y^{\frac{1}{3}}}\frac{dy}{dx} = \frac{4-x}{x}$$
   to show that
   $$y = \pm\frac{(k-x+4\ln x)^{\frac{3}{2}}}{216}$$
   where $x > 0$ and k is a constant.

16) The height above ground, H metres, of a mechanical arm can be modelled by the differential equation
   $$\frac{dH}{dt} = \frac{2H\cos 0.2t}{15}$$
   where t is the time in seconds since the mechanical arm was switched on. The arm is initially 2 m above the ground.
   a) Show that $H = 2e^{\frac{2}{3}\sin 0.2t}$
   b) Determine the first time after being switched on that the mechanical arm reaches its maximum height.

17) The rate of change of y with respect to x is directly proportional to $\frac{10-x}{y}$. When $x = 5$, both $y = 20$ and the rate of change of y with respect to x is 10. Show that
   $$y^2 = 800x - 40x^2 - 2600$$

18) A piece of paper has ink dropped onto it. Initially the ink blob has a radius of 4 mm and after 2 seconds it has a radius of 16 mm. In a simple model, the rate of increase of the radius of the ink blob is inversely proportional to the square root of the radius.
   a) Using this model and defining any variable you use, find an equation linking the radius and the time.
   b) Explain why this model may not be valid in the long-term.

19) A particle's acceleration, a ms^{-2}, at time t seconds, may be modelled in terms of its velocity v ms^{-1}, as $a = 1.02v^3$, where $v > 0$. The particle's initial velocity is 1 ms^{-1}. Show that
   $$v = \frac{5}{\sqrt{25-51t}}$$

20) A particle's acceleration, a ms^{-2}, at time t seconds, may be modelled in terms of its velocity v ms^{-1}, as $a = -0.02v^2$. The particle's initial velocity is 2 ms^{-1}. Show that
   $$v = \frac{50}{t+25}$$

2.40 Differential Equations (answers on page 537)

21) A particle's acceleration, a ms^{-2}, at time t seconds, may be modelled in terms of its velocity v ms^{-1}, as $a = -1.05v$. The particle's initial velocity is 6 ms^{-1}. Determine the speed of the particle when $t = 2$, giving your answer to 3 significant figures.

22) Dan measures the temperature, $T°C$, of a freshly made cup of coffee after t minutes. He assumes that the temperature decreases at a rate that is inversely proportional to the square root of coffee's temperature.
    a) Construct a differential equation involving T, t and a constant k, where $k > 0$, to model this situation.
    b) Explain why this model may not be valid.

23) A water tank contains a large amount of water with volume V m^3. There is a small hole at the bottom of the tank such that water flows out at a constant rate of 0.02 m^3 per minute. At time $t = 0$, the volume of the water in the tank is 20 m^3. A tap is turned on so that water flows into the tank at a rate proportional to the volume of the water already in the tank. Show that
$$50\frac{dV}{dt} = kV - 1$$
where k is a positive constant.

24) £500 is invested in a savings account. The amount of money in the account after t days is £P. The rate at which P increases is kP, where k is a positive constant.
    a) Construct a differential equation involving P, t and k.
    b) It is given that initially, the rate of increase of the money in the savings account is £0.02 per day. Determine how much money is in the account when $t = 365$

25) The rate of increase of bacteria in a petri dish is modelled as being directly proportional to the number of bacteria, N, at any time. When the bacteria are increasing at 5000 a day, there are 60 000 bacteria. Write down a differential equation to model this.

26) The rate at which a tree's height, h m, increases is known to be directly proportion to the natural logarithm of the tree's height at any time, t years. When the tree is 2 m tall, the rate at which the tree is growing is 0.2 m per year. Write down a differential equation to model this.

27) The diagram below shows a cuboid water tank, which has a width of 6 m, a depth of 10 m and a height of 6 m. The depth of the water in the tank is d metres.

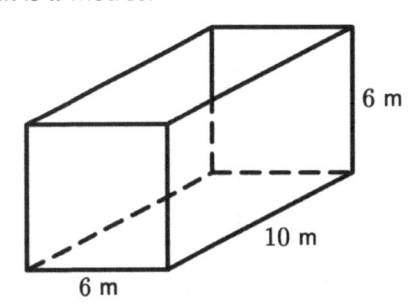

There is a small hole at the bottom of the tank such that water flows out at a rate of $0.05d$ m^3 per minute. A tap is turned on so that water flows into the tank at a rate of 0.1 m^3 per minute. Show that
$$\frac{dV}{dt} = \frac{120 - V}{1200}$$

28) The diagram below shows a cylindrical water tank, which has a height of 5 m and a diameter of 8 m. The depth of the water in the tank is d metres at time t minutes.

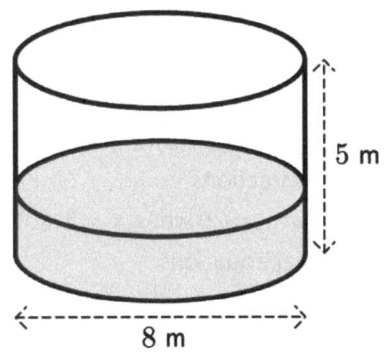

There is a small hole at the bottom of the tank such that water flows out at a rate of $0.2d^2$ m^3 per minute. A tap is turned on so that water flows into the tank at a rate of 0.8 m^3 per minute. Show that
$$\frac{dV}{dt} = \frac{(32\pi - V)(32\pi + V)}{1280\pi^2}$$

2.40 Differential Equations (answers on page 537)

**Prerequisite knowledge for Q29-Q46:
Connected Rates of Change, Double Angle
Formulae, Integration by Substitution,
Integration by Parts, Partial Fractions:**

29) Solve the following differential equations:
   a)
   $$\frac{dy}{dx} = \frac{(x+2)^3}{y}$$
   b)
   $$\frac{dy}{dx} = y(2x-5)^4$$
   c)
   $$\frac{dy}{dx} = \frac{\sqrt{4x+7}}{\sqrt{y}}$$
   d)
   $$\frac{dy}{dx} = \frac{(3-x)^{\frac{2}{3}}}{e^y}$$

30) Solve the following differential equations:
   a)
   $$\frac{dy}{dx} = \frac{xe^x}{y}$$
   b)
   $$\frac{dy}{dx} = xy \sin 10x$$
   c)
   $$\frac{20}{x}\frac{dy}{dx} = \frac{e^{2x}}{y^2}$$
   d)
   $$\frac{3}{x \ln x}\frac{dy}{dx} = \frac{5}{\ln y}$$

31)
   a) Express
   $$\frac{8x-10}{(x-3)(x+4)}$$
   as partial fractions
   b) Given that $y = 20$ when $x = 4$, solve the differential equation
   $$(x-3)(x+4)\frac{dy}{dx} = 2y(4x-5)$$
   writing your answer in the form $y = f(x)$

32) Solve the differential equation
$$\frac{dy}{dx} = 8y \cos x \sin^3 x$$
such that $y = 2$ when $x = 0$, writing y in terms of x

33) A spherical balloon is deflating so that at time t seconds, the balloon has radius r cm and volume V cm³. The volume of the balloon is modelled as decreasing at a constant rate.
   a) Using this model, show that
   $$\frac{dr}{dt} = -\frac{m}{r^2}$$
   where m is a positive constant.
   b) It is given that the balloon's initial radius is 30 cm, and after 2 seconds the radius is 29 cm. The volume of the balloon continues to decrease at a constant rate until the balloon is completely deflated. Solve the differential equation, writing your answer in the form $r = f(t)$

34) A large cylindrical drum contains a volume of water. The area of the base of the drum is 3600 cm². The depth of the water after t seconds is h cm. A tap is turned on and water pours into the drum at a constant rate of 2400 cm²s⁻¹. There is a hole at the bottom of the tank such that water flows out at a rate proportional to the square of the height of the water already in the cylinder. Show that at time t seconds, the height of the water in the drum satisfies the differential equation
$$\frac{dh}{dt} = \frac{2}{3} - kh^2$$
where k is a positive constant.

35) It is given that $P > 0$ and that
$$\frac{dP}{dt} = \frac{\sin^2 4t}{10P}$$
Given also that $P = 20$ when $t = 0$, show that
$$P = \frac{1}{20}\sqrt{160000 + 40t - 5\sin 8t}$$

36)
   a) Solve the differential equation
   $$\frac{dy}{dx} = \frac{y}{10}\left(1 - \frac{y}{100}\right)$$
   where $0 < y < 100$.
   It is also given that $y = 10$ when $x = 0$.
   b) Find the value of y as x tends to infinity.

2.40 Differential Equations (answers on page 537)

37)
  a) Express
  $$\frac{1}{x(200-x)}$$
  as partial fractions.
  b) It is given that $\frac{dx}{dt} = \frac{x(200-x)}{1000}$
  where $0 < x < 200$ and $t \geq 0$.
  It is also given that $x = 100$ when $t = 0$.
  Show that
  $$x = \frac{200e^{\frac{1}{5}t}}{1 + e^{\frac{1}{5}t}}$$
  c) As $t \to \infty$, $x \to k$. Find the value of k.

38) A study into the population, P thousands, of a certain type of insect can be modelled by the differential equation
$$15000\frac{dP}{dt} = P(1000 - P)$$
where t is the time measured in weeks since the study began. It is given that initially there are 4000 insects.
  a) Solve the differential equation to show that
  $$P = \frac{4000}{a + be^{-\frac{1}{15}x}}$$
  where a and b are constants to be determined.
  b) Hence show that the population will not drop below 1000.

39)
  a) Use the substitution $u = 2\sqrt{x} - 1$ to show that
  $$\int \frac{1}{2\sqrt{x} - 1}dx = \sqrt{x} + \frac{1}{2}\ln|1 - 2\sqrt{x}| + c$$
  where c is a constant.
  b) Hence solve the differential equation
  $$\frac{dx}{dt} = \frac{2\sqrt{x} - 1}{50t^{\frac{2}{5}}}$$
  given that $x = 0$ when $t = 0$ to show that
  $$t = (30\sqrt{x} + 15\ln|1 - 2\sqrt{x}|)^{\frac{5}{3}}$$
  c) Find, according to the model, the range of values of x.

40) The gradient function of a curve is given by
$$\frac{dy}{dx} = \frac{4xe^{2x}}{\cos 4y}$$
Solve the differential equation to show that
$$y = \frac{1}{4}\arcsin(4e^{2x}(2x - 1) + c)$$

41) The gradient function of a curve is given by
$$\frac{dy}{dx} = \frac{(3x - 1)^5}{y \sin 4y}$$
Find an equation for the curve in the form $f(y) = g(x)$.

42) The gradient function of a curve is given by
$$\frac{dy}{dx} = \frac{6\cos^2 2x}{y^2 \sin 3y}$$
Find an equation for the curve in the form $f(y) = g(x)$.

43) A colony of 2000 insects is introduced to an area. The rate of increase of the number of insects, N, after t weeks is modelled by
$$\frac{dN}{dt} = \frac{9\,000\,000 - N^2}{500}$$
Determine the long-term number of insects in the colony.

44) A cuboid water tank has base dimensions of 5 m by 3 m. The depth of the water after t seconds is h m. Initially the tank is empty. There is a hole at the bottom of the tank such that water flows out at a rate of $0.1h^2$ m³ per second. At time $t = 0$, a tap is turned on so that water flows into the tank at a constant rate of 0.4 m³ per second. Find the time taken for the water to reach a depth of 10 cm.

45) Given $y = 1$ when $x = 0$, solve:
$$\frac{dy}{dx} = \frac{y}{8 - 2\sqrt{x}}$$

46) The gradient function of a curve is given by the differential equation
$$(3x + 1)^3 \frac{dy}{dx} - 16y^2 = 0$$
Given that $y = 2$ when $x = 1$, show that
$$y = \frac{ax^2 + bx + c}{px^2 + qx + r}$$
where a, b, c, p, q, r are integers to be found.

2.41 Parametric Equations: Introduction (answers on page 538)

Prerequisite knowledge: Those questions that involve trigonometry use radians

1) For each of the following, complete the table, plot the coordinates and sketch the curve for $-2 \leq t \leq 2$

t	-2	-1	0	1	2
x					
y					

   a) $x = t^2$ and $y = 2t$
   b) $x = t + 2$ and $y = -2t^2 + 1$
   c) $x = t(t+2)$ and $y = t^2(t-1)$
   d) $x = t(1-t)$ and $y = 2t^2$
   e) $x = \frac{4}{2+t^2}$ and $y = t(t+1)$

2) Given that $x = t$, find y in terms of t for each of the following pairs of parametric equations:
   a) $y = 2x^2$
   b) $y = x^2 + 4$
   c) $y = e^x + 3$
   d) $y - 2x = 8$
   e) $\frac{x^2}{y} = 1$
   f) $x^3 y = -10$
   g) $x^2 + y^2 = 1$

3) Find the cartesian equation (with y as the subject) for each of the following curves:
   a) $x = 2t$ and $y = t^2$
   b) $x = t - 1$ and $y = 2t + 3$
   c) $x = t - 3$ and $y = e^t - t$
   d) $x = -2t$ and $y = t^2 \ln t$

4) Find the cartesian equation (with x as the subject) for each of the following curves:
   a) $x = t^2$ and $y = t - 1$
   b) $x = t^3 - t$ and $y = \frac{t}{2}$
   c) $x = t^5$ and $y = 3t$
   d) $x = e^{t-1}$ and $y = t + 5$

5) For each of the following curves, find the coordinates of intersections with the axes:
   a) $x = t + 1$ and $y = t - 3$
   b) $x = \cos t$ and $y = 2 \sin t$ for $0 \leq t < 2\pi$
   c) $x = 2 \cos t$ and $y = \sin 3t$ for $0 \leq t < 2\pi$
   d) $x = t + 5$ and $y = \ln(t + 1)$
   e) $x = t^3 - 3t^2 - 6t + 8$ and $y = e^{-2t}$

6) A curve is given by parametric equations
   $$x = t^2 \text{ and } y = t^3$$
   Given the curve may be written in the form $y^2 = x^k$, state the value of k

7) Find the cartesian equation for each of:
   a) $x = t + \frac{1}{t}$ and $y = t - \frac{1}{t}$
   b) $x = t - \frac{2}{t}$ and $y = t + \frac{2}{t}$
   c) $x = 7t + \frac{4}{t}$ and $y = 7t - \frac{4}{t}$

8) A curve is given by parametric equations
   $$x = \frac{1}{t} - 1 \text{ and } y = \frac{2+3t}{1+3t}$$
   Show that the cartesian equation of the curve can be written in the form $y = \frac{ax+b}{x+4}, x \neq -4$
   where a and b are integers to be found.

9) Find a pair of parametric equations for each of the following cartesian equations:
   a) $4x^2 + 9y^2 = 1$
   b) $3x^2 + 2y^2 = 1$
   c) $5x^2 + \frac{1}{2}y^2 = 1$
   d) $4x^2 + \frac{2}{3}y^2 = 2$

10) Show that each of the following may be written in the form
    $$(x-p)^2 + (y-q)^2 = k$$
    where p, q and k are integers and $0 \leq \theta < 2\pi$:
    a) $x = 3 \cos \theta$ and $y = 3 \sin \theta$
    b) $x = 5 \sin \theta$ and $y = -5 \cos \theta$
    c) $x = \cos \theta + 1$ and $y = \sin \theta - 3$
    d) $x = \sin \theta - 4$ and $y = \cos \theta + 8$

11) A fireworks display is planned so that a sequence of firework bursts follow a trajectory in the sky, in metres, such that
    $$x = 10 \cos t \text{ and } y = 15 + 5 \sin t$$
    for $0 < t \leq 2\pi$ where t represents the position of a burst of a firework at time t seconds. Find
    a) the cartesian equation of the trajectory traced by the fireworks.
    b) the maximum height reached by the fireworks in one cycle of motion.
    c) the value of t which provides the maximum height in part b).
    d) the period of the fireworks sequence.

2.41 Parametric Equations: Introduction (answers on page 538)

12) A curve is defined for $0 \leq \theta < 2\pi$ by
$$x = 2k + 2k\cos\theta \text{ and } y = 8 - 3\sin\theta$$
for $k > 0$
   a) Find a cartesian equation for the curve.

   It is given that the curve is a circle. Find
   b) the value of k
   c) the centre of the circle.
   d) the radius of the circle.

Prerequisite knowledge for Q13: Trig Identities

13) Write the following
$$x = 2\tan\theta \text{ and } y = 2 + \sec\theta, 0 \leq \theta < 2\pi$$
as a cartesian equation.

14) A curve is defined by the parametric equations
$$x = \sqrt{3}\sin\theta, y = \sin 2\theta$$
for $0 \leq \theta \leq \frac{\pi}{2}$

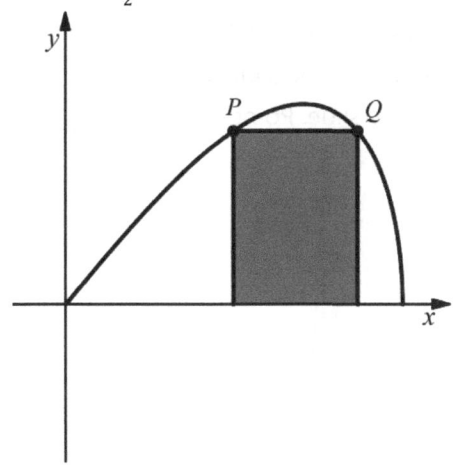

The points P and Q lie on the curve.
The line $x = \frac{3}{2}$ meets the curve at the point Q.
   a) Find the coordinates of P and Q.
   b) Hence find the exact area of the rectangle bounded by the x-axis and the points P and Q.

15) Find the cartesian form of the curve which satisfies all the points on:
   a) $x = \cos\theta$ and $y = \sin^2\theta$ for $0 \leq \theta \leq \pi$
   b) $x = \sin\theta$ and $y = \tan\theta$ for $-\frac{\pi}{2} \leq \theta \leq \frac{\pi}{2}$

16) A curve is defined by
$$x = \frac{t^2+7}{t^2+1} \text{ and } y = \frac{6t}{t^2+1}$$
   a) Show that the curve defined by for $t \in \mathbb{R}$ is a circle with centre $(4, 0)$
   b) Find the radius of the circle.

17) For each of the following parametric curves, find all the points of intersection with the circle given by
$$x^2 + y^2 = 100$$
where $0 \leq \theta \leq 2\pi$. Give your answers in exact form.
   a) $x = 10\cos\theta, y = 4\sin\theta$
   b) $x = 2\cos\theta, y = 10\sin\theta$
   c) $x = 6\sqrt{2}\cos\theta, y = 6\sqrt{5}\sin\theta$

Prerequisite knowledge for Q18: Double Angle Formulae

18) A curve is given by parametric equations
$$x = 5 + 2\cos\theta \text{ and } y = -1 + 2\cos 2\theta$$
for $0 \leq \theta \leq \pi$
   a) Find the cartesian equation of the curve which satisfies all the points on the curve.
   b) State the domain and range of the curve.

   The line $y + x = k$ where k is a constant, intersects the curve at two distinct points.
   c) State the range of the values of k giving your answer in interval notation.

Prerequisite knowledge for Q19-Q21: Domain and Range of Functions

19) Find the largest possible domain and range of:
   a) $x = t^2 + 1$ and $y = \sqrt{t}$
   b) $x = 2t^2$ and $y = 4 - \sqrt{t}$
   c) $x = \sin t$ and $y = \cos t$
   d) $x = \sin 3t$ and $y = \cos\left(t + \frac{\pi}{2}\right)$
   e) $x = e^t$ and $y = e^{-2t}$
   f) $x = \ln t$ and $y = t^2$

20) By first considering valid values for t, find the largest possible domain and range of the curve defined by $x = \sin\sqrt{t}$ and $y = \frac{1}{t}$

21) For $0 \leq \theta < 2\pi$ a curve is defined by
$$x = \sin 2\theta \text{ and } y = 4\cos\left(\theta - \frac{\pi}{6}\right)$$
   a) Find the domain and range of the curve.
   b) Find the coordinates where the curve meets the coordinate axes.
   c) Hence, sketch the curve.

2.42 Parametric Equations: Differentiation (answers on page 539)

1) By finding the cartesian equation of the curve, given by
$$x = t - 1$$
and
$$y = 2t^2$$
find $\frac{dy}{dx}$ in terms of x

2) Find $\frac{dy}{dx}$ in terms of t for each of the following:
   a) $x = t + 1$ and $y = -2t^2 + 2$
   b) $x = t(t + 3)$ and $y = t^2(t + 1)$
   c) $x = e^{2t}$ and $y = 2e^t$
   d) $x = 4e^{2t}$ and $y = 3e^{-t} + 1$
   e) $x = t - \cos t$ and $y = t - \sin t$

Prerequisite knowledge for Q3: Quotient Rule

3) Find $\frac{dy}{dx}$ in terms of t for $x = \frac{1}{1+t^2}$ and $y = t^2$

4) It is given that $x = t + 1$, $y = t^2$, where $t \geq 0$. Find the equation of the tangent to the curve when $t = 2$

5) It is given that $x = t^3$, $y = 2t$, where $t \geq 0$. Find the equation of the normal to the curve when $t = 3$

6) It is given that $x = e^t$, $y = \frac{1}{t}$, where $t > 0$. Find the equation of the tangent to the curve when $t = 1$

7) For each of the following curves defined parametrically, find the coordinates of the stationary point(s) of the curve:
   a) $x = t^4 + t$ and $y = 10t - 5t^2$ where $t \geq 0$
   b) $x = 6t$ and $y = t^3 - 5t^2 + 3t$ where $t \geq 0$
   c) $x = e^{2t}$ and $y = 6t^2 - \frac{4}{3}t^3 - 5t$ where $t \geq 0$

8) A curve is defined by the parametric equations
$$x = 3 - t^3$$
and
$$y = (t + 2)^2$$
for $-1 \leq t \leq 1$. Find
   a) the coordinate of the point with the maximum y-value.
   b) the coordinate of the point with the minimum y-value.
   c) $\frac{dy}{dx}$ in terms of t

9) A curve is defined by the parametric equations
$$x = \frac{4}{t}$$
and
$$y = 3t - 1$$
   a) Find the equation of the tangent line to the curve at the point $(2, 5)$ giving your answer in the form $y = mx + c$
   b) Hence find the area bounded by the tangent line at $(2, 5)$ and the coordinate axes.

10) Given that $t > 0$, find the equation of the normal to the curve given by
$$x = 1 - \frac{1}{t} \text{ and } y = 2t(t - 1)$$
when $x = -2$, giving your answer in the form $ax + by = c$ where a, b and c are integers to be found.

11) A drop slide is shown in the diagram. Point A is at the start of the slide and point is B at the end of the slide. Point O lies on the ground.

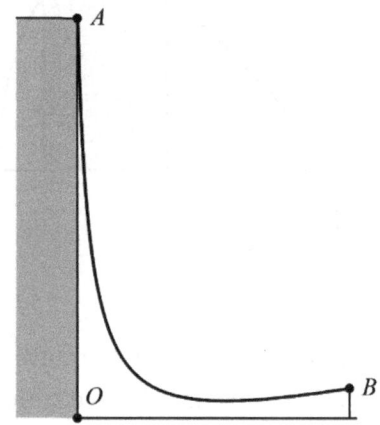

It may be modelled by parametric equations
$$x = \frac{1}{2}t^2 - \frac{1}{8} \text{ and } y = \frac{5}{t} + t - \frac{21}{5}$$
for $0.5 \leq t \leq 3$

The horizontal distance from O is x metres and the vertical distance from O is y metres. According to the model, find
   a) the height at the start of the slide.
   b) the height at the end of the slide.
   c) how far horizontally point B is from the start of the slide.
   d) the difference in height between the start of the slide and the lowest point of the slide, giving your answer to the nearest centimetre.

2.42 Parametric Equations: Differentiation (answers on page 539)

12) A curve is defined by parametric equations
$$x = t + a$$
and
$$y = t^2 + 3$$
where $-3 \leq t \leq 3$ and a is an integer.
Find
   a) the cartesian curve which satisfies all the points on the curve.
   b) the equation of the normal to the curve when $t = 1$ in the form $y = mx + c$, giving your answer in terms of a

13) A curve is defined by the parametric equations
$$x = \sqrt{3} \sin \theta$$
and
$$y = 2 \sin 2\theta$$
for $0 \leq \theta \leq \frac{\pi}{2}$
   a) Find $\frac{dy}{dx}$ in terms of θ
   b) Show that the equation for the normal line to the curve at $\theta = \frac{\pi}{3}$ may be written in the form
$$a\sqrt{3}y - bx = c$$
   where a, b and c are integers to be found.
   c) Find the area bound by the coordinate axes and the normal line to the curve when $\theta = \frac{\pi}{3}$

14) A curve is defined by the parametric equations
$$x = 2 \times 3^{-t} + 4$$
and
$$y = 4 \times 3^t - 6$$
for $t \in \mathbb{R}$
   a) Show that
$$\frac{dy}{dx} = -2 \times 3^{2t}$$
   b) Show that the curve does not cross the y-axis.
   c) Show that the cartesian equation which satisfies all the points on the curve may be written in the form
$$y = \frac{a}{x - b} - c$$
   where a, b and c are integers and $x > b$
   d) As $t \to \infty$, $x \to \alpha$. Find the value of α
   e) As $t \to -\infty$, $y \to \beta$. Find the value of β
   f) Find when the curve crosses the x-axis.
   g) Hence sketch the curve.

Prerequisite knowledge for Q15: Reciprocal Trigonometric Functions and Chain Rule

15) A curve is defined by
$$x = \sin 4\theta$$
and
$$y = \sec^2 \theta$$
for $0 \leq \theta \leq \frac{\pi}{2}$
   a) Find $\frac{dy}{dx}$ in terms of θ
   b) Hence find the exact value of the gradient of the tangent of the curve when $y = 2$

16) A curve is defined by parametric equations
$$x = 5t^3 \text{ and } y = t^4$$
for $t \geq 0$
   a) Show that the equation of the tangent at the point t on the curve is given by
$$15y - 4tx + 5t^4 = 0$$

A triangle is bound by the x-axis, the tangent line to the curve at $x = 40$ and the line $x = 40$
   b) Find the area of the triangle.

17) A curve is defined by parametric equations
$$x = \tan \theta + 1$$
and
$$y = \sec \theta$$
for $-\frac{\pi}{3} \leq \theta \leq \frac{\pi}{3}$
The point P lies on the curve when $\theta = a$ where $-\frac{\pi}{3} \leq a \leq \frac{\pi}{3}$
   a) Find the equation for the normal to the curve at the point P in the form
$$y = mx + c$$
   giving your answer in terms of a
   b) Hence find the minimum point on the curve.

18) A curve is defined by
$$x = t^4 \text{ and } y = 2t^3(1 + t)$$
where $t \in \mathbb{R}$
   a) Find $\frac{dy}{dx}$ in terms of t
   b) Find the coordinates of the minimum point of the curve.
   c) Find where the curve crosses the y-axis.
   d) Find where the curve crosses the x-axis.
   e) Hence sketch the curve.

2.42 Parametric Equations: Differentiation (answers on page 539)

Prerequisite knowledge for Q19-Q21: Harmonic forms

19) Two springs are attached to a ceiling with masses attached. They oscillate up and down, between two heights above the floor, p and q metres as shown in the diagram.

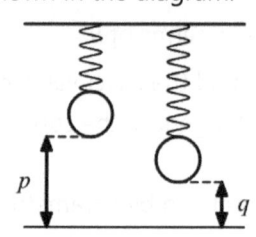

They may be modelled using the parametric equations
$$p = \frac{1}{2} - \frac{1}{2}\cos t \text{ and } q = \frac{1}{2} - \frac{1}{4}\sin t$$
where $0 \leq t \leq 8\pi$

a) Find the initial height difference between the two springs.
b) Write the difference in height between them in the form $R\sin(t - \alpha)$ where R and α are positive constants to be found and $0 < \alpha < \frac{\pi}{2}$
c) Find the maximum distance between the two springs.

20) A curve is defined by
$$x = 4\cos\theta \text{ and } y = 5\sin\theta$$
for $0 \leq \theta \leq 2\pi$
The tangent lines to the curve at two points, A and B, pass through the point $(4, 12)$
a) Find $\frac{dy}{dx}$ in terms of θ
b) State the equation of the tangent line in terms of θ
c) Hence show that values of θ at A and B satisfy the equation
$$12\sin\theta + 5\cos\theta = 5$$
d) Hence find the two possible values of θ

21) A curve is defined by
$$x = a\cos\theta \text{ and } y = 4a\sin\theta$$
a) Show that the equation of the tangent line is
$$y\sin\theta + 4x\cos\theta = 4a$$

The tangent line passes through the point $(a, 3a)$

b) For $0 < \alpha < \frac{\pi}{2}$, show that θ satisfies
$$5\sin(\theta + \alpha) = 4$$
c) Hence find the two possible values of θ when $0 \leq \theta < 2\pi$

Prerequisite knowledge for Q22: The Newton-Raphson Method and Double Angle Formulae

22) The curve C shown below is defined by parametric equations
$$x = \tan t \text{ and } y = \cos 2t$$
for $0 \leq t \leq \frac{\pi}{4}$

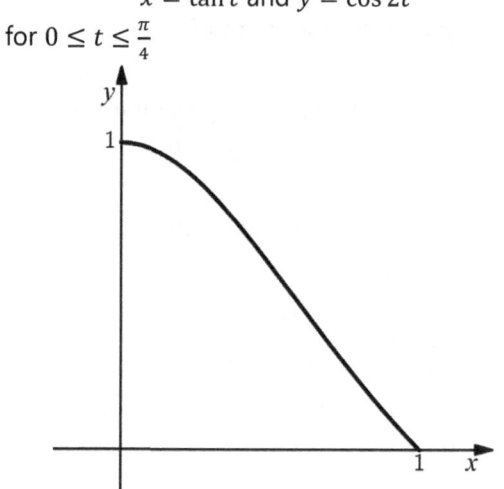

a) Find the equation of the line that connects the points on the curve C when $t = 0$ and $t = \frac{\pi}{4}$
b) Show that
$$\frac{dy}{dx} = -4\sin t \cos^3 t$$
c) Use Newton-Raphson's method starting at $x_0 = 0$ to show that a solution to
$$4\sin x \cos^3 x = 1$$
occurs when $x = 0.2874$ (to 4 significant figures).
d) Hence, find the range of values of k such that the line $y + x = k$ intersects the curve at least once.

Prerequisite knowledge for Q23: Product Rule and Small Angle Approximations

23) A curve is defined by the parametric equations
$$x = \frac{1}{8}\sin t \text{ and } y = t\cos t$$
where $0 \leq t \leq \pi$
a) Find $\frac{dy}{dx}$ in terms of t
b) Given the stationary point is close to the origin, find the stationary point of the curve by using small angle approximations for $\cos t$ and $\sin t$

2.43 Parametric Equations: Integration (answers on page 540)

Prerequisite knowledge: The following questions may involve integrating standard functions or reversing the chain rule.

1) For each of the following curves defined by the parametric equations, find $\int y \, dx$:
   a) $x = 3t^3 - t$ and $y = t^2 + 1$
   b) $x = e^t$ and $y = e^{-t} + 1$
   c) $x = 3^t$ and $y = 9^t$

2) For each of the following curves, where the lower and upper limit (t_1 and t_2 respectively) of the definite integral have already been found in terms of t, find $\int_a^b y \, dx$ giving your answers in exact form:
   a) $x = 2t^3 + t$ and $y = t^2 + 4$
      between $t_1 = 1$ and $t_2 = 2$
   b) $x = t^2$ and $y = e^t$
      between $t_1 = 0$ and $t_2 = 2$
   c) $x = 2^t$ and $y = 4^t$
      between $t_1 = 0$ and $t_2 = 1$
   d) $x = -\ln t$ and $y = t^3$
      between $t_1 = 0.5$ and $t_2 = 1$

3) A curve is defined by parametric equations
   $x = t^2$ and $y = \frac{1}{t}$
   for $1 \leq t \leq 2$
   a) Find the coordinates of the curve when $t = 1$ and $t = 2$
   b) Sketch the curve.
   c) Find the area enclosed by the curve, the coordinate axes, the lines $x = 1$ and $x = 4$

4) The diagram shows part of a curve defined by
   $x = 2t - 3$ and $y = \frac{4}{t}$

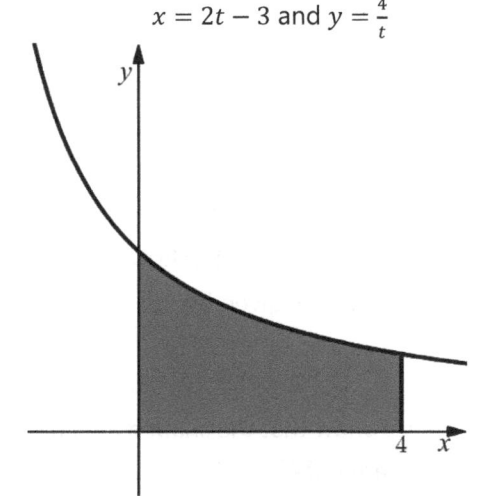

   The shaded region is bounded by the curve, the coordinate axes and the line $x = 4$
   a) Find the value of t when $x = 0$ and $x = 4$
   Find the area of the shaded region
   b) by using parametric integration.
   c) by first finding a cartesian equation for the curve.

5) A curve is defined by parametric equations
   $x = t^2 + 2t$ and $y = 2t^2 - t^3$ for $t \geq 0$

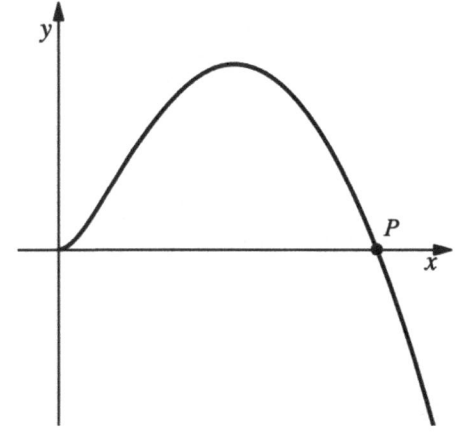

   The point P is where the curve meets the x-axis at $x = k$. The area enclosed by the curve and the x-axis is given by
   $$A = \int_0^k y \, dx$$
   a) Find the value of k
   b) Find the area A.

6) The diagram shows the curve given by
   $x = t^2(t - 1)$ and $y = t^2$ for $t \geq 0$
   The point P lies on the y-axis. The curve meets the y-axis when $t = 0$ and at point P.

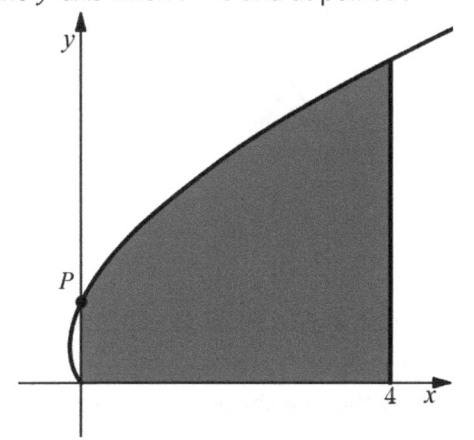

   a) Find the value of t at P.
   b) Find the area of the shaded region, which is bounded by the curve, the positive x-axis, the lines $x = 0$ and $x = 4$

2.43 Parametric Equations: Integration (answers on page 540)

7) A curve is defined by parametric equations
$$x = \frac{t}{t+4}$$
and
$$y = \frac{8}{(t+4)^2}$$
where $0 \leq t \leq 5$
   a) Find the coordinates of the curve when $t = 0$ and when $t = 5$
   b) Show that the area enclosed by the curve, the axes and the line $x = \frac{5}{9}$ is given by
$$\int_0^5 \frac{32}{(t+4)^4} \, dt$$
   c) Hence, find the exact area.

8) The parametric equations of a curve are
$$x = \tan\theta \text{ and } y = \cos^3\theta$$
where $0 \leq \theta < \frac{\pi}{2}$
   a) Find the coordinates of the curve when $x = 0$ and when $x = \sqrt{3}$
   b) Show that the area enclosed by the curve, the x-axis and the line $x = \sqrt{3}$ is
$$\frac{\sqrt{3}}{2}$$

9) The diagram shows part of the curve defined by parametric equations
$$x = 3\sin t \text{ and } y = 2\sin 2t$$
for $0 \leq t \leq \frac{\pi}{2}$

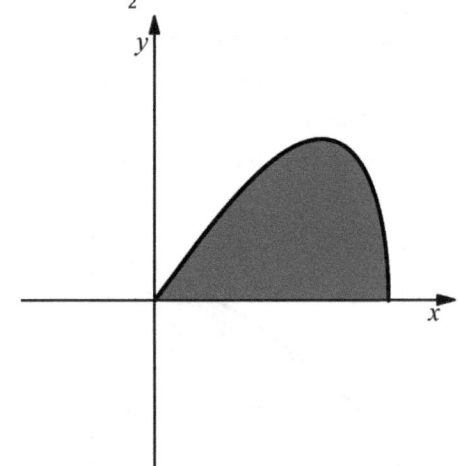

The shaded region is bounded by the curve and the x-axis.
   a) Show that the shaded area is given by
$$\int_0^{\frac{\pi}{2}} 12\sin t \cos^2 t \, dt$$
   b) Hence show that the shaded area has an exact area of 4

10) A curve is defined by the parametric equations
$$x = t^3 \text{ and } y = k\sqrt{t}$$
where $0 \leq t \leq 1$ and k is a positive constant.
It is known that the area between the curve, the x-axis and the lines $x = 0$ and $x = 1$ is 3
   a) Show that
$$k \int_0^1 3t^{\frac{5}{2}} \, dt = 3$$
   b) Find the value of k

11) A curve is defined by parametric equations
$$x = t + \sin t$$
and
$$y = \frac{1}{2}\cos^2 t$$
where $0 \leq t \leq \pi$
   a) Find the coordinates of the curve when $t = 0$ and when $t = \pi$
   b) Show that the area enclosed by the curve, the axes and the line $x = \pi$ is given by
$$\frac{1}{2}\int_0^\pi (1 - \sin^2 t + \cos t - \cos t \sin^2 t) \, dt$$
   c) Hence, find the exact area.

12) The diagram shows part of the curve defined by parametric equations
$$x = t^2 \text{ and } y = \frac{6}{t} - 2$$
for $0 \leq t \leq 3$

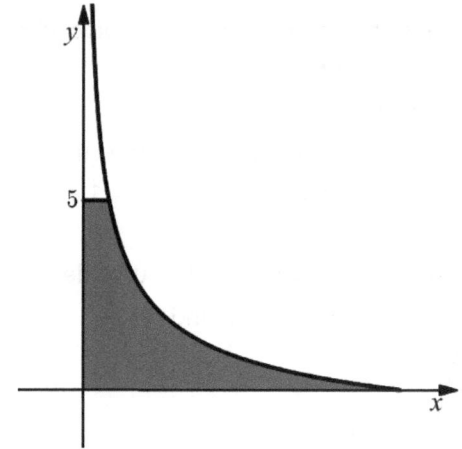

The shaded region is bounded by the curve, the coordinate axes and the line $y = 5$
   a) By considering $\int_0^5 y \, dx$, show that the shaded area is given by
$$\int_{\frac{6}{7}}^3 6 \, dt$$
   b) Hence show that the shaded area has an exact area of $\frac{90}{7}$

2.43 Parametric Equations: Integration (answers on page 540)

13) The diagram shows part of the curve defined by parametric equations
$$x = 5t^3 \text{ and } y = \frac{1}{t} - 1$$
for $0 \leq t \leq 1$

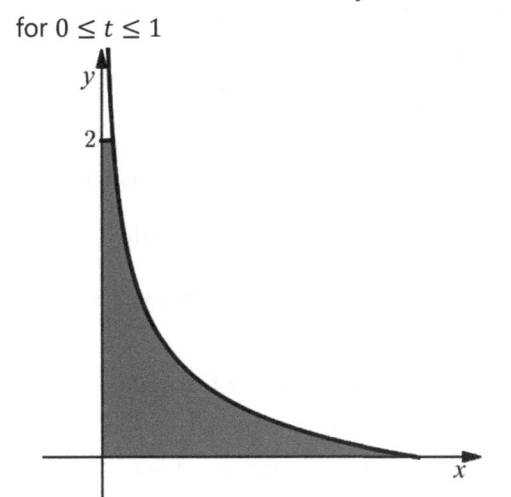

The shaded region is bounded by the curve, the coordinate axes and the line $y = 2$

a) Show that the shaded area is given by
$$-\int_{\frac{1}{3}}^{1} 5t \, dt$$

b) Hence show that the shaded area has an exact area of $\frac{20}{9}$

Prerequisite knowledge for Q14-Q22: Double Angle Formulae, Integration by Parts, Product Rule, Partial Fractions, Integration by Substitution

14) The diagram shows the curve given by
$$x = \frac{6}{t^2} \text{ and } y = t^3 e^{-t}$$
where $t > 0$
The maximum of the curve is at the point P with $x = a$ where a is a constant to be found.

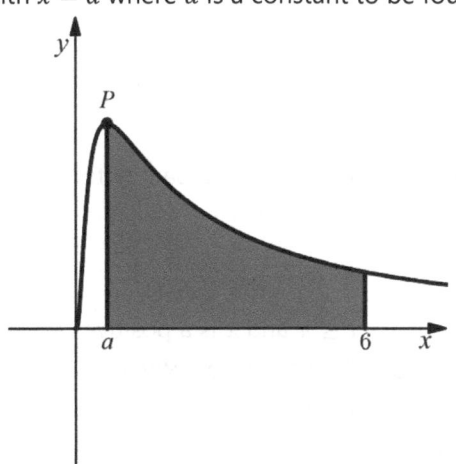

a) Find the coordinates of P.

b) Hence, show that the area enclosed by the curve, the x-axis and the line $x = a$ and $x = 6$ is
$$\frac{12}{e}\left(1 - \frac{1}{e^2}\right)$$

15) Find the area under the curve defined parametrically by
$$y = \cos\theta, \; x = \sin\theta$$
between $\theta = 0$ and $\theta = \frac{\pi}{2}$

16) Given
$$x = 2\sin t \text{ and } y = 4\cos t$$
find
$$\int_0^2 y \, dx$$
giving your answer in exact form.

17) The diagram shows the curve defined by parametric equations
$$x = 9\cos\theta \text{ and } y = 5\sin\theta$$
for $0 \leq \theta < 2\pi$

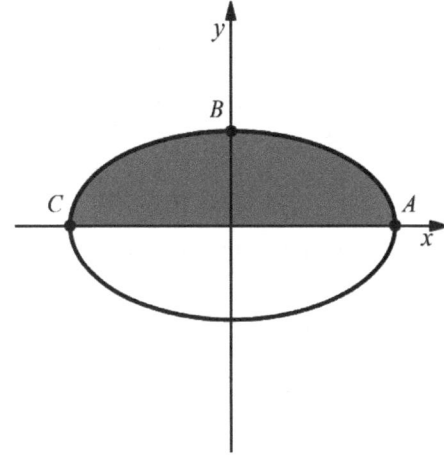

The curve meets the x-axis at points A and C. The curve meets the positive y-axis at the point B

a) Find the coordinates of A, B and C and hence the value of θ at each of these points.

b) Show that the shaded region enclosed by the ellipse is given by
$$\int_0^\pi -45\sin^2\theta \, d\theta$$

c) Hence show that the shaded region has area
$$\frac{45\pi}{2}$$

2.43 Parametric Equations: Integration (answers on page 540)

18) The diagram shows the curve defined by the parametric equations
$$x = \ln(t-2)$$
and
$$y = \frac{6}{t}$$
for $t \geq 2$

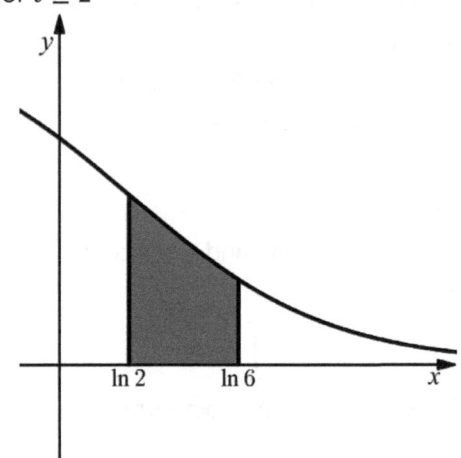

The shaded region is closed by the curve, the x-axis and the lines $x = \ln 2$ and $x = \ln 6$
a) Show that the shaded area may be given by
$$\int_{t_1}^{t_2} \frac{6}{t(t-2)} \, dt$$
where t_1 and t_2 are constants to be found.
b) Show that
$$\frac{6}{t(t-2)} \equiv \frac{A}{t-2} + \frac{B}{t}$$
for some constants A and B
c) Hence find the exact value of the shaded area in the form $\ln\left(\frac{a}{b}\right)$ where a and b are positive integers.

19) The parametric equations of a curve are
$$x = \frac{2t^2}{2+t}$$
and
$$y = \frac{1}{2(2+t)}$$
where $t \geq 0$
a) Find the coordinates of the curve when $x = 0$ and when $x = \frac{18}{5}$
b) Show that the area enclosed by the curve, the x-axis and the line $x = 9$ is given by
$$\int_0^3 \frac{t(4+t)}{(2+t)^3} \, dt$$
c) Hence, by using the substitution $u = 2 + t$ find the exact area.

20) A curve is defined by parametric equations
$$x = t^2 \text{ and } y = 1 - e^{-t}$$
for $t \geq 0$
a) Find the y-value of the curve when $x = 0$ and when $x = 4$
b) Sketch the curve for $0 \leq x \leq 4$
c) Show that the area enclosed by the curve, the x-axis and the line $x = 4$ is given by
$$\int_0^2 (2t - 2te^{-t}) \, dt$$
d) Hence, show the exact area is $2 + \frac{6}{e^2}$

21) The diagram shows part of the curve defined by parametric equations
$$x = \frac{t}{t+1} \text{ and } y = 1 - t^2$$
for $0 \leq t \leq 1$

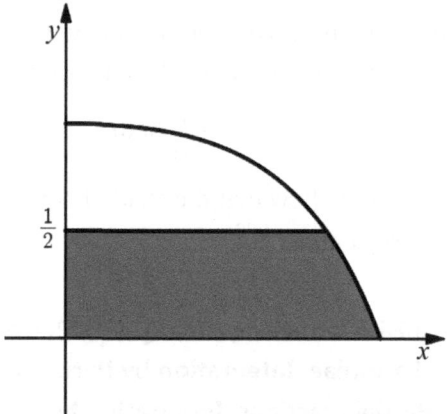

The shaded region is bounded by the curve, the coordinate axes and the line $y = \frac{1}{2}$
a) Show that the shaded area is given by
$$\int_{\frac{\sqrt{2}}{2}}^{1} \frac{2t^2}{t+1} \, dt$$
b) Hence, by using the substitution $u = t + 1$ show that the shaded area has an exact area of
$$-\frac{3}{2} + \sqrt{2} + 2\ln(4 - 2\sqrt{2})$$

22) A curve is defined by the parametric equations
$$x = t + t^2$$
and
$$y = k \sin t$$
where $0 \leq t \leq \pi$ and k is a positive constant.
It is known that the area between the curve, the x-axis and the lines $x = 0$ and $x = 12$ is 10
Show that
$$k = \frac{10}{1 + 2\sin 3 - 7\cos 3}$$

2.44 Proof by Contradiction (answers on page 541)

1) For each of the following statements, write the negation:
   a) The cat is black.
   b) Some cats are black.
   c) Some cats are not black.
   d) Line A is parallel to line B.
   e) The probability has increased.
   f) $x \geq 0$
   g) $x = 2$
   h) a is irrational.
   i) n is even and prime.

2) For each of the following statements, write the negation:
   a) For all integers, n, either n is positive or n is negative.
   b) n is a number with at least 3 distinct prime factors.
   c) Every prime number is odd.
   d) The sum of a rational number and an irrational number is always irrational.
   e) If n^2 is even, then n is even.

3) A student is attempting to prove by contradiction that if p and q are positive integers, that $\frac{p}{q} + \frac{q}{p} \geq 2$

   Their first line of working states:

   Given that p and q are negative numbers, $\frac{p}{q} + \frac{q}{p} \geq 2$

   Correct their first line.

4) Prove that the following numbers are irrational:
   a) $\sqrt{2}$
   b) $\sqrt{3}$
   c) $\sqrt{5}$

5) Prove that $\sqrt[3]{5}$ is an irrational number.

6) Prove that $1 + \sqrt{2}$ is irrational.

7) Prove that if $x^2 - 5x + 6 < 0$ then $2 < x < 3$

8) Prove that there is no greatest multiple of 3

9) Prove that there are infinitely many prime numbers.

10) Prove that the difference between a rational number and an irrational number is always irrational.

11)
   a) Prove that the product of a positive rational number and a positive irrational number is irrational
   b) Hence, state the values for which it is not true that the product of a rational and an irrational number is irrational.

12) Lynn has sketched the graph of
$$f(x) = kx^2 - 3kx + 8$$
which can be seen below.

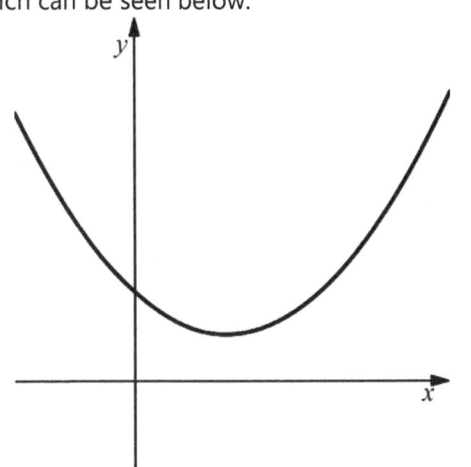

She claims that the values of k for which this is the correct sketch are $0 < k \leq \frac{32}{9}$

Use proof by contradiction to prove her claim.

13) Prove that for all $x > 0$, $x + \frac{1}{x} \geq 2$

14) Prove that for any real value of a and b then
$$a^2 + b^2 > 2ab$$

15) Prove that $\log_{10} 3$ is an irrational number.

16)
   a) Show that if $p^2 - 6q = 5$ then p^2 is odd
   b) Hence, prove by contradiction that there are no integers a and b such that
$$a^2 - 6b = 5$$

17) Prove that 3 is the only prime number which is 1 less than a square number.

Pre-requisite knowledge for Q18: Double Angle Formula

18) Prove that for all real θ,
$$\cos\theta + \sin\theta \leq \sqrt{2}$$

2.45 3D Vectors (answers on page 542)

1) It is given that A, B and C have the coordinates $(-2, 3, 5)$, $(4, -1, -3)$ and $(-1, 0, -8)$ respectively. Find
   a) $\overrightarrow{AB}$
   b) $\overrightarrow{BC}$
   c) $\overrightarrow{AC}$
   d) $\overrightarrow{BA}$

2) Find the magnitude of the following vectors:
   a) $4\mathbf{i} + 5\mathbf{j} + 7\mathbf{k}$
   b) $\begin{pmatrix} 6 \\ 7 \\ 1 \end{pmatrix}$
   c) $5\mathbf{i} - 2\mathbf{j} + 12\mathbf{k}$
   d) $\begin{pmatrix} a \\ b \\ 1 \end{pmatrix}$

3) Given each of the following position vectors, find the magnitude of the vector $\overrightarrow{AB}$
   a) $\overrightarrow{OA} = \begin{pmatrix} 2 \\ -1 \\ 5 \end{pmatrix}$ and $\overrightarrow{OB} = \begin{pmatrix} 3 \\ 4 \\ -1 \end{pmatrix}$
   b) $\overrightarrow{OA} = \mathbf{i} + 4\mathbf{j} - 5\mathbf{k}$ and $\overrightarrow{OB} = 3\mathbf{i} - \mathbf{j} + \mathbf{k}$
   c) $\overrightarrow{OA} = \begin{pmatrix} 0 \\ 1 \\ 10 \end{pmatrix}$ and $\overrightarrow{OB} = \begin{pmatrix} 0 \\ -2 \\ 13 \end{pmatrix}$

4) Find the unit vector in the direction of the following vectors:
   a) $2\mathbf{i} + 3\mathbf{j} - 7\mathbf{k}$
   b) $\begin{pmatrix} 6 \\ -1 \\ 4 \end{pmatrix}$
   c) $8\mathbf{i} - 6\mathbf{j} + 10\mathbf{k}$
   d) $\frac{1}{\sqrt{6}}\begin{pmatrix} 1 \\ -1 \\ 2 \end{pmatrix}$

5) Given each of the following position vectors, find the unit vector in the direction of the vector $\overrightarrow{AB}$:
   a) $\overrightarrow{OA} = \begin{pmatrix} 4 \\ 8 \\ -2 \end{pmatrix}$ and $\overrightarrow{OB} = \begin{pmatrix} 6 \\ 0 \\ -2 \end{pmatrix}$
   b) $\overrightarrow{OA} = 2\mathbf{i} - 10\mathbf{j} - \mathbf{k}$ and $\overrightarrow{OB} = 7\mathbf{i} - 6\mathbf{j} + 2\mathbf{k}$

6) It is given that $\overrightarrow{AB} = \begin{pmatrix} -1 \\ 2 \\ 8 \end{pmatrix}$ and $\overrightarrow{BC} = \begin{pmatrix} 0 \\ 3 \\ -2 \end{pmatrix}$ and the angle between $\overrightarrow{AB}$ and $\overrightarrow{BC}$ is θ. Show that
$$\cos\theta = \frac{10}{\sqrt{897}}$$

7) The diagram shows the triangle ABC.

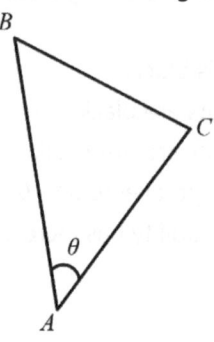

In each case, show that the value of the angle θ is the given value to the nearest degree:
   a) $\theta = 18.3°$, $\overrightarrow{AB} = 2\mathbf{i} - 10\mathbf{j} + \mathbf{k}$ and $\overrightarrow{BC} = 7\mathbf{i} - 6\mathbf{j} + 2\mathbf{k}$
   b) $\theta = 135.5°$, $\overrightarrow{AB} = \begin{pmatrix} 1 \\ 3 \\ -7 \end{pmatrix}$ and $\overrightarrow{AC} = \begin{pmatrix} 2 \\ 1 \\ 5 \end{pmatrix}$

8) Position vectors of the points A, B and C are given as $\overrightarrow{OA} = 3\mathbf{i} - 2\mathbf{j} + \mathbf{k}$, $\overrightarrow{OB} = \mathbf{i} + 3\mathbf{j} + 5\mathbf{k}$ and $\overrightarrow{OC} = 8\mathbf{i} - 7\mathbf{j} - \mathbf{k}$. Given that $\overrightarrow{BC} = \overrightarrow{AD}$, find the position vector of the point D.

9) Three position vectors are given to be
$$\overrightarrow{OA} = \begin{pmatrix} 3 \\ 6 \\ -1 \end{pmatrix}, \overrightarrow{OB} = \begin{pmatrix} 4 \\ 9 \\ 1 \end{pmatrix} \text{ and } \overrightarrow{OC} = \begin{pmatrix} 7 \\ 12 \\ p \end{pmatrix}$$
There is another point D and $\overrightarrow{AB} = \overrightarrow{BD}$
   a) Find the position vector of D
   b) Given $|\overrightarrow{AC}| = 9$, find the possible values of p

10) Show that the coordinates $(7, 4, 0)$, $(12, 6, -1)$ and $(-13, -4, 4)$ are collinear.

11) Show that the coordinates $(6, 4, -3)$, $(8, 5, -3)$ and $(12, 7, -15)$ are not collinear.

12) Three points A, B and C lie on a line. It is known that $\overrightarrow{OA} = \begin{pmatrix} 1 \\ 4 \\ -2 \end{pmatrix}$ and $\overrightarrow{OB} = \begin{pmatrix} 5 \\ 12 \\ 2 \end{pmatrix}$ and furthermore that $|\overrightarrow{AC}| = 3|\overrightarrow{BC}|$. Find the possible position vectors of the point C

13) The vector $\mathbf{a}$ is parallel to the given vector $\mathbf{p}$. In each case, find the vector $\mathbf{a}$:
   a) $\mathbf{p} = 4\mathbf{i} - \mathbf{j} + 2\mathbf{k}$ and $|\mathbf{a}| = \sqrt{84}$
   b) $\mathbf{p} = -\mathbf{i} + 7\mathbf{j} + 2\mathbf{k}$ and $|\mathbf{a}| = \sqrt{6}$
   c) $\mathbf{p} = \mathbf{i} - 8\mathbf{j} - \mathbf{k}$ and $|\mathbf{a}| = 3\sqrt{66}$

2.45 3D Vectors (answers on page 542)

14) Three position vectors are given as
$$\vec{OA} = \begin{pmatrix} 1 \\ 3 \\ -1 \end{pmatrix}, \vec{OB} = \begin{pmatrix} 24 \\ -3 \\ 7 \end{pmatrix} \text{ and } \vec{OC} = \begin{pmatrix} 5 \\ 1 \\ 2 \end{pmatrix}$$
Find
   a) the midpoint of AB, m_{AB}
   b) the midpoint of BC, m_{BC}
   c) the distance between m_{AB} and m_{BC}
   d) a unit vector in the direction of the line passing through m_{AB} and m_{BC}

The point D lies a distance of 1 unit away from A parallel to the line passing through the midpoints.
   e) Find the possible coordinates of D

15) For each of the following position vectors, state on which axes they lie:
   a) $\begin{pmatrix} 0 \\ 0 \\ 2 \end{pmatrix}$
   b) $-12\mathbf{i}$

16) Points A and B have position vectors $\mathbf{a}$ and $\mathbf{b}$ where
$$\mathbf{a} = \begin{pmatrix} 1 \\ -4 \\ 2 \end{pmatrix} \text{ and } \mathbf{b} = \begin{pmatrix} 3 \\ 8 \\ -4 \end{pmatrix}$$
A third point C has position vector $k(\mathbf{a} + k\mathbf{b})$ where $k \neq 0$ and it is known that C lies on the x-axis.
   a) Find the value of k
   b) Write down the position vector of C

17) Points A and B have position vectors $\mathbf{a}$ and $\mathbf{b}$
A point C lies on AB such that $AC:CB$ is $6:1$
Show that the position vector of C is
$$\frac{1}{7}(\mathbf{a} + 6\mathbf{b})$$

18) Points A, B and C have position vectors $\mathbf{a}$, $\mathbf{b}$ and $\mathbf{c}$ respectively, where
$$\mathbf{a} = \begin{pmatrix} 3 \\ -6 \\ 1 \end{pmatrix}, \mathbf{b} = \begin{pmatrix} 1 \\ 5 \\ 4 \end{pmatrix} \text{ and } \mathbf{c} = \begin{pmatrix} 14 \\ 7 \\ k \end{pmatrix}$$
where k is a positive integer.
It is known that $p\mathbf{a} + q\mathbf{b} = \mathbf{c}$ where p and q are positive integers.
Find the values of p, q and k

19) The diagram shows the triangle OAB with vertices labelled. The midpoint of $\vec{OA}$ is M_1 and the midpoint of $\vec{AB}$ is M_2.

By considering O as the origin, so that $\vec{OA} = \mathbf{a}$ and $\vec{OB} = \mathbf{b}$, show that the line joining M_1 and M_2 is parallel to $\mathbf{b}$ and is half the length of $\mathbf{b}$

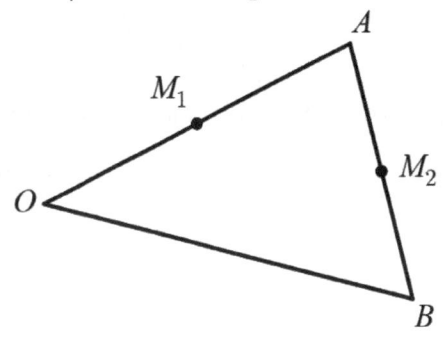

20) The diagram shows the cuboid $OAPBCQRS$, with vertices labelled. The midpoint of $\vec{PS}$ is M.

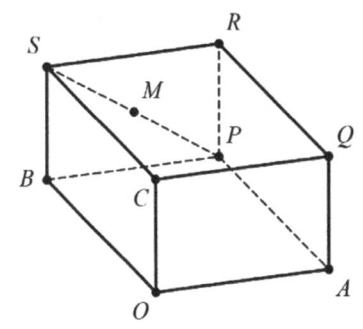

By considering O as the origin, so that $\vec{OA} = \mathbf{a}, \vec{OB} = \mathbf{b}$ and $\vec{OC} = \mathbf{c}$
   a) Show that $\vec{AM} = -\frac{1}{2}\mathbf{a} + \mathbf{b} + \frac{1}{2}\mathbf{c}$
   b) Find $\vec{PC}$ in terms of $\mathbf{a}, \mathbf{b}$ and $\mathbf{c}$
   c) Given that the lines $\vec{AM}$ and $\vec{PC}$ intersect, find the position vector of the point of intersection in terms of $\mathbf{a}, \mathbf{b}$ and $\mathbf{c}$

Consider now, that the point X lies on AM such that $XM = 2AX$

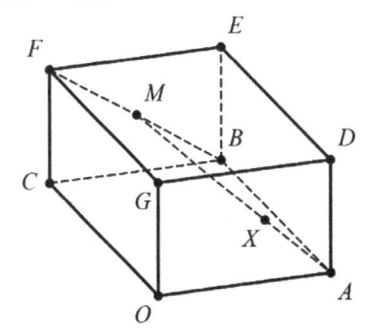

   d) Find the position vector of X in terms of $\mathbf{a}, \mathbf{b}$ and $\mathbf{c}$

2.45 3D Vectors (answers on page 542)

21) Three points ABC form an isosceles triangle such that $|\overrightarrow{AB}| = |\overrightarrow{BC}|$, and it is known that the position vectors of A and B are
$$\mathbf{a} = \begin{pmatrix} 1 \\ 8 \\ 4 \end{pmatrix} \text{ and } \mathbf{b} = \begin{pmatrix} 1 \\ 5 \\ 7 \end{pmatrix}$$
respectively. The third point C has position vector $\mathbf{c} = \begin{pmatrix} 5 \\ 4 \\ k \end{pmatrix}$ where k is a positive constant to be found.
   a) Find the possible values of k
   b) Find the exact area of ABC for both your values of k

22) Four points A, B, C and D have position vectors
$$\mathbf{a} = 2\mathbf{i} + \mathbf{j} - 3\mathbf{k}, \mathbf{b} = 6\mathbf{i} + 5\mathbf{k},$$
$$\mathbf{c} = 5\mathbf{i} + 5\mathbf{j} + 7\mathbf{k} \text{ and } \mathbf{d} = \mathbf{i} + 6\mathbf{j} - \mathbf{k}$$
Show that $ABCD$ is a parallelogram.

23) Three position vectors of points A, B and C are
$$\mathbf{a} = \begin{pmatrix} -1 \\ 5 \\ 8 \end{pmatrix}, \mathbf{b} = \begin{pmatrix} 1 \\ 4 \\ 13 \end{pmatrix} \text{ and } \mathbf{c} = \begin{pmatrix} -2 \\ 5 \\ 19 \end{pmatrix}$$
There is another point D such that $ABCD$ forms a parallelogram and $\overrightarrow{AB} = \overrightarrow{DC}$.
   a) Find the position vector of D.
   b) Find the exact length of $\overrightarrow{AC}$

24) Three position vectors of points A, B and C are
$$\mathbf{a} = 2\mathbf{i} - 4\mathbf{j} + 3\mathbf{k}, \mathbf{b} = 6\mathbf{i} + 12\mathbf{j} + \mathbf{k}$$
$$\text{and } \mathbf{c} = -\mathbf{i} + \mathbf{j} + 4\mathbf{k}$$
There is another point D such that $ABCD$ forms a parallelogram and $\overrightarrow{AB} = \overrightarrow{DC}$.
   a) Find the position vector of D.
   b) Find the length of $\overrightarrow{AB}$
   c) Find the length of $\overrightarrow{AC}$

The angle between $\overrightarrow{AB}$ and $\overrightarrow{BC}$ is θ
   d) Show that
$$\cos \theta = \frac{105}{\sqrt{12351}}$$
   e) Find the area of the parallelogram $ABCD$.

25) Four points A, B, C and D have position vectors
$$\mathbf{a} = \mathbf{i} - 4\mathbf{k}, \mathbf{b} = 3\mathbf{i} + 3\mathbf{j} - 8\mathbf{k},$$
$$\mathbf{c} = 8\mathbf{i} + 3\mathbf{j} - 10\mathbf{k} \text{ and } \mathbf{d} = 6\mathbf{i} - 6\mathbf{k}$$
   a) Show that $ABCD$ is a rhombus.
   b) Find the area of the rhombus.

26) The position vectors of points A and B are given by
$$\mathbf{a} = \mathbf{i} + 5\mathbf{j} - 4\mathbf{k} \text{ and } \mathbf{b} = 7\mathbf{i} - 3\mathbf{j} + 2\mathbf{k}$$
The points A and B lie on a circle of radius $\sqrt{66}$
A further point $C\ (-1, -3, p)$ exists such that AC is the diameter of the circle.

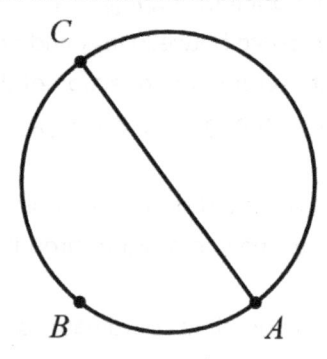

   a) Find the length of $\overrightarrow{BC}$
   b) Find the possible values of p

27) Four position vectors of points A, B, C and D are
$$\mathbf{a} = -\mathbf{i} + 4\mathbf{j} + 7\mathbf{k}, \mathbf{b} = \mathbf{i} + 5\mathbf{j} + 12\mathbf{k},$$
$$\mathbf{c} = -2\mathbf{i} + 10\mathbf{j} + 13\mathbf{k} \text{ and } \mathbf{d} = -7\mathbf{i} + 14\mathbf{j} + 9\mathbf{k}$$
Show that $ABCD$ is a trapezium.

28) Three position vectors of points A, B and C are
$$\mathbf{a} = \begin{pmatrix} 3 \\ -2 \\ 1 \end{pmatrix}, \mathbf{b} = \begin{pmatrix} 8 \\ -3 \\ 3 \end{pmatrix} \text{ and } \mathbf{c} = \begin{pmatrix} 9 \\ -1 \\ 7 \end{pmatrix}$$
There is an additional point D such that $ABCD$ forms a trapezium where $\overrightarrow{BC} = \frac{1}{3}\overrightarrow{AD}$

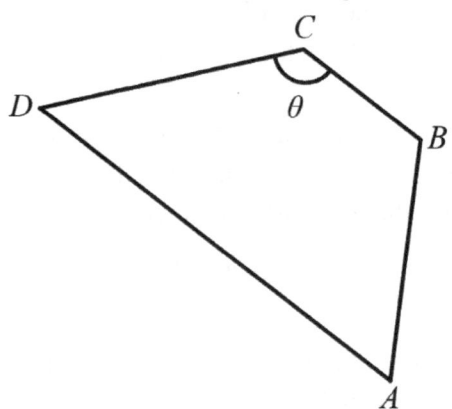

   a) Find the position vector of D.

The angle between BC and CD, θ, is given by
$$\cos \theta = -\frac{31}{7\sqrt{30}}$$
   b) Show that the exact area of the trapezium $ABCD$ is $2\sqrt{509}$ units^2

2.46 Bivariate Data: PMCC Hypothesis Testing (answers on page 543)

Tables of critical values are on page 356

1) Describe what is meant by the significance level.

2) In each of the following cases, r is the sample correlation coefficient. Determine whether you reject H_0 or fail to reject H_0
   a) $H_0: \rho = 0$, $H_1: \rho > 0$
      The test statistic is $r = 0.638$ and the p-value is 0.0444, where $n = 8$ and the significance level is 5%
   b) $H_0: \rho = 0$, $H_1: \rho < 0$
      The test statistic is $r = -0.456$ and the p-value is 0.0681, where $n = 12$ and the significance level is 5%

3) In each of the following cases, r is the sample correlation coefficient. Determine whether you reject H_0 or fail to reject H_0
   a) $H_0: \rho = 0$, $H_1: \rho > 0$
      $r = 0.487$
      Critical value for $n = 14$ for a one-tailed test at a 5% significance level is 0.4575
   b) $H_0: \rho = 0$, $H_1: \rho > 0$
      $r = 0.628$
      Critical value for $n = 7$ for a one-tailed test at a 5% significance level is 0.6694
   c) $H_0: \rho = 0$, $H_1: \rho > 0$
      $r = 0.593$
      Critical value for $n = 20$ for a one-tailed test at a 1% significance level is 0.5155
   d) $H_0: \rho = 0$, $H_1: \rho > 0$
      $r = 0.213$
      Critical value for $n = 60$ for a one-tailed test at a 10% significance level is 0.1678

4) In each of the following cases, r is the sample correlation coefficient. Determine whether you reject H_0 or fail to reject H_0
   a) $H_0: \rho = 0$, $H_1: \rho < 0$
      $r = -0.453$
      Critical value for $n = 15$ for a one-tailed test at a 5% significance level is 0.4409
   b) $H_0: \rho = 0$, $H_1: \rho < 0$
      $r = -0.829$
      Critical value for $n = 7$ for a one-tailed test at a 5% significance level is 0.6694
   c) $H_0: \rho = 0$, $H_1: \rho < 0$
      $r = -0.306$
      Critical value for $n = 25$ for a one-tailed test at a 1% significance level is 0.4622
   d) $H_0: \rho = 0$, $H_1: \rho < 0$
      $r = -0.197$
      Critical value for $n = 40$ for a one-tailed test at a 10% significance level is 0.2070

5) State the critical region for the following test:
   $H_0: \rho = 0$, $H_1: \rho > 0$
   Critical value for $n = 19$ for a one-tailed test at a 0.5% significance level is 0.5751

6) State the critical region for the following test:
   $H_0: \rho = 0$, $H_1: \rho < 0$
   Critical value for $n = 16$ for a one-tailed test at a 1% significance level is 0.5742

7) In each of the following cases, r is the sample correlation coefficient. Determine whether you reject H_0 or fail to reject H_0
   a) $H_0: \rho = 0$, $H_1: \rho \neq 0$
      $r = 0.658$
      Critical value for $n = 9$ for a two-tailed test at a 5% significance level is 0.6664
   b) $H_0: \rho = 0$, $H_1: \rho \neq 0$
      $r = -0.567$
      Critical value for $n = 11$ for a two-tailed test at a 5% significance level is 0.6021
   c) $H_0: \rho = 0$, $H_1: \rho \neq 0$
      $r = 0.621$
      Critical value for $n = 16$ for a two-tailed test at a 1% significance level is 0.6226
   d) $H_0: \rho = 0$, $H_1: \rho \neq 0$
      $r = -0.248$
      Critical value for $n = 50$ for a two-tailed test at a 10% significance level is 0.2353

8) State the critical region for the following test:
   $H_0: \rho = 0$, $H_1: \rho \neq 0$
   Critical value for $n = 7$ for a two-tailed test at a 2.5% significance level is 0.7545

9) State the critical region for the following test:
   $H_0: \rho = 0$, $H_1: \rho \neq 0$
   Critical value for $n = 80$ for a two-tailed test at a 5% significance level is 0.2199

2.46 Bivariate Data: PMCC Hypothesis Testing (answers on page 543)

10) Hannah believes there is a positive correlation between the amount of sugar consumed daily by adults and their blood glucose levels. She collects a random sample of 28 adults and finds the sample's correlation coefficient to be 0.382. Carry out a hypothesis test at the 5% significance level to investigate whether Hannah's belief is supported.

11) Kyle is a social scientist who wants to explore if there is a negative correlation between the number of hours spent on social media each day and self-reported feelings of loneliness in teenagers. He collects a random sample of 70 teenagers and finds the sample's correlation coefficient to be −0.297. Carry out a hypothesis test at the 0.5% significance level to investigate whether there is evidence to support a negative correlation.

12) Hamza is a biologist who is investigating whether there is a negative correlation between the water temperature they grow up in and their growth rate. Hamza wants to conduct a hypothesis test at the 1% significance level. He collects a random sample of 40 of a certain type of fish and finds the sample's correlation coefficient to be −0.364. Hamza's hypothesis test is shown below.

Let ρ be the sample correlation coefficient between the water temperature and the growth rate of the fish.
$H_0: \rho = 0$
$H_1: \rho < 0$
$r = -0.364$
The critical value for $n = 40$ for a one-tailed test at a 1% significance level is 0.3665
−0.364 < 0.3665 so we reject H_0
There is sufficient evidence to suggest that there is a negative correlation between the water temperature and the growth rate of the fish.

   a) Identify the first two errors made by Alan in his hypothesis test.
   b) Correct his test.

13) Aida is an environmental scientist who wants to determine if there is some correlation between the levels of air pollution and numbers of adults with respiratory illnesses. She collects data from 50 randomly selected neighbourhoods and finds the sample correlation coefficient to be 0.209. Carry out a hypothesis test at the 5% significance level to investigate whether there is any evidence to support there being some correlation.

14) Kaylee is a nutritionist who wants to determine if there is any correlation between the daily intake of fruits and vegetables in adults and their body mass index (BMI). She collects data from 100 randomly selected adults and finds the sample correlation coefficient to be −0.301. Carry out a hypothesis test at the 1% significance level to investigate whether there is any evidence to suggest a correlation.

15) Jo is a psychologist who is investigating whether there is any correlation between the number of hours of sleep college students get each night and their stress levels, as measured by a standardised test. Jo wants to conduct a hypothesis test at the 5% significance level. She collects data from 28 randomly selected college students and finds the sample correlation coefficient to be 0.482. Jo's hypothesis test is shown below.

Let ρ be the population correlation coefficient between the number of hours of sleep college students get each night and their stress levels.
$H_0: \rho = 0$
$H_1: \rho \neq 0$
$r = 0.482$
The critical value for $n = 28$ for a two-tailed test at a 5% significance level is 0.3172
0.3172 < 0.482 so we fail to reject H_0
There is insufficient evidence to suggest that there is some correlation between the number of hours of sleep college students get each night and their stress levels.

   a) Identify the first two errors made by Jodie in her hypothesis test.
   b) Correct her test.

2.47 Bivariate Data: Spearman's Rank Hypothesis Testing (answers on page 543)

Tables of critical values are on page 356

1) Richard is investigating whether there is an association between the scores students achieve in a Mathematics test and in an English test. He takes a random sample of 11 students and records their scores in each subject. The table below shows the data, and below that is a scatter graph to illustrate the data.

Mathematics	88	85	90	78	92	87	84	80	93	76	89
English	97	89	95	84	91	88	82	79	96	81	90

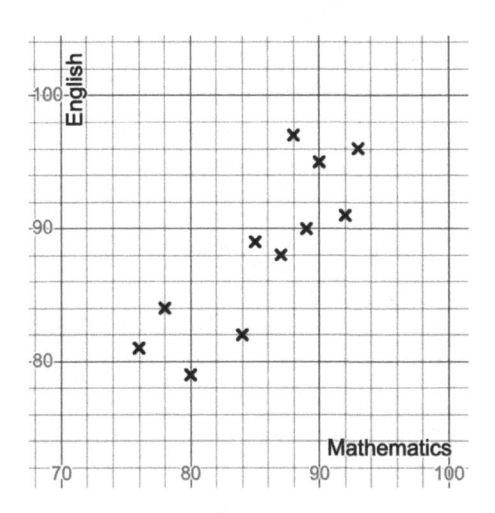

Richard decides to carry out a Spearman's Rank hypothesis test to investigate whether there is a positive association between the scores in the two subjects, using a 5% significance level.

$$r_s = 1 - \frac{6 \sum d_i^2}{n(n^2 - 1)}$$

a) Use the formula above to show that $r_s = 0.845$ to 3sf

Richard writes out the first part of his hypothesis test correctly, as shown below:

H_0: *There is no association between Mathematics test scores and English test scores in the population*
H_1: *There is a positive association between Mathematics test scores and English test scores in the population*
$r_s = 0.845$
The critical value for $n = 11$ *for a one-tailed test at a 5% significance level is* 0.5364
$0.845 > 0.5364$

b) Select Richard's correct conclusion:

Reject H_0. *There is sufficient evidence to suggest that there is a positive association between Mathematics and English test scores in the population.*	*Fail to reject* H_0. *There is insufficient evidence to suggest that there is a positive association between Mathematics and English test scores in the population.*

2) Beth is investigating whether there is a positive association between the length of a bird's wingspan and its tendency to migrate long distances. She measures the wingspans, in cm, and the maximum migration distances, in km, of a random sample of 12 different bird species. Beth correctly calculates the Spearman's Rank correlation coefficient to be 0.479. Conduct a test at the 5% level to investigate whether there is a positive association between the wingspan and migration distance.

3) Si is investigating whether there is a negative association between the air quality index (AQI) and the average life expectancy, in years, of people in different cities. He collects a random sample of 15 cities and correctly calculates the Spearman's Rank correlation coefficient to be -0.654. Conduct a test at the 1% level to investigate whether there is a negative association between the air quality index and average life expectancy.

4) Claire is investigating whether there is any association between the number of hours of sleep a person gets per night and their cognitive performance score on a standardised test. She collects a random sample of 22 individuals who were tested and correctly calculates the Spearman's Rank correlation coefficient to be 0.321. Conduct a test at the 10% level to investigate whether there is any association between the number of hours of sleep and cognitive performance score.

2.48 PMCC & Spearman's Rank Critical Values Tables

Critical Values for the Product Moment Correlation Coefficient

2-Tail	20%	10%	5%	2%	1%	Sample
1-Tail	10%	5%	2.5%	1%	0.5%	Size, n
	0.8000	0.9000	0.9500	0.9800	0.9900	4
	0.6870	0.8054	0.8783	0.9343	0.9587	5
	0.6084	0.7293	0.8114	0.8822	0.9172	6
	0.5509	0.6694	0.7545	0.8329	0.8745	7
	0.5067	0.6215	0.7067	0.7887	0.8343	8
	0.4716	0.5822	0.6664	0.7498	0.7977	9
	0.4428	0.5494	0.6319	0.7155	0.7646	10
	0.4187	0.5214	0.6021	0.6851	0.7348	11
	0.3981	0.4973	0.5760	0.6581	0.7079	12
	0.3802	0.4762	0.5529	0.6339	0.6835	13
	0.3646	0.4575	0.5324	0.6120	0.6614	14
	0.3507	0.4409	0.5140	0.5923	0.6411	15
	0.3383	0.4259	0.4973	0.5742	0.6226	16
	0.3271	0.4124	0.4821	0.5577	0.6055	17
	0.3170	0.4000	0.4683	0.5425	0.5897	18
	0.3077	0.3887	0.4555	0.5285	0.5751	19
	0.2992	0.3783	0.4438	0.5155	0.5614	20
	0.2914	0.3687	0.4329	0.5034	0.5487	21
	0.2841	0.3598	0.4227	0.4921	0.5368	22
	0.2774	0.3515	0.4133	0.4815	0.5256	23
	0.2711	0.3438	0.4044	0.4716	0.5151	24
	0.2653	0.3365	0.3961	0.4622	0.5052	25
	0.2598	0.3297	0.3882	0.4534	0.4958	26
	0.2546	0.3233	0.3809	0.4451	0.4869	27
	0.2497	0.3172	0.3739	0.4372	0.4785	28
	0.2451	0.3115	0.3673	0.4297	0.4705	29
	0.2407	0.3061	0.3610	0.4226	0.4629	30
	0.2070	0.2638	0.3120	0.3665	0.4026	40
	0.1843	0.2353	0.2787	0.3281	0.3610	50
	0.1678	0.2144	0.2542	0.2997	0.3301	60
	0.1550	0.1982	0.2352	0.2776	0.3060	70
	0.1448	0.1852	0.2199	0.2597	0.2864	80
	0.1364	0.1745	0.2072	0.2449	0.2702	90
	0.1292	0.1654	0.1966	0.2324	0.2565	100

Critical Values for Spearman's Rank Correlation Coefficient

2-Tail	10%	5%	2%	1%	Sample
1-Tail	5%	2.5%	1%	0.5%	Size, n
	1.0000	-	-	-	4
	0.9000	1.0000	1.0000	-	5
	0.8286	0.8857	0.9429	1.0000	6
	0.7143	0.7857	0.8929	0.9286	7
	0.6429	0.7381	0.8333	0.8810	8
	0.6000	0.7000	0.7833	0.8333	9
	0.5636	0.6485	0.7455	0.7939	10
	0.5364	0.6182	0.7091	0.7545	11
	0.5035	0.5874	0.6783	0.7273	12
	0.4835	0.5604	0.6484	0.7033	13
	0.4637	0.5385	0.6264	0.6791	14
	0.4464	0.5214	0.6036	0.6536	15
	0.4294	0.5029	0.5824	0.6353	16
	0.4142	0.4877	0.5662	0.6176	17
	0.4014	0.4716	0.5501	0.5996	18
	0.3912	0.4596	0.5351	0.5842	19
	0.3805	0.4466	0.5218	0.5699	20
	0.3701	0.4364	0.5091	0.5558	21
	0.3608	0.4252	0.4975	0.5438	22
	0.3528	0.4160	0.4862	0.5316	23
	0.3443	0.4070	0.4757	0.5209	24
	0.3369	0.3977	0.4662	0.5108	25
	0.3306	0.3901	0.4571	0.5009	26
	0.3242	0.3828	0.4487	0.4915	27
	0.3180	0.3755	0.4401	0.4828	28
	0.3118	0.3685	0.4325	0.4749	29
	0.3063	0.3624	0.4251	0.4670	30

2.49 Probability: Conditional Probability (answers on page 543)

1) It is given that A and B are two events such that $P(A) = \frac{1}{8}$, $P(B) = \frac{1}{2}$ and $P(A \cap B) = \frac{1}{10}$. Find $P(A|B)$

2) It is given that A and B are two events such that $P(A) = \frac{1}{4}$, $P(B) = \frac{2}{5}$ and $P(A \cup B) = \frac{1}{2}$. Find $P(B|A)$

3) Events A and B are shown in the Venn diagram below. It is given that $P(A|B) = 0.5$. Find the value of x and the value of y

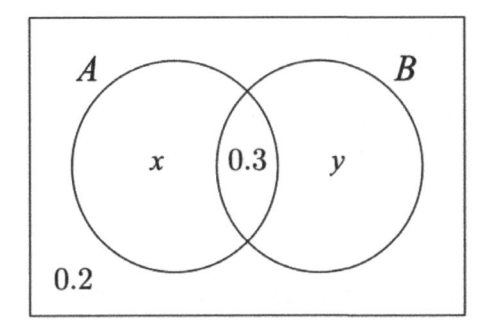

4) Events A, B, C and D are shown in the Venn diagram below.

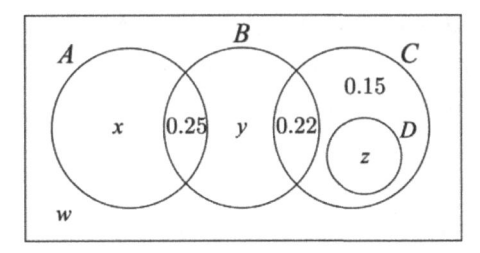

   a) Given $P(C) = 0.4$, find z
   b) Given B and C are independent, find y
   c) Given $P(B|A) = 0.625$, find x
   d) Find w

5) At a school, students must choose their GCSE options in Year 9. As part of their choices, they must choose to study History, or Geography, or both, or neither. The probability of a student choosing History is $\frac{1}{4}$ and the probability of a student choosing Geography is $\frac{2}{5}$. The probability of a student choosing History, given that they have chosen Geography, is $\frac{1}{3}$.
   Find the probability that a student picked at random chose neither History or Geography.

6) It is given that A and B are two events such that $P(A|B) = 0.3$, $P(B|A) = 0.4$ and $P(A' \cap B') = 0.1$. Find $P(A \cap B)$

7) Using the following probability tree, find:
   a) $P(A|B)$
   b) $P(B|A)$
   c) $P(A|B')$
   d) $P(B'|A)$
   e) $P(A'|B')$

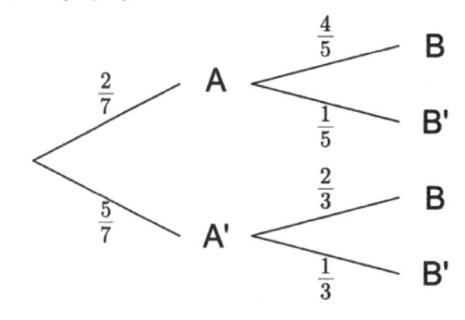

8) In a pack of ten playing cards, 8 are blue and 2 are red. One card is randomly chosen from the pack and is not replaced. Another card is then randomly chosen from the pack and not replaced. Find the probability that the first card was blue, given that the second card was red.

9) The Maxi box of chocolates has 60 chocolates with nine different varieties.

	Milk	Dark	White	Total
Plain	16	6	6	28
Caramel	7	4	4	15
Hazelnut	7	5	5	17
Total	30	15	15	60

Find the probability that a randomly selected chocolate
   a) is a plain milk chocolate;
   b) is a plain chocolate;
   c) is a white chocolate, given that it is a plain chocolate;
   d) is a hazelnut chocolate, given that it is a white chocolate.
   e) Write down two events from the table that are mutually exclusive.

2.49 Probability: Conditional Probability (answers on page 543)

10) It is given that A and B are two events such that $P(A \cap B) = 0.25$, $P(A' \cap B) = 0.3$, and $P(A) = P(B')$.
   a) Find $P(B)$
   b) Find $P(A|B)$
   c) Determine if A and B are independent events.

11) It is given that $P(A) = \frac{4}{5}$, $P(B) = \frac{2}{3}$, and $P(A \cap B) = \frac{11}{20}$
   a) Find $P(A|B)$
   b) Find $P(B|A)$
   c) Find $P(A|B')$
   d) Find $P(B'|A')$
   e) Find $P([A \cup B]')$

12) Using the following probability tree, find:
   a) $P(A|D)$
   b) $P(B|E)$
   c) $P(C|D)$
   d) $P(A'|E)$

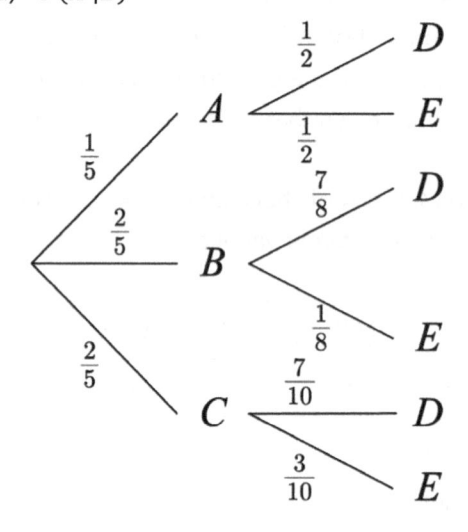

13) A large pottery makes bowls from two types of clay: terracotta and porcelain. The proportion that are made with terracotta is 73% and the proportion that are made with porcelain is 27%. During the firing process, the probability that a terracotta bowl cracks is 18% and the probability that a porcelain bowl cracks is 32%. One bowl is chosen at random.
   a) Find the probability that it is terracotta, given that it is cracked.
   b) Find the probability that it is porcelain, given that it is not cracked.

14) It is given that A and B are two events such that $P(A) = \frac{7}{10}$, $P(B) = \frac{2}{3}$ and $P(A \cap B) = \frac{3}{5}$
   a) Find $P(A|B)$
   b) Use your answer to a) to determine whether the events A and B are independent.

15) It is given that A and B are two events such that $P(A) = \frac{11}{15}$, $P(B') = \frac{5}{11}$ and $P(A' \cap B) = \frac{8}{55}$
   a) Find $P(A'|B)$
   b) Use your answer to a) to determine whether the events A and B are independent.

16) It is given that A and B are two events such that $P([A \cup B]') = \frac{1}{25}$, $P(A' \cap B) = \frac{2}{5}$ and $P(B) = \frac{2}{3}$. Find $P(A|[A \cap B]')$

17) At an auction house, two sets of cutlery, one set that is silver and one set that is silver-plated, are accidentally mixed into the same box. The silver set has four forks, four knives and four spoons. The silver-plated set has six forks, six knives and six spoons. A piece of cutlery is selected at random from the box. Find the probability that it is silver, given that it is a fork or a knife.

18) Tim made 2 cups of tea and 3 cups of coffee in five identical mugs and forgot which is which. He decides to randomly choose a mug and smell whether it is coffee or not, which he can identify with 100% accuracy. He will continue to randomly select mugs until he is sure he has identified 2 mugs of tea or 3 mugs of coffee.
   a) Draw a probability tree to display this information.
   b) Find the probability that Tim must check four mugs.
   c) It is given that Tim only checks three mugs. Find the probability that the last mug he checks is tea.
   d) Given that the second mug he checks is coffee, find the probability that the first mug was tea.

2.49 Probability: Conditional Probability (answers on page 543)

19) Events A, B, C and D are shown in the Venn diagram below.

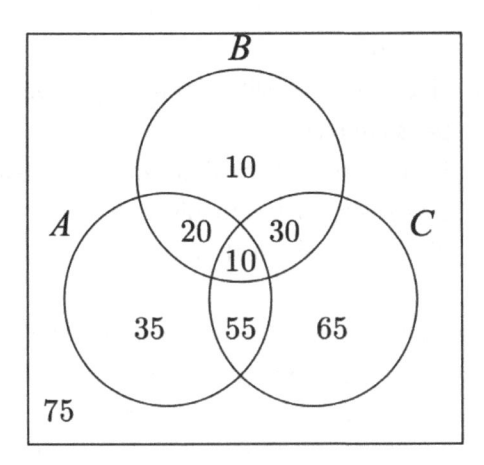

[Venn diagram with three circles A, B, C: values 10 (B only), 20 ($A\cap B$ only), 30 ($B\cap C$ only), 10 ($A\cap B\cap C$), 35 (A only), 55 ($A\cap C$ only), 65 (C only), 75 (outside)]

Find:
a) $P(A \cap B)$
b) $P(A|B)$
c) $P(B|A)$
d) $P(C|B)$
e) $P(B|C)$
f) $P(A|[B \cup C])$
g) $P(B|[A \cup C])$
h) $P(B|[A \cap C])$
i) $P([B \cup C]|A)$
j) $P([B \cap C]|A)$
k) $P([A \cap B]|[A \cup B])$

20) In a sixth form college, the numbers of students studying Psychology, Physics and Chemistry in Year 12 and 13 are shown in the table below.

	Year 12	Year 13
Psychology	250	200
Physics	100	80
Chemistry	60	40

All students at the college take at least one of Psychology, Physics, Chemistry and Maths. It is given that 58% of Psychology students, in both years, study A-Level Maths. 95% of Physics students, in both years, study A-Level Maths. 85% of Chemistry students, in both years, study A-Level Maths.
Event A is that the student studies Physics.
Event B is that the student studies Maths.
Event C is that the student is in Year 13.
a) Use this information to complete the Venn diagram below.

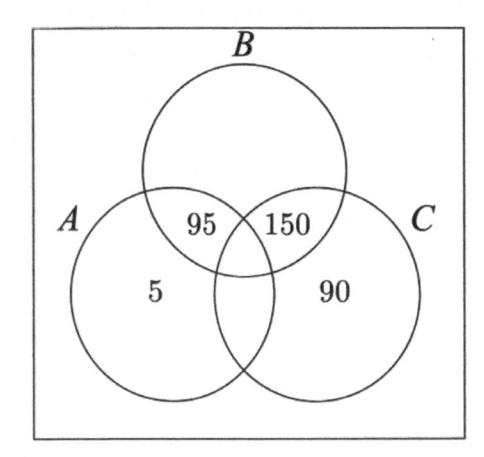

[Venn diagram with three circles A, B, C: values 95 ($A\cap B$ region), 150 ($B\cap C$ region), 5, 90]

b) Find $P(A \cap B \cap C)$
c) Find $P(A' \cap B \cap C)$
d) Find $P([A \cup B]|[B \cup C])$
e) Find $P([A \cap B]|[B \cap C])$

21) It is given that event A is mutually exclusive with both of the events B and C. It is also given that events B and C are independent. It is given that $P(A) = \frac{1}{12}$, $P(B) = \frac{1}{14}$ and $P(C) = \frac{1}{16}$. Find $P([A \cup B]|[B \cup C])$

22) It is given that A and B are two events such that $P([A \cap B]') = \frac{8}{9}$, $P(A' \cap B') = \frac{1}{10}$ and $P(A) = \frac{1}{5}$. Find $P([A \cup B]|[A \cup B'])$

23) It is given that A, B and C are three events such that $P(A \cap B \cap C) = \frac{1}{13}$, $P(A \cap B) = \frac{18}{65}$, $P(A \cap C) = \frac{3}{13}$ and $P(B \cap C) = \frac{11}{130}$.
It is also given that
$$P(B|A) = \frac{36}{71}, \quad P(A|B) = \frac{36}{41}$$
and
$$4P(B \cap [A \cup C]') = P(C \cap [A \cup B]')$$
Draw a fully labelled Venn diagram for the events A, B and C.

24) It is given that A and B are two events such that $P(A|B) = \frac{4}{11}$ and $P(B|A) = \frac{40}{49}$. It is also given that $P([A \cup B]') = \frac{3}{20}$. Draw a fully labelled Venn diagram for the events A and B.

2.49 Probability: Conditional Probability (answers on page 543)

25) Ellen is throwing darts at a darts board. The probability that she hits the bullseye with a single dart at a horizontal distance of 200 cm is $\frac{1}{6}$. Every 10 cm that she steps away from the darts board, the probability of her hitting the bullseye reduces by 15%. She plays a game where only if she hits the bullseye can she move 10 cm further away from the darts board to take the next shot. Given that she reaches 230 cm, find the probability that she hits the bullseye with her next shot.

26) On Sunday, Sarah gets to choose between three different activities: hiking, watching a movie, or attending a music concert. The probability of her choosing hiking is 0.35, the probability of choosing a movie is 0.4, and the probability of attending a concert is 0.25. If she goes hiking, the probability of her feeling tired the next day is 0.8. If she watches a movie, the probability of her feeling tired the next day is 0.2. If she attends the music concert, the probability of her feeling tired the next day is 0.6. On Monday, Sarah felt tired. Determine the probability that she chose to go hiking for her on Sunday.

27) On a mysterious island, 7.9% of the population have developed a unique psychic ability known as "Mindlink". 89.6% of the community has participated in a special meditation practice associated with developing Mindlink, while 9.2% of the community has neither engaged in the meditation practice nor possess the psychic ability. Emily has participated in the meditation practice. Find the probability that she will develop the Mindlink ability.

28) Olena's pencil case contains several gel pens and felt tip pens of varying colours, as shown in the table below.

	Red	Green	Blue
Gel	3	1	1
Felt-tip	4	2	2

Olena takes four pens out of the pencil case at random.
a) Find the probability that all four colours are red.
b) Find the probability that exactly three gel pens are picked.
c) Find the probability that exactly two felt-tip pens are selected, given that all three blue pens are selected.

29) In a television game show, a contestant is asked a series of multiple-choice questions, where each question has four possible answers. Valerie is playing the game and the probability that she knows the answer to the first question is $\frac{3}{8}$. The more questions she answers, the harder they become, so that the probability that she knows the answer is 10% less than the probability before. If she does not know the answer, she guesses. Given that Valerie reaches Question 4, find the probability that she guessed the answer to Question 3.

30) Priya has a bag that contains 28 blue marbles and n red marbles. There are fewer red marbles than blue marbles in the bag. Priya picks a marble at random from the bag and does not replace it. She then randomly selects another marble from the bag. The probability that the two marbles are the same colour is 0.5. Find the probability that the 1st marble is blue, given that the 2nd marble is red.

31) Issy has a box of assorted chocolates that contains 63 milk chocolates and n dark chocolates. There are more dark chocolates than milk chocolates in the box. Issy picks a chocolate at random from the box and eats it. She than randomly selects another chocolate from the box. The probability that the two chocolates are of different types is 0.35. Find the probability that 1st chocolate is milk, given that the 2nd chocolate is milk.

2.50 The Normal Distribution (answers on page 545)

1) The four diagrams below show $X \sim N(10,1)$, $X \sim N(10,3)$, $X \sim N(20,3)$ and $X \sim N(20,6)$. Which is which?

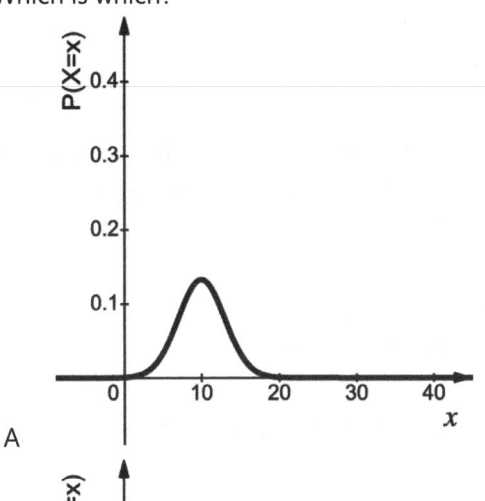

A

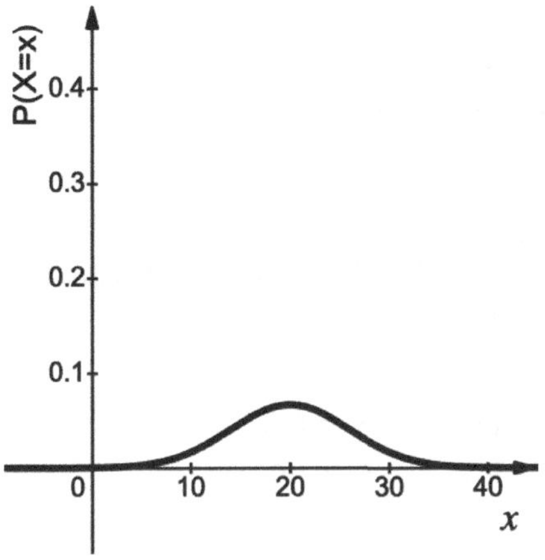

B

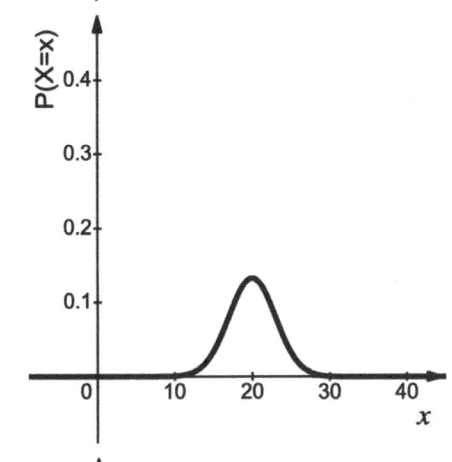

C

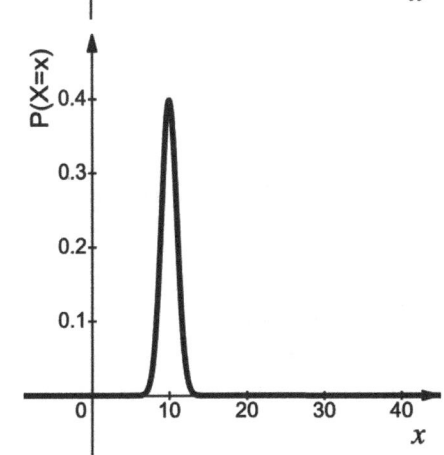

D

2) It is given that the distribution shown below is normal. Use the diagram to estimate the standard deviation.

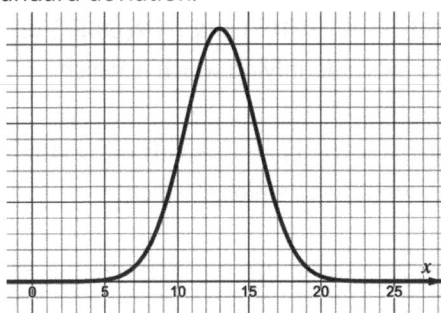

3) It is given that the distribution shown below is normal. Use the diagram to estimate the standard deviation.

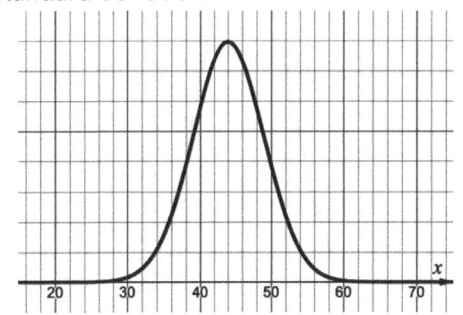

4) Find the standard deviation of each of these normal distributions:
   a) $X \sim N(72, 49)$
   b) $X \sim N(12, 5)$
   c) $X \sim N(53, \sqrt{2})$

5) Approximately 99.73% of the area under a normal distribution is within three standard deviations of the mean. For each of these normal distributions find the values of a and b such that $P(a < X < b) \approx 0.9973$:
   a) $X \sim N(30, 25)$
   b) $X \sim N(18, 12.25)$
   c) $X \sim N(605, 1024)$

6) Sketch the following normal distributions. You do not need to consider their heights.
   a) $X \sim N(25, 5^2)$
   b) $X \sim N(100, 100)$
   c) $X \sim N(50, 5)$

7) Sketch $X \sim N(60, 5^2)$ and $Y \sim N(100, 5^2)$ on the same axes, identifying which is which.

2.50 The Normal Distribution (answers on page 545)

8) Sketch $X \sim N(20, 5^2)$ and $Y \sim N(30, 5^2)$ on the same axes, identifying which is which.

9) Sketch $X \sim N(60, 10^2)$ and $Y \sim N(60, 5^2)$ on the same axes, identifying which is which.

10) Sketch $X \sim N(8, 1)$ and $Y \sim N(10, 4)$ on the same axes, identifying which is which.

11) It is given that $X \sim N(64, 4^2)$. The graph of X is shown below and it is given that
$$P(X > 68) = 0.158655$$

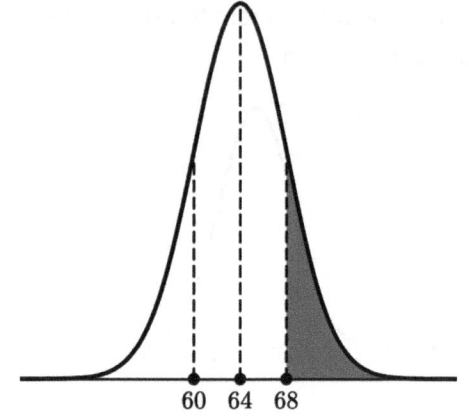

Use the symmetry of the graph to find:
a) $P(X < 68)$
b) $P(X < 60)$
c) $P(60 < X < 68)$

12) It is given that $X \sim N(100, 400)$. The graph of X is shown below and it is given that
$$P(80 < X < 120) = 0.682690$$

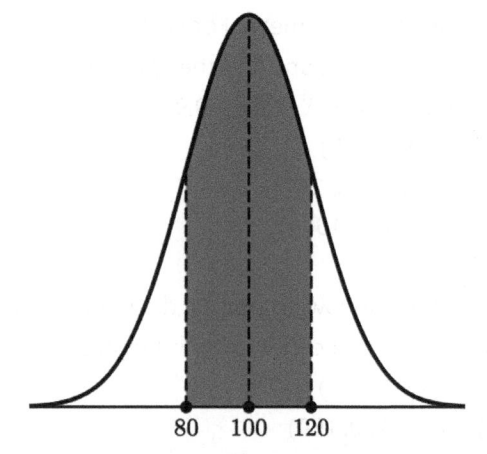

Use the symmetry of the graph to find:
a) $P(X < 80)$
b) $P(X > 120)$
c) $P(80 < X < 100)$

13) It is given that $X \sim N(8, 4)$ and $Z \sim N(0, 1)$. Find the corresponding values of Z when:
a) $X = 7$
b) $X = 9$
c) $X = 10.2$

14) It is given that $X \sim N(16, 16)$. By converting to the standard normal distribution, find:
a) $P(X < 15)$
b) $P(X > 17)$
c) $P(13 < X < 15)$
d) $P(18 < X < 19)$

15) Given $X \sim N(40, 8^2)$, use a calculator to find:
a) $P(X \leq 37)$
b) $P(X < 37)$
c) $P(X \geq 45)$
d) $P(X > 45)$

16) Given $X \sim N(800, 45^2)$, use a calculator to find:
a) $P(X < 860)$
b) $P(X > 810)$
c) $P(780 < X < 850)$
d) $P(860 < X < 880)$

17) Given $X \sim N(45, 3^2)$, find:
a) $P(X = 46)$
b) $P(X \neq 40)$

18) Given $X \sim N(80, 30)$, find:
a) $P(X < 80)$
b) $P(X < 75)$
c) $P(X > 83)$
d) $P(75 < X < 83)$

19) Given $X \sim N(125, 88)$, find:
a) $P(X < 110)$
b) $P(125 < X < 130)$
c) $P(130 < X < 135)$

20) Use the standard normal distribution, $Z \sim N(0, 1)$ to find:
a) $P(-1 < Z < 1)$
b) $P(-2 < Z < 2)$
c) $P(-3 < Z < 3)$
d) $P(-4 < Z < 4)$

2.50 The Normal Distribution (answers on page 545)

21) Given $X \sim N(200, 15^2)$ and your answers to the previous question, find:
    a) $P(185 < X < 215)$
    b) $P(170 < X < 230)$
    c) $P(155 < X < 245)$
    d) $P(140 < X < 260)$

22) Given $X \sim N\left(\mu, \frac{1}{4}\mu^2\right)$, find $P(X < 0.9\mu)$

23) Given $X \sim N(5\mu, 16\mu^2)$, find $P\left(X > \frac{11}{2}\mu\right)$

24) Given $X \sim N(2\mu, 81\mu^2)$, find $P(0.1\mu < X < 0.5\mu)$

25) 200 university students are surveyed and it is found that the monthly rent spent on student accommodation, £X, had the following results, where $\bar{x}$ denotes the sample mean:
    $\sum x = 105000$ and $\sum (x - \bar{x})^2 = 440000$
    a) Find the sample mean and sample standard deviation of X
    b) Using a normal distribution to model X, find the probability that a university student will pay more than £535 each month.

26) 100 sixth form students are surveyed, and it is found that the number of hours of driving lessons, X, had the following results:
    $\sum x = 4650$ and $\sum x^2 = 219000$
    a) Find the sample mean, $\bar{x}$, of X
    b) It is given that the sample variance,
    $$s^2 = \frac{\sum x^2}{n-1} - \frac{(\sum x)^2}{n(n-1)}$$
    Find s^2
    c) Assuming that X be modelled by $N(\bar{x}, s^2)$, find $P(43 < X < 46)$

27) The heights of male NBA basketball players have a mean of 2.01 metres and a standard deviation of 10 cm. It is known that 99% of the basketball players are between 1.79 m and 2.23 m. Comment on whether it would be suitable to model the heights of male NBA basketball players using a normal distribution.

28) Using the standard normal distribution, $Z \sim N(0, 1)$,
    a) Find a such that $P(Z < a) = 0.25$
    b) Find b such that $P(Z < b) = 0.9$
    c) Find c such that $P(Z > c) = 0.01$
    d) Find d such that $P(Z > d) = 0.8$

29) Given $X \sim N(27, 4^2)$,
    a) Find a such that $P(X < a) = 0.02$
    b) Find b such that $P(X < b) = 0.7$
    c) Find c such that $P(X > c) = 0.21$
    d) Find d such that $P(X > d) = 0.6$

30) It is given that $X \sim N(800, 900)$ and $P(a < X < b) = 0.9$. It is also given that $P(a < X < 800) = P(800 < X < b)$. Find the value of a and the value of b

31) It is given that $X \sim N(65, 12)$ and $P(64 < X < a) = 0.25$. Find the value of a

32) It is given that $X \sim N(18, 10)$.
    a) Find a given that $P(X > a) = 0.3$
    b) Find b given that $P(18 - b < X < 18 + b) = 0.3$

33) It is given that $X \sim N(\mu, 10^2)$. Find the value of μ in each of these cases:
    a) $P(X < 40) = 0.5$
    b) $P(X < 50) = 0.3$
    c) $P(X > 200) = 0.7$
    d) $P(X > 70) = 0.01$

34) It is given that $X \sim N(33, \sigma^2)$. Find the value of σ in each of these cases:
    a) $P(X < 30) = 0.2$
    b) $P(X < 50) = 0.6$
    c) $P(X > 1) = 0.99$
    d) $P(X > 34) = 0.49$

35) The average monthly food bill, £X, for a family of 4 in the UK can be modelled by a normal distribution with mean £725. It is known that $P(X < 700) = 0.35$. Find the standard deviation of X to one decimal place.

2.50 The Normal Distribution (answers on page 545)

36) Loose leaf tea is packaged in sachets. The mass, X, of loose leaf tea in each sachet is known to be normally distributed with a mean of 3 grams. It is given that 50% of the masses are between 2.6 g and 3.4 g. Determine the standard deviation of the masses.

37) Category 2 ambulance calls are classified as an emergency that may require immediate assessment or on-scene intervention. Response times should have an average of 18 minutes and 90% must be responded to within 40 minutes. Explain why response times may not be normally distributed.

38) The distribution $X \sim N(\mu, \sigma^2)$ is shown below. The two shaded regions add to an area of 0.2.

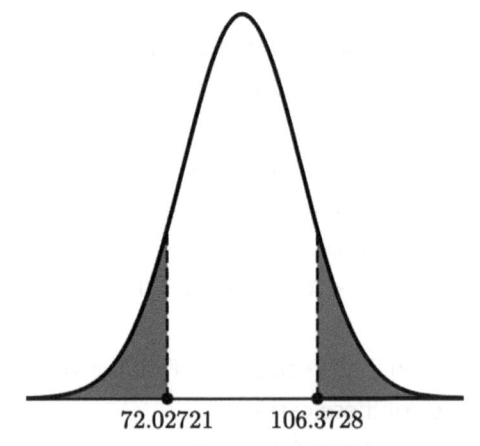

    a) Find μ and σ^2, given that the lower boundary is 72.02721 and the upper boundary is 106.3728
    b) It is given that Y is the random variable such that $Y = 3X - 2$. Write down the distribution of Y.

39) It is given that $X \sim N(84, 5.29)$ can be transformed to $Y \sim N(326.8, 72.4201)$ by the transformation $y = ax + b$. Find the values of a and b.

40) It is given that $X \sim N(\mu, \sigma^2)$. It is also given that $P(X < 99) = 0.5$ and $P(X > 90) = 0.6$. Find the value of μ and the value of σ

41) It is given that $X \sim N(\mu, \sigma^2)$. It is also given that $P(X > 11) = 0.5$ and $P(X > 12) = 0.05$. Find the value of μ and the value of σ

42) It is given that $X \sim N(\mu, \sigma^2)$. Find the value μ and σ in each of these cases:
    a) $P(X < 10) = 0.3$ and $P(X < 11) = 0.35$
    b) $P(X < 50) = 0.4$ and $P(X < 90) = 0.65$
    c) $P(X < 3) = 0.01$ and $P(X > 9) = 0.02$
    d) $P(X > 67) = 0.44$ and $P(X > 73) = 0.33$

43) A banana is considered small if it weighs less than 140 grams, medium if it weighs between 140 and 215 grams, and large if it weighs over 215 grams. 14% of bananas are found to be small and 14% are found to be large. By writing down an appropriate distribution, estimate the weight of the lightest banana.

Prerequisite knowledge for Q44-Q45: Points of Inflection, The Chain Rule & The Product Rule

44) The normal distribution with mean 0 and standard deviation 1 has equation
$$y = \frac{1}{\sqrt{2\pi}} e^{-\frac{1}{2}x^2}$$
Use differentiation to find the x-coordinates of the points of inflection.

45) The normal distribution with mean μ and standard deviation σ has equation
$$y = \frac{1}{\sigma\sqrt{2\pi}} e^{-\frac{1}{2}\left(\frac{x-\mu}{\sigma}\right)^2}$$
Use differentiation to find the the x-coordinates of the points of inflection.

Prerequisite knowledge: Conditional Probability

46) Given $X \sim N(75, 10)$, find:
    a) $P(X < 70 | X < 80)$
    b) $P(X > 70 | X < 76)$
    c) $P(X < 77 | X > 73)$
    d) $P(X > 81 | X > 80)$

47) A nursery grows yucca plants. When they are ready to be sold, their heights are normally distributed with a mean of 75 cm and standard deviation 5 cm. One yucca plant is selected at random. Find the probability that it is taller than 80 cm given that it is taller than 76 cm.

2.50 The Normal Distribution (answers on page 545)

48) The weight of sandbags is known to be normally distributed with a mean of 33 kg and standard deviation of 3.5 kg. A sandbag is selected at random. Given that it weighs more than 34 kg, find the probability that it weighs less than 36 kg.

Prerequisite knowledge for Q49: The Binomial Distribution

49) The weight, X, of a farm's eggs may be modelled by a normal distribution with mean 49 grams and variance 9 grams.
    a) A single egg is randomly selected. Find the probability that the egg has a weight of less than 47 grams.
    b) Eggs are put in boxes of twelve before they are transported to the local supermarket. A box of twelve eggs is randomly selected. Find the probability that at least four of the eggs has a weight of less than 47 grams.

50) The lifetime, X, of a discount battery may be modelled by a normal distribution with mean 10 hours and standard deviation 1.2 hours.
    a) A single discount battery is randomly selected. Find the probability that the battery has a lifetime:
       i) equal to 11 hours;
       ii) greater than 11 hours.
    b) Discount batteries are sold in packs of eight. A pack is randomly selected. Find the probability that at least two of the batteries have a lifetime of more than 11 hours.

51) The width, W, of a type of potato is known to be normally distributed with a mean of 5.7 cm and standard deviation 0.9 cm. A farmer is able to sell their potatoes to a large supermarket if $4.8 \leq W \leq 6.2$ for £0.25 per kilogram. If $W < 4.8$, the farmer sells the potatoes to a local farmer for £0.11 per kilogram. If $W > 6.2$, the farmer sells the potatoes to a local greengrocer for £0.21 per kilogram. Calculate how much the farmer will be able to sell a crop of 150 kg of potatoes for.

52) The heights of 180 students were measured to the nearest cm and recorded in the histogram below.

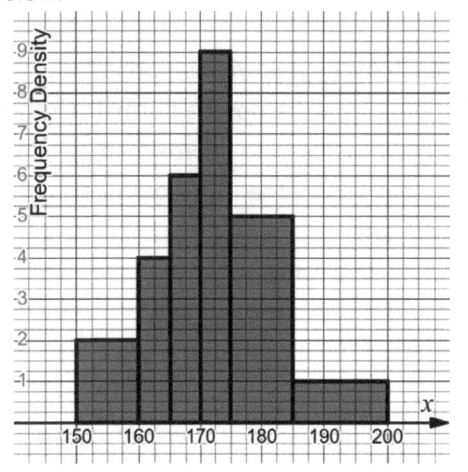

    a) Find $P(x < 170)$
    b) Find estimates for the mean and sample standard deviation of the heights.
    c) Using your answers to b), model the data as a normal distribution, $Y \sim N(\mu, \sigma^2)$. Find $P(Y < 170)$ and hence comment on whether the normal distribution is an appropriate model for the data.

53) The heights of a different group of 225 students were measured to the nearest cm and recorded in the histogram below.

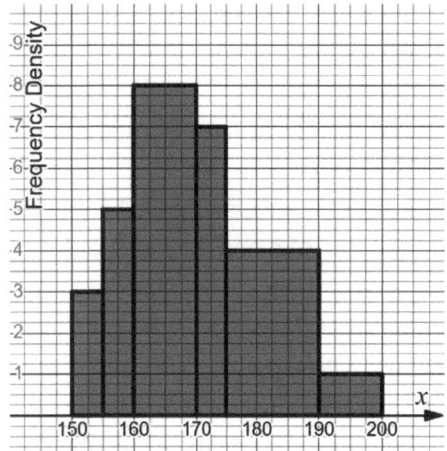

    a) Find $P(x < 170)$
    b) Find estimates for the mean and sample standard deviation of the heights.
    c) Using your answers to b), model the data as a normal distribution, $Y \sim N(\mu, \sigma^2)$. Find $P(Y < 170)$ and hence comment on whether the normal distribution is an appropriate model for the data.

2.50 The Normal Distribution (answers on page 545)

54) Describe the conditions for which the binomial distribution $X \sim B(n,p)$ can be approximated using the normal distribution $Y \sim N(\mu, \sigma^2)$

55) For each of the following binomial distributions, write down a normal distribution that could be used to approximate it, and comment on whether this would be appropriate.
    a) $X \sim B(500, 0.48)$
    b) $X \sim B(4, 0.5)$
    c) $X \sim B(150000, 0.001)$

56) A circular spinner has five sections labelled 1, 2, 3, 4 and 5. The section labelled 2 is a sector with an angle of 160°. The spinner is spun 2000 times and the number, X, of times 2 is chosen by the spinner is recorded.
    a) Write down the distribution of X and explain why it would be appropriate to approximate the distribution of X using a normal distribution.
    b) Using a normal distribution to approximate the distribution of X, find a such that $P(X > a) \approx 0.2$
    c) Using the distribution of X and your value of a, find the actual value of $P(X > a)$ to 3sf.

57) It is given that $X \sim N(300, 20^2)$ and $Y \sim N(400, 30^2)$ where
    $$P(X > a) = P(Y < a) = b$$
    Find the value of a and the value of b

58) It is given that $X \sim N(16.5, 5.29)$ and $Y \sim N(20.2, 6.76)$ where
    $$P(X > a) = P(Y < a) = b$$
    Find the value of a and the value of b

59) It is given that $X \sim N(50, 4)$ and $Y \sim N(60, 9)$ where
    $$P(X < a) = P(Y > a) = b$$
    Find the value of a and the value of b

60) It is given that $X \sim N(200, 25)$ and $Y \sim N(250, 49)$ where
    $$P(X < a - 15) = P(Y > a + 15) = b$$
    Find the value of a and the value of b

Continuity Corrections

61) The binomial distribution $X \sim B(600, 0.52)$ is to be approximated using the normal distribution $Y \sim N(312, 149.76)$. Approximate each of these probabilities.
    a) $P(X \leq 300)$
    b) $P(X > 320)$
    c) $P(X < 311)$
    d) $P(X \geq 295)$
    e) $P(X = 312)$

62) Given that $X \sim B(350, 0.6)$:
    a) Find $P(X \leq 200)$
    b) Use a normal distribution, Y, to approximate your answer to a).
    c) Find the percentage error between your answers to a) and b) to 3 significant figures.

63) Given that $X \sim B(540, 0.47)$:
    a) Find $P(X < 260)$
    b) Use a normal distribution, Y, to approximate your answer to a).
    c) Find the percentage error between your answers to a) and b) to 3 significant figures.

64) Given that $X \sim B(2000, 0.7)$:
    a) Find $P(X \geq 1440)$
    b) Use a normal distribution, Y, to approximate your answer to a).
    c) Find the percentage error between your answers to a) and b) to 3 significant figures.

65) It is known that 4.5% of adults aged 18 and over use e-cigarettes.
    a) In a random sample of 5000 adults, find the probability that:
        i) More than 230 use e-cigarettes.
        ii) More than 230 use e-cigarettes, given that less than 250 do.
    b) Use a normal approximation to estimate your answers to a) i) and a) ii).

2.51 Sample Means Hypothesis Testing (answers on page 547)

1) In each of these situations, find the p-value and hence determine whether you reject H_0 or fail to reject H_0
   a) $H_0: \mu = 30$, $H_1: \mu < 30$
      $X \sim N(30, 2^2)$
      $n = 4$, $\bar{x} = 27.1$, 5% significance level
   b) $H_0: \mu = 45$, $H_1: \mu < 45$
      $X \sim N(45, 16)$
      $n = 10$, $\bar{x} = 43.2$, 5% significance level
   c) $H_0: \mu = 50$, $H_1: \mu < 50$
      $X \sim N(50, 20)$
      $n = 30$, $\bar{x} = 49.2$, 10% significance level
   d) $H_0: \mu = 85$, $H_1: \mu < 85$
      $X \sim N(85, 37)$
      $n = 15$, $\bar{x} = 81.2$, 1% significance level

2) In each of these situations, find the p-value and hence determine whether you reject H_0 or fail to reject H_0
   a) $H_0: \mu = 60$, $H_1: \mu > 60$
      $X \sim N(60, 3^2)$
      $n = 5$, $\bar{x} = 62.1$, 5% significance level
   b) $H_0: \mu = 23$, $H_1: \mu > 23$
      $X \sim N(23, 25)$
      $n = 9$, $\bar{x} = 25.9$, 5% significance level
   c) $H_0: \mu = 104$, $H_1: \mu > 104$
      $X \sim N(104, 35)$
      $n = 20$, $\bar{x} = 107$, 10% significance level
   d) $H_0: \mu = 162$, $H_1: \mu > 162$
      $X \sim N(162, 54)$
      $n = 25$, $\bar{x} = 166.3$, 1% significance level

3) In each of these situations, find the p-value and hence determine whether you reject H_0 or fail to reject H_0
   a) $H_0: \mu = 34$, $H_1: \mu \neq 34$
      $X \sim N(34, 6^2)$
      $n = 16$, $\bar{x} = 30$, 5% significance level
   b) $H_0: \mu = 70$, $H_1: \mu \neq 70$
      $X \sim N(70, 2.25)$
      $n = 7$, $\bar{x} = 70.8$, 5% significance level
   c) $H_0: \mu = 125$, $H_1: \mu \neq 125$
      $X \sim N(125, 80)$
      $n = 100$, $\bar{x} = 126.6$, 10% significance level
   d) $H_0: \mu = 216$, $H_1: \mu \neq 216$
      $X \sim N(216, 17)$
      $n = 30$, $\bar{x} = 214$, 1% significance level

4) In each of these situations, find the critical region and hence determine whether you reject H_0 or fail to reject H_0
   a) $H_0: \mu = 28$, $H_1: \mu < 28$
      $X \sim N(28, 2.5^2)$
      $n = 18$, $\bar{x} = 26.7$, 5% significance level
   b) $H_0: \mu = 32$, $H_1: \mu < 32$
      $X \sim N(32, 25)$
      $n = 26$, $\bar{x} = 30.5$, 5% significance level
   c) $H_0: \mu = 106$, $H_1: \mu < 106$
      $X \sim N(106, 60)$
      $n = 33$, $\bar{x} = 104.4$, 10% significance level
   d) $H_0: \mu = 340$, $H_1: \mu < 340$
      $X \sim N(340, 129)$
      $n = 80$, $\bar{x} = 336.6$, 1% significance level

5) In each of these situations, find the critical region and hence determine whether you reject H_0 or fail to reject H_0
   a) $H_0: \mu = 75$, $H_1: \mu > 75$
      $X \sim N(75, 10^2)$
      $n = 19$, $\bar{x} = 79.3$, 5% significance level
   b) $H_0: \mu = 160$, $H_1: \mu > 160$
      $X \sim N(160, 121)$
      $n = 33$, $\bar{x} = 162.7$, 5% significance level
   c) $H_0: \mu = 1250$, $H_1: \mu > 1250$
      $X \sim N(1250, 3.2)$
      $n = 4$, $\bar{x} = 1252.3$, 10% significance level
   d) $H_0: \mu = 670$, $H_1: \mu > 670$
      $X \sim N(670, 163)$
      $n = 60$, $\bar{x} = 674.1$, 1% significance level

6) In each of these situations, find the critical region and hence determine whether you reject H_0 or fail to reject H_0
   a) $H_0: \mu = 85$, $H_1: \mu \neq 85$
      $X \sim N(85, 7^2)$
      $n = 54$, $\bar{x} = 82.9$, 5% significance level
   b) $H_0: \mu = 1.6$, $H_1: \mu \neq 1.6$
      $X \sim N(1.6, 0.04)$
      $n = 43$, $\bar{x} = 1.68$, 5% significance level
   c) $H_0: \mu = 57$, $H_1: \mu \neq 57$
      $X \sim N(57, 29)$
      $n = 10$, $\bar{x} = 54.04$, 10% significance level
   d) $H_0: \mu = 890$, $H_1: \mu \neq 890$
      $X \sim N(890, 250)$
      $n = 9$, $\bar{x} = 902.9$, 1% significance level

2.51 Sample Means Hypothesis Testing (answers on page 547)

7) In each of these situations, find the test statistic and the critical value of Z, and hence determine whether you reject H_0 or fail to reject H_0
   a) $H_0: \mu = 35$, $H_1: \mu < 35$
      $X \sim N(35, 8^2)$
      $n = 6$, $\bar{x} = 29.7$, 5% significance level
   b) $H_0: \mu = 60$, $H_1: \mu < 60$
      $X \sim N(60, 50)$
      $n = 10$, $\bar{x} = 55.9$, 5% significance level
   c) $H_0: \mu = 12.5$, $H_1: \mu < 12.5$
      $X \sim N(12.5, 1.4)$
      $n = 4$, $\bar{x} = 11.9$, 10% significance level
   d) $H_0: \mu = 496$, $H_1: \mu < 496$
      $X \sim N(496, 244)$
      $n = 45$, $\bar{x} = 490.1$, 1% significance level

8) In each of these situations, find the test statistic and the critical value of Z, and hence determine whether you reject H_0 or fail to reject H_0
   a) $H_0: \mu = 300$, $H_1: \mu > 300$
      $X \sim N(300, 20^2)$
      $n = 3$, $\bar{x} = 319.3$, 5% significance level
   b) $H_0: \mu = 133$, $H_1: \mu > 133$
      $X \sim N(133, 1.21)$
      $n = 5$, $\bar{x} = 133.7$, 5% significance level
   c) $H_0: \mu = 8.8$, $H_1: \mu > 8.8$
      $X \sim N(8.8, 0.56)$
      $n = 12$, $\bar{x} = 9.085$, 10% significance level
   d) $H_0: \mu = 8550$, $H_1: \mu > 8550$
      $X \sim N(8550, 560)$
      $n = 35$, $\bar{x} = 8560.3$, 1% significance level

9) In each of these situations, find the test statistic and the critical values of Z, and hence determine whether you reject H_0 or fail to reject H_0
   a) $H_0: \mu = 11$, $H_1: \mu \neq 11$
      $X \sim N(11, 1)$
      $n = 8$, $\bar{x} = 10.25$, 5% significance level
   b) $H_0: \mu = 77$, $H_1: \mu \neq 77$
      $X \sim N(77, 35)$
      $n = 22$, $\bar{x} = 78.9$, 5% significance level
   c) $H_0: \mu = 170$, $H_1: \mu \neq 170$
      $X \sim N(170, 560)$
      $n = 6$, $\bar{x} = 154.8$, 10% significance level
   d) $H_0: \mu = 35600$, $H_1: \mu \neq 35600$
      $X \sim N(35600, 19500)$
      $n = 85$, $\bar{x} = 35552$, 1% significance level

10) Molly owns a 1950s inspired café with a fizzy drink machine that dispenses soda into cups to a volume V, where V can be modelled as normally distributed with a mean of 380 ml and standard deviation 3 ml. She believes that the machine is not working properly, and the amount of soda being dispensed is less than it should be. Molly carries out a hypothesis test, at the 10% significance level, to test her belief. She collects a random sample of 15 soda drinks and measures their mean volume to be 378.7 ml. Molly's hypothesis test is shown below.

    Let X be the volume of soda dispensed by the machine in millilitres.
    $H_0: \mu = 380$
    $H_1: \mu > 380$
    $\bar{X} \sim N\left(380, \frac{3^2}{15}\right)$
    $P(\bar{X} < 378.7) = 0.0466$
    $0.0466 < 0.1$ so we fail to reject H_0
    There is insufficient evidence to suggest that the population mean amount of soda is less than expected.

    a) Identify the first two errors made by Molly in her hypothesis test.
    b) Write out a fully correct hypothesis test for Molly.

11) The heights of a certain type of plant over the age of 2 years can be modelled by a normal distribution with a mean of 130 cm and a standard deviation of 20 cm. Trisha works at a garden centre that sells this type of plant, and believes that the mean height is actually greater than 130 cm. She collects a random sample of 10 of these plants and finds their mean height to be 138 cm. Carry out a test at the 3% significance level to determine whether there is evidence to support Trisha's belief.

2.51 Sample Means Hypothesis Testing (answers on page 547)

12) Between April 2017 and March 2018, the weekly household spending in the UK can be modelled by a normal distribution with a mean of £589.60 (adjusting for inflation to compare with April 2018 to March 2019), and a standard deviation of £120.30. A random sample of size 150 was taken from April 2018 to March 2019, which had a mean of £605. Carry out a hypothesis test, at the 10% significance level, to investigate whether the weekly household spending in the UK significantly changed from the 2017-2018 to 2018-2019 tax years.

13) A confectionary company sells chocolate bars that are claimed to have a mean weight of 100 grams and a standard deviation of 5 grams. A quality control team is tasked with verifying if the mean weight of the chocolate bars is different from the claim. A random sample of 16 bars are selected from the production line, and their weights measured. The weights are shown in the table below.

97.7	97.1	97.2	97.2
98.8	97.6	100.2	96.5
96.5	97.0	96.5	96.9
98.4	99.7	100.0	99.5

Using a significance level of 5%, conduct a hypothesis test to determine if there is evidence to suggest that the mean weight of the chocolate bars produced by the company differs from the 100 grams claim.

14) A bakery bakes loaves of bread. The mass of the loaves can be modelled by a normal distribution with a mean weight of 400 grams and a standard deviation of 6 grams. The length of time the loaves are in the oven is shortened and the bakery wants to test whether this will reduce the weight of the loaves. A large number of loaves are baked using this new method, and a random sample of 50 are selected and weighed. The mean weight of these 50 loaves is 398 grams. The bakery carries out a hypothesis test, at the 5% significance level, to test their belief. The bakery's hypothesis test is shown below.

Let X be the weight of baked loaves in grams.
$H_0: \mu = 400$
$H_1: \mu \neq 400$
$\bar{X} \sim N\left(400, \frac{6^2}{\sqrt{50}}\right)$
$P(X < a) = 0.05 \Rightarrow a = 396.3$
$398 > 396.3$ *so we fail to reject* H_0
There is insufficient evidence to suggest that the population mean weight of loaves of bread has been reduced.

a) Identify the first two errors made by the bakery in their hypothesis test.
b) Write out a fully correct hypothesis test for the bakery.

15) The heights of female adults in the UK in 1998 can be modelled by a normal distribution with a mean of 161 cm and a standard deviation 9 cm. Gemma believes that the mean height of females in the UK is different to that in 1998. She decides to conduct a hypothesis test using a 5% significance level by collecting a random sample of 60 female adults.
a) Find the critical region for this test.

The mean height of the sample is 162.4 cm.
b) Write out and complete Gemma's hypothesis test.

16) It is known that the waiting times at a veterinary surgery for those without an appointment can be modelled by a normal distribution with a mean of 45 minutes and a standard deviation of 6 minutes. A new manager, Dave, takes over the surgery and introduces some new processes that he hopes will reduce the waiting time for those without an appointment. Dave decides to conduct a hypothesis test at the 5% significance level to test his belief and will collect a random sample of how long 12 people with their pets wait.
a) Find the critical region for the test.

2.51 Sample Means Hypothesis Testing (answers on page 547)

Dave collects the following data, where the times are in minutes.

41	39	49	46	38	58
33	42	36	38	45	44

b) Write out and complete Dave's hypothesis test.

17) In 2019, a large study was undertaken to look into adults' sleeping habits, and it was found that the amount of time an adult sleeps can be modelled by a normal distribution with a mean of 7 hours and a standard deviation of 1.1 hours. In 2023, Stuart wanted to determine if there is any evidence to suggest that the average amount of sleep an adult gets has changed since 2019 and decides to carry out a hypothesis test at the 1% significance level to test his belief. Stuart collects a random sample of 80 adults and finds their average amount of sleep to be 7.3 hours. Stuart's hypothesis test is shown below.

Let X be the number of hours of sleep an adult gets.
$H_0: \mu = 7$
$H_1: \mu \neq 7$
$\bar{X} \sim N\left(7.3, \frac{1.1^2}{80}\right)$
$Z = \frac{\bar{x} - \mu}{\frac{\sigma}{\sqrt{n}}} = \frac{7.3 - 7}{\frac{1.1}{\sqrt{80}}} = 2.439$
$P(Z > a) = 0.995 \Rightarrow a = 2.576$
2.439 < 2.576 so we fail to reject H_0
There is insufficient evidence to suggest that the population mean number of hours that adults sleep has changed.

a) Identify the first two errors made by Stuart in his hypothesis test.
b) Write out a fully correct hypothesis test for Stuart.

18) A pharmaceutical company produces a cold and flu medication claiming that the mean time for their medication to relieve symptoms is 30 minutes, with a standard deviation of 6 minutes. Assume that these times are normally distributed. A team of medical researchers believes that it takes longer for the medication to take effect than the average time claimed. The team wishes to conduct a hypothesis test at the 1% significance level to determine if there is any evidence to suggest the mean time is longer than 30 minutes. A random sample of ten patients suffering with cold and flu symptoms is given the medication and their response times are recorded. The sample has a mean response time of 35 minutes.
a) Find the test statistic and the critical values of Z.
b) Write out and complete the test.

19) A car manufacturer has just unveiled their latest model, and the manufacturer claims that the mean fuel efficiency is 40 miles to the gallon with a standard deviation of 2 miles to the gallon. An environmentalist watchdog wants to verify their claim and decides to conduct a hypothesis test at the 2% significance level to determine if there is evidence to suggest that the 40 miles to the gallon average is too high, and that actually it is less than the advertised figure. A random sample of 15 of the new cars are tested, and the sample has a mean fuel efficiency of 38.75 miles to the gallon.
a) Find the test statistic and the critical values of Z.
b) Write out and complete the test.

20) An established fitness club owns several gyms across the country and has a large membership of adults aged 18 and over. All members who join the gym are enrolled on a twelve-week course that is focused on weight-loss and maintaining this level. The fitness club claim that the mean weight-loss for their membership is 8.5 kilograms, with a standard deviation of 1.8 kilograms. It may be assumed that the weight-loss for the membership is normally distributed. A group of nutritionists decide to investigate whether the mean weight loss achieved by the membership differs from 8.5 kilograms, and they decide to conduct a hypothesis test using a significance level of 3%. The nutritionists track a random sample of 15 members of the fitness club and the sample's mean weight loss is 9.2 kilograms. Write out and complete the nutritionists' hypothesis test.

2.52 Forces: Equilibrium (answers on page 549)

1) Find the horizontal and vertical component x and y of the force P as shown in the diagram below, for each of the given values of P and θ:

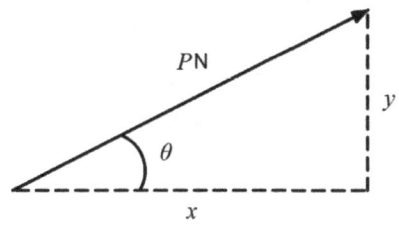

   a) $P = 5$ N, $\theta = 30°$
   b) $P = 6$ N, $\theta = 45°$
   c) $P = 8$ N, $\theta = 40°$
   d) $P = 3$ N, $\theta = 22°$

2) Find the horizontal and vertical component of each of the forces shown:

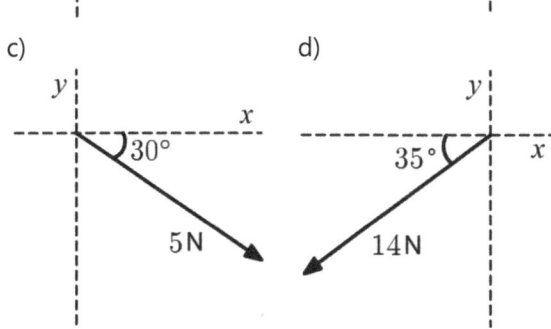

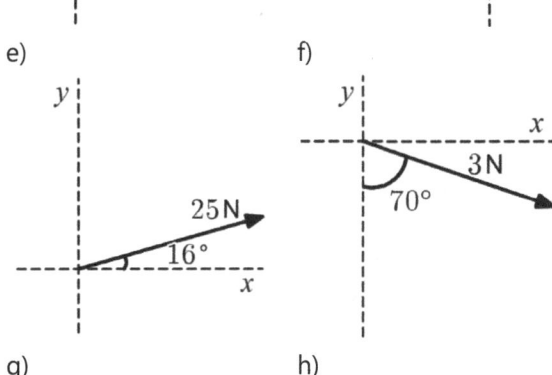

   a) 10 N at 45°
   b) 12 N at 20°
   c) 5 N at 30°
   d) 14 N at 35°
   e) 25 N at 16°
   f) 3 N at 70°
   g) 10 N at 64°
   h) 7 N at 28°

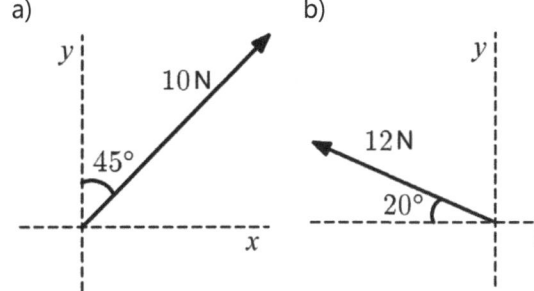

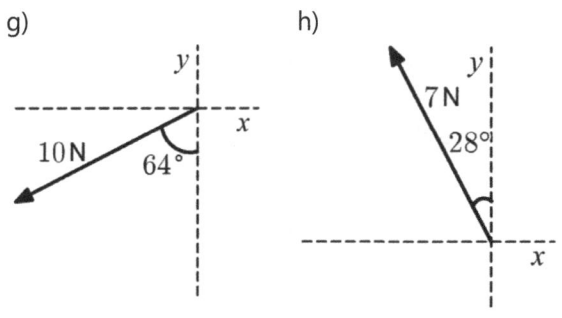

3) For each of the diagrams below, a particle is acted upon by the forces shown. Find the magnitude of the resultant force and the angle the resultant makes with Ox:

   a) 2 N up, 4 N at 34° above negative x-axis, 10 N along positive x-axis
   b) 6 N along negative x-axis, 8 N at 60° above positive x-axis, 3 N down
   c) 7 N at 45° above negative x-axis, 7 N at 45° above positive x-axis, 5 N down

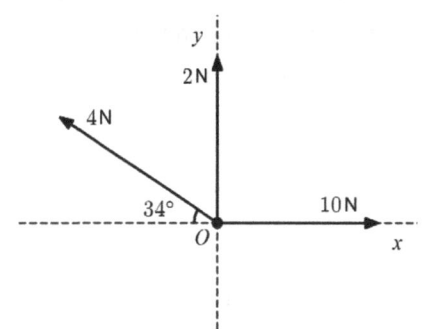

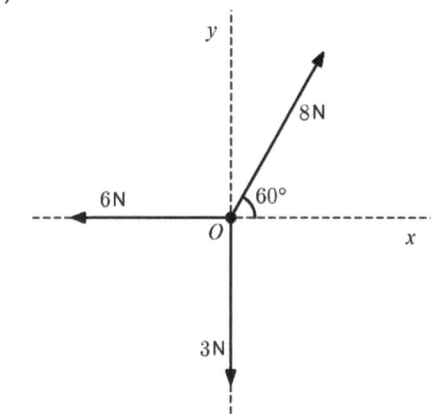

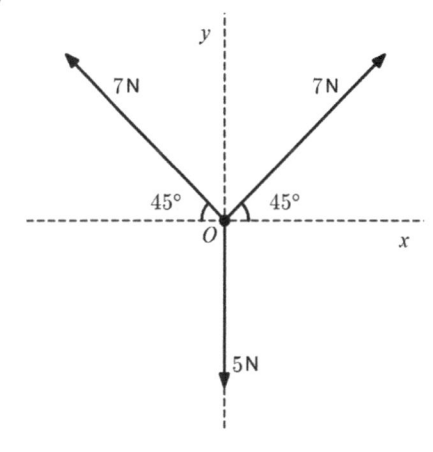

2.52 Forces: Equilibrium (answers on page 549)

4) In each of the diagrams below, a particle is at rest, under the action of the forces shown. In each case, by resolving in the x-direction and y-direction, find the magnitude of any unknown forces (labelled P, Q, R) and the size of any unknown angles (labelled θ):

a)

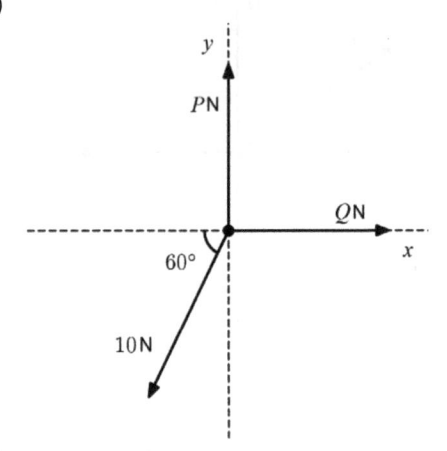

b)

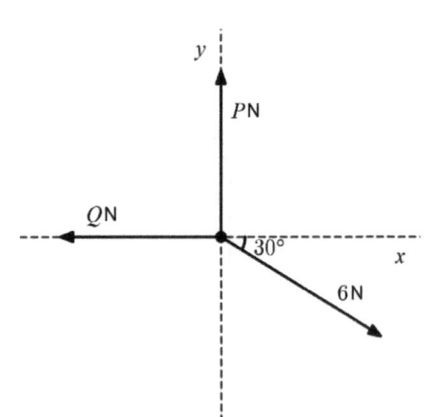

c)

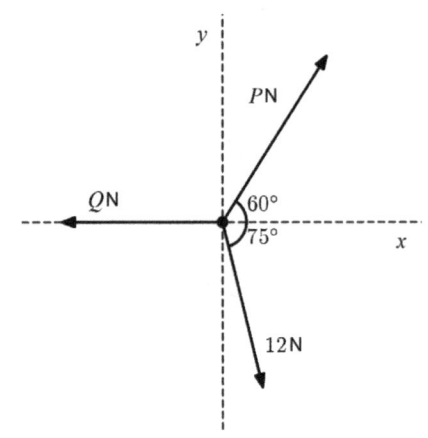

d)

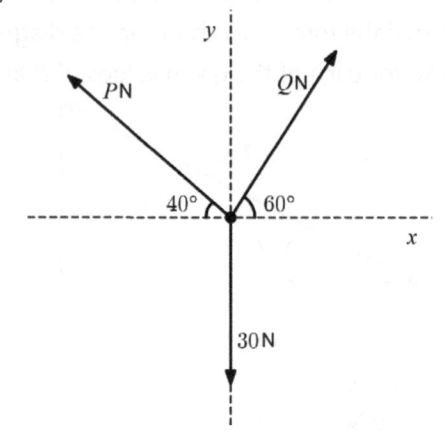

e)

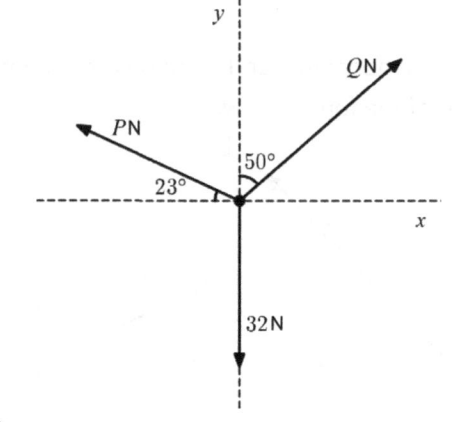

f)

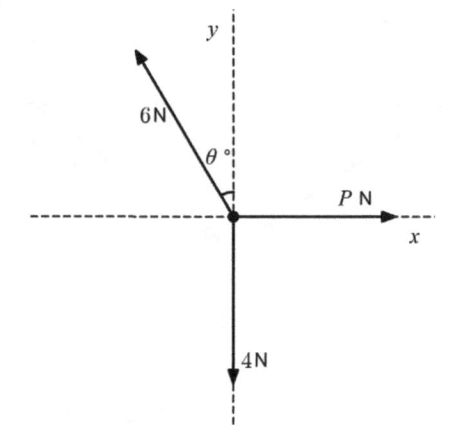

g)

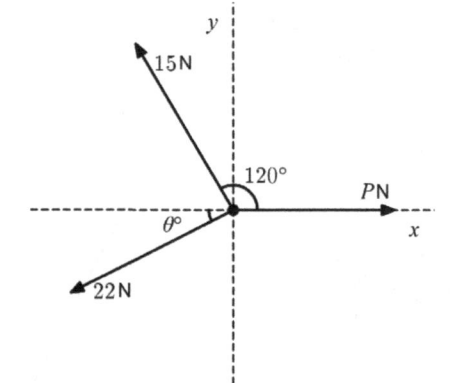

2.52 Forces: Equilibrium (answers on page 549)

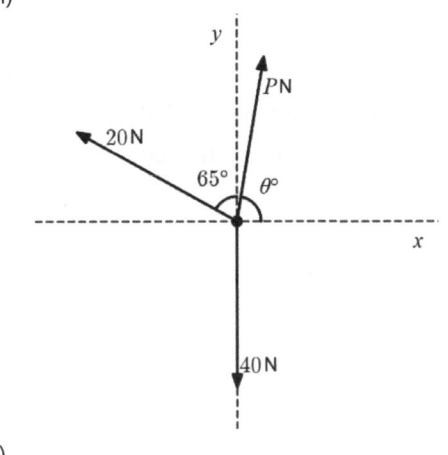

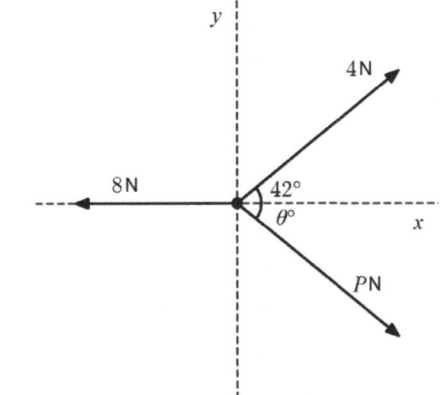

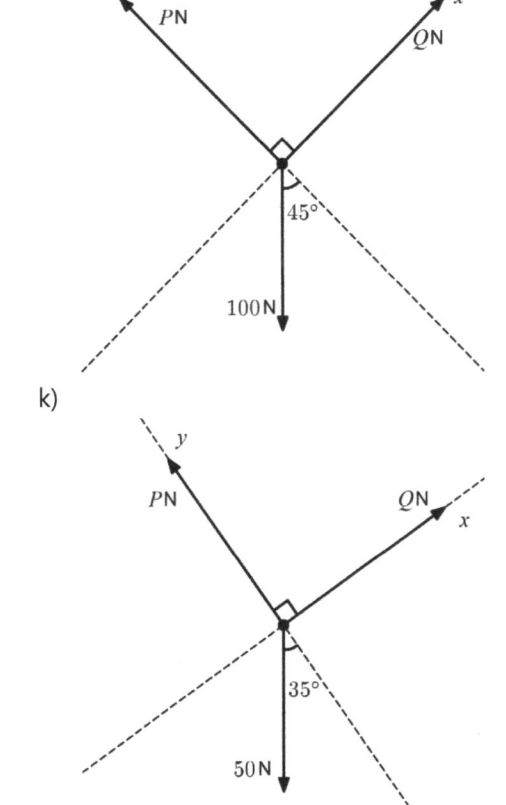

h) [Diagram: forces at origin — 20N at 65° above negative x-axis (measured from y-axis), PN at angle θ° from y-axis in first quadrant (65° shown between 20N and y-axis), 40N downward along negative y-axis]

i) [Diagram: 4N in first quadrant, 8N along negative x-axis, PN in fourth quadrant at angle θ° below x-axis, with 42° between 4N and x-axis]

j) [Diagram: PN along y-axis direction, QN along x-axis direction (perpendicular), 45° between QN and the downward 100N force]

k) [Diagram: PN and QN perpendicular, 35° between y-axis and downward 50N]

The following questions use $g = 9.8$ ms⁻²

5) An object of mass 2 kg is held in equilibrium, suspended by two light, inextensible strings as shown in each of the diagrams. The tensions in the string are given by T_1 N and T_2 N.
In each case, find the value of T_1 and T_2

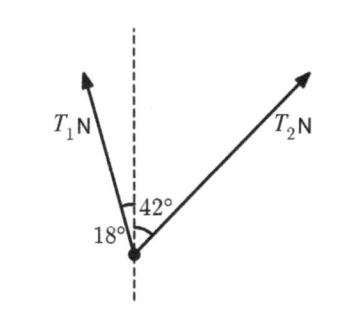

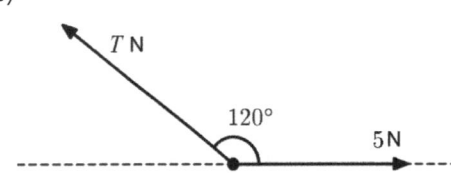

a) [Diagram: T_1N at 35° above horizontal (left), T_2N at 45° above horizontal (right)]

b) [Diagram: T_1N at 18° left of vertical, T_2N at 42° right of vertical]

c) [Diagram: T_1N at 50° below horizontal (left), T_2N at 50° below horizontal (right)]

6) A particle of mass m kg is held in equilibrium, suspended by two light, inextensible strings as shown in each of the diagrams. One of the strings has tension T
In each case, find the value of T and m

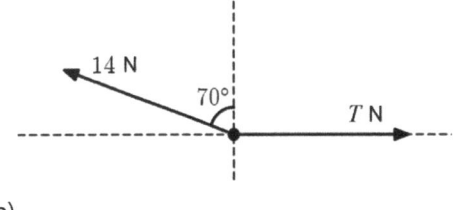

a) [Diagram: 14 N at 70° above horizontal (to the left of vertical), T N horizontal to the right]

b) [Diagram: T N up and to the left, 5N horizontal to the right, 120° between them]

2.52 Forces: Equilibrium (answers on page 549)

7) A particle of weight 10 N is held in equilibrium, suspended by two light, inextensible strings as shown in the following diagram. The tension in the horizontal string is T N. The second string is pulling with a tension of 15 N at $\theta°$ to the horizontal in the opposite direction to T

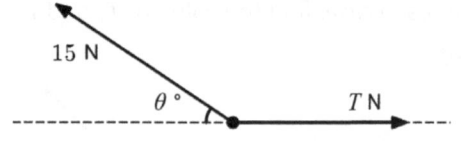

Find
a) the value of θ
b) the value of T

8) A particle of weight 4 N is held in equilibrium, suspended by two light, inextensible strings as shown in the following diagram. The tension in the horizontal string is 18 N. The second string is pulling with a tension of T N at $\theta°$ to the horizontal in the opposite direction to the first string.

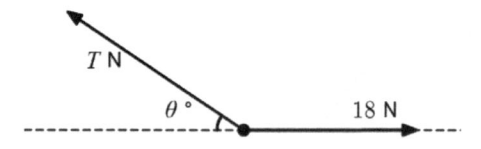

a) Find the value of θ
b) Find the value of T

The following questions use $g = 10ms^{-2}$

9) A box of mass 4 kg is suspended by a cable, which makes an angle $\theta°$ with a vertical wall. A horizontal rope keeps the box in equilibrium.

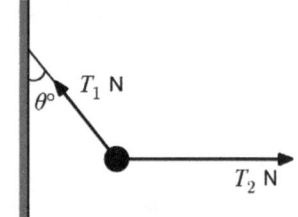

a) Calculate the tension in the cable giving your answer in terms of θ
b) Calculate the tension in the rope giving your answer in terms of θ
c) Find the angle at which the tension in the cable and the rope is the same.
d) Find the range of angles at which the tension in the rope is less than the weight of the box.

The following questions use exact values of g

10) A particle of mass m kg is held in equilibrium by two strings, each inclined at $\theta°$ to the horizontal. The tension in each string is T N.

a) Show that
$$T = \frac{mg}{2 \sin \theta°}$$

Find T when the particle has weight 5 N and
b) θ is 45°
c) θ is 60°
d) θ approaches 90°

11) Two ropes are hung from a ceiling, 2 metres apart and are connected at their other ends. The length of rope A and rope B are 1.4 metres and 1.6 metres respectively.

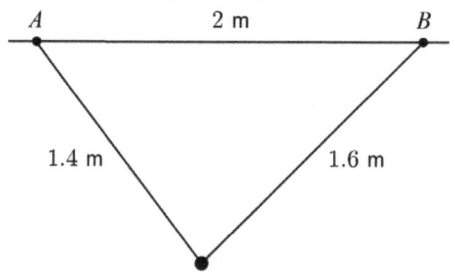

A crate of mass m kg is attached at the end of both rope A and rope B. The system is in equilibrium. Show that
a) the acute angle α between rope A and the ceiling is given by
$$\cos \alpha = \frac{17}{28}$$
b) the acute angle β between rope B and the ceiling is given by
$$\cos \beta = \frac{23}{32}$$

The tensions in rope A and rope B are given by T_A and T_B respectively.
c) Show that $\frac{T_B}{T_A} = \frac{136}{161}$

Given that $T_A = 322$ N,
d) find the exact value of m in terms of g

374

2.52 Forces: Equilibrium (answers on page 549)

The following questions use Triangles of Forces and g = 9.8 ms⁻²

12) An object of weight 50 N is suspended, in equilibrium, by two light inextensible strings A and B which make angles of 30° and 45° with the horizontal as shown in the diagram below:

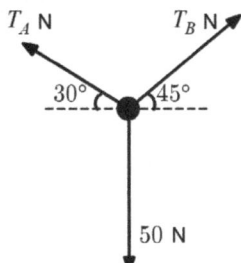

The tension in string A is given by T_A N and the tension in string B is given by T_B N.
a) Sketch the triangle of forces for these forces.
b) Hence determine the value of T_A and T_B

13) The diagram below shows a system of forces in equilibrium:

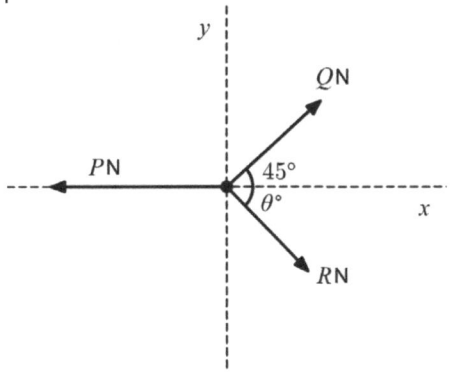

a) Sketch the triangle of forces for these forces.
b) Find the range of possible values of θ for the system to remain in equilibrium.

14) A particle of mass 2 kg is suspended by two light inextensible strings A and B which make an angle of 35° and 30° to the vertical respectively, as shown in the diagram below:

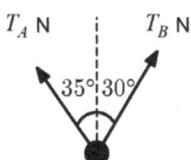

The tension in string A is given by T_A N and the tension in string B is given by T_B N.
a) Sketch the triangle of forces for the forces acting on the object.
b) Hence determine the value of T_A and T_B

15) An object of weight w N is attached to one end of a light inextensible string of length 8 cm. The other end is attached to a vertical wall at the point A. A horizontal force of magnitude 10 N is applied to the object. The particle is in equilibrium with the string taut, where it lies 5 cm from the wall.

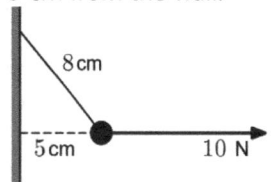

a) Sketch a triangle of forces for the forces acting on the object.

Hence find
b) the tension in the string.
c) the value of w
d) the mass of the object.

The following questions use exact values g

Prerequisite knowledge for Q16: Compound Angle Formulae

16) A small box B of mass m kg is held in equilibrium by two light strings, one attached to a vertical wall which has a tension of T_W N and one to the horizontal floor which has a tension of T_F N. They make an angle of $\alpha°$ with the wall and 30° with the floor respectively.

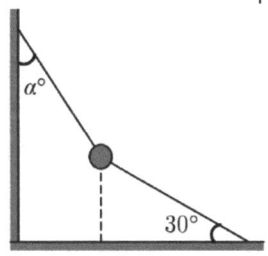

a) Sketch a triangle of forces for the forces acting on the box.

Hence or otherwise, given $\alpha = 30°$, show that
b) $T_W = \sqrt{3}mg$
c) $T_F = mg$

Angle α is increased and the angle of the string with the floor becomes 45°. The system remains in equilibrium.
d) Show that
$$T_W = \frac{mg}{\cos\alpha° - \sin\alpha°}$$

2.53 Forces: Coefficient of Friction (answers on page 550)

1) A particle of weight w N lies on a rough horizontal surface. A horizontal force 10 N is applied to the particle. The coefficient of friction between the particle and the surface is given by μ

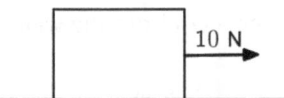

   The frictional force F N is the only resistance force, and the particle is at the point of moving. Find the value of F given
   a) $w = 50, \mu = 0.2$
   b) $w = 20, \mu = 0.45$
   c) $w = 15, \mu = 0.85$
   d) $w = 3, \mu = 1.1$

2) A crate of weight w N lies on a rough horizontal surface. A horizontal force P N is applied to the crate. The frictional force F N is the only resistance against the crate. The coefficient of friction between the crate and the surface is given by μ

   Given the crate is at the point of moving and
   a) $w = 12, \mu = 0.5$, find the value of P
   b) $w = 12, \mu = 0.9$, find the value of P
   c) $P = 20, \mu = 0.4$, find the value of w
   d) $P = 20, \mu = 0.8$, find the value of w
   e) $w = 50, P = 40$, find the value of μ
   f) $w = 300, P = 12$, find the value of μ
   g) $w = 12, P = 15$, find the value of μ

   The crate has objects added to it with weight M N as shown in the diagram below. The other forces on the crate remain the same.

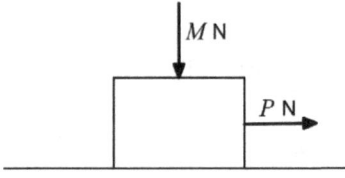

   Given the crate is at the point of moving and
   h) $w = 12, \mu = 0.7, M = 3$, find the value of P
   i) $P = 8, \mu = 0.8, M = 2$, find the value of w
   j) $w = 60, P = 10, M = 20$ find the value of μ

3) A box of weight w N lies on a rough horizontal surface. It is being pulled by a horizontal force T N at an angle $\theta°$ to the horizontal. The frictional force F N is the only resistance against the box. The coefficient of friction between the box and the surface is given by μ

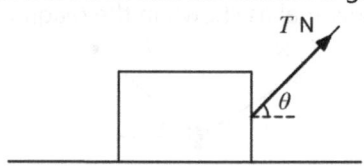

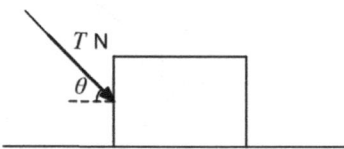

   Given the box is at the point of moving and
   a) $w = 20, T = 15, \theta = 30, \mu = 0.5$, find the value of F
   b) $w = 8, \theta = 45, \mu = 0.2$, find the value of T
   c) $T = 10, \theta = 60, \mu = 0.1$, find the value of w
   d) $w = 15, T = 4, \theta = 30$, find the value of μ

4) A box of weight w N lies on a rough horizontal surface. It is being pushed by a force T N at an angle $\theta°$ to the horizontal. The frictional force F N is the only resistance against the box. The coefficient of friction between the box and the surface is given by μ

   Given the box is at the point of moving and
   a) $w = 2, T = 4, \theta = 30, \mu = \frac{1}{2}\sqrt{3}$, find the value of F
   b) $w = 4, \theta = 45, \mu = 0.3$, find the value of T
   c) $T = 12, \theta = 60, \mu = 0.2$, find the value of w
   d) $w = 20, T = 5, \theta = 30$, find the value of μ

5) Will has drawn a correct force diagram for a particle in equilibrium, and some workings:

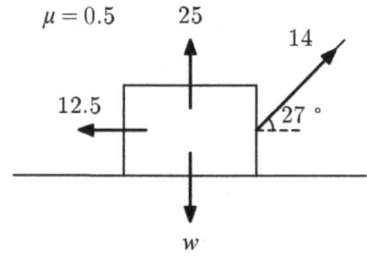

   $\mu = 0.5$ 25 14 12.5 27° w

   R(↑): $25 + 14 \sin 27° - w = 0$ ∴ mass is 31.4 kg

   Explain why his claim that the mass of the particle is 31.4 kg is incorrect.

2.53 Forces: Coefficient of Friction (answers on page 550)

The following questions use g = 9.8 ms⁻²

6) A box of mass m kg lies on a rough horizontal surface. It is being pulled by a force T N at an angle $\theta°$ to the horizontal. The frictional force F N is the only resistance against the box. The coefficient of friction between the box and the surface is given by μ

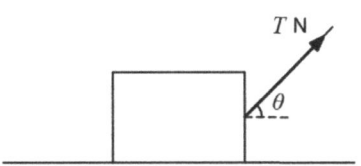

   Given the box is at the point of moving and
   a) $m = 5, \theta = 10, \mu = 0.5$, find the value of T
   b) $T = 20, \theta = 50, \mu = 0.1$, find the value of m
   c) $m = 2, T = 5, \theta = 70$, find the value of μ

7) The coefficient of friction between a particle of mass 2 kg, on a rough horizontal surface is 0.4 A horizontal force of P N acts on the particle.

   In each case, identify if the particle is moving under the given value of P and whether, when an additional mass of 100 g is added to the particle, it continues to move or remain stationary.
   a) $P = 8.4$
   b) $P = 8$
   c) $P = 7.6$

8) A crate of mass m kg lies on a rough floor. The coefficient of friction between the crate and the floor is 0.3. The crate is pushed by a horizontal force P N and it does not move.

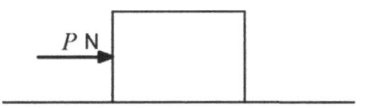

   For each of the following, find the greatest possible values of P
   a) $m = 5$ kg
   b) $m = 10$ kg
   c) $m = 20$ kg

The following questions use exact values of g

9) A box of mass m kg lies on a rough floor. The coefficient of friction between the crate and the floor is μ. The crate is pushed by a horizontal force P N and it does not move.

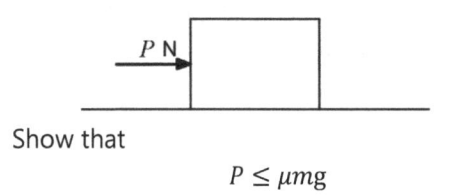

   Show that
   $$P \leq \mu mg$$

The following questions use g = 10 ms⁻²

10) A block of mass 15 kg lies on a rough horizontal surface. The block is being pulled by a force of 10 N at an angle $\theta°$ to the horizontal. The coefficient of friction between the block and the mass is 0.1

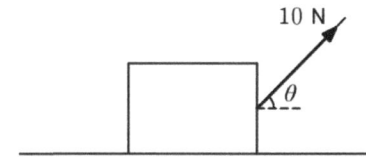

    Given the block is moving, show that
    $$10 \cos \theta + \sin \theta > 15$$

11) A block of mass 15 kg lies on a rough horizontal surface. The coefficient of friction between the block and the mass is 0.85

    The block is pushed by a horizontal force of 120 N.
    a) Does the block move, and if so, what is it acceleration?

    The pushing force is removed, and instead a pulling force of 120 N at angle of 30° to the horizontal is applied.

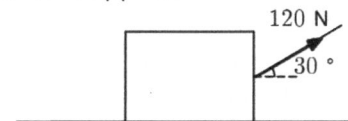

    b) Find the acceleration of the block.
    c) Find the maximum extra mass m kg which may be added to the block, for it still be moving.

2.53 Forces: Coefficient of Friction (answers on page 550)

The following questions use exact values of g

12) A box of mass m kg lies on a rough horizontal surface. It is being pulled by a horizontal force T N at an angle $\theta°$ to the horizontal. The coefficient of friction between the box and the surface is μ. The frictional force F N is the only resistance against the box.

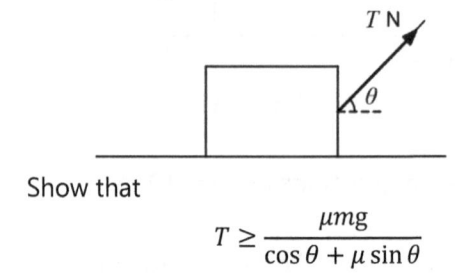

Show that
$$T \geq \frac{\mu mg}{\cos\theta + \mu\sin\theta}$$

13) For each of the following, find the exact values of $\sin\theta$ and $\cos\theta$:
   a) $\tan\theta = \frac{3}{4}$
   b) $\tan\theta = \frac{5}{12}$
   c) $\tan\theta = \frac{7}{16}$
   d) $\tan\theta = \frac{10}{21}$

The following questions use g = 9.8 ms⁻²

14) A box of mass m kg lies on a rough horizontal surface. It is being pulled by a force T N at an angle $\theta°$ to the horizontal where $\tan\theta = \frac{3}{4}$. The coefficient of friction between the box and the surface is μ. The frictional force F N is the only resistance against the box.

Given the box is moving and the acceleration of the box is a ms⁻², and
   a) $m = 10, T = 20, \mu = 0.1$, find the value of a
   b) $m = 3, T = 20, \mu = 0.2$, find the value of a
   c) $m = 5, T = 15, a = 1.5$, find the value of μ
   d) $T = 16, \mu = 0.25, a = 3$, find the value of m
   e) in general, show that
   $$a = \left(\frac{4}{5m} + \frac{3}{5m}\mu\right)T - 9.8\mu$$

15) A box of mass 4 kg lies on a rough horizontal surface. It is being pushed by a force 10 N at an angle $\theta°$ to the horizontal where $\tan\theta = \frac{5}{12}$. The coefficient of friction between the box and the surface is 0.2.

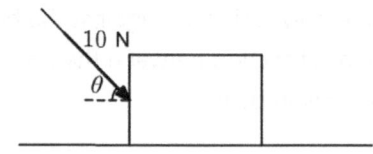

   a) Find the acceleration of the box.

The box has additional mass of m kg added to it, given
   b) $m = 0.1$ find the value of a
   c) $m = 0.5$ find the value of a

16) A trolley of mass 8 kg and is being pushed with a horizontal force of P N along a rough horizontal floor. It is at the point of moving. In doubling the horizontal pushing force, the trolley moves with acceleration 0.25 ms⁻². Find
   a) the value of P
   b) the coefficient of friction between the trolley and the floor.

The trolley is loaded with an extra m kg of products. The trolley is pushed at $2P$ N.
   c) Find the maximum value of m such that the trolley is moving or on the point of moving.

17) A sledge of mass 1.2 kg lies on a rough horizontal ground and is being pulled by a force 8 N at an angle 30° to the horizontal, having initially started at rest. The coefficient of friction between the sledge and the ground is 0.2

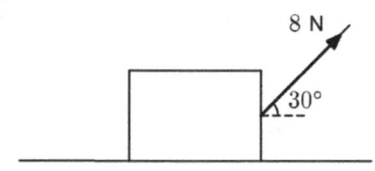

   a) Find the acceleration of the sledge.

After 5 seconds, the rope snaps. The sledge continues to move until it comes to a stop.
   b) Find the distance the sledge travels after the rope breaks.

2.53 Forces: Coefficient of Friction (answers on page 550)

The following questions use g = 9.81 ms⁻²

18) A crate of mass 5 kg lies on a rough horizontal surface. It is being pulled by a force 20 N at an angle 30° to the horizontal. The coefficient of friction between the crate and the surface is 0.4

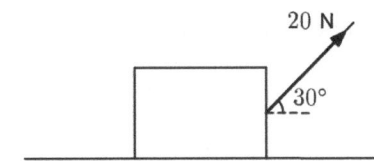

    a) Find the acceleration of the crate.

    The 20 N force needs to be replaced by a horizontal pushing force P N such that it moves at the same acceleration.
    b) Find the value of P

19) A car of mass 2000 kg is towing a trailer of mass 500 kg along a straight flat road. The driving force produced by the engine is 6000 N. The coefficient of friction between the road and the vehicles is 0.1

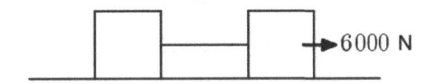

    a) Draw a force diagram.
    b) Find the acceleration of the car.
    c) Find the tension in the tow bar.
    d) State any modelling assumptions you have made.

20) A van of mass 3500 kg is towing a trailer of mass 750 kg along a rough straight flat road. The driving force produced by the engine is 10000 N. The tow bar is modelled as both light and inextensible.

    The van is accelerating at 1.5 ms⁻².
    The coefficient of friction between both the van and the road, and the trailer and the road is μ
    a) Find the value of μ.
    b) Find the tension in the tow bar.
    c) Explain how the modelling assumptions that the string is light and inextensible have been used.

The following questions use g = 9.8 ms⁻²

21) A toy train is made up of an engine A of mass 0.8 kg connected by a horizontal, light, inextensible string to truck B of mass 0.3 kg. The engine is being pulled by a light inextensible string with tension 8 N which is at an angle of 45° to the horizontal.

    The coefficient of friction between both the engine and the truck and the floor is 0.95
    a) Draw a force diagram.
    b) Find the friction force acting on the engine.
    c) Find the friction force acting on the truck.
    d) Find the acceleration of the engine.
    e) Find the tension in the string connecting the engine and the truck.
    f) State a modelling assumption you have used.

22) A particle A of mass 10 kg and a particle B of mass 5 kg are on a horizontal rough surface and are connected by a light, inextensible string which remains taut.
    Particle A is being pulled by a light, inextensible string, at an angle of $\theta°$ where $\tan\theta = \frac{5}{12}$. The tension in this string is 39 N.

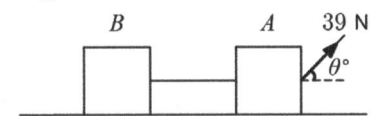

    The coefficient of friction between the particles and the surface is μ.
    The deceleration of the particles is 3.5 ms⁻².
    a) Draw a force diagram.
    b) Find the coefficient of friction between the particles and the surface.
    c) Find the tension in the string connecting the two particles.
    d) State a modelling assumption that you have used.

    It takes 10 seconds before coming to rest.
    e) Find the distance they have travelled.

2.53 Forces: Coefficient of Friction (answers on page 550)

The following questions use g = 10 ms⁻²

23) A particle A of mass 8 kg and a particle B of mass 3 kg are on a horizontal rough surface and are connected by a light, horizontal, inextensible string which remains taut. Particle A is being pulled by a light, inextensible string, at an angle of $\theta°$ where $\tan \theta = \frac{7}{24}$. The tension in this string is P N.

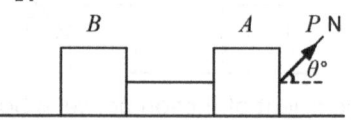

The coefficient of friction between the particles and the surface is μ
The system is initially at rest, then the acceleration of the particles is 2.5 ms⁻².
a) Draw a force diagram.

It is given that the friction acting against the motion of particle B is 6 N. Find
b) the value of μ
c) the tension in the string connecting particle A and particle B.
d) the value of P.
e) the friction acting against particle A.
f) How long it takes for the particles to travel 20 metres.

24) A van A of mass 4000 kg is towing a trailer B of mass 1000 kg along a straight flat road. The driving force produced by the engine is 12000 N.
The tow bar is angled upwards from B to A with angle 10° to the horizontal. The tow bar may be modelled as a rod.

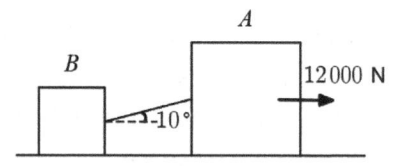

The coefficient of friction between the van and trailer and the surface is μ
The van is accelerating at 0.5 ms⁻².
a) Draw force diagram.
b) Find the value of μ
c) Find the tension in the tow bar.

25) A particle A of mass m kg and a particle B of mass $3m$ kg are on a horizontal rough surface and are connected by a light, horizontal, inextensible string which remains taut. Particle A is pulled forward by a force of $18m$ N. The tow bar is angled upwards from B to A with angle $\theta°$ to the horizontal. The tow bar may be modelled as a rod.

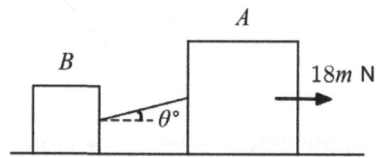

The coefficient of friction between the van and trailer and the surface is μ
The van is accelerating at 2 ms⁻².
a) Show that
$$\mu = 0.25$$
b) Show that
$$T = \frac{54m}{\sin \theta + 4 \cos \theta}$$

The following questions use exact values of g

26) A particle A of mass $4m$ kg and a particle B of mass m kg are on a horizontal rough surface and are connected by a light, inextensible string. Particle A is being pulled by a force of P N at an angle of $\theta°$ to the horizontal where $\tan \theta = \frac{3}{4}$. The coefficient of friction between the particles and the surface is μ.

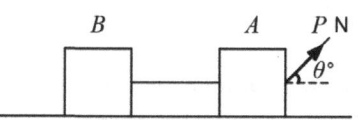

a) Draw a force diagram.

Given that $P = 25$ and the total resistance force is 10 N.
b) Show that
$$\mu = \frac{2}{mg - 3}$$
c) Show that the acceleration of the system is given by
$$a = \frac{2}{m} \text{ ms}^{-2}$$
d) Find the tension in the string between the two particles in terms of m

2.54 Forces: Connected Particles (answers on page 551)

The following questions use g = 9.8 ms^{-2}

1) Two particles, A and B of mass 5 kg and 3 kg respectively, are attached by a light inextensible string over a small smooth peg. Particle A rests on a rough table. Particle B hangs freely at the end of the string.

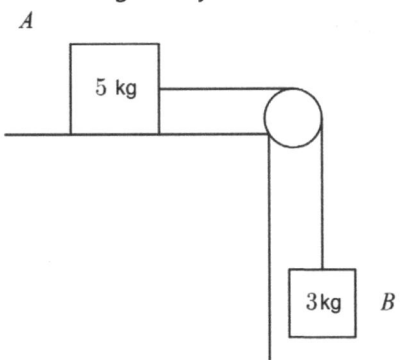

   The coefficient of friction between A and the table is 0.25
   a) Find the acceleration of the system.
   b) Find the tension in the string.

2) Two particles, A and B of mass 12 kg and 5 kg respectively, are attached by a light inextensible string over a small smooth peg. Particle A rests on a rough table. Particle B hangs freely at the end of the string.

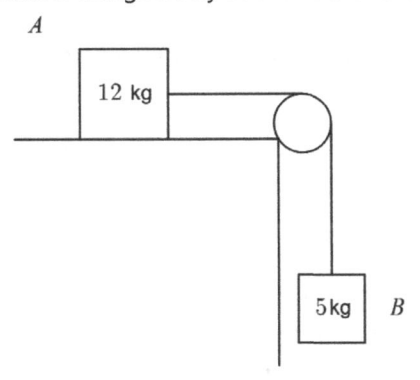

   The two particles are accelerating at 0.02 ms^{-2}. The system is released from rest.
   Find
   a) the tension in the string.
   b) the coefficient of friction between A and the table.

   Particle B is 0.5 m from the ground. Assuming that particle A does not reach the pulley,
   c) find the time it takes for B to reach the ground.
   d) state the distance A travels in this time.

3) Two particles, A and B of mass 4 kg and 5 kg respectively, are attached by a light inextensible string over a small smooth peg. Particle A rests on a rough table. Particle B hangs freely at the end of the string.

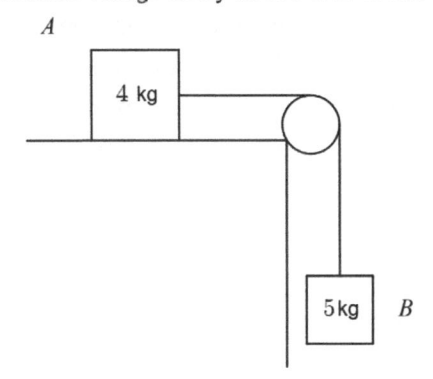

   The coefficient of friction between A and the table is 0.4
   a) Find the acceleration of the system.
   b) Find the tension in the string.

   The system is released from rest. It takes 1.05 seconds for Particle B to reach the ground and A does not reach the pulley.
   c) Determine the initial height of particle B above the ground.

The following questions use exact values of g

4) Two particles, A and B of mass 2 kg and m kg respectively, are attached by a light inextensible string over a small smooth peg. Particle A rests on a rough table. Particle B hangs freely at the end of the string.

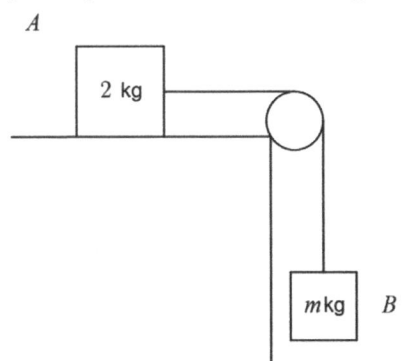

   The coefficient of friction between A and the table is 0.3
   a) Find the acceleration of the system in terms of m and g
   b) Find the minimum value of m such that Particle B is moving.

2.54 Forces: Connected Particles (answers on page 551)

The following questions use g = 9.8 ms⁻²

5) Two particles A and B of mass $3m$ kg and m kg respectively are attached by a light inextensible string over a small smooth peg. Particle A rests on a rough table. The coefficient of friction between A and the table is μ. Particle B hangs freely with the string taut. Particle A is pulled away from the pulley with horizontal force P N with acceleration a ms⁻².

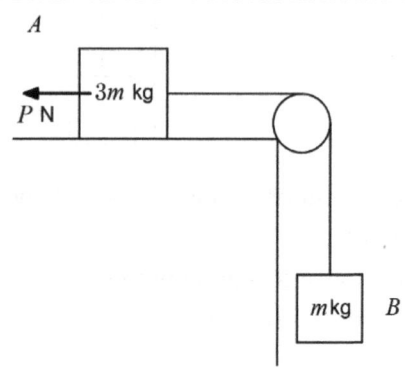

   Show that
   $$P = (29.4\mu + 4a + 9.8)m$$

6) Two particles, A and B of mass 2 kg and 0.5 kg respectively, are attached by a light inextensible string over a small smooth peg. Particle A rests on a rough table. Particle B hangs freely at the end of the string. Particle A is being pulled by a force of 10 N at an angle 30° to the horizontal away from the pulley.

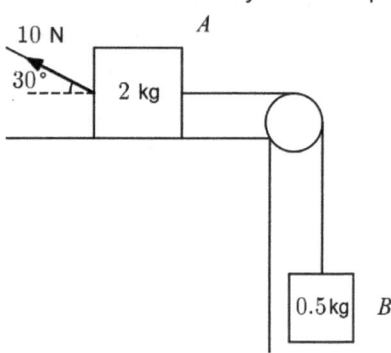

   The coefficient of friction between A and the table is 0.2. Find
   a) the frictional force acting on particle A
   b) the acceleration of the system.
   c) the tension in the string.

   Particle B is swapped for particle C of mass m kg.
   d) Find the maximum value of m such that C accelerate upwards.

7) Three boxes, A, B and C have masses 2 kg, 8 kg and 0.5 kg respectively. Box A is hanging freely from a string, which passes over a small, fixed, smooth pulley and connects to box B. Box B is on a rough horizontal table. Box C is hanging freely from a string, which passes over a small, fixed, smooth pulley and connects to the other side of box B.

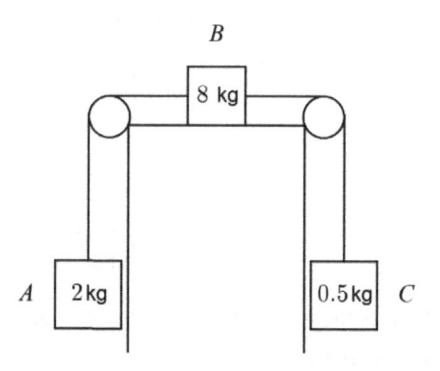

   The coefficient of friction between box B and the table is 0.1. The system is moving.
   a) Find the tension in the strings.
   b) Find the acceleration of the system and describe the motion of the boxes.
   c) State the modelling assumptions you have used.

The following questions use exact values of g

8) Two particles A and B of mass M kg and m kg respectively are attached by a light inextensible string over a small smooth peg. Particle A rests on a rough table. The coefficient of friction between A and the table is μ. Particle B hangs freely with the string taut.

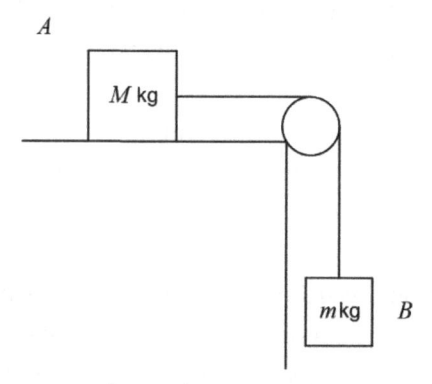

   The system is moving. Show that
   $$m \geq \mu M$$

2.55 Forces: Inclined Planes (answers on page 551)

The following question use $g = 9.8$ ms^{-2}

1) A box of mass 5 kg rests on a rough slope which is inclined with an angle of $\theta°$.

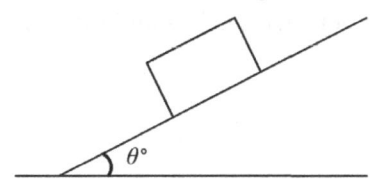

   The box is at the point of slipping.
   a) Draw a force diagram.

   Find the coefficient of friction between the box and the slope in each case when
   b) $\theta = 30$
   c) $\theta = 45$
   d) $\theta = 55$

   The box is replaced with a different box of mass m kg.
   e) Show that
   $$\mu = \tan\theta$$

2) A box of mass m kg rests on a rough slope which is inclined with an angle of $\theta°$. A pulling force T N acts upwards and parallel to the slope. A frictional force F N acts parallel to the slope.

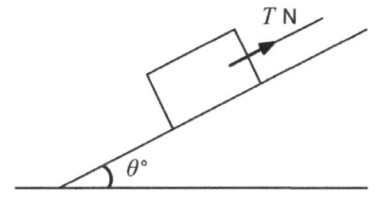

   The box is at the point of slipping up the slope.
   a) Draw a force diagram.

   Given
   b) $m = 4, \mu = 0.8, \theta = 30$, find the value of F and the value of T
   c) $m = 5, \mu = 0.6, \theta = 60$, find the value of F and the value of T
   d) $T = 10, \theta = 20, F = 5$, find the value of m
   e) $T = 12, m = 2, F = 8$, find the value of θ
   f) $T = 10, \mu = 0.6, \theta = 45$, find the value of m

   Now using exact values of g
   g) show that
   $$T = mg(\mu\cos\theta + \sin\theta)$$

3) A particle has mass 20 kg and is held at rest on a rough slope which is inclined with an angle of $\theta°$ where $\tan\theta = \frac{3}{4}$

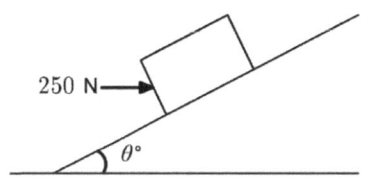

   It is being held at rest by a horizontal pushing force 250 N.
   a) Draw a force diagram.
   b) Show that
   $$\sin\theta = \frac{3}{5}$$
   and
   $$\cos\theta = \frac{4}{5}$$
   c) Find the normal reaction force on the box.
   d) Find the frictional force acting on the box.
   e) Find the coefficient of friction acting between the box and the slope.

4) A box of mass 100 N rests on a rough slope which is inclined with an angle of $\theta°$ where $\tan\theta = \frac{7}{24}$

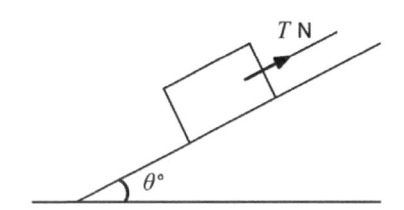

   A pulling force T N acts upwards and parallel to the slope. The box is at the point of slipping down the slope.
   a) Draw a force diagram.
   b) Show that
   $$\sin\theta = \frac{7}{25}$$
   and
   $$\cos\theta = \frac{24}{25}$$

   The coefficient of friction between the box and the slope is 0.25
   c) Find the value of T

2.55 Forces: Inclined Planes (answers on page 551)

The following question use g = 10 ms⁻²

5) A box of mass 5 kg rests on a rough slope which is inclined with an angle of $\theta°$ where $\tan\theta = \frac{2}{5}$

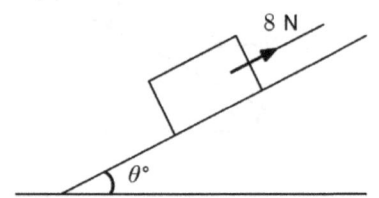

A pulling force 8 N acts upwards and parallel to the slope. The box is at the point of slipping down the slope.
a) Draw a force diagram.
b) Show that
$$\sin\theta = \frac{2}{\sqrt{29}}$$
and
$$\cos\theta = \frac{5}{\sqrt{29}}$$
c) Find the value of the coefficient of friction between the box and the slope.

6) A particle has mass m kg and is held at rest on a rough slope which is inclined with an angle of $\theta°$.

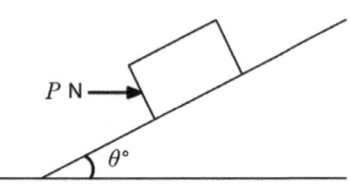

It is being held at rest by a horizontal pushing force P N.
a) Draw a force diagram.

Given that $m = 2, \mu = 0.4, \theta = 30$, find
b) the normal reaction force on the box in terms of P
c) the frictional force in terms of P
d) the value of P

7) A sledge is released from rest on a smooth slope which is at angle 10° to the horizontal. The mass of the sledge is m kg.
a) Draw a force diagram.
b) Find the acceleration of the sledge.
c) Find the speed once it has travelled 5 metres.

8) A box of mass m kg is held at rest before being released. It lies on a rough slope which is inclined with an angle of $\theta°$. A frictional force F N acts parallel to the slope. The coefficient of friction between the box and the slope is μ

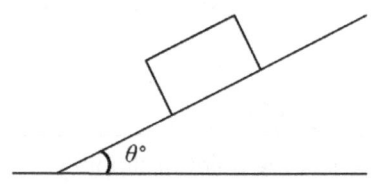

The acceleration of the box is given by a ms⁻².
a) Draw a force diagram.

Given
b) $\theta = 60, m = 4, \mu = 0.2$, find the value of a and state its direction.
c) $\theta = 45, m = 12, a = 1.5$, find the value of μ

Now, using exact values of g,
d) show that the acceleration down the slope is given by
$$a = g(\sin\theta - \mu\cos\theta)$$
e) State the acceleration down the slope if it were modelled as smooth.

9) A box has weight 39 N, initially at rest and on a rough slope which is inclined with an angle of $\theta°$ where $\tan\theta = \frac{5}{12}$. It is being pulled up the slope by a rope which may be modelled as a light, inextensible string which is parallel to the slope with the tension 50 N.

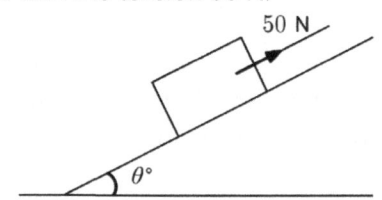

The coefficient of friction between the box and the slope is given by 0.5. The acceleration of the system is a ms⁻².
a) Draw a force diagram.
b) Show that
$$a = \frac{170}{39}$$

The box is pulled up the slope for 1.5 seconds.
c) Find the distance travelled by the box in this time.
d) Find the speed it is moving at 1.5 seconds.

2.55 Forces: Inclined Planes (answers on page 551)

The following questions use g = 9.8 ms⁻²

10) A particle has mass m kg and is held at rest on a rough slope which is inclined with an angle of $\theta°$ where $\tan\theta = \frac{3}{7}$. It is being pushed up the slope by a horizontal force of 20 N which is horizontal.

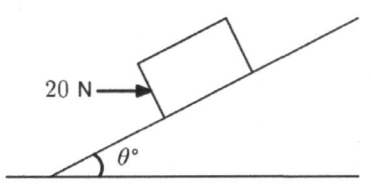

The coefficient of friction between the box and the slope is given by μ. The acceleration of the system is a ms⁻².
   a) Draw a force diagram.
   b) Find the exact values of $\cos\theta$ and $\sin\theta$

It is given that $m = 10$ and $\mu = 0.25$
   c) Find the value of a
   d) Describe the type and direction of the motion.

11) A particle has mass m kg is on a rough slope which is inclined with an angle of $\theta°$ where $\tan\theta = \frac{7}{24}$.
It is being pulled up the slope by a light, inextensible string which is at angle 45° to the the slope with the tension 12 N.

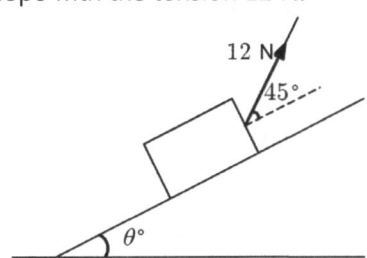

The coefficient of friction between the box and the slope is given by μ. The acceleration of the particle up the slope is 1.2 ms⁻².
   a) Draw a force diagram.
   b) Find the exact values of $\cos\theta$ and $\sin\theta$

Given
   c) $m = 2$, find the value of μ
   d) $\mu = 0.1$, find the value of m

12) A crate has mass 6 kg on a rough slope which is inclined with an angle of 40°. It is being pulled up the slope by a rope with the tension 25 N which is parallel to the slope. The rope may be modelled as a light, inextensible string.

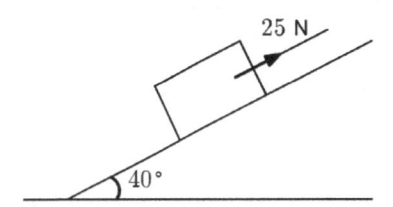

The coefficient of friction between the crate and the slope is 0.45.
   a) Find the acceleration of the crate.
   b) Describe the motion of the crate.

After coming to a stop, the rope is removed, and the crate slides down the slope.
   c) Find the value of the acceleration down the slope.

13) A crate has mass 4 kg, initially at rest and on a rough slope which is inclined with an angle of 24°. It is being pulled up the slope by a rope with the tension 35 N which is parallel to the slope. The rope may be modelled as a light, inextensible string.

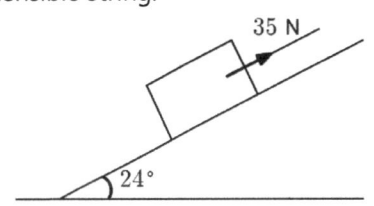

The acceleration of the system is 0.5 ms⁻².
   a) Show that the coefficient of friction between the box and the crate is approximately 0.48
   b) Find how long it takes for the crate to move 2 metres up the slope.

Upon travelling 2 metres, the end of the rope which is not attached to the crate is instantaneously adjusted upwards so that it creates an angle of 20° with the slope but remains at 35 N.
   c) Find how long it takes for the crate to move a further 2 metres up the slope.

2.55 Forces: Inclined Planes (answers on page 551)

14) A toy car is placed at the end of a straight flat track of length 5 metres, held at an incline of $\theta°$ where $\tan\theta = \frac{3}{4}$. The toy car is released from rest and takes 2 seconds to reach the bottom of the track. The coefficient of friction between the car and the track is μ
   a) Find the acceleration of the toy car
   b) Find the value of μ

15) A particle of mass m kg is projected up a rough slope which has an incline of $\theta°$ where $\tan\theta = \frac{12}{5}$
   The coefficient of friction between the particle and the slope is 0.8. The particle initially has speed 20 ms^{-1}.
   Find the time it takes for the particle to return to its original position.

16) A particle of mass m kg is projected up a rough slope which has an incline of $\theta°$ where $\tan\theta = \frac{3}{4}$.
   It starts with speed 15 ms^{-1} and takes 3 s to return to its start position.
   Find the coefficient of friction between the particle and the slope.

17) A block has weight 20 N and is on a rough slope which is inclined with an angle of 30°. It is being pushed up the slope by a horizontal force of P N.

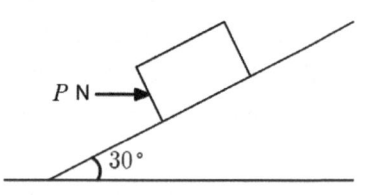

   The normal reaction on the block by the slope is $15\sqrt{3}$ N. The block is decelerating up the slope at 2.45 ms^{-2}.
   a) Find the friction force acting against the particle.

   The horizontal force is removed.
   b) Find the acceleration of the block.
   c) Suggest improvements to the model in order to make the model more realistic.

18) A particle has weight 49 N is on a rough slope which is inclined with an angle of $\theta°$ where $\tan\theta = \frac{5}{12}$

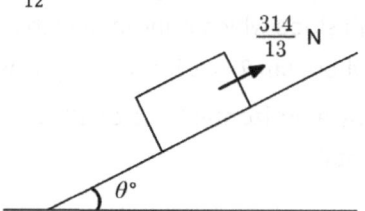

   The particle is pulled up the slope by a force parallel to the slope of $\frac{314}{13}$ N. The particle has deceleration 1.2 ms^{-2}.
   a) Find the coefficient of friction between the slope and the particle.
   b) Given it takes 4 seconds to come to a stop, find the distance it travels up the slope.

   The pulling force is then removed.
   c) Find how long it takes for the particle to return to its original position from the point at which it came to a stop.

The following questions use exact values of g

19) A particle has mass m kg and is on a rough slope which is inclined with an angle of $\theta°$ where $\tan\theta = \frac{4}{3}$. It is being pushed up the slope by a horizontal force of P N.

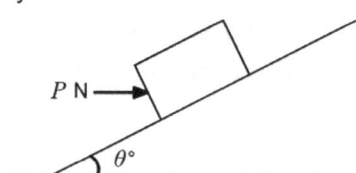

   The coefficient of friction between the particle and the slope is 0.6
   a) Find the values of P such that the particle is moving up the slope, giving your answer in terms of m and g

   The particle is placed on a rough slope which is inclined at an angle of $\theta°$ where $\tan\theta = \frac{4}{3}$. with coefficient of friction μ
   It is being pushed up the slope by a horizontal force of Q N.
   b) Find the values of μ in terms of Q, m and g such that the particle will not accelerate up the slope.

2.55 Forces: Inclined Planes (answers on page 551)

20) Particle A of weight W N is on a rough slope which is inclined with an angle of $\theta°$ where $\tan\theta = \frac{2}{3}$. The coefficient of friction between the particle and the slope is μ. The block is at the point of sliding down the slope.
   a) Show that $\mu = \frac{2}{3}$

   Particle A is replaced with particle B which has double the mass of A. The coefficient of friction remains the same.
   b) Describe the motion of particle B.

The following questions use g = 10 ms⁻²

21) A car of mass 2200 kg is towing a trailer of mass 400 kg along a straight flat road which is inclined at 10° to the horizontal. The driving force produced by the engine of the car is 6000 N.

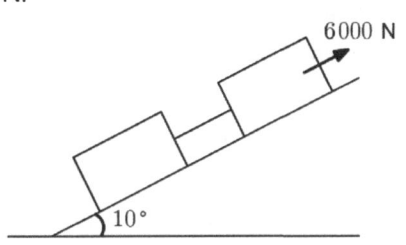

   The coefficient of friction between the road and the vehicles is 0.1.
   a) Find the deceleration of the car.
   b) Find the tension in the tow bar.

22) A van of mass 2400 kg is towing a trailer of mass 600 kg along a straight flat road which is inclined at 15° to the horizontal. The van is driving at 8 ms⁻¹ and after 5 seconds has reached 12 ms⁻¹ due to a driving force of D N.

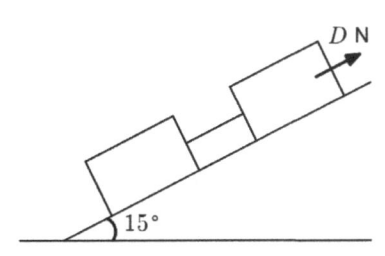

   The coefficient of friction between the road and the vehicles is 0.15.
   a) Find the value of D
   b) Find the tension in the tow bar.

The following questions use exact values of g

23) A block A has mass m kg and is on a rough slope which is inclined with an angle of 45°. It is being pushed up the slope by a force of P N which is parallel to the slope.
   The coefficient of friction between the block and the slope is μ. The block is at the point of sliding.
   Show that
   $$\mu = \frac{\sqrt{2}P}{mg} - 1$$

24) A particle of mass m kg lies on a rough slope which is inclined with an angle of $\theta°$.
   It is being pulled upwards by a force of km N which is parallel to the slope, where k is a positive constant.
   The coefficient of friction between the particle and the slope is μ
   Show that
   $$a = k - \mu g\cos\theta - g\sin\theta$$

25) A particle has weight w N and is on a rough slope which is inclined with an angle of $\theta°$ where $\tan\theta = \frac{7}{24}$. It is being pushed up the slope by a horizontal force of P N.
   The coefficient of friction between the box and the slope is given by μ
   The particle is at the point of sliding down the slope.

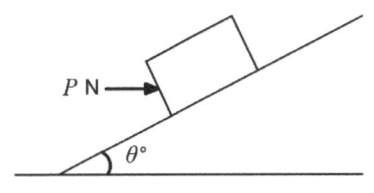

   a) Draw a force diagram.
   b) Show that
   $$\mu = \frac{7w - 24P}{24w + 7P}$$

   The force P is removed and consequently the block moves down the slope with acceleration $\frac{1}{5}g$ ms⁻².
   c) Show that
   $$\mu = \frac{1}{12}$$

2.56 Forces: Connected Particles on Inclined Planes (answers on page 553)

The following question use g = 9.8 ms⁻²

1) A particle A has mass M kg and is on a smooth slope which is inclined with an angle of $\theta°$. It is being pulled up the slope by a light, inextensible string which is parallel to the slope and passes over a small, smooth, fixed pulley. A second particle B of mass m kg hangs from the pulley freely with the string taut. The system is released from rest.

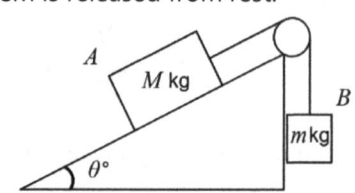

Find the tension in the string and the acceleration in the system, and state the direction that particle A is moving, given:
   a) $M = 1.5, m = 3.9, \theta = 30$
   b) $M = 2.1, m = 0.5, \theta = 60$
   c) $M = 2, m = 2, \theta = 40$
   d) $M = 40, m = 20, \theta = 30$

2) A particle A has mass M kg and is on a rough slope which is inclined with an angle of $\theta°$. It is being pulled up the slope by a light, inextensible string which is parallel to the slope and passes over a small, smooth, fixed pulley. A second particle B of mass m kg hangs from the pulley freely with the string taut. The coefficient of friction between the slope and the particle is μ. The tension in the string is T N.

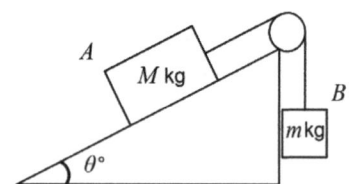

Given that the system is released from rest:
   a) $M = 2, m = 5, \theta = 30$ and $\mu = 0.85$ find T and the acceleration (and direction) of A
   b) $M = 7, \theta = 40, \mu = 0.7, a = 0.4$ (up the slope), find m and T
   c) $M = 7, m = 3, \theta = 60, a = 0.1$ (down the slope), find T and μ
   d) $M = 2k, m = k$, where k is a positive constant, $\theta = 30, \mu = 0.5$, find a and give the direction of motion.

3) A box A has mass 5 kg and is on a rough slope which is inclined with an angle of $\theta°$ where $\tan \theta = \frac{4}{3}$. A light, inextensible string connects box A to box B which has mass 3 kg. The string is parallel to the slope and passes over a small, smooth, fixed pulley. Box B hangs from the pulley freely with the string taut.

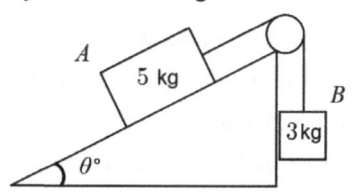

The system is released from rest and the box moves down the slope at 0.8 ms⁻².
   a) Find the coefficient of friction between the slope and the box.
   b) Find the distance travelled by the system in the first 5 seconds, stating any assumptions you have made.
   c) Explain what it means for the string to be modelled as light and inextensible.

4) A box A has mass m kg and is on a rough slope which is inclined with an angle of $\theta°$ where $\tan \theta = \frac{3}{4}$. It is being pulled up the slope by a light, inextensible string which is parallel to the slope and passes over a small, smooth, fixed pulley. A second box B of mass $2m$ kg hangs from the pulley freely with the string taut.

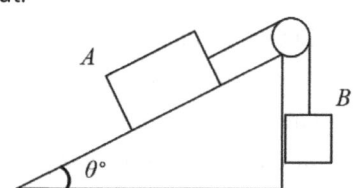

The coefficient of friction between the slope and the box is 0.4. The system is released from rest.
   a) Find the tension in the string in terms of m

Given that box B starts 1.5 m above the ground and assuming box A does not reach the pulley. Find
   b) the time it takes for box B to reach the ground.
   c) the total time box A spends moving from its initial position to when it stops.

2.56 Forces: Connected Particles on Inclined Planes (answers on page 553)

The following questions use g = 10 ms⁻²

5) A block A has mass M kg and is held at rest on a horizontal rough plane. A second block B of mass m kg lies on a smooth inclined plane with an angle of 30°. Blocks A and B are connected by a light, inextensible string which passes over a small, smooth, fixed pulley and is parallel to the surface. The coefficient of friction between block A and the plane is μ. The tension in the string is given by T N.

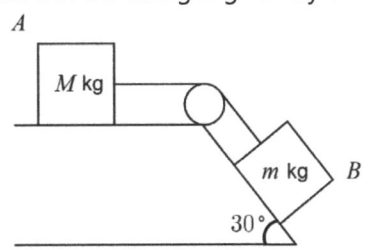

The system is released from rest. The acceleration of the system is a ms⁻². Given

a) $M = 10, m = 3, \mu = 0.2$, find the value of a and T
b) $M = 5, m = 4, a = 1.2$, find the value of T and μ
c) $M = 0.5, m = 2, T = 3$, find the value of a and μ
d) $a = 2, \mu = 0.4, T = 12$ find the value of m and M

6) A block A has mass 10 kg and lies on a horizontal rough plane. A second block B of mass 8 kg lies on a rough inclined plane with an angle of $\theta°$ where $\tan\theta = \frac{4}{3}$. Blocks A and B are connected by a light, inextensible string which passes over a small, smooth, fixed pulley. The coefficient of friction between block A and the plane is 0.2. The coefficient of friction between block B and the plane is 0.4

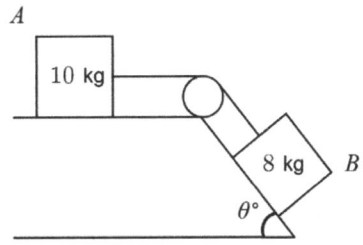

The system is released from rest and A travels towards the pulley. Find
a) the acceleration of the system.
b) the tension in each of the strings.

7) A block A has mass 8 kg and is on a rough slope which is inclined with an angle of 30°. A second block B has mass 3 kg and is on a smooth slope which is inclined with an angle of 45°. Blocks A and B are connected by a light, inextensible string which passes over a small, smooth, fixed pulley.

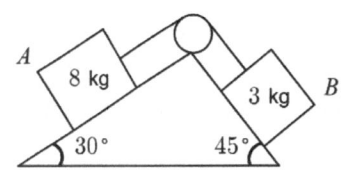

The coefficient of friction between block A and the plane is 0.05. The system is released from rest.
a) Find the acceleration of the system.
b) State the direction in which the blocks are moving.
c) Find the tension in the string.

Block B is replaced by block C with mass m kg and when released it remains in equilibrium.
d) Find the value of m

8) A block A has mass 2 kg and is on a rough slope which is inclined with an angle of $\theta°$ where $\tan\theta = \frac{3}{4}$. A second block B has mass 0.5 kg and is on a rough slope which is inclined with an angle of $\alpha°$ where $\tan\alpha = \frac{4}{3}$
Blocks A and B are connected by a light, inextensible string which passes over a small, smooth, fixed pulley. The rope is parallel to the slopes as shown in the diagram below.

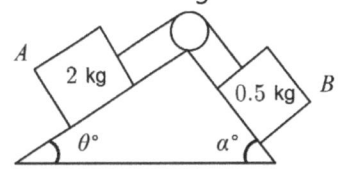

The coefficient of friction between block A and the plane on which it lies is twice the size of the coefficient of friction between block B and the plane on which it lies.
The system is released from rest and A travels with acceleration 0.1 ms⁻² away from the pulley. Find
a) the coefficient of friction between block A and the plane.
b) the tension in the string.

2.56 Forces: Connected Particles on Inclined Planes (answers on page 553)

The following questions use $g = 9.8$ ms^{-2}

9) A box has mass 10 kg is being pulled down a rough slope which is inclined with an angle of $\theta°$ where $\tan \theta = \frac{5}{12}$ by a light, inextensible, horizontal string which has tension 30 N. The particle is also attached to another light, inextensible string which passes over a small, smooth, fixed pulley to a second box of mass 2 kg which hangs from the pulley freely with the string taut.

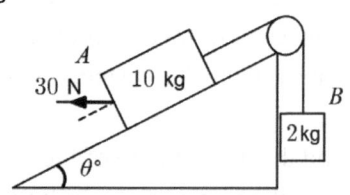

The acceleration of the system is 0.4 ms^{-2}.
   a) Find the tension in the string connecting the two boxes.
   b) Show that the value of the coefficient of friction is approximately 0.519
   c) State any modelling assumptions you have used.

10) Box A with mass 10 kg is on a rough plane inclined with an angle of 30° and is connected by a light inextensible string which passes over a small, smooth, fixed pulley to box B. Box B with mass 5 kg is on a rough horizontal plane slope and is connected by a light inextensible string which passes over a small, smooth, fixed pulley to box C. Box C has mass 2 kg lies on a rough plane which is inclined with an angle of 45° to the horizontal.

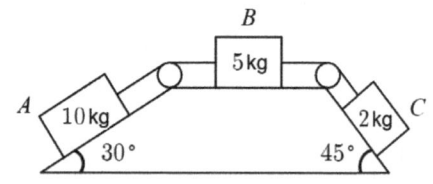

The coefficient of friction between each of the boxes and the planes are $\mu_A = 0.1$, $\mu_B = 0.4$ and $\mu_C = 0.2$. The system is released from rest and box A slides down the plane. Find
   a) the acceleration of the system
   b) the tension in the string between box A and B
   c) the tension in the string between box B and C

The following questions use $g = 10$ ms^{-2}

11) A box A has mass 2 kg and is on a rough slope which is inclined with an angle of $\theta°$ where $\tan \theta = \frac{3}{4}$. The box is being pulled by a horizontal force of 20 N. Box A is connected to a light, inextensible string which is parallel to the slope and passes over a small, smooth, fixed pulley to a light container. The container has two boxes stacked inside, with box B sitting directly on top of box C. Box B has mass 1 kg and box C has mass 3 kg.

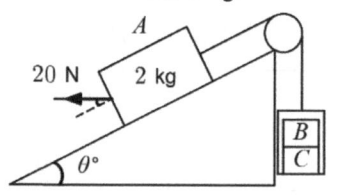

The coefficient of friction between the slope and the box is 0.5
   a) Find the acceleration of the system.
   b) Find the tension in the string.
   c) Find the force exerted on C by B
   d) Explain how the system changes if the container is no longer modelled as light.

The following questions use exact values of g

12) A crate A has mass km kg where k is a positive constant. It lies on a rough slope which is inclined with an angle of $\theta°$ where $\tan \theta = \frac{5}{12}$. It is connected by a rope which is parallel to the slope and passes over a small, smooth, fixed pulley. A second box B of mass m kg hangs with the string taut from the same rope.

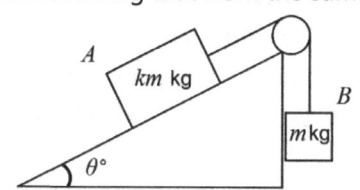

The coefficient of friction between the slope and the box is $\frac{1}{12}$. The system is released from rest and crate A moves down the slope.
   a) Show that
   $$a = \left(\frac{4k - 13}{13k + 13}\right)g$$
   b) State any modelling assumptions you have made.

2.56 Forces: Connected Particles on Inclined Planes (answers on page 553)

13) A particle A has mass km kg and is on a rough slope which is inclined with an angle of $\theta°$ where $\tan\theta = \frac{7}{24}$. It is being pulled up the slope by a light, inextensible string which passes over a small, smooth, fixed pulley. A second box B of mass m kg hangs from the pulley freely with the string taut. The coefficient of friction between the slope and the box is μ

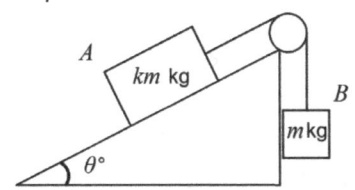

The system is released from rest and particle A moves 0.5 metres up the slope in 2 seconds.
a) Find the acceleration of particle A.
b) Show that the tension in the string is
$$\left(g - \frac{1}{4}\right)m$$

Given that $k = 0.05$
c) Find the value of μ in terms of g
d) State any modelling assumptions you have made.

14) A block A has mass M kg and is held at rest on a horizontal rough plane. A second block B of mass m kg lies on a smooth inclined plane with an angle of $\theta°$ where $\tan\theta = \frac{3}{4}$. Blocks A and B are connected by a light, inextensible string which passes over a small, smooth, fixed pulley and is parallel to the surface. The coefficient of friction between block A and the plane is μ

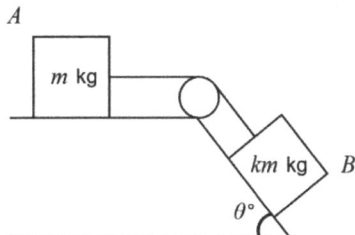

The system is released from rest. The acceleration of the system is a ms^{-2}.
a) Show that
$$a = \frac{(3k - 5\mu)g}{5k + 5}$$
b) Find the range of values of k for which the system will be moving.

15) A block A has mass km kg and is on a rough slope which is inclined with an angle of $\theta°$. A second block B has mass m kg and is on a rough slope which is inclined with an angle of $\alpha°$. Blocks A and B are connected by a light, inextensible string which passes over a small, smooth, fixed pulley.

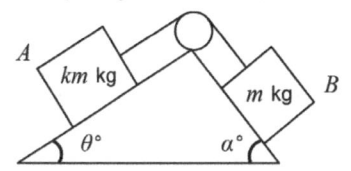

The coefficient of friction between block B and the plane on which it lies is μ. The coefficient of friction between block A and the plane on which it lies is 3 times larger.
The system is released and is at the point of block B sliding down the slope.
a) Show that
$$k = \frac{\sin\alpha - \mu\cos\alpha}{\sin\theta + 3\mu\cos\theta}$$

Block A is 10% heavier than block B.
b) In the case that $\theta = 30$ and $\alpha = 40$, find the value of μ giving your answer to 3 significant figures.

16) A box A has mass $4m$ kg and is being pulled along a rough slope which is inclined with an angle of 30° by a force of P N which is horizontal, where $P > 0$
The particle is attached to a light, inextensible string which passes over a small, smooth, fixed pulley to a second box of mass m kg which hangs from the pulley freely with the string taut.

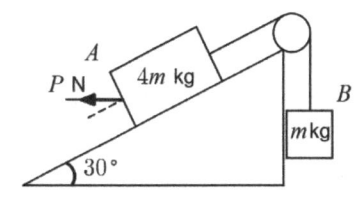

The coefficient of friction between box A and the slope is 0.5. The tension in the string which passes over the pulley is T N.
Find the values of P such that the B is moving upwards.

2.57 Constant Acceleration: Vectors (answers on page 554)

1) The points A and B have position vectors of $\begin{pmatrix}-2\\4\end{pmatrix}$ metres and $\begin{pmatrix}8\\11\end{pmatrix}$ metres respectively.
   Find the distance between A and B.

2) Given that a particle travels at the constant velocity given, find the speed of the particle:
   a) $(-2\mathbf{i} + 4\mathbf{j})$ ms^{-1}
   b) $\begin{pmatrix}5\\8\end{pmatrix}$ ms^{-1}
   c) $(2.5k\mathbf{i} - 6k\mathbf{j})$ ms^{-1}, where $k > 0$

3) A particle travels with a constant acceleration of $(-2\mathbf{i} + 3\mathbf{j})$ ms^{-2} from point A to point B. Its total displacement is $(-400\mathbf{i} + 250\mathbf{j})$ m. Its final velocity is $(12\mathbf{i} + 10\mathbf{j})$ ms^{-1}. Find
   a) the magnitude of the acceleration of the particle.
   b) the distance from A to B.
   c) the speed of the particle when it passes through B.

4) A particle moves with constant acceleration $(4\mathbf{i} + 7\mathbf{j})$ ms^{-2}, and initially has velocity $(\mathbf{i} + 2\mathbf{j})$ ms^{-1}.
   Find the velocity of the particle at $t = 3$ seconds.

5) A particle moves with constant acceleration $(-\mathbf{i} + 3\mathbf{j})$ ms^{-2}. It has velocity $(-2\mathbf{i} + 5\mathbf{j})$ ms^{-1} at $t = 7$ seconds. Find
   a) the initial velocity of the particle.
   b) the initial speed of the particle.

6) An object is moving with constant acceleration. Initially it has velocity $\begin{pmatrix}4\\3\end{pmatrix}$ ms^{-1} and 7 seconds later it is has velocity $\begin{pmatrix}8\\18\end{pmatrix}$ ms^{-1}.
   Find the magnitude of its acceleration.

7) Find the velocity vector of a particle in each case, given the speed v and direction in which the particle is travelling, leaving your answers in exact form:
   a) $v = 10$ ms^{-1} parallel to $\begin{pmatrix}1\\0\end{pmatrix}$
   b) $v = 4$ ms^{-1} parallel to $\begin{pmatrix}1\\1\end{pmatrix}$
   c) $v = 0.3$ ms^{-1} parallel to $\begin{pmatrix}2\\-3\end{pmatrix}$

8) The position vector, $\mathbf{r}$ metres, of an object at time t seconds is given by
   $$\mathbf{r} = t^2\mathbf{i} + (3 - 2t)\mathbf{j}$$
   Find its displacement between $t = 2$ and $t = 5$

9) An object is initially at the origin moving with velocity $(2\mathbf{i} + 5\mathbf{j})$ ms^{-1}. It has a constant acceleration of $(6\mathbf{i} - \mathbf{j})$ ms^{-2}. Find
   a) the position vector of the object at time t
   b) the distance it is from the origin after 10 seconds.

10) A particle P moves with constant acceleration $(5\mathbf{i} + 4\mathbf{j})$ ms^{-2}, having been initially at rest at initial position $(2\mathbf{i} - 7\mathbf{j})$ metres.
    a) Show that the expression for the position vector of P at time t seconds is given by
    $$\mathbf{r} = \left(2 + \frac{5}{2}t^2\right)\mathbf{i} + (2t^2 - 7)\mathbf{j}$$
    b) Hence, find the distance the particle has travelled after 2 seconds, giving your answer to 3 significant figures.

11) Initially a particle position is $(6\mathbf{i} + 2\mathbf{j})$ m relative to a fixed origin O and is moving with velocity $(\mathbf{i} - 3\mathbf{j})$ ms^{-1} and constant acceleration $(8\mathbf{i} - 5\mathbf{j})$ ms^{-2}.
    Find the position vector of the particle at $t = 4$ seconds.

12) A particle moves with constant acceleration $\begin{pmatrix}2\\-1\end{pmatrix}$ ms^{-2} and passes through the point A with velocity $\begin{pmatrix}-1\\3\end{pmatrix}$ ms^{-1}.
    It then reaches point B 4 seconds after leaving A, where point B has position vector $\begin{pmatrix}8\\-2\end{pmatrix}$ metres relative to the origin.
    Find the position vector of A.

13) A particle moves with constant acceleration $(2\mathbf{i} - 4\mathbf{j})$ ms^{-2}. At $t = 0$ it passes through the origin and T seconds later it passes through a point with position vector $(-12\mathbf{i} + k\mathbf{j})$ metres at velocity $(4\mathbf{i} + 3\mathbf{j})$ ms^{-1}. Find
    a) the value of T
    b) the value of k

2.57 Constant Acceleration: Vectors (answers on page 554)

14) A particle P moves in a straight line from position $\begin{pmatrix} 5 \\ -2 \end{pmatrix}$ metres to $\begin{pmatrix} k \\ 78 \end{pmatrix}$ metres, moving parallel to the vector $\begin{pmatrix} 7 \\ 10 \end{pmatrix}$. Find the value of k

15) For each of the following, the two vectors represent the velocity of two particles which are moving with constant velocity and parallel to each other. In each case, find the value of k for this to hold:
    a) $(2\mathbf{i} - 4\mathbf{j})$ ms^{-1} and $(4\mathbf{i} - k\mathbf{j})$ ms^{-1}
    b) $(5\mathbf{i} + 3\mathbf{j})$ ms^{-1} and $(-10\mathbf{i} + k\mathbf{j})$ ms^{-1}
    c) $(-3\mathbf{i} + 10\mathbf{j})$ ms^{-1} and $(k\mathbf{i} + 30\mathbf{j})$ ms^{-1}
    d) $(1.5\mathbf{i} - 9\mathbf{j})$ ms^{-1} and $k(2\mathbf{i} - 12\mathbf{j})$ ms^{-1}

16) A particle passes through the origin at $t = 0$ with velocity $\begin{pmatrix} 9 \\ 2 \end{pmatrix}$ ms^{-1} and constant acceleration $\begin{pmatrix} 2 \\ 3 \end{pmatrix}$ ms^{-2}. At $t = T$ the particle is moving in the direction of $\begin{pmatrix} 4 \\ 1 \end{pmatrix}$
    Find the value of T

17) Emily is cycling with constant velocity $(4.2\mathbf{i} + 14.4\mathbf{j})$ ms^{-1} along a straight road. Henry is cycling parallel to Emily, on the same path. Find the velocity for Henry in each of the following scenarios:
    a) Henry is cycling twice as fast as Emily in the same direction.
    b) Henry is cycling with speed 5 ms^{-1} in the same direction.
    c) Henry is cycling with speed 12 ms^{-1} in the same direction.
    d) Henry is cycling with speed 7.5 ms^{-1} in the opposite direction.

18) Two particles A and B are moving parallel to each other. Particle A is moving with constant velocity $\begin{pmatrix} 9 \\ 21.6 \end{pmatrix}$ ms^{-1}. Particle B begins at a position of $\begin{pmatrix} 2 \\ -3 \end{pmatrix}$ m relative to a fixed origin and moves with constant speed 19.5 ms^{-1}.
    Find
    a) the speed of Particle A.
    b) the velocity of Particle B.
    c) the position vector of B after 5 seconds.

19) Two particles A and B are moving parallel to each other, starting 4 metres apart and then at a fixed time later, are 4 metres apart again. Particle A travels a distance of 16 metres and particle B travels a distance of 20 metres. Find the shortest distance between their two lines of motion, giving your answer to 3 significant figures.

20) Two squirrels are running along parallel branches of trees in the woods. At time $t = 0$ seconds, the first squirrel starts at position $(-\mathbf{i} + 3\mathbf{j})$ metres and is moving with constant velocity 5.2 ms^{-1}. The second squirrel runs with constant velocity $(2.4\mathbf{i} + 5.76\mathbf{j})$ ms^{-1}.
    By modelling the branches as flat straight paths, find
    a) the speed of the second squirrel.
    b) the position vector of the first squirrel when 2.5 seconds have passed.

    Given that the second squirrel starts at position $(11\mathbf{i} - 2\mathbf{j})$ metres,
    c) determine the shortest distance between their two branches.

21) Particle A starts with position vector $\begin{pmatrix} -7 \\ -1 \end{pmatrix}$ m and constant velocity $\begin{pmatrix} 6 \\ 3 \end{pmatrix}$ ms^{-1}. Particle B starts with position vector $\begin{pmatrix} 0 \\ 2 \end{pmatrix}$ m and constant velocity $\begin{pmatrix} 2.5 \\ 1.5 \end{pmatrix}$ ms^{-1}.
    Show that the particles collide.

22) Particles A and B are both moving in straight lines. Particle A has a constant velocity of $\begin{pmatrix} 3.5 \\ 12 \end{pmatrix}$ ms^{-1} and Particle B moves with a speed of 6.25 ms^{-1}.
    Both particle A and particle B pass through point P. Particle B reaches point P first, with particle A passing through it 2 seconds later. Given that the two lines of motion are perpendicular to each other, find the distance between particle A and particle B 6 seconds after particle B passed through point P.

2.57 Constant Acceleration: Vectors (answers on page 554)

The following questions that use i and j are unit vectors pointing east and north respectively.

23) For each of the following velocity vectors, state the direction that the particle is moving in:
    a) $(2\mathbf{i})$ ms^{-1}
    b) $(-5\mathbf{j})$ ms^{-1}
    c) $(4\mathbf{i} + 4\mathbf{j})$ ms^{-1}
    d) $(-5\mathbf{i} + 5\mathbf{j})$ ms^{-1}

24) A ship, moving with constant acceleration, is travelling due east at a speed of 4 ms^{-1}. After 40 seconds it is travelling due south at 2 ms^{-1}. Find
    a) the displacement from its initial position after 40 seconds.
    b) the distance the ship has travelled from its initial position after 40 seconds.

25) The velocity, in ms^{-1}, of an object at time t seconds is given by
    $$\mathbf{v} = (t-5)\mathbf{i} + (2t-13)\mathbf{j}$$
    a) Show the object is never instantaneously at rest.
    b) Find the time at which the object is moving directly to the south.
    c) Find the time at which the object is moving to the north-east.

26) An object moves so that its position vector, in metres, relative to the origin O at time t seconds is given by
    $$\mathbf{r} = (105 + 8t)\mathbf{i} + t^2\mathbf{j}$$
    a) Determine the time at which the object is east of the origin.
    b) Determine the time at which the object is north-east of the origin.
    c) Explain why the object never moves in the westwards direction.

27) A particle passes point A with velocity $\binom{3}{8}$ ms^{-1} and has acceleration $\binom{2}{-20}$ ms^{-2}.
    At the instant it reaches point B it is travelling in a due east direction.
    Find the time at which it reaches point B.

28) A particle passes point A with velocity $\binom{1}{3}$ ms^{-1} and has acceleration $\binom{15}{5}$ ms^{-2}.
    At the instant it reaches point B it is travelling in a due north-east direction.
    Find the time at which it reaches point B.

29) A particle is travelling with constant acceleration. It passes point A with velocity $\binom{6}{-12}$ ms^{-1} and then passes point B with velocity $\binom{-2}{2}$ ms^{-1}
    Find
    a) the acceleration of the particle when $t = 2$
    b) the time at which it is travelling due south.
    c) the speed at which it is travelling at the exact point it is travelling due south.

30) A particle is travelling with constant acceleration $(-\mathbf{i} - 3\mathbf{j})$ ms^{-2} and passes the origin with velocity $(13\mathbf{i} - 5\mathbf{j})$ ms^{-1}. After a further 5 seconds have passes, the particle reaches point P. Find
    a) the speed the particle is travelling when it passes through point P giving your answer to 3 significant figures.
    b) Find the bearing at which the particle is travelling when it passes through point P

31) A helicopter is initially hovering above a field. It then sets off so that its acceleration is $(0.4\mathbf{i} + 0.2\mathbf{j})$ ms^{-2}.
    The helicopter does not change its height as it moves.
    Find
    a) the speed of the helicopter 15 seconds after it leaves its position above the field.
    b) the bearing on which the helicopter is travelling, 15 seconds after it leaves its position above the field, giving your answer correct to the nearest degree.

    The helicopter stops accelerating when it is 240 metres from its initial position.
    c) Find the time that it takes for the helicopter to travel from its initial position to the point where it stops accelerating.

2.58 Projectiles (answers on page 554)

The following questions use g = 9.8 ms⁻²

1) A particle is projected from point O on horizontal ground with a velocity of 25 ms⁻¹ at an angle of 30° above the horizontal before hitting the ground.

   Find
   a) the velocity of the particle after 0.5 seconds.
   b) the greatest height the particle reaches.
   c) the time the particle spends in the air.
   d) the total horizontal distance travelled by the particle.

2) A particle is projected from point O on horizontal ground with a velocity of 10 ms⁻¹ at an angle of 30° above the horizontal before hitting the ground.

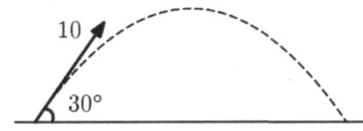

   Find
   a) the velocity of the particle after 0.4 seconds.
   b) the greatest height the particle reaches.
   c) the time the particle spends in the air.
   d) the total horizontal distance travelled by the particle.

3) A particle is projected from point O on horizontal ground with a velocity of 25 ms⁻¹ at an angle of 45° above the horizontal before hitting the ground.

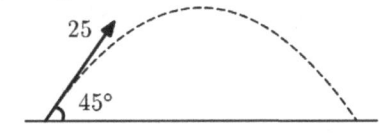

   Find
   a) the velocity of the particle after 3 seconds.
   b) the greatest height the particle reaches.
   c) the time the particle spends in the air.
   d) the total horizontal distance travelled by the particle.

4) A ball is projected at an angle of 20° above the horizontal from a tower of 10 metres high at a speed of 20 ms⁻¹ before landing on the floor.

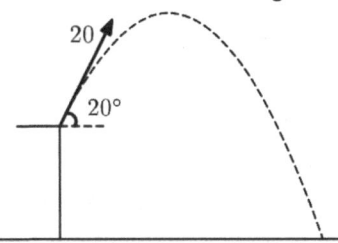

   Modelling the ball as a particle, find
   a) the initial velocity of the ball.
   b) the time the ball spends in the air.
   c) the velocity of the ball as it hits the ground.

5) A stone is thrown from a cliff of height 80m into a lake below. It is projected at 10 ms⁻¹ at an angle of α above the horizontal where $\tan \alpha = \frac{4}{3}$

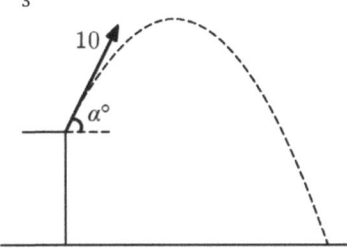

   By modelling the stone as a particle, find
   a) the time the stone spends in the air
   b) the vertical and horizontal components of the displacement of the stone in terms of t and hence show that the equation of the trajectory of the stone is
   $$y = 80 + \frac{4}{3}x - \frac{49}{360}x^2$$
   c) the maximum height of the stone above the ground.

6) A ball is projected at an angle of 35° above the horizontal from a height of 4.5 metres and at a speed of 22 ms⁻¹. A wall of height 1.5 metres stands 50 metres away.

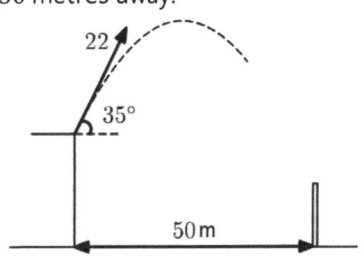

   Does the ball hit the wall?

2.58 Projectiles (answers on page 554)

7) A tennis ball is hit horizontally with a speed of 18 ms⁻¹ from a height of 2.5 metres. The net is 1.2 metres high and is 10 metres away.

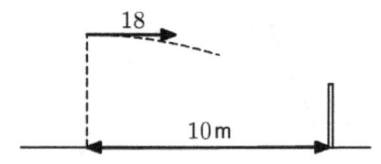

   a) Will the ball pass over the net?
   b) What modelling assumptions have you made?

8) A stone is thrown at speed 18 ms⁻¹ at an angle of $\alpha°$ where $\tan \alpha = \frac{5}{12}$ above the horizontal from a cliff of height 25 metres.

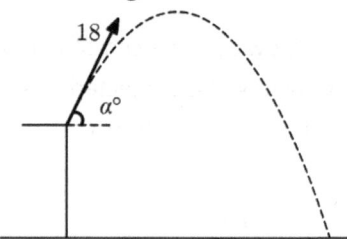

   Modelling the stone as a particle, find the amount of time that the ball is more than 27 metres above ground level.

9) A stone is thrown at a speed of u ms⁻¹ with angle of α above the horizontal where $\tan \alpha = \frac{3}{4}$ from a cliff of height 40 m. It travels 50 metres horizontally before hitting the water below t seconds later.
   a) Find the exact value of t
   b) Find the exact value of u
   c) State any modelling assumptions you have made.

   A stone of twice the mass is thrown from the cliff, otherwise the conditions remain the same.
   d) Explain how that would change the results in the previous part of the question.

10) A particle is thrown at 10 ms⁻¹ at an angle of α below the horizontal, where $\tan \alpha = \frac{2}{5}$, from a height of 2 metres. Show that the equation of the trajectory of the particle is
$$y = 2 - \frac{2}{5}x - \frac{1421}{25000}x^2$$

The following questions use g = 9.81 ms⁻²

11) Dean kicked a football on a horizontal field with speed 20 ms⁻¹ at an angle of θ above the horizontal where $\tan \theta = \frac{3}{4}$. The football goal is 36 metres away, the top of the goal post is 2.4 metres above the ground.
    a) Does the ball go into the goal?
    b) State any modelling assumptions you have made.

12) A stone is thrown towards some water at speed 12 ms⁻¹ at an angle of $\alpha°$ below the horizontal where $\tan \alpha = \frac{1}{10}$.

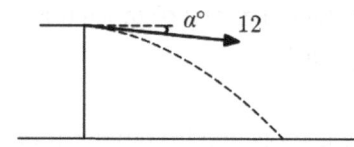

    It lands on the surface of the water 0.4 seconds later. By modelling the stone as a particle, find
    a) the direction of the motion when 0.2 seconds have passed.
    b) the height above the water surface area that the stone was thrown.
    c) the horizontal distance the stone travelled before hitting the water.

The following questions use g = 10 ms⁻²

13) A ball is projected at an angle of $\theta°$ above the horizontal from a height of 2.5 metres above the ground. The ball hits a target which is 1.5 metres above the ground and 7 metres away from the point of projection, 2 seconds after projection. Modelling the ball as a particle, find
    a) the speed at which the ball is projected.
    b) the angle at which the ball is projected.

14) A ball is projected at angle of 30° above the horizontal from a height of 2 metres at a speed of 18 ms⁻¹. It passes over a wall of height 3 metres stands d metres away. Find
    a) the maximum value of d such that the ball still passes over the wall.
    b) the direction at which the ball impacts the ground.

2.58 Projectiles (answers on page 554)

15) Two balls A and B are projected from a bridge at the same point. Ball A is projected with speed 10 ms^{-1} at an angle of 30° to the horizontal. Ball B is projected 1 second after ball A with speed 40 ms^{-1} at an angle of 60°. By modelling the balls as particles, identify the vertical distance between the balls when one passes over the other, assuming they have not reached the ground.

Prerequisite knowledge for Q16: Reciprocal Trigonometric Functions

16) A child shoots a water pistol, with the water firing at 10 ms^{-1} at an acute angle of θ° above the horizontal.
    a) Show that the equation of the trajectory of the water fired from a height of a metres above the ground is
    $$y = a + x\tan\theta - \frac{x^2}{20\cos^2\theta}$$

    The water is fired from a height of 0.7 metres above the ground and reaches a horizontal distance of 9.2 metres from the child.
    b) Find the angle(s) above the horizontal at which the water pistol was fired.

The following questions use exact values of g

17) Two balls A and B are projected at the same time towards each other from points P and Q respectively, where P and Q lie 5 metres apart. Ball A is projected with speed 10 ms^{-1} at 30° to the horizontal.
Ball B is projected with speed u ms^{-1} at 45° to the horizontal.

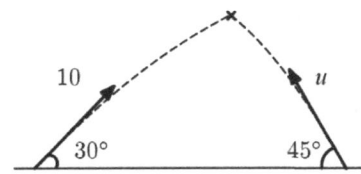

The two balls collide.
a) Show that the value of u is $5\sqrt{2}$
b) Find the exact time at which they collide giving your answer to 2 significant figures.
c) Find the horizontal distance ball A is from its start point when they collide, giving your answer to 3 significant figures.

18) A particle is projected with speed u ms^{-1} from the ground at an angle of θ above the horizontal. Show that the equation of the trajectory of the particle is
$$y = \tan\theta\, x - \frac{g}{2u^2}\sec^2\theta\, x^2$$

19) A particle is projected with speed u ms^{-1} at an angle of θ above the horizontal.
The particle reaches a maximum vertical height of h metres.
a) Show that
$$h = \frac{u^2}{2g}\sin^2\theta$$

It is known that $u = 10$ ms^{-1}.
b) Given θ is acute show that
$$0 \leq h \leq \frac{50}{g}$$

Given that a projectile needs to have a maximum height of 5 metres above its start point and is projected at 12 ms^{-1}.
Using g = 10 ms^{-2},
c) find the acute angle θ at which it must be projected, giving your answer to 1 decimal place.

Prerequisite knowledge for Q20: Double Angle Formulae

20) A particle is projected from the ground with speed u ms^{-1} at an angle of θ° above the horizontal.
The particle lands x metres from its original start location after t seconds have passed.
a) Show that
$$t = \frac{2u}{g}\sin\theta$$
b) Show that
$$x = \frac{u^2}{g}\sin 2\theta$$

It is known that the projectile needs to land at least 20 metres away from the start point and is projected at 15 ms^{-1}. Using g = 10 ms^{-2},
c) find the range of possible angles θ at which it must be projected, giving your answers to 1 decimal place.

2.59 Variable Acceleration: Vectors (answers on page 555)

1) A particle's displacement at time t seconds is given by $\mathbf{s}$ metres. In each case find the magnitude of the acceleration of the particle at the given value of t:
   a) $\mathbf{s} = (4t^3)\mathbf{i} + (t-3)\mathbf{j}$ at $t = 0.5$
   b) $\mathbf{s} = (3t^3 - 5)\mathbf{i} + (5 - t^4)\mathbf{j}$ at $t = 2$
   c) $\mathbf{s} = (2e^{0.5t})\mathbf{i} + (t^2)\mathbf{j}$ at $t = 0.1$

2) A particle's acceleration at time t seconds is given by $\mathbf{a}$ ms^{-2} and initial velocity $\mathbf{u}$ ms^{-1}. In each case find the speed of the particle at the given value of t:
   a) $\mathbf{a} = (4t - 5)\mathbf{i} + (9t^2)\mathbf{j}$, $\mathbf{u} = 3\mathbf{i} - \mathbf{j}$ at $t = 5$
   b) $\mathbf{a} = (12t + 2)\mathbf{i} - 18\mathbf{j}$, $\mathbf{u} = -2\mathbf{i} + 5\mathbf{j}$ at $t = 0.5$

3) A particle's velocity at time t seconds is given by $\mathbf{v}$ metres per second and initial position $\mathbf{s}$ metres. In each case find the distance of the particle from the origin at the given value of t:
   a) $\mathbf{v} = (t - 5)\mathbf{i} + 3\mathbf{j}$, $\mathbf{s} = -4\mathbf{i} + 10\mathbf{j}$ at $t = 0.4$
   b) $\mathbf{v} = (3t)\mathbf{i} + (2e^{-0.1t})\mathbf{j}$, $\mathbf{s} = 2\mathbf{i} + 0.5\mathbf{j}$ at $t = 0.1$

4) A particle P of mass 1.2 kg is moving under the action of a single force $\mathbf{F}$ newtons. At time t seconds, the velocity of P, $\mathbf{v}$ ms^{-1} is given by
$$\mathbf{v} = (5t + 2)\mathbf{i} + (t^2 - 2t)\mathbf{j}$$
When $t = 0$, P is at the point with position vector $(2\mathbf{i} + 3\mathbf{j})$ m. Find
   a) the magnitude of $\mathbf{F}$ when $t = 5$
   b) the distance from its initial position when $t = 5$

5) The velocity of a particle at time t seconds is
$$\mathbf{v} = (6t^2)\mathbf{i} + (4t - 1)\mathbf{j}$$
metres per second. When $t = 1$ it passes a point A and when $t = 5$ it passes the point B. Find the distance AB.

6) A particle's position vector at time t seconds is
$$\mathbf{r} = (2t^2 - t)\mathbf{i} + (2t + 5)\mathbf{j}$$
metres.
Find the time at the instant at which the speed of the particle is 2 ms^{-1}.

7) The acceleration, ms^{-2}, of a particle at time t seconds is
$$\mathbf{a} = (2t + 9)\mathbf{i} + (e^{-0.1t})\mathbf{j}$$
When $t = 0$ it passes a point A and when $t = 1$ it passes the point B. Show that the velocity between A and B is $10\mathbf{i} + (10 - 10e^{-0.1})\mathbf{j}$ m.

8) A particle's position vector at time t seconds is
$$\mathbf{r} = (t^2 + t)\mathbf{i} + (5t + 4)\mathbf{j}$$
metres.
   a) Find the speed of the particle when $t = 4$ seconds.
   b) Show that the acceleration of the particle is never zero.

9) A particle's velocity at time t seconds is
$$\mathbf{v} = (t - 4)\mathbf{i} + (t^3 + 2t)\mathbf{j}$$
metres per second.
The particle is initially at the origin. Find
   a) the position vector of the particle in terms of t
   b) the vertical displacement when the particle has travelled 10 metres horizontally.

10) A particle's velocity at time t seconds is
$$\mathbf{v} = \left(2t^{\frac{1}{2}}\right)\mathbf{i} + (3t - 1)\mathbf{j}$$
metres per second.
In each case find the values of t at the instant that the particle is moving in the given direction:
   a) $\mathbf{i} + \mathbf{j}$
   b) $\mathbf{i} - \mathbf{j}$
   c) $3\mathbf{i} + 13\mathbf{j}$

11) A particle's velocity at time t s is
$$\mathbf{v} = (2t - 36)\mathbf{i} + \left(6t^{\frac{1}{2}}\right)\mathbf{j}$$
metres per second. It is known that when 1 second has passed, the position vector of the particle is $-20\mathbf{i} + 6\mathbf{j}$. Find
   a) the value of t at the instant when the particle is moving in the direction $\mathbf{i} + \mathbf{j}$
   b) the position vector $\mathbf{r}$ in terms of t
   c) the exact position of the particle at the first instant it is moving with speed 24 ms^{-1}.

2.59 Variable Acceleration: Vectors (answers on page 555)

12) The mass of an object is 0.5 kg. The object is moving with velocity
$$\mathbf{v} = \left(6 - pt^{\frac{1}{2}}\right)\mathbf{i} + (5t - q)\mathbf{j}$$
metres per second, where t is the number of seconds passed.
Initially, the object has position vector $4\mathbf{i} + \mathbf{j}$
After 4 seconds it passes through $12\mathbf{i} + 21\mathbf{j}$
Find
   a) the values of p and q
   b) the magnitude of the force acting on the object when $t = 9$ seconds.

13) A particle's position $\mathbf{r}$ metres is
$$\mathbf{r} = \left(t + \frac{1}{8}k\right)\mathbf{i} + (kt^2 - 8t)\mathbf{j}$$
where t is time in seconds.
At the instant that $t = 0.5$, the particle is moving parallel to $\mathbf{i}$. Find
   a) the value of k
   b) the value of t when the particle is 1 metre from the origin.

14) The position vectors of particles P and Q are
$$\mathbf{r}_P = (3 - t^2)\mathbf{i} + (3t^2 + 4)\mathbf{j}$$
$$\mathbf{r}_Q = (12 - 2t^2)\mathbf{i} + (4t^2 + 5)\mathbf{j}$$
respectively, where t is time in seconds.
   a) Show that the distance between P and Q at any given time satisfies
   $$d^2 = 2t^4 - 16t^2 + 82$$
   b) Find the time at which the distance between P and Q is minimised during the motion.

15) The position vectors of particles P and Q are
$$\mathbf{r}_P = (t + 1)\mathbf{i} + (6t)\mathbf{j}$$
$$\mathbf{r}_Q = (4t^2)\mathbf{i} + (5t + 1)\mathbf{j}$$
respectively, where t is the time in seconds.
Find the exact value of t when both particles are moving at the same speed.

16) The velocity vectors of particles P and Q are
$$\mathbf{v}_P = (4t)\mathbf{i} + \left(t^2 + \frac{2}{3}\right)\mathbf{j}$$
$$\mathbf{v}_Q = (2t + 7)\mathbf{i} + (-4)\mathbf{j}$$
respectively, where t is the time in seconds.
They are both initially at the origin.
Show that particle Q is four times particle P's distance from the origin when $t = 1$

The following questions use that $\mathbf{i}$ and $\mathbf{j}$ are unit vectors pointing east and north respectively.

17) A particle's position $\mathbf{r}$ metres is
$$\mathbf{r} = (5e^{0.4t} + 1)\mathbf{i} + (t^2 - 4t)\mathbf{j}$$
where t is the time in seconds.
   a) Find an expression for the velocity of the particle at time t
   b) State the direction in which the particle is travelling when $t = 2$
   c) The mass of the particle is 10 kg. Find the magnitude of the resultant force on the particle when $t = 2$

18) The mass of an object is 2 kg. The object is moving with velocity
$$\mathbf{v} = (t^2 + pt)\mathbf{i} + (4t^3)\mathbf{j}$$
metres per second, where t is the number of seconds passed since passing through the origin. The magnitude of the force acting on the object is 120 N when $t = 2$ seconds.
   a) Show that p satisfies
   $$p^2 + 8p - 1280 = 0$$
   b) Hence, find the possibles value of p
   c) Show that there is no instantaneous time for which the object moves eastwards or westwards.

19) A particle's velocity at time t seconds is
$$\mathbf{v} = (2t^2)\mathbf{i} + (-5t)\mathbf{j}$$
metres per second.
Find the bearing the particle is travelling on when $t = 0.1$ seconds to the nearest degree.

20) An object's velocity $\mathbf{v}$ metres per second is
$$\mathbf{v} = (-t^2 - 4)\mathbf{i} + (-t^3)\mathbf{j}$$
where t is the time in seconds.
Find the time(s) at which the object is moving at a bearing of 225°

21) An object's position $\mathbf{r}$ metres is given by
$$\mathbf{r} = \left(t + \frac{7}{4}\right)\mathbf{i} + (2 - t^2)\mathbf{j}$$
where t is the time in seconds. Find
   a) the time(s) at which the object is at a bearing of 135° to the origin.
   b) the speed of the object when it is directly east of the origin.

2.59 Variable Acceleration: Vectors (answers on page 555)

22) A particle's position **r** metres is
$$\mathbf{r} = (3 + t^2)\mathbf{i} + (8t - 5)\mathbf{j}$$
metres where t is the time passed in seconds. At time T, in that instant the object is moving with a bearing 045°.
Find
a) an expression for the velocity in terms of t
b) the value of T and the speed of the object at your value of T

23) An object of mass 0.5 kg is initially at rest. It is being pushed by forces
$$\mathbf{F}_1 = -3\mathbf{i} + (t^2 - 3t + 7)\mathbf{j} \text{ N}$$
and $\mathbf{F}_2 = (t^2 - 1)\mathbf{i} - 5\mathbf{j}$ N
where t is the number of seconds passed.
Find
a) the value of t when the object is in equilibrium.
b) the instantaneous time at which the object is moving due north-east.
c) the speed of the object when $t = 1$

The following questions use Calculus from non-AS-Level content

24) A particle's displacement at time t seconds is
$$\mathbf{s} = (2^{2t})\mathbf{i} - (2^t)\mathbf{j}$$
metres. Find
a) the speed of the particle when $t = 0.4$
b) the time at which the particle is at a bearing of 135° to the origin.

25) A particle's acceleration in ms^{-2} at time t seconds is
$$\mathbf{a} = 8\mathbf{i} - \left(\frac{2}{t}\right)\mathbf{j}$$
When $t = 1$, the particle is moving with velocity $(3\mathbf{i} - 5\mathbf{j})$ ms^{-1}. Find
a) an expression for the velocity in terms of t
b) the speed of the particle at $t = 0.1$

26) A particle's velocity at time t seconds is
$$\mathbf{v} = (e^t)\mathbf{i} + (\sin t)\mathbf{j}$$
metres per second. When $t = 0$, the particle is $(2\mathbf{i} + 5\mathbf{j})$ m from the origin.
Find the distance of the particle from the origin when $t = 0.5$

27) A particle's position in relation to the origin is
$$\mathbf{r} = (4\cos(2t))\mathbf{i} + (t^2 \sin t)\mathbf{j}$$
metres where t is the time in seconds. The mass of the particle is 0.5 kg.
a) State the position of the particle in relation to the origin when $t = 3\pi$
b) Find the direction in which the particle is travelling when $t = 3\pi$
c) Find the magnitude of the resultant force on the particle when $t = 3\pi$

28) A particle's acceleration at time t is
$$\mathbf{a} = 6\mathbf{i} + (te^{2t})\mathbf{j}$$
metres per second.
The initial velocity of the particle is $\left(\frac{3}{4}\mathbf{j}\right)$ ms^{-1}.
Find the speed of the particle when $t = 0.2$ seconds.

29) The velocity of a particle at time t seconds is
$$\mathbf{v} = (\sin 2t)\mathbf{i} + (\cos t)\mathbf{j}$$
metres per second. When $t = 0$ it passes the point A and when $t = \frac{\pi}{2}$ it passes the point B. Find the distance AB.

30) A particle's position at time t seconds is
$$\mathbf{r} = (\cos t - e^t)\mathbf{i} + (\sin t - \ln t)\mathbf{j}$$
metres.
Find the bearing the particle is travelling on when $t = 0.5$ seconds to the nearest degree.

Prerequisite knowledge for Q31-Q32: Parametric Equations

31) For each of the following, write a cartesian equation for the path of the motion:
a) $\mathbf{r} = (8t - 5)\mathbf{i} + (3 + t^2)\mathbf{j}$
b) $\mathbf{r} = \left(t + \frac{7}{4}\right)\mathbf{i} + (2 - t^2)\mathbf{j}$
c) $\mathbf{r} = (t^2)\mathbf{i} + (5t + 4)\mathbf{j}$

32) A particle P has mass 2 kg and its position is given by **r** metres
$$\mathbf{r} = (\sin t)\mathbf{i} + (\cos t)\mathbf{j}$$
where t is the time in seconds.
a) Find the magnitude of the resultant force on the particle when $t = \frac{\pi}{2}$
b) Write a cartesian equation for the path of the motion.

2.60 Moments (answers on page 556)

1) Find the resultant moment about O in each case, stating whether it is anticlockwise or clockwise:

   a) [Diagram: 8 N force upward, 3 m from O]

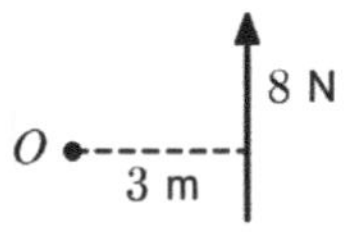

   b) [Diagram: 10 N force upward, 5 m from O]

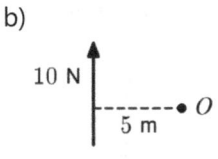

   c) [Diagram: 1.6 N force, 2 m from O]

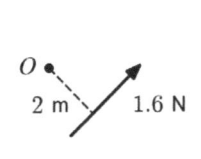

   d) [Diagram: 20 N force, 2 m from O]

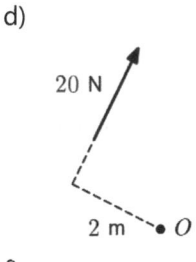

   e) [Diagram: 6 N upward at 2 m, 4 N downward at 3 m from O]

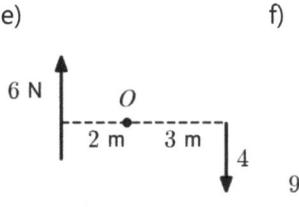

   f) [Diagram: 7 N and 9 N forces, 1.2 m and 0.5 m from O]

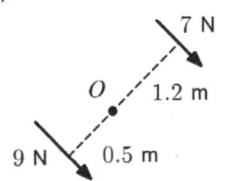

2) A uniform lamina $ABCD$ has height 2 metres and width 3 metres. It has a centre of mass at O which lies exactly where the lines AC and BD intersect. A force of 5 N acts on the point D parallel to DC as shown in the diagram.

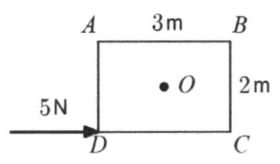

   Find the moment about
   a) A b) B c) C d) D e) O

3) A uniform lamina $ABCD$ has height 4 metres and width 6 metres. The point O lies exactly where the lines AC and BD intersect. A force of 12 N acts on the point A parallel to AB as shown in the diagram.

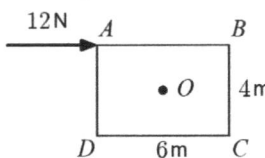

   Find the moment about
   a) A b) B c) C d) D e) O

4) A uniform lamina $ABCD$ has height 6 metres and width 5 metres. The point O lies exactly where the lines AC and BD intersect. A force of 12 N acts on the point A parallel to AB and a force of 10 N acts on the points B parallel to BC as shown in the diagram.

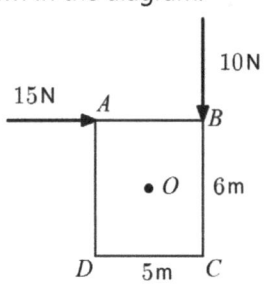

   Take moments about
   a) A b) B c) C d) D e) O

5) A uniform rectangular lamina $ABCD$ is such that $BC = 0.2$ m and $AB = 0.08$ m. Forces of magnitude 10 N, 20 N, X N and Y N act on it as shown in the diagram, where the force of Y N acts at the midpoint of AB.

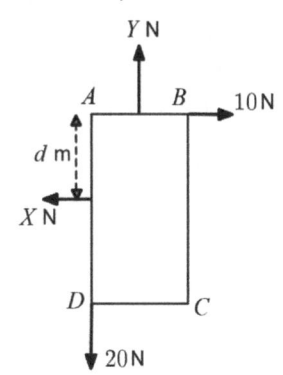

   The lamina is in equilibrium. Find the value of
   a) X b) Y c) d

   The force acting at B is moved down towards point C.
   d) Describe what happens to the lamina.

6) PQ is a light horizontal rod of length 16 m. Forces act on the rod as shown in the diagram

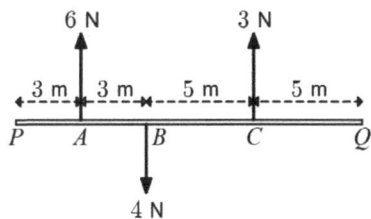

   Find the resultant moment about
   a) P b) A c) B d) C e) Q

2.60 Moments (answers on page 556)

7) Each of the diagrams show a horizontal beam AB which is in equilibrium. Two forces acting on the beam are given by P N and Q N. In each case, find the values of P and Q:

   a) [Diagram: beam AB with P N upward at 1 m from A, wait — P N downward at 1 m from A... forces: P N down, 50 N down at 3.5 m from A, Q N up at 6 m from A; segments: 1 m, 2.5 m, 2.5 m, 2 m]

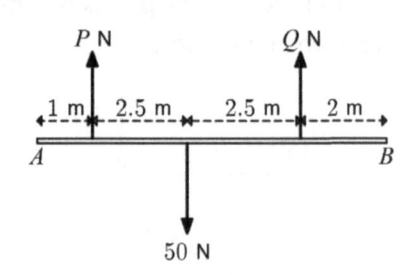

   b) [Diagram: beam AB with P N up at 4 m from A, 110 N down at 4 m, Q N up; segments: 4 m, 7 m, 2 m]

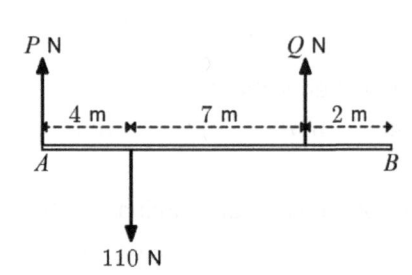

8) Each of the diagrams show a horizontal beam AB which is in equilibrium, being held in place by two inextensible cables, with tension P N and Q N respectively. In each case, find the values of P and Q:

   a) [Diagram: P N up, 30 N down at 5 m, Q N up at 3 m further; segments 5 m, 3 m]

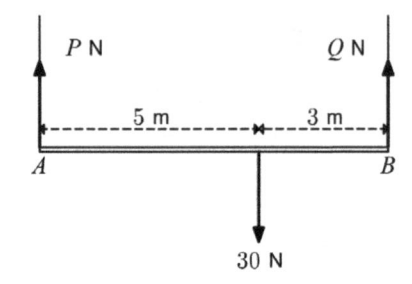

   b) [Diagram: P N up, 5 N down at 4 m, 64 N down at 3 m further, Q N up at 6 m further; segments 4 m, 3 m, 6 m]

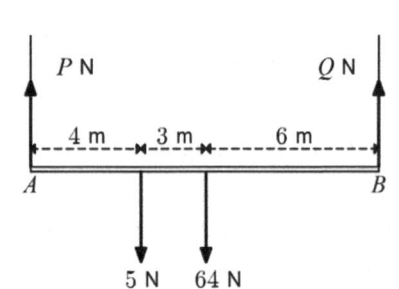

   c) [Diagram: P N up, 5 N down, Q N up; segments 15 cm, 25 cm, 40 cm]

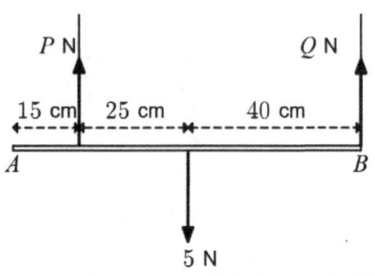

9) The diagrams below show a light horizontal rod PQ of length 0.5 m resting in equilibrium. Find

   a) the value of x and W
   [Diagram: 15 N up at A, 25 N up at B, W N down at x m from A]

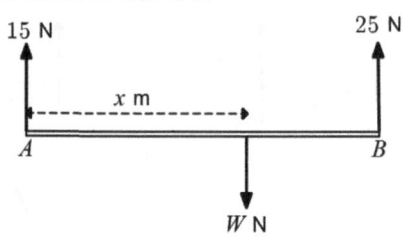

   b) the value of P and Q
   [Diagram: P N up at A, Q N up at B, 120 N down at 0.2 m from B]

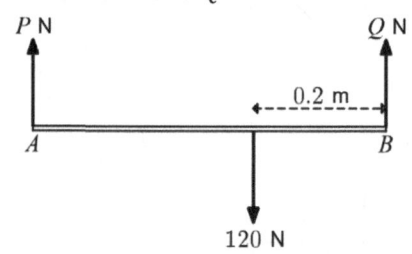

10) In each of the following cases, sketch a diagram, labelling the forces and distances:
    a) 3 m uniform rod AB of weight 20 N, held up by two smooth supports at A and B.
    b) 9 m uniform rod AB of weight 100 N, held by vertical ropes at A and B, with a object of weight 10 N which is 1m along from A.
    c) x m uniform rod AB of weight W N, held by vertical ropes at A and B, with an object of weight F N hanging from the rod d m along from A.

11) A uniform rod of weight 30 N rests horizontally on two smooth supports at A and B as shown.

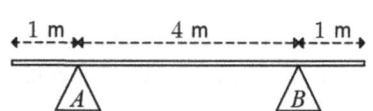

    [Diagram: segments 1 m, 4 m, 1 m with supports at A and B]

    Find the reaction at A and the reaction at B.

12) A uniform rod AB of length 3 m and weight 60 N. It rests horizontally on two smooth supports at A and B. A particle of weight 12 N is positioned on the rod 1 metre from B

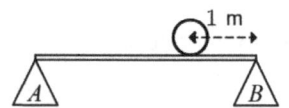

    [Diagram: particle at 1 m from B, supports at A and B]

    Find the reaction at A and the reaction at B.

2.60 Moments (answers on page 556)

The following questions use g = 9.8 ms⁻²

13) A uniform rod AB of length 10 m and mass 5 kg is held in place by two inextensible cables at A and B respectively. A particle of mass 0.5 kg is positioned on the rod 2.5 metres from A

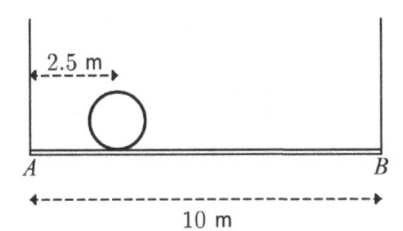

Given the rod is horizontal and in equilibrium, find the tension in the cables at A and B.

14) A plank of weight 500 N rests horizontally across two crates. An object of weight 200 N is placed 1 metre from the left-hand side as shown in the diagram.

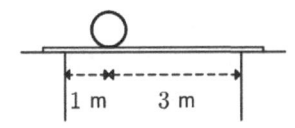

   a) Find the reaction forces exerted on each end of the plank. You may assume they act on the corners of the crates.
   b) State the modelling assumptions you have made.

15) A uniform rod AB of length 4 m and mass 2 kg rests horizontally on a smooth support at C, which is 3.5 metres from A.
A particle of weight W N is positioned at B so that the rod is held in equilibrium.

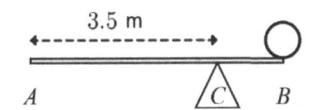

Find
   a) the reaction at C.
   b) the value of W
   c) the mass of the particle.

An extra particle of weight W is added 0.5 metres to the left of C and the system remains in equilibrium.
   d) Find the new value of R

16) Alice and Bob are 22 kg and 18 kg respectively. They sit on a seesaw of length 5 metres with mass 40 kg, pivoted at its midpoint. Bob sits at the right-hand side of the seesaw.

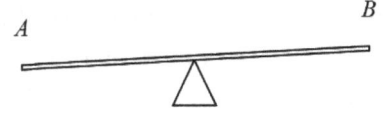

By modelling Alice and Bob as particles and the seesaw as a uniform rod, find how far Alice must sit from the left-hand end of the seesaw for the seesaw to be in equilibrium.

17) A uniform rod AB of length 0.5 m and mass 5 kg rests horizontally on two smooth supports at A and B as shown.

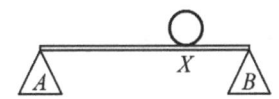

A particle of mass 8 kg is placed on the rod at X. The reaction on the rod at B is three times the reaction at A. Find the distance AX.

18) A plank of mass 10 kg rests horizontally on two crates which lie 8 metres apart. A particle of weight 1.5 kg is placed x metres from the left-hand side where $x < 4$ as shown.

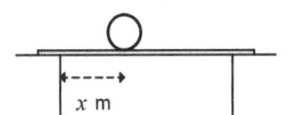

One normal reaction is 20% larger than the other. By modelling the plank as a uniform rod, find the value of x

19) A uniform rod AB of weight 300 N is held horizontally in position by a smooth support at A and a vertical string at C

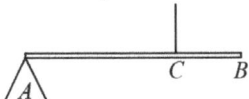

The string is 80% of the way along AB. Find the reaction at A and the tension in the string.

20) A non-uniform rod AB of length 3 metres and weight W N rests horizontally on two smooth supports at A and B. The reaction at A and B have magnitude 12 N and 18 N respectively. Calculate the distance of the centre of mass of the rod from A.

2.60 Moments (answers on page 556)

21) A non-uniform rod AB of weight W newtons and length 10 metres rests horizontally on two smooth supports at A and B. The reaction at A and B have magnitude R N and 40 N respectively. Calculate the distance from A of the centre of mass of the rod giving your answer in terms of W

The following questions use exact values of g

22) A uniform rod AB of length 5 metres and mass 5 kg. A particle of mass m kg is placed at A. A particle of mass $2m$ kg is placed at B. The rod is horizontal and at rest, balancing on a smooth support at C.

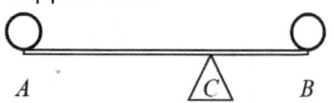

The distance between A and C is x metres. Show that
$$x = \frac{20m + 25}{6m + 10}$$

23) A uniform rod AB of mass 0.4 kg and length 0.7 m is resting on a table, with 20 cm of it overhanging the table. An object of mass m kg hangs from the rod x cm away from the table, as shown in the diagram, so that the rod is on the point of tilting.

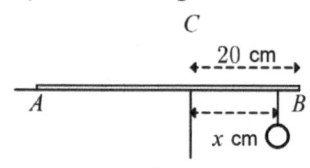

a) Show that $x = \frac{3}{50m}$
b) Find the range of values that m may be so that the rod remains horizontal.

24) A uniform rod of length 8 m and weight 60 N rests horizontally in equilibrium on two smooth supports at A and B as shown on the diagram. A particle of weight W N is positioned on the rod 1 m from B as shown. The rod is on the point of tilting.

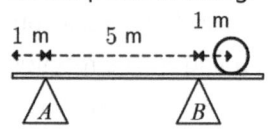

Find the reaction at B and the weight W

The following questions use g = 10 ms⁻²

25) A uniform rod AB of length 4 m and mass 5 kg rests horizontally in equilibrium on two smooth supports at C and D as shown on the diagram. A particle of mass m kg is placed on the rod positioned on the rod 1 m from A as shown. The rod is on the point of tilting.

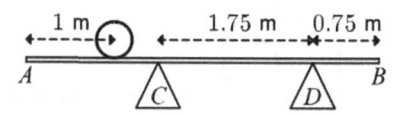

Find the reaction at C and the mass m

26) A uniform rod AB of length 1.6 m and weight W N rests horizontally in equilibrium on two smooth supports at C and D as shown on the diagram. A particle of mass $4W$ N is placed on the rod.

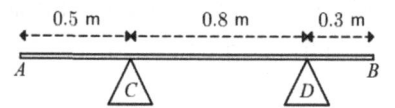

By taking moments about A, find the range of distances from A that the particle may be without it tipping.

27) A non-uniform plank AB of length 3 metres and mass 2 kg is held horizontally in equilibrium by two ropes that are positioned at C and D where $AC = 0.3$ m and $DB = 0.2$ m.

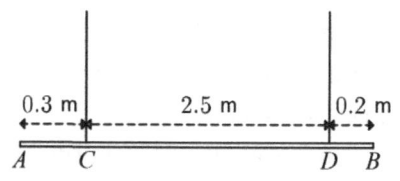

The tension in the rope at C is twice the tension in the rope at D.
a) Calculate the exact distance of the centre of mass from A.
b) State the modelling assumptions you have used.

A child of mass 25 kg wants to walk across the plank. The ropes will break if either of the tensions exceed 200 N.
c) Find the start and end point where the child can walk without the ropes breaking, giving your answer to 2 significant figures.

2.61 Moments: Ladders and Hinges (answers on page 557)

1) Find the resultant moment about O in each case, giving your answer to 3 significant figures and stating whether it is anticlockwise or clockwise:

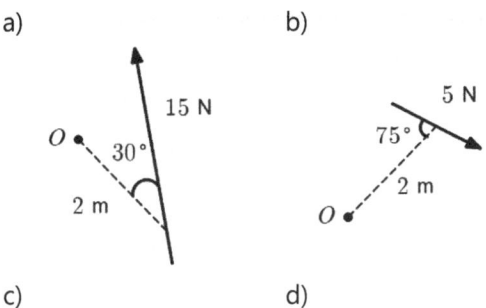

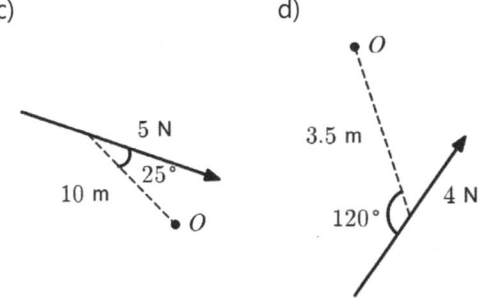

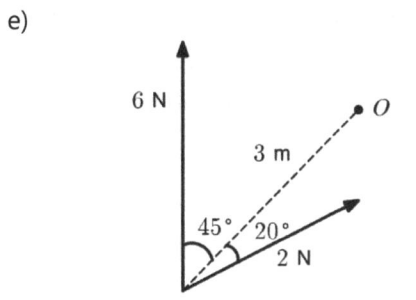

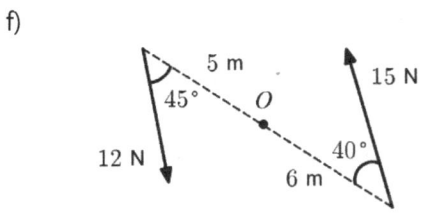

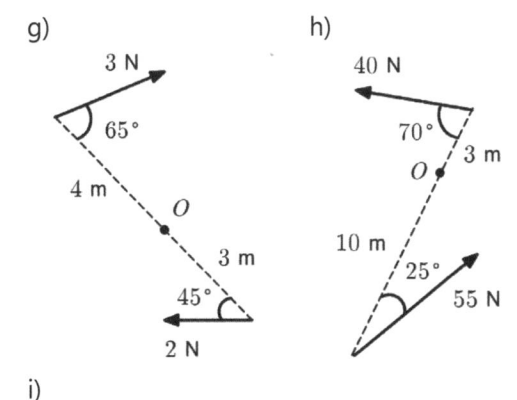

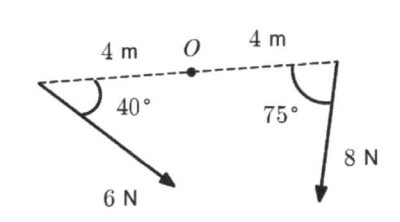

   a)

   b)

   c)

   d)

   e)

   f)

   g)

   h)

   i)

The following questions use $g = 9.8$ ms^{-2}

2) A uniform rod AB has mass m kg and length x metres. It rests with one end A on rough horizontal ground and is held in limiting equilibrium at angle θ by a light inextensible string at B with tension T N.

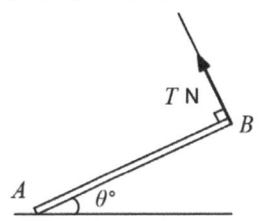

   a) Draw a force diagram.

   For each of the following, given

   b) $m = 30, x = 10, \theta = 30$, find the value of T and the value of μ

   c) $x = 8, \theta = 20, T = 180$, find the value of m and the value of μ

3) A ladder modelled as a uniform rod AB with length x metres and mass m kg rests against a wall. One end, A is resting on rough horizontal ground and the other end B rests on a smooth vertical wall. The ladder is held in limiting equilibrium at an angle of $\theta°$ to the horizontal.

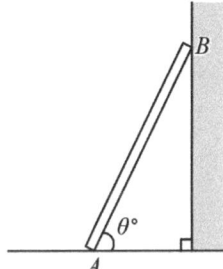

   The coefficient of friction between the ladder and the floor is μ

   a) Draw a force diagram.

   For each of the following find μ, given

   b) $\theta = 60, x = 5, m = 10$

   c) $\theta = 45, x = 5, m = 10$

   d) $\theta = 60, x = a, m = M$

   A mass of $\frac{1}{2}m$ kg is loaded onto the ladder $\frac{1}{4}$ of the way up. The ladder is a metres long and has mass m kg. Find the value of μ when

   e) $\theta = 60$

   f) $\theta = 45$

2.61 Moments: Ladders and Hinges (answers on page 557)

4) A uniform lever AB of mass m kg and length x metres is attached at A to a vertical wall by a smooth hinge. The acute angle between the lever and the wall is $\theta°$. The lever is held in equilibrium by a horizontal force of magnitude P N at B.

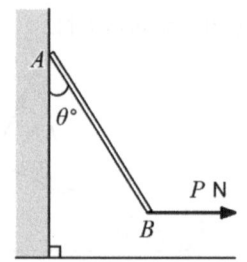

   a) Draw a force diagram.

   For each of the following, find P and the magnitude of the force exerted on the rod by the wall, given
   b) $\theta = 30, m = 2, x = 10$
   c) $\theta = 45, m = 5, x = a$

5) A uniform rod AB of mass m kg and length x metres is attached to a vertical wall by a smooth hinge. It is held in place horizontally by a light inextensible string which is attached to the wall directly above A and the other end is attached to the rod at B.

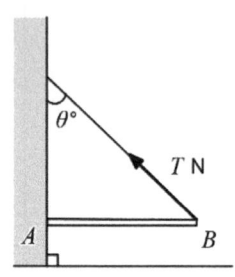

   The tension in the string is T N. The acute angle between the string and the wall is $\theta°$.
   a) Draw a force diagram.

   For each of the following, find the value of T and the magnitude of the force exerted on the rod by the wall, given
   b) $\theta = 30, m = 3, x = 10$
   c) $\theta = 60, m = 15, x = a$
   d) $\theta = 45, m = 10, x = a$

   A box of mass 3 kg is added to the rod 2 metres from A. Given
   e) $\theta = 40, m = 5, x = 8$, find the value of T

6) A uniform rod AB has weight 100 N and length $5a$ metres. It rests with one end A on rough horizontal ground and is resting on a smooth cylindrical drum at C where AC is $4a$ metres. The rod is angled at θ degrees to the horizontal where $\tan \theta = \frac{7}{24}$.

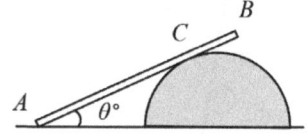

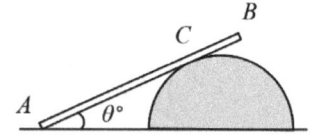

   a) Draw a force diagram.
   b) Find the normal and frictional components of the contact force at A.
   c) Find the least possible value of the coefficient of friction between the rod and the ground.

7) A non-uniform rod AB has weight 80 N which acts $2a$ metres from A. It rests with one end A on rough horizontal ground and is resting on a smooth cylindrical drum at C where AC is $4a$ metres. The rod is angled at θ degrees to the horizontal where $\tan \theta = \frac{7}{24}$.

   a) Draw a force diagram.
   b) Find the normal and frictional components of the contact force at A.
   c) Find the least possible value of the coefficient of friction between the rod and the ground.

8) A uniform ladder AB of length a metres rests against a wall in limiting equilibrium. One end, A rests on rough horizontal ground at an angle of θ and the other end B rests on a smooth vertical wall. The coefficient of friction is μ

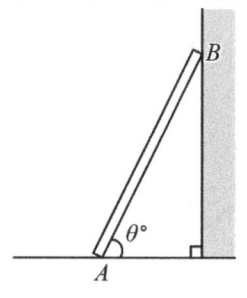

   Show that
   $$2\mu \tan \theta = 1$$

2.61 Moments: Ladders and Hinges (answers on page 557)

9) A uniform rod AB of mass 2 kg and length 4a metres with its end A against a rough vertical wall, with B being held in place horizontally by a light inextensible string which is attached to the wall directly above A

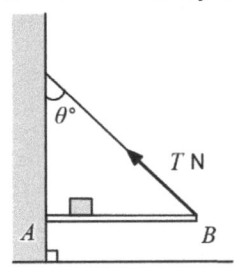

The tension in the string is T N. The acute angle between the string and the wall is $\theta°$ where $\tan\theta = \frac{3}{4}$

An object of mass 0.5 kg is placed on the rod, a metres from A
  a) Find the value of T
  b) Find the normal and frictional components of the contact force at A
  c) Find the value of the coefficient of friction.

The following question use exact values of g

10) A ladder modelled as a uniform rod AB of length 2a metres and mass m kg rests against a wall. One end, A is resting on rough horizontal ground and the other end B rests on a rough vertical wall. The ladder is held in limiting equilibrium at an angle of $\theta°$ to the horizontal

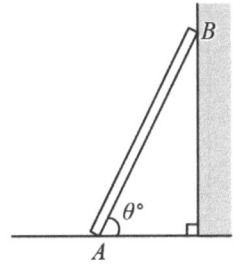

The coefficient of friction between the ladder and the floor is 0.4 and the coefficient of friction between the ladder and the wall is 0.1
  a) Show that the normal reaction at the floor is 2.5 times larger than the normal reaction at the wall.

The normal reaction at the floor is 15 N, find
  b) m in terms of g
  c) the value of θ to 3 significant figures.

11) A uniform ladder AB of length 6a metres and mass m kg rests against a wall. One end, A is resting on rough horizontal ground and the other end B rests on a smooth vertical wall. The ladder is held in limiting equilibrium at an angle of $\theta°$ to the horizontal where $\tan\theta = \frac{24}{7}$

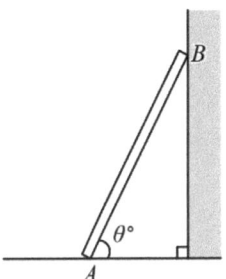

The coefficient of friction between the ladder and the floor is 0.25
A person of mass 6m kg stands at the top of the ladder, whilst another person holds the ladder in place, pushing with a force of magnitude P N horizontally at A.
  a) Show that the reaction of the wall on the ladder is
  $$\frac{91}{48}mg$$
  b) Find in terms of m and g the values of P for the ladder to remain in equilibrium.

12) A uniform plank AB of mass m kg and length a metres with its end A against a rough vertical wall, with B being held in place horizontally by a light inextensible string which is attached to the wall directly above A. The acute angle between the string and the wall is $\theta°$ where $\tan\theta = \frac{4}{3}$. The tension in the string is T N.

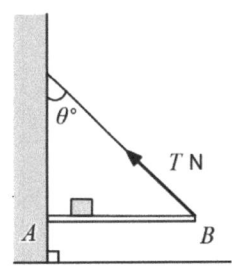

An object of mass 5m kg is placed on the rod, x metres from A
  a) Show that
  $$T = \frac{5mg(10x + a)}{6a}$$
  b) Find the % along the plank from A that the object may be placed given the maximum tension the rope can take is 8mg N.

1.1 Indices
Questions on page 1

1) a) 2

1) b) $\frac{1}{3}$

1) c) 49

1) d) 6

1) e) 729

1) f) $\frac{1}{8}$

1) g) 4

1) h) $\frac{2}{3}$

1) i) $\frac{1}{2}$

2) a) $x^{\frac{3}{2}}$

2) b) $x^{\frac{5}{2}}$

2) c) $x^{\frac{1}{2}}$

2) d) $x^{\frac{3}{2}}$

2) e) $x^{\frac{3}{2}}$

2) f) $x^{\frac{2}{3}}$

2) g) $x^{-\frac{1}{2}}$

2) h) $4x^{-\frac{3}{2}}$

2) i) $x^{\frac{7}{6}}$

3) a) $\sqrt{x}$

3) b) $\frac{1}{\sqrt{x}}$

3) c) $\frac{1}{x^5}$

3) d) $\sqrt[3]{x^2}$

3) e) $\frac{1}{\sqrt[4]{x^3}}$

3) f) $x^{\frac{3}{5}} = \sqrt[5]{x^3}$

4) a) $8a^{15}$

4) b) $2a^3 b^{\frac{5}{3}} c^{\frac{5}{6}}$

4) c) $\frac{4}{3}a^{-\frac{3}{2}}$

4) d) $4a^6 b^{-1} c^2$

4) e) $3a^3 b^7$

5) a) $54a^3$

5) b) $54a^{-5}$

5) c) a^9

5) d) $2a^{\frac{11}{5}}$

6) a) x^4

6) b) $x^{\frac{2}{5}}$

6) c) $\frac{7}{3}x^2 y^{-1}$

6) d) $\frac{4}{3}xy^{\frac{3}{2}}$

7) a) $2x^{\frac{3}{2}} + x^{\frac{1}{2}}$

7) b) $x - x^{\frac{1}{2}} - 2$

7) c) $x^{\frac{1}{2}} + x^{-\frac{1}{2}}$

7) d) $x^{\frac{3}{2}} - x^{-\frac{1}{2}}$

7) e) $4x^{-1} + \frac{1}{3}x^{-\frac{1}{2}}$

7) f) $x^{-1} - 3x^{-\frac{3}{2}}$

8) a) $25\sqrt{a}$

8) b) $84a$

9) a) 4

9) b) $x = \frac{1}{9}$

9) c) $x = \frac{1}{8}$

9) d) $x = \frac{1}{121}$

9) e) $x = \frac{9}{4}$

9) f) $x = \frac{8}{27}$

9) g) $x = \frac{27}{125}$

10) In finding the 7th root, he identifies -3 as an answer.

$x^{\frac{1}{4}} = \sqrt[7]{2187} = 3 \Rightarrow x = 3^4 \Rightarrow x = 81$

11) a) $8^{\frac{1}{2}}$

11) b) 8^{-2}

11) c) 8^{-2}

11) d) $8^{\frac{2}{3}}$

11) e) $8^{-\frac{1}{6}}$

11) f) $\frac{\left(8^{\frac{1}{3}}\right)^6}{(8)^{\frac{1}{2}}} = 8^{\frac{3}{2}}$

11) g) $8^{\frac{7}{9}}$

11) h) $8^{\frac{16}{9}}$

12) a) $x = \frac{1}{2}$

12) b) $x = -\frac{5}{2}$

12) c) $x = 1$

12) d) $x = \frac{1}{8}$

12) e) $x = 2$

12) f) $x = 8$

12) g) $x = \frac{3}{8}$

12) h) $x = 5$

12) i) $x = -2$ or $x = 3$

13) a) 4^{1+x}

13) b) 2^{2x+1}

13) c) 2^{3x+3}

13) d) 2^{1+8x}

13) e) 4×3^{2x}

13) f) 14×2^x or $7 \times 2^{x+1}$

13) g) 17×2^{2x}

14) a) $\frac{x^3+1}{x^4 y+x}$

14) b) $\frac{x^4+xy}{x^3 y^3+y}$

14) c) $\frac{x^6 z - yz^4}{x^4 z + x^2 y^{\frac{1}{2}}}$

14) d)
$\frac{x^8 y^{\frac{1}{2}} - x^2 y^{\frac{3}{2}} z^4}{y^{\frac{1}{2}} + x^3 z^2}$

15) a)
$x = \frac{1}{2}$

15) b)
$x = 1$

15) c)
$x = \frac{1}{2}$

15) d)
$x = 8$

16)
$x = \frac{1}{2}$

17)
$k = \frac{248}{3}$

18)
$x = 9$

19)
$x = -1$
(or $x = -0.2988475575\ldots$)

1.02 Surds
Questions on page 3

1) a)
$4\sqrt{2}$

1) b)
$4\sqrt{3}$

1) c)
$3\sqrt{5}$

1) d)
$10\sqrt{2}$

1) e)
$-20\sqrt{2}$

2) a)
$\sqrt{10}$

2) b)
$5\sqrt{3}$

2) c)
12

2) d)
$\frac{1}{3}$

2) e)
$\sqrt{2}$

2) f)
$5x$

3) a)
$\sqrt{5}$

3) b)
$4\sqrt{2}$

3) c)
$\sqrt{3}$

3) d)
$2\sqrt{2}$

3) e)
$9\sqrt{3}$

3) f)
$\sqrt{6}$

3) g)
$6\sqrt{3}$

3) h)
$5\sqrt{14} + 2\sqrt{6}$

3) i)
$4\sqrt{2a} + 2\sqrt{a} - 2\sqrt{7a}$

4) a)
$\sqrt{42}$

4) b)
6

4) c)
12

4) d)
$10\sqrt{15}$

4) e)
-18

4) f)
24

4) g)
72

4) h)
$40\sqrt{5}$

5) a)
$\sqrt{20}$

5) b)
$\sqrt{48}$

5) c)
$\sqrt{18}$

5) d)
$\sqrt{28}$

5) e)
$\sqrt{172}$

6) a)
$\sqrt{a^2 b}$

6) b)
$\sqrt{4a}$

6) c)
$\sqrt{a^5}$

6) d)
$\sqrt{a^6 b^3}$

7)
$5\sqrt{3}\sqrt{7}$

8)
$10:4:3$

9) a)
$\frac{3\sqrt{2}}{2}$

9) b)
$2\sqrt{3}$

9) c)
$\frac{3\sqrt{10}}{5}$

9) d)
$\frac{\sqrt{6}}{3}$

9) e)
$\sqrt{10}$

9) f)
2

10) a)
$\sqrt{5} + 5$

10) b)
$2\sqrt{11} + 11$

10) c)
$4\sqrt{3} + 8\sqrt{5}$

10) d)
2

10) e)
$3\sqrt{6} + 8$

10) f)
$2\sqrt{2} + 6$

10) g)
$4\sqrt{3} - 6\sqrt{6} + \sqrt{2} - 3$

11) a)
$\sqrt{2} - 1$

11) b)
$3\sqrt{5} + 6$

11) c)
$-\frac{2}{5} - \frac{2}{5}\sqrt{6}$

11) d)
$\frac{a - \sqrt{b}}{a^2 - b}$

11) e)
$\frac{a + \sqrt{2b}}{a^2 - 2b}$

12) a)
$8\sqrt{3}$

12) b)
$-\frac{1}{6}\sqrt{6}$

12) c)
$\frac{7}{2}\sqrt{6}$

13)
$4 + \sqrt{6}$

14) a)
$-\frac{1}{2}\sqrt{6} - \sqrt{3}$

14) b)
$-\frac{14}{11} + \frac{9}{11}\sqrt{5}$

14) c)
$-\frac{10\sqrt{3}}{3} - \frac{5\sqrt{21}}{3} + \frac{4}{3} + \frac{2\sqrt{7}}{3}$

14) d)
$-3 - \frac{8}{3}\sqrt{3}$

14) e)
$\frac{2 - (\sqrt{2} + 2)\sqrt{k} + \sqrt{2}k}{4 - 2k}$

15)
$x = -3 - 4\sqrt{2}$

16)
$5\sqrt{2} - 2\sqrt{5}$

17)
$-\frac{5}{2} - \frac{1}{2}\sqrt{21}$

18)
$p = -\frac{a+6}{5}$ and $q = -\frac{a+1}{5}$

19)
$3\sqrt{3}$

20)
Show that

21)
$a = \frac{k+18}{26}$ and $b = \frac{-3k-2}{26}$

22)
$a = \frac{k+2}{1-k}$ and $b = -\frac{3}{1-k}$

23) a)
$2a + 1 - 2\sqrt{a}\sqrt{a+1}$

23) b)
$13 - 2\sqrt{42}$

24)
9

25)
Show that

26)
Show that

1.03 Coordinate Geometry: Linear Graphs
Questions on page 5

1) a)
$2x + y = 9$

1) b)
$-3x + y = 1$

1) c)
$-2x + 3y = -7$

1) d)
$11x + 2y = -56$

1) e)
$3qx - py = 2pq$

1) f)
$x + y = \frac{p^2+1}{p}$

2) a)
$m_1 = -\frac{1}{2}$ and $m_2 = 2$

2) b)
$m_1 = \frac{2}{3}$ and $m_2 = -\frac{3}{2}$

2) c)
$m_1 = -\frac{1}{4}$ and $m_2 = 4$

2) d)
$m_1 = \frac{5}{7}$ and $m_2 = -\frac{7}{5}$

2) e)
$m_1 = -\frac{20}{17}$ and $m_2 = \frac{17}{20}$

2) f)
$m_1 = -\frac{a}{b}$ and $m_2 = \frac{b}{a}$

3) a)
$-\frac{2}{3} \times -\frac{3}{2} \neq -1$ ∴ not perpendicular.

3) b)
$\frac{1}{5} \times 5 \neq -1$ ∴ not perpendicular.

3) c)
$-2 \times -2 \neq -1$ ∴ not perpendicular (they are parallel).

3) d)
$-4 \times \frac{1}{4} = -1$ ∴ perpendicular.

3) e)
$-\frac{a}{b} \times \frac{b}{a} = -1$ ∴ perpendicular.

4) a)
$\frac{a}{b} = \frac{p}{q}$ or $aq = bp$

4) b)
$\frac{ap}{bq} = -1$ or $ap = -bq$

5) a)
$(0, 6)$

5) b)
$\left(\frac{7}{2}, 1\right)$

5) c)
$\left(-5, \frac{9}{2}\right)$

5) d)
$\left(-\frac{7}{4}, 5\right)$

5) e)
$\left(\frac{3}{2}p, 2q\right)$

5) f)
$\left(\frac{1}{2}p + \frac{1}{2p}, \frac{1}{2}p + \frac{1}{2p}\right)$

6) a)
$k = -\frac{8}{5}$

6) b)
$k = -4$

7) a)
$2x - y = 1$

7) b)
$x + y = 11$

7) c)
$6x + 16y = 111$

7) d)
$5x - 7y = -7$

7) e)
$16x - 26y = -197$

7) f)
$2px + 4qy = 8q^2 + 3p^2$

8)
$x - 2y = -9$

9)
$3x - y = 3$

10) a)
63.4°

10) b)
4.4°

10) c)
15.3°

11)
$\frac{121}{4}$ units²

12)
$\frac{2}{3k}$ units²

13)
Show that

14)
Show that

15)
$k = 21$

16)
$p = -1$

17)
$p = 1, q = 10$

18)
Show that

19) a)
$k = 27$

19) b)
$k = \frac{13}{6}$

20) a)
$y = x + 5$

20) b)
$y = -x + 11$

20) c)
$x = 3, y = 8$

20) d)
AP: $2\sqrt{10}$
BP: $2\sqrt{10}$
CP: $2\sqrt{10}$

21)
5

22) a)
Show that

22) b)
$y = -\frac{1}{5}x + 5$

22) c)
$y = -\frac{1}{5}x + 12$

22) d)
$k = 9$

22) e)
AD: $\sqrt{26}$
BC: $3\sqrt{26}$

22) f)
$y = 5x - 40$

22) g)
$\sqrt{26}$

22) h)
70 units^2

23) a)
Show that

23) b)
Show that

24)
A kite.

1.04 Graph Sketching: Linear Graphs
Questions on page 7

1) a)
$x = 3$

1) b)
$y = -4$

2) a)
$-3x + y = 6$

2) b)
$2x + 5y = -5$

2) c)
$-4x + 2y = -5$

2) d)
$x + 3y = 15$

3) a)

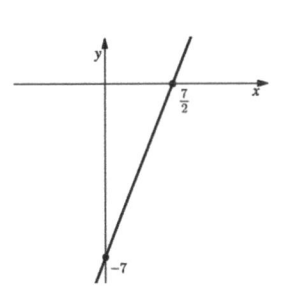

3) b)

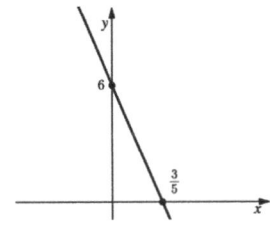

3) c)

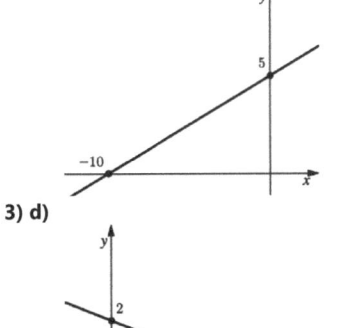

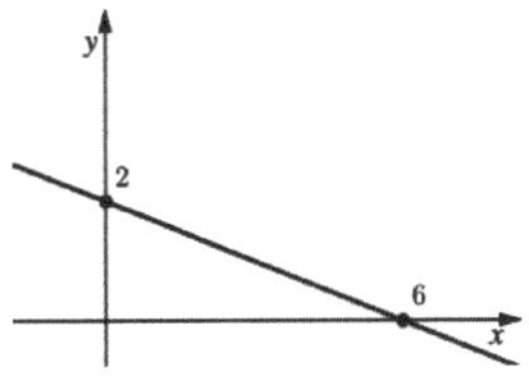

3) d)

4) a)
$8x + 4y = 1$

4) b)
$-25x + 10y = 12$

4) c)
$-28x + 8y = -1$

4) d)
$5x + 20y = -32$

5) a)

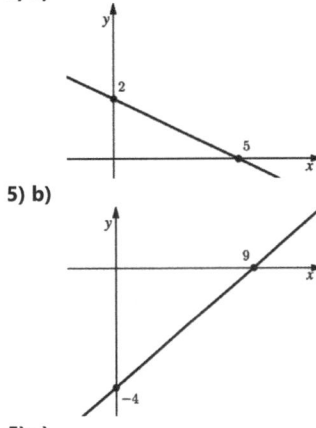

5) b)

5) c)

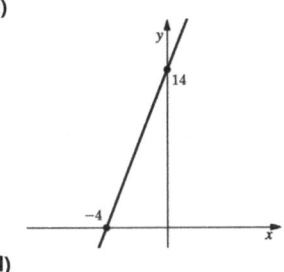

5) d)

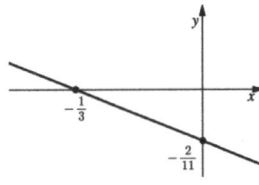

6) a)

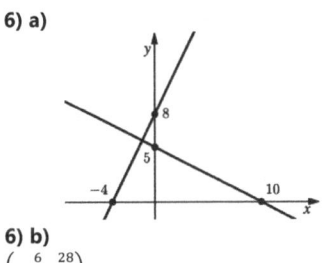

6) b)
$\left(-\frac{6}{5}, \frac{28}{5}\right)$

6) c)
$\frac{196}{5}$

6) d)
$\frac{9}{5}$

7) a)
$x + \frac{2}{3}y = -5$

7) b)
$x - \frac{1}{5}y = \frac{4}{3}$

7) c)
$x + \frac{7}{8}y = -\frac{2}{5}$

7) d)
$x - \frac{20}{3}y = \frac{17}{6}$

8) a)

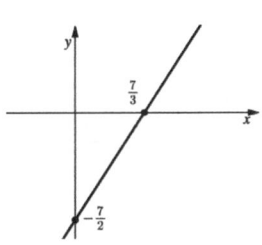

8) b)

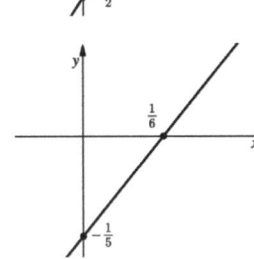

8) c)

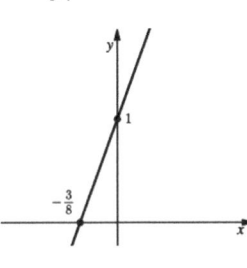

8) d)

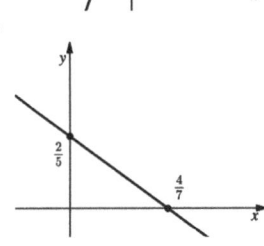

9)
$y = \frac{2k-2}{3k}x + \frac{5k+1}{3}$

10) a)

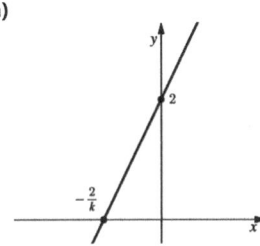

10) b)

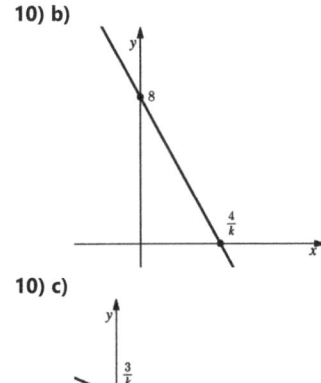

10) c)

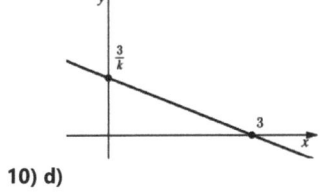

10) d)

10) e)

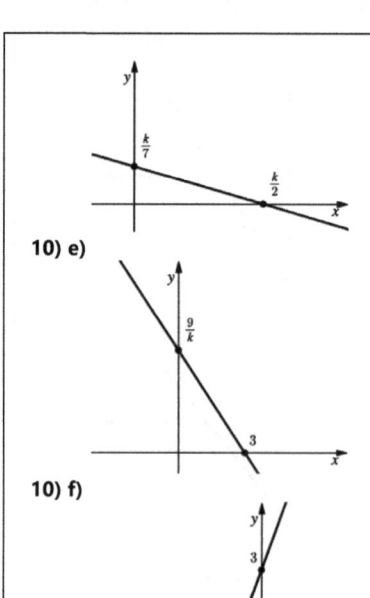

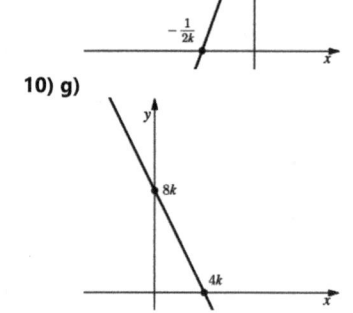

10) f)

10) g)

11) a)
$l_1: y - (12k^2 - 5k) = 3k^2(x - 4)$
11) b)
$l_2: y - \left(\frac{3}{k} + \frac{1}{k^2}\right) = -\frac{1}{2k^2}(x + 2)$
11) c)
$\left(\frac{10k^3 + 6k}{6k^4 + 1}, \frac{18k^3 - 5k}{6k^4 + 1}\right)$

1.05 Linear Modelling
Questions on page 77

1) a)
$y = \frac{1}{2}x$
1) b)
$y = 1.2x$
1) c)
$y = 1.29x$

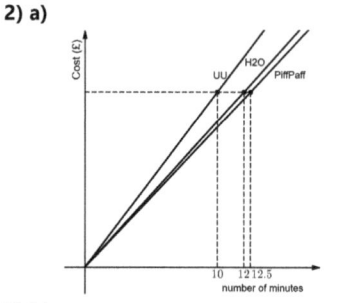

2) a)

2) b)
UU: $C = 0.3t$, H$_2$O: $C = 0.25t$
Piffpaff: $C = 0.24t$

3) a)
$y = x + 8$

3) b)
$y = \frac{2}{5}x + 2$
3) c)
$y = \frac{2}{3}x + 5$

4) a)
$y = 3x + 4$
4) b)
$y = 0.5x + 2$
4) c)
$y = \frac{5}{12}x + \frac{23}{4}$

5) a)
Answers similar to $d = 2.8t + 0.2$
5) b)
Answers close to 7.2 feet
5) c)
Depth per minute (feet per minute)
5) d)
Vertical intercept is 0.2 which is the theoretical time that the penguins will dive in 0 minutes
5) e)
This is not possible, so the model is not valid for low values.

6) a)
9.6 seconds
6) b)
The 10.2 represents a theoretical time for how long the gold medallist ran for in 1960.
The -0.0082 represents the rate of change in time per year, that is, for every 1 year that passes, the time to complete the 100m race goes down by 0.0082 seconds.
6) c)
As the years pass, the times are getting smaller by 0.0082 a year, so eventually it will become too quick for it to be possible for a human to achieve it. Eventually (in 3204) the model will predict a negative race time!

7) a)
Zelacoach: £800 cost for hiring the coach
Stagebus: £500 cost for hiring the coach
7) b)
Zelacoach: $C = \frac{45}{11}m + 800$
Stagebus: $C = \frac{75}{11}m + 500$
7) c)
P is $(110, 1250)$
When the number of miles is 110, both companies charge the same amount, £1250

8)
$(100, 2.5)$
When ice creams are sold at £2.50, 100 ice creams are sold. For a higher price point, the demand goes down as does the number sold. For a lower price point, the demand goes up as does the number sold.

9) a)
£50 call-out charge
9) b)
40
9) c)
£10

10) a)
Show that
10) b)
158

1.06 Quadratics: Introduction
Questions on page 12

1) a)
$(x + 9)(x + 5)$
1) b)
$(x - 3)(x + 6)$
1) c)
$x(x - 7)$
1) d)
$(x + 6)(2x - 1)$
1) e)
$(x - 5)(2x - 7)$
1) f)
$(2x - 3)(4x - 1)$
1) g)
$(3x + 1)(2x + 5)$
1) h)
$(2x + 3)(1 - x)$
1) i)
$(x - 3)(x + 3)$
1) j)
$(2x - \sqrt{5})(2x + \sqrt{5})$

2) a)
$(x - a)(x + a)$
2) b)
$(ax - b)(ax + b)$
2) c)
$(x - \sqrt{a})(x + \sqrt{a})$

3) a)
$(x + 2)(8 - x)$
3) b)
$(x - 2)(2x - 1)$
3) c)
$(x - 5)(12x - 59)$
3) d)
$(x - 5)(x - 9)$
3) e)
$2(x - 2)(3x - 14)$

4) a)
$x = 0$ or $x = 1$

412

4) b)
$x = 2$ or $x = -2$
4) c)
$x = \sqrt{8} = 2\sqrt{2}$ or $x = -\sqrt{8} = -2\sqrt{2}$
4) d)
$x = 0$ or $x = -2$
4) e)
$x = 5$
4) f)
$x = 6$
4) g)
$x = 5$ or $x = -4$
4) h)
$x = -3 \pm \sqrt{6}$
4) i)
$x = 1 \pm \sqrt{6}$

5) a)
$x = 4 \pm \sqrt{11}$
5) b)
$x = \pm 2\sqrt{5}$
5) c)
$x = -6$ or $x = -2$
5) d)
$x = \frac{5}{6} \pm \frac{\sqrt{37}}{6}$
5) e)
$x = \pm\sqrt{\frac{b}{a}}$
5) f)
$x = \frac{b}{a}$ or $x = 0$

6)
Student 1 spotted a solution but did not spot the second one.
Student 2 removed the solution $x = 5$ when dividing through by $(x - 5)$.
$(x-5)^2 - (x-5) = 0$
$\Rightarrow (x-5)[(x-5) - 1] = 0$
$\Rightarrow (x-5)(x-6) = 0$
$\therefore x = 5$ or $x = 6$

7)
$x = 9$

8)
$x = 4$ or $x = \frac{1}{4}$

9)
$x = \pm\sqrt{5}$ or $x = \pm\frac{1}{\sqrt{2}}$

10)
$x = \sqrt[3]{\frac{5}{3}}$ or $x = -2$

11)
$x = \frac{1}{4} \pm \frac{\sqrt{57}}{12}$

12) a)
$(x-4)^2 - 16$
12) b)
$(x+3)^2 - 9$
12) c)
$(x-5)^2 + 15$
12) d)
$(x-4)^2 - 36$
12) e)
$\left(x+\frac{5}{2}\right)^2 + \frac{3}{4}$
12) f)
$2(x-4)^2 - 52$
12) g)
$4(x-3)^2 - 11$
12) h)
$3\left(x-\frac{7}{6}\right)^2 - \frac{37}{12}$

13) a)
$1 - (x+1)^2$
13) b)
$\left(x+\frac{5}{2}\right)^2 - \frac{1}{4}$
13) c)
$-2(x+1)^2 + 7$
13) d)
$-2\left(x+\frac{7}{4}\right)^2 + \frac{57}{8}$
13) e)
$-3\left(x-\frac{5}{6}\right)^2 + \frac{145}{12}$
13) f)
$-4x^2 + 25$ is already in completed square form
13) g)
$\left(x+\frac{p}{2}\right)^2 - \frac{p^2}{4} + r$

14)
(a, b) is the coordinate of the minimum of the curve

15) a)
Minimum at $(2, -12)$
15) b)
Minimum at $(-2, -8)$
15) c)
Maximum at $(1, 12)$
15) d)
Maximum at $(-2, 0)$
15) e)
Minimum at $(-4, 5)$
15) f)
Minimum at $\left(-\frac{5}{2}, \frac{1}{2}\right)$
15) g)
Maximum at $(4, -100)$
15) h)
Minimum at $(3, -11)$
15) i)
Minimum at $(-1, 2)$
15) j)
Maximum at $(-1, 5)$
15) k)
Minimum at $(0, -5)$
15) l)
Maximum at $(0, 12)$
15) m)
Minimum at $(b, 0)$
15) n)
Minimum at $\left(-\frac{p}{2}, r - \frac{p^2}{4}\right)$

16) a)
$x = 4$
16) b)
$x = -3$
16) c)
$x = 1$
16) d)
$x = -1$
16) e)
$x = \frac{7}{2}$
16) f)
$x = 3$
16) g)
$x = \frac{a+b}{2}$
16) h)
$x = p$

17) a)
$y = (x-4)(x+1)$
Vertex at $\left(\frac{3}{2}, -\frac{25}{4}\right)$
17) b)
$y = -(x+2)(x-5)$
Vertex at $\left(\frac{3}{2}, \frac{49}{4}\right)$

18) a)
$y = (x-2)^2 + 2$
18) b)
$y = -(x+1)^2 + 8$
18) c)
$y = 2(x-3)^2 + 5$
18) d)
$y = \left(x-\frac{1}{2}\right)^2 - \frac{9}{4}$
18) e)
$y = -(x+2)^2 + 1$

19) a)
$y = x^2 - 2x + 6$
19) b)
$y = 2x^2 - 12x + 17$
19) c)
$y = -20x^2 - 80x - 76$

20) a)
$(2, 7)$ and $(-1, 1)$
20) b)
$(1, 1)$ and $(-1, 5)$
20) c)
$(2, 0)$ and $(-2, 4)$
20) d)
$(1, 3)$ and $(-1, 1)$
20) e)
$(1, -1)$ and $(-4, 24)$

21) a)
$(0, 1)$ and $\left(\frac{1}{2}, \frac{5}{4}\right)$
21) b)
$(2, 1)$ and $(-1, 4)$

21) c)
$(3, 2)$
21) d)
$(2, 3)$

22)
$(-2, -10)$ and $(4, 8)$

23)
$(-1 + \sqrt{3}, 3 + 5\sqrt{3})$ and
$(-1 - \sqrt{3}, 3 - 5\sqrt{3})$

1.07 Algebra: Simultaneous Equations
Questions on page 15

1) a)
$x = 1, y = 1$
1) b)
$x = 1, y = -\frac{1}{2}$
1) c)
$x = -3, y = 4$
1) d)
$x = 4, y = 8$

2) a)
$x = \frac{25}{2}$
2) b)
$k = \frac{2}{5}$

3) a)
$b = 8$
3) b)
$a = \frac{11}{10}$

4)
$4y + x = 32$ and $3y - x = 3$

5) a)
$(3, 10), (-1, 2)$
5) b)
$(-1, -5), (8, 22)$
5) c)
$(0, 0), \left(\frac{1}{4}, \frac{1}{8}\right)$

6) a)
Show that
6) b)
$(1, -3)$

7) a)
$(4, 3), (7, -6)$
7) b)
$(-3, 1), (2, 6)$
7) c)
$\left(\frac{1}{5}, \frac{62}{5}\right), (-5, 2)$

8) a)
$(0, 4), (4, 0)$
8) b)
$(1, 2)$

8) c)
$(4, 2), \left(\frac{4}{7}, \frac{26}{7}\right)$
8) d)
$(3, 1), \left(-\frac{11}{5}, -\frac{8}{5}\right)$

9)
$y = \frac{-52 \pm \sqrt{2704 - 2880}}{10}$ which has no real solutions.

10)
$p = 2, q = 3$

11)
$p = \frac{8}{3}, q = \frac{22}{3}$

12)
$\frac{10}{3}$ and 6

13)
$x = \frac{19}{7}, y = \frac{3}{7}$

14)
$x = \frac{5}{2}, y = 2$

1.08 Graph Sketching: Quadratic Graphs
Questions on page 16

1) a)
$y = (x + 3)(x - 3)$
1) b)
$y = (x + 2)(x - 4)$
1) c)
$y = (x + 3)(x + 1)$

2) a)

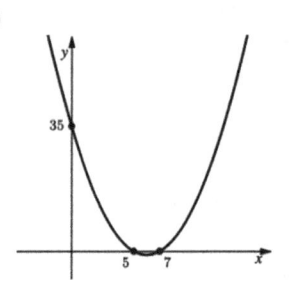

2) b)

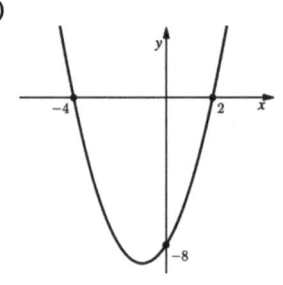

3) a)
$y = -(x + 4)(x - 4)$
3) b)
$y = -(x + 5)(x - 1)$
3) c)
$y = -(x - 2)(x - 6)$

4) a)

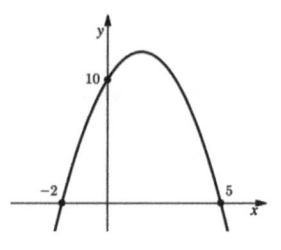

4) b)

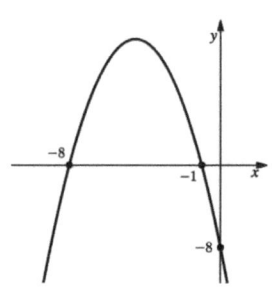

5) a)
$y = (x - 2)^2 + 1$
5) b)
$y = (x + 2)^2 + 3$
5) c)
$y = (x + 3)^2 - 5$

6) a)

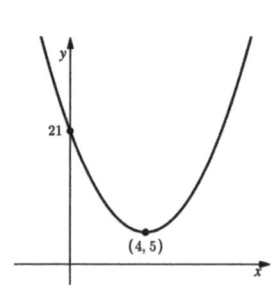

6) b)

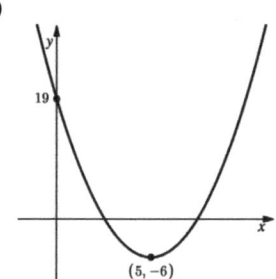

7) a)
$y = -(x - 2)^2 - 3$
7) b)
$y = -(x - 4)^2 + 3$
7) c)
$y = -(x + 2)^2 + 6$

8) a)

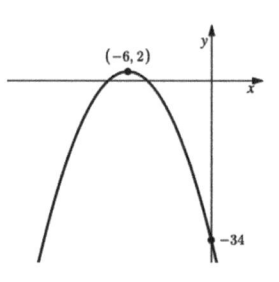

8) b) [graph: max at (3, −4), y-intercept −13]

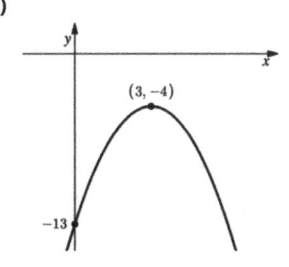

9) a) $y = 2(x+3)(x-2)$
9) b) $y = 4(x+4)(x+1)$
9) c) $y = -\frac{3}{2}(x+3)(x-2)$

10) a) [graph: roots −2 and 1, y-intercept −6]

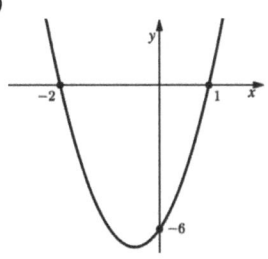

10) b) [graph: roots −4 and −3, y-intercept −24]

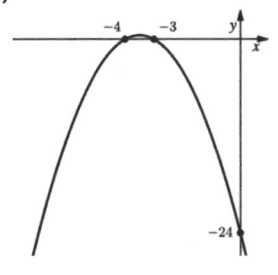

11) a) $y = 3(x-3)^2 + 4$
11) b) $y = \frac{1}{3}(x+6)^2 - 5$
11) c) $y = -\frac{5}{4}(x-4)^2 + 4$

12) a) [graph: min (−1, −2), y-intercept 3]

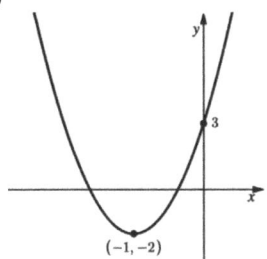

12) b) [graph: max (6, −12), y-intercept −66]

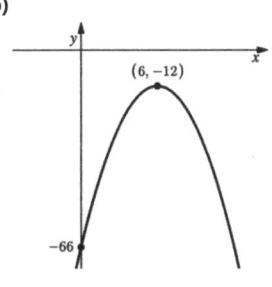

13) a) $y = 2x^2 - 4x - 4$
13) b) $y = -\frac{5}{2}x^2 - 20x - 25$
13) c) $y = -\frac{3}{4}x^2 + 6x$

14) [graph: roots −16 and 0, min (−8, −32)]

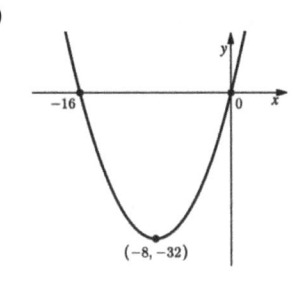

15) a) [graph: roots −2 and 8, y-intercept −16, min (3, −25)]

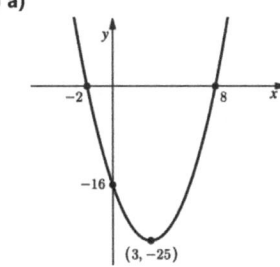

15) b) [graph: roots −4 and 6, y-intercept −24, min (1, −25)]

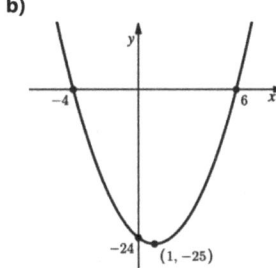

15) c) [graph: roots −6 and −2, y-intercept 36, min (−4, −12)]

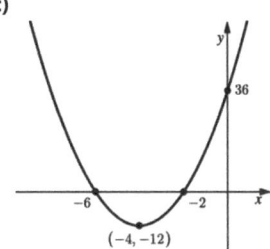

15) d) [graph: roots 2 and 4, max (3, 4), y-intercept −32]

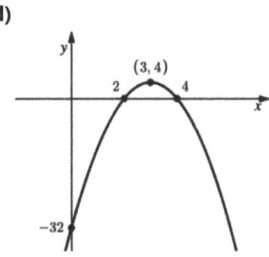

16) a) [graph: roots 3 and 7, y-intercept 21, min (5, −4)]

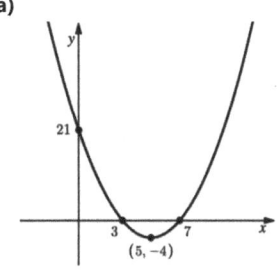

16) b) [graph: roots −6 and −2, y-intercept 4, min $(-4, -\frac{4}{3})$]

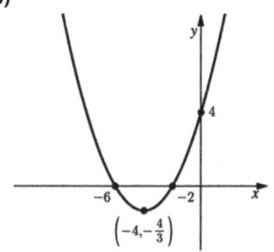

16) c) [graph: roots −9 and −5, max (−7, 4), y-intercept −45]

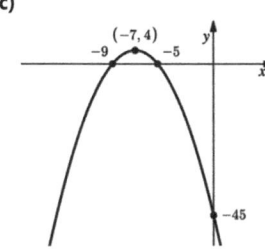

16) d) [graph: roots −4 and 8, y-intercept 56, max (2, 63)]

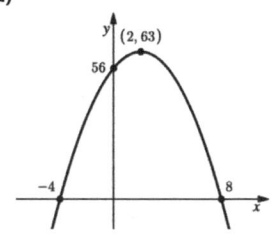

17) a) [graph: roots $6-\sqrt{2}$ and $6+\sqrt{2}$, y-intercept 34, min (6, −2)]

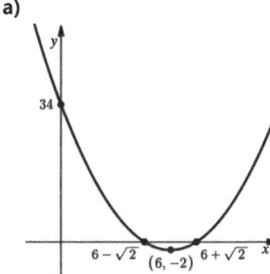

17) b) [graph: roots $-2-2\sqrt{2}$ and $-2+2\sqrt{2}$, max (−2, 8), y-intercept 4]

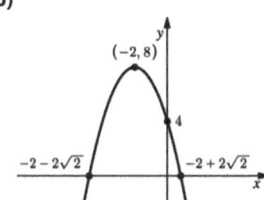

17) c) [graph: roots $\frac{-12-3\sqrt{3}}{2}$ and $\frac{-12+3\sqrt{3}}{2}$, y-intercept 39, min (−6, −9)]

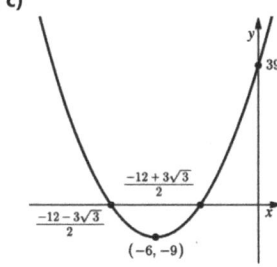

17) d) [graph: roots $\frac{6-3\sqrt{2}}{2}$ and $\frac{6+3\sqrt{2}}{2}$, max (3, 12), y-intercept −12]

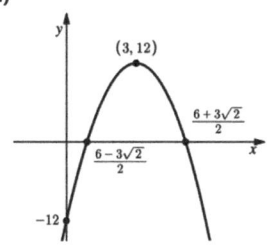

18) a)

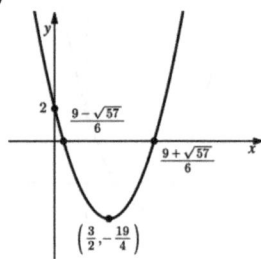

Graph with points 2, $\frac{9-\sqrt{57}}{6}$, $\frac{9+\sqrt{57}}{6}$, minimum $\left(\frac{3}{2}, -\frac{19}{4}\right)$

18) b)

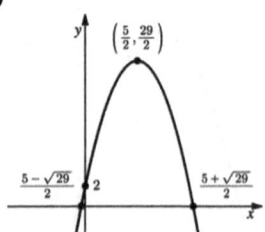

Graph with vertex $\left(\frac{5}{2}, \frac{29}{2}\right)$, points $\frac{5-\sqrt{29}}{2}$, 2, $\frac{5+\sqrt{29}}{2}$

18) c)

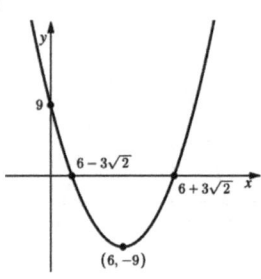

Graph with point 9, $6-3\sqrt{2}$, $6+3\sqrt{2}$, minimum $(6, -9)$

18) d)

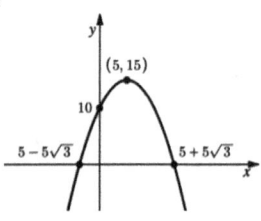

Graph with vertex $(5, 15)$, point 10, $5-5\sqrt{3}$, $5+5\sqrt{3}$

19) a)

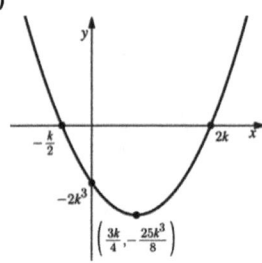

Graph with points $-\frac{k}{2}$, $2k$, $-2k^3$, minimum $\left(\frac{3k}{4}, -\frac{25k^3}{8}\right)$

19) b)

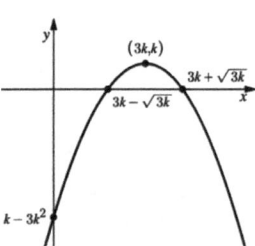

Graph with vertex $(3k, k)$, points $3k-\sqrt{3k}$, $3k+\sqrt{3k}$, $k-3k^2$

19) c)

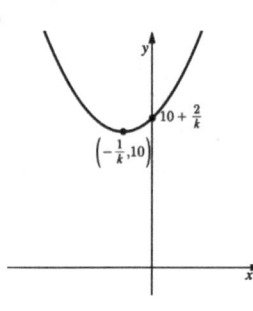

Graph with vertex $\left(-\frac{1}{k}, 10\right)$, point $10+\frac{2}{k}$

19) d)

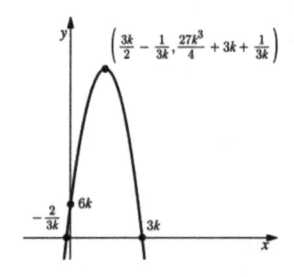

Graph with vertex $\left(\frac{3k}{2}-\frac{1}{3k}, \frac{27k^3}{4}+3k+\frac{1}{3k}\right)$, points $-\frac{2}{3k}$, $6k$, $3k$

20)

Graph with points $\left(\frac{1-k}{3}, \frac{4k-1}{3}\right)$, k^2, k, $-\frac{k}{3}$

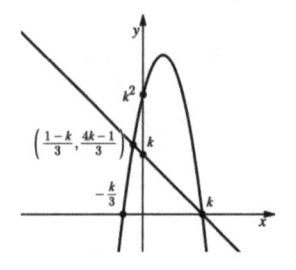

1.09 Quadratics: Problem Solving
Questions on page 20

1)
Length: 24m, Width: 14m

2)
$x = 2$

3) a)
Minimum at $(-1, 2)$ so no real roots
3) b)
Maximum at $(-1, 2)$ so 2 real roots
3) c)
Minimum at $(4, -3)$ so 2 real roots
3) d)
Maximum at $(4, -3)$ so no real roots
3) e)
Minimum at $(-2, 0)$ so 1 (repeated) real root

4) a)

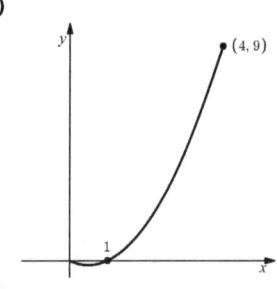

Graph with points $(4, 9)$, 1

4) b)
Minimum at $x = \frac{1}{2}$, $y = -\frac{3}{16}$

5) a)

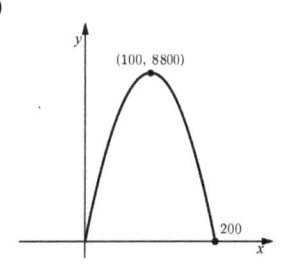

Graph with vertex $(100, 8800)$, point 200

5) b)
Maximum at $x = 100, y = 8800$

6) a)

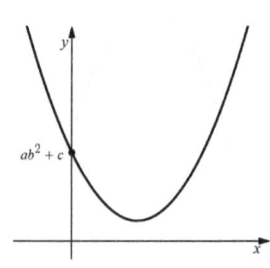

Graph with point $ab^2 + c$

6) b)

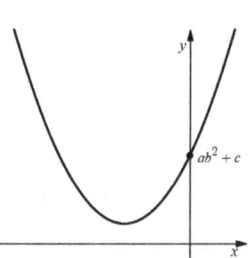

Graph with point $ab^2 + c$

6) c)

Graph with point $ab^2 + c$

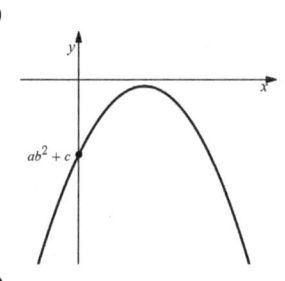

6) d)

Graph with point $ab^2 + c$

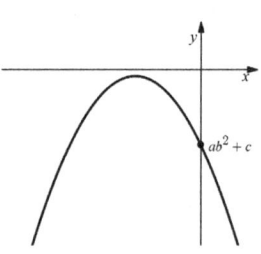

7)
$a > 0, b < 0$ and $c > 0$

8)
$a < 0, b > 0, c > 0$ and $ab^2 + c < 0$

9) a)
$\left(0, \frac{7}{4}\right)$
9) b)
$\left(-\frac{3}{2} + \frac{\sqrt{2}}{2}, 0\right)$ and $\left(-\frac{3}{2} - \frac{\sqrt{2}}{2}, 0\right)$

10) a)
$(0, 1)$
10) b)
$(2 - \sqrt{5}, 0)$ and $(2 + \sqrt{5}, 0)$

11)
Show that

12)
$k = -3$

13)
$k = \pm 2$

1.10 Quadratics: The Discriminant
Questions on page 21

1) a) 28
1) b) 36
1) c) 21
1) d) 24
1) e) 0
1) f) -7
1) g) -3
1) h) $k^2 - 8$

2) a) no roots
2) b) 2 roots
2) c) 2 roots
2) d) 0 roots
2) e) 1 repeated root
2) f) 2 roots
2) g) 1 repeated root

3) Show that

4) a) $k = -12$ or $k = 12$
4) b) $k = -12$ or $k = 12$
4) c) $k = 4$
4) d) $k = -2$
4) e) $k = \frac{1}{2}$ or $k = -\frac{1}{2}$

5) $p = \frac{4}{9}$

6) Show that

7) Show that

8) $k = -2$ or $k = 2$

9) a) $k = \frac{51}{20}$
9) b) $\left(0, \frac{51}{100}\right)$ and $\left(\frac{51}{20}, 0\right)$

10)

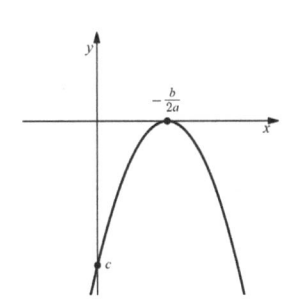

11) Show that

12) a) Show that
12) b) $k = \frac{1}{16}$
12) c) $\left(-\frac{1}{4}, \frac{3}{2}\right)$

13) $p = q$

14) a) $k = -1$ or $k = -33$
14) b) $\therefore (0, -4)$ and $(-8, -204)$

15) $k = 9$

1.11 Quadratics: Modelling
Questions on page 22

1) a) 19.6 m
1) b) $t = 1s, t = 3s$
1) c) $t = 1$ is when it reaches a height 14.7m
$t = 3$ is when it drops back to the height of 14.7m on the way back down.

2) a) 5s
2) b) $-2\left(t - \frac{5}{2}\right)^2 + \frac{25}{2}$
2) c) Maximum height is 12.5m after 2.5s has passed.
2) d) $-2t^2 + 10t + 1$

2) e) $\left(\frac{5}{2}, \frac{27}{2}\right)$

3) a) The posts are 0.5m high
3) b) $\frac{1}{100}\left(x - \frac{5}{4}\right)^2 + \frac{31}{64}$
3) c) Horizontal position of the minimum height of the rope is 1.25m from the left post
Vertical position of the minimum height of the rope $\frac{31}{64}$m ≈ 0.48m (2 SF)
3) d) The rope ends at the top of the right post which is 2.5m from the left post. So $x > 2.5$ would be modelling beyond the length of the rope.

4) a)

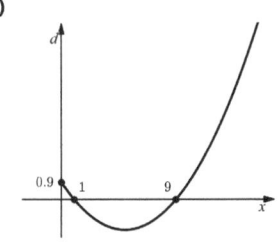

4) b) 1.6 metres below the surface.
4) c) 8 metres
4) d) 1 m
4) e) $x > 9$

5) a)

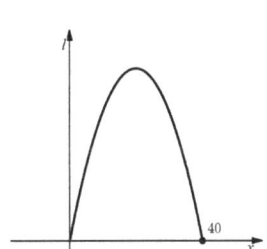

5) b) $I = 1000 - \frac{5}{2}(x - 20)^2$
5) c) £1000
5) d) £20
5) e) $x = £7.35$ or $x = £32.65$

6) a) £6 per book
6) b) Max profit of £0.10 per book
6) c) No profit when charging £2 or £10 per book

6) d)
12.5p loss per book

7) a)
15 days
7) b)
30 cm
7) c)
The quadratics reach a maximum then reduce, whilst the plants will either keep growing or reach a fixed height.

8)
$h = -2.5t^2 + 10t$

9) a)
$a = 1.5$ m.
9) b)
8 m
9) c)
2 m above the ground
9) d)
It doesn't hit it.

10)
$y = \frac{4}{875}x(350 - x)$

1.12 Coordinate Geometry: Circles
Questions on page 24

1) a)
Centre $(2, 7)$, Radius $= 3$
1) b)
Centre $(0, -2)$, Radius $= 3.5$
1) c)
Centre $(-8, 5)$, Radius $= \sqrt{16} = 4$
1) d)
Centre $(-12, -9)$, Radius $= \sqrt{19}$
1) e)
Centre $(3, -16)$, Radius $= \sqrt{28} = 2\sqrt{7}$
1) f)
Centre $\left(\frac{1}{2}, 0\right)$, Radius $= \sqrt{\frac{9}{4}} = \frac{3}{2}$
1) g)
Centre $(-2k, k - 1)$,
Radius $= \sqrt{100k} = 10\sqrt{k}$

2) a)
$(x + 3)^2 + (y + 8)^2 = 81$
2) b)
$(x - 4)^2 + (y + 3)^2 = 25$
2) c)
$x^2 + y^2 = 10\,000$
2) d)
$(x - 6)^2 + (y - 1)^2 = 17$
2) e)
$(x + 11)^2 + (y - 6)^2 = 8$
2) f)
$\left(x + \frac{2}{3}\right)^2 + \left(y + \frac{3}{2}\right)^2 = \frac{16}{81}$
2) g)
$(x - k + 5)^2 + (y - 7 + k)^2 = k^2 + 4k + 4$

3) a)
Area $= 25\pi$ units^2
3) b)
Area $= \frac{49\pi}{9}$ units^2
3) c)
Area $= 121\pi$ units^2

4) a)
$x^2 + y^2 + 12y + 34 = 0$
4) b)
$x^2 - 20x + y^2 - 30 = 0$
4) c)
$x^2 + 4x + y^2 - 16y + 2 = 0$
4) d)
$x^2 - 22x + y^2 - 30y - 1254 = 0$
4) e)
$x^2 + 8kx + y^2 - 4ky = 0$

5) a)
$(x + 1)^2 + (y + 3)^2 = 18$
5) b)
$(x - 4)^2 + (y - 10)^2 = 900$
5) c)
$(x + 3)^2 + (y - 7)^2 = 140$
5) d)
$\left(x - \frac{5}{2}\right)^2 + \left(y - \frac{9}{2}\right)^2 = 55$
5) e)
$\left(x - \frac{1}{3}\right)^2 + \left(y + \frac{2}{9}\right)^2 = 3$
5) f)
$\left(x - \frac{3}{4}\right)^2 + \left(y - \frac{5}{8}\right)^2 = 18$
5) g)
$\left(x + \frac{7}{5}\right)^2 + \left(y + \frac{5}{4}\right)^2 = 1$
5) h)
$\left(x + \frac{1}{2}k\right)^2 + (y - 3k)^2 = \frac{37}{4}k^2 - 3k$

6) a)
This does not represent a circle.
6) b)
This does not represent a circle.
6) c)
This does not represent a circle.
6) d)
This does not represent a circle.
6) e)
This is a circle, centre $\left(-\frac{1}{3}, \frac{2}{3}\right)$ with radius $\sqrt{\frac{8}{9}} = \frac{2\sqrt{2}}{3}$

7)
$k > -52$

8)
$k < 135$

9) a)
Does not intersect the x or y axes
9) b)
Intersects the y-axis but not the x-axis
9) c)
Intersects the x-axis but not the y-axis

9) d)
Intersects the x-axis but not the y-axis

10) a)
$0 < k < 6$
10) b)
$0 < k < \sqrt{10}$
10) c)
$k > 4$
10) d)
$k > 7$

11) a)
$0 < k < 7$
11) b)
$0 < k < 3$
11) c)
$k > 6$

12) a)

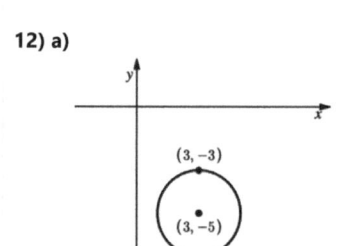

Circle with points $(3, -3)$ and $(3, -5)$ (centre).

12) b)

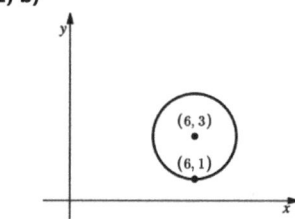

Circle with points $(6, 3)$ and $(6, 1)$.

12) c)

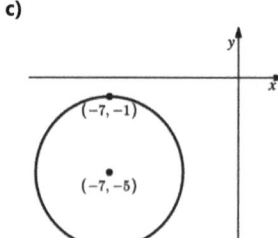

Circle with points $(-7, -1)$ and $(-7, -5)$ (centre).

12) d)

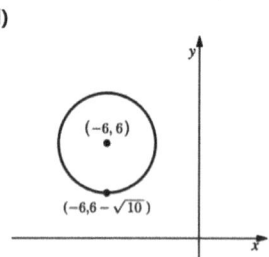

Circle with centre $(-6, 6)$ and point $(-6, 6 - \sqrt{10})$.

13) a)

Circle passing through origin, with points at 6 on y-axis and centre at 3.

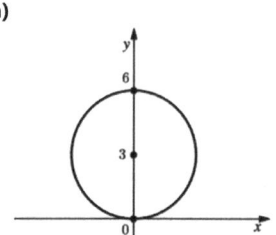

13) b) *[circle diagram: centre (0, −2), radius on x-axis ±2√3, y-intercepts 2 and −6]*

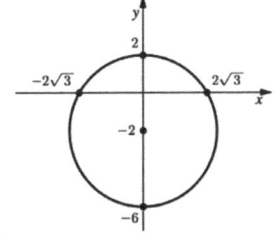

13) c) *[circle diagram: centre (5, 0), x-intercepts −4 and 14, y-intercepts ±2√14]*

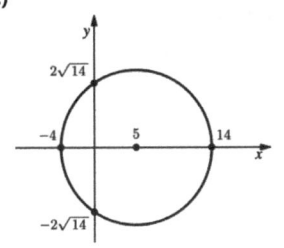

13) d) *[circle diagram: centre (−5, 0), x-intercepts −10 and 0]*

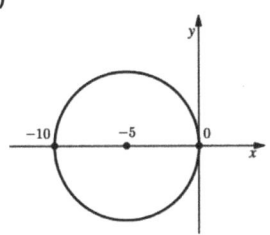

14) a) *[circle diagram: centre (4, 4), tangent to axes at 4 and 4]*

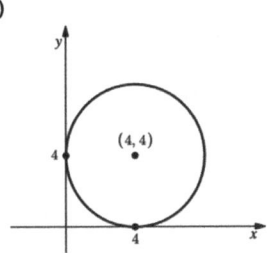

14) b) *[circle diagram: centre (−2, −2), x-intercepts −5 and 1, y-intercepts 1 and −5]*

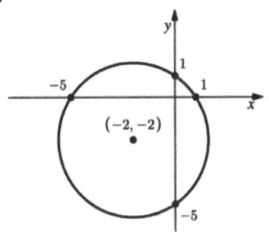

14) c) *[circle diagram: centre (1, −3), x-intercepts −2√2+1 and 2√2+1, y-intercept −7]*

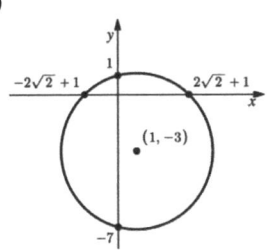

14) d) *[circle diagram: centre (−7, 2), x-intercepts −11 and −3]*

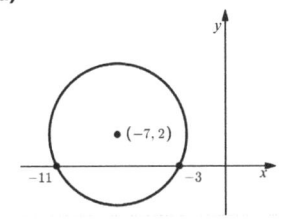

15) a) *[circle diagram: centre (−3, 8), y-intercepts √7+8 and −√7+8]*

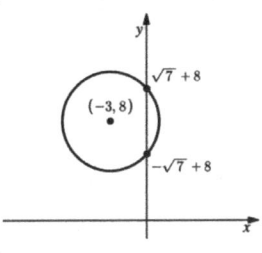

15) b) *[circle diagram: centre (−7, −5), x-intercepts −√11−7 and √11−7]*

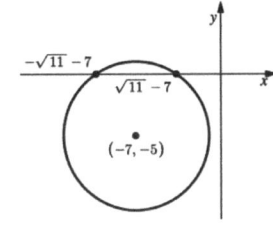

15) c) *[circle diagram: centre (8, 2), x-intercepts −√77+8 and √77+8, y-intercepts √17+2 and −√17+2]*

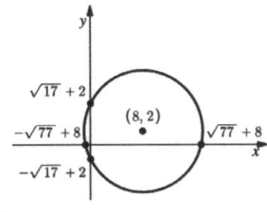

15) d) *[circle diagram: centre (1, 11/16), x-intercepts −1/2 and 5/2, y-intercept −5/8]*

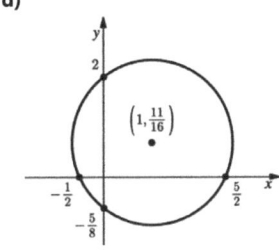

16) a)
Inside the circle

16) b)
On the circle

16) c)
Outside the circle

17)
$(x-2)^2 + (y+2)^2 = 400$

18) a)
$(x-2)^2 + (y+6)^2 = 50$

18) b)
$(x+1)^2 + (y-3)^2 = 65$

18) c)
$x^2 + (y+1)^2 = 80$

18) d)
$(x-5)^2 + (y-6)^2 = 98$

19)
$(x-10)^2 + y^2 = 20$

20)
$x^2 + \left(y - \frac{26}{5}\right)^2 = \frac{1066}{25}$

21)
$\left(x - \frac{44}{3}\right)^2 + \left(y - \frac{44}{3}\right)^2 = \frac{5525}{9}$

22)
Show that

23)
$(x-5)^2 + \left(y - 2\sqrt{21}\right)^2 = 100$
and $(x-5)^2 + \left(y + 2\sqrt{21}\right)^2 = 100$

24)
$(x-5)^2 + (y+2)^2 = 20$

25)
$(x-3)^2 + (y+9)^2 = 25$

26)
$(x+4)^2 + (y-6)^2 = 41$

27)
$x^2 + \left(y - \frac{19}{5}\right)^2 = \frac{221}{25}$

28)
Area = 16 units^2

29) a)
$AB = \sqrt{160}$, $BC = \sqrt{40}$, $AC = \sqrt{200}$

29) b)
Show that

29) c)
Area = 40 units^2

30)
$(x-4)^2 + \left(y - \frac{19}{2}\right)^2 = \frac{655}{4}$

31) a)
The line does not intersect the circle

31) b)
The line intersects the circle at two distinct points.

31) c)
The line is a tangent to the circle.

32)
$k = -\frac{17}{4}$ or $k = -\frac{23}{4}$

33)
Area = $\frac{100\pi}{3}$ units^2

34)
$\frac{3}{5}$

35)
$y = 3x - 1$

36)
$y = -2x + \frac{9}{2}$

37)
$AB = \sqrt{70}$

38)
Tangent at A: $y = \frac{4}{3}x + \frac{50}{3}$
Tangent at B: $y = -\frac{4}{3}x - \frac{26}{3}$

39)
Tangent at A: $y = -3x + 4$
Tangent at B: $y = 3x + 34$

40)
$\left(x + \frac{5}{2}\right)^2 + \left(y - \frac{17}{2}\right)^2 = \frac{25}{2}$

41) a)
Show that
41) b)
$k = 0$ or $k = -\frac{88}{105}$

42) a)
Show that
42) b)
$k = 29$ or $k = -11$

43)
Area = 22.7 units²

44)
Show that

45)
Distance = $\frac{6\sqrt{5}}{5} - \sqrt{3}$

46)
Arc Length = 14.1 units

47)
$\left(x - \sqrt{55}\right)^2 + y^2 = 16$
or $\left(x + \sqrt{55}\right)^2 + y^2 = 16$

48)
$(x - 8)^2 + (y - 3)^2 = \frac{36}{13}$

49)
$k = \frac{75 \pm 25\sqrt{5}}{2}$

50)
$\left(x - 10 - 5\sqrt{5}\right)^2 + (y - 5)^2 = 25$

51)
Percentage shaded = 9.08% (3 SF)

1.13 Algebra: Algebraic Fractions
Questions on page 28

1) a) $x + 4$
1) b) $x + 6$
1) c) $\frac{x}{x+2}$
1) d) $\frac{x-5}{x-3}$
1) e) $\frac{x+1}{x+8}$
1) f) $\frac{x+1}{3x-4}$
1) g) $\frac{x+4}{x(2x-5)}$
1) h) $\frac{x+6}{x(x+3)}$

2) a) $\frac{5}{2x}$
2) b) $\frac{3x^2+5x+5}{5x^2}$
2) c) $\frac{2x+1}{x(x+1)}$
2) d) $\frac{3x+8}{(x+2)(x+4)}$
2) e) $\frac{2(x-5)}{(x+1)(x-3)}$
2) f) $\frac{17x+9}{(2x-1)(x+2)}$
2) g) $\frac{7x-13}{4(x+5)(x-3)}$
2) h) $\frac{2(3x-5)}{(x+2)(x-2)}$

3) a) $\frac{2x+3}{5x}$
3) b) $\frac{x+2}{x(x+3)}$
3) c) $2(x-1)(x+5)$
3) d) $\frac{2(x-3)(x+2)}{(x+3)(x+1)}$

4) a) $\frac{x+1}{x}$
4) b) $\frac{2x}{x-2}$
4) c) $\frac{2(x-1)}{(x-5)^2}$
4) d) $\frac{2x+3}{(x+2)(x+4)}$
4) e) $(3x+5)(3x+1)$

5) a) Prove
5) b) Prove
5) c) Prove
5) d) Prove

1.14 Polynomials
Questions on page 29

1) a) $14x^3 - x^2 - 10x - 3$
1) b) $30x^3 - 67x^2 + 42x - 8$
1) c) $24x^3 + 14x^2 - 11x - 6$

2) a) $4x(2x^2 - 3x - 7)$
2) b) $10x^2(x+1)(x-3)$
2) c) $3x^2(2x+1)(x-8)$

3) a) $2xy^3z(x - 3yz)$
3) b) $2x^2y^2z^4(2y + 7xz)$
3) c) $3(x-1)^2yz(xy - y - 3z^2)$
3) d) $3x^2(2y-1)z(3(2y-1)^2 - 5xz)$
3) e) $4x^2(3y-2)(z+1)(2(3y-2)^2 - x(z+1))$

4) a) 4
4) b) 5
4) c) 6

5) a) $x^3 - x^2 - 5x + 2$
5) b) $2x^4 + 3x^3 - 4x^2 + 7x + 3$
5) c) $3x^4 + 5x^3 + 9x^2 + x + 6$
5) d) $2x^4 + 2x^3 - 15x^2 + 11x - 2$

6) a) $3x^2 + 5x$
6) b) $\frac{7}{3}x^2 + 3x^9$
6) c) $x - 5$
6) d) $(x-2)^2$

6) e)
$\frac{1}{3x-1}$

6) f)
$\frac{x-5}{(2x+1)(x-5)} = \frac{1}{2x+1}$

6) g)
$\frac{(2x-1)(x-7)}{(2x-1)(x+3)} = \frac{x-7}{x+3}$

7) a)
$2x^2 - 7x + 3$

7) b)
$x^3 + 4x^2 - x + 3$

7) c)
$2x^4 - x^3 + x^2 - 10x + 4$

7) d)
$8x^3 - x^2 + 5$

7) e)
$x^5 - x^4 + 2$

7) f)
$x^3 + 2x^2 + x - 5$

7) g)
$x^5 - 5x^4 + x^2 - 8x + 12$

8) a)
$x^2 + 8x + 11 - \frac{5}{x+1}$

8) b)
$2x^2 - 4x - 2 - \frac{14}{2x-1}$

8) c)
$x^3 - x + 3 + \frac{10}{x+2}$

8) d)
$x^3 - 3x + 1 + \frac{-3x+11}{x^2-1}$

9) a)
The only solution is $x = 5$

9) b)
Show that

10) a)
$(x+2)(2x^2 - 11x + 12)$

10) b)
$(x+2)(x-4)(2x-3)$

10) c)
$x = -2, x = 4, x = \frac{3}{2}$

10) d)
$x = 0, x = 6, x = \frac{7}{2}$

11) a)
$(2x+1)(x^2 - x + 18)$

11) b)
Show that

11) c)
$x = -\frac{1}{2}$

11) d)
$x = \frac{7}{2}$

12) a)
$f(x) = x(x-1)(x-5)$
$g(x) = (x-1)(x-1)(x+2)$

12) b)
Show that

13) a)
Prove

13) b)
Prove

13) c)
Prove

14) a)
$(x-6)(x-1)(2x+3)$

14) b)
$(2x+5)(3x-1)(x+1)$

15)
$k = 5$

16) a)
$p = 3$

16) b)
$(x+3)(2x-3)(x+2)$

17) a)
$p = 164$

17) b)
$(8x+15)(4x+1)(x+3)$

18) a)
$p = 9$

18) b)
$(x+1)(2x-3)(x+6)$

19) a)
$p = 2, q = 10$

19) b)
$(x+2)(x-5)(2x+1)$

20)
$p = 2, q = 13, r = 7$

21) a)
$p = 331$

21) b)
$(x+9)(x+2)(3x-1)(x+9)$

22)
Show that

23)
Show that

24) a)
Show that

24) b)
$(2, 0)$

24) c)

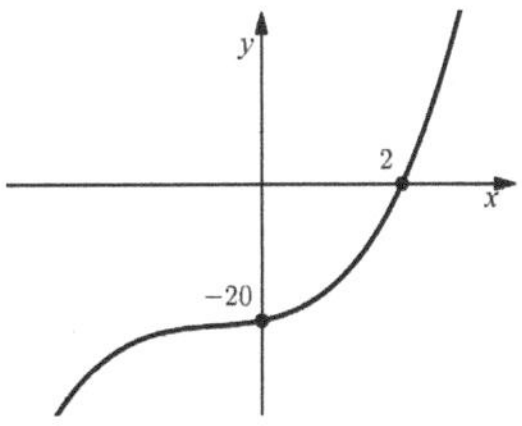

25) a)
$p = 5$

25) b)
$x = -3, x = 5$

25) c)

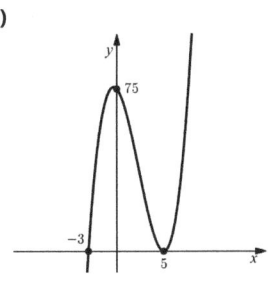

26)

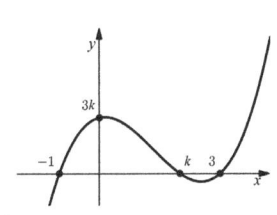

27) a)
Prove

27) b)

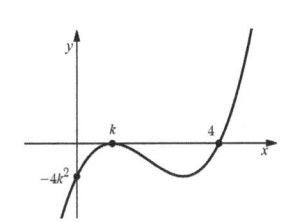

28) a)
$x^2 + \frac{5}{x-1}$

28) b)

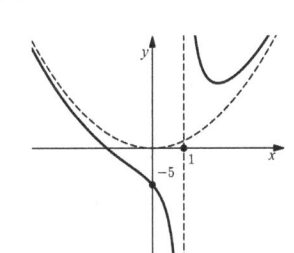

1.15 Graph Sketching: Cubic Graphs
Questions on page 31

1) a)
$y = x^3 - 4$

1) b)
$y = x^3 - 6x^2 + 12x - 7$

1) c)
$y = x^3 + 9x^2 + 27x + 7$

2) a)

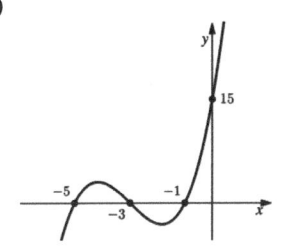

2) b)

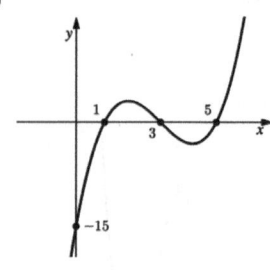

3) a)
$y = -x^3 - 3$
3) b)
$y = -x^3 - 12x^2 - 48x - 64$
3) c)
$y = -x^3 + 6x^2 - 12x + 28$

4) a)

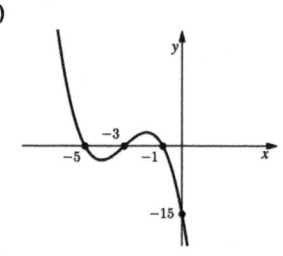

4) b)

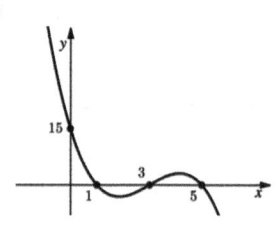

5) a)
$y = (x+3)(x+1)(x-2)$
5) b)
$y = (x+2)(x-1)(x-4)$
5) c)
$y = (x-2)(x-3)(x-5)$

6) a)

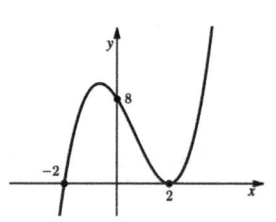

6) b)

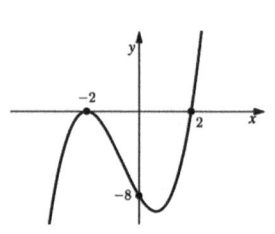

7) a)
$y = (x+4)(x+2)^2$
7) b)
$y = (x+3)^2(x-1)$
7) c)
$y = (x+4)x^2$

8) a)

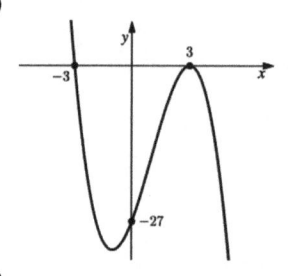

8) b)

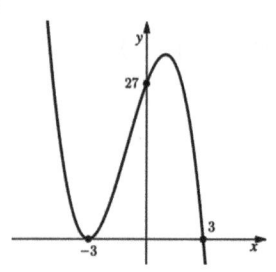

9) a)
$y = -2(x+3)(x-2)(x-4)$
9) b)
$y = -\frac{2}{3}(x+6)(x+3)(x-2)$
9) c)
$y = \frac{1}{2}x(x+2)(x-4)$

10) a)

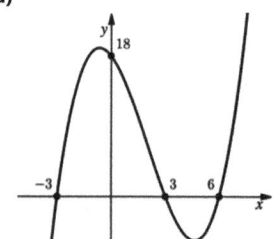

10) b)

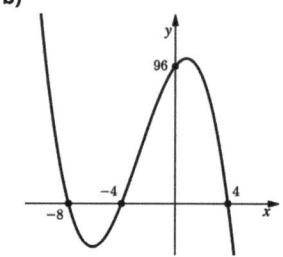

11) a)

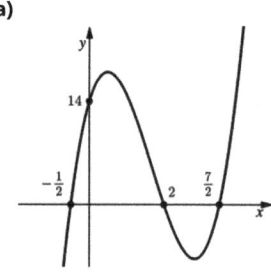

11) b)

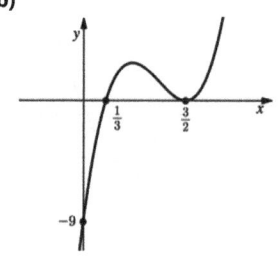

11) c)

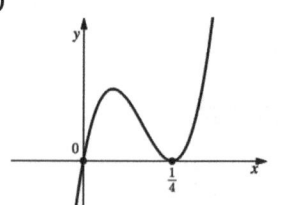

11) d)

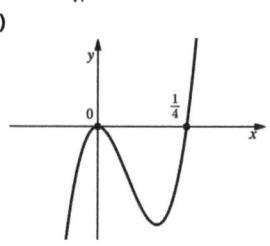

12) a)

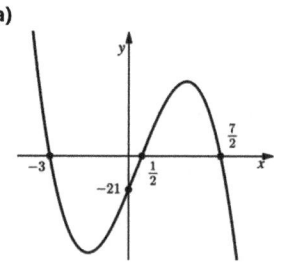

12) b)

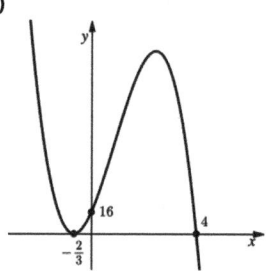

12) c)

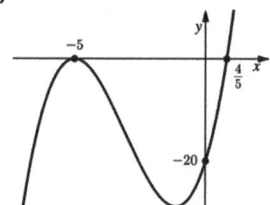

12) d)

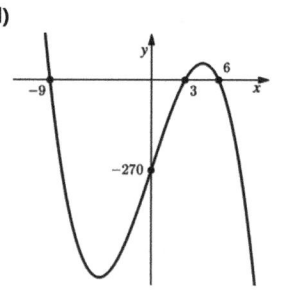

13) a)

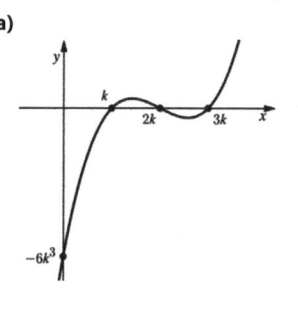

13) b)

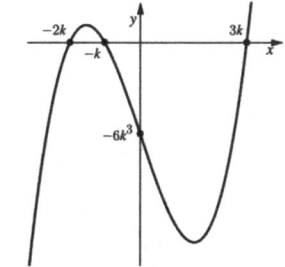

13) c)

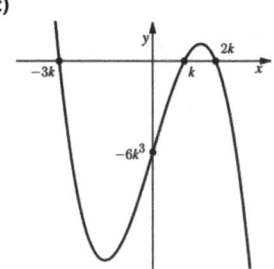

13) d)

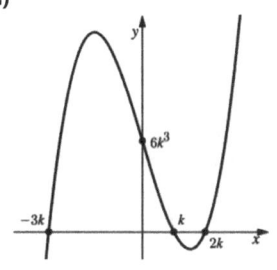

14)
$y = \frac{2}{3}\left(x + \frac{5k}{2}\right)\left(x - \frac{6}{k}\right)(x - 6k)$

15)

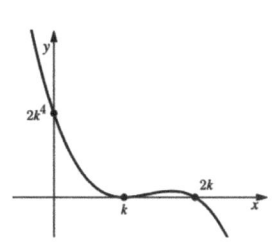

16)

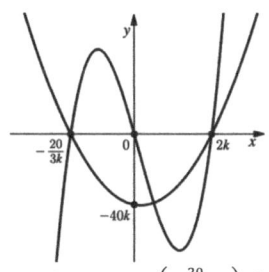

The curves intersect at $\left(-\frac{20}{3k}, 0\right)$, $(2k, 0)$ and $(1, -6k^2 - 37k + 20)$.

1.16 Graph Sketching: Quartic Graphs
Questions on page 34

1) a)
$y = x^4 - 4x^3 + 6x^2 - 4x + 1$
1) b)
$y = x^4 + 8x^3 + 24x^2 + 32x + 18$
1) c)
$y = x^4 - 12x^3 + 54x^2 - 108x$

2) a)

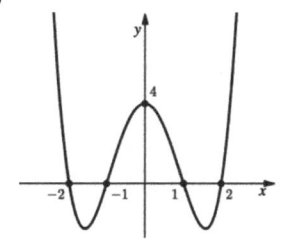

2) b)

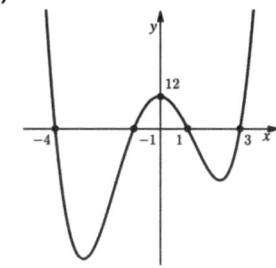

3) a)
$y = -x^4 - 4x^3 - 6x^2 - 4x$
3) b)
$y = -x^4 + 4x^3 - 6x^2 + 4x + 1$
3) c)
$y = -x^4 - 12x^3 - 54x^2 - 108x - 100$

4) a)

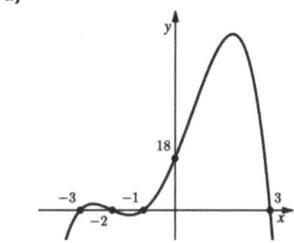

4) b)

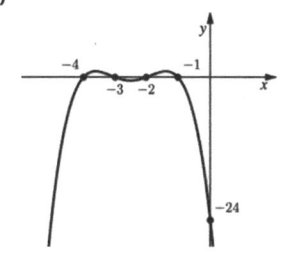

5) a)
$y = (x + 1)(x - 1)(x - 2)(x - 3)$
5) b)
$y = (x + 3)(x + 2)(x - 1)(x - 2)$
5) c)
$y = (x + 1)(x - 2)(x - 3)(x - 4)$

6) a)

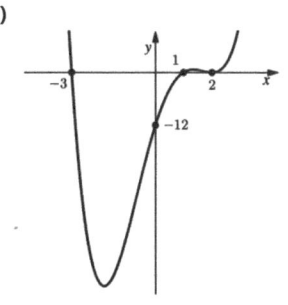

6) b)

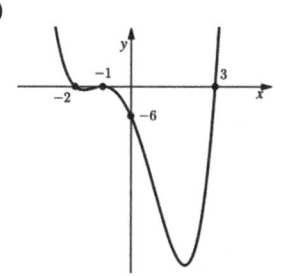

6) c)

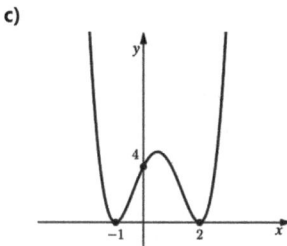

7) a)
$y = (x - 1)(x - 2)(x + 1)^2$
7) b)
$y = (x + 3)(x + 1)(x - 2)^2$
7) c)
$y = (x + 3)(x - 3)(x + 1)^2$

8) a)

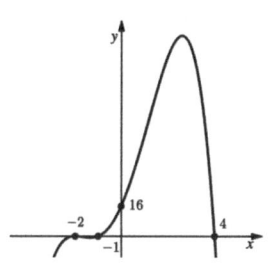

8) b)

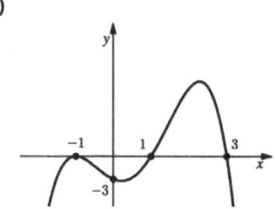

8) c)

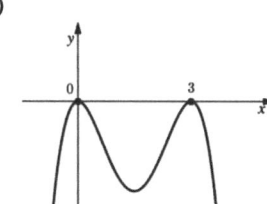

9) a)
$y = 2(x + 2)(x + 1)(x - 2)(x - 3)$
9) b)
$y = -3(x - 1)(x - 2)(x - 3)(x - 4)$
9) c)
$y = -\frac{1}{4}(x + 4)(x + 1)(x - 3)(x - 5)$

10) a) [graph: roots at −2, 2, 4, 6; minimum −48]

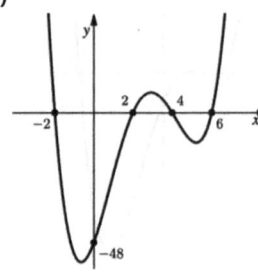

10) b) [graph: y-intercept 216; roots at −3, 3, 6]

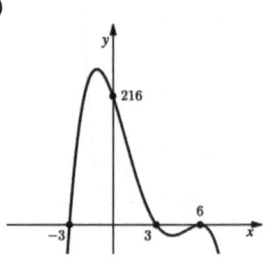

11) a) [graph: roots at −1, −1/2, 1/3, 1/2; y-intercept 1]

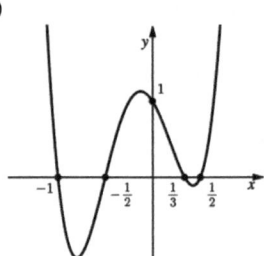

11) b) [graph: roots at −2, 1/2, 5/3; y-intercept −50]

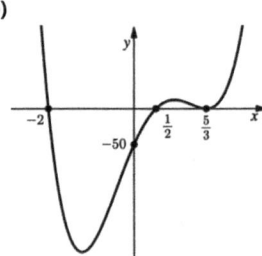

11) c) [graph: roots at −4/3, −2/5, 0]

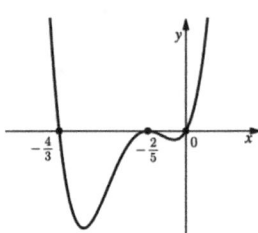

11) d) [graph: roots at 0, 3/4, 4/3]

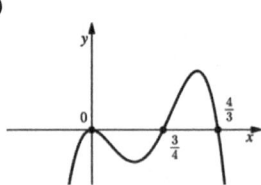

11) e) [graph: roots at −3/2, 1/2; point (+ve x, 9)]

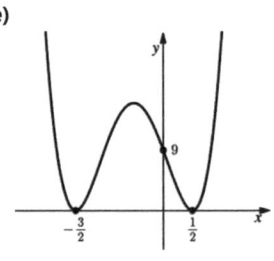

11) f) [graph: roots at −5/12, 1/6; min −25/6]

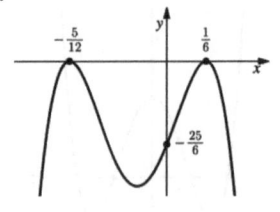

12) a) [graph: roots at −3, 2; min −54]

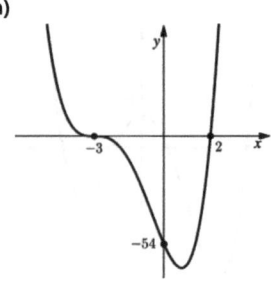

12) b) [graph: roots at −3/2, 3/4; y-intercept −81]

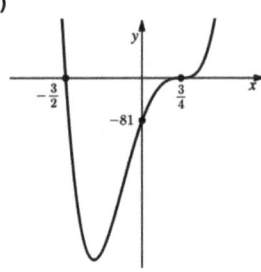

12) c) [graph: roots at 0, 3/2]

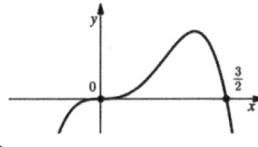

12) d) [graph: roots at −4/3, 1, 1/3]

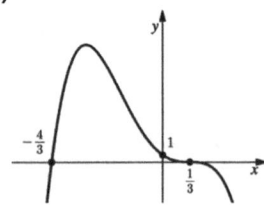

13) a) [graph: roots at 0, k, 2k, 3k]

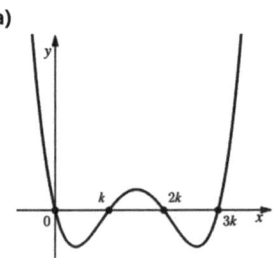

13) b) [graph: roots at −3k, −2k, −k, 2k; min −12k⁴]

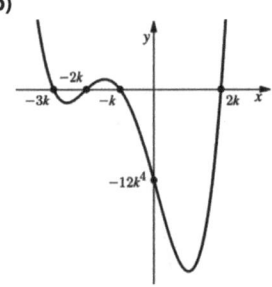

13) c) [graph: roots at −2k, k, 3k; y-intercept 18k⁴]

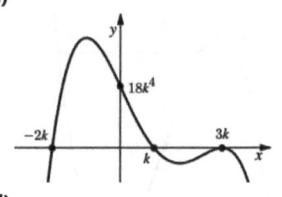

13) d) [graph: roots at −2k, 4k; y-intercept 32k⁴]

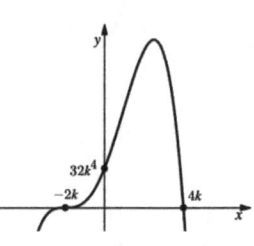

14)
$y = \frac{1}{k^2}(2x+k)(3x-k)(kx-2)(x-4k)$

15) [graph: intersections shown]

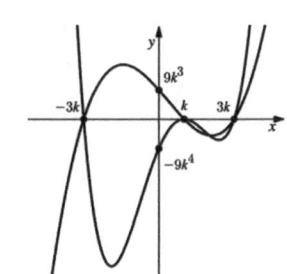

The curves intersect at $(-3k, 0)$, $(k, 0)$, $(3k, 0)$ and $(k+1, -8k^2 + 2k + 1)$

1.17 Proportion
Questions on page 37

1)
C appears to be a reciprocal curve, and could represent $y \propto \frac{1}{x^3}$

2) a)
$y = kx^2$

2) b)
$y = \frac{k}{x}$

2) c)
$y = kx^{\frac{1}{3}}$

2) d)
$y = \frac{k}{(x+1)^3}$

3)
$y = 400$

4)
$x = \frac{81}{10000}$

5)
$V = 32$ litres

6)
$n = 10$ workers

7)
$C = 1.5$ mol/L

8)
$d = 4$ metres

9)
$x = 8$

1.18 Graph Sketching: Rational Functions
Questions on page 38

1)
$\frac{5}{6} = \frac{k}{3} \Rightarrow k = \frac{5}{2}$

2) a)
$y = \frac{1}{x-2}$

2) b)
$y = \frac{1}{x} - 1$

2) c)
$y = \frac{1}{x-3} - 2$

2) d)
$y = \frac{1}{x+2} + 4$

3)
$x = 1$ and $y = 3$

4) a)

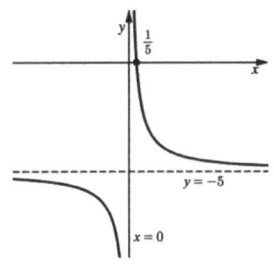

4) b)

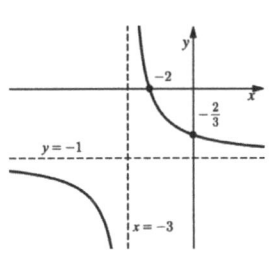

4) c)

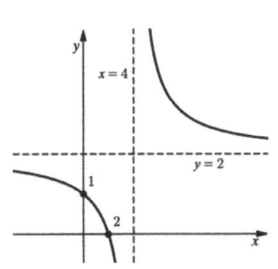

4) d)

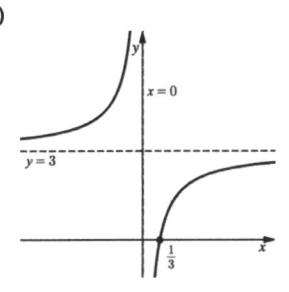

4) e)

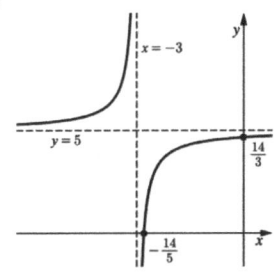

5) a)

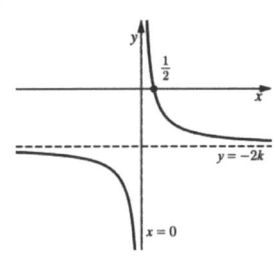

5) b)

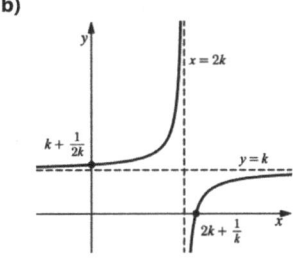

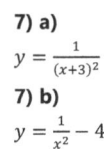

6)
$k = \frac{3}{10}$

7) a)
$y = \frac{1}{(x+3)^2}$

7) b)
$y = \frac{1}{x^2} - 4$

7) c)
$y = \frac{1}{(x+2)^2} - 1$

7) d)
$y = \frac{1}{(x-3)^2} + \frac{3}{4}$

8)
$x = -5$ and $y = -2$

9) a)

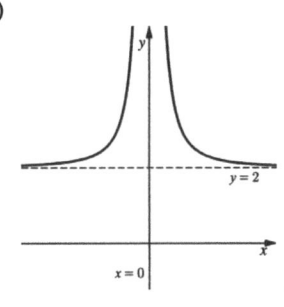

9) b)

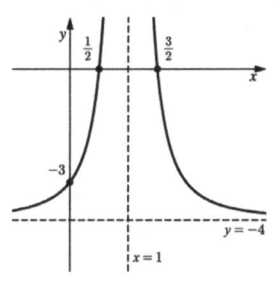

9) c)

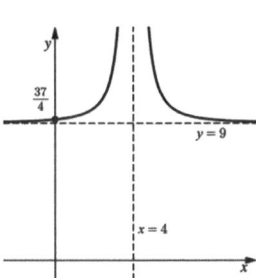

9) d)

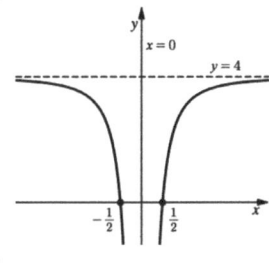

9) e)

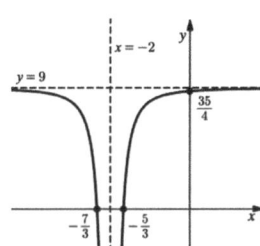

10) a)

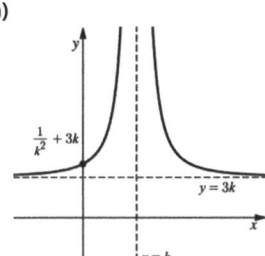

10) b)

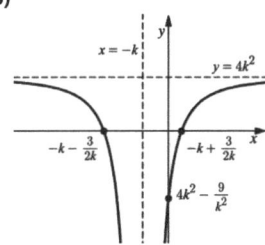

1.19 Graph Sketching: Radical Functions
Questions on page 40

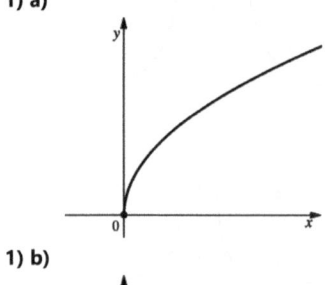

1) a) [graph of radical function starting at origin, increasing]

1) b) [graph of radical function, decreasing from y-axis]

2) $y = \frac{7}{2}\sqrt{x}$

3) a) $y = \sqrt{x-2}$

3) b) $y = \sqrt{x+3}$

4) a) $y = \sqrt{x+5} - 3$

4) b) $y = \sqrt{x+2} - \sqrt{2}$

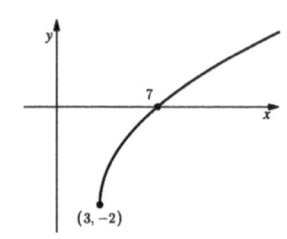

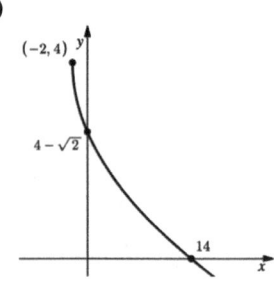

5) a) [graph with point (3, −2), passing through x = 7]

5) b) [graph with point (−2, 4), passing through $4 - \sqrt{2}$ on y-axis and 14 on x-axis]

6) $\left(\frac{9}{4}, \frac{5}{2}\right)$

1.20 Inequalities
Questions on page 41

1) a) $\{x : x \in \mathbb{R}, 1 \leq x < 3\}$

1) b) $\{x : x \in \mathbb{R}, -2 < x \leq 1\}$

1) c) $\{x : x \in \mathbb{R}, x \leq -1\} \cup \{x : x \in \mathbb{R}, x \geq 2\}$

1) d) $\{x : x \in \mathbb{R}, x \leq 2\} \cup \{x : x \in \mathbb{R}, x > 4\}$

2) a) $x \in (-3, 0]$

2) b) $x \in [-2, \infty)$

2) c) $x \in (-\infty, -3) \cup (3, \infty)$

2) d) $x \in (-\infty, 0) \cup [4, \infty)$

3) a) [number line]

3) b) [number line]

3) c) [number line]

3) d) [number line]

3) e) [number line] $\{x : x > 1\} \cap \{x : x < -3\}$ is the empty set, ϕ

4) a) $\{x : x \in \mathbb{R}, x < 4\}$

4) b) $\{x : x \in \mathbb{R}, x \geq -\frac{9}{2}\}$

4) c) $\{x : x \in \mathbb{R}, x > -\frac{1}{2}\}$

4) d) $\{x : x \in \mathbb{R}, x \leq \frac{1}{2}\}$

4) e) $\{x : x \in \mathbb{R}, -2 \leq x \leq 2\}$

4) f) $\{x : x \in \mathbb{R}, \frac{3}{5} < x < 6\}$

4) g) $\{x : x \in \mathbb{R}, -1 < x < 0\}$

4) h) $\{x : x \in \mathbb{R}, \frac{-1-b}{a} < x < \frac{1-b}{a}\}$

4) i) $\{x : x \in \mathbb{R}, \frac{1-b}{a} < x < \frac{-1-b}{a}\}$

5) a) $x \in (4, \infty)$

5) b) $x \in \left[-\frac{26}{5}, \infty\right)$

5) c) $x \in \left(-\infty, \frac{1}{2}\right]$

6) a) $\{x : x \in \mathbb{R}, -2 < x < 4\}$

6) b) $\{x : x \in \mathbb{R}, x \leq -7\} \cup \{x : x \in \mathbb{R}, x \geq -1\}$

6) c) $\{x : x \in \mathbb{R}, -5 \leq x \leq 2\}$

6) d) $\{x : x \in \mathbb{R}, x < -3\} \cup \{x : x \in \mathbb{R}, x > -2\}$

6) e) $\{x : x \in \mathbb{R}, x < -3\} \cup \{x : x \in \mathbb{R}, x > 8\}$

6) f) $\{x : x \in \mathbb{R}, -4 \leq x \leq 2\}$

6) g) $\{x : x \in \mathbb{R}, x \leq -\frac{3}{2} - \frac{1}{2}\sqrt{13}\} \cup \{x : x \in \mathbb{R}, x \geq -\frac{3}{2} + \frac{1}{2}\sqrt{13}\}$

6) h) $\{x : x \in \mathbb{R}\}$

6) i) $x \in \phi$

7) a) $x \in (-\infty, -4) \cup (5, \infty)$

7) b) $x \in (-5, 1)$

7) c) $x \in \left(-\infty, -\frac{7}{2}\right] \cup \left[\frac{5}{2}, \infty\right)$

7) d) $x \in \left[-\frac{1}{2}, \frac{5}{3}\right]$

8) a)
L1: he didn't change the inequality sign.
L2: he didn't factorise correctly
L3: he chose incorrect solutions
LF: he used strict inequalities rather than inclusive ones.

8) b)
$x^2 - 2x - 8 \geq 0$
$(x-4)(x+2) \geq 0$
$x = -2, x = 4$
$\{x : x \leq -2\} \cup \{x : x \geq 4\}$

9) $\{x : x \in \mathbb{R}, x < -1\} \cup \{x : x \in \mathbb{R}, x > 3\}$

10) $x \in \left(-\frac{9}{2}, 1\right)$

11) $\{x : x \in \mathbb{R}, x < -\frac{1}{23}\} \cup \{x : x \in \mathbb{R}, x > \frac{1}{23}\}$

12) $\{x : x \in \mathbb{R}, -\frac{1}{a} \leq x \leq \frac{1}{a}\}$

13) a) $x > 12$

13) b) $x < 100$

13) c) $x \leq 5$

13) d) $x \geq 10$

13) e) $x \geq 1.5y$

13) f) $x \leq 1.2y$

14) a)
Show that
14) b)
$0 < x \leq 8$
14) c)
112 cm^2

15)
$6 \leq x < 8$

16)
Prove

17) a)
Show that
17) b)
$\{k: k \in \mathbb{R}, k \leq -6\} \cup \{k: k \in \mathbb{R}, k \geq 6\}$

18) a)
$\left\{k: k \in \mathbb{R}, -\frac{2}{3} < k < \frac{2}{3}\right\}$
18) b)
$\{k: k \in \mathbb{R}, -10 < k < 2\}$
18) c)
$\left\{k: k \in \mathbb{R}, k < -\frac{1}{2}\right\} \cup \left\{k: k \in \mathbb{R}, k > \frac{9}{2}\right\}$

19)
$\left\{k: k \in \mathbb{R}, 0 < k < \frac{32}{9}\right\}$

20)
$k \in \left(-\frac{16}{9}, 0\right)$

21)
$\left\{k: k \in \mathbb{R}, k < -\frac{4}{3}\right\}$

22)
$\{k: k \in \mathbb{R}, -4\sqrt{2} \leq k \leq 4\sqrt{2}\}$

23) a)
$\{x: x \in \mathbb{R}, -2 < x < -1\} \cup \{x: x \in \mathbb{R}, x > 3\}$
23) b)
$\{x: x \in \mathbb{R}, x < -1\} \cup \{x: x \in \mathbb{R}, 1 < x < 5\}$
23) c)
$\{x: x \in \mathbb{R}, 1 \leq x \leq 2\} \cup \{x: x \in \mathbb{R}, x \geq 6\}$
23) d)
$\{x: x \in \mathbb{R}, x \leq -3\} \cup \{x: x \in \mathbb{R}, 0 \leq x \leq 5\}$

24) a)

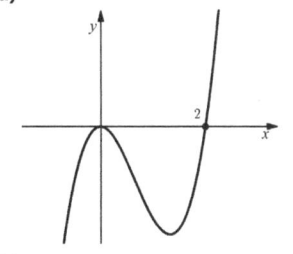

24) b)
$\{x: x \in \mathbb{R}, x \leq 2\}$

25) a)

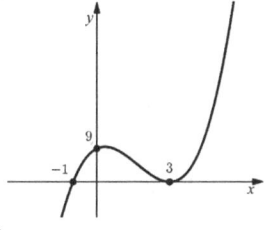

25) b)
$\{x: x \in \mathbb{R}, -1 < x < 3\} \cup \{x: x \in \mathbb{R}, x > 3\}$

26) a)

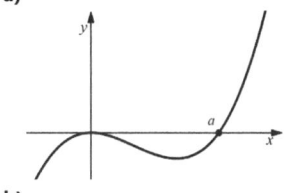

26) b)
$\{x: x \in \mathbb{R}, x > a\}$

27) a)

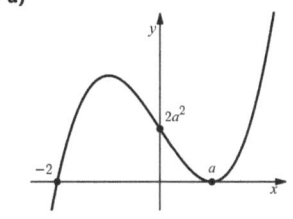

27) b)
$\{x: x \in \mathbb{R}, x \leq -2\} \cup \{x: x \in \mathbb{R}, x = a\}$

28) a)

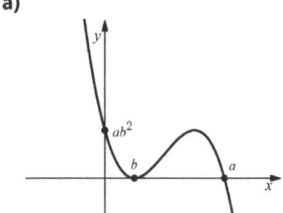

28) b)
$\{x: x \in \mathbb{R}, x \leq a\}$
28) c)
$x \in [b] \cup [a, \infty)$

29) a)
Show that
29) b)
$(x-5)(2x-1)^2$
29) c)

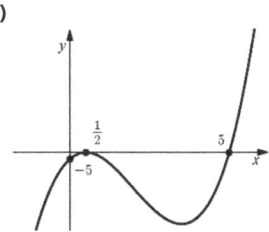

29) d)
$x \in (-\infty, 5]$

30) a)
Show that
30) b)
$(x-2)(x+3)^2$

30) c)

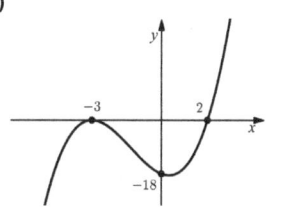

30) d)
$x = -\frac{3}{2}$ and $x = 1$
30) e)
$\{x: x \in \mathbb{R}, x > 1\}$

31) a)
$-2(x+1)(x-4)(x-6)$
31) b)
$\{x: x \in \mathbb{R}, x < -2\} \cup \{x: x \in \mathbb{R}, 8 < x < 12\}$

32) a)
$\{x: x \in \mathbb{R}, x \leq \ln 10\}$
32) b)
$\{x: x \in \mathbb{R}, x \geq \log_4 5\}$
32) c)
$\{x: x \in \mathbb{R}, x < -2\}$
32) d)
$\{x: x \in \mathbb{R}, x < 0\}$
32) e)
$\{x: x \in \mathbb{R}, x < -2\}$

33)
$x < \log_3 4$

34)
$x > 0$

1.21 Inequalities: Identifying Regions
Questions on page 44

1) a)

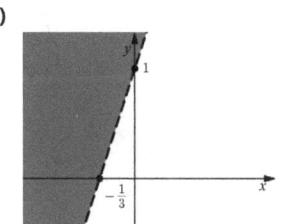

1) b)

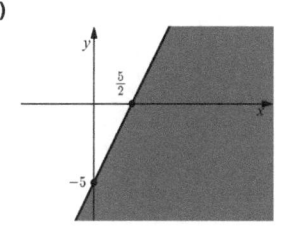

2) [graph]

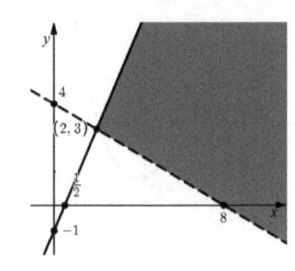

3) [graph]

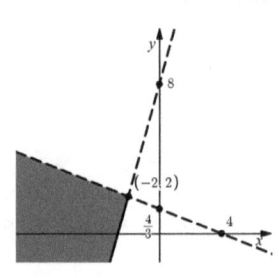

4) a) [graph]

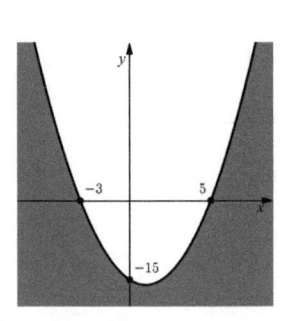

4) b) [graph]

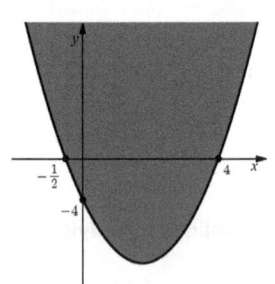

4) c) [graph]

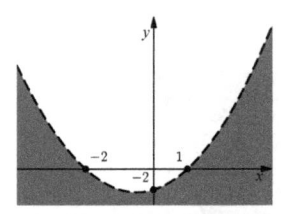

4) d) [graph]

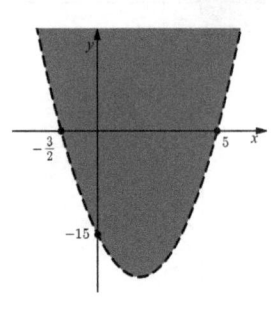

5) a) [graph]

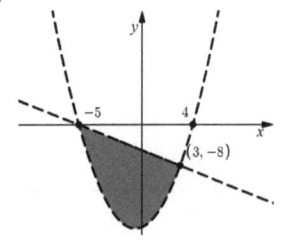

5) b) [graph]

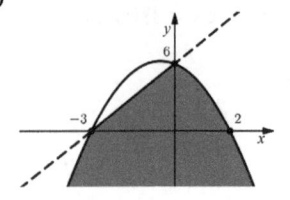

6) [graph]

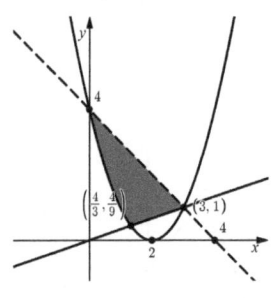

7) [graph]

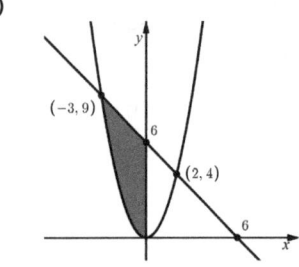

8) a)
$y > x + 1$ and $y \leq 6 + 5x - x^2$

8) b)
P is $(-1, 0)$ and Q is $(5, 6)$

8) c)
$\{x : x \in \mathbb{R}, -1 < x < 5\}$

9) a)
$y \geq x^2 - 7x + 12$ and $y < -x^2 + 2x + 8$

9) b)
$\left(\frac{1}{2}, \frac{35}{4}\right)$

10) a) [graph]

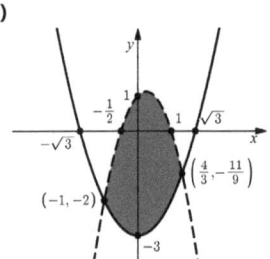

10) b)
$\left\{x : x \in \mathbb{R}, -1 < x < \frac{4}{3}\right\}$

11) [graph]

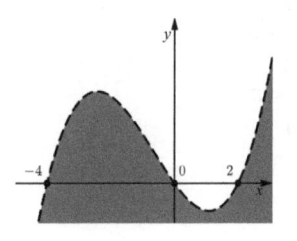

12) [graph]

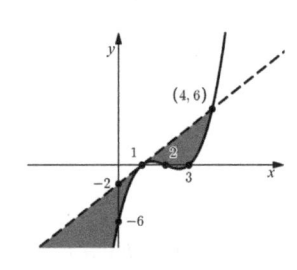

13) a)
centre $(-1, 0)$ and radius 2

13) b) [graph]

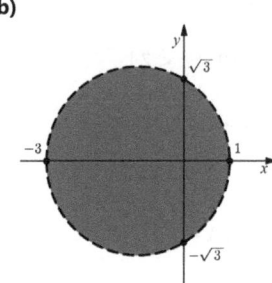

14) [graph]

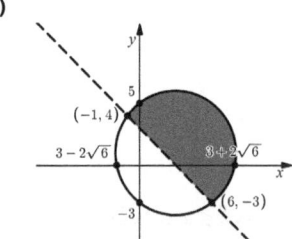

1.22 Exponentials and Logarithms
Questions on page 45

1) a)
$x = \log_4 3$

1) b)
$x = \log_3 4$

1) c)
$x = \log_2 \frac{1}{3} = -\log_2 3$

1) d)
$x = \log_9 \frac{1}{100} = -\log_9 100$

2) a)
$x = \log_2 3 - 1$

2) b)
$x = \log_3 \frac{1}{5} + 5$

2) c)
$x = 250 \log_5 7 + \frac{1}{2}$

2) d)
$x = \frac{250}{3} \log_6 9 - \frac{4}{3}$

3)
$x = \pm\sqrt{2}$

4)
$x = \pm\sqrt{5}$

5) a)
$x = 3^8$
5) b)
$x = 8^3$
5) c)
$x = 5^7$
5) d)
$x = 7^5$

6) a)
$\log_3 20$
6) b)
$\log_5 18$
6) c)
$\log_6 5$
6) d)
$\log_{10} \frac{1}{2}$

7) a)
$\log_{10} 6$
7) b)
$\log_2 10$
7) c)
$\log_5 \frac{3}{10}$
7) d)
$\log_9 \frac{16}{5}$

8)
$x = 40$

9)
$x = \frac{1}{2}$

10) a)
1
10) b)
1
10) c)
2
10) d)
-4

11) a)
$\log_3 25$
11) b)
$\log_4 27$
11) c)
$\log_9 16$
11) d)
$\log_7 10000$

12) a)
$\log_2 \frac{2}{3}$
12) b)
$\log_3 \frac{1}{2}$

12) c)
$\log_5 \frac{1}{4}$
12) d)
$\log_8 \frac{1}{27}$

13) a)
$\log_b 50$
13) b)
$\log_b 135$
13) c)
$\log_b \frac{5}{8}$
13) d)
$\log_b \frac{1}{25}$

14) a)
$\log_2 kx^3$
14) b)
$x = \sqrt[3]{\frac{8}{k}} = \frac{2}{\sqrt[3]{k}}$

15) a)
$\log_3 \frac{x^5}{m}$
15) b)
$x = \sqrt[5]{9m}$

16) a)
$\log_5 \frac{15}{2}$
16) b)
$\log_2 \frac{1}{3}$
16) c)
$\log_9 4$
16) d)
$= \log_3 20000$

17)
$y = 25x^{\frac{3}{5}}$

18) a)
$\log_3 256$
18) b)
$\log_5 16$
18) c)
$= \log_8 12500$
18) d)
$\log_4 \frac{1}{9}$

19) a)
$\log_7 4$

19) b)
$\log_{10} \frac{9}{5}$
19) c)
$\log_2 63$
19) d)
$\log_9 \frac{1}{2}$

20) a)
$\log_3 48$
20) b)
$\log_2 \frac{125}{2}$

20) c)
$\log_5 900$
20) d)
$\log_4 \frac{16}{25}$

21) a)
$\log_b 9b^2$
21) b)
$\log_b \frac{16}{b}$
21) c)
$\log_b 64b$
21 d)
$\log_b \frac{b}{400}$

22)
Show that

23)
Show that

24) a)
7
24) b)
5
24 c)
-1
24 d)
2

25)
Show that

26) a)
$\frac{1}{2}m$
26) b)
$m - 1$
26) c)
$2 - 2m$

27)
$y = \frac{40}{b^{\frac{3}{2}}} = 40b^{-\frac{3}{2}}$

28)
$y = 32k^{\frac{3}{4}}$

29) a)
$\log_b(p^2 q^3 r^2)$
29) b)
$\log_b \left(\frac{p^3 q^5}{r^3}\right)$
29) c)
$\log_b \left(\frac{p^5 r^4}{q^2}\right)$
29) d)
$\log_b \left(\frac{p^8}{q^6 r^9}\right)$

30) a)
$\log_7 5$
30) b)
$\log_8 \frac{5}{3}$

30) c)
$\log_6 \frac{3}{50}$

30) d)
$\log_{10} 2$

31) a)
$\log_b \left(\frac{p}{qr}\right)$

31) b)
$\log_b \left(\frac{pr^2}{q^2}\right)$

31) c)
$\log_b \left(\frac{p}{q^3 r^3}\right)$

31) d)
$\log_b \left(\frac{p^2 r^4}{q^2}\right)$

32) a)
$x + y$

32) b)
$2x$

32) c)
$x + z$

32) d)
$2z - 2y$

33)
Show that

34) a)
$\frac{5}{2}$

34) b)
$\frac{8}{3}$

34) c)
1

34) d)
2

35)
Show that

36)
$x = \log_2 9$ or $x = \log_2 5$.

37)
Show that

38)
$x = \log_2 3$

39) a)
$x = \log_2 5$ or $x = \log_2 3$

39) b)
$x = \log_3 4$ or $x = \log_3 5$

39) c)
$x = \log_4 7$

39) d)
$x = \log_3 10$

40) a)
$x = \log_2 3$ or $x = \log_2 7$

40) b)
$x = 1$ or $x = \log_5 10$

40) c)
$x = \frac{1}{2} \log_2 3$

40) d)
$x = \frac{1}{2} \log_2 10$

41)
$x = \log_2 3$ or $x = 0$

42) a)
$x = \log_3 5$

42) b)
$x = \log_4 9$

42) c)
$x = \frac{1}{2} \log_2 \frac{9}{2}$

42) d)
$x = \frac{1}{2} \log_2 \frac{3}{8}$

43)
Show that

44)
Show that

45)
Show that

46) a)
$x = \frac{\log_3 4}{\log_3 4 - 1}$

46) b)
$x = \frac{2\log_3 5 - 5}{1 + \log_3 5}$

46) c)
$x = \frac{3\log_3 8 + 8}{\log_3 8 - 1}$

46) d)
$x = \frac{5\log_3 10 + 6}{\log_3 10 + 1}$

47) a)
$x = \frac{2\log_2 9}{\log_2 5 - \log_2 9}$

47) b)
$x = \frac{2\log_2 3 + 3\log_2 7}{\log_2 3 + \log_2 7}$

47) c)
$x = \frac{4\log_2 10 - 4\log_2 9}{\log_2 9 + \log_2 10}$

47) d)
$x = \frac{4\log_2 6 - 8\log_2 5}{\log_2 6 - \log_2 5}$

48) a)
Show that

48) b)
$x = -8$ is the only solution.

49)
$x = \frac{51}{33}$

50) a)
The first error is using the power rule incorrectly from line 1 to line 2.
$3(\log_8 x)^2 \neq 6\log_8 x$

The second error is 'unlogging' from line 4 to line 5. If $\log_8 x = \frac{20}{17}$ then $x = 8^{\frac{20}{17}}$, not $x = \left(\frac{20}{17}\right)^8$.

50) b)
$x = 8^{\frac{4}{3}}$ or $x = 8^{-5}$

51) a)
The first error is dividing the 1 by the x from line 1 to line 2. $\frac{1}{\log_3 x} \neq \log_3 \frac{1}{x}$

The second error is from mistaking $\log_3 3 = 1$ for $\log_3 1 = 0$ line 4 to line 5. If George's working had been correct up to this point, the final answer would be $x = 1$ and not $x = 3$.

51) b)
$x = \sqrt{3}$ or $x = \frac{1}{\sqrt{3}} = \frac{\sqrt{3}}{3}$

52) a)
Show that

52) b)
$m = \frac{4}{9}$

53)
$x = \log_2(\log_2 5)$

54)
$A\left(\frac{38}{15}, 0\right)$
$B(4,0)$
$C\left(18 + 4\sqrt{15}, 2\log_5(15 + 4\sqrt{15})\right)$

55)
$A(1, \log_4 49)$
$B(-3, \log_4 9)$
$C(-5, 0)$

1.23 e and Natural Logarithms
Questions on page 49

1) a)
$x = \ln 8$

1) b)
$x = \frac{1}{2} \ln 4$

1) c)
$x = \frac{1}{3} \ln 8$

1) d)
$x = \frac{1}{5} \ln \frac{1}{100}$

2) a)
$x = \frac{1}{4}(\ln 9 - 3)$

2) b)
$x = \frac{1}{5}(\ln 10 + 2)$

2) c)
$x = \frac{9}{2} - \frac{1}{2} \ln 4$

2) d)
$x = \frac{1}{5} - \frac{1}{5} \ln \frac{1}{5} = \frac{1}{5} + \frac{1}{5} \ln 5$

3) a)
$x = e^7$

3) b) $x = e^{-5}$
3) c) $x = e^{\frac{1}{3}}$
3) d) $x = e^{-\frac{2}{5}}$

4) a) $\ln 120$
4) b) $\ln 6$
4) c) $\ln \frac{1}{2}$
4) d) $\ln 25$

5) a) $x = 4$
5) b) $x = 15$
5) c) $x = \frac{1}{24}$
5) d) $x = \frac{100}{3}$

6) a) 1
6) b) -4
6) c) -5
6) d) $\frac{1}{3}$

7) a) $\ln 64$
7) b) $\ln \frac{8}{9}$
7) c) $\ln \frac{243}{4}$
7) d) $\ln \frac{1}{4}$

8) $\ln\left(\frac{x^3}{y^4}\right)$

9) a) $\ln(p^4 q^2 r^3)$
9) b) $\ln\left(\frac{p^3 q^5}{r^4}\right)$
9) c) $\ln\left(\frac{p^6}{q^2 r^8}\right)$
9) d) $\ln\left(\frac{p^9 r^7}{q^5}\right)$

10) a) $\ln\left(\frac{p^2}{qr^3}\right)$
10) b) $\ln\left(\frac{pr}{q^2}\right)$
10) c) $\ln\left(\frac{p^3}{q^2 r^4}\right)$
10) d) $\ln\left(\frac{p^4 r^9}{q^6}\right)$

11) a) $xy = e^5$
11) b) $\frac{y}{x} = e^{12}$
11) c) $\frac{x^2}{y} = e^4$
11) d) $x^4 y^3 = e^8$

12) $(4, 2)$ and $(25, 5)$

13) a) $x = \ln 16$
13) b) $x = \ln 4$
13) c) $x = \ln 16$
13) d) $x = \ln 4$

14) a) $x = -1$
14) b) $x = 5$
14) c) $x = -\frac{1}{3}\ln 2$
14) d) $x = \ln 3 - 1$

15) a) Show that
15) b) $x = \ln \frac{3}{4}$ or $x = \ln 9$

16) a) $x = \ln 2$ or $x = \ln 4$
16) b) $x = \ln 5$ or $x = \ln 8$
16) c) $x = \ln 5$
16) d) $x = \ln 10$

17) a) $x = \ln \frac{3}{2}$ or $x = \ln 2$
17) b) $x = \ln \frac{5}{2}$ or $x = \ln \frac{9}{2}$
17) c) $x = \ln \frac{9}{2}$
17) d) $x = \ln \frac{3}{4}$

18) Show that

19) Show that

20) a) $x = \frac{4\ln 5}{\ln 2 - \ln 5}$
20) b) $x = \frac{2\ln 6 + \ln 4}{\ln 4 - \ln 6}$
20) c) $x = \frac{3\ln 5 - \ln 7}{\ln 5 + \ln 7}$
20) d) $x = \frac{8\ln 6 - 9\ln 8}{\ln 6 - \ln 8}$

21) a) $x = \frac{e^2 + 5}{3}$
21) b) $x = \frac{3 - e^{\frac{9}{2}}}{2}$
21) c) $x = -4 + \sqrt{5}$
21) d) $x = \sqrt{\frac{27}{32}e}$

22) $x = 4$

23) a) $x = \frac{3e^2}{1 - e^2}$
23) b) $x = \frac{e^3 + 1}{1 - e^3}$
23) c) $x = \sqrt{\frac{e^4}{2}} = \frac{e^2}{\sqrt{2}}$
23) d) $x = \frac{-4 + \sqrt{16 + 16e^5}}{8} = \frac{-1 + \sqrt{1 + e^5}}{2}$

24) $(\sqrt{2}, 4\ln\sqrt{2}) = (\sqrt{2}, \ln 4)$

25) a) $x = 2e^2$
25) b) $x = \frac{e}{2}$
25) c) $x = \frac{4}{e}$
25) d) $x = \frac{10}{e^4}$

26) $x = \frac{1}{2}e^{\frac{3}{4}}$

27) a) $x = e^8$ or $x = e^4$
27) b) $x = e^{\frac{3}{2}}$ or $x = e^2$

27) c)
$x = e^{-\frac{5}{2}}$ or $x = e^{\frac{2}{3}}$

27) d)
$x = e^{-\frac{1}{5}}$ or $x = e^{\frac{7}{2}}$

28)
$x = 25$

29)
$x = \frac{18}{13}$

30)
$x = -\frac{3}{4}$

31) a)
Show that

31) b)
$x = \ln 2$

32)
$x = e^2$ or $x = e$

33)
$x = e^{\frac{3}{2}}$ or $x = e^{\frac{5}{3}}$

34)
$x = \frac{1}{\sqrt{2}} e$

35)
$x = \frac{1}{1 - \ln 3}$

36)
$x = \frac{-3 + \ln 2}{2 + \ln 3}$

37)
$x = \frac{10}{8 + \ln 5}$

38)
$x = \frac{1 + 5e}{3 - e}$

39)
$(\ln 8, 5)$

40)
$a = \frac{7}{2}$ and $b = 5$

41)
$p = -\frac{8}{3}$ and $q = \frac{9}{2}$

42)
$x = \ln 2$ and $y = \sqrt{2}$

43)
$(4 - e^2, 2)$

44) a)
20 metres

44) b)
16.5 metres (3 SF)

44) c)
6.39 hours (3 SF)

45) a)
500

45) b)
598

45) c)
20 years

46) a)
6.11°C (3 SF)

46) b)
26 days

1.24 Graph Sketching: Exponentials and Logarithms
Questions on page 52

1) a)
$y = 2^x$

1) b)
$y = 2^x + 2$

1) c)
$y = 2^x - 8$

2) a)

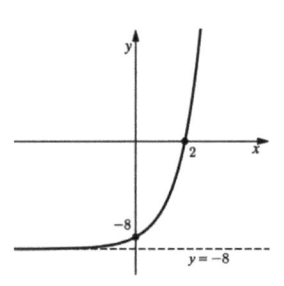

2) b)

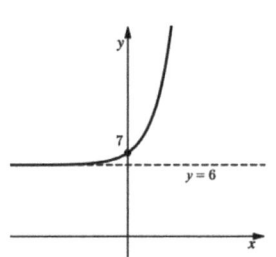

3) a)
$y = 2^{x-2} + 4$

3) b)
$y = 2^{x+1} - 4$

3) c)
$y = 2^{x+3} - 1$

4) a)

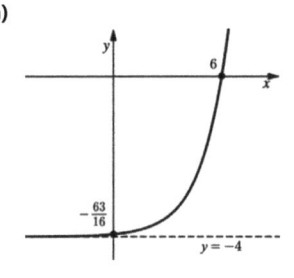

4) b)

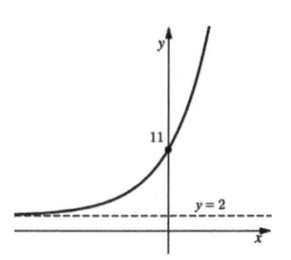

5) a)
$y = 2^{2-x} + 2$

5) b)
$y = 2^{3-x} - 2$

5) c)
$y = 2^{4-x} - 1$

6) a)

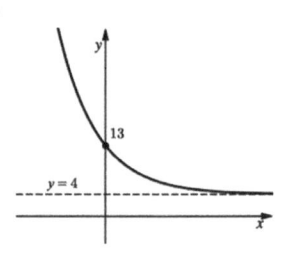

6) b)

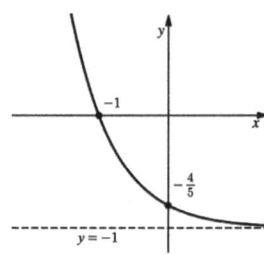

7) a)
$y = 2 - 2^{x-2}$

7) b)
$y = -4 - 2^{x+3}$

7) c)
$y = 3 - 2^{x+1}$

8) a)

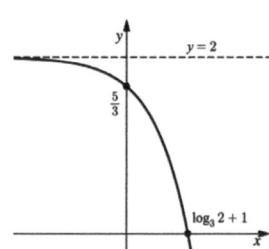

8) b)

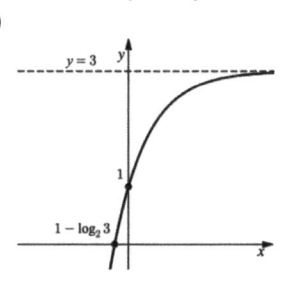

9) a)
$y = \log_2(x - 2)$

9) b)
$y = \log_2(x + 4)$

9) c)
$y = \log_2(x - 10)$

10) a)

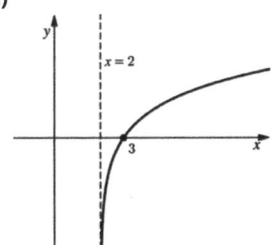

10) b)

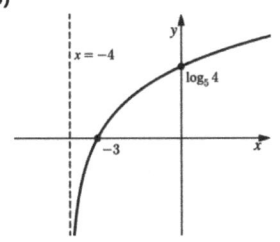

11) a)
$y = \log_2 x + 1$
11) b)
$y = \log_2 x - 2$
11) c)
$y = \log_2 x + \frac{1}{2}$

12) a)

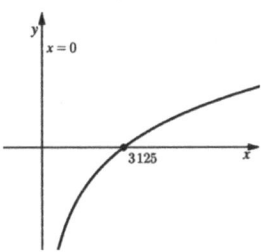

12) b)

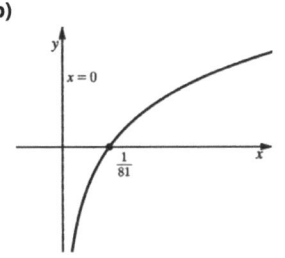

13) a)
$y = \log_2(x - 2) - 1$
13) b)
$y = \log_2(x + 4) + 3$
13) c)
$y = \log_2(x + 1) - 3$

14) a)

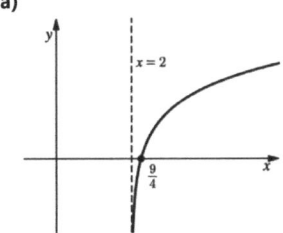

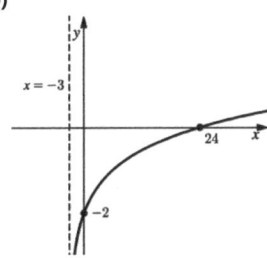

14) b)

15)
$y = \log_2(4 - x) + 3$

16) a)

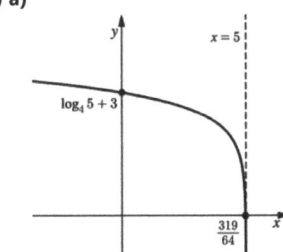

16) b)

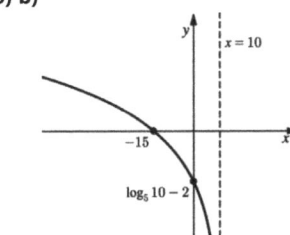

16) c)

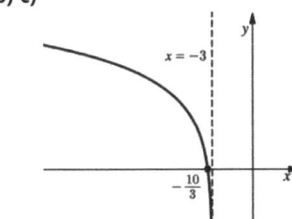

16) d)

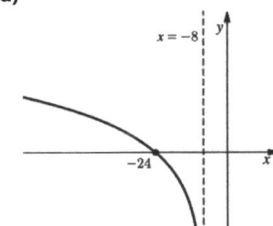

17)
$y = 5 - \log_2(x - 5)$

18) a)

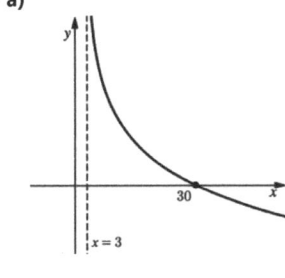

18) b)

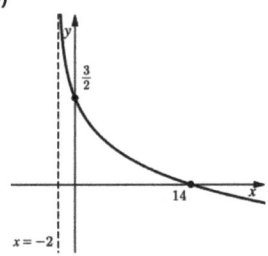

18) c)

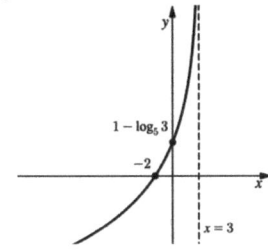

18) d)

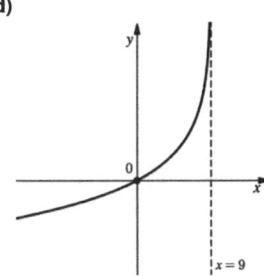

19) a)

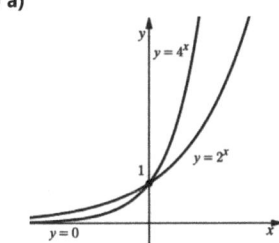

19) b)

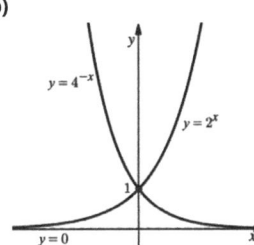

19) c)

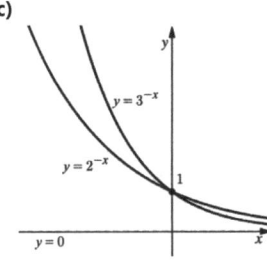

20) a)

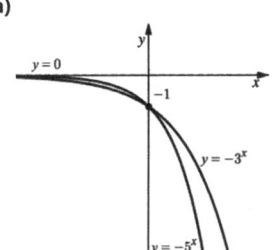

20) b)

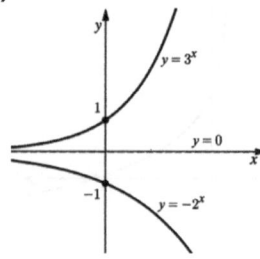

20) c)

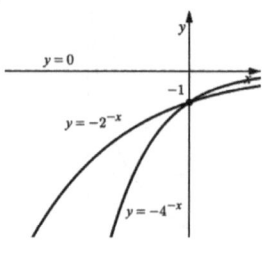

21) a)

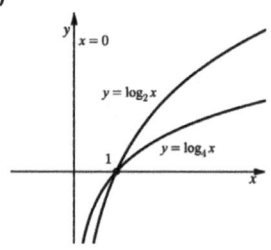

21) b)

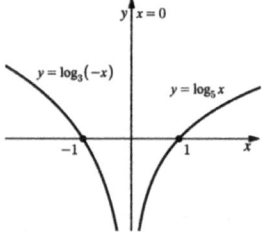

21) c)

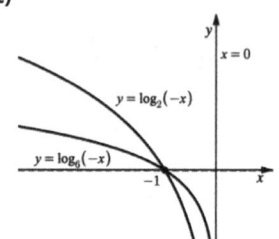

22) a)

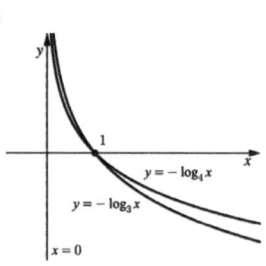

22) b)

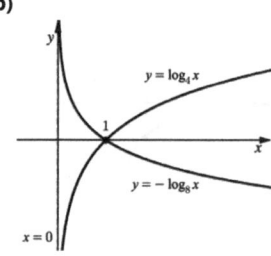

22) c)

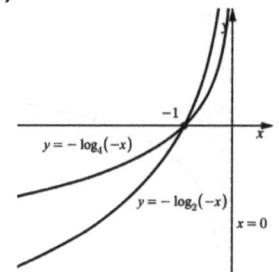

23) a)

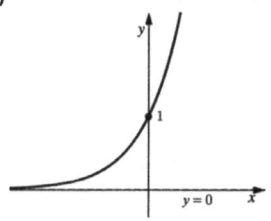

23) b)

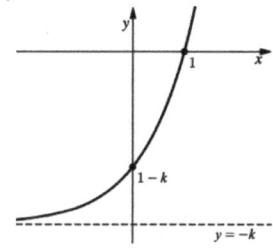

23) c)

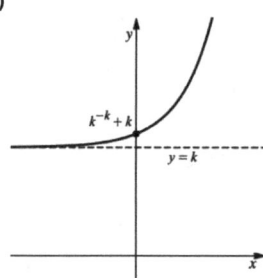

24) a)
$y = k^{k-x} - k$

24) b)
$y = 2k - k^{k-x}$

25) a)

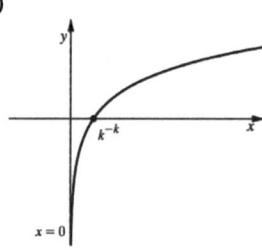

25) b)

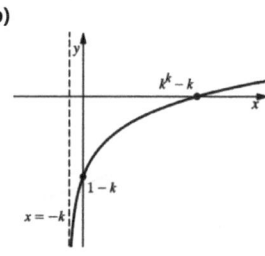

26) a)
$y = k - \log_k(x+k)$

26) b)
$y = k - \log_k(-k-x)$

1.25 Graph Sketching: e and Natural Logarithms
Questions on page 57

1) a)
$y = e^x$
1) b)
$y = e^x + 5$
1) c)
$y = e^x - 4$

2) a)

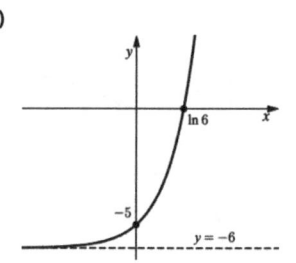

2) b)

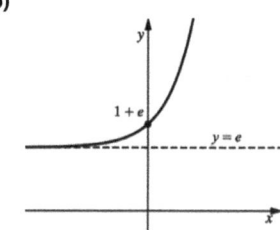

3) a)
$y = e^{x-1} + 3$
3) b)
$y = e^{x+2} - 5$
3) c)
$y = e^{x-3} - 1$

4) a)

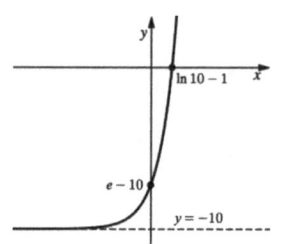

4) b)

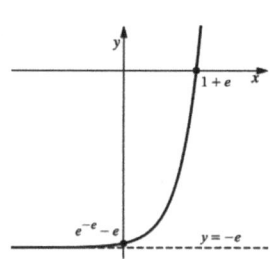

5) a)
$y = e^{1-x} + 2$
5) b)
$y = e^{2-x} - 3$
5) c)
$y = e^{-3-x} - 9$

6) a)

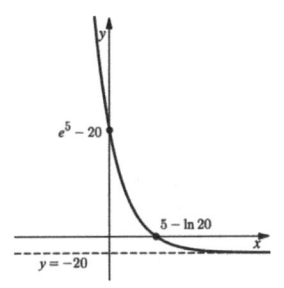

6) b)

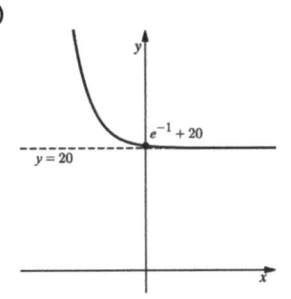

7) a)
$y = 1 - e^x$
7) b)
$y = -2 - e^{x-1}$
7) c)
$y = 4 - e^{x+2}$

8) a)

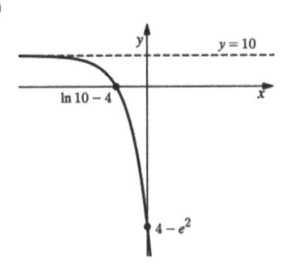

8) b)

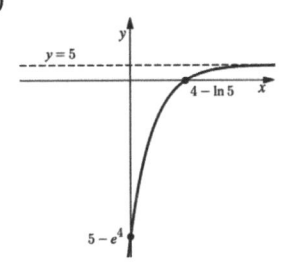

9) a)
$y = \ln(x-3)$
9) b)
$y = \ln(x+5)$
9) c)
$y = \ln(x+1)$

10) a)

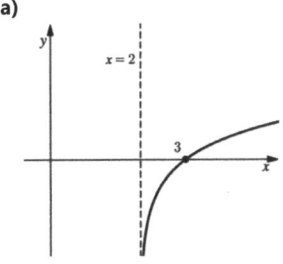

10) b)

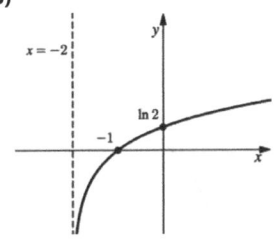

11) a)
$y = \ln x + 1$
11) b)
$y = \ln x - 2$
11) c)
$y = \ln x + \frac{1}{2}$

12) a)

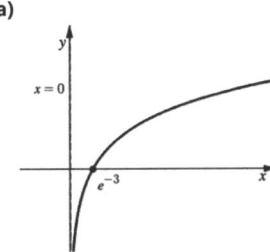

12) b)

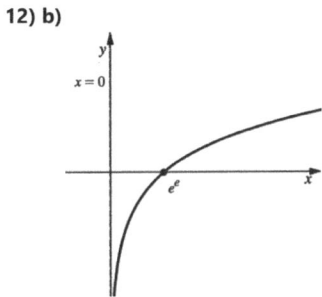

13) a)
$y = \ln(x-1) + 1$
13) b)
$y = \ln(x+2) + 3$
13) c)
$y = \ln(x+1) - 2$

14) a)

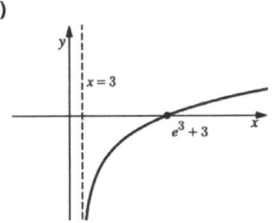

14) b)

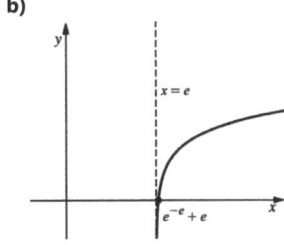

15)
$y = \ln(5-x) + 2$

16) a)

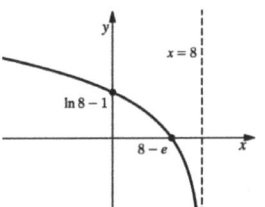

16) b)

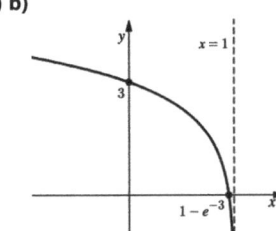

16) c)

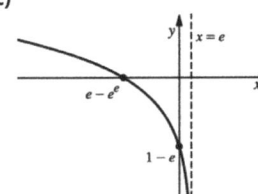

16) d)

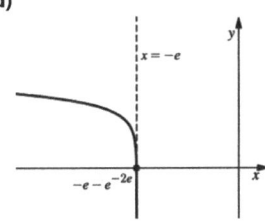

17)
$y = 3 - \ln(x-2)$

18) a)

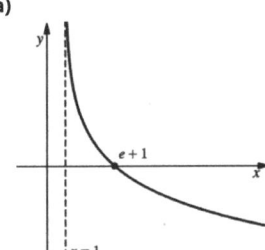

18) b)

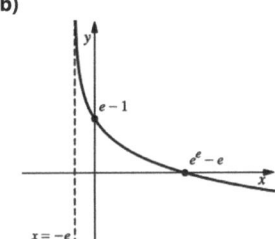

18) c)

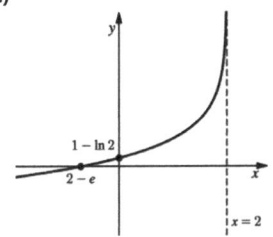

18) d)

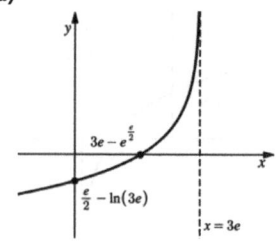

19) a)

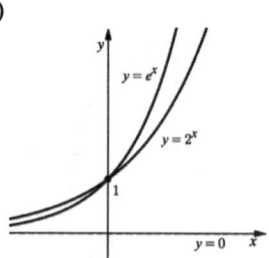

19) b)

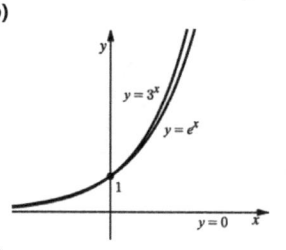

19) c)

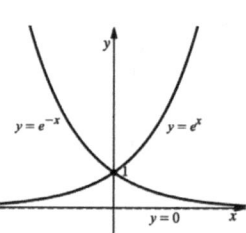

20) a)

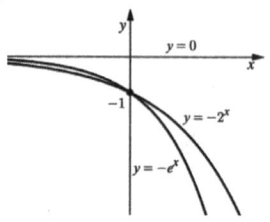

20) b)

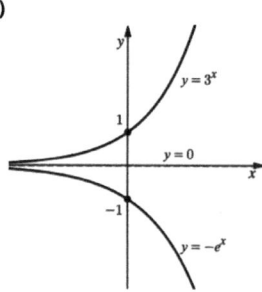

20) c)

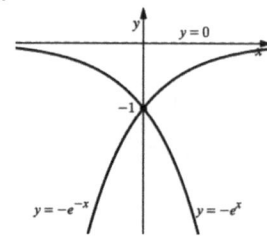

21) a)

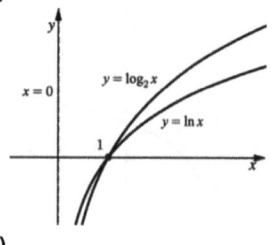

21) b)

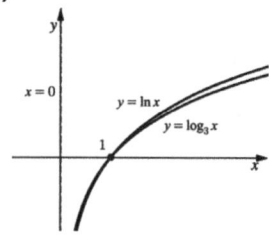

21) c)

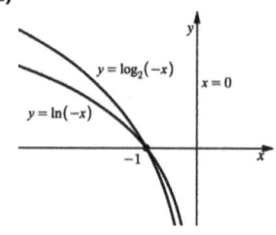

22) a)

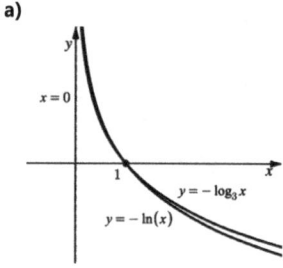

22) b)

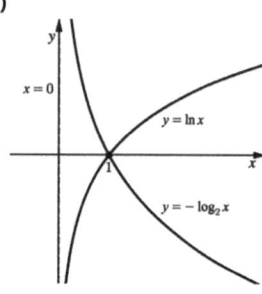

22) c)

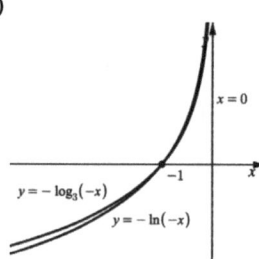

23) a)

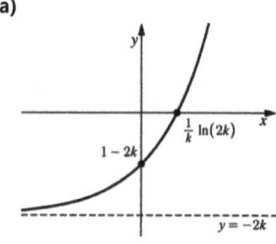

23) b)

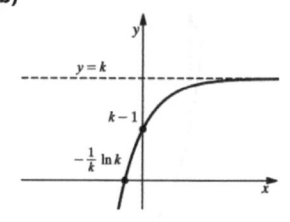

24) a)
$y = e^{-\frac{x}{k}} - k$

24) b)
$y = 2k - e^{-\frac{x}{2k}}$

25) a)

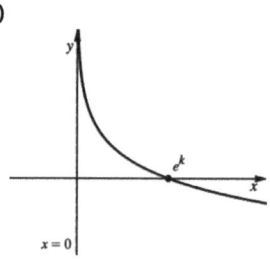

25) b)

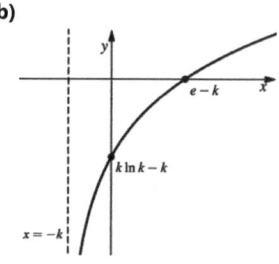

26) a)
$y = k - \ln(x + k)$

26) b)
$y = \ln(k - x) - \ln k$

1.26 Exponential Growth and Decay
Questions on page 62

1) a)
2
1) b)
19
1) c)
11
1) d)
4
1) e)
26
1) f)
-2

2) a)
50
2) b)
9

3) a)
$P = 250$
3) b)
$P = 5021$
3) c)
$t = 9.24$ hours (3 SF)

4) a)
1 hour
4) b)
$t = \frac{2}{3}\ln 2$ hours
4) c)
$t = \log_{1.6} 2$ hours
4) d)
$t = 10\ln 7$ hours

5) a)
1000 grams
5) b)
Show that

6) a)
$y_0 = 2$, as $x \to \infty, y \to 10$

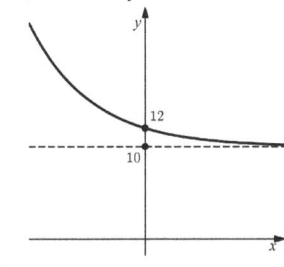

6) b)
$y_0 = 2$, as $x \to \infty, y \to -5$

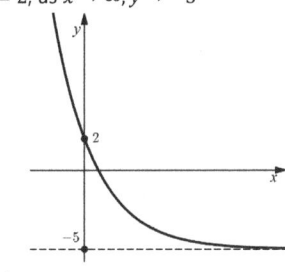

6) c)
$y_0 = 0.6$, as $x \to \infty, y \to 1.2$

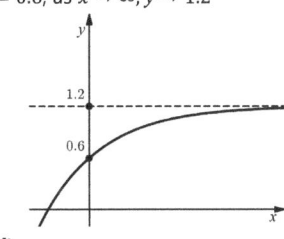

6) d)
$y_0 = 2$, as $x \to \infty, y \to \infty$

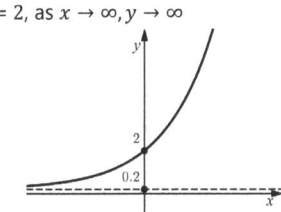

7) a)
98°C
7) b)
$y = 22$
7) c)
$A = 22, B = 76$
7) d)
Change the value of B to be 78
7) e)
$A = 18$ and $B = 79$

8) a)
10 cm
8) b)
50 cm

9) a)
50 cm
9) b)
84.2 cm
9) c)
32 months
9) d)
As $x \to \infty, h \to 130$ so it does not go over a height of 1.3m.

10) a)
$y = 1200 \times 1.05^t$
10) b)

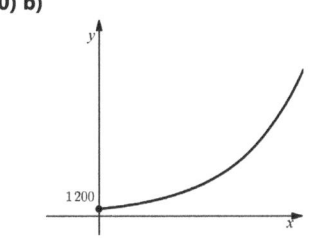

10) c)
£1458.61
10) d)
End of year 11

11) a)
£100
11) b)
$C = 100 \times 1.5^t$
11) c)
$C = £506.25$

12) a)
225 million active users initially.
12) b)
$P = 225 \times 0.8^{\frac{1}{3}t}$
12) c)
$P = 106.9$ million active users

13) a)
$P = 640 \times 1.5^t$, so $A = 640, r = 1.5$
13) b)
640 is the initial population of insects, and 1.5 shows that the population 150% of its previous size as each month passes.

14) a)
$P = 12000e^{\frac{1}{5}\ln\frac{5}{24}t}$
14) b)
6407 items
14) c)
10.1 months (3 SF)
14) d)
As $t \to \infty, P \to 0$ which does not allow for any changes in patterns such as ordering more stock.

15) a)
$C = 200e^{-\frac{1}{4}\ln(2)t}$, $A = 200$ and $k = \frac{1}{4}\ln 2$
15) b)
8 hours

16) a)
Show that
16) b)
$P = \frac{5000}{7}e^{(\ln 1.4)n}$, $A = \frac{5000}{7}$ and $k = \ln 1.4$
16) c)
2744 restaurants
16) d)
15 years
16) e)
As $t \to \infty, P \to \infty$ There cannot be infinite restaurants.

17)
$\frac{1}{k}\ln 100$

18) a)
$k = \frac{1}{5}\ln\frac{125}{22}$
18) b)
$A = 1001.8$

19) a)
$a = 49.3, b = 35.7$
19) b)
The approximate initial number of each type of fish
19) c)
After 8.07 months (3 SF)
19) d)
Both models predict infinite exponential growth but the fish populations will be limited by the size of the lake or may be affected by disease or predators.

20) a)
$T = 17 + 48e^{-\frac{1}{3}\ln\left(\frac{8}{3}\right)t}$, $a = 48, k = \frac{1}{3}\ln\left(\frac{8}{3}\right)$
20) b)
26.4°C (3 SF)
20) c)
17°C
20) d)
The value "17" in the model would be reduced and the "48" would increase by the same amount.
20) e)
$t = 1.66$ minutes (3SF)

21) a)
-0.329 (3 SF)
21) b)
1.19 (3 SF)
21) c)
9.69 (3 SF)

22)
Show that

23) a)
330ml
23) b)
20.0 ml/s (3 SF)

24) a)

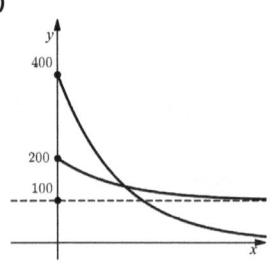

24) b)
Amount of product A → 100
Amount of product B → 0
24) c)
In both cases the models do not actually reach these values, and as the amount of each product is discrete the model only approximates at each point.
24) d)
Show that
24) e)
$k = \frac{1}{2}\ln 3$

25) a)
4.5 m
25) b)
Show that
25) c)
18 m
25) d)
$a = 16$

25) e)
0.575 metres per year (3 sf)
25) f)
As $t \to \infty, h \to 16$ m. So the first tree is expected to be taller.

26)
$k > \frac{1}{18}$

1.27 Reduction to Linear Form
Questions on page 66

1)
Show that

2)
Show that

3)
Show that

4)
Show that

5)
Show that

6)
Show that

7) a)
$y = 3.0x^5$
7) b)
$y = 2.3x^3$
7) c)
$y = 0.045x^{-0.89}$
7) d)
$y = 0.32x^{0.51}$

8) a)
$y = 0.029 \times 1.6^x$
8) b)
$y = 1.6 \times 1.7^x$
8) c)
$y = 1.2 \times 0.79^x$
8) d)
$y = 4.1 \times 1.9^x$

9) a)
$\ln y = 3.7 - 0.2x$
9) b)
$\ln y = -0.69 + 0.020x$
9) c)
$\ln y = 0.48 - 0.15x$

10) a)
$y = 3.2 \times 1000^x$
10) b)
$y = 1.2 \times 1.8^x$

11) a)
$y = 7.4e^{4x}$
11) b)
$y = 1.1e^{0.75x}$

12) a)
$y = 1.3x^4$
12) b)
$y = 10000x^{0.5}$

13) a)
$y = 1.3x^{-5}$
13) b)
$y = 15x^{-0.4}$

14)
$y = 7.4x^{0.5}$

15)
$y = 1000x^{0.44}$

16)
$y = 3.2 \times 2.0^x$

17)
$y = 8.70 \times 2.26^x$

18) a)

$\log_{10} x$	$\log_{10} y$
0.301	0.699
1.477	1.176
1.778	1.380
2	1.477
2.176	1.602

18) b)
See worked solutions
18) c)
y-intercept ≈ 0.54, Gradient ≈ 0.47
18) d)
Show that
18) e)
$a = 3.5, b = 0.5$

19) a)

$\log_{10} t$	$\log_{10} N$
0.477	1.083
0.699	1.393
0.903	1.702
1	1.819
1.255	2.231
1.380	2.396

19) b)
See worked solutions
19) c)
$a = 2.3, b = 1.5$
19) d)
95.6 million downloads
19) e)
16.4 months

20) a)

$\ln m$
3.912
3.040
2.186
1.281
0.531

20) b)
See worked solutions
20) c)
$a = 49.4, b = -0.0857$
20) d)
0.68 grams

21) a)
See worked solutions
21) b)
$a = 89, b = -0.1, T = 89 \times t^{-0.1}$
21) c)
a is the initial amount of time spent exercising (that is when 0 weeks have passed)
21) d)
80 minutes
21) e)
When $t = 0$ there are no solutions so the model cannot be used to predict the amount of exercise done at this point.

21) f)
Morgana may well continue to exercise regularly as she is still doing 10 weeks in, and so has formed a habit. But it is unlikely it will continue to decrease in this manner consistently for a long period of time.

22) a)
Since $y = ax^b \Rightarrow \log_{10} y = \log_{10} a + b \log_{10} x$
He should have plotted $\log_{10} y$ against $\log_{10} x$

22) b)
$y = 251 \times 4^x$

23) a)
She raised the right-hand side of her equation by 10 to the power rather than e

23) b)
$y = e^{-0.0395x + 7.82}$

24) a)
$\ln N = 7 - 0.8 \ln t$

24) b)
$N = 1096.6 t^{-0.8}$

24) c)
$t = 66$
38.4 million cinema admissions

24) d)
Due to the pandemic, there was a huge drop in cinema attendance in reality.

25) a)
$\log_2 y = 0.467x + 13.185$

25) b)
$y = 9312.8 \times 2^{0.467x}$

25) c)
a is the number of transistors in a microprocessor in 1975, according to the model.

25) d)
5.0×10^{11} transistors

26) a)
$N = 99.41 \times 1.73^t$

26) b)
a is the initial number of subscribers according to the model.
b is the exponential rate, ie the number of subscribers is increasing by 73% per year

26) c)
22646 subscribers

27) a)
98.0°C

27) b)
$T = 98.0 e^{-0.0701 t}$

28) a)
$T = 0.203 d^{1.48}$

28) b)
624 days

28) c)
9.17 %

29) a)
Show that

29) b)
$a = 0.196, b = 0.66$

29) c)

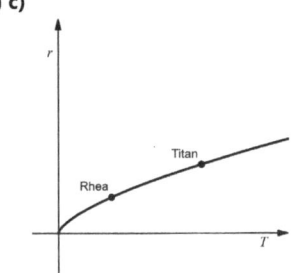

29) d)
0.378 million km

29) e)
0.53 %

30) a)
$a = 0.0217$

30) b)
$b = 79.1$

30) c)
33 years

30) d)
53 years

31)
$y = 0.7 x^2 + 5.2$

32)
$y = \frac{25}{\pi} \sin(x) - \frac{36}{5}$

1.28 Graph Transformations: Single Transformations
Questions on page 71

1) a)
Translation by the vector $\begin{pmatrix} 6 \\ 0 \end{pmatrix}$

1) b)
Translation by the vector $\begin{pmatrix} 0 \\ 8 \end{pmatrix}$

1) c)
Translation by the vector $\begin{pmatrix} -4 \\ 0 \end{pmatrix}$

1) d)
Translation by the vector $\begin{pmatrix} 0 \\ -11 \end{pmatrix}$

2) a)
Translation by the vector $\begin{pmatrix} 9 \\ 0 \end{pmatrix}$

2) b)
Translation by the vector $\begin{pmatrix} -5 \\ 0 \end{pmatrix}$

2) c)
Translation by the vector $\begin{pmatrix} 0 \\ 2 \end{pmatrix}$

2) d)
Translation by the vector $\begin{pmatrix} 0 \\ -10 \end{pmatrix}$

2) e)
Translation by the vector $\begin{pmatrix} 0 \\ -4 \end{pmatrix}$

2) f)
Translation by the vector $\begin{pmatrix} 0 \\ 8 \end{pmatrix}$

2) g)
Translation by the vector $\begin{pmatrix} -7 \\ -3 \end{pmatrix}$

2) h)
Translation by the vector $\begin{pmatrix} 1 \\ 7 \end{pmatrix}$

3) a)
Translation by the vector $\begin{pmatrix} -4 \\ 0 \end{pmatrix}$

3) b)
Translation by the vector $\begin{pmatrix} 2 \\ 0 \end{pmatrix}$

3) c)
Translation by the vector $\begin{pmatrix} 0 \\ -7 \end{pmatrix}$

3) d)
Translation by the vector $\begin{pmatrix} 0 \\ 15 \end{pmatrix}$

3) e)
Translation by the vector $\begin{pmatrix} 0 \\ -9 \end{pmatrix}$

3) f)
Translation by the vector $\begin{pmatrix} 3 \\ 8 \end{pmatrix}$

4)
$y = 9x + 25$

5) a)
$y = 21 - 6x$

5) b)
$y = (x-4)^2 - 7$

5) c)
$y = (x-4)^2 - 4$

5) d)
$y = (x-2)(x-12) - 7$

5) e)
$y = (x-13)(x-3) - 7$

5) f)
$y = 3^{x-4} - 7$

5) g)
$y = e^{x-4} - 12$

5) h)
$y = \frac{1}{x-4} - 7$

5) i)
$y = \frac{1}{(x-4)^2} - 7$

5) j)
$y = \sqrt{x-4} - 7$

5) k)
$y = \cos(x-4) - 7$

6) a)
$P'(15, -5)$

6) b)
$P'(7, -2)$

6) c)
$P'(-5, -5)$

6) d)
$P'(6, -10)$

7)
$y = f(x+2) + 2$

8)
Translation by the vector $\begin{pmatrix} -4 \\ -10 \end{pmatrix}$

9)
Translation by the vector $\begin{pmatrix} 8 \\ -7 \end{pmatrix}$

10)
Translation by the vector $\begin{pmatrix} -2 \\ -2 \end{pmatrix}$

11)
Translation by the vector $\begin{pmatrix} 4 \\ -2 \end{pmatrix}$

12)
Translation by the vector $\begin{pmatrix} 2 \\ 2 \end{pmatrix}$

13)
Translation by the vector $\begin{pmatrix} -2e \\ -5 \end{pmatrix}$

14)
$(x+7)^2 + (y-5)^2 = 9$

15)
$(x-2)^2 + (y+8)^2 = 3$

16)
$(x+7)^2 + (y+3)^2 = 10$

17)
$(x+8)^2 + (y-2)^2 = 16$

18)
$(x+3)(y-4) + (x+3)^2 - (y-4) = 10$

19)
$y = (x+2)^3 - 2(x+2)^2 + 8(x+2) + 5$

20)
$y = 8(x-4)^4 - 4(x-4)^8 - 8$

21)
$y = \frac{3}{(x+2)^2} - 6$

22)
$y = -\frac{4}{x-14}$

23)
Translation by the vector $\begin{pmatrix} 6 \\ 0 \end{pmatrix}$

24) a)
[graph with points (−2,0) and (0,2)]

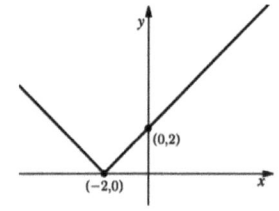

24) b)
[graph with points (−3,0), (0,−1), (1,0), (−1,−2)]

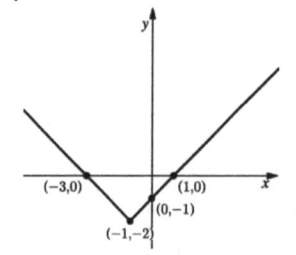

24) c)
[graph with points (0,4) and (3,1)]

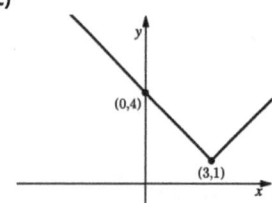

24) d)
[graph with points (0,5) and (−3,2)]

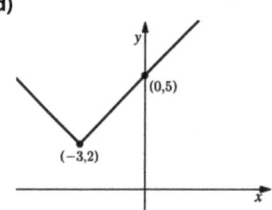

25) a)
Stretch, parallel to x-axis, scale factor $\frac{1}{5}$

25) b)
Stretch, parallel to y-axis, scale factor $\frac{1}{6}$

25) c)
Stretch, parallel to x-axis, scale factor $\frac{3}{2}$

26) a)
Stretch, parallel to y-axis, scale factor 3

26) b)
Stretch, parallel to x-axis, scale factor $\frac{1}{3}$

27) a)
Stretch, parallel to x-axis, scale factor $\frac{1}{2}$

27) b)
Stretch, parallel to y-axis, scale factor 8

27) c)
Stretch, parallel to y-axis, scale factor $\frac{1}{8}$

27) d)
Stretch, parallel to x-axis, scale factor 2

28) a)
$y = 20x + 3$

28) b)
$y = x + \frac{3}{5}$

29) a)
$y = 4 - 2x$

29) b)
$y = \frac{1}{9}x^2$

29) c)
$y = \frac{1}{9}x^2 + 3$

29) d)
$y = \frac{1}{9}(x+6)(x-24)$

29) e)
$y = \frac{1}{9}(x-27)(x+3)$

29) f)
$y = 3^{\frac{1}{3}x}$

29) g)
$y = e^{\frac{1}{3}x} - 5$

29) h)
$y = \frac{3}{x}$

29) i)
$y = \frac{9}{x^2}$

29) j)
$y = \frac{\sqrt{3x}}{3}$

29) k)
$y = \cos\left(\frac{1}{3}x\right)$

30) a)
$y = 12 - 18x$

30) b)
$y = 3x^2$

30) c)
$y = 3x^2 + 9$

30) d)
$y = 3(x+2)(x-8)$

30) e)
$y = 3(x-9)(x+1)$

30) f)
$y = 3^{x+1}$

30) g)
$y = 3e^x - 15$

30) h)
$y = \frac{3}{x}$

30) i)
$y = \frac{3}{x^2}$

30) j)
$y = 3\sqrt{x}$

30) k)
$y = 3\cos x$

31) a)
$(-2, 60)$

31) b)
$(-4, 10)$

31) c)
$(-2, 4)$

31) d)
$\left(-\frac{1}{3}, 10\right)$

32)
Stretch, parallel to x-axis, scale factor $\frac{1}{5}$

33)
Stretch, parallel to y-axis, scale factor 4

34)
$y = 2f(x)$

35)
Stretch, parallel to x-axis, scale factor 4

36) a)
Stretch, parallel to y-axis, scale factor 27

36) b)
Stretch, parallel to x-axis, scale factor $\frac{1}{3}$

37) a)
Stretch, parallel to y-axis, scale factor 8
37) b)
Stretch, parallel to x-axis, scale factor 8

38) a)
Stretch, parallel to y-axis, scale factor 100
38) b)
Stretch, parallel to x-axis, scale factor 10

39)
Stretch, parallel to y-axis, scale factor $\frac{1}{3}$

40)
Stretch, parallel to x-axis, scale factor $\frac{\sqrt{3}}{6}$

41)
$y = f\left(\frac{1}{2}x\right)$

42)
Stretch, parallel to x-axis, scale factor $\frac{1}{2}$

43)
Stretch, parallel to x-axis, scale factor $\frac{1}{4}$

44)
Stretch, parallel to x-axis, scale factor 8

45) a)
[Graph: V-shape with vertex at $(-\frac{1}{2}, 0)$ and y-intercept $(0,1)$]

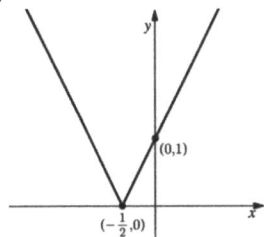

45) b)
[Graph: V-shape with vertex at $(-1,0)$ and y-intercept $(0,4)$]

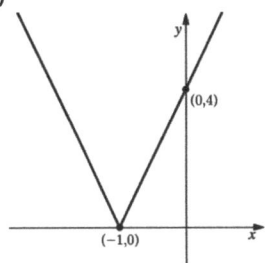

45) c)
[Graph: V-shape with vertex at $(-1,0)$ and y-intercept $(0,\frac{1}{3})$]

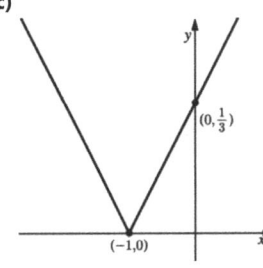

45) d)
[Graph: V-shape with vertex at $(-3,0)$ and y-intercept $(0,1)$]

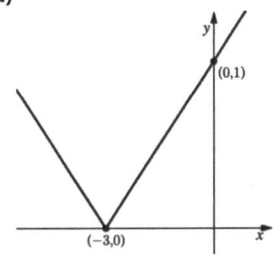

46) a)
Reflection in the y-axis
46) b)
Reflection in the x-axis

47) a)
Reflection in the x-axis
47) b)
Reflection in the y-axis

48) a)
Reflection in the y-axis
48) b)
Reflection in the x-axis
48) c)
Reflection in the y-axis

49) a)
$y = 9 - 8x$
49) b)
$y = -8x - 9$

50) a)
$y = 6x - 3$
50) b)
$y = 3 + 6x$

51) a)
$y = 5 - (-x) \Rightarrow y = 5 + x$
This function **will not** remain unchanged
51) b)
$y = (-x)^2 + 2 \Rightarrow y = x^2 + 2$
This function **will** remain unchanged
51) c)
$y = (-x)^3 \Rightarrow y = -x^3$
This function **will not** remain unchanged
51) d)
$y = 4^{-x}$
This function **will not** remain unchanged
51) e)
$y = \frac{1}{(-x)} \Rightarrow y = -\frac{1}{x}$
This function **will not** remain unchanged
51) f)
$y = \frac{1}{(-x)^2} \Rightarrow y = \frac{1}{x^2}$
This function **will** remain unchanged
51) g)
$y = \sin(-x)$
$(\Rightarrow y = -\sin x)$
This function **will not** remain unchanged
51) h)
$y = \cos(-x) \Rightarrow y = \cos x$
This function **will** remain unchanged

51) i)
$y = \frac{e^{-x} + e^{-(-x)}}{2} \Rightarrow y = \frac{e^{-x} + e^x}{2} \Rightarrow y = \frac{e^x + e^{-x}}{2}$
This function **will** remain unchanged

52) a)
$-y = 5 - x \Rightarrow y = -5 + x$
$\Rightarrow y = x - 5$
This function **will not** remain unchanged
52) b)
$-y = x^2 + 2 \Rightarrow y = -x^2 - 2$
This function **will not** remain unchanged
52) c)
$-y = x^3 \Rightarrow y = -x^3$
This function **will not** remain unchanged
52) d)
$-y = 4^x \Rightarrow y = -4^x$
This function **will not** remain unchanged
52) e)
$-y = \frac{1}{x} \Rightarrow y = -\frac{1}{x}$
This function **will not** remain unchanged
52) f)
$-y = \frac{1}{x^2} \Rightarrow y = -\frac{1}{x^2}$
This function **will not** remain unchanged
52) g)
$-y = \sin x \Rightarrow y = -\sin x$
This function **will not** remain unchanged
52) h)
$-y = \cos x \Rightarrow y = -\cos x$
This function **will not** remain unchanged
52) i)
$-y = \frac{e^x + e^{-x}}{2} \Rightarrow y = -\frac{e^x + e^{-x}}{2}$
This function **will not** remain unchanged

53) a)
[Graph: V-shape with vertex at $(1,0)$ and y-intercept $(0,1)$]

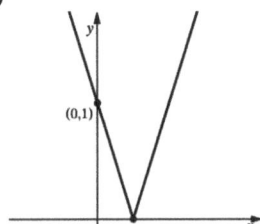

53) b)
[Graph: inverted V-shape with vertex at $(-1,0)$ and y-intercept $(0,-1)$]

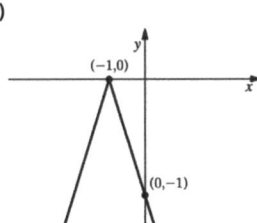

54) a)
$(4, 1)$
54) b)
$(-4, -1)$

55) a)
Translation by the vector $\begin{pmatrix} 0 \\ -3 \end{pmatrix}$
55) b)
Stretch, parallel to x-axis, scale factor $\frac{1}{2}$
Alternatively:
Stretch, parallel to y-axis, scale factor $\sqrt{2}$

55) c)
Translation by the vector $\begin{pmatrix} -9 \\ 0 \end{pmatrix}$

55) d)
Reflection in the x-axis

55) e)
Stretch, parallel to y-axis, scale factor 8
Alternatively:
Stretch, parallel to x-axis, scale factor $\frac{1}{64}$

56)
Show that

57)
Show that

58)
Stretch, parallel to y-axis, scale factor $\frac{9}{16}$

59) a)
$f(x+2) = \frac{1}{(x+2)^2}$
Asymptotes at $x = -2$ and $y = 0$

59) b)
$-f(x) = -\frac{1}{x^2}$
Asymptotes at $x = 0$ and $y = 0$

59) c)
$f(x-5) + 4 = \frac{1}{(x-5)^2} + 4$
Asymptotes at $x = 5$ and $y = 4$

59) d)
$f\left(\frac{1}{3}x\right) = \frac{1}{\left(\frac{1}{3}x\right)^2} = \frac{1}{\frac{1}{9}x^2} = \frac{9}{x^2}$
Asymptotes at $x = 0$ and $y = 0$

60) a)
$a = 0, b = 7, c = 9$

60) b)
$a = -12, b = -5, c = -3$

60) c)
$a = -2, b = 5, c = 7$

60) d)
$a = -1, b = \frac{5}{2}, c = \frac{7}{2}$

60) e)
$a = -6, b = 15, c = 21$

60) f)
$a = 2, b = -5, c = -7$

61) a)
$x = 2$ or $x = -4$ or $x = -5$

61) b)
$x = 4$ or $x = -2$ or $x = -3$

61) c)
$x = 1$ or $x = -2$ or $x = -\frac{5}{2}$

61) d)
$x = 6$ or $x = -12$ or $x = -15$

61) e)
$x = -2$ or $x = 4$ or $x = 5$

61) f)
$x = -13$ or $x = -19$ or $x = -20$

62) a)
$y = \frac{8}{x-3}$

62) b)
$y = \frac{4}{x-6}$

62) c)
$y = -\frac{8}{x+6}$

62) d)
$y = \frac{4}{2x-3}$

63)
$k = 4, k = -5,$ or $k = -7$

64)
$k = 8, k = 2,$ or $k = -3$

65) a)
Translation by the vector $\begin{pmatrix} 3 \\ 0 \end{pmatrix}$

65) b)
Stretch, parallel to y-axis, scale factor $\frac{1}{64}$

66)
Translation by the vector $\begin{pmatrix} -4 \\ -19 \end{pmatrix}$

67)
Translation by the vector $\begin{pmatrix} 2 \\ -13 \end{pmatrix}$

68) a)
$y = 3x^2 + 2x + 1$

68) b)
$y = 3x^2 - 20x + 28$

68) c)
$y = 27x^2 - 6x + 1$

69) a)
$x = -\frac{1}{2}$ or $x = 4$ or $x = 0$

69) b)
$x = -\frac{7}{2}$ or $x = 1$ or $x = -3$

69) c)
$x = -\frac{1}{6}$ or $x = \frac{4}{3}$ or $x = 0$

69) d)
$x = \frac{1}{2}$ or $x = -4$ or $x = 0$

70) a)

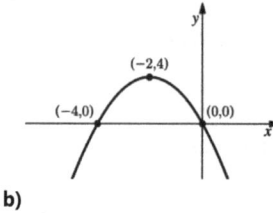

70) b)

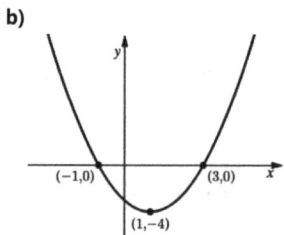

70) c)

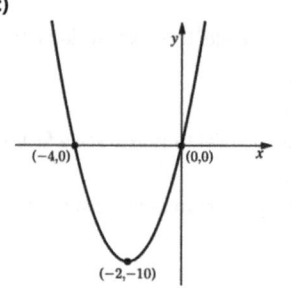

1.29 Binomial Expansion
Questions on page 77

1) a)
56

1) b)
19600

1) c)
4900

1) d)
7

1) e)
4950

1) f)
100

1) g)
50

1) h)
1

2) a)
n

2) b)
$\frac{1}{2}n(n-1)$

2) c)
n

2) d)
$n+1$

2) e)
1

2) f)
$(2n-1)(n-1)$

2) g)
$(2n-5)(n-2)$

3)
$n = 10$

4) a)
$n = 15$

4) b)
$n = 12$

4) c)
$n = 13$

5) a)
$1 + 6x + 15x^2 + 20x^3 + 15x^4 + 6x^5 + x^6$

5) b)
$32 + 80x + 80x^2 + 40x^3 + 10x^4 + x^5$

5) c)
$729 + 1458x + 1215x^2 + 540x^3 + 135x^4 + 18x^5 + x^6$

5) d)
$1 - 4x + 6x^2 - 4x^3 + x^4$

6) a)
$x^5 - 15x^4 + 90x^3 - 270x^2 + 405x - 243$

6) b)
$x^6 - 30x^5 + 375x^4 - 2500x^3 + 9375x^2 - 18750x + 15625$

6) c)
$16x^4 + 32x^3 + 24x^2 + 8x + 1$

6) d)
$243x^5 - 1215x^4 + 2430x^3 - 2430x^2 + 1215x - 243$

6) e)
$\frac{1}{16}x^4 + \frac{1}{2}x^3 + \frac{3}{2}x^2 + 2x + 1$

6) f)
$-\frac{1}{1024}x^5 + \frac{5}{128}x^4 - \frac{5}{8}x^3 + 5x^2 - 20x + 32$

6) g)
$16x^4 - 32x^2 + 24 - 8x^{-2} + x^{-4}$

7)
$-405 + 1080x - 1080x^2 + 480x^3 - 80x^4$

8)
$16 - 40x^2 + 40x^3 - 15x^4 + 2x^5$

9) a)
$28 + 16\sqrt{3}$

9) b)
$89 + 28\sqrt{10}$

9) c)
$209\sqrt{7} - 389\sqrt{2}$

10) a)
150

10) b)
-40

10) c)
5120

10) d)
135

10) e)
12012

11) a)
-4

11) b)
4860

11) c)
43758

12)
Show that

13)
Show that

14)
Each term has $(-3x)^k$ as a factor. When k is odd, there will be an odd number of negative terms multiplying together, so the result will be negative.

15)
$n = 7$

16)
$n = 15$

17)
$n = 11$

18)
$n = 9$

19) a)
924

19) b)
680

19) c)
196830

20) a)
$243 + 405x + 270x^2 + 90x^3 + 15x^4 + x^5$

20) b)
$843 - 589\sqrt{2}$

21) a)
$16 + 160x + 600x^2 + 1000x^3 + 625x^4$

21) b)
$16 - 160x + 600x^2 - 1000x^3 + 625x^4$

21) c)
14882

22) a)
$1 + 8ax + 28a^2x^2 + 56a^3x^3 + \cdots$

22) b)
$a = \frac{3}{2}$

23)
$a = \pm 3$

24)
$a = 5, n = 8$

25)
$a = 16$

26) a)
$32 - 560x + 3920x^2 - 13720x^3 + 24010x^4 - 16807x^5$

26) b)
$a = 3$

27) a)
$1 - 18x + 144x^2 + \cdots$

27) b)
0.834 (3 DP)

28)
Set $3 + \frac{2x}{5} = 2.96$ so that $x = -0.1$
Fully expanding the expression and using the value of $x = -0.1$ in the result gives the value of 2.96^5

29) a)
$\left(1 - \frac{x}{2}\right)^5 = 1 - \frac{5}{2}x + \frac{5}{2}x^2 + \cdots$
$\left(2 - \frac{3x}{4}\right)^6 = 64 - 144x + 135x^2 + \cdots$

29) b)
63.549

30)
$\binom{12}{r} \times (x^2)^{12-r} \times \left(\frac{1}{x^4}\right)^r$
gives $x^{24} \times x^{-2r} \times x^{-4r} = x^{24}x^{-6r} = x^{24-6r}$
$24 - 6r = 6(4 - r)$ so powers are always multiples of 6.
Or
$\binom{12}{r} \times (x^2)^r \times \left(\frac{1}{x^4}\right)^{12-r}$
gives $x^{2r} \times (x^{-4})^{12-r} = x^{2r} \times x^{-48+4r} = x^{6r-48}$
$64 - 48 = 6(r - 8)$ so powers are always multiples of 6.

31) a)
$1 + 12x + 54x^2 + 108x^3 + 81x^4$

31) b)
$1 - 12x + 54x^2 - 108x^3 + 81x^4$

31) c)
2

32) a)
$15625 - 37500x + 37500x^2 - \cdots$

32) b)
-75000

32) c)
46 875

1.30 Differentiation from First Principles
Questions on page 79

1) a)
3

1) b)
3

1) c)
19

1) d)
33

1) e)
$\frac{3}{2}$

2) a)
4

2) b)
1

2) c)
$-\frac{1}{2}$

2) d)
$\frac{1}{2}$

3)
28.028007, 2.0001, 28.00280007
Hence 28

4) a)
2, 1.43547, 1.39111, 1.38678, 1.38634
4) b)
1.39

5) a)
0.48
5) b)
0.700 (3 SF)
5) c)
Use a third point that is on the curve which is even closer to point P, so that its x-coordinate is
$2 < x < 2.1$

6) a)
-2
6) b)
At $x = 3$ the function is undefined (discontinuous)
6) c)
2. However there is a discontinuity between the two points, so it does not approximate the gradient at $x = 2$.

7) a)
$1 + 2h + h^2$
7) b)
$5 + 10h + 5h^2$
7) c)
$1 + 3h + 3h^2 + h^3$
7) d)
$-3 - 9h - 9h^2 - 3h^3$

8) a)
$4 + 4h + h^2$
8) b)
$63 + 42h + 7h^2$
8) c)
$-8 + 12h - 6h^2 + h^3$
8) d)
$-3000 - 900h - 90h^2 - 3h^3$

9) a)
$6h + h^2$
9) b)
$12h + 6h^2$
9) c)
$3h - 3h^2 + h^3$
9) d)
$-96h - 24h^2 - 2h^3$

10) a)
$f(2) = 4$
$f(2 + h) = 4 + 4h + h^2$
10) b)
$f(2) = 2$
$f(2 + h) = 2 + 3h + h^2$
10) c)
$f(2) = 5$
$f(2 + h) = 5 + 2h + h^2$
10) d)
$f(2) = -10$
$f(2 + h) = -10 - 11h - 3h^2$

10) e)
$f(2) = 0$
$f(2 + h) = -9h - 2h^2$

11) a)
2
11) b)
-5
11) c)
3
11) d)
1
11) e)
-4
12)
$\frac{h^2 - 2}{h} = 1 - \frac{2}{h}$, undefined when $h = 0$
As $h \to 0$, $\frac{2}{h}$ diverges.

13) a)
$1 + 3h + 3h^2 + h^3$
13) b)
$24 + 24h + 8h^2$
13) c)
24

14) a)
4
14) b)
6
14) c)
-7

15) a)
8.1
15) b)
$8 + h$
15) c)
8

16) a)
Show that
16) b)
$m = 30$

17)
Show that

18) a)
$4x$
18) b)
$10x + 1$
18) c)
$1 - 8x$
18) d)
$2x - 2$
18) e)
$3x^2$

19) a)
12
19) b)
-8

20)
Show that

21) a)
$-6h - h^2$
21) b)
0

22)
The expansion of $2(3 + h)^3$ and missing $\lim_{h \to 0}$ on penultimate line
$f(3) = 2 \times 3^3 - 9 = 45$
$f(3 + h) = 2(3 + h)^3 - 9 = 2(27 + 27h + 9h^2 + h^3) - 9$
$= 2h^3 + 18h^2 + 54h + 45$
$\frac{dy}{dx} = \lim_{h \to 0} \frac{2h^3 + 18h^2 + 54h + 45 - 45}{h}$
$= \lim_{h \to 0}(2h^2 + 18h + 54)$
$= 54$

23)
a

24)
$2ax + b$

25)
$a = -1 + 2\sqrt{6}$

26)
$p = 2, q = \frac{17}{2}$

27) a)
16
27) b)
86.4°

28) a)
7
28) b)
$x = 2$

29) a)
$-\frac{1}{x^2}$
29) b)
$\frac{1}{(1-x)^2}$
29) c)
$\frac{1}{2\sqrt{x+1}}$

30)
Show that

31)
Show that

32) a)
Show that
32) b)
as $h \to 0$, $h^2 + 6h \to 0$ so the gradient is zero, hence a stationary point.

33) a)
$5x^4$
33) b)
$8x^3$

34) a)
Show that
34) b)
Show that
34) c)
Show that

1.31 Differentiation: Introduction
Questions on page 82

1) a)
$\frac{dy}{dx} = 12x^2 + 10x$
1) b)
$\frac{dy}{dx} = 2 - 18x$
1) c)
$\frac{dy}{dx} = 6x^5 - 36x$
1) d)
$\frac{dy}{dx} = -x^{-2} + 4x^3$
1) e)
$\frac{dy}{dx} = 3x^2 + 2x^{-3}$
1) f)
$\frac{dy}{dx} = 2x - 2x^{-3} + 1$
1) g)
$\frac{dy}{dx} = 15x^4 + 4x^{-3} - 1$
1) h)
$\frac{dy}{dx} = 2ax^{a-1} - bx^{b-1}$

2) a)
$f'(t) = t - 3t^{-6}$
2) b)
$f'(t) = \frac{1}{4} - 3t^3$
2) c)
$f'(t) = \frac{1}{2}t^5 + 20t$
2) d)
$f'(t) = -\frac{1}{3}t^{-2} + 4t^5$
2) e)
$f'(t) = \frac{7}{4}t + \frac{3}{4}t^{-4}$
2) f)
$f'(t) = \frac{16}{7}t^7 - \frac{3}{7}t^{-4} + \frac{1}{5}$
2) g)
$f'(t) = -\frac{3}{2}t^5 + t^{-3} + 2$
2) h)
$f'(t) = -\frac{a}{2b}t^{\frac{a}{b}-1}$

3) a)
$\frac{dy}{dt} = 4t^{-\frac{1}{2}} + 2t^{-\frac{2}{3}}$
3) b)
$\frac{dy}{dt} = 5 - 3t^{-\frac{3}{4}}$
3) c)
$\frac{dy}{dt} = \frac{2}{3}t^{-\frac{1}{3}} + \frac{21}{5}t^{-\frac{2}{5}}$
3) d)
$\frac{dy}{dt} = -\frac{1}{2}t^{-\frac{3}{2}} + \frac{1}{4}t^{-\frac{3}{4}}$

3) e)
$\frac{dy}{dt} = -\frac{5}{8}t^{-\frac{3}{8}} + \frac{3}{8}t^{-\frac{11}{8}} - \frac{1}{8}$
3) f)
$\frac{dy}{dt} = \frac{1}{3}t^{-\frac{2}{3}} - \frac{1}{3}t^{-\frac{4}{3}} + 3$
3) g)
$\frac{dy}{dt} = -\frac{3}{4}t^{-\frac{1}{4}} - \frac{8}{9}t^{-\frac{17}{9}} - \frac{1}{12}$

4) a)
$f'(x) = 2x + 3$
4) b)
$f'(x) = 2x + 6$
4) c)
$f'(x) = 4x + 1$
4) d)
$f'(x) = 2x + 8$
4) e)
$f'(x) = 3x^2 - 2x$
4) f)
$f'(x) = 15x^2 + 38x + 16$
4) g)
$f'(x) = 14x - 4x^3$
4) h)
$f'(x) = 27x^2 - 1$
4) i)
$f'(x) = x^{-2} + 8$
4) j)
$f'(x) = 3x^2 + 3 - 3x^{-2} - 3x^{-4}$
4) k)
$f'(x) = 2x + p + q$
4) l)
$f'(x) = 2x + \frac{1-a^2}{a}$

5) a)
$x^{\frac{1}{2}}$
5) b)
$x^{\frac{5}{2}}$
5) c)
x^{-4}
5) d)
$x^{\frac{9}{2}}$
5) e)
$x^{\frac{1}{3}}$
5) f)
$x^{-\frac{1}{2}}$
5) g)
$x^{\frac{1}{2}}$
5) h)
$x^{-\frac{3}{2}}$
5) i)
$x^{\frac{a+1}{a}}$

6) a)
$5x^{\frac{1}{3}}$
6) b)
$2x^{\frac{3}{2}}$
6) c)
$3x^{-2}$
6) d)
$32x^{\frac{15}{2}}$

6) e)
$\frac{1}{3}x^{\frac{1}{8}}$
6) f)
$4x^{-\frac{1}{2}}$
6) g)
x^{-1}
6) h)
$\frac{5}{4}x^{-\frac{3}{2}}$
6) i)
$\frac{2}{5\sqrt{a}}x^{-\frac{1}{2}}$

7) a)
$\frac{dy}{dx} = \frac{1}{4}x^{-\frac{3}{4}}$
7) b)
$\frac{dy}{dx} = \frac{15}{2}x^{\frac{3}{2}}$
7) c)
$\frac{dy}{dx} = -15x^{-4}$
7) d)
$\frac{dy}{dx} = \frac{9}{2}x^{\frac{7}{2}}$
7) e)
$\frac{dy}{dx} = \frac{1}{15}x^{-\frac{2}{3}}$
7) f)
$\frac{dy}{dx} = -\frac{3}{2}x^{-\frac{3}{2}}$
7) g)
$\frac{dy}{dx} = -x^{-2}$
7) h)
$\frac{dy}{dx} = -\frac{1}{2}x^{-2}$

8) a)
$f'(x) = \frac{3}{2}x^{\frac{1}{2}} + x^{-\frac{1}{2}}$
8) b)
$f'(x) = \frac{5}{2}x^{\frac{3}{2}} - \frac{1}{2}x^{-\frac{1}{2}}$
8) c)
$f'(x) = 2x$
8) d)
$f'(x) = 4x - 12x^{-4}$
8) e)
$f'(x) = -9x^{-2} + 2x^{-3}$
8) f)
$f'(x) = \frac{3}{2}x^{\frac{1}{2}} + x^{-\frac{1}{2}} + \frac{1}{2}x^{-\frac{3}{2}}$
8) g)
$f'(x) = \frac{3}{2}x^{\frac{1}{2}} + 10x^{-3} - \frac{1}{2}x^{-\frac{3}{2}}$

9) a)
$\frac{dy}{dx} = 3x^2 - 3x^{-4} + 4x - 5$
$\frac{d^2y}{dx^2} = 6x + 12x^{-5} + 4$
9) b)
$\frac{dy}{dx} = \frac{1}{3}x^3 + \frac{1}{4}x + 7, \frac{d^2y}{dx^2} = x^2 + \frac{1}{4}$
9) c)
$\frac{dy}{dx} = -3x^2 + x^{-\frac{3}{2}} + 15, \frac{d^2y}{dx^2} = -6x - \frac{3}{2}x^{-\frac{5}{2}}$
9) d)
$\frac{dy}{dx} = -\frac{1}{7}x^{-\frac{6}{7}} + \frac{1}{7}x^{-\frac{8}{7}} - \frac{1}{9}$
$\frac{d^2y}{dx^2} = \frac{6}{49}x^{-\frac{13}{7}} - \frac{8}{49}x^{-\frac{15}{7}}$

9) e)
$y = 2x^4 + x$ so $\frac{dy}{dx} = 8x^3 + 1$
$\frac{d^2y}{dx^2} = 24x^2$

9) f)
$\frac{dy}{dx} = 6x^{\frac{1}{2}}, \frac{d^2y}{dx^2} = 3x^{-\frac{1}{2}}$

9) g)
$\frac{dy}{dx} = -x^{-\frac{3}{2}}, \frac{d^2y}{dx^2} = \frac{3}{2}x^{-\frac{5}{2}}$

9) h)
$\frac{dy}{dx} = 4x^3 - 12x^{-4}, \frac{d^2y}{dx^2} = 12x^2 + 48x^{-5}$

10) a)
7

10) b)
10

10) c)
$-\frac{257}{16}$

10) d)
$-\frac{3}{4}$

10) e)
0

11) a)
-4

11) b)
4

11) c)
$\frac{15}{4}$

11) d)
$\frac{4}{9}$

11) e)
$-\frac{1}{4}p - \frac{8}{q}$

12) a)
$f'(x) = 2e^{2x}$

12) b)
$f'(x) = 13e^{13x}$

12) c)
$f'(x) = -8e^{-8x}$

12) d)
$f'(x) = \frac{1}{5}e^{\frac{1}{5}x}$

12) e)
$f'(x) = -\frac{4}{7}e^{-\frac{4}{7}x}$

12) f)
$f'(x) = 6e^{3x}$

12) g)
$f'(x) = -5e^{-x}$

12) h)
$f'(x) = -\frac{1}{2}e^{-2x}$

12) i)
$f'(x) = -5e^{-5x}$

12) j)
$f'(x) = \frac{3}{2}e^{\frac{3}{2}x}$

12) k)
$f'(x) = \frac{1}{2}e^{\frac{1}{2}x}$

12) l)
$f'(x) = -2e^{-2x}$

12) m)
$f'(x) = \frac{a}{2}e^{\frac{a}{2}x}$

12) n)
$f'(x) = \left(1 - \frac{a}{2}\right)e^{\left(1-\frac{a}{2}\right)x}$

13) a)
$\frac{dy}{dx} = 4e^{2x} - 15e^{3x}$

13) b)
$\frac{dy}{dx} = 2 - 3e^{3x}$

13) c)
$\frac{dy}{dx} = -e^{-x} + 5x^4$

13) d)
$\frac{dy}{dx} = 5x^4 - x^3 - \frac{3}{5}e^{-\frac{1}{5}x}$

13) e)
$\frac{dy}{dx} = \frac{1}{2}x^{-\frac{1}{2}} - 6e^{-2x}$

13) f)
$\frac{dy}{dx} = \frac{\sqrt{a}}{2}x^{-\frac{1}{2}} - 5be^{-bx}$

14) a)
$f'(x) = 3e^{3x} - 8e^{-x}, f''(x) = 9e^{3x} + 8e^{-x}$

14) b)
$f'(x) = 10x - 5e^{5x}, f''(x) = 10 - 25e^{5x}$

14) c)
$f'(x) = -2e^{-x} + 2x, f''(x) = 2e^{-x} + 2$

14) d)
$f'(x) = 4x^3 + 4e^{-\frac{1}{2}x} + \frac{3}{2}x^{-4}$
$f''(x) = 12x^2 - 2e^{-\frac{1}{2}x} - 6x^{-5}$

15) a)
e^2

15) b)
$5e^{-1}$

15) c)
-1

15) d)
$1 - e$

15) e)
$2 + 4e$

15) f)
$-3 - 3ae^a$

16)
$\left.\frac{dy}{dx}\right|_{x=-1.423} = 2.28$ (2 DP)
$\left.\frac{dy}{dx}\right|_{x=0} = 3 - 3e^0 = 0$
$\left.\frac{dy}{dx}\right|_{x=0.969} = -4.91$ (2 DP)

17)
$\frac{d^2y}{dx^2} > 0$ when $k > 0$ and $p \neq 0$

1.32 Differentiation: Tangents and Normals
Questions on page 84

1) a)
$\frac{25}{2}$

1) b)
$\frac{\sqrt{2}}{4} + 64$

1) c)
$\frac{49}{4}$

1) d)
$-\frac{5\sqrt{2}}{4}$

2) a)
$4x - y - 4 = 0$

2) b)
$2x - y + 3 = 0$

2) c)
$x + 4y + 4 = 0$

2) d)
$x - y + 1 = 0$

3) a)
$y - 3 = 0$

3) b)
It has a gradient of 0 at this point, so the curve is neither increasing nor decreasing at $x = 0$ (it is the maximum point of the quadratic)

4) a)
$x + 6y - 19 = 0$

4) b)
$x - 18y - 73 = 0$

4) c)
$16x - 4y - 31 = 0$

4) d)
$2x + y - 3 = 0$

4) e)
$x + 5y + 11 = 0$

5)
$x = 3$

6)
$x = -2$ or $x = 2$

7)
$x = 8$

8) a)
$y = -x + 4$

8) b)
$y = x + 2$

8) c)
Tangent line meets axes at $(0, 4)$ and $(4, 0)$
Normal line meets axes at $(0, -2)$ and $(-2, 0)$

9) a)
$x + 9y = 21$

9) b)
$\left(0, \frac{7}{3}\right)$ and $(21, 0)$

9) c)
24.5 units^2

10) a)
$y = 2x + 1$

10) b)
$(0, 1)$ and $\left(-\frac{1}{2}, 0\right)$

10) c)
$(-1, -1)$

11) a)
$(-1, 6)$
11) b)
$y = -9x - 3$
11) c)
$\left(\frac{1}{5}, -\frac{24}{5}\right)$

12)
$-4k$

13)
$p = 2, q = -10$

14)
54 units^2

15)
$p = 2, q = 1$

16)
$y = -16x + 15$

17)
Show that

18) a)
Show that
18) b)
$x^2 \geq 0$ for all real values of x, so $3x^2 + 2 > 0$ for all real values of x, so the gradient of the curve is always positive.

19)
$2 - 12k$

20) a)
$6e^6$
20) b)
$\frac{1}{2} - 2e^2$
20) c)
$\frac{4}{3} + 2e^2$

21)
$y = x + 1$

22)
$y = \frac{1}{2}x - 1$

23) a)
$y = (e - 4)x + 2$
23) b)
$(0, 2)$ and $\left(\frac{2}{4-e}, 0\right)$
23) c)
$\frac{2}{4-e}$

24) a)
$-4ke^{-2k}$
24) b)
Show that

25)
$k = 2$

26) a)
$y = \frac{1}{e^2-5}x - \frac{2}{e^2-5} + 10 - e^2$
26) b)
$(e^4 - 15e^2 + 52, 0)$

27)
$k = -11, p = 26.5$

1.33 Differentiation: Graphs of Gradient Functions
Questions on page 87

1) a)

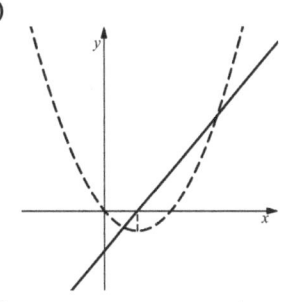

1) b)

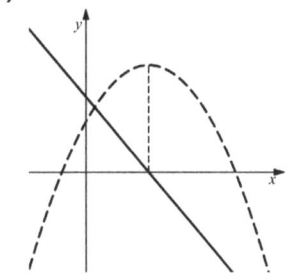

1) c)

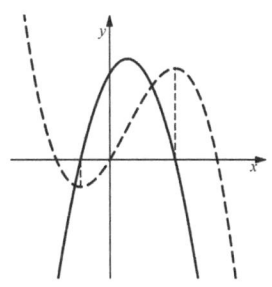

1) d)

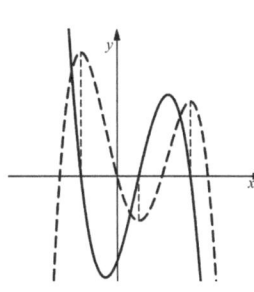

1) e)

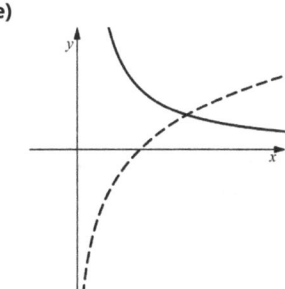

1) f)

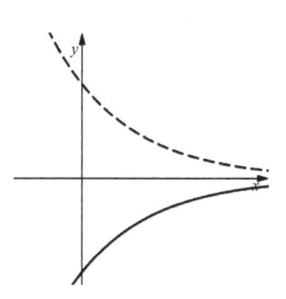

2) a)

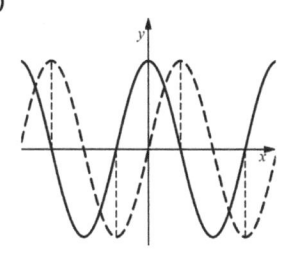

2) b)

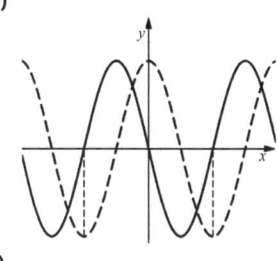

2) c)

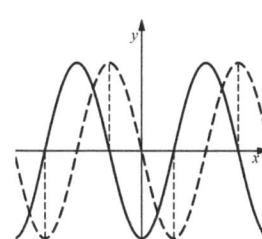

3) a)
$a = 0, b > 0, c = 0$
3) b)
$a = 0, b < 0, c > 0$
3) c)
$a = 0, b < 0, c < 0$
3) d)
$a > 0, b = 0, c > 0$
3) e)
$a < 0, b > 0, c = 0$.
3) f)
$a > 0, b > 0, c > 0$
3) g)
$a < 0, b > 0, c < 0$
3) h)
$a > 0, b < 0, c < 0$

4)

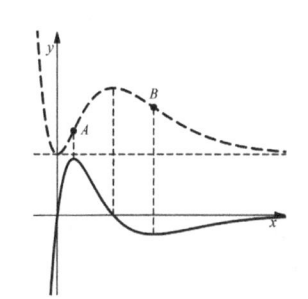

1.34 Differentiation: Stationary Points
Questions on page 90

1) a) $x = 1$
1) b) $x = \frac{9}{2}$
1) c) $x = 1$
1) d) $x = -2$
1) e) $x = \pm 2$
1) f) $x = 1$
1) g) $x = \pm\sqrt{3}$

2) a) $\{x : x \in \mathbb{R}, x < -2\}$
2) b) $\{x : x \in \mathbb{R}, x < \frac{5}{6}\}$
2) c) $\{x : x \in \mathbb{R}, -1 < x < 1\}$
2) d) $\{x : x \in \mathbb{R}, -3 < x < 2\}$
2) e) $\{x : x \in \mathbb{R}, x < -4\} \cup \{x : x \in \mathbb{R}, x > -2\}$

3) $x \in (-5, 1)$

4) Show that

5) Show that

6) a) $\left(\frac{5}{2}, \frac{1}{4}\right)$ is a maximum
6) b) $(1, 11)$ is a maximum
$(4, -16)$ is a minimum
6) c) $(-1, 14)$ is a maximum
$(2, -13)$ is a minimum
6) d) $(-4, -176)$ is a minimum
$(6, 324)$ is a maximum
6) e) $(-1, -2)$ is a minimum, $(0, 0)$ is a maximum and $(1, -2)$ is a minimum
6) f) $\left(-\frac{1}{2}, -4\right)$ is a maximum
$\left(\frac{1}{2}, 4\right)$ is a minimum
6) g) $(1, 3)$ is a minimum
6) j) $(0, 0)$ is a stationary point of inflection
$(1, -3)$ is a minimum

7) $\left(-\frac{2}{9}, \frac{227}{243}\right)$

8) a) $(-3, 81)$ is a maximum
$(5, -175)$ is a minimum

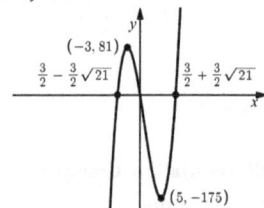

8) b) $(-1, -9)$ is a minimum, $(0, 7)$ is a maximum and $(1, -9)$ is a minimum

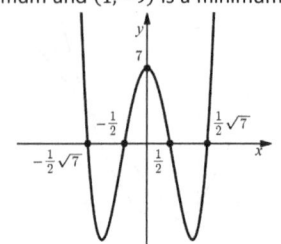

8) c) $\left(\frac{1}{25}, -\frac{1}{5}\right)$ is a minimum

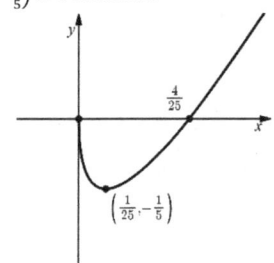

9) $a = 6, b = -9, c = 6$

10) $a = -1, b = 4, c = -3$
Maximum at $(-\sqrt{2}, 1)$, minimum at $(0, -3)$ and maximum at $(\sqrt{2}, 1)$

11) $x \in \left(-\frac{1}{2}\ln 2, \infty\right)$

12) a) $(0, 4)$ and $(\ln 5, 0)$
12) b) $\left(\ln\left(\frac{5}{2}\right), \frac{25}{4}\right)$
12) c) Show that

13) Minimum at $(0, -1)$

1.35 Differentiation: Optimisation
Questions on page 91

1) a) $A = x^2, P = 4x$
1) b) $A = 2x^2, P = 6x$
1) c) $A = \pi x^2, P = 2\pi x$
1) d) $A = x^2, P = (1 + \sqrt{17})x$

2) a) $V = x^3, A = 6x^2$
2) b) $V = 3x^3, A = 14x^2$
2) c) $V = \pi x^2 h, A = 2\pi x^2 + 2\pi xh$
2) d) $V = \frac{1}{2}x^2 h, A = x^2 + (1 + \sqrt{5})xh$

3) a) $\frac{dV}{dr} = 20 - \frac{1}{2}r$
3) b) $\frac{dP}{dt} = -t^{-2} + \frac{1}{2}$
3) c) $\frac{dA}{dr} = 2 + 20r^{-3}$
3) d) $\frac{dC}{dx} = -10x^{-2} + \frac{1}{2}x$

4) a) $r = \frac{5}{2}$
4) b) $x = 24$
4) c) $x = \sqrt[3]{20}$
4) d) $r = \sqrt[3]{108\pi}$
4) e) $r = \sqrt[3]{\frac{15}{\pi}}$

5) a) Show that
5) b) 112.5 m^2

6) a) $v = 102 \text{ kmh}^{-1}$
6) b) £72.67

7) a) Show that
7) b) 54.4 cm^3 (1 DP)
7) c) $\frac{5}{\sqrt{6}}$ by $\frac{20}{\sqrt{6}}$ by $\frac{4}{3}\sqrt{6}$

8) a) Show that
8) b) 30.2 cm^3 (1 DP)
8) c) $2\sqrt{2}$ by $4\sqrt{2}$ by $\frac{4}{3}\sqrt{2}$

9) a)
Show that
9) b)
64π cm^3

10)
$V = 4740.7$ cm^3 (1 DP)

11) a)
$r = 6$ cm, $h = 12$ cm
11) b)
£0.48

12)
24.3 units^2

13) a)
Show that
13 b)
$r = 6$ cm
13) c)
$h = 0$ so that means that the least amount of material needed would only be a semi-sphere.

14) a)
Show that
14) b)
$r = 3.24$ cm (3 SF)
$h = 15.1$ cm (3 SF)
14) c)
Minimum cost is 14 pence (to the nearest pence)

15) a)
Show that
15) b)
Show that
15) c)
£6

16) a)
Show that
16) b)
Show that
16) c)
Quarter-circle area is 9.39 m^2

17) a)
Show that
17) b)
$r = 12.4$ cm (3 SF),
$A = 325$ cm^2 (3 SF)
$h = 6.45$ cm (3 SF)

1.35 Integration: Indefinite Integrals
Questions on page 94

1) a)
$4x^2 + x + c$
1) b)
$2x^3 - 3x + c$
1) c)
$\frac{5}{2}x^4 + 9x + c$
1 d)
$3x^4 + \frac{20}{3}x^3 - \frac{3}{2}x^2 - 5x + c$

2) a)
$\frac{1}{3}x^2 + \frac{1}{3}x^5 + c$
2) b)
$\frac{1}{4}x^3 - \frac{2}{3}x^7 + c$
2) c)
$\frac{1}{20}x^8 - \frac{1}{16}x^6 + c$
2) d)
$\frac{2}{45}x^5 + \frac{14}{9}x^3 + 24x + c$

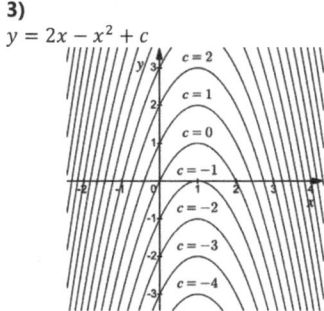

3)
$y = 2x - x^2 + c$

4)
$y = \frac{1}{18}x^3 + \frac{1}{12}x^2 - x - 8$

5) a)
$y = \frac{3}{5}x^{\frac{5}{3}} + c$
5) b)
$y = 5x^{\frac{6}{5}} + c$
5) c)
$y = 12x^{\frac{7}{4}} + c$
5) d)
$y = 3x^{\frac{4}{3}} + \frac{72}{7}x^{\frac{7}{6}} + 9x + c$

6) a)
$\frac{1}{4}x^4 + \frac{2}{5}x^{\frac{5}{2}} + c$
6) b)
$-5x^{-1} - \frac{6}{5}x^{\frac{5}{3}} + c$
6) c)
$-4x^{-2} - \frac{20}{11}x^{\frac{11}{5}} + c$
6) d)
$-\frac{4}{7}x^{-7} + \frac{36}{5}x^{-\frac{5}{3}} + \frac{27}{11}x^{\frac{11}{3}} + c$

7)
$x + \frac{1}{3}x^{-3} + x^{-1} - \frac{1}{5}x^{-5} + c$

8)
$y = 30x^{\frac{4}{3}} - 2x^2 - 29$

9) a)
$y = 5x + \frac{2}{3}x^{-3} + c$
9) b)
$y = -6x^{-1} + 6x^{-2} + c$
9) c)
$y = \frac{16}{5}x^{\frac{5}{2}} - 2x^{\frac{3}{2}} + c$
9) d)
$y = -4x^{-1} + \frac{28}{5}x^{\frac{5}{2}} + \frac{12}{5}x^{-\frac{5}{2}} - 21x + c$

10) a)
$\frac{8}{5}x^{\frac{5}{2}} - \frac{8}{3}x^{\frac{3}{2}} + c$
10) b)
$\frac{4}{5}x^{\frac{5}{4}} + c$
10) c)
$\frac{5}{3}x^{\frac{6}{5}} + c$
10) d)
$\frac{48}{17}x^{\frac{17}{16}} + c$

11)
$y = 2x^{\frac{7}{2}} - x^3 - 5$

12) a)
$y = -4x^{-1} + c$
12) b)
$y = -3x^{-2} + c$
12) c)
$y = -\frac{3}{5}x^{-3} + c$
12) d)
$y = -\frac{100}{81}x^{-9} + c$

13) a)
$\frac{8}{3}x^{\frac{1}{2}} + c$
13) b)
$9x^{\frac{2}{3}} + c$
13) c)
$10x^{\frac{4}{5}} - 75x^{\frac{2}{3}} + c$
13) d)
$3x^{\frac{1}{3}} - \frac{36}{5}x^{-\frac{5}{6}} - \frac{9}{2}x^{-2} + c$

14)
$y = 8(2\sqrt{x} - 9)$

15)
$y = 250\left(1 - \frac{1}{x^3}\right)$

16)
$y = 3\left(\sqrt{2} - \frac{4}{\sqrt{x}}\right)$

17)
$y = 18\left(x^{\frac{2}{3}} - 9\right)$

18)
$y = 1 - \frac{2}{5x^2\sqrt{x}}$

19)
$-\frac{8}{3}x^{-\frac{3}{2}} + \frac{1}{2}x^{-\frac{1}{2}} + c$

20)
$-\frac{12}{25}x^{-\frac{5}{2}} + \frac{1}{10}x^{-\frac{5}{3}} + c$

21)
$-\frac{3}{10}x^{-\frac{5}{2}} - \frac{6}{13}x^{\frac{13}{6}} + c$

22)
$f(x) = 3x^3 - 24x - \frac{16}{x} + 30$

23) a)
$y = \frac{1}{5}kx^5 + c$
23) b)
$y = \frac{1}{5}k^4x^5 + c$
23) c)
$y = -\frac{1}{3k}x^{-3} + c$
23) d)
$y = -\frac{9}{64k^5}x^{-4} + c$

24) a)
$-5x^{-1} + \frac{1}{2}x^2 + c$
24) b)
$-x^{-3} + 2x^{-1} + c$
24) c)
$-8x^{-1} + 3x^{-3} + c$
24) d)
$-\frac{15}{7}x^{-1} + \frac{1}{8}x^{-4} + c$

25) a)
$y = \frac{1}{2}x^2 + \frac{7}{10}x^{-2} + c$
25) b)
$y = -\frac{3}{16}x^{-4} + 2x^{-1} + c$
25) c)
$y = \frac{1}{8}x^4 + \frac{2}{9}x^{-3} + c$
25) d)
$y = -\frac{1}{9}x^{-3} + \frac{1}{16}x^{-2} + \frac{1}{16}x^{-4} + c$

26)
$f(x) = \frac{-3x^2 - x + 4}{x^2}$

27)
$f(x) = 21x^{\frac{1}{3}}(x^2 - 7x + 28)$

28)
$(-7, 250)$

29)
$f(x) = 6x^3 + 37x^2 + 11x - 120$

30)
$f(x) = 144x^4 - 96x^3 - 104x^2 + 40x + 25$

31) a)
$10x^4 + 500x^2 + c$
31) b)
$\frac{32}{3}x^6 + 360x^4 + 810x^2 + c$
31) c)
$-486x^6 - 270x^4 - 18x^2 + c$
31) d)
$32x^8 + 224x^6 + 140x^4 + 14x^2 + c$

32)
$y = 4(x^2 + 12)^2$

33) a)
$y = \frac{3}{2}x^2 + c_1 x + c_2$
33) b)
$y = \frac{2}{3}x^3 - \frac{5}{2}x^2 + c_1 x + c_2$
33) c)
$y = \frac{1}{2}x^4 - \frac{4}{3}x^3 + \frac{9}{2}x^2 + c_1 x + c_2$
33) d)
$y = \frac{2}{5}x^5 - \frac{1}{6}x^7 + c_1 x + c_2$

34) a)
$y = \frac{4}{15}x^{\frac{5}{2}} + c_1 x + c_2$
34) b)
$y = \frac{20}{3}x^{\frac{3}{2}} + c_1 x + c_2$
34) c)
$y = \frac{3}{2}x^{-1} - \frac{1}{60}x^5 + c_1 x + c_2$
34) d)
$y = \frac{32}{35}x^{-\frac{5}{2}} + c_1 x + c_2$

35)
$y = -\frac{80}{3}x^{\frac{3}{2}} + \frac{1}{2}c_1 x^2 + c_2 x + c_3$

36)
$f(x) = 7x^2 - x^3 - 13x - 127$

37)
$f(x) = 3x^4 + 4x^3 + 6x^2 - 31x + 21$

38)
$f(x) = 9x^{-1} + 7x - \frac{27}{2}$

39)
$f(x) = 35x^3 - 56x^{\frac{5}{2}} + 24x^{\frac{7}{2}} + 8$

40)
$f(x) = -4x^3 - 6x^2 + 9x + 1$

1.37 Integration: Definite Integrals and Areas
Questions on page 97

1) a)
132
1) b)
73
1) c)
1136
1) d)
0

2)
$\frac{44}{5}$

3)
$\frac{392}{15}$

4)
46

5)
10

6) a)

6) b)
Part of the area is below the x-axis. The integral of a region below the x-axis is negative, so $\int_1^5 (x-1)(4-x)\,dx$ is the sum of a positive part and a negative part, and is not the sum of two positive values, which is want we want for the area.

7)
$k = \sqrt{3}$

8)
$\frac{5120}{3}$

9)
$k = 28$

10)
$\frac{64}{3}$

11)
32

12)
Show that

13)
Show that

14)
Show that

15) a)
$\frac{112}{3}$
15) b)
3
15) c)
16
15) d)
$\frac{143}{6} - \frac{64}{3}\sqrt{2}$

16)
$\frac{4}{5}$

17)
54

18)
$\frac{6}{5}$

19)
$\frac{71}{6}$

20)
$\frac{7}{6}k^3 + 3k$

21)
$\frac{32}{3}$

22)
32

23)
$k = 8$

24)
$k = 2$

25)
$k = -5$ or $k = 19$

26)
$\frac{1}{216}$

27)
$\frac{3}{4}$

28)
$k = \frac{20}{3}$

29)
$k = 484$

30)
$\frac{318}{5}$

31)
3

32)
$\frac{71}{1296}$

33)
$k = 204\sqrt{3}$

34)
$8k^{\frac{1}{2}} + 40k^2$

35)
$\frac{-9+6\sqrt{3}}{2}$

36) a)
$\frac{14}{9}$

36) b)
$\frac{55}{96}$

36) c)
$-\frac{23}{9}$

36) d)
$\frac{65}{576}$

37)
$\frac{8}{3}$

38)
1

39)
$\frac{2}{75}$

40)
$k = \frac{5}{3}$

41)
$k = \frac{2}{5}$

42)
$k = 2$

43)
$\frac{71}{5400}$

44)
$192\sqrt{2}$

45)
Show that

46)
$\frac{875}{324}$

47)
$\frac{4096}{3}$

48) a)
$\frac{1}{2}$

48) b)
$\frac{1}{12}$

48) c)
$\frac{1}{12}$

48) d)
$\frac{1}{30}$

49)
32.8 cm^2

50)
$\frac{\pi}{2} + \frac{1}{3}$

51)
$\frac{4}{3}$

52)
$\frac{50}{3}$

53)
$\frac{20295}{2}$

1.38 Applications of Calculus
Questions on page 104

1) a)
$\frac{dC}{dx} = 5.8$
For each additional mile travelled, the cost increases by £5.80
Units: £/mile

1) b)
$\frac{dh}{dt} = -2t + 6$ so $\frac{dh}{dt}\big|_{t=10} = -14$
At $t = 10$ exactly, the height of the ball is decreasing at a rate of 14 ms^{-1}
Units: ms^{-1}

1) c)
$\frac{dT}{dt} = 0.03t^2 - 0.4t$ so $\frac{dT}{dt}\big|_{t=5} = -1.25$
At $t = 5$ exactly, the temperature of the substance is decreasing at a rate of 1.25°C per hour.
Units: °C per hour

2) a)
$\frac{dN}{dt}\big|_{t=5} = 5.1$
On day 5 there is 5.1 people per day gaining the virus

2) b)
$\frac{dN}{dt}\big|_{t=15} = -4.7$
On day 15 there is 4.7 fewer people per day gaining the virus

2) c)
$(10.2, 54.6)$

2) d)
The model says that at $t = 10.2$ days, the maximum amount of people with the virus is 54.6 people.

3) a)
$\frac{dh}{dn} = -0.056n + 1.95$
It represents the rate in change (per day) of the height of the seedling (in mm)

3) b)
During the 34th day it reaches its maximum height of 40.2 mm

3) c)
The model is no longer valid after reaching its maximum because it then models the height as decreasing.

4) a)
$(2, 4.67), (0.286, 10.5)$

4) b)
When $t = 0$, $v = 10$
When $t = 3$, $v = 13$
When $t = 2$ seconds the object has its lowest velocity, and when $t = 0.286$ seconds, the object has a local maximum velocity (the actual max is 13 ms^{-1})

5) a)
56.2 cm

5) b)
$t = \frac{8}{3} = 2.67$ seconds (3 SF)
$\frac{dy}{dt}\big|_{t=\frac{8}{3}} = -\frac{19}{3} = -6.33$ (3 SF)
$y = 50.1$ cm (3 SF)

5) c)
The boat is moving downwards as this point happens after the local maximum.

6) a)
94

6) b)
43.2°

6) c)
The maximum height occurs 194 feet horizontally from the beginning of the rollercoaster and reaches a height of 234 feet above the ground.

7)
0.167 metres

8) a)
£1274

8) b)
£362

1.39 Trigonometry: Introduction
Questions on page 105

1) a)
$A = 40°$
$AC = 24.8$ cm (3 SF)
$BC = 15.9$ cm (3 SF)
1) b)
$AC = 15.0$ cm (3 SF)
$B = 29.9°$ (3 SF)
$C = 60.1°$ (3 SF)

2)
$\sqrt{3}$ m²

3)
$AB = \frac{\sqrt{3}}{3}$

4)
$OA = \sqrt{3}$

5)
489 m² (3 SF)

6)
14.2 cm (3 SF)

7) a)
$A = 59.0°$ (1 DP) or $A = 121.0°$ (1 DP)
7) b)

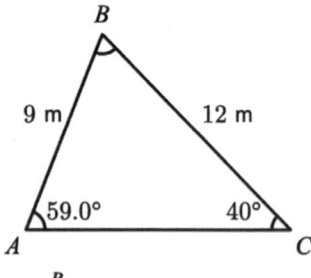

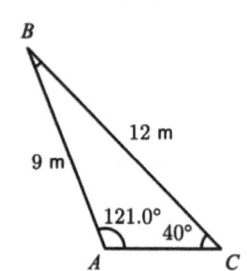

8) a)
$A = 44.6$ (3 SF)
$C = 60.4°$ (3 SF)
$AB = 9.90$ (3 SF)
8) b)
$A = 73.0°$ (3 SF)
$B = 61.2°$ (3 SF)
$C = 45.8°$ (3 SF)
8) c)
$B = 57.9°$ (3 SF)
$A = 89.1°$ (3 SF)
$BC = 16.5$ m (3 SF)
Or
$B = 122°$ (3 SF)
$A = 24.9°$ (3 SF)
$BC = 6.96$ m (3 SF)

9)
7.02 cm² (3 SF)

10)
$x = 20.3°$ (1 DP) or $x = 159.7°$ (1 DP)

11)
$XC = 2.67$ m (3 SF)

12)
$k = 4$

13) a)
Show that
13) b)
$x = \frac{11}{4}$
13) c)
$A = 129°$ (3 SF)

14)
$\frac{301}{500}\sqrt{3}$ m²

15)
Show that

16)
32.6 m (3 SF)

17) a)
180°
17) b)
255°
17) c)
230°

18)
$p = \frac{-5+5\sqrt{13}}{2}$

19)
5.91 cm (3 SF)

20)
9.65 cm (3 SF)

21) a)
$p = 8$ and $q = 6$
21) b)
$0 < r \leq \frac{4}{3}$

1.40 Trigonometry: $\sin x$, $\cos x$, $\tan x$ in Degrees
Questions on page 108

1) a)

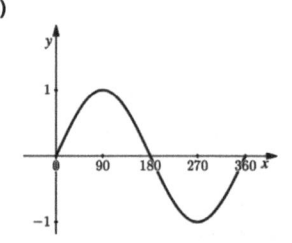

1) b)

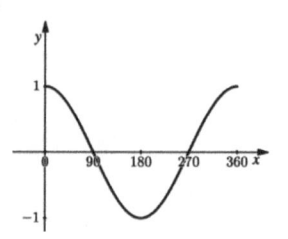

1) c)

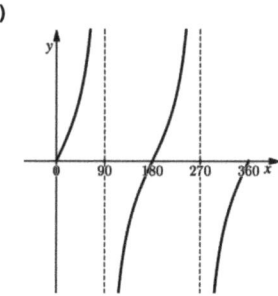

2) a)
$\theta = \{0°, 180°\}$
2) b)
$\theta = \{90°\}$
2) c)
$\theta = \{270°\}$

3) a)
$\theta = \{90°, 270°\}$
3) b)
$\theta = \{0°\}$
3) c)
$\theta = \{180°\}$

4) a)
$\theta = \{0°, 180°\}$
4) b)
$\theta = \{45°, 225°\}$
4) c)
$\theta = \{135°, 315°\}$

5)
$0° < x < 90°$ or $270° < x < 360°$

6)
$0° < x < 180°$

7) a)
$\theta = \{19.5°, 161°\}$ (3 SF)
7) b)
$\theta = \{48.2°, 312°\}$ (3 SF)
7) c)
$\theta = \{53.1°, 233°\}$ (3 SF)

8) a)
$\theta = \{-336°, -204°\}$ (3 SF)
8) b)
$\theta = \{-295°, -64.6°\}$ (3 SF)
8) c)
$\theta = \{-336°, -156°\}$ (3 SF)

9) a)
$\theta = \{192°, 348°\}$ (3 SF)
9) b)
$\theta = \{125°, 235°\}$ (3 SF)

9) c)
$\theta = \{104°, 284°\}$ (3 SF)

10) a)
$\theta = \{-116°, -64.2°\}$ (3 SF)
10) b)
$\theta = \{-233°, -127°\}$ (3 SF)
10) c)
$\theta = \{-257°, -77.5°\}$ (3 SF)

11) a)
$\theta = \{-338°, -202°, 22.0°, 158°\}$ (3 SF)
11) b)
$\theta = \{-314°, -45.6°, 45.6°, 314°\}$ (3 SF)
11) c)
$\theta = \{-279°, -99.5°, 80.5°, 261°\}$ (3 SF)
11) d)
$\theta = \{-121°, -59.0°, 239°, 301°\}$ (3 SF)
11) e)
$\theta = \{-257°, -103°, 103°, 257°\}$ (3 SF)
11) f)
$\theta = \{-232°, -52.4°, 128°, 308°\}$ (3 SF)

12)
$-630° < x < -450°$

13)
$180° < x < 360°$

14) a)
$\theta = \{34.8°, 145°, 395°, 505°\}$ (3 SF)
14) b)
$\theta = \{873°, 927°\}$ (3 SF)
14) c)
$\theta = \{-624°\}$ (3 SF)

15) a)
$\theta = \{3.18°, 177°\}$ (3 SF)
15) b)
$\theta = \{201°, 381°\}$ (3 SF)
15) c)
$\theta = \{-87.1°, 87.1°\}$ (3 SF)

16)
$A = \{24.2°, 156°\}$ (3 SF)

17)
$A: x = p,\ B: x = 180° - p$
$C: x = -180° - p,\ D: x = p - 360°$

18)
$\sin 60° = \frac{\sqrt{3}}{2}$
$\sin(180° - 60°) = \sin 120° = \frac{\sqrt{3}}{2}$
$\sin(120° + 360°) = \sin 480° = \frac{\sqrt{3}}{2}$

19)
$\cos 45° = \frac{\sqrt{2}}{2}$
$\cos(90° + 45°) = \cos 135° = -\frac{\sqrt{2}}{2}$
$\cos(135° + 360°) = \cos 495° = -\frac{\sqrt{2}}{2}$

20)
$\tan(-210°) = -\frac{\sqrt{3}}{3}$
$\tan 210° = \frac{\sqrt{3}}{3}$
$\tan(210° + 180°) = \tan 390° = \frac{\sqrt{3}}{3}$

21)
$A: x = p,\ B: x = 180° - p$
$C: x = p - 360°,\ D: x = -180° - p$

22)
$A: x = -p,\ B: x = -180° + p$
$C: x = 180° + p,\ D: x = 360° - p$

23) a)
$\theta = \{75.5°, 104°, 256°, 284°\}$ (3 SF)
23) b)
$\theta = \{-153°, -27.3°, 27.3°, 153°\}$ (3 SF)
23) c)
$\theta = \{56.8°, 123°, 237°, 303°, 417°, 483°\}$ (3 SF)

24) a)
$\theta = \{90°, 270°\}$
24) b)
$\theta = \{0°, 150°, 180°, 330°\}$ (3 SF)
24) c)
$\theta = \{0°, 23.6°, 156°, 180°\}$ (3 SF)

25)
$A: x = p,\ B: x = 360° - p$
$C: x = -p,\ D: x = -360° + p$

26)
$A: x = p,\ B: x = 360° - p$
$C: x = p - 360°,\ D: x = -p$

27)
$A: x = p,\ B: x = p + 180°$
$C: x = p - 180°,\ D: x = p - 360°$

28) a)
$\theta = \{19.5°, 90°, 161°, 270°\}$ (3 SF)
28) b)
$\theta = \{0°, 48.4°, 180°, 228°\}$ (3 SF)
28) c)
$\theta = \{90°, 270°\}$

29) a)
$\theta = \{30°, 150°, 270°\}$
29) b)
$\theta = \{0°, 132°, 228°\}$ (3 SF)
29) c)
$\theta = \{36.9°, 117°, 217°, 297°\}$ (3 SF)

30) a)
$\theta = \{30°, 150°, 199°, 341°\}$ (3 SF)
30) b)
$\theta = \{63.6°, 296°\}$ (3 SF)
30) c)
$\theta = \{58.0°, 140°, 238°, 320°\}$ (3 SF)

31) a)
$3(x-2)^2 + 5$
31) b)
$(2, 5)$
31) c)
Minimum is 8 when $\theta = 90°$

32) a)
$360°$
32) b)
$360°$
32) c)
$180°$
32) d)
$180°$
32) e)
$120°$
32) f)
$45°$
32) g)
$1080°$
32) h)
$540°$
32) i)
$225°$

33) a)
4 solutions
33) b)
2 solutions
33) c)
8 solutions

34) a)
2 solutions
34) b)
6 solutions
34) c)
16 solutions

35) a)
12 solutions
35) b)
12 solutions
35) c)
14 solutions

36) a)
$360°$
36) b)
$360°$
36) c)
$180°$
36) d)
$360°$
36) e)
$360°$
36) f)
$180°$
36) g)
$180°$

36) h) 90°
36) i) 20°
36) j) 1800°
36) k) 2880°
36) l) 324°

37) a) $\theta = \{7.46°, 153°\}$ (3 SF)
37) b) $\theta = \{60.5°, 220°\}$ (3 SF)
37) c) $\theta = \{131°, 359°\}$ (3 SF)

38) a) $\theta = \{129°, 351°\}$ (3 SF)
38) b) $\theta = \{6.79°, 303°\}$ (3 SF)
38) c) $\theta = \{69.3°, 321°\}$ (3 SF)

39) a) $\theta = \{113°, 293°\}$ (3 SF)
39) b) $\theta = \{140°, 320°\}$ (3 SF)
39) c) $\theta = \{138°, 318°\}$ (3 SF)

40) $0° \leq x < 120°$ or $300° < x \leq 360°$

41) a) $\theta = \{22.5°, 67.5°, 202.5°, 247.5°\}$
41) b) $\theta = \{34.6°, 145°, 215°, 325°\}$ (3 SF)
41) c) $\theta = \{15°, 105°, 195°, 285°\}$

42) a) $\theta = \{17.7°, 42.3°, 138°, 162°, 258°, 282°\}$ (3 SF)
42) b) $\theta = \{35.8°, 84.2°, 156°, 204°, 276°, 324°\}$ (3 SF)
42) c) $\theta = \{36.1°, 96.1°, 156°, 216°, 276°, 336°\}$ (3 SF)

43) a) $\theta = \{23.1°, 337°\}$ (3 SF)
43) b) $\theta = \{573°, 867°, 2010°, 2310°, 3450°\}$ (3 SF)
43) c) $\theta = \{152°\}$ (3 SF)

44) a) $\theta = \{27.5°, 87.5°, 207.5°, 267.5°\}$

44) b) $\theta = \{3.33°, 83.3°, 123°, 203°, 243°, 323°\}$ (3 SF)
44) c) $\theta = \{80°, 170°, 260°, 350°\}$

45) a) $\theta = \{63.9°, 80.6°, 154°, 171°\}$ (3 SF)
45) b) $\theta = \left\{ \begin{array}{c} -89.6°, -59.0°, -29.6°, \\ 0.953°, 30.4°, 61.0° \end{array} \right\}$ (3 SF)
45) c) $\theta = \{131°, 167°\}$ (3 SF)

46) $0° < x < 90°$

47) $0° < x < 30°$ or $90° < x < 150°$ or $210° < x < 270°$ or $330° < x < 360°$

48) a) $\theta = \{18°, 98°, 138°, 218°, 258°, 338°\}$
48) b) $\theta = \{27.5°, 87.5°, 207.5°, 267.5°\}$
48) c) $\theta = \left\{ \begin{array}{c} 35°, 80°, 125°, 170°, \\ 215°, 260°, 305°, 350° \end{array} \right\}$

49) a) $\theta = \{4.45°, 42.7°, 76.5°, 115°\}$ (3 SF)
49) b) $\theta = \{23.0°, 28.3°, 143°, 148°, 263°, 268°\}$ (3 SF)
49) c) $\theta = \{2.91°, 25.4°, 47.9°\}$ (3 SF)

50) a) $\theta = \left\{ \begin{array}{c} 15°, 75°, 105°, 165°, \\ 195°, 255°, 285°, 345° \end{array} \right\}$
50) b) $\theta = \{25.2°, 34.8°, 85.2°, 94.8°, 145°, 155°\}$ (3 SF)
50) c) $\theta = \{-19.0°, -17.0°, 17.0°, 19.0°\}$ (3 SF)

51) $A: x = p$, $B: x = p - 180°$
$C: x = 90° - p$, $D: x = p - 540°$

52) a) $\theta = \{120°, 140°\}$ (3 SF)
52) b) $\theta = \{225°, 245°\}$
52) c) $\theta = \{35°, 295°\}$
52) d) $\theta = \{90°, 270°\}$

53) a) $\theta = \{290°, 320°\}$ (3 SF)
53) b) $\theta = \{112.5°, 167.5°, 292.5°, 347.5°\}$

53) c) $\theta = \{20°, 80°, 140°, 200°, 260°, 320°\}$

54) a) Show that
54) b) $\theta = \{66.4°, 109°, 251°, 294°\}$ (3 SF)

1.41 Graph Sketching: $\sin x$, $\cos x$, $\tan x$ in Degrees
Questions on page 108

1) a) $y = 2\sin x$
1) b) $y = 4\sin x$

2) a) $y = \frac{1}{2}\cos x$
2) b) $y = \frac{3}{2}\cos x$

3) a)

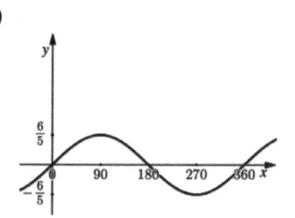

3) b)

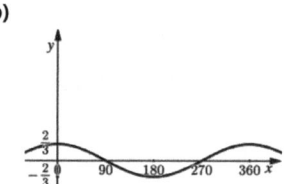

4) a) $y = \sin 2x$
4) b) $y = \sin 3x$
4) c) $y = \sin \frac{3}{2}x$
4) d) $y = \sin \frac{1}{2}x$

5)

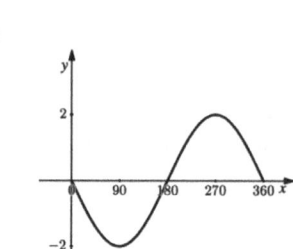

6) a) $y = \cos 3x$
6) b) $y = \cos 4x$
6) c) $y = \cos \frac{4}{3}x$

6) d)
$y = \cos\frac{1}{3}x$

7)

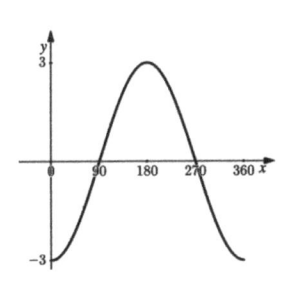

8)

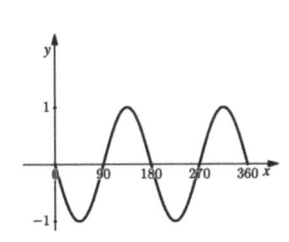

9) a)
$y = \tan 2x$
9) b)
$y = \tan\frac{1}{2}x$
9) c)
$y = \tan\frac{1}{3}x$
9) d)
$y = \tan\frac{5}{3}x$

10) a)

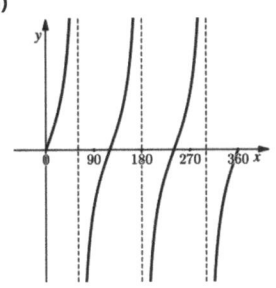

10) b)

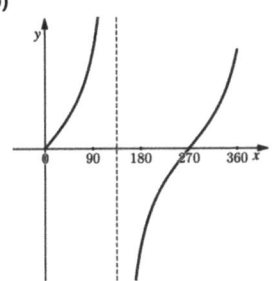

11) a)
$y = 2 + \sin x$
11) b)
$y = 1 + \cos x$
11) c)
$y = -1 + \sin x$
11) d)
$y = -3 + \cos x$

12) a)
$y = 1 + 2\sin x$

12) b)
$y = 2 + 3\cos x$
12) c)
$y = -1 + 3\sin x$
12) d)
$y = -2 + 2\cos x$

13) a)

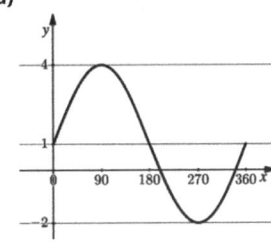

13) b)

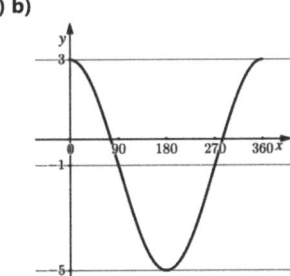

13) c)

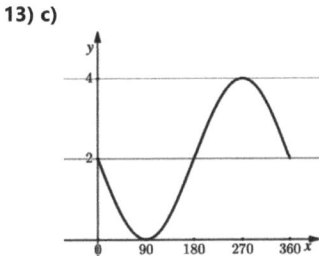

13) d)

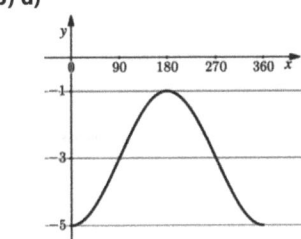

14) a)
$y = \sin(x + 30°)$
14) b)
$y = \sin(x - 60°)$

15) a)
$y = \cos(x - 60°)$
15) b)
$y = \cos(x - 150°)$

16) a)
$y = \tan(x + 30°)$
16) b)
$y = \tan(x - 45°)$

17) a)

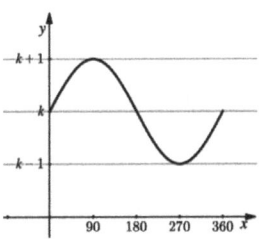

17) b)

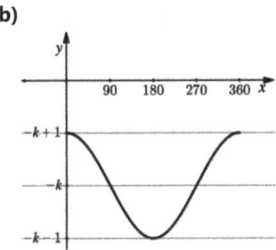

18) a)

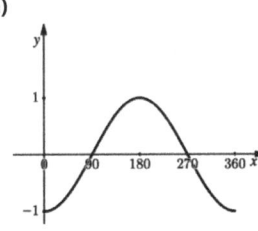

18) b)

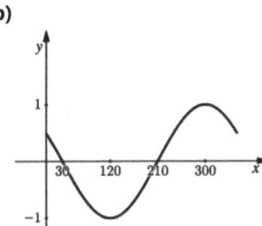

18) c)

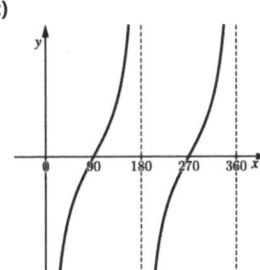

18) d)

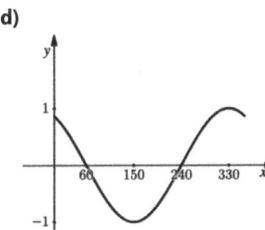

18) e)

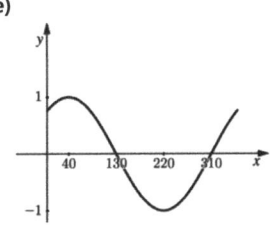

18) f)

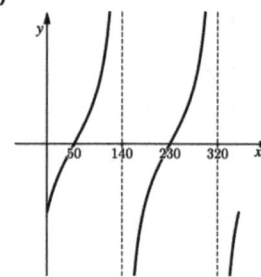

19) a)

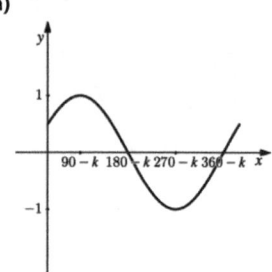

19) b)

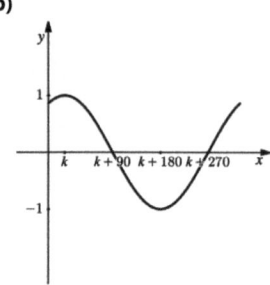

20) a)
$y = \sin(2x + 90°)$
20) b)
$y = \sin\left(\frac{1}{2}x - 30°\right)$
20) c)
$y = \sin\left(\frac{3}{2}x + 60°\right)$

21) a)
$y = \cos(3x - 90°)$
21) b)
$y = \cos\left(\frac{3}{4}x + 66°\right)$
21) c)
$y = \cos\left(\frac{5}{2}x - 100°\right)$

22) a)
$y = \tan(2x + 120°)$
22) b)
$y = \tan\left(\frac{1}{2}x - 60°\right)$
22) c)
$y = \tan\left(\frac{2}{3}x + 36°\right)$

23) a)
$y = 1 + \sin\left(\frac{1}{2}x + 40°\right)$
23) b)
$y = -1 + 2\sin(x - 80°)$
23) c)
$y = -2 + 2\sin(2x - 30°)$
23) d)
$y = 1 + 3\sin\left(\frac{3}{2}x - 96°\right)$

24) a)
$y = 1 + 2\cos(x - 60°)$
24) b)
$y = -1 + 3\cos 2x$
24) c)
$y = -2 + \cos\left(\frac{6}{5}x + 30°\right)$
24) d)
$y = 2 + 4\cos\left(\frac{3}{5}x - 120°\right)$

25) a)

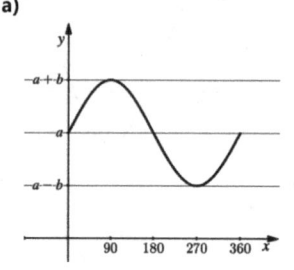

25) b)

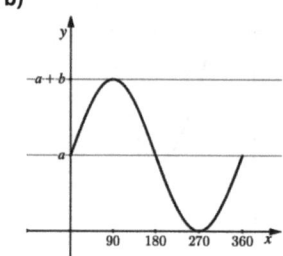

25) c)

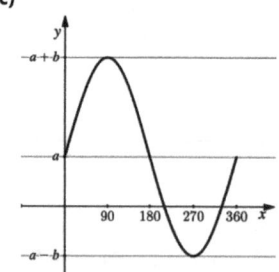

1.42 Trigonometry: Trigonometric Identities in Degrees
Questions on page 120

1) a)
$-\sin x$
1) b)
$-\sin x$
1) c)
$\sin x$
1) d)
$-\sin x$
1) e)
$\sin x$
1) f)
$\sin x$
1) g)
$-\sin x$
1) h)
$\sin x$
1) i)
$-\sin x$
1) j)
$-\sin x$
1) k)
$\sin x$
1) l)
$\sin x$
1) m)
$-\sin x$

2) a)
$-\cos x$
2) b)
$\cos x$
2) c)
$-\cos x$
2) d)
$-\cos x$
2) e)
$\cos x$
2) f)
$\cos x$
2) g)
$\cos x$
2) h)
$-\cos x$
2) i)
$-\cos x$
2) j)
$-\cos x$
2) k)
$\cos x$
2) l)
$-\cos x$
2) m)
$\cos x$

3) a)
$\tan x$
3) b)
$-\tan x$
3) c)
$\tan x$
3) d)
$\tan x$
3) e)
$-\tan x$
3) f)
$-\tan x$
3) g)
$\tan x$
3) h)
$-\tan x$
3) i)
$\tan x$
3) j)
$\tan x$
3) k)
$\tan x$
3) l)
$-\tan x$
3) m)
$-\tan x$

4) a)
$\cos x$
4) b)
$-\cos x$
4) c)
$-\cos x$

4) d) $\cos x$
4) e) $-\cos x$
4) f) $\cos x$
4) g) $-\cos x$
4) h) $\cos x$
4) i) $\cos x$
4) j) $-\cos x$
4) k) $\cos x$
4) l) $-\cos x$
4) m) $\cos x$

5) a) $-\sin x$
5) b) $\sin x$
5) c) $\sin x$
5) d) $\sin x$
5) e) $-\sin x$
5) f) $-\sin x$
5) g) $\sin x$
5) h) $\sin x$
5) i) $-\sin x$
5) j) $-\sin x$
5) k) $-\sin x$
5) l) $-\sin x$
5) m) $\sin x$

6) a) $\cos x = \frac{2\sqrt{2}}{3}$
6) b) $\tan x = \frac{1}{2\sqrt{2}} = \frac{\sqrt{2}}{4}$

7) a) $\sin x = \frac{3\sqrt{5}}{7}$
7) b) $\tan x = \frac{3\sqrt{5}}{2}$

8) a) $\sin x = \frac{3}{\sqrt{109}} = \frac{3\sqrt{109}}{109}$
8) b) $\cos x = \frac{10}{\sqrt{109}} = \frac{10\sqrt{109}}{109}$

9) a) $\cos x = -\frac{\sqrt{77}}{9}$
9) b) $\tan x = -\frac{2}{\sqrt{77}} = -\frac{2\sqrt{77}}{77}$

10) a) $\sin x = \frac{2\sqrt{10}}{7}$
10) b) $\tan x = -\frac{2\sqrt{10}}{3}$

11) a) $\sin x = \frac{11}{\sqrt{521}} = \frac{11\sqrt{521}}{521}$
11) b) $\cos x = -\frac{20}{\sqrt{521}} = -\frac{20\sqrt{521}}{521}$

12) a) $\cos x = -\frac{\sqrt{11}}{6}$
12) b) $\tan x = \frac{5}{\sqrt{11}} = \frac{5\sqrt{11}}{11}$

13) a) $\sin x = -\frac{2\sqrt{2}}{3}$
13) b) $\tan x = 2\sqrt{2}$

14) a) $\sin x = -\frac{7}{\sqrt{53}} = -\frac{7\sqrt{53}}{53}$
14) b) $\cos x = -\frac{2}{\sqrt{53}} = -\frac{2\sqrt{53}}{53}$

15) a) $\cos x = \frac{3\sqrt{11}}{10}$
15) b) $\tan x = -\frac{1}{3\sqrt{11}} = -\frac{\sqrt{11}}{33}$

16) a) $\sin x = -\frac{\sqrt{29}}{15}$
16) b) $\tan x = -\frac{\sqrt{29}}{14}$

17) a) $\sin x = -\frac{20}{\sqrt{409}} = -\frac{20\sqrt{409}}{409}$
17) b) $\cos x = \frac{3}{\sqrt{409}} = \frac{3\sqrt{409}}{409}$

18) a) $\sin 75° = \frac{\sqrt{2+\sqrt{3}}}{2}$
18) b) Show that

19) a) $\sin x$
19) b) $\frac{\sin^2 x}{\cos x}$
19) c) $\frac{1}{\cos x}$
19) d) $\frac{\sin x}{\cos^2 x}$

20) a) $\sin^2 x$
20) b) $\sin^3 x$
20) c) $\frac{\cos x}{\sin x}$
20) d) $\frac{\sin x}{\cos^3 x}$

21) a) $\sin^2 x$
21) b) $|\cos x|$
21) c) $-9\sin^2 x$
21) d) 8

22) a) 1
22) b) 1
22) c) $\frac{4}{5}\tan^2 x$
22) d) $-\frac{1}{6\tan x}$

23) a) $\theta = \{45°, 225°\}$
23) b) $\theta = \{56.3°, 236°\}$ (3 SF)
23) c) $\theta = \{148°, 328°\}$ (3 SF)
23) d) $\theta = \{74.1°, 254°\}$ (3 SF)

24) a)

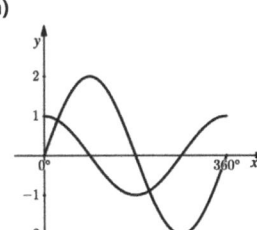

24) b) $(26.6°, 0.894)$ and $(207°, -0.894)$ (3 SF)

25) a) $\theta = \{13.3°, 73.3°, 133°, 193°\}$ (3 SF)
25) b) $\theta = \{104°, 464°\}$ (3 SF)
25) c) $\theta = \{71.2°, 611°, 1150°, 1690°\}$ (3 SF)

26) a)
$\theta = \{194°, 346°\}$ (3 SF)
26) b)
$\theta = \{99.3°, 261°\}$ (3 SF)
26) c)
$\theta = \{93.5°, 176°, 274°, 356°\}$ (3 SF)

27) a)
$\theta = \{30°, 150°\}$
27) b)
$\theta = \{75.5°, 284°, 436°\}$ (3 SF)
27) c)
$\theta = \{8.21°, 172°\}$ (3 SF)
27) d)
$\theta = \{-84.3°, 84.3°, 276°\}$ (3 SF)

28)
$2\sin^4 x - 5\sin^2 x + 3$

29) a)
$\theta = \{30°, 90°, 150°\}$
29) b)
$\theta = \{132°, 180°, 228°\}$ (3 SF)
29) c)
$\theta = \{41.8°, 138°, 210°, 330°\}$ (3 SF)
29) d)
$\theta = \{66.4°, 139°, 221°, 294°\}$ (3 SF)

30) a)
$\theta = \{19.5°, 161°, 210°, 330°\}$ (3 SF)
30) b)
$\theta = \{161°, 199°\}$ (3 SF)
30) c)
$\theta = \{239°, 301°\}$ (3 SF)

31) a)
$x = \{70.5°, 132°, 228°, 289°\}$ (3 SF)
31) b)
$\theta = 10.3°$ (3 SF)

32)
$\theta = \{15.7°, 51.7°\}$ (3 SF)

33) a)
$\theta = \{114°, 246°\}$ (3 SF)
33) b)
$\theta = \{99.6°, 132°, 228°, 260°\}$ (3 SF)
33) c)
$\theta = \{23.6°, 156°, 194°, 346°\}$ (3 SF)

34)
(68.2°, 0.690), (171°, 4.86), (248°, 0.690), (351°, 4.86) (3 SF)

35) a)
$\theta = \{45°, 76.0°, 225°, 256°\}$ (3 SF)
35) b)
$\theta = \{71.6°, 169°, 252°, 349°\}$ (3 SF)
35) c)
$\theta = \{50.2°, 124°, 230°, 304°\}$ (3 SF)

36) a)
$\theta = \{117°, 171°, 297°, 351°\}$ (3 SF)
36) b)
$\theta = \{56.3°, 59.0°, 236°, 239°\}$ (3 SF)

37) a)
Prove
37) b)
Prove
37) c)
Prove
37) d)
Prove

38) a)
Prove
38) b)
Prove
38) c)
Prove
38) d)
Prove

39) a)
Prove
39) b)
Prove
39) c)
Prove
39) d)
Prove

40) a)
Prove
40) b)
Prove
40) c)
Prove
40) d)
Prove

41) a)
Prove
41) b)
Prove
41) c)
Prove
41) d)
Prove

42) a)
Prove
42) b)
Prove
42) c)
Prove
42) d)
Prove

43) a)
Prove
43) b)
Prove
43) c)
Prove
43) d)
Prove

44) a)
Prove
44) b)
Prove
44) c)
Prove

45) a)
Prove
45) b)
Prove

46) a)
Prove
46) b)
Prove

47)
Prove

48)
Prove

49)
The first error is replacing $\cos^2 x$ with $\sin^2 x - 1$. She should instead have replaced $\cos^2 x$ with $1 - \sin^2 x$.
Note that Isla *can* divide through by $\sin x$ when she goes from
$\sqrt{2}\sin^2 x + \sin x = 0$ to $\sqrt{2}\sin x + 1 = 0$
without losing solutions as $\sin x$ does not equal zero between 180° and 360° (exclusive), so this is not an error.
Isla's second mistake is in not finding solutions between 180° and 360°, as she only finds −45° which is not in the range we are looking for.

50) a)
Show that
50) b)
$\theta = \{2.83°, 29.9°\}$ (3 SF)

51) a)
Show that
51) b)
$x = \{9.59°, 170°\}$ (3 SF)
51) c)
$\theta = \{32.3°, 113°\}$ (3 SF)

52) a)
Show that

52) b)
$x = \{87.1°, 267°\}$ (3 SF)
52) c)
$\theta = \{39.1°, 129°\}$ (3 SF)

53)
$\theta = \{60°, 120°, 240°, 300°\}$

54)
$\theta = \begin{Bmatrix} 0°, 30°, 90°, 150°, \\ 180°, 270°, 360° \end{Bmatrix}$

1.43 Trigonometry: Modelling in Degrees
Questions on page 125

1) a) 180°
1) b) 60°
1) c) 180°
1) d) 720°
1) e) 1440°
1) f) 90°
1) g) 7°
1) h) 52°
1) i) 365°
1) j) 24°

2) a) $y = \sin(2x)$
2) b) $y = \sin(360x)$
2) c) $y = \sin\left(\frac{360}{7}x\right)$
2) d) $y = \sin\left(\frac{90}{13}x\right)$

3) a) $y = 3\sin\left(\frac{360}{7}x\right)$
3) b) $y = 2\sin(360x)$
3) c) $y = 10\sin(18x)$
3) d) $y = -5\sin(12x)$

4) a) 1
4) b) 1
4) c) 18
4) d) 5

5) a) Max is 1, Min is -1
5) b) Max is 5, Min is 3
5) c) Max is 40, Min is 10
5) d) Max is 11, Min is -5

6) a) $y = 3 + 2\sin(36x)$
6) b) $y = 2 + 0.5\sin(12x)$
6) c) $y = -1 + 3\sin(18x)$
6) d) $y = 20 - 5\sin\left(\frac{360}{7}x\right)$

7) a) $a = 6.25$, $b = 1.75$
7) b) Change the period to match the yearly cycle: $y = 6.25 + 1.75\cos\left(\frac{360}{365}x\right)$

8) a) $y = 100 + 20\sin(360t)$
8) b) Blood pressure will change depending on how much a person is moving.

9) a) 21 mice
9) b) Max is 26 mice, Min is 14 mice
9) c) Max occurs between day 33 and day 34
Min occurs between day 93 and day 94

10) a) £100
10) b) 365 days later or 501 days after 1st January

11) a) 3.3 metres
11) b) 4.3 hours.
11) c) 4: 34 am
11) d) $h = 2.5 + 1.65\sin(25t - 330)°$

1.44 Hidden Polynomials
Questions on page 128

1) a) $x = \pm\sqrt{\frac{3}{5}}$
1) b) $x = 49$
1) c) $x = \frac{27}{8}$
1) d) $x = 2, x = -\frac{3}{5}$
1) e) $x = 0$
1) f) $x = 0, x = -1$
1) g) $x = -1$ is the only solution.
1) h) $x = -1$

2) $2^6 \times 17 \times 353$

3) $x = \pm\frac{1}{\sqrt{2}}$ or $x = \pm\sqrt{2}$

4) $x = -\frac{1}{2}, x = -1$

5) $x = \frac{1}{\sqrt[3]{2}}, x = -1, x = 1$

6) $x = 2 \pm \sqrt{2}$ or $x = 1 \pm \sqrt{5}$

7) $x = 1, x = 16$

8) $x = -1, x = 1$

9) a) $x = \log_2 \frac{2}{3}$
9) b) $x = 0, x = \ln\frac{3}{2}$
9) c) $x = 0$

10) a) $x = \frac{5}{2}, x = -2$
10) b) $x = \log_5\left(\frac{5}{2}\right)$

11) a) $x = 6, x = -2$
11) b) $x = 1, x = \frac{3}{5}$
11) c) $x = \frac{1}{3}, x = \frac{3}{2}$

12) $x = e^2$

13) $x = 30°, x = 150°$

14) $x = 90°, x = 270°$

1.45 Proof
Questions on page 129

1) a) $A \Leftarrow B$
1) b) $A \Leftarrow B$
1) c) $A \Leftrightarrow B$
1) d) $A \Rightarrow B$

2) a) $A \Rightarrow B$
2) b) $A \Leftarrow B$
2) c) $A \Rightarrow B$
2) d) $A \Leftrightarrow B$

3) a) True
3) b) True
3) c) False, not true for $x = 0$
3) d) True
3) e) False. It is always odd.
3) f) False as x could be another value such as 300 for example

4) At the point $\frac{(10-10)(10+10)}{10(10-10)}$ there is a zero in the denominator

5) a) 2 is a prime number and its even.
5) b) $3^2 = 9$ which is odd.
5) c) 3 is a multiple of 3 but not a multiple of 6
5) d) -1 does not have a real square root.
5) e) $4 = 2^2$ ends in an even number.

6) a) $x = -2$
6) b) $n = 13$
6) c) $n = 16$

7) a) Prove
7) b) Prove
7) c) Prove

8) Prove

9) a) Show that
9) b) Show that
9) c) Show that
9) d) Show that

10) a) Prove
10) b) Prove

11) Prove

12) Prove

13) a) Prove
13) b) Prove

14) Prove

1.46 2D Vectors
Questions on page 130

1) a) $6\mathbf{i} + 3\mathbf{j}$
1) b) $-\mathbf{i} + 2\mathbf{j}$
1) c) $4\mathbf{i} + 9\mathbf{j}$

2) a) $-\mathbf{a} + \mathbf{b}$
2) b) $-\mathbf{c} + \mathbf{b}$
2) c) $-\mathbf{a} + \mathbf{c}$

3) a) $\frac{1}{2}\mathbf{a}$
3) b) $\frac{1}{2}\mathbf{b}$
3) c) $\mathbf{b} + \frac{1}{2}\mathbf{a}$
3) d) $\mathbf{a} + \frac{1}{2}\mathbf{b}$
3) e) $-\frac{1}{2}\mathbf{b}$
3) f) $\frac{1}{2}\mathbf{b} - \frac{1}{2}\mathbf{a}$
3) g) $-\frac{1}{2}\mathbf{a} - \mathbf{b} + \frac{1}{2}\mathbf{a} = -\mathbf{b}$
3) h) $-\mathbf{a} + \frac{1}{2}\mathbf{b}$

4) a) Parallel
4) b) Parallel
4) c) Not parallel
4) d) Not parallel

5) a) $\sqrt{41}$
5) b) $5\sqrt{2}$
5) c) $\sqrt{10}$
5) d) $\sqrt{p^2 + q^2}$

6) a) $2\sqrt{37}$ on bearing 009°
6) b) $\sqrt{34}$ on bearing 301°
6) c) $5\sqrt{5}$ on bearing 243°
6) d) 17 on bearing 152°

7) a) $9.19\mathbf{i} + 7.71\mathbf{j}$
7) b) $-4.57\mathbf{i} + 2.03\mathbf{j}$
7) c) $-10.1\mathbf{i} - 14.9\mathbf{j}$
7) d) $18.8\mathbf{i} - 51.7\mathbf{j}$

8) a) $\frac{1}{\sqrt{13}}(2\mathbf{i} + 3\mathbf{j})$
8) b) $\frac{1}{\sqrt{17}}\begin{pmatrix}-1\\4\end{pmatrix}$
8) c) $\frac{1}{\sqrt{89}}(5\mathbf{i} - 8\mathbf{j})$

9) a) $\mathbf{a} = -3\mathbf{i} + 9\mathbf{j}$
9) b) $\mathbf{a} = \mathbf{i} + \frac{7}{2}\mathbf{j}$
9) c) $\mathbf{a} = 3\mathbf{i} - 24\mathbf{j}$

10) a)
$\binom{1}{2}$

10) b)
$\binom{-1}{6}$

10) c)
$\binom{p}{q}$

11) a)
$\overrightarrow{AB} = \binom{-3}{4}$

11) b)
$\overrightarrow{BC} = \binom{5}{-16}$

11) c)
$\overrightarrow{AC} = \binom{2}{-12}$

11) d)
$\overrightarrow{BA} = \binom{3}{-4}$

12) a)
$\sqrt{73}$

12) b)
$\sqrt{170}$

13) a)
$\frac{1}{5\sqrt{5}}\binom{2}{11}$

13) b)
$\frac{1}{\sqrt{73}}(-8\mathbf{i} + 3\mathbf{j})$

14) a)
90°

14) b)
29.1°

14) c)
58.1°

14) d)
149.0°

14) e)
161.6°

15)
Show that

16)
$\overrightarrow{OD} = \binom{-6}{-5}$

17) a)
Collinear

17) b)
Collinear

17) c)
Not collinear

18)
$\overrightarrow{OB} = 3\mathbf{i} - 5\mathbf{j}$

19) a)
$p = 8, k = 12$

19) b)
AB is parallel to CD and has same length

$\overrightarrow{AC} = \binom{4}{5}$, $\overrightarrow{BD} = \binom{4}{5}$

AC is parallel to BD and has same length but different to AB
Therefore, it is a parallelogram.

20) a) i)
$\overrightarrow{OC} = \mathbf{b} - \mathbf{a}$

20) a) ii)
$\overrightarrow{OP} = \frac{1}{2}\mathbf{b} + \frac{1}{2}\mathbf{a}$

20) a) iii)
$\overrightarrow{OQ} = \frac{1}{2}\mathbf{b} - \frac{1}{2}\mathbf{a}$

20) b)
Show that

21) a)
$k = 1$

21) b)
$A(-4, 0)$ and $B(0, 5)$

22) a)
$P(3, 4)$ and $Q(-2, 5)$

22) b)
Show that

22) c)
$(-4, -1)$ or $\left(\frac{46}{5}, \frac{17}{5}\right)$

23)
$(5.5, 5.5)$ or $(-2.5, 6.5)$

1.47 Sampling Methods
Questions on page 132

1)
Opportunity

2)
Simple random

3)
Systematic

4)
Cluster

5)
Quota

6) a)
The 7200 employees

6) b)
Method B will collect responses from 5 of the 24 stores, so it will less likely be biased based on location. Method C will only consider the London store, and there could be job satisfaction issues based on the location, which is not reflected elsewhere.

6) c)
Method B will collect responses from just 5 stores, but Method A could potentially seek responses from all 24 stores. Method B is therefore more likely to be quicker to implement. Method A also has the potential to collect more responses from some stores than others, introducing potential bias.

7)
A census can be very expensive and time-consuming to both collect and analyse.

8)
There is no sampling frame i.e. we do not have a list of all of the African penguins (the number of breeding pairs is only an estimate, and will regularly fluctuate anyway).

9)
Attendees to a national union conference will mostly be made up of union representatives and so will already have strong views on the conditions in their place of work. So it is unlikely that Chloe will gather many moderate views. She is also relying on attendees visiting her stall and then deciding to fill in the questionnaire, so she will only get the views of people who are willing to do this (even if her stall is close to the entrance, there is no guarantee that everyone will visit it).

10)
Clive should first number all of the customers from 001 to 680. He should then use a random number generator to generate three-digit numbers from 001 to 680. He should generate 40 numbers from 001 to 680, ignoring repeats, and select the customers given those numbers.

11)
86 should be surveyed from the shops and 64 from the warehouses.

12)
A simple random sample would ask the opinion of a number of people who would not be affected and would be a waste of their time. An opportunity sample would be more appropriate as it would gather the opinions of those who work near the kitchen who are the most likely people to own the mugs. It would also be very easy to gain opinions quickly.

13)
Susan should first number all the classes from 01 to 50. She should then use a random number generator to generate two-digit numbers from 01 to 50. She should generate 3 numbers from 01 to 50, ignoring repeats, and survey all the students in the classes given those numbers.

14)
Some members of staff may not check their emails within the 24-hour window as they may be ill or only work part-time. Of those that do read the email, the responses may only come from those with a strong interest or motivation to take part, and hence may lead to a biased sample.

15) a)
$k = 50$

15) b)
15

16) a)
Systematic

16) b)
There are only 6 sets of 120 employees that can be selected using this method. For example: {1, 7, 13, 19, ...}, {2, 8, 14, 20, ...} can be selected, but {1, 2, 3, 4, ...} cannot. So because not **all** samples of size 20 can be selected, the sampling method is not random.

1.48 Measures of Central Tendency and Variation
Questions on page 134

1)
Set A: Mean = 13, Median = 12, Mode = 3, Range = 25, Lower Quartile = 6, Upper Quartile = 20, Interquartile Range = 14
Set B: Mean = 13, Median = 12, Mode = 3, Range = 25, Lower Quartile = 6, Upper Quartile = 20, Interquartile Range = 14
The statistics are the same for both sets of data despite coming from different data.

2) a)
The mean uses all values in its calculation and is good for roughly symmetric data.

2) b)
The median is unaffected by extreme values and is good for skewed data.

3) a)
Mean = 10.9 (3 SF)

Standard Deviation = 6.10 (3 SF)

3) b)
25 is an outlier

3) c)
Median = 10, Interquartile Range = 7

3) d)
There are no outliers

4) a)
Mean = 20.5
Standard Deviation = 13.3 (3 SF)

4) b)
55 is an outlier

4) c)
Median = 19, Interquartile Range = 17.5

4) d)
55 is an outlier

5) a)
$\bar{x} = 45.8$ (3 SF)

5) b)
If 59 was removed, the value of $\sum x$ would decrease, hence $\bar{x}$ would decrease.

6) a)
$\bar{x} = 56.5$

6) b)
If 40 is changed to 56.5, the value of $\sum x$ would increase, hence $\bar{x}$ would increase.

7) a)
$\bar{x} = 172$

7) b)
$\bar{x} = 2480$ (3 SF)

8)
$n = 987$

9)
$\sigma = 44.0$ (3 SF)

10)
$\sigma = 4.61$ (3 SF)

11)
$\sigma^2 = 17.1$ (3 SF)

12)
$\sigma^2 = 942$ (3 SF)

13) a)
Standard deviation is 5.15 (3 SF)
(Sample standard deviation is 5.65 (3 SF))

13) b)
If 125 was removed, then the data is less spread out, hence the standard deviation would decrease.

14) a)
Standard deviation is 1.99 (3 SF)
(Sample standard deviation is 2.18 (3 SF))

14) b)
If 43.6 was changed to 48.6, then the data is more spread out, hence the standard deviation would increase.

15) a)
$\bar{x} = 14.5$

15) b)
Standard deviation is 4.35 (3 SF)
(Sample standard deviation is 4.76 (3 SF))

15) c)
e.g. $x = 14$ and $y = 15$. x and y must be the same distance from the mean to keep the mean the same (so the mean of x and y must be 14.5), but less than 1 standard deviation away from the mean to reduce the standard deviation.

16) a)
$\bar{x} = 32$

16) b)
Median = 33

16) c)
e.g. $x = 31$ and $y = 33$. x and y must be the same distance from the mean to keep the mean the same (so the mean of x and y must be 32), but y must be less than 34 to reduce the median.

17) a)
$x = 22$

17) b)
$x = 17.3$

17) c)
$x = 17.6$

18) a)
$\bar{x} = 138$ cm (3 SF)

18) b)
$s = 29.2$ cm (3 SF)

18) c)
196 is an outlier

19)
Mean is 1.975
Sample standard deviation is 1.46 (3 SF)
(Standard deviation is 1.45 (3 SF))

20)
Mean is 1.68 (3 SF)
Sample standard deviation is 0.704 (3 SF)
(Standard deviation is 0.700 (3 SF))

21) a)
$\bar{x} = 2.6$

21) b)
Median = 2.5

22) a)
$\bar{x} = 22.75$

22) b)
Sample standard deviation is 4.50 (3 SF)

(Standard deviation is 4.49 (3 SF))
22) c)
Using the midpoints assumes that the data is evenly distributed within the class intervals.

23) a)
11.5
23) b)
$\bar{x} = 10.4$ kg (3 SF)
23) c)
Sample standard deviation is 2.25 kg (3 SF)
(Standard deviation is 2.22 kg (3 SF))

24) a)
27
24) b)
66 (as of writing) is England's state retirement age
24) c)
Mean is 38.9 years (3 SF)
Sample standard deviation is 10.8 years (3 SF)
(Standard deviation is 10.7 years (3 SF))

25) a)
$\bar{x} = £46{,}000$ (3 SF)
25) b)
Sample standard deviation is £8,200 (3 SF)
(Standard deviation is £7,950 (3 SF))
25) c)
These are only estimates as we have used the midpoints of the class intervals. This assumes that the raw data is evenly distributed within the class intervals, which may not actually be the case.
25) d)
This data only comes from one publishing house, so is only a very small sample. Also, the salaries in London would be expected to be higher than other places in the UK.

26)
$\sum x = 297$, $\sum x^2 = 2739$

27)
$\sum x = 9020$, $\sum x^2 = 814700$

28)
£61.61 to the nearest penny

29)
$n = 40000$

30)
The two pieces of data that can be replaced are 19 and 21. If 19 is removed it must be replaced with 35. If 21 is removed it must be replaced with 37.

31)
$\sigma = \sqrt{21}$

32)
$k = 66$

33)
$\bar{x} = \frac{725}{24} = 30.2$ (3 SF), $\sigma = 4.996$ to 4sf

34)
$\bar{x} = 39.5$ (3 SF), $\sigma = 7.77$ (3 SF)

35)

| y | 5 | $\frac{45}{2}$ | 40 | 75 | 115 |

36)
$\bar{y} = 850$, $\sigma_y = 120$

37)
$\bar{y} = 410$, $\sigma_y = 15$

38)
$\bar{x} = 63.75$, $\sigma_x = 15$

39)
$\bar{x} = -743$ (3 SF), $\sigma_x = 135$

40) a)
Mean is 25.4°C (3 SF)
Sample standard deviation is 2.07°C (3 SF)
(Standard deviation is 1.92°C (3 SF))
40) b)
Mean is 77.8°F (3 SF)
Sample standard deviation is 3.73°F (3 SF)
(Standard deviation is 3.45°F (3 SF))

41) a)
Mean is 61.4°F (3 SF)
Sample standard deviation is 1.99°F (3 SF)
(Standard deviation 1.84°F (3 SF))
41) b)
Mean is 16.3°C (3 SF)
Sample standard deviation is 1.10°C (3 SF)
(Standard deviation is 1.02°C (3 SF))

42)
Mid-range = 51.5

43) a)
Mode = 1
43) b)
Mid-range = 3
43) c)
Median = 1

44)
$Q_2 = 7.08$ (3 SF)

45)
$Q_2 = 10.6$ (3 SF)

46)
$Q_2 = 73.3$ (3 SF)

47)
$IQR = 24.375$

48) a)
20th to 80th percentile range = 63.0 (3 SF)
48) b)
480

49)
$Q_3 = 6.64$ (3 SF)

50)
$Q_1 = 16.9$

51)
We expect 70% of the fish to be longer than 8.94 cm.

52)
$Q_2 = 45.2$ (3 SF)

1.49 Representing Data
Questions on page 140

1)
$\bar{x} = \frac{75}{14} = 5.36$ (3 SF)

2) a)
Unimodal, discrete, symmetric
2) b)
Unimodal, continuous, symmetric
2) c)
Unimodal, discrete, negatively skewed
2) d)
Bimodal, continuous, symmetric
2) e)
Unimodal, continuous, positively skewed
2) f)
Bimodal, discrete, negatively skewed
2) g)
Uniform, unimodal, discrete, symmetric

3) a)
Positive skew
3) b)
$Q_2 = 40$
3) c)
$IQR = 19$

4) a)
$Q_2 = 86$ cm
$IQR = 18$ cm

4) b)
Mean is 88 cm (3 SF)
Sample standard deviation is 15.1 cm (3 SF)
(Standard deviation is 14.5 cm (3 SF))

4) c)
The median would be a better measure of central tendency as it is not affected by the tallest kangaroo at 129 cm, which is considerably taller than the next largest at 100 cm.

5) a)
Negative skew

5) b)
36.0 is the mode

5) c)
34.8 is the median

5) d)

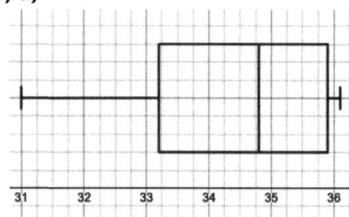

6)
Any example where $Q_3 - Q_2 < Q_2 - Q_1$
e.g.

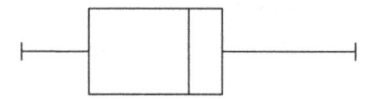

7) a)
Positive skew

7) b)
Negative skew

7) c)
Roughly symmetric

7) d)
Positive skew

8) a)
Median = 15

8) b)
Positive skew

9) a)
$Q_2 = 14.5$

9) b)
Range = 23

9) c)
$IQR = 6$

10) a)
Median of Class A = 66
Median of Class B = 62

10) b)
Class A: $IQR = 54.5$,
Class B: $IQR = 44$

10) c)
Class A has a higher median at 66 than Class B at 62, so on average Class A performed better than Class B.
Class A has a larger interquartile range at 54.5 than Class B at 44, so the scores of Class A are more spread out.

11)
$\frac{41}{76} = 53.9\%$ (3 SF)

12) a)

Meeting Length (t)	Frequency
$0 \leq t < 30$	30
$30 \leq t < 60$	45
$60 \leq t < 120$	20
$120 \leq t < 150$	15
$150 \leq t < 180$	5

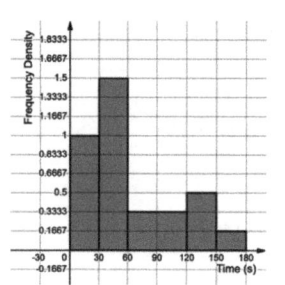

12) b)
43

13) a)
The median is in the 30 to 40 class interval.

13) b)
Mean is 33.9 (3 SF)

13) c)
Sample standard deviation is 11.4 (3 SF)
(Standard deviation is 11.3 (3 SF))

13) d)
$\frac{14}{95}$

14)
Width: 3 cm. Height: $\frac{8}{3}$ cm

15)
$Q_1 = 2.94$ (3 SF), $Q_2 = 4.17$ (3 SF),
$Q_3 = 5.23$ (3 SF)

16) a)
Graph B should be used as the data should be plotted at the top end of each class interval.

16) b)
The data is cumulative, so an increasing gradient would show that sales week on week are increasing. However, at the end of the 5th week, the gradient starts to decrease and so the number of sales here are decreasing.

16) c)
Amber only sold ice creams on a Sunday. Estimating from the graph would assume that the sales of ice creams were evenly spread across the week, which is not the case.

1.50 Bivariate Data: PMCC & Linear Regression
Questions on page 144

1)
Strong Positive Correlation

2)
Moderate Negative Correlation

3)
$r = 0.864$

4)
$r = -0.391$

5)
$-\frac{7}{3}$ or $\frac{8}{7}$

6)
Explanatory (independent) variable = Hours of exercise per week
Response (dependent) variable = Levels of stress

7)
Probably correct.

8)
Definitely incorrect.

9)
Definitely correct.

10)
$r = 0.995$ (3 SF)

11) a)
$r = 0.934$ (3 SF)

11) b)
There is a strong positive correlation between the number of hours the students studied in a week and the students' test scores.

12)
A PMCC of -0.76 shows a moderately strong negative correlation between the frequency of smoking and respiratory function. So as the amount of smoking increases, the respiratory function decreases.

13) a)
Winchester appears to be an outlier. It may have an unusually high average house price due to its high quality of life / historical significance / transport links to London, etc.

13) b)
Correlation is how close the data is to a straight line. So the data does not appear to be well-correlated, but there appears to be a negative association (non-linear as the data is curved).

13) c)
As the distance from London increases, the average house price decreases.

14) a)
Data point A is unlikely to be correct as the runner did zero exercise the week before but has still managed to run a long distance. It may well be the case, however, that the runner has considerable prior experience of running long distance and does not require as much exercise in the lead-up.

14) b)
Data point B could represent a new member of the running group. As they are already on week 10, it may well be that B has only just joined the group and hence needs to put in more exercise the week before to reach a similar standard in the run as others in the group.

15) a)
Data point A has a zero score for the Spelling test. This may be accurate as the student may have been absent.

15) b)
Data point B has a much higher Arithmetic test score than any of the other students. All of the other students scored under 100, so it may well be that the tests had a maximum score of 100. It may well have been inputted incorrectly when the graph was drawn.

16)
The claim is incorrect as the negative correlation in the scatter diagram does not imply that amount of sleep and the task completion time are causally linked (correlation does not imply causation).

17)
The data in the diagram is close to a straight line, and the PMCC of -0.979 is very strong (negative) correlation.

18)
$y = 2.88x + 2.84$

19)
$y = 773 - 3.90x$

20)
For every £1 spent each month in advertising, the sales revenue increases by £1.96.

21)
For every 1 penny increase in price, the number of cans sold is reduced by 2.22 cans.

22)
The fastest time will be recorded by the person that sleeps the longest. If a runner sleeps for 82 hours, they will finish the 100 m race instantaneously.

23) a)
Marks scored in the final exam per hour of lunchtime workshop attended.

23) b)
This will increase the average student's final score on the exam by $1.12 \times 2.5 = 2.8$.

23) c)
$1 \times 5 \times 38 = 190$. So there is a maximum number of 190 hours of lunchtime workshop. So $t = 200$ would be beyond the number of hours available, and it would also give $98 + 1.12 \times 200 = 322$ which is a higher score than the maximum possible score for the exam (which was 300).

24) a)
3 days is 6×12 hours, so $t = 6$. $0.67 \times 6 = 4.02$, so on average the fungus will have grown by 4.02 mm.

24) b)
35 days

25) a)
6.5327 tonnes per hectare

25) b)
5.986 tonnes per hectare

25) c)
The estimate calculated for Farm A uses interpolation as 132 is within the data collected, and so is more reliable than the estimate calculated for Farm B as that uses extrapolation as 121 is outside of the data collected.

1.51 Bivariate Data: Association & Spearman's Rank
Questions on page 149

1)
Positive nonlinear association

2)
Positive nonlinear association

3) a)
$r_s = \frac{6}{7} = 0.857$ (3 SF)

3) b)
Strong Positive

4)
$r_s = -\frac{41}{42} = -0.976$ (3 SF)

5)
$r_s = \frac{3}{7} = 0.429$ (3 SF)

1.52 Probability: Introduction
Questions on page 150

1) a)

		2nd coin	
		H	T
1st coin	H	HH	HT
	T	TH	TT

1) b)
$\frac{3}{4}$

2) a)

			2nd die		
		1	2	3	4
	1	1	2	3	4
1st die	2	2	4	6	8
	3	3	6	9	12
	4	4	8	12	16

2) b)
$\frac{3}{16}$

3) a)

			2nd spinner	
		1	4	9
	2	3	6	11
1st spinner	3	4	7	12
	5	6	9	14
	7	8	11	16

3) b)
$\frac{7}{12}$

4) a)

		2nd pack		
		A	B	C
	A	AA	AB	AC
1st pack	B	BA	BB	BC
	C	CA	CB	CC

4) b)
$\frac{4}{9}$

5) a)
RR, RG, RB, GR, GG, GB, BR, BG

5) b)
$\frac{2}{3}$

6) a)
RR, RB, BR, BB
6) b)
$\frac{1}{4}$
6) c)
$\frac{1}{2}$
6) d)
$\frac{1}{2}$

7) a)

		2nd bucket					
		0	1	2	3	4	5
1st bucket	0	00	01	02	03	04	05
	1	10	11	12	13	14	15
	2	20	21	22	23	24	25
	3	30	31	32	33	34	35
	4	40	41	42	43	44	45
	5	50	51	52	53	54	55

7) b)
$\frac{5}{36}$
7) c)
$\frac{2}{9}$

8) a)
RRB, RRG, RBR, RBB, RBG, RGR, RGB, BRR, BRB, BRG, BBR, BBG, BGR, BGB, GRR, GRB, GBR, GBB
8) b)
$\frac{3}{10}$
8) c)
$\frac{2}{5}$
8) d)
$\frac{1}{5}$

9) a)
0.55
9) b)
0.3
9) c)
0.35
9) d)
0.9
9) e)
0.25
9) f)
0.65

10) a)
0.83
10) b)
0.23
10) c)
0.97
10) d)
0.14
10) e)
0.4
10) f)
0.4

11)

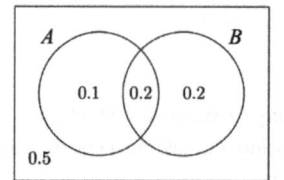

Venn diagram: A and B overlapping. A only: 0.1, intersection: 0.2, B only: 0.2, outside: 0.5

12)

Venn diagram: C and D overlapping. C only: 0.05, intersection: 0.7, D only: 0.15, outside: 0.1

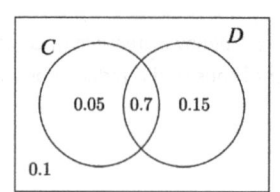

13) a)
$k = \frac{7}{30}$
13) b)
$P(A) = \frac{17}{30}$

14)
$p = \frac{19}{45}$ and $q = \frac{1}{90}$

15)
$r = \frac{2}{9}$ and $s = \frac{1}{18}$

16) a)
$v = \frac{1}{5}$ and $w = \frac{7}{10}$

16) b)

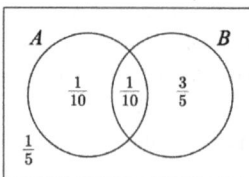

Venn diagram: A and B overlapping. A only: $\frac{1}{10}$, intersection: $\frac{1}{10}$, B only: $\frac{3}{5}$, outside: $\frac{1}{5}$

17)

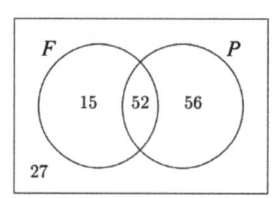

Venn diagram: F and P overlapping. F only: 15, intersection: 52, P only: 56, outside: 27

18)
$k = 0.13$

19)
$q = 0.2$

20)

Venn diagram with three sets A, V, E. V only: 328, $A \cap V$ only: 59, $V \cap E$ only: 19, centre: 4, A only: 16, $A \cap E$ only: 1, E only: 6, outside: 147

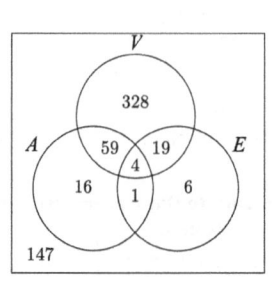

21) a)
$\frac{7}{12}$

21) b)
$\frac{5}{6}$
21) c)
$\frac{23}{36}$
21) d)
$\frac{1}{6}$
21) e)
0
21) f)
$\frac{8}{9}$
21) g)
$\frac{31}{36}$
21) h)
$\frac{5}{12}$
21) i)
$\frac{1}{2}$
21) j)
$\frac{1}{12}$

22) a)
$\frac{19}{50}$
22) b)
$\frac{3}{10}$
22) c)
$\frac{27}{50}$
22) d)
$\frac{17}{100}$
22) e)
$\frac{1}{5}$
22) f)
$\frac{3}{5}$
22) g)
$\frac{7}{10}$
22) h)
$\frac{6}{25}$
22) i)
$\frac{49}{100}$
22) j)
$\frac{4}{5}$

23)
0.687 (3 SF)

24)
0.215 (3 SF)

25)
0.0533 (3 SF)

26) a)
$\frac{11}{1190} = 0.00924$ (3 SF)
26) b)
0.0115 (3 SF)
26) c)
0.0468 (3 SF)

27) a) $\frac{2}{5}$

27) b) $\frac{2}{15}$

27) c) $\frac{19}{30}$

27) d) $\frac{19}{30}$

27) e) $= \frac{4}{15}$

27) f) $= \frac{23}{30}$

28) a) $\frac{7}{12}$

28) b) $\frac{31}{60}$

28) c) $\frac{19}{20}$

28) d) $\frac{11}{30}$

28) e) $\frac{14}{15}$

28) f) $\frac{14}{15}$

29) a) i)
$A \cap B'$: The employee travels to and from work by bicycle and has worked at the company for under 5 years.

29) a) ii)
$A' \cap B$: The employee does not travel to and from work by bicycle and has worked at the company for over 5 years.

29) a) iii)
$A' \cup B'$: The employee does not travel to and from work by bicycle or has worked at the company for under 5 years, or both.

29) b) i) $\frac{5}{38}$

29) b) ii) $\frac{9}{19}$

29) b) iii) $\frac{2}{19}$

29) b) iv) $\frac{1}{2}$

29) b) v) $\frac{1}{2}$

29) b) vi) $\frac{17}{19}$

30) a)
$P((A' \cup B)') = \frac{7}{50} = 14\%$

30) b)
$P((A \cap B')') = \frac{43}{50} = 86\%$

31)

	A	A'
B	0.19	0.03
B'	0.47	0.31

32) a) 0.05

32) b) 0.75

32) c) 0.65

32) d) 0.25

32) e) 0.45

33)
2.5% of the onions bought from farm B are spoilt.

34)
95% of the students from class C are right-handed.

35)
$p = \frac{1}{38}$ and $q = \frac{17}{114}$

36)
$r = 0.1, s = 0.08, t = 0.06$

37) a) $\frac{1}{3}$

37) b) $\frac{2}{3}$

37) c) $\frac{29}{60}$

37) d) $\frac{31}{60}$

38) a) $\frac{1}{60}$

38) b) $\frac{3}{50}$

38) c) $\frac{143}{300}$

39) a)

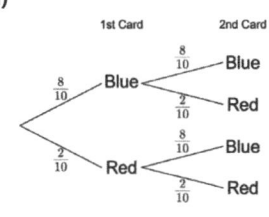

Tree diagram: 1st Card — Blue ($\frac{8}{10}$) / Red ($\frac{2}{10}$); 2nd Card from Blue — Blue ($\frac{8}{10}$) / Red ($\frac{2}{10}$); from Red — Blue ($\frac{8}{10}$) / Red ($\frac{2}{10}$).

39) b) $\frac{17}{25}$

40) a)

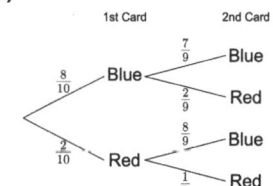

Tree diagram: 1st Card — Blue ($\frac{8}{10}$) / Red ($\frac{2}{10}$); 2nd Card from Blue — Blue ($\frac{7}{9}$) / Red ($\frac{2}{9}$); from Red — Blue ($\frac{8}{9}$) / Red ($\frac{1}{9}$).

40) b) $\frac{29}{45}$

41) a) $\frac{1}{24}$

41) b) $\frac{29}{72}$

41) c) $\frac{3}{4}$

42) a)

Tree diagram: Rain (0.2) — Before (0.75) / After (0.25); No Rain (0.8) — Before (0.95) / After (0.05).

42) b) 0.09

42) c) 0.624 (3 SF)

43) a) 0.0367 (3 SF)

43) b) 0.0670 (3 SF)

44) a) 0.135 (3 SF)

44) b) 0.089 (3 SF)

45) a) $\frac{1}{720}$

45) b) $\frac{1}{30}$

45) c) $\frac{1}{30}$

45) d) $\frac{1}{120}$

46) a) $\frac{1}{56}$

46) b) $\frac{1}{56}$

46) c) $\frac{1}{56}$

47) a) $\frac{2}{5}$

47) b) $\frac{4}{45}$

47) c)
$\frac{1}{210}$

47) d)
$\frac{17}{45}$

47) e)
$\frac{136}{315} = 0.432$ (3 SF)

47) f)
$\frac{179}{315} = 0.568$ (3 SF)

48)
$\frac{1}{2}$

49)
$\frac{2}{3}$

50) a)
$\frac{31}{81}$

50) b)
$\frac{59}{76}$

51)
$k = 18$

52)
$\frac{31}{78}$

53)
$\frac{4}{33}$

54) a)

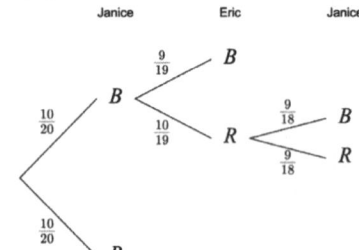

Janice — Eric — Janice

Tree diagram with branches: $\frac{10}{20}$ to B, then $\frac{10}{19}$ to R and $\frac{9}{19}$ to B; from R: $\frac{9}{18}$ to B and $\frac{9}{18}$ to R; $\frac{10}{20}$ to R.

54) b)
$\frac{6}{323}$

55) a)
$\frac{225}{1081}$

55) b)
$\frac{285}{4324}$

55) c)
$\frac{255}{1081}$

55) d)
$\frac{765}{8648}$

56) a)
0.809 (3 SF)

56) b)
0.0703 (3 SF)

57) a)
0.137 (3 SF)

57) b)
0.410 (3 SF)

1.53 Probability: Regions of Venn Diagrams
Questions on page 40

1) a)
A

1) b)
$A \cap B$

1) c)
B'

1) d)
$A \cap B'$

1) e)
$(A \cup B)'$ or $A' \cap B'$

2) a)
$B \cap C$

2) b)
$B \cup C$

2) c)
$B \cap C'$

2) d)
$A' \cap C'$ or $(A \cup C)'$

2) e)
$(B \cup C)'$ or $B' \cap C'$

2) f)
$A' \cap B \cap C'$

2) g)
$B \cup C$

2) h)
$B \cap C$

2) i)
$A \cap B \cap C$

2) j)
$(A \cup B \cup C)'$ or $A' \cap B' \cap C'$

2) k)
$A \cap B'$

2) l)
$A' \cap B \cap C'$

2) m)
$A \cap B' \cap C$

2) n)
$(A \cup B) \cap C$

2) o)
$A \cap (B \cup C)$

2) p)
$(A \cup C) \cap B$

2) q)
$(A \cap C) \cup ((A \cup C) \cap B)$

2) r)
$(A \cap B) \cup (A' \cap B \cap C')$

2) s)
$A \cap B \cap C'$

2) t)
$(A \cap B \cap C') \cup (A' \cap B' \cap C)$

2) u)
$(A \cap B' \cap C') \cup (A' \cap B \cap C') \cup (A' \cap B' \cap C)$

1.54 Probability: Independence and Mutually Exclusive
Questions on page 162

1)
A and C, A and D, B and D

2)

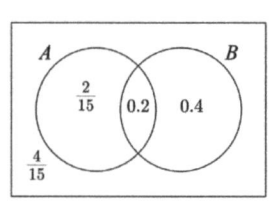

Venn diagram with rectangle containing two overlapping circles A and B; A only: $\frac{2}{15}$; intersection: 0.2; B only: 0.4; outside: $\frac{4}{15}$.

3)
$0.95 = 0.4 + 0.55 - P(A \cap B)$
$\Rightarrow P(A \cap B) = 0$
$\therefore A$ and B are mutually exclusive events

4)
$0.7 = 0.32 + 0.46 - P(A \cap B)$
$\Rightarrow P(A \cap B) = 0.08 \neq 0$
$\therefore A$ and B are not mutually exclusive events

5)
$P(A) \times P(B) = \frac{3}{7} \times \frac{2}{3} = \frac{2}{7}$, $P(A \cap B) = \frac{2}{7}$
$P(A \cap B) = P(A) \times P(B)$ so A and B are independent events

6)
$P(A) \times P(B) = \frac{3}{10} \times \frac{5}{6} = \frac{1}{4}$, $P(A \cap B) = \frac{1}{2}$
$P(A \cap B) \neq P(A) \times P(B)$ so A and B are not independent events

7)
$P(A) \times P(B) = 0.45 \times 0.52 = 0.234$
$P(A \cap B) = 0.23$
$P(A \cap B) \neq P(A) \times P(B)$ so A and B are not independent events

8)
$P(\text{Bathroom}) \times P(\text{Balcony})$
$= \frac{16}{30} \times \frac{8}{30} = \frac{32}{225} = 0.142\ldots$
$P(\text{Bathroom} \cap \text{Balcony}) = \frac{6}{30} = 0.2$
$P(\text{Bathroom} \cap \text{Balcony}) \neq P(\text{Bathroom}) \times P(\text{Balcony})$ so A and B are not independent events

9)
$P(B) = \frac{9}{14}$

10)
$P(A \cup B) = \frac{189}{190}$

11)
$P(A) \times P(B) = \frac{33}{56} \times \frac{4}{7} = \frac{33}{98} = 0.336\ldots$
$P(A \cap B) = \frac{3}{10} = 0.3$
$P(A \cap B) \neq P(A) \times P(B)$ so A and B are not independent events

12)
$k = \frac{1}{8}$, $p = \frac{33}{56}$

13)
$x = 0.1, y = 0.2, z = 0$
or $x = 0, y = 0.3, z = 0$

14)
$P(A) = \frac{1}{4}$ and $P(B) = \frac{2}{3}$
or $P(A) = \frac{2}{3}$ and $P(B) = \frac{1}{4}$

15)

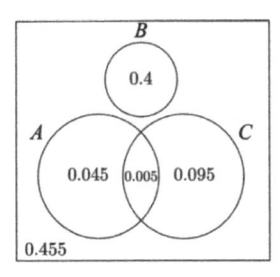

Venn diagram: B contains 0.4; A and C intersect with A=0.045, A∩C=0.005, C=0.095; outside 0.455

16)

Venn diagram: A=0.045, A∩B=0.005, B=0.055, B∩C=0.04, C=0.36; outside 0.495

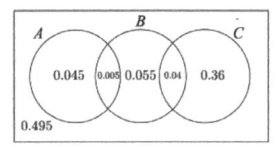

17)
$P(A) \times P(B) = 0.36 \times 0.145$
$= 0.0522 = P(A \cap B)$
$\therefore A$ and B are independent events.

18) a)
$P(B) = \frac{24}{75}$

18) b)
$P(A \cup B) = \frac{62}{75}$

18) c)
$P(A \cap B') = \frac{38}{75}$

18) d)
$P(A) \times P(B) = \frac{52}{75} \times \frac{24}{75} = \frac{416}{1875} \neq \frac{14}{75} = P(A \cap B)$
so A and B are not independent events

19)
$P(A' \cap B) \times P(C) = 0.16 \times 0.25$
$= 0.04 = P((A' \cap B) \cap C)$
so $A' \cap B$ and C are independent events

20)
$k = 0.8$

1.55 Discrete Random Variables
Questions on page 164

1)
0.72

2)

x	2	4	6	8
$P(X=x)$	$\frac{11}{50}$	$\frac{6}{25}$	$\frac{13}{50}$	$\frac{7}{25}$

3)
$\frac{3}{5}$

4)
$k = 0.18$

5)
$k = \frac{1}{11}$

6) a)
Show that

6) b)
$\frac{3}{10}$

7)
$k = \frac{97}{13}$

8)
0.65

9)
0.2025

10)
0.5502

11)
0.0909

12)
0.4864

13)
0.34

14)
0.03

15)
0.5511

16)
$\frac{11}{40}$

17)
$\frac{12}{343}$

18)

x	1	2	4	5
$P(X=x)$	0.2	0.3	0.2	0.3

19)

x	4	5	9	10
$P(X=x)$	0.32	0.18	0.32	0.18

20)

x	2	3	4	5	6
$P(X=x)$	0.11	$\frac{67}{300}$	$\frac{1}{3}$	0.11	$\frac{67}{300}$

21) a)
$\frac{2}{15}$

21) b)
$k = \frac{1}{3}$

21) c)
$\frac{7}{15}$

22) a)
$4a$

22) b)
$a = \frac{1}{9}$ and $b = \frac{31}{630}$

23) a)
Show that

23) b)
$a = \frac{1}{36}$ and $b = \frac{1}{66}$

24)
$k = \frac{1}{5}$ and $P(X > 1) = \frac{22}{25}$

25)
Show that

26)
$k = \frac{125}{281}$ and $P(X \leq 3) = \frac{31}{281}$

27)
0.0567

28)
$\frac{1}{6}$

29)
0.89

30)
0.3

31) a)
$\frac{49}{300}$

31) b)
$\frac{29}{4500}$

31) c)
$\frac{1}{30}$

32)
0.75

33)
$\frac{275}{343}$

34)
$\frac{73}{324}$

35)
$\frac{337}{675}$

36)
$\frac{1}{5}$

37)
$\frac{1}{n}$

38)
0.45

39) a)
$\frac{19}{81}$

39) b)
$\frac{10}{19}$

40) a)
$\frac{16}{675}$
40) b)
$\frac{3}{5}$

1.56 The Binomial Distribution
Questions on page 168

1)
$X \sim B(6, 0.2)$ is diagram B
$X \sim B(6, 0.5)$ is diagram C
$X \sim B(6, 0.7)$ is diagram A
$X \sim B(6, 0.89)$ is diagram D

2) a)
Positive skew
2) b)
Symmetric
2) c)
Negative skew
2) d)
Negative skew

3)

x	0	1	2	3
$P(X=x)$	0.512	0.384	0.096	0.008

4)

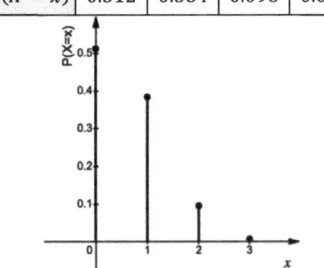

x	0	1	2	3
$P(X=x)$	0.064	0.288	0.432	0.216

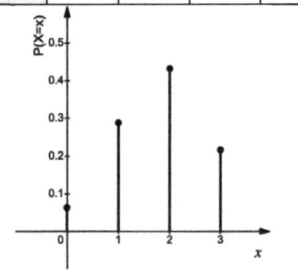

5)

x	0	1	2	3	4
$P(X=x)$	0.0625	0.25	0.375	0.25	0.0625

6)
$X \sim B(10, 0.5)$

7) a)
0.1700 (4 DP)
7) b)
0.2186 (4 DP)
7) c)
0.2061 (4 DP)
7) d)
0.0116 (4 DP)

8) a)
0.2508 (4 DP)
8) b)
0.1867 (4 DP)
8) c)
0.0844 (4 DP)
8) d)
0.0320 (4 DP)

9) a)
0.1029 (4 DP)
9) b)
0.2668 (4 DP)
9) c)
0.0282 (4 DP)

10) a)
0.0576 (4 DP)
10) b)
0.2054 (4 DP)
10) c)
0.1746 (4 DP)

11) a)
0.2187 (4 DP)
11) b)
0.4734 (4 DP)
11) c)
0.7159 (4 DP)

12) a)
0.0061 (4 DP)
12) b)
0.1629 (4 DP)
12) c)
0.6466 (4 DP)

13) a)
0.7512 (4 DP)
13) b)
0.9884 (4 DP)
13) c)
0.6720 (4 DP)
13) d)
0.1713 (4 DP)

14) a)
0.0329 (4 DP)
14) b)
0.1878 (4 DP)
14) c)
0.7309 (4 DP)

15) a)
0.6552 (4 DP)
15) b)
0.1462 (4 DP)
15) c)
0.0611 (4 DP)

16) a)
0.8982 (4 DP)
16) b)
0.1760 (4 DP)
16) c)
0.8158 (4 DP)

17) a)
0.1181 (4 DP)
17) b)
0.3669 (4 DP)
17) c)
0.3190 (4 DP)

18) a)
0.1839 (4 DP)
18) b)
0.6133 (4 DP)
18) c)
0.4405 (4 DP)

19) a)
0.4057 (4 DP)
19) b)
0.6261 (4 DP)
19) c)
0.2243 (4 DP)

20) a)
0.1989 (4 DP)
20) b)
0.5529 (4 DP)
20) c)
0.2864 (4 DP)

21)
0.1723 (4 DP)

22)
0.7838 (4 DP)

23) a)
$E(X) = 4$, $Var(X) = 2$
23) b)
$E(X) = 6$, $Var(X) = 4.2$
23) c)
$E(X) = 14.4$, $Var(X) = 1.44$
23) d)
$E(X) = 5$, $Var(X) = 4$

23) e)
E(X) = 24, Var(X) = 9.6

24) a)
0.5460 (4 DP)

24) b)
0.4540 (4 DP)

25)
E(X) = 11.25, Var(X) = 8.4375

26)
$p = \frac{2}{9}$

27)
$p = \frac{1}{13}$

28)
$n = 200$

29)
$n = 2700$

30)
$n = 11200$

31)
$n = 47040$

32)
$n = 28, p = 0.9$,

33)
$n = 36, p = 0.4$

34)
$n = 65, p = \frac{1}{13}$

35)
$k = 0.8$

36)
$p = \frac{1}{3}$

37)
$p = \frac{14}{15}$

38)
There are fixed number of independent trials. There are exactly two outcomes of each trial, with the probability of success (or failure) being constant.

39)
There is no fixed number of trials, n.

40)
The sixteen days are consecutive, so this is not a random sample of days throughout the year. If it rains on one day, then this will affect the likelihood of it raining on the next, so the trials are not independent.

41) a)
0.4616 (4 DP)

41) b)
0.1625 (4 DP)

42) a)
Var(X) = 3.828125

42) b)
0.8605 (4 DP)

42) c)
0.3855 (4 DP)

43) a)
E(X) = 5

43) b)
Var(X) = 4.5

43) c)
0.1849 (4 DP)

43) d)
0.7702 (4 DP)

43) e)
0.1608 (4 DP)

44) a)
0.2642 (4 DP)

44) b)
0.4291 (4 DP)

44) c)
Curtis either hits the bullseye or does not hit the bullseye. The event that Curtis hits the bullseye is independent of every other time Curtis hits the bullseye. The probability of hitting the bullseye remains constant at 0.15.

45) a)
0.1644 (4 DP)

45) b)
The probability of having a graphical calculator in a single class may not be constant at 64% as students may influence each other's choice to purchase one.

46) a)
0.4679 (4 DP)

46) b)
0.0384 (4 DP)

46) c)
A tile being faulty is independent of any other tile being faulty. The tiles are either faulty or they are not faulty. The probability of each tile being faulty is fixed at 0.025.

47) a)
0.1901 (4 DP)

47) b)
0.8296 (4 DP)

47) c)
0.036

47) d)
0.3446 (4 DP)

48)
Show that

49) a)
0.0441 (4 DP)

49) b)
0.9648 (4 DP)

50) a)
0.1117 (4 DP)

50) b)
0.3103 (4 DP)

50) c)
0.2949 (4 DP)

51) a)
0.3685 (4 DP)

51) b)
0.4278 (4 DP)

51) c)
0.0166 (4 DP)

51) d)
0.1887 (4 DP)

52)
$k = 30$

53)
$k = 65$

54)
$k = 19$

55)
$k = 156$

56)
$n = 74$

57)
$n = 116$

58)
$n = 643$

59)
$n = 158$

60)
$n = 89$

61)
$n = 509$

62)
44

63)
321

1.57 Binomial Hypothesis Testing
Questions on page 173

1)
The significance level is the probability of rejecting the null hypothesis when in fact it is true.

2) a)
$P(X \leq 10) = 0.0480 < 0.05$
Reject H_0
2) b)
$P(X \leq 5) = 0.0766 > 0.05$
Fail to reject H_0
2) c)
$P(X \leq 12) = 0.0743 < 0.1$
Reject H_0
2) d)
$P(X \leq 40) = 0.0045 < 0.01$
Reject H_0

3) a)
$P(X \geq 9) = 0.0950 > 0.05$
Fail to reject H_0
3) b)
$P(X \geq 41) = 0.1748 > 0.05$
Fail to reject H_0
3) c)
$P(X \geq 26) = 0.0263 < 0.1$
Reject H_0
3) d)
$P(X \geq 18) = 0.0199 > 0.01$
Fail to reject H_0

4) a)
p-value $= 0.0173 \ldots < 0.05$
Reject H_0
4) b)
p-value $= 0.0265 > 0.01$
Fail to reject H_0
4) c)
p-value $= 0.0791 < 0.1$
Reject H_0
4) d)
p-value $= 0.0382 > 0.01$
Fail to reject H_0

5) a)
The critical region is the range of values of X that will reject the null hypothesis.
5) b)
The acceptance region is the range of values of X that will fail to reject the null hypothesis.

6) a)
CR is $\{0 \leq X \leq 5\}$
Actual Significance Level = 4.81%
6) b)
CR is $\{0 \leq X \leq 15\}$
Actual Significance Level = 3.94%
6) c)
CR is $\{0 \leq X \leq 6\}$
Actual Significance Level = 7.38%
6) d)
CR is $\{0 \leq X \leq 48\}$
Actual Significance Level = 0.70%

7) a)
CR is $\{13 \leq X \leq 40\}$
Actual Significance Level = 4.32%
7) b)
CR is $\{23 \leq X \leq 49\}$
Actual Significance Level = 2.95%
7) c)
CR is $\{11 \leq X \leq 49\}$
Actual Significance Level = 9.69%
7) d)
CR is $\{105 \leq X \leq 125\}$
Actual Significance Level = 0.54%

8) a)
CR is $\{0 \leq X \leq 16\} \cup \{33 \leq X \leq 60\}$
Actual Significance Level = 3.54%
8) b)
CR is $\{0 \leq X \leq 34\} \cup \{X = 44\}$
Actual Significance Level = 2.00%
8) c)
CR is $\{0 \leq X \leq 9\} \cup \{22 \leq X \leq 55\}$
Actual Significance Level = 7.00%
8) d)
CR is $\{0 \leq X \leq 3\} \cup \{22 \leq X \leq 135\}$
Actual Significance Level = 0.611%

9) a)
Firstly, the alternative hypothesis should be $H_1: p < 0.25$ and not $H_1: p \leq 0.25$ (alternative hypothesis should be a strict inequality). Secondly, the p-value to be found should be $P(X \leq 5)$ and not $P(X = 5)$ (it should be a cumulative probability).
9) b)
Let X be the number customers purchasing a chrysanthemum bouquet.
$H_0: p = 0.25$
$H_1: p < 0.25$
$X \sim B(34, 0.25)$
$P(X \leq 5) = 0.1138$
$0.1138 > 0.1$ so we fail to reject H_0
There is insufficient evidence to suggest that the proportion of customers purchasing a chrysanthemum bouquet has decreased.

10)
$0.0250 < 0.05$ so we reject H_0
There is sufficient evidence to suggest that the proportion of games won by Chaisai has increased.

11)
$0.0593 > 0.05$ so we fail to reject H_0
There is insufficient evidence to suggest that the proportion of adults who met the fruit recommendation of 2 servings a day is less than 44.1%.

12)
$0.0029 < 0.01$ so we reject H_0
There is sufficient evidence to suggest that the proportion of unemployed adults in Brighton is lower than the UK average.

13)
$0.0229 < 0.05$ so we reject H_0
There is sufficient evidence to suggest that the proportion of batteries lasting longer than 20 hours has increased.

14)
$0.0325 > 0.025$ so we fail reject H_0
There is insufficient evidence to suggest that the proportion of TV remotes that have a fault has changed.

15)
$0.0019 < 0.015$ so we reject H_0
There is sufficient evidence to suggest that the proportion of roses that suffer from powdery mildew has changed.

16) a)
$H_0: p = \frac{1}{6}$, $H_1: p > \frac{1}{6}$
16) b)
CR is $\{10 \leq X \leq 35\}$
16) c)
7 is not in the critical region, so we fail to reject H_0. There is insufficient evidence to suggest that the proportion of times the die rolls a 1 is greater than $\frac{1}{6}$.

17) a)
$H_0: p = 0.76$
$H_1: p < 0.76$
17) b)
CR is $\{0 \leq X \leq 44\}$
17) c)
43 is in the critical region, so we reject H_0
There is sufficient evidence to suggest that the proportion of patients who wait longer than 3 hours has reduced.

18) a)
$H_0: p = 0.32$, $H_1: p \neq 0.32$
18) b)
CR is $\{0 \leq X \leq 17\} \cup \{35 \leq X \leq 80\}$
18) c)
4.14%

18) d)
45 is in the critical region, so we reject H_0
There is sufficient evidence to suggest that the proportion of customers purchasing single cans of soup has changed.

19) a)
$H_0: p = 0.05$, $H_1: p < 0.05$
19) b)
CR is $\{0 \leq X \leq 20\}$
19) c)
13 is in the critical region, so we reject H_0
There is sufficient evidence to suggest that the proportion of spoilt bags of grain from Baker Farms is less than that of Aaron Farms.

20) a) i)
0.2704
20) a) ii)
0.6768
20) b) i)
Let X be the number of gold dinosaurs.
$H_0: p = 0.08$, $H_1: p > 0.08$
20) b) ii)
CR is $\{6 \leq X \leq 30\}$
20) b) iii)
4 is not in the critical region, so we fail to reject H_0
There is insufficient evidence to suggest that the proportion of gold dinosaurs found is higher when shopping through the website.

1.58 Large Data Set: Edexcel
Questions on page 176

1) a)
Camborne, Heathrow, Hurn, Leeming and Leuchars
1) b)
Beijing, Jacksonville and Perth
1) c)
1987 and 2015
1) d)
May to October inclusive – which is 184 days

2) a)
Camborne, Hurn, Leuchars, Jacksonville and Perth
2) b)
Perth

3)
Perth is experiencing winter months due to being in the southern hemisphere

4)
Leuchars

5) a)
°C

5) b)
mm
5) c)
Hours
5) d)
None (it's a %)
5) e)
Knots or the Beaufort Scale
5) f)
Oktas (number out of 8)
5) g)
decametres
5) h)
hPa (hectopascals)

6) a)
3.45 mph
6) b)
5 knots
6) c)
26.1 knots

7) a)
$\frac{1}{8}$ of the sky has could coverage
7) b)
$\frac{1}{2}$ of the sky has cloud coverage
7) c)
There is full cloud coverage

8)
Note that: The daily mean wind direction is rounded to the nearest 10 degrees. The original data point can be found in the Large Data Set. WNW

9)
The cloud coverage is negatively skewed across the data set so it is more likely to be high cloud coverage.

10)
She is ignoring all the data points with the letters "tr" which represents a small amount of rain, missing 6 days which she should have put as 0 mm of rainfall.
By including those, the mean and the standard deviation will both go down.

11) a)
$\bar{x} = 5.18$ hours, $s_x = 4.11$
11) b)
The average amount of hours of daily sunshine has gone down since 1987 but the spread of the hours has roughly stayed the same (there has been a slight decrease in spread).

12) a)
September & October due to the coldest days
12) b)
Invalid claim as it is only a comparison of two months and only in one location.

13) a)
Systematic sampling
13) b)
He did not choose a random start point, between 1 and 6, he started with the first date.

14) a)
$P(c = 1) = 0.01$ and $P(c = 8) = 0.1$
14) b)
26 days
14) c)
25 days
14) d)
For this distribution there are more than the 0 to 8 oktas,
$P(X \geq 9) = 1 - 0.8597 = 0.1403$
meaning that around 26 days are 9 or more Oktas which does not exist!

15) a)
6.926 knots
15) b)
There is a high correlation coefficient, and whilst there should be a positive correlation, the magnitude of PMCC value of this data set is too strong.

16) a)
Mean is 8.162
Standard deviation is 2.47 (3 SF)
16) b)
16 days
16) c)
0.0601
16) d)
For $X \sim N(8.162, 2.47^2)$, $2.47 \times 3 = 7.41$
So distribution lies approximately between 0.752 and 15.572 which is reasonable. The mean is 8.162 and the median is ≈ 7.95, and the mode ≈ 8 which are similar to each other, this indicates the symmetry needed for a normal distribution.

17) a)
$-0.3613 > -0.3783$ so we fail to reject H_0. There is insufficient evidence to suggest that there is a negative correlation between the windspeed in Leeming on a given day and the relative humidity in Leeming on that same day.
17) b)
$-0.6234 < -0.3783$ so we reject H_0. There is sufficient evidence to suggest that there is a negative correlation between the windspeed in Leeming on a given day and the relative humidity in Leeming on that same day.

1.59 Large Data Set: AQA
Questions on page 178

1) a)
BMW, Ford, Toyota, Vauxhall, Volkswagen
1) b)
2002, 2016
1) c)
London, North West, South West
1) d)
2002, 2016
1) e)
Petrol, Diesel, Electric, Gas/Petrol, Electric/Petrol

1) f)
2 door saloon, 4 door saloon, Saloon, Convertible, Coupe, Estate, 3 Door Hatchback, 5 Door Hatchback, Multi Purpose Vehicle

2) a)
g/km

2) b)
g/km

2) c)
g/km

2) d)
g/km

2) e)
g/km

2) f)
kg

2) g)
cubic cm

3)
There is data missing from the LDS for carbon monoxide, nox, particulate and hydrocarbon emissions.

4)
Mass: 0 kg is a mistake as every car has a mass as it should include the 75 kg for driver. Particulate emissions: The LDS does have blanks in particulate emissions where it has not been recorded but there are no 0. Engine size: There is one car with a 0 for engine size in the LDS

5) a)
Systematic sampling

5) b)
Quota sampling

5) c)
Stratified sampling

6) a)
Incorrect as 75kg is included in the mass, not the mean mass of the population at the time

6) b)
Incorrect as the LDS doesn't include:
All car manufacturers
All regions of the UK
Vehicles other than cars

7) a)
Mean is 1405.4 cm^3,
Sample standard deviation is 334.4 cm^3
(Standard deviation is 317.2 cm^3)

7) b)
There are significantly more cars in the sample from London than expected as London is the smallest number of cars by region in the LDS (998 from London, 1164 from the North West and 1665 from the South West)

7) c)
The BMW with engine size 2171cm^3 is an outlier.

7) d)
The engine size is larger on average in the whole data set than the sample.

7) e)
The engine size is more spread out on average in the whole data set than the sample.

8) a)
2002: 153 g/km
2016: 116 g/km

8) b)
124 g/km

8) c)
There's only one electric vehicle in the large data set, and its CO2 emissions are zero.

9) a)
Total of 50 cars were sampled.

9) b)
1414.14 kg

9) c)
$\frac{33}{50} = 0.66$

10) a)
$k = \frac{26}{125}$

10) b)
Male: 100
Female: 52
Unknown: 8
Company: 90

10) c)
The proportions of Male / female / unknown and company are very similar to that of the large data set.

11)
Engine sizes vary between 647 and 4951 so the data lies within this, but the middle 50% of the Large Data Set lies between 1395 and 1968. The box plot has a much lower middle 50% so it is likely that there has been a higher selection of lower engines selected which indicates it may not have been sampled randomly.

12) a)
Mean: 139.8 g/km
Standard deviation: 23.25 g/km

12) b)
0.3304

12) c)
0

1.60 Large Data Set: OCR MEI (Health)
Questions on page 180

1) a)
17 years old

1) b)
85 years old

1) c)
Divorced, living with partner, married, never married, separated, widowed.

1) d)
Food30 records whether they have eaten within the last 30 minutes

1) e)
Which arm blood pressure is taken on

1) f)
USA

2) a)
kg

2) b)
cm

2) c)
No units

2) d)
cm

2) e)
Beats per second

2) f)
mm Hg

3)
The National Health and Nutrition Examination Survey in the united states.

4)
The systolic blood pressure is the pressure at the time when the heart beats. The diastolic blood pressure is the pressure between heart beats.

5)
Interviewers are instructed to get three measurements of blood pressure (with at least one minute between readings). This might take up to four attempts. They are instructed not to take more than four attempts to get the blood pressure readings.

6)
The average of the 3 readings for systolic and diastolic blood pressure.

7) a)
Simple random sampling

7) b)
There are missing data points in the spreadsheet labelled with a #N/A which might mean that he cannot use that line of data and therefore he ends up with less than 20 pieces of data.

8)
Any person who has a detail missing in the weight, height, BMI, thigh length, upper arm length, waist, Food30, arm, pulse would no longer be available to sample from. Furthermore, she has actively decide to only include people who had their blood pressure readings measured in the first 3 attempts out of 4.

9) a)
She should use stratified sampling.
9) b)
She should remove any lines of data that contain a #N/A in the BMI column.
She would need to find the total number in each of the categories n then use the formula $\frac{n}{200-\text{number of any removed lines}} \times 50$ to find out how many to sample from the category. She could then randomly sample the amount she needs from the large data set, using either simple random sampling or systematic sampling, dismissing.

10) a)
Mean is 177.53 cm
Sample standard deviation is 8.81 cm
(Standard deviation is 8.36 cm)
10b)
Whilst there are slightly more males than females, the sample of heights is consistently high except for one shorter male!

10) c)
The male with height 159.4 cm is an outlier.
10) d)
The height is larger on average in the sample than in whole data set.
10) e)
The height is less spread out in the sample than in the whole data set.

11) a)
$t = 14.2$ cm
11) b)
The data in the diagram is close to a straight line, and the PMCC of 0.7518 is strong (positive) correlation.

12) a)
They are both approximately 120 mm Hg
12) b)
Male: ~17.5 mm Hg
Female: ~30 mm Hg
12) c)
Males and females have the same systolic blood pressure on average, but the systolic blood pressure of males is more consistent than females.

13) a)
The median is in the age 35 to 45 class interval.
13) b)
$\bar{x} = 46$
13) c)
0.531

14) a)
$0.8159 > 0.3783$ so we reject H_0
There is sufficient evidence to suggest that there is a positive correlation between average systolic blood pressure.
14) b)
The data on the teacher's graph is distributed across the whole area, which means that it is a closer to zero corelation, so the only way that such a high correlation could have been found is if the data points were picked to be close to a straight line.

1.60 Large Data Set: OCR MEI (World Bank)
Questions on page 182

1) a)
USA
1) b)
North Africa, South Africa, Central America, North America, South America, Asia. Asia/Europe, Caribbean, Europe. Oceania
1) c)
It is based on statistics from population censuses, vital statistics registration systems, or sample surveys pertaining to the recent past and on assumptions about future trends.
1) d)
Square kilometres

2)
It is the average number of births (or deaths) per year per 1000 population at midyear.

3) a)
GDP is a measure of all the goods and services produced within a nation in a year.
3) b)
It is the purchasing power parity basis divided by population in 1 July for that year. Purchasing power measures the GDP in dollars based on comparing prices between the country and the US rather than using the official exchange rate between currencies.

4)
the number of medical doctors (physicians), including generalist and specialist medical practitioners, per 1000 of the population. Medical doctors are defined as doctors that study, diagnose, treat, and prevent illness, disease, injury, and other physical and mental impairments in humans through the application of modern medicine. They also plan, supervise, and evaluate care and treatment plans by other health care providers.

5) a)
physicians, nurses, and midwives only
5) b)
The World Health Organization estimates that fewer than 2.3 health workers per 1,000 would be insufficient to achieve coverage of primary healthcare needs.
5) c)
Albania, Bosnia and Herzegovina or Cyprus
5) d)
North Africa

6)
Current Health Expenditure (CHE) describes the share of spending on health in each country relative to the size of its economy. It includes expenditures corresponding to the final consumption of health care goods and services and excludes investment, exports, and intermediate consumption. CHE shows the importance of the health sector in the economy and indicates the priority given to health in monetary terms.

7) a)
Sometimes the number in 'mobile phone subscribers per 100 population' is greater than 100 which shows that there are countries in which people have more than one phone per person on average.
7) b)
Oceania
7) c)
There are 236 countries in total so he approximately wants a $\frac{1}{5}$th of the data set.
He should randomly sample 1 per 5 countries in each of the regions.
Eg. Select randomly the number below from each of the regions
North Africa: 1, South Africa: 11, Central America: 1, North America: 1, South America: 2, Asia: 10, Asia/Europe: 0, Caribbean: 5, Europe: 10, Oceania: 5

8) a)
No
8) b)
Yes – Lesotho

9) a)
Kosovo and Monaco are the outliers in the Europe boxplot
Japan is the outlier in the Asia boxplot

9) b)
The median age is not available for the Holy See (Vatican City) and this is identifiable with a #N/A

9) c)
The median age of a person in Europe is higher on average and more consistent.

9) d)
Russia is categorised as Asia/Europe

10) a)
Mean is 5.88 %
Sample standard deviation is 1.81 %
(Standard deviation is 1.74 %)

10) b)
Cuba with CHE 11.3 % is an outlier.

10) c)
Countries on average spend more per capita in Oceania then they do in the Caribbean, but there is less consistency in this in Oceania.

10) d)
CHE is a measure of what % of the countries expenditure is
Both regions contain lots of smaller islands so may face similar challenges in providing healthcare.

10) e)
Oceania includes Australia and New Zealand which are commonly known as first world countries due to being highly industrialized, possessing a low poverty rate, and/or high accessibility to modern resources and infrastructure. Countries such as these tend to have higher CHE. If you were wanting to only compare small islands, you may want to choose to not include higher population countries.

11) a)
Botswana: 57.135
Chilie: 69.15

11) b)
Botswana: 50.1 is the life expectancy of someone in 1960. 2.01 is the rate at which life expectancy increases per decade. Chile: 55.5 is the life expectancy of someone in 1960. 3.90 is the rate at which life expectancy increases per decade.

11) c)
According to the models, Chile has a higher life expectancy in 1960 and has a higher rate of improvement in life expectancy than Botswana.

12) a)
The value of r indicates a strong positive linear relationship. However, median age will be limited by the age humans can reach, so it cannot be linear for higher values of physician density.

12) b)
The right most point, identified with a circle below:

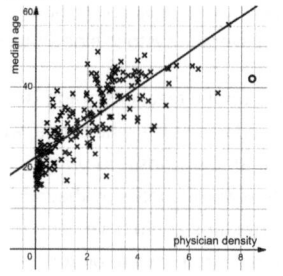

12) c)
$y = 53.19$ years

12) d)
Whilst it is interpolation (just) having a median age of 53.19 is high, which indicates a much older population. Therefore it is unlikely that physician density is the only factor they need to change to have a sustainable older median age.

13) a)
0.7157 (4 DP)

13) b)
0.9545 (4 DP)

13) c)
10.9 (3 SF)

1.62 Large Data Set: OCR MEI (London)
Questions on page 184

1) a)
32 plus the City of London

1) b)
9 areas: North East, North West, Yorkshire and The Humber, East Midlands, West Midlands, East of England, London, South East and South West

1) c)
The definitions of Inner and Outer London are those used by the Office of National Statistics and the Census.

2) a)
The percentage of people employed in the region with the age 16 to 64

2) b)
The amount of personal income per year.

2) c)
The percentage of all pupils in state schools at the end of key stage 4 who achieve five GCSEs at grade A* to C, including mathematics and English or better.

2) d)
The percentage of collected household waste which is recycled or composted.

3)
Houses which are sold for less than market value (such as right to buy houses).

4)
This data is estimated based on the Survey of Personal Incomes (SPI) which is an annual sample survey of HMRC records for individuals who could be liable to UK Income Tax.

5) a)
Simple random sampling

5) b)
Systematic sampling

5) c)
Convenience/opportunity sampling

5) d)
Stratified sampling (with simple random sampling).

6) a)
Mean is £686 867.80
Sample standard deviation is £221 761.90
(Standard deviation is £210 381.81)

6) b)
Whilst there are a reasonable amount of areas selected from inner and outer London, the areas are predominantly from the top half of the data within Inner or Outer London (i.e. more expensive areas).

6) c)
There are no areas with median house price which are outliers.

6) d)
The median house price is much higher on average and more spread out in 2021 than in 2007.

7)
In both cases the median is less than the mean, which indicates a positive skew (more people with lower income). Kensington and Chelsea has a highly skewed income distribution with significant income inequality, characterised by a few extremely high earners that raise the average income far above the median. Lewisham has a more balanced income distribution with less income inequality and a median income closer to the mean.

8) a)
The top graph is Hackney.

8) b)
Hackney: £49 800
North West of England: £33062.50

8) c)
Hackney: 26.3 ⇒ £26 300 is the mean income of someone in 2009/2010 . 1.88 is the rate at which mean income increases per year. North West of England: 23.3 ⇒ £23 300 is the mean income of someone in 2009/2010 . 0.781 is the rate at which mean income increases per year.

8) d)
According to the models, Hackney has a higher mean income in 2009/10 and has a higher rate of improvement in mean income than the North West of England.

9) a)
The City of London

9) b)
On average, outer London schools achieve the highest rates of percentage of all pupils getting 5 or more A*- C grades, with the rest of England performing slightly higher on average than Inner London. The rates of percentage of all pupils getting 5 or more A*- C grades are least consistent in Outer London, with similar consistency for Inner London and the Rest of England.

10) a)
The value of r indicates a moderate linear relationship between the recycling rate and the employment rate in 2011.

10) b)
38.9 % (3 SF)

10) c)
It would lower the value of r
The gradient of the regression line would be less steep and the y-intercept would translate upwards.

11) a)
0.7580 (4 DP)

11) b)
0.0139 (4 DP)

11) c)
0.9545 (4 DP)

1.63 Large Data Set: OCR A
Questions on page 186

1) a)
Work from home; Underground, metro, light rail, tram; Train; Bus, minibus or coach; Motorcycle, scooter or moped; Driving a car or van; Passenger in a car or van; Taxi; Bicycle; On foot

1) b)
Not in employment

1) c)
2001 and 2011

2)
New unitary authorities were made in 2009, meaning that when creating the data set for 2011 some of the data was summed to create the new unitary authorities.

3)
People who predominantly drive (which is likely to be the case for those using park and ride) will be in the "Driving a car or van" or "Passenger in a car or van" category.

4)
Not in employment as the Census was based on the week before the Census.

5)
Exeter has a big university which explains a noticeably larger number of people aged 20-24.

6) a)
$\bar{x} = 53.2$ (3 SF), $s_x = 51.9$ (3 SF)

6) b)
He didn't include the Isles of Scilly which should have 0 in the large data set rather than the dash.
$\bar{x} = 51.8$ (3 SF), $s_x = 51.9$ (3 SF)

7)
A baby less than 1 year old is categorised as 0 years old (as it is either based on age at last birthday for the 2001 data or their age on 27 March 2011 for the 2011 data.)

8) a)
Systematic sampling

8) b)
Quota sampling

8) c)
Stratified sampling

8) d)
Cluster sampling

9) a)
No. Whilst the number of 25-44 year olds has gone up, the percentage of this category has only gone from 35.3% to 35.5% so it has only marginally increased.

9) b)
As 10 years has passed the 5-14 years olds have gone into the 15-24 year old category, which has increased by 214 441. However, this does not account for people leaving the category such as moving to a different area.

9) c)
The category includes people over 94 and up to 110 and 115 for 2001 and 2011 respectively.

10) a)
This point is where a lower proportion of people drive to work but a higher proportion do not necessarily take public transport. This point is a rural area (it is Eden which is the Lake District) and demonstrates a high proportion of people who must work from home.

10) b)
45.15%

11)
On average, the proportions of people using bus, minibus or coach across the country hasn't changed. Both data sets have a (negative) skew but the middle 50% of people using bus, minibus or coach across the country has gone down from 2001 to 2011. The range is similar, but by 2011 the proportions of people using bus, minibus or coach across the country has become more consistent.

12)
Category 5 will be other. London will be really high public transport use, rural will be significantly lower which would suggest category 4. Identifying A as London and B as the rural authority. Rural will be high car use, with London lower so 'car as a passenger or a driver' will be category 2. Category 1&3 are not significantly different enough, but will be the remaining categories which are working from home and bicycle or on foot, which suggests 1 as being working from home which was much more common in rural areas and 3 for on bicycle or by foot.

13) a)
Mean is 39.3 years old (3 SF)
Standard deviation is 23.1 (3 SF)

13) b)
1 281 285 people. This is an estimation as the distribution of the data in the category 59 to 74 years is unknown and it assumes each age has an equal likelihood.

13) c)
The data is not symmetrical around the mean, so it would not be. Using the mean and standard deviation from part a), the normal distribution would be

centred around 39.4 years. With a standard deviation of 23.1, the majority of the data would have to lie between −29.9 years and 108.7 years, which does not work as you cannot have people less than 0 years old.

13) d)
The final category finishes at 90.5 which is not a fair representation of people over the age of 90

1.64 Graphs of Motion
Questions on page 188

1) a)
2 metres from the start of the track
1) b)
4 metres from the start of the track
1) c)
0.25 ms^{-1}
1) d)
3.25 metres (1.25 metres from where the car started)

2) a)
40 m
2) b)
−8 m
2) c)
2 ms^{-1}
2) d)
4 ms^{-1}
2) e)
$-\frac{3}{2}$ ms^{-1}
2) f)
$-\frac{2}{5}$ ms^{-1}

3) a)
18 m
3) b)
6 m
3) c)
1.8 ms^{-1}
3) d)
0.6 ms^{-1}

4)
[Graph: s-t graph with values 180, 60, 80, 125]

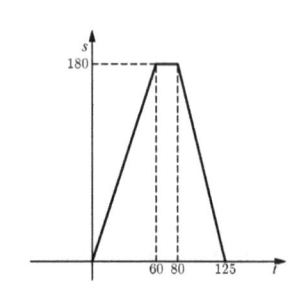

5) a)
[Graph: s-t graph with values 12, 6, 11, 20]

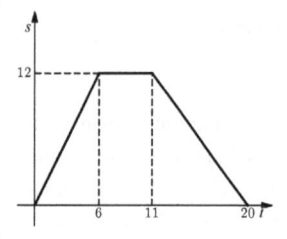

5) b)
$-\frac{4}{3}$ ms^{-1}

6) a)
[Graph: s-t graph with values 20, t, $t+5$, $\frac{3}{2}t+5$]

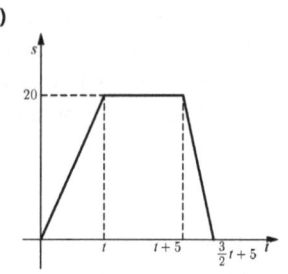

6) b)
$t = 4$ seconds

7) a)
The object starts at 2 ms^{-1} and it accelerates uniformly until it reaches 4 ms^{-1}. It remains at a constant speed until a total of 6 seconds have passed since the start.

7) b)
The object starts from rest, accelerating at 2 ms^{-2} for the first 3 seconds and reaches 6 ms^{-1}. The object travels at 6 ms^{-1} for 4 seconds. The object decelerates at 4 ms^{-2} uniformly, before reaching a speed of 0 ms^{-1} in 1.5 seconds. It then accelerates in the negative direction at 4 ms^{-2} for a further 1.5 seconds.

8) a)
[Graph: a-t graph with values $\frac{4}{3}$, 3, 5, 10, $-\frac{4}{5}$]

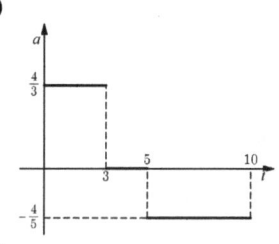

8) b)
[Graph: a-t graph with values 2.5, 5, 12, 16, $-\frac{6}{5}$]

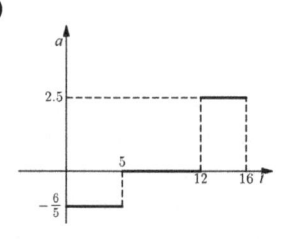

9) a)
[Graph: v-t graph with values 3, 5, 6.4, −4.5, −8]

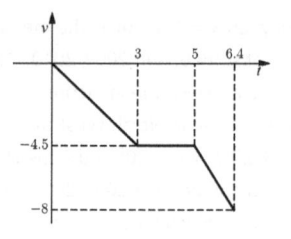

9) b)
[Graph: v-t graph with values 17.5, 15.5, 10, 5, 8, 10]

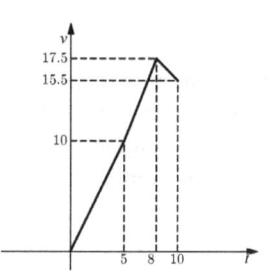

10) a)
[Graph: v-t graph with values 13.5, 3, 10]

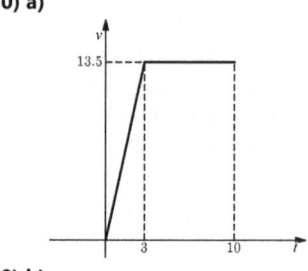

10) b)
114.75 m

11) a)
0.5 ms^{-2}, −0.2 ms^{-2}, $\frac{2}{3}$ ms^{-2}
11) b)
80.6 m
11) c)
8 s
11) d)
6.2 ms^{-1}

12) a)
[Graph: v-t graph with values 10, 6, 6+T]

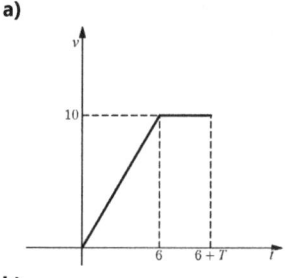

12) b)
[Graph: v-t graph with values 10, 2, 5, 6, 6+T]

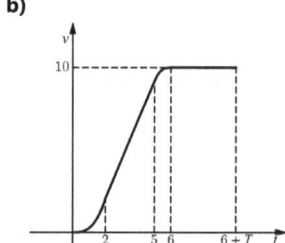

13) a)
0.6 ms^{-2}
13) b)
57 s
13) c)
1236 m

14) a)

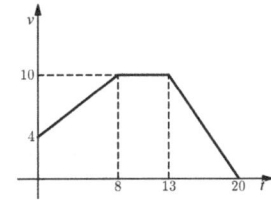

14) b)
141 m

14) c)
7.05 ms⁻¹

15) a)

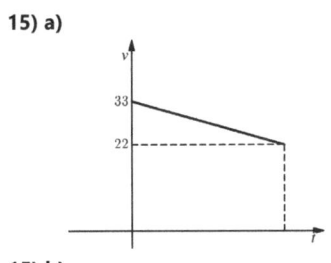

15) b)
40 seconds

15) c)
25.8 ms⁻¹

15) d)
27.5 ms⁻¹

16) a)
1.2 ms⁻², 0 ms⁻², −0.5 ms⁻², 1 ms⁻²

16) b)
69 m

16) c)
37.5 m

16) d)
69 m

16) e)
31.5 m

16) f)
106.5 m

16) g)
3.04 ms⁻¹ (3 SF)

17) a)

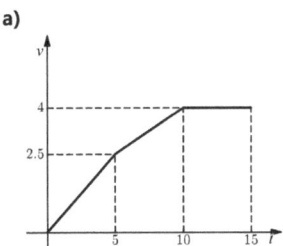

17) b)
4 ms⁻¹

17) c)
42.5 m

18) a)
0 ms⁻², −3 ms⁻², 0.5 ms⁻²

18) b)
$t = 10$ s and $t = 24$ s

18) c)
154.5 m

18) d)
70.5 m

19) a)
5 ms⁻²

19) b)
$t = 0$ s, $t = 12.4$ s and $t = 35$ s

19) c)
68 m before the initial start point.

20) a)
13.5 ms⁻¹

20) b)

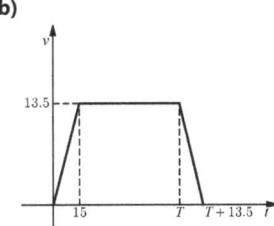

20) c)
82.25 s

21) a)
$t = 35\frac{1}{6}$ seconds

21) b)
209 metres

22) a)

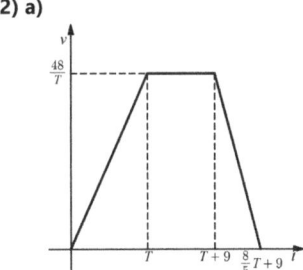

22) b)
$T = 4$

23) a)
0.25 ms⁻²

23) b)
$v = 0.25t + 12$

23) c)
24 s

23) d)
360 m

23) e)
The time at which both cars are instantaneously driving at the same speed of 15 ms⁻¹

24) a)

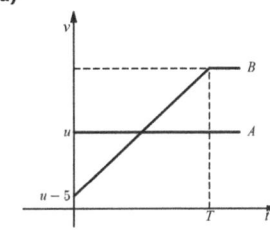

24) b)
Overtake at time T when they have travelled the same distance, so by symmetry $v = (u + 5)$ ms⁻¹

24) c)
$T = 40, u = 25$ ms⁻¹

1.65 Constant Acceleration
Questions on page 193

1) a)
Show that
1) b)
Show that
1) c)
Show that
1) d)
Show that

2)
Show that

3) a)
$t = 1$
3) b)
$u = \frac{14}{3}$
3) c)
$a = 1$
3) d)
$s = \frac{5}{3}$
3) e)
$v = 0$

4)
$u = \frac{25}{6}, a = -\frac{5}{18}$

5)
$a = \frac{3}{5}, t = 5$

6) a)
20 ms⁻¹
6) b)
91 m

7)
45.0 s (3 SF)

8) a)
55 m
8) b)
6 ms⁻¹

9) a)
$a = \frac{4}{3}$ ms⁻², $s = 6$ m
9) b)
$a = 0.007$ ms⁻², $s = 0.014$ m
9) c)
$a = \frac{10}{t}$ ms⁻², $s = 5t$ m

479

10)
$a = \frac{u}{6}$ ms^{-2}

11) a)
3.42 s

11) b)
Stopping times are affected by the type of vehicle, the condition of the brakes, the weather and condition of the road. Also the driver needs 'thinking time' to apply the brakes.

12)
0.00463 ms^{-2} (3SF)

13) a)
0.267 ms^{-2} (3 SF)

13) b)
8 ms^{-1}

14) a)
−1.2 ms^{-2}

14) b)
12.2 ms^{-1}

14) c)
0.0167 m (3 SF) beyond C

15) a)
3 : 4

15) b)
44.5 m (3SF)

16) a)
2.71 ms^{-2} (3SF)

16) b)
9.79 ms^{-1} (3SF)

17)
6.38 s (3 SF)

18) a)
$a = \frac{96 - 12k}{k^2}$

18) b)
$u = \frac{96}{k} - 6$

18) c)
$k = 5$

18) d)
$a = 1.44$ ms^{-2}

19)
1.24 ms^{-2} (3 SF)

20) a)
Height is 3.1392 m, 3.14 m (3 SF – AQA)

20) b)
$v = -9.81t$

20) c)
Speed is 7.848 ms^{-1}.
7.85 ms^{-1} (3 SF – AQA)

21) a)
$v = 12.5$ ms^{-1} (3 SF)
$v = 13$ ms^{-1} (2 SF – AQA)

21) b)
$t = 1.28$ s (3 SF), $t = 1.3$ s (2 SF – AQA)

21) c)
A more accurate value of the acceleration of gravity (including considering where on the planet the object is dropped as g is variable). Calculating for the object having a shape rather than being modelled as a particle.

22) a)
3 s

22) b)
12.25 metres

23) a)
2.5 s

23) b)
47.8 m (3 SF), 48 m (2 SF – AQA)

23) c)
109 m (3 SF), 110 m (2 SF – AQA)

24) a)
1.5 s

24) b)
1.22 s, 0.0587 s (3 SF)
1.2 s, 0.059 s (2 SF – AQA)

24) c)
1.15 s (3 SF), 1.2 s (2 SF – AQA)

25) a)
13.6 ms^{-1} (3 SF), 10 ms^{-1} (1 SF – AQA)

25) b)
24.2m (3 SF), 20 m (1 SF – AQA)

25) c)
3.56 s (3 SF), 4 s (1 SF – AQA)

25) d)
Consider reducing the acceleration due to gravity (as resistance would change the total acceleration). The initial speed would be larger to give the same final speed, this would lead to a higher greatest height reached, and it would take longer to hit the ground.

26) a)
40 ms^{-1}

26) b)
670.5 seconds
700 s (1 SF – AQA)

27) a)
$t = 0.4$ s

27) b)
$s = 7.2$ m
$s = 7$ m (1 SF – AQA)

28) a)
2.13 s (3 SF), 2 seconds (1 SF – AQA)

28) b)
0.625 s (3 SF), 0.6 s (1 SF – AQA)

28) c)
$s = 10.5$ m (3 SF), $s = 10$ m (1 SF – AQA)

29)
Show that

30)
Show that

31) a)
$s = \frac{u^2}{2g}$

31) b)
$t = \frac{2u}{g}$

32)
$t = 0.731$ s

1.66 Forces: Equilibrium
Questions on page 196

1) a)
15 N upwards

1) b)
7 N left

2) a)
$P = 2$

2) b)
$P = 6$

2) c)
$P = 34$

2) d)
$P = 20$

3) a)
$P = 62, Q = 31$

3) b)
$P = 27, Q = 8$

3) c)
$P = 12, Q = 4$

4) a)
$P = 8, Q = 2$

4) b)
$P = 3, Q = 2$

5) a)
$T = 50$N

5) b)
5 kg

5) c)
50 N downwards

6) a)
490.5 N, 491 N (3 SF – AQA)

6) b)
81.5 N

7)
653 N (3 SF), 650 N (2 SF – AQA)

8) a)
3.92 N
8) b)
$P = 3.92$
8) c)
3.92 N downwards

9) a)

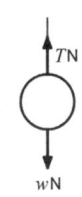

[Diagram: circle with arrow up labelled T N and arrow down labelled w N]

9) b)
19.6 N
9) c)
3.5 kg
9) d)
4.9 N
9) e)
14.7 N

10) a)

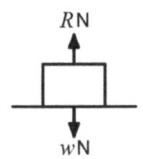

[Diagram: box with arrow up labelled R N and arrow down labelled w N]

10) b)
15 N
10) c)
6 kg
10) d)
0.5 kg
10) e)
12 N
10) f)
25 N

11)

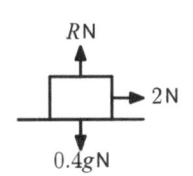

[Diagram: box with arrow up labelled R N, arrow right labelled 2 N, arrow down labelled $0.4g$ N]

1.67 Forces: Particles in Motion
Questions on page 198

1) a)
$F = 1.28$ N
1) b)
$m = 25$ kg
1) c)
$a = 3$ ms^{-2}
1) d)
$F = 0.16$ N

2) a)
$m = 5$, $P = 20$, $Q = 49$
2) b)
$m = 3$, $P = 12.5$, $Q = 29.4$
$Q = 29$ (2 SF – AQA)
2) c)
$m = 6.5$, $P = 20.5$, $Q = 63.7$
$Q = 64$ (2 SF – AQA)

3) a)
$m = 3$, $P = 36$,
$P = 40$ (1 SF – AQA)
3) b)
$m = 7.5$, $P = 86.25$
$P = 90$ (1 SF – AQA)
3) c)
$m = 1.2$, $P = 11.04$
$P = 10$ (1 SF – AQA)

4) a)
$T = 26$ N, $T = 30$ N (1 SF – AQA)
4) b)
$\Rightarrow T = 17$ N, $T = 20$ N (1 SF – AQA)
4) c)
$T = 15$ N, $T = 20$ N (1 SF – AQA)
4) d)
$T = 20$ N

5)
$T = 4.9$ N, $T = 5$ N (1 SF – AQA)

6)
$R = 11.2$ N

7)
$R = 100$ N

8) a)
$w = 156.8$ N, $w = 160$ N (2 SF – AQA)
8) b)
The box is a particle OR the box is rigid OR the force pushing the box is parallel to the floor

9)
0.4 ms^{-2}

10)
$F = 0.365$ N, $F = 0.37$ N (2 SF – AQA)

11) a)
Breaking force of 7200 N
11) b)
There are no other external forces acting on the car. The car is modelled as a particle. The acceleration is uniform.

12)
Breaking force of 2550 N

13)
5.4 m

14) a)
2.04 kg (3 SF), 2.0 kg (2 SF – AQA)
14) b)
7.35 ms^{-1}, 7.4 ms^{-1} (2 SF – AQA)

15)
4 metres

16)
$F = 6160$ N

17)
$s = 17.0$ m so the car stops in time.

18) a)
7.8125 ms^{-2}
18) b)
$F = 17\,578.125$ N $\Rightarrow F = 17\,600$ N (3 SF)
This calculates the total force in the direction of motion so the assumption is that there is no resistance to motion such as friction from the road and air resistance. The van is also modelled as particle and is assumed to be rigid.
18) c)
$R = 22050$ N, $R = 22000$ (2 SF – AQA)

1.68 Forces: Vectors
Questions on page 200

1) a)
3.16 N (3 SF), 71.6° (3 SF)
1) b)
8.25 N (3 SF), 346° (3 SF)
1) c)
9.22 N (3 SF), 103° (3 SF)
1) d)
1.41 N (3 SF), 225° (3 SF)
1) e)
16.2 N (3 SF), 279° (3 SF)

2) a)
5.39 N (3 SF), Bearing: 068°
2) b)
5.83 N (3 SF), Bearing: 301°
2) c)
13 N, Bearing: 157°
2) d)
1.02 (3 SF), Bearing: 259°
2) e)
2.69 N (3 SF), Bearing: 248°

3)
$k = 2$

4)
$k = -3\sqrt{2}$

5) a)
5**i** − 12**j**
5) b)
$\begin{pmatrix} 5 \\ -\sqrt{3} \end{pmatrix}$
5) c)
$-\sqrt{2}\mathbf{i} + 4\mathbf{j}$
5) d)
$\begin{pmatrix} 0 \\ -2 \end{pmatrix}$
5) e)
$-3b\mathbf{j}$
5) f)
$\begin{pmatrix} -2a - b \\ -6a - 6b \end{pmatrix}$

6)
$a = 4, b = 1$

7) a)
$-2\mathbf{j}$
7) b)
2 N, 180°
7) c)
The particle is accelerating due south.
7) d)
$2\mathbf{j}$

8)
$b = 4$ and $a > -15$

9) a)
14.3°
9) b)
45°

10)
$k = \frac{21}{19}$

11) a)
$a = 2, b > -10$
11) b)
$b = -4$

12) a)
$\frac{1}{5}\begin{pmatrix} 3 \\ 4 \end{pmatrix}$
12) b)
$\frac{1}{\sqrt{10}}(3\mathbf{i} - \mathbf{j})$
12) c)
$\frac{1}{\sqrt{17}}\begin{pmatrix} -1 \\ 4 \end{pmatrix}$
12) d)
$\frac{1}{10}(-8\mathbf{i} - 6\mathbf{j})$

13) a)
$\mathbf{F} = \frac{5}{\sqrt{2}}\begin{pmatrix} -1 \\ -1 \end{pmatrix}$
13) b)
$\mathbf{F} = \frac{4}{\sqrt{13}}\begin{pmatrix} 3 \\ 2 \end{pmatrix}$
13) c)
$\mathbf{F} = \frac{1}{\sqrt{2}}\begin{pmatrix} 8 \\ 8 \end{pmatrix}$

14) a)
$a = 10$ ms^{-2}
14) b)
$t = 4$ seconds

15) a)
$a = \frac{\sqrt{5}}{2} = 1.19$ ms^{-2} (3 SF)
15) b)
$v = 2\sqrt{5} = 4.47$ ms^{-1} (3 SF)

16) a)
$a = 1.25$ ms^{-2}
16) b)
$t = 4$ seconds

17)
$a = 4.72$ ms^{-2} (3 SF)

18) a)
3.75 ms^{-2}
18) b)
0.25 ms^{-2}

19)
$\mathbf{F} = \begin{pmatrix} 8 \\ -6 \end{pmatrix}$

20)
$\mathbf{F} = \frac{9\sqrt{5}}{25}\begin{pmatrix} 2 \\ 1 \end{pmatrix}$

21) a)
Show that
21) b)
22.4 m (3 SF)

22) a)
Show that
22) b)
5.10 ms^{-1} (3 SF)

23) a)
Show that
23) b)
9.5 ms^{-1}

1.69 Forces: Connected Particles
Questions on page 202

1) a) 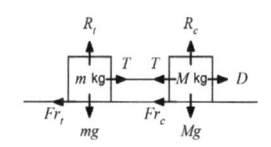
1) b) 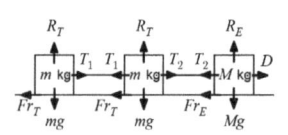
1) c)

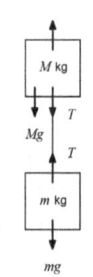

2) a)
$a = 0.24$ ms^{-2}

2) b)
$T = 120$ N

3) a)
$a = 0.389$ ms^{-2} (3 SF)
3) b)
$T = 172$ N (3 SF)

4) a)
$a = 6$ ms^{-2}
4) b)
$T = 12$ N
4) c)
Inextensible: the acceleration of the two particles is the same.
Light: the mass is negligible.

5)
$T = 5$ N

6) a)
$m = 30$
6) b)
$T = 90$ N

7) a)
$a = \frac{4.2}{m}$
7) b)
$T = 10.8$ N
7) c)
Inextensible: the acceleration of the two particles is the same.
Light: the mass is negligible.

8) a)
$R = 1100$ N
8) b)
$T_1 = 10\,600$ N, $T_2 = 5300$ N

9)
$R = 120$ N

10) a)
$a = 0.625$ ms^{-2}
10) b)
$F = 1755$ N
10) c)
$T = 540$ N

11) a)
$a = 2.6$ ms^{-2}
11) b)
32.5 metres
11) c)
$T = 33.6$ N

12) a)
Show that
12) b)
$T = 2800$ N

12) c)
1600 N
12) d)
5600 N

13) a)
0.8 ms^{-2}
13) b)
6.25 seconds
13) c)
$T = 740$ N
13) d)
The tow bar is inextensible OR it has no mass OR it is horizontal.
13) e)
382 metres (3 SF)
13) f)
No tension would mean that that it would be a higher overall force leading to an increase in acceleration.

14) a)
$a = -2.3$ ms^{-2}
14) b)
$T = 6$ N
14) c)
An increase of 4.6 N

15) a)
103 N, 100 N (2 SF – AQA)
15) b)
1.05 N, 1.1 N (2 SF – AQA)
15) c)
17.2 N, 17 N (2 SF – AQA)
15) d)
4.05 N, 4.1 N (2 SF – AQA)

16) a)
$T = 8370$ N
16) b)
$T = 9270$ N
16) c)
$T = 9180$ N
16) d)
$T = 8460$ N
16) e)
$a = -6.8$ ms^{-2}
16) f)
$a = -0.8$ ms^{-2}
16) g)
$a = 0.7$ ms^{-2}

17) a)
$a = 0.0889$ ms^{-2} (3 SF)
17) b)
$T = 6560$ N (3 SF),
$T = 7000$ N (1 SF – AQA)
17) c)
$R = 504$ N (3 SF),
$R = 500$ N (1 SF – AQA)

18) a)
$T = 1196$ N, $T = 1000$ N (1 SF – AQA)
18) b)
$R = 52$ N, $R = 50$ N (1 SF – AQA)
18) c)
$F = 156$ N, $F = 200$ N (1 SF – AQA)

19) a)
$T = 1345.5$ N, $T = 1000$ N (1 SF – AQA)
19) b)
$R = 29.25$ N, $R = 30$ N (1 SF – AQA)
19) c)
$F = 175.5$ N, $F = 200$ N (1 SF – AQA)

20) a)
$T = 1272.6$ N, $T = 1000$ N (1 SF – AQA)
20) b)
$F = 60.6$ N, $F = 60$ N (1 SF – AQA)

21) a)
2 ms^{-2}
21) b)
$T = 10$ N
21) b)
$D = 26$ N

1.70 Forces: Pulleys
Questions on page 205

1) a)
$T = 44.1$ N, $a = 4.9$ ms^{-2}
$T = 44$ N (2 SF – AQA),
$a = 4.9$ ms^{-2} (2 SF – AQA)
1) b)
$T = 31.36$ N, $a = 5.88$ ms^{-2}
$T = 32$ N (2 SF – AQA),
$a = 5.9$ ms^{-2} (2 SF – AQA)
1) c)
$T = 2.8$ N, $a = 4.2$ ms^{-2}

2) a)
$a = 2.45$ ms^{-2}, $a = 2.5$ ms^{-2} (2 SF – AQA)
2) b)
$T = 36.75m$ N
2) c)
$2T = 73.5m$ N

3)
Show that

4)
Show that

5) a)
$a = 0.1$ ms^{-2}
5) b)
$T = 19.42$ N, $T = 19.4$ N (3 SF – AQA)
5) c)
$m = 1.96$ kg (3 SF)

6) a)
$a = 1.40$ ms^{-2}
6) b)
$T = 33.6$ N
6) c)
$t = 1.46$ s
6) d)
1.71 metres above the floor.
Modelling assumptions: there are no resistances to motion (such as friction or air resistance), the pulley is small, particle B does not hit the pulley.

7) a)
$a = 1.6$ ms^{-2}
7) b)
$T = 57$ N
7) c)
Show that
7) d)
1.26 ms^{-1}
7) e)
1.58 metres above the floor.
7) f)
0.258 s, 0.26 s (2 SF – AQA)
7) g)
Box A will ascend with a higher acceleration and there will be a higher tension in the string.
7) h)
Box A will not move as $a = 0$ and so the tension will be equal to the weight of box A
7) i)
Box A will descend and there will be a lower tension in the string.

8) a)
$a = 7$ ms^{-2}
8) b)
$T = 8.4$ N
8) c)
Show that

9) a)
$78.4 < T \leq 94.08$
$78 < T \leq 94$ (2 SF – AQA)
9) b)
$6 < m \leq 10$

10) a)
$a = 2.88$ ms^{-2}, $a = 2.9$ ms^{-2} (2 SF – AQA)
10) b)
$T = 3.46$ N, $T = 3.5$ N (2 SF – AQA)

11) a)
$a = 1.88$ ms^{-2}, $T = 23.76$ N
$a = 1.9$ ms^{-2} (2 SF – AQA),
$T = 24$ N (2 SF – AQA)
11) b)
$a = 4.72$ ms^{-2}, $T = 35.6$ N
$a = 4.7$ ms^{-2} (2 SF – AQA),

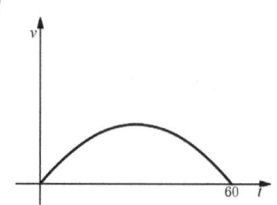

$T = 36$ N (2 SF – AQA)
11) c)
$a = 4.41$ ms^{-2}, $T = 2.83$ N
$a = 4.4$ ms^{-2} (2 SF – AQA),
$T = 2.8$ N (2 SF – AQA)
11) d)
$a = 0$ ms^{-2}, $T = 58.8$ N
$a = 0$ ms^{-2} (2 SF – AQA),
$T = 59$ N (2 SF – AQA)

12) a)
$t = 1.03$ s, $t = 1.0$ s (2 SF – AQA)
12) b)
$t = 0.651$ s, $t = 0.65$ s (2 SF – AQA)
12) c)
$t = 0.695$ s, $t = 0.69$ s (2 SF – AQA)
12) d)
The system does not move

13) a)
$a = 0.5$ ms^{-2}
13) b)
$T = 1050$ N, $T = 1000$ N (1 SF – AQA)
13) c)
The cable is light and inextensible. The pulley is small.
13) d)
25 metres
13) e)
No resistances acting against the pallet of building materials.

14)
Show that

15) a)
$a = \frac{9.8m - 5}{m + 0.3}$
15) b)
$m = 0.563$ kg, $m = 0.56$ kg (2 SF – AQA)
15) c)
There are no resistances to motion on block B

16) a)
$a = 3.77$ ms^{-2}, $a = 3.8$ ms^{-2} (2 SF – AQA)
16) b)
$T_1 = 48.2$ N, $T_2 = 40.7$ N,
$T_1 = 48$ N (2 SF – AQA),
$T_2 = 41$ N (2 SF – AQA)

17)
Show that

18) a)
$a = 0.01$ ms^{-2}
18) b)
Show that

1.71 Variable Acceleration
Questions on page 209

1) a)
$v = \frac{ds}{dt} = 12t^2 - 5$, $a = \frac{dv}{dt} = 24t$
1) b)
$v = \frac{ds}{dt} = 5t^4 + 4t$, $a = \frac{dv}{dt} = 20t^3 + 4$
1) c)
$v = \frac{ds}{dt} = 2t - 1$, $a = \frac{dv}{dt} = 2$
1) d)
$v = \frac{ds}{dt} = 2 + \frac{1}{t^2}$, $a = \frac{dv}{dt} = -\frac{2}{t^3}$

2) a)
$s = 2t^4 - 3t^2$
2) b)
$s = 2t^3 + \frac{1}{2}t^4$
2) c)
$s = \frac{1}{40}t^4 - \frac{1}{6}t^3$
2) d)
$s = \frac{1}{2}t^2 + \frac{1}{t}$

3) a)
$v = 3t^5 - 2t^2 + 1$
3) b)
$v = 3t^3 + t^5 - 16$
3) c)
$v = -0.03t^3 + 1.35t^2 + 1.68$
3) d)
$v = \frac{2}{3}t^{\frac{3}{2}} + \frac{1}{2}t^{-2} + \frac{5}{6}$

4) a)
$t = 0$ and $t = 60$
4) b)

4) c)
720m
4) d)
$a = -\frac{1}{25}t + \frac{6}{5}$
4) e)
The model suggest that the car will continue with only a negative velocity after 60 seconds which would mean the car reversing from that point on.

5) a)
$v = 10t^{\frac{3}{2}} + 2t - 11$
5) b)
$s = 4t^{\frac{5}{2}} + t^2 - 11t + 12$

6) a)
$v = -10t^3 + \frac{25}{4}t^4 + 5$
6) b)
10 m

7) a)
Show that
7) b)
6 m
7) c)
8 seconds
7) d)
$v = -\frac{3}{4}t + 3$
7) e)
4.5 ms^{-1}

8) a)
$s = 4t^2 - t^3$
8) b)
$a = \frac{dv}{dt} = 8 - 6t$
8) c)
$F = 3.2 - 2.4t$

9) a)
$k = 2$
9) b)
73.15 m

10) a)
Show that
10) b)
2.4 ms^{-1}
10) c)
0 ms^{-1}
10) d)
Ellis has reached a speed of 0 ms^{-1} so he will have a negative velocity which means he has changed direction.

11)
20 s

12) a)
$v = 4t + \frac{3}{2}t^2 - 2t^3$
12) b)
$s = 2t^2 + \frac{1}{2}t^3 - \frac{1}{2}t^4 + 4$
12) c)
$t = 2.80$ s (3 SF)

13) a)
520 m
13) b)
$t = 2$
13) c)
$v = 3t^2 - 12t + 12 = 3(t - 2)^2 \geq 0$ for all t. So it never reverses direction.
13) d)
$a = \frac{dv}{dt} = 6t - 12$ which is not constant.

14)
10 seconds

15) a)
360 m

15) b)
$a = \frac{dv}{dt} = 2 - 0.003t^2$
a has max of 2 when $t = 0$,
When $t = 25, a = 0.125$ so $a \geq 0$ for $0 \leq t \leq 25$ so always accelerating.

16) a)
$v = \frac{ds}{dt} = \frac{8}{15}t - \frac{1}{225}t^2$

16) b)
16 ms⁻¹

17) a)
$a = 0.75, b = -0.05$

17) b)
Show that

18) a)
$v = \frac{1}{2}t$

18) b)
12.25 metres

18) c)
$k = 14$

18) d)
16.3 metres

18) e)

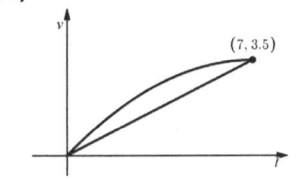

The area under the curve between $t = 0$ and $t = 7$ is larger for the second (quadratic) model.

19) a)
20.8 m (3 SF)

19) b)
$t = \frac{5}{3}$ s: $v = 7.41$ ms⁻¹ (3 SF)

19) c)
$t = \frac{10}{3}$: $a = -\frac{10}{3}$ so minimum acceleration is -3.33 ms⁻²
This is a deceleration of 3.33 ms⁻²

19) d)
As $t \to \infty, v \to \infty$ which is not possible. That is, the cyclist will not be able to increase in velocity (with increasing acceleration) permanently.

20) a)
Show that

20) b)
77.1 m (3 SF)

20) c)
For smaller values of t, the velocity is very large which is unrealistic.

21)
$p = 0.3, q = -2.4, r = 7.8$

22)
$k = 2$ which gives $v = t^2 - 4t + 8$

23)
8 metres

24) a)
Show that

24) b)
Show that

24) c)
18 metres

24) d)
0 metres

24) e)
$t = 6$

24) f)
72 metres

25) a)
0 ms⁻¹

25) b)
The cyclist has increasing velocity for the first 6 seconds before cycling with constant velocity of 9.72 ms⁻¹ for the next 14 seconds.

25) c)
2.43 ms⁻²

26) a)
400 seconds.

26) b)
650 seconds

26) c)
3 ms⁻¹

26) d)
3.2 ms⁻¹

26) e)
It is unlikely that Filip ran with a fixed velocity instantaneously from leaving the shop to arriving exactly back at his house.

27) a)
Show that

27) b)
$v = 6 - 6e^{-0.4t}$

27) c)
0 ms⁻¹

27) d)
$a = 2.4$ ms⁻²

28) a)
8.8 m above ground level

28) b)
9.96 seconds (3 SF)

28) c)
3.26 metres (3 SF)

28) d)
$v = 0.08e^{0.1t} - 1.6e^{-0.2t}$

28) e)
4.37 ms⁻¹

28) f)
19.6 metres

29)
$F = -\frac{3}{25}t + \frac{4}{25}$

2.01 Trigonometry: Radians, Arcs and Sectors
Questions on page 213

1)
$\pi, \frac{\pi}{2}, 60°, 45°, 36°, \frac{\pi}{6}, \frac{\pi}{9}, 160°, \frac{9\pi}{1000}, 92.8°$ (3SF)

2) a)
5 cm

2) b)
5π m

2) c)
11.25 cm

2) d)
45π m

3) a)
80 cm²

3) b)
$\frac{27\pi}{2}$ m²

3) c)
2355.2 cm²

3) d)
$\frac{1369\pi}{160}$ m²

4) a)
Area = 25.6 cm², Perimeter = 22.4 cm

4) b)
Area = $\frac{98\pi}{5}$ m², Perimeter = $\left(28 + \frac{14\pi}{5}\right)$ m = 36.8 m (3 SF)

4) c)
Area = $\frac{9971}{160}$ cm² = 62.31875 cm², Perimeter = $\frac{1287}{40}$ cm = 32.175 cm

4) d)
Area = 39600π m², Perimeter = $(720 + 220\pi)$ m = 1410 m (3 SF)

5)
$\theta = \frac{9}{8}$

6)
$r = \sqrt{\frac{100}{\theta}}$

7)
$\theta = \frac{84 - 2r}{r}$

8)
$OA = 3\sqrt{3}$ cm

9) a)
2.10 cm² (3 SF)

9) b)
10.6 cm (3 SF)
10) a)
25.0 units² (3 SF)
10) b)
23.4% (3 SF)

11)
25.3 m² (3 SF)

12) a)
Show that
12) b)
Show that
12) c)
14.5 cm² (3 SF)

13)
31.3 cm (3 SF)

14) a)
Show that
14) b)
$\theta = 1$ or $\theta = 4$

15)
$\theta = \frac{2}{5}, r = 15$

16)
10.2 cm (3 SF)

17)
$r = 4$ and $\theta = \frac{1}{5}$

18)
$\left(\frac{1}{4}\sqrt{3} - \frac{41}{400}\pi\right)r^2$

19)
$\frac{25}{2}(\pi - \sqrt{3})$ cm²

20)
$AB = 20\sqrt{6}$ m

21)
40.1 units² (3 SF)

22)
3.82 m (3 SF)

23) a)
Show that
23) b)
10.2 units² (3 SF)

24)
44.0 units (3 SF)

25)
$k = 4\sqrt{2 - \sqrt{2}}$

2.02 Trigonometry: $\sin x$, $\cos x$, $\tan x$ in Radians
Questions on page 217

1) a)

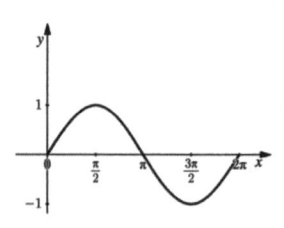

1) b)

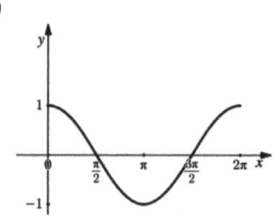

1) c)

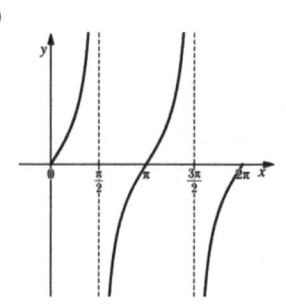

2) a)
$\theta = \{0, \pi\}$
2) b)
$\theta = \left\{\frac{\pi}{2}\right\}$
2) c)
$\theta = \left\{\frac{3\pi}{2}\right\}$

3) a)
$\theta = \left\{\frac{\pi}{2}, \frac{3\pi}{2}\right\}$
3) b)
$\theta = \{0\}$
3) c)
$\theta = \{\pi\}$

4) a)
$\theta = \{0, \pi\}$
4) b)
$\theta = \left\{\frac{\pi}{4}, \frac{5\pi}{4}\right\}$
4) c)
$\theta = \left\{\frac{3\pi}{4}, \frac{7\pi}{4}\right\}$

5)
$\frac{\pi}{2} < x < \frac{3\pi}{2}$

6)
$\pi < x < 2\pi$

7) a)
$\theta = \{0.253, 2.89\}$ (3 SF)
7) b)
$\theta = \{0.723, 5.56\}$ (3 SF)

7) c)
$\theta = \{0.896, 4.04\}$ (3 SF)

8) a)
$\theta = \{-5.16, -4.26\}$ (3 SF)
8) b)
$\theta = \{-4.82, -1.46\}$ (3 SF)
8) c)
$\theta = \{-5.41, -2.27\}$ (3 SF)

9) a)
$\theta = \{3.31, 6.12\}$ (3 SF)
9) b)
$\theta = \{2.67, 3.62\}$ (3 SF)
9) c)
$\theta = \{2.82, 5.96\}$ (3 SF)

10) a)
$\theta = \{-2.42, -0.721\}$ (3 SF)
10) b)
$\theta = \{-4.13, -2.15\}$ (3 SF)
10) c)
$\theta = \{-4.60, -1.46\}$ (3 SF)

11) a)
$\theta = \{-5.82, -3.60, 0.461, 2.68\}$ (3 SF)
11) b)
$\theta = \{-4.82, -1.46, 1.46, 4.82\}$ (3 SF)
11) c)
$\theta = \{-5.02, -1.87, 1.27, 4.41\}$ (3 SF)
11) d)
$\theta = \{-2.16, -0.979, 4.12, 5.30\}$ (3 SF)
11) e)
$\theta = \{-4.61, -1.67, 1.67, 4.61\}$ (3 SF)
11) f)
$\theta = \{-4.56, -1.42, 1.72, 4.86\}$ (3 SF)

12)
$-\frac{9\pi}{2} < x < -\frac{7\pi}{2}$ or $-\frac{5\pi}{2} < x < -\frac{3\pi}{2}$ or $-\frac{\pi}{2} < x < 0$

13)
$8\pi < x < 9\pi$

14) a)
$\theta = \{0.189, 2.95, 6.47, 9.24\}$ (3 SF)
14) b)
$\Rightarrow \theta = \{10.5, 14.6, 16.8\}$ (3 SF)
14) c)
$\theta = \{-19.6, -16.5\}$ (3 SF)

15) a)
$\theta = \{0.0606, 3.08\}$ (3 SF)
15) b)
$\theta = \{3.78, 6.92, 10.1\}$ (3 SF)
15) c)
$\theta = \{-7.79, -4.78\}$ (3 SF)

16)
$A = \{1.21, 1.93\}$ (3 SF)

17)
$A: x = -p, B: x = p - 3\pi$
$C: x = p - \pi, D: x = 2\pi - p$

18)
$\sin \frac{\pi}{6} = \frac{1}{2}$
$\sin\left(\pi - \frac{\pi}{6}\right) = \sin \frac{5\pi}{6} = \frac{1}{2}$
$\sin\left(\frac{5\pi}{6} + 2\pi\right) = \sin \frac{17\pi}{6} = \frac{1}{2}$

19)
$\cos \frac{\pi}{4} = \frac{\sqrt{2}}{2}$
$\cos\left(\frac{\pi}{2} + \frac{\pi}{4}\right) = \cos \frac{3\pi}{4} = -\frac{\sqrt{2}}{2}$
$\cos\left(\frac{3\pi}{4} - 8\pi\right) = \cos\left(-\frac{29\pi}{4}\right) = -\frac{\sqrt{2}}{2}$

20)
$\tan\left(-\frac{39\pi}{4}\right) = 1$, $\tan\left(\frac{39\pi}{4}\right) = -1$
$\tan\left(\frac{39\pi}{4} - 5\pi\right) = \tan \frac{19\pi}{4} = -1$

21)
$A: x = -p, B: x = p - 3\pi$
$C: x = \pi - p, D: x = p$

22)
$A: x = p, B: x = 2\pi - p$
$C: x = \pi - p, D: x = -\pi - p$

23) a)
$\theta = \{0.201, 2.94, 3.34, 6.08, 6.48, 9.22\}$ (3 SF)
23) b)
$\theta = \{-8.60, -7.11, -5.45, -3.97\}$ (3 SF)
23) c)
$\theta = \{4.61, 4.81, 7.75, 7.95, 10.9, 11.1\}$ (3 SF)

24) a)
$\theta = \{0, \pi\}$
24) b)
$\theta = \{0, 0.507, \pi, 3.65\}$ (3 SF)
24) c)
$\theta = \left\{\frac{\pi}{2}, 2.15, 4.14, \frac{3\pi}{2}\right\}$ (3 SF)

25)
$A: x = p, B: x = p + \pi$
$C: x = p - \pi, D: x = p - 2\pi$

26)
$A: x = -p, B: x = p - 3\pi$
$C: x = \pi - p, D: x = p$

27)
$A: x = -p, B: x = p - 2\pi$
$C: x = p - \pi, D: x = 3\pi - p$

28) a)
$\theta = \left\{0.644, \frac{\pi}{2}, 2.50, \frac{3\pi}{2}\right\}$ (3 SF)
28) b)
$\theta = \{0, 1.70, \pi, 4.84\}$ (3 SF)
28) c)
$\theta = \{0, 0.361, \pi, 5.92\}$ (3 SF)

29) a)
$\theta = \left\{\frac{\pi}{6}, \frac{5\pi}{6}, 3.48, 5.94\right\}$ (3 SF)
29) b)
$\theta = \{1.23, 1.77, 4.51, 5.05\}$ (3 SF)
29) c)
$\theta = \{2.55, 2.98, 5.70, 6.12\}$ (3 SF)

30) a)
$\theta = \{0.384, 2.76, 3.99, 5.44\}$ (3 SF)
30) b)
$\theta = \{1.79, 4.49\}$ (3 SF)
30) c)
$\theta = \{1.29, 2.31, 4.43, 5.45\}$ (3 SF)

31) a)
$5(x + 7)^2 - 26$
31) b)
$(-7, -26)$
31) c)
Minimum is 154 when $\theta = \pi$

32) a)
2π
32) b)
2π
32) c)
π
32) d)
$\frac{2\pi}{3}$
32) e)
π
32) f)
$\frac{\pi}{6}$
32) g)
10π
32) h)
$\frac{3\pi}{4}$
32) i)
$\frac{16\pi}{9}$

33) a)
4 solutions
33) b)
6 solutions
33) c)
24 solutions

34) a)
4 solutions
34) b)
15 solutions
34) c)
30 solutions

35) a)
10 solutions
35) b)
45 solutions
35) c)
260 solutions

36) a)
2π
36) b)
2π
36) c)
π
36) d)
2π
36) e)
2π
36) f)
π
36) g)
$\frac{2\pi}{3}$
36) h)
$\frac{\pi}{3}$
36) i)
$\frac{\pi}{12}$
36) j)
12π
36) k)
$\frac{8\pi}{25}$
36) l)
$\frac{4\pi}{5}$

37) a)
$\theta = \{0.147, 2.35\}$ (3 SF)
37) b)
$\theta = \{3.82, 4.90\}$ (3 SF)
37) c)
$\theta = \{0.864, 4.68\}$ (3 SF)

38) a)
$\theta = \{2.34, 5.88\}$ (3 SF)
38) b)
$\theta = \{2.92, 5.04\}$ (3 SF)
38) c)
$\theta = \{1.62, 2.57\}$ (3 SF)

39) a)
$\theta = \{2.54, 5.68\}$ (3 SF)
39) b)
$\theta = \{0.987, 4.13\}$ (3 SF)
39) c)
$\theta = \{0.702, 3.84\}$ (3 SF)

40)
$\frac{3\pi}{4} < x < \frac{7\pi}{4}$

41) a)
$\theta = \left\{\frac{\pi}{8}, \frac{3\pi}{8}, \frac{9\pi}{8}, \frac{11\pi}{8}\right\}$
41) b)
$\theta = \left\{\frac{\pi}{12}, \frac{11\pi}{12}, \frac{13\pi}{12}, \frac{23\pi}{12}\right\}$
41) c)
$\theta = \left\{\frac{\pi}{6}, \frac{2\pi}{3}, \frac{7\pi}{6}, \frac{5\pi}{3}\right\}$

42) a)
$\theta = \{0.288, 0.759, 2.38, 2.85, 4.48, 4.95\}$ (3 SF)
42) b)
$\theta = \{0.941, 1.15, 3.04, 3.25, 5.13, 5.34\}$ (3 SF)
42) c)
$\theta = \{0.944, 1.99, 3.04, 4.09, 5.13, 6.18\}$ (3 SF)

43) a)
$\theta = \{6.85, 12.0\}$ (3 SF)
43) b)
$\theta = \{6.73, 24.7\}$ (3 SF)
43) c)
$\theta = \{1.40\}$ (3 SF)

44) a)
$\theta = \{0.200, 1.05, 2.29\}$ (3 SF)
44) b)
$\theta = \{0.00597, 1.080, 1.58, 2.65\}$ (3 SF)
44) c)
$\theta = \{0.482, 1.11\}$ (3 SF)

45) a)
$\theta = \{-2.49, -1.93, -1.24, -0.672\}$ (3 SF)
45) b)
$\theta = \{3.53, 3.95, 4.43\}$ (3 SF)
45) c)
$\theta = \{6.87, 7.92, 8.97, 10.0, 11.1, 12.1\}$ (3 SF)

46)
$0 < x < \frac{\pi}{3}$ or $\frac{2\pi}{3} < x < \pi$

47)
$0 < x < 2\pi$ or $6\pi < x < 10\pi$

48) a)
$\theta = \left\{\frac{11\pi}{16}, \frac{15\pi}{16}, \frac{27\pi}{16}, \frac{31\pi}{16}\right\}$
48) b)
$\theta = \left\{\frac{13\pi}{90}, \frac{23\pi}{90}, \frac{73\pi}{90}, \frac{83\pi}{90}, \frac{133\pi}{90}, \frac{143\pi}{90}\right\}$
48) c)
$\theta = \left\{\frac{\pi}{80}, \frac{21\pi}{80}, \frac{41\pi}{80}, \frac{61\pi}{80}, \frac{81\pi}{80}, \frac{101\pi}{80}, \frac{121\pi}{80}, \frac{141\pi}{80}\right\}$

49) a)
$\theta = \{0.318, 1.07, 1.89\}$ (3 SF)
49) b)
$\theta = \{0.345, 0.495\}$ (3 SF)
49) c)
$\theta = \{0.188, 0.637\}$ (3 SF)

50) a)
$\theta = \left\{\begin{array}{c}-0.836, -0.735, -0.0503, \\ 0.0503, 0.735, 0.836\end{array}\right\}$ (3 SF)
50) b)
$\theta = \left\{\begin{array}{c}-1.61, -1.53, -0.561, \\ -0.486, 0.486, 0.561, 1.53, 1.61\end{array}\right\}$ (3 SF)
50) c)
$\theta = \{-0.183, -0.166, 0.166, 0.183\}$ (3 SF)

51)
$A: x = -p, B: x = \pi - p$
$C: x = \frac{5\pi}{3} - p, D: x = p + \frac{2\pi}{3}$

52) a)
$\theta = \left\{\frac{\pi}{2}, \frac{7\pi}{6}\right\}$
52) b)
$\theta = \left\{\frac{7\pi}{10}, \frac{3\pi}{2}\right\}$
52) c)
$\theta = \left\{\frac{5\pi}{21}, \frac{19\pi}{21}\right\}$
52) d)
$\theta = \left\{\frac{11\pi}{16}, \frac{27\pi}{16}\right\}$

53) a)
$\theta = \left\{\frac{4\pi}{9}, \frac{11\pi}{9}\right\}$
53) b)
$\theta = \left\{\frac{7\pi}{8}, \frac{23\pi}{24}, \frac{15\pi}{8}, \frac{47\pi}{24}\right\}$
53) c)
$\theta = \left\{\frac{\pi}{8}, \frac{3\pi}{8}, \frac{5\pi}{8}, \frac{7\pi}{8}, \frac{9\pi}{8}, \frac{11\pi}{8}, \frac{13\pi}{8}, \frac{15\pi}{8}\right\}$

2.03 Graph Sketching: $\sin x$, $\cos x$, $\tan x$ in Radians
Questions on page 218

1) a)
$y = 3 \sin x$
1) b)
$y = \frac{4}{5} \sin x$

2) a)
$y = \frac{5}{2} \cos x$
2) b)
$y = \frac{9}{4} \cos x$

3) a)

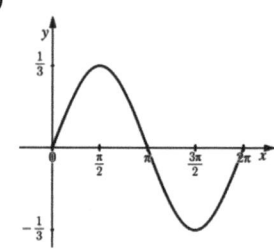

3) b)

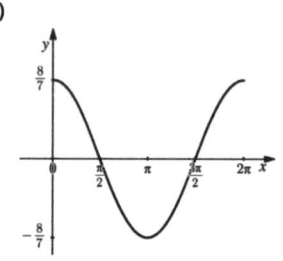

4) a)
$y = \sin 2x$
4) b)
$y = \sin 3x$
4) c)
$y = \sin \frac{5}{6} x$
4) d)
$y = \sin \frac{3}{5} x$

5)

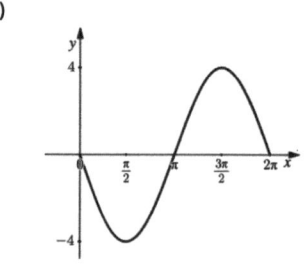

6) a)
$y = \cos \frac{5}{2} x$
6) b)
$y = \cos \frac{3}{4} x$
6) c)
$y = \cos \frac{6}{5} x$
6) d)
$y = \cos \frac{5}{8} x$

7)

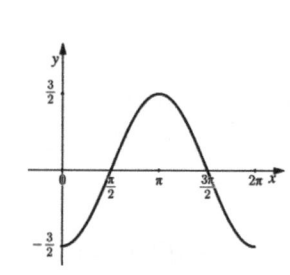

8)

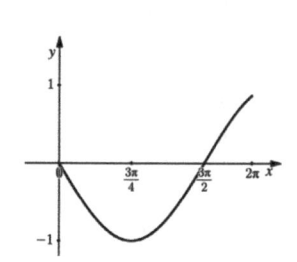

9) a)
$y = \tan 2x$
9) b)
$y = \tan \frac{3}{2} x$
9) c)
$y = \tan \frac{4}{3} x$
9) d)
$y = \tan \frac{8}{9} x$

10) a)

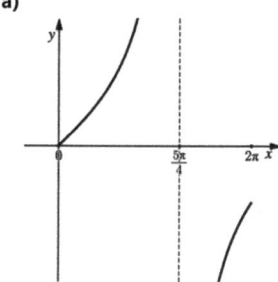

10) b)

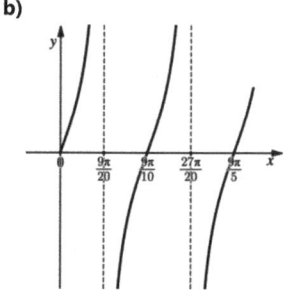

11) a)
$y = 3 + \sin x$
11) b)
$y = -4 + \cos x$

11) c)
$y = -1 + \sin x$
11) d)
$y = 2 + \cos x$

12) a)
$y = 3 + 2\sin x$
12) b)
$y = -2 + 2\cos x$
12) c)
$y = 1 + 4\cos x$
12) d)
$y = 2 - 3\sin x$

13) a)

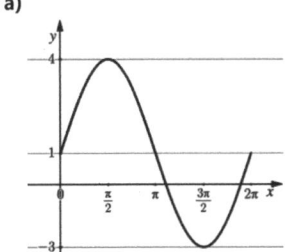

13) b)

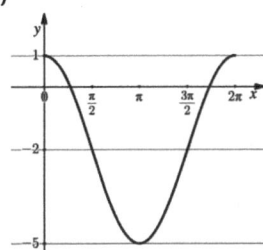

13) c)

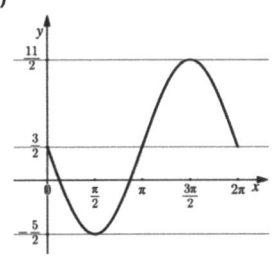

13) d)

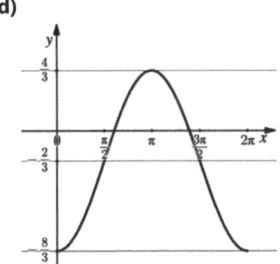

14) a)
$y = \sin\left(x - \frac{\pi}{2}\right)$
14) b)
$y = \sin\left(x + \frac{\pi}{3}\right)$

15) a)
$y = \cos\left(x + \frac{2\pi}{3}\right)$
15) b)
$y = \cos\left(x - \frac{3\pi}{5}\right)$

16) a)
$y = \tan\left(x + \frac{\pi}{6}\right)$
16) b)
$y = \tan\left(x - \frac{2\pi}{3}\right)$

17) a)

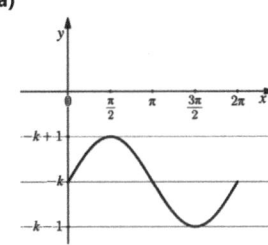

17) b)

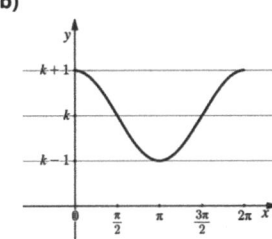

18) a)

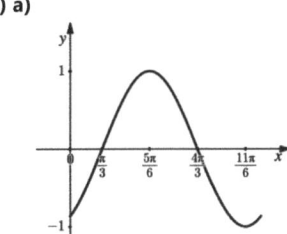

18) b)

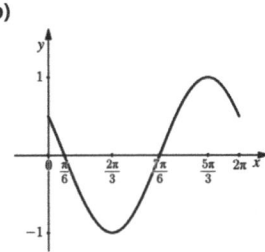

18) c)

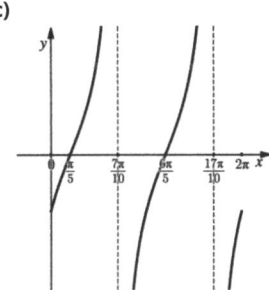

18) d)

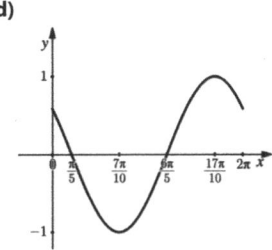

18) e)

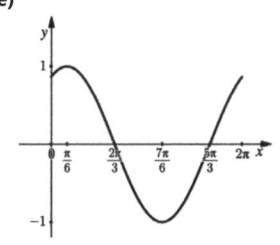

18) f)

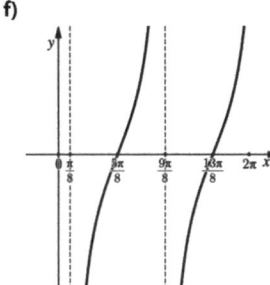

19) a)

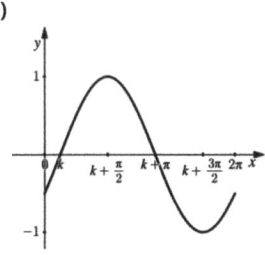

19) b)

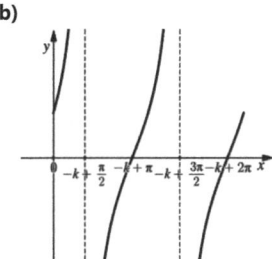

20) a)
$y = \sin\left(2x - \frac{\pi}{3}\right)$
20) b)
$y = \sin\left(\frac{1}{2}x + \frac{\pi}{4}\right)$
20) c)
$y = \sin\left(\frac{2}{3}x - \frac{\pi}{5}\right)$

21) a)
$y = \cos\left(\frac{1}{2}x - \frac{\pi}{8}\right)$
21) b)
$y = \cos\left(\frac{5}{3}x + \frac{\pi}{5}\right)$
21) c)
$y = \cos\left(\frac{5}{7}x - \frac{5\pi}{9}\right)$

22) a)
$y = \tan\left(\frac{5}{3}x + \frac{\pi}{3}\right)$
22) b)
$y = \tan\left(\frac{7}{8}\pi - \frac{\pi}{4}\right)$
22) c)
$y = \tan\left(\frac{4}{9}x + \frac{\pi}{6}\right)$

23) a)
$y = 2 + 3\sin\left(x + \frac{\pi}{8}\right)$

23) b)
$y = -3 + \sin\left(\frac{3}{2}x - \frac{\pi}{10}\right)$

23) c)
$y = -1 + 4\sin\left(\frac{4}{5}x + \frac{\pi}{4}\right)$

23) d)
$y = 2 - 3\sin\left(\frac{5}{4}x - \frac{\pi}{3}\right)$

24) a)
$y = 4\cos\left(\frac{3}{4}x - \frac{\pi}{4}\right)$

24) b)
$y = -2 + 2\cos\left(\frac{4}{5}x + \frac{2\pi}{5}\right)$

24) c)
$y = -1 - 4\cos\left(\frac{7}{9}x - \frac{\pi}{6}\right)$

24) d)
$y = 2 - 3\cos\left(\frac{7}{10}x + \frac{4\pi}{5}\right)$

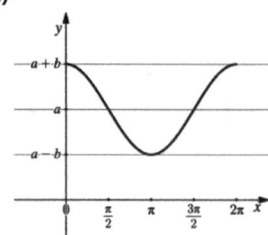

25) a)

25) b)

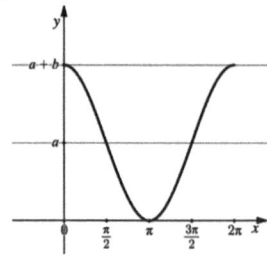

25) c)

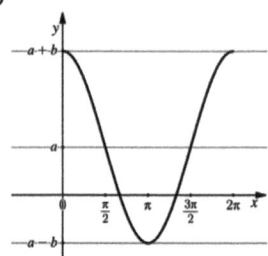

2.04 Trigonometry: Trigonometric Identities in Radians
Questions on page 229

1) a)
$\cos x = \frac{1}{\sqrt{5}}$
1) b)
$\tan x = 2$

2) a)
$\sin x = \frac{1}{3}$
2) b)
$\tan x = \frac{\sqrt{2}}{4}$

3) a)
$\sin x = \frac{5}{\sqrt{31}}$
3) b)
$\cos x = \frac{31\sqrt{6}}{31}$

4) a)
$\cos x = -\frac{\sqrt{21}}{6}$
4) b)
$\tan x = -\frac{\sqrt{35}}{7}$

5) a)
$\sin x = \frac{\sqrt{13}}{13}$
5) b)
$\tan x = -\frac{\sqrt{3}}{6}$

6) a)
$\sin x = \frac{\sqrt{1110}}{37}$
6) b)
$\cos x = -\frac{7}{\sqrt{259}}$

7) a)
$\cos x = -\frac{\sqrt{13}}{4}$
7) b)
$\tan x = \frac{\sqrt{39}}{13}$

8) a)
$\sin x = -\frac{\sqrt{3}}{6}$
8) b)
$\tan x = \frac{\sqrt{11}}{11}$

9) a)
$\sin x = \frac{3\sqrt{58}}{29}$
9) b)
$\cos x = -\frac{\sqrt{319}}{29}$

10) a)
$\cos x = \frac{1}{5}$
10) b)
$\tan x = -2\sqrt{6}$

11) a)
$\sin x = -\frac{5}{7}$
11) b)
$\tan x = -\frac{5\sqrt{6}}{12}$

12) a)
$\sin x = -\frac{2\sqrt{1155}}{77}$
12) b)
$\cos x = \frac{\sqrt{1309}}{77}$

13) a)
$\sin\frac{\pi}{10} = \frac{\sqrt{3-\sqrt{5}}}{2\sqrt{2}}$
13) b)
Show that

14)
The first error is going from
$3\sin x \cos x = \sin x$ to $3\cos x = 1$. Serkan has divided his equation by $\sin x$, but this will lose the solutions when $\sin x = 0$. He should factorise rearrange and factorise instead:
$3\sin x \cos x - \sin x = 0$
$\Rightarrow \sin x (3\cos x - 1) = 0$
The second error is that he has solved $\cos x = \frac{1}{3}$ in radians when the question said to solve between 0° and 360°.

15) a)
$\theta = \left\{\frac{\pi}{4}, \frac{5\pi}{4}, \frac{9\pi}{4}, \frac{13\pi}{4}\right\}$
15) b)
$\theta = \{2.99, 6.13\}$ (3 SF)
15) c)
$\theta = \{1.22, 4.36\}$ (3 SF)

16) a)
$\theta = \left\{\frac{\pi}{8}, \frac{5\pi}{8}, \frac{9\pi}{8}, \frac{13\pi}{8}, \frac{17\pi}{8}, \frac{21\pi}{8}\right\}$
16) b)
$\theta = \{0.0696, 0.855\}$ (3 SF)
16) c)
$\theta = \{5.70, 12.0\}$ (3 SF)
16) d)
$\theta = \{3.48, 16.0, 28.6, 41.2, 53.7\}$ (3 SF)

17) a)
$\theta = \{3.30, 6.12\}$ (3 SF)
17) b)
$\theta = \{1.63, 4.65\}$ (3 SF)
17) c)
$\theta = \{6.45, 12.4, 19.0, 25.0\}$ (3 SF)

18) a)
$\theta = \{0.340, 2.80\}$ (3 SF)
18) b)
$\theta = \{1.46, 4.82, 7.74\}$ (3 SF)
18) c)
$\theta = \{0.0834, 3.06\}$ (3 SF)
18) d)
$\theta = \{-1.55, 1.55\}$ (3 SF)

19) a)
$\theta = \{0.253, 2.89\}$ (3 SF)
19) b)
$\theta = \{0.841, 1.91, 4.37, 5.44\}$ (3 SF)
19) c)
$\theta = \{3.55, 5.87\}$ (3 SF)
19) d)
$\theta = \{2.30, 3.98\}$ (3 SF)

20) a)
$\theta = \{0.253, 0.985, 2.16, 2.89\}$
20) b)
$\theta = \{1.13, 2.50, 3.79, 5.16\}$ (3 SF)
20) c)
$\theta = \{1.16, 1.98\}$ (3 SF)

20) d)
$\theta = \{0.622, 5.66\}$ (3 SF)

21) a)
$x = \begin{Bmatrix} -5.30, -4.13, -2.89, -0.253, \\ 0.985, 2.16, 3.39, 6.03 \end{Bmatrix}$ (3 SF)

21) b)
$\theta = 0.370$ (3 SF)

22) a)
$\theta = \{0.430, 0.841, 2.71, 5.44\}$ (3 SF)

22) b)
$\theta = \{1.40, 2.42, 3.86, 4.88\}$ (3 SF)

22) c)
$\theta = \{0.443, 2.70, 3.55, 5.87\}$ (3 SF)

22) d)
$\theta = \left\{1.16, \frac{7\pi}{6}, 5.12, \frac{11\pi}{6}\right\}$ (3 SF)

23) a)
$\theta = \left\{\frac{\pi}{4}, 2.82, \frac{5\pi}{4}, 5.96\right\}$ (3 SF)

23) b)
$\theta = \{0.124, 1.77, 3.27, 4.91\}$ (3 SF)

23) c)
$\theta = \{0.418, 0.876, 3.56, 4.02\}$ (3 SF)

23) d)
$\theta = \{0.896, 2.85, 4.04, 5.99\}$ (3 SF)

24) a)
$x = \begin{Bmatrix} 0.586, 2.30, 3.98, 5.70, \\ 6.87, 8.58, 10.3, 12.0 \end{Bmatrix}$ (3 SF)

24) b)
$\theta = \begin{Bmatrix} 1.76, 6.90, 11.9, 17.1, \\ 20.6, 25.8, 30.8, 35.9 \end{Bmatrix}$ (3 SF)

25) a)
Show that

25) b)
$\theta = \{0.424, 3.57\}$ (3 SF)

26) a)
Show that

26) b)
$\theta = \{1.23, 2.80\}$ (3 SF)

27)
$\theta = \{0.927, 6.12\}$ (3 SF)

28) a)
Prove

28) b)
Prove

2.05 Trigonometry: Modelling in Radians
Questions on page 231

1) a)
2π

1) b)
$\frac{1}{2}$

1) c)
1

1) d)
4

1) e)
8

1) f)
2

1) g)
52

1) h)
7

1) i)
365

1) j)
52

2) a)
$y = \sin\left(\frac{2\pi}{7}x\right)$

2) b)
$y = \sin(2\pi x)$

2) c)
$y = \sin\left(\frac{\pi}{300}x\right)$

2) d)
$y = \sin\left(\frac{\pi}{12}x\right)$

3) a)
$y = 3\sin\left(\frac{2\pi}{13}x\right)$

3) b)
$y = 2\sin(2\pi x)$

3) c)
$y = 10\sin\left(\frac{\pi}{10}x\right)$

3) d)
$y = -5\sin\left(\frac{\pi}{15}x\right)$

4) a)
1

4) b)
1

4) c)
12

4) d)
2

5) a)
Max is 1, Min is -1

5) b)
Max is 4, Min is 2

5) c)
Max is 20, Min is 8

5) d)
Max is 9, Min is -5

6) a)
$y = 3 + 2\sin\left(\frac{\pi}{5}x\right)$

6) b)
$y = 2 + \frac{1}{2}\sin\left(\frac{\pi}{15}x\right)$

6) c)
$y = -1 + 3\sin\left(\frac{\pi}{10}x\right)$

6) d)
$y = 20 - 5\sin\left(\frac{2\pi}{7}x\right)$

7) a)
$a = 2.8, b = 0.8$

7) b)
Period is 5

7) c)
2 hours 20 minutes

8) a)
37.44°C

8) b)
16:00

8) c)
$T = 38.75 - 0.45\sin\left(\frac{\pi}{12}(h+2)\right)$

9) a)
Stretch parallel to t-axis, scale factor 2.4
Followed by Translation by vector $\begin{pmatrix} 0 \\ 18.5 \end{pmatrix}$

9) b)
16:07

10) a)
104 passengers

10) b)
04:00 to 08:00, and 16:00 to 20:00

10) c)
The number of passengers is a discrete number which is then being represented by a continuous curve. It also assumes the same pattern outside of the data collection times, so for example, the model says the platform is never empty and that there are symmetries a39round the times when there are a maximum number of passengers, so it predicts the same number of passengers at both 2am and 10am which seems unrealistic!

2.06 Sequences and Series: Introduction
Questions on page 234

1) a)
Divergent as the distances between the terms are not decreasing.

1) b)
Periodic as the numbers 1 and 5 are repeating. This sequence has period 2.

1) c)
Convergent as the distances between the terms is decreasing.

1) d)
Convergent as the distances between the terms is decreasing.

2) a)
46

2) b)
53

2) c)
$\frac{137}{30}$

2) d)
$\frac{15}{2}$

3) a)
46

3) b)
$\frac{199}{3300}$

3) c)
7750

4) a)
$15 + 5k$

4) b)
$4 - 100k$

4) c)
$\frac{47}{60}k$

4) d)
$\frac{18k}{k+1}$

5)
$\frac{11}{10}$

6) a)
25

6) b)
135

6) c)
81

6) d)
$(k + 3)^2$

7)
212

8)
Show that

9)
Show that

10)
Show that

11)
Show that

12)
Show that

13)
$k = 98$

14)
$u_2 = 154, u_3 = 466, u_4 = 1402$

15)
$u_2 = \frac{501}{5}, u_3 = \frac{511}{50}, u_4 = \frac{611}{500}$

16)
$k = -\frac{23}{3}$

17)
$\frac{201}{55}$

18)
$k = 3$ or $k = \frac{7}{2}$

19) a)
Converges

19) b)
Diverges

19) c)
Diverges

19) d)
Diverges

19) e)
Converges

20) a)
Diverges

20) b)
Diverges

20) c)
Converges

20) d)
Converges

21) a)
$L = 75$

21) b)
$L = -40$

21) c)
$L = \frac{15}{29}$

21) d)
$L = -\frac{125}{31}$

22) a)
One quarter of the painkiller already in Arnold's system is broken down over the four-hour period.

22) b)
Show that

22) c)
$\frac{1600}{3} = 533.3333 \dots$ mg

23) a)
$u_1 = -40$

23) b)
$u_1 = \frac{10}{16} = \frac{5}{8}$

23) c)
$u_1 = -\frac{7}{2}$ or $u_1 = 6$

23) d)
$u_1 = -\frac{11}{6}$ or $u_1 = \frac{5}{3}$

24) a)
0

24) b)
1

24) c)
1

24) d)
2

24) e)
0

25)
$u_{155} = 1$

26)
$u_{88} = 7$

27) a)
$u_{100} = -1$

27) b)
0

28) a)
$u_{500} = -3$

28) b)
250

29) a)
$u_{99} = \frac{5}{3}$

29) b)
$\frac{809}{15}$

30) a)
$k = 1$

30) b)
$\frac{215}{2}$

31)
994

32)
-498

33)
$k = 301$

34)
2

35)
4

36)
3

37)
6

38)
8

39)
4

40)
4

41)
6

42)
Show that

43)
Show that

44)
Show that

45)
Show that

46)
Prove

47)
Prove

48)
Prove

49)
Prove

50)
Prove

51)
Prove

52)
Prove

53)
Prove

2.07 Sequences and Series: Arithmetic Sequences
Questions on page 238

1) a)
$u_n = 3 + (n-1) \times 5$
1) b)
$u_n = 25 + (n-1) \times 10$
1) c)
$u_n = 12 + (n-1) \times -6$
1) d)
$u_n = 150 + (n-1) \times -15$
1) e)
$u_n = -44 + (n-1) \times 3$
1) f)
$u_n = -100 + (n-1) \times -21$
1) g)
$u_n = \frac{5}{6} + (n-1) \times \frac{1}{5}$
1) h)
$u_n = \frac{11}{4} + (n-1) \times -\frac{9}{2}$

2)
1506

3)
$\frac{487}{6}$

4)
164

5)
289

6)
120

7)
$a = 6, d = 8$

8)
$a = 28, d = \frac{5}{2}$

9)
$a = -9, d = 80$

10)
$u_{33} = 101$

11)
$d = \frac{1}{6} - \frac{1}{6}\sqrt{3} + \frac{1}{6}\sqrt{5}$

12)
23 cm

13)
$p = \frac{4}{9}k$

14)
$k = -\frac{7}{2}$ or $k = \frac{5}{3}$

15)
$k = -\frac{2}{5}$ or $k = \frac{1}{9}$

16)
$k = \ln\frac{1}{2}$ or $k = \ln 2$

17)
$k = 3$ or $k = \log_2 20$

18) a)
$k = -4$
18) b)
911

19)
1580

20)
128330

21)
−16590

22)
−2160

23) a)
$a = \frac{83}{2}, d = 6$

23) b)
34

24)
Show that

25)
Show that

26)
20512.5

27)
3808

28)
4884

29)
−18739.5

30)
£5220

31)
1855

32)
$k = 17$

33)
$k = 38$

34) a)
66 weeks
34) b)
£9.50

35)
6880

36)
$d = -\frac{16}{275}$

37)
$n = 88$

38) a)
Show that
38) b)
$a = 11, d = -6$

39) a)
Show that
39) b)
$a = 5, d = 8$
39) c)
3630

40) a)
Show that
40) b)
$a = 11, d = 3$

41) a)
Show that
41) b)
$a = 8, d = -2$

42) a)
Show that
42) b)
$a = 1, d = 3$

43)
$637 - 13\sqrt{2}$

44)
$24 \leq n \leq 34$

45)
$p = 2, q = -100$

46)
$k + \frac{1}{2}$

47)
$k = 12$

48)
$k = 250$

49)
Show that

50)
60300

51)
77280

52)
$p = 2, u_5 = \frac{460}{3}$

2.08 Sequences and Series: Geometric Sequences
Questions on page 241

1) a)
$u_n = 3 \times 4^{n-1}$
1) b)
$u_n = 80 \times \left(\frac{3}{2}\right)^{n-1}$
1) c)
$u_n = 240 \times \left(\frac{1}{2}\right)^{n-1}$
1) d)
$u_n = 10 \times \left(\frac{9}{10}\right)^{n-1}$
1) e)
$u_n = 22 \times (-2)^{n-1}$
1) f)
$u_n = \frac{1}{3} \times (-4)^{n-1}$
1) g)
$u_n = \frac{9}{10} \times \left(-\frac{10}{9}\right)^{n-1}$

1) h)
$u_n = 88 \times \left(-\frac{3}{4}\right)^{n-1}$

2)
$u_{15} = 24576$

3)
$u_6 = \frac{2048}{81}$

4)
16

5)
11

6)
9

7)
$u_{14} = 1638.4$

8)
$a = \frac{7}{2}, r = 2$

9)
$a = \frac{2}{81}, r = \pm 9$

10)
$a = -\frac{3}{5}, r = -25$

11)
$a = 0.008, r = 6$

12)
$u_{15} = 10\,522\,531.8$

13) a)
Show that
13) b)
$-\frac{22}{19} + \frac{13}{19}\sqrt{5}$

14)
$k = \frac{2}{5}$

15)
$k = -\frac{18}{37}$ or $k = \frac{8}{3}$

16)
$k = \frac{14}{3}$

17)
$k = \frac{3}{8}$

18)
5115

19)
2839.85

20)
17.4762

21)
-4901.0949

22)
$\frac{35839}{1536}$

23) a)
9.10 miles (2 DP)
23) b)
82.90 miles (2 DP)

24) a)
$a = \frac{128}{9375}, r = \frac{2}{5}$
24) b)
20 terms

25)
1779.77 (2 DP)

26)
22510.34 (2 DP)

27)
Show that

28)
Show that

29)
$r = \sqrt{7}$

30) a)
Show that
30) b)
$k = 20$

31) a)
Show that
31) b)
$k = 78$

32) a)
Show that
32) b)
$k = 3$

33)
$k = 9$

34)
$k = 10$

35) a)
£281.38 to the nearest penny.
35) b)
48 weeks
35) c)
£30,424.05 to the nearest penny

36)
As $r > 1$, the series is not convergent, and hence will have no sum to infinity.

37) a) Converges
37) b) Converges
37) c) Diverges
37) d) Converges
37) e) Converges
37) f) Converges

38) a) $S_\infty = \frac{100}{3}$
38) b) $S_\infty = \frac{4000}{3}$
38) c) $S_\infty = 9900$
38) d) $S_\infty = -12800$
38) e) $S_\infty = \frac{2460}{11}$
38) f) $S_\infty = -\frac{108}{23}$

39) $r = \frac{1}{9}$

40) $S_\infty = \frac{3}{2}$

41) $S_\infty = 500$

42) $S_\infty = \frac{3\sqrt{2}+\sqrt{6}}{4}$

43) $\{r : r \in \mathbb{R}, -1 < r < \frac{1}{2}\}$

44) $\{r : r \in \mathbb{R}, -1 < r < \frac{31}{40}\}$

45) $\{r : r \in \mathbb{R}, \frac{4}{5} < r < 1\}$

46) $\{r : r \in \mathbb{R}, \frac{1}{5} < r < \frac{3}{5}\}$

47) 9000

48) 99825

49) $\frac{16}{25}$

50) a) $S_\infty = 72$
50) b) $v_1 = 576, v_2 = 432, v_3 = 324, v_4 = 243$
50) c) 2304

51) $a = 20, r = \frac{1}{2}$

52) $a = 18000$ or $a = 27000$

53) 494.59 (2 DP)

54) $r = \pm\frac{3\sqrt{10}}{10}$

55) $\frac{500}{199}\sqrt{2} + \frac{50}{199}$

56) $k > \frac{1}{2}$

57) 1

58) $\frac{2\sqrt{6}+3}{5}$

59) $-\frac{1}{6}$

60) $-\frac{107}{191} + \frac{48}{191}\sqrt{15} - \frac{6}{191}\sqrt{5} + \frac{92}{191}\sqrt{3}$

61) $k < \frac{\ln\left(\frac{81}{25000}\right)}{\ln\frac{3}{5}} = 11.22$ (2 DP)

62) Show that

63) $k > \frac{\ln\frac{5}{16}}{\ln\frac{1}{2}} = 1.68$ (2 DP)

64) a) $u_5 = \frac{1}{16}k$
64) b) Show that
64) c) $\sum_{n=1}^{\infty} v_n = \sum_{n=1}^{\infty}(0.5 + k \times 0.5^{n-1})$
$= \sum_{n=1}^{\infty} 0.5 + \sum_{n=1}^{\infty} k \times 0.5^{n-1} = \left(\sum_{n=1}^{\infty} 0.5\right) + 2k$

$\sum_{n=1}^{\infty} 0.5$ diverges as it is adding together an infinite number of 0.5s. Therefore, the sum to infinity of v_n does not converge.

65) $k = 4000\pi$

66) Show that

67) Show that

68) $k = \frac{1}{2}$

69) $\frac{1330}{729}$

70) a) This is an arithmetic sequence, with first term 5 and common difference 3
70) b) This is a geometric sequence, with first term 4 and common ratio 2
70) c) Not an arithmetic or geometric sequence

71) a) Not an arithmetic or geometric sequence
71) b) 6696

72) a) Show that
72) b) $k = 8$

2.09 Binomial Series
Questions on page 245

1) a) $1 + \frac{1}{4}x - \frac{3}{32}x^2 + \frac{7}{128}x^3 + \cdots, |x| < 1$
1) b) $1 - 2x + 4x^2 - 8x^3 + \cdots, |x| < \frac{1}{2}$
1) c) $1 - \frac{3}{2}x - \frac{9}{8}x^2 - \frac{27}{16}x^3 + \cdots, |x| < \frac{1}{3}$
1) d) $\frac{1}{8} - \frac{3}{16}x + \frac{3}{16}x^2 - \frac{5}{32}x^3 + \cdots, |x| < 2$
1) e) $3 + \frac{1}{2}x - \frac{1}{24}x^2 + \frac{1}{144}x^3 + \cdots, |x| < 3$
1) f) $4 - \frac{9}{2}x - \frac{81}{32}x^2 - \frac{729}{256}x^3 + \cdots, |x| < \frac{4}{9}$
1) g) $2 - \frac{1}{6}x - \frac{1}{72}x^2 - \frac{5}{2592}x^3 + \cdots, |x| < 4$
1) h) $\frac{\sqrt{3}}{3} - \frac{\sqrt{3}}{9}x + \frac{\sqrt{3}}{18}x^2 - \frac{5\sqrt{3}}{162}x^3 + \cdots, |x| < \frac{3}{2}$

2) a)
$-\frac{6237}{2048}$

2) b)
$-\frac{112}{243}$

2) c)
$\frac{1155}{32768}$

3) a)
$\frac{1}{3} - \frac{1}{54}x + \frac{1}{648}x^2 + \cdots$

3) b)
$\frac{1}{3} + \frac{1}{54}x^2 + \frac{1}{648}x^4 + \cdots$

4) a)
$1 - \frac{1}{2}x + \frac{3}{8}x^2 + \cdots$

4) b)
$\left(1 + \frac{2}{5}x\right)^{-\frac{1}{2}} \approx 1 - \frac{1}{5}x + \frac{3}{50}x^2 + \cdots$
$\sqrt{10}(5 + 2x)^{-\frac{1}{2}} = \sqrt{2} - \frac{\sqrt{2}}{5}x + \frac{3\sqrt{2}}{50}x^2 + \cdots$

5) a)
$3 - \frac{1}{6}x - \frac{1}{216}x^2 + \cdots$

5) b)
$\frac{611}{216}$

6) a)
$1 + \frac{5}{3}x - \frac{25}{9}x^2 + \frac{625}{81}x^3 + \cdots$

6) b)
1.0362

7)
$(2 - 3x)^{\frac{1}{2}} \approx \sqrt{2} - \frac{3\sqrt{2}}{4}x - \frac{9\sqrt{2}}{32}x^2 + \cdots$
$\sqrt{170} \approx \frac{2951}{320}\sqrt{2}$

8) a)
$(1 + 3x)^{\frac{1}{2}} \approx 1 + \frac{3}{2}x - \frac{9}{8}x^2 + \cdots$
$(1 + x)^{-\frac{1}{2}} \approx 1 - \frac{1}{2}x + \frac{3}{8}x^2 + \cdots$

8) b)
Show that

8) c)
$|x| < \frac{1}{3}$

9) a)
$1 - x - \frac{1}{2}x^2 + \cdots$

9) b)
Show that

9) c)
$|x| < \frac{1}{2}$

10) a)
$p = -\frac{1}{2}, q = 4$

10) b)
$|x| < \frac{1}{4}$

11) a)
$k = -\frac{1}{288}$

11) b)
1.913 (3 DP)

12) a)
$\frac{59}{72}\sqrt{3}$

12) b)
Overestimate since all the terms in the expansion except the first are negative.

13) a)
$\frac{\sqrt{2}}{2} + \frac{\sqrt{2}}{8}x + \frac{3\sqrt{2}}{64}x^2 + \cdots$

13) b)
Let the expansion be f(x)
At $x = -3$: $\frac{1}{\sqrt{2+3}} = \frac{1}{\sqrt{5}}$ so $\sqrt{5} = 5f(-3)$
At $x = -\frac{6}{5}$: $\frac{1}{\sqrt{2-\frac{6}{5}}} = \frac{\sqrt{5}}{2}$ so $\sqrt{5} = 2f\left(-\frac{6}{5}\right)$
At $x = \frac{1}{5}$: $\frac{1}{\sqrt{2-\frac{1}{5}}} = \frac{\sqrt{5}}{3}$ so $\sqrt{5} = 3f\left(\frac{1}{5}\right)$

13) c)
Since it is valid for $\left|\frac{1}{2}x\right| < 1 \Rightarrow |x| < 2$
$x = -3$ may not be used.

14)
$\frac{49\sqrt{7}}{69}$

15)
With each additional term the approximation should get closer to the exact result within the range of validity. $y = 1 + \frac{1}{2}x^2$ is a positive quadratic graph so this means the second term increases in value and is further away from the graph of $\sqrt{1-x^2}$ than $y = 1$

16)
$a = 3, b = -2$

17) a)
$1 + \frac{1}{2}px - \frac{1}{8}p^2x^2 + \frac{1}{16}p^3x^3 + \cdots$

17) b)
$p = 4, q = -6$

18) a)
$1 + 2x + 3x^2 + 4x^3 + \cdots$

18) b)
$\frac{2-x}{(1-x)^2}$

19) a)
$A = -3, B = 1$

19) b)
$2 + x - x^2 - \cdots$

20) a)
$\frac{5}{3} - \frac{47}{9}x + \frac{677}{27}x^2 - \cdots$

20) b)
$|x| < \frac{1}{5}$

21) a)
$2 - 19x + 112x^2 - \cdots$

21) b)
$|x| < \frac{1}{3}$

22) a)
$\frac{1}{3} + \frac{1}{54}x + \frac{1}{648}x^2 + \cdots$

22) b)
$\frac{367}{1080}$

23)
$2 + \frac{1}{2}x - \frac{1}{16}x^2 + \cdots$

2.10 Domain, Range and Composite Functions
Questions on page 247

1) a)
A function

1) b)
Not a function

1) c)
Not a function

1) d)
A function

2) a)
Many-to-one, a function

2) b)
One-to-one, a function

2) c)
Many-to-one, not a function due to asymptote

2) d)
One-to-many, not a function

2) e)
Many-to-many, not a function

2) f)
Many-to-one, a function

3) a)

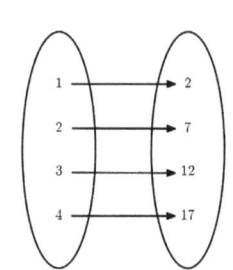

3) b)
One-to-one function

4) a)

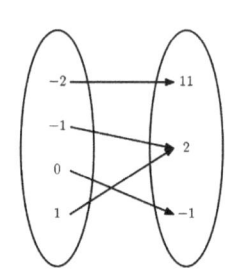

4) b)
A function, many-to-one

5) a)

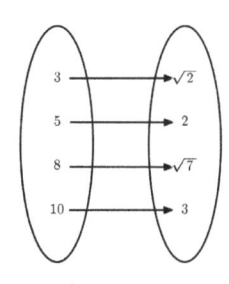

(Mapping diagram: 3 → √2, 5 → 2, 8 → √7, 10 → 3)

5) b)
A function, one-to-one

6) a) One-to-one
6) b) Many-to-one
6) c) One-to-one
6) d) One-to-one
6) e) Many-to-one
6) f) One-to-one
6) g) One-to-one
6) h) Many-to-one
6) i) One-to-one
6) j) One-to-one

7) a)
$\{f(x): f(x) \in \mathbb{R}, f(x) > 1\}$
7) b)
$\{y: y \in \mathbb{R}, y \geq -5\}$
7) c)
$\{f(x): f(x) \in \mathbb{R}\}$
7) d)
$\{y: y \in \mathbb{R}, y < 0\}$
7) e)
$\{f(x): f(x) \in \mathbb{R}, f(x) < 5\}$

8) a)
Domain: $\{x: x \in \mathbb{R}\}$
Range: $\{y: y \in \mathbb{R}, -1 \leq y \leq 1\}$
8) b)
Domain: $\{x: x \in \mathbb{R}, x \geq 0\}$
Range: $\{y: y \in \mathbb{R}, y \geq 0\}$
8) c)
Maximum at $(-4, -9)$
Domain: $\{x: x \in \mathbb{R}\}$
Range: $\{y: y \in \mathbb{R}, y \leq -9\}$
8) d)
Minimum at $\left(\frac{3}{2}, \frac{9}{2}\right)$
Domain: $\{x: x \in \mathbb{R}\}$
Range: $\left\{y: y \in \mathbb{R}, y \geq \frac{9}{2}\right\}$
8) e)
Domain: $\{x: x \in \mathbb{R}, x \neq 3\}$
Range: $\{y: y \in \mathbb{R}, y \neq 0\}$

8) f)
Domain: $\{x: x \in \mathbb{R}, x \neq 0\}$
Range: $\{y: y \in \mathbb{R}, y \neq 4\}$
8) g)
Domain: $\{x: x \in \mathbb{R}\}$
Range: $\{y: y \in \mathbb{R}, y > 2\}$
8) h)
Domain: $\{x: x \in \mathbb{R}, x > 0\}$
Range: $\{y: y \in \mathbb{R}\}$
8) i)
Domain: $\{x: x \in \mathbb{R}, x > 2\}$
Range: $\{y: y \in \mathbb{R}\}$

9) a)
Domain: $\{x: x \in \mathbb{R}, -5 \leq x \leq 3\}$
Range: $\{y: y \in \mathbb{R}, -2 \leq y \leq 6\}$
9) b)
Domain: $\{x: x \in \mathbb{R}, -3 < x < 4\}$
Range: $\{y: y \in \mathbb{R}, -5 < y < 9\}$
9) c)
Domain: $\{x: x \in \mathbb{R}, -2 \leq x < 2\}$
Range: $\{y: y \in \mathbb{R}, -8 \leq y < 8\}$
9) d)
Domain: $\{x: x \in \mathbb{R}, -3 \leq x < 2\}$
Range: $\{y: y \in \mathbb{R}, 0 \leq y \leq 9\}$
9) e)
Domain: $\{x: x \in \mathbb{R}, -3 \leq x \leq 4\}$
Range: $\{y: y \in \mathbb{R}, -\sqrt{7} \leq y \leq \sqrt{7}\}$
9) f)
Domain: $\{x: x \in \mathbb{R}, -2 \leq x \leq 3\}$
Range: $\{y: y \in \mathbb{R}, -6.5 < y \leq 7\}$
9) g)
Domain: $\{x: x \in \mathbb{R}\}$
Range: $\{y: y \in \mathbb{R}, -2\pi < y < 2\pi\}$
9) h)
Domain: $\left\{x: x \in \mathbb{R}, -\pi \leq x < -\frac{\pi}{2}\right\}$
$\cup \left\{x: x \in \mathbb{R}, -\frac{\pi}{2} < x < \frac{\pi}{2}\right\}$
$\cup \left\{x: x \in \mathbb{R}, \frac{\pi}{2} < x \leq \pi\right\}$
Range: $\{y: y \in \mathbb{R}, y \geq 4\}$
$\cup \{y: y \in \mathbb{R}, y \leq 0\}$

10)
(Graph with maximum point $(p, 3)$)

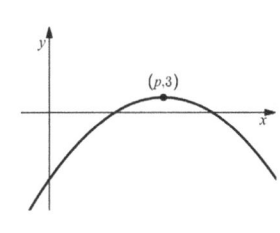

11)
(Graph with point of inflection at $x = p$)

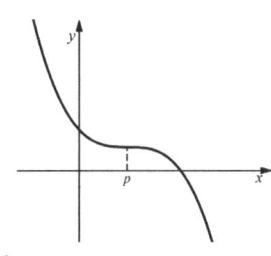

12) a)
Range: $\{f(x): f(x) \in \mathbb{R}, f(x) \geq 0\}$

12) b)
Range: $\left\{f(x): f(x) \in \mathbb{R}, f(x) \geq -\frac{3}{2}\right\}$
12) c)
Range: $\left\{f(x): f(x) \in \mathbb{R}, 2 \leq f(x) \leq \frac{5}{2}\right\}$

13) a)
Range: $\{f(x): f(x) \in \mathbb{R}, f(x) \leq 18\}$
13) b)
Range: $\{f(x): f(x) \in \mathbb{R}, f(x) \leq 13.5\}$
13) c)
Range: $\{f(x): f(x) \in \mathbb{R}, 5.5 < f(x) < 17.5\}$
13) d)
$f(3) = -0.5 \times 3^2 + 5 \times 3 + 5.5 = 16$
$= f(7)$
Range: $\{f(x): f(x) \in \mathbb{R}, 16 \leq f(x) \leq 18\}$

14) a)
Range: $\left\{f(x): f(x) \in \mathbb{R}, 0 < f(x) \leq \frac{1}{3}e^3\right\}$
14) b)
Range: $\left\{f(x): f(x) \in \mathbb{R}, 0 \leq f(x) \leq \frac{1}{3}e^3\right\}$
14) c)
Range: $\{f(x): f(x) \in \mathbb{R}, f(x) \geq 0\}$
14) d)
Range: $\left\{f(x): f(x) \in \mathbb{R}, \frac{25}{12} \leq f(x) < \frac{1}{3}e^3\right\}$

15) a)
(Parabola with intercepts $-4 - 2\sqrt{3}$ and $-4 + 2\sqrt{3}$, y-intercept 4, minimum at $(-4, -12)$)

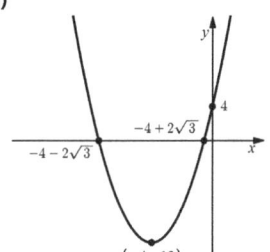

15) b)
Range: $\{f(x): f(x) \in \mathbb{R}, f(x) \geq -12\}$

16) a)
Range: $\{f(x): f(x) \in \mathbb{R}, f(x) \geq 5\}$
16) b)
Range: $\left\{f(x): f(x) \in \mathbb{R}, f(x) < -\frac{8}{3}\right\}$
16) c)
Range: $\{f(x): f(x) \in \mathbb{R}, -27 < f(x) \leq 21\}$
16) d)
Range: $\{f(x): f(x) \in \mathbb{R}, 4 < f(x) \leq 8\}$
16) e)
Range: $\{f(x): f(x) \in \mathbb{R}, 0 < f(x) \leq 2.25\}$
16) f)
Range: $\{f(x): f(x) \in \mathbb{R}, 0 < f(x) \leq 1\}$

17)
Range: $\left\{f(x): f(x) \in \mathbb{R}, 0 < f(x) \leq \frac{1}{7}\right\}$

18)
$k = \pm 12$

19)
$p = \frac{1}{2}(a + b)$

20) a)
Domain: $\{x : x \in \mathbb{R}\}$
Range: $\{f(x) : f(x) \in \mathbb{R}, f(x) \leq 3\}$

20) b)
Domain: $\{x : x \in \mathbb{R}\}$
Range: $\{f(x) : f(x) \in \mathbb{R}, f(x) \leq 2\ln(2) - 2\}$

21)
Domain: $\{x : x \in \mathbb{R}, x \geq 0\}$
Range: $\left\{f(x) : f(x) \in \mathbb{R}, f(x) \geq -\frac{1}{e}\right\}$

22) a)
Range: $\{f(x) : f(x) \in \mathbb{R}, -80 \leq f(x) < 28\}$

22) b)
Range: $\{g(x) : g(x) \in \mathbb{R}, -160 \leq g(x) < 56\}$

22) c)
Range: $\{h(x) : h(x) \in \mathbb{R}, -244 \leq h(x) < 80\}$

23) a)
$fg(-1) = 14$

23) b)
$gf(2) = 83$

23) c)
$ff(4) = 94$

24)
$fg(4) = \frac{1}{4}$

25) a)
$fg(x) = 2x^2 + 1$

25) b)
$gf(x) = (2x+1)^2$

25) c)
$ff(x) = 4x + 3$

25) d)
$a = 4$

25) e)
$x = 0$ or $x = -2$

26) a)
$f^{89}(x) = \frac{1}{x}$

26) b)
$f^{100}(x) = x$

27)
$x = 4$

28) a)
$fg(x) = 5x$

28) b)
Show that

29)
$k = 3$ or $k = \frac{7}{4}$

30) a)
-3

30) b)
$ff\left(\frac{1}{2}\right) = -2$

30) c)
$\{x : x \in \mathbb{R}, x < -6\} \cup \left\{x : x \in \mathbb{R}, x > \frac{9}{2}\right\}$

31) a)
1

31) b)
$\{f(x) : f(x) \in \mathbb{R}, f(x) \geq -4\}$

31) c)
$ff(2) = 0$

31) d)
$\{x : x \in \mathbb{R}, -3 < x < 3\}$

32) a)
$a = 4$

32) b)
$ff\left(\frac{7}{2}\right) = 6 - e^{\frac{1}{2}}$

32) c)
$\{x : x \in \mathbb{R}, -1 \leq x \leq 4 + \ln 6\}$

33) a)
$\{f(x) : f(x) \in \mathbb{R}, f(x) \neq 2\}$

33) b)
$ff(x) = \frac{5x-3}{26-3x}$

34) a)
$\{f(x) : f(x) \in \mathbb{R}, f(x) \neq a\}$

34) b)
$ff(x) = \frac{(a^2+1)x + a - 4}{(a-4)x + 17}$

35) a)
$fg(x) = (a^2 - x)^2 - a(a^2 - x)$

35) b)
$a = 2, a = -1, a = \sqrt{2}, a = -\sqrt{2}$

36) a)
$f(x) \in (0, \infty)$

36) b)
$f(x) \in [0, 8e]$

36) c)
$f(x) \in (0, 18]$

36) d)
$f(x) \in [0, 8e]$

37) a)

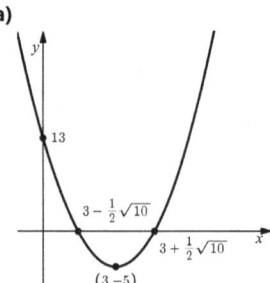

37) b)
$f(x) \in [-5, \infty)$

38) a)
Range: $f(x) \in \left[-\frac{49}{32}, \frac{113}{32}\right]$

38) b)
Range: $f(x) \in [-98, 226]$

38) c)
Range: $f(x) \in \left[-\frac{89}{4}, \frac{73}{4}\right]$

38) d)
Domain: $x \in \left(-\frac{1}{2}, \frac{1}{2}\right)$

39) a)
$a = 2$

39) b)
$k = 2$

39) c)
$ff(3) = 7 - 2e$

39) d)
$x \in \left[\frac{3}{2}, 2 + \ln 2\right]$

2.11 Graph Transformations: Combinations
Questions on page 253

1) a)
Stretch, parallel to x-axis, scale factor $\frac{1}{4}$
Stretch, parallel to y-axis, scale factor 3
(either order)

1) b)
Stretch, parallel to x-axis, scale factor 3
Stretch, parallel to y-axis, scale factor $\frac{1}{2}$
(either order)

2) a)
Translation by the vector $\begin{pmatrix} -6 \\ 0 \end{pmatrix}$
Reflection in the x-axis (either order)

2) b)
Reflection in the y-axis
Translation by the vector $\begin{pmatrix} 0 \\ 6 \end{pmatrix}$
(either order)

3) a)
Translation by the vector $\begin{pmatrix} 4 \\ 0 \end{pmatrix}$
Stretch, parallel to y-axis, scale factor $\frac{1}{6}$
(either order)

3) b)
Stretch, parallel to x-axis, scale factor $\frac{1}{8}$
Translation by the vector $\begin{pmatrix} 0 \\ -5 \end{pmatrix}$
(either order)

4) a)
Translation by the vector $\begin{pmatrix} \frac{\pi}{3} \\ 0 \end{pmatrix}$
Stretch, parallel to y-axis, scale factor 2
(either order)

4) b)
Stretch, parallel to x-axis, scale factor 5
Translation by the vector $\begin{pmatrix} 0 \\ 2 \end{pmatrix}$
(either order)

5) a)
Stretch, parallel to x-axis, scale factor 3
Stretch, parallel to y-axis, scale factor 5
(either order)

5) b)
Reflection in the y-axis

Translation by the vector $\begin{pmatrix} 0 \\ -9 \end{pmatrix}$
(either order)

6) a)
Translation by the vector $\begin{pmatrix} 4 \\ 0 \end{pmatrix}$
Stretch, parallel to y-axis, scale factor 2
(either order)

6) b)
Reflection in the y-axis
Translation by the vector $\begin{pmatrix} 0 \\ 5 \end{pmatrix}$
(either order)
Or
Reflection in the x-axis followed by a
Translation by the vector $\begin{pmatrix} 0 \\ 5 \end{pmatrix}$

7) a)
Stretch, parallel to x-axis, scale factor $\frac{1}{2}$
Translation by the vector $\begin{pmatrix} 0 \\ -3 \end{pmatrix}$
(either order)

7) b)
Stretch, parallel to x-axis, scale factor $\frac{e^3}{2}$

8) a)
$(-2, 24)$
8) b)
$(-1, -12)$

9) a)
$(24, -1)$
9) b)
$(-4, 6)$

10) a)

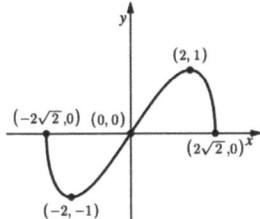

Graph with points $(-2\sqrt{2},0)$, $(0,0)$, $(2\sqrt{2},0)$, $(2,1)$, $(-2,-1)$

10) b)

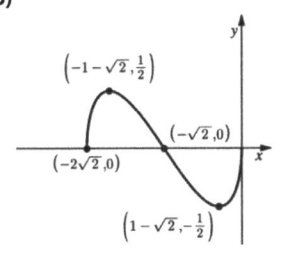

Graph with points $(-1-\sqrt{2}, \frac{1}{2})$, $(-\sqrt{2},0)$, $(-2\sqrt{2},0)$, $(1-\sqrt{2},-\frac{1}{2})$

10) c)

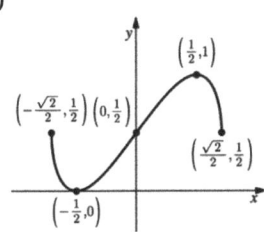

Graph with points $(-\frac{\sqrt{2}}{2}, \frac{1}{2})$, $(0, \frac{1}{2})$, $(\frac{1}{2}, 1)$, $(\frac{\sqrt{2}}{2}, \frac{1}{2})$, $(-\frac{1}{2}, 0)$

10) d)

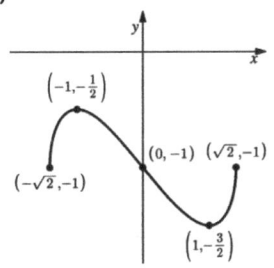

Graph with points $(-1, -\frac{1}{2})$, $(-\sqrt{2}, -1)$, $(0, -1)$, $(\sqrt{2}, -1)$, $(1, -\frac{3}{2})$

11) a)
Stretch, parallel to y-axis, scale factor 2
followed by a
Translation by the vector $\begin{pmatrix} 0 \\ 3 \end{pmatrix}$

11) b)
Stretch, parallel to y-axis, scale factor 9
followed by a
Translation by the vector $\begin{pmatrix} 0 \\ -8 \end{pmatrix}$

11) c)
Stretch, parallel to y-axis, scale factor $\frac{1}{5}$
followed by a
Translation by the vector $\begin{pmatrix} 0 \\ -10 \end{pmatrix}$

11) d)
Stretch, parallel to y-axis, scale factor $\frac{3}{2}$
followed by a
Translation by the vector $\begin{pmatrix} 0 \\ 20 \end{pmatrix}$

12) a)
Stretch, parallel to y-axis, scale factor $\frac{5}{4}$
followed by a
Translation by the vector $\begin{pmatrix} 0 \\ \frac{13}{4} \end{pmatrix}$

12) b)
Stretch, parallel to y-axis, scale factor $\frac{1}{2}$
followed by a
Translation by the vector $\begin{pmatrix} 0 \\ \frac{19}{2} \end{pmatrix}$

13) a)
Translation by the vector $\begin{pmatrix} -1 \\ 0 \end{pmatrix}$
followed by a
Stretch, parallel to x-axis, scale factor $\frac{1}{3}$

13) b)
Translation by the vector $\begin{pmatrix} 9 \\ 0 \end{pmatrix}$
followed by a
Stretch, parallel to x-axis, scale factor $\frac{1}{2}$

13) c)
Translation by the vector $\begin{pmatrix} -2 \\ 0 \end{pmatrix}$
followed by a Reflection in the y-axis

13) d)
Translation by the vector $\begin{pmatrix} 3 \\ 0 \end{pmatrix}$
followed by a Reflection in the y-axis

14) a)
Translation by the vector $\begin{pmatrix} 4 \\ 0 \end{pmatrix}$
followed by a
Stretch, parallel to x-axis, scale factor $\frac{1}{2}$

14) b)
Translation by the vector $\begin{pmatrix} 1 \\ 0 \end{pmatrix}$
followed by a
Stretch, parallel to x-axis, scale factor 5

15) a)
$(15, -3)$
15) b)
$(10, -6)$
15) c)
$\left(\frac{5}{3}, -3\right)$
15) d)
$(48, -3)$

16) a)
Stretch, parallel to y-axis, scale factor 4
followed by a
Translation by the vector $\begin{pmatrix} 0 \\ -6 \end{pmatrix}$

16) b)
Translation by the vector $\begin{pmatrix} \frac{\pi}{9} \\ 0 \end{pmatrix}$
followed by a
Stretch, parallel to x-axis, scale factor 6

16) c)
Translation by the vector $\begin{pmatrix} -\frac{\pi}{2} \\ 0 \end{pmatrix}$
followed by a
Stretch, parallel to x-axis, scale factor $\frac{1}{2}$

16) d)
Translation by the vector $\begin{pmatrix} -\frac{\pi}{2} \\ 0 \end{pmatrix}$
Stretch, parallel to y-axis, scale factor 3
(either order)

17) a)
Translation by the vector $\begin{pmatrix} 10 \\ 0 \end{pmatrix}$
followed by a
Stretch, parallel to x-axis, scale factor $\frac{1}{6}$

17) b)
Translation by the vector $\begin{pmatrix} -9 \\ 0 \end{pmatrix}$
followed by a Reflection in the y-axis

17) c)
Stretch, parallel to y-axis, scale factor 10
followed by a
Translation by the vector $\begin{pmatrix} 0 \\ 20 \end{pmatrix}$

17) d)
Reflection in the x-axis followed by a
Translation by the vector $\begin{pmatrix} 0 \\ 18 \end{pmatrix}$

18) a)
Translation by the vector $\begin{pmatrix} 2 \\ 0 \end{pmatrix}$
followed by a
Stretch, parallel to x-axis, scale factor $\frac{1}{5}$

18) b)
Stretch, parallel to y-axis, scale factor 8
followed by a

Translation by the vector $\begin{pmatrix} 0 \\ -9 \end{pmatrix}$

18) c)
Stretch, parallel to y-axis, scale factor 6
followed by a
Translation by the vector $\begin{pmatrix} 0 \\ -5 \end{pmatrix}$

18) d)
Stretch, parallel to y-axis, scale factor $\frac{7}{20}$
followed by a
Translation by the vector $\begin{pmatrix} 0 \\ \frac{1}{10} \end{pmatrix}$

19) a)
Stretch, parallel to y-axis, scale factor $\frac{1}{5}$
Reflection in the x-axis (either order)

19) b)
Stretch, parallel to y-axis, scale factor 2
followed by a
Translation by the vector $\begin{pmatrix} 0 \\ 1 \end{pmatrix}$

19) c)
Stretch, parallel to y-axis, scale factor $\frac{1}{3}$
followed by a
Translation by the vector $\begin{pmatrix} 0 \\ \frac{1}{2} \end{pmatrix}$

19) d)
Stretch, parallel to y-axis, scale factor $\frac{1}{8}$
followed by a
Translation by the vector $\begin{pmatrix} 0 \\ -\frac{3}{4} \end{pmatrix}$

20) a)
Translation by the vector $\begin{pmatrix} -2 \\ 0 \end{pmatrix}$
followed by a
Stretch, parallel to x-axis, scale factor $\frac{1}{3}$

20) b)
Translation by the vector $\begin{pmatrix} 0 \\ \ln 3 \end{pmatrix}$

21) a)
Translation by the vector $\begin{pmatrix} -1 \\ 0 \end{pmatrix}$
followed by a
Stretch, parallel to x-axis, scale factor $\frac{1}{2}$

21) b)
Translation by the vector $\begin{pmatrix} -\frac{3}{4} \\ 0 \end{pmatrix}$
followed by a
Stretch, parallel to x-axis, scale factor 2

22) a)
Translation by the vector $\begin{pmatrix} -\frac{12}{5} \\ 0 \end{pmatrix}$
followed by a
Stretch, parallel to x-axis, scale factor $\frac{1}{2}$

22) b)
Reflection in the x-axis
Stretch, parallel to y-axis, scale factor 10
(either order)

23)
$y = \frac{5}{2}x - 11$

24)
$y = \frac{8}{3} - 2x$

25) a)
Stretch, parallel to y-axis, scale factor 4
followed by a
Translation by the vector $\begin{pmatrix} 0 \\ 9 \end{pmatrix}$

25) b)
Reflection in the x-axis followed by a
Translation by the vector $\begin{pmatrix} 0 \\ 5 \end{pmatrix}$

25) c)
Stretch, parallel to y-axis, scale factor $\frac{1}{4}$
followed by a
Translation by the vector $\begin{pmatrix} 6 \\ 3 \end{pmatrix}$

25) d)
Stretch, parallel to y-axis, scale factor 8
followed by a
Translation by the vector $\begin{pmatrix} -1 \\ -2 \end{pmatrix}$

25) e)
Translation by the vector $\begin{pmatrix} -1 \\ 0 \end{pmatrix}$
followed by a
Stretch, parallel to x-axis, scale factor $\frac{1}{4}$

25) f)
Translation by the vector $\begin{pmatrix} 9 \\ 0 \end{pmatrix}$
followed by a
Stretch, parallel to x-axis, scale factor 3

25) g)
Stretch, parallel to y-axis, scale factor 2
followed by a
Translation by the vector $\begin{pmatrix} 1 \\ -9 \end{pmatrix}$

25) h)
Stretch, parallel to y-axis, scale factor 3
followed by a
Translation by the vector $\begin{pmatrix} -2 \\ -11 \end{pmatrix}$

25) i)
Reflection in the x-axis followed by a
Translation by the vector $\begin{pmatrix} 4 \\ 25 \end{pmatrix}$

25) j)
Stretch, parallel to y-axis, scale factor 18
followed by a
Translation by the vector $\begin{pmatrix} -\frac{4}{3} \\ -72 \end{pmatrix}$

26) a)
Stretch, parallel to y-axis, scale factor 9
followed by a
Translation by the vector $\begin{pmatrix} 0 \\ -8 \end{pmatrix}$

26) b)
Stretch, parallel to y-axis, scale factor 8
followed by a
Translation by the vector $\begin{pmatrix} 0 \\ -16 \end{pmatrix}$

26) c)
Reflection in the x-axis followed by a
Translation by the vector $\begin{pmatrix} 0 \\ 9 \end{pmatrix}$

27)
$y = 8(x + 1)^3 + 2$

28)
$y = \ln\left(\frac{1}{27}x - \frac{2}{9}\right)$

29) a)
$y = 4 - x^2$

29) b)
$y = (x - 6)^2$

30) a)
$y = -\cos x - 6$

30) b)
$y = \cos\left(x + \frac{2\pi}{3}\right)$

31) a)

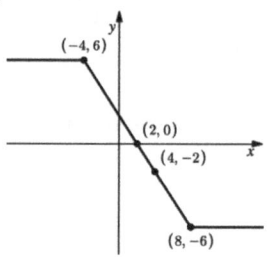

Points: $(-4, 6)$, $(2, 0)$, $(4, -2)$, $(8, -6)$

31) b)

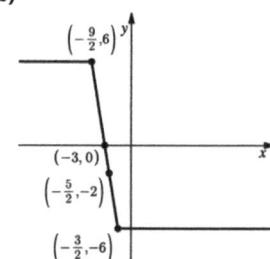

Points: $\left(-\frac{9}{2}, 6\right)$, $(-3, 0)$, $\left(-\frac{5}{2}, -2\right)$, $\left(-\frac{3}{2}, -6\right)$

31) c)

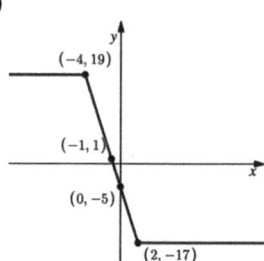

Points: $(-4, 19)$, $(-1, 1)$, $(0, -5)$, $(2, -17)$

31) d)

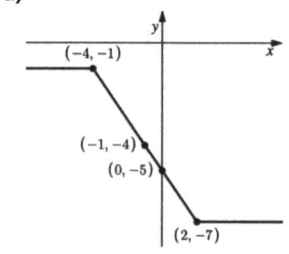

Points: $(-4, -1)$, $(-1, -4)$, $(0, -5)$, $(2, -7)$

32)
Reflection in the line $y = x$

33)
Reflection in the line $y = x$

34)

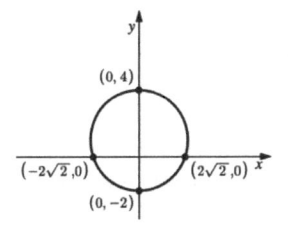

(0, 4), (−2√2, 0), (2√2, 0), (0, −2)

35)
Reflection in the line $y = x$
followed by a
Stretch, parallel to x-axis, scale factor $\frac{1}{3}$

36)
Translation by the vector $\begin{pmatrix} -5 \\ 0 \end{pmatrix}$
followed by a Reflection in the line $y = x$

37)
Reflection in the line $y = x$ followed by a
Stretch, parallel to y-axis, scale factor 2
followed by a
Translation by the vector $\begin{pmatrix} 0 \\ -7 \end{pmatrix}$

38)

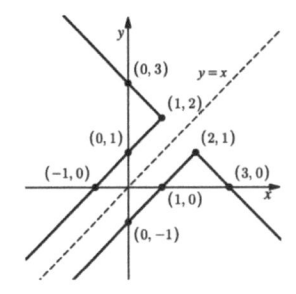

(0, 3), (1, 2), (0, 1), (2, 1), (−1, 0), (3, 0), (1, 0), (0, −1), $y = x$

39)
Translation by the vector $\begin{pmatrix} 9 \\ 0 \end{pmatrix}$
followed by a
Stretch, parallel to x-axis, scale factor 3
followed by a
Stretch, parallel to y-axis, scale factor 2
followed by a
Translation by the vector $\begin{pmatrix} 0 \\ -3 \end{pmatrix}$

40)
Translation by the vector $\begin{pmatrix} -4 \\ 0 \end{pmatrix}$
followed by a
Stretch, parallel to x-axis, scale factor $\frac{5}{2}$
followed by a
Stretch, parallel to y-axis, scale factor $\frac{3}{5}$
followed by a
Translation by the vector $\begin{pmatrix} 0 \\ 2 \end{pmatrix}$

41)
$-x = y^2$

2.12 Inverse Functions
Questions on page 256

1) a) f^{-1} exists
1) b) f^{-1} doesn't exist
1) c) f^{-1} exists
1) d) f^{-1} doesn't exist
1) e) f^{-1} exists

2) a) $f^{-1}(x) = \frac{1-x}{2}, x \in \mathbb{R}$
2) b) $f^{-1}(x) = 4x - 3, x \in \mathbb{R}$
2) c) $f^{-1}(x) = \frac{7}{2x}, x \neq 0$
2) d) $f^{-1}(x) = \frac{3x+1}{x-2}, x \neq 2$
2) e) $f^{-1}(x) = \frac{x+9}{4-2x}, x \neq 2$

3) a) $f^{-1}(x) = 2x + 10, x \in \mathbb{R}$
3) b)

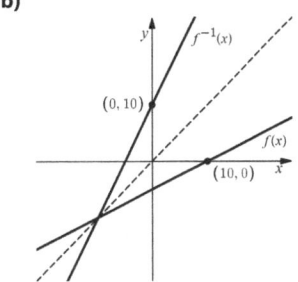

(0, 10), (10, 0), $f(x)$, $f^{-1}(x)$

3) c)
$f(12) = 1$
$f^{-1}(1) = 12$
The point (12, 1) on f reflected across the line $y = x$ is the point (1, 12) on f^{-1}

4)
A reflection across the line $y = x$

5) a) (3, 1)
5) b) (4, −5)
5) c) $(-3e^2, 3)$
5) d) (b, a)

6) a) $x = \frac{1}{2}$
6) b) $x = 2$
6) c) $x = 3$
6) d) $x = 1$
6) e) $x = 0$ or $x = 1$
6) f) $x = 1$

7) It is not one-to-one

8) a) Show that
8) b) $f^{-1}(x) = -x + 12, x \neq 12$

9) a) Show that
9) b) $f^{-1}(x) = \frac{1}{x}, x \neq 0$

10) a) Show that
10) b) $f^{-1}(x) = \frac{7x-48}{x-7}, x \neq 7$

11) a) Show that
11) b) $f^{-1}(x) = \frac{ax-1}{x-a}, x \neq a$

12) a) $f^{-1}(x) = \sqrt{\frac{1}{2}x - \frac{1}{2}}, x \geq 1$
12) b) Show that
12) c)

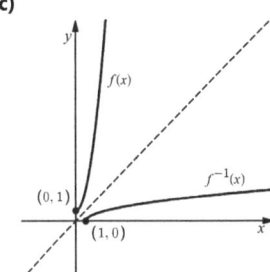

(0, 1), (1, 0), $f(x)$, $f^{-1}(x)$

12) d)
Domain: $\{x : x \in \mathbb{R}, x \geq 1\}$
Range: $\{f^{-1}(x) : f^{-1}(x) \in \mathbb{R}, f^{-1}(x) \geq 0\}$

13) a) $f^{-1}(x) = \frac{1}{3}\ln\left(\frac{1}{2}x + \frac{1}{2}\right), x > -1$
13) b)

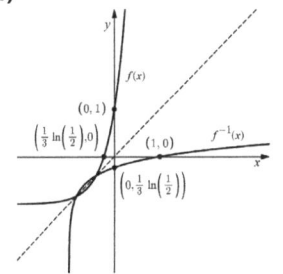

(0, 1), $\left(\frac{1}{3}\ln\left(\frac{1}{2}\right), 0\right)$, (1, 0), $\left(0, \frac{1}{3}\ln\left(\frac{1}{2}\right)\right)$, $f(x)$, $f^{-1}(x)$

13) c)
Domain: $\{x : x \in \mathbb{R}, x > -1\}$
Range: $\{f^{-1}(x) : f^{-1}(x) \in \mathbb{R}\}$

14) a)
$f^{-1}(x) = -\log_5(x-3), x > 3$

14) b)

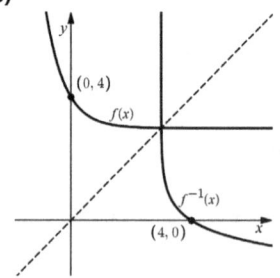

14) c)
Domain: $\{x : x \in \mathbb{R}, x > 3\}$
Range: $\{f^{-1}(x) : f^{-1}(x) \in \mathbb{R}\}$

15) a)
$f^{-1}(x) = \frac{5x}{2-x}, -8 \leq x \leq \frac{1}{3}$

15) b)
$x = 0$ or $x = -3$

15) c)

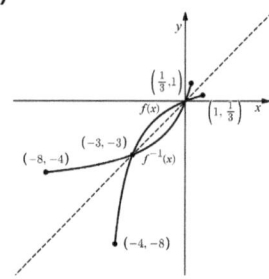

15) d)
Domain: $\left\{x : x \in \mathbb{R}, -8 \leq x \leq \frac{1}{3}\right\}$
Range: $\{f^{-1}(x) : f^{-1}(x) \in \mathbb{R}, -4 \leq f^{-1}(x) \leq 1\}$

16) a)
Domain: $\{x : x \in \mathbb{R}, -7 \leq x \leq 23\}$
Range: $\{f^{-1}(x) : f^{-1}(x) \in \mathbb{R}, -1 \leq f^{-1}(x) \leq 5\}$

16) b)
Domain: $\{x : x \in \mathbb{R}\}$
Range: $\{f^{-1}(x) : f^{-1}(x) \in \mathbb{R}, f^{-1}(x) > 0\}$

16) c)
Domain: $\{x : x \in \mathbb{R}, x > 0\}$
Range: $\{f^{-1}(x) : f^{-1}(x) \in \mathbb{R}\}$

16) d)
Domain: $\{x : x \in \mathbb{R}, -1 \leq x \leq 1\}$
Range: $\left\{f^{-1}(x) : f^{-1}(x) \in \mathbb{R}, -\frac{\pi}{2} \leq f^{-1}(x) \leq \frac{\pi}{2}\right\}$

16) e)
Domain: $\{x : x \in \mathbb{R}, -1 \leq x \leq 1\}$
Range: $\{f^{-1}(x) : f^{-1}(x) \in \mathbb{R}, 0 \leq f^{-1}(x) \leq \pi\}$

16) f)
Domain: $\{x : x \in \mathbb{R}\}$
Range: $\left\{f^{-1}(x) : f^{-1}(x) \in \mathbb{R}, -\frac{\pi}{2} < f^{-1}(x) < \frac{\pi}{2}\right\}$

17) a)
Show that

17) b)
Show that

17) c)
The function is not one-to-one.

17) d)
Domain: $\{x : x \in \mathbb{R}, 0 \leq x \leq 12\}$
Range: $\{f^{-1}(x) : f^{-1}(x) \in \mathbb{R}, 0 \leq f^{-1}(x) \leq 2\}$

18) a)

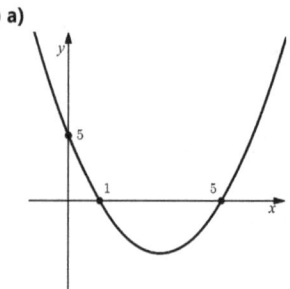

18) b)
$p = 3$

18) c)
Range: $\{f(x) : f(x) \in \mathbb{R}, f(x) \geq -4\}$

18) d)
Domain: $\{x : x \in \mathbb{R}, x \geq -4\}$
Range: $\{f^{-1}(x) : f^{-1}(x) \in \mathbb{R}, f^{-1}(x) \geq 3\}$

18) e)
Show that

18) f)

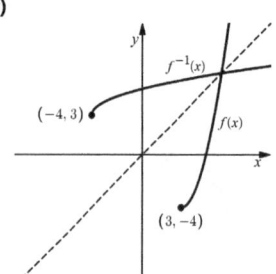

19) a)
$p = 3$

19) b)
$\{f(x) : f(x) \in \mathbb{R}, f(x) \in \mathbb{R}\}$

19) c)
$f^{-1}(x) = e^x + 3, x \in \mathbb{R}$

19) d)
Domain: $\{x : x \in \mathbb{R}, x \in \mathbb{R}\}$
Range: $\{f^{-1}(x) : f^{-1}(x) \in \mathbb{R}, f^{-1}(x) > 3\}$

19) e)

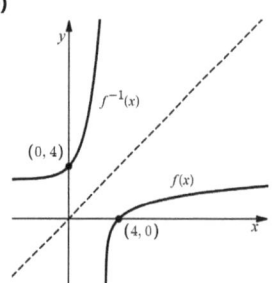

20) a)
$p = 5$

20) b)
$\{f(x) : f(x) \in \mathbb{R}, f(x) \geq 0\}$

20) c)
$f^{-1}(x) = x^2 + 5, x \geq 0$

20) d)
Domain: $\{x : x \in \mathbb{R}, x \geq 0\}$
Range: $\{f^{-1}(x) : f^{-1}(x) \in \mathbb{R}, f^{-1}(x) \geq 5\}$

20) e)
Show that

20) f)

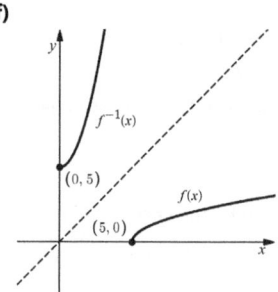

21) a)
$p = -2$

21) b)
$\{f(x) : f(x) \in \mathbb{R}, f(x) < 3\}$

21) c)
$f^{-1}(x) = \frac{1}{3-x} - 2, x \neq 3$

21) d)
Domain: $\{x : x \in \mathbb{R}, x < 3\}$
Range: $\{f^{-1}(x) : f^{-1}(x) \in \mathbb{R}, f^{-1}(x) > -2\}$

21) e)
$x = \frac{1}{2} \pm \frac{1}{2}\sqrt{21}$

21) f)

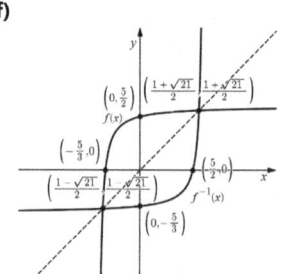

22) a)
$p = 5$

22) b)
$\{f(x) : f(x) \in \mathbb{R}, f(x) \neq 4\}$

22) c)
Domain: $\{x : x \in \mathbb{R}, x \neq 4\}$
Range: $\{f^{-1}(x) : f^{-1}(x) \in \mathbb{R}, f^{-1}(x) \neq 5\}$

23) a)
$x = -\frac{2}{5}$

23) b)
$x = 3$

23) c)
$x = -4$

24) a)
$f'(x) = a$

24) b)
$(f^{-1})'(x) = \frac{1}{a}$

24) c)
$a = \pm 1$

25) a)
$f'(2) = 4$
25) b)
$(f^{-1})'(2) = \frac{\sqrt{2}}{4}$
25) c)
$x = \frac{1}{2\sqrt[3]{2}}$

26) a)
$fg(x) = -2x + 6$
26) b)
$k = -9$

27) a)
$fg(x) = 50x^2 - 20x - 6$
27) b)
$k = 771$

28) a)
$gf(x) = -6x + 8$
28) b)
$\{x : x \in \mathbb{R}, x < 0\}$
28) c)
$\{gf(x) : gf(x) \in \mathbb{R}, gf(x) > 8\}$
28) d)
$(gf)^{-1}(x) = -\frac{1}{6}x + \frac{4}{3}, x > 8$
28) e)
Domain: $\{x : x \in \mathbb{R}, x > 8\}$
Range: $\{(gf)^{-1}(x) : (gf)^{-1}(x) \in \mathbb{R}, (gf)^{-1}(x) < 0\}$

29) a)
$gf(x) = x^2 + 7x + 12$
29) b)
$\{x : x \in \mathbb{R}, -3 < x < 0\}$
29) c)
$\{gf(x) : gf(x) \in \mathbb{R}, 0 < gf(x) < 12\}$
29) d)
$(gf)^{-1}(x) = -3.5 + \sqrt{x + 0.25}, 0 < x < 12$
29) e)
Domain: $\{x : x \in \mathbb{R}, 0 < x < 12\}$
Range: $\{(gf)^{-1}(x) : (gf)^{-1}(x) \in \mathbb{R}, -3 < (gf)^{-1}(x) < 0\}$

30) a)
$fg(x) = 7 - \frac{6}{x+2}$
30) b)
$\{x : x \in \mathbb{R}, x \neq -2\}$
30) c)
$\{fg(x) : fg(x) \in \mathbb{R}, fg(x) \neq 7\}$
30) d)
$f^{-1}(x) = \frac{2x-8}{7-x}, x \neq 7$
30) e)
Domain: $\{x : x \in \mathbb{R}, x \neq -7\}$
Range: $\{(fg)^{-1}(x) : (fg)^{-1}(x) \in \mathbb{R}, (fg)^{-1}(x) \neq -2\}$

31) a)
$f(x) \in [-4, \infty)$
31) b)
$f^{-1}(x) = 3 - \sqrt{x+8}, x \in [-4, \infty)$
31) c)
$x = \frac{7}{2} - \frac{3}{2}\sqrt{5}$

31) d)

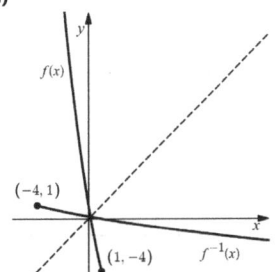

31) e)
Domain: $x \in (-4, \infty)$
Range: $f^{-1}(x) \in (-\infty, 1)$

32) a)
$f^{-1}(x) = \frac{1}{2}\ln(x-2), x \in (2, \infty)$
32) b)

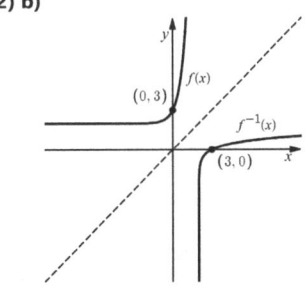

32) c)
Domain: $x \in (2, \infty)$
Range: $f^{-1}(x) \in \mathbb{R}$

33) a)
$f^{-1}(x) = \frac{4x}{x-3}, x \in \left[-3, \frac{5}{3}\right]$
33) b)
$x = 0$
33) c)

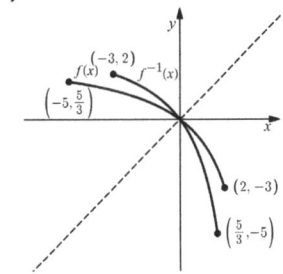

33) d)
Domain: $x \in \left[-3, \frac{5}{3}\right]$
Range: $f^{-1}(x) \in [-5, 2]$

34) a)
Domain: $x \in [-11, 7]$
Range: $f^{-1}(x) \in [-2, 7]$
34) b)
Domain: $x \in (-\infty, \infty)$
Range: $f^{-1}(x) \in (0, \infty)$
34) c)
Domain: $x \in (2, \infty)$
Range $f^{-1}(x) \in (-\infty, \infty)$

35) a)
$p = -2$
35) b)
$f(x) \in (-\infty, \infty)$

35) c)
$f^{-1}(x) = e^x - 2, x \in \mathbb{R}$
35) d)
Domain: $x \in (-\infty, \infty)$
Range: $f^{-1}(x) \in (-2, \infty)$
35) e)

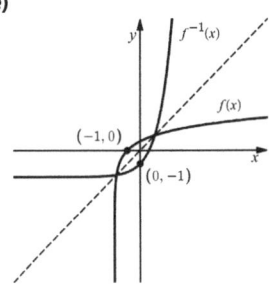

36) a)
$p = 9$
36) b)
$f(x) \in (-\infty, 5) \cup (5, \infty)$
36) c)
Domain: $x \in (-\infty, 5) \cup (5, \infty)$
Range: $f^{-1}(x) \in (-\infty, 9) \cup (9, \infty)$

37) a)
$p = -1$
37) b)
Domain of f^{-1}: $x \in [-7, \infty)$
Range of f^{-1}: $f^{-1}(x) \in [-1, \infty)$
37) c)

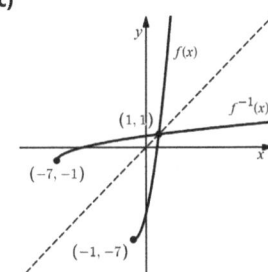

38) a)
$gf(x) = 4x^2 + 4x + 1$
38) b)
$x \in \left(\frac{1}{2}, 1\right)$
38) c)
$gf(x) \in (4, 9)$
38) d)
$(gf)^{-1}(x) = -\frac{1}{2} + \frac{1}{2}\sqrt{x}, x \in (4, 9)$
38) e)
Domain: $x \in (4, 9)$
Range: $(gf)^{-1}(x) \in \left(\frac{1}{2}, 1\right)$

39) a)
$fg(x) = -13 + \frac{70}{x+7}$
39) b)
$x \in (-\infty, -7) \cup (-7, \infty)$
39) c)
$fg(x) \in (-\infty, -13) \cup (-13, \infty)$
39) d)
$(fg)^{-1}(x) = \frac{-21-7x}{x+13}, x \neq -13$
39) e)
Domain: $x \in (-\infty, -13) \cup (-13, \infty)$
Range: $(fg)^{-1}(x) \in (-\infty, -7) \cup (-7, \infty)$

2.13 Modulus Functions
Questions on page 260

1) a) 1
1) b) 5
1) c) 2
1) d) −1
1) e) −2
1) f) 2

2) 3 or 13

3) 7 or 9

4) a) 2
4) b) 1
4) c) 8
4) d) 16
4) e) −9
4) f) 4

5) a) $(-3, 0)$
5) b) $(2, 0)$
5) c) $(1, 5)$
5) d) $(-3, -8)$
5) e) $(-4, 13)$
5) f) $(1, 9)$
5) g) $\left(-\frac{1}{2}, 2\right)$

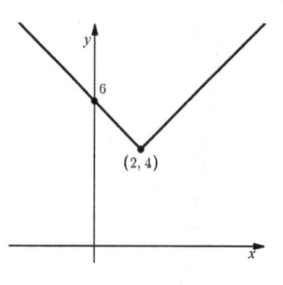

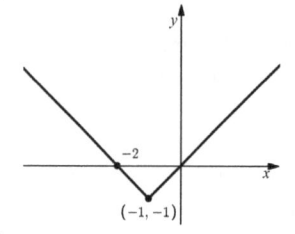

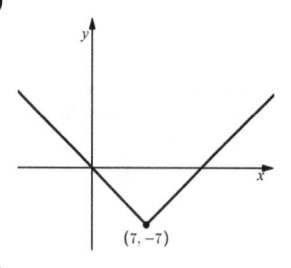

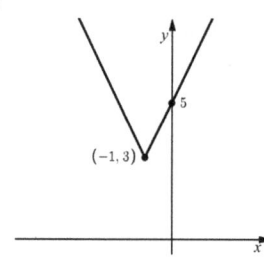

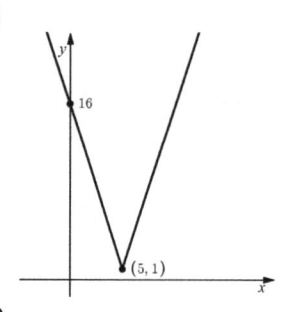

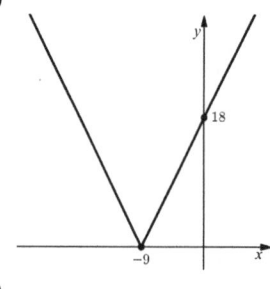

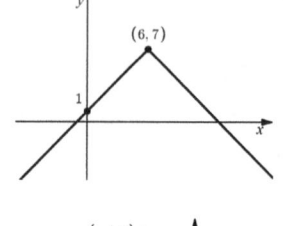

6) a) graph with vertex at -3, y-intercept 3
6) b) graph with vertex at $(2, 4)$, y-intercept 6
6) c) graph with vertex at $(-1, -1)$, x-intercept -2
6) d) graph with vertex at $(7, -7)$
6) e) graph with vertex at $(-1, 3)$, y-intercept 5
6) f) graph with vertex at $(5, 1)$, y-intercept 16
6) g) graph with vertex at -9, y-intercept 18
6) h) graph with vertex at 1, y-intercept -4
6) i) graph with vertex at $(6, 7)$, y-intercept 1
6) j) graph with vertex at $(-4, 9)$, y-intercept 1

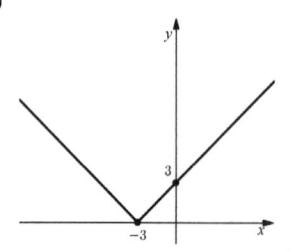

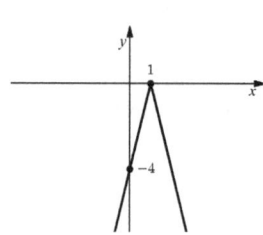

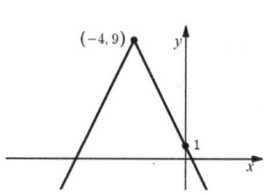

7) a) $y = |x - 5|$
7) b) $y = 2|x - 3|$
7) c) $y = 3|x + 2|$
7) d) $y = |x + 3| + 1$
7) e) $y = |x - 2| - 2$
7) f) $y = 2|x - 2| - 6$
7) g) $y = -|x - 2| + 4$
7) h) $y = -|x - 5| - 1$
7) i) $y = -4|x + 7| + 8$

8) a) $x = 2$ or $x = -8$
8) b) $x = 9$ or $x = -5$
8) c) $x = \frac{3}{2}$ or $x = -\frac{5}{2}$
8) d) $x = -2$ or $x = -8$
8) e) No solutions

9) a) $k = 10$
9) b) $\left(-\frac{9}{2}, 0\right)$ and $\left(\frac{1}{2}, 0\right)$

10) a) $x = -\frac{7}{2}$
10) b) $x = \frac{1}{2}$
10) c) $x = \frac{11}{2}$

11) a) $x = -1$ or $x = -\frac{11}{3}$
11) b) $x = \frac{15}{2}$ or $x = \frac{3}{4}$

11) c)
$x = -\frac{19}{2}$ or $x = -7$

11) d)
$x = \frac{13}{2}$ or $x = -\frac{7}{2}$

11) e)
$x = \frac{15}{2}$ or $x = -\frac{1}{2}$

11) f)
No Solutions

12) a)
$|x| < 6$

12) b)
$|x - 1| < 5$

12) c)
$|x - 11| < 14$

12) d)
$|x - 3| > 2$

13)
$|2x - 20| \leq 8$

14) 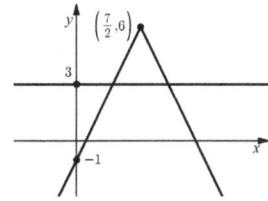
$2 < x < 5$

15) 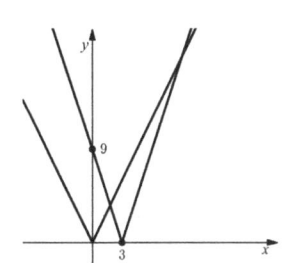
$\frac{9}{5} \leq x \leq 9$

16) a)
$x \in (-\infty, -13) \cup (7, \infty)$

16) b)
$x \in (-\infty, -1] \cup [3, \infty)$

16) c)
$x \in \left(-\infty, -\frac{9}{2}\right] \cup \left[\frac{3}{2}, \infty\right)$

17)

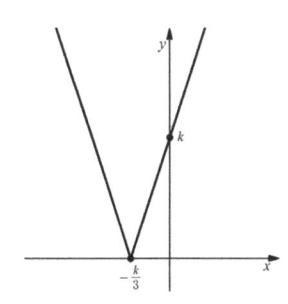

18) a)

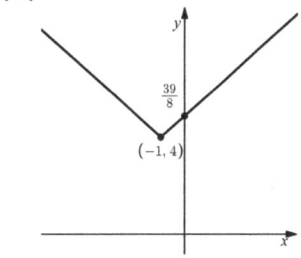

18) b)
$x = \frac{5}{3}$

19) a)
$k = 3$

19) b)
$x = 3$ or $x = -\frac{7}{3}$

20) a)
$x \in \mathbb{R}$

20) b)
$x > -4$

21) a)
$k > 6$

21) b)
$k > -4$

21) c)
$k < 8$

22) a)

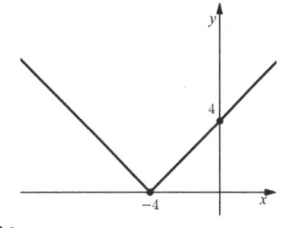

22) b)
$-1 < k < 0$

23) a)

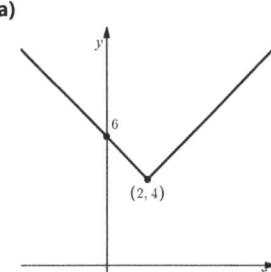

23) b)
$0 < k < 1$

24) a)
$(0, 5)$ and $\left(\frac{5}{2}, 0\right)$

24) b)
$-\frac{4}{5} < k < 2$

24) c)
$x = \frac{7}{2-k}$ or $x = \frac{3}{2+k}$

25) a)
Stretch, parallel to y-axis, scale factor 2

25) b)
Translation by the vector $\binom{5}{0}$

25) b)
Translation by the vector $\binom{-2}{0}$

25) d)
Stretch, parallel to y-axis, scale factor 2

25) e)
Stretch, parallel to x-axis, scale factor $\frac{1}{2}$

26) a)
Stretch, parallel to y-axis, scale factor 3

26) b)
Stretch, parallel to y-axis, scale factor 3

26) c)
Translation by the vector $\binom{1}{0}$

26) d)
Translation by the vector $\binom{-3}{4}$

26) e)
Translation by the vector $\binom{\frac{3}{2}}{-1}$

27) a)
Stretch, parallel to y-axis, scale factor $\frac{1}{2}$

27) b)
Stretch, parallel to x-axis, scale factor 3

27) c)
Translation by the vector $\binom{-4}{-5}$

28) a)
$(-2, -2)$

28) b)
$x = \frac{7}{3}$

28) c)
$k \leq 1$ or $k > 3$

29) a)

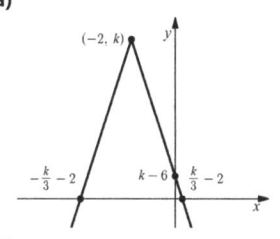

29) b)
$x = \frac{1}{4}k - \frac{5}{2}$ or $x = -\frac{1}{2}k - 1$

30) a)

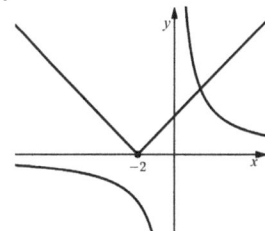

30) b)
The graphs of $y = |x + 2|$ and $y = \frac{k}{x}$ where k is positive only have one point of intersection.

30) c)
$x = -1 + \sqrt{1+k}$

31) a)

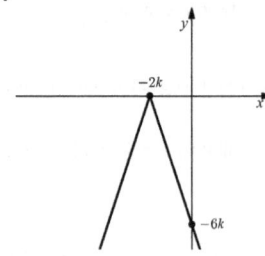

31) b)

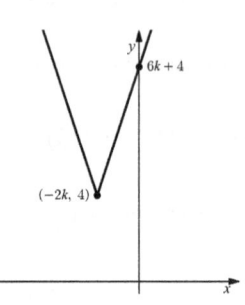

32) a)

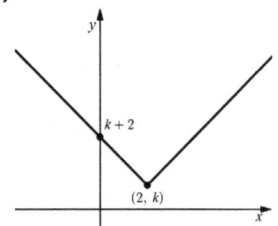

32) b)

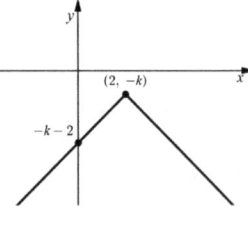

32) c)

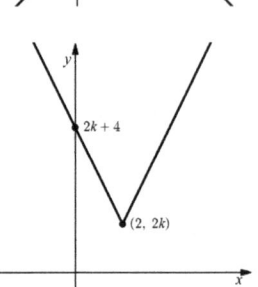

33)
$p = 4$ and $q = -9$

34)
No solutions

35) a)
$\{f(x): f(x) \in \mathbb{R}, f(x) \geq 0\}$
35) b)
$\{f(x): f(x) \in \mathbb{R}, f(x) \leq 0\}$
35) c)
$\{f(x): f(x) \in \mathbb{R}, f(x) \geq 3\}$
35) d)
$\{f(x): f(x) \in \mathbb{R}, f(x) \geq -1\}$

35) e)
$\{f(x): f(x) \in \mathbb{R}, f(x) \leq 5\}$
35) f)
$\{f(x): f(x) \in \mathbb{R}, f(x) \leq 1\}$

36) a)

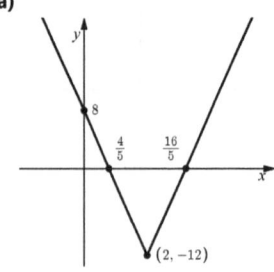

36) b)
Range: $\{f(x): f(x) \in \mathbb{R}, f(x) \geq -12\}$
Domain: $\{x: x \in \mathbb{R}\}$
36) c)
$\frac{1}{2} \leq x \leq \frac{47}{14}$

37) a)

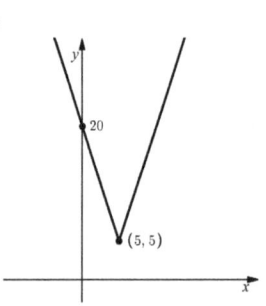

37) b)
$\{f(x): f(x) \in \mathbb{R}, f(x) \geq 5\}$
37) c)
$x = \frac{3}{2}$ or $x = \frac{75}{8}$

38) a)
$2|x| + 1$
38) b)
$|2x + 1|$
38) c)
$-|2x + 1|$
38) d)
$-|2|x| + 1|$

39) a)
$|x|^2 + 1 = x^2 + 1$
39) b)
$|x^2 + 1|$
39) c)
$-|x^2 + 1|$
39) d)
$|x|^2 + 1 = x^2 + 1$
39) e)
$-|x|$

40) a)

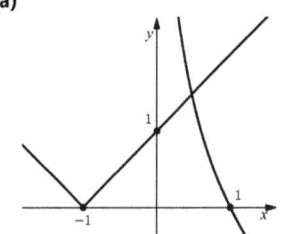

40) b)
The graphs of $y = |x + 1|$ and $y = k \ln x$ where k is negative only have one point of intersection.
40) c)
$x = 0.710$ (3 SF)

41) a)

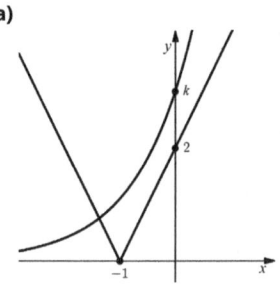

41) b)
The graphs only intersect once. If $k \leq 2$ then the graphs intersect twice.
41) c)
$x = -1.46$ (3 SF)

42) a)

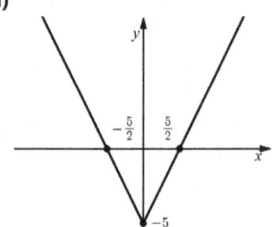

42) b)

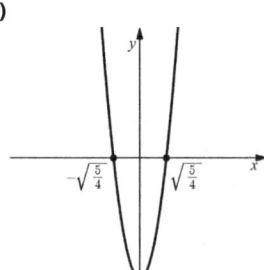

42) c)

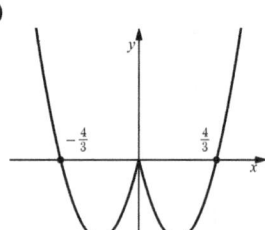

42) d)

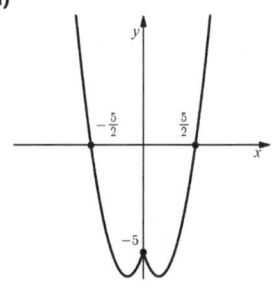

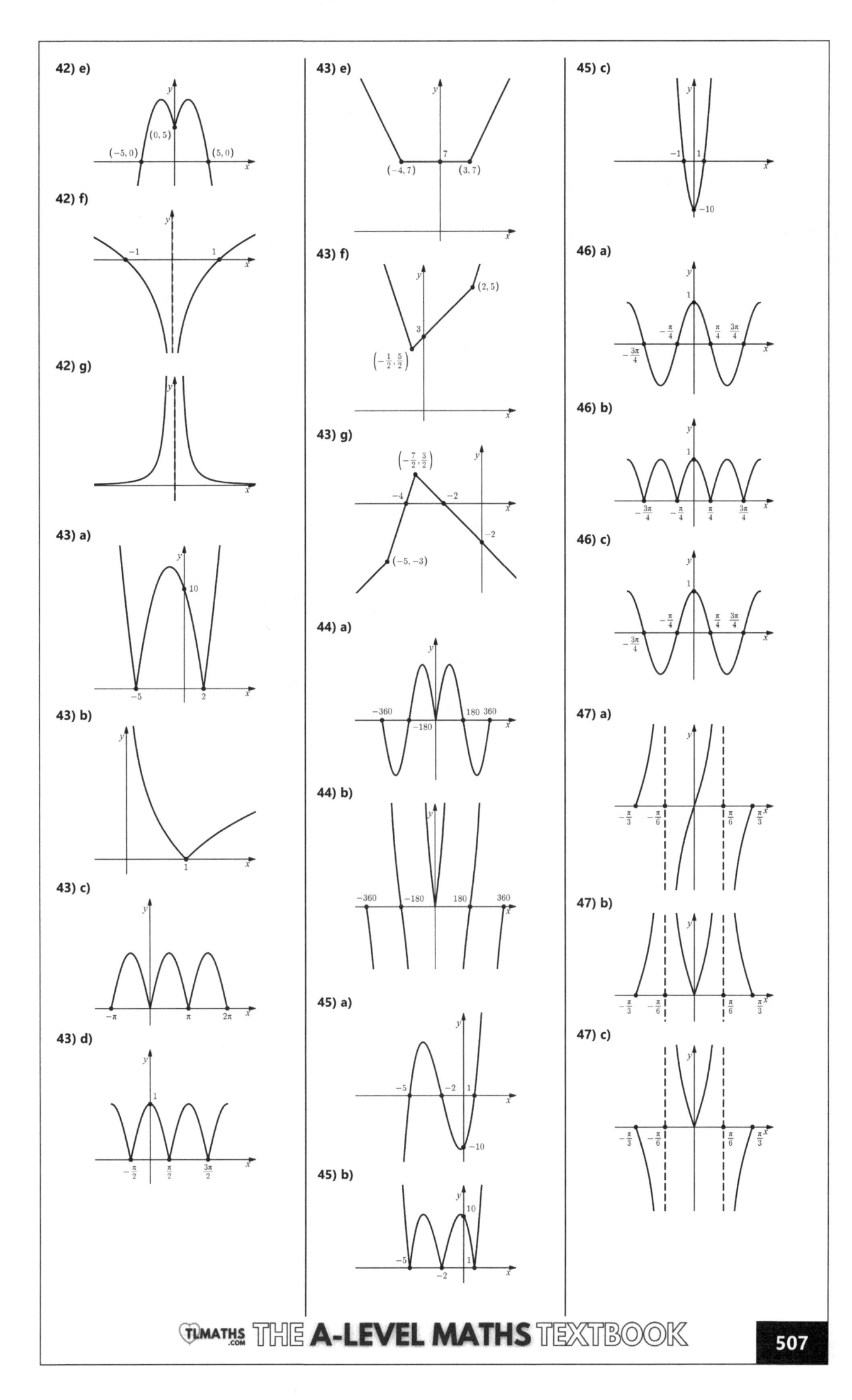

42) e)
42) f)
42) g)
43) a)
43) b)
43) c)
43) d)
43) e)
43) f)
43) g)
44) a)
44) b)
45) a)
45) b)
45) c)
46) a)
46) b)
46) c)
47) a)
47) b)
47) c)

48) a) [graph]

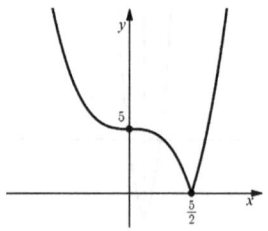

48) b) [graph]

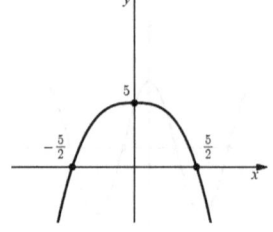

48) c) [graph]

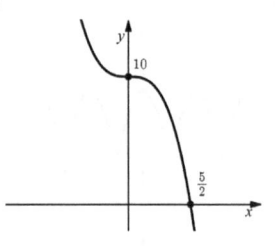

49) a) [graph]

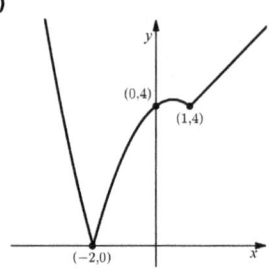

49) b) [graph]

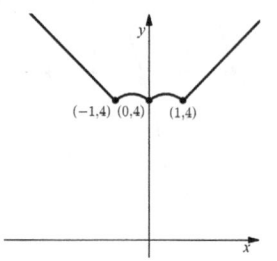

49) c) [graph]

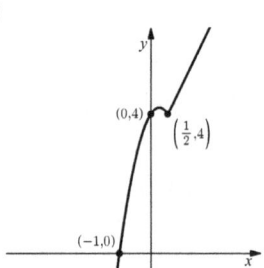

50) a) [graph]

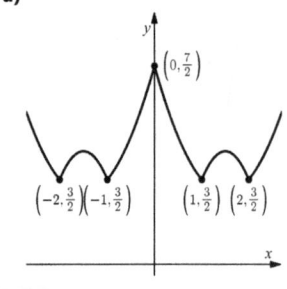

50) b) [graph]

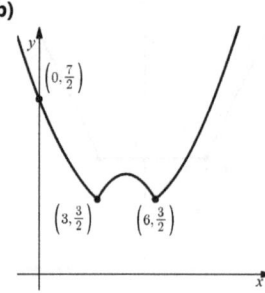

50) c) [graph]

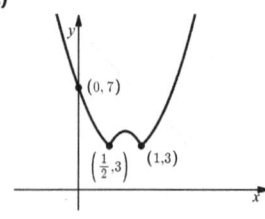

51) a) [graph]

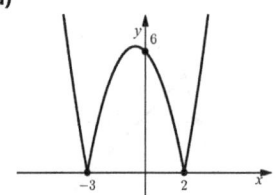

51) b) [graph]

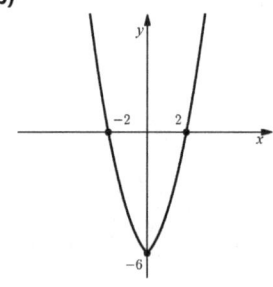

51) c)
$x \geq 2$ or $x = -\sqrt{6}$

52) a) [graph]

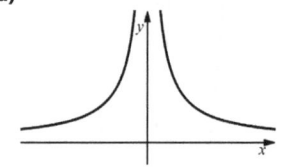

52) b) [graph]

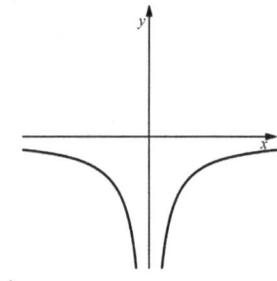

52) c)
$x \in \mathbb{R}, x \neq 0$

53) a) [graph]

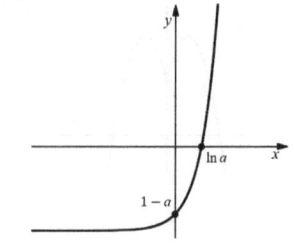

53) b) [graph]

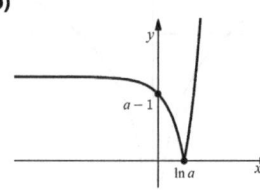

53) c) [graph]

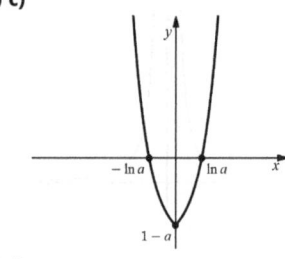

53) d)
$x \geq \ln a$ or $x = \ln(a - \sqrt{a^2 - 1})$

54) a)
$x \geq 0$

54) b)
$-2 \leq x \leq 0$ or $x \geq 2$

54) c)
$x \geq 1$

54) d)
$x \geq 1 + \ln 4$

55) a)
$-x^2 - 4x - 5$

55) b) [graph]

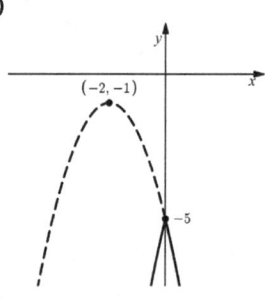

55) c)
$x \geq 0$

56)
$x = \frac{\sqrt{2}}{2}$ or $x = 2 - \frac{3\sqrt{2}}{2}$

2.14 Trigonometry: Inverse Trigonometric Functions
Questions on page 267

1) a) Graph through $(0,0)$, $(1, \frac{\pi}{2})$, $(-1, -\frac{\pi}{2})$

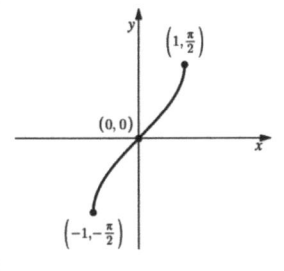

1) b) Graph through $(-1, \pi)$, $(0, \frac{\pi}{2})$, $(1, 0)$

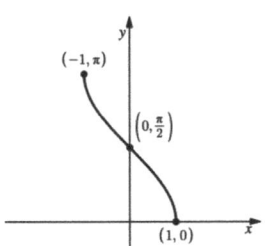

1) c) Graph with asymptotes $y = \frac{\pi}{2}$ and $y = -\frac{\pi}{2}$, through $(0,0)$

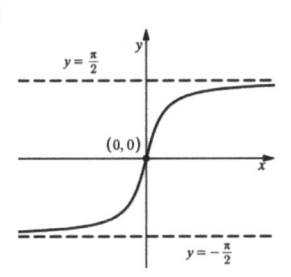

2) a) Graph through $(0,0)$, $(1, \frac{3\pi}{2})$, $(-1, -\frac{3\pi}{2})$

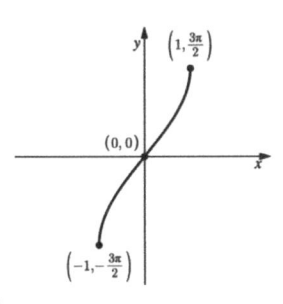

2) b) Graph through $(0,0)$, $(1, \frac{\pi}{8})$, $(-1, -\frac{\pi}{8})$

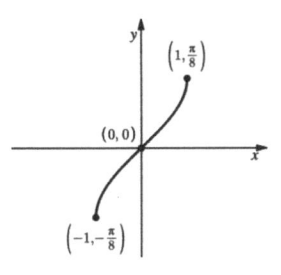

3) a) Graph through $(0,0)$, $(\frac{1}{3}, \frac{\pi}{2})$, $(-\frac{1}{3}, -\frac{\pi}{2})$

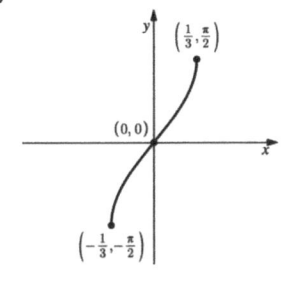

3) b) Graph through $(0,0)$, $(\frac{5}{2}, \frac{\pi}{2})$, $(-\frac{5}{2}, -\frac{\pi}{2})$

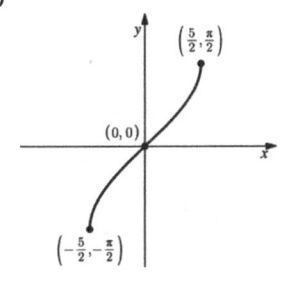

4) a) $y = \arcsin x + \frac{\pi}{2}$

4) b) $y = -\arcsin x$ or $y = \arcsin(-x)$

4) c) $y = \arcsin(x-1)$

4) d) $y = \arcsin(x+1) - \frac{\pi}{2}$

5) Graph through $(-2, \frac{\pi}{2})$, $(-1, 0)$, $(0, -\frac{\pi}{2})$

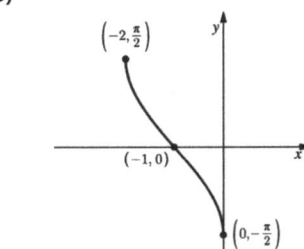

6) Graph through $(-1, \pi)$, $(0, \frac{\pi}{2})$, $(1, 0)$

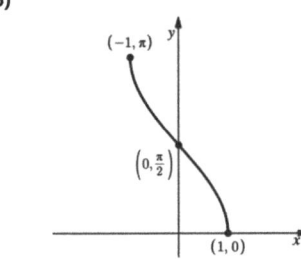

7) $x = \frac{\sqrt{3}}{4} + \frac{3}{2}$

8) a) Show that

8) b) Domain of $f^{-1}(x)$: $\{x : x \in \mathbb{R}, -1 \leq x \leq 5\}$
Range of $f^{-1}(x)$: $\{f^{-1}(x) : f^{-1}(x) \in \mathbb{R}, -\frac{\pi}{2} \leq f^{-1}(x) \leq \frac{\pi}{2}\}$

9) a) Graph through $(-1, \frac{5\pi}{2})$, $(0, \frac{5\pi}{4})$, $(1, 0)$

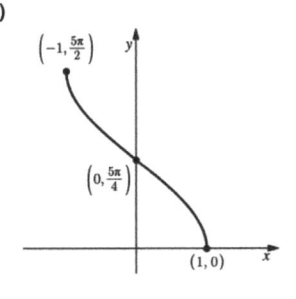

9) b) Graph through $(-1, \frac{\pi}{5})$, $(0, \frac{\pi}{10})$, $(1, 0)$

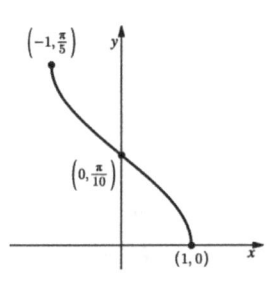

10) a) Graph through $(-\frac{1}{4}, \pi)$, $(0, \frac{\pi}{2})$, $(\frac{1}{4}, 0)$

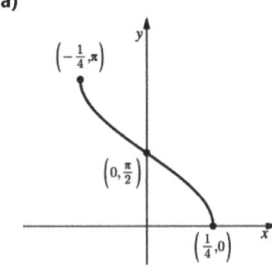

10) b) Graph through $(-\frac{5}{4}, \pi)$, $(0, \frac{\pi}{2})$, $(\frac{5}{4}, 0)$

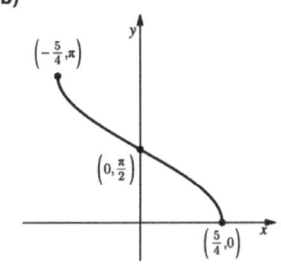

11) a) $y = -\arccos x$

11) b) $y = \arccos(-x)$

11) c) $y = \arccos x - \frac{\pi}{2}$

11) d) $y = \arccos(x+1) - \pi$

12) Graph through $(0, -\pi)$, $(1, -\frac{\pi}{2})$, $(2, 0)$

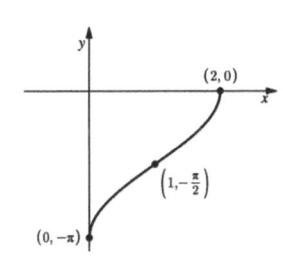

13) Graph through $(-\frac{1}{2}, 0)$, $(0, -\frac{\pi}{2})$, $(\frac{1}{2}, -\pi)$

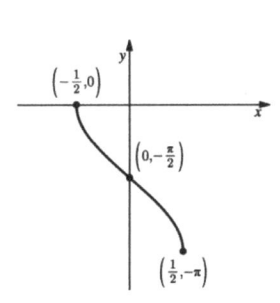

14) $x = -\frac{35}{2}$

15) a) $x = \frac{1}{2}$

15) b)
$\arcsin x = \frac{\pi}{6}$

16) a)

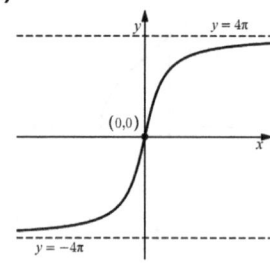

16) b)

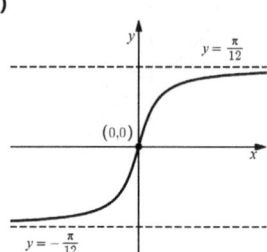

17) a)

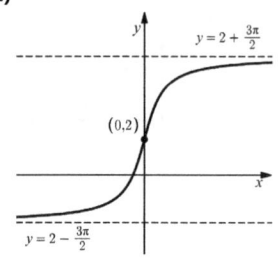

17) b)

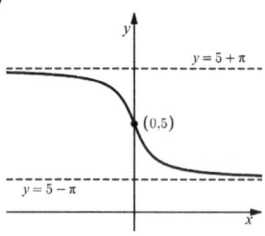

18) a)

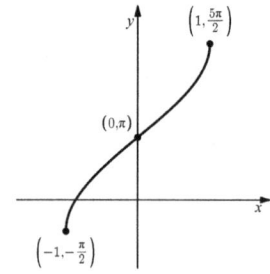

18) b)

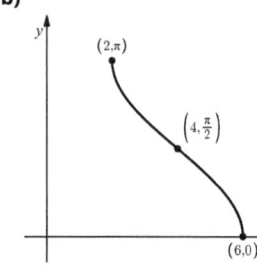

18) c)

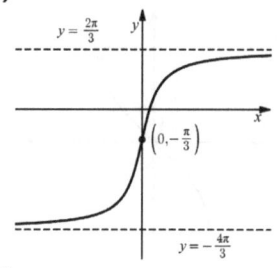

18) d)

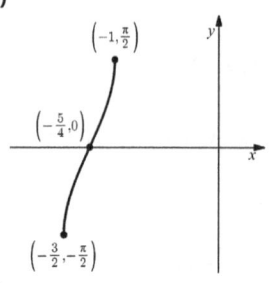

18) e)

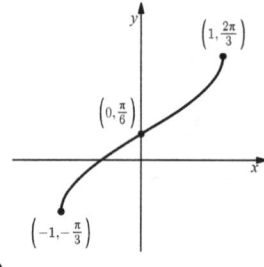

18) f)

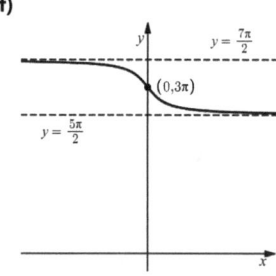

19) a)
$y = 2\arcsin x + \frac{\pi}{4}$
$y = \frac{5\pi}{4} - 2\arccos x$

19) b)
$y = \frac{\pi}{2} + \arcsin\left(\frac{1}{4} - 2x\right)$
$y = \arccos\left(2x - \frac{1}{4}\right)$

19) c)
$y = \arcsin\left(3x + \frac{5}{6}\right)$
$y = \frac{\pi}{2} - \arccos\left(3x + \frac{5}{6}\right)$

20)
$x = \frac{1}{\sqrt{2}} = \frac{\sqrt{2}}{2}$

21) a)
$x = \frac{\sqrt{3}}{2}$

21) b)
$x = \frac{1}{2}$

22)
Show that

23) a)
$\frac{\pi}{2}$

23) b)
π

23) c)
0

23) d)
0

24) a)
$\frac{\pi}{4}$

24) b)
$\frac{\pi}{4}$

24) c)
$-\frac{\pi}{4}$

24) d)
$\frac{3\pi}{4}$

25) a)
$\frac{\pi}{4}$

25) b)
$-\frac{\pi}{4}$

25) c)
$\frac{\pi}{3}$

25) d)
$-\frac{\pi}{6}$

26) a)
$\frac{\pi}{2}$

26) b)
$\frac{\pi}{2}$

26) c)
$-\frac{\pi}{2}$

26) d)
$=\frac{\pi}{2}$

26) e)
$\frac{\pi}{3}$

26) f)
$\frac{2\pi}{3}$

26) g)
$-\frac{\pi}{3}$

26) h)
$\frac{\pi}{3}$

26) i)
$\frac{\pi}{4}$

26) j)
$-\frac{\pi}{3}$

27) a)
$\frac{\sqrt{2}}{2}$

27) b)
$-\frac{\sqrt{2}}{2}$

27) c)
$\frac{\sqrt{3}}{2}$

27) d)
$\frac{\sqrt{3}}{2}$

27) e)
$\frac{1}{2}$

27) f)
$\frac{1}{2}$

27) g)
$-\frac{\sqrt{3}}{3}$

2.15 Trigonometry: Reciprocal Trigonometric Functions
Questions on page 270

1) a) 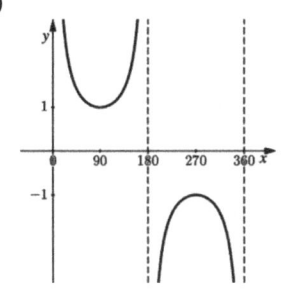
1) b) 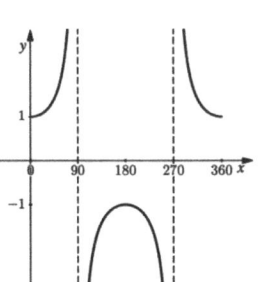
1) c) 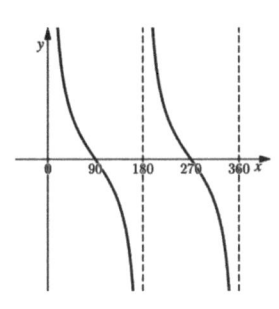
2) a) 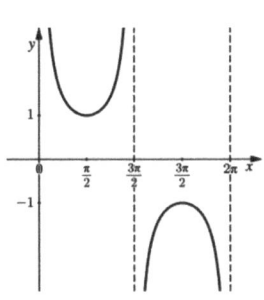
2) b) 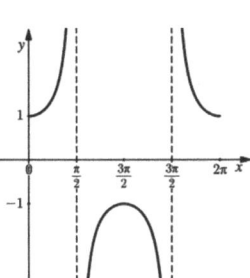
2) c) 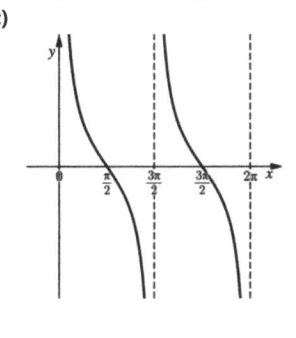
3) a) 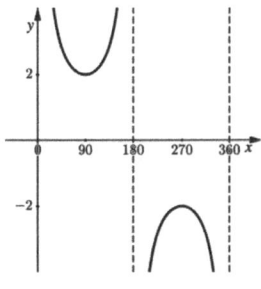
3) b) 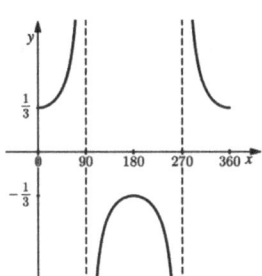
3) c) 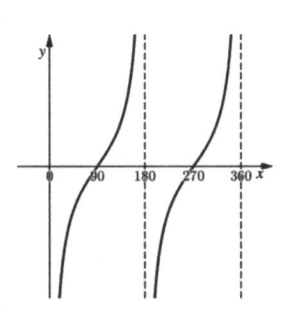
3) d) 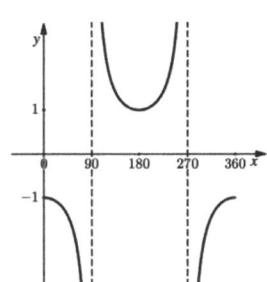
3) e) 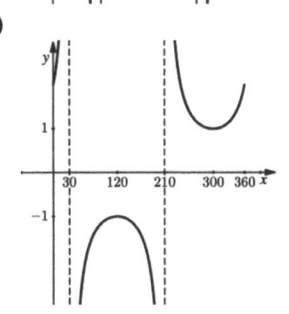
3) f) 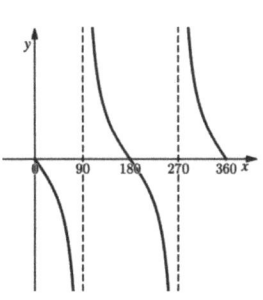
4) a) 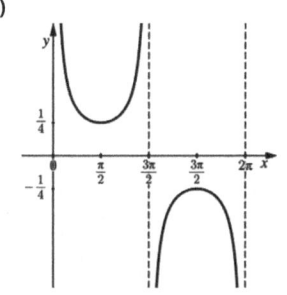
4) b) 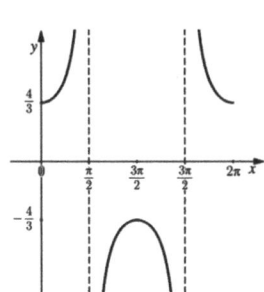
4) c) 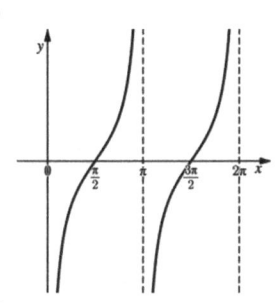
4) d) 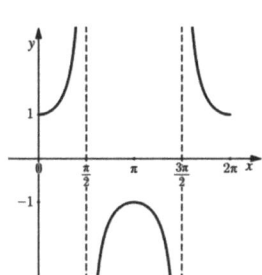
4) e) 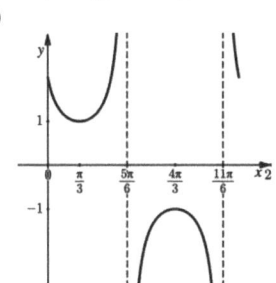
4) f)

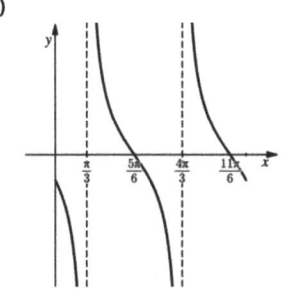

5) a) [graph]

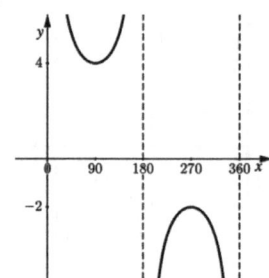

5) b) [graph]

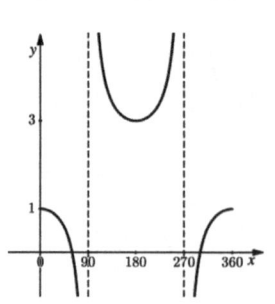

5) c) [graph]

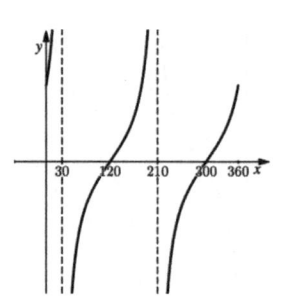

5) d) [graph]

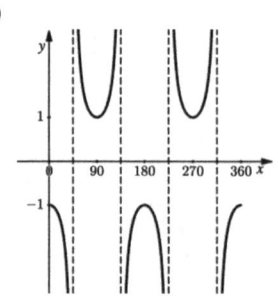

5) e) [graph]

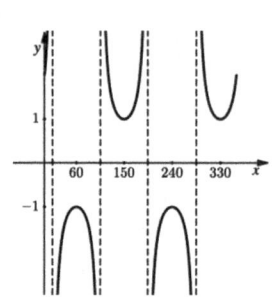

5) f) [graph]

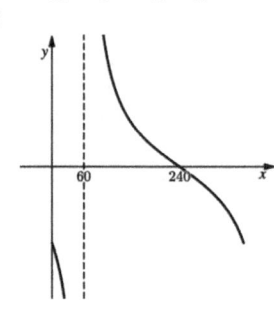

6) a) [graph]

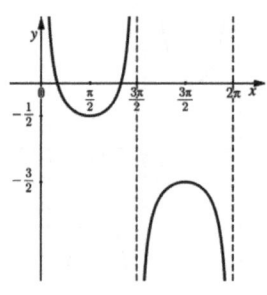

6) b) [graph]

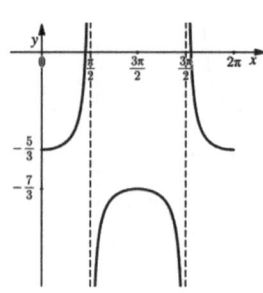

6) c) [graph]

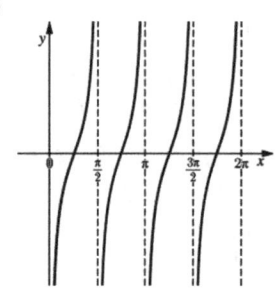

6) d) [graph]

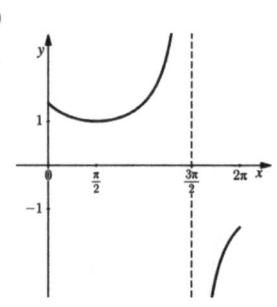

6) e) [graph]

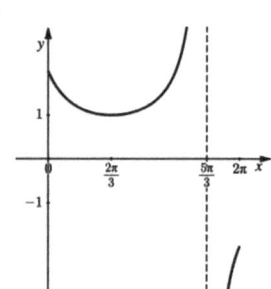

6) f) [graph]

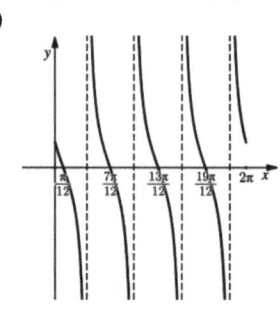

7) a)
$\theta = \{30°, 150°\}$

7) b)
$\theta = \{1.23, 5.05\}$ (3 SF)

7) c)
$\theta = \{14.0°, 194°\}$ (3 SF)

7) d)
$\theta = \{4.07, 5.36\}$ (3 SF)

7) e)
$\theta = \{103°, 257°\}$ (3 SF)

7) f)
$\theta = \{2.87, 6.01\}$ (3 SF)

8)
$\frac{\pi}{2} < x < \pi$ or $\pi < x < \frac{3\pi}{2}$

9)
$180° < x < 270°$ or $270° < x < 360°$

10) a)
$\sin x = \frac{2}{7}$

10) b)
$\cos x = \frac{3\sqrt{5}}{7}$

10) c)
$\tan x = \frac{2}{3\sqrt{5}} = \frac{2\sqrt{5}}{15}$

10) d)
$\sec x = \frac{7}{3\sqrt{5}} = \frac{7\sqrt{5}}{15}$

10) e)
$\cot x = \frac{3\sqrt{5}}{2}$

11) a)
$\cos x = \frac{4}{9}$

11) b)
$\sin x = \frac{\sqrt{65}}{9}$

11) c)
$\tan x = \frac{\sqrt{65}}{4}$

11) d)
$\cosec x = \frac{9}{\sqrt{65}}$

11) e)
$\cot x = \frac{4}{\sqrt{65}} = \frac{4\sqrt{65}}{65}$

12) a)
$\tan x = \frac{11}{12}$

12) b)
$\sin x = \frac{11}{\sqrt{265}} = \frac{11\sqrt{265}}{265}$

12) c)
$\cos x = \frac{12}{\sqrt{265}} = \frac{12\sqrt{265}}{265}$

12) d)
$\cosec x = \frac{\sqrt{265}}{11}$

12) e)
$\sec x = \frac{\sqrt{265}}{12}$

13) a)
$\sin x = \frac{9}{10}$

13) b)
$\cos x = -\frac{\sqrt{19}}{10}$

13) c)
$\tan x = -\frac{9}{\sqrt{19}} = -\frac{9\sqrt{19}}{19}$

13) d)
$\sec x = -\frac{10}{\sqrt{19}} = -\frac{10\sqrt{19}}{19}$

13) e)
$\cot x = -\frac{\sqrt{19}}{9}$

14) a)
$\cos x = -\frac{5}{12}$

14) b)
$\sin x = -\frac{\sqrt{119}}{12}$

14) c)
$\tan x = \frac{\sqrt{119}}{5}$

14) d)
$\operatorname{cosec} x = -\frac{12}{\sqrt{119}} = -\frac{12\sqrt{119}}{119}$

14) e)
$\cot x = \frac{5}{\sqrt{119}} = \frac{5\sqrt{119}}{119}$

15) a)
$\tan x = -\frac{7}{2}$

15) b)
$\sin x = -\frac{7}{\sqrt{53}} = -\frac{7\sqrt{53}}{53}$

15) c)
$\cos x = \frac{2}{\sqrt{53}} = \frac{2\sqrt{53}}{53}$

15) d)
$\operatorname{cosec} x = -\frac{\sqrt{53}}{7}$

15) e)
$\sec x = \frac{\sqrt{53}}{2}$

16) a)
$\sec x = \frac{3}{2}$

16) b)
$\cos x = \frac{2}{3}$

16) c)
$\sin x = \frac{\sqrt{5}}{3}$

16) d)
$\tan x = \frac{\sqrt{5}}{2}$

16) e)
$\operatorname{cosec} x = \frac{3}{\sqrt{5}} = \frac{3\sqrt{5}}{5}$

16) f)
$\cot x = \frac{2}{\sqrt{5}} = \frac{2\sqrt{5}}{5}$

17)
$\frac{2}{3}$

18) a)
$\tan x \sec x$

18) b)
$\cot x \operatorname{cosec} x$

19) a)
1

19) b)
1

19) c)
1

20) a)
$\frac{1}{\sin x \cos x}$

20) b)
$\frac{\cos x}{\sin^2 x}$

20) c)
$\frac{1}{\sin x}$

20) d)
$\frac{1}{\sin^2 x}$

21) a)
$\cot x$

21) b)
$\sec x$

21) c)
$\tan x$

21) d)
$\sec x \tan x$

21) e)
$\cot^2 x \cos x$

21) f)
$\cos^3 x \sin^2 x$

22) a)
$\theta = \{45°, 225°\}$

22) b)
$\theta = \{3.81, 5.62\}$ (3 SF)

22) c)
$\theta = \{38.2°, 142°\}$ (3 SF)

22) d)
$\theta = \{\frac{\pi}{4}, \frac{3\pi}{4}, \frac{5\pi}{4}, \frac{7\pi}{4}\}$

23) a)

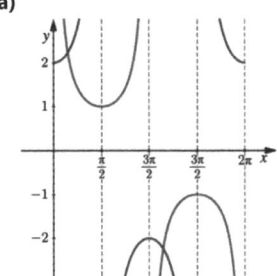

23) b)
$(0.464, 2.24)$ and $(3.61, -2.24)$ (3 SF)

24)
$k \leq 2$ or $k \geq 4$

25) a)
$1 + \cot^2 x \equiv \operatorname{cosec}^2 x$

25) b)
$\tan^2 x + 1 \equiv \sec^2 x$

26) a)
1

26) b)
1

26) c)
$\tan^2 x$

26) d)
$\frac{2}{5} \cot^2 x$

27) a)
$\theta = \{90°, 194°, 346°\}$ (3 SF)

27) b)
$\theta = \{1.32, 1.91, 4.37, 4.97\}$ (3 SF)

27) c)
$\theta = \{26.6°, 172°, 207°, 352°\}$ (3 SF)

27) d)
$\theta = \{1.25, 1.68, 4.39, 4.82\}$ (3 SF)

28) a)
$\theta = \{71.6°, 135°, 252°, 315°\}$ (3 SF)

28) b)
$\theta = \{1.19, 2.16, 4.33, 5.30\}$ (3 SF)

28) c)
$\theta = \{197°, 343°\}$ (3 SF)

28) d)
$\theta = \{2.42, 3.86\}$ (3 SF)

29)
Show that

30) a)
$\sec x = -\frac{6}{5}$

30) b)
$\cos x = -\frac{5}{6}$

30) c)
$\sin x = \frac{\sqrt{11}}{6}$

30) d)
$\tan x = -\frac{\sqrt{11}}{5}$

30) e)
$\operatorname{cosec} x = \frac{6}{\sqrt{11}} = \frac{6\sqrt{11}}{11}$

30) f)
$\cot x = -\frac{5}{\sqrt{11}} = -\frac{5\sqrt{11}}{11}$

31) a)
$\operatorname{cosec} x = -\frac{5}{4}$

31) b)
$\sin x = \frac{1}{-\frac{5}{4}} = -\frac{4}{5}$

31) c)
$\cos x = \frac{3}{5}$

31) d)
$\tan x = -\frac{4}{3}$

31) e)
$\sec x = \frac{5}{3}$

31) f)
$\cot x = -\frac{3}{4}$

32)
Prove

33) a)
Show that

33) b)
$p(x) = (3x+2)(2x-1)(2x+1)$

33) c)
Prove

34) a)
Prove
34) b)
Prove
34) c)
Prove
34) d)
Prove

35) a)
Prove
35) b)
Prove
35) c)
Prove
35) d)
Prove

36) a)
Prove
36) b)
Prove
36) c)
Prove
36) d)
Prove

37) a)
Prove
37) b)
Prove
37) c)
Prove
37) d)
Prove
37) e)
Prove

38) a)
Prove
38) b)
$\theta = \{0.100, 3.04\}$ (3 SF)

39)
$\theta = \{0.101, 1.47, 3.24, 4.61\}$ (3 SF)

2.16 Trigonometry: Compound Angle Formulae
Questions on page 273

1) a)
$h = b \cos x$
1) b)
$h = a \cos y$
1) c)
Area of triangle $ACM = \frac{1}{2}bh \sin x$
Area of triangle $BCM = \frac{1}{2}ah \sin y$
1) d)
Area of triangle $ABC = \frac{1}{2}ab \sin(x+y)$
1) e)
Show that

2) a)
$AE = \cos x, CE = \sin x$
2) b)
$\angle ACE = 90° - x, \angle AEF = 90° - y,$
$\angle CED = y, \angle ECD = 90° - y,$
$\angle BCA = x + y, \angle BAC = 90° - (x+y)$
2) c)
$BC = \cos(x+y), AB = \sin(x+y)$
2) d)
$AF = \cos x \cos y, EF = \cos x \sin y$
2) e)
$DE = \sin x \cos y, CD = \sin x \sin y$
2) f)
$\sin(x+y) = \sin x \cos y + \cos x \sin y$
$\cos(x+y) = \cos x \cos y - \sin x \sin y$

3)
Show that

4)
Show that

5)
Show that

6)
Show that

7) a)
$\sin 75°$
7) b)
$\sin 15°$
7) c)
$\sin 20°$
7) d)
$\sin 60°$

8) a)
$\cos 20°$
8) b)
$\cos 55°$
8) c)
$\cos 64°$
8) d)
$\cos 10°$

9) a) i)
$\sin 30° = \frac{1}{2}$
$\sin 45° = \frac{1}{\sqrt{2}} = \frac{\sqrt{2}}{2}$
$\sin 60° = \frac{\sqrt{3}}{2}$
9) a) ii)
$\cos 30° = \frac{\sqrt{3}}{2}$
$\cos 45° = \frac{1}{\sqrt{2}} = \frac{\sqrt{2}}{2}$
$\cos 60° = \frac{1}{2}$
9) b) i)
$\sin 75° \equiv \frac{\sqrt{2}+\sqrt{6}}{4}$
9) b) ii)
$\sin 15° \equiv \frac{\sqrt{6}-\sqrt{2}}{4}$
9) b) iii)
$\cos 105° \equiv \frac{\sqrt{2}-\sqrt{6}}{4}$
9) b) iv)
$\cos 15° \equiv \frac{\sqrt{2}+\sqrt{6}}{4}$

10) a)
$\tan 30° = \frac{\sqrt{3}}{3}, \tan 45° = 1, \tan 60° = \sqrt{3}$
10) b) i)
$\tan 75° \equiv 2 + \sqrt{3}$
10) b) ii)
$\tan 15° \equiv 2 - \sqrt{3}$

11) a)
$\cos x$
11) b)
$\sin x$
11) c)
$-\cos x$
11) d)
$\cos x$

12) a)
$\sin 18°$
12) b)
$\sin 87°$

13) a)
$\cos 1°$
13) b)
$\cos 23°$

14) a)
$\frac{12+\sqrt{1001}}{63}$
14) b)
$\frac{6\sqrt{77}-2\sqrt{13}}{63}$

15)
Show that

16)
Show that

17) a)
Minimum $= -1$ when $\theta = 210°$
17) b)
Minimum $= -2$ when $\theta = 215°$
17) c)
Minimum $= -4$ when $\theta = 312°$

18) a)
Minimum $= -1$ when $\theta = \frac{5\pi}{4}$
18) b)
Minimum $= -3$ when $\theta = \frac{\pi}{4}$
18) c)
Minimum $= -2$ when $\theta = 0.0292$

19) a)
Maximum $= 1$ when $\theta = \frac{\pi}{10}$
19) b)
Maximum $= 9$ when $\theta = \frac{2\pi}{3}$
19) c)
Maximum $= \frac{2}{3}$ when $\theta = \frac{\pi}{12}$

20) a)
$\theta = \{10°, 130°\}$
20) b)
$\theta = \{110°, 330°\}$ (3 SF)

21) a)
$\tan 54°$
21) b)
$\tan 68°$

22) a)
$\theta = \{56.0°, 236°\}$ (3 SF)
22) b)
$\theta = \{44.2°, 104°\}$ (3 SF)

23) a)
$\sqrt{2} \cos x$
23) b)
$\sqrt{2} \cos x$
23) c)
$-\frac{\sqrt{2}}{2} \sin x - \frac{\sqrt{6}}{2} \cos x$

24) a)
$\frac{1+\sqrt{3}}{2} \sin x + \frac{1-\sqrt{3}}{2} \cos x$
24) b)
$\frac{\sqrt{3}+1}{2} \cos x + \frac{\sqrt{3}-1}{2} \sin x$
24) c)
$\frac{\sqrt{2}+\sqrt{3}}{2} \sin x - \frac{1+\sqrt{2}}{2} \cos x$

25) a)
$\theta = \{30°, 210°\}$
25) b)
$\theta = \{75°, 255°\}$
25) c)
$\theta = \{30°, 210°\}$

26) a)
$\theta = \{37.5°, 217.5°\}$
26) b)
$\theta = \{52.5°, 233°\}$ (3 SF)
26) c)
$\theta = \{75°, 255°\}$

27)
Show that

28)
Prove

29)
$\theta = \{93.4°, 273°\}$ (3 SF)

30) a)
$\theta = \{161°, 341°\}$ (3 SF)
30) b)
$\theta = \{53.6°, 114°\}$ (3 SF)

31)
Show that

32)
$\tan y = \frac{5}{3}$

33)
Prove

34)
Show that

35)
Prove

36)
Prove

2.17 Trigonometry: Double Angle Formulae
Questions on page 276

1)
$\sin 2x \equiv 2 \sin x \cos x$

2)
$\cos 2x \equiv \cos^2 x - \sin^2 x$

3)
$\cos 2x \equiv 2 \cos^2 x - 1$

4)
$\cos 2x \equiv 1 - 2 \sin^2 x$

5)
$\tan 2x \equiv \frac{2 \tan x}{1 - \tan^2 x}$

6)
$\tan 2x \equiv \frac{2 \tan x}{1 - \tan^2 x}$

7) a)
$\sin 4x$
7) b)
$\cos 6x$
7) c)
$\cos 8x$
7) d)
$2 \sin x$
7) e)
$6 \cos \frac{2}{3} x$
7) f)
$3 \cos 3x$

8) a)
$\theta = \{0°, 180°\}$
8) b)
$\theta = \{0, \pi\}$
8) c)
$\theta = \{19.5°, 90°, 161°, 270°\}$ (3 SF)
8) d)
$\theta = \{\frac{\pi}{2}, 3.46, \frac{3\pi}{2}, 5.97\}$ (3 SF)

9) a)
$\theta = \{60°, 180°, 300°\}$

9) b)
$\theta = \{1.91, 4.37\}$ (3 SF)
9) c)
$\theta = \{90°, 210°, 330°\}$
9) d)
$\theta = \{0.205, 2.94\}$ (3 SF)

10) a)
$\theta = \{75.5°, 284°\}$ (3 SF)
10) b)
$\theta = \{0.201, 2.94\}$ (3 SF)
10) c)
$\theta = \{16.3°, 164°\}$ (3 SF)
10) d)
$\theta = \{0.597, 2.22, 4.06, 5.69\}$ (3 SF)

11)
$\sin 2x = \frac{4\sqrt{21}}{25}$

12)
$\cos 2x = \frac{23}{32}$

13)
$\tan 2x = \frac{4\sqrt{77}}{73}$

14)
Show that

15) a)
$\theta = \{71.6°, 112°, 252°, 292°\}$ (3 SF)
15) b)
$\theta = \{0.381, 1.89, 3.52, 5.03\}$ (3 SF)

16) a)
Prove
16) b)
Prove
16) c)
Prove

17) a)
Prove
17) b)
Prove
17) c)
Prove

18)
Prove

19) a)
Prove
19) b)
Prove
19) c)
Prove

20) a)
Prove
20) b)
Prove

20) c)
Prove

21) a)
Prove
21) b)
$\theta = \{0.714, 2.43, 3.86, 5.57\}$ (3 SF)

22) a)
$x = \{0.340, \frac{\pi}{2}, 2.80, \frac{3\pi}{2}\}$ (3 SF)
22) b)
$\theta = \{0.375, \frac{\pi}{4}, 1.20, \frac{7\pi}{12}\}$ (3 SF)

23) a)
Prove
23) b)
Prove
23) c)
Prove
23) d)
Prove

24)
Prove

25)
Prove

26)
Show that

27)
Show that

28)
Prove

29)
$f(x)$ has period π (or 180°)

30)
Prove

31)
$\tan 3x \equiv \frac{3\tan x - \tan^3 x}{1 - 3\tan^2 x}$

32)
$\theta = \{45°, 135°, 180°, 225°, 315°\}$

33)
$\theta = \{\frac{3\pi}{8}, \frac{15\pi}{8}\}$

34)
$\theta = \{23.5°, 96.5°, 144°, 216°, 264°, 336°\}$ (3 SF)

35) a)
Show that
35) b)
Distance $= \sqrt{3}k$

36)
Prove

37)
Prove

38)
Prove

2.18 Trigonometry: Harmonic Forms
Questions on page 278

1) a)
$R = 17$
1) b)
$R = 37$
1) c)
$R = 2\sqrt{149}$
1) d)
$R = \sqrt{58}$

2) a)
Max $= 41$, Min $= -41$
2) b)
Max $= 26$, Min -26
2) c)
Max $= 2\sqrt{39}$, Min $= -2\sqrt{39}$
2) d)
Max $= 2\sqrt{30}$, Min $= -2\sqrt{30}$

3) a)
Max $= 233$, Min $= -209$
3) b)
Max $= 50 + 5\sqrt{13}$, Min $= 50 - 5\sqrt{13}$
3) c)
Max $= 65^2 = 4225$, Min $= 0$
3) d)
Max $= 122 + 10\sqrt{97}$, Min $= 0$

4) a)
Max $= 1$, Min $= \frac{1}{27}$
4) b)
Max $= 3$, Min $= \frac{3}{35}$
4) c)
Max $= \frac{950 + 300\sqrt{2}}{289}$, Min $= \frac{950 - 300\sqrt{2}}{289}$
4) d)
Max $= \frac{125}{18}$, Min $= \frac{125}{863}$

5)
$k = \pm 3$

6) a)
$3\sqrt{2}\sin(x + 45°)$
6) b)
$10\sin(x + 60°)$
6) c)
$18\sin(x + 30°)$
6) d)
$2\sqrt{10}\sin(x + 26.6°)$

7) a)
$4\sin\left(x - \frac{\pi}{4}\right)$
7) b)
$5\sin(x - 0.927)$
7) c)
$13\sin(x - 0.395)$
7) d)
$\sqrt{122}\sin(x - 0.0907)$

8) a)
$8\cos(x + 60°)$
8) b)
$\sqrt{26}\cos(x + 11.3°)$
8) c)
$10\sqrt{13}\cos(x + 56.3°)$
8) d)
$\sqrt{205}\cos(x + 24.8°)$

9) a)
$101\cos(x - 1.37)$
9) b)
$65\cos(x - 0.533)$
9) c)
$50\sqrt{3}\cos\left(x - \frac{\pi}{3}\right)$
9) d)
$\sqrt{145}\cos(x - 1.19)$

10) a)
$10\cos(\theta + 53.1°)$
10) b)
$25\sin(\theta + 16.3°)$ or $25\cos(x - 73.7°)$
10) c)
$5\sqrt{2}\sin(\theta - 45°)$
10) d)
$2\sqrt{82}\sin(\theta + 51.3°)$
or $2\sqrt{82}\cos(x - 38.7°)$

11) a)
Maximum $= \sqrt{26}$ when $\theta = 11.3°$
11) b)
Maximum $= \sqrt{83}$ when $\theta = 351°$
11) c)
Maximum $= 2\sqrt{13}$ when $\theta = 124°$
11) d)
Maximum $= 10\sqrt{5}$ when $\theta = 63.4°$

12) a)
$\theta = \{106°, 164°\}$ (3 SF)
12) b)
$\theta = \{0.272, 1.94\}$ (3 SF)
12) c)
$\theta = \{36.9°, 100°, 157°, 220°\}$ (3 SF)
12) d)
$\theta = \{0.276, 1.99, 3.42, 5.13\}$ (3 SF)

13)
Stretch, parallel to the y-axis, scale factor $\sqrt{102}$, followed by a Translation by the vector $\begin{bmatrix} 0.140 \\ 4 \end{bmatrix}$

14)
$58\cos(5x + 0.810)$

15)
$\sqrt{73}\cos\left(\frac{1}{2}x - 0.359\right)$

16)
$p = 2\sqrt{3}$ and $q = 2$

17) a)
$k = 89$
17) b)
$t = 0.273$ seconds (3 SF)

18) a)
$\left(4 - \frac{\sqrt{13}}{2}\right)$ metres
18) b)
Derek must wait 2 hours and 4 minutes, to the nearest minute.

19) a)
100 mm
19) b)
$T = 0.0927$ seconds (3 SF)
19) c)
Currently, the model has the piston making $\frac{20}{2\pi} \approx 3.183$ cycles per second. 15 cycles per second corresponds to replacing $20t$ with $15 \times 2\pi t = 30\pi t$:
$h = 50 + 40\cos(30\pi t) - 30\sin(30\pi t)$

20) a)
$23\cos(\theta + 60.40\ldots°)$
20) b)
$C = 2$
20) c)
$T = 24.2°C$ (3 SF)

2.19 Trigonometry: Small Angle Approximations
Questions on page 280

1)
Show that

2)
Show that

3) a)
θ^2
3) b)
$-\frac{\theta^2}{2}$
3) c)
$1 - 18\theta^2$
3) d)
$-\theta$
3) e)
$1 - \frac{3}{2}\theta^2$

4) a)
2

4) b)
$1 - 5\theta^2$
4) c)
$1 - \theta^2$
4) d)
1
4) e)
$\frac{1}{2}\theta$
4) f)
-6θ
4) g)
$\frac{1}{3}\theta$

5) a)
0.9210610 …
5) b)
0.115%

6) a)
1
6) b)
$\frac{1}{2\theta}$
6) c)
$-\frac{1}{4\theta}$

7) a)
Show that
7) b)
Show that
7) c)
0.0304%

8) a)
Show that
8) b)
$\theta \approx -0.243$

9)
$x \approx 0.199$

10) a)
$f(x) \approx 2 - \frac{2}{3}x$
10) b)

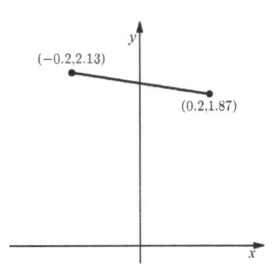

11)
$x \approx -0.345$

12)
Show that

13) a)
$\frac{3}{2}\theta^2$
13) b)
$\frac{3}{2}$

14) a)
0
14) b)
1

15) a)
$\cos\theta$
15) b)
$-\sin\theta$

16)
$-\frac{\sqrt{3}}{2}$

17)
Prove

18)
Prove

19) a)
$1 - 4x - 8x^2$
19) b)
$1 + 2x - 4x^2$

20) a)
$1 - \frac{1}{6}x^2$
20) b)
0.1996 (4 DP)

21) a)
$\frac{1}{2}$
21) b)
$\frac{1}{2}$

22) a)
$-\frac{1}{48}$
22) b)
$-\frac{1}{24\pi}$

23) a)
0.167
23) b)
0.169

24) a)
-0.169
24) b)
-0.169

2.20 Differentiation: Points of Inflection and Applications
Questions on page 282

1) a)

From convex (concave upwards) to concave (concave downwards)

1) b)

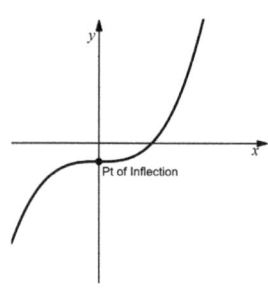

From concave (concave downwards) to convex (concave upwards)

1) c)

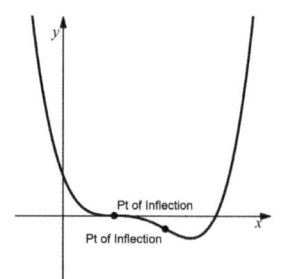

From convex (concave upwards) to concave (concave downwards) to convex (concave upwards)

2)

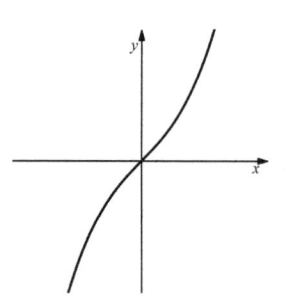

3)

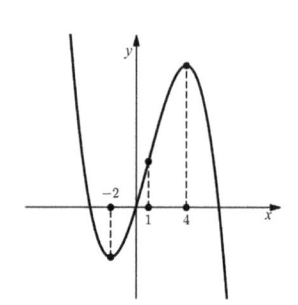

4) a)
concave (concave downwards) at $x = 0$
4) b)
convex (concave upwards) at $x = 1$
4) c)
concave (concave downwards) at $x = -1$
4) d)
neither at $x = 4$

5)
$(1, 43), (3, -135)$

6)
$\{x : x \in \mathbb{R}, -1 < x < \frac{1}{2}\} \cup \{x : x \in \mathbb{R}, x > \frac{3}{2}\}$

7) a)
$(-5, -871)$ is not stationary
$\left(\frac{1}{2}, \frac{23}{16}\right)$ is not stationary
7) b)
$(0, -1)$ is not stationary
$(-1, 0)$ is stationary

8)
$t = 1.5$ seconds
The rate of change in height (vertical velocity) is at its steepest between the local maximum value and the local minimum value when $t = 1.5$ seconds.

9) a)
£104
9) b)
$\therefore t = \frac{13}{3} = 4.33$ seconds (3 SF)
The rate of change in the value of the shares is at its steepest between the local maximum value and the local minimum value when $t = 4.33$ months.

10)
Show that

11) a)
$(-1, 1)$
11) b)
Show that
11) c)

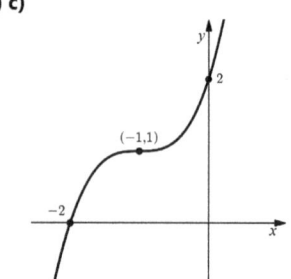

12) a)
$(1, -3)$ is a stationary point of inflection
$\left(-\frac{1}{2}, \frac{15}{4}\right)$ is a minimum
12) b)
Show that

13) a)
Show that
13) b)
Show that
13) c)

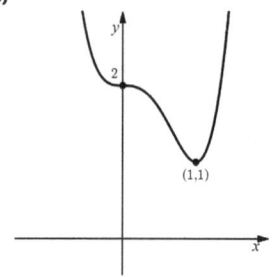

14) a)
$(0, 0)$ is a stationary point of inflection.
$\left(-\frac{3}{2}, -\frac{27}{8}\right)$ is a minimum
14) b)
$(-1, -2)$
14) c)

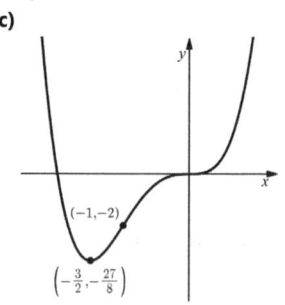

15)
$a = -15, b = 75, c = -123$

16)
$a = 3, b = -3, c = 7$

17) a)
convex at $x = 2$
17) b)
concave at $x = 0$
17) c)
convex at $x = 1$
17) d)
convex at $x = -1$
17) e)
concave at $x = 0$

18) a)
Show that
18) b)
$(\ln 2, 2 - (\ln 2)^2)$

19) a)
Show that
19) b)
Show that

20) a)
140.7 pence, $t = 1$ month
20) b)
Show that
According to the model, there is no point in the 12 months where there is a change in convexity of the curve. This means that the rate in change in the price of petrol decreases in the first month, before increasing for the remainder of the year.

2.21 Differentiation: Standard Functions
Questions on page 284

1) a)
$\frac{dy}{dx} = 5(x + 3)^4$

1) b)
$\frac{dy}{dx} = 24(4x+1)^5$
1) c)
$\frac{dy}{dx} = -20(6-5x)^3$
1) d)
$\frac{dy}{dx} = 14(2x-3)^6$
1) e)
$\frac{dy}{dx} = -(7-4x)^{-\frac{3}{4}}$
1) f)
$\frac{dy}{dx} = -(x+2)^{-2}$
1) g)
$\frac{dy}{dx} = -6(x-5)^{-3}$
1) h)
$\frac{dy}{dx} = 12(8-3x)^{-5}$
1) i)
$\frac{dy}{dx} = \frac{1}{2}(2+x)^{-\frac{1}{2}}$
1) j)
$\frac{dy}{dx} = -\frac{3}{2}(3x-1)^{-\frac{3}{2}}$

2) a)
$f''(x) = 980(7x-1)^3$
2) b)
$f''(x) = -\frac{1}{4}(3-x)^{-\frac{3}{2}}$
2) c)
$f''(x) = \frac{16}{9}(2x+9)^{-\frac{7}{3}}$

3) a)
$\left(\frac{7}{2}, 0\right)$
3) b)
$\left(\frac{1}{2}, 0\right)$ and $\left(\frac{3}{2}, 8\right)$

4) a)
$f'(x) = 12(3x+1)^3 = 324x^3 + 324x^2 + 108x + 12$
4) b)
$f(x) = 81x^4 + 108x^3 + 54x^2 + 12x + 1$
$f'(x) = 324x^3 + 324x^2 + 108x + 12$

5)
$6y - x = 16$

6) a)
Minimum height is 200 m
6) b)
$\frac{d^2h}{dt^2} = 32 > 0$ so it is a minimum

7) a)
$\frac{dy}{dx} = 3\cos 3x$
7) b)
$\frac{dy}{dx} = -5\sin 5x$
7) c)
$\frac{dy}{dx} = -14\cos 2x$
7) d)
$\frac{dy}{dx} = -3\sin 6x$
7) e)
$\frac{dy}{dx} = \frac{1}{4}\sin\frac{1}{4}x$
7) f)
$\frac{dy}{dx} = -2\cos 10x$
7) g)
$\frac{dy}{dx} = \pi \sin \pi x$
7) h)
$\frac{dy}{dx} = -\frac{1}{\pi}\sin\frac{x}{\pi}$

8)
$y = 12x - 12\pi$

9)
$y = 3x + 1 - \frac{3}{2}\pi$

10)
$y = \frac{1}{\pi}x + 3 - \frac{2}{\pi}$

11)
Show that

12)
$\left(\frac{3\pi}{2}, -\frac{3\pi}{2}\right)$

13) a)
$p = 260$
13) b)
$\frac{dp}{dt} = \frac{20\pi}{3}\sin\left(\frac{\pi}{6}t\right)$
13) c)
$\frac{dp}{dt}\bigg|_{t=6} = 0$ so the attraction has either a maximum or minimum amount of visitors
13) d)
$\frac{d^2p}{dt^2} = \frac{10\pi^2}{9}\cos\left(\frac{\pi}{6}t\right)$ so $\frac{d^2p}{dt^2}\bigg|_{t=6} = -\frac{10\pi^2}{9} < 0$
so it is a maximum amount of visitors.

14) a)
$\frac{ds}{dt} = -\pi\sin\left(\frac{\pi}{3}t\right)$
14) b)
$\frac{ds}{dt} = \pi$ when $t = \frac{9}{2}$
It is a point of inflection, so $\frac{ds}{dt}$ is at its most steep. Ie. Acceleration is maximum.

15)
$x = \frac{1}{12}$ or $x = \frac{5}{12}$

16)
$\frac{dy}{dx} = \frac{\pi}{180}\cos\left(\frac{\pi}{180}x\right) = \frac{\pi}{180}\cos(x°)$

17) a)
$\frac{dy}{dx} = \frac{1}{x}$
17) b)
$\frac{dy}{dx} = \ln 3 \times 3^x$
17) c)
$\frac{dy}{dx} = 5e^{5x}$
17) d)
$\frac{dy}{dx} = \frac{2}{x}$
17) e)
$\frac{dy}{dx} = -\ln 6 \times 6^x$
17) f)
$\frac{dy}{dx} = 2e^{2x} - 2e^x$

18) a)
Minimum at $(\ln 2, 4 - 4\ln 2)$
18) b)
Maximum at $\left(\frac{1}{8}, \ln\frac{1}{8} - 1\right)$
18) c)
Minimum at $\left(\frac{1}{4}, \frac{39}{8} - 5\ln\frac{1}{4}\right)$

19) a)
$x = 1$
19) b)
$x = -\frac{1}{2}\ln(2)$
19) c)
$x = \log_3 \frac{1}{\ln 3}$

20)
Show that

21)
$y = \frac{5}{e}x$

22)
$25\ln 5\, y + x = 2 + 625\ln 5$

23) a)
37 hours
23) b)
-0.0209 bugs per day

24)
Minimum at $\left(\log_3 \frac{1}{2}, -\frac{1}{4}\right)$

25)
$\frac{2}{\ln 2} + 32\ln 2$

26)
Show that

27) a)
$\frac{dy}{dx} = 3\sec^2 3x$
27) b)
$\frac{dy}{dx} = 8\sec^2 4x$
27) c)
$\frac{dy}{dx} = -3\sec^2 9x$
27) d)
$\frac{dy}{dx} = -\pi\sec^2 \pi x$
27) e)
$\frac{dy}{dx} = -\frac{1}{2\pi}\sec^2\left(\frac{6}{\pi}x\right)$

28)
$y = \frac{4\pi}{9}x + 6 - \pi$

29)
Show that

30) a)
$\frac{dy}{dx} = 21\cos 7x + 4\sec^2 2x$

30) b)
$\frac{dy}{dx} = \ln 15 \times 15^x + \frac{15}{2}\sin\left(\frac{5}{2}x\right)$

30) c)
$\frac{dy}{dx} = \frac{1}{x} + \sqrt{2}\sec^2(\sqrt{2}x)$

30) d)
$\frac{dy}{dx} = \frac{1}{2}x^{-\frac{1}{2}} - 6\cos(2x)$

30) e)
$\frac{dy}{dx} = 3e^{3x} + 8\cos 8x$

30) f)
$\frac{dy}{dx} = -\frac{1}{4}\sin\left(\frac{1}{4}x\right) + 2\pi\sin(2\pi x)$

30) g)
$\frac{dy}{dx} = \ln 4 \times 4^x + \pi\sec^2(\pi x)$

31) a)
$\frac{d^2y}{dx^2} = 60x^3 - 25e^{5x}$

31) b)
$\frac{d^2y}{dx^2} = 2 + 4\sin(2x)$

31) c)
$\frac{d^2y}{dx^2} = -\frac{1}{4}x^{-\frac{3}{2}} - x^{-2}$

31) d)
$\frac{d^2y}{dx^2} = 2x^{-3} + \frac{1}{4}\cos\left(\frac{1}{2}x\right)$

31) e)
$\frac{d^2y}{dx^2} = 4e^{2x} + \pi^2\sin \pi x$

31) f)
$\frac{d^2y}{dx^2} = \frac{1}{4}e^{-\frac{1}{2}x} - 9\sin 3x + 9\cos 3x$

31) g)
$\frac{d^2y}{dx^2} = \frac{3}{4}x^{-\frac{5}{2}} + x^{-2}$

32) a)
$2e^2 + \frac{1}{4}$

32) b)
$\frac{9}{4}$

32) c)
$\sqrt{2}\ln 2 - 2\pi$

32) d)
$-\frac{\pi}{2}$

33)
$\left.\frac{dy}{dx}\right|_{x=\pi} = -2e^{-2\pi} + 15 > 0$ so increasing

34)
$y = \left(\frac{1}{3} - \pi\right)x - 1 + 3\pi + \ln 6$

35) a)
$y = -2x + 4$

35) b)
$y = \frac{1}{2}x + 4$

35) c)
20 units²

36) a)
Show that

36) b)
Show that

37)
$\left(\frac{\pi}{3}, 0\right)$

2.22 Differentiation: The Chain Rule
Questions on page 287

1) a)
$\frac{dy}{dx} = 12x(2x^2 + 1)^2$

1) b)
$\frac{dy}{dx} = 4(6x^2 - 5)(2x^3 - 5x)^3$

1) c)
$\frac{dy}{dx} = 6x^3(3x^4 - 1)^{-\frac{1}{2}}$

1) d)
$\frac{dy}{dx} = 3x(1 - x^2)^{-\frac{5}{2}}$

1) e)
$\frac{dy}{dx} = -\left(\frac{1}{2}x^{-\frac{1}{2}}\right)\left(5 + x^{\frac{1}{2}}\right)^{-2}$

1) f)
$\frac{dy}{dx} = 7(-x^{-2} - 1)(x^{-1} - x)^6$

2)
$\frac{9}{\sqrt{17}}$

3)
$8y - 21x = 1$

4)
$3y - x = 4$

5) a)
$\left(-\frac{1}{\sqrt{3}}, 0\right)$, $(0, 1)$ and $\left(\frac{1}{\sqrt{3}}, 0\right)$

5) b)
$(0, 1)$

6) a)
$\frac{dy}{dx} = \frac{1}{6y - 1}$

6) b)
$\frac{dy}{dx} = \frac{1}{8y^2}$

6) c)
$\frac{dy}{dx} = \frac{2}{3y}(3y^2 - 1)^{\frac{3}{4}}$

7) a)
$-\frac{1}{5}$

7) b)
1

7) c)
-4

8)
$y = 6x - 3$

9) a)
$\frac{dx}{dy} = 6x^{\frac{5}{6}}$

9) b)
$\frac{dy}{dx} = \frac{1}{6}x^{-\frac{5}{6}}$

10)
$x = \frac{3}{4}$

11) a)
$\frac{dy}{dx} = 2\cos x \sin x$

11) b)
$\frac{dy}{dx} = -2\sin x \cos x$

11) c)
$\frac{dy}{dx} = 6\cos(3x - 7)$

11) d)
$\frac{dy}{dx} = \frac{\pi}{5}\sin\left(\pi x - \frac{\pi}{2}\right)$

11) e)
$\frac{dy}{dx} = -\frac{1}{2}x^{-\frac{1}{2}}\sin(\sqrt{x})$

11) f)
$\frac{dy}{dx} = -6x^2\cos(x^3)$

11) g)
$\frac{dy}{dx} = -\cos x \sin(\sin x)$

11) h)
$\frac{dy}{dx} = \cos x \cos(\sin x)$

11) i)
$\frac{dy}{dx} = \frac{1}{2}\cos x\, (\sin x)^{-\frac{1}{2}}$

11) j)
$\frac{dy}{dx} = -\frac{1}{3}\sin x\, (\cos x)^{-\frac{2}{3}}$

11) k)
$\frac{dy}{dx} = 15\sin(5x)\cos^2(5x)$

12)
Show that

13) a)
Max = 3.6, Min = 2.1

13) b)
Maximums at 07:04 and 21:11

14)
Show that

15)
The rate at which the number of daylight hours is increasing is higher on day 85 compared to day 17

16)
Show that

17)
Show that

18) a)
$\frac{dy}{dx} = 10e^{2x}(e^{2x} - 1)^4$

18) b)
$\frac{dy}{dx} = 4\ln 5 \times 5^x(5^x - 1)^3$

18) c)
$\frac{dy}{dx} = \frac{4}{x}(\ln x)^3$

18) d)
$\frac{dy}{dx} = 6xe^{3x^2 - 1}$

18) e)
$\frac{dy}{dx} = (2x + 5)\ln 4 \times 4^{x^2 + 5x + 1}$

18) f)
$\frac{dy}{dx} = -\ln 5 \times 4x \times 5^{2x^2 + 1}$

19) a)
$\left(\frac{1}{2}\ln 3, 3\ln 3 - 3\right)$

19) b)
$\left(\frac{1}{2}, -2\ln 4\right)$

20) a)
$\frac{dy}{dx} = \frac{6x^2+2x}{2x^3+x^2-1}$

20) b)
$\frac{dy}{dx} = \frac{1}{2}x^{-1}$

20) c)
$\frac{dy}{dx} = -\frac{1}{2}x^{-1}$

20) d)
$\frac{dy}{dx} = \frac{\cos x}{\sin x} = \cot x$

20) e)
$\frac{dy}{dx} = \frac{e^x}{e^x+1}$

20) f)
$\frac{dy}{dx} = \frac{1}{x\ln x}$

21) a)
$\frac{dy}{dx} = \frac{1}{2\sqrt{x}}\sec^2 \sqrt{x}$

21) b)
$\frac{dy}{dx} = (\cos x - 2x)\sec^2(\sin x - x^2)$

21) c)
$\frac{dy}{dx} = \sec^2 x \; e^{\tan x}$

21) d)
$\frac{dy}{dx} = \left(\frac{1}{x} + 2e^{2x}\right)\sec^2(\ln x + e^{2x})$

21) e)
$\frac{dy}{dx} = \sec^2 x \, (\sec^2(\tan x))$

22)
Show that

23)
Show that

24)
$a = \frac{2}{3}\pi$ or $a = \frac{4}{3}\pi$ or $a = 2\pi$

25) a)
Show that

25) b)
107 cm³ (3 SF)

2.23 Differentiation: Connected Rates of Change
Questions on page 289

1) a)
$A = \pi r^2$

1) b)
$A = \frac{\sqrt{3}}{4}x^2$

1) c)
$V = \pi r^2 h$

1) d)
$V = \frac{4}{3}\pi r^3$

1) e)
$A = 4\pi r^2$

1) f)
$V = \frac{1}{3}\pi r^2 h$

2) a)
$\frac{dA}{dt} = 2\pi r$

2) b)
$\frac{dy}{dt} = -10x$

2) c)
$\frac{dP}{dt} = 18x^2 - 3$

2) d)
$\frac{dM}{dt} = -5x^4 + x$

3)
$\left.\frac{dA}{dt}\right|_{x=4} = 80$ cm² s⁻¹

4)
$\left.\frac{dA}{dt}\right|_{r=5} = 30\pi$ cm² s⁻¹

5)
$\frac{dc}{dt} = -0.2\pi$

6)
$\left.\frac{dA}{dt}\right|_{x=2\sqrt{3}} = 36$ cm² s⁻¹

7)
$\left.\frac{dr}{dt}\right|_{r=3} = 0.5$ cm s⁻¹

8) a)
$V = x^3, A = 6x^2$

8) b)
Show that

8) c)
$\left.\frac{dA}{dt}\right|_{x=4} = 5$ cm² s⁻¹

9)
$\left.\frac{dp}{dt}\right|_{x=25} = -0.48$ atmospheric units per second.

10)
$\left.\frac{dA}{dt}\right|_{r=25} = 0.32$ cm² s⁻¹

11) a)
$\frac{dh}{dt} = \frac{10}{\pi r^2}$

11) b)
$\left.\frac{dh}{dt}\right|_{r=45} = 0.00157$ cm s⁻¹

12)
$\left.\frac{dh}{dt}\right|_{r=45} = -0.0159$ cm per minute

13) a)
Show that

13) b)
$\frac{dV}{dh} = \frac{1}{3}\pi h^2$

13) c)
$\left.\frac{dh}{dt}\right|_{h=10} = 0.0573$ cm s⁻¹

14)
$\left.\frac{dh}{dt}\right|_{h=4} = 0.239$ cm s⁻¹

15)
Show that

16) a)
$\frac{dr}{dt} = \frac{2}{r^2}$

16) b)
Show that

2.24 Differentiation: The Product Rule
Questions on page 291

1) a)
$\frac{dy}{dx} = 2x(x-1) + (x-1)^2$

1) b)
$\frac{dy}{dx} = 4x^2(2x+7) + 2x(2x+7)^2$

1) c)
$\frac{dy}{dx} = -16x(7-x)^3 + 4(7-x)^4$

1) d)
$\frac{dy}{dx} = 9(x+2)^2(3x+1)^2 + 2(x+2)(3x+1)^3$

1) e)
$\frac{dy}{dx} = 6(3-x)^2(2x+9)^2 - 2(3-x)(2x+9)^3$

1) f)
$\frac{dy}{dx} = 8\sqrt{2x+5}(4x+1) + (2x+5)^{-\frac{1}{2}}(4x+1)^2$

1) g)
$\frac{dy}{dx} = \left(\frac{1}{x} - 2x\right)(-3x^{-4} - 10x^{-6}) + (-x^{-2} - 2)\left(\frac{1}{x^3} + \frac{2}{x^5}\right)$

2)
Student A: incorrectly multiplied out the brackets and they incorrectly differentiated the x^5 term.
Student B: Correct method.
Student C: incorrectly multiplied the two differentials of the two brackets together without applying the product rule.

3) a)
$\frac{dy}{dx} = x \times \frac{1}{x} + \ln x = 1 + \ln x$

3) b)
$\frac{dy}{dx} = x^{\frac{1}{2}}\sec^2 x + \frac{1}{2}x^{-\frac{1}{2}}\tan x$

3) c)
$\frac{dy}{dx} = \cos^2 x - \sin^2 x$

3) d)
$\frac{dy}{dx} = \frac{1}{x}\cos x - \sin x \ln x$

3) e)
$\frac{dy}{dx} = 2^x \cos x + \ln 2 \times 2^x \sin x$

3) f)
$\frac{dy}{dx} = \frac{(x^2+1)}{x} + 2x\ln x$

4) a)
$\frac{dy}{dx} = e^x(x+1)$
4) b)
$\frac{dy}{dx} = 4^x x(x \ln 4 + 2)$
4) c)
$\frac{dy}{dx} = e^x(\sec^2 x + \tan x)$
4) d)
$\frac{dy}{dx} = 2^x(\cos x + \ln 2 \sin x)$
4) e)
$\frac{dy}{dx} = -2e^{-2x}(x-4)(x+3)$
4) f)
$\frac{dy}{dx} = 2e^{4x}(\cos 2x + 2 \sin 2x)$

5) a)
$y = 33x - 24$
5) b)
$y = 332x - 536$
5) c)
$y = 3x + 4$

6) a)
$2x^2(x+3)$
6) b)
$(2x-1)^4(5x-4)$
6) c)
$(3x+1)^2(13x+6)$
6) d)
$\frac{1}{3}(6x+5)^2(9x+2)$
6) e)
$\frac{9-x}{2(1-x)^6}$
6) f)
$(3x-1)(4x+1)^2(43x-5)$

7) a)
$x = -1, x = -\frac{1}{3}$
7) b)
$x = -2$
7) c)
$x = \frac{1}{7}, 1, \frac{1}{16}$

8) a)
$(1, 4)$
8) b)
$(0, 0)$ and $\left(\frac{5}{3}, 0\right)$
8) c)
$0 < y < 4$

9) a)
$\frac{dy}{dx} = (x+3)^6(8x+3)$
9) b)
$\frac{dy}{dx} = (2x+1)^3(10x-23)$
9) c)
$\frac{dy}{dx} = 5(2x+1)^3(10x+1)$
9) d)
$\frac{dy}{dx} = 2e^{2x}(x^2+x+1)$

10) a)
$\left(\frac{1}{2}, \frac{2000}{e}\right)$

10) b)
Maximum

11) a)
150 employees
11) b)
$\frac{dy}{dt} = -\frac{\pi}{2}t \cos\frac{\pi}{6}t - 3\sin\frac{\pi}{6}t$
11) c)
$\left.\frac{dy}{dt}\right|_{t=3} = -3$
11) d)
On exactly 1st April the company is losing 3 employees per day

12)
$y - 6e^2x + 5e^2 = 0$

13)
Minimum at $(e^{-1}, -e^{-1})$

14) a)
Show that
14) b)
Maximum

15) a)
Show that
15) b)
$r = -e^\pi$

16) a)
$f'(x) = 6x^2 e^{10x^2} \cos(2x^3) + 20xe^{10x^2} \sin(2x^3)$
16) b)
$f'(x) = 2x5^{x^2} \sin(1-x^2) + 2x(\ln 5)5^{x^2} \cos(1-x^2)$
16) c)
$f'(x) = \frac{2}{x}(x^2+1)^3 + 6x(x^2+1)^2 \ln(x^2)$

17)
Show that

18)
Show that

19) a)
$(0, 0)$ and $(-2, 4e^{-2})$
19) b)
$\left(\frac{1}{3}, 0\right)$
19) c)
$(1, 0)$ and $\left(\frac{1}{5}, \frac{16}{25\sqrt{5}}\right)$

20) a)
$\frac{dx}{dy} = \frac{3(y+2)}{3y-5} + \ln(3y-5)$
20) b)
$\frac{1}{12}$

21)
Show that

22) a)
60 m
22) b)
$\frac{dh}{dt} = -\frac{25\pi}{3}e^{-0.05t} \sin\left(\frac{\pi}{3}t\right) - \frac{5}{4}e^{-0.05t}\left(1 + \cos\left(\frac{\pi}{3}t\right)\right)$
$\left.\frac{dh}{dt}\right|_{t=3} = 0$
22) c)
When $t = 3, h = 10$ so the bungee jumper has reached the minimum height which is 10m above the ground.

23) a)
$f'(x) = 2x(4x+3)^{-\frac{1}{2}} + \sqrt{4x+3}$
23) b)
Show that
23) c)
Minimum at $x = -\frac{1}{2}$

2.25 Differentiation: The Quotient Rule
Questions on page 293

1) a)
$f'(x) = -\frac{5}{(2x-1)^2}$
1) b)
$f'(x) = \frac{1}{(1-4x)^2}$
1) c)
$f'(x) = \frac{6}{(3x-1)^2}$
1) d)
$f'(x) = \frac{e^x(x-4)}{(x-3)^2}$
1) e)
$f'(x) = \frac{1 - 3\ln x}{x^4}$
1) f)
$f'(x) = \frac{10x \cos(10x) - 3\sin(10x)}{2x^4}$
1) g)
$f'(x) = \frac{5^x(x \ln 5 - 2)}{x^3}$
1) h)
$f'(x) = \frac{e^{2x}(2\cos x + \sin x)}{\cos^2 x}$
1) i)
$f'(x) = \frac{2x + 1 - 2x \ln 5x}{x(2x+1)^2}$
1) j)
$f'(x) = \frac{2x \sin x - x^2 \cos x + \cos x}{\sin^2 x}$

2)
$\frac{dy}{dx} = \frac{x+5}{2x\sqrt{x}}$

3)
$\frac{dy}{dx} = \frac{3^x(x \ln 3 - 2)}{x^3 \ln 3}$

4) a)
$\frac{1-2x}{(x+1)^3}$
4) b)
$-\frac{x+6}{(x-4)^3}$
4) c)
$-\frac{26x}{(x+2)^4}$

4) d)
$\frac{(x+3)^3(17x-13)}{(5x-1)^4}$

4) e)
$-\frac{24(2x+5)^2(x+1)}{(1-2x)^5}$

5)
Show that

6) a)
$f'(x) = \frac{4(1-x)}{(x+1)^3}$

6) b)
$f'(x) = \frac{21-36x}{(1+2x)^4}$

6) c)
$f'(x) = \frac{2e^{4x}(2x+1)}{(x+1)^3}$

7) a)
$x = \frac{25}{2}$

7) b)
$x = \frac{3}{2}$ and $x = \frac{13}{2}$

7) c)
$x = 6$

7) d)
$x = \frac{2}{a} - a$

8) a)
Show that

8) b)
Show that

8) c)
Show that

9) a)
$\frac{dy}{dx} = 3\sec^2(3x)$

9) b)
$\frac{dy}{dx} = -2\csc 2x \cot 2x$

9) c)
$\frac{dy}{dx} = \frac{1}{3}\sec\frac{1}{3}x \tan\frac{1}{3}x$

9) d)
$\frac{dy}{dx} = -2\pi \sec^2(\pi x)$

9) e)
$\frac{dy}{dx} = -\frac{1}{5}\sec^2\left(-\frac{1}{5}x\right)$

9) f)
$\frac{dy}{dx} = -9 \sec 9x \tan 9x$

9) g)
$\frac{dy}{dx} = \frac{1}{2}\csc\frac{1}{2}x \cot\frac{1}{2}x$

10)
$x \in \left(-\infty, -\frac{1}{2}\right) \cup \left(\frac{3}{2}, \infty\right)$

11)
Maximum at $\left(\frac{9}{4}, \frac{1}{8}\right)$

12) a)
$\left(1, \frac{1}{2}\right)$

12) b)
Minimum

13)
Show that

14) a)
$P(1,4)$ and $Q\left(4, \frac{7}{2}\right)$

14) b)
$y - 4 = -\frac{1}{6}(x - 1)$

14) c)
$\frac{625}{12}$ units^2

15)
Show that

16) a)
Show that

16) b)
$k = 4$

17) a)
$f'(x) = \frac{-(10x^2+20x+k)}{e^x(k+10x^2)^2}$

17) b)
$k \leq 10$

18) a)
$\frac{dy}{dx} = 2x \sec^2(x^2 - 1)$

18) b)
$\frac{dy}{dx} = -2e^{2x} \cot(e^{2x}) \csc(e^{2x})$

18) c)
$\frac{dy}{dx} = \cos x \tan(\sin x) \sec(\sin x)$

19)
Show that

20) a)
$k = \frac{1}{2}$

20) b)
Show that

20) c)
$\left(-\frac{1}{3}, \frac{2}{27}\right)$ and $\left(1, -\frac{2}{5}\right)$

20) d)

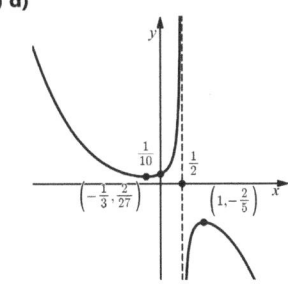

21) a)
$\frac{dy}{dx} = \frac{2xe^{x^2}(x^2-2)}{(x^2-1)^2}$

21) b)
$\frac{dy}{dx} = \frac{3x^2 \cos(x^3) - 2\sin(x^3)}{e^{2x+1}}$

21) c)
$\frac{dy}{dx} = -\frac{3^{\cos x}[x(\ln 3)\sin x + 2]}{x^3}$

22) a)
$\frac{dy}{dx} = 4\left(\frac{x^2-1}{x-3}\right)^3 \times \frac{2x(x-3)-(x^2-1)}{(x-3)^2}$

22) b)
Show that

23) a)
$\frac{dy}{dx} = \frac{2\ln x - 4(\ln x)^2}{x^5}$

23) b)
Show that

24) a)
Show that

24) b)
$x = \frac{\pi}{9}$ and $x = \frac{4\pi}{9}$

25) a)
Show that

25) b)
Show that

26)
$a = -2, b = 3$

27) a)
$f'(x) = \frac{xe^x(2x^2+x-2)}{(2x-1)^2}$

27) b)
$f'(x) = \frac{(2^x \cos x + \ln 2 \times 2^x \sin x)(\sqrt{x}) - 2^x \sin x \times \frac{1}{2}x^{-\frac{1}{2}}}{x}$

27) c)
$f'(x) = \frac{(1+\ln x)(x-1)^2 - 2x(x-1)\ln x}{(x-1)^4}$

28)
Show that

29) a)
$x = 3$

29) b)
Show that

29) c)
Show that

30)
Show that

31)
$\frac{2}{5(1+\sin 2x)}$

32) a)
$k = 3$

32) b)
$B\left(0, -\frac{1}{9}\right), C(1,0)$

32) c)
$A\left(-1, -\frac{1}{8}\right)$

523

32) d) i)

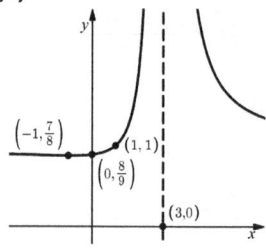

32) d) ii)

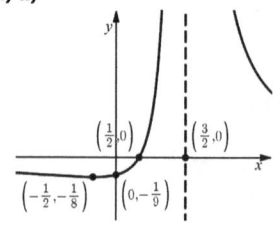

33) a)
Maximum domain is $x > 0$ and $x \neq e$
33) b)
Show that
33) c)
$f(x) < 0$ when $e < x < e^{\frac{5}{2}}$

34) a)
$\{x : x \in \mathbb{R}, x \neq 1\}$
34) b)
$f'(x) = \frac{2^x(x\ln 2 - \ln 2 - 1)}{(x-1)^2}$
$(2.44, 3.77)$
34) c)
$\{f(x) : f(x) \in \mathbb{R}, f(x) < 0\} \cup \{f(x) : f(x) \in \mathbb{R}, f(x) > 3.77\}$

2.26 Differentiation: Implicit Differentiation
Questions on page 297

1) a)
$2y\frac{dy}{dx} + 1$
1) b)
$x\frac{dy}{dx} + y$
1) c)
$x^2\frac{dy}{dx} + 2xy$
1) d)
$e^x \cos y \frac{dy}{dx} + e^x \sin y$
1) e)
$-\sin x \, e^y + e^y \frac{dy}{dx} \cos y$
1) f)
$\frac{x}{y}\frac{dy}{dx} + \ln y$
1) g)
$2^x \frac{dy}{dx} + (\ln 2) \times 2^x y$

2) a)
$\frac{dy}{dx} = \frac{8}{2y} = \frac{4}{y}$
2) b)
$\frac{dy}{dx} = -\frac{y}{x}$

2) c)
$\frac{dy}{dx} = \frac{-3x^2 - y^2}{2xy - 2y}$
2) d)
$\frac{dy}{dx} = \frac{2x-1}{2xy}$
2) e)
$\frac{dy}{dx} = \frac{y - 5x}{5y - x}$
2) f)
$\frac{dy}{dx} = \frac{-\cos x - e^y}{xe^y}$
2) g)
$\frac{dy}{dx} = \frac{\cos y - 1}{x \sin y}$
2) h)
$\frac{dy}{dx} = -\frac{y^2 + xy\ln y}{x^2 + xy \ln x}$

3)
$3y + 4x - 25 = 0$

4)
Show that

5)
Show that

6)
Show that

7)
Show that

8)
Show that

9) a)
$\frac{1}{2}$
9) b)
$-\frac{1}{4}$
9) c)
$\frac{5}{3}$ or $\frac{5}{12}$

10)
Show that

11)
Show that

12) a)
$(0, 1)$
12) b)
$(-\sqrt{2}, 0)$ and $(\sqrt{2}, 0)$

13) a)

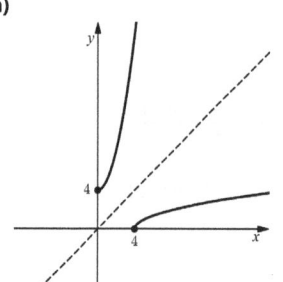

13) b)
Show that

14)
$\frac{1}{7}$

15)
$(2, \ln 2)$

16)
Show that

17) a)
For $\left(\frac{1}{4}, 0\right)$: $8y + 4x = 1$
For $\left(\frac{1}{4}, -2\right)$: $8y + 28x = -9$
17) b)
Show that
17) c)
$\frac{2}{21}$ units^2

18)
Show that

19) a)
Show that
19) b)
Show that

20) a)
Show that
20) b)
$3y - 4x = 2$
20) c)
Show that

21) a)
Show that
21) b)
Show that
21) c)
$y = \frac{1}{2}x + \frac{3\pi}{2}$

22) a)
Show that
22) b)
Show that
22) c)
$y = -x + 2$

23) a)
P is $(13, 16)$
23) b)
$8\sqrt{5}$

24) a)
$k = 2$
24) b)
$(e^2, 0)$

25) a)
Show that

25) b)
Tangent line: $y - 4\ln 2 = -\frac{1}{2}(x-2)$
Normal line: $y - 4\ln 2 = 2(x-2)$
25) c)
Tangent: $x = 2 + 8\ln 2$
Normal: $x = 2 - 2\ln 2$
25) d)
Show that

26)
$A(\sqrt{5}, \sqrt{5})$, $B(5,1)$, $C(-\sqrt{5}, -\sqrt{5})$, $D(-5, -1)$

27) a)
Show that
27) b)
Show that
27) c)
$k = -6$

28) a)
$a = 2, b = -1, k = 5$
28) b)
$(2, -1)$ and $(10, -5)$

29) a)
Show that
29) b)
$2x + 5y = 11$
29) c)
Prove

30) a)
Show that
30) b)
$\frac{dP}{dt} = \frac{1}{(t-31)^2 + 1}$
30) c)
During the year 2004

2.27 Numerical Methods: Change of Sign Method
Questions on page 300

1) a)
Show that
1) b)
Show that
1) c)
Show that
1) d)
Show that

2) a)
$f(2) = -1, f(4) = 1$
2) b)
[graph]

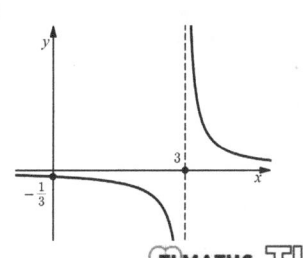

2) c)
Whilst there is a change in sign between the two values of x, the graph is not continuous for the interval, at $x = 3$ there is a vertical asymptote so we cannot determine if there is a root or not.

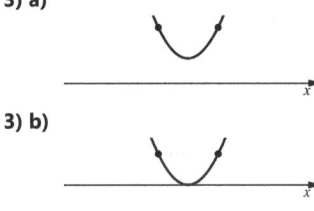

3) a)
[graph]
3) b)
[graph]
3) c)
[graph]

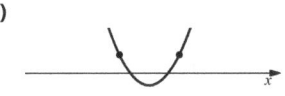

4) a) 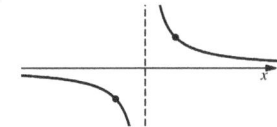
[graph]
4) b) 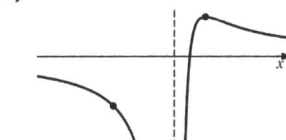
[graph]
4) c) 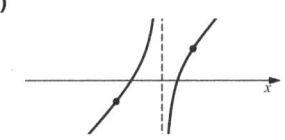
[graph]
4) d) 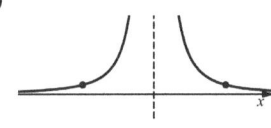
[graph]
4) e) 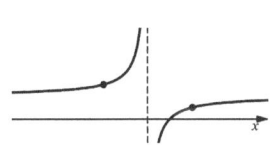
[graph]
4) f)
[graph]

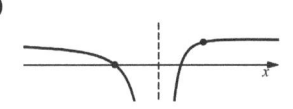

5) a)
$f(-1) = 1.0574\ldots, f(1) = -1.0574\ldots,$
$f(2) = 2.4350\ldots$
5) b)
There is an asymptote at $x = 0$ and $x = \frac{\pi}{2}$ so the curve is discontinuous between each of the pairs of points, therefore we cannot determine if there are any roots.

6) a)
Show that

6) b)
Whilst there is a change in sign between the two values of x, the graph is not continuous for the interval, at $x = 0$ there is a vertical asymptote so we cannot determine if there is a root or not.

7)
[graph showing $y = e^x$ and $y = 4 - x^2$]

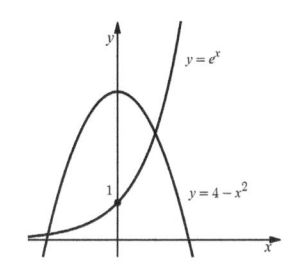

Two points of intersection ⇒ two solutions

8)
[graph showing $y = \frac{1}{2}x$ and $y = \sqrt{x-1}$ with point $(2, 1)$]

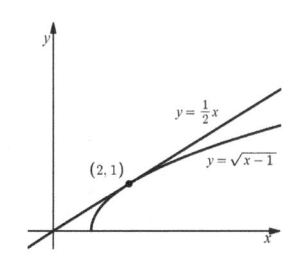

One point of intersection ⇒ one solution

9) a)
Show that
9) b)
Show that
9) c)
Show that

10) a)
Show that
10) b)
Show that
10) c)
1.6786

2.28 Numerical Methods: $x = g(x)$ Method
Questions on page 301

1) a)
Show that
1) b)
Show that

2) a)
Show that
2) b)
Show that

3) a)
Show that
3) b)
$x_{n+1} = \sqrt{\frac{3 - 2x_n}{x_n - 1}}$

525

Solution is $x = 2.3744$ to 4 decimal places.

4) a)
Show that
4) b)
Show that
4) c)
$x_{n+1} = \sqrt{\dfrac{5x_n}{\ln(x_n)}}$
Solution is $x = 3.7687$ to 4 decimal places.

5) a)
$x_2 = 2.8284, x_3 = 2.4142, x_4 = 2.3268$
5) b)
$x_2 = 1.6829, x_3 = 1.9874, x_4 = 1.8289$
5) c)
$x_2 = 7.9000, x_3 = 5.6429, x_4 = 7.0444$
5) d)
$x_2 = 1.8000, x_3 = 1.6832, x_4 = 1.6353$

6) a) 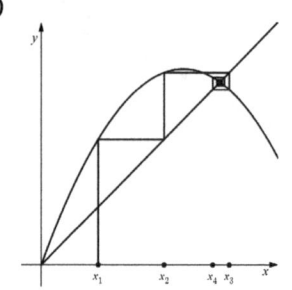
6) b) 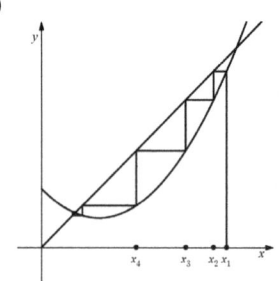
6) c) 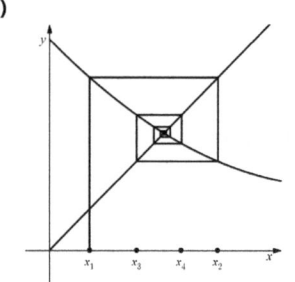
6) d)

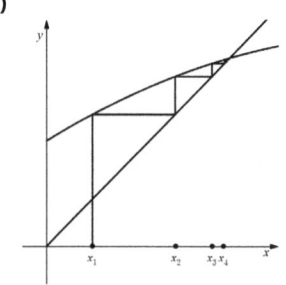

7) a)
It will staircase and then cobweb to the intersection at $x = \frac{1}{4}$
7) b)
It directly gives the intersection at $x = \frac{1}{4}$
7) c)
It will cobweb or staircase/cobweb to the intersection at $x = \frac{1}{4}$
7) d)
It directly gives the intersection at $x = 1$
7) e)
It diverges away (as a staircase) from the solution at $x = 1$

8) a)
$\alpha \in [3.146, 3.147]$
8) b)
$\alpha \in [1.159, 1.160]$
8) c)
$\alpha \in [1.709, 1.710]$

9) a)
$g'(2.273) = 39.51 > 1$ so it diverges from the root.
9) b)
$g'(1.146) = 3.146 > 1$ so it diverges from the root.
9) c)
$g'(-1.841) = 0.1587 < 1$ so it converges to the root.

10) a)
Iterative formula B: $x = 2.86$ (2 DP)
Iterative formula C: $x = 1.25$ (2 DP)
Iterative formula D: $x = -1.11$ (2 DP)
10) b)
$g'(-1.11) = 9.3563 > 1$
Convergence requires $-1 < g'(\alpha) < 1$ which does not hold for any of the roots.

11) a) 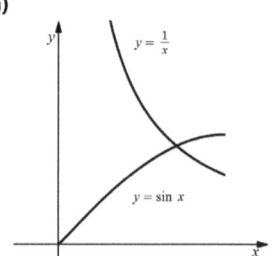
11) b)
Show that
11) c)
$x = 1.1142$ (4 DP)

12) a) 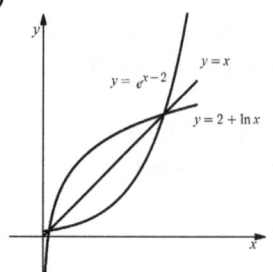
12) b)
All 3 graphs intercept at the same points.

13) a)
Show that
13) b)
$T = 50.133$ hours
13) c)
50 hours

14) a)
Show that
14) b)
$\theta = 1.97$ radians (2 DP)

15) a)
$x = 0.739$ (3 DP)
15) b)
$x = -1.686$ (3 DP)

16) a) 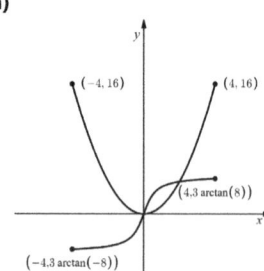
16) b)
Show that
16) c)
$x = 1.994$ (3 DP)

17) a)
Show that
17) b)
$x = 0.74$ (2 DP)

18) a)
$x_2 = 0.3165, x_3 = 0.3122, x_4 = 0.3133$
18) b)
Show that

2.29 Numerical Methods: Newton-Raphson Method
Questions on page 304

1) a)
$x_2 = 2.91$ (2 DP)
1) b)
$x_2 = 1.87$ (2 DP)

1) c)
$x_2 = 1.14$ (2 DP)

2) a)
$x_2 = 1.15$ (2 DP)
2) b)
$x_2 = 2.39$ (2 DP)
2) c)
$x_2 = 1.26$ (2 DP)

3) a)

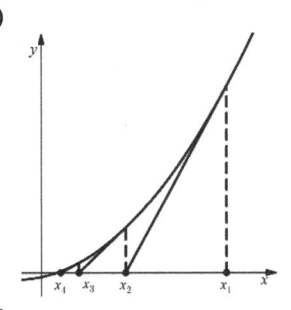

3) b)

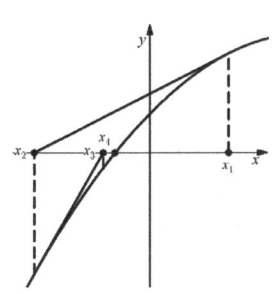

3) c)

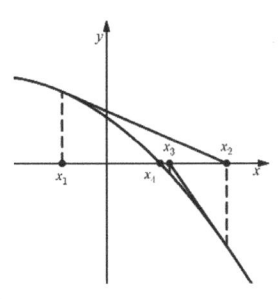

3) d)

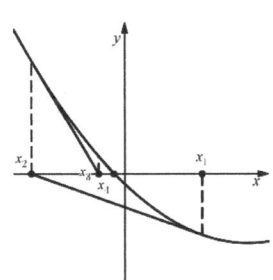

4) a)
It reaches the root at $-\frac{9}{40}$
4) b)
It coincides with a stationary point, therefore fails to reach a root
4) c)
It reaches the root at $\frac{89}{40}$

5) a)
It is already at the root at A

5) b)
It reaches the root at A

5) c)
It coincides with a stationary point, therefore fails to reach a root
5) d)
Closer to B it will reach the root at E, before switching to reach the root at C
5) e)
It is already at the root at C
5) f)
It coincides with a stationary point, therefore fails to reach a root
5) g)
It reaches the root at E
5) h)
It is already at the root at E

6)
2.714 (3 DP)

7) a)
Show that
7) b)
Show that
7) c)
$x_2 = 0.7647$, $x_3 = 0.7049$
7) d)
There is a zero on the denominator of the formula This is caused by a stationary point on the curve at $x = 0$ and $x = -\frac{8}{9}$

8) a)
Show that
8) b)
$n = 1.895$ (4 SF)

9) a)
Show that
9) b)
$x = 1.504$ (4 SF)
9) c)
-3.97 (3 SF)

10) a)

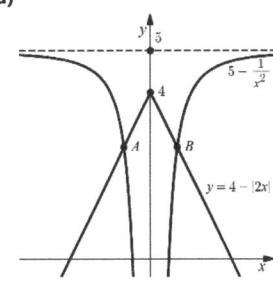

10) b)
Show that
10) c)
Show that
10) d)
$x = -0.657$ (3 DP)

10) e)
$x_1 = \frac{1}{3}$ or $x_1 = 0$

11) a)
Show that
11) b)
$x = 1.11$ (3 SF)

12) a)

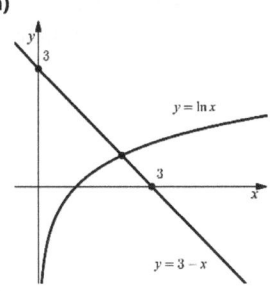

12) b)
Show that
12) c)
$x = 2.21$ (3 SF)

13) a)
Show that
13) b)
Show that
13) c)
$x_2 = 0.1670$, $x_3 = 0.1911$, $x_4 = 0.1927$
13) d)
$x^2 \geq 0 \therefore 2x^2 + 1 \geq 1$ so there is no fail case

14) a)
Show that
14) b)
Show that
14) c)
$x = 7.292$ (3 DP)

15) a)
Show that
15) b)
$x = 1.11$ (3 SF)

16) a)
$f'(x) = 2^x x(x\ln 2 + 2)$
By using $x_0 = 0$ she will get a fail case caused by the stationary point of the curve at $(0, -3)$
16) b)
$x_3 = 1.16$ (3 SF)

17) a)
$(1, 0)$
17) b)
Show that
17) c)
$x = 0.717$ (3 SF)

18) a)
Show that
18) b)
Show that

18) c)
$x_1 = 0.4$, $x_2 = 0.425$, $x_3 = 0.426$
18) d)
$x = 0.618$ (3 SF)
18) e)
$x \approx 0.62$ is approximately the x-coordinate of the minimum of the curve

19) a)
Show that
19) b)
Show that
19) c)
$x = 2.17$ (3 SF)

20) a)
Show that
20) b)
16 days and 14 hours
20) c)
When $t = 0$ there are $N = -1481$ which does not make sense as the initial number of views
20) d)
The graph is negative after 16 days and 14 hours so the model would no longer be valid.

21) a)
Show that
21) b)
$x = 3.714$ (3 DP)
21) c)
1.2 units^2

2.30 Integration: The Trapezium Rule
Questions on page 308

1)
A will produce an overestimate as the curve is convex (concave up). The top of each trapezium will be above the curve.
B will produce an underestimate as the curve is concave (concave down). The top of each trapezium will be below the curve.
C will produce the exact result as the function is linear.
D will produce an underestimate as the curve is concave (concave down). The top of each trapezium will be below the curve.

2)
$h = 0.625$

3) a)
$A = \frac{19}{2}$
3) b)
$A = 26$

3) c)
$A = \frac{17}{4}$
3) d)
$A = \frac{429}{16}$

4) a)
$A = \frac{111}{2}$
4) b)
$A = \frac{117}{32}$
4) c)
$A = \frac{311}{8}$
4) d)
$A = \frac{75}{2}$

5) a)
$A = 38$
5) b)
$A = \frac{45}{2}$
5) c)
$A = \frac{116}{25}$
5) d)
$A = \frac{1132}{75}$

6) a)
$A = 342$
6) b)
$A = 24616$
6) c)
$A = \frac{262}{63}$
6) d)
$A = \frac{922}{105}$

7)
$A = 0.153$ (3 SF)

8) a)
Show that
8) b)
Show that
8) c)
Show that
8) d)
Show that

9)
$A = 7.362$ (3 DP). This will be an underestimate as the curve is concave between $x = 2$ and $x = 8$.

10)
$s = 225$ m

11) a)
$A = 0.481$ (3 SF)
11) b)
Increase the number of strips.

12)
$A = 0.427$ (3 SF)

13)
$s = 106.65$ m

14) a)
$A = 2.521$
14) b)
$\int_{2.5}^{4.5} \sqrt{\frac{2+x}{2x}}\, dx \approx 1.78$ (3 SF)

15) a)
$A = 23000$ (3 SF)
15) b)
$\int_{1}^{2} e^{3x^2+1}\, dx \approx 62600$ (3 SF)

16)
$A = 2660$ (3 SF)

17) a)
$A = 9.233$
17) b)
$\int_{8}^{24} \log_5 \frac{x^2}{36}\, dx \approx 18.466$

18) a)
$A = 5.38$ (3 SF)
18) b)
This will be an underestimate as the curve is concave between $x = -2.5$ and $x = -0.5$.

19)
$A = 1.23$ (3 SF)

20)
$A = 1.57$ (3 SF)

21) a)
$A = \frac{3}{5}p + \frac{92}{5}$
21) b)
$p = 11$

22) a)
Show that
22) b)
Show that

23) a)
Show that
23) b)
$A_2 = 33.81$ (4 SF)
23) c)
$A_T = 25.48$ (4 SF)
23) d)
$\int_{-5}^{5} \frac{25}{5+x^2}\, dx = 25.72$ (4 SF)

2.31 Integration: Areas Between Curves
Questions on page 312

1)
$\frac{1}{12}$

2)
$\frac{8}{3}$

3)
9

4)
$\frac{8}{3}$

5)
$\frac{3}{2}$

6)
$\frac{1}{3}$

7)
$\frac{4}{9}$

8)
$\frac{1}{8}$

9)
$\frac{1}{2}$

10)
$\frac{5581}{96}$

11)
$\frac{16875}{64}$

12)
$6 - 4\sqrt{2}$

2.32 Integration: Standard Functions
Questions on page 314

1) a)
$e^x + \frac{1}{2}x^2 + c$

1) b)
$\frac{1}{2}e^{2x} + c$

1) c)
$\frac{2}{3}e^{3x} - \frac{1}{3}x^3 + c$

1) d)
$2e^{\frac{1}{2}x} + \frac{9}{x} + c$

1) e)
$-\frac{1}{4}e^{-8x} - \frac{1}{7}e^{-7x} + c$

2) a)
$y = -\frac{1}{3}e^{-3x} + c$

2) b)
$y = 4x + e^{-2x} + c$

2) c)
$y = \frac{2}{3}x^{\frac{3}{2}} - 2e^{\frac{1}{6}x} + c$

2) d)
$y = \frac{1}{2}x^2 + \frac{3}{2}e^{-4x} + c$

3)
$y = 2e^{6x}$

4)
$y = 3 - 16e^{-\frac{3}{2}x}$

5) a)
$e^3 + 5$

5) b)
$100 \ln \frac{3}{2} - 1$

5) c)
$36\left(e^{\frac{2}{3}} - e^{-\frac{2}{3}}\right)$

5) d)
$\frac{7}{4}$

6) a)
$\frac{1}{6}(e^{18} - 1)$

6) b)
$-\frac{3}{2} + 3e^{-\frac{2}{3}}$

7)
$\frac{4\sqrt{3}}{3}$

8)
$6 \ln 2 - \frac{5}{2}$

9)
$4e^2 - e^3 + e$

10) a)
$-\cos x + \sin x + c$

10) b)
$-2\cos x - \sin x + c$

10) c)
$-\cos 2x + \frac{1}{4}\sin 4x + c$

10) d)
$-6\cos\frac{1}{2}x - 9\sin\frac{2}{3}x + c$

11) a)
$y = \frac{1}{2}\sin 6x + c$

11) b)
$y = -\frac{3}{x} + 16\cos\frac{1}{4}x + c$

11) c)
$y = \frac{3}{4}x^{\frac{4}{3}} - \frac{4}{5}\sin\frac{5}{6}x + c$

12)
$y = \frac{4}{3}\cos 3x + 3$

13)
$y = -6\cos\frac{1}{3}x - 9\sin\frac{1}{3}x$

14) a)
1

14) b)
$\frac{1}{2}$

14) c)
$\frac{15}{2}(\sqrt{3} - 1)$

14) d)
$\frac{11}{4}$

15) a)
$1 - 2\sqrt{3}$

15) b)
$\frac{3\sqrt{2}}{8} - \frac{1}{4}$

16)
$9\pi - 4$

17)
$16\pi - \frac{3\sqrt{3}}{2} - \frac{9}{2}$

18)
4π

19) a)
$\frac{1}{\ln 3}3^x + c$

19) b)
$\frac{1}{\ln 9}9^x + \frac{1}{\ln 64}64^x + c$

19) c)
$-\frac{1}{(\ln 5)5^x} + \frac{2}{(\ln 2)2^x} + c$

19) d)
$-\frac{1}{(\ln 3)3^x} + \frac{4}{(\ln 9)9^x} + c$

20) a)
$y = \frac{1}{\ln 8}8^x + c$

20) b)
$y = \frac{1}{2}x^2 - 3^x + c$

20) c)
$y = \frac{1}{3\ln 6}6^{3x} + \frac{1}{3\ln 6}6^{-3x}$

20) d)
$y = \frac{1}{3\ln 2}2^{3x} + \frac{4}{\ln 2}2^{-x}$

21)
Show that

22)
Show that

23) a)
$\frac{8}{3\ln 3}$

23) b)
$-\frac{9}{2\ln 2}$

23) c)
$\frac{8}{5\ln 5}$

23) d)
$\frac{3}{8\ln 4} - 1$

24) a)
$\frac{100}{\ln 5}$

24) b)
$\frac{\sqrt{2}-1}{\ln 2}$

25)
$\frac{3}{2} - \frac{1}{\ln 2}$

26)
$\frac{2}{\ln 2} - \frac{8}{\ln 4} + 6\log_2 3$

27)
$\frac{5e-10}{2e\ln 5}$

28) a)
$\ln|x| + c$
28) b)
$3\ln|x| - \frac{1}{6}x^2 + c$
28) c)
$\frac{5}{3}\ln|x| - \frac{2}{5x} + c$
28) d)
$\frac{3}{4}\ln|x| + 2x + c$

29) a)
$y = 4\ln|x| + \frac{4}{x} + c$
29) b)
$y = 2x - \frac{7}{4}\ln|x| + c$
29) c)
$y = 3\ln|x| - \frac{2}{\sqrt{x}} + c$
29) d)
$y = \frac{8}{5}\ln|x| + \frac{3}{5x} + c$

30)
$y = -24 + 5\ln|x|$

31)
$y = -10x + 11 + 3\ln|x|$

32) a)
$9\ln 2$
32) b)
$8\ln\frac{1}{3}$
32) c)
$\frac{1}{5}\ln 4 + \frac{3}{10}$
32) d)
$\frac{3}{5}\ln\frac{2}{3} - \frac{1}{5}$

33) a)
$4\ln\frac{1}{2} - 4$
33) b)
$\frac{3}{2} + 6e^4 - 6e^2$

34)
$10\ln 3$

35)
$6\ln 3$

36)
$\frac{3}{4} + 4\ln 2$

37) a)
$\tan x + c$
37) b)
$-\csc x + c$
37) c)
$\sec x + c$
37) d)
$-\cot x + c$
37) e)
$\frac{1}{4}\tan 4x + c$

37) f)
$-3\cot\frac{1}{3}x + c$

38) a)
$y = -\frac{1}{2}\csc 2x + c$
38) b)
$y = -32\csc\frac{1}{4}x + c$
38) c)
$y = \frac{1}{5}\sec\frac{10}{3}x + c$

39)
$y = \frac{5}{3}\tan 3x + \frac{10}{3}$

40)
$y = 2\sec\frac{1}{2}x - \frac{4\sqrt{3}}{3}$

41) a)
$2\sqrt{3}$
41) b)
27
41) c)
$\frac{1}{9}$
41) d)
$\sqrt{3} - \frac{3}{2}$

42) a)
$3(2 + \sqrt{3})$
42) b)
$\frac{1}{2} - \frac{\sqrt{3}}{6}$

43)
$\frac{\pi}{4} - \frac{1}{2}$

44)
$-2\sqrt{2} + 2 + \frac{2\sqrt{3}}{3}$

45)
$\frac{\pi^2 + 4\pi + 8}{8}$

2.33 Integration: Reversing the Chain Rule Part 1
Questions on page 319

1) a)
$\frac{1}{4}(x+1)^4 + c$
1) b)
$\frac{1}{7}(x-4)^7 + c$
1) c)
$\frac{3}{4}(2+x)^8 + c$
1) d)
$2(x-1)^5 - \frac{2}{3}(x-3)^6 + c$

2) a)
$y = \frac{4}{7}(x-3)^{\frac{7}{4}} + c$
2) b)
$y = -\frac{1}{2}(9+x)^{-2} + c$
2) c)
$y = \frac{2}{3}(x+7)^{\frac{3}{2}} + c$

2) d)
$y = 8(5+x)^{\frac{1}{2}} + c$

3)
$y = 4 - \frac{3}{2(x-2)^4}$

4)
$y = \frac{3}{4}(4+x)^{\frac{4}{3}} - 9$

5) a)
$\frac{1}{9}$
5) b)
2916
5) c)
$\frac{16}{3} - \frac{4\sqrt{2}}{3}$
5) d)
198

6) a)
$\frac{1275}{8}$
6) b)
$\frac{304}{3}$

7) a)
$\frac{1}{10}(2x+1)^5 + c$
7) b)
$-\frac{1}{18}(2-3x)^6 + c$
7) c)
$\frac{1}{15}(5+6x)^{10} + c$
7) d)
$\frac{3}{8}(2x-3)^4 + \frac{4}{25}(4-5x)^5 + c$

8) a)
$y = \frac{3}{50}(5x+9)^{\frac{10}{3}} + c$
8) b)
$y = \frac{1}{4}(6-x)^{-4} + c$
8) c)
$y = \frac{2}{9}(10+3x)^{\frac{3}{2}} + c$
8) d)
$y = -30(1-x)^{\frac{2}{3}} + c$

9)
$y = \frac{3x-4}{3x-2}$

10)
$y = 3 - \frac{1}{(4x+1)^{\frac{7}{2}}}$

11) a)
0
11) b)
$\frac{7}{12}$
11) c)
$\frac{15}{7}$
11) d)
$\frac{15}{4}$

12) a)
$\frac{85}{4}$

12) b)
$\frac{26}{3}$

13) a)
$\frac{1}{6}(x^2+3)^6 + c$

13) b)
$\frac{1}{4}(4x^3-5)^4 + c$

13) c)
$\frac{1}{20}(2x^2-5)^5 + c$

13) d)
$\frac{1}{3}(x^4-3)^{\frac{3}{2}} + c$

14) a)
$y = -(3x^3+2)^{-1} + c$

14) b)
$y = \frac{1}{7}(2-5x^4)^{\frac{7}{2}} + c$

14) c)
$y = \frac{3}{32}(3+4x^2)^{\frac{4}{3}} + c$

14) d)
$y = -\frac{10}{9}(2x^9+3)^{-5} + c$

15)
$y = \frac{1}{16}(5x^2-1)^8 + \frac{15}{16}$

16)
$y = \frac{2}{9}(27x^3-1)^{\frac{1}{2}}$

17) a)
1158562

17) b)
$\frac{93}{40}$

17) c)
$\frac{19}{145}$

17) d)
$\frac{124}{9}$

18) a)
$3375\sqrt{15} - 1$

18) b)
$\frac{28}{3}$

19)
$\frac{1}{12}$

20)
$\frac{1}{6}$

21)
$\frac{4128}{3481}$

2.34 Integration: Reversing the Chain Rule Part 2
Questions on page 321

1) a)
$\frac{1}{2}e^{2x+1} + c$

1) b)
$-\frac{1}{3}e^{5-3x} + c$

1) c)
$\frac{1}{2}e^{x^2} + c$

1) d)
$\frac{1}{6}e^{6x^4} + c$

2) a)
$e^{\sin x} + c$

2) b)
$-\frac{1}{18}e^{6\cos 3x} + c$

2) c)
$8e^{\sin\frac{1}{2}x} + c$

2) d)
$-9e^{3\cos\frac{1}{3}x} + c$

3) a)
$e^{2^x} + c$

3) b)
$\frac{1}{\ln 3}e^{3^x} + c$

3) c)
$\frac{1}{(\ln 5)^2}e^{5^x+1} + c$

3) d)
$\frac{1}{4\ln 4}e^{4^{x+1}} + c$

4) a)
$e^{\tan x} + c$

4) b)
$-\frac{1}{2}e^{4\cosec x} + c$

4) c)
$\frac{1}{2}e^{\sec 6x} + c$

4) d)
$e^{16\cot\frac{1}{2}x} + c$

5)
$y = e^{\frac{1}{x}} - 2e$

6) a)
$\frac{1}{3}(e-1)$

6) b)
$\frac{1}{2}(e-1)$

6) c)
$3e^2(e^2-1)$

6) d)
$5(e-1)$

7)
$e^2(e^2-1)$

8) a)
$-\frac{1}{5}\cos\left(5x + \frac{\pi}{7}\right) + c$

8) b)
$2\sin 2x^2 + c$

8) c)
$-\cos(x^2+x-1) + c$

8) d)
$-\frac{1}{2}\sin\left(6x - \frac{2}{3}x^3\right) + c$

9) a)
$-\cos(\sin x) + c$

9) b)
$\cos(\cos x) + c$

9) c)
$\sin(\sin x) + c$

9) d)
$-\sin(\cos x) + c$

10) a)
$-\cos(5^x) + c$

10) b)
$\frac{1}{\ln 9}\sin(9^x) + c$

10) c)
$-\cos(\ln x) + c$

10) d)
$\frac{1}{2}\sin(2\ln x) + c$

11) a)
$y = -\frac{1}{2}\cos(\cot 2x) + c$

11) b)
$y = -\frac{1}{6}\sin(8\cosec 3x) + c$

11) c)
$y = -\frac{3}{10}\cos\left(90\tan\frac{1}{3}x\right) + c$

11) d)
$y = \frac{1}{12}\sin\left(18\sec\frac{2}{3}x\right) + c$

12)
$y = -6\cos x^2 + 24 + 3\sqrt{2}$

13) a)
1

13) b)
$\frac{5}{2}$

13) c)
18

13) d)
0

14)
$-\frac{\sqrt{2}}{64}$

15) a)
$\ln|x+1| + c$

15) b)
$\ln|2x-5| + c$

15) c)
$2\ln|4x+3| + c$

15) d)
$\frac{9}{5}\ln|5x-7| + c$

15) e)
$-\frac{3}{4}\ln|1-4x| + c$

15) f)
$-5\ln|5-3x| + c$

16) a)
$\ln|x^2 + 3| + c$
16) b)
$\ln|5 + x^3| + c$
16) c)
$\frac{3}{2}\ln|5x^2 - 4| + c$
16) d)
$-\frac{5}{2}\ln|6 - x^4| + c$

17) a)
$\ln|x^2 + 4x - 3| + c$
17) b)
$\ln|x^3 - 3x + 9| + c$
17) c)
$\frac{1}{8}\ln|4x^2 + 8x + 5| + c$
17) d)
$2\ln|x^3 - 6x^2 - 1| + c$

18) a)
$\ln|e^x + 1| + c$
18) b)
$\ln|2^x - x| + c$
18) c)
$\ln|\sec x| + c$
18) d)
$\ln|\sin x| + c$

19) a)
$\frac{1}{2}\ln|\sin 2x + 1| + c$
19) b)
$2\ln|4 - \cos 3x| + c$
19) c)
$\frac{4}{3}\ln|3\sin 6x - 5| + c$
19) d)
$-\frac{2}{15}\ln|15\cos 9x + 4| + c$

20) a)
$y = \ln|3^x + e^x| + c$
20) b)
$y = 2\ln|4x + \tan 2x| + c$
20) c)
$y = \ln\left|\sin x + x^{\frac{1}{2}}\right| + c$
20) d)
$y = -10\ln|e^{2x} - e^{3x}| + c$

21)
$y = \ln|x^2 + x - 2|$

22) a)
$4\ln 4$
22) b)
$\frac{3}{2}\ln 2$
22) c)
$\ln\frac{1}{3}$
22) d)
$4\ln\frac{4}{3}$

23)
$\frac{1}{2}\ln\frac{61}{9}$

24) a)
$\frac{1}{4}(e^x + 5)^4 + c$
24) b)
$\frac{1}{15}(e^{3x} - 3)^4 + c$
24) c)
$\frac{1}{48}(4e^{4x} - 1)^6 + c$
24) d)
$\frac{40}{3}\left(e^{\frac{1}{2}x} + 2\right)^{\frac{3}{2}} + c$

25) a)
$\frac{1}{3}\sin^3 x + c$
25) b)
$\frac{1}{4}\sin^4 x + c$
25) c)
$-\frac{1}{5}\cos^5 x + c$
25) d)
$-\frac{1}{12}\cos^6 2x + c$

26) a)
$\frac{1}{6}(\tan x - 4)^6 + c$
26) b)
$\frac{1}{12}\tan^4 3x + c$
26) c)
$\frac{1}{5\ln 2}(2^x + 4)^5 + c$
26) d)
$\frac{1}{2\ln 3}(3^x - 2)^6 + c$

27) a)
$y = \frac{1}{4}(\ln x + 3)^4 + c$
27) b)
$y = \frac{1}{144}(2\ln x + 5)^6 + c$
27) c)
$y = \frac{1}{4}(\sec x + 2)^4 + c$
27) d)
$y = -\frac{1}{8}\text{cosec}^8 x + c$

28) a)
910
28) b)
$\frac{1}{5}$
28) c)
$\frac{1}{3}$
28) d)
$\frac{3843}{4}$

29)
$2\ln\frac{9}{5}$

30)
$\ln\frac{7}{6}$

31)
$\frac{1}{2}\ln\frac{7}{2}$

32)
Show that

33)
$1 + \frac{\sqrt{2}}{2}$

34)
$\frac{1}{6}\ln 33$

35)
$\ln 4 - \frac{3}{4}$

36)
$\frac{37}{768} - \frac{3\sqrt{3}\pi}{512}$

37)
$\frac{4\pi\sqrt{3}}{27}$

2.35 Integration by Substitution
Questions on page 324

1) a)
$\frac{1}{15}(3x + 1)^5 + c$
1) b)
$-\frac{1}{(2x-3)^2} + c$
1) c)
$\frac{1}{12}(8x + 5)^{\frac{3}{2}} + c$
1) d)
$-\frac{9}{2}(5 - 2x)^{\frac{2}{3}} + c$

2) a)
Translation by the vector $\begin{pmatrix} -3 \\ 0 \end{pmatrix}$
2) b)
Show that

3) a)
Translation by the vector $\begin{pmatrix} 4 \\ 0 \end{pmatrix}$
3) b)
Show that

4) a)
$\frac{2}{5}(x + 1)^{\frac{5}{2}} - \frac{2}{3}(x + 1)^{\frac{3}{2}} + c$
4) b)
$\frac{1}{5}(2x + 3)^{\frac{5}{2}} - (2x + 3)^{\frac{3}{2}} + c$
4) c)
$\frac{1}{5}(x^2 + 1)^{\frac{5}{2}} - \frac{1}{3}(x^2 + 1)^{\frac{3}{2}} + c$

5) a)
$(x + 1) - \ln|x + 1| + c$
5) b)
$\frac{1}{2}(2x + 3) - \frac{3}{2}\ln|2x + 3| + c$
5) c)
$\frac{1}{2}(x^2 + 1) - \frac{1}{2}\ln|x^2 + 1| + c$

6) a)
$\frac{1}{2}(x + 1)^2 - 2(x + 1) + \ln|x + 1| + c$
6) b)
$\frac{1}{4}(2x + 3)^2 - 3(2x + 3) + \frac{9}{2}\ln|2x + 3| + c$

6) c)
$\frac{1}{2}(x^2+1) - \frac{1}{2}\ln|x^2+1| + c$

7) a)
$\frac{2}{7}(x+1)^{\frac{7}{2}} - \frac{4}{5}(x+1)^{\frac{5}{2}} + \frac{2}{3}(x+1)^{\frac{3}{2}} + c$

7) b)
$\frac{1}{7}(2x+3)^{\frac{7}{2}} - \frac{6}{5}(2x+3)^{\frac{5}{2}} + 3(2x+3)^{\frac{3}{2}} + c$

7) c)
$\frac{1}{5}(x^2+1)^{\frac{5}{2}} - \frac{1}{3}(x^2+1)^{\frac{3}{2}} + c$

8)
$-7(2-3x)^6 + 3(2-3x)^7 + c$

9)
$4 - 2\sqrt{2}$

10)
Show that

11)
$1 + \ln\frac{9}{16}$

12)
Show that

13)
$\frac{62}{1225}$

14)
Show that

15)
Show that

16)
$(x+2)^{\frac{3}{2}}(12x - 46) + c$

17)
$-\frac{11}{14}$

18)
$\frac{2}{3}(x^2+5)^{\frac{3}{2}} - 11(x^2+5)^{\frac{1}{2}} + c$

19)
$\ln\frac{4}{3} - \frac{1}{4}$

20)
$\ln\frac{256}{81} - 1$

21)
$(2\sqrt{x} - 1) + \ln|2\sqrt{x} - 1| + c$

22)
$4 + 4\ln\frac{2}{3}$

23)
$\frac{712}{15}$

24)
$2(1+\sqrt{x})^3 - 9(1+\sqrt{x})^2 + 18(1+\sqrt{x}) - 6\ln|1+\sqrt{x}| + c$

25)
$2\ln\frac{4}{3}$

26)
$5 - \frac{1}{4}\ln\frac{7}{3}$

27)
$\frac{13}{3}$

28) a)
$32 - 8\ln 5$

28) b)
$\frac{44}{3}$

29) a)
$8\ln\frac{10}{7}$

29) b)
$8\ln\frac{10}{7}$

30)
$\frac{5}{3}$

31) a)
$\frac{1}{5}\sin^5 x + c$

31) b)
$4e^{x^2+1} + c$

31) c)
$-e^{\cos x} + c$

31) d)
$\ln|\tan x| + c$

32)
$6\ln\frac{10}{9}$

33)
$5\ln|5 + e^x| + c$

34) a)
$3 - 9\ln\frac{5}{4}$

34) b)
$\frac{2}{5}$

35) a)
$\frac{2}{3}$

35) b)
$\frac{7\sqrt{2}}{240}$

36) a)
$e^2 - e$

36) b)
$\frac{21}{20}$

37)
$-\frac{1}{1+\tan x} + c$

38) a)
$\frac{35}{1152\ln 5}$

38) b)
$\frac{35}{1152\ln 5}$

39) a)
$\frac{486 - 8\sqrt{2}}{5\ln 8}$

39) b)
$\frac{486 - 8\sqrt{2}}{5\ln 8}$

40)
$-6\ln|3 - \ln x| + c$

41) a)
$\frac{2}{25} - \frac{2}{(6+\ln 2)^2}$

41) b)
$\ln\frac{16}{9} - \frac{2}{3}$

42) a)
$\arcsin x + c$

42) b)
$\frac{x}{\sqrt{1-x^2}} + c$

43)
$\frac{\pi}{12}$

44)
$\frac{1}{2}\arcsin\left(\frac{2}{3}x\right) + c$

45) a)
$\arctan x + c$

45) b)
$\arctan\left(\frac{1}{5}x\right) + c$

46)
$\frac{1}{4} + \frac{\pi}{8}$

47)
$\frac{\pi}{2} - 1$

48) a)
$\frac{\pi}{4}$

48) b)
$\frac{\pi}{8}$

49)
$\frac{\pi}{12}$

50)
$\ln 64 + 5\pi$

51)
$3\sqrt{3} - 3$

52)
Show that

53)
$x(\ln 2x)^2 - 2x\ln 2x + 2x + c$

54)
$\tan^3(\arcsin x) + 3\tan(\arcsin x) + c$

55)
$-8(\cos x - 2) - 16\ln|\cos x - 2| + c$

56)
$\frac{1}{2}\arcsin x - \frac{1}{2}x\sqrt{1-x^2} + c$

57) a)
$\frac{9\sqrt{3}}{8} + \frac{3\pi}{4}$

57) b)
$\frac{\pi}{12}$

58) a)
$\frac{1}{2}$

58) b)
$\frac{5\pi}{4} - \frac{5}{2}$

59) a)
$\left(\frac{1}{2}(x^2+4)^2 - 4(x^2+4)\right)\ln(x^2+4) - \frac{1}{4}(x^2+4)^2 + 4(x^2+4) + c$

59) b)
$\frac{1}{\ln 3}\left((3^x+1)\ln(3^x+1) - (3^x+1)\right) + c$

60)
$\frac{1}{2}\sqrt{x+2}\,e^{4\sqrt{x+2}} - \frac{1}{8}e^{4\sqrt{x+2}} + c$

61)
$2e^3$

62) a)
$\frac{33}{40}\ln 11 - \frac{1}{3}$

62) b)
$\pi - 2$

63)
$\ln 9$

64)
$4\ln\frac{8}{7}$

65) a)
$\ln\left(\frac{3+2\sqrt{2}}{3}\right)$

65) b)
$\frac{1+3\sqrt{2}}{6} - \ln\left(\frac{3+2\sqrt{2}}{3}\right)$

66) a)
Show that

66) b)
$\frac{2\sqrt{3}}{3} - \frac{\pi}{3}$

2.36 Integration by Parts
Questions on page 328

1) a)
$xe^x - e^x + c$

1) b)
$xe^x - 2e^x + c$

1) c)
$2xe^x + 3e^x + c$

2)
$\frac{2}{9e}$

3)
$\frac{1}{3}e - \frac{2}{3}$

4)
$y = 4xe^{\frac{1}{2}x} - 8e^{\frac{1}{2}x} - 8\ln 4 + 18$

5) a)
$-x\cos x + \sin x + c$

5) b)
$-\left(x + \frac{\pi}{3}\right)\cos x + \sin x + c$

6)
$\frac{\pi}{2}$

7) a)
$x\sin x + \cos x + c$

7) b)
$\left(x - \frac{\pi}{6}\right)\sin x + \cos x + c$

8)
$\frac{8\pi}{3}$

9)
$\frac{3\pi}{8} - \frac{1}{2}$

10)
$\frac{\pi}{16}$

11)
$y = \sin 8x - 8x\cos 8x$

12)
$y = 18x\sin\frac{1}{3}x + 54\cos\frac{1}{3}x - 9\sqrt{3}\pi - 28$

13) a)
$\frac{1}{2}x^2\ln x - \frac{1}{4}x^2 + c$

13) b)
$\frac{1}{3}x^3\ln x - \frac{1}{9}x^3 + c$

13) c)
$\left(\frac{1}{2}x^2 + 4x\right)\ln x - \left(\frac{1}{4}x^2 + 4x\right) + c$

13) d)
$\frac{1}{4}x^2(x^2+6)\ln x - \left(\frac{1}{16}x^4 + \frac{3}{4}x^2\right) + c$

14)
$\frac{1}{3}\ln 8 - \frac{1}{24}\ln 16 + \frac{7}{72}$

15)
$y = -x^2 + 2x^2\ln\frac{1}{2}x - 4e^2 + 1$

16) a)
$\frac{1}{\ln 2}x2^x - \frac{1}{(\ln 2)^2}2^x + c$

16) b)
$\frac{1}{\ln 3}(3x+9)3^x - \frac{3}{(\ln 3)^2}3^x + c$

17)
$\frac{4}{\ln 2} - \frac{1}{(\ln 2)^2} = \frac{4\ln 2 - 1}{(\ln 2)^2}$

18)
$\frac{90}{\ln 5} - \frac{40}{(\ln 5)^2}$

19)
$y = -\frac{1}{\ln 4}x4^{-x} - \frac{1}{(\ln 4)^2}4^{-x} + 1 + 4^{-\ln 4}\left(1 + \frac{1}{(\ln 4)^2}\right)$

20)
Show that

21)
$\pi\sqrt{3} + 18\ln\sqrt{3}$

22) a)
$x\ln x - x + c$

22) b)
$x(\ln x)^2 - 2x\ln x + 2x + c$

23)
$e^2 + 1$

24) a)
$x^2 e^x - 2xe^x + 2e^x + c$

24) b)
$-x^2 e^{-x} - 2xe^{-x} - 2e^{-x} + c$

24) c)
$-x^2\cos x + 2x\sin x + 2\cos x + c$

24) d)
$\frac{1}{\ln 2}x^2 2^x - \frac{2}{(\ln 2)^2}x2^x + \frac{2}{(\ln 2)^3}2^x + c$

25)
$\frac{5}{8} - \frac{25}{16}e^{-1}$

26)
$e + 1$

27)
$6 - 2e$

28) a)
$\frac{1}{2}(e^x\sin x - e^x\cos x) + c$

28) b)
$\frac{1}{2}(e^x\cos x + e^x\sin x) + c$

29)
$\frac{2}{29}\left(1 + e^{\frac{5\pi}{2}}\right)$

30)
$\frac{1}{3}(\ln 3)^2 + \frac{3}{2}\ln 3 - 1$

2.37 Partial Fractions
Questions on page 330

1)
$\frac{5}{x+1} + \frac{2}{x+3}$

2)
$\frac{7}{x-2} + \frac{5}{x+1}$

3) a)
$\frac{1}{x-2} + \frac{1}{x+5}$

3) b)
$\frac{9}{(x+4)} - \frac{2}{(2x+1)}$

3) c)
$\frac{4}{5(x+2)} - \frac{4}{5(x+7)}$

3) d)
$\frac{7}{2(x+2)} - \frac{7}{2(x+6)}$

3) e)
$-\frac{2}{x} + \frac{2}{x-2}$

3) f)
$-\frac{2}{x+5} + \frac{3}{x+3}$

3) g)
$\frac{5}{2x-3} - \frac{3}{x+8}$

3) h)
$\frac{5}{4x+1} + \frac{3}{2x+1}$

3) i)
$-\frac{k}{9(x+5)} + \frac{k}{9(x-4)}$

4) a)
In his second line of working he incorrectly rearranged, putting $25 - 4x^2$ rather than 1 on the left hand side of the equation

4) b)
$\frac{1}{10(5-2x)} + \frac{1}{10(5+2x)}$

5)
$\frac{3}{x-5} + \frac{2}{x+1} - \frac{2}{x-3}$

6) a)
$\frac{2}{x-5} - \frac{3}{x-2} + \frac{5}{x+1}$

6) b)
$\frac{2}{49(x-4)} + \frac{2}{49(x+3)} - \frac{8}{49(2x-1)}$

7) a)
Show that

7) b)
$(2x-1)(2x+3)(x-3)$

7) c)
$-\frac{1}{2(2x-1)} + \frac{5}{18(2x+3)} + \frac{1}{9(x-3)}$

8) a)
Lisa has not multiplied both sides by $(x+2)(x^2-1)$

8) b)
$\frac{1}{3(x+2)} + \frac{1}{6(x-1)} - \frac{1}{2(x+1)}$

9)
$\frac{4}{x+1} - \frac{1}{(x+1)^2} + \frac{3}{x+3}$

10)
$-\frac{2}{x-2} + \frac{3}{(x-2)^2} + \frac{1}{x+4}$

11) a)
$\frac{1}{3(x-1)} - \frac{1}{3(x+2)} - \frac{1}{(x+2)^2}$

11) b)
$\frac{8}{9(2x-1)} - \frac{4}{9(x+1)} + \frac{1}{3(x+1)^2}$

11) c)
$-\frac{2}{x} - \frac{2}{x^2} + \frac{2}{x-1}$

11) d)
$\frac{1}{x-1} + \frac{4}{(x-1)^2}$

12) a)
A

12) b)
B

12) c)
A

12) d)
B

12) e)
C

13) a)
$\frac{1}{x+5} + \frac{1}{x-2}$

13) b)
$\frac{1}{x} - \frac{1}{x-2} + \frac{2}{(x-2)^2}$

13) c)
$-\frac{3}{5(x+2)} + \frac{7}{10(2x-1)} + \frac{1}{2(2x+1)}$

13) d)
$-\frac{4}{25(x+1)} - \frac{2}{5(x+1)^2} + \frac{8}{25(2x-3)}$

14) a)
$\frac{1}{x-2} + \frac{4}{x+7}$

14) b)
$\frac{4}{x+7} - \frac{1}{x-2} + \frac{2}{(x-2)^2}$

14) c)
$x = 3$

15)
$p = 5$

16) a)
$\frac{2}{3(x-4)} - \frac{1}{3(2x+1)}$

16) b)
$\left(1, -\frac{1}{3}\right)$ or $\left(-5, -\frac{1}{27}\right)$

17) a)
$\frac{5}{3(x-2)} + \frac{2}{(x-2)^2} + \frac{5}{3(5-x)}$

17) b)
Show that

18) a)
$-\frac{1}{x+3} + \frac{1}{2-x} = -\frac{1}{x+3} - \frac{1}{x+2}$

18) b)
Show that

19) a)
$\frac{5}{x+1} - \frac{5}{x+2}$

19) b)
Show that

20) a)
$\frac{2}{3(1-2x)} + \frac{1}{3(1+x)}$

20) b)
$1 + x + 3x^2 + 5x^3 + \cdots$

20) c)
$|x| < \frac{1}{2}$

21) a)
$-\frac{1}{3(x+1)} + \frac{1}{3(x-2)}$

21) b)
$-\frac{1}{2} + \frac{1}{4}x - \frac{3}{8}x^2 + \frac{5}{16}x^3 + \cdots$

21) c)
$|x| < 1$

22) a)
$\frac{1}{(1-x)^2} + \frac{3}{3+x}$

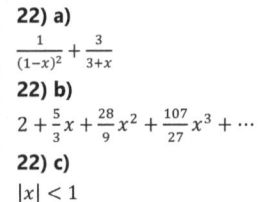

22) b)
$2 + \frac{5}{3}x + \frac{28}{9}x^2 + \frac{107}{27}x^3 + \cdots$

22) c)
$|x| < 1$

23)
$A = 4, B = 2, C = 3$

24) a)
$A = 3, B = 5$

24) b)

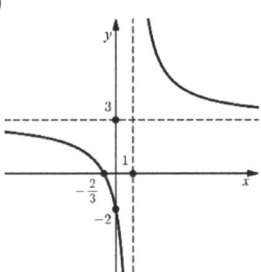

25) a)
$A = 5, B = -2$

25) b)

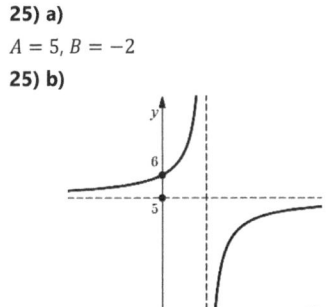

26) a)
$A = 1, B = 2, C = 8$

26) b)
$A = -2, B = -6, C = 33$
26) c)
$A = 3, B = 2, C = -3, D = -2$
26) d)
$A = 1, B = -1, C = 2, D = -1$

27)
$A = 2, B = 1, C = -3$

28)
$A = 5, B = 2$

29)
$A = -3, B = 1, C = \frac{1}{5}, D = \frac{64}{5}$

30)
$A = 1, B = -1, C = \frac{5}{2}, D = \frac{17}{2}$

31)
$A = 2, B = -1, C = -2, D = 6$

32) a)
$\frac{1}{p+1} + \frac{3}{p+2}$
32) b)
$\frac{5p+4}{(p+1)(p+2)}$
32) c)
$p = 7$
$u_1 = \frac{1}{8}, u_2 = \frac{1}{3}, u_3 = \frac{13}{24}$

2.38 Integration: Partial Fractions
Questions on page 333

1) a)
$8 + \frac{7}{x+2}$
1) b)
$8x + 7\ln|x+2| + c$

2) a)
$2x - 4 + \frac{3}{x+1}$
2) b)
$x^2 - 4x + 3\ln|x+1| + c$

3) a)
$\frac{1}{x-3} - \frac{1}{x+1}$
3) b)
$\ln\left|\frac{x-3}{x+1}\right| + c$

4) a)
$\ln\left|\frac{5+x}{5-x}\right| + c$
4) b)
$\ln\frac{9}{4}$

5) a)
$\ln|x+5| + \ln|x-6| + c$
5) b)
$k = 10$

6) a)
$\frac{\frac{k}{2}}{x} + \frac{\frac{k-8}{2}}{2-x} \equiv \frac{k}{2x} + \frac{k-8}{2(2-x)}$
6) b)
$k = 6$

7) a)
$\frac{2}{3-x} + \frac{1}{x+2} + \frac{3}{3x-1}$
7) b)
$\ln\frac{40}{3}$

8) a)
$-\frac{2}{x} + \frac{1}{x^2} + \frac{8}{2x+1}$
8) b)
$-\frac{1}{x} + \ln\left(\frac{(2x+1)^4}{x^2}\right) + c$

9) a)
$\frac{4}{x-2} + \frac{6}{(x-2)^2}$
9) b)
$3 + \ln 16$

10) a)
$\frac{2}{x+1} - \frac{4}{(x+1)^2} + \frac{4}{3x-1}$
10) b)
$-\frac{1}{2} + \frac{16}{3}\ln 2 - \ln 9$

11)
$\ln\frac{5184}{3125} - \frac{1}{2}$

12)
$\ln\frac{18}{11}$

13)
$\ln|\sqrt{x} - 2| + \ln|\sqrt{x} + 1| - 2\ln|\sqrt{x} + 4| + c$

14)
Show that

2.39 Integration: Double Angles
Questions on page 334

1) a)
$\cos 2x = 2\cos^2 x - 1$
1) b)
Show that

2) a)
$\cos 2x = 1 - 2\sin^2 x$
2) b)
Show that

3) a)
$\frac{3}{2}x - 2\cos x - \frac{1}{4}\sin 2x + c$
3) b)
$\frac{3}{2}x - 2\sin x + \frac{1}{4}\sin 2x + c$
3) c)
$x + \frac{1}{2}\cos 2x + c$
3) d)
$-\frac{1}{2}\sin 2x + c$

4) a)
$-\frac{1}{4}\cos 2x + c$
4) b)
$-\cos 4x + c$
4) c)
$-\frac{3}{5}\cos 10x + c$
4) d)
$-\frac{1}{24}\cos 12x + c$

5) a)
$\frac{41\pi}{4} - 24$
5) b)
$\frac{19\pi}{2} - 12$

6) a)
$\frac{33\pi}{4} - 8$
6) b)
179π

7) a)
$\frac{1}{6}\sin 6x + c$
7) b)
$\frac{1}{10}\sin 10x + c$
7) c)
$\frac{1}{4}\sin 4x + c$
7) d)
$\frac{1}{20}\sin 20x + c$

8)
$-\frac{5}{2}\ln|\cos 2x + 4| + c$

9)
$y = 2\sec x + 8$

10)
$1 - \frac{\sqrt{2}}{2}$

11)
$\sqrt{2} - 1$

12)
$\frac{2\sqrt{3}}{3}$

13) a)
Show that
13) b)
$-\frac{3}{4}\cos x + \frac{1}{12}\cos 3x + c$

14) a)
Show that
14) b)
$\frac{1}{12}\sin 3x + \frac{3}{4}\sin x + c$

15)
$\pi - 2$

16)
$\frac{9\pi}{2}$

2.40 Differential Equations
Questions on page 335

1) a) $y = Ae^x$
1) b) $y = Ax$
1) c) $y = Ae^{-\frac{1}{x}}$
1) d) $y = Ae^{-\frac{1}{18x^3}}$

2) a) $y = \pm\sqrt{x^2 + A}$
2) b) $y = (x^3 + A)^{\frac{1}{3}}$
2) c) $y = \pm\left(\frac{1}{4}x^4 + c\right)^{\frac{1}{2}}$
2) d) $y = \pm\left(\frac{2}{9}x^6 + A\right)^{\frac{1}{4}}$

3) $x = \left(\frac{10}{10 - 3\ln|t|}\right)^{\frac{1}{3}}$

4) $x = \left(\frac{3}{4}t^2 + 8\right)^{\frac{2}{3}}$
When $t = 2\sqrt{39}$: $x = 25$

5) a) $y = \frac{2}{B - \ln|x|}$
5) b) $y = \frac{x}{1 + Ax}$
5) c) $y = \pm\left(\frac{3x^2}{1 + Bx^2}\right)^{\frac{1}{2}}$

6) a) $y = Ae^{\frac{1}{2}x^2}$
6) b) $y = Ae^{\frac{1}{3}x^3}$
6) c) $y = Ae^{x^4}$
6) d) $y = Ae^{\frac{2}{3}x^{\frac{3}{2}}}$

7) $P = 500e^{-25t^4}$
When $t = 0.7$, $P = 1.24$ to 3sf

8) a) $y = \frac{2}{B - x^2}$
8) b) $y = \frac{3}{B - x^3}$
8) c) $y = \pm\left(\frac{3}{B - x^4}\right)^{\frac{1}{2}}$

9) $y = \pm\left(\frac{36}{7 - 6x^4}\right)^{\frac{1}{2}}$
When $x = 0$: $y = \pm\frac{6\sqrt{7}}{7}$

10) a) $y = \pm(A - 2\cos x)^{\frac{1}{2}}$
10) b) $y = \pm\left(\frac{1}{2}\sin x + A\right)^{\frac{1}{2}}$
10) c) $y = \left(A - \frac{1}{4}\cos 2x\right)^{\frac{2}{3}}$
10) d) $y = \left(\frac{15}{32}\sin 4x + A\right)^{\frac{2}{3}}$

11) a) $y = \left(4\sqrt{2}\sin 3x\right)^{\frac{2}{3}}$
11) b) $y = 32^{\frac{1}{3}}$

12) a) $y = \ln\left(\frac{1}{2}\sin 2x + c\right)$
12) b) $y = Ax + B$
12) c) $y = 5\ln\left(\frac{4}{5}x^3 + A\right)$
12) d) $y = Me^{-20e^{0.2x}} + \frac{3}{4}$

13) a) $y = \ln(e^x + c)$
13) b) $y = \ln\left(\frac{e^x}{1 + Ae^x}\right)$
13) c) $y = \frac{1}{3}\ln\left(\frac{2}{B - 3e^{4x}}\right)$
13) d) $y = \frac{1}{4}\ln\left(\frac{e^{\frac{1}{2}x}}{64 + Ae^{\frac{1}{2}x}}\right)$

14) $y = \ln\left(\frac{6}{e^{-6x} + 5}\right)$
When $x = 0.5$: $y = \ln\left(\frac{6}{e^{-3} + 5}\right)$

15) Show that

16) a) Show that
16) b) $t = \frac{5\pi}{2} \approx 7.85$ seconds

17) Show that

18) a) $\frac{dr}{dt} = \frac{k}{\sqrt{r}}$ where r is the radius in mm, t is the time in seconds, and k is a positive constant

$r = (28t + 8)^{\frac{2}{3}}$
18) b) $r = (28t + 8)^{\frac{2}{3}}$ may not be a valid equation to model the ink in the long term as in reality the ink will eventually stop spreading. Whereas in our model the radius grows indefinitely.

19) Show that

20) Show that

21) $v = 6e^{-1.05t}$
When $t = 2$, $v = 6e^{-1.05 \times 2} = 0.735$ ms⁻¹

22) a) $\frac{dT}{dt} = -\frac{k}{\sqrt{T}}$
22) b) The coffee cannot decrease in temperature indefinitely as it will eventually reach room temperature.

23) Show that

24) a) $\frac{dP}{dt} = kP$
24) b) $P = 500e^{0.00004t}$
When $t = 365$, $P = 500e^{0.00004 \times 365} \approx £507.35$

25) $\frac{dN}{dt} = \frac{N}{12}$

26) $\frac{dh}{dt} = \frac{\ln h}{\ln 32}$

27) Show that

28) Show that

29) a) $y = \pm\left(\frac{1}{2}(x + 2)^4 + A\right)^{\frac{1}{2}}$
29) b) $y = Ae^{\frac{1}{10}(2x - 5)^5}$
29) c) $y = \left(\frac{1}{4}(4x + 7)^{\frac{3}{2}} + A\right)^{\frac{2}{3}}$
29) d) $y = \ln\left(c - \frac{3}{5}(3 - x)^{\frac{5}{3}}\right)$

30) a) $y = \pm(2xe^x - 2e^x + A)^{\frac{1}{2}}$

30) b)
$y = Ae^{-\frac{1}{10}x}\cos 10x + \frac{1}{100}\sin 10x$

30) c)
$y = \left(\frac{3}{40}xe^{2x} - \frac{3}{80}e^{2x} + A\right)^{\frac{1}{3}}$

30) d)
$3y\ln y - 3y = \frac{5}{2}x^2 \ln x - \frac{5}{4}x^2 + c$

31) a)
$\frac{2}{x-3} + \frac{6}{x+4}$

31) b)
$y = \frac{5(x-3)^2(x+4)^6}{65536}$

32)
$y = \pm 2e^{2\sin^4 x}$

33) a)
Show that

33) b)
$r = \left(27000 - \frac{2611}{2}t\right)^{\frac{1}{3}}$

34)
Show that

35)
Show that

36) a)
$y = \frac{100e^{\frac{1}{10}x}}{9 + e^{\frac{1}{10}x}}$

36) b)
As $x \to \infty, y \to 100$

37) a)
$\frac{1}{200x} + \frac{1}{200(200-x)}$

37) b)
Show that

37) c)
As $t \to \infty, x \to 200, \therefore k = 200$

38) a)
$P = \frac{4000}{4 - 3e^{-\frac{1}{15}t}}$ so $a = 4$ and $b = -3$

38) b)
Show that

39) a)
Show that

39) b)
Show that

39) c)
$0 \leq x < \frac{1}{4}$

40)
Show that

41)
$-\frac{1}{4}y\cos 4y + \frac{1}{16}\sin 4y = \frac{1}{18}(3x-1)^6 + c$

42)
$-\frac{1}{3}y^2\cos 3y + \frac{2}{9}y\sin 3y + \frac{2}{27}\cos 3y = \frac{3}{4}\sin 4x + 3x + c$

43)
$N = \frac{15000 - 3000e^{-12t}}{e^{-12t} + 5}$
As $t \to \infty, N \to 3000$

44)
$\frac{dV}{dt} = 0.4 - 0.1h^2$
$\Rightarrow t = \frac{75}{2}\ln\left|\frac{2+h}{2-h}\right|$
When $h = 0.1, t = 3.75$ seconds

45)
$y = \frac{256e^{-\sqrt{x}}}{(4-\sqrt{x})^4}$

46)
$y = \frac{9x^2 + 6x + 1}{3x^2 + 2x + 3}$

2.41 Parametric Equations: Introduction
Questions on page 340

1) a)

t	-2	-1	0	1	2
x	4	1	0	1	4
y	-4	-2	0	2	4

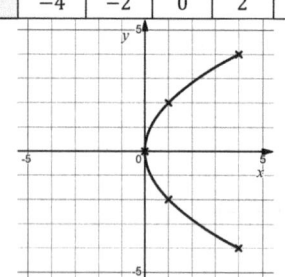

1) b)

t	-2	-1	0	1	2
x	0	1	2	3	4
y	-7	-1	1	-1	-7

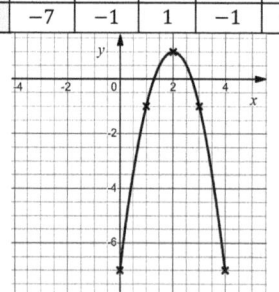

1) c)

t	-2	-1	0	1	2
x	0	-1	0	3	8
y	-12	-2	0	0	4

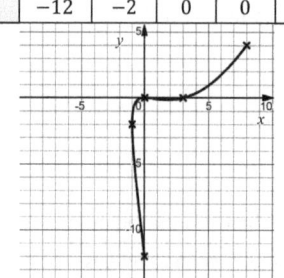

1) d)

t	-2	-1	0	1	2
x	-6	-2	0	0	-2
y	8	2	0	2	8

1) e)

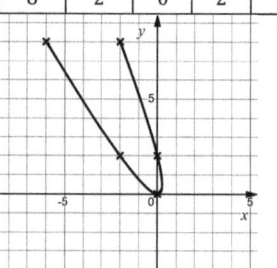

t	-2	-1	0	1	2
x	$\frac{2}{3}$	$\frac{4}{3}$	2	$\frac{4}{3}$	$\frac{2}{3}$
y	2	0	0	0	6

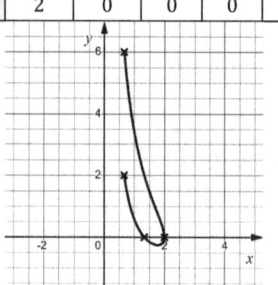

2) a)
$y = 2t^2$

2) b)
$y = t^2 + 4$

2) c)
$y = e^t + 3$

2) d)
$y = 8 + 2t$

2) e)
$y = t^2$

2) f)
$y = -\frac{10}{t^3}$

2) g)
$x = \cos t, y = \sin t$ or $x = \sin t, y = \cos t$

3) a)
$y = \left(\frac{x}{2}\right)^2 \Rightarrow y = \frac{1}{4}x^2$

3) b)
$y = 2(x+1) + 3 \Rightarrow y = 2x + 5$

3) c)
$y = e^{x+3} - x - 3$

3) d)
$y = \frac{1}{4}x^2 \ln\left(-\frac{x}{2}\right)$

4) a)
$x = (1+y)^2$

4) b)
$x = (2y)^3 - 2y \Rightarrow x = 8y^3 - 2y$

4) c)
$x = \left(\frac{y}{3}\right)^5 \Rightarrow x = \frac{1}{243}y^5$

4) d)
$x = e^{y-6}$

5) a)
$(0, -4)$ and $(4, 0)$

5) b)
$(0,2)$ and $(0,-2)$
$(1,0)$ and $(-1,0)$
5) c)
$(0,-1)$ and $(0,1)$
$(2,0), (1,0), (-1,0)$ and $(-2,0)$
5) d)
$(5,0)$
5) e)
$(0, e^{-2}), (0, e^4)$ and $(0, e^{-8})$

6)
$k = 3$

7) a)
$x^2 - y^2 = 4$
7) b)
$x^2 - y^2 = -8$
7) c)
$x^2 - y^2 = 112$

8)
$y = \frac{2x+5}{x+4}$, $a = 2, b = 5$

9) a)
$x = \frac{1}{2}\cos\theta, y = \frac{1}{3}\sin\theta$
or $x = \frac{1}{2}\sin\theta, y = \frac{1}{3}\cos\theta$
9) b)
$x = \frac{1}{\sqrt{3}}\cos\theta, y = \frac{1}{\sqrt{2}}\sin\theta$
or $x = \frac{1}{\sqrt{3}}\sin\theta, y = \frac{1}{\sqrt{2}}\cos\theta$
9) c)
$x = \frac{1}{\sqrt{5}}\cos\theta, y = \sqrt{2}\sin\theta$
or $x = \frac{1}{\sqrt{5}}\sin\theta, y = \sqrt{2}\cos\theta$
9) d)
$x = \frac{1}{\sqrt{2}}\cos\theta, y = \sqrt{3}\sin\theta$
or $x = \frac{1}{\sqrt{2}}\sin\theta, y = \sqrt{3}\cos\theta$

10) a)
$x^2 + y^2 = 9$
10) b)
$x^2 + y^2 = 25$
10) c)
$(y+3)^2 + (x-1)^2 = 1$
10) d)
$(x+4)^2 + (y-8)^2 = 1$

11) a)
$4(y-15)^2 + x^2 = 100$
11) b)
20 m
11) c)
$t = \frac{\pi}{2}$
11) d)
2π

12) a)
$\left(\frac{y-8}{-3}\right)^2 + \left(\frac{x-2k}{2k}\right)^2 = 1$
$\Rightarrow 9(x-2k)^2 + 4k^2(y-8)^2 = 36k^2$
12) b)
$k = \frac{3}{2}$

12) c)
$(3, 8)$
12) d)
$r = 3$

13)
$\left(\frac{x}{2}\right)^2 + 1 = (y-2)^2$

14) a)
$P\left(\frac{\sqrt{3}}{2}, \frac{\sqrt{3}}{2}\right), Q\left(\frac{3}{2}, \frac{\sqrt{3}}{2}\right)$
14) b)
$\frac{3}{4}(\sqrt{3} - 1)$ units²

15) a)
$y = 1 - x^2$
15) b)
$y = \frac{x}{\sqrt{1-x^2}}$

16) a)
Show that
16) b)
3

17) a)
$(10, 0)$ and $(-10, 0)$
17) b)
$(0, 10)$ and $(0, -10)$
17) c)
$(3\sqrt{6}, 3\sqrt{5}), (-3\sqrt{6}, 3\sqrt{5}), (3\sqrt{6}, -3\sqrt{5})$
and $(-3\sqrt{6}, -3\sqrt{5})$

18) a)
$y = (x - 5)^2 - 3$
18) b)
Domain is $\{x : x \in \mathbb{R}, 3 \leq x \leq 7\}$
Range is $\{y : y \in \mathbb{R}, -3 \leq y \leq 1\}$
18) c)
$\frac{7}{4} < k < 4$

19) a)
Domain: $x \geq 1$, Range: $y \geq 0$
19) b)
Domain: $x \geq 0$, Range: $y \leq 4$
19) c)
Domain: $-1 \leq x \leq 1$, Range: $-1 \leq y \leq 1$
19) d)
Domain: $-1 \leq x \leq 1$, Range: $-1 \leq y \leq 1$
19) e)
Domain: $x > 0$, Range: $y > 0$
19) f)
Domain: $x \in \mathbb{R}$, Range: $y \geq 0$

20)
Domain: $-1 \leq x \leq 1$, Range: $y > 0$

21) a)
Domain: $-1 \leq x \leq 1$, Range: $-4 \leq y \leq 4$
21) b)
$(0, 2\sqrt{3}), (0, 2), (0, -2\sqrt{3})$ and $(0, -2)$
$\left(-\frac{\sqrt{3}}{2}, 0\right)$

21) c)

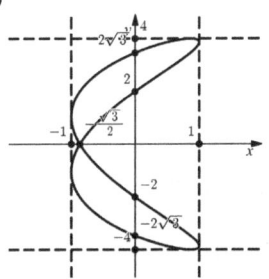

2.42 Parametric Equations: Differentiation
Questions on page 342

1)
$\frac{dy}{dx} = 4x + 4$

2) a)
$\frac{dy}{dx} = -4t$
2) b)
$\frac{dy}{dx} = \frac{3t^2 + 2t}{2t+3}$
2) d)
$\frac{dy}{dx} = \frac{1}{e^t}$
2) e)
$\frac{dy}{dx} = -\frac{3}{8e^{3t}}$
2) f)
$\frac{dy}{dx} = \frac{1 - \cos t}{1 + \sin t}$

3)
$\frac{dy}{dx} = -(1 + t^2)^2$

4)
$y = 4x - 8$

5)
$y = -\frac{27}{2}x + \frac{741}{2}$

6)
$y = ex - e^2 + 1$

7) a)
$(2, 5)$
7) b)
$\left(2, \frac{13}{27}\right)$ and $(18, -9)$
7) c)
$\left(e, -\frac{7}{6}\right)$ and $\left(e^5, \frac{25}{6}\right)$

8) a)
$(2, 9)$
8) b)
$(4, 1)$
8) c)
$\frac{dy}{dx} = -\frac{2(t+2)}{3t^2}$

9) a)
$y = -3x + 11$
9) b)
$\frac{121}{6}$ units²

10)
$18y - 243x = 478$

11) a)
6.3 metres
11) b)
0.467 metres (3 SF)
11) c)
4.375 metres
11) d)
6.03 metres (3 SF)

12) a)
$y = (x-a)^2 + 3$
12) b)
$y = -\frac{1}{2}x + \frac{a+9}{2}$

13) a)
$\frac{dy}{dx} = \frac{4\cos 2\theta}{\sqrt{3}\cos\theta}$
13) b)
$8\sqrt{3}y - 6x = 15$
13) c)
$\frac{25\sqrt{3}}{32}$ units²

14) a)
Show that
14) b)
Show that
14) c)
$y = \frac{8}{x-4} - 6$
14) d)
$\alpha = 4$
14) e)
$\beta = -6$
14) f)
$\left(\frac{16}{3}, 0\right)$
14) g)

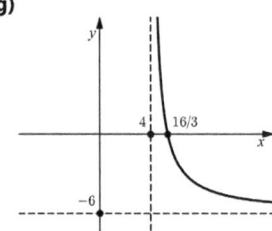

15) a)
$\frac{dy}{dx} = \frac{\sin\theta}{2\cos 4\theta \cos^3\theta}$
15) b)
-1

16) a)
Show that
16) b)
240 units²

17) a)
$y = -\frac{1}{\sin a}x + \csc a + 2\sec a$
17) b)
$(1, 1)$

18) a)
$\frac{dy}{dx} = \frac{3+4t}{4t}$
18) b)
$\left(\frac{81}{256}, -\frac{27}{128}\right)$
18) c)
$(0, 0)$
18) d)
$(0, 0)$ and $(1, 0)$
18) e)

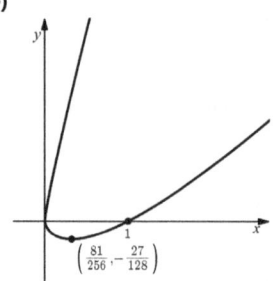

19) a)
0.5 metres
19) b)
$\frac{\sqrt{5}}{4}\sin(t - 1.11)$
19) c)
0.56 metres

20) a)
$\frac{dy}{dx} = -\frac{5\cos\theta}{4\sin\theta}$
20) b)
$y - 5\sin\theta = -\frac{5\cos\theta}{4\sin\theta}(x - 4\cos\theta)$
20) c)
Show that
20) d)
$\theta = \{0, 2.35\}$ (3SF)

21) a)
Show that
21) b)
$5\sin(\theta + 0.927\ldots) = 4$
21) c)
$\theta = \{0, 1.29\}$ (3 SF)

22) a)
$y = -x + 1$
22) b)
Show that
22) c)
Show that
22) d)
$1 \leq k \leq 1.135$

23) a)
$\frac{dy}{dx} = 8\left(\frac{\cos t - t\sin t}{\cos t}\right)$
23) b)
$(0.0911, 0.559)$ (3 SF)

2.43 Parametric Equations: Integration
Questions on page 345

1) a)
$\frac{9}{5}t^5 + \frac{8}{3}t^3 - t + c$
1) b)
$t + e^t + c$
1) c)
$\frac{1}{3} \times 3^{3t} + c$

2) a)
$\frac{1493}{15}$
2) b)
$2e^2 + 2$
2) c)
$\frac{7}{3}$
2) d)
$-\frac{7}{24}$

3) a)
$t = 1: (1, 1), t = 2: \left(4, \frac{1}{2}\right)$
3) b)

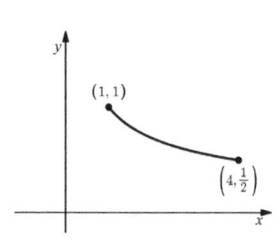

3) c)
2

4) a)
$x = 4: t = \frac{7}{2}, x = 0: t = \frac{3}{2}$
4) b)
$8\ln\frac{7}{3}$
4) c)
$8\ln\frac{7}{3}$

5) a)
$k = 8$
5) b)
$\frac{88}{15}$

6) a)
$t = 1$
6) b)
$\frac{111}{10}$

7) a)
$t = 0: \left(0, \frac{1}{2}\right), t = 5: \left(\frac{5}{9}, \frac{8}{81}\right)$
7) b)
Show that
7) c)
$\frac{665}{4374}$

8) a)
$x = 0: (0, 1)$

540

$x = \sqrt{3}$: $\left(\sqrt{3}, \frac{1}{8}\right)$

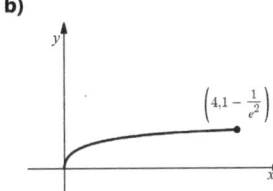

8) b)
Show that

9) a)
Show that
9) b)
Show that

10) a)
Show that
10) b)
$k = \frac{7}{2}$

11) a)
$t = 0$: $\left(0, \frac{1}{2}\right)$, $t = \pi$: $\left(\pi, \frac{1}{2}\right)$
11) b)
Show that
11) c)
$\frac{1}{4}\pi$

12) a)
Show that
12) b)
Show that

13) a)
Show that
13) b)
$\frac{20}{9}$

14) a)
$P\left(\frac{2}{3}, 27e^{-3}\right)$
14) b)
Show that

15)
$\frac{\pi}{4}$

16)
2π

17) a)
At $A(9,0)$, $\theta = 0$; At $B(0,5)$, $\theta = \frac{\pi}{2}$
At $C(-9,0)$, $\theta = \pi$
17) b)
Show that
17) c)
Show that

18) a)
Show that
18) b)
$-\frac{3}{t} + \frac{3}{t-2}$
18) c)
$\ln \frac{27}{8}$

19) a)
$\left(0, \frac{1}{4}\right)$ and $\left(\frac{18}{5}, \frac{1}{10}\right)$

19) b)
Show that
19) c)
$\ln\left(\frac{5}{2}\right) - \frac{21}{50}$

20) a)
$x = 0$: $y = 0$; $x = 4$: $y = 1 - e^{-2}$
20) b)

20) c)
Show that
20) d)
Show that

21) a)
Show that
21) b)
Show that

22)
Show that

2.44 Proof by Contradiction
Questions on page 349

1) a)
The cat is not black
1) b)
No cats are black
1) c)
All cats are black
1) d)
Line A is not parallel to line B
1) e)
The probability has not increased
1) f)
$x < 0$
1) g)
$x \neq 2$
1) h)
a is rational
1) i)
n is odd and not prime

2) a)
There exists an integer n that is neither positive nor negative (e.g. $n = 0$)
2) b)
n is a number with less than 3 distinct prime factors
2) c)
There exists at least one prime number that is not odd (e.g. 2)
2) d)
There exists a rational number and an irrational number whose sum is rational.

2) e)
If n^2 is even, then n is odd

3)
Given that p and q are positive integers, assume that $\frac{p}{q} + \frac{q}{p} < 2$

4) a)
Prove
4) b)
Prove
4) c)
Prove

5)
Prove

6)
Prove

7)
Prove

8)
Prove

9)
Prove

10)
Prove

11) a)
Prove
11) b)
It is true for all cases except 0

12)
Prove

13)
Prove

14)
Prove

15)
Prove

16) a)
Show that
16) b)
Prove

17)
Prove

18)
Prove

2.45 3D Vectors
Questions on page 350

1) a)
$\overrightarrow{AB} = \begin{pmatrix} 6 \\ -4 \\ -8 \end{pmatrix}$

1) b)
$\overrightarrow{BC} = \begin{pmatrix} -5 \\ 1 \\ -5 \end{pmatrix}$

1) c)
$\overrightarrow{AC} = \begin{pmatrix} 1 \\ -3 \\ -13 \end{pmatrix}$

1) d)
$\overrightarrow{BA} = \begin{pmatrix} -6 \\ 4 \\ 8 \end{pmatrix}$

2) a)
$3\sqrt{10}$

2) b)
$\sqrt{86}$

2) c)
$\sqrt{173}$

2) d)
$\sqrt{a^2 + b^2 + 1}$

3) a)
$|\overrightarrow{AB}| = \sqrt{62}$

3) b)
$|\overrightarrow{AB}| = \sqrt{65}$

3) c)
$|\overrightarrow{AB}| = 3\sqrt{2}$

4) a)
$\frac{1}{\sqrt{62}}(2\mathbf{i} + 3\mathbf{j} - 7\mathbf{k})$

4) b)
$\frac{1}{\sqrt{53}}\begin{pmatrix} 6 \\ -1 \\ 4 \end{pmatrix}$

4) c)
$\frac{1}{10\sqrt{2}}(8\mathbf{i} - 6\mathbf{j} + 10\mathbf{k})$

4) d)
$\frac{1}{\sqrt{6}}\begin{pmatrix} 1 \\ -1 \\ 2 \end{pmatrix}$ is already a unit vector.

5) a)
$\frac{1}{2\sqrt{17}}\begin{pmatrix} 2 \\ -8 \\ 0 \end{pmatrix}$ or $\frac{1}{\sqrt{17}}\begin{pmatrix} 1 \\ -4 \\ 0 \end{pmatrix}$

5) b)
$\frac{1}{5\sqrt{2}}(5\mathbf{i} + 4\mathbf{j} + 3\mathbf{k})$

6)
Show that

7) a)
Show that
7) b)
Show that

8)
$10\mathbf{i} - 12\mathbf{j} - 5\mathbf{k}$

9) a)
$\begin{pmatrix} 5 \\ 12 \\ 3 \end{pmatrix}$

9) b)
$p = -1 \pm \sqrt{29}$

10)
Show that

11)
Show that

12)
$\begin{pmatrix} \frac{16}{3} \\ \frac{32}{3} \\ \frac{16}{3} \end{pmatrix}$ or $\begin{pmatrix} 6 \\ 12 \\ 6 \end{pmatrix}$

13) a)
$8\mathbf{i} - 2\mathbf{j} + 4\mathbf{k}$
or $-8\mathbf{i} + 2\mathbf{j} - 4\mathbf{k}$

13) b)
$\frac{1}{3}(-\mathbf{i} + 7\mathbf{j} + 2\mathbf{k})$
or $\frac{1}{3}(\mathbf{i} - 7\mathbf{j} - 2\mathbf{k})$

13) c)
$3\mathbf{i} - 24\mathbf{j} - 3\mathbf{k}$
or $-3\mathbf{i} + 24\mathbf{j} + 3\mathbf{k}$

14) a)
$m_{AB} = \left(\frac{25}{2}, 0, 3\right)$

14) b)
$m_{BC} = \left(\frac{29}{2}, -1, \frac{9}{2}\right)$

14) c)
$\frac{1}{2}\sqrt{29}$

14) d)
$\frac{2}{\sqrt{29}}\begin{pmatrix} 2 \\ -1 \\ \frac{3}{2} \end{pmatrix}$

14) e)
$\left(1 + \frac{4}{\sqrt{29}}, 3 - \frac{2}{\sqrt{29}}, -1 + \frac{3}{\sqrt{29}}\right)$
or $\left(1 - \frac{4}{\sqrt{29}}, 3 + \frac{2}{\sqrt{29}}, -1 - \frac{3}{\sqrt{29}}\right)$

15) a)
z-axis

15) b)
x-axis

16) a)
$k = \frac{1}{2}$

16) b)
$\begin{pmatrix} \frac{5}{4} \\ 0 \\ 0 \end{pmatrix}$

17)
Show that

18)
$p = 3, q = 5, k = 23$

19)
Show that

20) a)
Show that

20) b)
$\overrightarrow{PC} = -\mathbf{a} - \mathbf{b} + \mathbf{c}$

20) c)
$\frac{2}{3}\mathbf{a} + \frac{2}{3}\mathbf{b} + \frac{1}{3}\mathbf{c}$

20) d)
$\frac{5}{6}\mathbf{a} + \frac{1}{3}\mathbf{b} + \frac{1}{6}\mathbf{c}$

21) a)
$k = 6$ or $k = 8$

21) b)
When $k = 6$, Area = 9 units^2
When $k = 8$, Area = $6\sqrt{2}$ units^2

22)
Show that

23) a)
$\overrightarrow{OD} = \begin{pmatrix} -4 \\ 6 \\ 14 \end{pmatrix}$

23) b)
$\sqrt{122}$

24) a)
$\overrightarrow{OD} = -5\mathbf{i} - 15\mathbf{j} + 6\mathbf{k}$

24) b)
$|\overrightarrow{AB}| = 2\sqrt{69}$

24) c)
$|\overrightarrow{AC}| = \sqrt{35}$

24) d)
Show that

24) e)
72.8 units^2 (3 SF)

25) a)
Show that

25) b)
$\sqrt{517}$

26) a)
$|\overrightarrow{BC}| = 8\sqrt{2}$

26) b)
$p = 10$ or $p = -6$

27)
Show that

28) a)
$\begin{pmatrix} 6 \\ 4 \\ 13 \end{pmatrix}$

28) b)
Show that

2.46 Bivariate Data: PMCC Hypothesis Testing
Questions on page 353

1)
The significance level is the probability of rejecting the null hypothesis when in fact it is true.

2) a)
$0.638 > 0.6215$ so Reject H_0
or $0.0444 < 0.05$ so Reject H_0

2) b)
$-0.456 > -0.4973$ so Fail to reject H_0
or $0.0681 > 0.05$ so Fail to reject H_0

3) a)
$0.487 > 0.4575$
Reject H_0

3) b)
$0.628 < 0.6694$
Fail to reject H_0

3) c)
$0.593 > 0.5155$
Reject H_0

3) d)
$0.213 > 0.1678$
Reject H_0

4) a)
$-0.453 < -0.4409$
Reject H_0

4) b)
$-0.829 < -0.6694$
Reject H_0

4) c)
$-0.306 > -0.4622$
Fail to reject H_0

4) d)
$-0.197 > -0.2070$
Fail to reject H_0

5)
CR is $[0.5751, 1]$

6)
CR is $[-1, -0.5742]$

7) a)
$0.658 < 0.6664$
Fail to reject H_0

7) b)
$-0.567 > -0.6021$
Fail to reject H_0

7) c)
$0.621 < 0.6226$
Fail to reject H_0

7) d)
$-0.248 < -0.2353$
Reject H_0

8)
CR is $[-1, -0.7545] \cup [0.7545, 1]$

9)
CR is $[-1, -0.2199] \cup [0.2199, 1]$

10)
$0.382 > 0.3172$ so we reject H_0
There is sufficient evidence to suggest that there is a positive correlation between the amount of sugar consumed daily by adults and their blood glucose levels.

11)
$-0.297 > -0.3060$ so we fail to reject H_0
There is insufficient evidence to suggest that there is a negative correlation between the number of hours spent on social media each day and self-reported feelings of loneliness in teenagers.

12) a)
In the opening statement, ρ is the **population** corelation coefficient and not the sample correlation coefficient (that is r).
r must be compared against -0.3665 and not 0.3665, as we are considering the bottom tail.

12) b)
Let ρ be the population correlation coefficient between the water temperature and the growth rate of the fish.
$H_0: \rho = 0$
$H_1: \rho < 0$
$r = -0.364$
The critical value for $n = 40$ for a one-tailed test at a 1% significance level is 0.3665
$-0.364 > -0.3665$ so we fail to reject H_0
There is insufficient evidence to suggest that there is a negative correlation between the water temperature and the growth rate of the fish.

13)
$0.209 < 0.2787$ so we fail to reject H_0
There is insufficient evidence to suggest that there is any correlation between the levels of air pollution and numbers of adults with respiratory illnesses.

14)
$-0.301 < -0.2565$ so we reject H_0
There is sufficient evidence to suggest that there is some correlation between the daily intake of fruits and vegetables in adults and their BMI.

15) a)
The critical value for $n = 28$ for a two-tailed test at a 5% significance level is 0.3739, and not 0.3172.
If the sample product moment correlation coefficient, r, is greater than the critical value, then we **reject** H_0, so Jo has come to the wrong conclusion as $0.482 > 0.3172$ (and 0.482 is greater than the correct critical value of 0.3739, so the conclusion holds).

15) b)
Let ρ be the population correlation coefficient between the number of hours of sleep college students get each night and their stress levels.
$H_0: \rho = 0$
$H_1: \rho \neq 0$
$r = 0.482$
The critical value for $n = 28$ for a two-tailed test at a 5% significance level is 0.3739
$0.482 > 0.3739$ so we reject H_0
There is sufficient evidence to suggest that there is some correlation between the number of hours of sleep college students get each night and their stress levels.

2.47 Bivariate Data: Spearman's Rank Hypothesis Testing
Questions on page 355

1) a)
Show that

1) b)
Reject H_0. There is sufficient evidence to suggest that there is a positive association between Mathematics and English test scores in the population.

2)
$0.479 < 0.5035$. Fail to reject H_0. There is insufficient evidence to suggest that there is a positive association between the length of a bird's wingspan and its tendency to migrate long distances in the population.

3)
$-0.654 < -0.6036$. Reject H_0. There is sufficient evidence to suggest that there is a negative association between the air quality index and the average life expectancy of people in different cities in the population.

4)
$0.321 < 0.3608$. Fail to reject H_0. There is insufficient evidence to suggest that there is some association between the number of hours of sleep a person gets per night and their cognitive performance score on a standardised test in the population.

2.49 Probability: Conditional Probability
Questions on page 357

1)
$\frac{1}{5}$

2)
$\frac{3}{5}$

3)
$x = 0.2, y = 0.3$

4) a)
$z = 0.03$
4) b)
$y = 0.08$
4) c)
$x = 0.15$
4) d)
$w = 0.12$

5)
$\frac{29}{60}$

6)
$\frac{27}{145}$

7) a)
$\frac{12}{37}$
7) b)
$\frac{4}{5}$
7) c)
$\frac{6}{31}$
7) d)
$\frac{1}{5}$
7) e)
$\frac{25}{31}$

8)
$\frac{8}{9}$

9) a)
$\frac{4}{15}$
9) b)
$\frac{7}{15}$
9) c)
$\frac{3}{14}$
9) d)
$\frac{1}{3}$
9) e)
Any one of: Milk chocolate and Dark chocolate, Milk Chocolate and White Chocolate, Plain and Caramel, Plain and Hazelnut, Caramel and Hazelnut

10) a)
0.55
10) b)
$\frac{5}{11}$
10) c)
$P(A) = 0.45 \neq \frac{5}{11} = P(A|B)$
So A and B are not independent events

11) a)
$\frac{33}{40}$
11) b)
$\frac{11}{16}$
11) c)
$\frac{3}{4}$
11) d)
$\frac{5}{12}$
11) e)
$\frac{1}{12}$

12) a)
$\frac{10}{73}$
12) b)
$\frac{5}{27}$
12) c)
$\frac{28}{73}$
12) d)
$\frac{17}{27}$

13) a)
$\frac{73}{121} = 0.6033$ (4 DP)
13) b)
$\frac{918}{3911} = 0.2347$ (4 DP)

14) a)
$\frac{9}{10}$
14) b)
$P(A) = \frac{7}{10} \neq \frac{9}{10} = P(A|B)$
So A and B are not independent events.

15) a)
$\frac{4}{15}$
15) b)
$P(A') = \frac{4}{15} = P(A'|B)$
So A and B are independent events.

16)
$\frac{2}{5}$

17)
$\frac{2}{5}$

18) a)

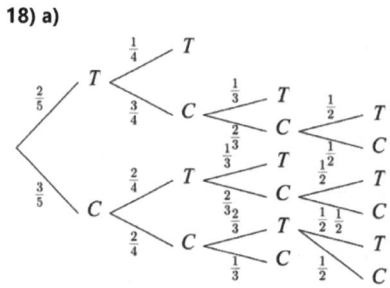

18) b)
$\frac{4}{5}$
18) c)
$\frac{2}{3}$

18) d)
$\frac{1}{2}$

19) a)
$\frac{1}{10}$
19) b)
$\frac{3}{7}$
19) c)
$\frac{1}{4}$
19) d)
$\frac{4}{7}$
19) e)
$\frac{1}{4}$
19) f)
$\frac{17}{38}$
19) g)
$\frac{12}{43}$
19) h)
$\frac{2}{13}$
19) i)
$\frac{17}{24}$
19) j)
$\frac{1}{12}$
19) k)
$\frac{3}{16}$

20) a)

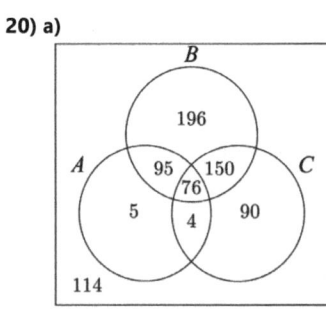

20) b)
$\frac{38}{365} = 0.1041$ (4 DP)
20) c)
$\frac{15}{73} = 0.2055$ (4 DP)
20) d)
$\frac{171}{611} = 0.2799$ (4 DP)
20) e)
$\frac{38}{113} = 0.3363$ (4 DP)

21)

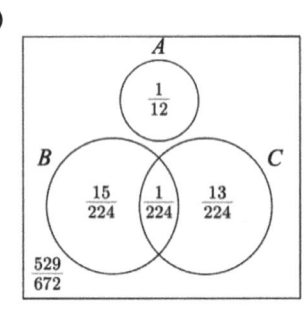

$\frac{16}{29}$

22)
$\frac{2}{3}$

23)

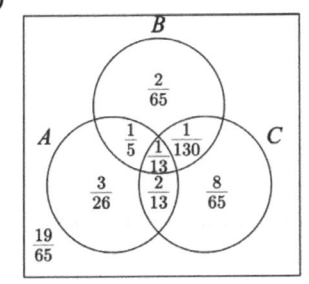

Venn diagram with three circles A, B, C:
- B only: $\frac{2}{65}$
- A ∩ B only: $\frac{1}{5}$
- B ∩ C only: $\frac{1}{130}$
- A ∩ B ∩ C: $\frac{1}{13}$
- A only: $\frac{3}{26}$
- A ∩ C only: $\frac{2}{13}$
- C only: $\frac{8}{65}$
- Outside: $\frac{19}{65}$

24)

Venn diagram with two circles A, B:
- A only: $\frac{9}{140}$
- A ∩ B: $\frac{2}{7}$
- B only: $\frac{1}{2}$
- Outside: $\frac{3}{20}$

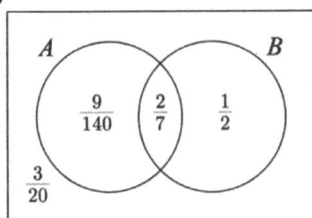

25)
$= 0.1024$ (4 DP)

26)
$\frac{28}{51} = 0.5490$ (4 DP)

27)
$\frac{67}{896} = 0.0748$ (4 DP)

28) a)
$\frac{7}{143}$

28) b)
$\frac{16}{143}$

28) c)
$\frac{2}{15}$

29)
$\frac{557}{1529} = 0.3643$ (4 DP)

30)
$\frac{7}{12}$

31)
$\frac{2}{9}$

2.50 The Normal Distribution
Questions on page 361

1)
A is $X \sim N(10, 3)$, B is $X \sim N(20, 6)$
C is $X \sim N(20, 3)$, D is $X \sim N(10, 1)$

2)
Mean is 13, Standard Deviation is 2.5

3)
Mean is 44, Standard Deviation is 5

4) a)
$\sigma = 7$

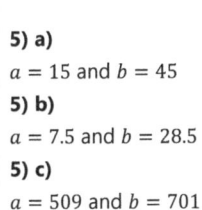

4) b)
$\sigma = \sqrt{5}$

4) c)
$\sigma = \sqrt[4]{2}$

5) a)
$a = 15$ and $b = 45$

5) b)
$a = 7.5$ and $b = 28.5$

5) c)
$a = 509$ and $b = 701$

6) a)
Normal curve centred at 25, x-axis 5 to 45.

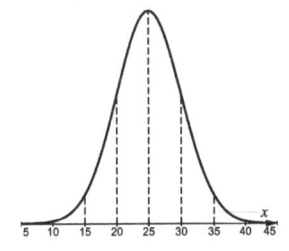

6) b)
Normal curve centred at 100, x-axis 70 to 130.

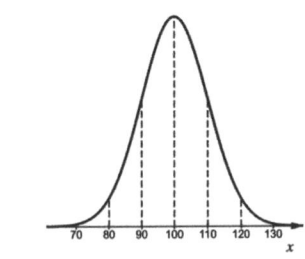

6) c)
Normal curve centred at 50, x-axis 40 to 60.

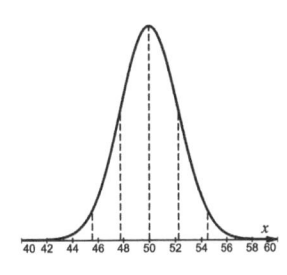

7)
Two normal curves: X centred at 60, Y centred at 100.

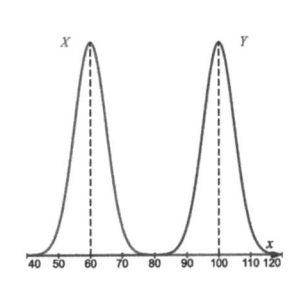

8)
Two overlapping normal curves: X centred at 20, Y centred at 30.

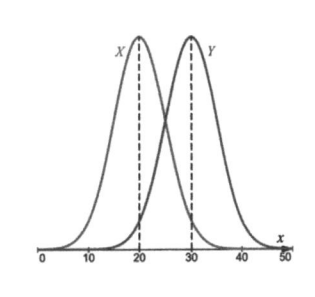

9)
Two normal curves centred at 60, Y taller/narrower than X.

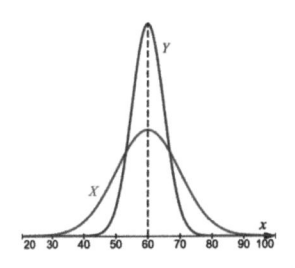

10)
Two normal curves: X centred at 8 (taller), Y centred at 10 (shorter).

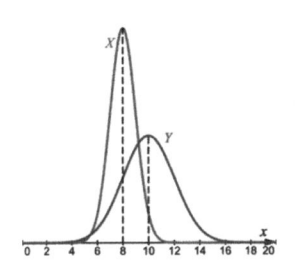

11) a)
0.8413 (4 DP)

11) b)
0.1587 (4 DP)

11) c)
0.6826 (4 DP)

12) a)
0.1587 (4 DP)

12) b)
0.1587 (4 DP)

12) c)
0.3413 (4 DP)

13) a)
$Z = -0.5$

13) b)
$Z = 0.5$

13) c)
$Z = 1.1$

14) a)
0.4013 (4 DP)

14) b)
0.4013 (4 DP)

14) c)
0.1747 (4 DP)

14) d)
0.0819 (4 DP)

15) a)
0.3538 (4 DP)

15) b)
0.3538 (4 DP)

15) c)
0.2660 (4 DP)

15) d)
0.2660 (4 DP)

16) a)
0.9088 (4 DP)

16) b)
0.4121 (4 DP)
16) c)
0.5384 (4 DP)
16) d)
0.0535 (4 DP)

17) a)
0
17) b)
1

18) a)
0.5
18) b)
0.1807 (4 DP)
18) c)
0.2919 (4 DP)
18) d)
0.5274 (4 DP)

19) a)
0.0549 (4 DP)
19) b)
0.2030 (4 DP)
19) c)
0.1538 (4 DP)

20) a)
0.6827 (4 DP)
20) b)
0.9545 (4 DP)
20) c)
0.9973 (4 DP)
20) d)
0.9999 (4 DP)

21) a)
0.6827 (4 DP)
21) b)
0.9545 (4 DP)
21) c)
0.9973 (4 DP)
21) d)
0.9999 (4 DP)

22)
0.4207 (4 DP)

23)
0.1303 (4 DP)

24)
0.0174 (4 DP)

25) a)
$\bar{x} = 525$
$s^2 = 2211.055276$
25) b)
0.4158 (4 DP)

26) a)
$\bar{x} = 46.5$
26) b)
$s^2 = \frac{925}{33} \approx 28.03$
26) c)
0.2081 (4 DP)

27)
$2.01 - 3 \times 0.1 = 1.71$,
$2.01 + 3 \times 0.1 = 2.31$
If a normal distribution was appropriate, we would expect 99% to be approximately between 1.71 m and 2.31 m. If we use $X \sim N(2.01, 0.1^2)$ then $P(1.83 < X < 2.19) \approx 93\%$ so it does not appear to be suitable.

28) a)
$a = -0.674$
28) b)
$b = 1.28$
28) c)
$c = 2.33$
28) d)
$d = -0.842$

29) a)
$a = 18.8$
29) b)
$b = 29.1$
29) c)
$c = 30.2$
29) d)
$d = 26.0$

30)
$a = 751, b = 849$

31)
$a = 66.2$

32) a)
$a = 19.7$
32) b)
$b = 1.21849 \approx 1.22$

33) a)
$\mu = 40$
33) b)
$\mu = 55.2$
33) c)
$\mu = 205$
33) d)
$\mu = 46.7$

34) a)
$\sigma = 3.56$
34) b)
$\sigma = 67.1$
34) c)
$\sigma = 13.8$
34) d)
$\sigma = 39.9$

35)
$\sigma = 64.9$ (1 DP)

36)
$\sigma = 0.593$ g

37)
$18 - 3 \times 17.166685 = -33.500055$
∴ a large proportion of ambulance response times would be negative, so they are unlikely to be normally distributed.

38) a)
$\mu = 89.200005 \approx 89.2$
$\sigma^2 = 179.5599624 \approx 179.56$
38) b)
$Y \sim N(265.6, 1616.04)$

39)
$a = 3.7, b = 16$

40)
$\mu = 99, \sigma = 35.5$

41)
$\mu = 11, \sigma = 0.608$ (3 SF)

42) a)
$\mu = 13.8, \sigma = 7.19$
42) b)
$\mu = 65.9, \sigma = 62.6$
42) c)
$\mu = 6.19, \sigma = 1.37$
42) d)
$\mu = 63.9, \sigma = 20.8$

43)
$X \sim N(177.5, 34.71196933^2)$
$177.5 - 3 \times 34.71196933 \approx 73.4$ grams

44)
$x = \pm 1$

45)
$x = \mu \pm \sigma$

46) a)
0.0604 (4 DP)
46) b)
0.9088 (4 DP)
46) c)
0.6421 (4 DP)
46) d)
0.5075 (4 DP)

47)
0.3771 (4 DP)

48)
0.4951 (4 DP)

49) a)
0.2525 (4 DP)
49) b)
0.3590 (4 DP)

50) a) i)
0
50) a) ii)
0.2023 (4 DP)
50) b)
0.5034 (4 DP)

51)
£32.43

52) a)
0.3889 (4 DP)
52) b)
$\bar{x} = 172.3611$
$s = 10.05244$
52) c)
$Y \sim N(172.3611, 10.05244^2)$
$P(Y < 170) = 0.4072$
The data in the histogram appears fairly symmetric, and the percentage difference in the probabilities is $\frac{0.4072 - 0.3889}{0.3889} \times 100 = 4.71\%$. So modelling the data as a normal distribution is not very accurate but might be suitable for rough calculations.

53) a)
0.5333 (4 DP)
53) b)
$\bar{x} = 170.5$
$s = 10.83305$
53) c)
$Y \sim N(170.5, 10.83305^2)$
$P(Y < 170) = 0.4816$
The data in the histogram appears positively skewed and the percentage difference in the probabilities is $\frac{0.5333 - 0.4816}{0.5333} \times 100 = 9.69\%$. So it appears that modelling the data as a normal distribution would not be appropriate.

54)
For a binomial distribution $X \sim B(n, p)$ to be approximated by a normal distribution $Y \sim N(\mu, \sigma^2)$, n must be large and p must be close to 0.5. It can also be shown that a normal distribution can approximate a binomial distribution if both $np > 5$ and $n(1-p) > 5$

55) a)
$Y \sim N(240, 124.8)$
500 is large and 0.48 is close to 0.5
Check: $500 \times 0.48 = 240 > 5$ and $500 \times 0.52 = 260 > 5$
So a normal approximation would be appropriate.

55) b)
$Y \sim N(2, 1)$
2 is small but 0.5 is exactly what is required
Check: $2 \times 0.5 = 1 < 5$ and $2 \times 0.5 < 5$
So a normal approximation would not be appropriate.

55) c)
$Y \sim N(150, 149.85)$
150000 is large but 0.001 is not close to 0.5
Check: $150000 \times 0.001 = 150 > 5$ and $150000 \times 0.999 = 149850 > 5$
So a normal approximation might be appropriate, depending on the context, as a relatively large proportion (11%) would give negative values.

56) a)
$X \sim B\left(2000, \frac{4}{9}\right)$
2000 is large and $\frac{4}{9} = 0.4444$ to 4dp is relatively close to 0.5
Check: $2000 \times \frac{4}{9} = \frac{8000}{9} > 5$ and $2000 \times \frac{5}{9} = \frac{10000}{9} > 5$
So a normal approximation would be appropriate.
56) b)
$a = 908$
56) c)
0.1887 (4 DP)

57)
$a = 340, b = 0.0228$ (4 DP)

58)
$a = \frac{4468}{245} = 18.2$ (3 SF), $b = 0.2251$ (4 DP)

59)
$a = 54, b = 0.9772$ (4 DP)

60)
$a = \frac{670}{3} = 223$ (3 SF), $b = 0.9522$ (4 DP)

61) a)
0.1737 (4 DP)
61) b)
0.2437 (4 DP)
61) c)
0.4512 (4 DP)
61) d)
0.9236 (4 DP)
61) e)
0.0326 (4 DP)

62) a)
0.1500297
62) b)
0.1499763
62) c)
0.0356% (3 SF)

63) a)
0.6886616
63) b)
0.6884506
63) c)
0.0306% (3 SF)

64) a)
0.02639502
64) b)
0.02696474
64) c)
2.16% (3 SF)

65) a) i)
0.3505 (4 DP)
65) a) ii)
0.3169 (4 DP)
65) b)
0.3538 (4 DP) and 0.3217 (4 DP)

2.51 Sample Means Hypothesis Testing
Questions on page 367

1) a)
$P(\bar{X} < 27.1) = 0.0019 < 0.05$
Reject H_0
1) b)
$P(\bar{X} < 43.2) = 0.0774 > 0.05$
Fail to reject H_0
1) c)
$P(\bar{X} < 49.2) = 0.1640 > 0.1$
Fail to reject H_0
1) d)
$P(\bar{X} < 81.2) = 0.0078 < 0.01$
Reject H_0

2) a)
$P(\bar{X} > 62.1) = 0.0588 > 0.05$
Fail to reject H_0
2) b)
$P(\bar{X} > 25.9) = 0.0409 < 0.05$
Reject H_0
2) c)
$P(\bar{X} > 107) = 0.0117 < 0.1$
Reject H_0
2) d)
$P(\bar{X} > 166.3) = 0.0017 < 0.01$
Reject H_0

3) a)
p-value $= 2 \times 0.0038 \ldots = 0.0077 < 0.05$
Reject H_0
3) b)
p-value $= 2 \times 0.0791 \ldots = 0.1582 > 0.05$
Fail to reject H_0
3) c)
p-value $= 2 \times 0.0368 \ldots = 0.0736 < 0.1$
Reject H_0
3) d)
p-value $= 2 \times 0.0039 \ldots = 0.0079 < 0.01$
Reject H_0

4) a)
CR is $\{\bar{X} < 27.03076\}$
$26.7 < 27.03076$, Reject H_0
4) b)
CR is $\{\bar{X} < 30.38709\}$
$30.5 > 30.38709$, Fail to reject H_0
4) c)
CR is $\{\bar{X} < 104.272\}$
$104.4 > 104.272$, Fail to reject H_0
4) d)
CR is $\{\bar{X} < 337.0459\}$
$336.6 < 337.0459$, Reject H_0

5) a)
CR is $\{\bar{X} > 78.77355\}$
$79.3 > 78.77355$, Reject H_0
5) b)
CR is $\{\bar{X} > 163.1497\}$
$162.7 < 163.1497$, Fail to reject H_0
5) c)
CR is $\{\bar{X} > 1251.146\}$
$1252.3 > 1251.146$, Reject H_0
5) d)
CR is $\{\bar{X} > 673.8344\}$
$674.1 > 673.8344$, Reject H_0

6) a)
CR is $\{\bar{X} < 83.13298\} \cup \{\bar{X} > 86.86702\}$
$82.9 < 83.13298$, Reject H_0
6) b)
CR is $\{\bar{X} < 1.540222\} \cup \{\bar{X} > 1.659778\}$
$1.68 > 1.659778$, Reject H_0
6) c)
CR is $\{\bar{X} < 54.19892\} \cup \{\bar{X} > 59.80108\}$
$54.04 < 54.19892$, Reject H_0
6) d)
CR is $\{\bar{X} < 876.4242\} \cup \{\bar{X} > 903.5758\}$
$876.4242 < 902.9 < 903.5758$
Fail to reject H_0

7) a)
$Z = -1.622786955$
$-1.622786955 > -1.644854$
Fail to reject H_0
7) b)
$Z = -1.833575742$
$-1.833575742 < -1.644854$
Reject H_0
7) c)
$Z = -1.014185106$
$-1.014185106 > -1.281552$
Fail to reject H_0
7) d)
$Z = -2.533747629$
$-2.533747629 < -2.326348$
Reject H_0

8) a)
$Z = 1.671429029$
$1.671429029 > 1.644854$, Reject H_0
8) b)
$Z = 1.422952349$
$1.422952349 < 1.644854$
Fail to reject H_0
8) c)
$Z = 1.319293642$
$1.319293642 > 1.281552$, Reject H_0
8) d)
$Z = 2.575$
$2.575 > 2.326348$, Reject H_0

9) a)
$Z = -2.121320344$
$-2.121320344 < -1.959964$
Reject H_0
9) b)
$Z = 1.506367438$
$-1.959964 < 1.506367438 < 1.959964$
Fail to reject H_0
9) c)
$Z = -1.573349475$
$-1.644854 < -1.573349475 < 1.644854$
Fail to reject H_0
9) d)
$Z = -3.169081401$
$-3.169081401 < -2.575829$
Reject H_0

10) a)
Molly believes that the machine is dispensing less soda than it should be, but the alternative hypothesis is $\mu > 380$, when it should be $\mu < 380$. Molly finds the correct p-value and compares it with the significance level, but she has come to the wrong conclusion. As $0.0466 < 0.1$, she should reject H_0, rather than fail to reject it.

10) b)
Let X be the volume of soda dispensed by the machine in millilitres.
$H_0: \mu = 380$
$H_1: \mu < 380$
$\bar{X} \sim N\left(380, \frac{3^2}{15}\right)$
$P(\bar{X} < 378.7) = 0.0466$
$0.0466 < 0.1$ so we reject H_0
There is sufficient evidence to suggest that the population mean amount of soda is less than expected.

11)
$0.1030 > 0.03$ so we fail to reject H_0
There is insufficient evidence to suggest that the population mean height of plants over the age of 2 years is greater than 130 cm.

12)
$0.0585 > 0.05$ so we fail to reject H_0
There is insufficient evidence to suggest that the population mean weekly household spend has changed.

13)
$0.0485 > 0.025$ so we fail to reject H_0
There is insufficient evidence to suggest that the population mean weight of the chocolate bars is different to 100 grams.

14) a)
The bakery wants to test whether shortening the time in the oven will reduce the weight of the loaves, but their alternative hypothesis is $\mu \neq 400$ when it should be $\mu < 400$. The second error is that the sample variance is $\frac{6^2}{\sqrt{50}}$ when it should be $\frac{6^2}{50}$.

14) b)
Let X be the weight of baked loaves in grams
$H_0: \mu = 400$
$H_1: \mu < 400$
$\bar{X} \sim N\left(400, \frac{6^2}{50}\right)$
$P(X < a) = 0.05 \Rightarrow a = 398.6043$
$398 < 398.6043$ so we reject H_0
There is sufficient evidence to suggest that the population mean weight of loaves of bread has been reduced.

15) a)
CR is $\{\bar{X} < 158.723\} \cup \{\bar{X} > 163.277\}$
15) b)
$158.723 < 162.4 < 163.277$ so we fail to reject H_0
There is insufficient evidence to suggest that the population mean height of female adults in the UK is different from that in 1998.

16) a)
CR is $\{\bar{X} < 42.15103\}$
16) b)
$42.41667 > 42.15103$ so we fail to reject H_0. There is insufficient evidence to suggest that the population mean waiting time at the veterinary surgery for those without an appointment has decreased.

17) a)
The first mistake is that the sample means should be normally distributed with a mean of 7 hours and not 7.3 hours. The second mistake is that because we are looking at the top tail, we want to find the value of a such that there is 0.5% at the top, and not 99.5% at the top.

17) b)
Let X be the number of hours of sleep an adult gets.
$H_0: \mu = 7$
$H_1: \mu \neq 7$
$\bar{X} \sim N\left(7, \frac{1.1^2}{80}\right)$
$Z = \frac{\bar{x} - \mu}{\frac{\sigma}{\sqrt{n}}} = \frac{7.3 - 7}{\frac{1.1}{\sqrt{80}}} = 2.439$

$P(Z > a) = 0.005 \Rightarrow a = 2.576$
$2.439 < 2.576$ so we fail to reject H_0
There is insufficient evidence to suggest that the population mean number of hours that adults sleep has changed.

18) a)
$Z = 2.635231383$,
Critical Value $= 2.326348$
18) b)
$2.635231383 > 2.326348$ so we reject H_0
There is sufficient evidence to suggest that the population mean time for the medication to relieve symptoms is longer than 30 minutes.

19) a)
$Z = -2.420614591$,
Critical Value $= -2.053749$
19) b)
$-2.420614591 < -2.053749$ so we reject H_0. There is sufficient evidence to suggest that the population mean fuel efficiency of the car is less than 40 miles to the gallon.

20)
$0.0660 > 0.015$ so we fail to reject H_0. There is insufficient evidence to suggest that the population mean weight-loss of adult members of the fitness club is different to 8.5 kilograms.

2.52 Forces: Equilibrium
Questions on page 371

1) a)
x-comp: 4.33 N (3 SF), y-comp: 2.5 N
1) b)
x-comp: 4.24 N (3 SF),
y-comp: 4.24 N (3 SF)
1) c)
x-comp: 6.13 N (3 SF)
y-comp: 5.14 N (3 SF)
1) d)
x-comp: 2.78 N (3 SF)
y-comp: 1.12 N (3 SF)

2) a)
x-comp: 7.07 N (3 SF)
y-comp: 7.07 N (3 SF)
2) b)
x-comp: -11.3 N (3 SF)
y-comp: 4.10 N (3 SF)
2) c)
x-comp: 4.33 N (3 SF)
y-comp: -2.5 N
2) d)
x-comp: -11.5 N (3 SF)
y-comp: -8.03 N (3 SF)

2) e)
x-comp: 24.0 N (3 SF)
y-comp: 6.89 N (3 SF)
2) f)
x-comp: 2.82 N (3 SF)
y-comp: -1.03 N (3 SF)
2) g)
x-comp: -8.99 N (3 SF)
y-comp: -4.38 N (3 SF)
2) h)
x-comp: -3.29 N (3 SF)
y-comp: 6.18 N (3 SF)

3) a)
7.91 N (3 SF), 32.4° (3 SF)
3) b)
4.41 N (3 SF), 117° (3 SF)
3) c)
4.90 N (3 SF), 90°

4) a)
$P = 5\sqrt{3} = 8.66$ (3 SF), $Q = 5$
4) b)
$P = 3, Q = 3\sqrt{3} = 5.20$ (3 SF)
4) c)
$P = 2\sqrt{6} + 6\sqrt{2} = 13.4$ (3 SF)
$Q = 4\sqrt{6} = 9.80$ (3 SF)
4) d)
$P = 15.2$ (3 SF), $Q = 23.3$ (3 SF)
4) e)
$P = 27.5$ (3 SF), $Q = 33.1$ (3 SF)
4) f)
$\theta = 48.2°$ (3 SF), $P = 4.47$ (3 SF)
4) g)
$\theta = 36.2$ (3 SF), $P = 25.3$ (3 SF)
4) h)
$P = 36.4$ (3 SF), $\theta = 60.1$ (3 SF)
4) i)
$P = 5.70$ (3 SF), $\theta = 28.0$ (3 SF)
4) j)
$P = 50\sqrt{2} = 70.7$ N (3 SF)
$Q = 50\sqrt{2} = 70.7$ N (3 SF)
4) k)
$P = 41.0$ N (3 SF), $Q = 28.7$ N (3 SF)

5) a)
$T_1 = 14.1$ (3 SF), $T_2 = 16.3$ (3 SF)
5) b)
$T_1 = 15.1$ (3 SF), $T_2 = 6.99$ (3 SF)
5) c)
$T_1 = 12.8$ (3 SF), $T_2 = 12.8$ (3 SF)

6) a)
$T = 13.2$ (3 SF), $m = 0.487$ (3 SF)
6) b)
$T = 10$, $m = 0.884$ (3 SF)

7) a)
$\theta = 41.8$ (3 SF)
7) b)
$T = 11.2$ (3SF)

8) a)
$\theta = 12.5$ (3 SF)

8) b)
$T = 18.4$ (3 SF)

9) a)
$T_1 = \frac{40}{\cos\theta°}$ N
9) b)
$T_2 = 40\tan\theta°$ N
9) c)
90°
9) d)
$0° \le \theta < 45°$

10) a)
Show that
10) b)
$T = \frac{5}{2}\sqrt{2} = 3.54$ (3SF)
10) c)
$T = \frac{5}{3}\sqrt{3} = 2.89$ (3SF)
10) d)
As $\theta \to 90°$, $T \to 2.5$

11) a)
Show that
11) b)
Show that
11) c)
Show that
11) d)
$m = \frac{60\sqrt{55}}{g}$

12) a)

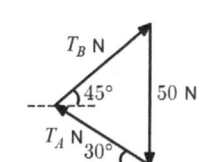

12) b)
$T_A = 36.6$ (3 SF), $T_B = 44.8$ (3 SF)

13) a)

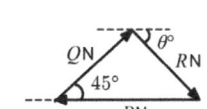

13) b)
$0 < \theta < 135$

14) a)

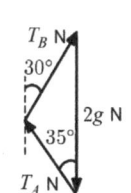

14) b)
$T_A = 10.8$ (3 SF), $T_B = 12.4$ (3 SF)

15) a)

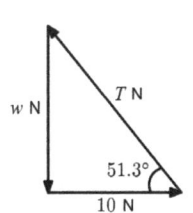

15) b)
$T = 16$ N

15) c)
$w = 2\sqrt{39} = 12.5$ (3 SF)
15) d)
Mass is 1.27 kg
1.3 kg (2 SF – AQA)

16) a)

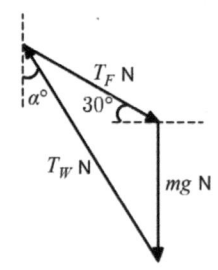

16) b)
Show that
16) c)
Show that
16) d)
Show that

2.53 Forces: Coefficient of Friction
Questions on page 376

1) a)
$F = 10$
1) b)
$F = 9$
1) c)
$F = 12.75$
1) d)
$F = 3.3$

2) a)
$P = 6$
2) b)
$P = 10.8$
2) c)
$w = 50$
2) d)
$w = 25$
2) e)
$\mu = 0.8$
2) f)
$\mu = 0.04$
2) g)
$\mu = 1.25$
2) h)
$P = 10.5$
2) i)
$w = 8$
2) j)
$\mu = 0.125$

3) a)
$F = 6.25$
3) b)
$T = \frac{4\sqrt{2}}{3} = 1.89$ (3 SF)
3) c)
$w = 50 + 5\sqrt{3} = 58.7$ (3 SF)
3) d)
$\mu = \frac{2\sqrt{3}}{13} = 0.266$ (3 SF)

4) a)
$F = 2\sqrt{3} = 3.46$ (3 SF)
4) b)
$T = \frac{12\sqrt{2}}{7} = 2.42$ (3 SF)
4) c)
$w = 30 - 6\sqrt{3} = 19.6$ (3 SF)
4) d)
$\mu = \frac{\sqrt{3}}{9} = 0.192$ (3 SF)

5) a)
$R(\uparrow): 25 + 14\sin 27° - w = 0 \Rightarrow w = 31.4$
This is the weight and not the mass of the particle.

6) a)
$T = 22.9$ (3 SF), $T = 23$ (2 sf – AQA)
6) b)
$m = 14.7$ (3 SF), $m = 15$ (2 sf – AQA)
6) c)
$\mu = 0.115$ (3 SF), $\mu = 0.11$ (2 sf – AQA)

7) a)
$R(\rightarrow): 0.56 > 0$ so the particle moves.
$R(\rightarrow): 0.168 > 0$ so the particle continues to move.
7) b)
$R(\rightarrow): 0.16 > 0$ so the particle moves.
$R(\rightarrow): -0.232 < 0$ so the particle no longer moves.
7) c)
$R(\rightarrow): -0.24 < 0$ so the particle doesn't move. Adding more mass, means it continues to not move.

8) a)
$P_{max} = 14.7$, $P_{max} = 15$ (2 SF – AQA)
8) b)
$P_{max} = 29.4$, $P_{max} = 29$ (2 SF – AQA)
8) c)
$P_{max} = 58.8$, $P_{max} = 60$ (2 SF – AQA)

9)
Show that

10)
Show that

11) a)
$R(\rightarrow): 120 - 127.5 < 0 \therefore$ does not move
11) b)
$\Rightarrow a = 1.83$ ms^{-2}, $a = 2$ ms^{-2} (1 SF – AQA)
11) c)
$m < 3.23$, $m < 3$ (1 SF – AQA)

12)
Show that

13) a)
$\sin\theta = \frac{3}{5}$, $\cos\theta = \frac{4}{5}$

13) b)
$\sin\theta = \frac{5}{13}$, $\cos\theta = \frac{12}{13}$
13) c)
$\sin\theta = \frac{7}{\sqrt{305}}$, $\cos\theta = \frac{16}{\sqrt{305}}$
13) d)
$\sin\theta = \frac{10}{\sqrt{541}}$, $\cos\theta = \frac{21}{\sqrt{541}}$

14) a)
$a = 0.74$
14) b)
$a = 4.17$ (3 SF), $a = 4.2$ (2 SF – AQA)
14) c)
$\mu = 0.1125$, $\mu = 0.11$ (2 SF – AQA)
14) d)
$\Rightarrow m = \frac{304}{109} = 2.79$ (3 SF)
$m = 2.8$ (2 SF – AQA)
14) e)
Show that

15) a)
$a = 0.155$ (3 SF), $a = 0.16$ (2 SF – AQA)
15) b)
$a = 0.104$ (3 SF), $a = 0.10$ (2 SF – AQA)
15) c)
$a = 0$

16) a)
$P = 2$
16) b)
$\mu = 0.0255$ (3 SF), $\mu = 0.026$ (2 SF – AQA)
16) c)
$m \leq 8$

17) a)
$a = 4.48$ ms^{-2} (3 SF)
$a = 4.5$ ms^{-2} (2 SF – AQA)
17) b)
$s = 128$ m (3 SF), $s = 130$ m (2 SF – AQA)

18) a)
$a = 0.340$ ms^{-2} (3 SF)
18) b)
$P = 21.3$ (3 SF)

19) a)

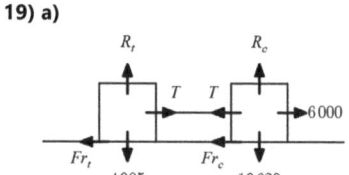

19) b)
$a = 1.419$ ms^{-2}
$a = 1.42$ ms^{-2} (3 SF – AQA)
19) c)
$T = 1200$ N
19) d)
The tow bar is modelled as a light inextensible string. There are no other resistances other than friction. The tow bar is horizontal.

20) a)
$\mu = 0.0869$ (3 SF)
20) b)
$T = 1770$ N (3 SF)
20) c)
Light: the mass of the towbar is negligible. Inextensible: acceleration of the masses is the same.

21) a)

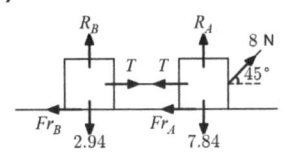

21) b)
$Fr_A = 2.07$ N, $Fr_A = 2.1$ N (2 SF – AQA)
21) c)
$Fr_B = 2.793$ N, $Fr_B = 2.8$ N (2 SF – AQA)
21) d)
$a = 0.718$ ms^{-2}
$a = 0.72$ ms^{-2} (2 SF – AQA)
21) e)
$T = 3.01$ N, $T = 3.0$ N (2 SF – AQA)
21) f)
Both strings are taut/rigid.

22) a)

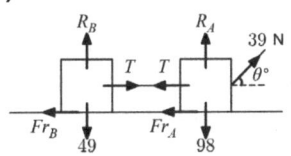

22) b)
$\mu = 0.670$ (3 SF), $\mu = 0.67$ (2 SF – AQA)
22) c)
$T = 15.4$ N (3 SF), $T = 15$ N (2 SF – AQA)
22) d)
The string connecting A and B is horizontal.
22) e)
$s = 175$ m

23) a)

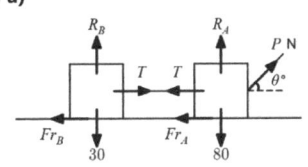

23) b)
$\mu = 0.2$
23) c)
$T = 13.5$
23) d)
$P = 48.7$ N (3 SF), $P = 50$ N (1 SF – AQA)
23) e)
$Fr_A = 13.3$ N (3 SF)
$Fr_A = 10$ N (1 SF – AQA)
23) f)
$t = 4$ s

24) a)

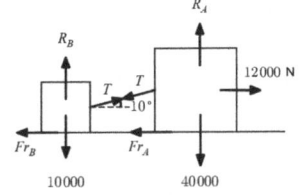

24) b)
$\mu = 0.19$
24) c)
$T = 2360$ N (3 SF)
$T = 2000$ N (1 SF – AQA)

25) a)
Show that
25) b)
Show that

26) a)

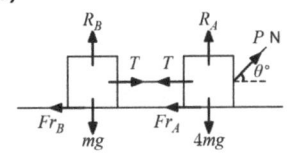

26) b)
Show that
26) c)
Show that
26) d)
$T = \frac{4mg-6}{mg-3}$

2.54 Forces: Connected Particles
Questions on page 381

1) a)
$a = 2.14$ ms^{-2} (3 SF)
$a = 2.1$ ms^{-2} (2 SF – AQA)
1) b)
$T = 23.0$ N (3 SF), $T = 23$ N (2 SF – AQA)

2) a)
$T = 48.9$ N, $T = 50$ N (2 SF – AQA)
2) b)
$\mu = 0.414$ (3 SF), $\mu = 0.41$ (2 SF – AQA)
2) c)
$t = 7.07$ s (3 SF), $t = 7.1$ s (2 SF – AQA)
2) d)
0.5 m

3) a)
$a = 3.70$ ms^{-2} (3 SF)
$a = 3.7$ ms^{-2} (2 SF – AQA)
3) b)
$T = 30.5$ N (3 SF), $T = 30$ N (2 SF – AQA)
3) c)
$s = 2.04$ metres (3 SF)
$s = 2.0$ metres (2 SF – AQA)

4) a)
$a = \frac{mg - 0.6g}{2+m}$ ms^{-2}

4) b)
Minimum value of m is 0.6

5)
Show that

6) a)
$F = 2.92$ N, $F = 3.0$ N (2 SF – AQA)
6) b)
$a = 0.336$ ms^{-2} (3 SF)
$a = 0.34$ ms^{-2} (2 SF – AQA)
6) c)
$T = 5.07$ N (3 SF), $T = 5.1$ N (2 SF – AQA)
6) d)
m is less than 0.586 kg (3 SF)
m is less than 0.59 kg (2 SF – AQA)

7) a)
$T_A = 18.3$ N (3 SF)
$T_A = 18$ N (2 SF – AQA)
$T_B = 5.23$ N (3 SF)
$T_B = 5.2$ N (2 SF – AQA)
7) b)
$a = 0.653$ ms^{-2} (3 SF)
$a = 0.65$ ms^{-2} (2 SF – AQA)
A is accelerating downwards
B is accelerating to the left
C is accelerating upwards
7) c)
The string is light, inextensible and taut.
The boxes are modelled as particles.

8)
Show that

2.55 Forces: Inclined Planes
Questions on page 383

1) a)

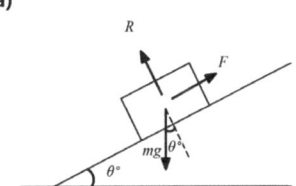

1) b)
$\mu = 0.577$ (3 SF), $\mu = 0.58$ (2 SF – AQA)
1) c)
$\mu = 1$
1) d)
$\mu = 1.43$ (3 SF), $\mu = 1.4$ (2 SF – AQA)
1) e)
Show that

2) a)

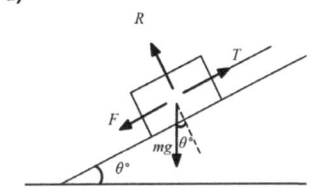

2) b)
$F = 27.2$ (3 SF), $F = 27$ (2 SF – AQA)
$T = 46.8$ (3 SF), $T = 47$ (2 SF – AQA)
2) c)
$F = 14.7$, $F = 15$ (2 SF – AQA)
$T = 57.1$ (3 SF), $T = 57$ (2 SF – AQA)
2) d)
$m = 1.49$ (3 SF), $m = 1.5$ (2 SF – AQA)
2) e)
$\theta = 11.8$ (3 SF), $\theta = 12$ (2 SF – AQA)
2) f)
$m = 0.902$ (3 SF), $m = 0.90$ (2 SF – AQA)
2) g)
Show that

3) a)

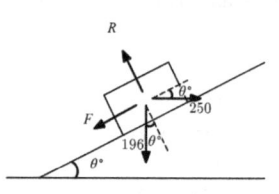

3) b)
Show that
3) c)
$R = 306.8$ N, $R = 310$ N (2 SF – AQA)
3) d)
$F = 82.4$ N, $F = 82$ N (2 SF – AQA)
3) e)
$\mu = 0.269$ (3 SF), $\mu = 0.27$ (2 SF – AQA)

4) a)

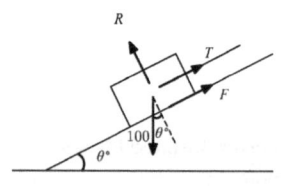

4) b)
Show that
4) c)
$T = 4$

5) a)

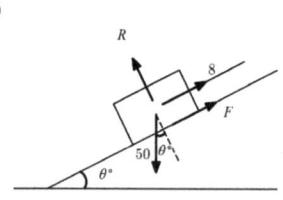

5) b)
Show that
5) c)
$\mu = 0.228$ (3 SF), $\mu = 0.2$ (1 SF – AQA)

6) a)

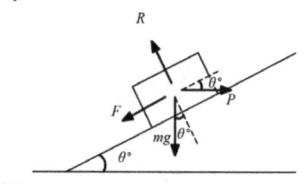

6) b)
$R = \frac{1}{2}P + 10\sqrt{3}$ N

6) c)
$F = \frac{1}{5}P + 4\sqrt{3}$
6) d)
$P = 25.1$ (3 SF), $P = 30$ (1 SF – AQA)

7) a)

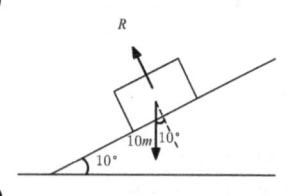

7) b)
$a = 1.74$ ms^{-2} down the slope (3 SF)
$a = 2$ ms^{-2} down the slope (1 SF – AQA)
7) c)
$v = 4.17$ ms^{-1} (3 SF)
$v = 4$ ms^{-1} (1 SF – AQA)

8) a)

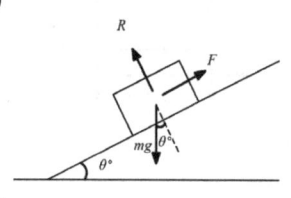

8) b)
$a = 7.66$ ms^{-2} down the slope (3 SF)
$a = 8$ ms^{-2} down the slope (1 SF – AQA)
8) c)
$\mu = 0.789$ (3 SF)
8) d)
Show that
8) e)
$a = g\sin\theta$

9) a)

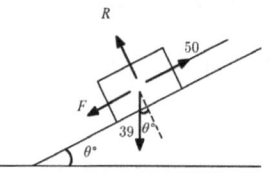

9 b)
Show that
9) c)
$s = 4.90$ m (3 SF), $s = 5$ m (1 SF – AQA)
9) d)
$v = 6.54$ ms^{-1} (3 SF)
$v = 7$ ms^{-1} (1 SF – AQA)

10) a)

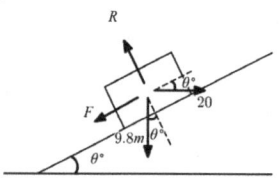

10) b)
$\cos\theta = \frac{7}{\sqrt{58}}$ and $\sin\theta = \frac{3}{\sqrt{58}}$
10) c)
$a = -4.47$ ms^{-2} (3 SF)
$a = -4.8$ ms^{-2} (2 SF – AQA)

10) d)
Deceleration up the slope.

11) a)

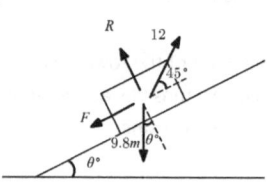

11) b)
$\cos\theta = \frac{24}{25}$ and $\sin\theta = \frac{7}{25}$
11) c)
$\mu = 0.0578$ (3 SF), $\mu = 0.058$ (2 SF – AQA)
11) d)
$m = 1.91$ (3 SF), $m = 1.9$ (2 SF – AQA)

12) a)
$a = -5.51$ ms^{-2} (3 SF)
$a = -5.5$ ms^{-2} (2 SF – AQA)
12) b)
It is decelerating up the slope.
12) c)
$a = 2.92$ ms^{-2} (3 SF)
$a = 2.9$ ms^{-2} (2 SF – AQA)

13) a)
Show that
13) b)
$t = 2.83$ s (3 SF), $t = 2.8$ s (2 SF – AQA)
13) c)
$t = 0.963$ seconds (3 SF)
$t = 0.96$ seconds (2 SF – AQA)

14) a)
$a = 2.5$ ms^{-2}
14) b)
$\mu = 0.431$ (3 SF), $\mu = 0.43$ (2 SF – AQA)

15)
$t = 3.32$ s (3 SF), $t = 3.3$ s (2 SF – AQA)

16)
$\mu = 0.526$ (3 SF), $\mu = 0.53$ (2 SF – AQA)

17) a)
$F = 10$
17) b)
$a = 1.63$ ms^{-2} down the slope (3 SF)
$a = 1.6$ ms^{-2} down the slope (2 SF – AQA)
17) c)
Do not model the block as a particle, i.e. consider dimensions. Include air resistance. Consider any rotational effects of forces such as spin.

18) a)
$\mu = 0.25$
18) b)
$s = 9.6$ m
18) c)
$t = 3.57$s (3 SF), $t = 3.6$s (2 SF – AQA)

19) a)
$P \geq \frac{29}{3}mg$

19) b)
$\mu \geq \frac{3Q-4mg}{3mg+4Q}$

20) a)
Show that

20) b)
The particle remains at rest.

21) a)
Deceleration is 0.414 ms^{-2} (3 SF)
0.4 ms^{-2} (1 SF – AQA)

21) b)
$T = 923$ N (3 SF), $T = 900$ N (1 SF – AQA)

22) a)
$D = 14\,500$ N (3 SF)
$D = 10\,000$ N (1 SF – AQA)

22) b)
$T = 2900$ N (3 SF)
$T = 3000$ N (1 SF – AQA)

23)
Show that

24)
Show that

25) a)

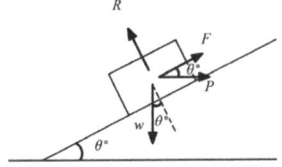

25) b)
Show that

25) c)
Show that

2.56 Forces: Connected Particles on Inclined Planes
Questions on page 388

1) a)
$T = 15.925$ N, $a = \frac{343}{60} = 5.72$ ms^{-2} (3 SF)
$T = 15$ N, $a = 5.7$ ms^{-2} (2 SF – AQA)
Particle A is moving up the slope.

1) b)
$T = 7.39$ N, $a = -4.97$ ms^{-2} (3 SF)
$T = 7.4$ N, $a = -5.0$ ms^{-2} (2 SF – AQA)
Particle A is moving down the slope.

1) c)
$T = 16.1$ N, $a = 1.75$ ms^{-2} (3 SF)
$T = 16$ N, $a = 1.8$ ms^{-2} (2 SF – AQA)
Particle A is moving up the slope.

1) d)
$T = 196$ N, $a = 0$ ms^{-2}
$T = 200$ N, $a = 0$ ms^{-2} (2 SF – AQA)

Particle A is not moving in either direction since a is zero.

2) a)
$T = 31.3$ N, $a = 3.35$ ms^{-2} (3 SF)
$T = 31$ N, $a = 3.4$ ms^{-2} (2 SF – AQA)
Particle A is moving up the slope.

2) b)
$T = 83.7$ N (3 SF), $T = 84$ N (2 SF – AQA)
$m = 8.90$ kg (3 SF)
$m = 8.9$ kg (2 SF – AQA)

2) c)
$T = 29.7$ N
$\mu = 0.846$ (3 SF), $\mu = 0.85$ (2 SF – AQA)

2) d)
$a = 2.83$ ms^{-2} (decelerating up the slope) (3 SF)
$a = 2.8$ ms^{-2} (decelerating up the slope) (2 SF – AQA)

3) a)
$\mu = 0.116$ (3 SF), $\mu = 0.12$ (2 SF – AQA)

3) b)
$s = 10$ m
This assumes that box B does not reach the pulley, and that the boxes may be modelled as particles.

3) c)
Light means the string has zero mass. Inextensible means that the acceleration is constant throughout.

4) a)
$T = 12.544m$

4) b)
$t = 0.922$ seconds (3 SF)
$t = 0.92$ (2 SF – AQA)

4) c)
1.28 seconds (3 SF)
1.3 seconds (2 SF – AQA)

5) a)
$a = -0.385$ ms^{-2} (3 SF)
$a = -0.3$ ms^{-2} (1 SF – AQA)
$T = 16.2$ N (3 SF), $T = 20$ N (1 SF – AQA)

5) b)
$T = 15.2$ N, $T = 20$ N (1 SF – AQA)
$\mu = 0.184$, $\mu = 0.2$ (1 SF – AQA)

5) c)
$a = 3.5$ ms^{-2}
$\mu = 0.25$, $\mu = 0.3$ (1 SF – AQA)

5) d)
$m = 4$ kg, $M = 2$ kg

6) a)
$a = 1.38$ ms^{-2} (3 SF)
$a = 1$ ms^{-2} (1 SF – AQA)

6) b)
$T = 33.8$ N (3 SF), $T = 30$ N (1 SF – AQA)

7) a)
$a = 1.39$ ms^{-2} (3 SF)
$a = 1$ ms^{-2} (1 SF – AQA)

7) b)
Block A is going down the slope and block B is going up the slope.

7) c)
$T = 25.4$ N (3 SF), $T = 30$ N (1 SF – AQA)

7) d)
$m = 5.17$ kg (3 SF)
$m = 5$ kg (1 SF – AQA)

8) a)
$\mu = 0.443$ (3 SF), $\mu = 0.4$ (1 SF – AQA)

8) b)
$T = 4.71$ N (3 SF), $T = 5$ N (1 SF – AQA)

9) a)
$T = 20.4$ N, $T = 20$ N (2 SF – AQA)

9) b)
Show that

9) c)
The second box is does not have any resistances acting upon it. The string which passes over the pulley is parallel to the slope.

10) a)
$a = 0.252$ ms^{-2} (3 SF)
$a = 0.25$ ms^{-2} (2 SF – AQA)

10) b)
$T_1 = 38.0$ N (3 SF)
$T_1 = 38$ N (2 SF – AQA)

10) c)
$T_2 = 17.1$ N (3 SF)
$T_2 = 17$ N (2 SF – AQA)

11) a)
$a = 1.67$ ms^{-2} (3 SF) up the slope
$a = 2$ ms^{-2} (1 SF – AQA) up the slope

11) b)
$T = 33.3$ N (3 SF), $T = 30$ N (1 SF – AQA)

11) c)
8.33 N (3 SF), 8 N (1 SF – AQA)

11) d)
The force exerted on the container by box C would decrease.

12) a)
Show that

12) b)
The rope is modelled as light and inextensible, so that it has no mass and acceleration is constant through the system. The boxes are modelled as particles.

13) a)
$a = \frac{1}{4}$ ms^{-2}

13) b)
Show that

13) c)
$\mu = \frac{493}{24} - \frac{173}{32g}$
13) d)
The string is parallel to the slope.

14) a)
Show that
14) b)
$k > \frac{5}{3}\mu$

15) a)
Show that
15) b)
$\mu = 0.0256$ (3 SF)

16)
$P > \frac{4}{11}(7 - 3\sqrt{3})mg$

2.57 Constant Acceleration: Vectors
Questions on page 392

1)
12.2 m (3 SF)

2) a)
4.47 ms^{-1} (3 SF)
2) b)
9.43 ms^{-1} (3 SF)
2) c)
6.5k ms^{-1} (3 SF)

3) a)
3.61 ms^{-2} (3 SF)
3) b)
472 m (3 SF)
3) c)
15.6 ms^{-1}

4)
(13$\mathbf{i}$ + 23$\mathbf{j}$) ms^{-1}

5) a)
(5$\mathbf{i}$ − 16$\mathbf{j}$) ms^{-1}
5) b)
16.8 ms^{-2} (3 SF)

6)
2.22 ms^{-2} (3 SF)

7) a)
$\binom{10}{0}$ ms^{-1}
7) b)
$\frac{1}{\sqrt{2}}\binom{4}{4}$ ms^{-1}
7) c)
$\frac{3\sqrt{13}}{130}\binom{2}{-3}$ ms^{-1}

8)
(21$\mathbf{i}$ − 6$\mathbf{j}$) metres

9) a)
$(2t + 3t^2)\mathbf{i} + \left(5t - \frac{1}{2}t^2\right)\mathbf{j}$
9) b)
320 metres

10) a)
Show that
10) b)
12.0 metres (3 SF)

11)
(74$\mathbf{i}$ − 50$\mathbf{j}$) metres

12)
$\binom{-4}{-6}$ metres

13) a)
$T = 6$
13) b)
$k = 90$

14)
$k = 61$

15) a)
$k = 8$
15) b)
$k = -6$
15) c)
$k = -9$
15) d)
Any real value of k

16)
$T = 0.1$ seconds

17) a)
8.4$\mathbf{i}$ + 28.8$\mathbf{j}$ ms^{-1}
17) b)
1.4$\mathbf{i}$ + 4.8$\mathbf{j}$ ms^{-1}
17) c)
3.36$\mathbf{i}$ + 11.52$\mathbf{j}$ ms^{-1}
17) d)
−2.1$\mathbf{i}$ − 7.2$\mathbf{j}$ ms^{-1}

18) a)
23.4 ms^{-1}
18) b)
$\binom{7.5}{18}$ ms^{-1}
18) c)
$\binom{39.5}{87}$ metres

19)
3.46 metres (3 SF)

20) a)
6.24 ms^{-1}
20) b)
(4$\mathbf{i}$ + 15$\mathbf{j}$) metres
20) c)
13 metres

21)
Show that

22)
62.5 m

23) a)
East
23) b)
South
23) c)
North-East

23) d)
North-West

24) a)
$\binom{80}{-40}$ m
24) b)
89.4 m (3 SF)

25) a)
Show that
25) b)
$t = 5$ s
25) c)
$t = 8$ s

26) a)
$t = 0$ s
26) b)
$t = 15$ s
26) c)
Since $t \geq 0$, $105 + 8t \geq 105$ so cannot move West

27)
$t = 0.4$ s

28)
$t = 0.2$ s

29) a)
$\binom{-4}{7}$ ms^{-1}
29) b)
$t = 1.5$ s
29) c)
1.5 ms^{-1}

30) a)
21.5 ms^{-1} (3 SF)
30) b)
158°

31) a)
6.71 ms^{-1} (3 SF)
31) b)
063°
31) c)
$t = 32.8$ s (3 SF)

2.58 Projectiles
Questions on page 395

1) a)
(21.7$\mathbf{i}$ + 7.6$\mathbf{j}$) ms^{-1} (3 SF)
(22$\mathbf{i}$ + 7.6$\mathbf{j}$) ms^{-1} (2 SF – AQA)
1) b)
$s = \frac{3125}{392} = 7.97$ m (3 SF)
$s = 8.0$ m (2 SF – AQA)
1) c)
$t = 2.55$ s (3 SF), $t = 2.6$ s (2 SF – AQA)
1) d)
$s = 25 \cos 30° \times \frac{125}{49} + 0 = 55.2$ m (3 SF)
$s = 55$ m (2 SF – AQA)

2) a)
(8.66$\mathbf{i}$ + 1.08$\mathbf{j}$) ms^{-1} (3 SF)
(8.7$\mathbf{i}$ + 1.1$\mathbf{j}$) ms^{-1} (2 SF – AQA)

2) b)
$s = 1.28$ m (3 SF)
$s = 1.3$ m (2 SF – AQA)
2) c)
$t = 1.02$ s (3 SF), $t = 1.0$ s (2 SF – AQA)
2) d)
$s = 8.84$ m (3 SF), $s = 8.9$ m (2 SF – AQA)

3) a)
$(17.7\mathbf{i} - 11.7\mathbf{j})$ ms^{-1} (3 SF)
$(18\mathbf{i} - 12\mathbf{j})$ ms^{-1} (2 SF – AQA)
3) b)
$s = 15.9$ m (3 SF), $s = 16$ m (2 SF – AQA)
3) c)
$t = 3.61$ s (3 SF), $t = 3.7$ s (2 SF – AQA)
3) d)
$s = 63.8$ m (3 SF), $s = 64$ m (2 SF – AQA)

4) a)
$(18.8\mathbf{i} + 6.84\mathbf{j})$ ms^{-1} (3 SF)
$(19\mathbf{i} + 6.9\mathbf{j})$ ms^{-1} (2 SF – AQA)
4) b)
$t = 2.28$ s (3 SF), $t = 2.3$ s (2 SF – AQA)
4) c)
$(18.8\mathbf{i} - 15.6\mathbf{j})$ ms^{-1} (3 SF)
$(19\mathbf{i} - 16\mathbf{j})$ ms^{-1} (2 SF – AQA)

5) a)
$t = 4.93$ s (3 SF), $t = 4.9$ s (2 SF – AQA)
5) b)
$y - 80 = 8t - 4.9t^2$, $x = 6t$
Show that
5) c)
83.3 m (3SF), 83 m (2SF – AQA)

6)
1.79 m (3 SF). Since the wall is 1.5 metres high, the ball passes over the wall.

7) a)
0.988 m (3 SF). Since the net is 1.2 metres high, the ball hits the net.
7) b)
Modelling the ball as a particle (dimensions of ball not allowed for, no spin on ball), no air resistance.

8)
0.603 seconds (3 SF)
0.60 seconds (2 SF – AQA)

9) a)
$t = \frac{5}{7}\sqrt{31}$
9) b)
$u = \frac{175}{2\sqrt{31}}$
9) c)
That g is exactly 9.8 ms^{-2}, there are no resistances to motion (like air resistance or wind), the dimension and shape of stone is ignored, any spin on the stone is ignored.

9) d)
It would not change the results as the modelling assumptions were that the original stone was modelled as a particle.

10)
Show that

11) a)
$s = 2.17$ m (3 SF)
Yes the ball goes into the goal.
11) b)
No resistances to motion (air resistance / wind). Dimensions of ball and goal post not included so treating as a particle/rod. That g is 9.81 ms^{-2} exactly. No spin on the ball.

12) a)
14.8° below the horizontal.
12) b)
1.26 metres (3 SF)
12) c)
4.77 metres (3 SF)

13) a)
$u = 10.1$ ms^{-1} (3 SF)
$u = 10$ ms^{-1} (1 SF – AQA)
13) b)
$\theta = 69.8°$ (3 SF), $\theta = 70°$ (1 SF – AQA)

14) a)
$s = 26.2$ m (3 SF), $s = 30$ m (1 SF – AQA)
14) b)
35.2° below the horizontal

15)
Difference in height of 30.3 m (3 SF)
30 m (1 SF – AQA)

16) a)
Show that
16) b)
$\theta = 59.2°$ or $\theta = 26.5°$ (3 SF) above the horizontal

17) a)
Show that
17) b)
$t = \frac{-1+\sqrt{3}}{2}$ seconds
17) c)
3.17 m

18)
Show that

19) a)
Show that
19) b)
Show that

20) a)
Show that
20) b)
Show that
20) c)
$31.4° < \theta < 58.6°$

2.59 Variable Acceleration: Vectors
Questions on page 398

1) a)
12 ms^{-2}
1) b)
60 ms^{-2}
1) c)
2.07 ms^{-2} (3 SF)

2) a)
375 ms^{-1} (3 SF)
2) b)
4.03 ms^{-1} (3 SF)

3) a)
12.7 m (3 SF)
3) b)
2.14 m (3 SF)

4) a)
11.3 N (3 SF)
4) b)
77.1 m (3 SF)

5)
252 m (3 SF)

6)
$t = 0.25$ s

7)
Show that

8) a)
10.3 ms^{-1} (3 SF)
8) b)
$\mathbf{a} = \frac{d\mathbf{v}}{dt} = 2\mathbf{i} + 0\mathbf{j}$ which will always have a non-zero horizontal component, hence the acceleration can never be zero.

9) a)
$\mathbf{s} = \left(\frac{1}{2}t^2 - 4t\right)\mathbf{i} + \left(\frac{1}{4}t^4 + t^2\right)\mathbf{j}$
9) b)
2600 m

10) a)
$t = 1$ s
10) b)
$t = \frac{1}{9}$ s
10) c)
$t = 9$ s

11) a)
$t = 36$ s
11) b)
$\mathbf{r} = (t^2 - 36t + 15)\mathbf{i} + \left(4t^{\frac{3}{2}} + 2\right)\mathbf{j}$
11) c)
$\mathbf{r} = -273\mathbf{i} + \left(96\sqrt{3} + 2\right)\mathbf{j}$

12) a)
$p = 3$, $q = 5$
12) b)
2.51 N (3 SF)

13) a)
$k = 8$
13) b)
$t = 0$ s

14) a)
Show that
14) b)
$t = 2$ s

15)
$t = \frac{\sqrt{3}}{4}$ s

16)
Show that

17) a)
$\mathbf{v} = 2e^{0.4t}\mathbf{i} + (2t - 4)\mathbf{j}$
17) b)
East
17) c)
26.8 N (3SF)

18) a)
Show that
18) b)
$p = 32$ or $p = -40$
18) c)
No vertical component $\Rightarrow t = 0$
But $t^2 + pt$ would also be 0 so never moves exactly westwards or eastwards.

19)
178°

20)
$t = 2$ s

21) a)
$t = 2.5$ s
21) b)
3 ms⁻¹

22) a)
$\mathbf{v} = (2t)\mathbf{i} + 8\mathbf{j}$
22) b)
$T = 4$, Speed = 11.3 ms⁻¹ (3 SF)

23) a)
$t = 2$ s
23) b)
$t = 4$ s
23) c)
7.52 ms⁻¹ (3 SF)

24) a)
2.58 ms⁻¹ (3 SF)
24) b)
$t = 0$ s

25) a)
$\mathbf{v} = (8t - 5)\mathbf{i} + (-2\ln t - 5)\mathbf{j}$
25) b)
4.22 ms⁻¹ (3 SF)

26)
5.77 m (3 SF)

27) a)
East of the origin
27) b)
South
27) c)
20.5 N (3 SF)

28)
1.43 ms⁻¹ (3 SF)

29)
1.41 m (3 SF)

30)
242°

31) a)
$y = 3 + \left(\frac{x+5}{8}\right)^2$
31) b)
$y = 2 - \left(x - \frac{7}{4}\right)^2$
31) c)
$x = \left(\frac{y-4}{5}\right)^2$

32) a)
2 N
32) b)
$x^2 + y^2 = 1$

2.60 Moments
Questions on page 401

1) a)
24 Nm anti-clockwise
1) b)
50 Nm clockwise
1) c)
3.2 Nm anti-clockwise
1) d)
40 Nm clockwise
1) e)
24 Nm clockwise
1) f)
3.9 Nm clockwise

2) a)
10 Nm anti-clockwise
2) b)
10 Nm anti-clockwise
2) c)
0 Nm
2) d)
0 Nm
2) e)
5 Nm anti-clockwise

3) a)
0 Nm
3) b)
0 Nm
3) c)
48 Nm clockwise
3) d)
48 Nm clockwise
3) e)
24 Nm clockwise

4) a)
50 Nm clockwise
4) b)
0 Nm
4) c)
90 Nm clockwise
4) d)
140 Nm clockwise
4) e)
70 Nm clockwise

5) a)
$X = 10$
5) b)
$Y = 20$
5) c)
$d = 0.08$ m
5) d)
The forces have a resultant moment which have an anti-clockwise turning effect on the lamina

6) a)
27 Nm anti-clockwise
6) b)
12 Nm anti-clockwise
6) c)
3 Nm clockwise
6) d)
28 Nm clockwise
6) e)
53 Nm clockwise

7) a)
$Q = 25$ N and $P = 25$ N
7) b)
$Q = 40$ N and $P = 70$ N

8) a)
$Q = 18.75$ N and $P = 11.25$ N
8) b)
$Q = 36$ N and $P = 33$ N
8) c)
$Q = 1.92$ N (3 SF) and $P = 3.08$ N (3 SF)

9) a)
$W = 40$ and $x = 0.3125$
9) b)
$Q = 72$ and $P = 48$

10) a)

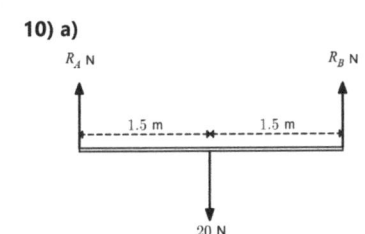

[Diagram: beam with R_A N up at left, R_B N up at right, 1.5 m each side of centre, 20 N down at centre]

10) b)

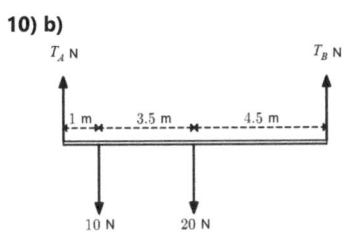

[Diagram: beam with T_A N up at left, T_B N up at right, 1 m, 3.5 m, 4.5 m segments, 10 N and 20 N down]

10) c)

[Diagram: beam A to B with T_A N up at A, T_B N up at B, 0.5x m each side of centre, F N and W N down, d m from A]

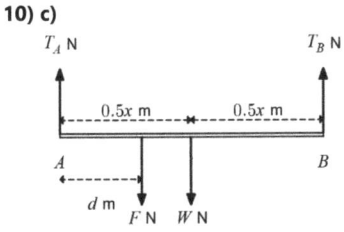

11)
$R_B = 15$ and $R_A = 15$

12)
$R_B = 38$ N and $R_A = 34$ N

13)
$T_B = 25.7$ N, $T_A = 28.2$ N (3 SF)
$T_B = 26$ N, $T_A = 28$ N (2 SF – AQA)

14) a)
$R_B = 300$ N and $R_A = 400$ N

14) b)
The plank is modelled as a uniform rod. The object is modelled as a particle.

15) a)
$R = 78.4$ N

15) b)
$W = 58.8$ N (3 SF)
$W = 59$ N (2 SF – AQA)

15) c)
$m = 6$ kg

15) d)
$R = 137.2$ N (3 SF)
$R = 140$ N (2 SF – AQA)

16)
$x = 0.455$ m (3 SF)
$x = 0.50$ m (2 SF – AQA)

17)
$x = 0.453$ m (3 SF)
$x = 0.45$ m (2 SF – AQA)

18)
$x = 1.21$ m (3 SF)
$x = 1.2$ m (2 SF – AQA)

19)
$T = 187.5$ N and $R = 112.5$ N

20)
$x = 1.8$ m

21)
$x = \frac{400}{W}$ m

22)
Show that

23 a)
Show that

23) b)
$0 < m \leq 0.3$

24)
$W = 120$ N and $R_B = 180$ N

25)
$m = 5$ kg and $R_C = 100$ N

26)
$\frac{17}{40} \leq x \leq \frac{57}{40}$

27) a)
$x = \frac{17}{15}$

27) b)
The plank is modelled as a rod. The ropes are modelled as light, inextensible string and are vertical.

27) c)
From 0.93 m from A to 2.2 m from A

2.61 Moments: Ladders and Hinges
Questions on page 405

1) a)
15 Nm anti-clockwise

1) b)
9.66 Nm clockwise (3 SF)

1) c)
21.1 Nm clockwise (3 SF)

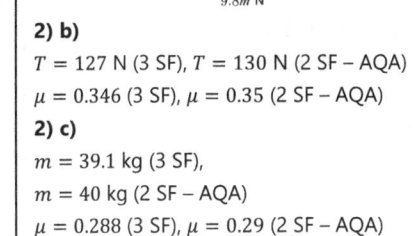

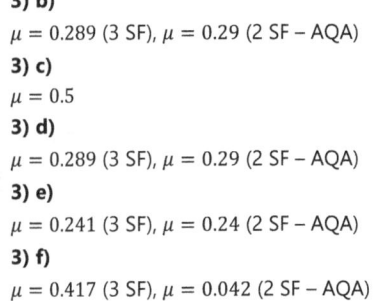

1) d)
12.1 Nm anti-clockwise (3 SF)

1) e)
10.7 Nm clockwise (3 SF)

1) f)
100 Nm anti-clockwise (3 SF)

1) g)
15.1 Nm clockwise (3 SF)

1) h)
345 Nm anti-clockwise (3 SF)

1) i)
15.5 Nm clockwise (3 SF)

2) a)

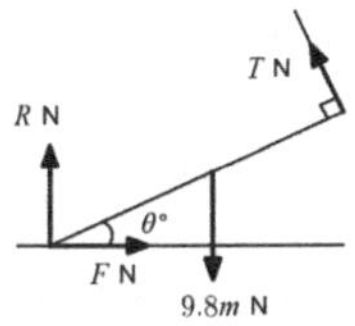

[Diagram: inclined rod with R N up, F N right, 9.8m N down, T N at angle at top, angle $\theta°$ at base]

2) b)
$T = 127$ N (3 SF), $T = 130$ N (2 SF – AQA)
$\mu = 0.346$ (3 SF), $\mu = 0.35$ (2 SF – AQA)

2) c)
$m = 39.1$ kg (3 SF),
$m = 40$ kg (2 SF – AQA)
$\mu = 0.288$ (3 SF), $\mu = 0.29$ (2 SF – AQA)

3) a)

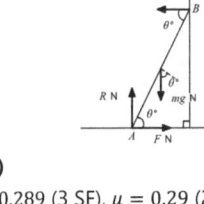

[Diagram: ladder AB with R N, F N at A, P N and mg N at B, angles $\theta°$]

3) b)
$\mu = 0.289$ (3 SF), $\mu = 0.29$ (2 SF – AQA)

3) c)
$\mu = 0.5$

3) d)
$\mu = 0.289$ (3 SF), $\mu = 0.29$ (2 SF – AQA)

3) e)
$\mu = 0.241$ (3 SF), $\mu = 0.24$ (2 SF – AQA)

3) f)
$\mu = 0.417$ (3 SF), $\mu = 0.042$ (2 SF – AQA)

4) a)

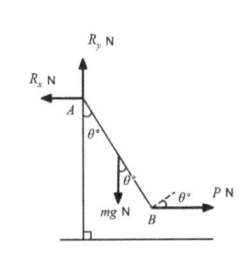

[Diagram: rod AB hinged at A with R_x N, R_y N at A, mg N at middle, P N at B, angles $\theta°$]

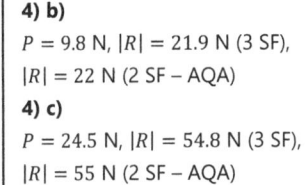

4) b)
$P = 9.8$ N, $|R| = 21.9$ N (3 SF),
$|R| = 22$ N (2 SF – AQA)

4) c)
$P = 24.5$ N, $|R| = 54.8$ N (3 SF),
$|R| = 55$ N (2 SF – AQA)

5) a)

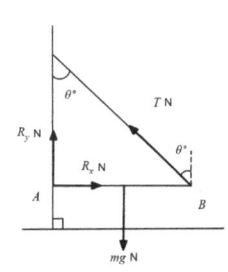

[Diagram: rod AB with R_x N, R_y N at A, mg N, T N at angle $\theta°$ at B]

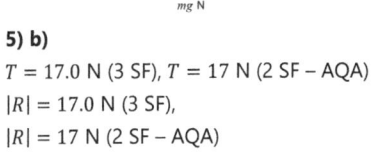

5) b)
$T = 17.0$ N (3 SF), $T = 17$ N (2 SF – AQA)
$|R| = 17.0$ N (3 SF),
$|R| = 17$ N (2 SF – AQA)

5) c)
$T = 147$ N, $T = 150$ N (2 SF – AQA)
$|R| = 147$ N (3 SF),
$|R| = 150$ N (2 SF – AQA)

5) d)
$T = 69.3$ N (3 SF), $T = 69$ N (2 SF – AQA)
$|R| = 69.3$ N (3 SF),
$|R| = 69$ N (2 SF – AQA)

5) e)
$T = 41.6$ N (3 SF), $T = 42$ N (2 SF – AQA)

6) a)

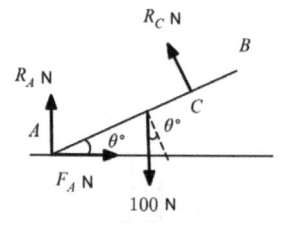

[Diagram: inclined plane from A to B with point C, showing R_A N upward at A, F_A N horizontal at A, R_C N perpendicular to slope at C, 100 N downward at C, angles $\theta°$]

6) b)
$R_A = 42.4$ N, $R_A = 40$ N (3 SF – AQA)
$F_A = 16.8$ N (3 SF),
$F_A = 17$ N (2 SF – AQA)

6) c)
$\mu = 0.396$ (3 SF), $\mu = 0.40$ (2 SF – AQA)

7) a)

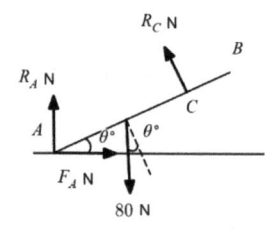

[Diagram: inclined plane from A to B with point C, showing R_A N upward at A, F_A N horizontal at A, R_C N perpendicular to slope at C, 80 N downward at C, angles $\theta°$]

7) b)
$R_A = 43.1$ N (3 SF),
$R_A = 43$ N (2 SF – AQA)
$F_A = 10.8$ N (3 SF),
$F_A = 11$ N (3 SF – AQA)

7) c)
$\mu = 0.249$ (3 SF), $\mu = 0.25$ (2 SF – AQA)

8)
Show that

9) a)
$T = 13.8$ N (3 SF), $T = 14$ N (2 SF – AQA)

9) b)
$F = 13.5$ N (3 SF), $F = 13$ N (2 SF – AQA)
$R = 82.7$ N (3 SF), $R = 22$ N (2 SF – AQA)

9) c)
$\mu = 1.63$ (3 SF), $\mu = 1.6$ (2 SF – AQA)

10) a)
Show that

10) b)
$m = \frac{78}{5g}$

10) c)
$\theta = 50.2°$ (3 SF)

11) a)
Show that

11) b)
$\frac{7}{48}mg \leq P \leq \frac{175}{48}mg$

12) a)
Show that

12) b)
$x = \frac{43}{50}a$ so 86% across from A

Index

Algebra
    Algebraic Fractions p28
    Simultaneous Equations p15

Applications of Calculus p104

Binomial Distribution
    Binomial Distribution p168 – 172
    The Binomial Distribution p365

Binomial Expansion
    Binomial Expansion p77 – 78
    Binomial Expansion p81, p96

Binomial Hypothesis Testing p173 – 175

Binomial Series
    Binomial Series p245 - 246
    Binomial Series p281, p332

Bivariate Data
    Association and Spearman's Rank p149
    PMCC and Linear Regression p144 - 148
    PMCC and Spearman's Critical Tables p356
    PMCC Hypothesis Testing p353 - 354
    PMCC Hypothesis Testing p177, p181
    Spearman's Rank Hypothesis Testing p355

Constant Acceleration
    Constant Acceleration p193 - 195
    Constant Acceleration p199, p201, p211
    Graphs of Motion p188 – 192
    Vectors p392 – 394

Coordinate Geometry
    Circles p24 - 27
    Circles p42, p44
    Linear Graphs p5 - 6

Differential Equations p335 - 339

Differentiation
    Applications of differentiation p103
    Chain Rule p287 - 288
    Chain Rule p343, p364
    Differentiation of $(ax+k)^n$ p331
    Connected Rates of Change p289 - 290
    Connected Rates of Change p338
    Differentiation of ae^{kx} p65, p83, p86, p90, p212, p283
    Differentiation from First Principles p79 – 81
    Differentiation from First Principles p281
    Differentiation of $\ln x$ p303
    Graphs of Gradient Functions p87 – 89
    Implicit Differentiation p297 - 299
    Implicit Differentiation p327
    Introduction p82 - 83
    Optimisation p91 - 93
    Points of Inflection and Applications p282 - 283
    Points of Inflection p285, p286, p287, p292, p294, p364
    Product Rule p291 - 292
    Product Rule p250, p283, p296, p306, p344, p347, p364
    Quotient Rule p293 - 296
    Quotient Rule p323, p342
    Standard Functions p284 – 286
    Standard Functions p250, p283, p306
    Stationary Points p90
    Stationary Points p81, p96
    Tangents and Normals p84 - 86
    Trigonometric Functions p281

Discrete Random Variables p164 - 167

Exponentials and Logarithms
    e and Natural Logarithms p49 - 51
    e^x and $\ln(x)$ p74
    Exponential Growth and Decay p62 - 65
    Exponentials and Logarithms p45 – 48
    Exponentials and Logarithms p43, p128
    Logarithms p72, p172
    Reduction to Linear Form p66 - 70

Forces
    Coefficient of friction p376 - 380
    Connected Particles p202 – 204
    Connected Particles p381 - 382
    Connected Particles on Inclined Planes p388 - 391
    Equilibrium p196 - 197
    Equilibrium p371 - 375
    $F_{net}=ma$ p210, p212
    Inclined Planes p383 - 387
    Particles in Motion p198 - 199
    Pulleys p205 – 208
    Vectors p200 - 201

Functions
    Composite Functions p264
    Domain, Range and Composite Functions p247 - 252
    Domain and Range p264, p267, p296, p299, p341
    Inverse Functions p256 - 259
    Inverse Functions p297, p267
    Modulus Functions p260 - 266
    Modulus Functions p305

Graph Sketching
    Cubic Graphs p31 - 33
    Cubic Graphs p42, p44, p102
    e and Natural Logarithms p57 - 61
    Exponentials and Logarithms p52 - 56
    Graph Sketching p30
    Linear Graphs p7 - 9
    Quadratic Graphs p16 - 19
    Quartic Graphs p34 - 36
    Quartic Graphs p102
    Radical Functions p40
    Radical Functions p102
    Rational Functions p38 - 39
    $\sin(x)$, $\cos(x)$, $\tan(x)$ in Degrees p112 - 119
    $\sin(x)$, $\cos(x)$, $\tan(x)$ in Radians p221 – 228

Graph Transformations
    Combinations p253 – 255
    Combinations p115, p117, p223, p226, p278, p232, p270
    Single Transformations p71 - 76
    Graph Transformations p43

Hidden Polynomials p128

Indices p1 - 2

Inequalities
    Inequalities p41 - 43
    Identifying Regions p44

Index

Integration
    Areas between Curves p312 - 313
    Definite Integrals and Areas p97 – p103
    Double Angles p334
    Indefinite Integrals p94 – 96
    Integration by Parts p328 – 329
    Integration by Parts p326, p338, p347
    Integration by Substitution p324 – 327
    Integration by Substitution p311, p329, p333, p338, p348
    Partial Fractions p333
    Polynomials p246
    Reversing the Chain Rule Part 1 p319 - 320
    Reversing the Chain Rule Part 2 p321 - 323
    Second Derivatives p85, p96
    Standard Functions p314 - 318

Large Data Set
    AQA p178 - 179
    Edexcel p176 – 177
    OCR A p186 - 187
    OCR MEI (Health) p180 - 181
    OCR MEI (London) p184 - 185
    OCR MEI (World Bank) p182 – 183

Linear Modelling p10 - 11

Measures of Central Tendency and Variation p134 – 139

Moments
    Ladders and Hinges p405 - 407
    Moments p401 - 404

Normal Distribution
    Normal Distribution p361 - 366
    Normal Distribution p177, p179, p183, p185, p187

Numerical Methods
    Change of Sign Method p300
    Newton-Raphson Method p304 - 307
    Newton-Raphson Method p265, p281, p299, p344
    x = g(x) Method p301 - 303
    Trapezium Rule p308 - 311
    Trapezium Rule p307

Parametric Equations
    Differentiation p342 - 344
    Equations p340 - 341
    Parametric Equations p400
    Integration p345 - 348

Partial Fractions
    Partial Fractions p330 - 332
    Partial Fractions p246, p330, p338, p348

Polynomials
    Factor Theorem p43, p96
    Polynomials p29 - 30
    Polynomials p111

Probability
    Conditional Probability p357 – 360
    Conditional Probability p167, p172, p364
    Independence and Mutually Exclusive p162 - 163
    Introduction p150 - 158
    Regions of Venn Diagrams p159 - 161

Projectiles p395 - 397

Proof
    Proof p129
    Proof by Contradiction p349

Proportion p37

Quadratics
    Modelling p22 – 23
    The Discriminant p21
    Introduction p12 - 14
    Problem Solving p20
    Quadratics p110

Representing Data p140 - 143

Sample Means Hypothesis Testing p367 - 370

Sampling Methods p132 - 133

Sequences and Series
    Arithmetic Sequences p238 - 240
    Arithmetic Sequences p244
    Arithmetic Series p305, p332
    Geometric Sequences p241 - 244
    Geometric Sequences p292
    Geometric Series p305
    Introduction p234 – 237
    Recurrence Relations p244

Surds p3 – 4

Trigonometry
    Arc Length and Sector Area in radians p93
    Compound Angle Formulae p273 - 275
    Compound Angle Formulae p281, p375
    Double Angle Formulae p276 - 277
    Double Angle Formulae p269, p272, p296, p303, p327, p338, p341, p344, p347, p349, p397
    Harmonic Forms p278 - 279
    Harmonic Forms p344
    Introduction p105 - 107
    Inverse Trigonometric Functions p267 - 269
    Modelling in Degrees p125 – 127
    Modelling in Radians p231 - 233
    Radians, Arcs and Sectors p213 – 216
    Radians p246, p253, p254, p255, p274, p278, p300, p301, p302, p303, p310
    Reciprocal Trigonometric Functions p270 - p272
    Reciprocal Trigonometric Functions p276, p277, p343, p397
    sin(x), cos(x), tan(x) in Degrees p108 - 111
    sin(x), cos(x), tan(x) in Radians p217 - 220
    Small Angle Approximations p280 - 281
    Small Angle Approximations p275, p344
    Solving Trigonometric Equations p128
    Trigonometric Identities in Degrees p120 - 124
    Trigonometric Identities in Radians p229 - 230
    Trigonometric Identities p341

Variable Acceleration
    Variable Acceleration p209 - 212
    Variable Acceleration p204
    Vectors p398 – 400

Vectors
    2D Vectors p130 - 131
    3D Vectors p350 - 352